APR 28 1980

QD
501
.L59

Phase Diagrams for Ceramists

Ernest M. Levin,
Carl R. Robbins and
Howard F. McMurdie

Compiled at the National Bureau of Standards

Margie K. Reser, *Editor*

FOURTH PRINTING 1979

© Copyright, 1964, by
The American Ceramic Society
65 Ceramic Drive, Columbus, Ohio 43214

Printed in U.S.A.
ISBN 0-916094-04-9

TABLE OF CONTENTS

	Page
Preface	3
Temperature Scales	4
General Discussion of Phase Diagrams	5
I. Glossary	5
II. The Phase Rule	8
(1) Statement	8
(2) Limitations	8
III. Interpretation of Diagrams	9
(1) One-Component Systems	9
(2) Two-Component Systems	9
(A) Binary Systems Without Solid Solutions	9
(a) No Compounds Present	9
(b) Compounds Present	11
(B) Binary Systems With Solid Solutions	11
(a) Mechanics of Crystallization	11
(b) Types of Systems	12
(C) Binary Systems With Immiscible Liquids	12
(D) Binary Systems of a Complex Nature	13
(a) Crystallization Paths	13
(3) Three-Component Systems	14
(A) Ternary Systems Without Solid Solutions	15
(a) Typical Cases	15
(b) Crystallization Paths	17
(c) Summary Relating to Crystallization	19
(d) Alternate Method for Determining Phase Composition	20
(e) Summary of Alternate Method	21
(B) Ternary Systems With Solid Solutions	21
(a) The Solid Solution Diagram	21
(b) Mechanics of Crystallization	22
(c) Determining Three-Phase Boundaries	24
(4) Multicomponent Systems	25
(A) General	25
(B) Graphical Representation	28
(a) Joins	28
(b) Sections	28
(c) Other Specializing Conditions	29
IV. Experimental Methods for High-Temperature Heterogeneous Equilibrium	29
V. Selected Bibliography	32
(1) Theory	32
(2) Interpretation	32
(3) Methods and Techniques	33
(A) General	33
(B) Optical Mineralogy	33
(C) Differential Thermal Analysis	33
(D) X-ray and Crystal Chemistry	34
(E) Hydrothermal	34
(F) High Pressure	34
(4) Mathematical Treatment	35
(5) Thermodynamic Calculations	35
(6) Silicate Chemistry	35
(7) Special Collections of Phase Diagrams	36
(8) Phase Diagrams in Related Fields	36
Specific Diagrams*	
A. Metal-Oxygen Systems, including those containing valence changes	
I. One Metal with Oxygen	37
II. Two Metals with Oxygen	43
III. Three Metals with Oxygen	72

	Page
B. Metal Oxide Systems	
I. One Oxide	83
II. Two Oxides	86
III. Three Oxides	145
IV. Four Oxides	265
V. Five Oxides	313
C. Systems with Oxygen-Containing Radicals	
I. Carbonates Only	
(a) One Carbonate	321
(b) Two Carbonates	322
(c) Three Carbonates	324
II. Perchlorates Only	324
III. Hydroxides	
(a) Two Substances	325
(b) Three Substances	327
(c) Four Substances	328
IV. Nitrates Only	
(a) Two Nitrates	329
(b) Three Nitrates	334
V. Nitrites Only	337
VI. Nitrites-Nitrates, mixed	
(a) Two Substances	337
(b) Four Substances	338
VII. Nitrates-Carbonates, mixed	339
VIII. Sulfates Only	
(a) Two Sulfates	340
(b) Three Sulfates	348
IX. Sulfates with Metal Oxides	
(a) Two Substances	351
(b) Three Substances	351
(c) Four Substances	352
X. Sulfates with Other Oxygen-Containing Radicals	
(a) Two Substances	352
(b) Three Substances	355
(c) Four Substances	356
D. Systems Containing Halides Only	
I. Bromides Only	
(a) Two Bromides	361
(b) Three Bromides	366
II. Chlorides Only	
(a) One Chloride	367
(b) Two Chlorides	368
(c) Three Chlorides	403
III. Fluorides Only	
(a) One Fluoride	413
(b) Two Fluorides	413
(c) Three Fluorides	432
IV. Iodides Only	
(a) One Iodide	448
(b) Two Iodides	448
(c) Three Iodides	451
V. Mixed Halides Only	
(a) Two Halides	451
(b) Three Halides	460
(c) Four Halides	461
E. Systems Containing Halides with Other Substances	
I. Halides with Metals	
(a) Two Substances	467
(b) Three Substances	469
(c) Four Substances	469
II. Fluorides with Metal Oxides	
(a) Two Substances	470
(b) Three Substances	470
(c) Four Substances	479
III. Halides with Hydroxides	483
IV. Halides with Nitrates	
(a) Two Substances	485
(b) Three Substances	486
(c) Four Substances	487

Table of Contents

		Page
V.	Halides with Sulfates	
	(a) Two Substances	491
	(b) Three Substances	492
	(c) Four Substances	494
VI.	Halides with Miscellaneous Oxides	
	(a) Two Substances	500
	(b) Three Substances	503
	(c) Four Substances	509
	(d) Six Substances	512
VII.	Halides with Miscellaneous Substances	
	(a) Two Substances	513
	(b) Four Substances	513
F.	Systems Containing Cyanides, Sulfides, etc.	
I.	Cyanides Only	515
II.	Metal with Sulfur	
	(a) Two Substances	516
	(b) Three Substances	518
	(c) Four Substances	519
III.	Sulfides Only	
	(a) Two Sulfides	519
	(b) Three Sulfides	521

		Page
IV.	Sulfides with Metal Oxides	
	(a) Two Substances	522
	(b) Three Substances	522
V.	Miscellaneous	523
G.	Systems Containing Water	
I.	Water	525
II.	One Metal Oxide with Water	526
III.	Two Metal Oxides with Water	532
IV.	Three Metal Oxides with Water	554
V.	Four Metal Oxides with Water	563
VI.	Five Metal Oxides with Water	564
VII.	Miscellaneous Substances with Water	
	(a) One Substance with Water	564
	(b) Two Substances with Water	565
	(c) Three Substances with Water	568
	(d) Four Substances with Water	568
Melting Points of the Metal Oxides		569
Molecular Weights of Oxides		574
Author Index		575
System Index		581

---

*System of Arranging Diagrams

A diagram is classified according to one of the listed sections, such as, metal oxides, halides, water containing, etc. Within each section the systems are grouped according to the total number of different oxides (or materials) represented in all the components, such as, two oxides, three fluorides, four materials, etc. Since the systems are not arranged according to the number of components, quaternary, ternary, and binary systems will be grouped together when the same oxides are represented in each. For example the phase diagrams of $CaO-5CaO \cdot 3Al_2O_3-2CaO \cdot SiO_2-4CaO \cdot Al_2O_3 \cdot Fe_2O_3$, $CaO-CaO \cdot SiO_2-4CaO \cdot Al_2O_3 \cdot Fe_2O_3$, and $2CaO \cdot SiO_2-4CaO \cdot Al_2O_3 \cdot Fe_2O$, are all listed together under the heading $CaO-Al_2O_3-Fe_2O_3-SiO_2$, because each contains CaO, Al_2O_3, Fe_2O_3, and SiO_2. These oxide groups are shown in boldface type in the compilation.

The diagrams in all groupings are arranged according to a combination valence, alphabetical, and simplicity order. The order of writing oxides of a compound and the order of writing components of a system follow the same rule. In this method, the oxides are first grouped according to increasing valence of the cations, the R_2O's first and the RO_3's last, and then arranged in alphabetical order within each valence grouping. In listing compounds of systems containing complex components (compounds of two or more oxides are termed complex), the simplest compounds are given first, followed by the other compounds in order of increasing complexity, each compound being written in the valence-alphabetical order.

The following examples are given for clarification: The Cs_2O-MoO_3 category precedes the Cu_2O-SiO_2 one; the $CaO-FeO-Al_2O_3-SiO_2$ category precedes the $CaO-MgO-Al_2O_3-Fe_2O_3$ one; and the $NaF-Na_2O-SiO_2$ category precedes the $NaF-Na_2O-CrO_3$ one. The formula for the mineral diopside is written $CaO \cdot MgO \cdot 2SiO_2$ and anorthite is $CaO \cdot Al_2O_3 \cdot 2SiO_2$. The phase diagram of albite, diopside, and anorthite is written $Na_2O \cdot Al_2O_3 \cdot 6SiO_2-CaO \cdot MgO \cdot 2SiO_2-CaO \cdot Al_2O_3 \cdot 2SiO_2$.

In order to conserve space the listing of the title of each diagram according to figure sequence is not given; however, in the alphabetical index at the end of the compilation, a diagram is listed under each of its constituent oxides or materials. An author index has also been included for the convenience of the reader.

Preface

The compilation of phase diagrams dates back more than 30 years, to when F. P. Hall and H. Insley published 178 diagrams as the October 1933 issue of the JOURNAL OF THE AMERICAN CERAMIC SOCIETY. Five years later, the same authors published a supplement, as the September 1938 issue of the JOURNAL. In November 1947 a complete revision, containing 507 diagrams, was issued as Part II of the JOURNAL of that month. A supplement, compiled by H. F. McMurdie and F. P. Hall, appeared two years later as a part of the December 1949 JOURNAL.

Continuing activity in phase equilibria studies led to the publication in 1956 of a completely revised edition, under the title "Phase Diagrams for Ceramists." The volume contained 811 diagrams and was compiled by E. M. Levin, H. F. McMurdie, and F. P. Hall. Only three years later, a supplement of 462 diagrams was issued, in which increased emphasis was placed on non-metal oxide and salt systems. For convenience of the user, the supplement included author and subject indexes covering the 1956 compilation, as well as the supplement.

Once again extensive activity in the field has prompted the publication of this new and revised edition, containing over 2000 diagrams. Because of the technological importance of fused salt systems, their coverage has been expanded to include the very early work, done principally by the heating curve method. References to other compilations of related data are given in the bibliography.

Introductory material under "General Discussion of Phase Diagrams" remains unchanged except for an addition to Part IV on experimental methods and a revision of the bibliography. In the selection of diagrams for inclusion in this edition, the authors have been more critical than previously. In general, when several diagrams exist for identical regions of a given system, only the best diagram, in the opinion of the compilers, has been included. However, in a few cases involving important systems, such as SiO_2 and Al_2O_3-SiO_2, several diagrams representative of current interpretations have been chosen. In some cases composite diagrams have been constructed from the work of several investigators. Figure reproductions have been made from tracings of edited enlargements of the original work. Editing of diagrams was concerned chiefly with uniformity of statement and legibility.

Three tables have been added as appendixes. One is a compilation, made at the National Bureau of Standards, of published melting points of the metallic oxides. Preferred melting points are indicated for only a few oxides, and it is evident that reliable melting point data are extremely limited. The other two tables, Relative Atomic Weights 1961 (based on $^{12}C = 12$) and Molecular Weights of Oxides, are for convenience and uniformity in calculations.

This introduction would not be complete without acknowledgment to the many workers in the field who have contributed both diagrams and suggestions. In particular, the authors express their gratitude to Miss Margie K. Reser, of The American Ceramic Society, who was responsible for the editorial detail and complexities associated with this compilation.

Temperature Scales

Temperatures are given in degrees Celsius (Centigrade) and thse compositions are given in weight perceutage unles otherwise stated on the individual diagram. It might be emphasized, however, that several ceutigrade temperature scales have been used since the tnru of the century and that they are not exactly eqnivalent, especially at higher temperatures. Since 1914, the Geophysical Laboratory of the Carnegie Institntion of Washington has used a scale based on the nitrogen-thermometer calibration of Day and Sosman with certain minor conversions by L. H. Adams.[1] This scale is still used by the staff of the Geophysical Laboratory, for purposes of consistency, and is implicit, therefore, in any temperature measurements it reports. In 1927 the Seventh General Conference in Weights and Measure, representing 33 nations, adopted the *International Temperature Scale of 1927*,[2] which became widely accepted. The *International Temperature Scale of 1948*[3] adopted by the Ninth General Conference on Weights and Measures represented the first revision of the 1927 scale. *The International Practical Temperatrue Scale of 1948*,[4] adopted by the Eleventh General Conference on Weights and Measures, is only a text revision of the 1948 scale without any changes in the numerical values of the temperatures. As the IPTS of 1948 is known to differ from the Thermodynamic Temperature Scale, future revisions in both text and numerical values are probable. Figure I shows the differences between the three scales using the International 1948 scale as the basis of comparison. It can be seen that up to the gold point (1063°C.) the differences are insignificant. Even at the palladium point (1552°C.) the differences are well within the limit of experimental error of most high-temperature phase-equilibrium measurements. Above the palladium point, however, differences become more appreciable. The melting point of cristobalite, for example, originally reported as 1713° on the Geophysical scale, was adjusted to 1728° on the *International Temperature Scale of 1927*, and becomes 1723°C. on the *International Practical Temperature Scale of 1948*.

---

[1] R. B. Sosman, "Temperature Scales and Silicate Research," *Am. J. Sci.*—Bowen Volume, 517–528 (1952).

[2] G. K. Burgess, "The International Temperature Scale," *BS J. Research*, **1**, 635 (1928); RP 22.

[3] (*a*) H. F. Stimson, "The International Temperature Scale of 1948," *J. Research NBS*, **42**, 209 (1949); RP 1962.
(*b*) Robert J. Corruccini, "Differences Between the International Temperature Scales of 1948 and 1927," *ibid.*, **43**, 133–136 (1949).
(*c*) R. E. Wilson, "Standards of Temperature," *Physics Today*, **6** [1], 10–15 (1953).

[4] H. F. Stimson, "International Practical Temperature Scale of 1948 (Text Revision of 1960)," *J. Research Nat'l. Bureau of Standards*, **65A** [3] 139–145 (1961).

General Discussion of Phase Diagrams

I. GLOSSARY

(1) **Alkemade (Van Rijn van Alkemade) Theorem:** The direction of falling temperature on the boundary curve of two intersecting primary phase areas is always away from the Alkemade line (see (2)). If the Alkemade line intersects the boundary curve, the point of intersection represents a temperature maximum on the boundary curve. If the Alkemade line does not intersect the boundary curve, then the maximum on the boundary curve is represented by that end which if prolonged would intersect the Alkemade line.

(2) **Alkemade Line:** In a ternary phase diagram a straight line connecting the composition points of two primary phases whose areas are adjacent and the intersection of which forms a boundary curve.

(3) **Boundary Line (Curve):** The intersection of adjoining liquidus surfaces in a ternary phase diagram. The area enclosed by a series of boundary lines is termed a primary phase area.

(4) **Components (of a System):** The smallest number of independently variable chemical constituents necessary and sufficient to express the composition of each phase present in any state of equilibrium. Zero and negative quantities of the components are permissible in expressing the composition of a phase.

(5) **Composition (or Compatibility) Tetrahedron:** In the phase diagram of a condensed quaternary system, the four triangular planes connecting the compositions of four solid phases which can coexist in equilibrium with liquid. The composition of the liquid is represented by a quaternary invariant point, which may lie within the composition tetrahedron (eutectic point) or outside the tetrahedron (peritectic or reaction point).

(6) **Composition (or Compatibility) Triangle:** In the phase diagram of a condensed ternary system the three joins connecting the composition points of the three primary phases whose liquidus surfaces meet at a point.

If in the diagram of a ternary system, all of the Alkemade lines (see (2)) be constructed, the ternary diagram will be divided into a number of composition triangles. If the three substances designating the vertices of any of these triangles are not miscible in the solid state, they represent the final equilibrium products of crystallization at the solidus temperature for compositions within the triangle. When crystalline solid solutions exist between any of the three substances, the final products of crystallization may be reduced in number by one or two.

(7) **Condensed System:** One in which the vapor pressures of the solid and liquid phases present are negligible or small in comparison to the atmospheric pressure. For such systems, e.g., the refractory oxide ones, as the pressure may be considered constant, one degree of freedom is lost, and the phase rule may be modified accordingly: The sum of the number of phases plus the number of degrees of freedom equals the sum of the number of components plus one (instead of two).

(8) **Congruent Melting Point:** At a specified pressure, the temperature at which a solid substance changes to a liquid of identical chemical composition.

(9) **Conjugate Phase:** One of two phases in equilibrium with each other defining a conode (see (10)).

(10) **Conode (or Tie Line):** For a particular temperature, the straight line connecting the compositions of two (conjugate) phases in equilibrium with each other.

(11) **Critical Pressure:** In a one component system, the unique pressure at which the liquid and vapor phases become identical.

(12) **Critical Temperature:** In a one component system, the unique temperature at which the liquid and vapor phases become identical. At the critical temperature the system passes from a heterogeneous state to a homogeneous phase. Above the critical temperature no liquid phase can exist however great the pressure.

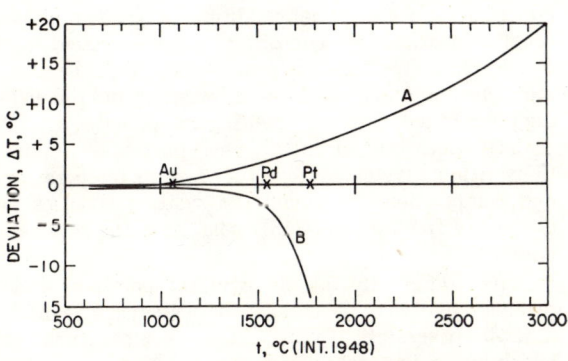

FIG. 1.—Comparison of Geophysical Laboratory Temperature Scale of 1914 and International Temperature Scale of 1927 with that of the International Temperature Scale of 1948. (A) ΔT = Int. 1927—Int. 1948; (B) ΔT = Geophysical 1914—Int. 1948.

(13) **Degrees of Freedom (or Variance):** "The number of intensive variables which can be altered independently and arbitrarily without bringing about the disappearance of a phase or the formation of a new one is called the number of degrees of freedom of a system."[4] Intensive variables are those which are independent of mass, such as pressure, temperature, and composition.

The number of degrees of freedom of a system may also be defined as the "number of variable factors, temperature, pressure, and concentration of the components, which must be arbitrarily fixed in order that the condition of the system may be perfectly defined."[5]

A system is termed invariant, mono-variant, bi-variant, tri-variant, and so on, according to whether it possesses, respectively, 0, 1, 2, 3, etc., degrees of freedom.

(14) **Devitrification:** The formation of crystalline material from glass.

(15) **Enantiotropic Forms:** Polymorphic forms (see (43)) which possess an inversion point at which they are in reversible equilibrium, that is, they are interconvertible; for example, α- and β-$2CaO \cdot SiO_2$ and α- and β-quartz. In such cases the vapor-pressure curves intersect below the melting point of the highest temperature polymorphic form.

(16) **Equilibrium:** From the theoretical, thermodynamic standpoint, the conditions for equilibrium can be exactly and precisely defined; because for any reversible process, no useful energy passes from or into the system.

From the practical, experimental standpoint, however, the actual attainment of an equilibrium state within a system may be very difficult to assess. Three criteria have been used variously either singly or together: (1) The time criterion, based on the constancy of phase properties with the passage of time; (2) the approach from two directions criterion, yielding under the same conditions phases of identical properties, e.g., from undersaturation and supersaturation, or from raising and lowering the temperature to the same value; and (3) the attainment by different procedures criterion, producing phases having the same properties when the same conditions, with respect to the variants, are reached.

None of these criteria are entirely adequate for excluding metastable relationships. In silicate systems, in particular, metastable equilibrium is common and may persist for long periods of time and at high temperatures. In the final analysis, interpretation and judgment by the investigator are of prime importance.

(17) **Eutectic:** A eutectic represents an invariant (unique temperature, pressure, composition) point for a system at which the phase reaction on the addition or removal of heat results in an increase or decrease, respectively, of the proportion of liquid to solid phases, without change of temperature. At a eutectic temperature the composition of the liquid phase in equilibrium with the solid phases can always be expressed in terms of positive quantities of the solid phases.

The eutectic composition is that combination of components in a simple system having the lowest melting temperature of any ratio of the components and is located at the intersection of the two solubility curves in a binary system and of the three solubility surfaces in a ternary system.

(18) **Eutectoid:** An invariant point (see (24)) composed solely of crystalline phases, at which the phase reaction on change of heat content at constant temperature results in a change in proportions of the solid phases exactly analogous to that at a eutectic point (see (17)), in which one of the phases is liquid.

(19) **Glass:** In ceramic phase equilibria studies glass refers to supercooled liquid.

(20) **Heterogeneous Equilibrium:** A system is heterogeneous and is in heterogeneous equilibrium when it consists of two or more homogeneous portions (phases) in equilibrium with each other. In the usual consideration of the phase rule, changes in equilibrium due to electrical, magnetic, capillary, and gravitational forces are not considered; but only those changes due to temperature, pressure, and concentration.

(21) **Homogeneous Equilibrium:** A system is homogeneous and is in homogeneous equilibrium when it consists of one phase and all processes or reactions occurring within it are in reversible equilibrium. A homogeneous phase need not consist of one atomic or molecular species, e.g., in the single phase system sodium chloride solution, Na^+, Cl^-, H_3O^+, OH^-, H_2O, and associated molecules may all be present but the reactions involving them are at equilibrium.

(22) **Incongruent Melting Point:** At a specified pressure the temperature at which one solid phase transforms into another solid phase plus a liquid phase both of different chemical compositions than the original substance.

(23) **Indifferent Point:** In a two or more component system the special conditions where two phases become identical in composition and the system loses one degree of freedom. Typical cases include the maximum or minimum in a solid solution series and the melting point of a congruently melting compound.

(24) **Invariant Point:** The particular conditions within a system, in terms of pressure, temperature, and composition, for which the system possesses no degrees of freedom (see (13)) constitute the invariant points.

Stated differently, at an invariant point, no independent changes in the state of the system can be made.

(25) **Inversion Point:** The temperature at which one polymorphic form of a substance (see (43)) changes into another under invariant conditions.

(26) **Isobar:** The locus of all points of constant pressure.

(27) **Isofract:** For compositions within a ternary system the locus of all glasses of constant index of refraction.

---

[4] F. H. MacDougall, "Thermodynamics and Chemistry," John Wiley and Sons, New York, 1939.

[5] See Alexander Findlay, A. N. Campbell, and N. O. Smith, under "Theory" in "Selected Bibliography," Part V.

(28) **Isopleth:** A line in a phase diagram of constant composition.

(29) **Isoplethal Study:** The method of considering the changes occurring in a system in which the composition variable is held constant and the temperature varied.

(30) **Isotherm:** In a ternary system the locus of all points on the liquidus of constant temperature.

(31) **Isothermal Study:** The method of considering the changes occurring in a system in which the temperature variable is held constant and the composition (or pressure) is varied.

(32) **Join:** The region of a phase diagram representing all mixtures that can be formed from a given number of selected compositions. A join may be binary (straight line), ternary (plane), etc., depending on the number of selected compositions, which need not be compounds. Each selected composition, however, must be incapable of formation from the others.

(33) **Le Chatelier's Theorem:** If a system in equilibrium is disturbed, a reaction tends to take place which opposes the effect of the disturbance, i.e., one by which the effect is partially annulled. The theorem of Le Chatelier is an important augment to the phase rule for it enables one to predict qualitatively the effect of external changes on the equilibrium of a system.

(34) **Lever Rule (or Center of Gravity Principle):** When a particular composition separates into only two phases, the given composition and that of the two phases are colinear; furthermore, the amounts of the two separated phases are inversely proportional to their distances from the given composition. Thus, in the adjacent figure, A and B represent the compositions of two phases formed from composition C: Amt. of $A \times$ length AC = Amt. of $B \times$ length BC or $A/B = BC/AC$

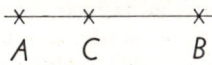

(35) **Liquidus:** The locus of temperature-composition points representing the maximum solubility (saturation) of a solid phase in the liquid phase. In a binary system, it is a line; and in a ternary system, it is a surface, usually curved. At temperatures above the liquidus, the system is completely liquid, and a point on the liquidus represents equilibrium between liquid and, in general, one crystalline phase (the primary one).

(36) **Metastable Phase:** A phase exists metastably in a system if it would not be present at final (thermodynamic) equilibrium, under unchanged conditions, and if the system is not approaching thermodynamic equilibrium at an observable rate.

(37) **Monotropic Forms:** In certain instances of polymorphism (see (43)), the vapor-pressure curves of the two forms do not meet below the melting point. They, therefore, lack a stable inversion point, and the form with the higher vapor pressure is metastable with respect to the other at all temperatures below the melting point. Such forms are called monotropic and are not interconvertible.

(38) **Peritectic Point:** An invariant point (see (24)) at which the composition of the liquid phase in equilibrium with the solid phases cannot be expressed in terms of positive quantities of the solid phases. Whereas the composition of a eutectic point always lies between or within the composition limits of the solid phases in equilibrium with liquid, the composition of a peritectic point always lies outside the composition limits.

At a peritectic point the intersecting univariant curves do not produce a minimum point on the liquidus curve as for a eutectic.

(39) **Peritectoid:** An invariant point (see (24)) composed entirely of crystalline phases, at which the phase reactions on change of heat content at constant temperature are exactly analogous to those at a peritectic point (see (38)), in which one of the phases is liquid.

(40) **Phase:** Any portion, including the whole, of a system which is physically homogeneous within itself and bounded by a surface so that it is mechanically separable from any other portions. A separable portion need not form a continuous body, as for example, one liquid dispersed in another.

A system composed of one phase is a homogeneous system; a system composed of more than one phase is heterogeneous; and in order for the phase rule to apply, each phase must be in homogeneous as well as heterogeneous equilibrium.

(41) **Phase Rule:** For a system in equilibrium, the sum of the number of phases plus the number of degrees of freedom must equal the sum of the number of components plus two, or $P + F = C + 2$.

(42) **Piercing Point:** In a quaternary system, the intersection of a univariant curve with a ternary joint (see (32)) at a point other than a ternary invariant point. The univariant curve represents the compositions of liquids that can exist in equilibrium with three particular solid phases. The composition of these solid phases usually all lie in the plane of the ternary join if the intersection is a ternary invariant point, but they cannot all lie in that plane if the intersection is a piercing point.

(43) **Polymorphism:** The property possessed by some substances of existing in more than one crystal form, all forms being of the same chemical composition but differing in crystalline structure and physical properties, and yielding identical liquid or gaseous phases on melting or evaporating.

(44) **Primary Phase:** The only crystalline phase which can exist in equilibrium with liquid of a given composition. The primary phase is the first crystalline phase to appear on cooling a composition from the liquid state; or conversely, it is the last crystalline phase to disappear on heating a composition to melting (see, also, Boundary Line (3) and Liquidus (35)).

(45) **Primary Phase Region:** The locus of all compositions in a phase diagram having a common primary phase.

(46) **Pseudo System:** It is frequently convenient or necessary to refer to portions of a binary or ternary, etc., system which are not (true) subsystems (see (49)). In such instances the term pseudo binary, or pseudo ternary, etc., is used.

For example, in Fig. XIV, $3CaO \cdot Al_2O_3\text{-}2CaO \cdot SiO_2$ is a pseudo binary system and $Al_2O_3\text{-}CaO \cdot Al_2O_3 \cdot 2SiO_2\text{-}3Al_2O_3 \cdot 2SiO_2$ is a pseudo ternary system. A ternary system must consist of components and be bounded by three binary systems. In the pseudo ternary system cited, one of the boundary lines, $3Al_2O_3 \cdot 2SiO_2\text{-}CaO \cdot Al_2O_3 \cdot 2SiO_2$, is only a pseudo binary system, as no combination of $CaO \cdot Al_2O_3 \cdot 2SiO_2$ and $3Al_2O_3 \cdot 2SiO_2$ can yield Al_2O_3 which appears as a primary phase.

(47) **Solid Solution:** A single crystalline phase which may be varied in composition within finite limits without the appearance of an additional phase.

(48) **Solidus:** The locus of temperature-composition points in a system at temperatures above which solid and liquid are in equilibrium and below which the system is completely solid. In binary diagrams without solid solutions, it is a straight line, representing constant temperature, and with solid solutions, it is a curved line or combination of curved and straight lines. Likewise, in ternary systems, the solidus is represented by a flat plane or a curved surface, respectively.

(49) **Subsystem:** Any portion of a binary, ternary, etc. system which can be treated as an independent binary or ternary, etc. system. The selected substances designating the subsystem must be components for the subsystem (see (4)). In the $CaO\text{-}Al_2O_3$ binary system (Fig. 231), the lime-alumina compounds with congruently melting points form binary systems with each other, for example, the $CaO \cdot 2Al_2O_3\text{-}CaO \cdot Al_2O_3$ and the $CaO\text{-}CaO \cdot Al_2O_3$ systems; and in the ternary system $CaO\text{-}Al_2O_3\text{-}SiO_2$ (Fig. XIV), the binary joins, such as $CaO \cdot SiO_2\text{-}CaO \cdot Al_2O_3 \cdot 2SiO_2$ are true binary systems, and the three congruently melting compounds, $CaO \cdot SiO_2$, $CaO \cdot Al_2O_3 \cdot 2SiO_2$, and $2CaO \cdot Al_2O_3 \cdot SiO_2$, whose common boundary lines meet in a eutectic, constitute a true ternary system.

(50) **System:** Any portion of the material universe which can be isolated completely and arbitrarily from the rest for consideration of the changes which may occur within it under varying conditions.

The term *system* is used in two senses: the general and the specific. In the general sense, one specifies a system by naming the chosen components, for example, the binary system $CaO\text{-}Al_2O_3$, or the ternary system $K_2O\text{-}B_2O_3\text{-}H_2O$. In the specific sense, one may designate restricted portions of a "general" system for study or discourse. Thus one may refer to an invariant system or bi-variant system, etc. in which the restriction is based on the degrees of freedom (see (13)). One may also refer to a one phase (homogeneous) system (for example, system water vapor) or to a two phase (heterogeneous) system (for example, system calcium disilicate-liquid), etc., in which case the restriction is based on the number of phases present. Finally, one may refer to a system in which the restriction is based on chemical composition, for example, the system 20 per cent $Li_2O \cdot SiO_2$-80 percent $Li_2O \cdot B_2O_3$. The use of the term system to designate a particular chemical composition is not necessary and should be avoided.

If the definition of system be kept in mind, the varied use of the word need not be confusing and can easily be interpreted in context.

(51) **Tie Line:** See Conode (10).
(52) **Variance:** See Degrees of Freedom (13).

II. THE PHASE RULE

(1) Statement

The basis of all work on equilibrium diagrams is, of course, the phase rule of Willard Gibbs.[6] Its use has been greatly facilitated by the interpretations of Roozeboom,[7] Schreinemakers,[8] and others. Extensive explanations are to be found in textbooks on physical chemistry or books devoted exclusively to the phase rule (see V. Selected Bibliography, (1)).

The diagram known variously as phase diagram, equilibrium diagram, etc., is essentially a graphical expression of the phase rule. Equation (1) gives the usual mathematical form of the phase rule.

$$P + F = C + 2 \qquad (1)$$

C = number of components of system.
P = number of phases present at equilibrium.
F = degrees of freedom (variance) of system.

The terms used in equation (1) as well as others necessary to an understanding and application of the phase rule are defined in the Glossary.

(2) Limitations

The phase rule applies only to equilibrium states of a system, which require both homogeneous equilibrium within each phase and heterogeneous equilibrium between co-existing phases. The phase rule does not depend on the nature of the components or on the nature and amounts of the phases present, but only on their numbers; nor does it give information concerning rates of reactions.

A system in equilibrium always obeys the phase rule, but conformance, in itself, is not a sufficient test for equilibrium, because of the possible existence of nonequilibrium phases and conditions. Non-conformance with the phase rule, however, is proof that equilibrium conditions do not exist.

The phase rule provides the basis for the classification, according to number of components, of the diverse cases of chemical equilibrium. If the number of components be known, which is usually the case for a specified system, the sum of the number of phases and the number of degrees of freedom is fixed at C plus 2.

The number of components plus two also represents the maximum number of phases that can coexist at equilibrium, as the degrees of freedom (F) can never be less than 0 (at invariant conditions).

---

[6] (a) J. W. Gibbs, "Equilibrium of Heterogeneous Substances," *Trans. Conn. Acad. Sci.*, **3**, 108–248, 343–524 (1874–78).
(b) J. W. Gibbs, The Collected Works of J. Willard Gibbs, Vol. I, pp. 54–371. Longmans, Green and Co., New York, 1928.
[7] H. W. B. Roozeboom, Die heterogenen Gleichgewichte, 6 vols. F. Vieweg Co. Sohn, Braunschweig, 1911.
[8] F. A. H. Schreinemakers, "Mischkristalle in Systemen dreier Stoffe," *Z. physik. Chem.*, **50** [2] 169–99: **51** [5] 547–76; **52** [5] 513–50 (1905).

III. INTERPRETATION OF DIAGRAMS

(1) One-Component Systems

The independent variables in a one-component system are limited to temperature and pressure because the composition is fixed. It follows from the phase rule that the system is bivariant if one phase is present, univariant if two phases are present and invariant if three are present. A diagram of a one-component system in which the independent variables, temperature and pressure, are the abscissa and ordinate, respectively, is shown in Fig. II. The following facts are observed:

(1) Curve F-A (univariant) is the sublimation curve for modification A.

(2) Curve A-B (univariant) is the sublimation curve for modification B.

(3) Curve B-C (univariant) is the vapor pressure curve for the liquid.

(4) Curve A-D (univariant) is the transition curve for modifications A and B and represents the change of the transition point with pressure.

(5) Curve B-E (univariant) is the melting curve for modification B and represents the change of the melting point with pressure.

(6) Point A (invariant) is the transition point for the two crystalline modifications. It is called a triple point since it is the point at which three phases (two solids and vapor) are in equilibrium.

(7) Point B is the triple point (invariant) for the equilibrium between crystalline modification B, liquid, and vapor.

(8) The system is bivariant in all parts not on these lines or their intersections.

In dealing with refractory substances it is at present impossible, with a few exceptions, to measure directly the vapor pressure of the solid and liquid phases. It is possible, however, to construct diagrams for refractory substances showing qualitatively the vapor pressure for the different phases if the temperature-stability relations are known because the unstable phase always has a higher vapor pressure than the stable phase even though the vapor pressure is infinitesimally small.

(2) Two-Component Systems

Two-component systems have three independent variables, namely, temperature, pressure, and composition. In systems of importance in ceramics where the vapor pressure remains very low for large variations in temperature, the pressure variable and the vapor phase may be eliminated from consideration.

Systems with the pressure variable eliminated are called condensed systems. The reduced phase rule by which such systems may be represented is shown in equation (2) in which P refers only to solid and liquid phases.

$$P + F = C + 1 \qquad (2)$$

In binary systems under these conditions three coexisting phases produce an invariant condition, two a univariant condition, and one a bivariant condition.

In representing condensed binary systems it is customary to make the ordinate the temperature scale and the abscissa the composition scale. In Fig. III, the intersection of the ordinate with the abscissa at T_0 represents 100 percent S_1 (and 0 percent S_2); the intersection at T_x represents 100 percent S_2 (and 0 percent S_1). The scale as labeled, reading from left to right, refers to percentage of S_2 present; the percentage of S_1 represented by any point equals $100 - S_2$. The composition variable is usually given in weight percent or mole percent, according to convenience; occasionally in weight fraction or mole fraction.

(A) Binary Systems Without Solid Solutions

(a) No Compounds Present

In Fig. III is shown a simple type of condensed binary system with (1) no intermediate compounds, (2) complete solubility in the liquid state, and (3) no solubility (i.e., no solid solution) between the solid phases. The point, C_1, is an invariant point of the eutectic type where two solid phases (components S_1 and S_2) and liquid (of composition 40% S_2, 60% S_1) are in equilibrium. A change in either variable will cause one or more phases to disappear. The curves F-C_1 and G-C_1 determine the position of univariant equilibrium. The coexistence of both phases (one solid and one liquid) can be maintained with change of one variable if a compensatory change is made in the other, the proportion of the two phases changing accordingly. The univariant boundary defines the quantitative relations of the variables.

In Fig. III, the vertical boundaries represent one-component systems. The solid phases, S_1 and S_2, representing these components have sharp melting points at the temperatures F and G, respectively. Elsewhere, the F-C_1 and G-C_1 (liquidus) curves separate the one phase (liquid) region from the areas where both solid and liquid are present.

The difference between a one-phase region and a region representing two or more phases in equilibrium is a basic concept in the interpretation of phase diagrams

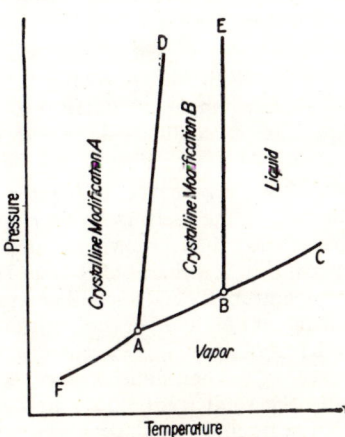

Fig. II.—Phase relations in a one-component system.

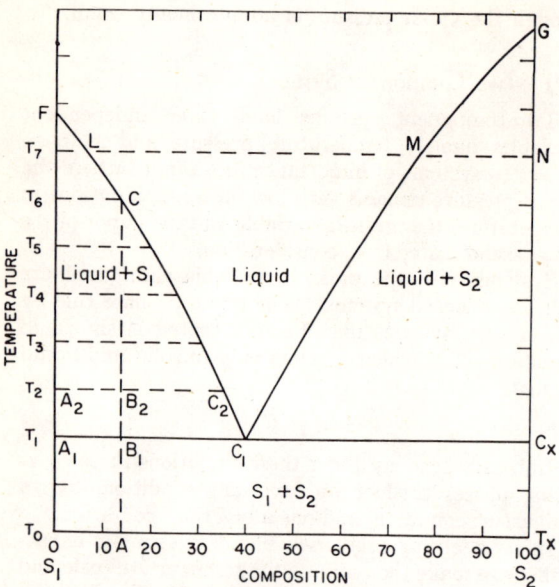

Fig. III.—Two-component system without intermediate compounds or solid solutions.

Every location within a one phase region, such as in the field designated *Liquid* in Fig. III, represents an actual state of the system, in terms of temperature and concentration variables. Locations within a two-phase region, however, do not represent actual states, but merely correspond to the overall chemical composition of two phases in equilibrium with each other, for example, points in the fields designated *Liquid* + S_1, *Liquid* + S_2, and $S_1 + S_2$, in Fig. III. The two-phase areas are, in effect, gaps or voids in which no single homogeneous phase can exist. To construct isobaric diagrams, it is necessary and sufficient to know the one-phase-region boundaries.[9]

Mechanics of Crystallization: Changes in the system illustrated in Fig. III may be followed by varying either temperature or composition. On heating, a mass of composition C_1 will show sharp melting at the temperature T_1. This is the lowest temperature in this system at which liquid is in equilibrium with solids (the eutectic temperature). Conversely, a homogeneous liquid of this composition will crystallize completely at this same temperature, on cooling, provided equilibrium conditions are maintained. Since neither of the variables, temperature or composition, can be changed without the complete disappearance of one or more phases, the point C_1 is an invariant point.

A liquid of composition A on cooling under equilibrium conditions from a temperature above T_6 will behave quite differently. The substance will remain a homogeneous liquid until temperature T_6 is reached, when the first infinitesimal amount of the solid with composition S_1 will crystallize. Upon further cooling, S_1 will continue to crystallize while the composition of the liquid follows curve C-C_1.

[9] See J. S. Marsh, under "Theory" in Selected Bibliography, Part V.

At any temperature the composition of the liquid coexisting with the solid is represented by the point of intersection of the horizontal line corresponding to that temperature with the liquidus curve. The relative amounts of solid and liquid coexisting are represented by the relative lengths of the temperature horizontal from the composition of the initial material to intersection with the liquidus and with the vertical, S_1-F, respectively. Thus, composition A at the temperature T_2 will consist of a solid of composition S_1 and a liquid of C_2 in the ratio of line lengths B_2-C_2/A_2-B_2.

With further cooling, solid S_1 will continue to crystallize until the temperature of the eutectic, C_1, is reached. The material will remain, because of the heat of crystallization, at that temperature until completely crystallized into a mixture of solids S_1 and S_2. The ratio of the amounts of the two solid substances can be found by the same lever rule $S_1/S_2 = A$-S_2/A-S_1.

If the two solids, S_1 and S_2 (Fig. III), are mixed in such proportion as represented by point A and are heated, no reaction will take place until the temperature T_1, of the intersection of the isopleth and the solidus is reached. At this temperature a liquid of composition C_1 which contains both S_1 and S_2 is formed. The quantity of liquid C_1 formed at T_1 is measured by the line A_1-B_1 and the ratio of solid to liquid is B_1-C_1/A_1-B_1. At temperature T_2 the ratio of solid to liquid is B_2-C_2/A_2-B_2, and the composition of the liquid is given at C_2, the intersection of the tie line A_2-C_2 with the liquidus F-C_1. As the temperature is raised, the amount of solid decreases and the liquid increases, the liquid becoming richer in S_1. At dT below T_6, only a very small amount of solid remains and the composition of the liquid has changed from C_1 to C. At T_6, the solid is completely melted and the liquid is of composition A.

The relative amounts of liquid and solid S_1 present at various temperatures from T_1 to T_6 for the substance of composition A are shown in Fig. IV. This figure serves fairly well to illustrate the use of the lever rule. The compositions, S_1 and S_2, and the composition corresponding to C_1 are special cases to which curves of the type of Fig. IV do not apply since melting of these compositions takes place at definite temperatures and not over temperature intervals.

The above statements regarding the crystallization of melts are true only if the rate of cooling is sufficiently slow to allow equilibrium to be attained at every instant. A rapid rate of cooling will lead to quite different results.

The liquidus boundary curve shows the effect of soluble impurities on the melting point of pure compounds. This effect can be shown by an isothermal study of the solid, S_1. If a charge of S_1 (Fig. III) is held constant at temperature T_7 and a small amount of S_2 is added, a small amount of liquid of composition L will be formed. As the amount of S_2 is increased, the amount of liquid L increases. When sufficient flux, S_2, has been added to bring the total composition to L, the solid phase, S_1, disappears and the charge becomes entirely liquid. Therefore, a small amount of soluble impurity lowers the melting temperature from F to T_7.

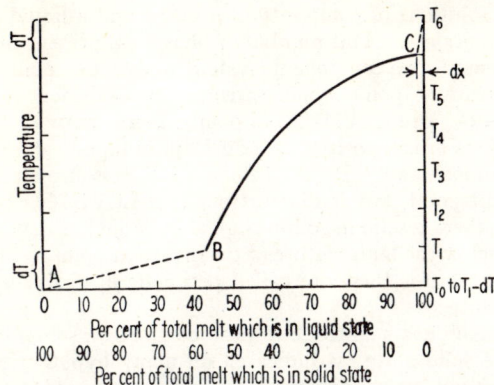

Fig. IV.—Variation in amounts of liquid and solid phase, S_1, upon heating composition A of Fig. III from the temperature T_0 to $T_6 + dT$ along the isopleth A-C; dT and dx are infinitesimal increments in temperature and amount of solid phase, S_1, respectively.

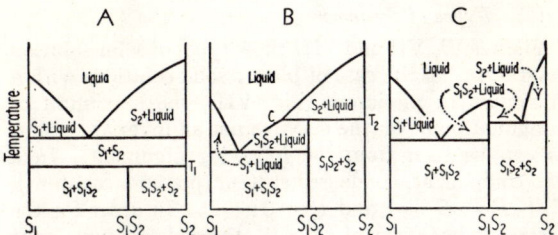

Fig. V.—Two-component systems with compounds present. (A), the compound, S_1S_2, decomposes at a temperature below the eutectic temperature; (B), the compound, S_1S_2, decomposes at a temperature above that of the eutectic; (C), the compound, S_1S_2, is stable at its melting point.

Further additions of S_2 completely dissolve and the composition of the liquid varies until point M is reached when the solid phase, S_2, no longer dissolves, that is, the liquid is saturated with S_2. Further additions of S_2 do not dissolve and the relative amount of liquid decreases; the mass approaches a complete solid as the total composition approaches S_2.

Since the curve F-C_1 represents solutions saturated with S_1, and the curve G-C_1, solutions saturated with S_2, the intersection C_1 (eutectic) must represent a solution saturated with both solids. The curves are, thus, solubility curves or freezing point curves.

(b) *Compounds Present*

Three types of binary diagrams with compounds, are shown in Fig. V. Figure V(A) shows a compound, S_1S_2, which decomposes at T_1 into the solids, S_1 and S_2. Figure V(B) shows a compound, S_1S_2, with an incongruent melting point; that is, it decomposes at T_2 into a solid, S_2, and liquid, C, neither of which has the composition of the original compound. Figure V(C) shows a compound, S_1S_2, which melts congruently, that is, the liquid resulting from the melting of S_1S_2 is of the same composition as the solid, S_1S_2.

The system represented in Fig. V(C) may conveniently be divided into two systems or subsystems, one containing S_1 and S_1S_2 as its components, the other S_1S_2 and S_2. Both may be studied in the same manner as Fig. III.

(B) Binary Systems with Solid Solutions

(a) *Mechanics of Crystallization*

Figure VI represents the case of simple solid solution where the melting point of A is depressed by B and that of B is raised by A. The composition X at the temperature, X_0, will consist entirely of liquid. If this melt is allowed to cool along the isopleth to the temperature, T_1, a solid of composition SS_1 crystallizes out of the solu-

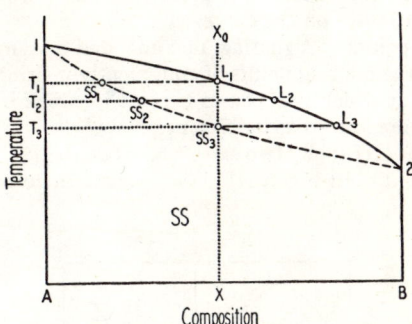

Fig. VI.—Complete solid solution without maximum or minimum in a binary system. Line 1-L_1-L_2-L_3-2 represents the composition of the liquid phase and is called the liquidus curve. Line 1-SS_1-SS_2-SS_3-2 represents the composition of the solid phase and is called the solidus curve. The tie lines, SS_1-L_1, SS_2-L_2, and SS_3-L_3, show the conjugate relation of liquid and solid phase for the three temperatures T_1, T_2, T_3.

tion. It is apparent that the solid is richer in A than is the liquid L. At temperature T_2 the liquid has a composition of L_2 and the solid solution a composition of SS_2; and the ratio of solid to liquid is L_2-X/SS_2-X.

As the temperature falls from T_1 to T_3 the isopleth crosses the tie lines joining compositions of solids from SS_1 to SS_3 with the compositions of liquid from L_1 to L_3. For the system to maintain equilibrium in cooling, each and every crystal of solid solution must change in composition continuously throughout its mass. There is a constant interchange of material between solid and liquid phases and a constant change in composition in all parts of the solid as cooling progresses.

The solid is thus increasing in concentration of B along SS_1-SS_3 while the liquid is increasing in concentration of B along L_1-L_3. Simultaneously the amount of solid is increasing and the amount of liquid decreasing, the total composition of the system, of course, remaining constant. The last drop of liquid has the composition L_3 and the total solid the composition SS_3.

Fractional crystallization can be obtained between T_1 and T_3 by removing the solid phase at any temperature between T_1 and T_3.

(b) Types of Systems

Figures VI, VII and VIII show types of solid-solution diagrams. In the case of binary solid solutions with a maximum or minimum (Fig. VII), the maximum or minimum point on the curve is not an invariant point, as can be seen from the following argument. In a two-component, condensed system (pressure constant), $P + F = C + 1$ and $F = 3 - P$. In order for the system to be invariant ($F = 0$), three phases must exist in equilibrium. Such a condition, however, can never exist in a solid-solution series as shown in Fig. VII, because there are never more than two phases present, i.e., solid solution and liquid solution. The system can at no point become invariant, and the equilibrium curves must be continuous, in contrast to a "true" invariant point (eutectic or peritectic) which is a point of discontinuity on the curve.

Such points designating maxima and minima, including the melting points of congruently melting compounds, at which two phases become identical in composition are known as indifferent points.[5]

Figure VIII shows two cases where two solid solutions are present. In Fig. VIII(A) at point c there are two solid solutions of compositions a and b and a liquid of composition c. This number of phases (3) present in a condensed two-component system makes the point c an invariant point which satisfies the definition of a eutectic. In Fig. VIII(B) at point c there are two solid solutions of compositions a and b and a liquid solution of composition c. Point c in Fig. VIII(B) is also an invariant point, but it differs from that in VIII(A) because there is solid in equilibrium with liquid both above and below the temperature of the invariant point. The relationship is therefore called peritectic in distinction from eutectic.

Marsh[9] has pointed out that complete insolubility in the solid state, as indicated on many hypothetical and actual binary diagrams, is highly improbable. It will be observed that this compilation shows many binary diagrams of the silicates and refractory oxides in which there is no indication of solid solubility, that is, no solid solution. In many cases solubility is so slight that evidences of it cannot be obtained or the scale of the diagram is insufficient to show it. In too many instances, however, the relations in the solid state have not been sufficiently explored. Data of this kind are important to the ceramic engineer and technologist as well as to the petrologist, and are becoming more prominent in literature.

(C) Binary Systems with Immiscible Liquids

In some systems two liquid solutions can exist in equilibrium as shown by the area a-b-c in Fig. IX. The extremities of the "tie line" L_1-L_2 represent the compositions of the two solutions which are in equilibrium at the temperature T_2. The point c at temperature T_1 is a triple point which, in a two-component system, is an invariant point since there are two liquid phases and one solid, A, present. The work of Greig has shown that there are often similar conditions present in certain SiO_2-rich regions of silicate systems.

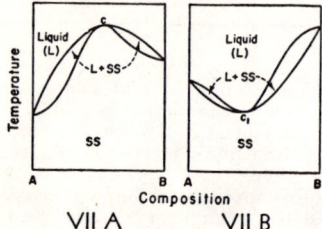

Fig. VII.—Binary systems with a single solid solution. (A), system with a maximum melting point at C, which is not a compound; (B), system with a minimum melting point at C_1, which is not a eutectic. Handbuch der Metallphysik. Edited by Georg Masing. Vol. II, Die heterogenen Gleichgewichte, by Rudolf Vogel, p. 241. Akademische Verlagsgesellschaft, Berlin, 1937.

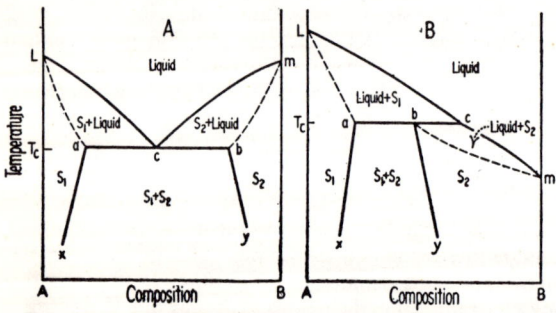

Fig. VIII.—Solid solutions showing conjugate relationships. The two solid solutions, S_1 and S_2, have a conjugate relation to each other in area a-b-x-y. The solid solutions in areas L-a-c and m-b-c have a conjugate relation to the corresponding liquid phases present in these areas.

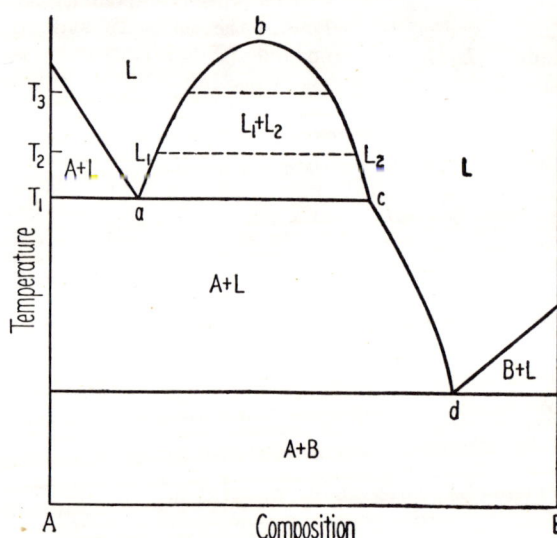

Fig. IX.—Conjugate liquid phases. The broken lines in area a-b-c show the conjugate relationship of the two liquid phases for the two temperatures, T_2 and T_3.

(D) Binary Systems of a Complex Nature

An example of a hypothetical binary diagram of considerable complexity is given in Fig. X to illustrate how the courses of crystallization under equilibrium conditions can be visualized. The two components are A and B. Component A, within the temperature range of the diagram, occurs in 5 enantiotropic forms: liquid, α_1, α_2, α_3, and α_4. Component B exists in only two forms within this temperature range, liquid and β_1. There is one intermediate compound, γ, which does not exist at the liquidus. In a part of the composition range two liquids, L_1 and L_2, coexist. The components, by definition, melt congruently to form liquids of their own composition.

(a) Crystallization Paths

Three vertical lines of constant composition (isopleths), C, D, and E will be considered.

(i) *Crystallization along isopleth C:* At temperatures above point k on the isopleth C, the substance is a homogeneous liquid. When the temperature drops to point k, separation into two liquids results, the two liquids changing in composition and in relative amounts as the temperature is lowered from k to n. At the temperature represented by point m, for example, the compositions of the two liquids correspond to the intersections of the horizontal line (tie line) through point m with the boundaries of the field (points d and e) and the relative amounts of the two liquids, L_1 and L_2, are pro-

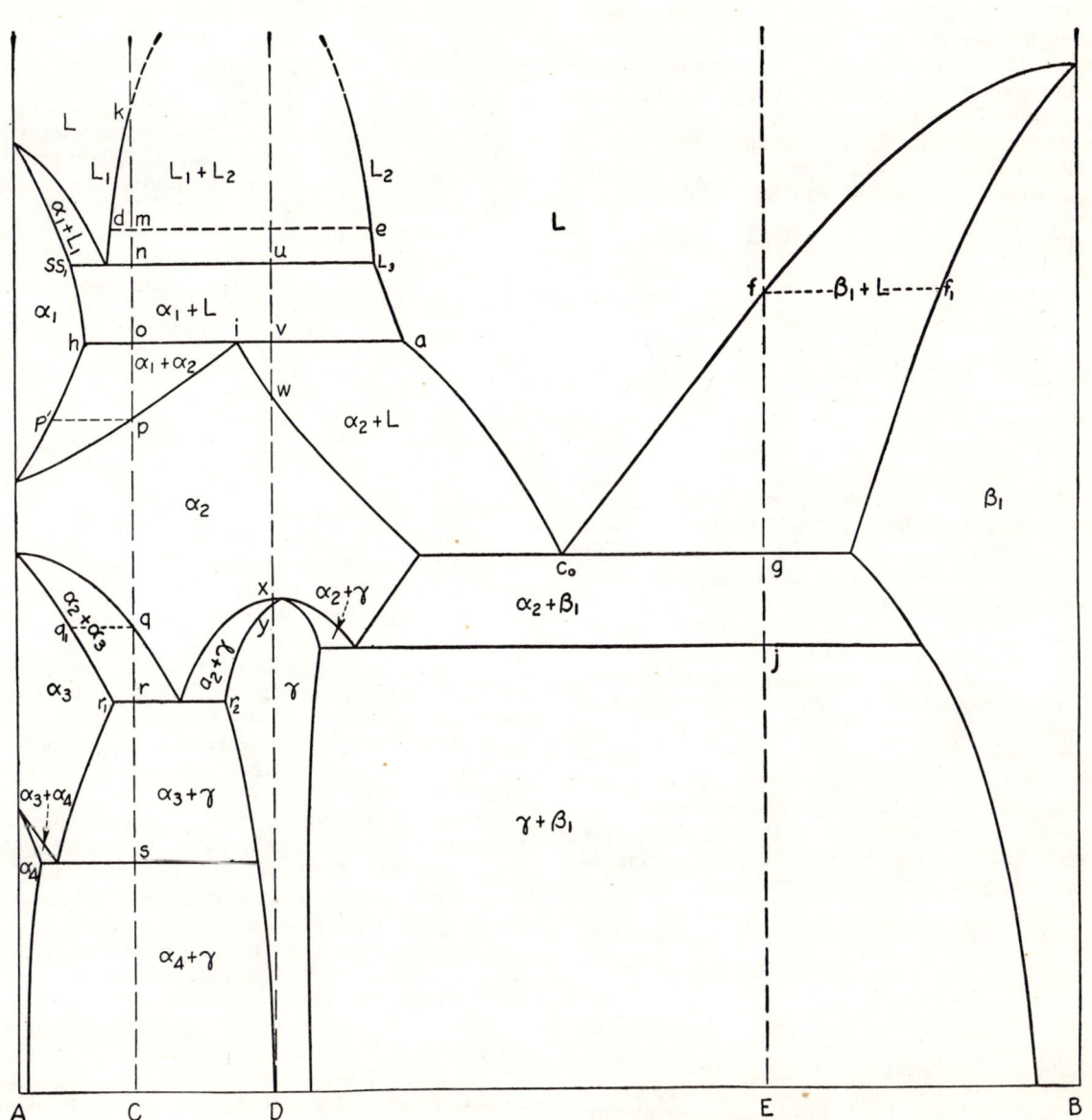

Fig. X.—Hypothetical binary diagram to illustrate possible phase changes.

portional to the lengths m-e and d-m, respectively.

At the temperature of point n, crystallization of the phase, α_1, a solid solution, occurs. The coexisting solid and liquid have the compositions of the left and right extremities, SS_1 and L_3, respectively, of the horizontal line through n. Further temperature drop results only in changes in compositions and in relative amounts of these two phases represented by horizontal lines through the composition at the proper temperature level until point o is reached, at which time the compositions of solid and liquid are h and a, respectively. At the temperature of point o, the cooling is arrested until all of the liquid of composition a has crystallized to a mixture of the solid solutions, α_1 and α_2, α_1 having the composition of point h, and α_2 having the composition of the point i.[10]

Further cooling through the region from o to p results in reaction of the two phases, α_1 and α_2, the quantitative relations between compositions and amounts being found as before by passing horizontals through the temperature levels of the reaction to the intersections with the phase-region boundaries. Solid α_1 changes in composition from h to p' and α_2 from i to p. At the temperature of point p, continued reaction between solids α_1 and α_2 results in the complete disappearance of α_1.

Solid solution α_2 of composition p persists, unchanged in composition, until the temperature of point q is reached. At this point, there is a partial decomposition of phase α_2 to form α_3 of composition q_1 (a eutectoid decomposition). These two phases co-exist, continually changing in relative amounts and in compositions, until point r is reached when another eutectoid decomposition takes place. Solid α_2 completely disappears, and the new solid solution, γ, of composition r_2 forms. At point s, α_3 is in turn replaced by α_4 and the two phases α_4 and γ exist together until the lowest temperature shown in the diagram is reached.

(ii) *Crystallization along isopleth D:* A mass of composition D cooled from above the temperature where it exists as a homogeneous liquid undergoes qualitatively the same changes as did sample C until the temperature represented by point v is reached. Here reaction between the solid α_1 and liquid L takes place, resulting in the re-solution of α_1 and precipitation of α_2. The phases α_2 and L coexist, changing in composition with decreasing temperature until the last bit of liquid disappears at the point w. From w to x only the solid phase α_2 is present. At x phase γ precipitates from α_2, but α_2 persists through only a very short cooling range when it reacts with (is resorbed by) phase γ, and γ persists alone throughout the remaining cooling range shown in the diagram.

The phase γ is a solid solution of an intermediate compound of the composition A_xB_y which can take either A or B into its structure in greater than stoichiometric proportions. The compound A_xB_y cannot be considered a component of a subsidiary system since it does not melt to a homogeneous liquid but instead inverts to the solid solution α_2 on heating.

(iii) *Crystallization along isopleth E:* Cooling of composition E results in the crystallization of solid solution β_1 of composition f_1 at the temperature of point f. Changes in composition of the coexistent solid and liquid phases take place until the temperature of the eutectic, C_0, is reached. Then an arrest in the cooling takes place until reaction has caused the complete disappearance of liquid by the crystallization of phases α_2 and β_1 in the proportion indicated by the lever principle about point g. Further reactions on cooling are not sufficiently different from previous descriptions to require comment.

(3) Three-Component Systems

There are four independent variables in a ternary system, namely, pressure, temperature, and two concentration variables since a ternary solution requires a statement of its composition with respect to two components before its total composition is fixed. Five coexisting phases (a quintuple point) produce an invariant system, four give an univariant system, three, a bivariant, etc. A complete graphical representation of the ternary system is a very difficult matter. If, however, the vapor pressure is so low as to be negligible, the ternary systems may be treated as condensed systems as was the case with binary systems, the phase rule again expressed as $P + F = C + 1$.

The compositions can then be represented by triangular coordinates. This method is illustrated in Fig. XI. In this figure, each side of the equilateral triangle is divided into 100 parts, each tenth division being intersected by lines parallel to each of the other two sides. A point at the apex, C, is composed wholly of component C. A point on the base line A-B is composed entirely of components A and B with none of C. The relative distance of a point, such as x, from each of the three apices may be expressed in percentage and it thus may represent a percentage composition of a ternary mixture or solution in terms of components A, B, C. Point X, for example, represents a composition of 45% A, 20% B, and 35% C, whereas y represents 15% A, 15% B, and 70% C.

By the appropriate construction the coordinates of a point in a triangular diagram can also be read off of any one of the sides. In Fig. XI if the lines XE and XF be constructed parallel to the sides BC and AB, respectively, the length CE represents the percentage of A (45) in composition X, the length AF the percentage of C (35), and the length FE the percentage of B (20). By a similar construction the composition of X, in terms of A, B, and C can be read off on the sides BC and AB. The two end segments of each line represent the proportional amounts (in terms of the whole line) of the substances designated at the opposite ends; the middle segment represents the proportion of the third substance, not located on the line.

---

[10] Crystallization of any solid solution in equilibrium with a liquid solution or with another solid solution under changing temperature conditions requires that continual reaction takes place not only between the materials crystallizing at that instant but with all the solids already crystallized. The difficulty of the attainment of equilibrium under such conditions is obvious.

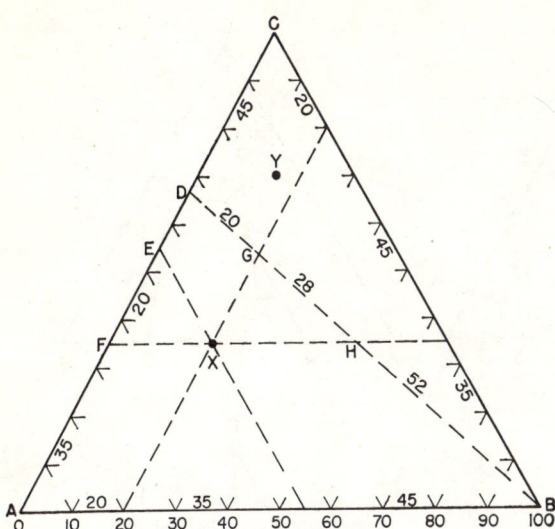

Fig. XI.—Representation of composition in a ternary system by means of triangular coordinates.

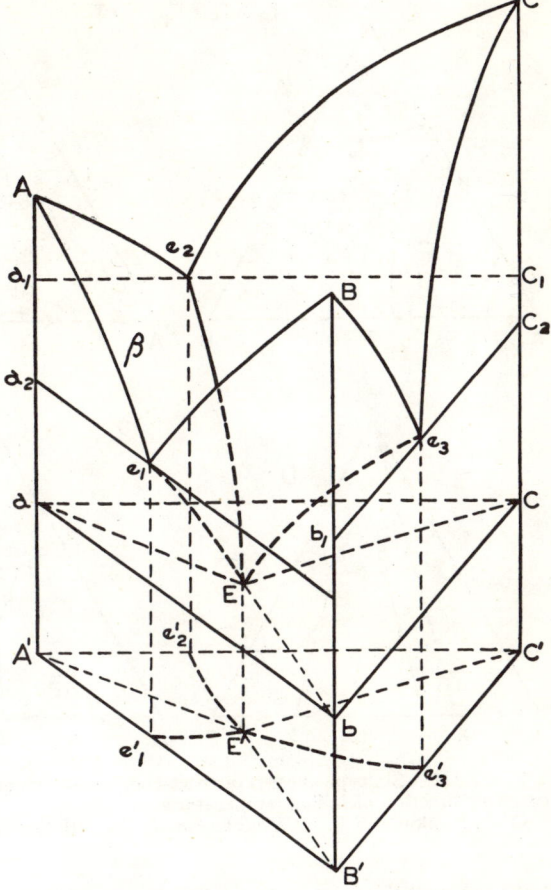

Fig. XII.—Perspective drawing of a space model of a ternary system with a simple eutectic and no ternary compound.
Modified from R. Vogel, Die heterogenen Gleichgewichte, in G. Masing, Handbuch der Metallphysik, Vol. II, Fig. 266, p. 370, 1937.

This method of expressing a composition in terms of three others by the appropriate construction, so as to designate the three proportions as segments of a line is not limited to equilateral triangles but is applicable, also, to scalene triangles. For example, point X in Fig. XI can be expressed in terms of A, B, and D on the line BD. The percentage of $B = DG \times 100/DB = 20$; the percentage of $D = HB \times 100/DB = 52$; and the percentage of $A = GH \times 100/DB = 28$. By rotating the line XE until it were parallel to BD, the proportions of A, D, and B in X could be determined, similarly, on the sides AD and AB.

As will be shown later, when dealing with subsystems or when tracing the course of crystallization of a liquid, it frequently becomes necessary to express an overall composition in terms of three others which do not form an equilateral triangle.

A triangular composition diagram also has the advantage that a series of additions of a third component to a mixture in any ratio of the other two components may be represented by a straight line from the apex of the third component. In Fig. XI for instance, additions of component B to a mixture of 33% A, 67% C (point D) all lie on the line D-B.

Temperatures can be represented by lengths perpendicular to the plane of the composition triangle and therefore cannot be shown directly on a two-dimensional surface, but the temperatures on one of the thermal surfaces (usually the liquidus surface) may be indicated for uniform temperature intervals by isotherms as are elevation contours on topographic maps. The actual solid diagram has an appearance like that of Fig. XII.

The liquidus surface is then a series of intersecting curved surfaces representing the primary phase fields of compounds in the system. A primary phase field of a congruently melting ternary compound is a domed surface, the highest elevation of which represents the melting point of the compound. Its field intersects that of an adjacent congruently melting compound in a sloping valley or boundary line. In the case of an incongruently melting compound the intersection of the primary phase field of the first with that of the second solid is a sloping terrace and not a valley.

(A) Ternary Systems Without Solid Solutions

(a) *Typical Cases*

In Fig. XIII, six typical cases of ternary systems are shown. Figure XIII(A) shows a system without either binary or ternary compounds present. The field 1-4-2-C represents the field of stability of component C in equilibrium with solution. Solid C is the primary phase for this area and is the last solid to disappear when any composition within this area is heated. It is also the first solid phase to appear when liquids represented by points in this area are cooled. Points 1, 2, and 3 are binary eutectics, while point 4 is a ternary eutectic. Lines 1-4, 2-4, and 3-4 are known as boundary curves, each of which represents a condition

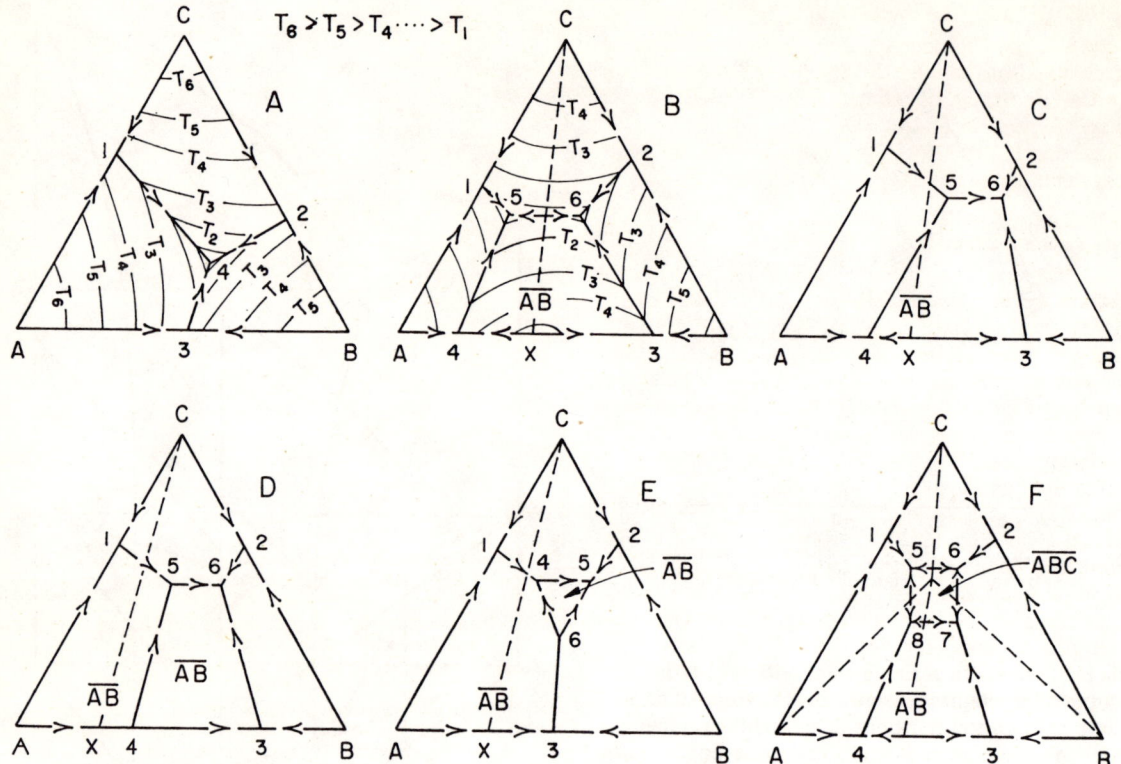

FIG. XIII.—Six typical cases of three-component systems; A and B show hypothetical isotherms. Arrows indicate the direction of falling temperatures.
G. A. Rankin and F. E. Wright, *Am. J. Sci.*, 4th Ser., **39**, 18 (1915).

of 3-phase equilibrium among two solid phases and liquid. The two solid phases at equilibrium along 1-4 are A and C, along 2-4 are C and B, and along 3-4 are A and B. Point 4 is a quadruple point in a condensed system at which solids A, and B, C are in equilibrium with solution.

In diagrams (B), (C), (D), and (E) of Fig. XIII an intermediate binary compound, $\overline{AB}$, is present. The straight line which joins this compound to the third component, C, of the ternary system is called an Alkemade line. Alkemade lines divide ternary systems into composition triangles. The final phases produced by equilibrium crystallization within one of these triangles are indicated by the apices of the triangle. For example, in diagrams of Figs. XIII(B), (C), (D), and (E), the final phases within the triangle A-$\overline{AB}$-C are crystalline A, $\overline{AB}$, and C. In XIII(F), where a ternary compound is present with the binary compound, conjugation lines form four subsidiary triangles.

Figure XIII(B) shows a ternary system where the binary compound $\overline{AB}$ is stable at its melting point. According to the theorem of Alkemade[11] if the line C-X crosses the line 5-6, the point of intersection will be a maximum on the boundary 5-6 and the points 5 and 6 will be eutectics and each composition triangle will behave as a true ternary system. However, if C-X does not cross 5-6 but intersects 1-5 and 4-5 (as in Fig. XIII-(C)), then only 6 will be the eutectic. In the case of

Fig. XIII(B), the line C-X forms a true binary system with components C and $\overline{AB}$. It divides the ternary system A-B-C into two ternary systems, each of which may be treated individually. In Fig. XIII(C), however the line C-X crosses the primary phase field of another compound, the composition of which does not lie on this line, and therefore the line does not describe a binary system.

In Fig. XIII(D) the composition of the binary compound, $\overline{AB}$, lies outside the field 4-5-6-3 because it has an incongruent melting point. In the binary system A-B it dissociates at a temperature corresponding to point 4 into solid A and liquid. In the ternary system A-B-C the compound $\overline{AB}$ is the primary phase in field 3-4-5-6 and is stable in this field.

In Fig. XIII(E), $\overline{AB}$ dissociates into solids A and B in the binary system A-B. In the ternary system compound $\overline{AB}$ has a stable field 4-5-6.

[11] "A theorem by Van Rijn Van Alkemade serves as a very effective guide in regard to temperature changes in the interior of the triangle. If the two points in the triangle which correspond to the composition of two solid phases be connected by a line, the temperature at which these same two phases can be in equilibrium with solutions and vapor rises as the boundary curve approaches this line, becoming a maximum at the intersection though the boundary curve often ceases to be stable before this point is reached." (W.D. Bancroft, "The Phase Rule," *J. Phys. Chem.*, **1**, 149 (1897))

In Fig. XIII(F), the system has a binary compound $\overline{AB}$ and a ternary compound, $\overline{ABC}$, each of which has a congruent melting point as their respective compositions are within or on the boundaries of the fields in which they are the primary phases.

These simple cases are also applicable to such complex systems as the one shown in Fig. XIV.

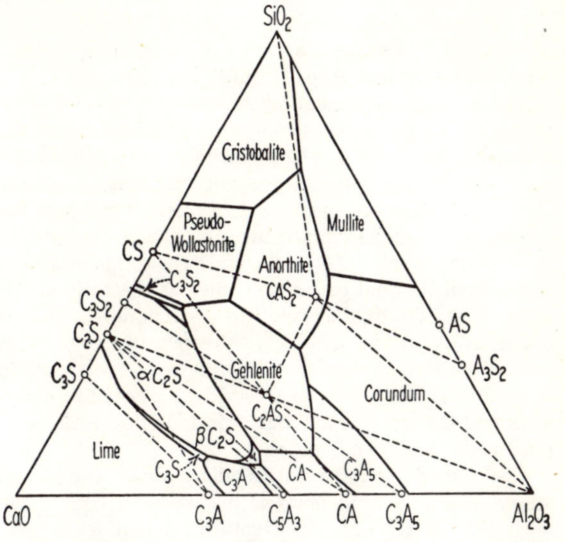

(b) *Crystallization Paths*[12]

(i) *Simple systems:* Geer[12(a)] states that "the crystallization curve denotes the locus of points which represent the compositions of the solutions formed on cooling any given solution from any given temperature to the temperature (quintuple point in case of ternary systems) at which it becomes solid, under the assumption that no phase is removed during the cooling." The relations of solid phases to liquid phases of any system that does not have solid solutions are known when the liquidus of the system is determined for all compositions. The liquidus is the temperature at which the first solid (primary phase) appears on cooling under equilibrium conditions. A knowledge of the crystallization curve or the melting curve (the reverse of the crystallization curve) for any particular melt is very valuable in the study of the firing of ceramic bodies. A few types of crystallization curves will be described using the diagrams and terms given by Andersen.[12(b)]

In Fig. XV, point m is the ternary eutectic and all crystallization curves of this system are terminated at this point. If a liquid of composition a is chosen and allowed to cool, the system remains liquid until the liquidus is reached, at which temperature the solid, A, begins to crystallize. The course of the crystallization curve from this point to boundary m-k follows a straight line drawn through A and a. This is true for all crystallization curves where solid solutions are not present. As the liquid changes in composition from a to b, solid A crystallizes. At b, a second phase appears and the crystallization curve follows boundary k-m with phases A and C crystallizing together. At point m, the temperature remains constant with solid phases A, B, and C crystallizing together until all liquid has disappeared. The final product will be a mixture of large crystals of A and C, and small crystals (eutectic mixture) of A, B, and C.

The composition of the solids crystallizing at any instant along b-m is given at the point where the tangent to the crystallization curve intersects that side of the composition triangle representing the two solid phases coexisting. For example, at b, it is indicated by the intersection of the tangent to the curve m-b-k at the point b with the line A-C at point b''. The ratio of A to C is given by the lever b''-C/b''-A. The mean composition of the two solid phases that have crystal-

---

[12] (a) W. C. Geer, "Crystallization in Three-Component Systems," *J. Phys. Chem.*, **8**, 257–87 (1904).
(b) Olaf Andersen, "System Anorthite-Forsterite-Silica," *Am. J. Sci.*, 4th Ser., **39**, 407 (1915).
(c) N. L. Bowen, "Ternary System, Diopside-Forsterite-Silica," *ibid.*, **38**, 207–64 (1914).

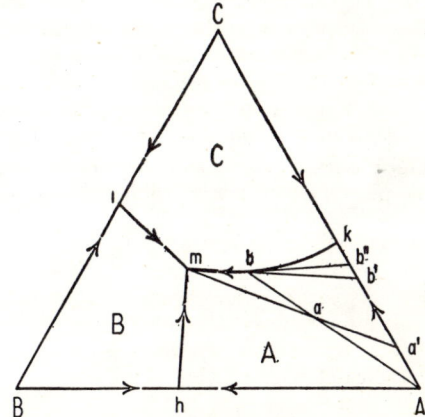

Fig. XIV.—Three-component system, Al_2O_3–CaO–SiO_2, showing (1) boundary curves as solid lines, (2) composition lines (Alkemade lines) as dashed lines. The final product of crystallization (on slow cooling) of ternary solutions of this system always consists of three solid phases whose fields of stability are adjacent. The same three solid phases will be the final product of crystallization from any solution whose composition lies within the triangle (composition triangle) formed by lines joining the compositions of these three phases. Note: In compound designations, $C = CaO$, $A = Al_2O_3$, $S = SiO_2$.
G. A. Rankin and F. E. Wright, *Am. J. Sci.*, 4th Ser., **39**, 52 (1915). (See Fig. 630 for revised and redrawn diagram.)

Fig. XV.—Course of crystallization in a simple ternary system
O. Anderson, *Am. J. Sci.*, 4th Ser., **39**, 427 (1915).

lized between points b and m is represented by the intersection of a line drawn through m and b and the side of the composition triangle at b'. In this case it is a mixture of A and C in the proportion b'-C/b'-A. The mean composition of the total solid which separates out between a and m (before B begins to crystallize) is determined by drawing a line through m and a to the side of the composition triangle at a'.

During eutectic crystallization at m, the composition of the total solids changes from a' to a, reaching the latter point as the last drop of liquid disappears.

The method of calculating the amounts of solid separating between various temperatures by the use of the lever rule as shown in the preceding paragraphs is the same as described in the case of the binary system. For example, in cooling a melt of original composition a from the liquidus temperature to point b, the ratio of the amount of solid A crystallized to the total amount of the system is equal to $a\text{-}b/A\text{-}b$ whereas the ratio of the amount of liquid of composition b remaining at b is equal to $a\text{-}A/A\text{-}b$. Similarly, the relative amount of solid of mean composition a' crystallized between a and m (before B has begun to crystallize) is equal to $a\text{-}m/a'\text{-}m$ and the relative amount of liquid of eutectic composition is equal to $a\text{-}a'/a'\text{-}m$.

The case of a ternary system with a binary compound stable at its melting point is discussed in connection with Fig. XIII(B). The crystallization in each subsystem would be treated in the same manner as the above case.

(ii) *System with ternary peritectic:* In Fig. XVI the ternary eutectic point m and the peritectic point, o, lie on the same side of the conjugation line, $C\text{-}\overline{AB}$, and the binary eutectic points, h and j, lie on opposite sides. The field of A ($A\text{-}j\text{-}o\text{-}k$) extends across the conjugation line, $C\text{-}\overline{AB}$, hence the system $C\text{-}\overline{AB}$ is not a true binary system. The temperature along line $o\text{-}m$ decreases according to the theorem of Alkemade toward m, and o is not a eutectic but a peritectic.

In cooling a melt of composition a, solid phase A crystallizes out along $a\text{-}b$. From b to o, A and $\overline{AB}$ crystallize together and as the total composition of the solid separated between a and o is given at a', there must be liquid left when o is reached.

The point a is within the composition triangle $A\text{-}\overline{AB}\text{-}C$, and the final products of crystallization must be these three phases. It is evident from the diagram that the three phases which are in equilibrium at o are A, $\overline{AB}$, and C. Therefore, the final solidification of composition a must take place at o and not at m since a lies in the composition triangle whose solid phase areas meet at o. During the final solidification at o, solid phases C and $\overline{AB}$ crystallize while some of the phase, A, is resorbed or dissolved.

The melt, c, on the conjugation line $C\text{-}\overline{AB}$ crystallizes as follows: Along $c\text{-}b$, A separates; along $b\text{-}o$, A and $\overline{AB}$ separate together; at o, C and $\overline{AB}$ separate and A completely dissolves, the final products of crystallization being only $\overline{AB}$ and C. Crystallization at invariant points frequently involves more than physical processes and may involve chemical reactions as well. At o, for example, with melt c the following chemical reaction must occur: A (solid) $+$ B (in liquid) $= AB$ (solid).

The melt, d, on the left of the conjugation line in the composition triangle $B\text{-}\overline{AB}\text{-}C$, crystallizes as follows: Along $d\text{-}e$, A separates; along $e\text{-}o$, C and A crystallize together; at o, A dissolves and $\overline{AB}$ is formed (temperature remains constant until all of A disappears); along $o\text{-}m$, $\overline{AB}$ and C crystallize together. At m, the final products of crystallization are $\overline{AB}$, B, and C, which is to be expected because the point d lies in the composition triangle $B\text{-}\overline{AB}\text{-}C$. The mean composition of the solid separating between d and o is represented by d'. At the moment at which all of A has disappeared at o and before crystallization begins to proceed along $o\text{-}m$, the mean composition of the solid has changed from d' to d'' and is composed of C and $\overline{AB}$. Along $o\text{-}m$, the mean composition of the solid changes from d'' to d''' and during the final crystallization at m from d''' to d.

(iii) *Systems with both binary and ternary peritectics:* In Fig. XVII, the quadruple points, h and j, and the quintuple points, o and m, lie on the same side of the Alkemade line, $C\text{-}\overline{AB}$. A melt of composition e crystallizes as follows: Along $e\text{-}b$, A separates; at b, the compound $\overline{AB}$ begins to crystallize and A to redissolve (the intersection

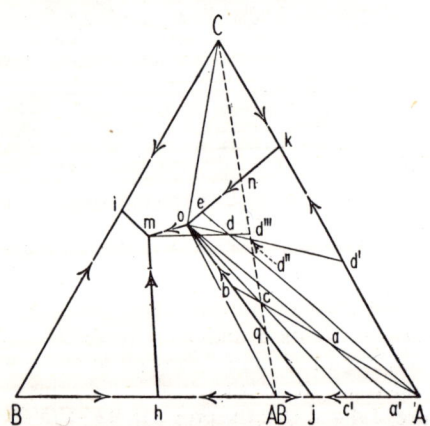

Fig. XVI.—Ternary system with a binary compound which does not form a binary system with the third component.
O. Andersen, *Am. J. Sci.*, 4th Ser., **39**, 431 (1915).

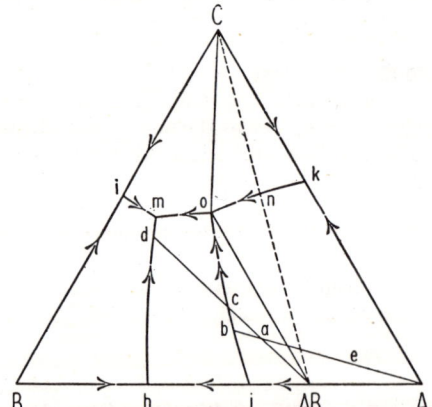

Fig. XVII.—Ternary system containing a binary compound unstable at its melting point.
O. Andersen, *Am. J. Sci.*, 4th Ser., **39**, 433 (1915).

of the tangent to line j-b-o intersects an extension of line A-$\overline{AB}$). This process continues until o is reached because all lines drawn through points on the curve b-o and point e intersect the line A-$\overline{AB}$. Final crystallization takes place when the composition of the liquid and the temperature of the system reaches o. In this case, A does not completely dissolve and the final products are A, $\overline{AB}$, and C. If the tangent to line j-o intersected the line A-$\overline{AB}$ and not its prolongation, the solid A would tend to increase in amount along j-o.

A melt of composition a which lies in the other composition triangle will crystallize as follows: From a to b, A separates; between b and c, $\overline{AB}$ separates and A dissolves (is resorbed), and at point c all of A has disappeared (the mean composition of total solid separated between a and c is represented by $\overline{AB}$). The point of mean composition moves from A to $\overline{AB}$ along line A-$\overline{AB}$ as the temperature falls from b to c. From c, the crystallization curve leaves the boundary j-o and continues across the field to d while $\overline{AB}$ is separating. From d to m, B and $\overline{AB}$ separate together, and at m, $\overline{AB}$, B, and C crystallize together. All melts in the field j-o-$\overline{AB}$ pass through the field h-m-o-j. The line j-o in this case is called an alteration curve and is indicated by double arrows.

(iiii) *Recurrent Crystallization:* Figure XVIII illustrates a case of recurrent crystallization. At point c in the crystallization of a melt of composition a, the phase A disappears and the crystallization curve follows the straight line to point d, at which A begins to separate again. Along d-o, A continues to crystallize and $\overline{AB}$ is slowly resorbed (the tangent to d-o intersects the extension of $\overline{AB}$-A). At point o, A dissolves with simultaneous separation of C and $\overline{AB}$, the temperature remaining constant until A is completely dissolved. Along o-m, C and $\overline{AB}$ separate together and the final products of crystallization at m are B, $\overline{AB}$, and C, as the original point lies within the composition triangle B-$\overline{AB}$-C.

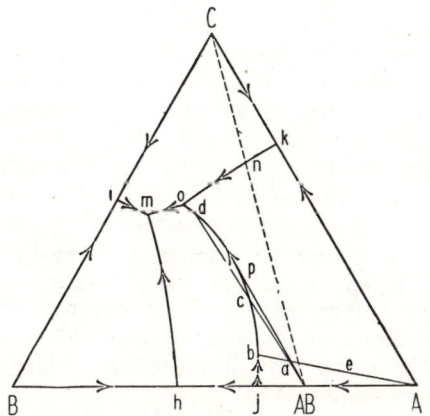

Fig. XVIII.—Recurrent crystallization in a ternary system.
O. Andersen, *Am. J. Sci.*, 4th Ser., **39**, 435 (1915).

A melt e crystallizes as follows: Along e-b, A separates; between b and p, $\overline{AB}$ crystallizes and A is partially resorbed; from p to d, the tangent to the crystallization curve, i.e., the tie line, intersects line A-$\overline{AB}$ and not an extension of it, therefore, A and $\overline{AB}$ separate together; from d to o, the tangent intersects an extension of $\overline{AB}$-A, hence A crystallizes again and $\overline{AB}$ is partially resorbed; at o, A is once more partially redissolved while $\overline{AB}$ and C crystallize out, and the final products of crystallization are A, $\overline{AB}$, and C.

(c) *Summary Relating to Crystallization*[13]

(1) When a liquid is cooled, the first phase to appear is the primary phase for that part of the system in which the composition of the melt is represented.

(2) The crystallization curve follows to the nearest boundary the extension of the straight line connecting the composition of the original liquid with that of the primary phase of that field. The composition of the liquid within the primary fields is represented by points on the crystallization curve. This curve is the intersection of a plane (passed perpendicular to the base triangle and passing through the compositions of original melt and the primary phase) with the liquidus surface.

(3) At the boundary line a new phase appears which is the primary phase of the adjacent field. The two phases separate together along this boundary as the temperature is lowered.

(4) Two conditions may appear that would alter the crystallization along the boundary. In one case the first phase will increase as is the case where the tangent to the boundary curve intersects the line connecting the compositions of the two phases separating or it will decrease (be resorbed) if this tangent intersects the prolongation of this line. In the other case the crystallization curve will leave the boundary curve when the first phase has become completely resorbed leaving only the second solid phase. This may be inferred from a study of the mean composition of the solid separating between successive points on the crystallization curve.

(5) The crystallization curve always ends at the invariant point which represents equilibrium of liquid with the three solid phases of the three components within whose composition triangle the composition of the original liquid was found.

(6) The mean composition of the solid which is crystallizing at any point on a boundary line is shown by the intersection of the tangent at that point with the line joining the composition of the two solid phases which are crystallizing at the given point.

(7) The mean composition of the total solid which has crystallized to any point on the crystallization curve is found by extending the line connecting the given point with the original liquid composition to the line connecting the compositions of the phases that have been separating.

---

[13] See G. A. Rankin and F. E. Wright, pp. 51–69, under "Methods and Techniques, General," of Selected Bibliography, Part V. See also Andersen, reference 12(b).

(8) The mean composition of the solid that has separated between two points on a boundary is found at the intersection of a line passing through these two points with the line connecting the compositions of the two solid phases separating along this boundary.

(d) Alternate Method for Determining Phase Composition

An alternate method for determining phase composition during the course of crystallization is based on the construction of similar triangles and has the advantage that the proportions of three phases in equilibrium with each other can be represented as segments of a straight line. The method has a further advantage in that it is simple and easily remembered, which is important to those who have only occasional recourse to the use of phase diagrams.

The method will be demonstrated for a system having a binary compound with a ternary field but without a binary field. Figure XIII(E) is such a ternary system, and the binary system containing a compound without a binary field is shown in Fig. V(A). In Fig. XIII(E), the binary compound $\overline{AB}$ decomposes into the components A and B below the eutectic (point 3); however, with addition of the third component, C, the liquidus values in the ternary system have been lowered to the point where the compound $\overline{AB}$ possesses a small primary field. Point 6 (Fig. XIII(E)) represents the unique temperature and composition at which ternary liquid can exist in equilibrium with substances A, B, and $\overline{AB}$. Point 6 is also the highest temperature at which $\overline{AB}$ can exist in equilibrium with a ternary solution.

The phase composition during the course of crystallization of the melt P_1 of Fig. XIX will now be considered. In Fig. XIX the path followed by the liquid during the process of crystallization is shown by the hatched line. From P_1 to a_1 solid B crystallizes; B, P_1, and a_1 are colinear, and the amounts of liquid and B at any point along the path P_1 to a_1 can be determined by the "Lever Rule," as described earlier. At point a_1, compound $\overline{AB}$ starts to crystallize and the composition of the liquid moves along the path a_1 to e. At any point along the path, e.g., at a_2, three phases are in equilibrium, namely, B, $\overline{AB}$, and liquid of composition a_2. These three phases, moreover, must be equivalent in composition to the original melt, P_1. Consequently, if the line P_1-p_1 be constructed parallel to the line a_2-$\overline{AB}$, and the line P_1-p_2 parallel to a_2-B, the proportions of the three phases are found along the line $\overline{AB}$-B, as $(\overline{AB}$-$p_1)/(\overline{AB}$-$B)$ of B, $(p_2$-$B)/(\overline{AB}$-$B)$ of $\overline{AB}$, and $(p_1$-$p_2)/(\overline{AB}$-$B)$ of liquid a_2.

At point e, at maximum heat content, that is, before any C has started to crystallize, the proportions of B, $\overline{AB}$, and liquid e, can be determined by constructing the line P_1-p_3 parallel to e-$\overline{AB}$ and the line P_1-p_4 parallel to e-B. The phase composition is then given along the line $\overline{AB}$-B as $(\overline{AB}$-$p_3)/(\overline{AB}$-$B)$ of B, $(p_4$-$B)/(\overline{AB}$-$B)$ of $\overline{AB}$, and $(p_3$-$p_4)/(\overline{AB}$-$B)$ of liquid e.

As P_1 lies within the compatibility triangle formed by $\overline{AB}$, C, and B, crystallization is completed at e, which point represents the equilibrium between the three named phases. At point e, at minimum heat content, that is, when the last trace of liquid has crystallized, the proportions of $\overline{AB}$, C, and B may be found by the appropriate construction of triangle p_5-P_1-p_6 similar to triangle $\overline{AB}$-C-B. The phase composition is then given as $(\overline{AB}$-$p_5)/(\overline{AB}$-$B)$ of B, $(p_6$-$B)/(\overline{AB}$-$B)$ of $\overline{AB}$, and $(p_5$-$p_6)/(\overline{AB}$-$B)$ of C.

The increase in amount of B during crystallization along the path from a to e can be seen both qualitatively and quantitatively by comparing the lengths of the segments $\overline{AB}$-p_1, $\overline{AB}$-p_3, and $\overline{AB}$-p_5. By a similar comparison of the segments p_2-B, p_4-B, and p_6-B the changes in amount of solid $\overline{AB}$ can be followed. The segments p_3-p_5, p_5-p_6, and p_6-p_4 represent the amounts of solids B, C, and $\overline{AB}$, respectively, formed from the eutectic liquid e during the change from maximum to minimum heat content at constant temperature.

The phase composition at the various selected points along the path of crystallization is given in the table associated with Fig. XIX.

The course of crystallization of the melt P_2, Fig. XX will next be considered. From P_2 to b_1, solid A crystallizes, and A, P_2, and b_1 are colinear. Along the path b_1 to b_2, compound $\overline{AB}$ crystallizes, and the two solid

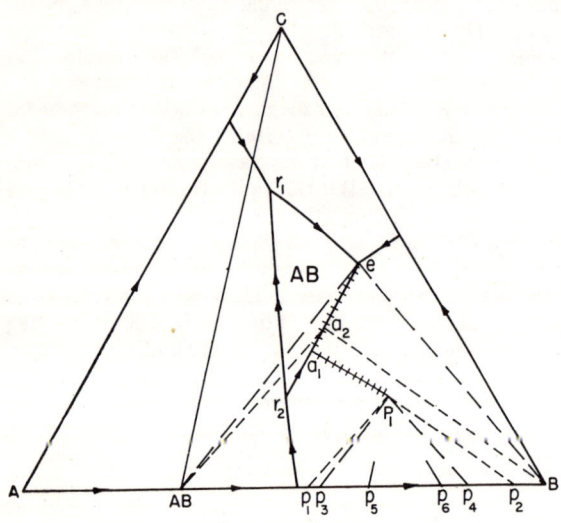

POINT	PHASE COMPOSITION, WT. FRACTION				
	A	B	C	AB	LIQ.
a_1		$\dfrac{P_1-a_1}{B-a_1}$			$\dfrac{B-P_1}{B-a_1}$
a_2		$\dfrac{AB-p_1}{AB-B}$		$\dfrac{p_2-B}{AB-B}$	$\dfrac{p_1-p_2}{AB-B}$
e(Max.H)		$\dfrac{AB-p_3}{AB-B}$		$\dfrac{p_4-B}{AB-B}$	$\dfrac{p_3-p_4}{AB-B}$
e(Min.H)		$\dfrac{AB-p_5}{AB-B}$	$\dfrac{p_5-p_6}{AB-B}$	$\dfrac{p_6-B}{AB-B}$	

Fig. XIX.—Composition of phase assemblages for a simple crystallization path in a system having a binary compound with a ternary field but without a binary field.

phases A and $\overline{AB}$ are in equilibrium with liquid of varying composition as represented along the line b_1-b_2. Determination of the amounts of the three phases present at any point along the crystallization path is similar to that for the melt P_1. At point b_2 the side b_2-$\overline{AB}$ of the triangle A-b_2-$\overline{AB}$ is coincident with the side P_2-$\overline{AB}$ of the similar triangle l_1-P_2-$\overline{AB}$. Thus, along the path b_1 to b_2, solid A has been completely resorbed, as the segment of line representing the amount of A has been reduced to zero. It is evident that crystallization along the A, $\overline{AB}$, boundary cannot proceed beyond b_2 (toward r_1), for then P_2 would lie outside the triangle formed by the equilibrium phases A, $\overline{AB}$, and liquid, and no positive combination of these phases could give the composition P_2. At b_2, therefore, the crystallization path cuts across the $\overline{AB}$ field as $\overline{AB}$ continues to crystallize, and $\overline{AB}$, P_2, b_2, and b_3 are colinear. At b_3 solid C starts to crystallize and the melt proceeds along the path b_3 to e. At any point along the path, b_4, e.g., C, $\overline{AB}$, and liquid of composition b_4 are in equilibrium. If the triangle l_2-P_2-l_3 be constructed similar to triangle $\overline{AB}$-b_4-C, then $(\overline{AB}$-$l_2)/(\overline{AB}$-$C)$ = fraction of C, $(l_3$-$C)/(\overline{AB}$-$C)$ = fraction of $\overline{AB}$, and $(l_2$-$l_3)/(\overline{AB}$-$C)$ = fraction of liquid. As P_2 lies within the crystallization triangle $\overline{AB}$-B-C, crystallization will be completed at e, as for melt P_1 (Fig. XIX), and the proportion of phases present at maximum and minimum heat content can be determined as has been illustrated. The table associated with Fig. XX gives the phase composition at the various points considered.

(e) Summary of Alternate Method

This alternate method for determining phase composition can be summarized as follows: The composition of three phases (usually but not necessarily, two solids and a liquid) in equilibrium at any point along the crystallization path are connected together with straight lines to form a triangle. If the composition of the original melt does not fall within this triangle, an inconsistency exists for it is not now possible to describe the original melt in terms of positive quantities of the separated phases. Using the composition of the original melt for one vertex and any convenient side of the phase triangle for a base, another triangle is constructed similar geometrically to the phase triangle. The side of the phase triangle selected as a base will then be divided into three segments proportional to the amounts of the three phases.

(B) Ternary Systems with Solid Solutions

In ternary systems it is necessary to locate boundary curves, isotherms, and the compositions of the phases crystallizing in order to determine the course of crystallization within the system. In systems without solid solutions where the phases are invariable in composition, the crystallization curve within the primary phase field of a compound follows the extension of the straight line connecting the composition of the mixture and the primary phase until the line intersects a boundary curve. In ternary systems containing solid solutions, however, the crystallization curve within the primary phase field of the solid solution is no longer a straight line because the composition of the primary phase continually changes as crystallization proceeds. In determining the crystallization curve in a ternary system containing solid solution, it is necessary to know the position of the boundary curves, the isotherms, and the tie lines[12(c)] for numerous points on the isotherms.[14]

(a) The Solid Solution Diagram

A simple case of solid solution in ternary systems as discussed by Bowen is shown in Figs. XXI(A) and (B). In the binary system A-B of Fig. XXI(A), the tie lines, 1'-5' and similar ones parallel to the base, give the composition of the solid phase in equilibrium with the liquid phase at the corresponding temperature. The line 5-4-6 (Fig. XXI(B)) is the isotherm for the temperature T_1, and line 5'-4'-6' is the isotherm for the temperature T_2.

[14] The tie lines have been referred to by some writers, particularly Schreinemakers,[8] as solid-phase indicating lines because they indicate the composition of the solid phase in equilibrium with a definite liquid phase. The tie lines must be determined experimentally; they cannot be constructed. In special cases where the extremities of the tie lines or solid-phase indicating lines are on the boundary lines of the three-phase system, they are designated as three-phase boundaries

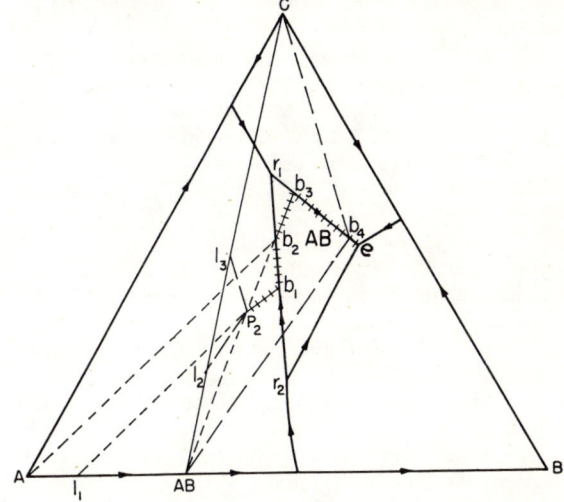

POINT	PHASE COMPOSITION, WT. FRACTION				
	A	B	C	AB	LIQ.
b_1	$\dfrac{P_2-b_1}{A-b_1}$				$\dfrac{A-P_2}{A-b_1}$
b_2				$\dfrac{A-l_1}{A-\overline{AB}}$	$\dfrac{l_1-\overline{AB}}{A-\overline{AB}}$
b_2 Alternate				$\dfrac{P_2-b_2}{\overline{AB}-b_2}$	$\dfrac{\overline{AB}-P_2}{\overline{AB}-b_2}$
b_3				$\dfrac{P_2-b_3}{\overline{AB}-b_3}$	$\dfrac{\overline{AB}-P_2}{\overline{AB}-b_3}$
b_4			$\dfrac{\overline{AB}-l_2}{\overline{AB}-C}$	$\dfrac{l_3-C}{\overline{AB}-C}$	$\dfrac{l_2-l_3}{\overline{AB}-C}$

FIG. XX.—Composition of phase assemblages for a crystallization path showing resorption of (A) in a system having a binary compound with a ternary field but without a binary field.

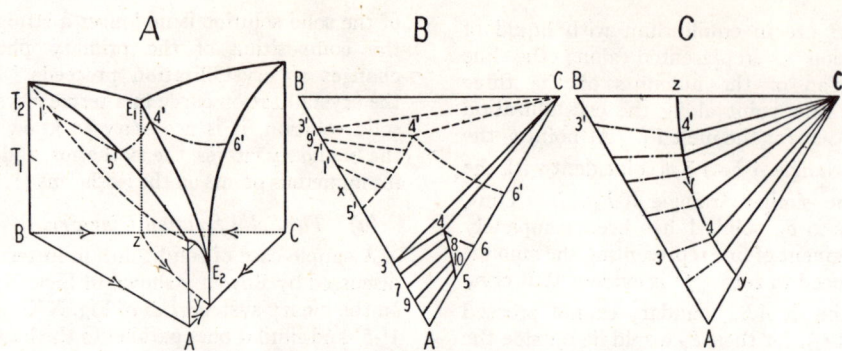

Fig. XXI.—(A) Melting surface of a three-component system in which one of the binary systems, A-B, has solid solutions present. In the ternary system these binary solid solutions exist as the primary phase to boundary E_1-E_2.

Modified from N. L. Bowen, *Am. J. Sci.*, 4th Ser., **38**, 221 (1914).

(B) Sections produced by planes passed parallel to base triangle at the temperatures, T_1 (solid lines) and T_2 (dotted lines), and tie lines at these temperatures.

(C) Series of three-phase boundaries (tie lines) at temperatures from E_1 to E_2; solid-solution tie lines are broken and single-crystal, C, tie lines are solid; tie lines 3-4-C and 3'-4'-C taken from Fig. XXIB.

The phases present at T_1 (Fig. XXI(B)) are as follows:

(1) Area 5-4-6, all liquid.

(2) Area A-5-4-3, liquid plus solid solution; composition of liquids on line 4-5 and solid solutions on line 3-A.

(3) Area 6-4-C, liquid and solid C; composition of liquid on 4-6.

(4) Area 3-4-C, liquid 4, solid solution 3, and solid C.

(5) Area 3-C-B, solid C and solid solution; compositions of solid solutions on line 3-B.

The phases present at T_2 (Fig. XXI(B)) are as follows:

(1) Area A-5'-4'-6', all liquid.

(2) Area 5'-4'-3', liquid plus solid solution; compositions of liquids on line 4'-5', compositions of solid solutions on line 3'-5'.

(3) Area 6'-4'-C, liquid and solid C; compositions of liquids on line 4'-6'.

(4) Area 3'-4'-C, liquid 4', solid solutions 3', and solid C.

(5) Area 3'-C-B, solid C and solid solution; compositions of solid solutions on line 3'-B.

Point 4 represents the composition of liquid in equilibrium at temperature T_1 with two solid phases, namely, solid solution 3 and solid C. Point 4' represents the composition of the liquid in equilibrium at temperature T_2 with the two solid phases, C and solid solution 3'. Both 4 and 4' lie on the boundary, E_1-E_2, of Fig. XXI(A). The extremities of lines 3-4 and 3'-4' lie on boundaries (lines E_2-E_1 and A-B) of the ternary system. These tie lines are two of the infinite number of possible three-phase boundaries, some of which are shown in Fig. XXI(C). Line 7-8 is the simple type of tie line showing the composition of solid solution 7 in equilibrium with liquid 8 at temperature T_1.

Melts represented by a composition in area z-y-C on crystallizing follow a radial line from C with phase C separating until boundary z-y is reached at which point of intersection a solid solution begins to separate; the composition of the solid solution is represented by the other extremity of the three-phase boundary at that temperature. For example, a melt whose composition lies on the line C-4 will reach the boundary curve at 4 and the composition of the first solid solution to appear is given at point 3. As the temperature drops the liquid changes in composition along y-z toward y while at the same time the solid solution is changing along B-A toward A and the solid C remains unchanged in composition. All solutions do not reach y before complete solidification takes place, as the original composition of the melt determines the final solidification temperature.

(b) *Mechanics of Crystallization*

(i) *No compounds present:* To aid discussion of crystallization paths in ternary systems with solid solutions, a hypothetical diagram is given in Fig. XXII. This diagram is the two-dimensional drawing of Fig. XXI(A) on which the following are shown: (1) isotherms, T_1, T_2, T_3, etc. (not related to designations in Fig. XXI), (2) representative tie lines for each of the isotherms in the solid-solution area B-E_1-E_2-A, and (3) representative crystallization paths (d-e, a-k-3, k-m, and f-4). It is to be emphasized again that in solid solution systems, the crystallization paths are curved lines. The compositions of the solid phase and the liquid existing in equilibrium with it for any respective temperature are joined by a tie line. For example, the solid in equilibrium with the liquid 1 on the isotherm T_5 is shown at 1' and the solid in equilibrium with a liquid of composition 2 on isotherm T_4 is shown at 2'.

As an example of crystallization, the composition a on isotherm T_5 is selected. The crystallization curve for this point is a-k-3. Upon cooling this melt, the first solid appears at temperature T_5 and is represented at the other extremity of the tie line passing through point a at temperature T_5. This tie line is shown as a dot-dash line and the composition of the solid solution is a'. As the temperature is lowered the trace of the compositions of the liquid phase follows the crystallization curve from a to 3 and thence the boundary line from 3 to c.

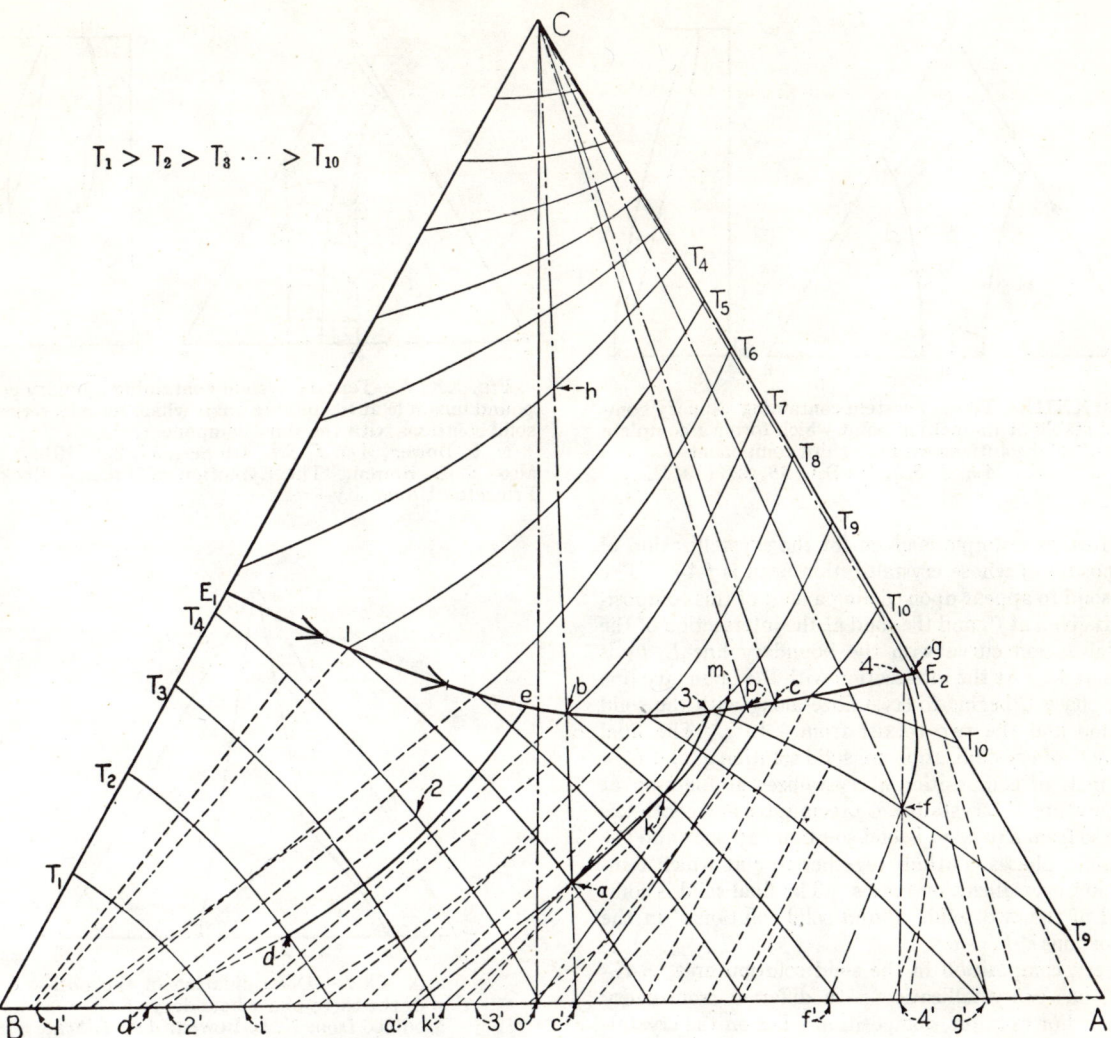

Fig. XXII.—Course of crystallization in a ternary system where one of the binary systems is a complete solid solution series.

The compositions of the solid solutions in equilibrium with the liquids during crystallization from a to 3 are represented by the series of tie lines connecting points on the crystallization curve with the corresponding solid solutions; all of these tie lines pass through a, for example, a-a', k-a-k', and 3-a-3'. But, during cooling from 3 to c, the tie lines do not pass through a; instead they pass to the right of a.[15]

---

[15] In any system, the summation of the phases present equals the composition of the mass being studied. During cooling of composition a from a to 3, the sum of the liquid plus the solid solution always must equal a; hence, the line (i.e., tie line) connecting the compositions of the two coexisting phases must pass through a. During cooling from 3 to c, a third phase (solid C) is present, therefore, the complete composition is expressed by a triangle rather than a straight line. For example, at temperature p which is between 3 and c, the two solid phases, C and solid solution o, and liquid p coexist; hence, the composition a is represented by corners of the triangle C-o-p, and the tie line o-p does not pass through a.

The crystallization will end when the temperature of tie line c-c' has been reached; this tie line being the one which passes through the point where the extension of line C-a intersects the boundary line, A-B; that is, line C-a-c' connects the compositions of the two final solid phases. The line c-c' is the tie line (three-phase boundary line) for temperature c and the triangle c'-c-C is of the same type as 3-4-C in Fig. XXI(B).

During the cooling period from 3 to c, crystalline C and solid solution are precipitated together, the solid solution changing in composition from 3' to c', and the liquid from 3 to c. The final products of crystallization are, therefore, crystalline C and the solid solution of composition c', with the ratio of C to solid solution being given by the lever reaction c'-a/C-a.

The above example shows that in solid solution systems the end point of the crystallization is not determined by a eutectic but stops at a point determined by the join passing through the composition of the original melt.

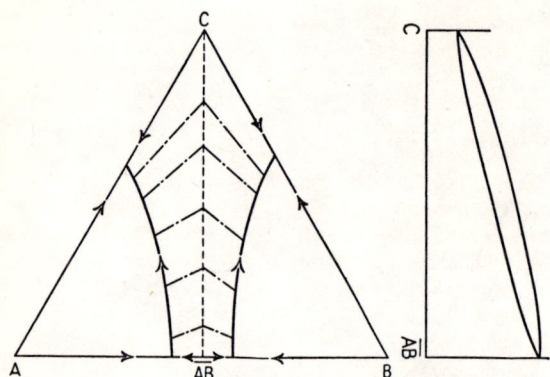

Fig. XXIII.—Ternary system containing a binary compound stable at its melting point which forms a complete series of solid solutions with the third component.
N. L. Bowen, *Am. J. Sci.*, 4th Ser., **38**, 223 (1914).

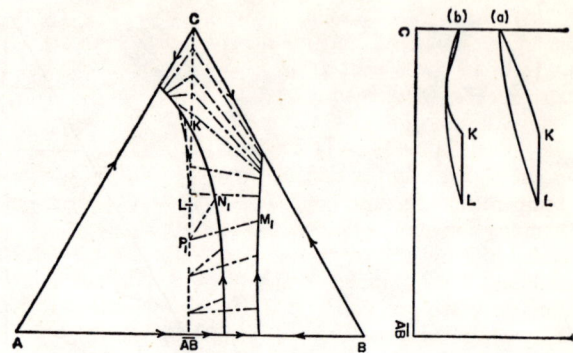

Fig. XXIV.—Ternary system containing a binary compound unstable at its melting point which forms a series of solid solutions with the third component.
N. L. Bowen, *Am. J. Sci.*, 4th Ser., **38**, 225 (1914); see also N. L. Bowen, "The Evolution of Igneous Rocks," Princeton University Press.

A further example is given for the crystallization of composition f whose crystallization path is f-4-g. The first solid to appear upon cooling a melt of this composition is given at f', and the solid at the intersection of the crystallization curve with the boundary line E_1-E_2 is given at $4'$. At the intersection with the boundary line at 4, phase C begins to crystallize along with the solid solution and the two coexist from 4 to g. The final products of crystallization are solid solution g' and C.

A melt of composition h crystallizes as follows: at temperature T_4, crystalline C precipitates and continues to do so from h to b; at b solid solution i appears and the two solid phases continue together to c at which temperature crystallization ceases. The final solid is composed of the two solids, C and solid solution c' in the proportions c'-h/C-h.

Every composition in the solid solution area, B-E_1-E_2-A, has a crystallization path different from every other. For example, composition k lies on the crystallization path of composition a, but its crystallization path is along k-m and not k-3.

(ii) *Congruent melting binary compound:* Figure XXIII illustrates a case where the compound, $\overline{AB}$, forms a complete series of solid solutions with the component C. This case can be considered as two ternary systems of the type represented in Fig. XXI(A), namely, systems C-$\overline{AB}$-A and C-$\overline{AB}$-B.

(iii) *Incongruent melting binary compound:* In the case where the compound, $\overline{AB}$, is unstable at its melting point (Fig. XXIV) the system cannot be treated in the same manner as the last case. Mixtures along C-K behave as true binary mixtures, but from K to $\overline{AB}$, the crystallization is ternary. There are two possibilities in this region of the ternary system as are shown in Fig. XXIV at (a) and (b), that is, there may be a simple solid solution or a solid solution with a minimum. Mixtures in the region K to L behave on melting as true binary mixtures until the temperature of the isotherm passing through K is reached, at which point solid A begins to separate from the solid solution. Mixtures between L and $\overline{AB}$ melt at temperatures above that of point K with the separation of pure A. The broken lines in Fig. XXIV are the three-phase boundaries for this type of system. The diopside-forsterite-silica system (Fig. 608) is of the type just described.

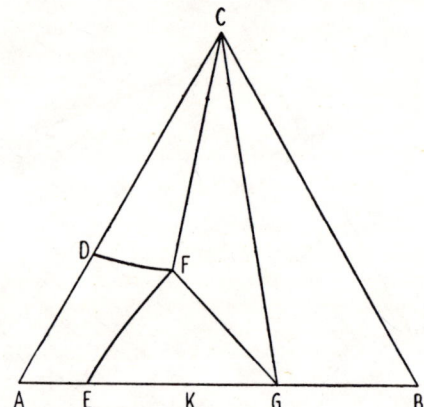

Fig. XXV.—Determination of the position of the three-phase boundary, F-G.
Modified from N. L. Bowen, *Am. J. Sci.*, 4th Ser., **40**, 171 (1916).

(c) *Determining Three-Phase Boundaries*

In determining the position of the three-phase boundaries shown on Fig. XXI(C), it is necessary to determine the physical properties of the solid solutions present. In Fig. XXV, the lines D-F and F-E are the isotherms for another system, A-B-C, at temperature T (similar to lines $4'$-$5'$ and $4'$-$6'$ of Fig. XXI(B)). The phases present in the areas are as follows:

(1) Area D-F-E-A, all liquid.
(2) Area E-F-G, liquid and solid solution; compositions of liquids on line E-F, and composition of solid solutions on line K-G (solid solution G is in equilibrium with liquid F at this temperature).
(3) Area C-D-F, liquid plus solid C; composition of liquids on line D-F.
(4) Area C-F-G, liquid F, solid solution G, and solid C.

(5) Area *B-G-C*, solid solution plus solid *C*; composition of solid solutions on line *B-G*.

In order to be able to predict the phases present at any specific temperature, such as *T*, it is necessary to know the position of the line, *F-G*. If the isotherms are known, this line can be fixed by determining only point *G*, the composition of the solid solution at that temperature, because point *F* is already known. This composition may be determined by optical methods provided the optical properties, especially the refractive indices, of the complete series of solid solutions in the binary system are known. X-Ray diffraction methods may be similarly used provided the diffraction constants of the binary solid solutions are known. Lines *C-F* and *C-G* also bound the three-phase field *C-F-G*, and they are merely straight lines radiating from *C*.

A three-phase boundary may also be determined by starting with a mixture of known composition and (1) determining the temperature at which the three-phase area is entered by cooling the melt from a higher temperature to a lower temperature or (2) determining the temperature of beginning of melting by starting with a mixture of solid *C* (Fig. XXV) and a solid solution of known composition lying between *A* and *B*.

(4) Multicomponent Systems

(A) General

The phase rule, of course, does not limit the number of components that may compose a system, but as the number of components increases, the complexity of the system mounts. Furthermore, for systems of more than three components a simple, convenient graphical representation of equilibrium relations is no longer possible.

The complete phase diagram of a condensed binary system may be represented on a plane surface as the two available dimensions of a plane suffice to permit the plotting of the two independent variables of composition and temperature.

A condensed ternary system is visualized usually as a triangular prism, in which the variables of composition (two independent and one dependent) are represented on an equilateral triangular base, and the independent variable of temperature on an axis vertical to the triangular variable of temperature on an axis vertical to the triangular base (Fig. XII). Three binary systems constitute the boundaries, or outside faces, of the triangular prism. From the binary invariant points (eutectics and peritectics) univariant curves, possessing one degree of freedom, penetrate the interior of the triangular prism. These univariant curves are formed by the intersection of two divariant primary phase surfaces, and along the univariant curves two solid phases are in equilibrium with liquids whose compositions lie on the boundary curves. The junction of three univariant curves forms a ternary invariant point corresponding to the single composition and temperature of a liquid in equilibrium with three solid phases.

In practice, the ternary system is drawn as a plane equilateral triangle, the base of a prism, upon which liquidus temperatures of the primary phase fields are projected as contours (isotherms). In such a representation, the unique liquidus temperature for every composition, a dependent variable, has been chosen as a specialized condition for projection on the composition plane. Fortunately, the physical nature of systems is such that by proper interpretation of the ternary "map," equilibrium relations at any other temperature within the melting range are easily obtainable, provided that solid solutions are not present. If, however, solid solutions exist within the system, the compositions of two coexisting phases must be indicated for every composition and temperature within the solid solution region, namely, that of the liquid solution and that of the coexisting solid solution. For the case of solid solutions, therefore, representation of equilibrium relations on a plane surface may be extremely complicated.

The complete graphical representation of a condensed quaternary system, or one showing the relationships between temperature and all possible mixtures of four components, requires four dimensional space. All possible mixtures of four components, however, may be depicted by a regular tetrahedron, each apex of which represents 100% of one component. The tetrahedron is subdivided into various primary phase volumes, which are separated, in general, by curved surfaces. Each primary phase volume represents the composition of all quaternary liquids that can exist in equilibrium with one particular solid phase. Boundaries between two primary phase volumes are quaternary divariant curved surfaces, in which lie the compositions of all quaternary liquids that can exist in equilibrium with two particular solid phases. The intersections of three divariant curved surfaces, the common boundaries of three intersecting primary phase volumes, form univariant curved lines, on which lie the compositions of quaternary liquids that can exist in equilibrium with three particular solid phases. Finally, quaternary invariant points,[16] each representing the composition of the liquid in equilibrium with four particular solids, are formed by the intersection of four univariant curves, each of which has been formed by three divariant surfaces, each of which, in turn, is the common boundary of two adjoining primary phase volumes. Six curved surfaces intersect at a quaternary invariant point.

Four ternary systems limit the quaternary system, and they are represented by the four equilateral triangles forming the faces of the tetrahedron. The quaternary univariant curves end (or originate) in the ternary invariant points; the quaternary divariant surfaces end (or originate) in the ternary univariant curved lines.

In a condensed ternary system, forming binary and/or ternary compounds, Alkemade lines (joins connecting the composition points of primary phases whose fields are adjacent) divide the equilateral triangle into smaller triangles. The substances indicated at the three vertices of these smaller triangles designate the final products of crystallization at the solidus for all

---

[16] An invariant point may be a eutectic point, an inversion point, or a reaction point, depending upon the nature of the solid phases and the temperature distribution along the univariant lines which meet at the point.

compositions within the triangular field. If a univariant boundary curve intersects its Alkemade line, it always does so at a maximum temperature on the boundary curve. If the boundary curve does not intersect the Alkemade line, then the maximum on the boundary curve is that end which if prolonged would intersect the Alkemade line.

A condensed quaternary system, similarly, may be divided into smaller tetrahedra by triangular planes connecting the compositions of all primary phases having adjoining liquidus regions. The final products of crystallization at the solidus for all compositions within each tetrahedron are designated by the four phases forming the apices of the tetrahedron. Analogous to the case for ternary systems, the intersection of a quaternary univariant curve with its composition plane always represents a temperature maximum on the univariant curve; or the maximum temperature on the univariant curve is that end which if extended would intersect the appropriate composition plane.

Inasmuch as no independent axis is available for the temperature variable in a condensed quaternary system represented by a compositional tetrahedron, isotherms are depicted as a series of intersecting curved surfaces through the various primary phase volumes. They cannot be represented in a simple manner on a plane diagram. An isothermal section may be represented as a tetrahedron, but mapping difficulties are still inherent in this approach. The above discussion has been based on fixed compositions of the crystallizing phases and excludes solid solution, which further complicates the determination and the mapping of the equilibrium relations.

A five component condensed system illustrates the increasing difficulties involved in studying multicomponent systems composed of more than four components, because for such cases the independent variables of composition can no longer be represented on a single model. Ten basic binary systems form the simplest boundary systems. These ten binary systems, in groups of three, form the boundaries for ten different ternary systems, which, in turn, in groups of four, form the boundaries for the five quaternary systems. Finally, the five quaternary systems define a region in five-dimensional space in which lie the primary phase regions and invariant points. Five-dimensional space can neither be portrayed or even visualized, and it becomes necessary to resort to mathematical concepts.[17]

Figure XXVI shows in tabular form the interrelations of sub-regions both within a given system and between boundary systems and is designed to clarify some of the relations already discussed, as well as to indicate those in more complicated systems. The figure gives (numbers in the squares) the number of solid phases in equilibrium with liquid for regions having different numbers of degrees of freedom, in 2, 3, 4, 5, or n component systems. The number of solid phases is derived simply from the phase rule formula, $P = C + 1 - F$, bearing in mind that the number of solid phases is $P-1$, inasmuch as a liquid phase has been specified as being present in all cases. The sub-regions designated as 0 solid phases are, of course, regions corresponding to the presence of liquid alone.

In the geometrical representation of degrees of freedom, an invariant condition, zero degrees of freedom, is always represented by a point, whether it be located in a 2, 3, 4, or n dimensional system. A univariant condition, one degree of freedom, is always depicted by a curve, which may, also, lie in a system having any number of dimensions. Similarly, a divariant condition, with two degrees of freedom, may be represented on a surface; and a trivariant condition, with three degrees of freedom, within a volume. Variance of greater than three degrees of freedom cannot be represented physically on a single model.

The number in parentheses at the base and to the left of each vertical arrow in Fig. XXVI indicates the number of regions designated below the arrow that intersect to form the region of next lower variance, indicated above the arrow. For example, in a quinary system the intersection of two, nonrepresentable, four-dimensional primary-phase regions forms a trivariant region wherein lie compositions for which two particular solid phases are in equilibrium with quinary liquids. Three of these trivariant regions intersect in a divariant surface in which three particular solid phases are in equilibrium with quinary liquids. Four divariant surfaces intersect in a univariant boundary curve, along which four solid phases are in equilibrium with quinary liquids. Finally, five univariant curves meet in an invariant point, corresponding to the composition and temperature of liquid in equilibrium with five solid phases.

Figure XXVI also shows the relations between regions in a system and its boundary systems, as indicated by the diagonal arrows. Thus, a curve in a quaternary system may end (or originate) in a ternary invariant point of the boundary system; a volume in a quinary system may end (or originate) in quaternary divariant surfaces, which, in turn, may end (or originate) in univariant curves of ternary boundary systems, etc. It should be noted at this point, however, that regions do not necessarily originate or terminate in boundary systems. As an example, a quaternary univariant curve may connect two quaternary invariant points, in which case, also, the three intersecting divariant surfaces (and volumes) do not contact boundary systems along the univariant curve in question.

From inspection in Fig. XXVI of the numbers (in parentheses) beside each vertical arrow, indicating the relation between two different subregions within a system, it may be observed that the values are prescribed by the mathematical law of combination. Each value is the number of sets of s_2 solid phases (in the region below the arrow) that can be formed from s_1 solid phases (in the region above the arrow). The

---

[17] See Mathematical Treatment in Selected Bibliography.

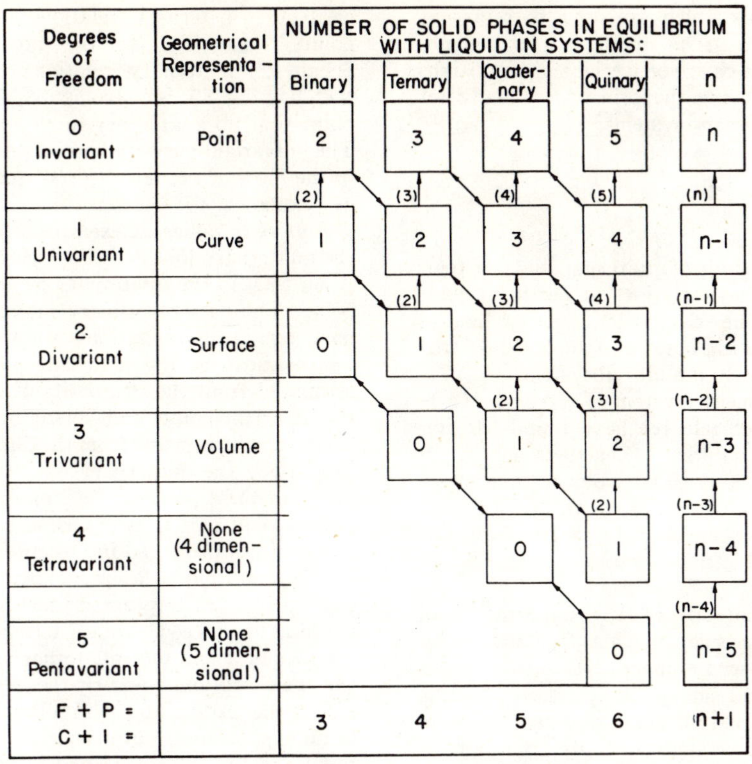

Degrees of Freedom	Geometrical Representation	NUMBER OF SOLID PHASES IN EQUILIBRIUM WITH LIQUID IN SYSTEMS:				
		Binary	Ternary	Quaternary	Quinary	n
0 Invariant	Point	2	3	4	5	n
		(2)	(3)	(4)	(5)	(n)
1 Univariant	Curve	1	2	3	4	n−1
			(2)	(3)	(4)	(n−1)
2 Divariant	Surface	0	1	2	3	n−2
				(2)	(3)	(n−2)
3 Trivariant	Volume		0	1	2	n−3
					(2)	(n−3)
4 Tetravariant	None (4 dimensional)			0	1	n−4
						(n−4)
5 Pentavariant	None (5 dimensional)				0	n−5
F + P = C + 1 =		3	4	5	6	n+1

Fig. XXVI.—The interrelationships of sub-regions of multicomponent condensed systems.

The number of regions of F_2 degrees of freedom that can be common to a region of lesser degrees of freedom, F_1, is given by the mathematical law of combination:

$$C_{s_1, s_2} = \frac{s_1!}{s_2!(s_1 - s_2)!},$$

where Cs_1, s_2 is the number of sets of s_2 solid phases that can be selected from s_1 solid phases, where s_2 and s_1 are the number of solid phases, respectively, for the two regions.

general formula, which can be found in most algebra texts, is given as:

$$C_{s_1, s_2} = s_1!/s_2!(s_1 - s_2)! \qquad (1)$$

where C_{s_1, s_2} is the number of sets or combinations of s_2 objects that can be chosen from s_1 objects. As an illustration, suppose the problem is to calculate for a quinary system the number of intersecting divariant surfaces (regions corresponding to equilibrium between three solids and a liquid) that form a univariant curve (regions corresponding to equilibrium between four solids and a liquid). Substituting in equation (1):

$$C_{s_1, s_2} = C_{4, 3} = \frac{4!}{3!(4-3)!} = 4$$

The four combinations of three solids chosen from four solids, designated A, B, C, D are ABC, ABD, ACD, BCD.

The combination formula can be further applied to obtain additional information not given in the figure. For example, in a quinary system how many trivariant regions participate in forming a univariant curve? It cannot be reasoned that because three trivariant regions intersect to form a surface, and four surfaces to form a univariant curve that, therefore, 4×3 or 12 regions meet at a univariant line, for trivariant regions may be common to more than one of the surfaces involved. Mathematically, the problem is to calculate the number of pairs of solid phases that can be chosen from four. Substituting in equation (1):

$$C_{4, 2} = \frac{4!}{2!(4-2)!} = 6$$

Half of the trivariant regions must be common to two surfaces.

Considering the complexity of even the simplest possible types of multicomponent systems, it is not surprising that the complete equilibrium relations for even a single quaternary system have never been established. Nevertheless, portions of multicomponent systems, especially those of geological and industrial importance, are being studied, and an ever increasing body of data is being accumulated in the literature. It suffices for the purposes of this Compilation to indicate some of the more common methods which are used to represent graphically such data. The methods, in general, depend on the judicious choice of specializing conditions and restrictions, employing the principles already set forth for binary and ternary systems.

(B) Graphic Representation

(a) Joins

One approach to the study of multicomponent systems is to determine the phase relations of pertinent joins within the system. The selected joins, either binary, ternary, quaternary, etc., may be found to be true subsystems or most likely pseudo-subsystems.

In the five-component system K_2O–CaO–MgO–Al_2O_3–SiO_2, Bowen and Schairer have found the important join $K_2O \cdot Al_2O_3 \cdot 4SiO_2$ (Leucite)-$CaO \cdot MgO \cdot 2SiO_2$ (Diopside) (Fig. 974) to be a binary subsystem. All phases appearing in the system can be expressed in terms of the end-members, and the course of crystallization of any liquid can be traced to completion. In the system Na_2O–CaO–MgO–Al_2O_3–SiO_2, however, Bowen has shown that the equally important join $CaO \cdot MgO \cdot 2SiO_2$ (Diopside)-$Na_2O \cdot Al_2O_3 \cdot 2SiO_2$ (Nephelite) (Fig. 978) is pseudo-binary. All of the phases appearing in the system cannot be expressed in terms of the end-members; and all areas representing more than two phases in equilibrium are non-binary. Courses of crystallization can be partially traced for the binary portions of the liquidus curve, but invariant points, showing final products of crystallization, lie outside the plane of the figure.

Similarly, in the quaternary system CaO–MgO–Al_2O_3–SiO_2 the join $MgO \cdot Al_2O_3$–$2CaO \cdot SiO_2$–$2CaO \cdot Al_2O_3 \cdot SiO_2$ (Fig. 919) is a ternary system. Within the same quaternary system, the join $CaO \cdot MgO \cdot 2SiO_2$–$2MgO \cdot SiO_2$–$CaO \cdot Al_2O_3 \cdot 2SiO_2$ (Fig. 909) is not a ternary system because of the appearance of spinel ($MgO \cdot Al_2O_3$), whose composition lies outside the join. Courses of crystallization which do not enter the spinel field can be traced.

Equilibrium relations in a single ternary join of a quaternary system can only provide limited information about the quaternary system. The study of several selected joins, of course, gives additional information. Such a procedure was used by Greene and Bogue in studying four composition planes (joins) in the system Na_2O–CaO–Al_2O_3–SiO_2 (Figs. 828 and 829). If, however, the significant portions of a system be divided by joins into lesser tetrahedra, and the equilibrium relations for the four ternary joins forming each tetrahedron be determined, then considerable information about the quaternary relations may be derived. The principle of this method is illustrated in Fig. 894 which shows relations in the quaternary join $2MgO \cdot SiO_2$ (Forsterite)-$CaO \cdot MgO \cdot 2SiO_2$ (Diopside)-$CaO \cdot Al_2O_3 \cdot 2SiO_2$ (Anorthite)-SiO_2 (Silica), part of the quaternary system CaO–MgO–Al_2O_3–SiO_2. Univariant curves, along which three crystalline phases and liquid are in equilibrium, intersect the ternary joins at points. Such points may be of two general types: ternary invariant points, either eutectic or peritectic, if ternary relations obtain; or "piercing points," according to the nomenclature of J. F. Schairer,[18] if ternary relations do not apply. The sketch in Fig. 894 is conveniently simple from the standpoint that no quaternary invariant points are present. The univariant curves all originate and end in the points in the faces of the tetrahedron (see lower part of figure and explanation).

A more complicated example is shown in Fig. 873, for the quaternary join FeO–$CaO \cdot SiO_2$–$2CaO \cdot SiO_2$–$2CaO \cdot Al_2O_3 \cdot SiO_2$ in the quaternary system CaO–FeO–Al_2O_3–SiO_2. The data of primary interest regarding quaternary equilibrium relations are the compositions and temperatures at the invariant points. These can be estimated from the compositions and temperatures of the invariant points and "piercing points" on the faces of the tetrahedron and from the direction of temperature drop along the quaternary univariant curves originating from these points. The maximum temperature on a univariant curve, as discussed earlier, is at its intersection (or a projected intersection) with the appropriate composition triangle. Therefore, with one exception the temperature decreases along a univariant curve as it proceeds from a ternary invariant point in the face of a join to a quaternary invariant point. In the one exception, however, the temperature may rise if within the tetrahedron solid solutions exist between one or more substances not present at the ternary invariant point and the three solid phases present along the univariant curve. The temperature may rise or fall along a univariant curve penetrating the interior of a tetrahedron from a piercing point. A piercing point by definition is not an invariant point and can only lie in a join other than the one defined by the compositions of the three solid phases in equilibrium with liquids represented by the univariant line. The diagram in the lower part of Fig. 873 indicates schematically, after the method of Schairer,[18] the relationships between univariant curves and invariant points existing in the system. It should be emphasized that the diagram does not depict spatial relations of the univariant curves within the tetrahedron and that the lengths of the univariant curves (shown as straight lines) are arbitrary and without significance. Additional accuracy in locating the compositions and temperatures of the invariant points may be secured by the study of selected quaternary mixtures.

(b) Sections

Another general approach which has been used in the study of multicomponent systems is the section method. Sections may be isothermal, in which case temperature is constant, or they may be planes through a system at constant percentages of one or more of the components.

(i) *Isothermal:* The phase relations at four different temperatures in a quaternary section $Na_2O \cdot SiO_2$–$CaO \cdot SiO_2$–$Na_2O \cdot Al_2O_3 \cdot 2SiO_2$–$CaO \cdot Al_2O_3 \cdot 2SiO_2$ of the

---

[18] J. F. Schairer, "The System CaO–FeO–Al_2O_3–SiO_2: I, Results of Quenching Experiments on Five Joins," *J. Am. Ceram. Soc.*, **25**, 241–73 (1942).

system Na_2O–CaO–Al_2O_3–SiO_2 (Fig. 840) were determined by Joseph Spivak. This method is especially convenient when studying complicated solid solution regions where tie lines between solid solutions and liquid solutions are to be indicated, as shown, for example, in Fig. 684. The method of isothermal sections is most useful for showing sub-liquidus relations in refractory oxide systems whose liquidus temperatures can not be determined with available equipment. For example, Fig. 581 shows the 1750°C isotherm for the system BeO–CeO_2–ZrO_2. Sub-solidus compatibility tetrahedra have been determined in the system CaO–MgO–SnO_2–TiO_2 (Figs. 939 and 940). In other words, for a given composition in this system, it is possible to state the final products of crystallization, although crystallization paths cannot be followed.

(ii) *Compositional Restraint.* A quaternary system can be represented by a series of ternary diagrams in which one of the composition variables is held at a constant but different value in each diagram. Thus, Lea and Parker (Figs. 943–947) determined equilibria relations in the system CaO–$5CaO \cdot 3Al_2O_3$–$2CaO \cdot SiO_2$–$4CaO \cdot Al_2O_3 \cdot Fe_2O_3$, of particular interest in cement technology, by studying planes through the base (CaO–$5CaO \cdot 3Al_2O_3$–$2CaO \cdot SiO_2$) of the tetrahedron at 0, 2, 5, 10, and 20 percent Fe_2O_3, or 0, 6.1, 15.2, 30.4, and 60.8 percent, respectively, of $4CaO \cdot Al_2O_3 \cdot Fe_2O_3$. Figure 943 is a perspective diagram of the system showing primary phase volumes and temperatures of some of the invariant points, derived from the data on the study of the sections together with the phase diagrams of the faces of the tetrahedron (Fig. 944).

A five component system could be studied by a similar technique, although it would be a long and tedious process. Three chosen components would form the apices of a series of ternary diagrams in a two dimensional array. The percentage of a fourth component would vary in the horizontal series, while the percentage of the fifth component would be held constant. In each vertical series the fourth component would remain constant while the fifth increased from row to row.

A five component system might be represented, also, as a series of tetrahedra in which the percentage of the fifth component varied systematically in the diagrams. It should be noted that each tetrahedron would correspond to a constant percentage of the fifth component and that the position of a point within a given tetrahedron would give only the ratio of percentages of the first four components. A primary phase region in any particular tetrahedron would identify the one phase which crystallized first from a liquid whose composition lay within the boundaries of that region.

(c) *Other Specializing Conditions*

In a study of the quaternary system CaO–MgO–$2CaO \cdot SiO_2$–$5CaO \cdot 3Al_2O_3$ (Fig. 891), McMurdie and Insley determined the divariant surface representing the compositions of all liquids that can exist in equilibrium with MgO and one other solid phase (the lower level of the primary phase volume of MgO). By making the MgO component a dependent variable by considering only the MgO content of the limiting surface, it becomes possible to project this surface, as contours of constant MgO content, onto the ternary base of the tetrahedron (Fig. 892); and the representation is greatly enhanced.

An interesting restricting condition was applied by Swayze in a study of a portion of the quinary system CaO–MgO–Al_2O_3–Fe_2O_3–SiO_2, namely, the system CaO–$5CaO \cdot 3Al_2O_3$–$2CaO \cdot Fe_2O_3$–$2CaO \cdot SiO_2$ modified by 5% MgO (Fig. 991). This figure requires careful interpretation. The addition of 5% MgO served to saturate the liquids at or near complete melting; to quote Swayze: "..., a small amount of periclase has been observed in the MgO-saturated glasses formed by complete or nearly complete melting of the compositions studied." Periclase, therefore, is the primary phase for the whole tetrahedron of compositions. The volumes labeled in the figure are the secondary phases, and the temperature values refer, not to liquidus temperatures, but to temperatures at which the secondary phases start to crystallize. The convenient and essential restricting condition in this case is not the 5% MgO content, per se, but that only the boundaries between the MgO primary phase region and the secondary crystallizing phases be considered. The 5% MgO designated in the figure as "Edge 95% tetrahedron" is of little significance, except to indicate the estimated maximum solubility of MgO in the compositions studied. Any higher percentage of MgO in the mixtures would give the same diagram.

In the graphic representation of multicomponent systems, no one method can be considered superior to another. To a large extent a chosen method will depend upon such factors as the exact region to be studied, the type of information sought, and the detail desired, as well as upon the actual data obtained. Any scheme of representation is acceptable which shows the data with the greatest clarity. It is of utmost importance, however, that the specializing or restricting conditions adopted be explicitly and carefully stated, both in the text and in the legends of the figures. As more and more phase diagrams of binary and ternary systems are made available, an increasing emphasis will be placed on the study of quaternary, quinary, and higher systems. It seems reasonable to suppose that mathematical expression of the relationships among many variables must accordingly assume greater importance in the future as an indispensable complement to the graphical representation of these complex relationships.

IV. EXPERIMENTAL METHODS FOR HIGH-TEMPERATURE HETEROGENOUS EQUILIBRIUM

Several methods of phase equilibrium determination are possible. All[19] have their definite fields of usefulness and several methods may be needed in different parts of the same system.

---

[19] Deformation of Seger or similar ceramic cones is not a method by which phase equilibrium can be determined. The temperature at which a cone deforms depends on a number of factors: amount and viscosity of liquid formed, and rigidity and strength of the cone at the particular temperature, etc.

Methods of determination may be divided into two general classes: the static and the dynamic. Static methods are those in which the temperature of the sample is held constant until equilibrium is attained. The fractionation method and the quenching method are two examples. The dynamic methods are those in which phase changes causing heat effects are indicated by an arrest in temperature change within the sample during uniform heating or cooling.

In systems where equilibrium is reached quickly, such as systems of metallic components, the dynamic methods are the most satisfactory. In systems where equilibrium is sluggish, because of high viscosity in the liquids or of low "crystallization potential" of the solids, the dynamic methods are unusable and it is necessary to resort to the static methods.

The most useful of the static methods is the quenching method since it permits the use of small samples. This procedure was developed for use in the silicate systems at the Geophysical Laboratory and the apparatus is described in several publications.[20] A small sample of homogeneous character having a known composition within the system under consideration is enclosed in a suitable container (usually an envelope of platinum foil or other chemically inert, high-melting metal) and heated at the desired temperature until equilibrium is established. The sample is then rapidly quenched from that temperature by dropping it instantaneously into a liquid at low temperature and the equilibrium conditions prevailing at the high temperature are "frozen." In other words, solid phases present at the high temperature are retained, perhaps metastably, at ordinary temperatures; phases which are liquids at high temperatures exist as glasses at low temperatures. Phases are identified by microscopic means or by crystal diffraction methods. The quenching method is applicable only to systems where changes are sufficiently sluggish to prevent transitions during quenching. The method has the great advantage that phases solid at the equilibrium temperature may be identified at room temperatures by their physical and optical properties.

If it is found that the sample contains more than one crystalline phase, a sample of the same composition is heated to a higher temperature, quenched and examined. If this quenched charge contains only one crystalline phase, which in this case is called the primary phase for that part of the system, the process is repeated until a temperature is found at which the primary phase disappears and leaves only liquid (glass at room temperature). This temperature and composition locate a point on the melting curve of the system investigated. Another static method of determining equilibrium diagrams is by fractionation in which the phases are separated mechanically at the equilibrium temperature and individually analyzed. In ternary systems containing solid solutions, Bowen[21] has shown how the results obtained under conditions of perfect fractionation, where the solids already crystallized have no opportunity of reacting with the liquid, can be used to determine the phase relations under conditions of perfect equilibrium. While the fractionation method is very useful in phase studies in aqueous systems or systems at ordinary temperatures, it presents great difficulties in nonaqueous systems at high temperatures where liquids are often of high viscosity and where it is difficult to prevent the erosion of filtering agencies such as screens.

However, even such difficulties have been partially overcome by the use of a high-temperature centrifuge developed by Newkirk.[22] With this equipment a charge is centrifuged within the furnace, after having been heated at a desired temperature sufficiently long for equilibrium conditions to prevail. Liquid is separated from solid by forced filtration through a platinum sponge filter contained in the small platinum sample holder. The liquid can then be analyzed chemically by micro methods. Applying this method to a multicomponent system, it is possible to determine the path of crystallization of a particular composition from liquidus to invariant point, without prior knowledge of the location of invariant points or primary phase boundaries.

Heating and cooling curves are the usual dynamic methods of determining phase diagrams. In these, time-temperature determinations for each sample tested are plotted. Every temperature arrest indicates a phase change, that is, the final disappearance or the first appearance of a phase during temperature change. The length of time of the arrest is an indication of the relative amount of the phase changed at that temperature.

A refinement of the time-temperature method called differential thermal analysis (DTA) is more sensitive to phase changes, as measured by the heat effects. In DTA the experimental sample and a standard inert sample, usually Al_2O_3 powder, are heated simultaneously according to a prescribed schedule, generally about 10°C. per minute. Thermocouples inserted into both samples are so connected that measurements can be made of both the actual temperature and the differential temperature (in mv.) between the standard and the experimental samples. As long as no phase change occurs, the differential in temperature remains relatively constant, but a temperature arrest or acceleration within the experimental sample, indicating a phase change, is reflected by a sharp change in differential.

Heating and cooling methods are suited to the study of systems which reach equilibrium rapidly and where the heat effect of transitions is large. Even in such systems, however, they have the disadvantage that the phases stable at high temperatures cannot be retained

---

[20] (a) H. S. Roberts, "Automatic Control of Laboratory Furnaces by Wheatstone Bridge Method," *J. Wash. Acad. Sci.*, 11, 401 (1921).
(b) G. W. Morey, "Comparison of Heating Curves and Quenching Methods of Melting-Point Determinations," *ibid.*, 13, 325–29 (1923).

[21] N. L. Bowen, "Certain Singular Points on Crystallization Curves of Solid Solutions," *Proc. Nat. Acad. Sci.*, 27, 301–309 (1941).
[22] See R. H., Bogue, under "Silicate Chemistry" in Selected Bibliography, Part V.

at ordinary temperatures for determination of their physical characteristics.

Another dynamic method, especially useful in studying oxidation and reduction reactions, volatility effects, or hydration and carbonation phenomena, depends on sensitive weight recording measurements during heating and cooling of a specimen. Equipment has been designed which simultaneously records weight changes and differential heat effects as a function of temperature.[23]

Definitive phase equilibria studies in oxide systems involving changes in oxidation state require careful control of partial pressures of the gas phase, in particular oxygen. When considering partial pressures, the condensed statement of the phase rule no longer applies, and the pressure variable introduces another degree of freedom in the system. Considerable progress has been made in the development of techniques for controlling partial pressures at high temperature and in the experimental determination and representation of phase diagrams, as for example, the iron-containing systems.[24]

In studies involving non-quenchable transformations it is necessary to identify phases at temperature. Two methods which have been successfully employed for this purpose are high-temperature X-ray diffractometry[25] and high-temperature microscopy.[26] Additional advantages of these methods, when applicable, are that rapid surveys, with small amounts of sample, can be made over a wide temperature range.

With increasing emphasis on refractory systems, methods are receiving attention for achieving, controlling, and measuring temperatures above 1700°C, about the maximum obtainable with the conventional platinum-rhodium (80%:20%) quench furnace. The strip furnace[27] provides an easy means of attaining high temperatures. It usually consists of a narrow, V-shaped, short strip of refractory metal (e.g., Pt-40% Rh, or Ir), which acts both as a sample holder and heating element. For phase studies up to 2400°C., an induction furnace,[28] having an iridium-crucible susceptor, also has been used successfully. Both methods have their limitations, and it cannot be emphasized too strongly that the accuracy of temperature measurements with an optical pyrometer depends in the final analysis on the degree of realization of blackbody conditions.[29]

The diagrams presented refer only to systems which are in equilibrium for the temperatures specified. They give no information as to the velocity of reactions, but they do define the relative amounts of crystalline and liquid phases present at any temperature and composition provided the time has been sufficient for equilibrium to be attained. Most ceramic and related processes are incomplete chemical reactions and the results obtained are dependent upon time as well as temperature. In applying equilibrium diagrams to manufacturing processes, consideration must be given to this difference between the ideal and the practical conditions.

For example, the composition of a porcelain body composed of $X\%$ of flint, $Y\%$ of clay (containing alumina, silica, and water only) and $Z\%$ of potash feldspar can be represented in the system K_2O–Al_2O_3–SiO_2 (Fig. 407) if the batch materials are pure and the body had been fired to a sufficiently high temperature to eliminate water. The composition of the porcelain in terms of the components is as follows:

$$SiO_2 = X + 0.541Y + 0.6472Z$$
$$Al_2O_3 = 0.459Y + 0.184Z$$
$$K_2O = 0.169Z$$

It is most improbable, however, that the phases as determined from the equilibrium diagram would be equivalent, quantitatively or qualitatively, to the constituents observed microscopically in the commercially manufactured porcelain of this composition. The high viscosity of the melted feldspar, the low rate of solution of the quartz, and the slow diffusion of the partially mixed and partially melted batch materials would hinder the attainment of equilibrium and of homogeneity so that the constituents would not only differ from the equilibrium phases but the composition (in terms of constituents) would vary widely from place to place in the body.

In the application of equilibrium diagrams to actual processes care must be taken that the actual and ideal are comparable. The presence of an impurity even in small amounts introduces an additional component which in many instances alters the solubility and melting relations so profoundly that no adequate interpretation on the basis of the simpler diagram is possible.

Although the phase diagram gives no information as to rates of reaction, it is often possible to infer from such diagrams what the products of reaction may be under certain conditions of disequilibrium and arrested reaction. The calculations of Lea and Parker[30] and of Dahl[31] on the phases existing in the system, CaO–Al_2O_3–Fe_2O_3–SiO_2, under certain conditions of arrested reaction have enabled the cement chemist to estimate phases present in cement clinkers under plant conditions of partial disequilibrium.

---

[23] See F. A. Mauer under "Methods and Techniques, (A) General" of Selected Bibliography.

[24] See A. Muan under "Theory" in Selected Bibliography.

[25] Ernest M. Levin and Floyd A. Mauer, "Improved Sample Holder for X-Ray Diffractometer Furnace," *J. Am. Ceram. Soc.*, **46** [1], 59–60(1963). This paper cites other references on the subejct.

[26] See F. Ordway, also J. H. Welch, under "Methods and Techniques, (A) General" of Selected Bibliography.

[27] H. S. Roberts and G. W. Morey, "Micro Furnace for Temperatures above 1000°C," *Rev. Sci. Inst.*, **1**, 576–579 (1930).

[28](a) S. J. Schneider, "Phase Equilibria in Systems Involving the Rare Earth Oxides. Part III. The Eu_2O_3–In_2O_3 System," *J. Research Natl. Bur. Standards*, **65A** [5], 429–434 (1961).

(b) S. J. Schneider and J. L. Waring, "Phase Equilibrium Relations in the Sc_2O_3–Ga_2O_3 System," *J. Research Natl. Bur. Standards*, **67A**, [1], 19–25 (1962).

[29] Samuel J. Schneider, "Compilation of the Melting Points of the Metal Oxides," National Bureau of Standards Monograph 68, Oct. 10, 1963. 31 pp.

[30] See section on "Interpretation" in Selected Bibliography, Part V.

[31] See section on "Mathematical Treatment" in Selected Bibliography, Part V.

V. SELECTED BIBLIOGRAPHY

It is neither the purpose nor intention of the authors to compile a complete set of references for the various topics included in the "Selected Bibliography." Neither is it claimed that all of the most important ones have been listed. Nevertheless, it is hoped that the selection will prove beneficial in introducing the novice to some of the aspects of phase equilibria study which perforce have not been discussed or only partially so in the text.

(1) Theory

Bowden, S. T., "Phase Rule and Phase Reactions," Macmillan and Co., London, 1938.

Dahl, L. A., "Equilibrium in Heterogeneous Systems of Two or More Components," *J. Chem. Education*, 26, 411 (1949).
A concise exposition of some simple types of phase equilibria diagrams.

Findlay, A., Campbell, A. N., and Smith, N. O., "The Phase Rule and Its Applications," Ninth Edition, Dover Publications, Inc., New York, 1951. 494 pp. + 236 figs.
A comprehensive introduction to the subject, including chapters on liquid-vapor equilibria, aqueous systems, practical applications, thermodynamic deductions, and experimental determination of binary diagrams.

"The Collected Works of J. Willard Gibbs," Volume I, Thermodynamics, Yale University Press, New Haven, Conn. Reprinted 1957. 434 pp.
This volume includes the well known memoir "On the Equilibrium of Heterogeneous Substances," which forms the theoretical basis for the Phase Rule and the graphical representation of equilibria.

Marsh, J. S., "Principles of Phase Diagrams," First Edition, McGraw-Hill Book Co., New York, 1935. xiv + 193 pp. + 180 figs.

Masing, G. (translated by B. A. Rogers), "Ternary Systems," Reinhold Publishing Corporation, New York, 1944. 173 pp. + 166 figs. Available as a paperback from Dover Publications, Inc., New York, N. Y., 1960.
Discussion of the fundamental theory underlying three component systems. Several alloy systems explained in detail.

Muan, Arnulf, "Phase Equilibria at High Temperatures in Oxide Systems Involving Changes in Oxidation States," *Am. J. Sci.*, 256, 171–206 (1958).
This paper discusses theoretical principles of controlling partial pressures of the gas phase and the representation of phase relationships in binary (Fe–O), ternary (FeO–Fe_2O_3–SiO_2) and quaternary (MgO–FeO–Fe_2O_3–SiO_2) systems. Four different idealized conditions are considered: (1) constant total composition of condensed phases, (2) constant O_2 pressure, (3) constant mixing ratio pCO_2/pH_2, and (4) crystallization in contact with metallic iron.

Ricci, John E., "The Phase Rule and Heterogeneous Equilibrium," D. Van Nostrand Co., Inc., New York, 1951. 505 pp.
A systematic study of the meaning and application of the Phase Rule. Good discussion of reciprocal ternary systems and aqueous quaternary and quinary systems. Book is profusely illustrated.

Ricci, John E., "Guide to the Phase Diagrams of the Fluoride Systems," Oak Ridge National Laboratory Report No. ORNL-2396, Oak Ridge, Tennessee, 1958. 106 pp. Available from the Office of Technical Services, Department of Commerce, Washington, D. C.
Detailed discussion of phase equilibria occurring in several complex ternary salt systems. First sections present general principles and explanations, as an aid in reading, interpreting, and using actual diagrams.

Tammann, G., "Lehrbuch der heterogenen Gleichgewichte," Vieweg, Braunschweig (1924).
Classical textbook on heterogeneous equilibria.

Vogel, R., "Die heterogenen Gleichgewichte," Vol. II: Handbuch der Metallphysik, Leipzig, 1937.

Wetmore, F. E. W., and LeRoy, D. J., "Principles of Phase Equilibria," First Edition, McGraw-Hill Book Company, Inc., New York, 1951. 200 pp.
Lucid explanation of phase equilibria principles, using a combination of the practical and theoretical approaches. Contains a section on thermodynamic considerations. Contrary to convention, arrows on boundary curves in the phase diagrams point in the direction of rising temperatures.

Wilson, A. J. C., "Binary Equilibrium," *J. Inst. Metals* 70, Part II, 543–560 (1944).
Thermodynamic derivations are given for several rules governing the construction of phase equilibrium diagrams for binary equilibrium.

Zernike, J., "Chemical Phase Theory," Kluwer's Publishing Co., Ltd., Deventer, Netherlands, 1955. 493 pp. (In English).
This book is a critical, comprehensive, and modern treatise on the deduction, the applications, and the limitations of the phase rule.

(2) Interpretation

Bowen, N. L., "Evolution of Igneous Rocks," p. 60, Princeton University Press, Princeton, N. J., 1928. 332 pp.; *Ceram. Abstracts*, 8 [8] 609 (1929).

Dahl, L. A., "Interpretation of Phase Diagrams of Ternary Systems," *J. Physical Chemistry*, 50 [3] March (1946).
Quantitative determination of phase composition during various stages of crystallization, by the graphical method, by construction of similar triangles.

Foster, Wilfrid R., "Contribution to the Interpretation of Phase Diagrams," *J. Am. Ceram. Soc.*, 34, 151–160 (1951).
A critical review and discussion involving compatibility triangles, and the relations of invariant points, melting intervals, and courses of melting to the firing behavior of ceramic bodies.

Foster, Wilfrid R., "Solid-State Reactions in Phase Equilibrium Research," I. *Bull. Am. Ceram. Soc.*, 30 [8] 267–270; II. *Ibid.*, [9] 291–296 (1951).
Outlines general principles for the systematic application of solid-state reactions to phase equilibrium studies.

Korzhinskiĭ, D. S., "Physicochemical Basis of the Analysis of the Paragenesis of Minerals," Translated from Russian by the Consultants Bureau, Inc., New York, 1959. 142 pp.
This book concentrates on the study of the dependence of mineralogical composition on known factors affecting phase equilibrium, namely, pressure, temperature, and composition. Projective geometry is used freely to represent the relationships.

Lea, F. M., and Parker, T. W., "The Quaternary System CaO–Al_2O_3–SiO_2–Fe_2O_3 in Relation to Cement Technology," *Bldg. Res. Tech. Paper*, 16, London (1935).
Contains section on the compound content in quarternary mixes in which equilibrium between solid and liquid is not maintained during cooling.

Levin, Ernest M., and Block, Stanley, "Structural Interpretation of Immiscibility in Oxide Systems: I. Analysis and Calculation of Immiscibility. II. Coordination

Principles Applied to Immiscibility. III. Effect of Alkalis and Alumina in Ternary Systems," *J. Am. Ceram. Soc.*, **40** [3] 95–106, [4] 113–118 (1957); **41** [2] 49–54 (1958).

These articles represent an attempt to apply crystal chemistry principles to the quantitative interpretation of immiscibility in borate and silicate systems.

Morey, George W., "The Interpretation of Phase Equilibrium Diagrams," *The Glass Industry*, **12** [4] 69–80 (1931).

Discusses the interpretation of binary and ternary diagrams as applied to several glass forming systems.

Niggli, Paul, "Das Magma und Seine Produkte," I Teil: Physikalisch-Chemische Grundlagen, Akademische Verlagsgesellchaft M.b.H., Leipsig, 1937. 379 pp.

This volume deals entirely with phase equilibrium principles and a review of experimental work. A considerable portion of the book is devoted to the construction and interpretation of diagrams of anhydrous systems; also included is a section dealing with systems with components of different volatility.

Osborn, E. F., "Segregation of Elements During the Crystallization of a Magma," *J. Am. Ceram. Soc.*, **33** [7] 219–224 (1950).

The principles and factors affecting fractional crystallization of a basalt, following Bowen's reaction series, are discussed.

(3) Methods and Techniques

(A) General

Bockris, J. O'M., White, J. L., and Mackenzie, J. D., "Physicochemical Measurements at High Temperatures," Academic Press Inc., New York, N. Y., 1959. 394 pp.

The book is a cooperative effort by leading workers in the field, which stresses the techniques of fundamental investigations. The first chapters summarize the general aspects of obtaining, controlling, and measuring high temperatures. In the latter chapters, techniques concerned with a specific type of property are discussed, e.g., calorimetry, liquid densitometry, surface tension, vapor pressure, ultrasonic velocity, to name a few.

Herzfeld, Charles M., Editor-in-Chief, "Temperature, Its Measurement and Control in Science and Industry," Volume 3: Part 1. "Basic Concepts, Standards, and Methods"; Part 2. "Applied Methods and Instruments," Reinhold Publishing Corp., New York, N. Y., 1962. 848 pp., 1094 pp.

These two books comprise the most comprehensive treatise on the subject.

Hume-Rothery, W., Christian, J. W., and Pearson, W. B., "Metallurgical Equilibrium Diagrams," The Institute of Physics, London, 1952.

Textbook dealing almost entirely with apparatus and experimental methods used in determining metallurgical equilibrium diagrams. Some general theory of binary and ternary systems.

Mauer, Floyd A., "An Analytical Balance for Recording Rapid Changes in Weight," *Rev. Sci. Inst.*, **25** [6] 598–602 (1954).

For adaptation of this equipment to the simultaneous recording of weight change and differential thermal analysis curves see Bogue, "The Chemistry of Portland Cement" (p. 317), under (6) Silicate Chemistry.

Ordway, Fred, "Techniques for Growing and Mounting Small Single Crystals of Refractory Compounds," *J. Research Natl. Bureau Standards*, **48** [2] 152–158 (1952).

The apparatus may be used for growing single crystals, noting melting temperatures, and studying phase changes. A small amount of sample is placed on a noble metal thermocouple, which serves both as the heating and the temperature measuring device.

For additional description see Bogue, "The Chemistry of Portland Cement" (p. 95), under (6) Silicate Chemistry.

Rankin, G. A., and Wright, Fred E., "The Ternary System $CaO-Al_2O_3-SiO_2$," *Am. J. Sci.*, **39**, 4th Ser., 1–79, (1915).

Classic study on a system, elucidating the "quenching" method and using the polarizing microscope for identifying phases.

Welch, J. H., "A Simple Microscope Attachment fo Observing High-Temperature Phenomena," *J. Sci. Inst.*, **31**, 458–462 (1954).

This article describes modifications of the hot wire apparatus developed by Ordway.

(B) Optical Mineralogy

Bulletin of the National Research Council, Number 118, June, 1949, "Data on Chemicals for Ceramic Use," published by the University of Pittsburgh for the National Academy of Sciences, Washington, D. C. New edition is being prepared.

A collection of the "best" values for selected physical properties of chemical substances that are of interest to ceramists, including data on density, melting point, transition point, boiling point, sublimation point, decomposition temperature, refractive index, crystal form, and color.

Hartshorne, N. H., and Stuart, A., "Crystals and the Polarizing Microscope," Third Edition, Edward Arnold and Co., London, 1960. 557 pp.

In an advanced manner, covers both theoretical and practical aspects of the use of the polarizing microscope for examination, identification, and characterization of substances.

Insley, Herbert, and Fréchette, Van Derck, "Microscopy of Ceramics and Cements," Academic Press Inc., New York, N. Y., 1955. 286 pp.

One of the few texts, if not the only one, dealing with the application of the polarizing microscope in the research, manufacture, and use of ceramics. The methods, results, and interpretations are applied to the fields of whitewares, refractories, glass, cements, porcelain enamels, structural clay products, abrasives, foundry sands, and metallurgical slags.

Larsen, Esper S., and Berman, Harry, "The Microscopic Determination of the Nonopaque Minerals," Second Edition, Geological Survey Bulletin 848, United States Government Printing Office, 1934.

Brief discussion of methods for determining optical constants; tables for the determination of minerals from their optical constants.

Winchell, Alexander N., "Elements of Optical Mineralogy, An Introduction to Microscopic Petrography," John Wiley and Sons, Inc., New York. In three volumes:

Part I. Principles and Methods, Fifth Edition, 1937. 263 pp. + 267 figs.

Part II. Descriptions of Minerals, With the Collaboration of Horace Winchell, Fourth Edition, 1951. 551 pp. + 427 figs.

Part III. Determinative Tables, Second Edition, 1929. 204 pp. + 3 charts.

Winchell, A. N., and Winchell, Horace, "Microscopical Characters of Artificial Inorganic Solid Substances," Third Edition, Academic Press, Inc., New York, N. Y., 1964. About 400 pp. (In press).

The determinative table constitutes an invaluable aid in the identification of inorganic substances with the polarizing microscope.

(C) Differential Thermal Analysis

Mackenzie, R. C., "Scifax Differential Thermal Analysis Data Index," 1962. Available from Cleaver-Hume Press Ltd., 31 Wright's Lane, Kensington, London, W.8.

The index contains a total of 1630 punched cards, coded according to 29 ranges of the principal and second peaks, respectively. Cards are grouped into Mineral, Inorganic, and Organic sections. Future supplementary sets and occasional replacement cards are planned.

Smothers, W. J., and Chiang, Yao, "Differential Thermal Analysis: Theory and Practice," Chemical Publishing Co., Inc., 212 Fifth Ave., New York, N. Y., 1958. 444 pp.

A concise book covering various aspects of differential thermal analysis (DTA), such as origins, equipment, theory, qualitative and quantitative results, and recent developments. About two-thirds of the book is devoted to four appendices: (1) Publications (covering the period 1877–1957). (2) Information on Equipment. (3) Index of Operators of DTA Equipment. (4) Reference List of Materials Studied.

A new edition of this book by the same authors and publisher titled "Handbook of Differential Thermal Analysis" is scheduled for publication in 1964.

(D) X-ray Powder Diffraction and Crystal Chemistry

Azároff, Leonid V., "Introduction to Solids," McGraw-Hill Book Co., Inc., New York, N. Y., 1960. 460 pp.

Primarily a college text for physical science and engineering students in which crystal chemistry is applied to the structure and properties of various classes of solids, e.g., metals, semiconductors, and insulators.

Azároff, Leonid V., and Buerger, Martin J., "The Powder Method in X-Ray Crystallography," McGraw-Hill Book Co., Inc., New York, N. Y., 1958. 342 pp.

Primarily a textbook on the theory and practice of the x-ray powder method, for the making and interpretation of powder photographs.

Bunn, C. W., "Chemical Crystallography," Second Edition, Oxford, at the Clarendon Press, 1961. 509 pp.

Part I of this book is a guide to the identification of solid substances by means of optical properties and x-ray powder photographs. Part II is concerned with structure determination. The Author uses a practical approach and avoids formal physical or mathematical treatment.

Donnay, J. D. H., Donnay, Gabrielle, Cox, E. G., Kennard, Olga, and King, Murray Vernon, "Crystal Data, Determinative Tables," Second Edition, ACA Monograph Number 5, American Crystallographic Association, 1963. 1302 pp. Available from Polycrystal Book Service, G.P.O. Box 620, Brooklyn 1, New York, N. Y., U. S. A.

This monograph contains an estimated 13,000 entries of inorganic compounds, organic compounds, and proteins, systematically arranged in groups according to the six crystal systems. Data given include unit cell dimensions, space point group, Z, crystal structure, and specific gravity.

Hey, Max H., "An Index of Mineral Species and Varieties," Printed by Order of the Trustees of the British Museum, Jarrold and Sons Ltd., Norwich, Great Britain, 1962. 728 pp.

Minerals are listed according to chemical classification and again according to alphabetical order. A pronouncing index of accepted mineral names is included.

Klug, Harold P., and Alexander, Leroy, E., "X-Ray Diffraction Procedures," John Wiley and Sons, Inc., New York, N. Y., 1954. 716 pp.

This book presents in detail the basic techniques, procedures, and applications of the powder method to polycrystalline and amorphous materials. Geiger-counter spectrometric techniques, small-angle scattering methods, and radial-distribution analysis are brought together in a single volume.

Peiser, H. S., Rooksby, H. P., and Wilson, A. J. C., Technical Editors, "X-Ray Diffraction by Polycrystalline Materials," John Wright and Sons Ltd., Bristol, 1955.

Thirty experts make authoritative contributions to the various aspects of crystal analysis through the study of polycrystalline materials. Theory, experimental techniques, and interpretation are covered in detail. An interesting feature of the book is a series of essays at the end, describing application of the polycrystalline method to diverse fields of research.

Wells, A. F., "Structural Inorganic Chemistry," Third Edition, Oxford University Press, Amen House, London, 1962. 1055 pp.

This book is both a textbook and a reference source for crystal chemistry and particular structures.

Wyckoff, Ralph W. G., "Crystal Structures," Second Edition, Vol. 1, John Wiley and Sons, Inc., New York, N. Y., 1963. 467 pp.

This monumental reference work, published in a hardcover edition, is the first of a series of several volumes. Contents include structures of the elements and of compounds with formula type RX and RX_2.

"Standard X-ray Diffraction Powder Patterns," National Bureau of Standards Circular 539, Vols. 1–10 (1953–1960). Continued in National Bureau of Standards Monograph 25, Sections 1–2 (1962–1963). Available from Superintendent of Documents, U. S. Government Printing Office, Washington, D. C. 20234

Standard patterns, suitable for identification of unknown crystalline materials, are presented. High purity materials are used. New data is published continually on groups of 30 to 60 substances.

"X-Ray Powder Data Card File and Index Book (XR-PDF)," issued by the American Society for Testing and Materials, X-Ray Department, 1916 Race St., Philadelphia Pa., U. S. A.

Contains X-ray powder data on some 7000 substances; supplements are being added continually.

(E) Hydrothermal

Ahrens, L. A., Rankama, Kalervo, and Runcorn, S. K., Editors, "Physics and Chemistry of the Earth," McGraw-Hill Book Co., Inc., New York, N. Y., 1956. 317 pp.

A collection of eight articles on geophysics and geochemistry. In the chapter "Investigations Under Hydrothermal Conditions," Rustum Roy and O. F. Tuttle present a brief historical review, a description of equipment, results of studies on mineral synthesis and analysis of phase equilibrium data.

Gilman, J. J., Editor, "The Art and Science of Growing Crystals," John Wiley and Sons, Inc., New York, 1963. 493 pp.

Chapter 13, "Hydrothermal Growth," by A. A. Ballman and R. A. Laudise is a concise discussion of the methods and techniques of growing large crystals under controlled conditions, with particular reference to the growth of quartz.

Morey, George W., "Hydrothermal Synthesis," *J. Am. Ceram. Soc.*, **36** [9] 279–285 (1953).

Presents underlying theory, describes apparatus, and gives illustration of the experimental method.

(F) High Pressure

Bridgman, P. W., "The Physics of High Pressure," G. Bell and Sons, Ltd., London, Reprinted 1952. 445 pp.

This book, by an outstanding investigator, surveys the important work in the field up to the time of publication. Chapters on melting phenomena and polymorphic transitions are of especial interest in phase studies.

Paul, William, and Warschauer, Douglas M., "Solids Under Pressure," McGraw-Hill Book Co., Inc., New York, N. Y., 1963. 478 pp.

This volume contains a collection of thirteen articles dealing with different aspects of research on the physics of crystalline solids at high pressures.

Wentorf, R. H., Jr., Editor, "Modern Very High Pressure Techniques," Butterworth Inc., Washington, D. C., 1962. 233 pp.

The book takes the form of a collection of papers written by experts. Main emphasis is on the design of apparatus for studying the properties of materials at pressures above 20,000 atmospheres.

(4) Mathematical Treatment

Dahl, L. A., "Analytical Treatment of Multicomponent Systems," *J. Phys. Colloid Chem.*, 52, 698–729 (1948).

Properties and applications of intrinsic equations in the analytical treatment of multicomponent systems, such as the conversion of compositions from one system of components to another, the classification of compositions, and the estimation of phase proportions at invariant points.

Dahl, L. A., "Parametric Equations in the Treatment of Multicomponent Systems," *J. Phys. Colloid Chem.*, 54, 547–564 (1950).

Application of parametric equations to phase equilibria problems; especially useful when dealing with a system of a small number of components within a multicomponent system.

Dahl, L. A., "Estimation of Phase Composition of Clinker," *Rock Products*, 41–42 (1938–39).

Paper deals essentially in principles and calculations involved in determining phase composition at temperatures at or above the temperature of liquid formation in the quaternary system $3CaO \cdot SiO_2$–$2CaO \cdot SiO_2$–$3CaO \cdot Al_2O_3$–$4CaO \cdot Al_2O_3 \cdot Fe_2O_3$.

Morey, George W., "Analytical Methods in Phase Rule Problems," *J. Phys. Chem.*, 34 [8] 1745-1750 (1930).

Application of linear equations to phase rule problems, and their solution by determinant notation.

Ordway, Fred, "Matrix Algebra for Calculating Multicomponent Mixtures," *J. Portland Cement Assoc. Research and Development Labs*, 2 [1] 28–36 (1960).

Pepper, Paul M., "Coexistence Relations of $n + 1$ Phases in n-Component Systems," *J. Appl. Phys.*, 20 [8] 754–60 (1949).

In isothermic-isobaric sections of n-component systems representing coexistence relationships of $n + 1$ phases, two coexistence patterns at most are possible. Formulas are derived (using vector methods) for calculating the composition of a "well-chosen sample" from which examination of the phases present clarifies the coexistence relationships of the $n + 1$ phases of known composition. Supplementary formulas are given for the range of coexistence, in the composition space, for any n phases known to coexist.

(5) Thermodynamic Calculations

Epstein, Leo F., and Howland, W. H., "Binary Mixtures of UO_2 and Other Oxides," *J. Am. Ceram. Soc.*, 36 [10] 334–335 (1953).

Phase diagrams of the binary systems containing UO_2 and the oxides MgO, Al_2O_3, and BeO are constructed using the laws of ideal solutions and an approximation for the entropy of fusion.

Knapp, W. J., "Use of Free Energy Data in the Construction of Phase Diagrams," *J. Am. Ceram. Soc.*, 36 [2] 43–47 (1953).

A review of the use of free energy-composition curves (at constant pressure and temperature and with the assumption of ideal solution behavior) for determining the range of composition over which a phase, or combination of phases, may be stable.

Kubaschewski, O., and Evans, E. LL., "Metallurgical Thermochemistry," Third Edition, Pergamon Press, New York, 1958. 425 pp.

This book serves as both an introduction to the subject and as a work of reference. Emphasis is on practical application of chemical thermodynamics, and extensive tables of thermochemical data are included.

Morey, George W., "The Application of Thermodynamics to Heterogeneous Equilibria," *J. Franklin Institute*, 194 [10] 425–484 (1922).

An exposition of the application of the first two laws of thermodynamics to the equilibrium of heterogeneous substances.

White, James, "Phase Relations in Ceramics," pp. 94–161, in "Ceramics: A Symposium," arranged and edited by A. T. Green and Gerald H. Stewart, published by the British Ceramic Society, Stoke-on-Trent (1953). 877 pp.

Elucidation of thermodynamic principles applied specifically to solids and solutions of interest in ceramics.

The general symposium is a survey of the practice, the technology, and the basic science of ceramics and constitutes a valuable contribution to the literature of ceramics.

Wygant, J. F. and Kingery, W. D., "Applications of Thermodynamics in Ceramics":

I. Energy and Heat Content, *Am. Ceram. Soc. Bull.*, 31 [5] 165–168 (1952).

II. Free Energy, Entropy, and Equilibrium, *ibid.*, 31 [6] 213–217 (1952).

III. Stability of Ceramic Materials, *ibid.*, 31 [7] 252–254 (1952).

IV. Crystal Chemistry, Physical Processes and Surface Effects, *ibid.*, 31 [8] 294–297 (1952).

V. Semiempirical Calculations, *ibid.*, 31 [9] 344–347 (1952).

VI. Summary, bibliography, and sources of data, *ibid.* 31 [10] 386–388 (1952).

(6) Silicate Chemistry

Bogue, Robert Herman, "The Chemistry of Portland Cement," Second Edition, Reinhold Publishing Corporation, New York, N. Y., 1955. 793 pp.

This comprehensive treatise is divided into three parts: I. The Chemistry of Clinker Formation; II. The Equilibria of Clinker Components; III. The Chemistry of Cement Utilization. Part II deals with the principles and techniques of high-temperature phase research, as well as with the detailed consideration of specific systems.

Eitel, Wilhelm, "Physical Chemistry of the Silicate," 5 volumes, Academic Press, New York, N. Y.

Volume 1, 1963, about 630 pp. Contents (tentative): Silicate Crystal Structures; Clay Mineral Structures; Silicate Dispersoids.

Volume 2, 1963, about 700 pp. Contents (tentative): Properties and Constitution of Silicate Glasses; Industrial Glass and Enamels; Industrial Slags.

Volumes 3–5, in preparation.

Eitel, Wilhelm, "Thermochemical Methods in Silicate Investigation," Rutgers University Press, New Brunswick, N. J., 1952. 132 pp.

An exposition of the science of calorimetry applied to silicate investigations. Calorimeters for determining heats of reaction and specific heats are described, along with methods of calculating the thermodynamic properties from the obtained data.

Eitel, Wilhelm, "Silicate Melt Equilibria," Rutgers University Press, New Brunswick, N. J., 1951. 159 pp. + 200 figs.

An exposition of phase equilibria principles applied to the understanding of anhydrous silicate systems.

Frondel, Clifford, "The System of Mineralogy" of James Dwight Dana and Edward Salisbury Dana, Seventh Edition, Vol. III, Silica Minerals, John Wiley and Sons, Inc., New York, 1962. 334 pp.

This book describes the mineralogy of the polymorphs of silica.

Greig, J. W., "Immiscibility in Silicate Melts," *Am. J. Sci.*, 13, 1–44, 133–154 (1927).

Classic study on silicate immiscibility.

Sosman, Robert B., "The Properties of Silica," The Chemical Catalog Company, Inc., J. J. Little and Ives Company, New York, 1927. 856 pp.

A comprehensive text on the properties of silica, with a section on industrial applications. A revised edition is in preparation.

(7) Special Collections of Phase Diagrams

Osborn, E. F. and Muan, A., "Phase Equilibrium Diagrams of Oxide Systems," The American Ceramic Society, 4055 N. High St., Columbus, Ohio.

Ten large-scale ternary phase equilibrium diagrams of fundamental importance are drawn from published data. Each diagram is on a 500 millimeter equilateral triangle, with a scale accurate enough to permit interpolation to within 0.1%.

Thoma, R. E., "Phase Diagrams of Nuclear Reactor Materials," Oak Ridge National Laboratory Report No. ORNL-2548, Oak Ridge, Tennessee, 1959. 205 pp. Available from the Office of Technical Services, Department of Commerce, Washington, D. C.

This compilation presents 145 phase diagrams for possible materials for use in nuclear reactors. Composition and temperature of invariant points, as well as the phase reaction, are listed separately.

(8) Phase Diagrams in Related Fields

American Society for Metals, "Metals Handbook," 1948 edition, 1332 pp., Cleveland 3, Ohio.

Sections on the constitution of binary alloys (pp. 1146 to 1240) and ternary alloys (pp. 1241 to 1268). Each diagram is annotated.

Hansen, Max, "Constitution of Binary Alloys," Second Edition, McGraw-Hill Book Co., Inc., New York, N. Y., 1958. 1305 pp. + 684 figs.

This comprehensive volume containing 1334 systems, 717 diagrams, and about 9800 references deals only with binary alloy systems, including the borides, carbides, nitrides, and silicides. All data are critically evaluated and incorporated into composite diagrams. Symmetry and lattice spacings of intermediate phases are given.

Gschneidner, Karl A., Jr., "Rare Earth Alloys," D. Van Nostrand Co., Inc., Princeton, N. J., 1961. 449 pp.

This book is a critical review of the alloy systems of the rare earth, scandium and yttrium metals. A hundred phase diagrams are included which have been constructed from the data of various references. The alloy systems, involving rare earth metals as one or more components, are divided into subgroups of binary, ternary and higher multicomponent systems.

Schwarzkopf, Paul, and Kieffer, Richard, "Refractory Hard Metals," Macmillan, New York, 1953. 447 pp. + 92 figs. + 100 tables.

Deals with the preparation, properties, and application to high temperature materials of the so-called hard metals, namely, the refractory and hard carbides, nitrides, borides, and silicides of transition metals. Phase diagrams given where available.

A. METAL-OXYGEN SYSTEMS, INCLUDING THOSE CONTAINING VALENCE CHANGES

I. One Metal with Oxygen

Al–O

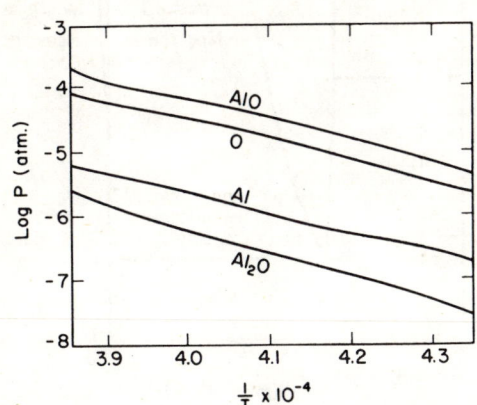

Fig. 1.—System Al–Al$_2$O$_3$; gaseous species over liquid Al$_2$O$_3$.

Leo Brewer and Alan W. Searcy, *J. Am. Chem. Soc.*, **73**, 5313 (1951).

Ce–O

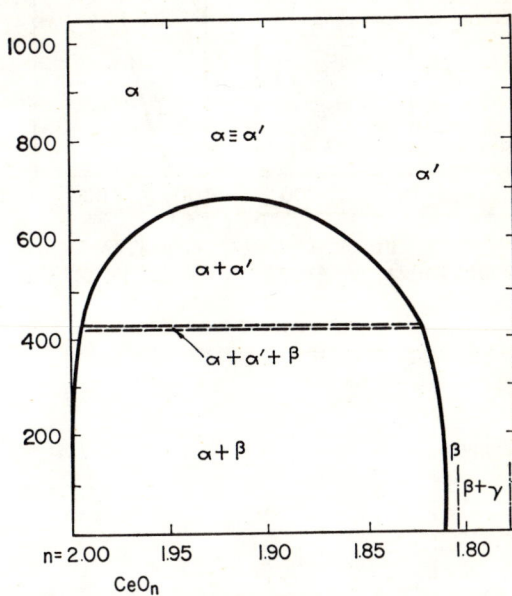

Fig. 3.—System CeO$_2$–CeO$_{1.78}$; immiscibility dome. Solid solution has defective fluorite structure type.

Georg Brauer and Karl Gingerich, "Rare Earth Research," edited by Eugene V. Kleber, The Macmillan Co., New York, p. 101 (1961).

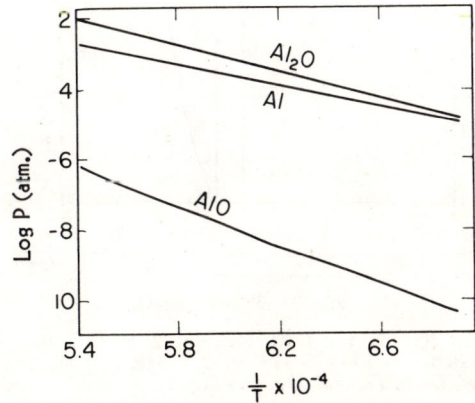

Fig. 2.—System Al–Al$_2$O$_3$; gaseous species over liquid Al and solid Al$_2$O$_3$ mixture.

Leo Brewer and Alan W. Searcy, *J. Am. Chem. Soc.*, **73**, 5313 (1951).

Co–O

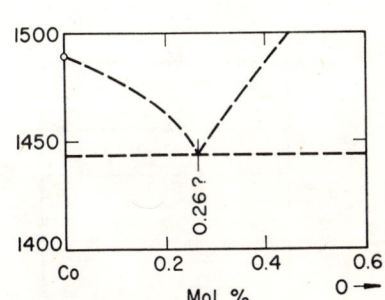

Fig. 4.—System Co–O$_2$.

P. Asanti and E. J. Kohlmeyer, *Z. anorg. Chem.*, **265**, 91 (1951).

Cr–O

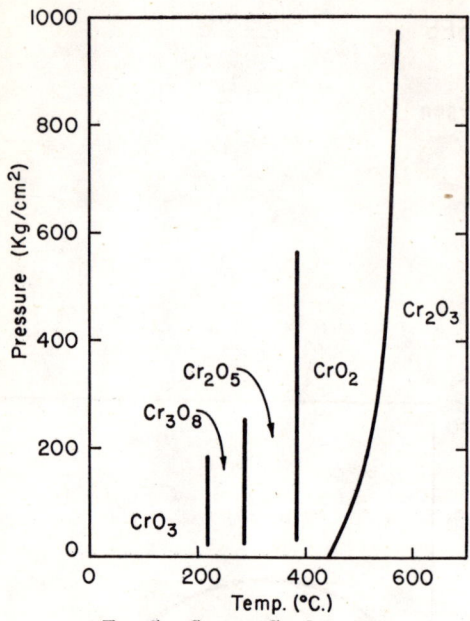

Fig. 5.—System Cr–O, p–T.
Buichi Kubota, *J. Am. Ceram. Soc.*, 44 [5] 247 (1961).

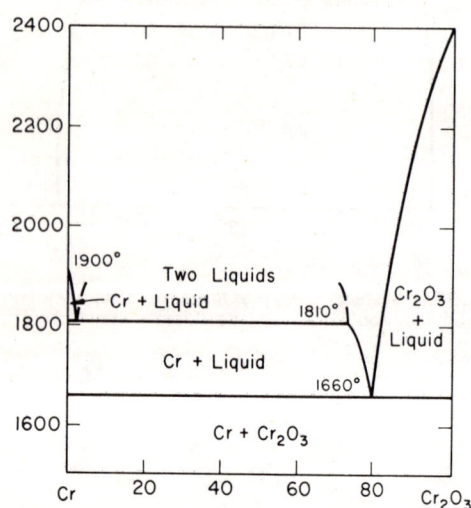

Fig. 6.—System Cr–Cr$_2$O$_3$.
Ya. I. Ol'shanskiĭ and V. K. Shlepov, *Doklady Akad. Nauk S.S.S.R.*, 91, 563 (1953).

Cu–O

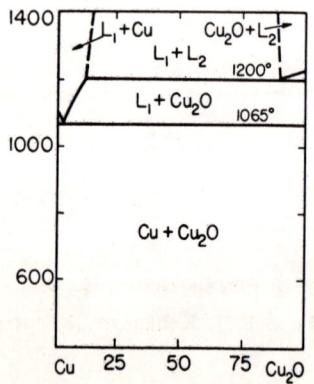

Fig. 7.—System Cu–Cu$_2$O.
Erich Gebhardt and Walter Obrowski, *Z. Metallk.*, 45, 333 (1954).

Fe–O

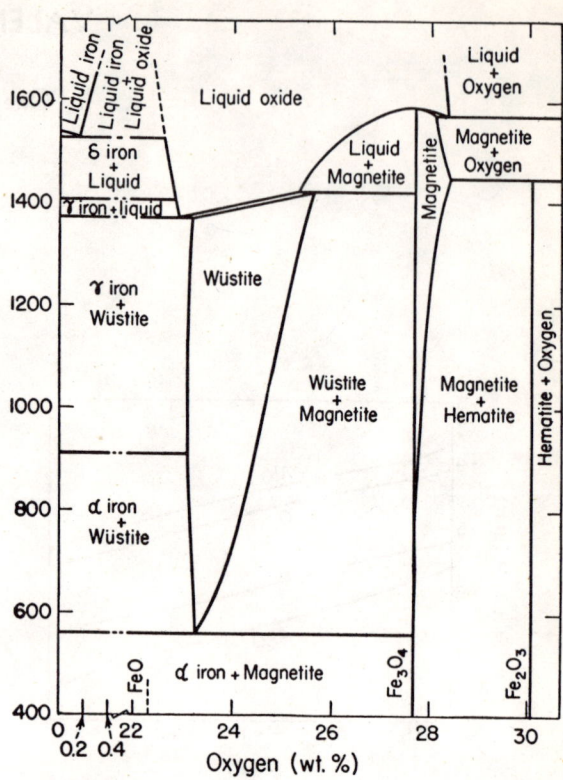

Fig. 8.—System Fe–Fe$_2$O$_3$; at a total pressure of one atmosphere.

L. S. Darken and R. W. Gurry, *J. Am. Chem. Soc.*, 68 [5] 799 (1946).

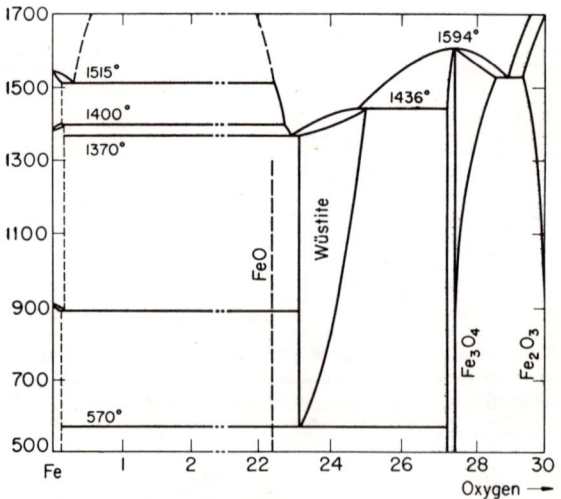

Fig. 9.—System Fe–O$_2$.

After R. Hay and J. M. McLeod, *J. West Scotland Iron Steel Inst.*, 52, 109 (1944). P. T. Carter and M. Ibrahim, *J. Soc. Glass Technol.*, 36, 144 (1952).

Fe–O (concl.)

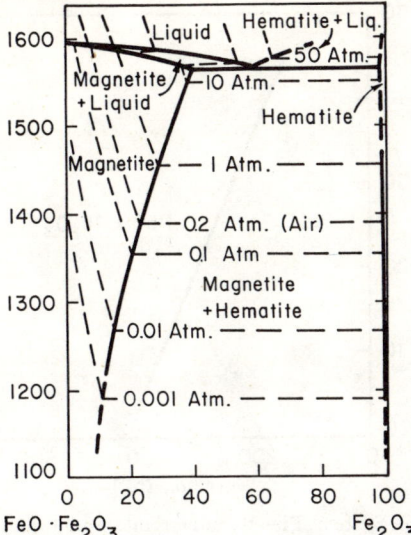

Fig. 10.—System FeO·Fe₂O₃–Fe₂O₃ in air; composite diagram. Based on data of: J. W. Greig, E. Posnjak, H. E. Merwin, and R. B. Sosman, *Am. J. Sci.* [5] **30** [9] 239–316 (1935); L. S. Darken and R. W. Gurry, *J. Am. Chem. Soc.*, **67** [8] 1398–1412 (1945); *ibid.*, **68** [5] 798–816 (1946); Bert Phillips and Arnulf Muan, *J. Phys. Chem.*, **64** [10] 1452 (1960).

Arnulf Muan and Shigeyuki Sōmiya, *J. Am. Ceram. Soc.*, **43** [4] 205 (1960).

Ge–O

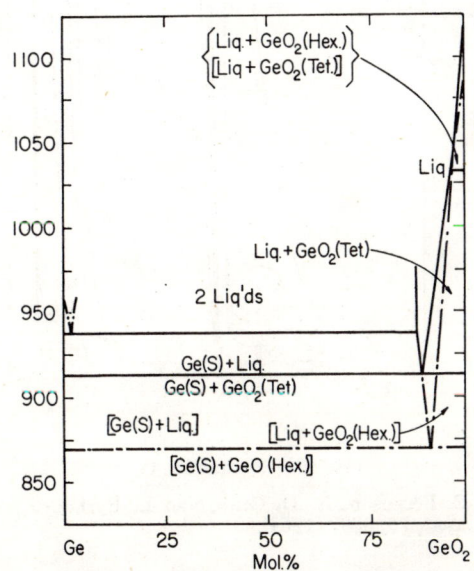

Fig. 11.—System Ge–GeO₂. ———stable equilibria– – – –metastable equilibria; phases in brackets are metastable. Melting point of Ge is 937°C, very close to the monotectic between Ge and 2 liquids. Private communication, F. A. Trumbore, Dec. 16, 1959.

F. A. Trumbore, C. D. Thurmond, and M. Kowalchik, *J. Chem. Phys.* **24**, 1112 (1956).

Mn–O

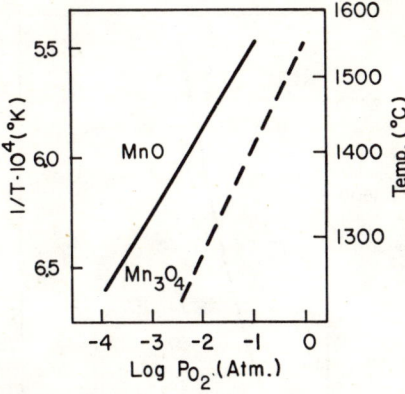

Fig. 12.—System MnO–Mn₃O₄ as a function of O₂ pressure and temp. Dashed curve after data of J. P. Coughlin, *U. S. Bur. Mines Bull.*, **1954**, No. 542, XII.

W. C. Hahn, Jr., and Arnulf Muan, *Am. J. Sci.*, **258**, 73 (1960).

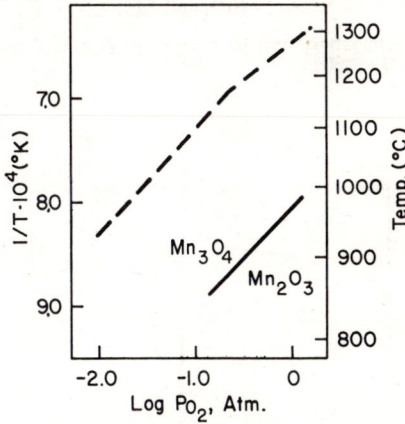

Fig. 13.—System Mn₂O₃–Mn₃O₄ as a function of O₂ pressure and temp. Dashed curve after data of J. P. Coughlin, *U. S. Bur. Mines Bull.*, **1954**, No. 542, XII.

W. C. Hahn, Jr., and Arnulf Muan, *Am. J. Sci.*, **258**, 69 (1960).

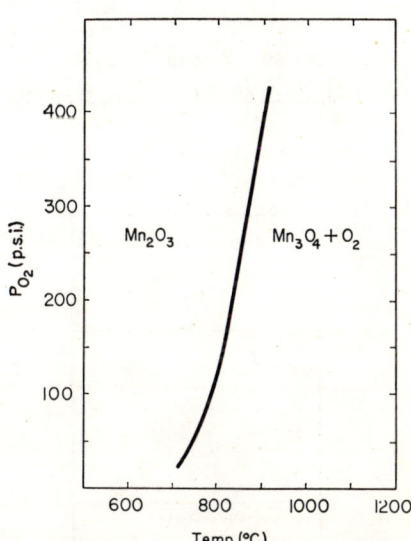

Fig. 14.—System Mn–O; univariant equilibrium curve between Mn₂O₃ and Mn₃O₄.

Cyrus Klingsberg and Rustum Roy, *J. Am. Ceram. Soc.*, **43** [12] 623 (1960).

Mn–O (concl.)

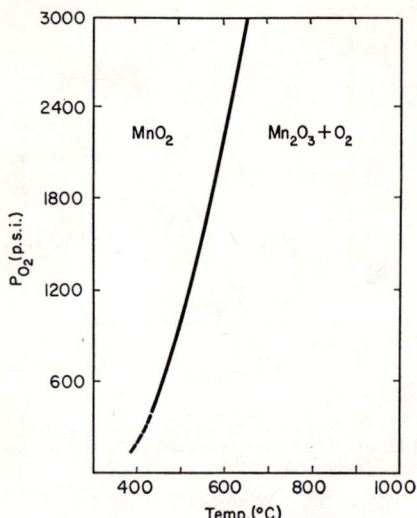

FIG. 15.—System Mn–O; univariant equilibrium curve between βMnO_2 and Mn_2O_3.

Cyrus Klingsberg and Rustum Roy, *J. Am. Ceram. Soc.*, **43** [12] 623 (1960).

Ni–O

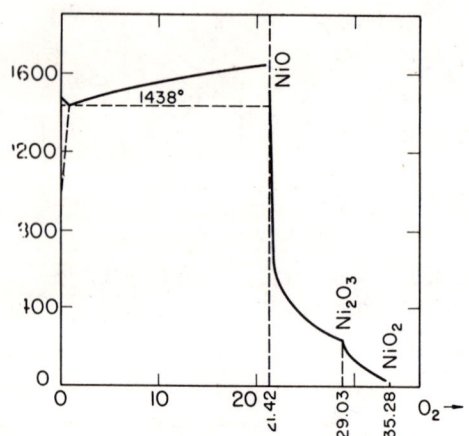

FIG. 16.—System Ni–O_2.

D. P. Bogatski, *Zhur. Obshcheĭ Khim.*, **21**, 9 (1951).

Pb–O

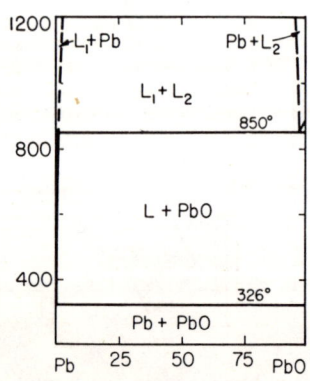

FIG. 17.—System Pb–PbO.

Erich Gebhardt and Walter Obrowski, *Z. Metallk.*, **45**, 333 (1954).

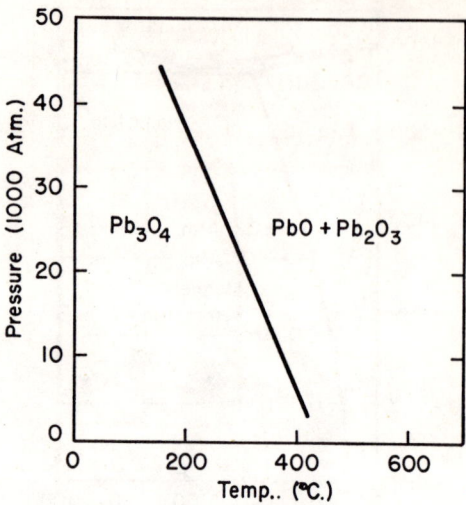

FIG. 18.—System Pb–O; univariant equilibrium curve between Pb_3O_4 and $PbO + Pb_2O_3$.

W. B. White, Frank Dachille, and Rustum Roy, *J. Am. Ceram. Soc.*, **44** [4] 174 (1961).

Pr–O

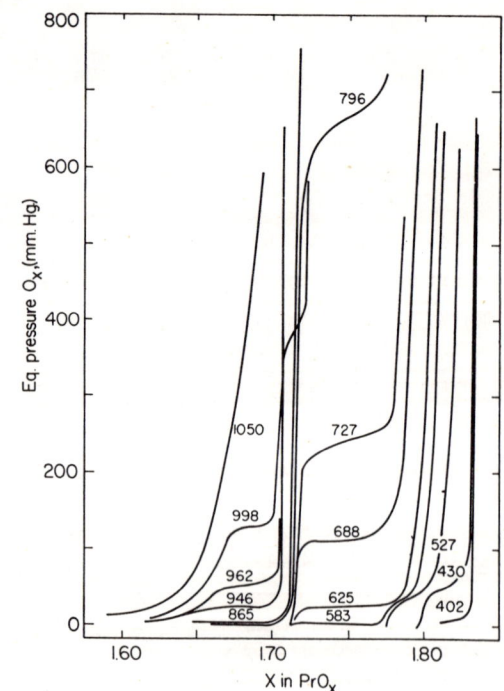

FIG. 19.—System Pr–O_2.

R. E. Ferguson, E. D. Guth, and L. Eyring, *J. Am. Chem. Soc.*, **76**, 3892 (1954).

Sb–O

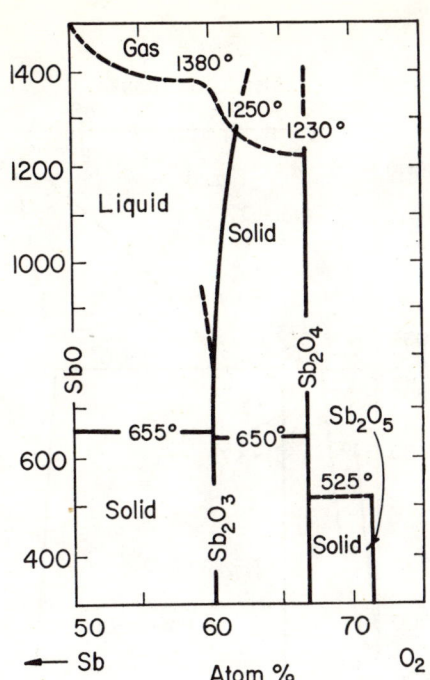

FIG. 20.—System SbO–O.
Helmut Hennig and E. J. Kohlmeyer, *Z. Erzbergbau, Metallhüettenw.*, **10**, 12 (1957).

Si–O

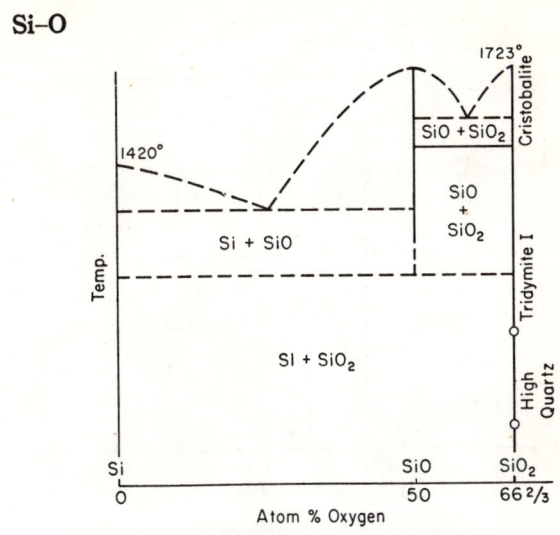

FIG. 21.—System Si–SiO$_2$, at low p, tentative.
R. B. Sosman, *Trans. Brit. Ceram. Soc.*, **54**, 657 (1955).

Ti–O

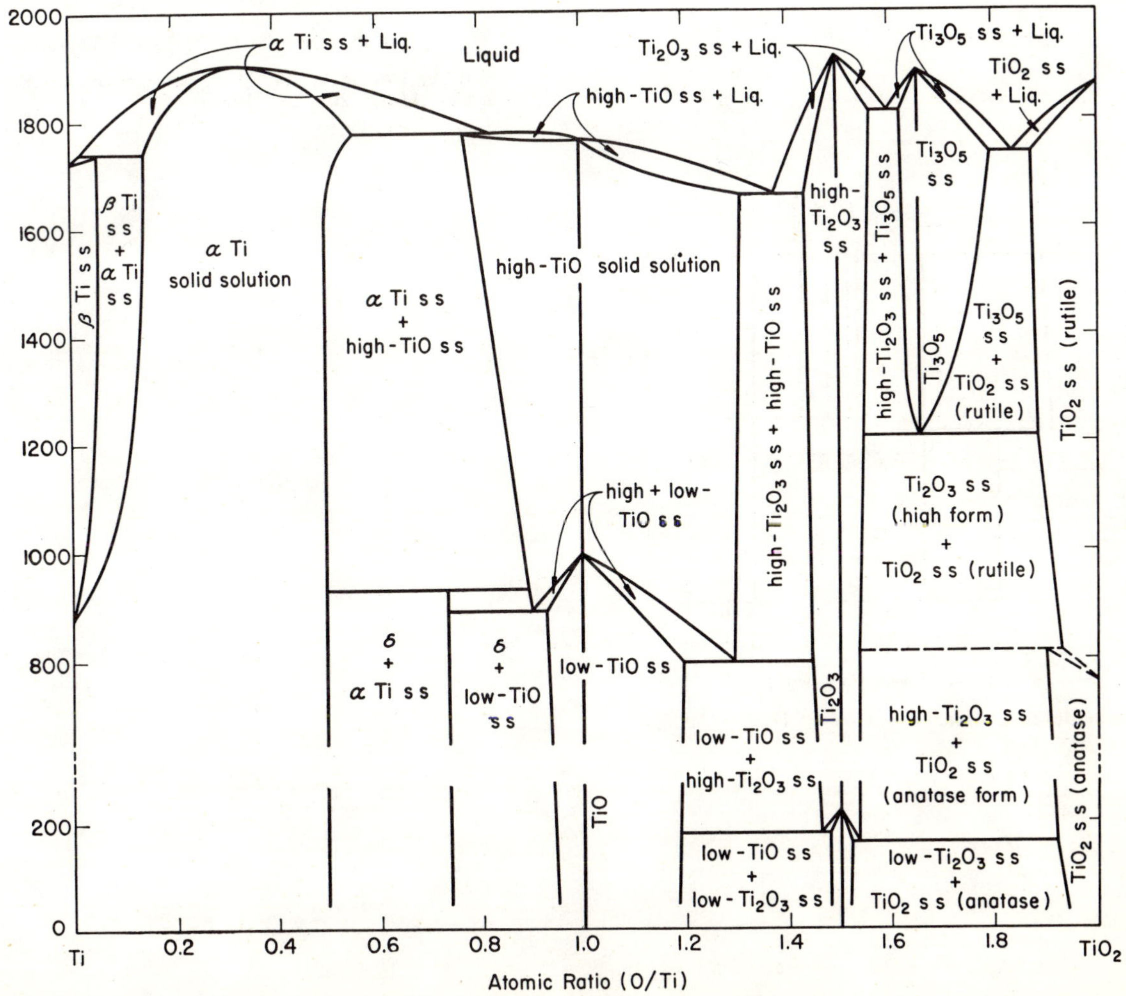

FIG. 22.—System Ti–TiO$_2$ from literature data. For Ti-rich region see also, E. S. Bumps, H. D. Kessler, and M. Hansen, *Trans. Am. Soc. Metals*, **45**, 1013 (1954).
R.C. DeVries and Rustum Roy, *Am. Ceram. Soc. Bull.*, **33** [12] 370–72 (1954).

U–O

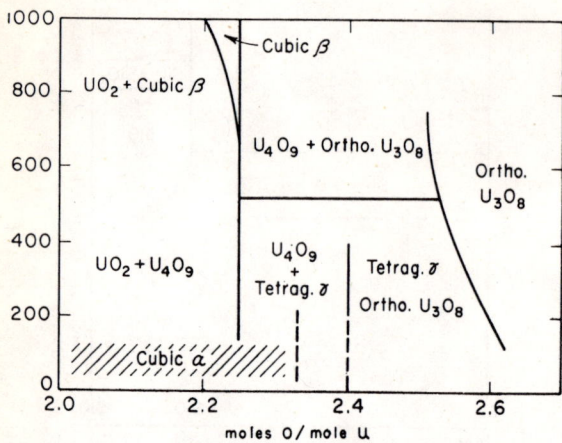

Fig. 23.—System UO_2–U_3O_8.

S. M. Lang, F. P. Knudsen, C. L. Fillmore, and R. S. Roth, *Natl. Bur. Standards Circ.*, No. 568, p. 3 (Feb. 20, 1956).

After P. Perio, *Bull. Soc. Chim. France*, p. 256 (1953).

Zr–O

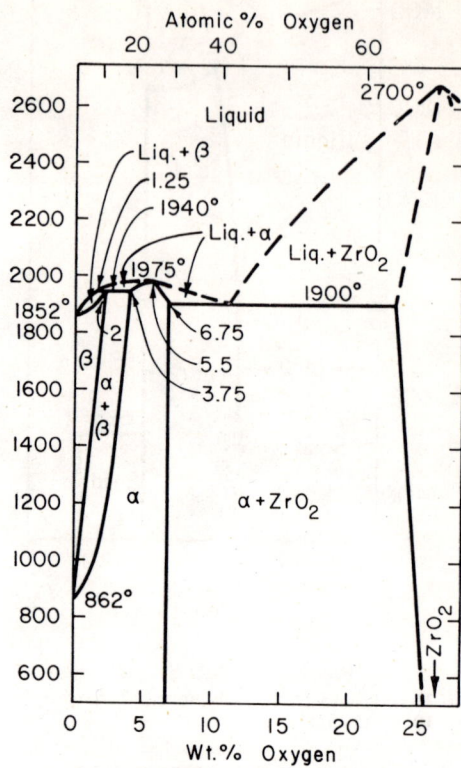

Fig. 25.—System Zr–O; partial diagram.

R. F. Domagala and D. J. McPherson, *J. Metals*, **6**; *Trans. AIME*, **200**, 241 (1954).

V–O

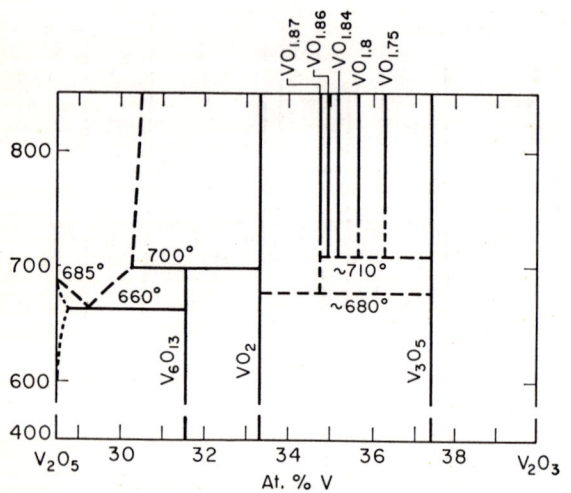

Fig. 24.—System V_2O_3–V_2O_5.

Aurelio Burdese, *Ann. Chim. (Rome)*, **47**, 795 (1957).

II. Two Metals with Oxygen

Al–Fe–O

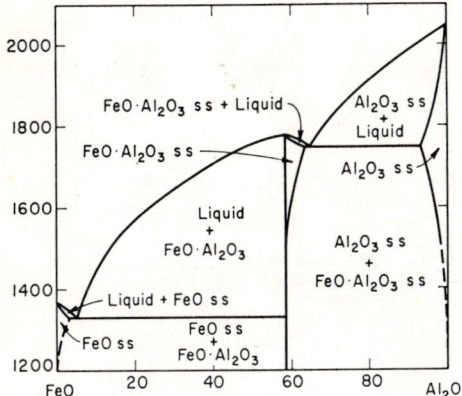

FIG. 26.—System $FeO-Al_2O_3$. ss = solid solution.
W. A. Fischer and Alfred Hoffmann, *Arch. Eisenhüttenw.*, **27** [5] 344 (1956).

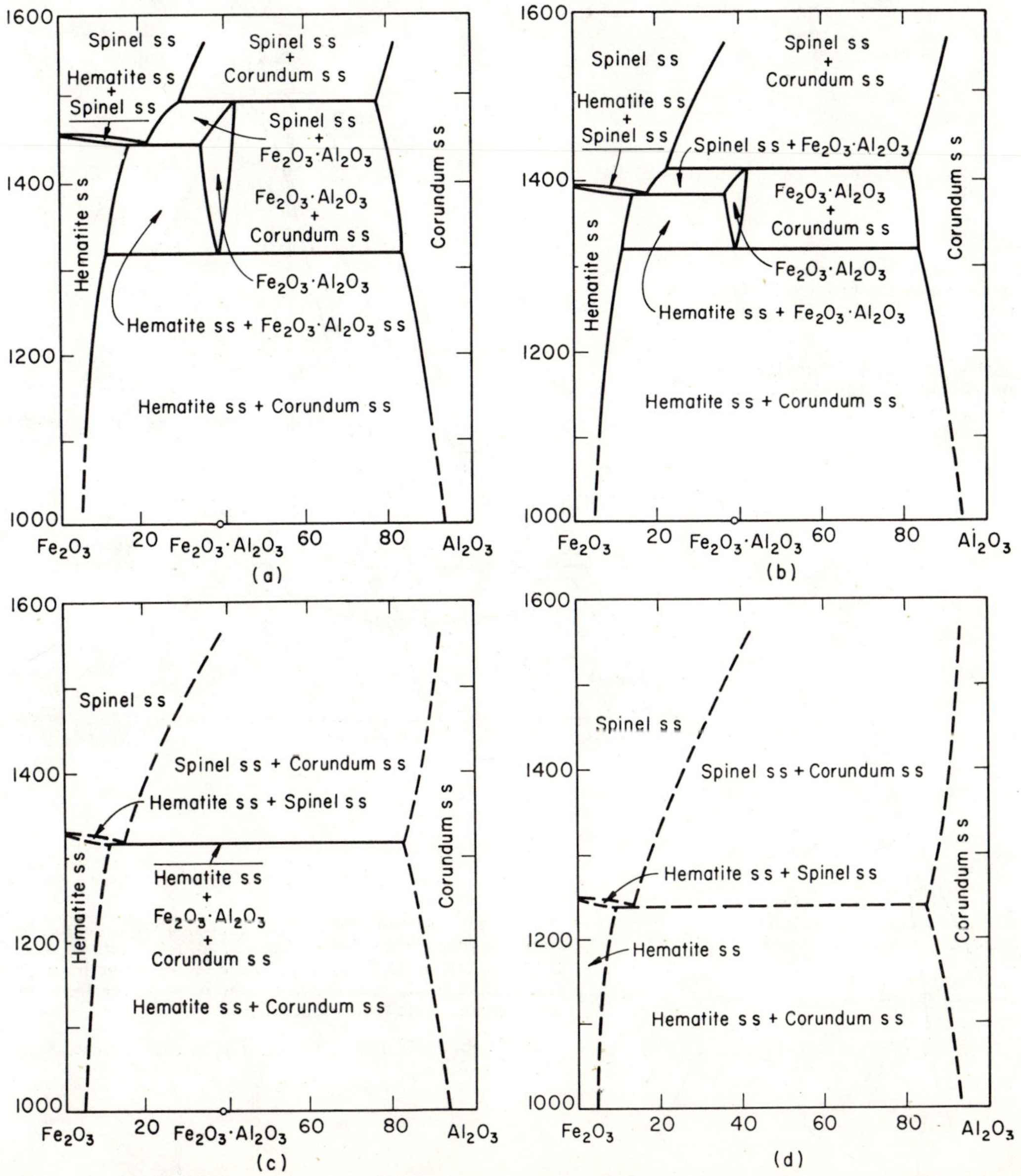

FIG. 27.—System $Fe_2O_3-Al_2O_3$; effect of O_2 partial pressures on subsolidus equilibria. (a), (b), (c), and (d) refer, respectively, to 1, 0.2, 0.03, and slightly below 0.03 atmosphere O_2 partial pressure.

Arnulf Muan, *Am. J. Sci.*, **256**, 420 (1958).

Al–Fe–O (cont.)

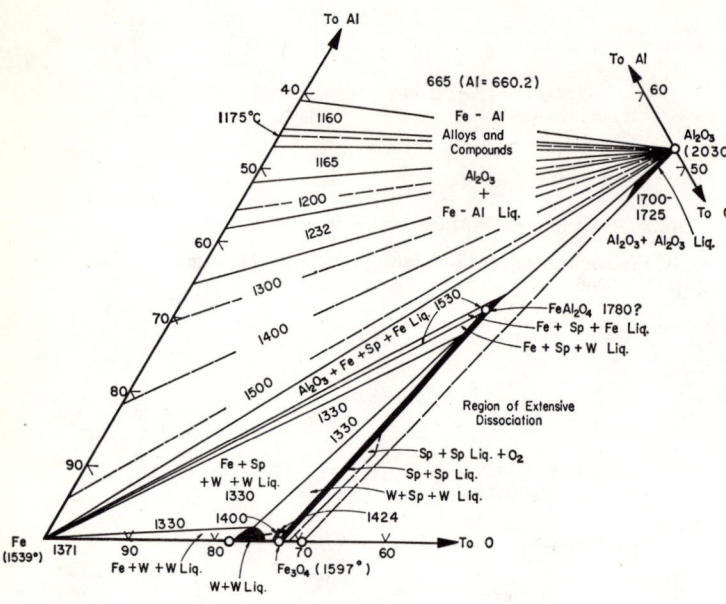

FIG. 28.—System Fe–Al–O; solidus surface. Sp = Fe(Fe, Al)$_2$O$_4$ spinel, W = wüstite, Liq. = liquid.

L. M. Atlas and W. K. Sumida, *J. Am. Ceram. Soc.*, 41 [5] 159 (1958).

FIG. 30.—System FeO–Al$_2$O$_3$–Fe$_2$O$_3$; changes in solid solution (ss) equilibria with temperature. R W = Richards and White.

L. M. Atlas and W. K. Sumida, *J. Am. Ceram. Soc.*, 41 [5] 158 (1958).

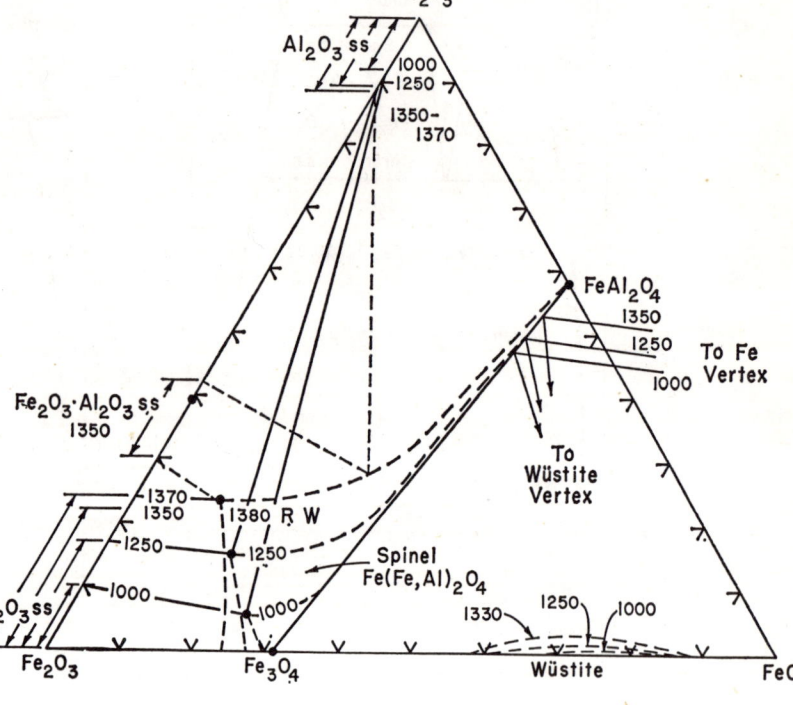

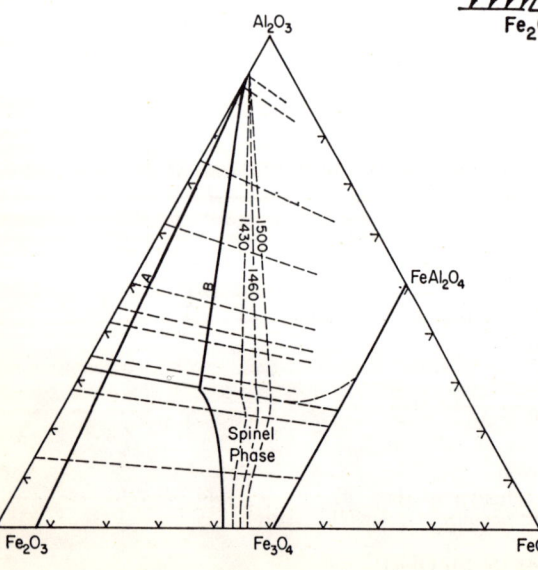

FIG. 29.—System FeO–Al$_2$O$_3$–Fe$_2$O$_3$, showing equilibrium between sesquioxide phase and spinel phases at oxygen pressure of 15.9 cm. Hg. Line A = upper limit of sesquioxide-spinel transition. Line B = lower limit of spinel-sesquioxide transition.

R. G. Richards and J. White, *Trans. Brit. Ceram. Soc.*, 53, 253 (1954).

Al–Fe–O (cont.)

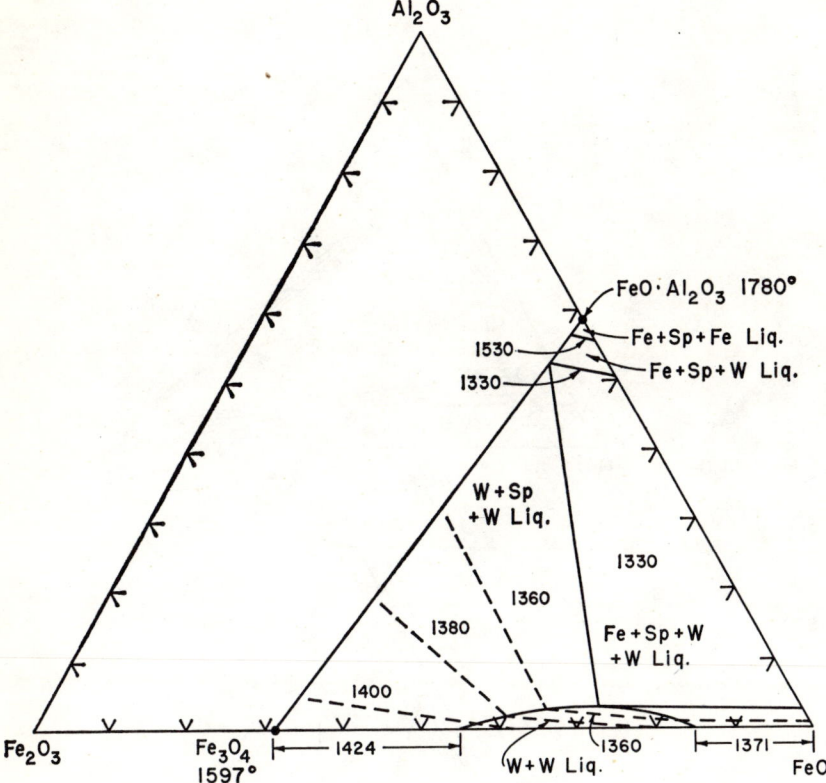

Fig. 31.—System FeO–Al$_2$O$_3$–Fe$_3$O$_4$; solidus equilibria. Sp = Fe(Fe, Al)$_2$O$_4$ spinel, W = wüstite, Liq. = liquid.

L. M. Atlas and W. K. Sumida, *J. Am. Ceram. Soc.*, 41 [5] 159 (1958).

Fig. 32. See page 46.

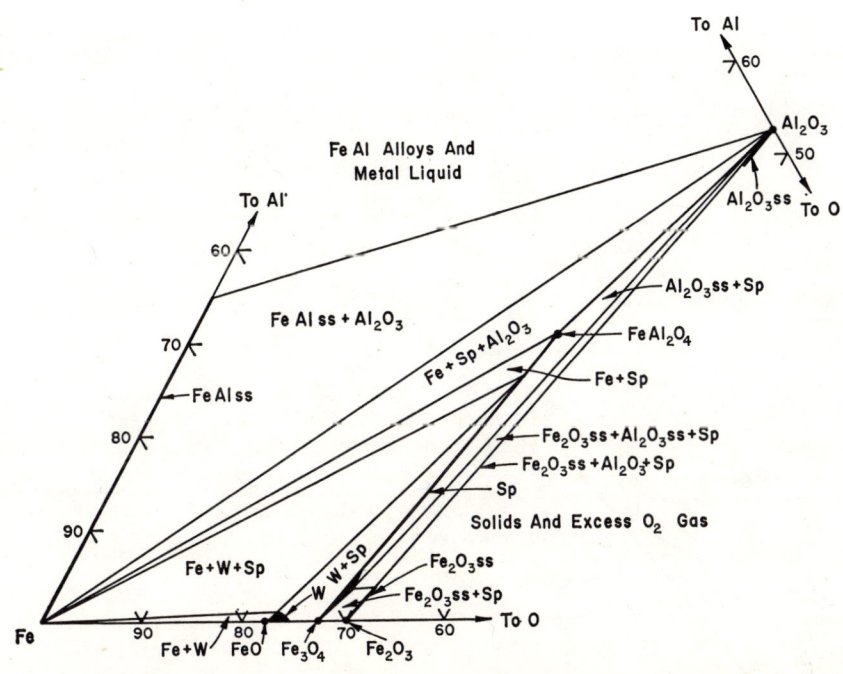

Fig. 33.—System Fe–Al–O at 1000°C. Sp = Fe(Fe, Al)$_2$O$_4$ spinel, W = wüstite, ss = solid solution.

L. M. Atlas and W. K. Sumida, *J. Am. Ceram. Soc.*, 41 [5] 157 (1958).

Al–Fe–O (cont.)

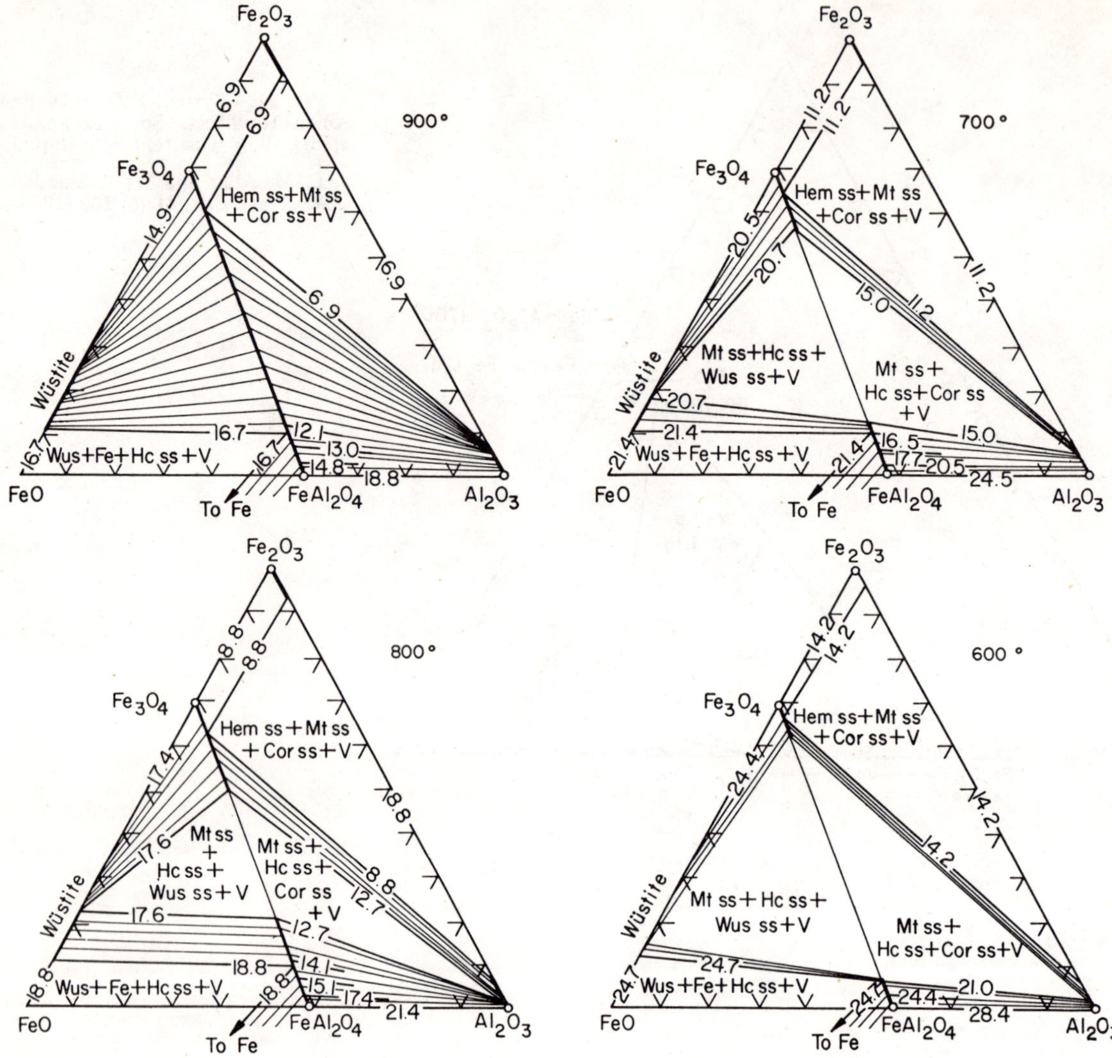

Fig. 32.—System FeO–Fe_2O_3–Al_2O_3; isothermal sections. Tie lines contoured for oxygen fugacity (in atms.), with values in $-\log f_{O_2}$. Cor = corundum; Fe = iron; Hc = hercynite; Hem = hematite, Mt = magnetite; V = vapor; Wus = wüstite.

A. C. Turnock and D. H. Lindsley, Ann. Rept. Director Geophys. Lab. 1960–61; in *Carnegie Inst. Washington Year Book*, **60**, 157 (1961).

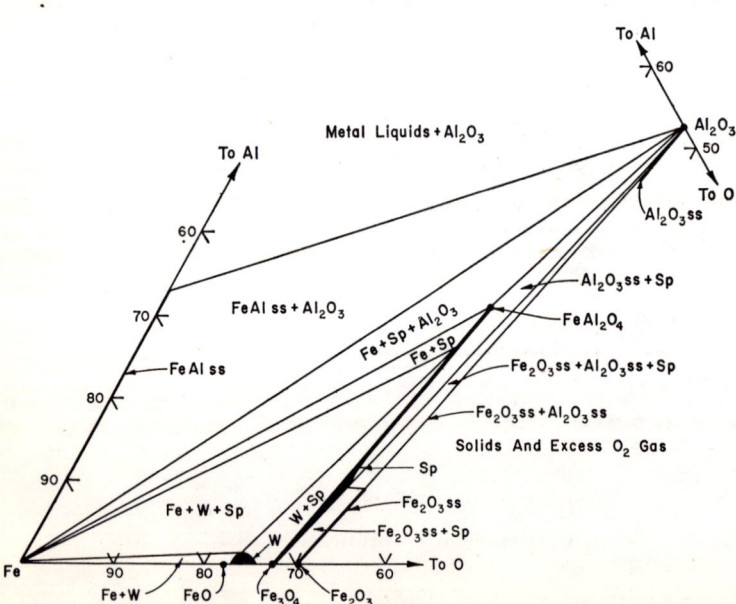

Fig. 34.—System Fe–Al–O at 1250°C. Sp = $Fe(Fe, Al)_2O_4$ spinel, W = wüstite, ss = solid solution.

L. M. Atlas and W. K. Sumida, *J. Am. Ceram. Soc.*, **41** [5] 157 (1958).

Al–Fe–O (concl.)

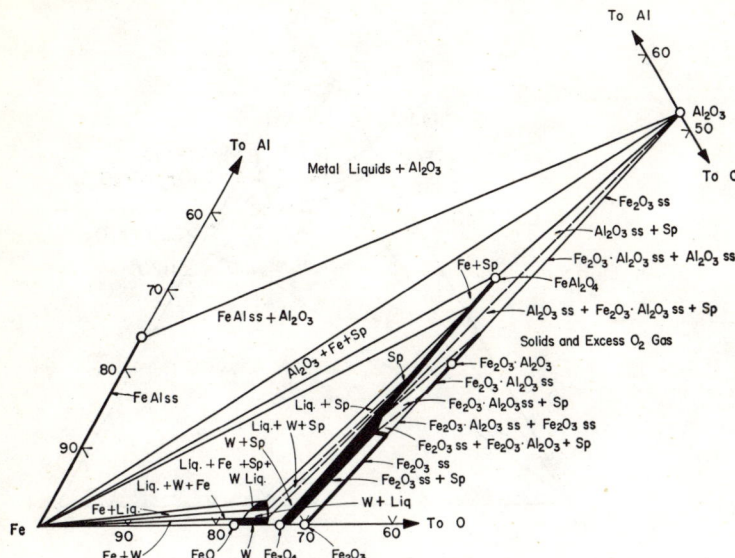

Fig. 35.—System Fe–Al–O at 1350°C. Sp = Fe(Fe, Al)$_2$O$_4$ spinel, W = wüstite, ss = solid solution, Liq. = liquid.

L. M. Atlas and W. K. Sumida, *J. Am. Ceram. Soc.*, **41** [5] 158 (1958).

C–Ti–O

Fig. 36.—System Ti–C–O; isothermal sections.

L. Stone and H. Margolin, *J. Metals*, **5**; *Trans. AIME*, **197**, 1500 (1953).

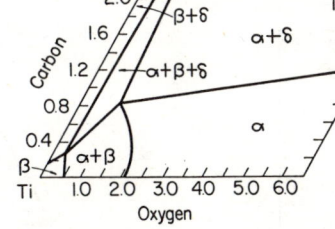
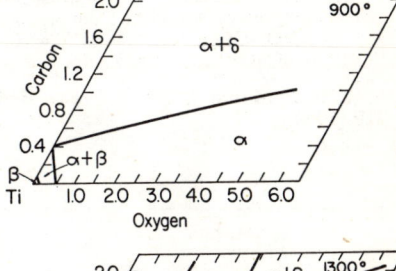
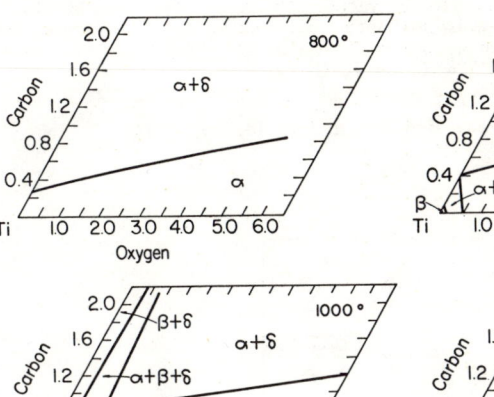
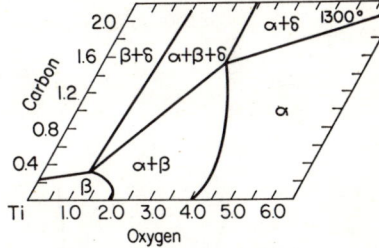

Ca–Cr–O

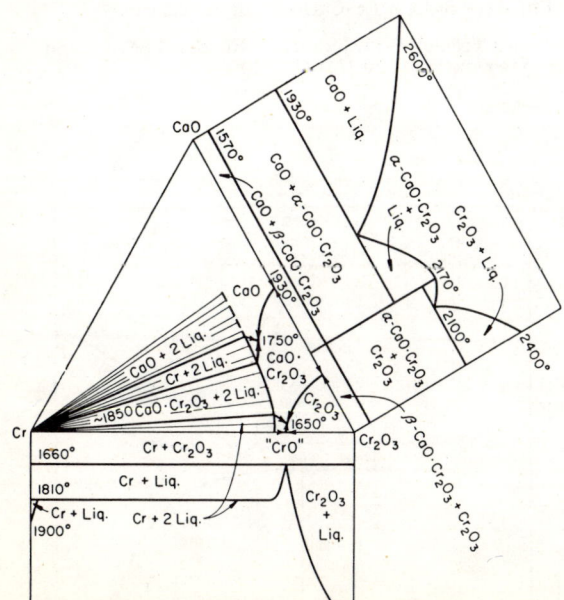

Fig. 37.—System Cr–CaO–Cr$_2$O$_3$.

Ya. I. Ol'shanskiĭ, A. I. Tsvetkov, and V. K. Shlepov, *Doklady Akad. Nauk S.S.S.R.*, **96**, 1008 (1954).

Ca–Cr–O (concl.)

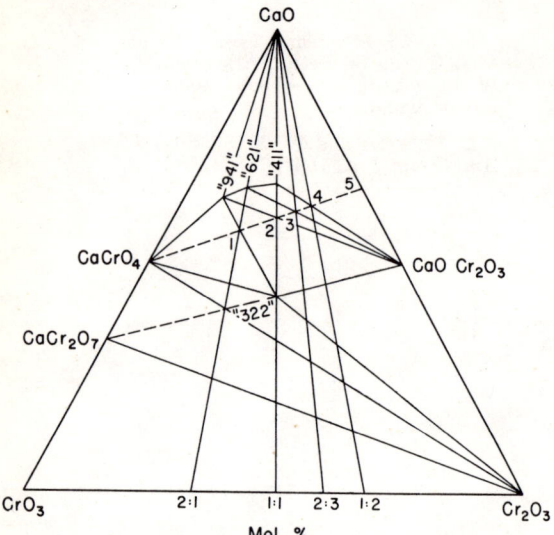

Fig. 38.—System CaO–CrO_3–Cr_2O_3; phase distribution in vacuo. Upper dotted straight line constitutes "dissociation path" of calcium chromate in vacuo. Under normal oxygen pressures, it is improbable that the "621" and "411" compounds would be formed; 2:1 = $2CrO_3 \cdot Cr_2O_3$; 1:1 = $CrO_3 \cdot Cr_2O_3$, etc. "941" = $9CaO \cdot 4CrO_3 \cdot Cr_2O_3$; "621" = $6CaO \cdot 2CrO_3 \cdot Cr_2O_3$, etc.

W. F. Ford and J. White, *Trans. Brit. Ceram. Soc.*, **48**, 419 (1949).

Ca–Fe–O

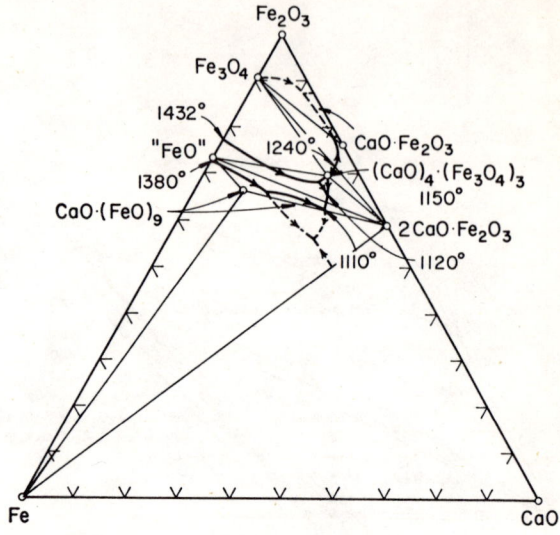

Fig. 40.—System CaO–Fe–Fe_2O_3.

E. Martin and R. Vogel, *Arch Eisenhüttenw.*, **6**, 109 (1932–33); *ibid.*, **8**, 249 (1934–35); from James White, *J. Iron Steel Inst.* (*London*) **148**, No. II, 636 (1943).

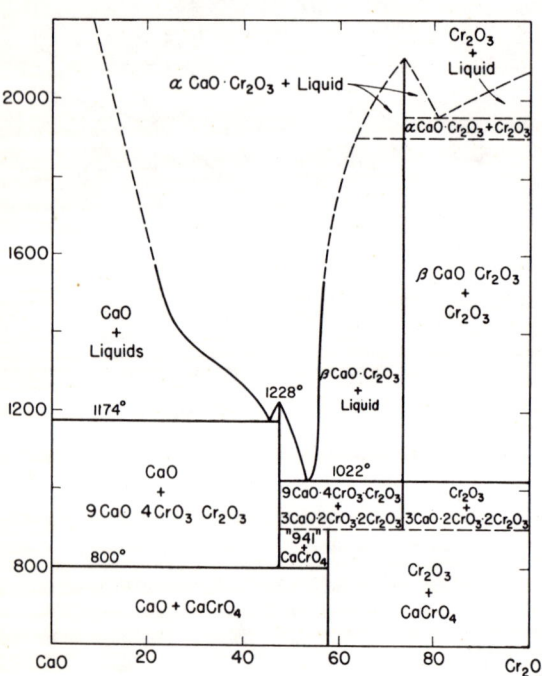

Fig. 39.—System CaO–Cr_2O_3; probable under atmospheric oxygen pressure. Some CrO_3 present because of oxygen reactions.

W. F. Ford and J. White, *Trans. Brit. Ceram. Soc.*, **48**, 423 (1949).

Fig. 42.—System CaO–FeO. Circles—estimated limits of solid solution.

W. C. Allen and R. B. Snow, *J. Am. Ceram. Soc.*, **38** [8] 268 (1955).

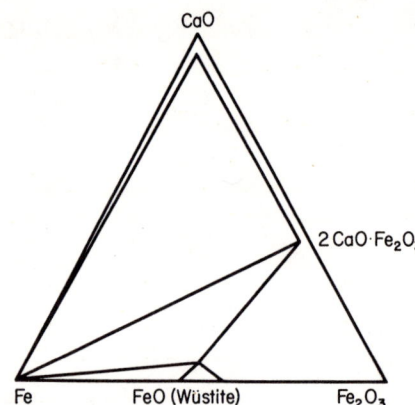

Fig. 41.—System Fe–CaO–Fe_2O_3; schematic.

Gerhard Trömel, Willi Jäger, and Eberhard Schürmann, *Arch. Eisenhüettenw.*, **26** [11] 695 (1955).

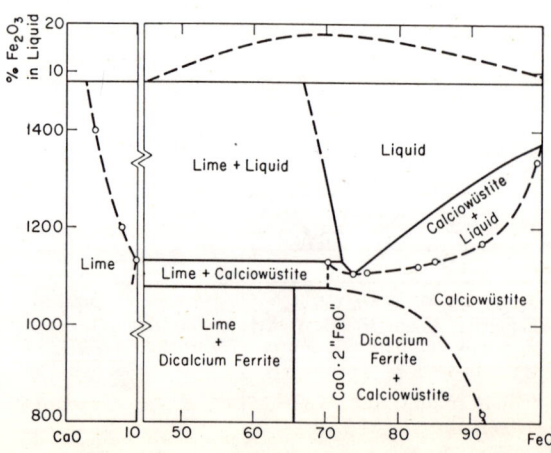

Ca–Fe–O (cont.)

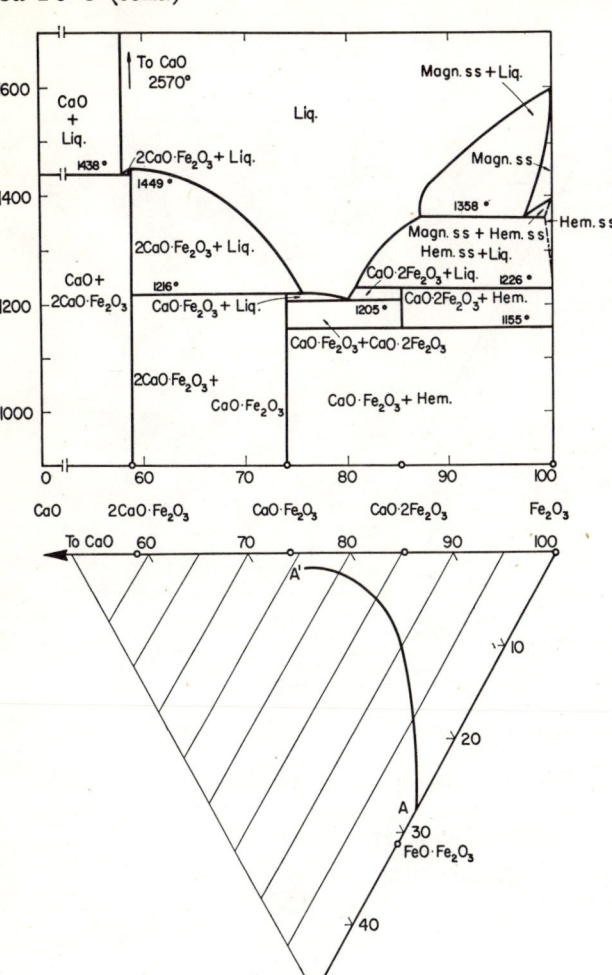

Fig. 43.—System CaO–Fe$_2$O$_3$ in air. Hem = hematite, Magn = magnetite, ss = solid solution. Curve A-A' in triangular diagram shows composition of liquids at liquidus temperatures.

Bert Phillips and Arnulf Muan, *J. Am. Ceram. Soc.*, **41** [11] 448 (1958).

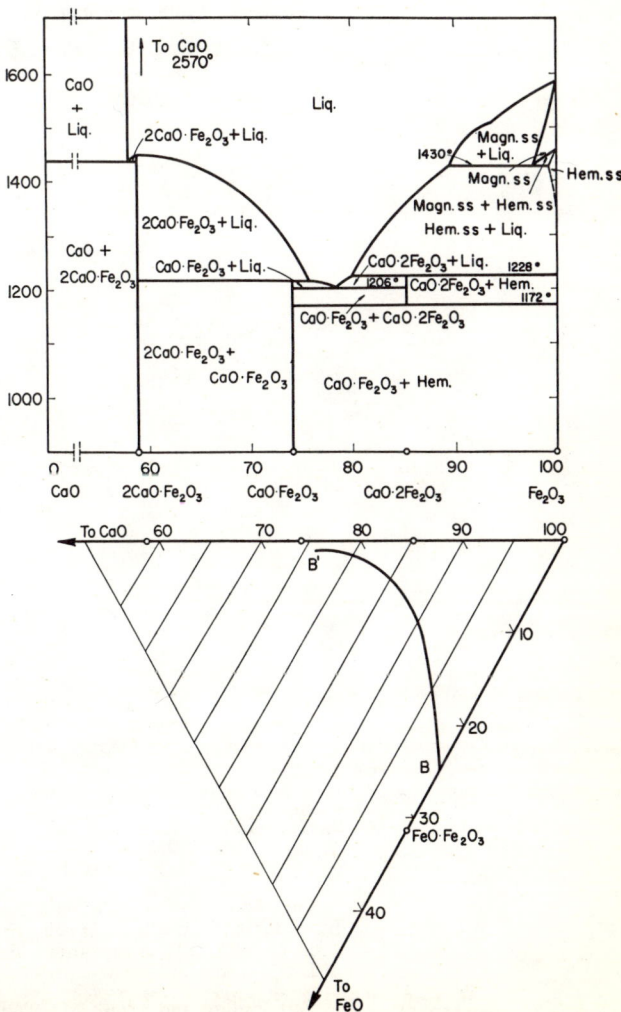

Fig. 44.—System CaO–Fe$_2$O$_3$ at 1 atm. O$_2$ pressure. Hem = hematite, Magn = magnetite, ss = solid solution. Curve B-B' in triangular diagram shows composition of liquids at liquidus temperatures.

Bert Phillips and Arnulf Muan, *J. Am. Ceram. Soc.*, **41** [11] 449 (1958).

Ca–Fe–O (cont.)

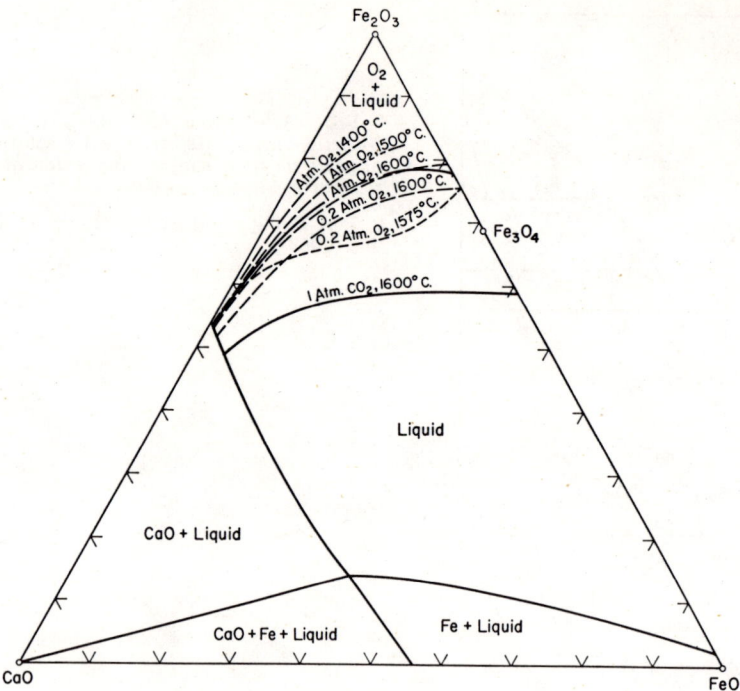

Fig. 45.—System CaO–FeO–Fe$_2$O$_3$ at 1600°C.
——— this investigation.
– – – Krings and Shackmann, *Z. Elektrochem. angew. physikal chem.*, **41**, 479 (1935).
— — J. White, *Iron and Steel Inst., Carnegie Schol. Mem.*, **27**, 1 (1938).
R. W. Gurry and L. S. Darken, *J. Am. Chem. Soc.*, **72**, 3908 (1950).

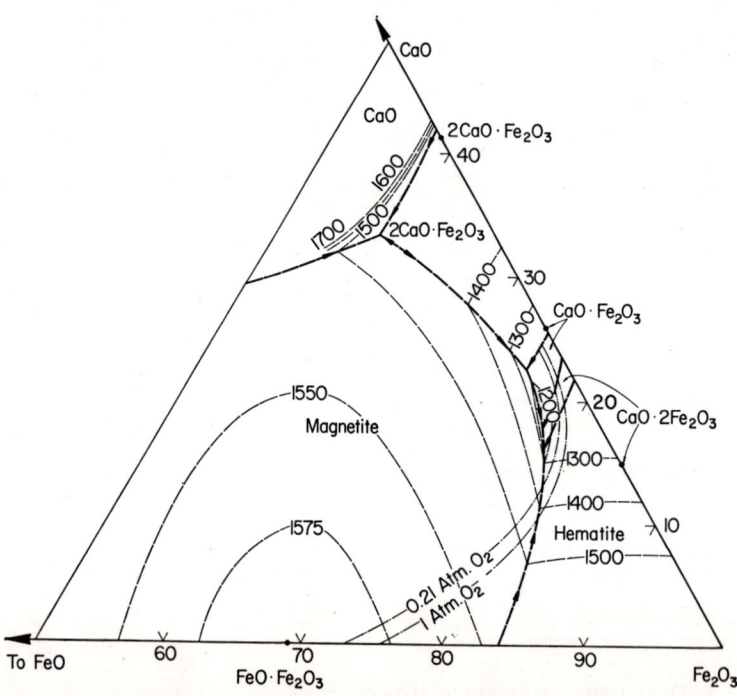

Fig. 46.—System 2CaO·Fe$_2$O$_3$–FeO·Fe$_2$O$_3$–Fe$_2$O$_3$; probable. Modified from original version of Aurelio Burdese and Cesare Brisi (*Ricerca Sci.*, **22**, 1564–67 (1952)) to incorporate a field of CaO·2Fe$_2$O$_3$ and to agree with data of L. S. Darken and R. W. Gurry (*J. Am. Chem. Soc.*, **68** [5] 798–816 (1946)).

Bert Phillips and Arnulf Muan, *J. Am. Ceram. Soc.*, **41** [11] 453 (1958).

Ca–Fe–O (concl.)

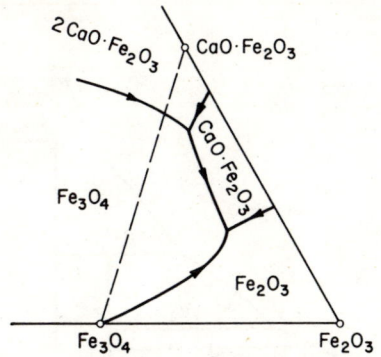

Fig. 47.—System Fe_2O_3–Fe_3O_4–$CaO \cdot Fe_2O_3$.

Aurelio Burdese and Cesare Brisi, *Ricerca Sci.*, 22, 1565 (1952).

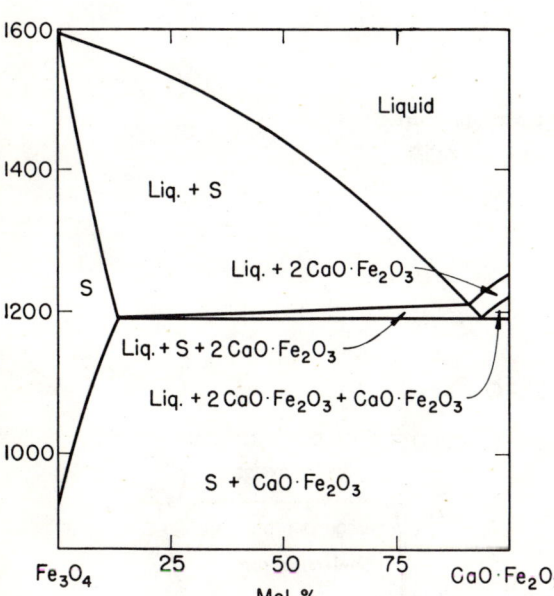

Fig. 48.—System Fe_3O_4–$CaO \cdot Fe_2O_3$.

Aurelio Burdese and Cesare Brisi, *Ricerca Sci.*, 22, 1566 (1952).

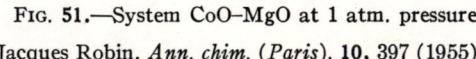

Fig. 51.—System CoO–MgO at 1 atm. pressure.

Jacques Robin, *Ann. chim.* (*Paris*), 10, 397 (1955).

Co–Fe–O

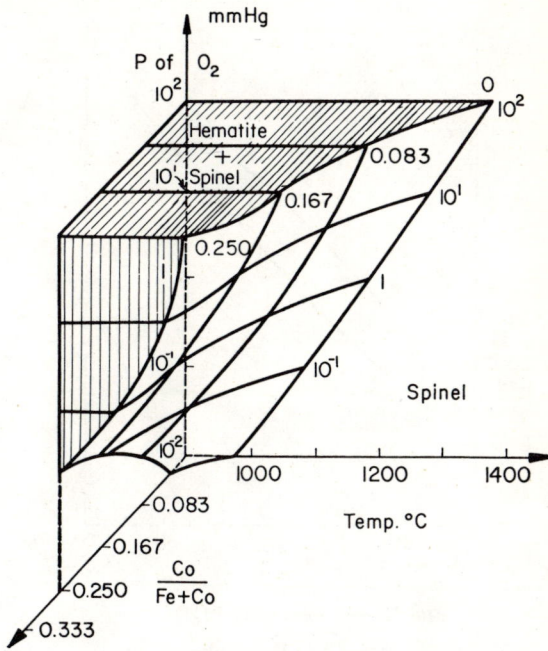

Fig. 49.—System Fe–Co–O; schematic. For isothermal diagrams at 1000°, 1100°, 1200°, and 1300°C, see B. D. Roiter and A. E. Paladino, *J. Am. Ceram. Soc.*, 45 [3] 132 (1962).

Shuichi Iida, *J. Phys. Soc. Japan*, 11, 847 (1956).

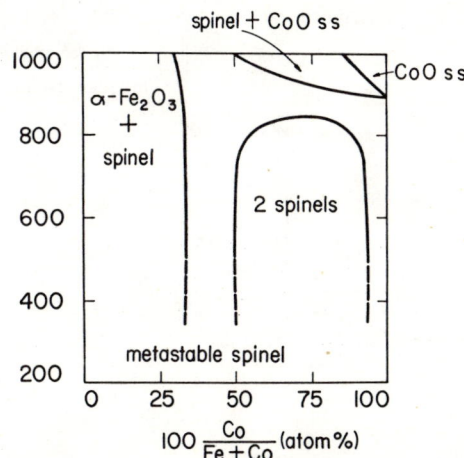

Fig. 50.—System CoO–Fe_2O_3 at 1 atm. pressure.

Jacques Robin, *Ann. chim.* (*Paris*), 10, 393 (1955).

Co–Mg–O

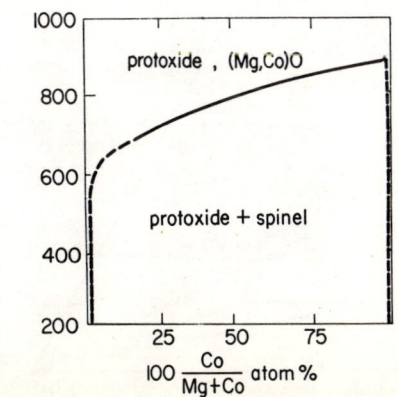

Co–Mg–O (concl.)

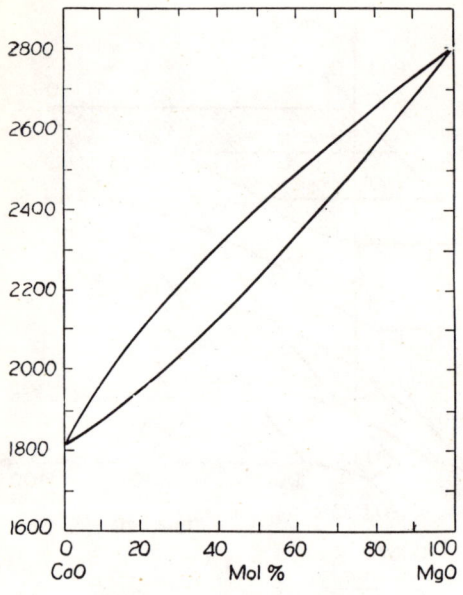

Fig. 52.—System CoO–MgO.

H. v. Wartenberg and E. Prophet, *Z. anorg. u. allgem. Chem.*, **208**, 379 (1932).

Co–Ni–O

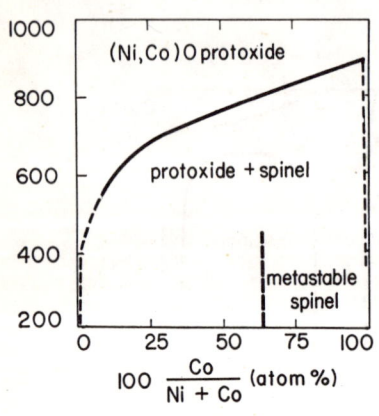

Fig. 53.—System CoO–NiO at 1 atm. pressure.

Jacques Robin, *Ann. chim.* (Paris), **10**, 395 (1955).

Co–V–O

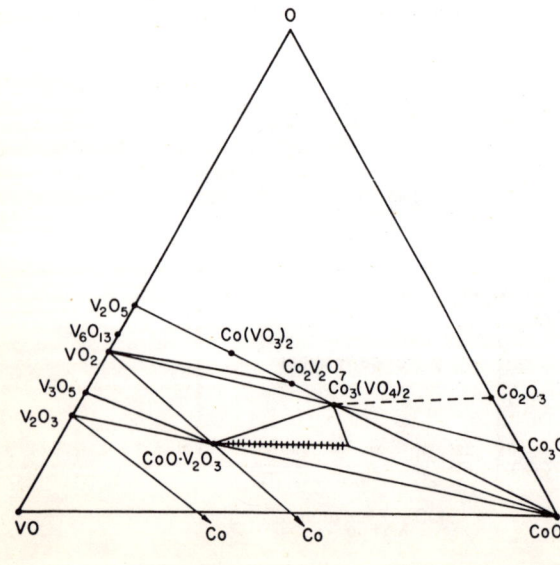

Fig. 54.—System CoO–VO–O.

Cesare Brisi, *Ann. chim.* (Rome), **47**, 830 (1957).

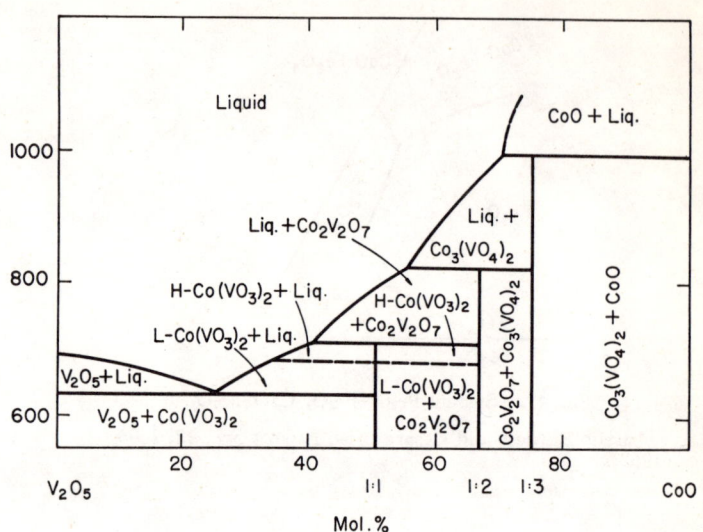

Fig. 55.—System CoO–V_2O_5.

Cesare Brisi, *Ann. chim.* (Rome), **47** [7,8] 815 (1957).

Co–Zn–O

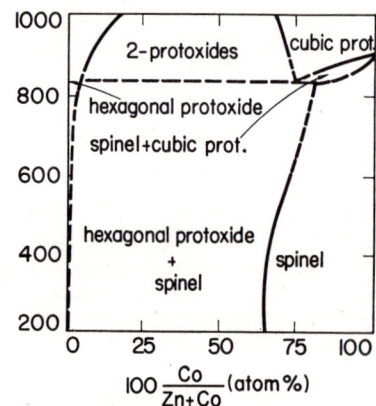

Fig. 56.—System CoO–ZnO at 1 atm. pressure.

Jacques Robin, *Ann. chim.* (Paris), **10**, 400 (1955).

Cr–Fe–O

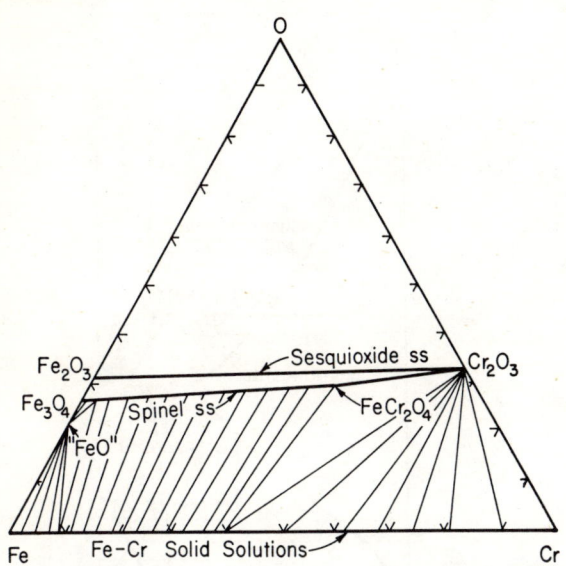

FIG. 57.—System Fe–Cr–O, above 575°C.; tentative.

D. Woodhouse and J. White, *Trans. Brit. Ceram. Soc.*, **54**, 365 (1955).

Cu–Fe–O

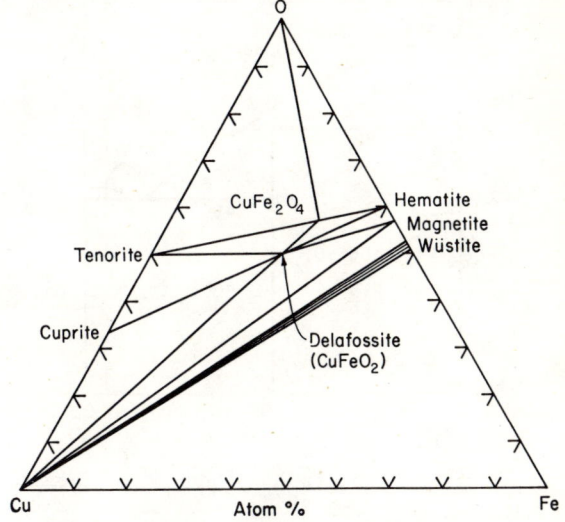

FIG. 59.—System Cu–Fe–O at 800°C.

R. A. Yund and G. Kullerud, Ann. Rept. Director Geophys. Lab. 1960–61; in *Carnegie Inst. Washington Year Book*, **60**, 181 (1961).

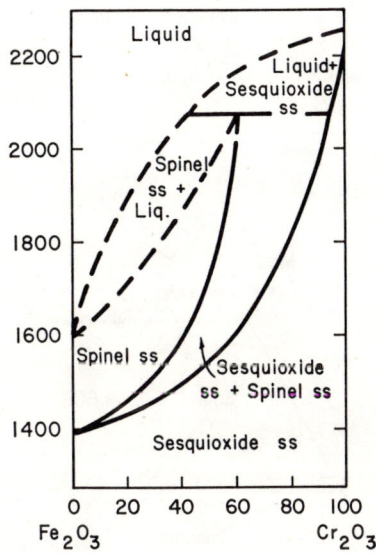

FIG. 58.—System Cr_2O_3–Fe_2O_3 in air. System is pseudo-binary.

Arnulf Muan and Shigeyuki Sōmiya, *J. Am. Ceram. Soc.*, **43** [4] 207 (1960).

Fe–La–O

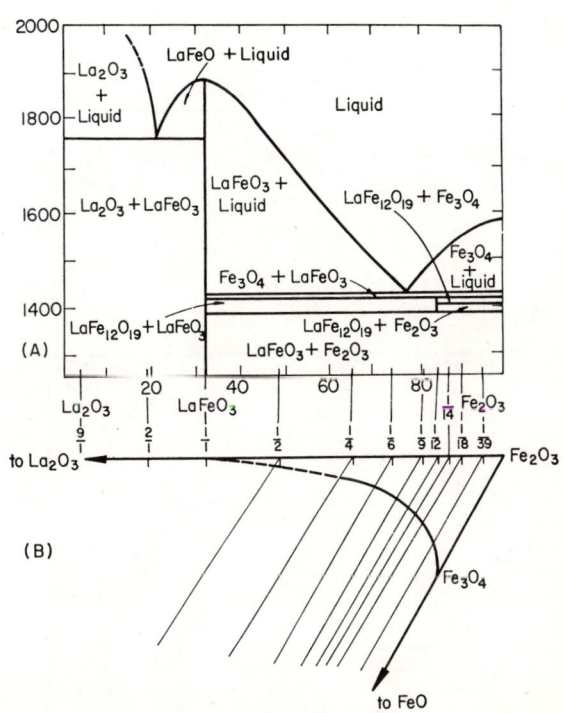

FIG. 60.—System Fe_2O_3–La_2O_3 in air. Light lines almost parallel to the FeO–Fe_2O_3 join of (B) represent constant lanthanum to iron ratios.

V. L. Moruzzi and M. W. Shafer, *J. Am. Ceram. Soc.*, **43** [7] 369 (1960).

Fe–La–O (concl.)

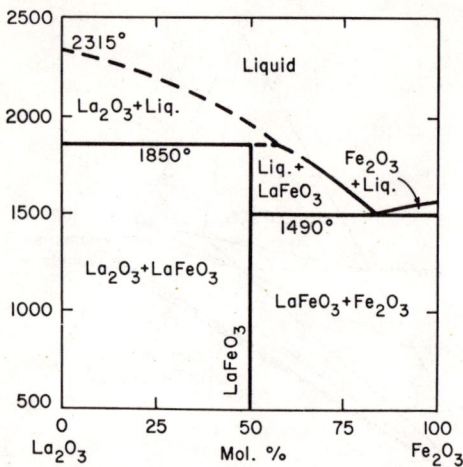

Fig. 61.—System Fe_2O_3–La_2O_3 at 1 atm. oxygen pressure above 1400°C.

Yasumasa Goto, Toshio Kitamura, Toshio Takada, and Sukeji Kachi, presented at the Annual Meeting of Japan Soc. of Powder Metallurgy, Tokyo, April 1960.

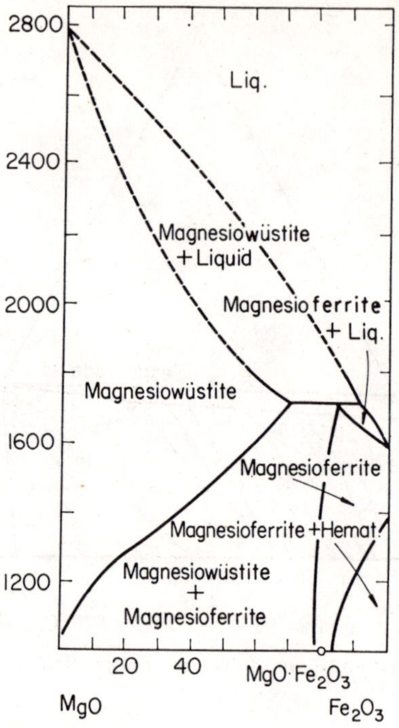

Fe–Li–O

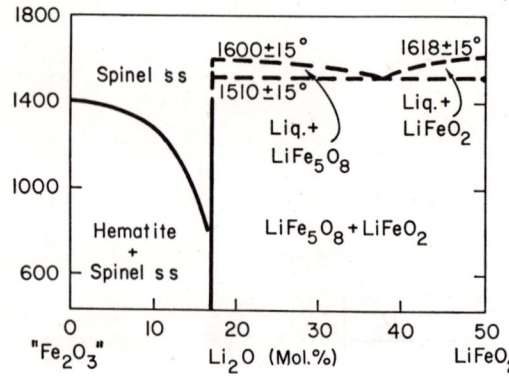

Fig. 62.—System $LiFeO_2$–Fe_2O_3; pseudobinary. Liquids throughout and solid phases in the Fe_2O_3-rich region are ternary and are only appproximately represented.

D. W. Strickler and Rustum Roy, J. Am. Ceram. Soc., 44 [5] 227 (1961).

Fig. 65. See page 55.

Fe–Mg–O

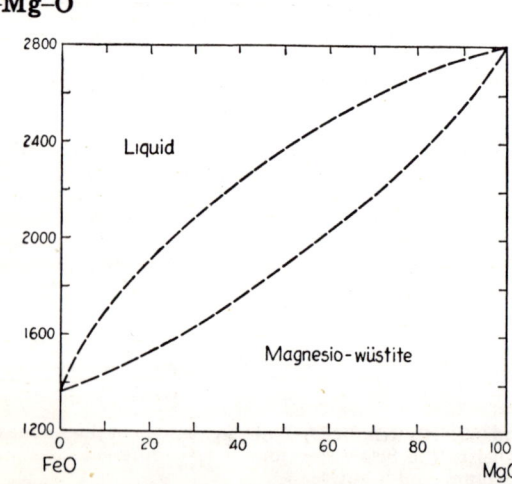

Fig. 63.—System FeO–MgO; this system has not been studied in detail, but the equilibrium diagram is believed to be as shown above if the incongruent melting of FeO is neglected.

N. L. Bowen and J. F. Schairer, Am. J. Sci., 5th Ser., 29, 153 (1935).

Fig. 64.—System MgO–Fe_2O_3 in air.

Bert Phillips, Shigeyuki Sōmiya, and Arnulf Muan, J. Am. Ceram. Soc., 44 [4] 169 (1961).

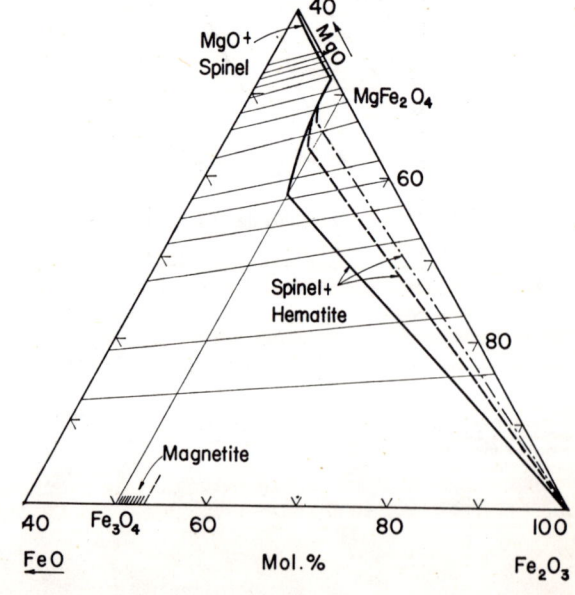

Fig. 66.—System FeO–MgO–Fe_2O_3 at 1100°C. Solid line = 1.0×10^{-2} isobar, dashed line = 2.1×10^{-1} isobar, dash dot line = 1.0×10^0 isobar.

A. E. Paladino, J. Am. Ceram. Soc., 43 [4] 187 (1960).

Fe–Mg–O (cont.)

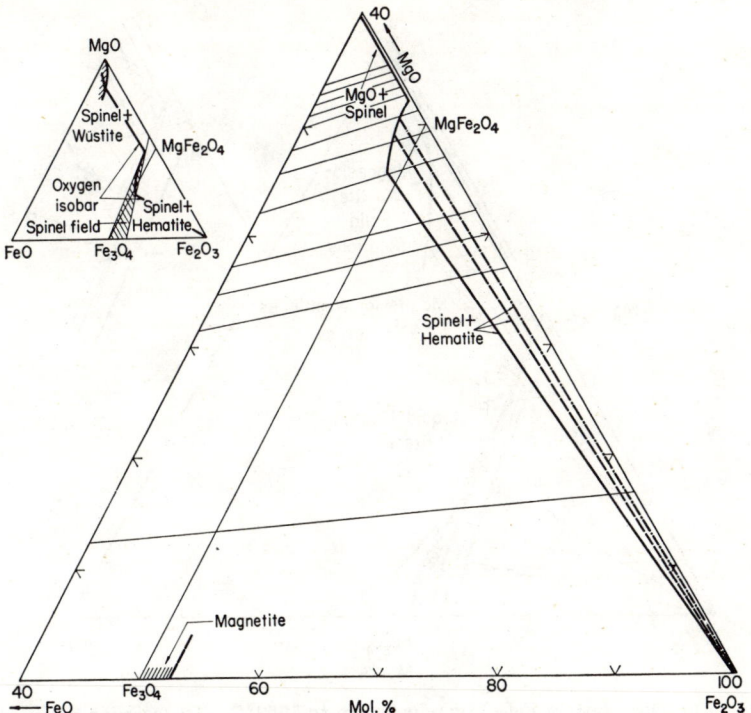

Fig. 65.—System FeO–MgO–Fe$_2$O$_3$ at 1000°C. Solid line = 1.0×10^{-2} isobar, dashed line = 2.1×10^{-1} isobar, dash dot line = 1.0×10^0 isobar.

A. E. Paladino, *J. Am. Ceram. Soc.*, **43**, [4] 186 (1960).

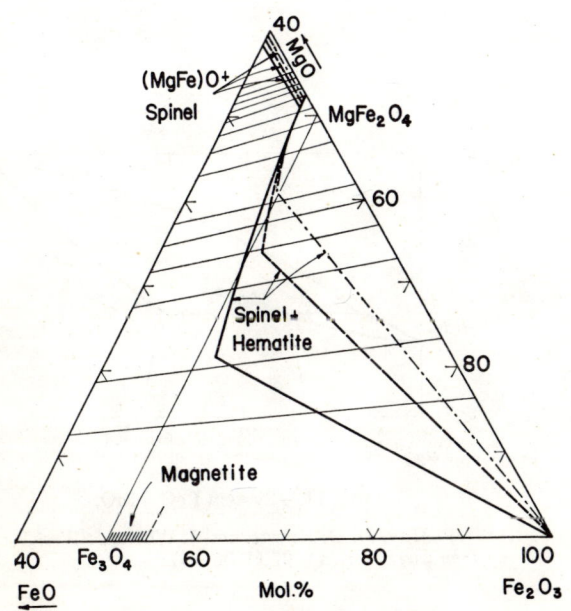

Fig. 67.—System FeO–MgO–Fe$_2$O$_3$ at 1200°C. Solid line = 1.0×10^{-2} isobar, dashed line = 2.1×10^{-1} isobar, dash dot line = 1.0×10^0 isobar.

A. E. Paladino, *J. Am. Ceram. Soc.*, **43** [4] 189 (1960).

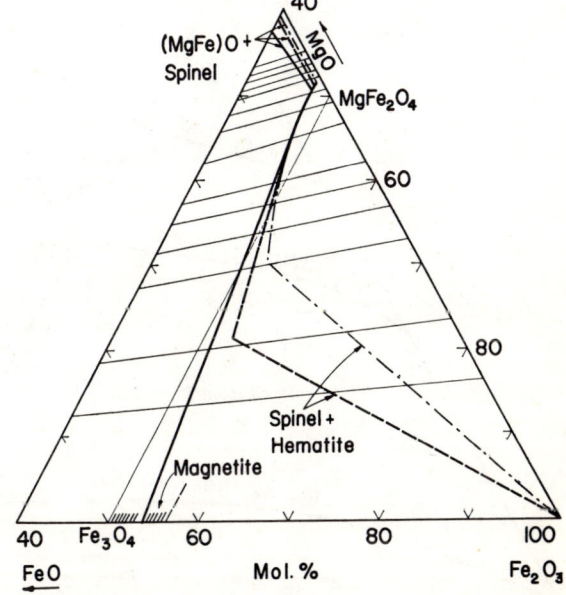

Fig. 68.—System FeO–MgO–Fe$_2$O$_3$ at 1300°C. Solid line = 1.0×10^{-2} isobar, dashed line = 2.1×10^{-1} isobar, dash dot line = 1.0×10^0 isobar.

A. E. Paladino, *J. Am. Ceram. Soc.*, **43** [4] 189 (1960).

Fe–Mg–O (concl.)

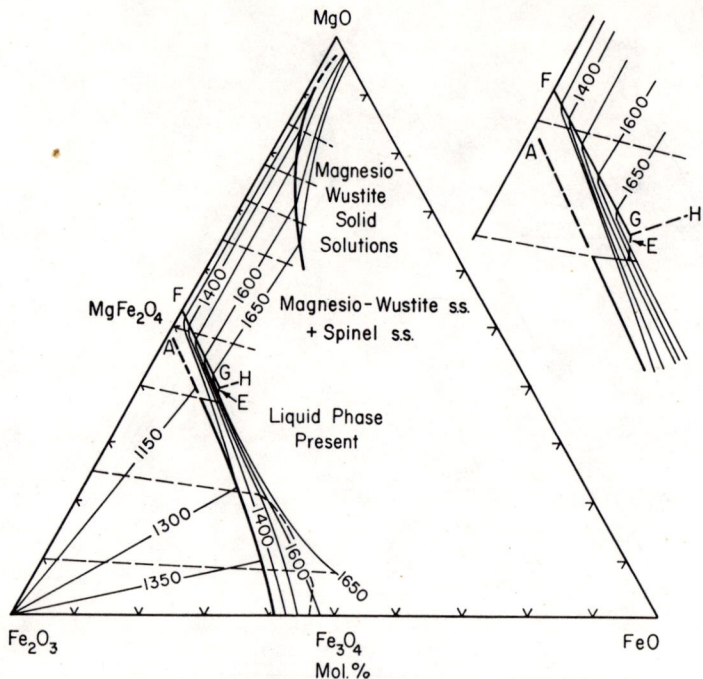

FIG. 69.—System FeO–MgO–Fe$_2$O$_3$, up to 1650°C. Dissociation path shown dashed.

D. Woodhouse and J. White, *Trans. Brit. Ceram. Soc.*, **54**, 339 (1955).

Fe–Mn–O

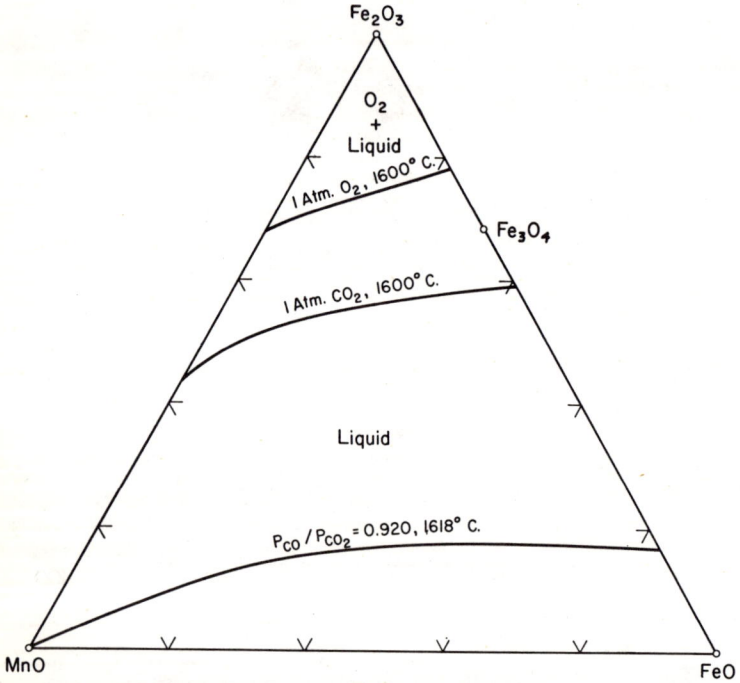

FIG. 70.—System MnO–FeO–Fe$_2$O$_3$ at 1600°C.

R. W. Gurry and L. S. Darken, *J. Am. Chem. Soc.*, **72**, 3909 (1950).

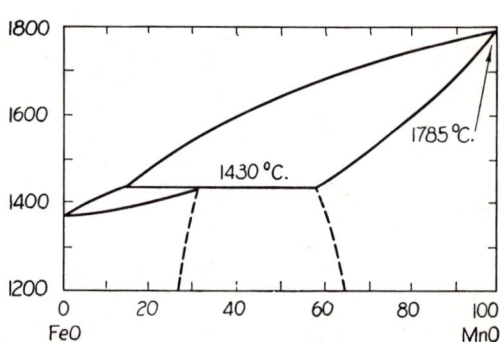

FIG. 71.—System FeO–MnO.

R. Hay, D. D. Howat, and J. White, *J. West Scot. Iron Steel Inst.*, **41**, 97 (1933–34).

Fe–Mn–O (concl.)

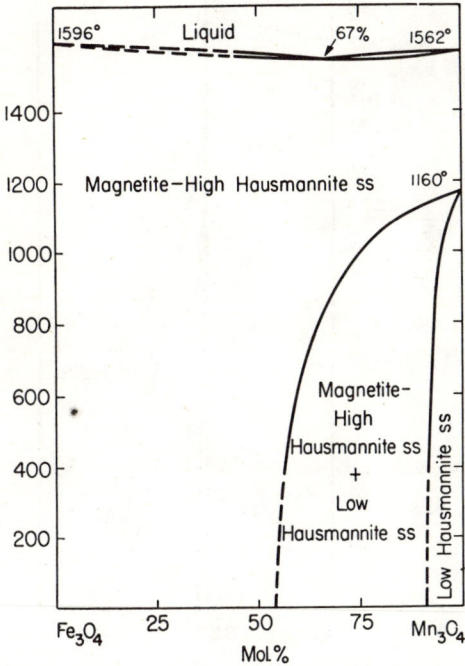

Fig. 72.—System Fe_3O_4–Mn_3O_4.
H. J. Van Hook and M. L. Keith, *Am. Mineral.*, **43** [1/2] 80 (1958).

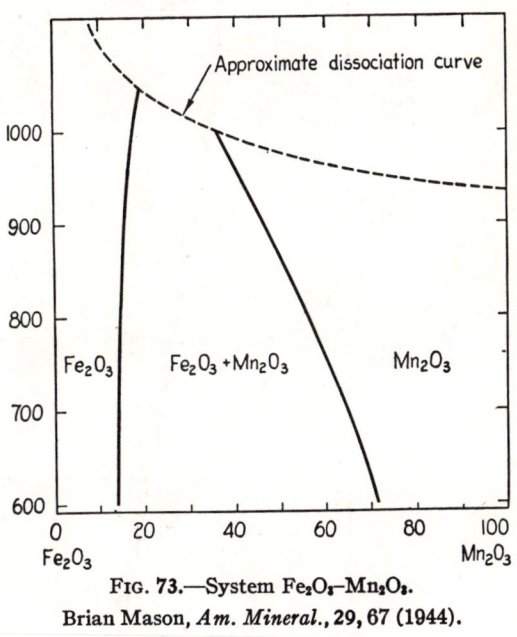

Fig. 73.—System Fe_2O_3–Mn_2O_3.
Brian Mason, *Am. Mineral.*, **29**, 67 (1944).

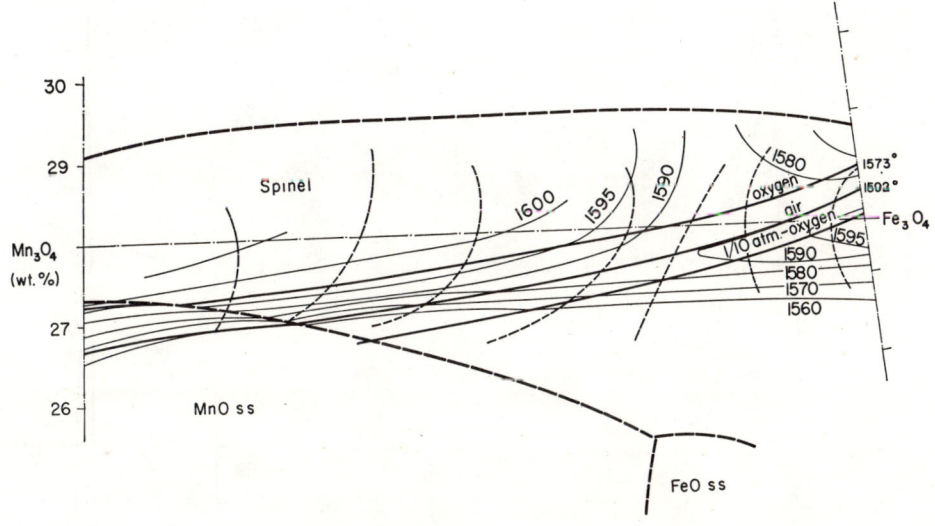

Fig. 74.—System Fe–Mn–O; ferrite region.
M. W. Shafer, *IBM J. Res. Develop.*, **2**, 195 (1958).

Fe–Ni–O

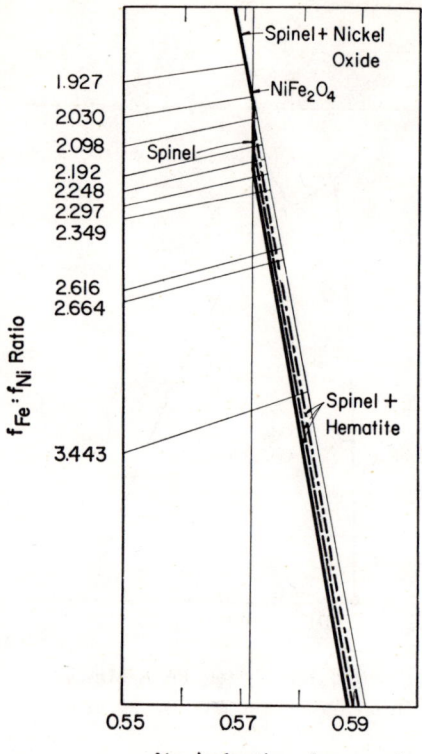

Fig. 75.—System Fe–Ni–O; ferrite region at 1000°C. Solid line = 1.0×10^{-2} isobar, dashed line = 2.1×10^{-1} isobar, dash dot line = 1.0×10^0 isobar.

A. E. Paladino, Jr., *J. Am. Ceram. Soc.*, **42** [4] 173 (1959).

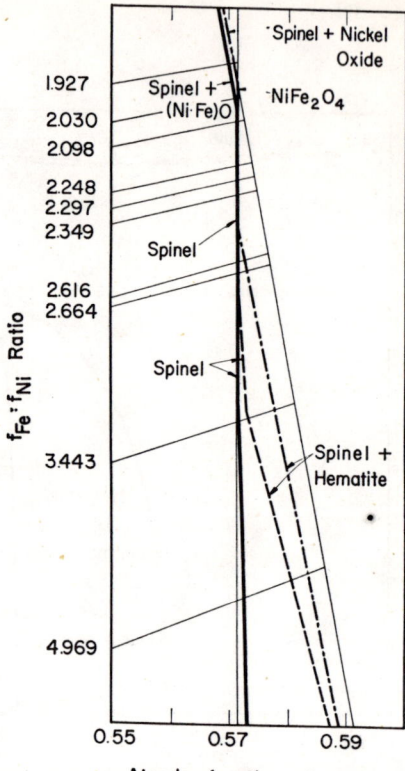

Fig. 77.—System Fe–Ni–O; ferrite region at 1200°C. Solid line = 1.0×10^{-2} isobar, dashed line = 2.1×10^{-1} isobar, dash dot line = 1.0×10^0 isobar.

A. E. Paladino, Jr., *J. Am. Ceram. Soc.*, **42** [4] 174 (1959).

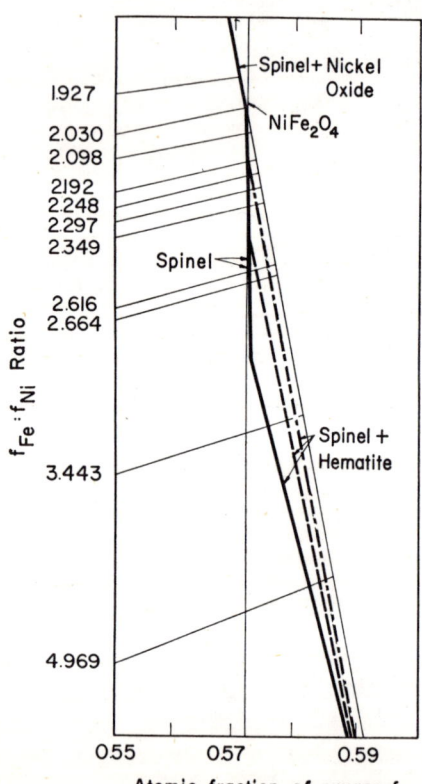

Fig. 76.—System Fe–Ni–O; ferrite region at 1100°C. Solid line = 1.0×10^{-2} isobar, dashed line = 2.1×10^{-1} isobar, dash dot line = 1.0×10^0 isobar.

A. E. Paladino, Jr., *J. Am. Ceram. Soc.*, **42** [4] 173 (1959).

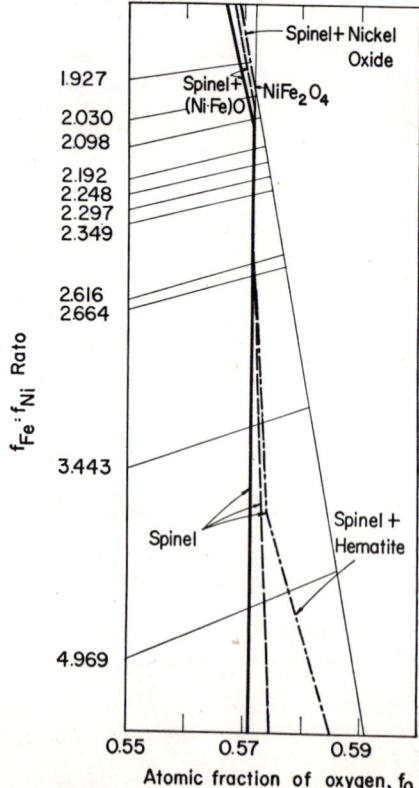

Fig. 78.—System Fe–Ni–O; ferrite region at 1300°C. Solid line = 1.0×10^{-2} isobar, dashed line = 2.1×10^{-1} isobar, dash dot line = 1.0×10^0 isobar.

A. E. Paladino, Jr., *J. Am. Ceram. Soc.*, **42** [4] 174 (1959).

Fe–Sc–O

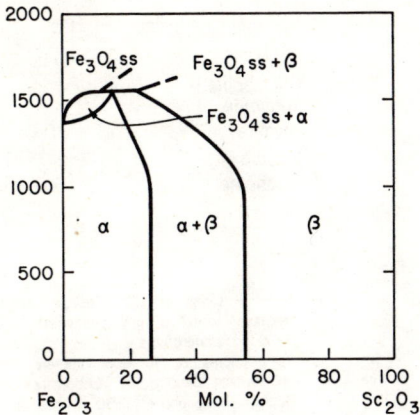

FIG. 79.—System Fe_2O_3–Sc_2O_3. α = rhombohedral Fe_2O_3 ss; β = cubic Sc_2O_3 ss.

Jeannine Cassedanne and Hubert Forestier, *Compt. rend.*, **245**, 2899 (1960).

Fe–Si–O

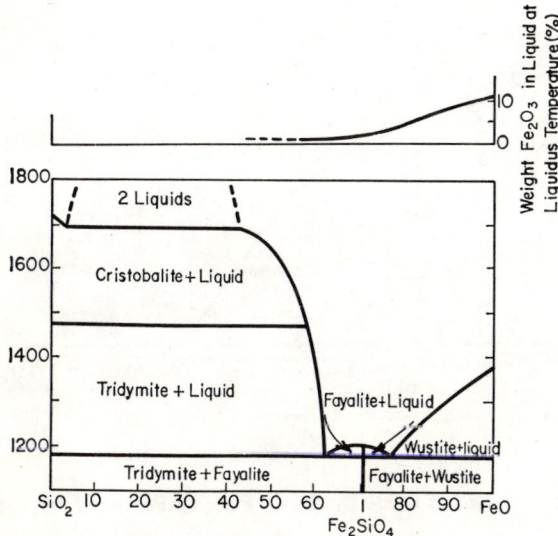

FIG. 80.—System FeO–SiO_2.

N. L. Bowen and J. F. Schairer, *Am. J. Sci.*, 5th Ser., **24**, 200 (1932).

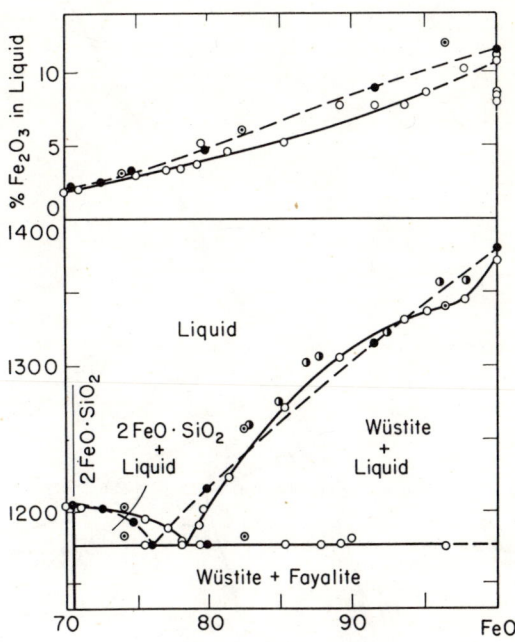

FIG. 81.—System $2FeO \cdot SiO_2$–FeO.

W. C. Allen and R. B. Snow, *J. Am. Ceram. Soc.*, **38** [8] 268 (1955).
○ = N_2 atmosphere; ☉ = CO/CO_2 atmosphere; ● = Bowen and Schairer, *Am. J. Sci.*, [5th Series], **24** [141] 177 (1932); ◐ = R. Schuhmann, Jr., and P. J. Ensio, *J. Metals*, **3** [5] 401 (1951).

Fe–Si–O (cont.)

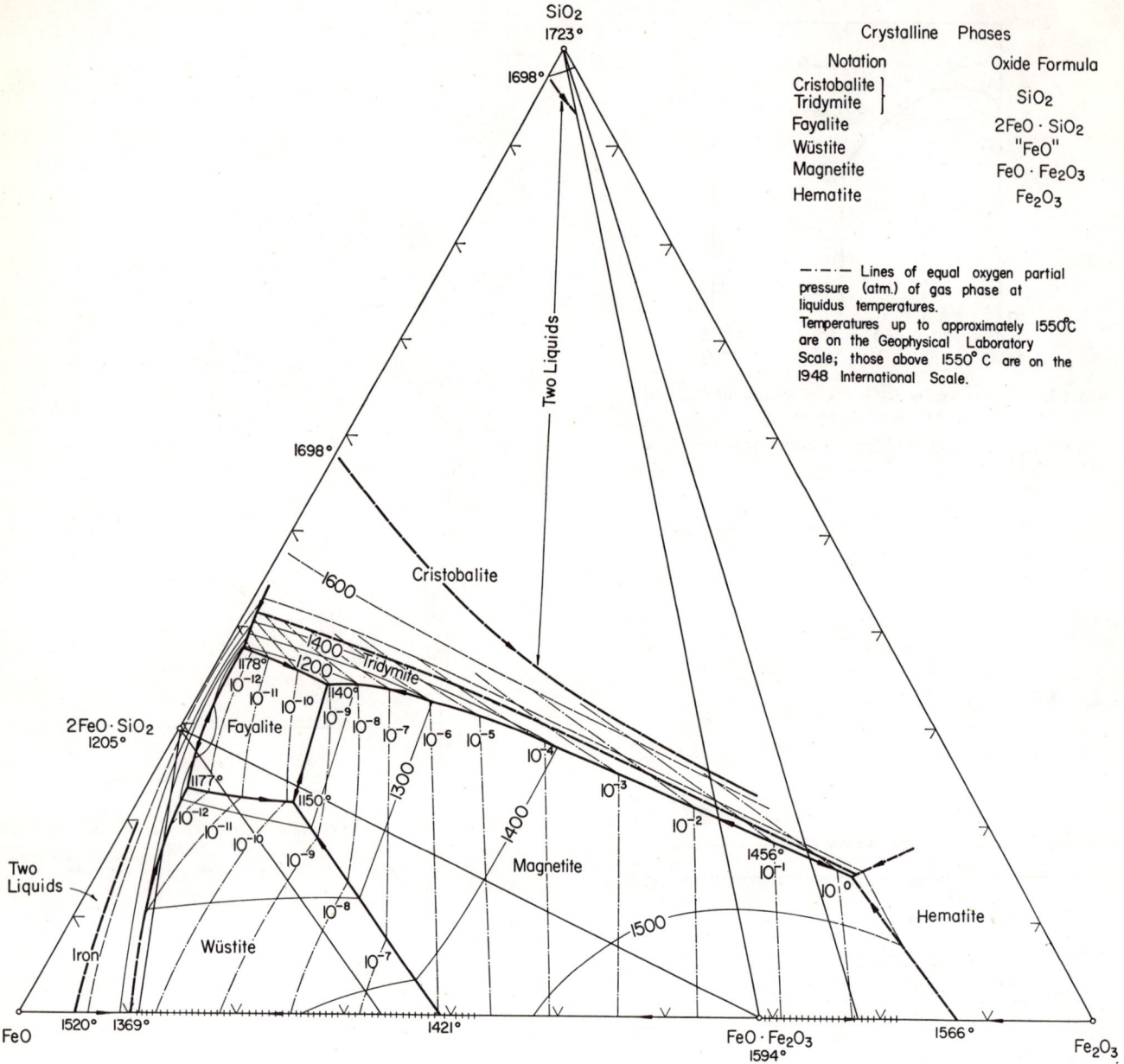

Fig. 82.—System FeO–Fe$_2$O$_3$–SiO$_2$; composite.

E. F. Osborn and Arnulf Muan, revised and redrawn "Phase Equilibrium Diagrams of Oxide Systems," Plate 6, Published by the American Ceramic Society and the Edward Orton, Jr., Ceramic Foundation, 1960.

Principal References

N. L. Bowen and J. F. Schairer, *Am. J. Sci.* (5th Series), **24**, 177–213 (1932).
L. S. Darken and R. W. Gurry, *J. Am. Chem. Soc.* **68**, 798–816 (1946).
L. S. Darken, *J. Am. Chem. Soc.*, **70**, 2046–53 (1948).
J. W. Greig, *Am. J. Sci.* (5th Series), **13**, 1–44; 133–54 (1927).
J. W. Greig, *Am. J. Sci.* (5th Series), **14**, 473–84 (1927).
R. Schuhmann, Jr., R. G. Powell and E. J. Michal. *J. Metals*, **5**, September 1953; *Trans. Am. Inst. Mining Met. Engrs.*, **197**, 1097–1104 (1953).
Arnulf Muan, *J. Metals*, **7**, September 1955; *Trans. Am. Inst. Mining Met. Engrs.*, **203**, 965–76 (1955).

Fe–Si–O (concl.)

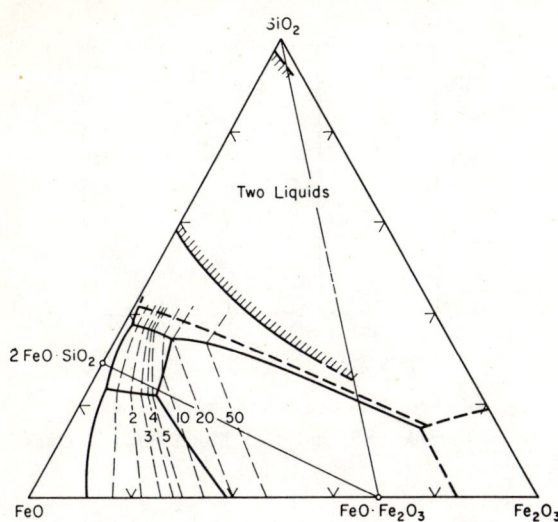

FIG. 83.—System FeO–Fe$_2$O$_3$–SiO$_2$, showing lines (dashed) of equal CO$_2$/H$_2$ mixing ratios for points on the liquidus surface.

Arnulf Muan, *J. Metals*, **7** (1955); *Trans. Am. Inst. Mining Met. Engrs.*, **203**, 972 (1955).

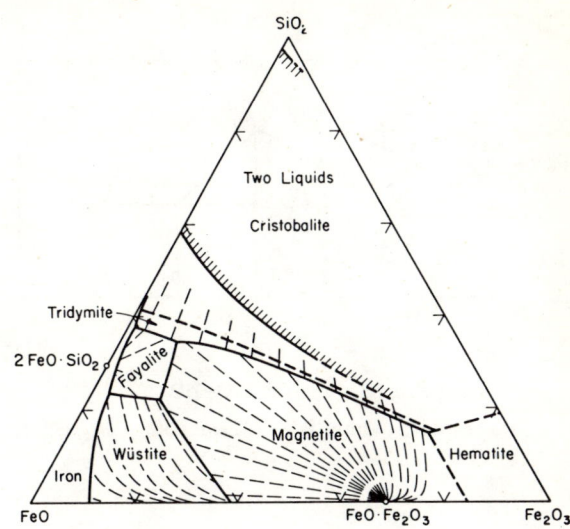

FIG. 84.—System FeO–Fe$_2$O$_3$–SiO$_2$, showing fractionation curves.

Arnulf Muan, *J. Metals*, **7**, (1955); *Trans. Am. Inst. Mining Met. Engrs.*, **203**, 972 (1955).

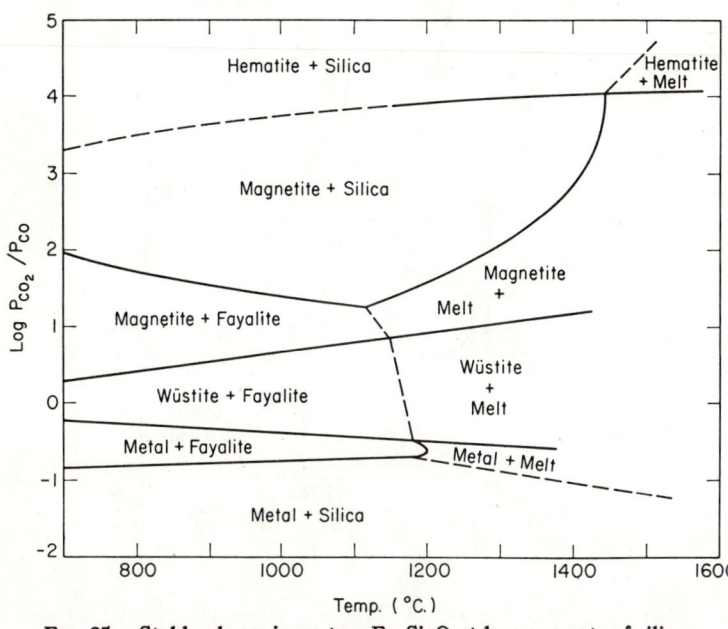

FIG. 85.—Stable phases in system Fe–Si–O at low amounts of silicon.
L. S. Darken, *J. Am. Chem. Soc.*, **70**, 2051 (1948).

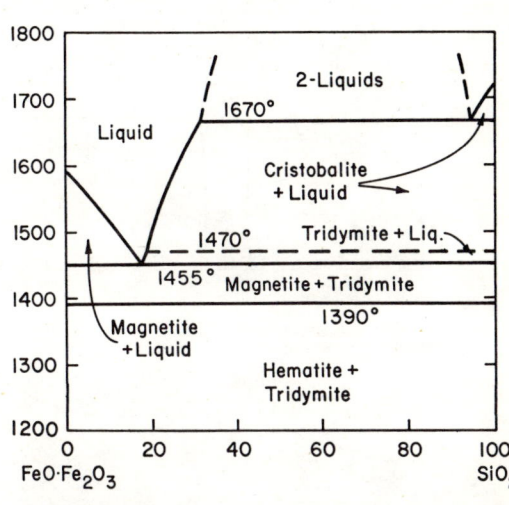

FIG. 87.—System FeO·Fe$_2$O$_3$–SiO$_2$ in air; isobaric.
Bert Phillips and Arnulf Muan, *J. Am. Ceram. Soc.*, **42** [9] 415 (1959).

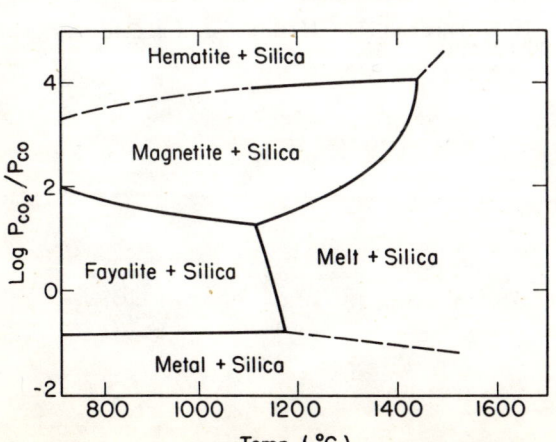

FIG. 86.—Stable phases in system Fe–Si–O in the presence of solid silica.
L. S. Darken, *J. Am. Chem. Soc.*, **70**, 2051 (1948).

Fe–Ti–O

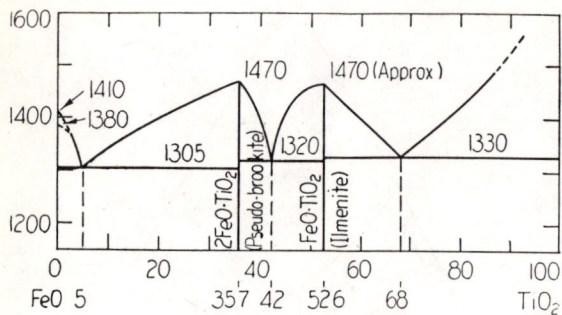

FIG. 88.—System FeO–TiO$_2$.

J. Grieve and J. White, *J. Roy. Tech. Coll. (Glasgow)*, **4**, 444 (1939).

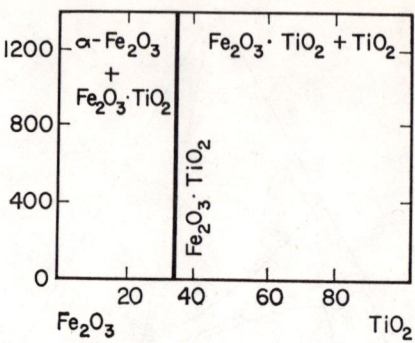

FIG. 89.—System Fe$_2$O$_3$–TiO$_2$; subsolidus.

M. D. Karkhanavala and A. C. Momin, *J. Am. Ceram. Soc.*, **42** [8] 400 (1959).

FIG. 91. See page 63.

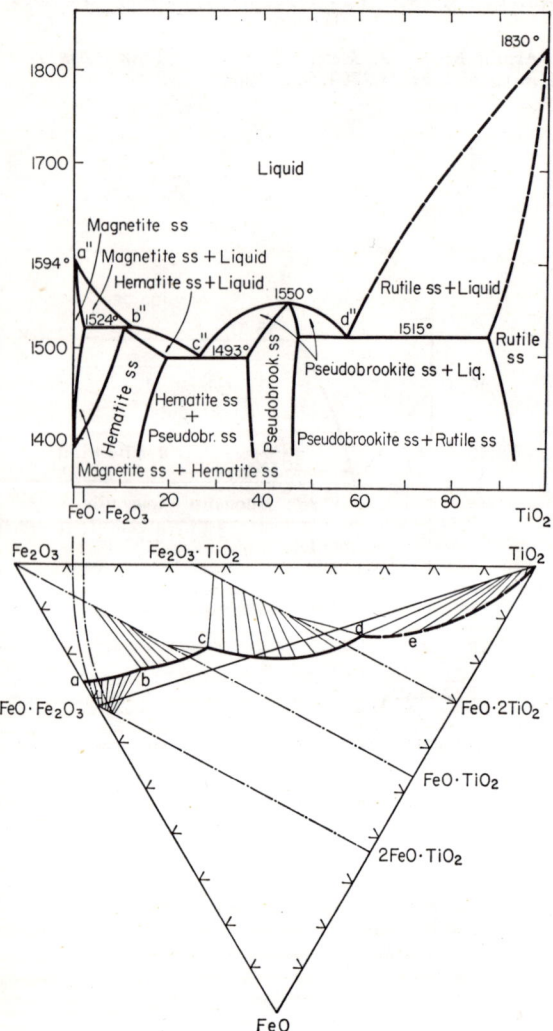

FIG. 90.—System FeO·Fe$_2$O$_3$–TiO$_2$ in air. Lower triangular diagram shows compositions of crystalline and liquid phases. The heavy dashed curve *abcde* represents compositions of liquids along the liquidus surface in air. Approximate compositions of crystalline phases lie on indicated joins.

J. B. MacChesney and Arnulf Muan, *Am. Mineral.*, **44**, 938 (1959).

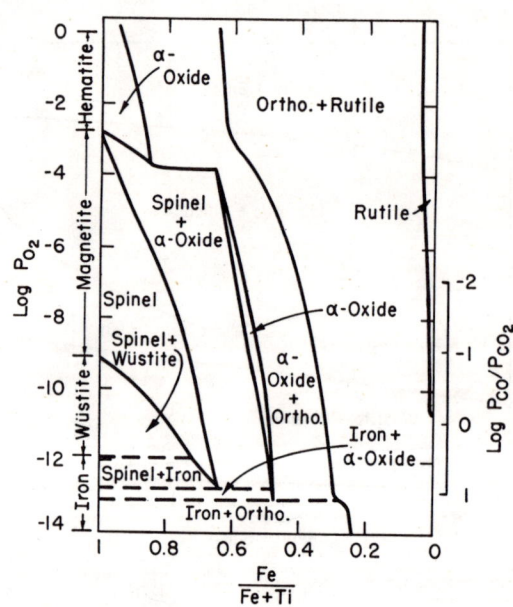

FIG. 92.—System Fe–Ti–O; phases present at 1200°C. The lines separating the (α-oxide + ortho.) field from the (ortho. + rutile) field and from the (spinel + α-oxide) field represent solid solutions of small, but finite width. Ortho. = orthorhombic oxide.

A. H. Webster and N. F. H. Bright, *J. Am. Ceram. Soc.*, **44** [3] 112 (1961).

Fe–Ti–O (concl.)

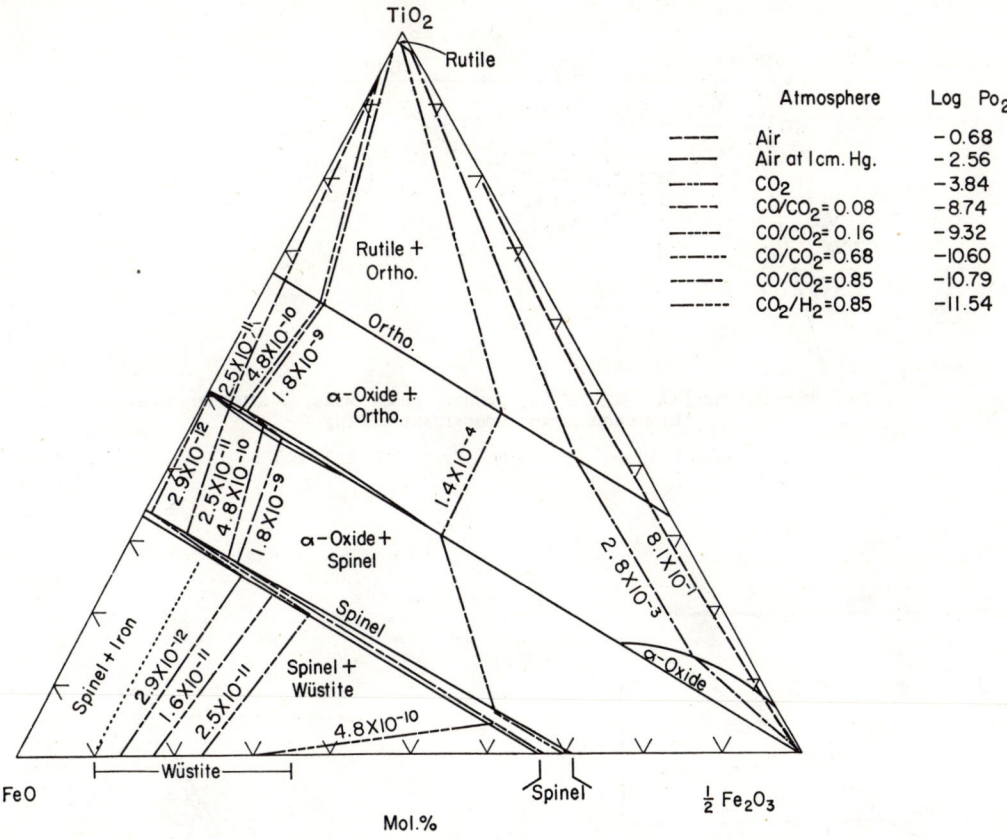

Fig. 91.—System FeO–Fe_2O_3–TiO_2 at 1200°C and showing oxygen isobars. Ortho. = orthorhombic oxide.

A. H. Webster and N. F. H. Bright, *J. Am. Ceram. Soc.*, **44** [3] 115 (1961).

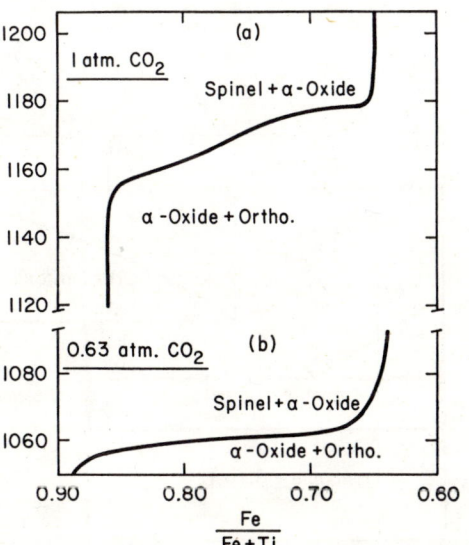

Fig. 93.—System Fe–Ti–O; phases in equilibrium with (a) CO_2 and (b) 63% CO_2 in argon. Ortho. = orthorhombic oxide.

A. H. Webster and N. F. H. Bright, *J. Am. Ceram. Soc.*, **44** [3] 112 (1961).

Fe–V–O

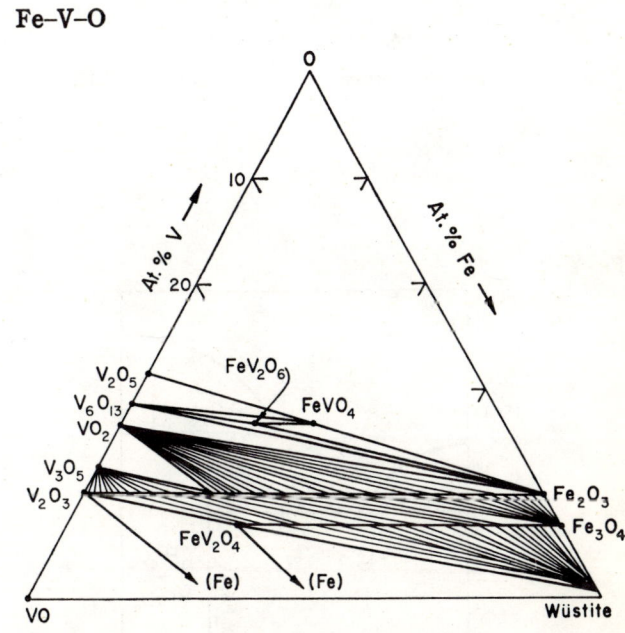

Fig. 94.—System FeO–VO–O at 600°C.

Aurelio Burdese, *Ann. chim. (Rome)*, **47**, 824 (1957).

Fe–V–O (concl.)

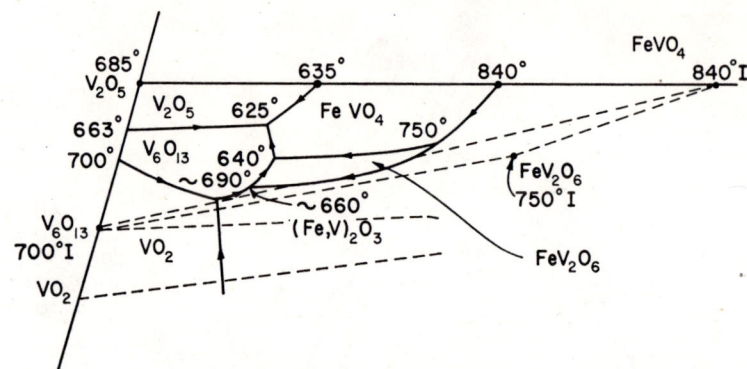

Fig. 95.—System VO_2–V_2O_5–$FeVO_4$; primary phase fields. I after temperature value means incongruent melting.

Aurelio Burdese, *Ann. chim. (Rome)*, **47**, 825 (1957).

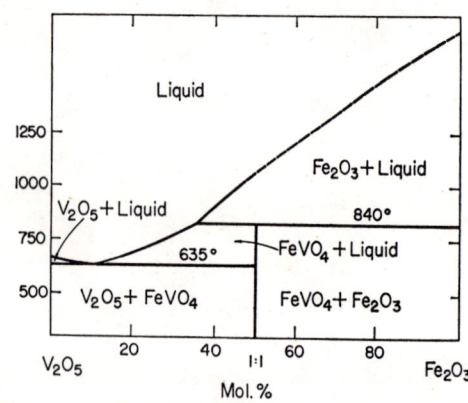

Fig. 96.—System Fe_2O_3–V_2O_5.

Aurelio Burdese, *Ann. chim. (Rome)*, **47** [7, 8] 804 (1957).

Fe–Y–O

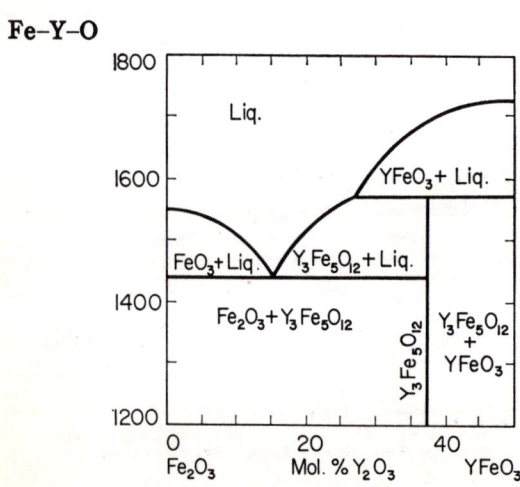

Fig. 97.—System Fe_2O_3–Y_2O_3.

J. W. Nielsen and E. F. Dearborn, *Phys. Chem. Solids*, **5** [3] 203 (1958).

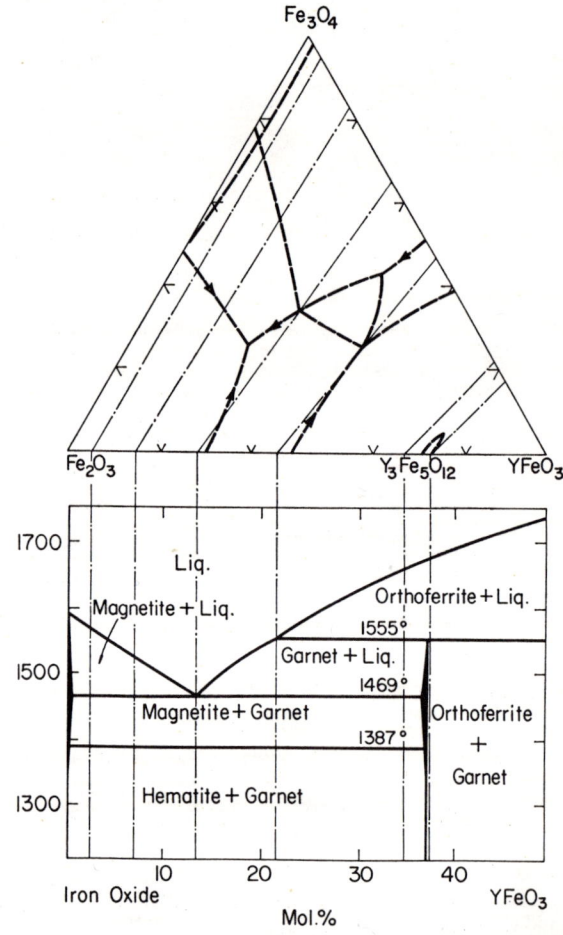

Fig. 98.—System Fe_3O_4–Fe_2O_3–$YFeO_3$. Relation of binary isobar to actual ternary compositions. Dashed line is air isobar; boundary curves are inferred; dash-dot lines are oxygen reaction lines.

H. J. Van Hook, *J. Am. Ceram. Soc.*, **44** [5] 213 (1961).

Mg–U–O

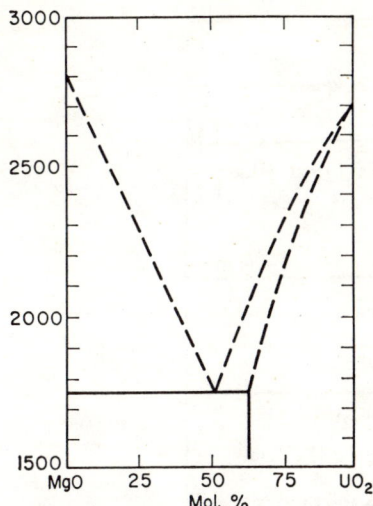

Fig. 99.—System MgO–UO$_{2+x}$ in air.

P. P. Budnikov, S. G. Tresvyatskiĭ, and V. I. Kushakovskiĭ, *Proc. U. N. Intern. Conf. Peaceful Uses At. Energy, 2nd, Geneva*, **6**, 130 (1958).

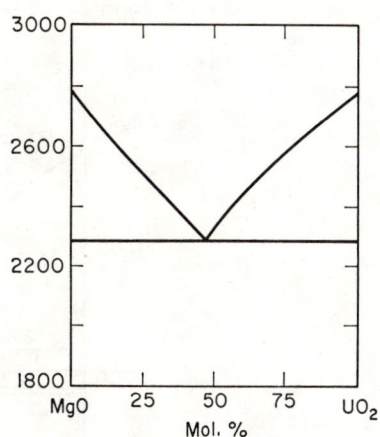

Fig. 100.—System MgO–UO$_2$; idealized.

P. P. Budnikov, S. G. Tresvyatskiĭ, and V. I. Kushakovskiĭ, *Proc. U. N. Intern. Conf. Peaceful Uses At. Energy, 2nd, Geneva*, **6**, 130 (1958).

Mn–Si–O

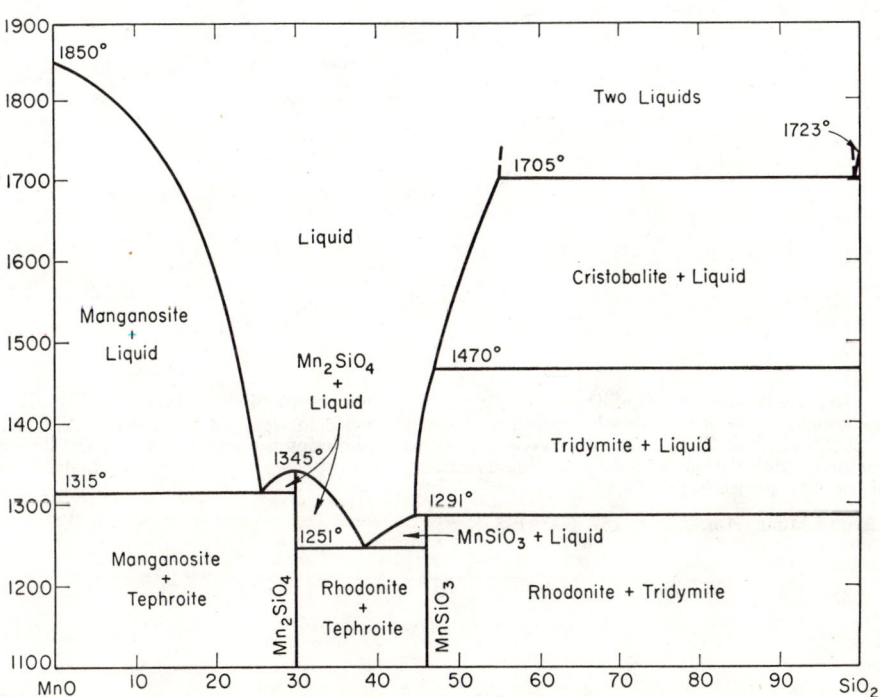

Fig. 101.—System MnO–SiO$_2$.

F. P. Glasser, *Am. J. Sci.*, **256** [6] 405 (1958).

See also J. White, D. D. Howatt, and R. Hay, *J. Roy. Tech. Coll.* (*Glasgow*), **3**, 239 (1933–36).

Mn–Si–O (cont.)

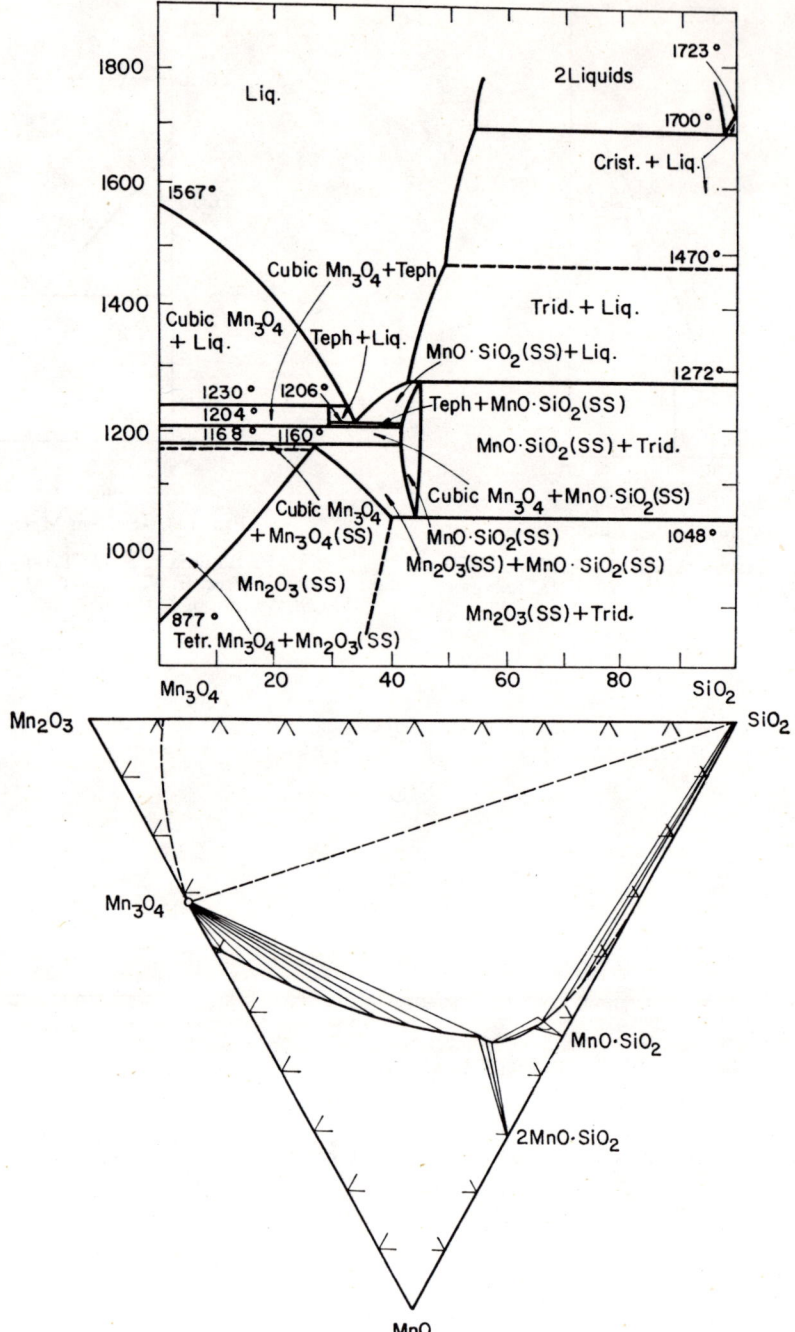

FIG. 102.—System Mn₃O₄–SiO₂ in air. Upper diagram is pseudobinary. Lower diagram shows true compositions of liquid and crystalline phases expressed in terms of the chosen components MnO, Mn₂O₃, and SiO₂. Heavy curve is the 0.21 atm. liquidus isobar. Light straight lines approximate conjugation lines between liquids and crystalline phases in equilibrium. Light dashed curves show projection method.

Arnulf Muan, *Am. J. Sci.*, **257**, 300 (1959).

Mn–Si–O (concl.)

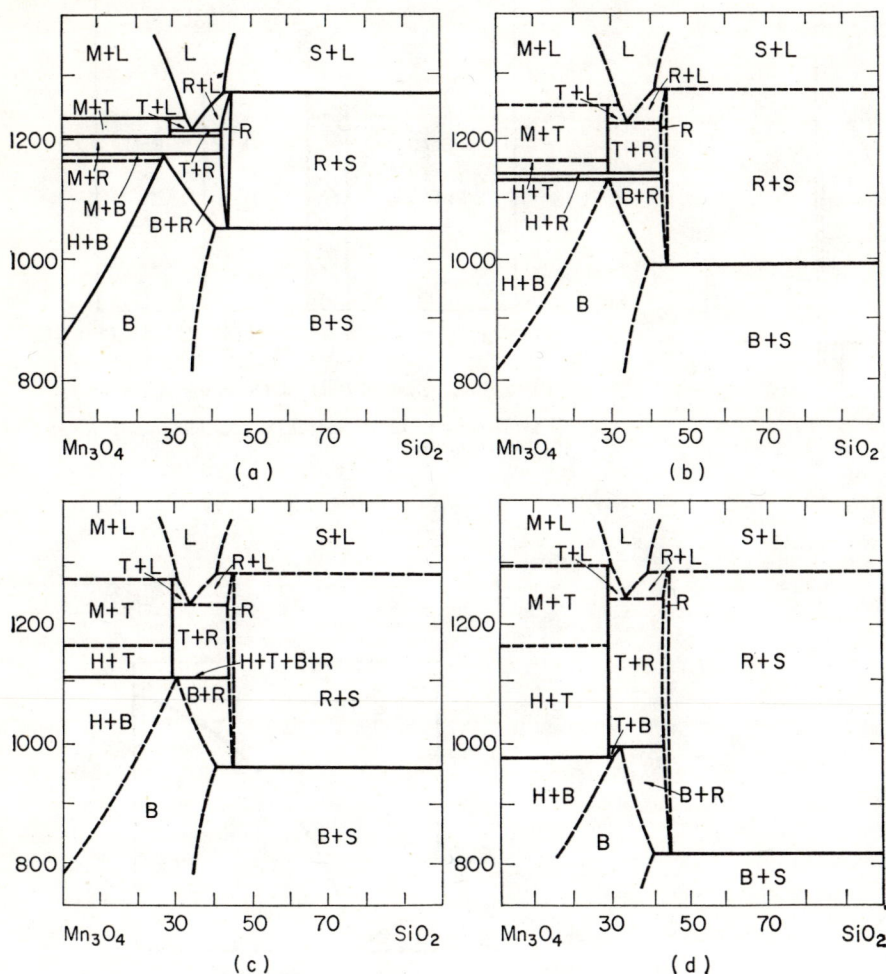

FIG. 103.—System Mn_3O_4–SiO_2 at various levels of O_2 partial pressure. Partial O_2 pressures for the diagrams labeled (a) through (d) are, respectively: $10^{-0.7}$, $10^{-1.1}$, $10^{-1.35}$, $10^{-2.5}$ atm. B - braunite (Mn_2O_3 ss); H - hausmannite (tetragonal Mn_3O_4); L - liquid; M - cubic Mn_3O_4; R - rhodonite (MnO·SiO$_2$ ss); S - silica (tridymite); T - tephroite ($2MnO·SiO_2$).

Arnulf Muan, *Am. Mineral.*, **44**, 954 1959).

N–Ti–O

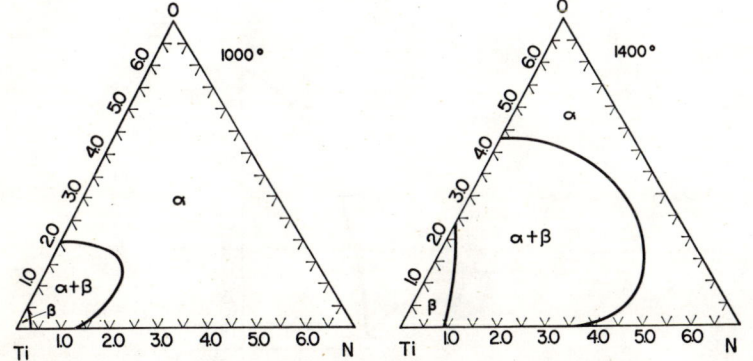

FIG. 104.—System Ti–N–O; isothermal sections.

L. Stone and H. Margolin, *J. Metals*, **5**, *Trans. AIME*, **197**, 1501 (1953).

Na–V–O

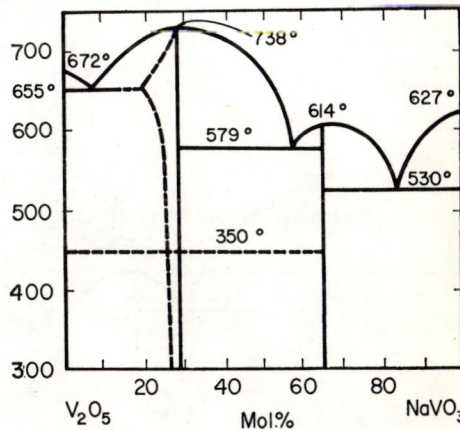

FIG. 105.—System $NaVO_3$–V_2O_5.

V. V. Illarionov, R. P. Ozerov, and E. V. Kil'disheva, *Zhur. Neorg. Khim.*, **2**, 884 (1957).

Na–V–O (concl.)

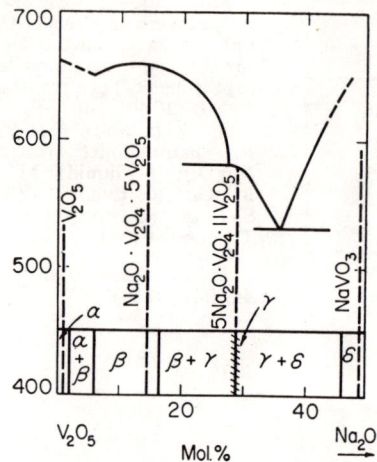

Fig. 106.—System Na_2O–V_2O_5.
Tidskr. Kjemi, Bergvesen Met., No. 5, p. 56 (1943).

Pb–Sb–O

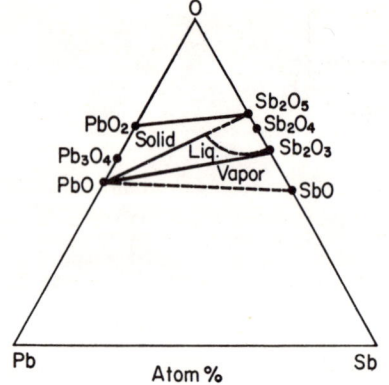

Fig. 107.—System Pb–Sb–O; oxide formation.
Helmut Hennig and E. J. Kohlmeyer, *Z. Erzbergbau Metallhüettenw.*, **10**, 8 (1957).

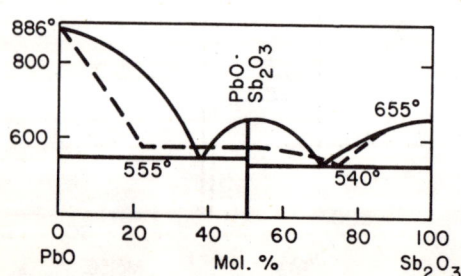

Fig. 108.—System PbO–Sb_2O_3.
Helmut Hennig and E. J. Kohlmeyer, *Z. Erzbergbau Metallhüettenw.*, **10**, 14 (1957).

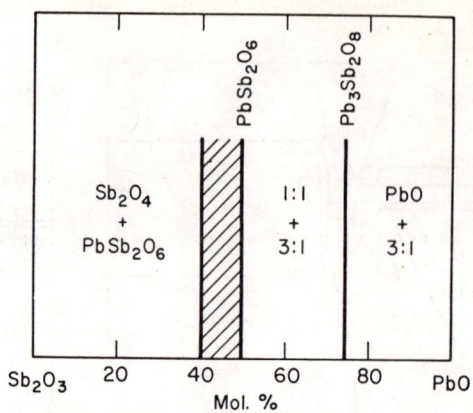

Fig. 109.—System PbO–Sb_2O_3, in air at 700°C.
G. G. Urazov and E. I. Speranskaya, *Zhur. Neorg. Khim.*, **1** [6] 1426 (1956).

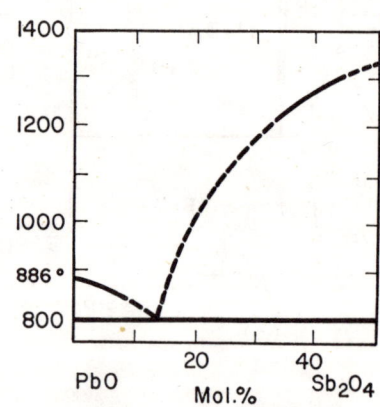

Fig. 110.—System PbO–Sb_2O_4.
Helmut Hennig and E. J. Kohlmeyer, *Z. Erzbergbau Metallhüettenw.*, **10**, 68 (1957).

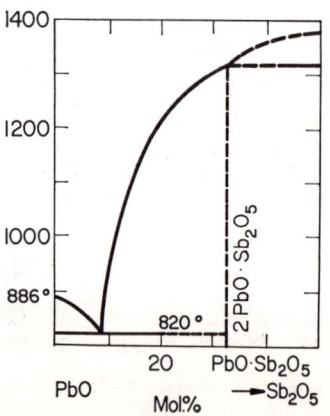

Fig. 111.—System PbO–$PbO \cdot Sb_2O_5$.
Helmut Hennig and E. J. Kohlmeyer, *Z. Erzbergbau Metallhüettenw.*, **10**, 65 (1957).

Pb–Sb–O (concl.)

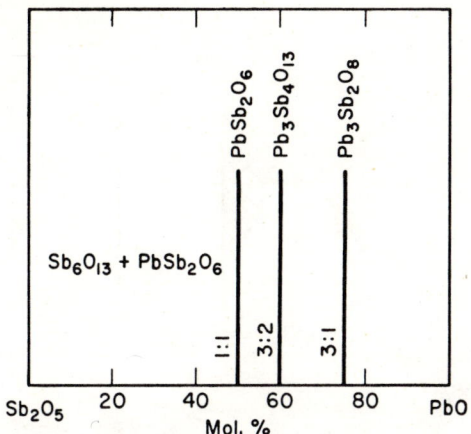

FIG. 112.—System PbO–Sb$_2$O$_5$ at 700°C.

G. G. Urazov and E. I. Speranskaya, *Zhur. Neorg. Khim.*, **1** [6] 1428 (1956).

Si–Ti–O

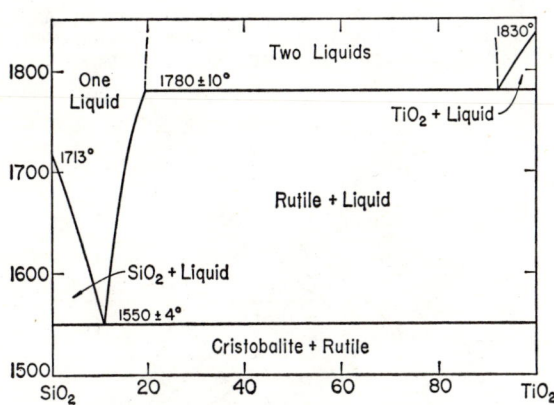

FIG. 113.—System SiO$_2$–TiO$_2$.

R. C. DeVries, R. Roy, and E. F. Osborn, *Trans. Brit. Ceram. Soc.*, **53** [9] 531 (1954).

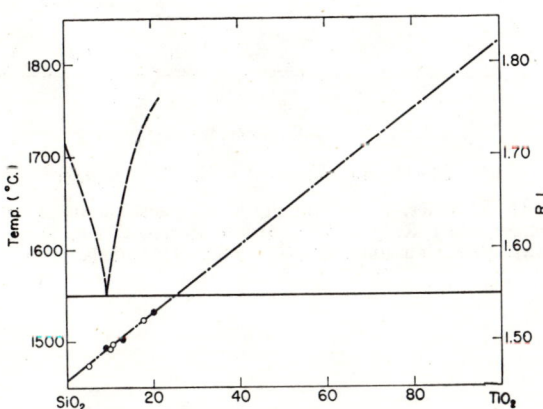

FIG. 114.—System SiO$_2$–TiO$_2$ (n of glasses).

R. C. DeVries, R. Roy, and E. F. Osborn, *Trans. Brit. Ceram. Soc.*, **53** [9] 533 (1954).

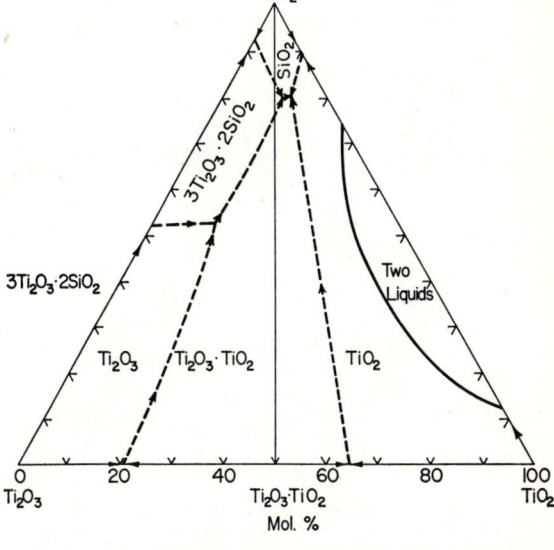

FIG. 115.—System Ti$_2$O$_3$–SiO$_2$–TiO$_2$; not impossible form.

R. Roy, R. C. DeVries, D. E. Rase, M. W. Shafer, and E. F. Osborn, Pennsylvania State College Quarterly Report on Contract No. DA36-039 sc-5594, Oct. 1–Dec. 31 (1952).

Th–U–O

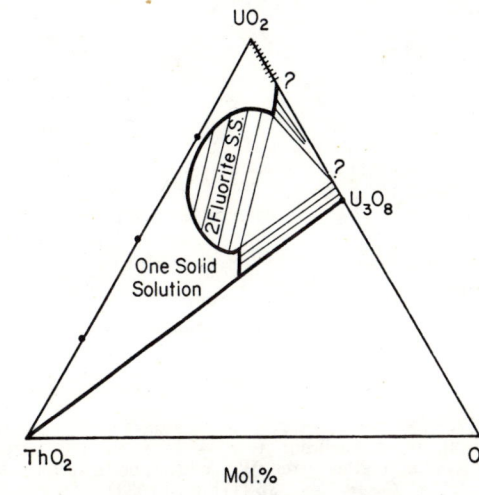

FIG. 116.—System ThO$_2$–UO$_2$–O below 800°C; hypothetical.

F. A. Mumpton and Rustum Roy, *J. Am. Ceram. Soc.*, **43** [5] 239 (1960).

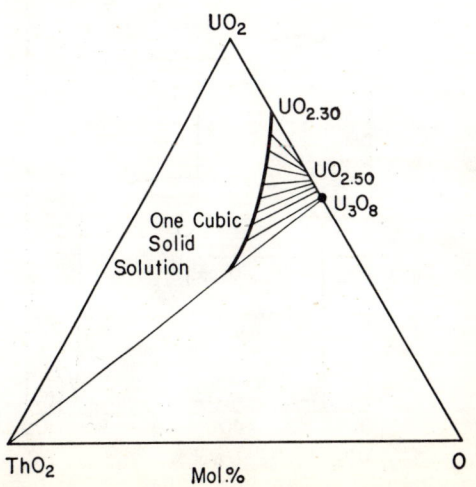

FIG. 117.—System ThO$_2$–UO$_2$–O near 1350°C; hypothetical.

F. A. Mumpton and Rustum Roy, *J. Am. Ceram. Soc.*, **43** [5] 239 (1960).

Th–U–O (concl.)

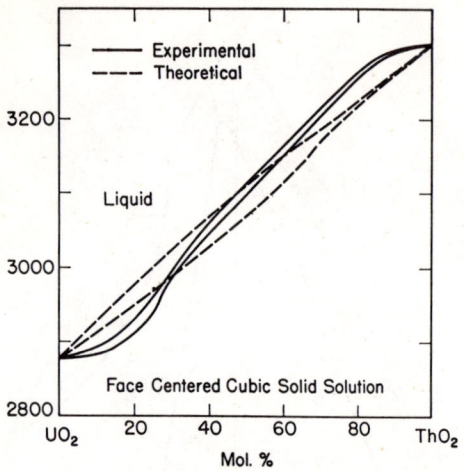

Fig. 118.—System ThO_2–UO_2.

W. A. Lambertson, M. H. Mueller, and F. H. Gunzel, Jr., *J. Am. Ceram. Soc.*, **36** [12] 399 (1953).

U–Zr–O

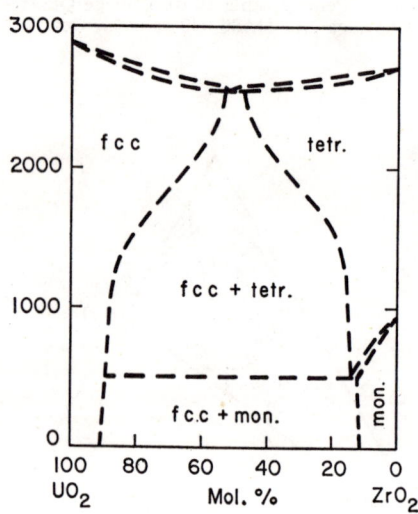

Fig. 119.—System UO_2–ZrO_2; proposed. fcc = face-centered cubic; mon. = monoclinic; tetr. = face-centered tetragonal. Liquidus-solidus region after W. A. Lambertson and M. H. Mueller, *J. Am. Ceram. Soc.*, **36** [11] 367 (1953).

P. E. Evans, *J. Am. Ceram. Soc.*, **43** [9] 446 (1960). Above about 2285°C, a cubic form of ZrO_2 was reported by D. K. Smith and C. F. Cline, *J. Am. Ceram. Soc.*, **45** [5] 249 (1962).

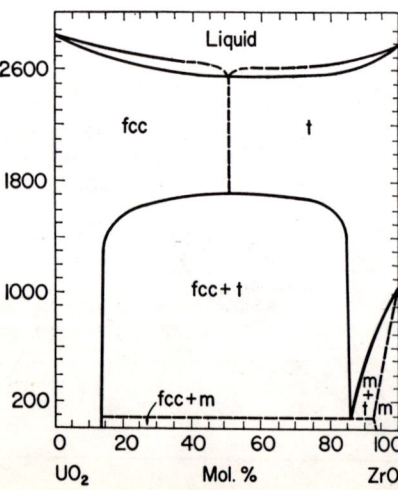

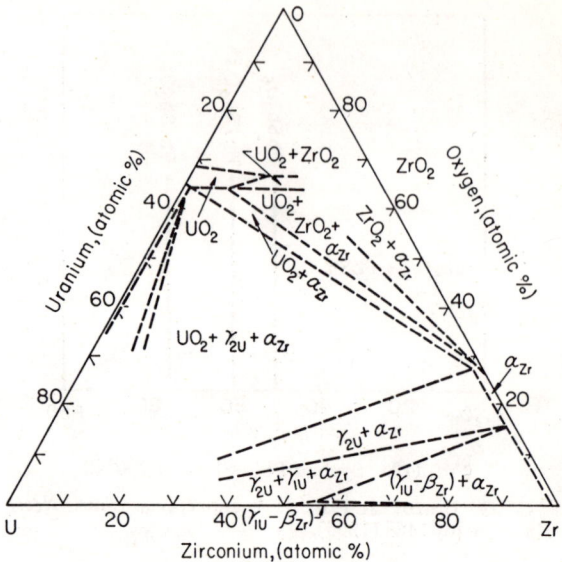

Fig. 121.—System Zr–U–O at 1000°F.; tentative.

H. A. Saller, F. A. Rough, J. M. Fackelmann, A. A. Bauer, and J. R. Doig, U. S. Atomic Energy Comm.-Rept., BMI-1023, Columbus, Ohio (July 28, 1955), unclassified.

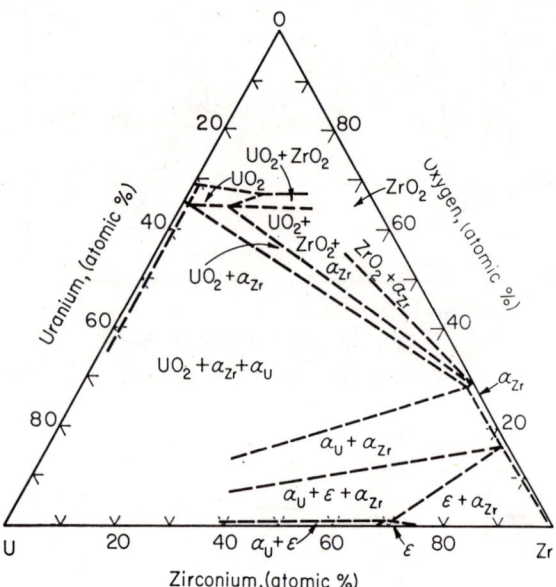

Fig. 122.—System Zr–U–O at 1300°F.; tentative.

H. A. Saller, F. A. Rough, J. M. Fackelmann, A. A. Bauer, and J. R. Doig, U. S. Atomic Energy Comm.-Rept., BMI-1023, Columbus, Ohio (July 28, 1955), unclassified.

Fig. 120.—System UO_2–ZrO_2. Solid solutions: fcc = face-centered cubic, m = monoclinic, t = tetragonal.

N. M. Voronov, E. A. Voitekhova, and A. S. Danilin, *Proc. U. N. Intern. Conf. Peaceful Uses At. Energy, 2nd, Geneva*, **6**, 223 (1958). Extent of immiscibility verified by E. Gebhardt and G. Elssner, Plansee Proc. 1961, pp. 133–39 (Publ. 1962) (in German).

For diagram showing smaller immiscibility gap in the subsolidus, see G. M. Wolten, *J. Am. Chem. Soc.*, **80**, 4774 (1958).

U–Zr–O (concl.)

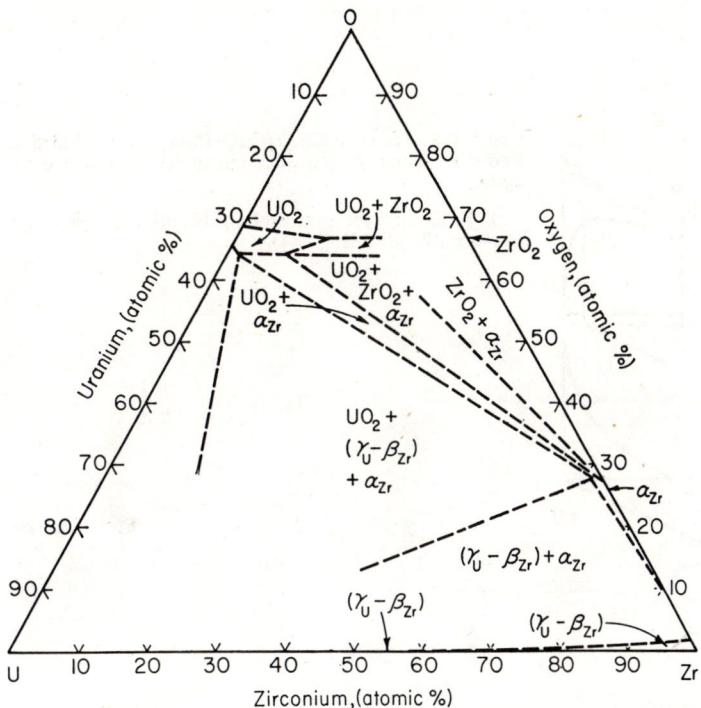

Fig. 123.—System Zr–U–O at 2000°F.; tentative.

H. A. Saller, F. A. Rough, J. M. Fackelmann, A. A. Bauer, and J. R. Doig, U. S. Atomic Energy Comm.-Rept., BMI-1023, Columbus, Ohio (July 28, 1955), unclassified.

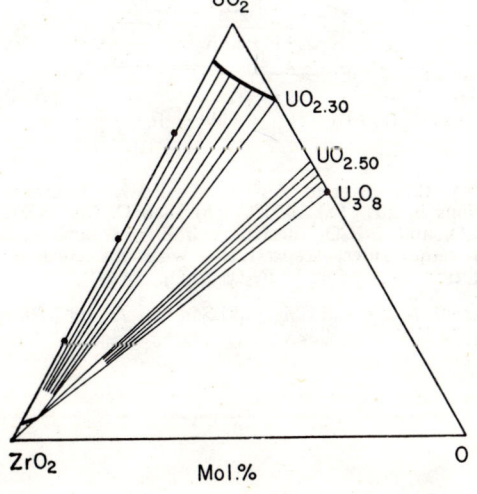

Fig. 124.—System UO_2–ZrO_2–O near 1350°C; hypothetical.

F. A. Mumpton and Rustum Roy, J. Am. Ceram. Soc., 43 [5] 238 (1960).

III. Three Metals with Oxygen

Ag–Cu–Pb–O

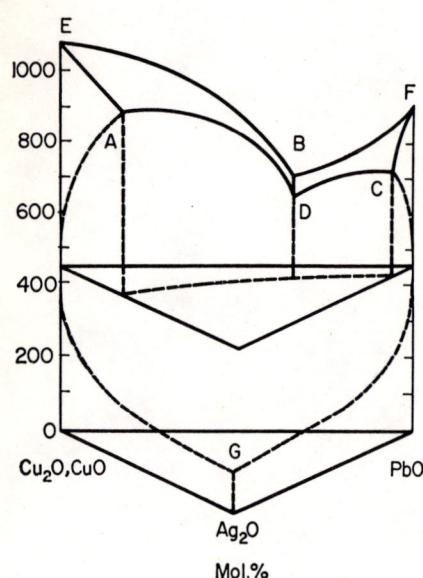

Fig. 125.—System Cu_2O/CuO–PbO–Ag_2O at 1 atm. O_2 pressure. A to F = primary crystallization surface; G = decomposition of Ag_2O.

E. J. Kohlmeyer and Helmit Hennig, *Z. Erzbergbau Metallhüettenw.*, **7**, [8] 331 (1954).

Fig. 126. See page 73.

Al–Cr–Fe–O

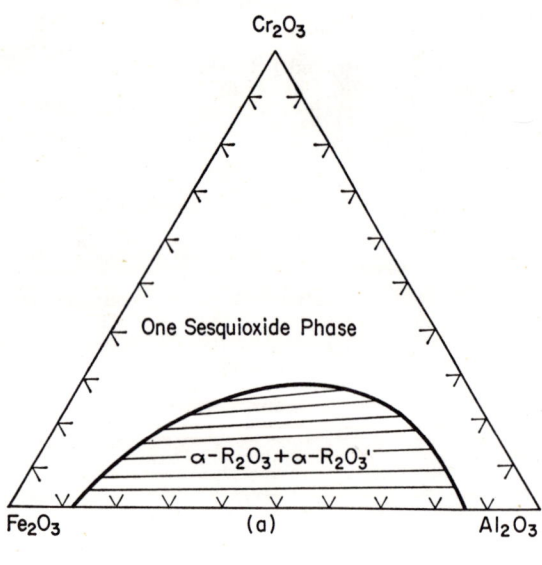

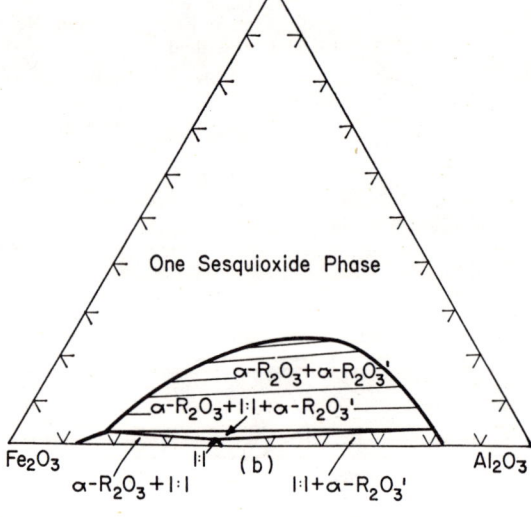

Fig. 127.—System Al_2O_3–Cr_2O_3–Fe_2O_3; isothermal sections in air. (a) 1250°C, (b) 1350°C, (c) 1500°C. α-R_2O_3 and α-R_2O_3' designate iron oxide and Al_2O_3-rich sesquioxides, respectively, with the corundum structure. 1:1 refers to $Fe_2O_3 \cdot Al_2O_3$.

Arnulf Muan and Shigeyuki Sōmiya, *J. Am. Ceram. Soc.*, **42** [12] 609 (1959).

Al–Cr–Fe–O (cont.)

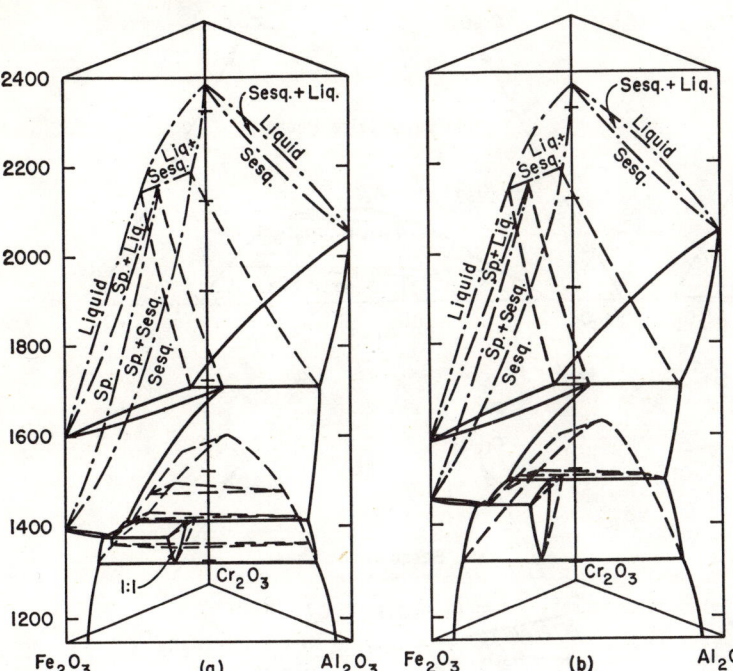

Fig. 126.—System Al_2O_3–Cr_2O_3–Fe_2O_3; perspective drawings. (a) in air; (b) at 1 atm. O_2 partial pressure.

Arnulf Muan and Shigeyuki Sōmiya, *J. Am. Ceram. Soc.*, **42** [12] 611 (1959).

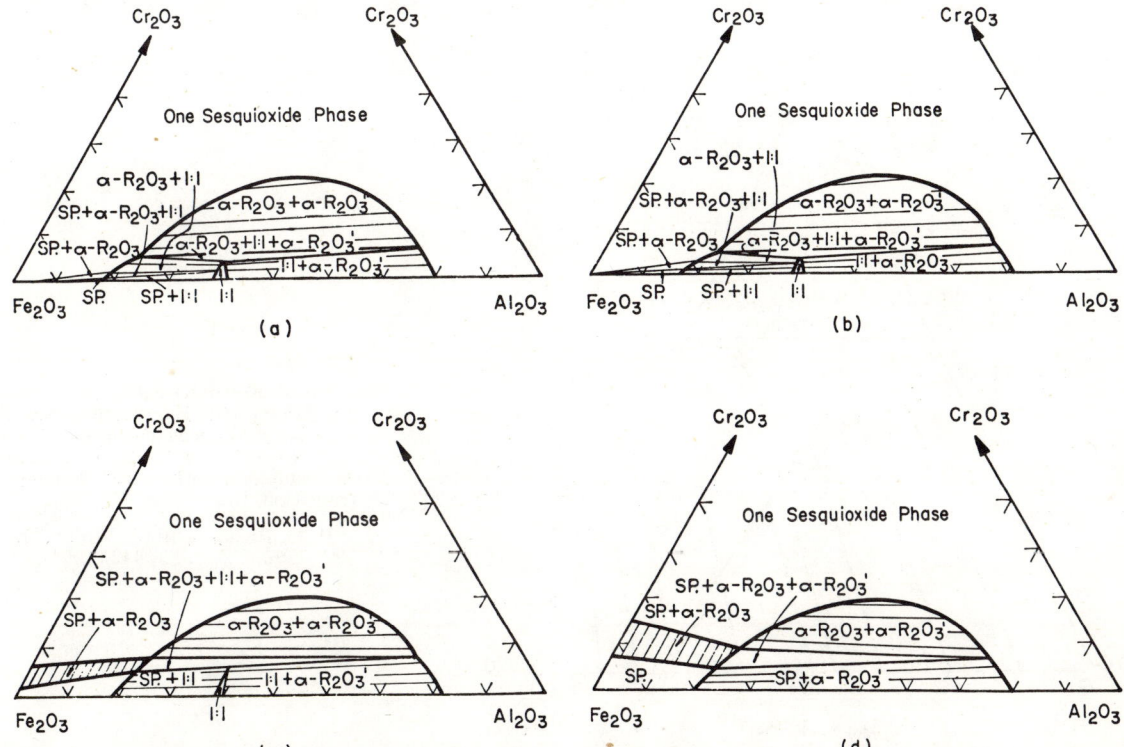

Fig. 128.—System Al_2O_3–Cr_2O_3–Fe_2O_3; isothermal sections in air. (a) 1385°C, (b) 1390°C, (c) 1410°C, (d) 1415°C. α-R_2O_3 and α-R_2O_3' designate iron oxide and Al_2O_3-rich sesquioxides, respectively, with the corundum structure. 1:1 refers to $Fe_2O_3 \cdot Al_2O_3$. SP designates a solid solution with spinel structure ($FeO \cdot Fe_2O_3$–$FeO \cdot Al_2O_3$–$FeO \cdot Cr_2O_3$).

Al–Cr–Fe–O (concl.)

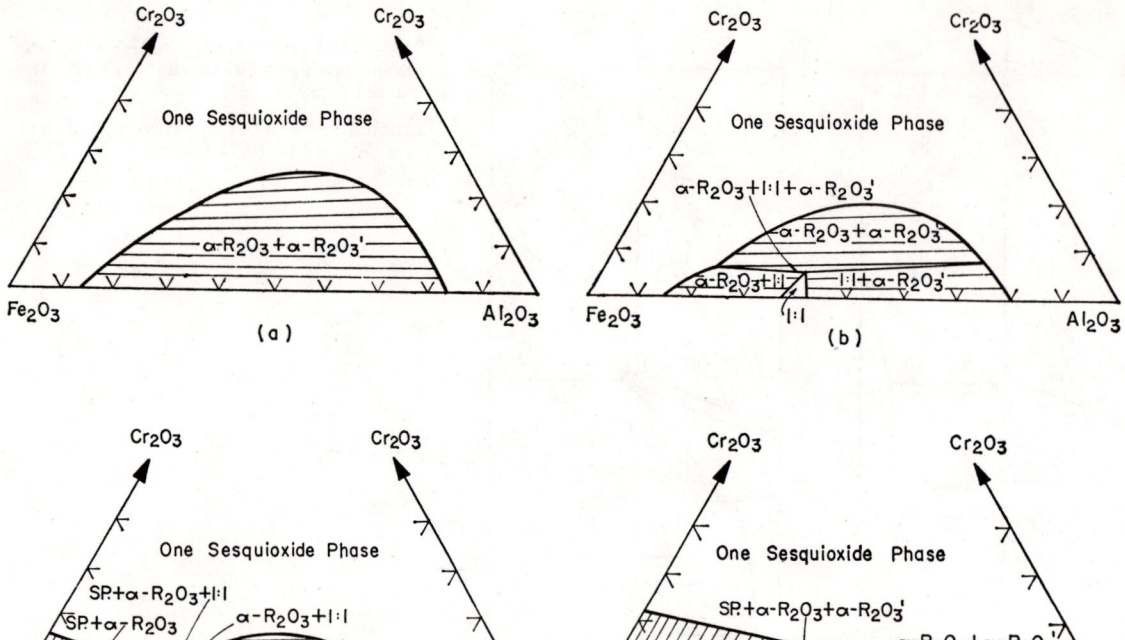

Fig. 129.—System Al_2O_3–Cr_2O_3–Fe_2O_3; isothermal sections at 1 atm. O_2 partial pressure. (a) 1300°C, (b) 1418°C, (c) 1483°C, (d) 1500°C. α-R_2O_3 and α-R_2O_3' designate iron oxide and Al_2O_3-rich sesquioxides, respectively, with the corundum structure. 1:1 refers to $Fe_2O_3 \cdot Al_2O_3$. SP designates a solid solution with spinel structure.

Arnulf Muan and Shigeyuki Sōmiya, *J. Am. Ceram. Soc.*, **42** [12] 611 (1959).

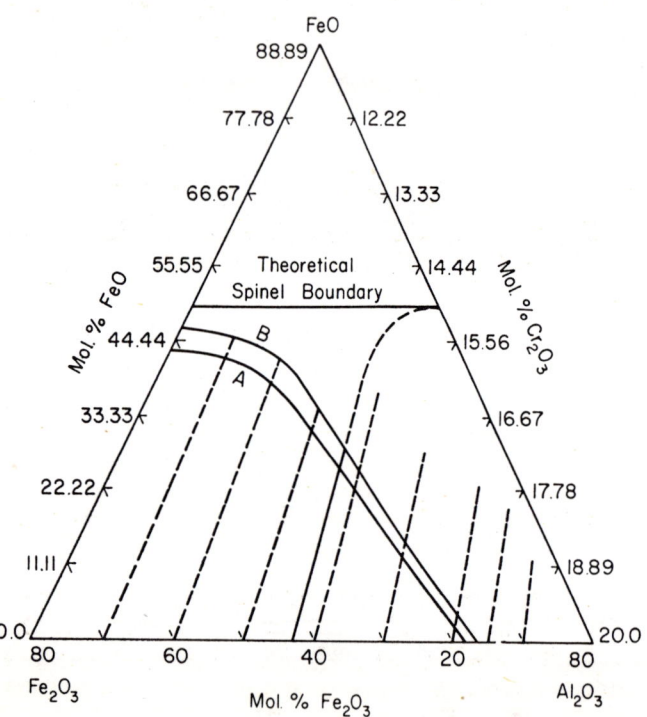

Fig. 130.—FeO–Al_2O_3–Cr_2O_3–Fe_2O_3; quaternary section containing dissociation paths (dashed) of sesquioxide mixtures with 20 mol. % Cr_2O_3. Curve A = completion of spinel sesquioxide transition, Curve B = 1550°C.

R. G. Richards and J. White, *Trans. Brit. Ceram. Soc.*, **53**, 442 (1954).

Al–Fe–Si–O

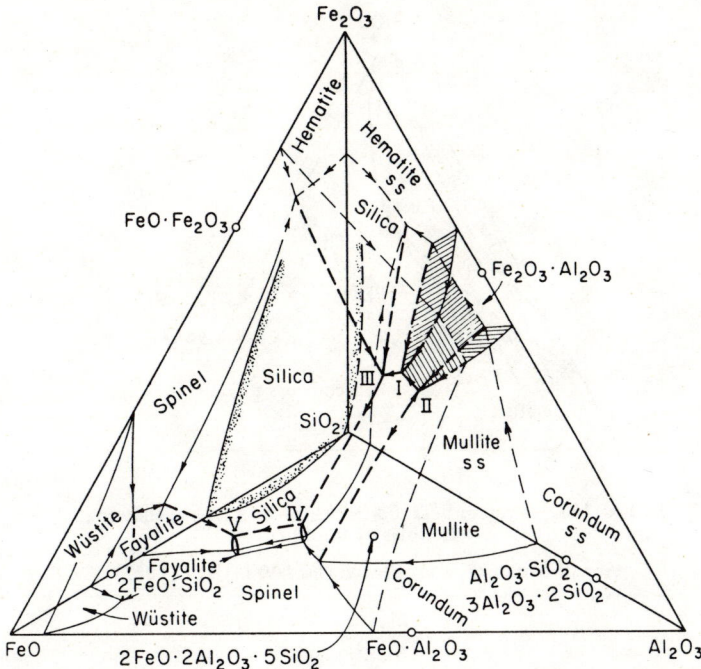

Fig. 131.—System $FeO-Al_2O_3-Fe_2O_3-SiO_2$; phase relationships at liquidus temperatures. Boundary curves are shown as light lines in bounding ternary systems; quaternary univariant lines are drawn as heavy lines. Lines with stippling indicate limits of two-liquid region.

Arnulf Muan, *J. Am. Ceram. Soc.* **40** [12] 425 (1957).

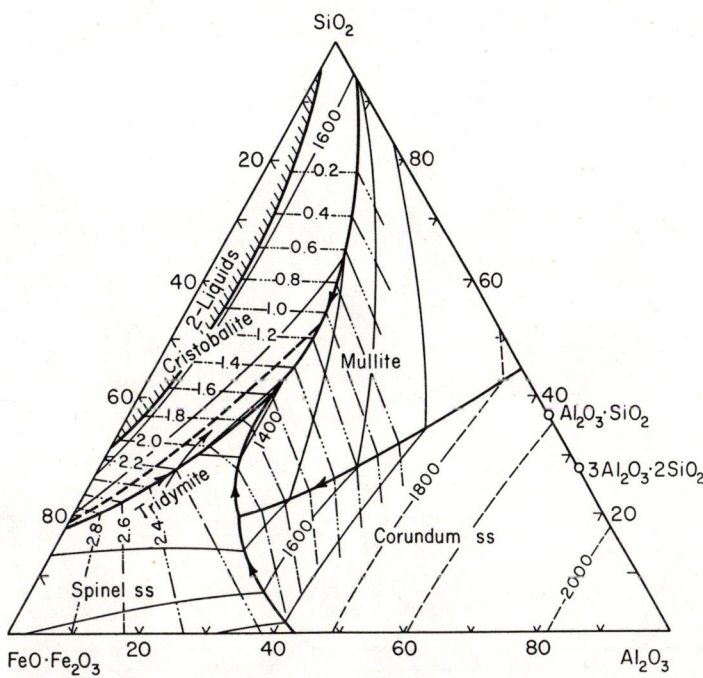

Fig. 132.—System $FeO \cdot Fe_2O_3-Al_2O_3-SiO_2$ in air. Light dash-triple dot lines are curves passing through points of equal Fe_2O_3/FeO ratios (wt.% basis) on liquidus surface.
Arnulf Muan, *J. Am. Ceram. Soc.*, **40** [4] 127 (1957).

Al–Fe–Si–O (cont.)

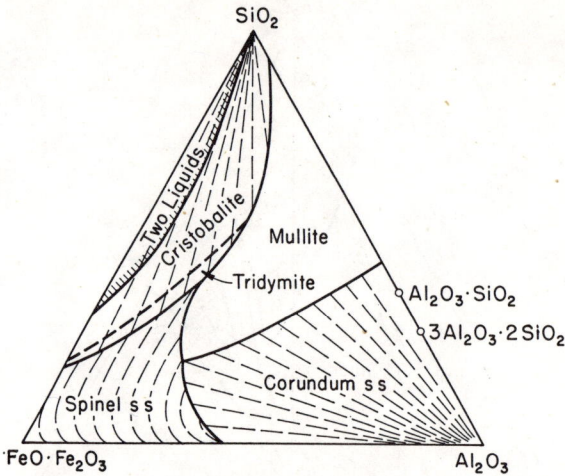

FIG. 133.—System FeO·Fe$_2$O$_3$–Al$_2$O$_3$–SiO$_2$ in air, showing tie lines at liquidus.

Arnulf Muan, *J. Am. Ceram. Soc.*, **40** [4] 127 (1957).

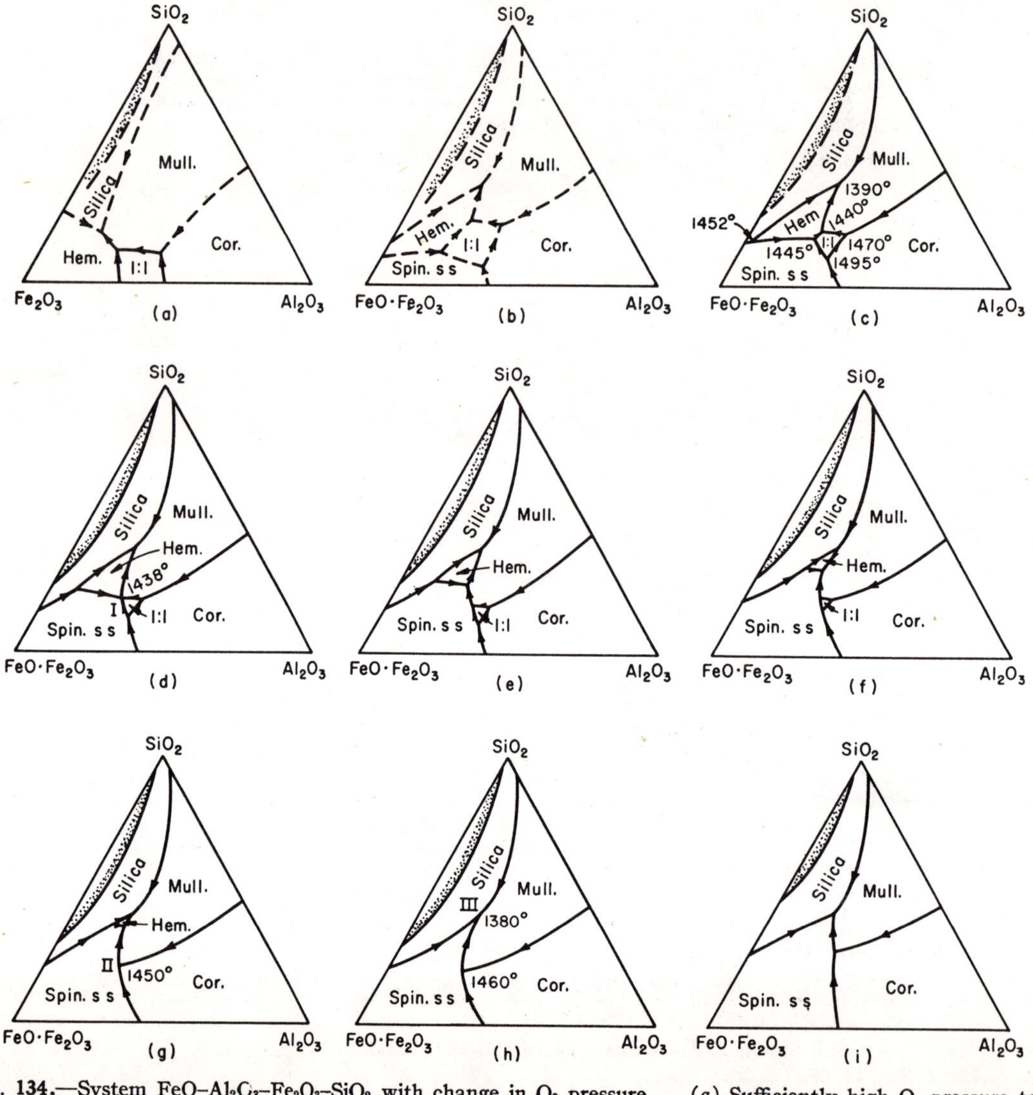

FIG. 134.—System FeO–Al$_2$O$_3$–Fe$_2$O$_3$–SiO$_2$ with change in O$_2$ pressure. (*a*) Sufficiently high O$_2$ pressure to keep essentially all the iron in the ferric state. (*b*) Lower O$_2$ pressure. (*c*) 1 atm. O$_2$. (*d*) 0.9 atm. (*e*) 0.8 atm. (*f*) 0.5 atm. (*g*) 0.4 atm. (*h*) 0.2 atm. (*i*) 10^{-3} atm. O$_2$.

Arnulf Muan, *J. Am. Ceram. Soc.*, **40** [12] 428 (1957).

Al–Fe–Si–O (concl.)

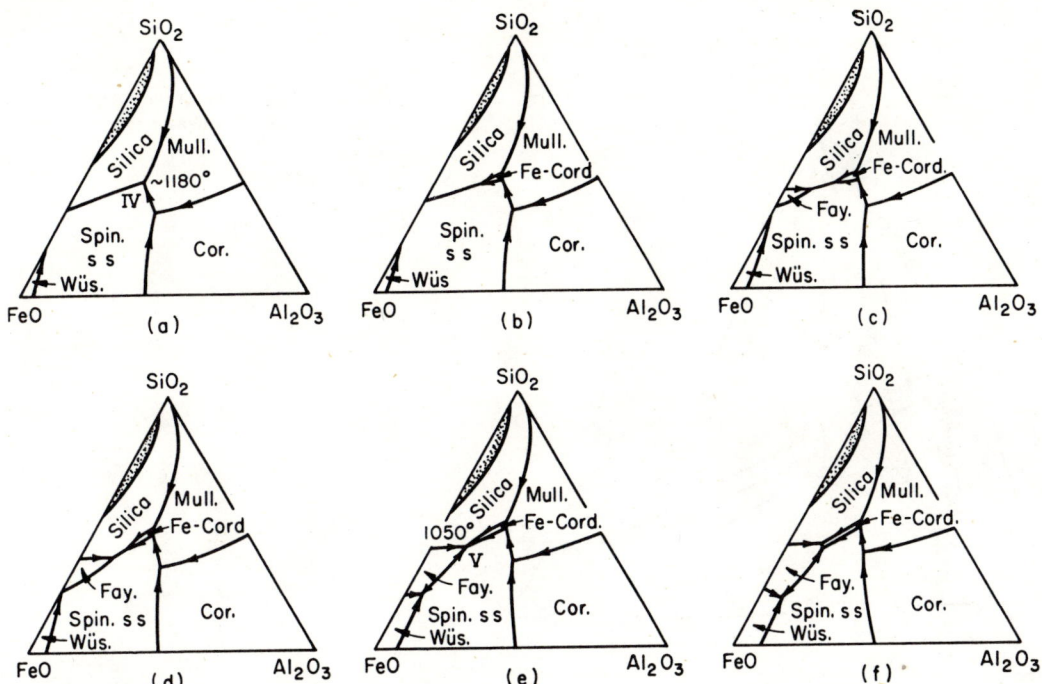

Fig. 135.—System FeO–Al$_2$O$_3$–Fe$_2$O$_3$–SiO$_2$; probable, with changes in oxygen pressure under strongly reducing conditions. Oxygen pressure decreases from approx. 10^{-10} atm. in (a) to 10^{-13} atm. in (f). Diagram (f) after Schairer and Yagi. Abbreviations apparent from Fig. 131.

Arnulf Muan, *J. Am. Ceram. Soc.*, **40** [12] 429 (1957).

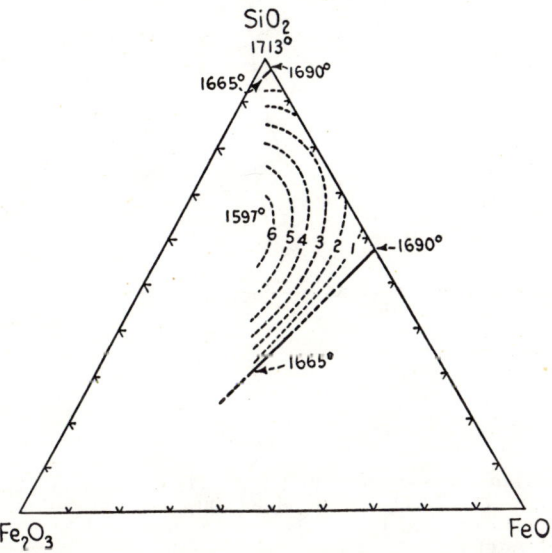

Fig. 136.—Projection onto the FeO–Fe$_2$O$_3$–SiO$_2$ face of the system FeO–Fe$_2$O$_3$–Al$_2$O$_3$–SiO$_2$ of the curved surface bounding the mixtures that form two liquids on melting; the Al$_2$O$_3$ apex of the tetrahedron is the pole of projection; dotted lines represent intersections of above surface by planes corresponding to the percentages of Al$_2$O$_3$ indicated on the lines; temperatures of equilibrium of cristobalite with liquid on the curved bounding surface are indicated for several compositions.

J. W. Greig, *Am. J. Sci.*, **14**, 480 (1927).

Cr–Fe–Mg–O

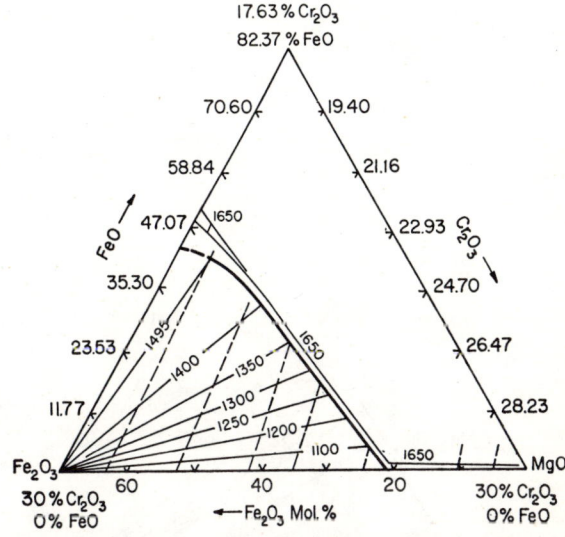

Fig. 137.—FeO–MgO–Cr$_2$O$_3$–Fe$_2$O$_3$, containing dissociation paths (dashed) of mixtures with initially 30 mol. % Cr$_2$O$_3$.

D. Woodhouse and J. White, *Trans. Brit. Ceram. Soc.*, **54**, 351 (1955).

Fe–Mg–Si–O

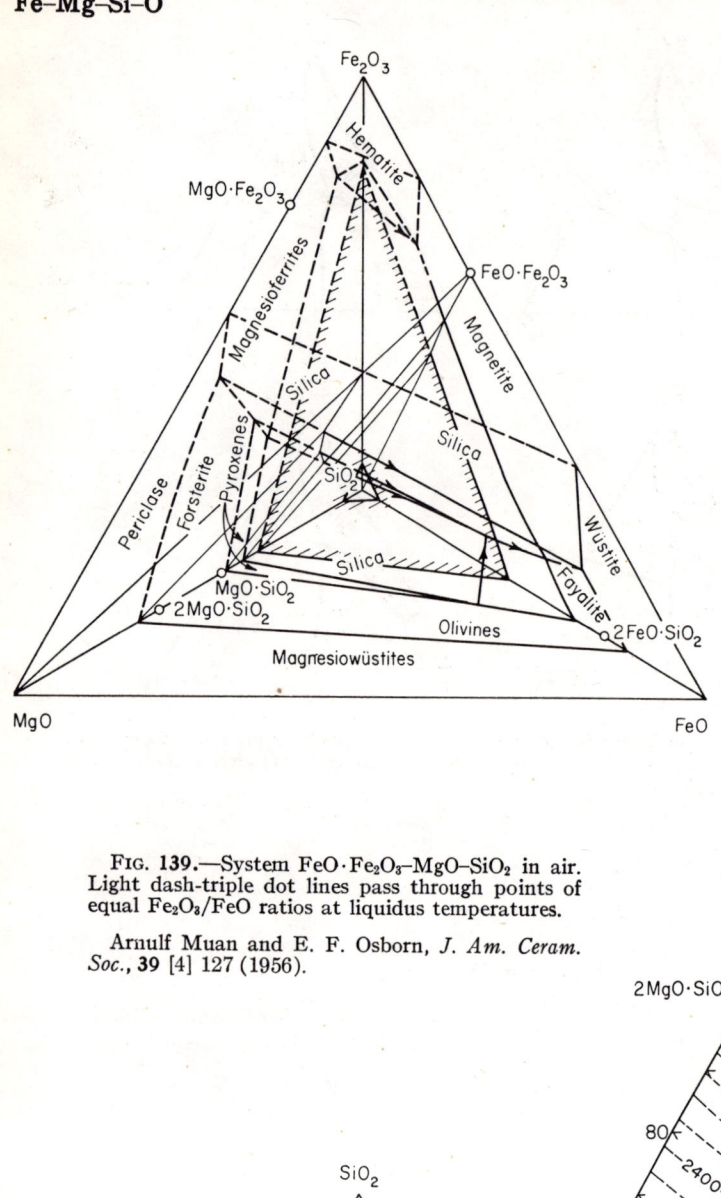

FIG. 138.—System FeO–MgO–Fe$_2$O$_3$–SiO$_2$; phase relationships.

Arnulf Muan and E. F. Osborn, *J. Am. Ceram. Soc.*, **39** [4] 134 (1956).

FIG. 139.—System FeO·Fe$_2$O$_3$–MgO–SiO$_2$ in air. Light dash-triple dot lines pass through points of equal Fe$_2$O$_3$/FeO ratios at liquidus temperatures.

Arnulf Muan and E. F. Osborn, *J. Am. Ceram. Soc.*, **39** [4] 127 (1956).

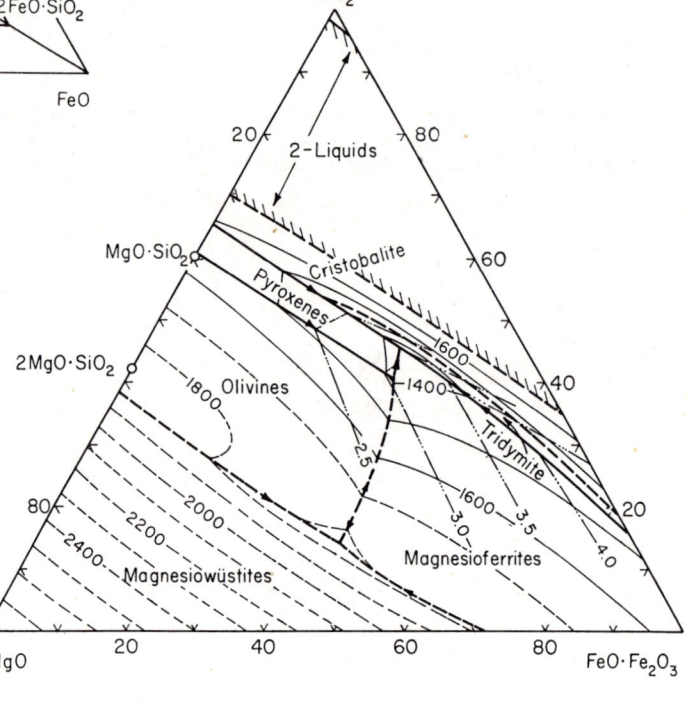

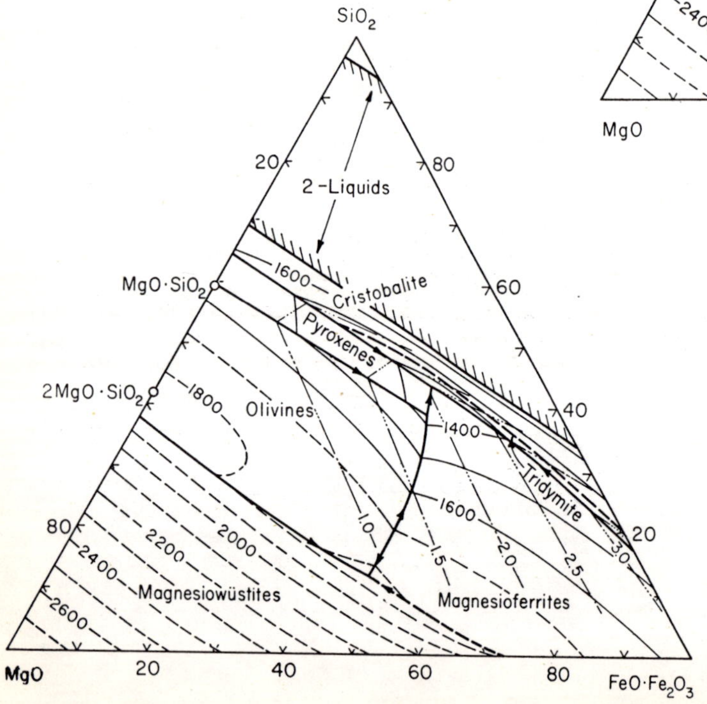

FIG. 140.—System FeO·Fe$_2$O$_3$–MgO–SiO$_2$ at 1 atm. O$_2$. Light dash-triple dot lines pass through points of equal Fe$_2$O$_3$/FeO ratios at liquidus temperatures.

Arnulf Muan and E. F. Osborn, *J. Am. Ceram. Soc.*, **39** [4] 129 (1956).

Fe–Mg–Si–O (cont.)

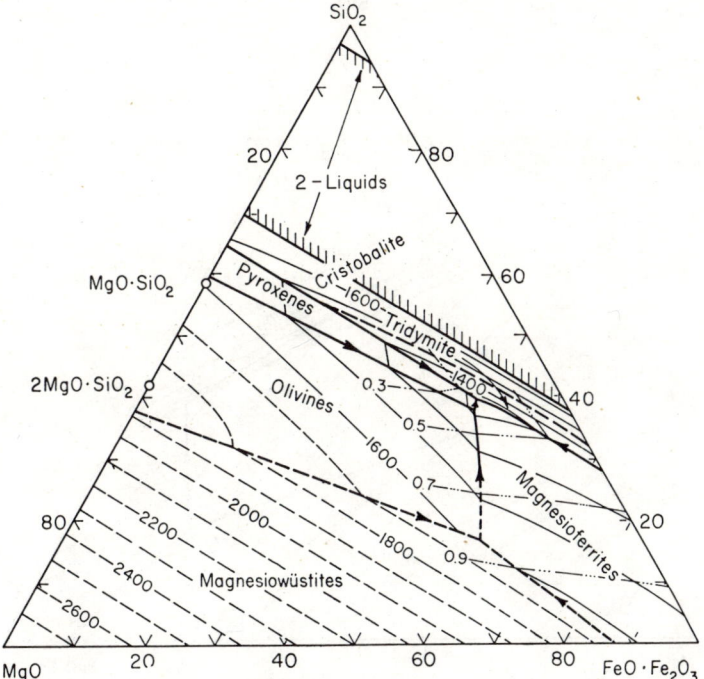

Fig. 141.—System $FeO \cdot Fe_2O_3$–MgO–SiO_2, for constant mixing ratio $CO_2/H_2 = 132$. Light dash-triple dot lines pass through points of equal Fe_2O_3/FeO ratios at liquidus temperatures.

Arnulf Muan and E. F. Osborn, *J. Am. Ceram. Soc.*, **39** [4] 129 (1956)

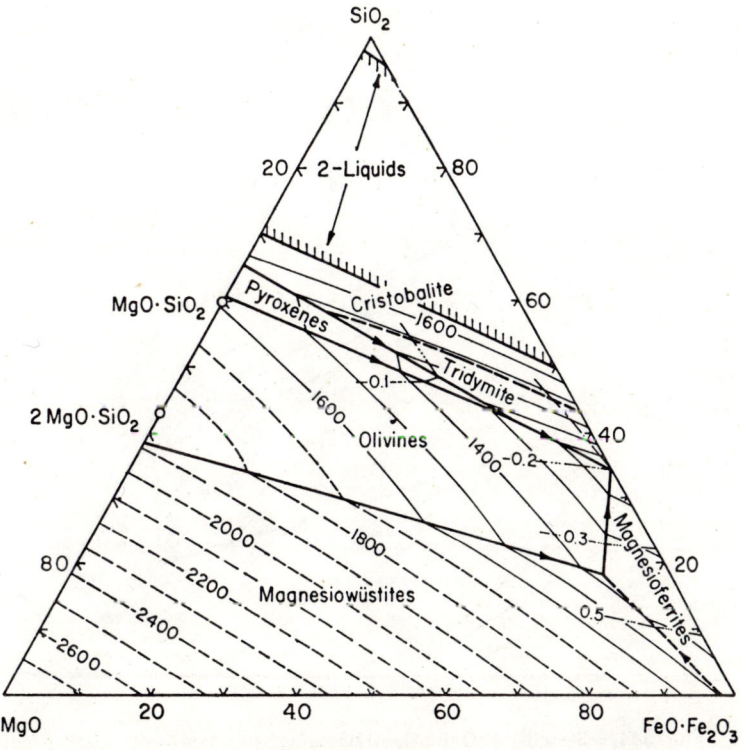

Fig. 142.—System $FeO \cdot Fe_2O_3$–MgO–SiO_2, for constant mixing ratio $CO_2/H_2 = 40$. Light dash-triple dot lines pass through points of equal Fe_2O_3/FeO ratios at liquidus temperatures.

Arnulf Muan and E. F. Osborn, *J. Am. Ceram. Soc.*, **39** [4] 130 (1956).

Fe–Mg–Si–O (cont.)

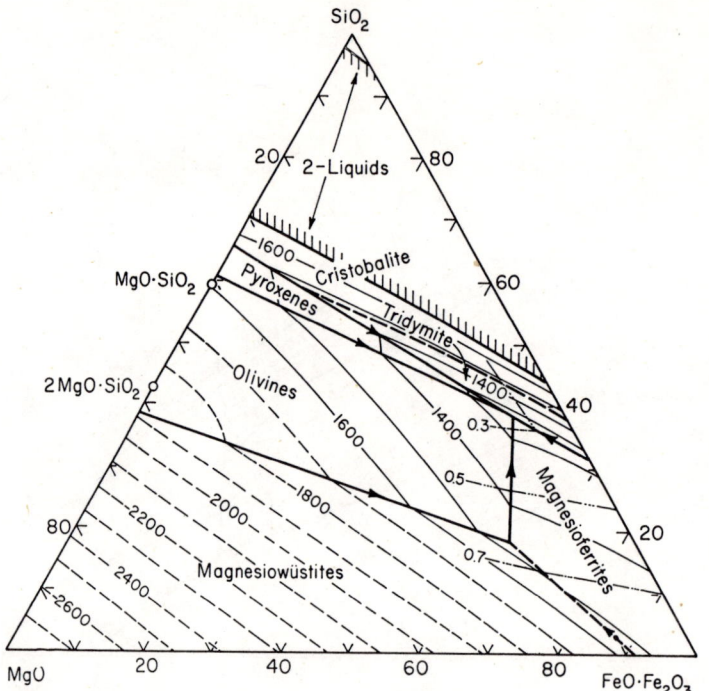

Fig. 143.—System $FeO \cdot Fe_2O_3$–MgO–SiO_2, for constant mixing ratio $CO_2/H_2 = 24$. Light dash-triple dot lines pass through points of equal Fe_2O_3/FeO ratios at liquidus temperatures.

Arnulf Muan and E. F. Osborn, *J. Am. Ceram. Soc.*, **39** [4] 130 (1956).

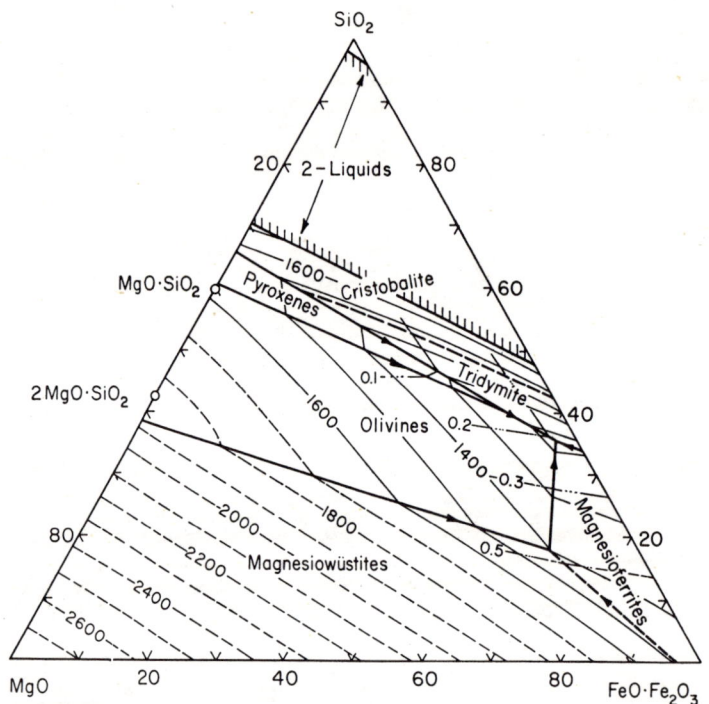

Fig. 144.—System $FeO \cdot Fe_2O_3$–MgO–SiO_2, for constant mixing ratio $CO_2/H_2 = 19$. Light dash-triple dot lines pass through points of equal Fe_2O_3/FeO ratios at liquidus temperatures.

Arnulf Muan and E. F. Osborn, *J. Am. Ceram. Soc.*, **39** [4] 132 (1956).

Fe–Mg–Si–O (concl.)

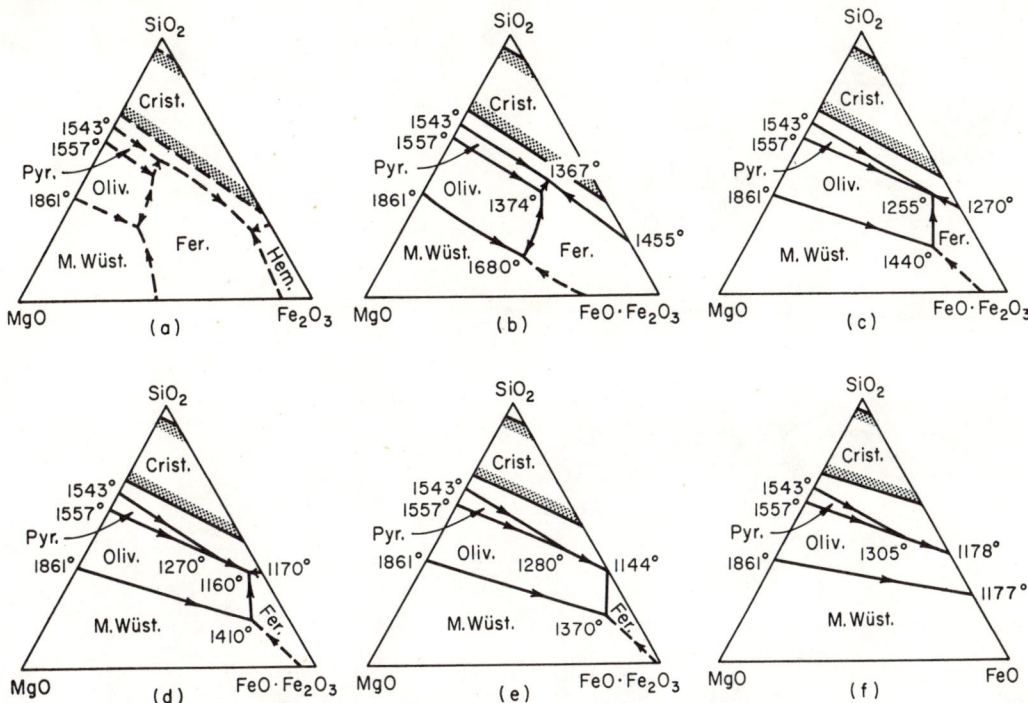

Fig. 145.—System FeO–MgO–Fe$_2$O$_3$–SiO$_2$, with decreasing O$_2$ pressure. (a) At sufficiently high O$_2$ pressure to keep essentially all of the iron in the ferric state. (b) in air, (c), (d), and (e) at constant CO$_2$/H$_2$ ratios of 40, 24, and 19, respectively. (f) extreme reducing conditions of melts in contact with metallic iron (after Bowen and Schairer). Oliv. = olivine; Fer. = magnesioferrite; M. Wüst. = magnesiowüsite; Crist. = cristobalite; Pyr. = pyroxene; Hem. = hematite.

Arnulf Muan and E. F. Osborn, *J. Am. Ceram. Soc.*, **39** [4] 133 (1956).

Fe–Si–Ti–O

Fig. 147.—System Fe$_3$O$_4$–SiO$_2$–TiO$_2$ showing compositions of liquids on liquidus surface as FeO/Fe$_2$O$_3$ ratios. The dash-dot curve 0.45 represents the FeO/Fe$_2$O$_3$ ratio of stoichiometric magnetite.

J. B. MacChesney and Arnulf Muan, *J. Am. Ceram. Soc.*, **43** [11] 589 (1960).

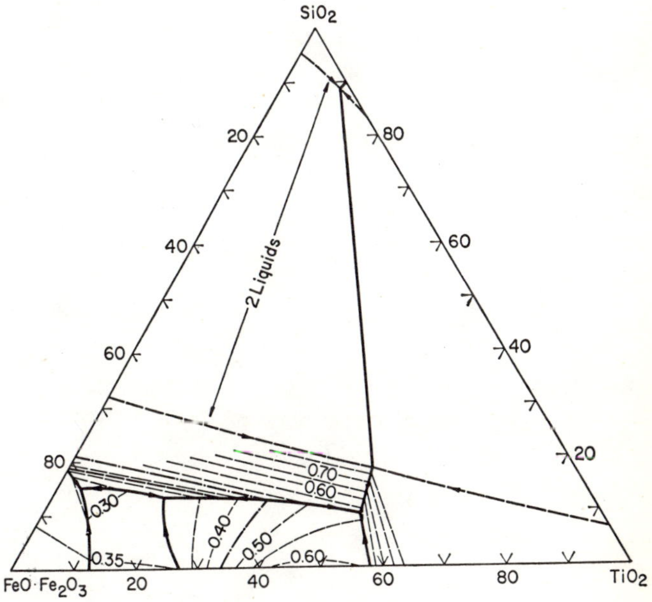

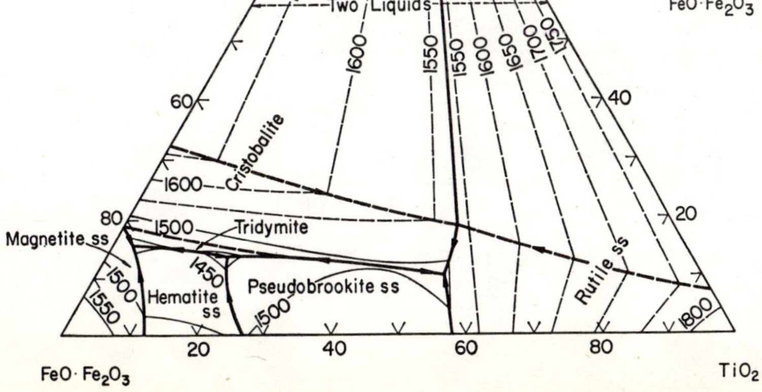

Fig. 146.—System Fe$_3$O$_4$–SiO$_2$–TiO$_2$ in air.

J. B. MacChesney and Arnulf Muan, *J. Am. Ceram. Soc.*, **43** [11] 588 (1960).

B. METAL OXIDE SYSTEMS

1. One Oxide

PbO

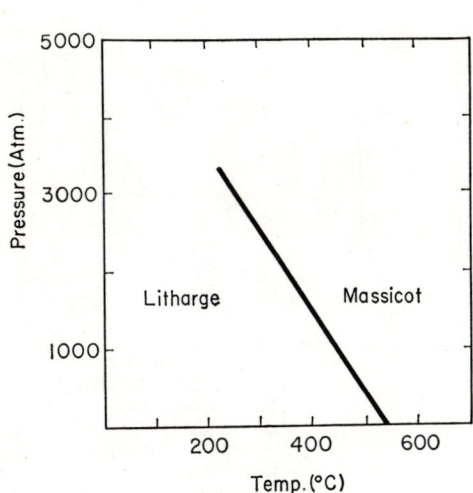

Fig. 148.—System PbO; p–T.

W. B. White, Frank Dachille, and Rustum Roy, *J. Am. Ceram. Soc.*, **44** [4] 172 (1961).

B_2O_3

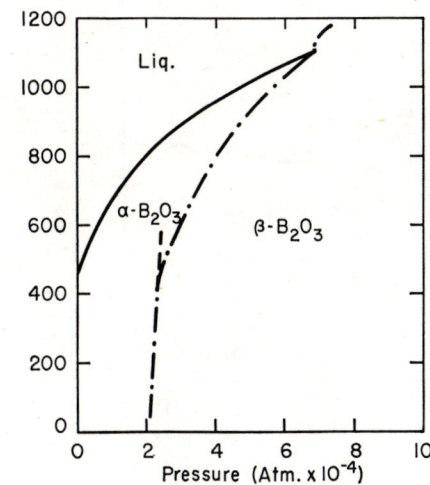

Fig. 150.—System B_2O_3; tentative p–T relations.

J. D. Mackenzie and W. F. Claussen, *J. Am. Ceram. Soc.*, **44** [2] 81 (1961).

As_2O_3

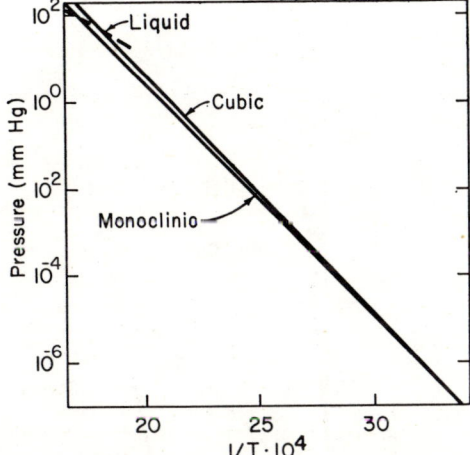

Fig. 149.—System As_2O_3; vapor pressure of arsenolite (cubic) and claudetite (monoclinic).

I. Karutz and I. N. Stranski, *Z. anorg. u. allgem. Chem.*, **292** [6] 332 (1957).

GeO_2

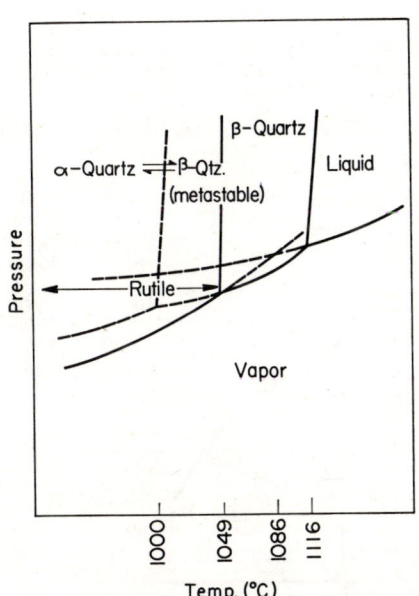

Fig. 151.—System GeO_2; p–T, schematic.

James F. Sarver and F. A. Hummel, *J. Am. Ceram. Soc.*, **43** [6] 336 (1960).

PbO₂

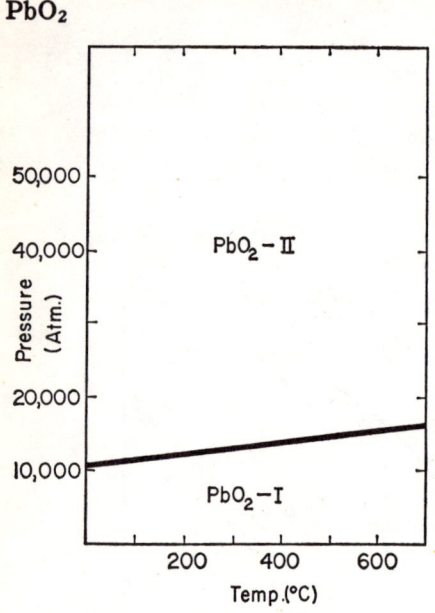

FIG. 152.—System PbO₂; p–T.

W. B. White, Frank Dachille, and Rustum Roy, *J. Am. Ceram. Soc.*, **44** [4] 173 (1961).

SiO₂

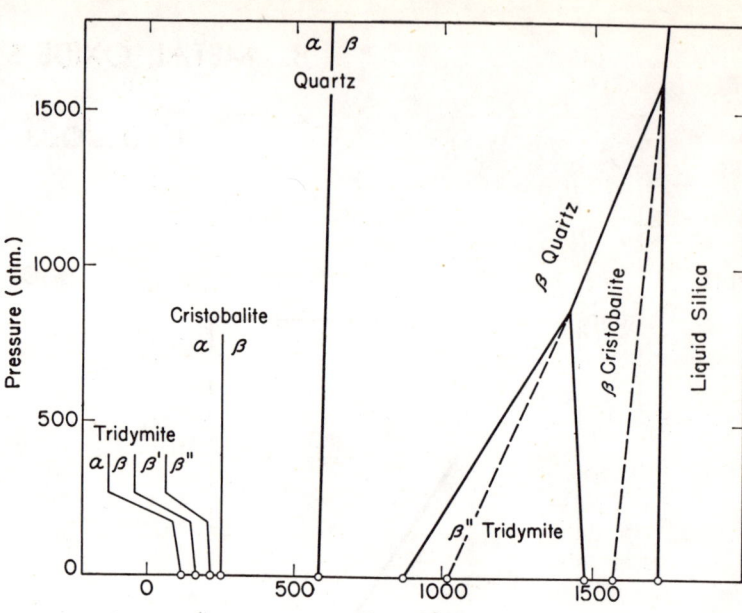

FIG. 155.—SiO₂ (p-t diag.). Solid lines represent transitions of phases stable in the respective adjoining regions. Broken lines represent transitions of metastable phases. Tridymite and cristobalite in the regions of their alpha–beta transitions are each metastable with respect to alpha quartz, which is here the stable phase.

Frank Charles Kracek, "Polymorphism," Encyclopaedia Britannica, Copyright 1953.

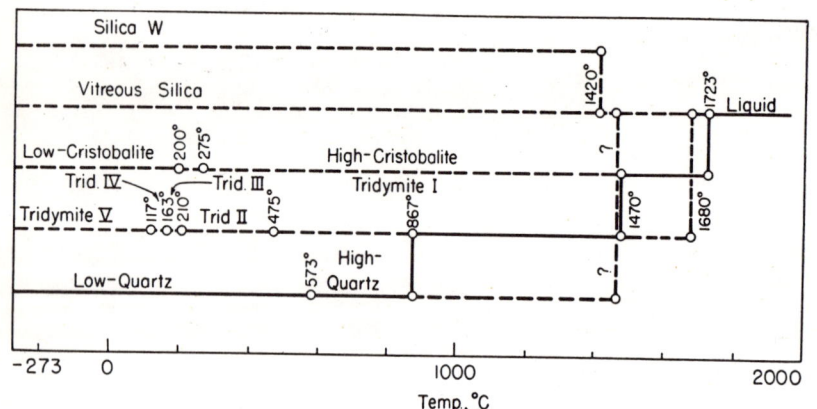

FIG. 153.—SiO₂; stability relations at one atmosphere pressure, as revised in 1955. Ordinate is undefined measure of stability. Solid lines represent stable state; dashed lines represent metastable state.

R. B. Sosman, *Trans. Brit. Ceram. Soc.*, **54**, 663 (1955).

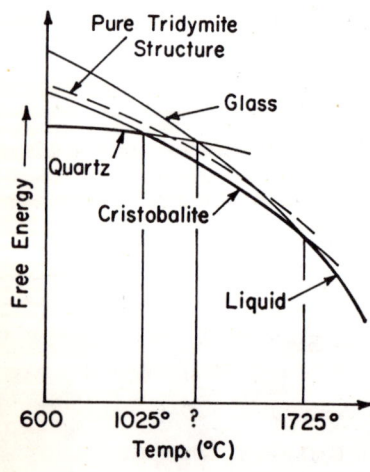

FIG. 154.—SiO₂, free energy for pure phases; schematic. Quartz, cristobalite, and liquid are assumed to be the only thermodynamically stable phases at ordinary pressures. See figures 168 and 184 for stability of tridymite.
S. B. Holmquist, *J. Am. Ceram. Soc.*, **44** [2] 86 (1961); see also, S. B. Holmquist, *Z. Krist.*, **111**, 71–76 (1958).

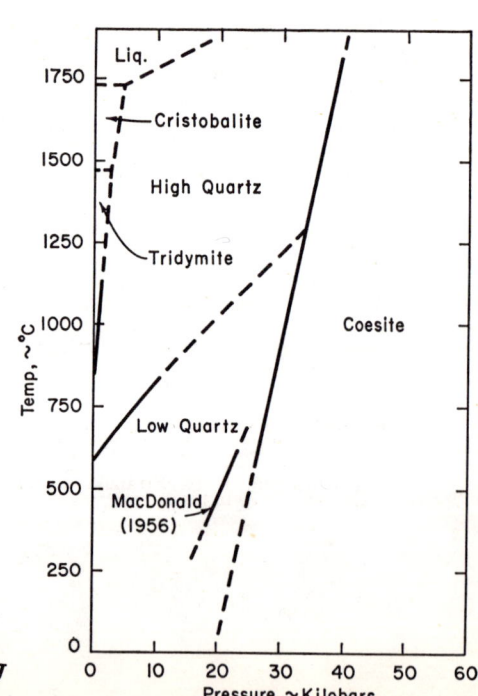

FIG. 156.—System SiO₂; quartz–coesite transition.

F. R. Boyd and J. L. England, *J. Geophys. Res.*, **65** [2] 752 (1960).

SiO₂ (concl.)

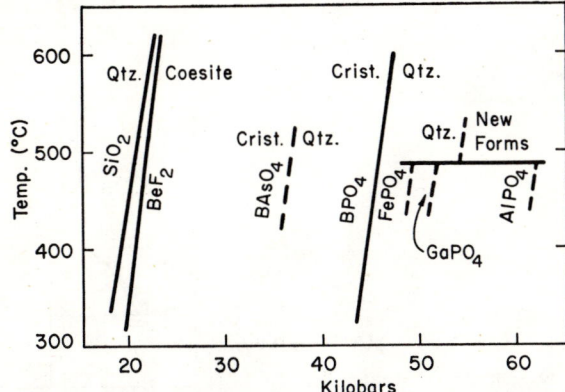

FIG. 157.—System SiO₂ isotypes. Univariant p–T lines for various compounds separate fields of pairs of polymorphs.

Frank Dachille and Rustum Roy, *Z. Krist.*, 111 [6] 455 (1959).

P₂O₅

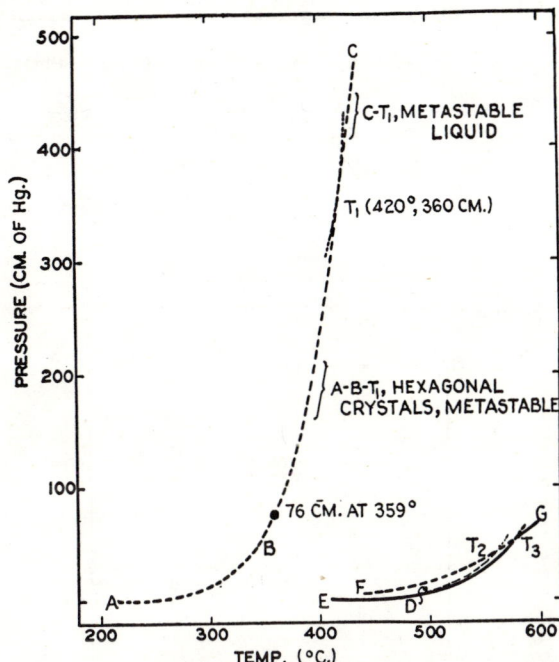

FIG. 158.—System P_2O_5; pressure-temperature diagram, 200° to 600°C.; see FIG. 159 for details of lower curves.

W. L. Hill, G. T. Faust, and S. B. Hendricks, *J. Am. Chem. Soc.*, 65 [5] 798 (1943).

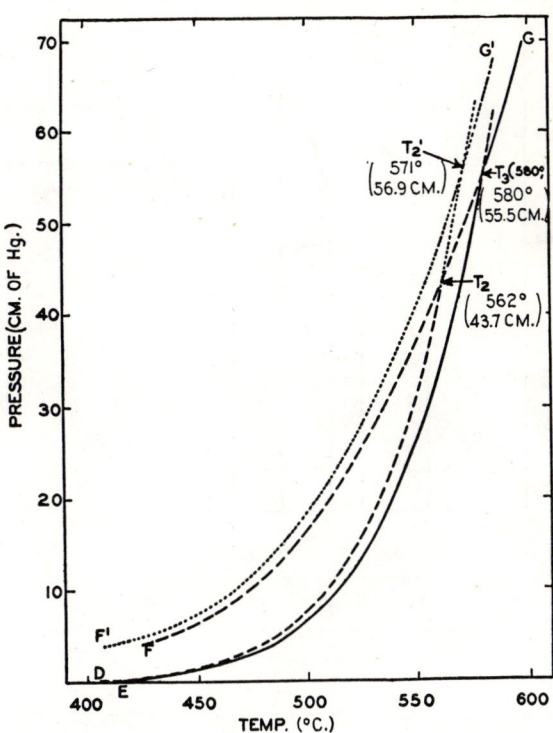

FIG. 159.—System P_2O_5; pressure-temperature diagram, 400° to 600°C.; *curve $D-T_2$*: orthorhombic crystals, metastable; *curve $E-T_2$*: tetragonal (?) crystals, stable; *curve F-G*: stable liquid; *curve F'-G'*: liquid according to data of Hoeflake and Scheffer.

W. L. Hill, G. T. Faust, and S. B. Hendricks, *J. Am. Chem. Soc.*, 65 [5] 798 (1943).

II. Two Oxides

Cs₂O–B₂O₃

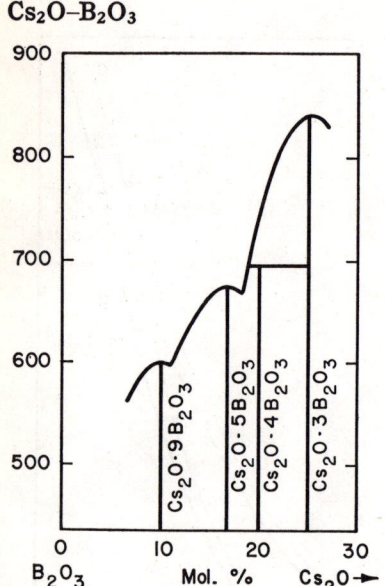

FIG. 160.—System Cs₂O–B₂O₃; B₂O₃ region.

J. Krogh-Moe, *Arkiv Kemi*, 12, 248 (1958).

Cs₂O–TiO₂

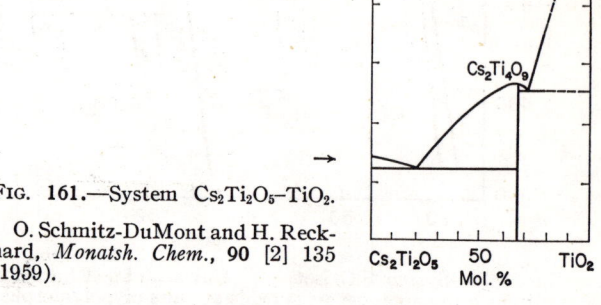

FIG. 161.—System Cs₂Ti₂O₅–TiO₂.

O. Schmitz-DuMont and H. Reckhard, *Monatsh. Chem.*, 90 [2] 135 (1959).

Cu₂O–PbO

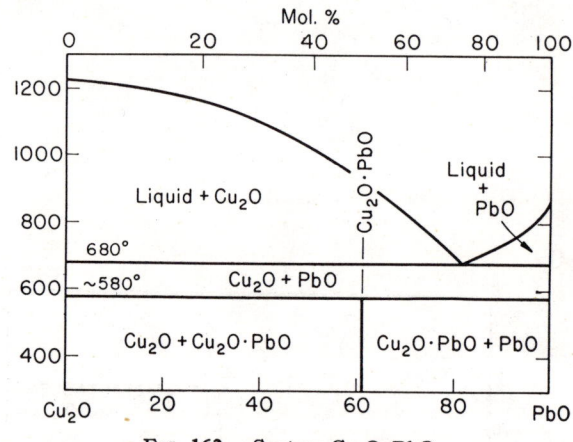

FIG. 163.—System Cu₂O–PbO.

Erich Gebhardt and Walter Obrowski, *Z. Metallk.*, 45, 333 (1954).

Cs₂O–MoO₃

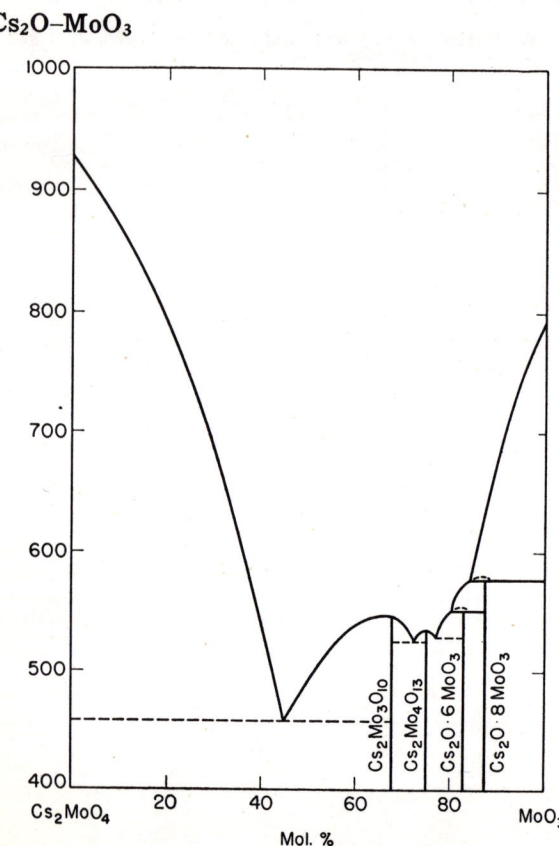

FIG. 162.—System Cs₂MoO₄–MoO₃.

Viktor Spitsyn and I. M. Kuleshov, *Zhur. Obshchei Khim.*, 21, 1372 (1951).

Cu₂O–SiO₂

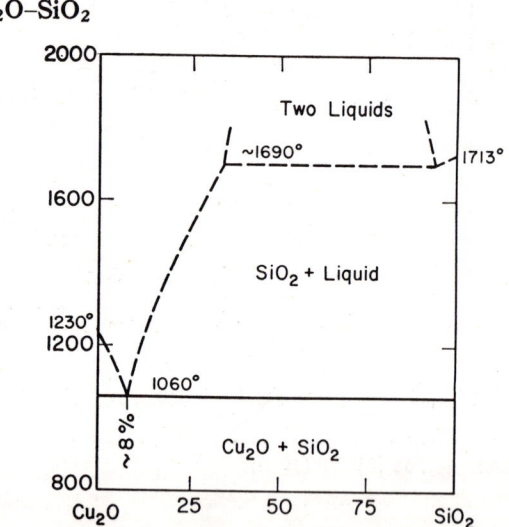

FIG. 164.—System Cu₂O–SiO₂. CuO content not greater than 5%.

A. S. Berezhnoi, L. I. Karyakin, and I. F. Dudavskiĭ, *Doklady Akad. Nauk S.S.S.R.*, 83 [3] 401 (1952).

K_2O–B_2O_3

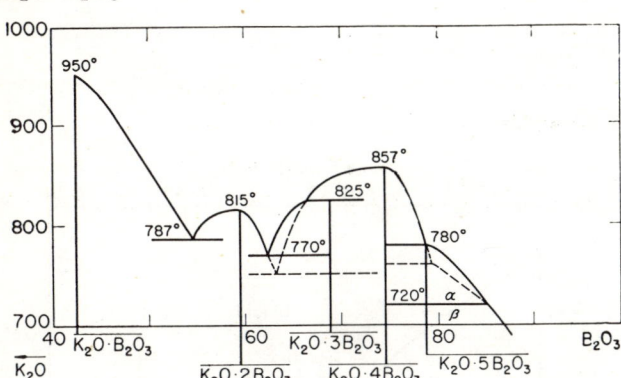

Fig. 165.—System K_2O–B_2O_3. (Dashed lines indicate metastable equilibrium.)

A. P. Rollet, *Compt. rend.*, 200, 1764 (1935); *ibid.*, 202, 1864 (1936).

K_2O–GeO_2

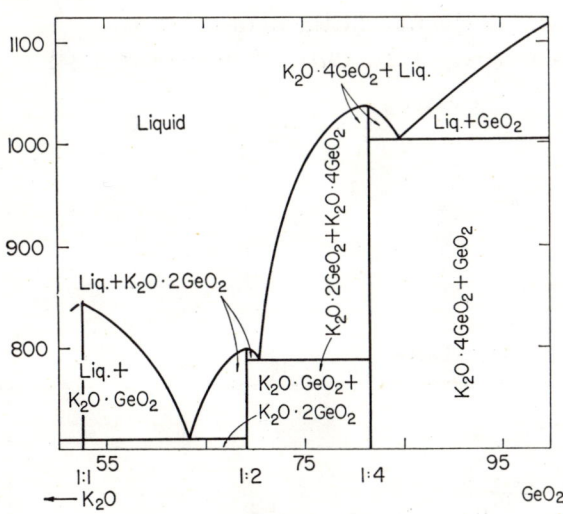

Fig. 166.—System K_2O–GeO_2.

Robert Schwarz and Fritz Heinrich, *Z. anorg. u. allgem. Chem.*, 205, 44 (1932).

K_2O–SiO_2

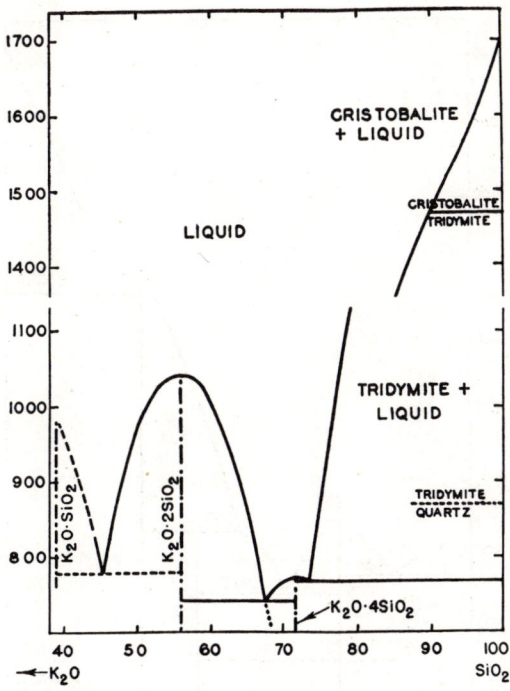

Fig. 167.—System SiO_2–$K_2O\cdot SiO_2$.

F. C. Kracek, N. L. Bowen, and G. W. Morey, *J. Phys. Chem.*, 41, 1188 (1937).

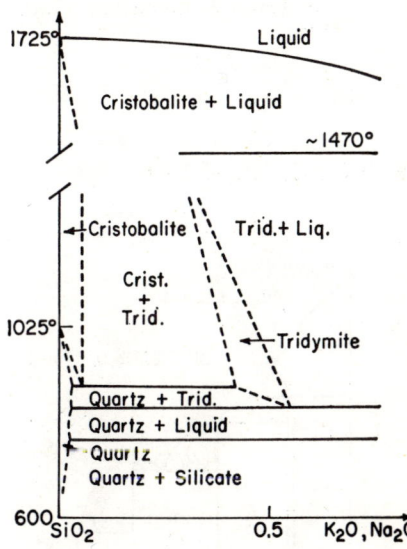

Fig. 168.—System SiO_2–K_2O, Na_2O; tentative for high-silica region of systems with eutectic below 870°C.

S. B. Holmquist, *J. Am. Ceram. Soc.* 44 [2] 85 (1961).

K_2O–TiO_2

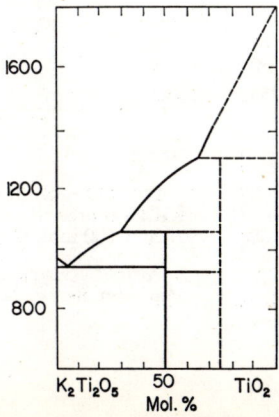

Fig. 169.—System $K_2Ti_2O_5$–TiO_2.

O. Schmitz-DuMont and H. Reckhard, *Monatsh. Chem.*, 90 [2] 135 (1959).

K_2O–As_2O_5

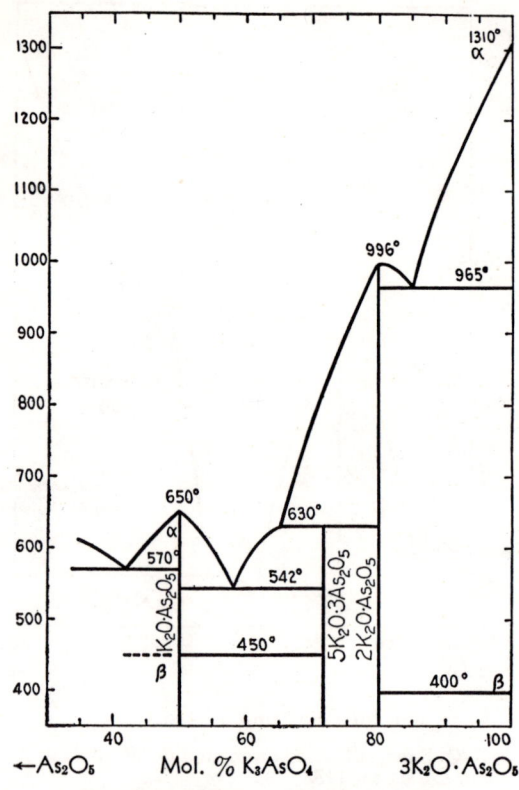

Fig. 170.—System As_2O_5–$3K_2O \cdot As_2O_5$.

M. Amadori, *Atti reale ist. Veneto sci.*, **73** [II] 1677 (1914).

K_2O–P_2O_5

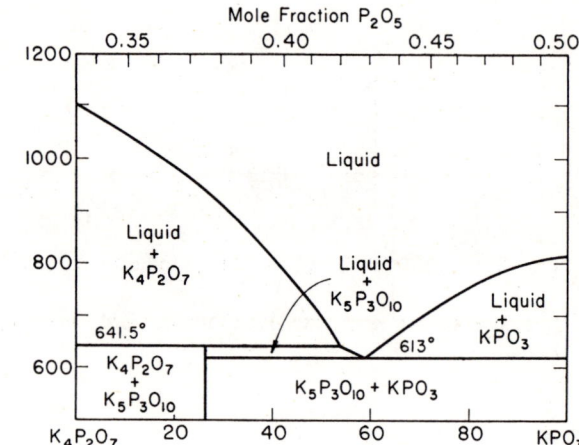

Fig. 172.—System KPO_3–$K_4P_2O_7$.

G. W. Morey, *J. Am. Chem. Soc.*, **76** [18] 4726 (1954).

K_2O (or K_2CO_3)–Nb_2O_5

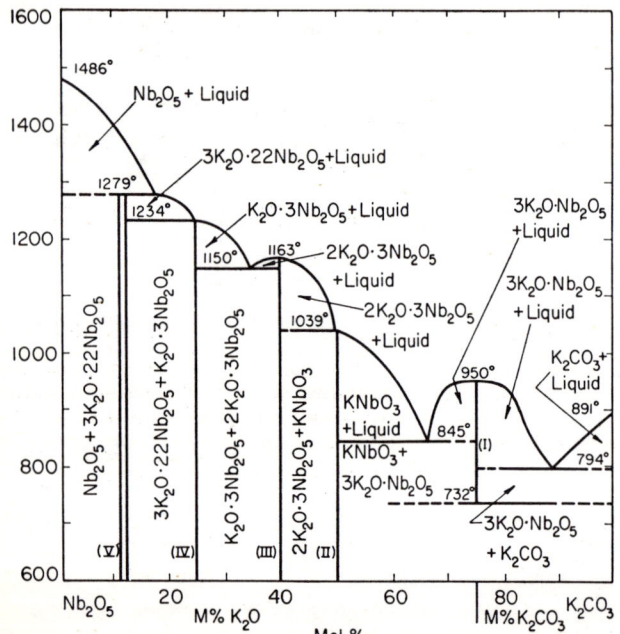

Fig. 171.—System K_2CO_3–Nb_2O_5. The region 0 = 75% K_2CO_3 represents equilibrium between K_2O and Nb_2O_5; between 75 and 100% K_2CO_3, K_2CO_3 is a binary component.

Arnold Reisman and Frederic Holtzberg, *J. Am. Chem. Soc.*, **77** [8] 2117 (1955).

K_2O (or K_2CO_3)–Ta_2O_5

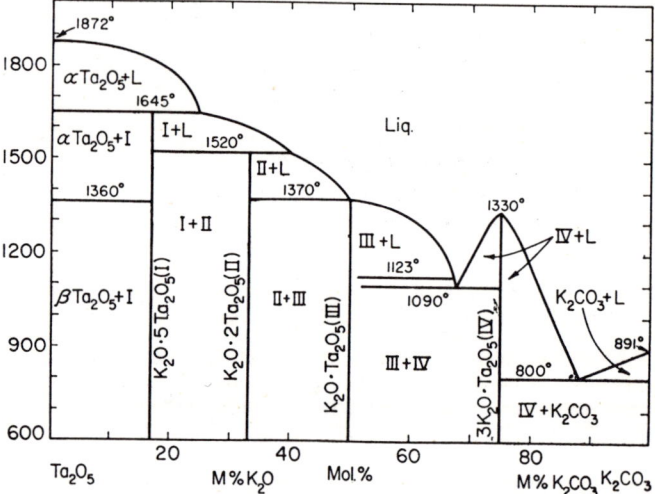

Fig. 173.—System K_2CO_3–Ta_2O_5. The region 0 to 75 mol. % K_2CO_3 represents equilibrium between K_2O and Ta_2O_5; between 75 and 100 mol. % K_2CO_3, K_2CO_3 is a binary component.

Arnold Reisman, Frederic Holtzberg, Melvin Berkenblit, and Margaret Berry, *J. Am. Chem. Soc.*, **78** [18] 4517 (1956).

$K_2O-(or\ K_2CO_3)-V_2O_5$

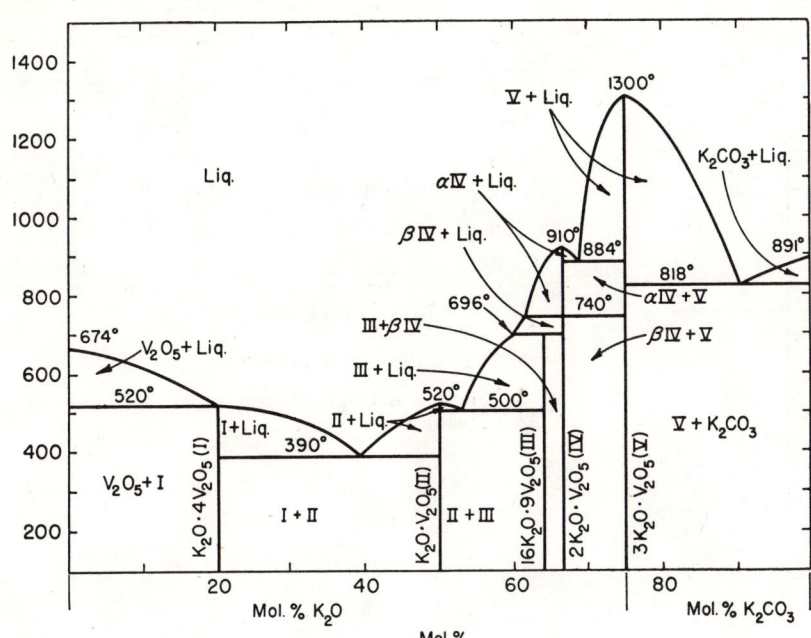

Fig. 174.—System K_2O (or K_2CO_3)–V_2O_5.

Frederic Holtzberg, Arnold Reisman, Margaret Berry, and Melvin Berkenblit, *J. Am. Chem. Soc.* **78** [8] 1538 (1956); see also, V. V. Illarionov, R. P. Ozerov, and E. V. Kil'disheva, *Zhur. Neorg. Khim.*, **1** [4] 779 (1956).

K_2O-CrO_3

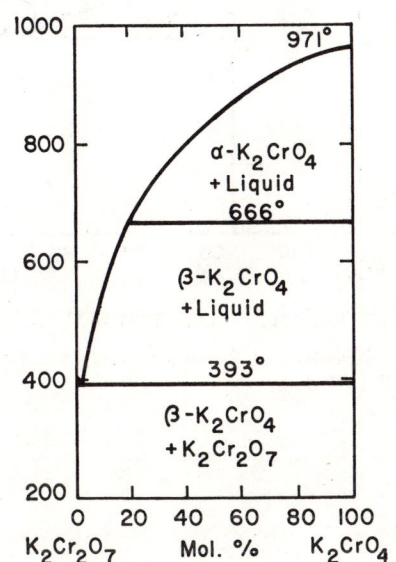

Fig. 175.—System K_2CrO_4–$K_2Cr_2O_7$

E. Groschuff, *Z. anorg. Chem.*, **58**, 111 (1908).

K_2O-MoO_3

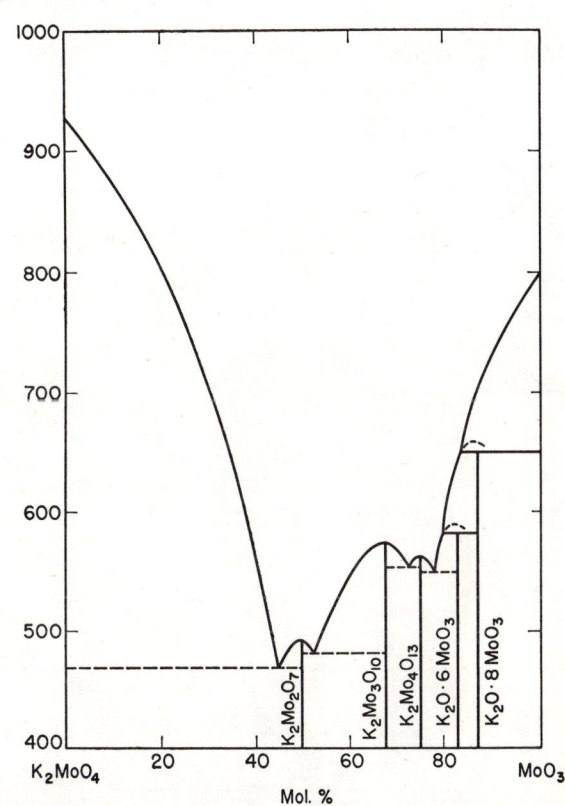

Fig. 176.—System K_2MoO_4–MoO_3.

Viktor Spitzyn and I. M. Kuleshov, *Zhur. Obshcheĭ. Khim.*, **21**, 1367 (1951); see also, M. Amadori, *Atti reale ist. Veneto sci.*, **72** [II] 893 (1913).

K₂O–WO₃

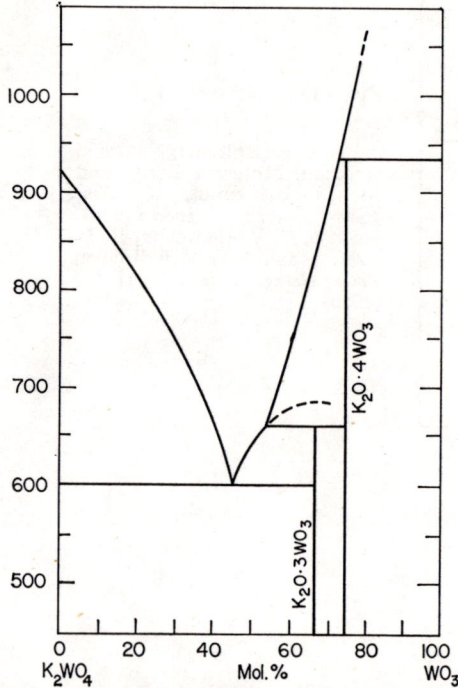

FIG. 177.—System K₂WO₄–WO₃.

F. Hoermann, *Z. anorg. u. allgem. Chem.*, **177**, 170 (1928–1929).

Li₂O–MnO

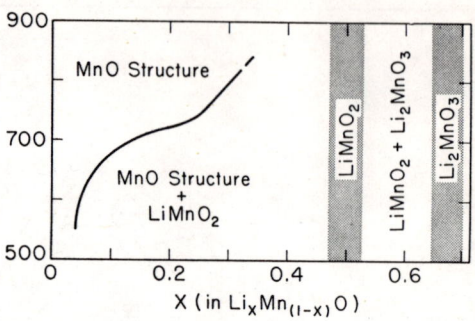

FIG. 178.—System $Li_xMn_{(1-x)}O$.

W. D. Johnston and R. R. Heides, *J. Am. Chem. Soc.*, **78** [14] 3257 (1956).

Li₂O–Al₂O₃

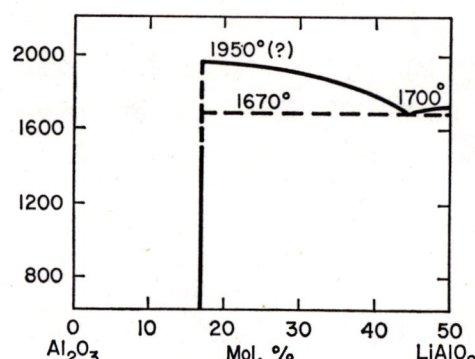

FIG. 179.—System LiAlO₂–Al₂O₃.

D. W. Strickler and Rustum Roy, *J. Am. Ceram. Soc.*, **44** [5] 228 (1961).

Li₂O–B₂O₃

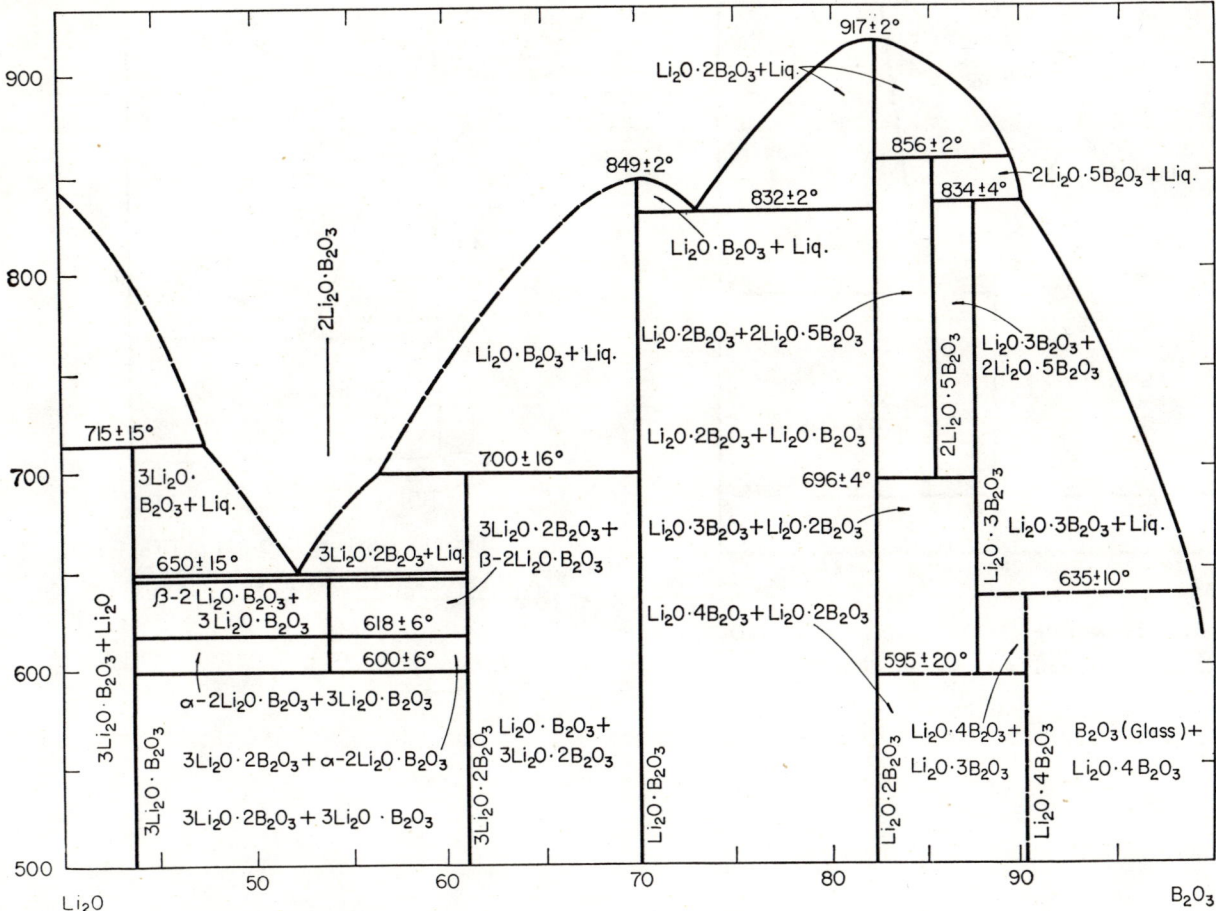

Fig. 180.—System Li₂O–B₂O₃.

B. S. R. Sastry and F. A. Hummel, *J. Am. Ceram. Soc.*, **42** [5] 218 (1959).

Li₂O–GeO₂

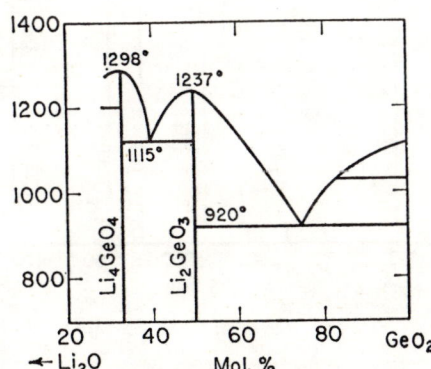

Fig. 181.—System Li₂O–GeO₂.

P. P. Budnikov, S. G. Tresvyatskiĭ, and R. I. Baĭkova, *Dokl. Akad. Nauk S.S.S.R.*, **99**, 761 (1954); see also, Robert Schwarz, *Chem. Ber.*, **62**, 2479 (1929).

Li₂O–SiO₂

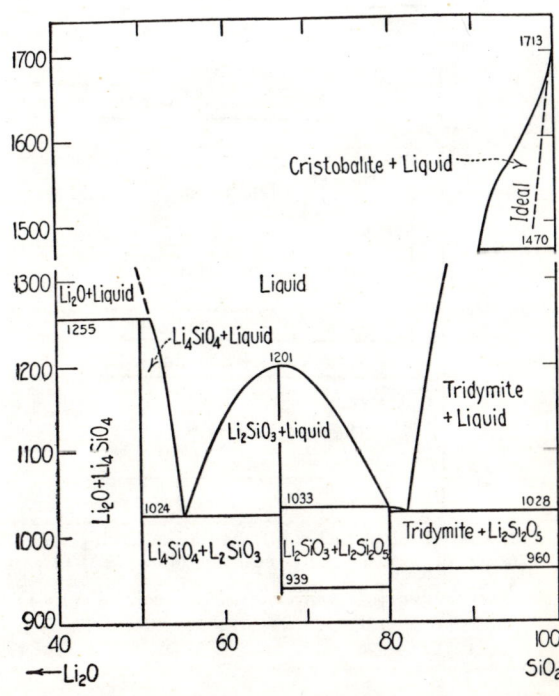

Fig. 182.—System SiO₂–2Li₂O·SiO₂.

F. C. Kracek, *J. Phys. Chem.*, **34**, Part II, 2645 (1930).

Li₂O–SiO₂ (concl.)

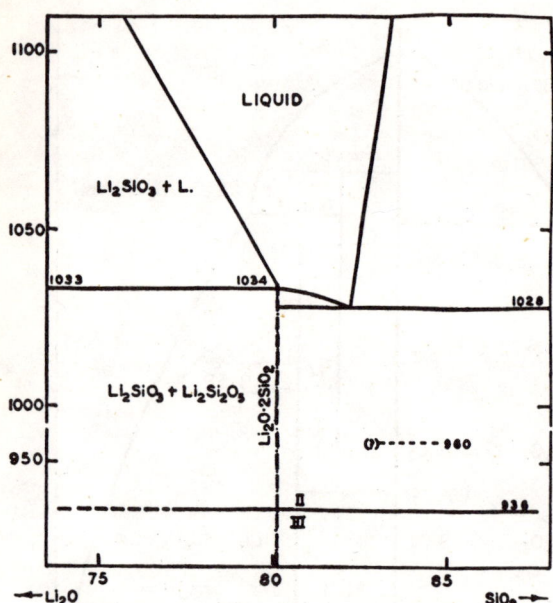

Fig. 183.—The disilicate region of system SiO₂–Li₂O·SiO₂.

F. C. Kracek, *J. Am. Chem. Soc.*, **61**, 2870 (1939).

Li₂O–TiO₂

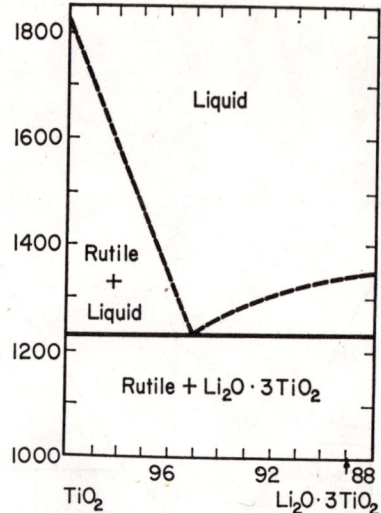

Fig. 185.—System Li₂O·3TiO₂–TiO₂.

F. A. Hummel and Tseng-Ying Tien, *J. Am. Ceram. Soc.*, **42** [4] 207 (1959).

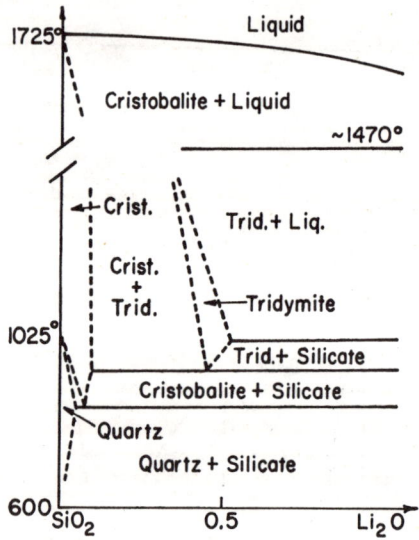

Fig. 184.—System SiO₂–Li₂O; tentative for high-silica region of systems with eutectic above 870°C.

S. B. Holmquist, *J. Am. Ceram. Soc.*, **44** [2] 85 (1961).

Li₂O–MoO₃

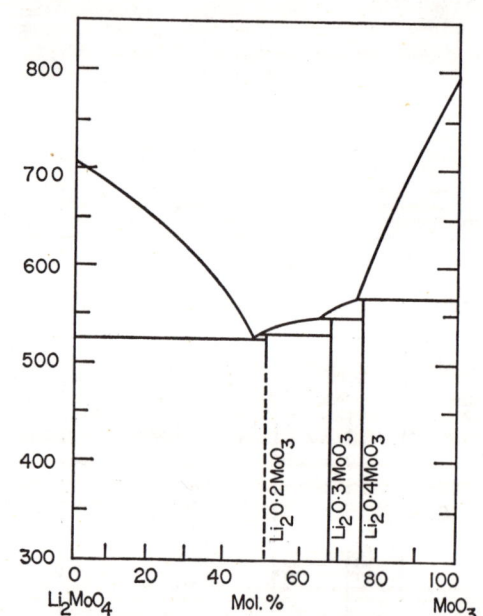

Fig. 186.—System Li₂MoO₄–MoO₃.

F. Hoermann, *Z. anorg. u. allgem. Chem.*, **177**, 154 (1928–1929).

Li_2O-WO_3

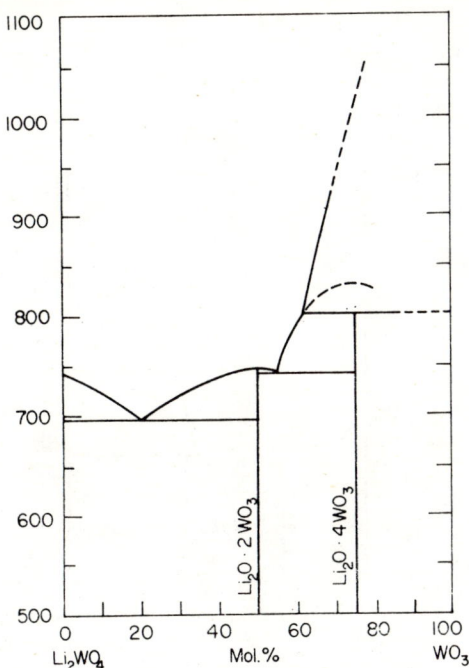

Fig. 187.—System $Li_2WO_4-WO_3$.

F. Hoermann, *Z. anorg. u. allgem. Chem.*, **177**, 163 (1928–1929).

$Na_2O-B_2O_3$

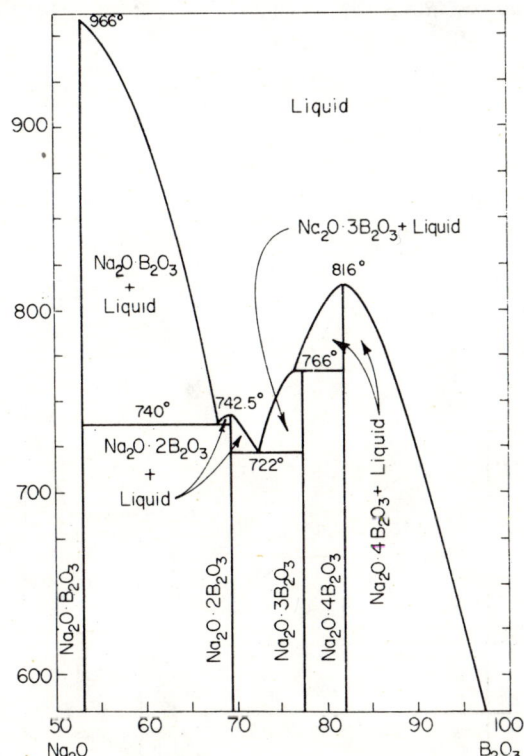

Fig. 188.—System $Na_2O \cdot B_2O_3-B_2O_3$.

G. W. Morey and H. E. Merwin, *J. Am. Chem. Soc.*, **58**, 2252 (1936). In addition, the authors describe the compound $2Na_2O \cdot B_2O_3$; melting point 625°C.

$Na_2O-Fe_2O_3$

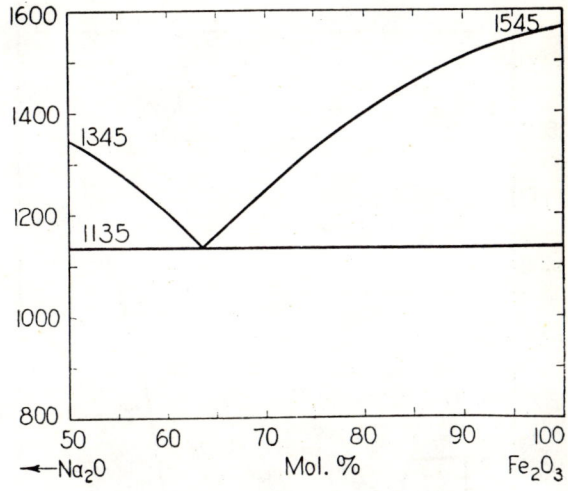

Fig. 189.—System $Fe_2O_3-Na_2O \cdot Fe_2O_3$.

R. Knick and E. J. Kohlmeyer, *Z. anorg. u. allgem. Chem.*, **244**, 77 (1940).

Na_2O-GeO_2

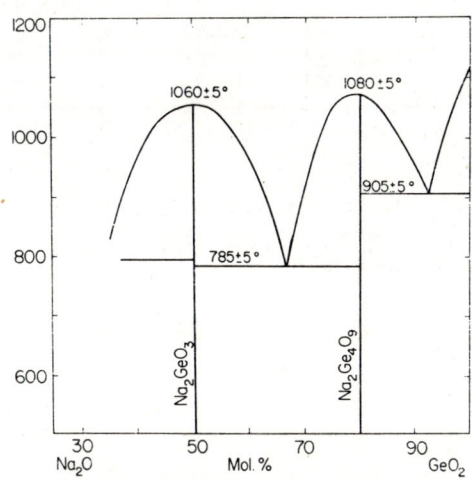

Fig. 190.—System Na_2O-GeO_2.

S. G. Tresvyatskiĭ, *Dopovidi Akad. Nauk Ukr. R.S.R.*, **3**, 295 (1958).

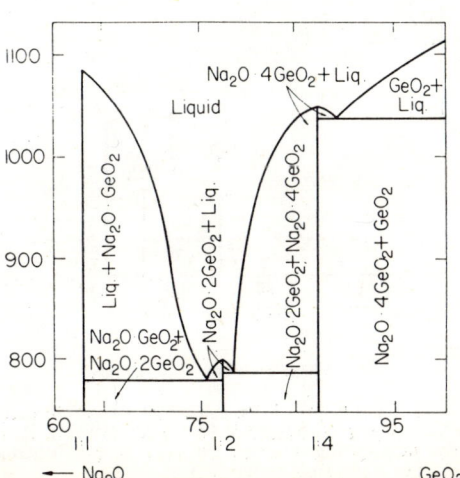

Fig. 191.—System Na_2O-GeO_2.

Robert Schwarz and Fritz Heinrich, *Z. anorg. u. allgem. Chem.*, **205**, 44 (1932).

Na$_2$O–SiO$_2$

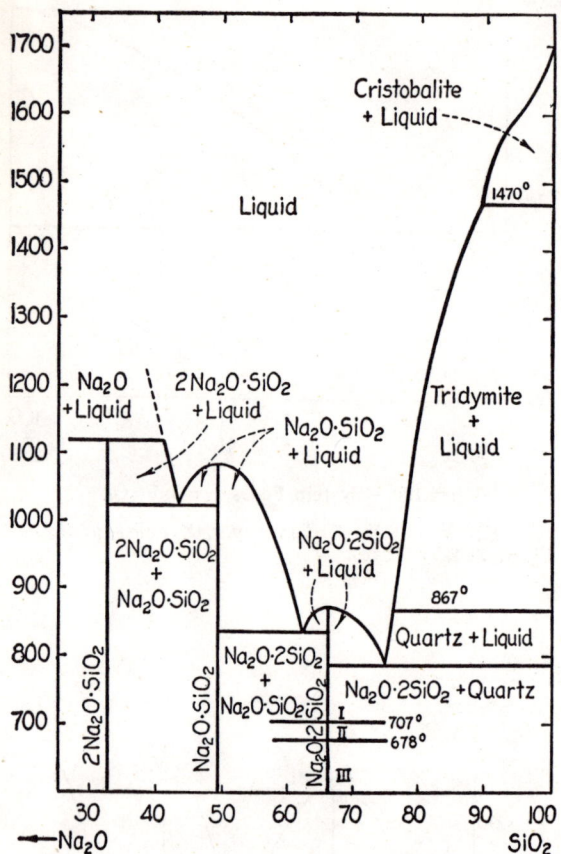

Fig. 192.—System SiO$_2$–2Na$_2$O·SiO$_2$.

F. C. Kracek, *J. Phys. Chem.*, **34**, 1588 (1930); *J. Am. Chem. Soc.*, **61**, 2869 (1939).

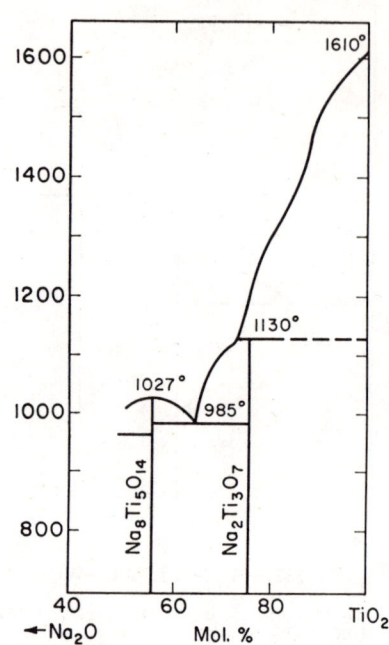

Fig. 194.—System Na$_2$O–TiO$_2$.

P. P. Budnikov and S. G. Tresvyatskiĭ, *Dopovidi Akad. Nauk Ukr. R.S.R.*, 374 (1956).

Na$_2$O–TiO$_2$

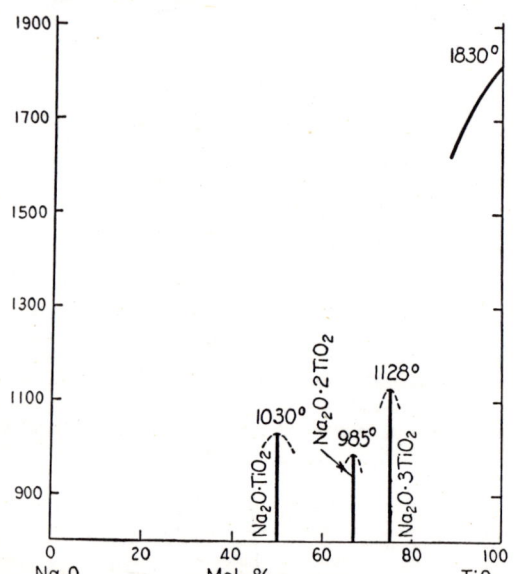

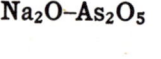

Fig. 193.—System Na$_2$O–TiO$_2$.

Data from E. W. Washburn and E. N. Bunting, *Bur. Standards J. Research*, **12** [2] 239 (1934); R. P. 648; for melting point of TiO$_2$ see E. N. Bunting, *ibid.*, **11**; 719 (1933). RP 619.

Na$_2$O–As$_2$O$_5$

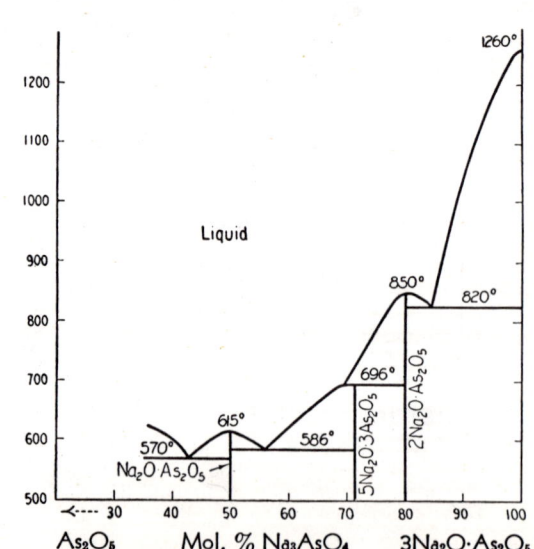

Fig. 195.—System As$_2$O$_5$–3Na$_2$O·As$_2$O$_5$.

M. Amadori, *Atti reale ist. Veneto sci.*, **73** [II] 1672 (1914).

Na₂O–Nb₂O₅

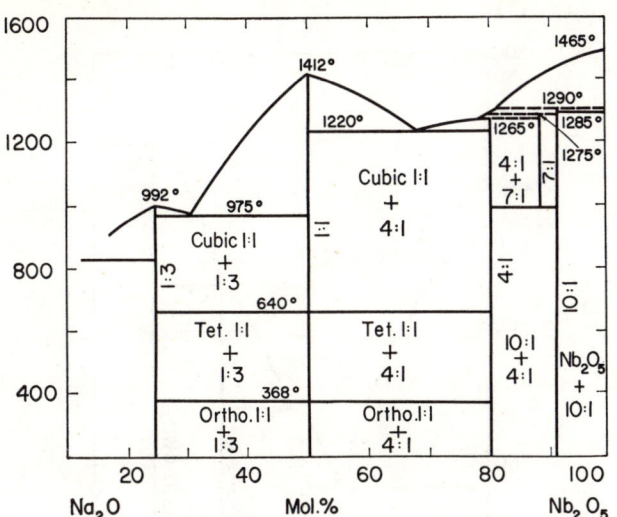

Fig. 196.—System Na₂O–Nb₂O₅. Compound ratios given as moles Na₂O to moles Nb₂O₅.

M. W. Shafer and Rustum Roy, *J. Am. Ceram. Soc.*, **42** [10] 485 (1959).

Na₂O–P₂O₅

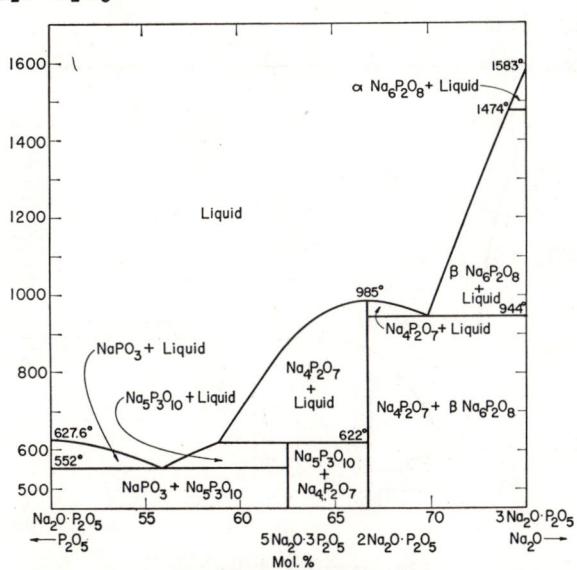

Fig. 197.—System Na₂O·P₂O₅–3Na₂O·P₂O₅; composite.

The Na₂O·P₂O₅–2Na₂O·P₂O₅ subsystem, after George W. Morey and Earl Ingerson, *Am. J. Sci.*, **242** [1] 4 (1944); remainder, after E. T. Turkdogan and W. R. Maddocks, *J. Iron Steel Inst.*, **172**, 6 (1952).

Na₂O–Ta₂O₅

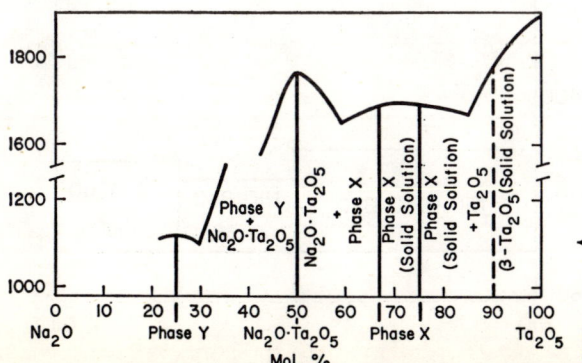

Fig. 198.—System Na₂O–Ta₂O₅.

B. W. King, John Schultz, E. A. Durbin, and W. H. Duckworth, U. S. Atomic Energy Comm., BMI-1106, 9 (1956).

Na₂O–MoO₃

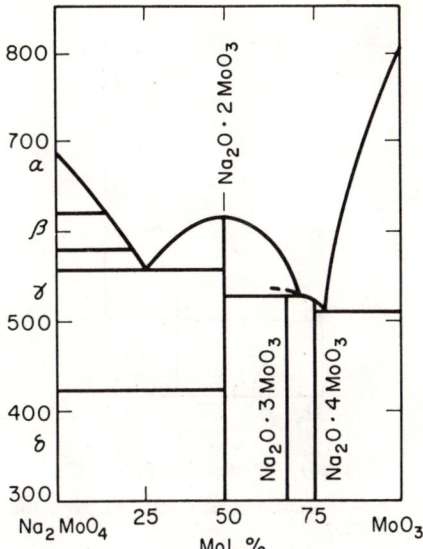

Fig. 199.—System Na₂MoO₄–MoO₃.

A. N. Zelikman and N. N. Gorovitz, *Zhur. Obshcheĭ Khim.*, **24**, 1920 (1954); see also, Fritz Hoermann, *Z. anorg. Chem.*, **177**, 157 (1929).

Na₂O–WO₃

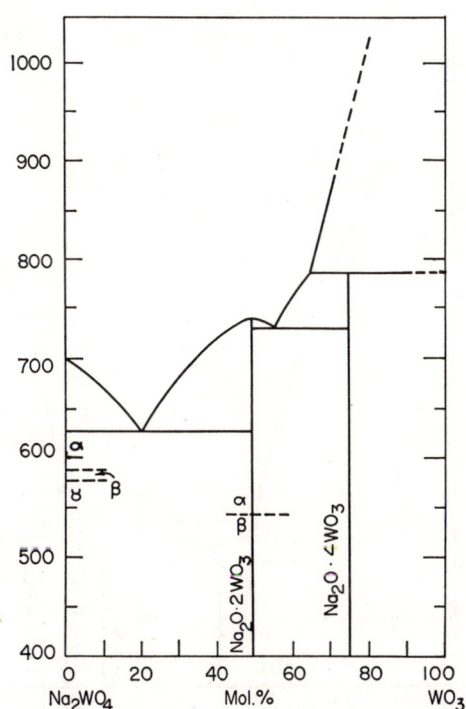

Fig. 200.—System Na₂WO₄–WO₃.

F. Hoermann, *Z. anorg. u. allgem. Chem.*, **177**, 167 (1928–1929).

Rb$_2$O–TiO$_2$

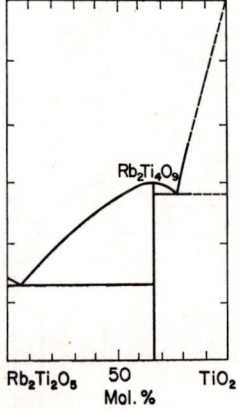

FIG. 201.—System Rb$_2$Ti$_2$O$_5$–TiO$_2$.

O. Schmitz-DuMont and H. Reckhard, *Monatsh. Chem.* **90** [2] 135 (1959).

Rb$_2$O–WO$_3$

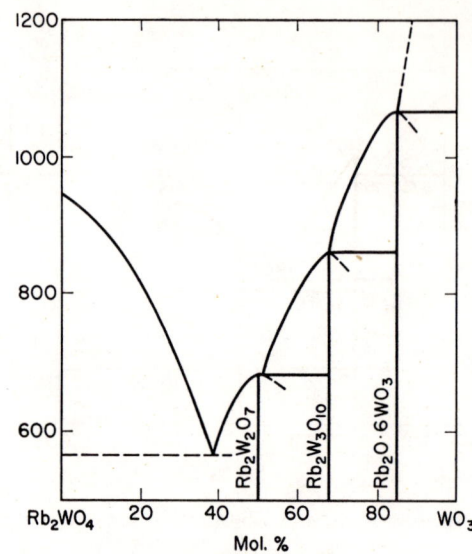

FIG. 203.—System Rb$_2$WO$_4$–WO$_3$.

V. I. Spitsyn and I. M. Kuleshov, *J. Phys. Chem. U.S.S.R.*, **24** [10] 1197 (1950).

Rb$_2$O–MoO$_3$

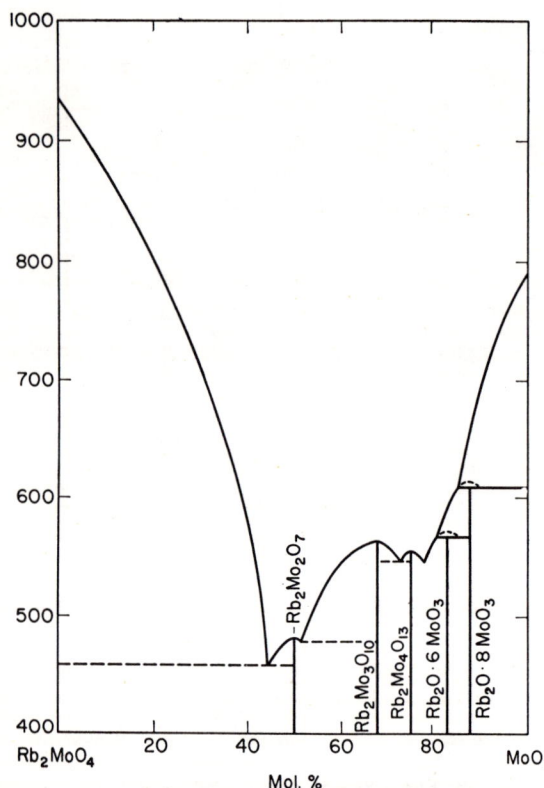

FIG. 202.—System Rb$_2$MoO$_4$–MoO$_3$.

Viktor Spitzyn and I. M. Kuleshov, *Zhur. Obshcheĭ Khim.*, **21**, 1370 (1951).

BaO–BeO

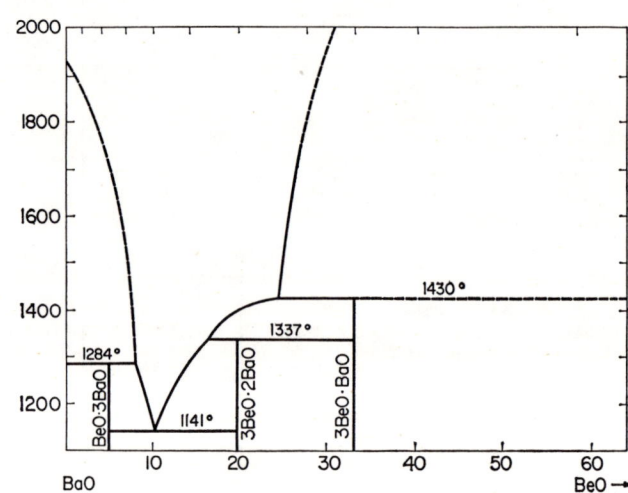

FIG. 204.—System BaO–BeO.

A. Auriol and J. G. Wurm, private communication, Nov. 19, 1961.

BaO–NiO

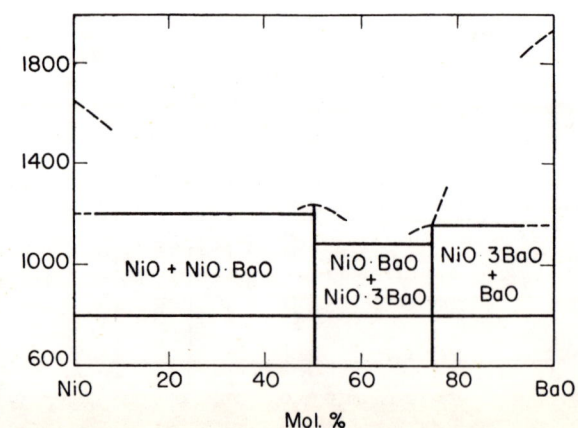

FIG. 205.—System BaO–NiO.

J. J. Lander, *J. Am. Chem. Soc.*, **73**, 2451 (1951).

BaO–Al₂O₃

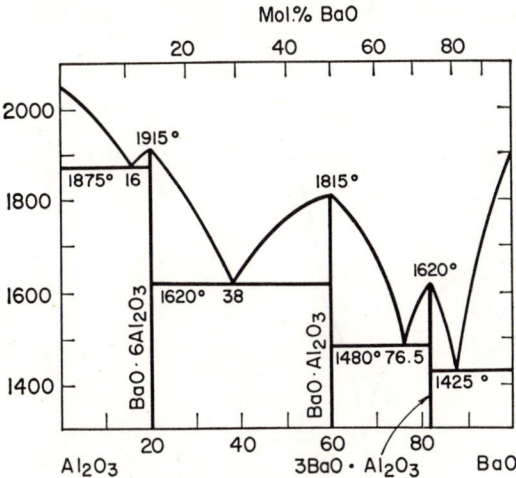

Fig. 206.—System BaO–Al₂O₃.

G. Purt, *Radex Rundschau*, 1960, No. 4, p. 201; see also, N. A. Toropov and F. Ya. Galakhov, *Doklady Akad. Nauk S.S.S.R.*, **82** [1] 70 (1952).

BaO–Fe₂O₃

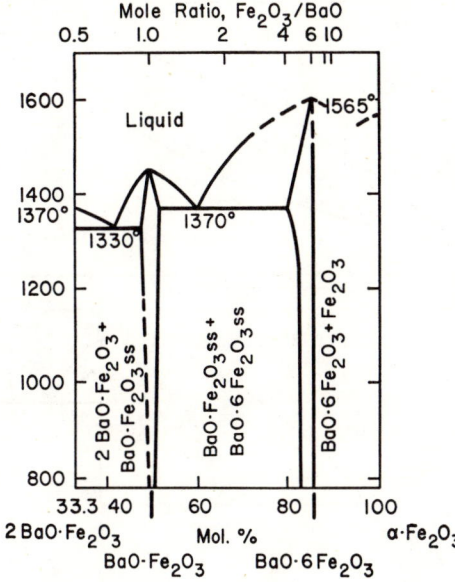

Fig. 208.—System 2BaO·Fe₂O₃–Fe₂O₃.

Yasumasa Goto and Toshio Takada, *J. Am. Ceram. Soc.*, **43** [3] 151 (1960).

BaO–B₂O₃

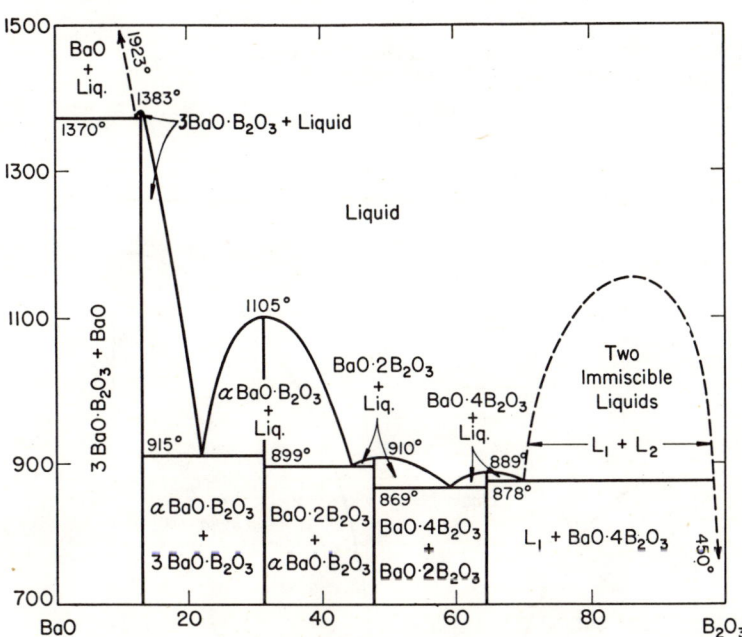

Fig. 207.—System BaO–B₂O₃.

After Ernest M. Levin and Howard F. McMurdie, *J. Research Natl. Bur. Standards*, **42** [2] 135 (1949); RP 1956; Modified by Ernest M. Levin and George Ugrinic, *ibid.*, **51** [1] 40 (1953); RP 2430; shape of immiscibility dome, by Ernest M. Levin, and G. W. Cleek, *J. Am. Ceram. Soc.*, **41** [5] 177 (1958). BaO·4B₂O₃ transforms reversibly to a high form at about 734°C., according to C. R. Robbins, private communication, Dec. 1962.

BaO–La₂O₃

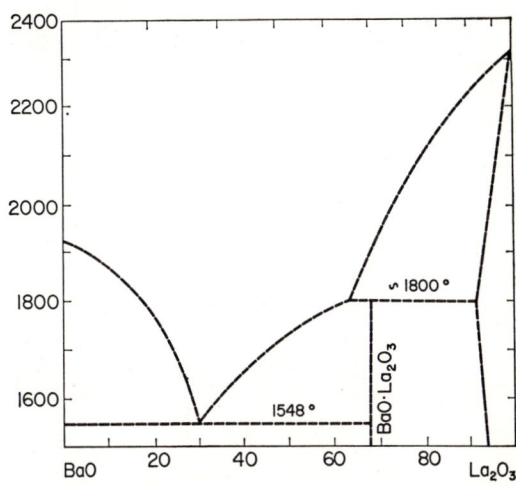

Fig. 209.—System BaO–La₂O₃; tentative.

A. Auriol and J. C. Wurm; solid solubility after M. Foex, *Bull. soc. chim. France*, p. 109, Jan., 1961. Private communication, Nov. 19, 1961.

BaO–SiO₂

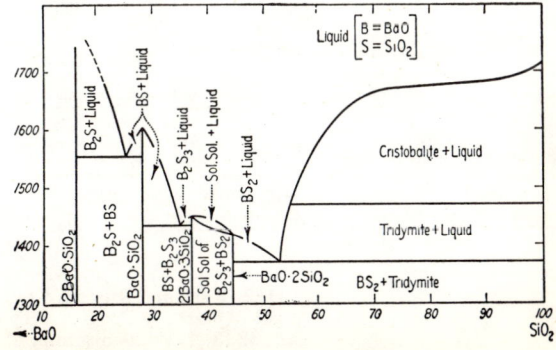

Fig. 210.—System BaO–SiO₂.

P. Eskola, *Am. J. Sci.*, 5th Ser., **4**, 345 (1922); modified by J. W. Greig, *Am. J. Sci.*, 5th Ser., **13**, 27 (1927).

BaO–SiO$_2$ (concl.)

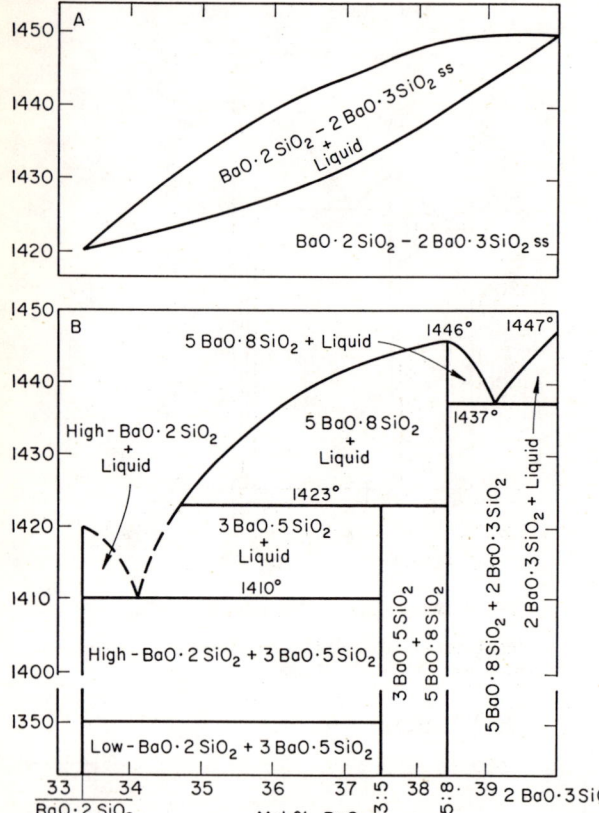

BaO–SnO$_2$

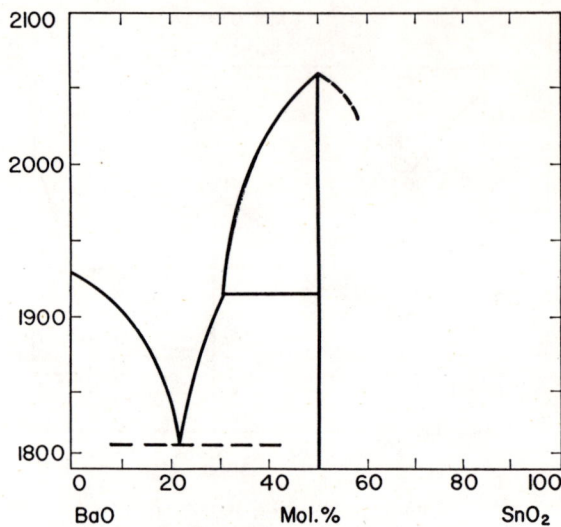

Fig. 212.—System BaO–SnO$_2$.

Gustav Wagner and Horst Binder, *Z. anorg. u. allgem. Chem.*, **297**, 334 (1958).

Fig. 211.—Subsystem BaO·2SiO$_2$–2BaO·3SiO$_2$; proposed. (*a*) Continuous solid solution diagram according to Eskola, with data converted to mole percent. (*b*) Without solid solution, showing two compounds and two eutectics.

R. S. Roth and E. M. Levin, *J. Research Natl. Bur. Standards*, **62** [5] 194 (1959); RP 2953.

BaO–TiO$_2$

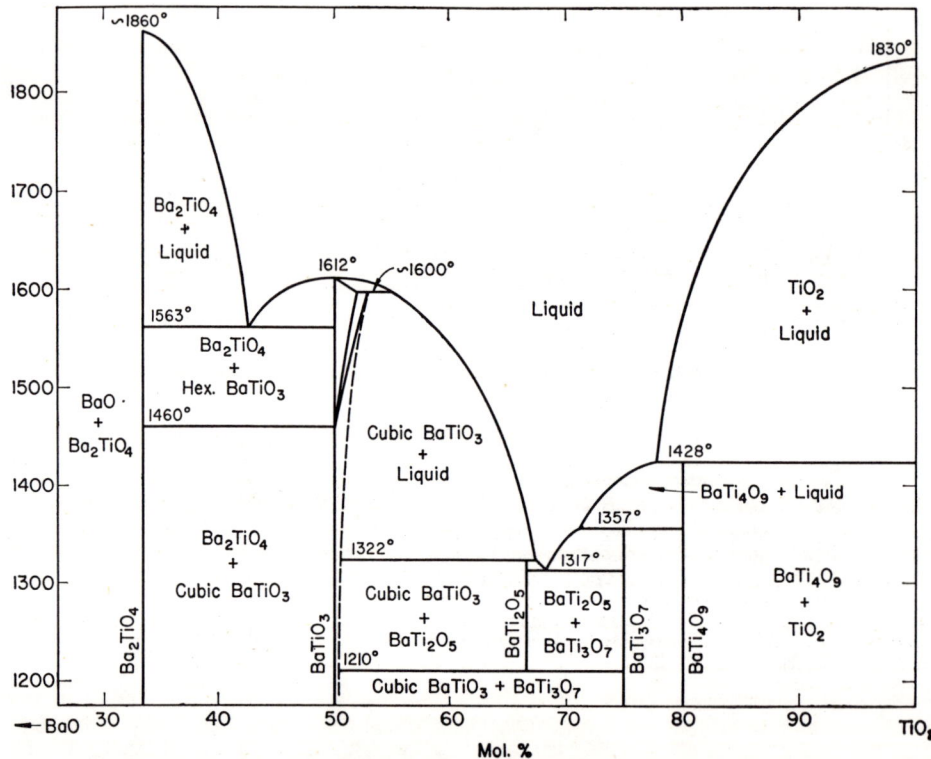

Fig. 213.—System BaO–TiO$_2$.

D. E. Rase and Rustum Roy, The Pennsylvania State University, College of Mineral Industries; Eighth Quarterly Progress Report, April 1 to June 30, Appendix II, p. 32 (1953); *J. Am. Ceram. Soc.*, **38** [3] 111 (1955); m.p. of BaTiO$_3$ given as 1618°C.

BaO–Nb₂O₅

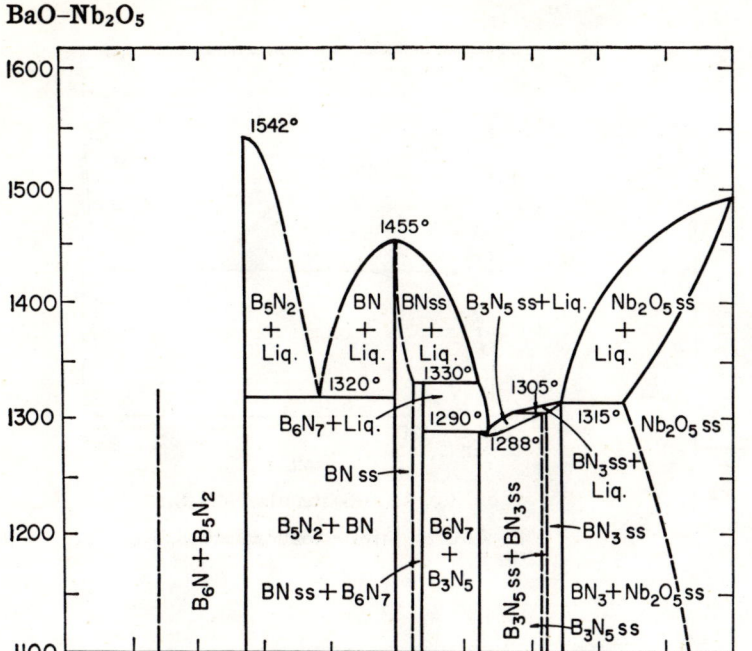

Fig. 214.—System BaO–Nb₂O₅; postulated. B = BaO, N = Nb₂O₅, ss = solid solution.

R. S. Roth and J. L. Waring, *J. Research Natl. Bur. Standards*, **65A** [4] 341 (1961).

BeO–MgO

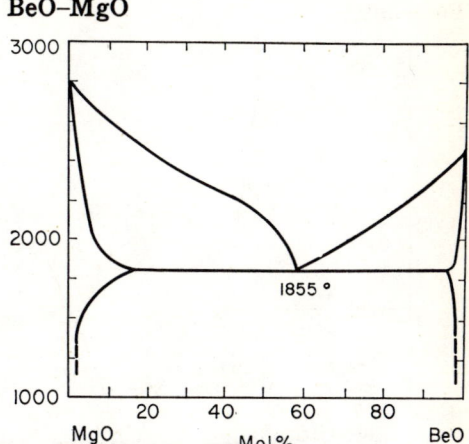

Fig. 216.—System BeO–MgO.

H. E. Otto, private communication, Dec. 27, 1961.

BeO–CaO

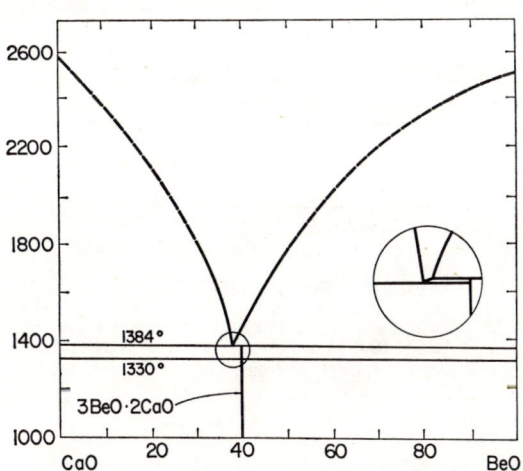

Fig. 215.—System BeO–CaO.

A. Auriol and J. G. Wurm, private communication, Nov. 19, 1961. R. A. Potter and L. A. Harris (*J. Am. Ceram. Soc.*, **45** [12] 615 (1962), consider the compound 3BeO·2CaO metastable. They place the eutectic at 1360 ± 5°C and 59.9 mol. % CaO (40.1 mol. % BeO).

BeO–SrO

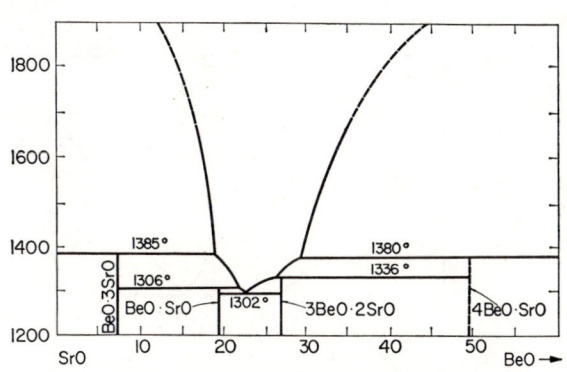

Fig. 217.—System BeO–SrO.

A. Auriol and J. G. Wurm, private communication, Nov. 19, 1961.

BeO–Al₂O₃

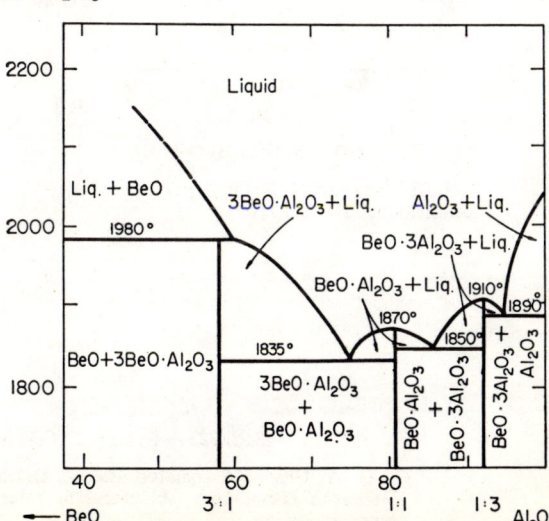

Fig. 218.—System BeO–Al₂O₃. Data from 70% to 100% Al₂O₃ after S. M. Lang, C. L. Fillmore, and L. H. Maxwell, *J. Research Natl. Bur. Standards*, **48** [4] 301 (1952); RP 2316.

F. Ya. Galakhov, *Izvest. Akad. Nauk S.S.S.R., Otdel. Khim. Nauk*, 1035 (1957).

BeO–La₂O₃

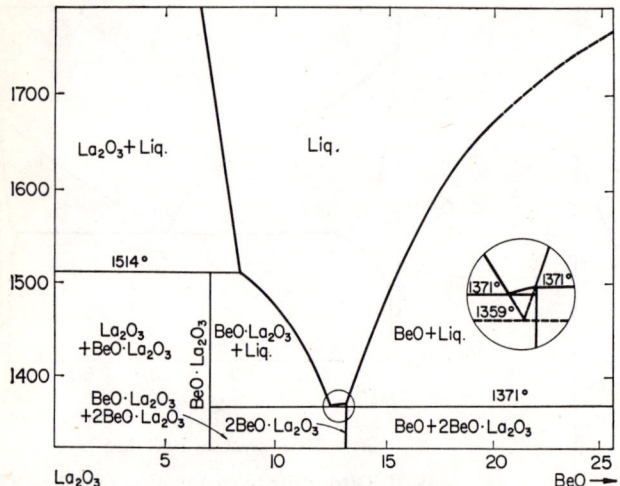

FIG. 219.—System BeO–La₂O₃.

A. Auriol and J. G. Wurm, private communication, Nov. 19, 1961.

BeO–Y₂O₃

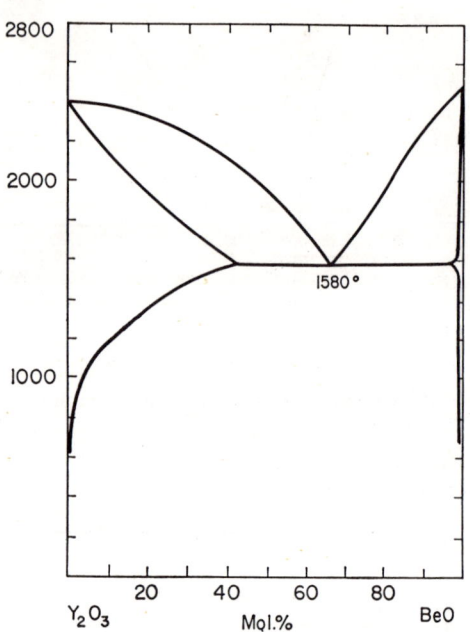

FIG. 220.—System BeO–Y₂O₃.

L. E. Olds and H. E. Otto, private communication, Dec. 27, 1961.

BeO–Yb₂O₃

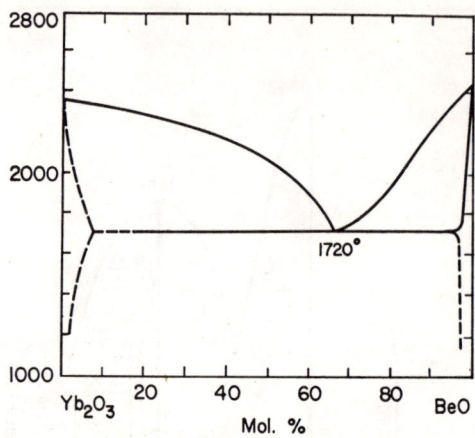

FIG. 221.—System BeO–Yb₂O₃.

H. E. Otto, private communication, Dec. 27, 1961.

BeO–SiO₂

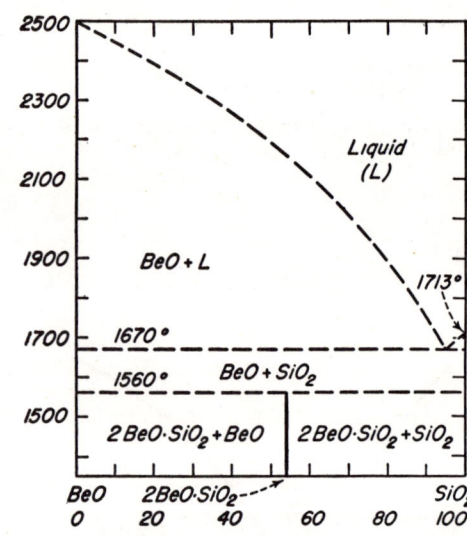

FIG. 222.—System BeO–SiO₂.

R. A. Morgan and F. A. Hummel, *J. Am. Ceram. Soc.*, **32** [8] 255 (1949); see also, P. P. Budnikov and A. M. Cherepanov, *Vopr. Petrogr. i Mineralog., Akad. Nauk S.S.S.R.*, **2**, 243 (1953).

BeO–ThO₂

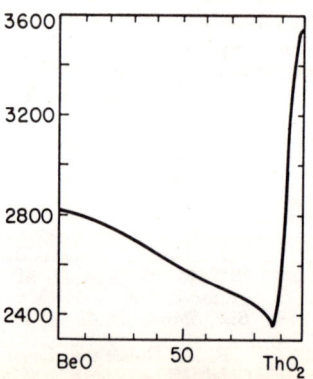

FIG. 223.—System BeO–ThO₂.

G. A. Geach, Associated Electrical Industries Limited Research Laboratory, Aldermaston, Berkshire; private communication.

BeO–TiO$_2$

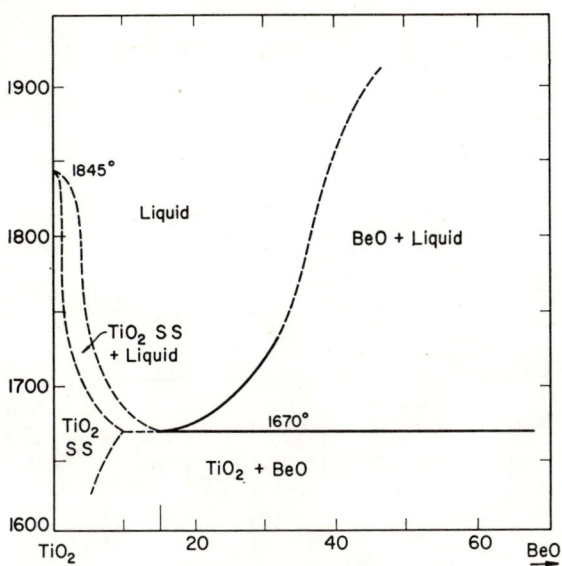

FIG. 224.—System BeO–TiO$_2$; revised.

S. M. Lang, C. L. Fillmore, and L. H. Maxwell, *J. Res. Natl. Bur. Std.*, **48** [4] 300 (1952); R. P. 2316; see also, Włodzimierz Trzebiatowski, Miroslawa Dryś, and Wladyslaw Baran, *Roczniki Chem.*, **27**, 440 (1953).

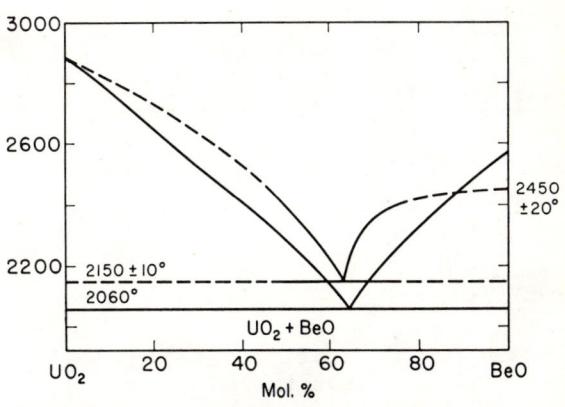

FIG. 226.—System BeO–UO$_2$.

S. M. Lang, F. P. Knudsen, C. L. Fillmore, and R. S. Roth, *Natl. Bur. Standards Circ.*, No. **568**, p. 5 (Feb. 20 1956).

Dashed line—Lang, *et al.*, solid line—L. F. Epstein and W. H. Howland (theoretical), *J. Am. Ceram. Soc.*, **36** [10] 334 (1953).

BeO–UO$_2$

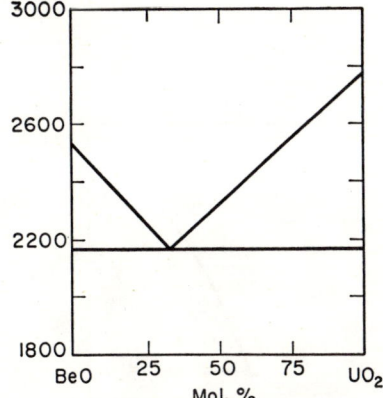

FIG. 225.—System BeO–UO$_2$.

P. P. Budnikov, S. G. Tresvyatskiĭ, and V. I. Kushakovskiĭ, *Proc. U. N. Intern. Conf. Peaceful Uses At. Energy, 2nd*, Geneva, **6**, 127 (1958).

BeO–ZrO$_2$

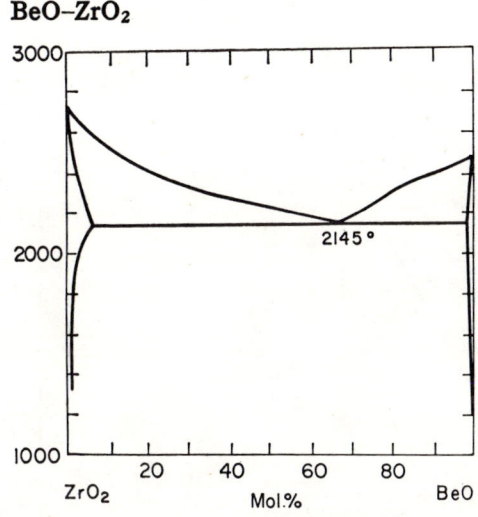

FIG. 227.—System BeO–ZrO$_2$.

H. E. Otto, private communication, Dec. 27, 1961.

CaO–CoO

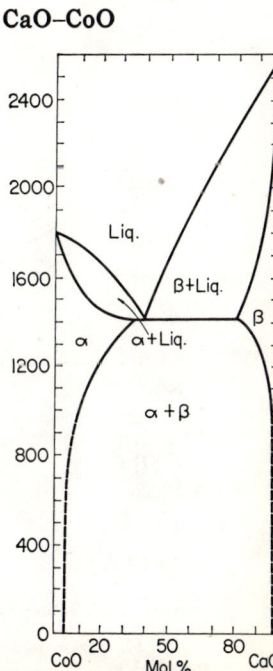

FIG. 228.—System CaO–CoO.

M. H. Tikkanen, private communication, Jan. 5, 1962.

CaO–MgO

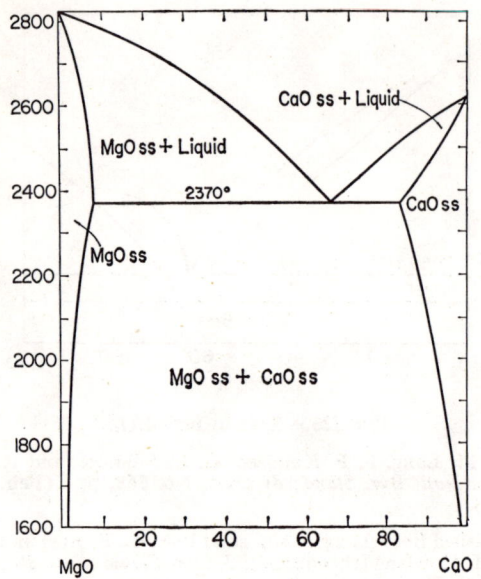

FIG. 229.—System CaO–MgO.

R. C. Doman, J. B. Barr, R. N. McNally, and A. M. Alper, *J. Am. Ceram. Soc.*, **46** [7] 314 (1963).

CaO–Al₂O₃

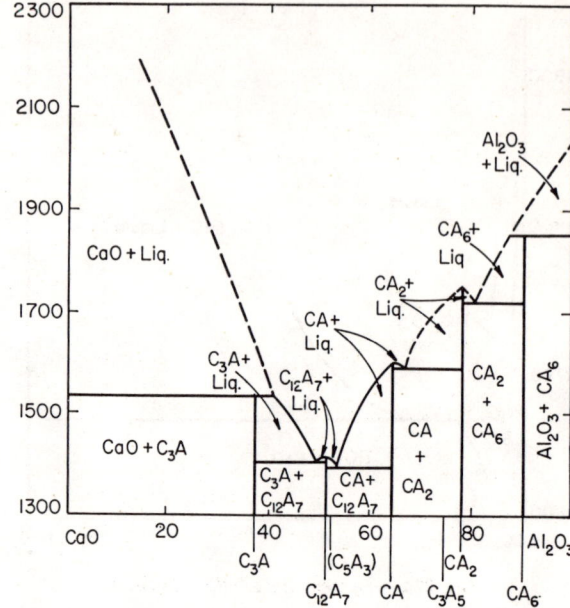

FIG. 231.—System CaO–Al$_2$O$_3$. C = CaO; A = Al$_2$O$_3$.

F. M. Lea and C. H. Desch, The Chemistry of Cement and Concrete, 2d ed., p. 52. Edward Arnold & Co., London, 1956.

CaO–NiO

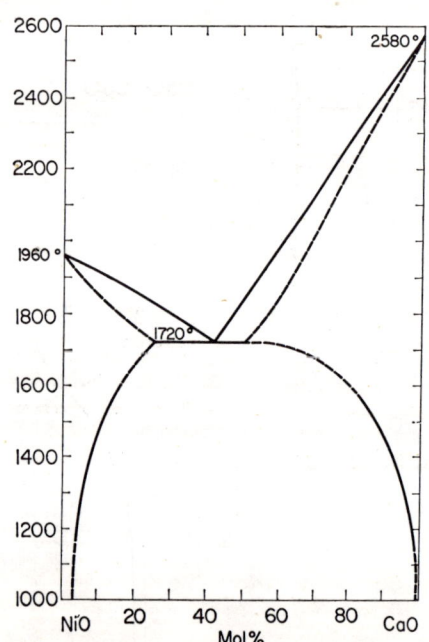

FIG. 230.—System CaO–NiO.

M. H. Tikkanen, private communication, Jan. 5, 1962.

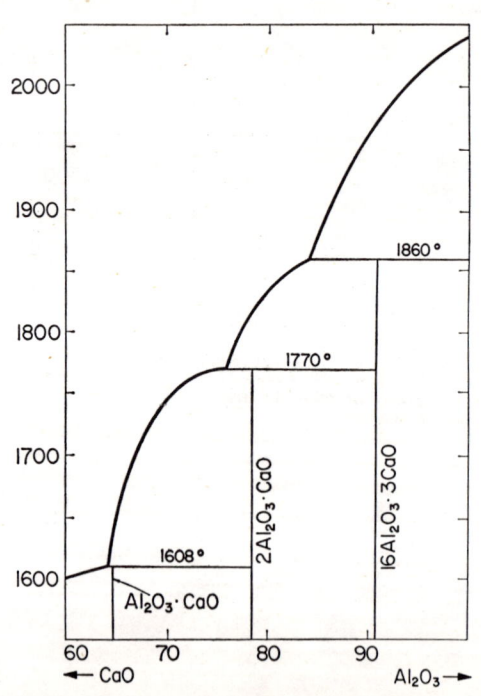

FIG. 232.—System CaO–Al$_2$O$_3$.

A. Auriol, G. Hauser, and J. G. Wurm, private communication, Nov. 19, 1961.

CaO–Al$_2$O$_3$ (concl.)

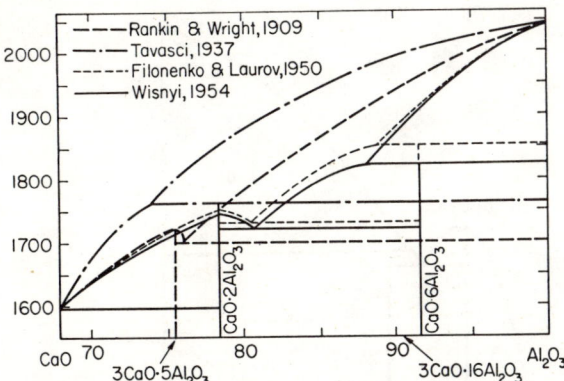

Fig. 233.—System CaO–Al$_2$O$_3$; high alumina portion.

L. G. Wisnyi, Doctor's thesis, Rutgers University, State University of New Jersey, New Brunswick, New Jersey, January, 1955.

CaO–Ga$_2$O$_3$

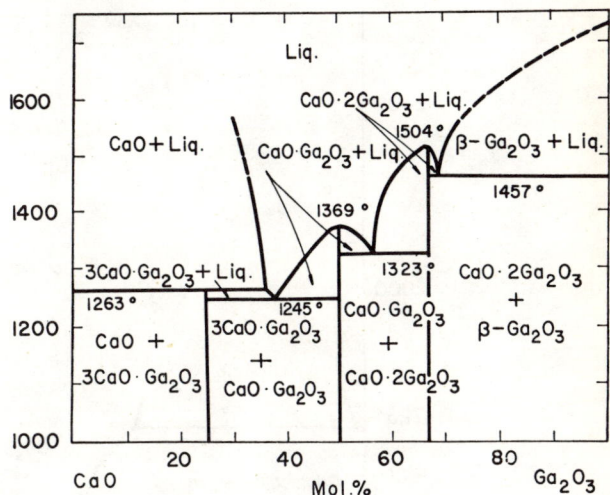

Fig. 235.—System CaO–Ga$_2$O$_3$.

J. Jeevaratnam and F. P. Glasser, *J. Am. Ceram. Soc.*, **44** [11] 564 (1961).

CaO–B$_2$O$_3$

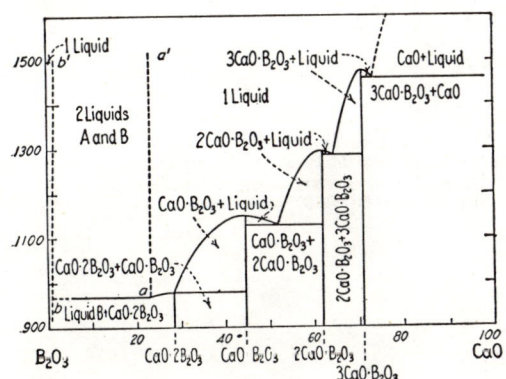

Fig. 234.—System CaO–B$_2$O$_3$.

E. T. Carlson, *Bur. Standards J. Research*, **9**, 830 (1932); R. P. 516.

CaO–HfO$_2$

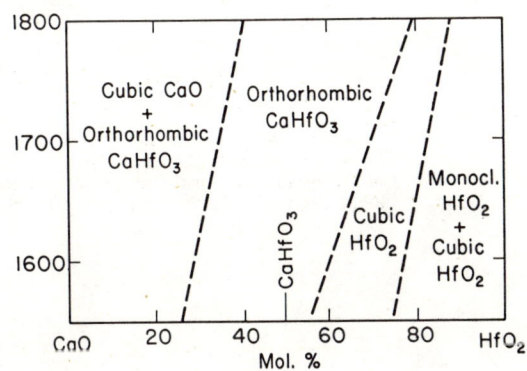

Fig. 236.—System CaO–HfO$_2$; results of solid-state reactions at 1550° and 1800°C. Diagram does not obey the phase rule.

C. E. Curtis, L. M. Doney, and J. R. Johnson, *J. Am. Ceram. Soc.*, **37** [10] 464 (1954)

CaO–SiO$_2$

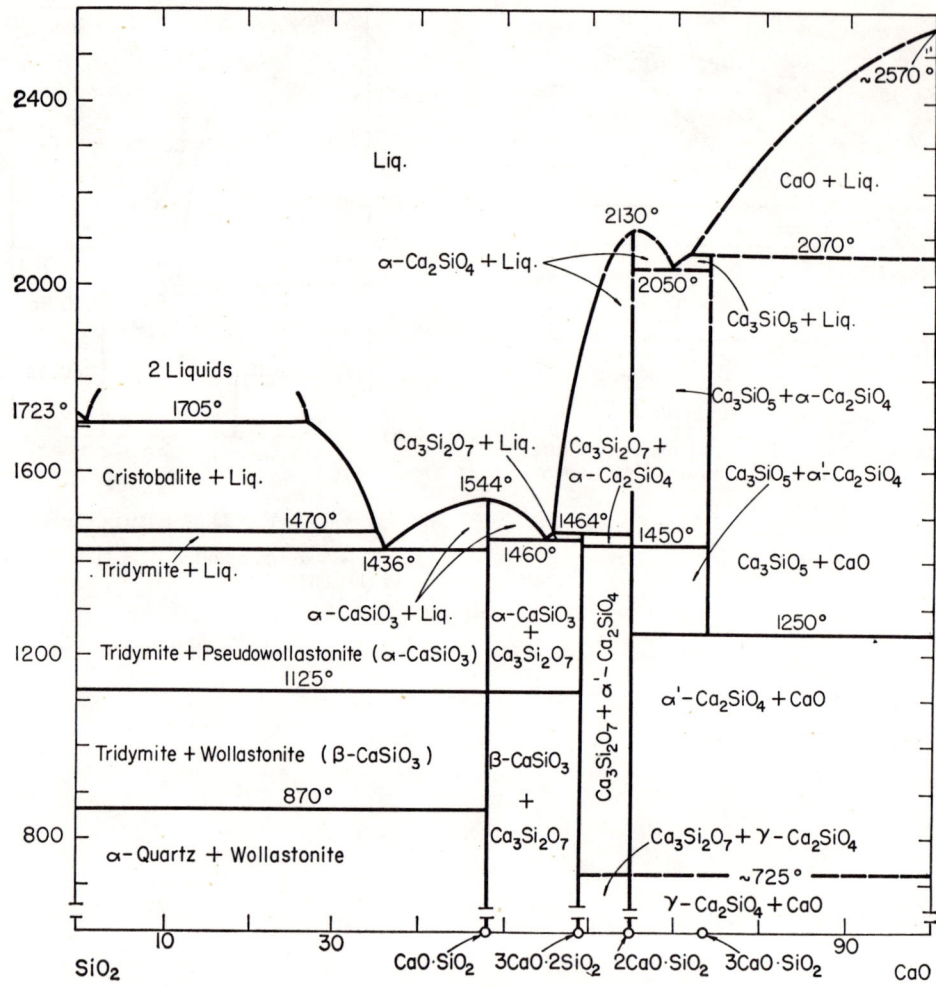

FIG. 237.—System CaO–SiO$_2$.

Bert Phillips and Arnulf Muan, *J. Am. Ceram. Soc.*, **42** [9] 414 (1959).

Based mainly on data of G. A. Rankin and F. E. Wright, *Am. J. Sci.* [4], **39**, 5 (1915) and J. W. Greig, *Am. J. Sci.*, [5], **13**, 1–44; [74] 133–54 (1927). Changes with respect to stability relations of tricalcium and dicalcium silicates based on data of D. M. Roy, *J. Am. Ceram. Soc.*, **41** [8] 293–99 (1958) and J. H. Welch and W. Gutt, *J. Am. Ceram. Soc.*, **42** [1] 11–15 (1959).

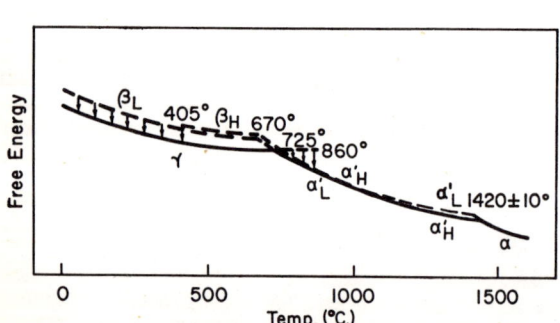

FIG. 238.—System Ca$_2$SiO$_4$; hypothetical, schematic free energy-temp. curves. Solid lines indicate stable phases; dashed lines, metastable phases.

Deane K. Smith, A. J. Majumdar, and Fred Ordway, *J. Am. Ceram. Soc.*, **44** [8] 411 (1961).

CaO–TiO$_2$

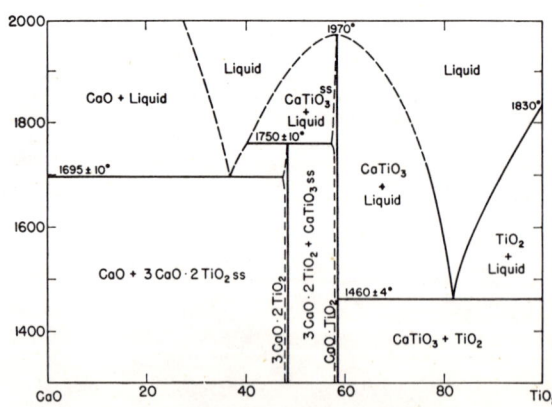

FIG. 239.—System CaO–TiO$_2$.

R. C. DeVries, R. Roy, and E. F. Osborn, *J. Phys. Chem.*, **58**, 1072 (1954).

CaO–TiO$_2$ (concl.)

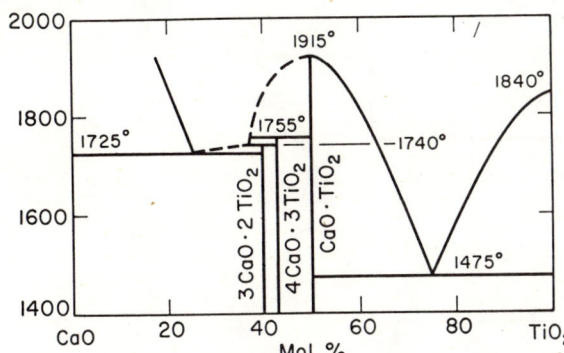

FIG. 240.—System CaO–TiO$_2$; revised to show compound Ca$_4$Ti$_3$O$_{10}$.

R. S. Roth, *J. Research Natl. Bur. Standards*, **61** [5] 440 (1958).

CaO–ZrO$_2$

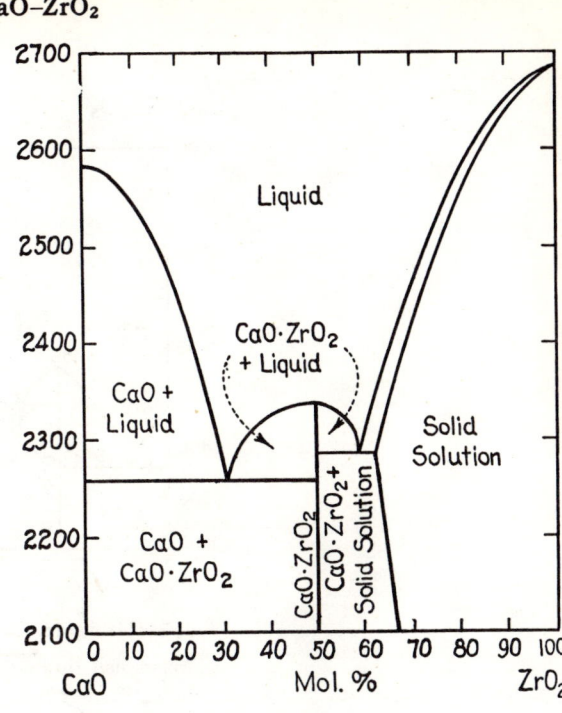

FIG. 242.—System CaO–ZrO$_2$.

O. Ruff, F. Ebert, and E. Stephan, *Z. anorg. u. allgem. Chem.*, **180**, 219 (1929).

CaO–UO$_2$

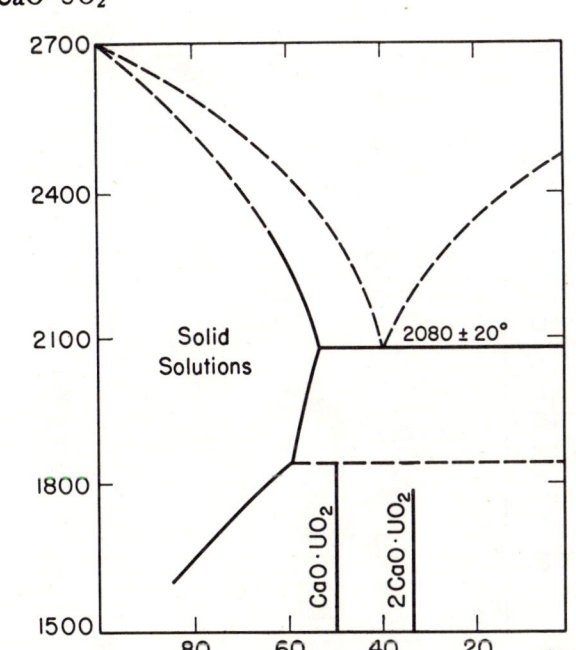

FIG. 241.—System CaO–UO$_2$.

K. B. Alberman, R. C. Blakey, and J. A. Anderson, *J. Chem. Soc.*, **26**, 1354 (1951).

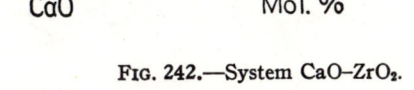

FIG. 243.—System CaO–ZrO$_2$; tentative.

Pol Duwez, Francis Odell, and Frank H. Brown, Jr., *J. Am. Ceram. Soc.*, **35** [5], 109 (1952).

CaO–Nb$_2$O$_5$

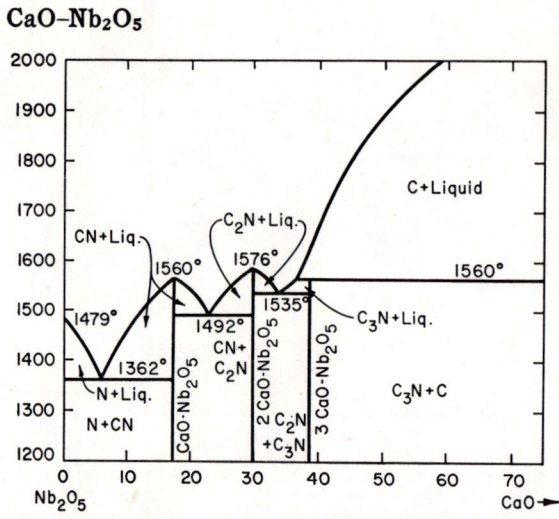

FIG. 244.—System CaO–Nb$_2$O$_5$. C = CaO; N = Nb$_2$O$_5$.

Mohammad Ibrahim, N. F. H. Bright, and J. F. Rowland, *J. Am. Ceram. Soc.*, **45** [7] 331 (1962).

CaO–P$_2$O$_5$

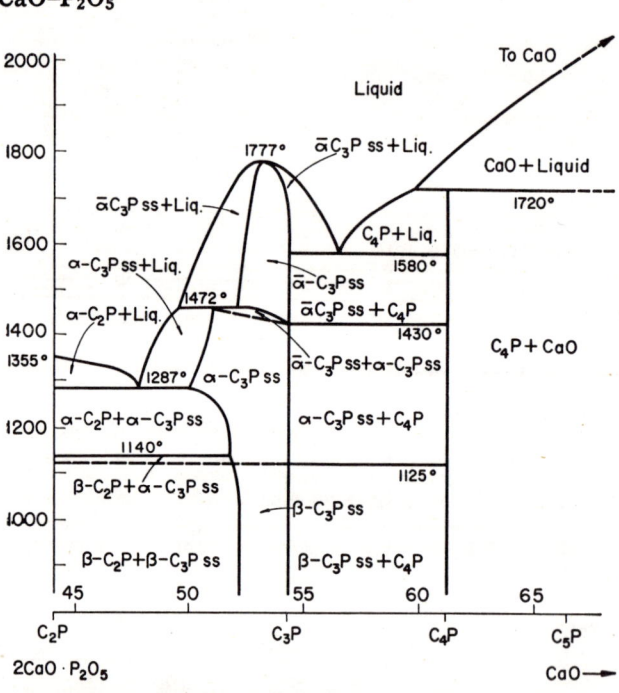

FIG. 245.—System CaO–2CaO·P$_2$O$_5$. C = CaO. P = P$_2$O$_5$.

J. H. Welch and W. Gutt, *J. Chem. Soc.*, **874**, 4442 (1961).

CaO–P$_2$O$_5$ (cont.)

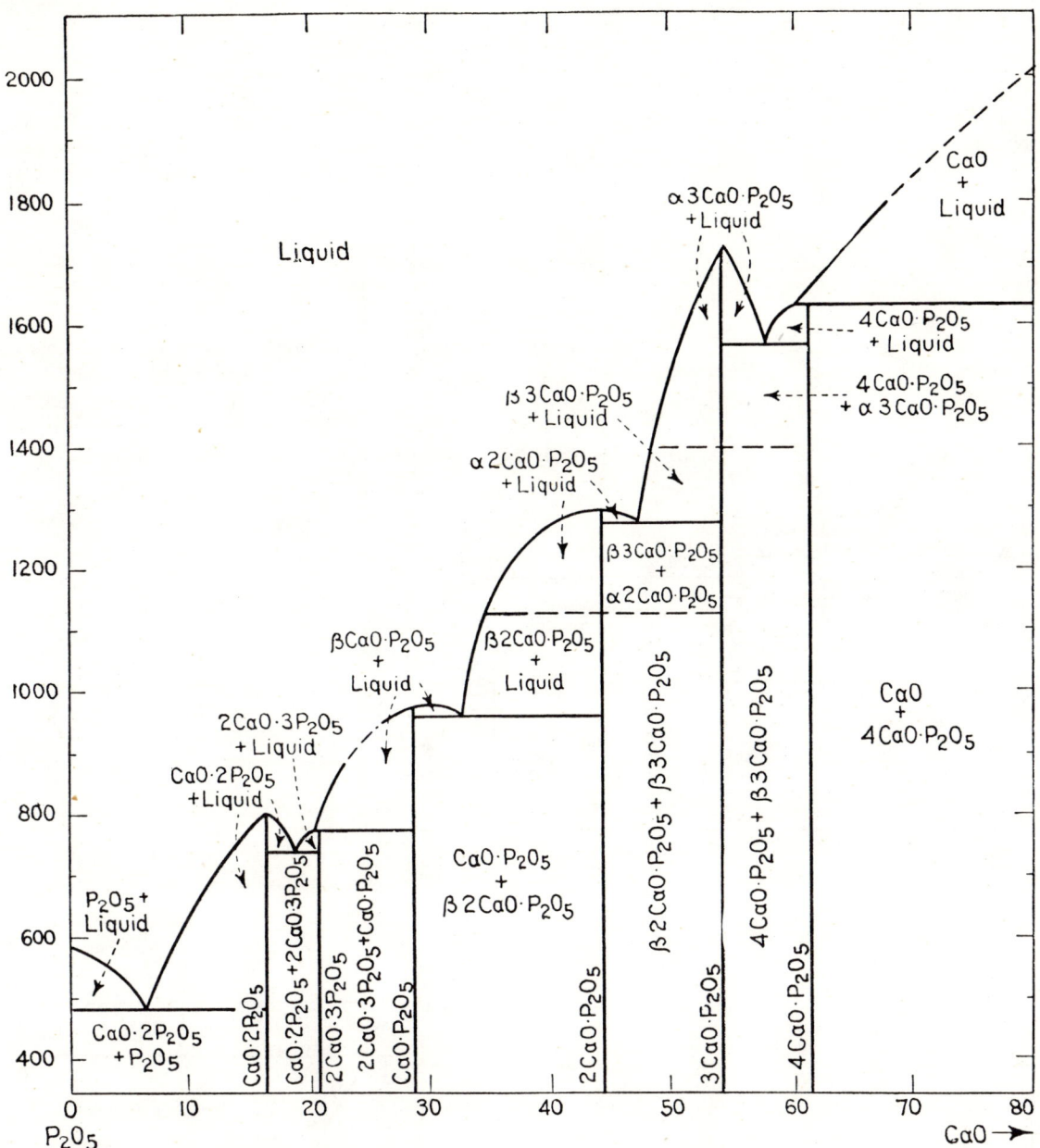

FIG. 246.—System CaO–P$_2$O$_5$.

Portion of system from 23 to 100% CaO by G. Trömel, *Stahl u. Eisen*, **63**, 21 (1943); portion of system from 0.0 to 23% CaO by W. L. Hill, G. T. Faust, and D. S. Reynolds, *Am. J. Sci.*, **242**, 469 (1944).

FIG. 247.—System CaO–P$_2$O$_5$. Modifies Fig. 246 in 4CaO·P$_2$O$_5$ plus beta 3CaO·P$_2$O$_5$ region.

G. Trömel, H.-J. Harkort and W. Hotop, *Z. anorg. Chem.*, **256** [5–6] 257 (1948).

CaO–P$_2$O$_5$ (concl.)

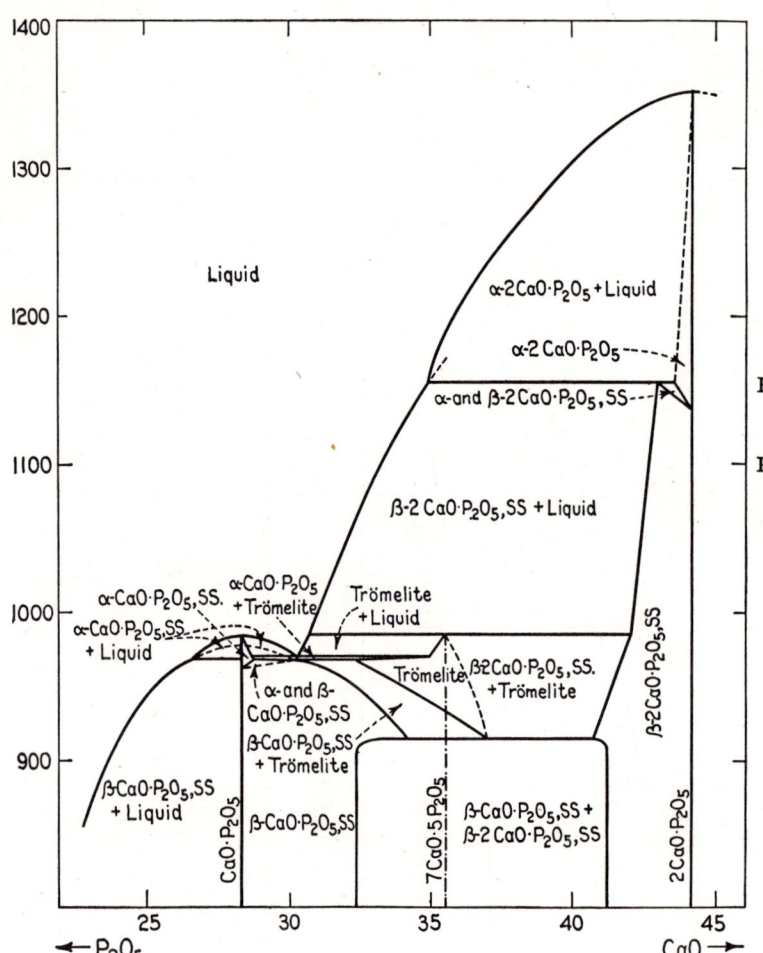

Fig. 248.—Portion of system CaO–P$_2$O$_5$ from 23 to 44.1% of CaO.

W. L. Hill, G. T. Faust, and D. S. Reynolds, *Am. J. Sci.*, **242**, 470 (1944).

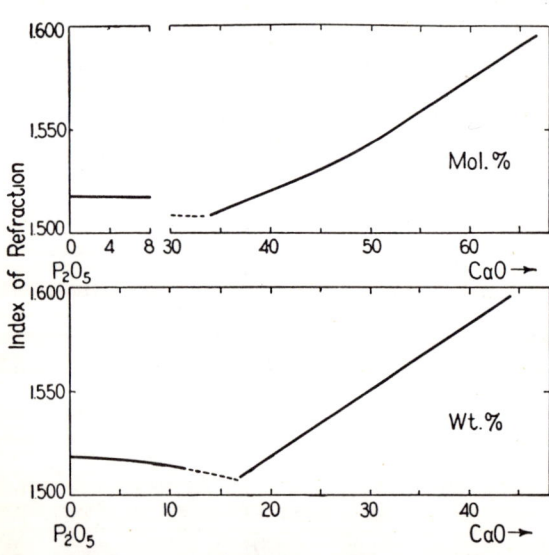

Fig. 249.—Refractive index-composition curves of glasses in system CaO–P$_2$O$_5$.

W. L. Hill, G. T. Faust, and D. S. Reynolds, *Am. J. Sci.*, **242**, 547 (1944).

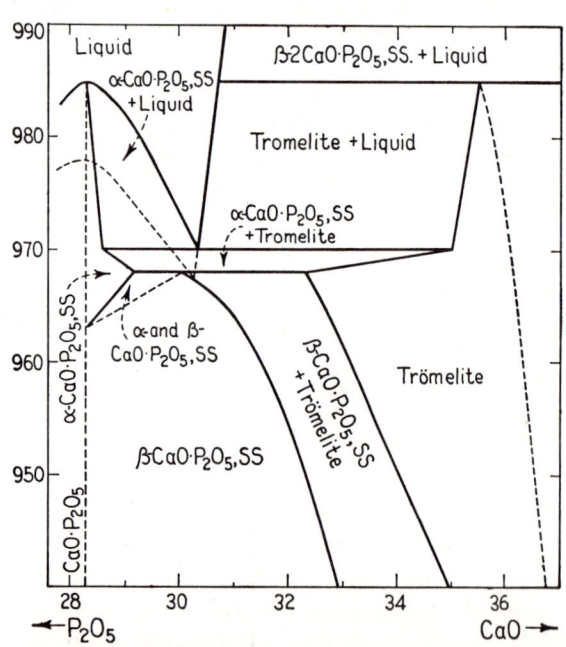

Fig. 250.—Relationships in vicinity of calcium metaphosphate and trömelite in the system CaO–P$_2$O$_5$.

W. L. Hill, G. T. Faust, and D. S. Reynolds, *Am. J. Sci.*, **242**, 473 (1944).

CaO–V$_2$O$_5$

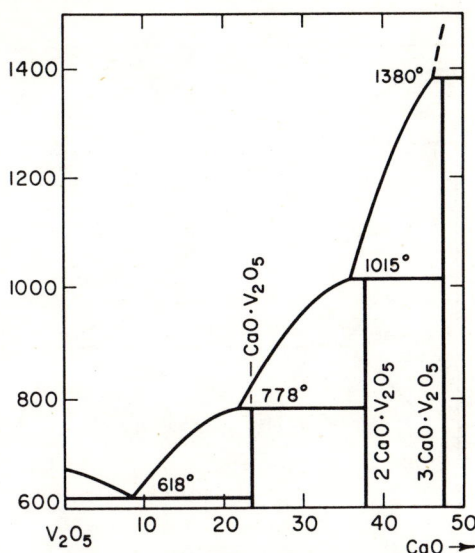

FIG. 251.—System CaO–V$_2$O$_5$.

A. N. Morozov, *Metallurg.*, **13** [12] 24 (1938).

CdO–B$_2$O$_3$

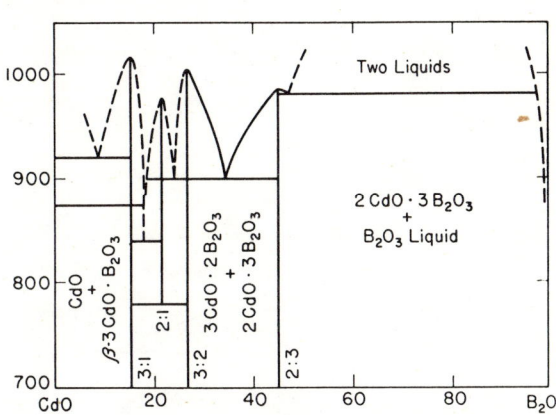

FIG. 252.—System CdO–B$_2$O$_3$.

E. C. Subbarao and F. A. Hummel, *J. Electrochem. Soc.* **103** [1] 30 (1956).

W. D. Hand and J. Krogh-Moe (*J. Am. Ceram. Soc.*, **45** [4] 197 (1962)) question the existence of the 3:2 and 2:3 compounds. They report, instead, a 1:2 compound and a low temperature form of the 2:1.

CdO–Nb$_2$O$_5$

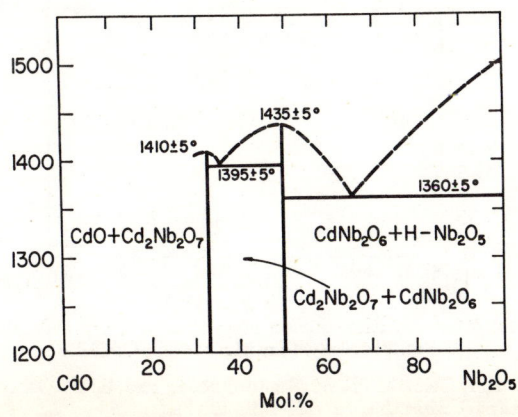

FIG. 253.—System CdO–Nb$_2$O$_5$; postulated liquidus.

R. S. Roth, *J. Am. Ceram. Soc.*, **44** [1] 49 (1961).

CoO–B$_2$O$_3$

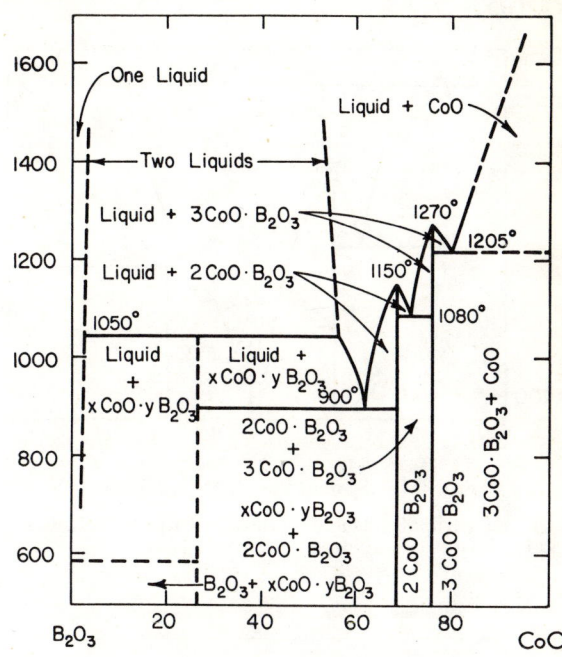

FIG. 254.—System CoO–B$_2$O$_3$; composite.

P. F. Konovalov, *Doklady Akad. Nauk S.S.S.R.*, **70** [5] 849 (1950). Compound $xCoO \cdot yB_2O_3$ after I. N. Belyaev, *Zhur. Fiz. Khim.*, **30**, 1419 (1956).

CoO–SiO$_2$

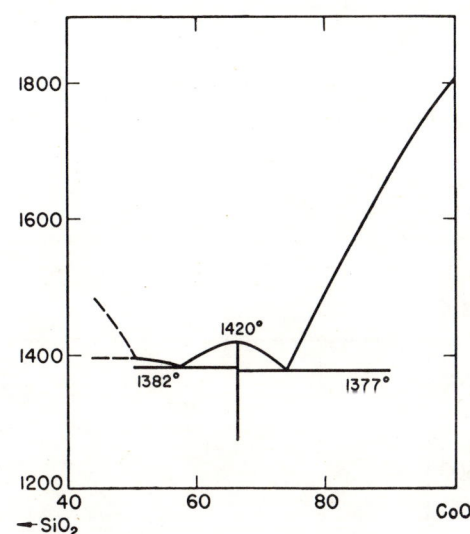

FIG. 255.—System CoO–SiO$_2$.

P. Asanti and E. J. Kohlmeyer, *Z. anorg. Chem.*, **265** 96 (1951).

CoO–TiO₂

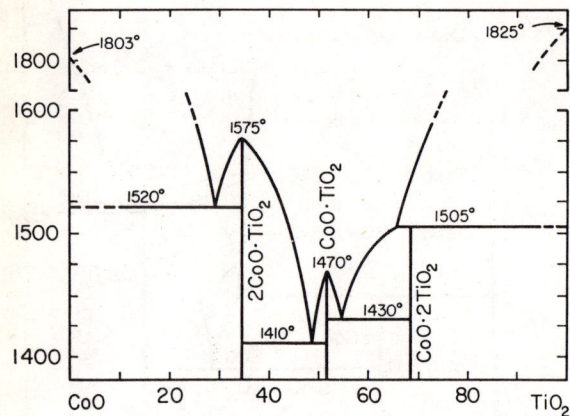

Fig. 256.—System CoO–TiO₂; proposed.

J. H. Strimple, Doctor's thesis, Rutgers University, State University of New Jersey, May, 1957.

FeO–ZrO₂

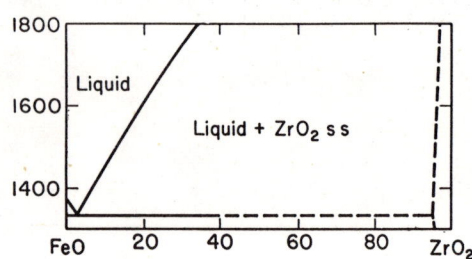

Fig. 257.—System FeO–ZrO₂.

W. A. Fischer and Alfred Hoffmann, *Arch. Eisenhüttenw.*, **28** [11] 743 (1957).

MgO–NiO

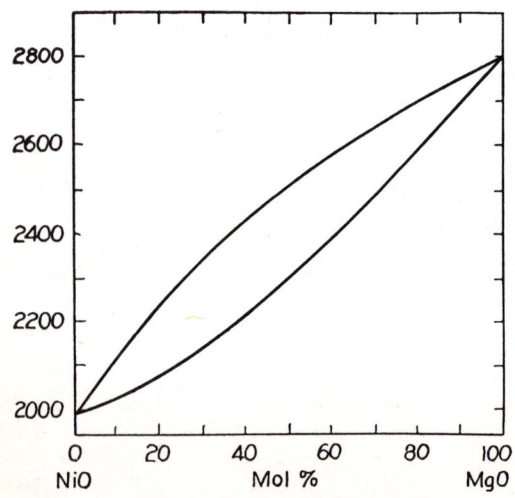

Fig. 258.—System MgO–NiO.

H. v. Wartenberg and E. Prophet, *Z. anorg. u. allgem. Chem.*, **208**, 379 (1932).

MgO–Al₂O₃

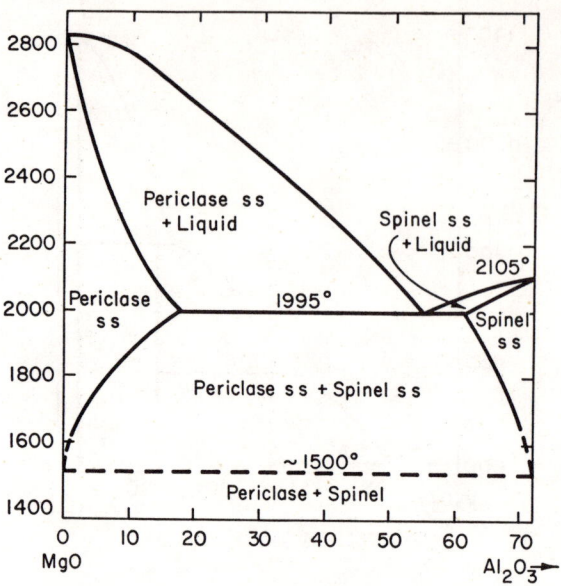

Fig. 259.—System MgO–MgO·Al₂O₃.

A. M. Alper, R. N. McNally, P. G. Ribbe, and R. C. Doman, *J. Am. Ceram. Soc.*, **45** [6] 264 (1962).

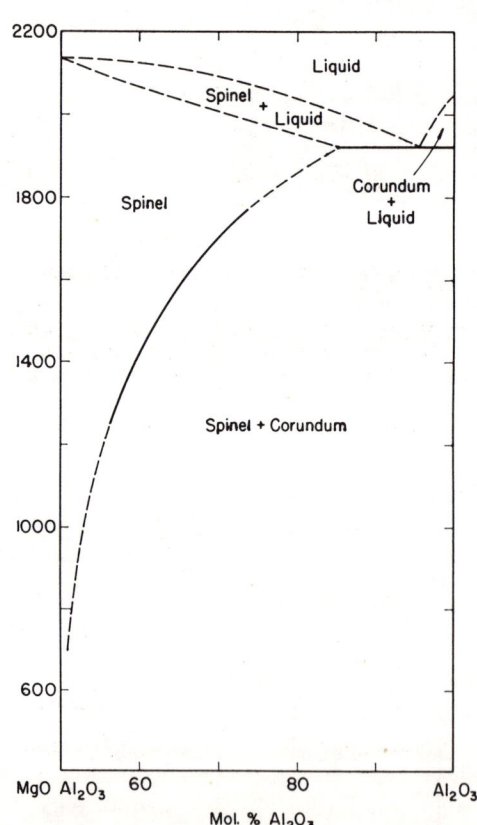

Fig. 260.—System MgO·Al₂O₃–Al₂O₃.

After Della M. Roy, Rustum Roy, and E. F. Osborn, *J. Am. Ceram. Soc.*, **36** [5] 149 (1953).

MgO–B₂O₃

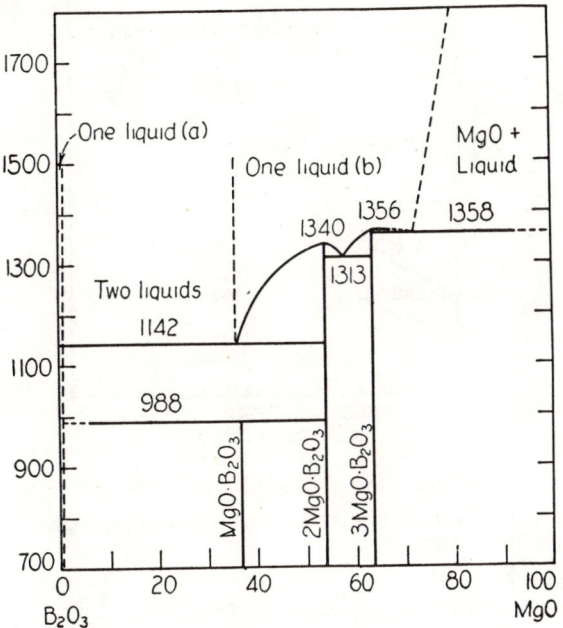

FIG. 261.—System MgO–B₂O₃.

H. M. Davis and M. A. Knight, *J. Am. Ceram. Soc.*, **28** [4] 100 (1945). Melting point of 3MgO·B₂O₃ should read 1366 and not 1356°C.

MgO–Y₂O₃

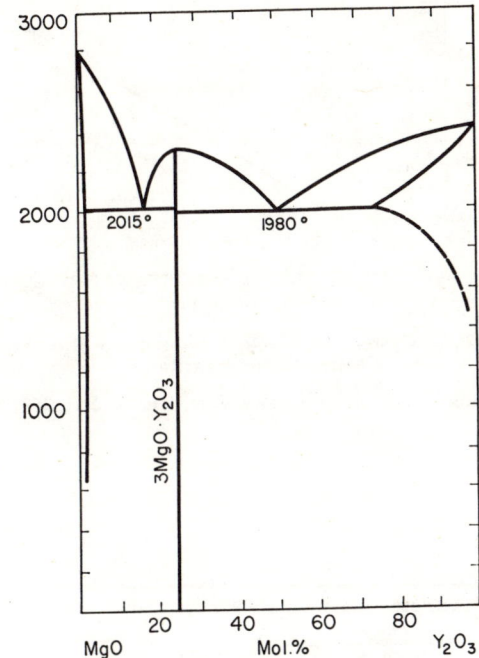

FIG. 263.—System MgO–Y₂O₃.

H. E. Otto, private communication, Dec. 27, 1961.

MgO–Cr₂O₃

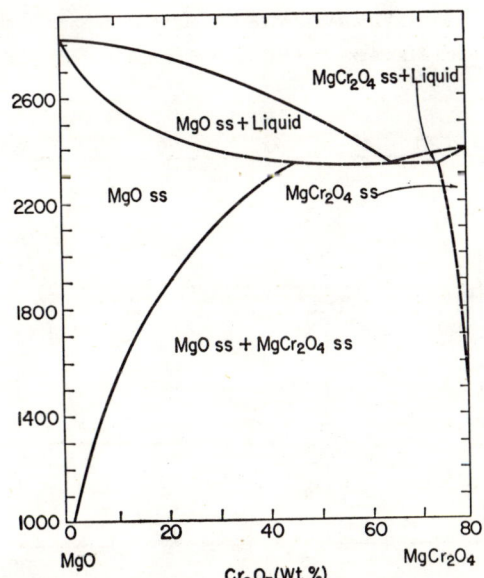

FIG. 262.—System MgO–MgCr₂O₄.

A. M. Alper, R. N. McNally, R. C. Doman, and F. G. Keihn, *J. Am. Ceram. Soc.*, **47** [1] 30 (1964).

MgO–GeO₂

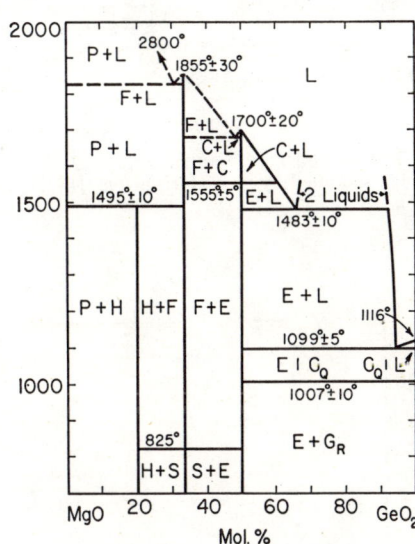

FIG. 264.—System MgO–GeO₂.

Symbol	Composition	Structure Type
C	MgO·GeO₂	Probably clinoenstatite
E	MgO·GeO₂	Enstatite
F	2MgO·GeO₂	Forsterite (olivine)
G_Q	GeO₂	Quartz
G_R	GeO₂	Rutile
H	4MgO·GeO₂	
L		Liquid
P	MgO	Periclase (NaCl)
S	2MgO·GeO₂	Spinel

Carl R. Robbins and E. M. Levin, *Am. J. Sci.*, **257** 65. (1959).

MgO–GeO₂ (concl.)

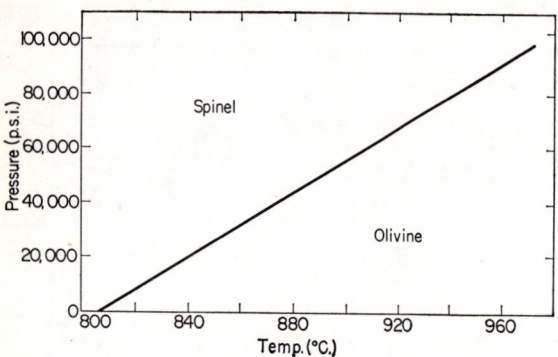

Fig. 265.—System Mg₂GeO₄; univariant p–T curve for spinel-olivine transition.

Frank Dachille and Rustum Roy, *Am. J. Sci.*, 258, 231 (1960).

MgO–SiO₂

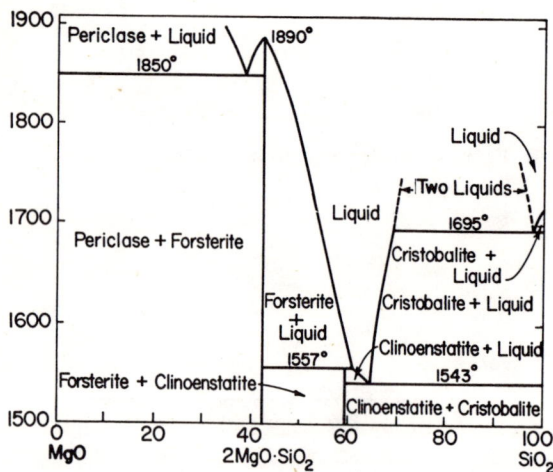

Fig. 266.—System MgO–SiO₂.

N. L. Bowen and Olaf Andersen, *Am. J. Sci.* [4], 37, 488 (1914); modified by J. W. Greig, *ibid.* [5] 13, 15, 133–54 (1927).

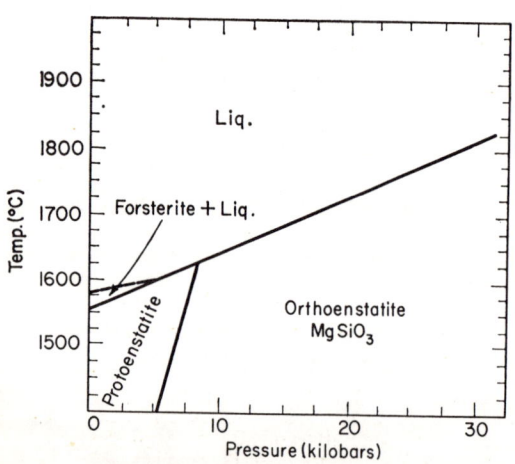

Fig. 267.—System MgSiO₃; p–T.

F. R. Boyd and J. L. England, Ann. Rept. Director Geophys. Lab. 1960–61; in *Carnegie Inst. Washington Year Book*, 60, 115 (1961).

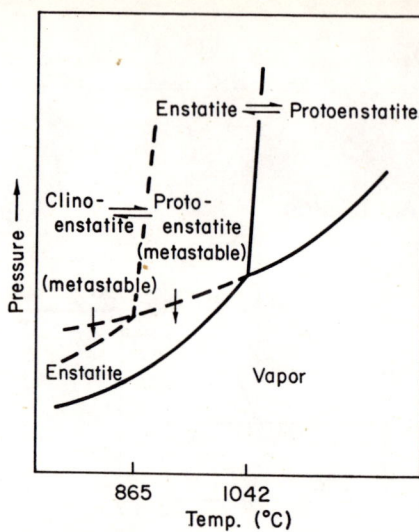

Fig. 268.—System MgSiO₃; p–T schematic.

J. F. Sarver and F. A. Hummel, *J. Am. Ceram. Soc.*, 45 [4] 156 (1962).

MgO–TiO₂

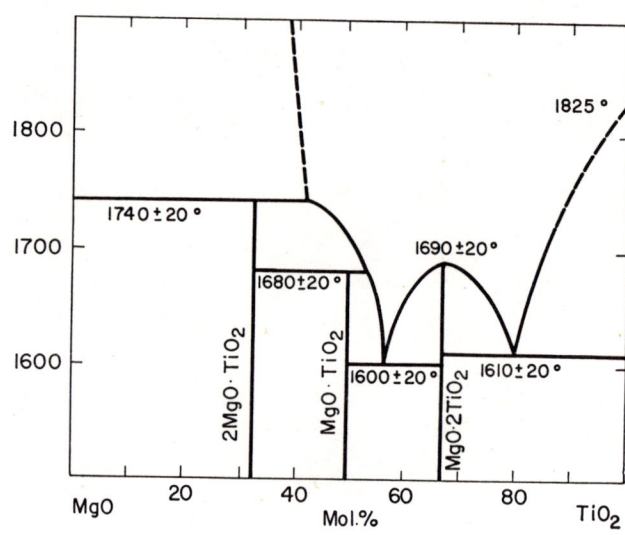

Fig. 269.—System MgO–TiO₂.

Franco Massazza and Efisia Sirchia, *Chim. Ind.* (Milan), 40, 378 (1958).

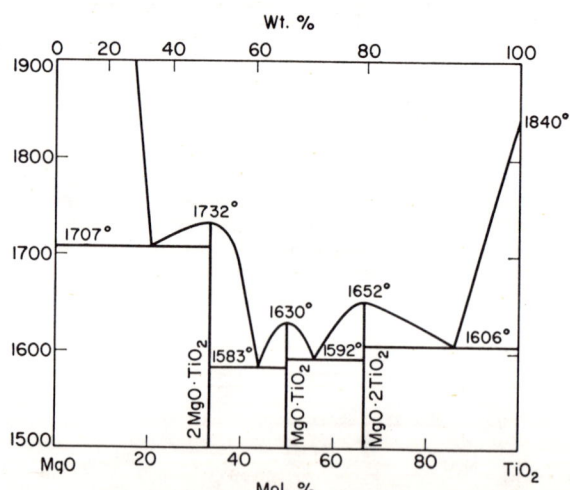

Fig. 270.—System MgO–TiO₂; proposed.

L. W. Coughanour and V. A. DeProsse, *J. Research Natl. Bur. Standards*, 51 [2] 87 (1953); RP 2435.

MgO–ZrO₂

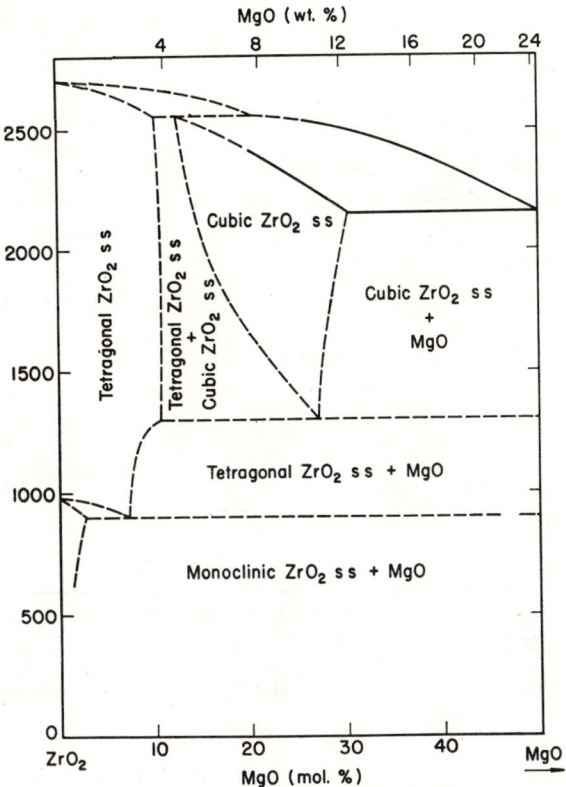

Fig. 271.—System MgO–ZrO₂; tentative.

Pol Duwez, Francis Odell, and Frank H. Brown, Jr., *J. Am. Ceram. Soc.*, **35** [5] 109 (1952).

MgO–Ta₂O₅

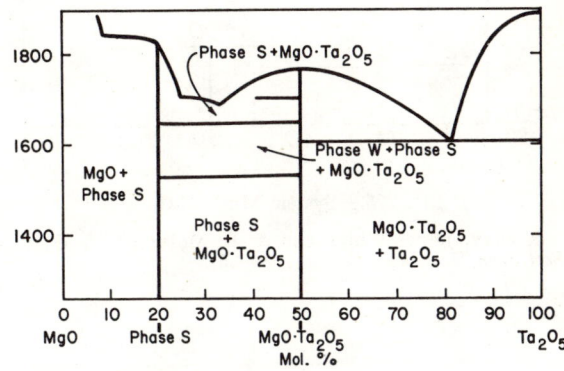

Fig. 273.—System MgO–Ta₂O₅.

B. W. King, John Schultz, E. A. Durbin, and W. H. Duckworth, U. S. Atomic Energy Comm., BMI-1106, 12 (1956).

MgO–P₂O₅

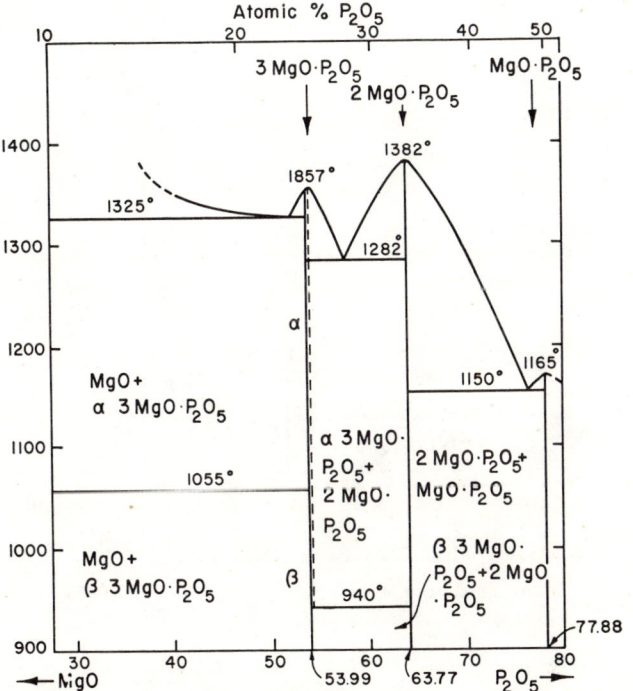

Fig. 272.—System MgO–P₂O₅.

Józef Berak, *Roczniki Chem.*, **32**, 19 (1958): (English summary).

MgO–R₂O,RO

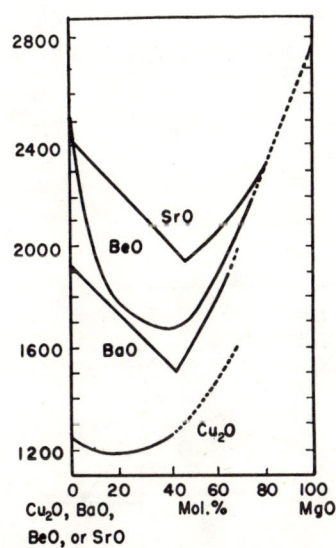

Fig. 274.—Liquidus curves of systems Cu₂O–MgO BaO–MgO, BeO–MgO, and MgO–SrO.
H. von Wartenberg and E. Prophet, Part V, *Z. anorg. u. allgem. Chem.*, **208**, 378 (1932).

MnO–Al$_2$O$_3$

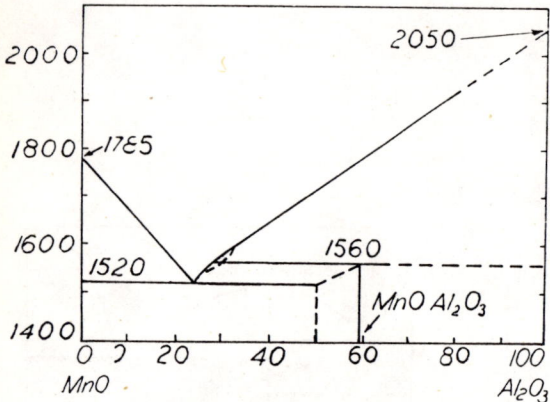

Fig. 275.—System MnO–Al$_2$O$_3$.

R. Hay, James White, and A. B. McIntosh, *J. West Scot. Iron Steel Inst.*, **42**, 99 (1934–35).

MnO–B$_2$O$_3$

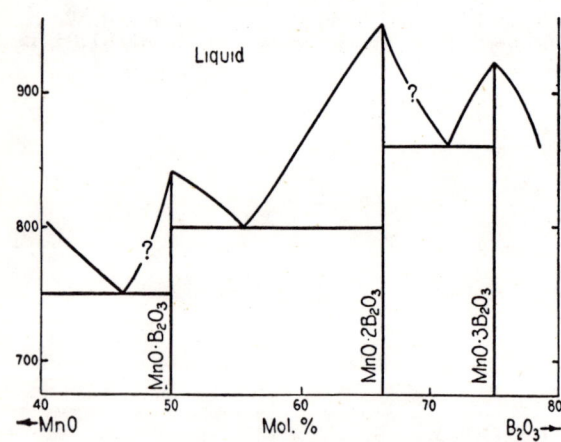

Fig. 276.—System MnO–B$_2$O$_3$.

Drawn from data of C. Mazzetti and F. de Carli, *Gazz. chim. ital.*, **56**, 27 (1926).

MnO–TiO$_2$

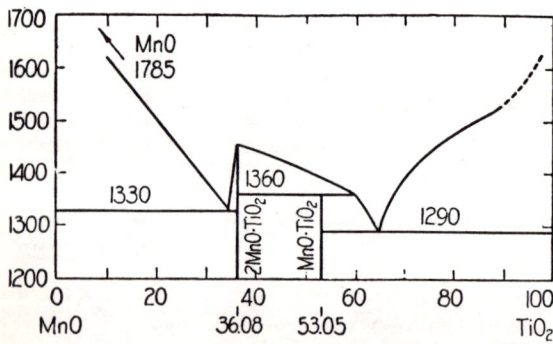

Fig. 277.—System MnO–TiO$_2$.

J. Grieve and J. White, *J. Roy. Tech. Coll. (Glasgow)*, **4**, 661 (1940).

MnO–P$_2$O$_5$

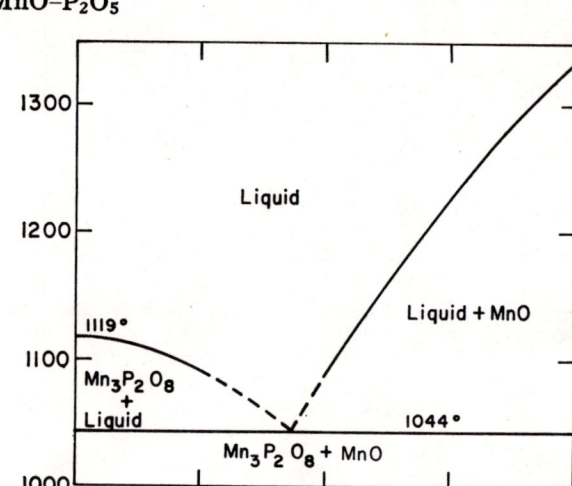

Fig. 278.—System MnO–MnO·P$_2$O$_5$.

J. Pearson, E. T. Turkdogan, and E. M. Fenn, *J. Iron Steel Inst.*, **176**, 442 (1954).

NiO–V$_2$O$_5$

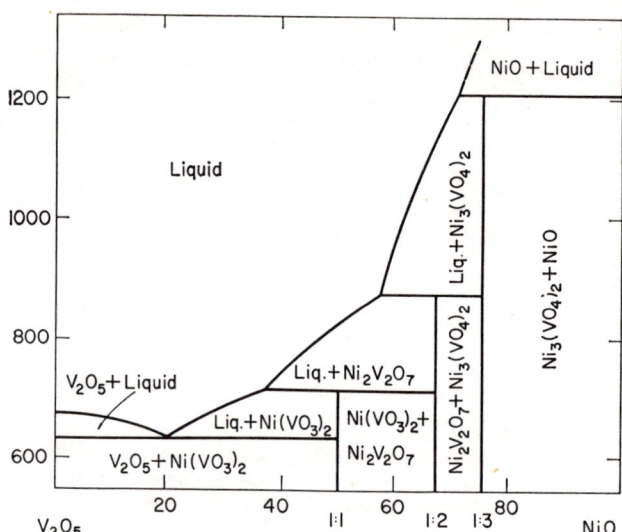

Fig. 279.—System NiO–V$_2$O$_5$.

Cesare Brisi, *Ann. chim. (Rome)*, **47** [7, 8] 806 (1957).

PbO–Al$_2$O$_3$

PbO–Fe$_2$O$_3$

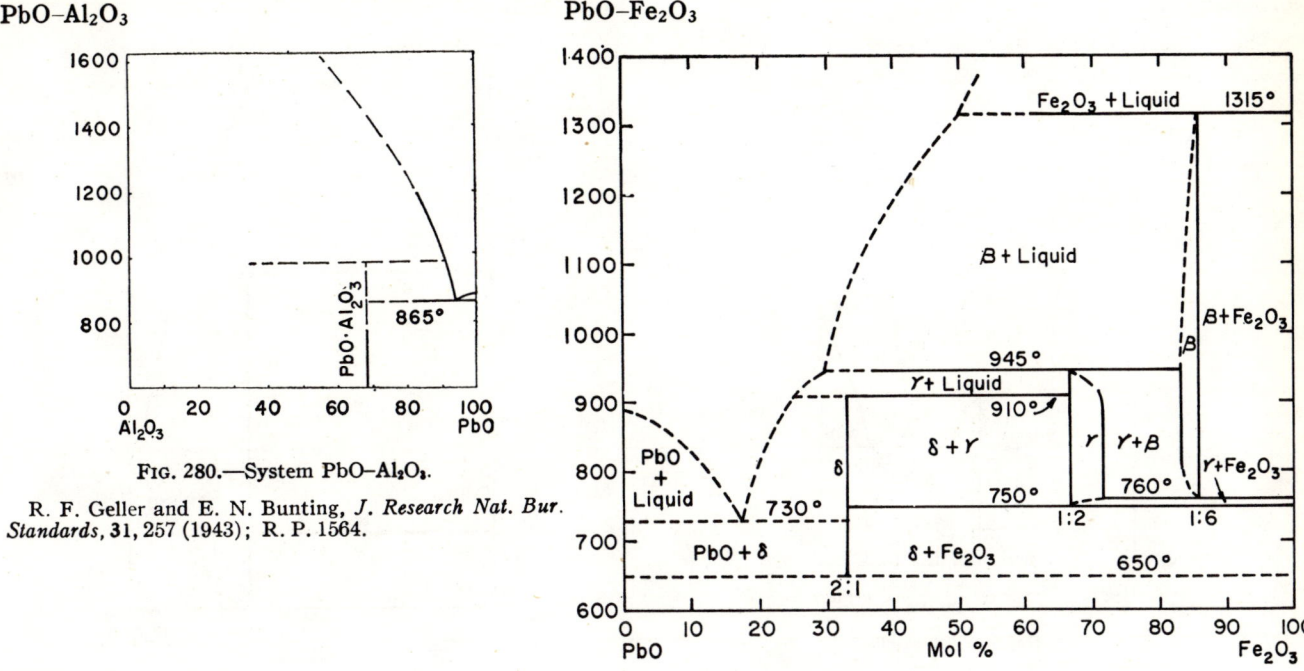

Fig. 280.—System PbO–Al$_2$O$_3$.

R. F. Geller and E. N. Bunting, *J. Research Nat. Bur. Standards*, 31, 257 (1943); R. P. 1564.

Fig. 282.—System PbO–Fe$_2$O$_3$.

A. J. Mountvala and S. F. Ravitz, *J. Am. Ceram Soc.*, 45 [6] 286 (1962).

PbO–B$_2$O$_3$

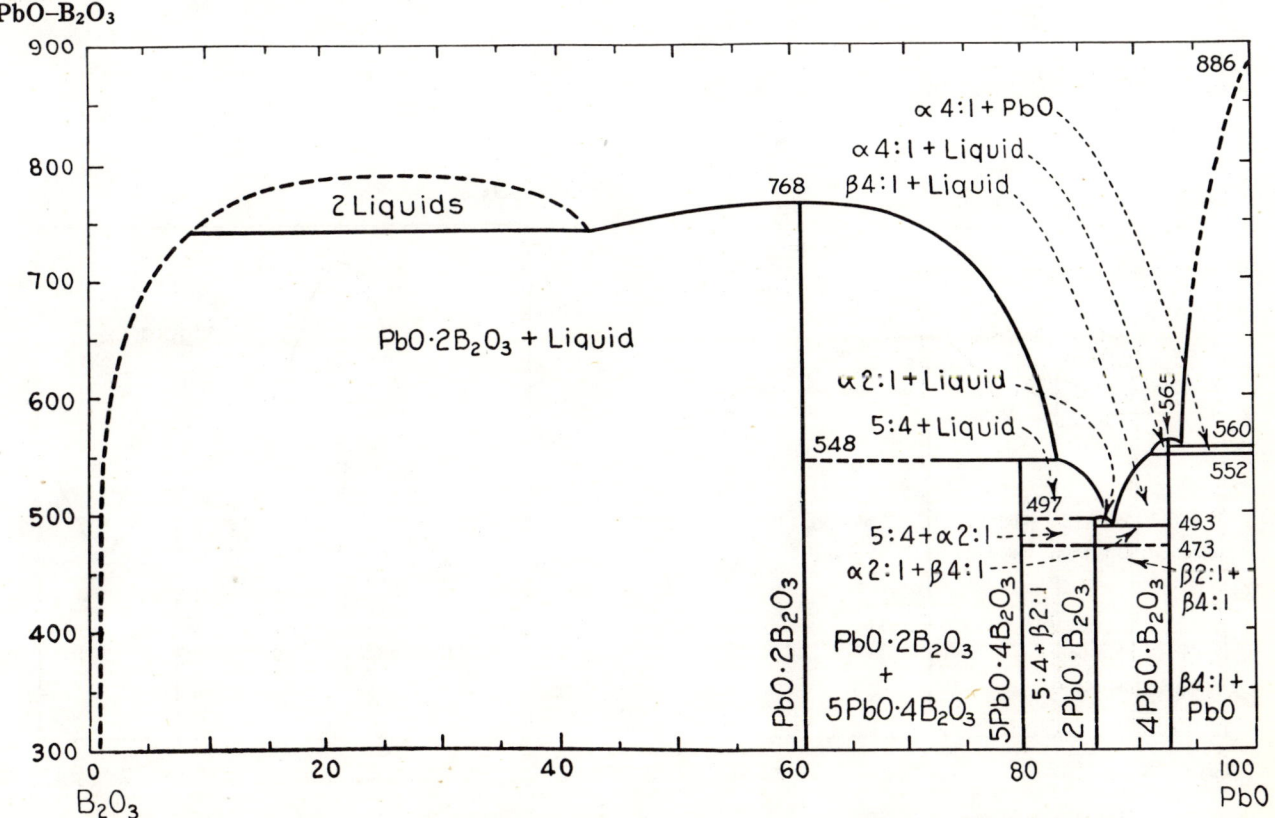

Fig. 281.—System PbO–B$_2$O$_3$.

R. F. Geller and E. N. Bunting, *J. Research Nat. Bur. Standards*, 18 [5] 585, 589 (1937); R. P. 995.

PbO–GeO₂

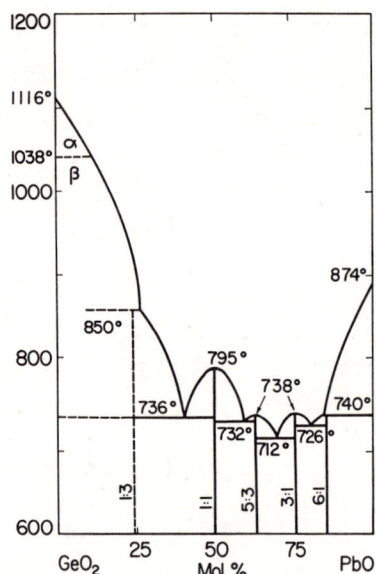

Fig. 283.—System PbO–GeO₂.

E. I. Speranskaya, *Izvest. Akad. Nauk S.S.S.R., Otdel Khim. Nauk*, 163 (1959).

PbO–SnO₂

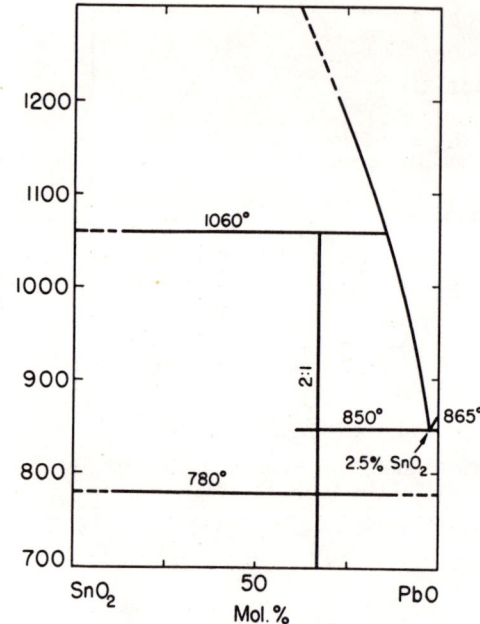

Fig. 285.—System PbO–SnO₂.

G. G. Urazov, E. I. Speranskaya, and Z. F. Gulyanitskaya, *Zhur. Neorg. Khim.*, 1 [6] 1414 (1956).

PbO–SiO₂

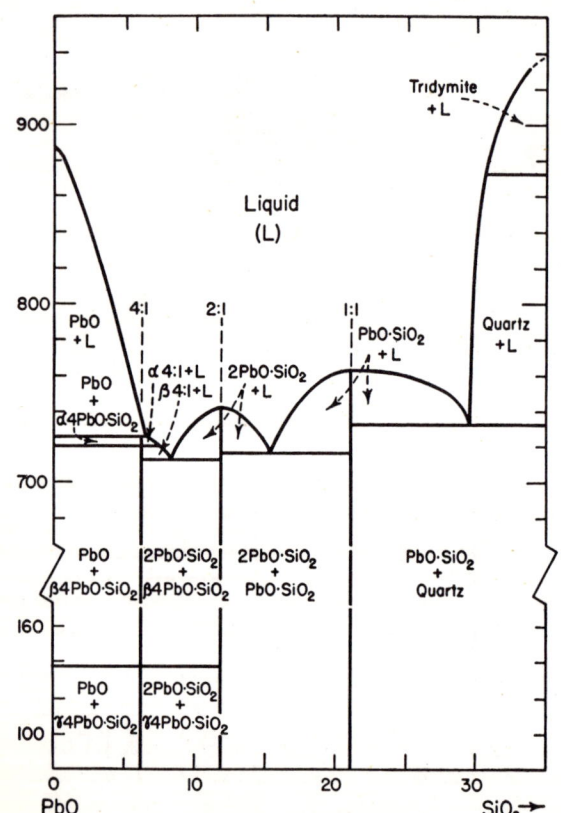

Fig. 284.—System PbO–SiO₂.

R. F. Geller, A. S. Creamer, and E. N. Bunting, *J. Research Natl. Bur. Standards*, 13 [2] 243 (1934); RP 705.

PbO–As₂O₅

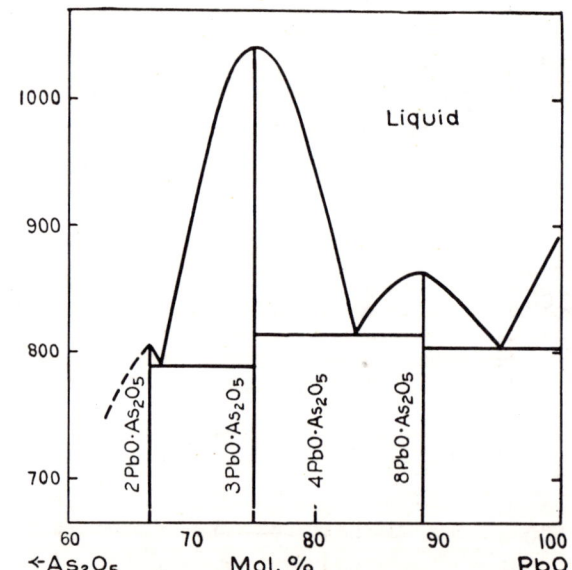

Fig. 286.—System PbO–2PbO·As₂O₅.

Modified from M. Amadori, *Atti reale ist. Veneto sci.*, 76 [II] 419 (1917).

PbO–Nb₂O₅

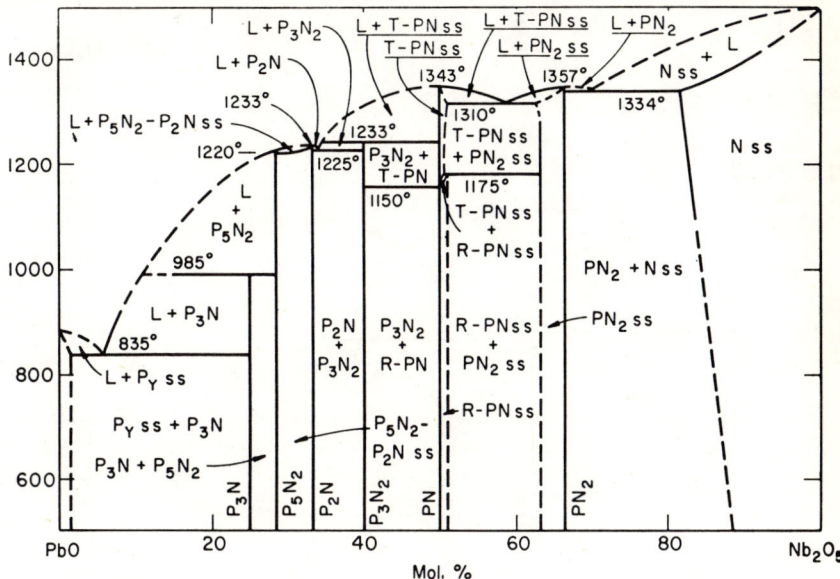

FIG. 287.—System PbO–Nb₂O₅. P_Y—yellow PbO, orthorhombic; P₃N—3PbO·Nb₂O₅; P₅N₂—5PbO·2Nb₂O₅; P₂N—2PbO·Nb₂O₅; P₃N₂—3PbO·2Nb₂O₅; T-PN—tetragonal PbO·Nb₂O₅; R-PN—rhombohedral PbO·Nb₂O₅; PN₂—PbO·2Nb₂O₅; N—Nb₂O₅; ss—solid solution; L—liquid.

R. S. Roth, *J. Research Natl. Bur. Standards*, **62**, [1] 34 (1959).

PbO–P₂O₅

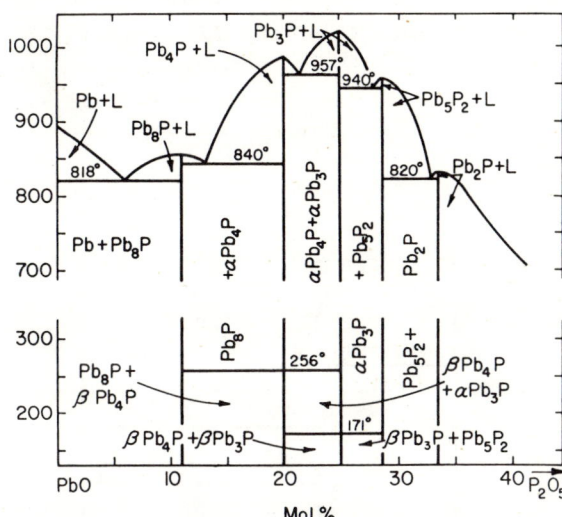

FIG. 288.—System PbO–P₂O₅. Pb = PbO; P = P₂O₅; L = liquid.

H. H. Paetsch and Adolf Dietzel, *Glastech. Ber.*, **29** [9] 348 (1956).

PbO–Ta₂O₅

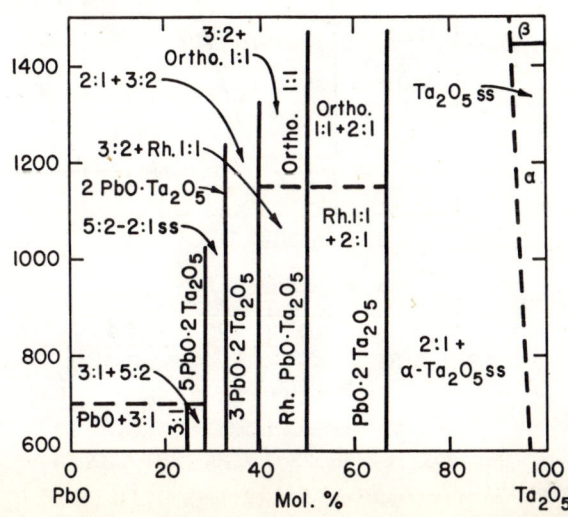

FIG. 289.—System PbO–Ta₂O₅; subsolidus. Ortho = orthorhombic modification; Rh = rhombohedral modification. Compounds in two-phase areas designated as mole ratios of components.

E. C. Subbarao, *J. Am. Ceram. Soc.*, **44** [2] 93 (1961).

PbO–V₂O₅

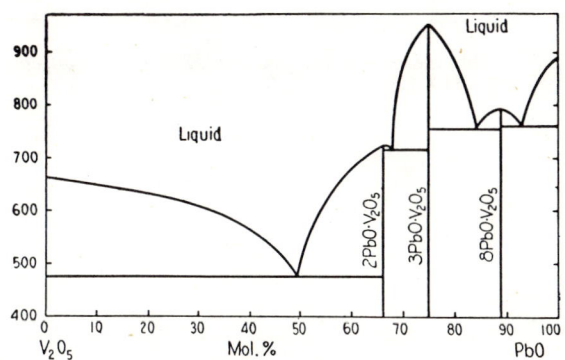

FIG. 290.—System PbO–V₂O₅.

M. Amadori, *Atti reale ist. Veneto sci.*, **76** [II] 419 (1917).

PbO–CrO₃

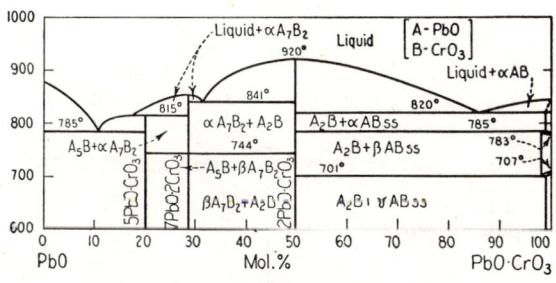

FIG. 291.—System PbO–PbO·CrO₃.

F. M. Jaeger and H. C. Germs, *Z. anorg. u. allgem Chem.*, **119**, 155 (1921).

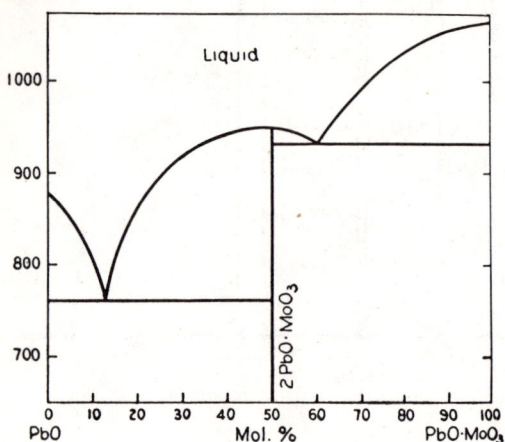

Fig. 292.—System PbO–PbO·MoO$_3$.

F. M. Jaeger and H. C. Germs, *Z. anorg. u. allgem. Chem.*, **119**, 159 (1921).

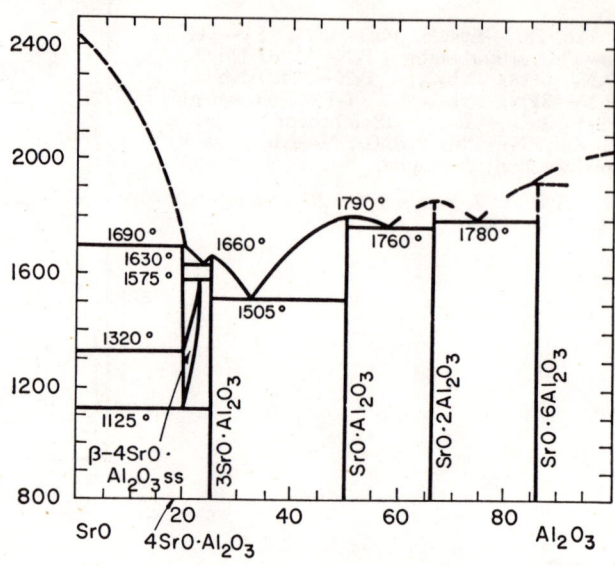

Fig. 294.—System SrO–Al$_2$O$_3$.

Franco Massazza, *Chim. Ind.* (*Milan*), **41**, 114 (1959).

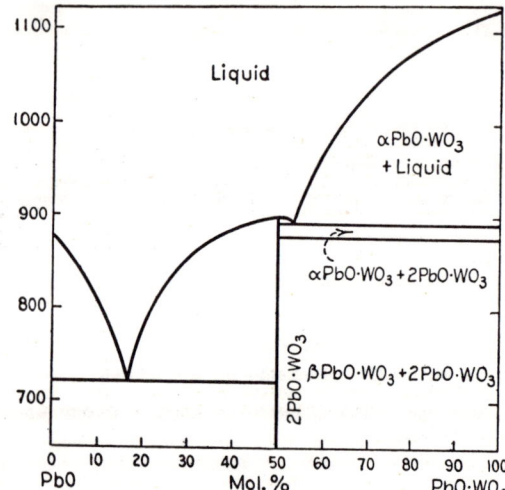

Fig. 293.—System PbO–PbO·WO$_3$.

F. M. Jaeger and H. C. Germs, *Z. anorg. u. allgem. Chem.*, **119**, 162 (1921).

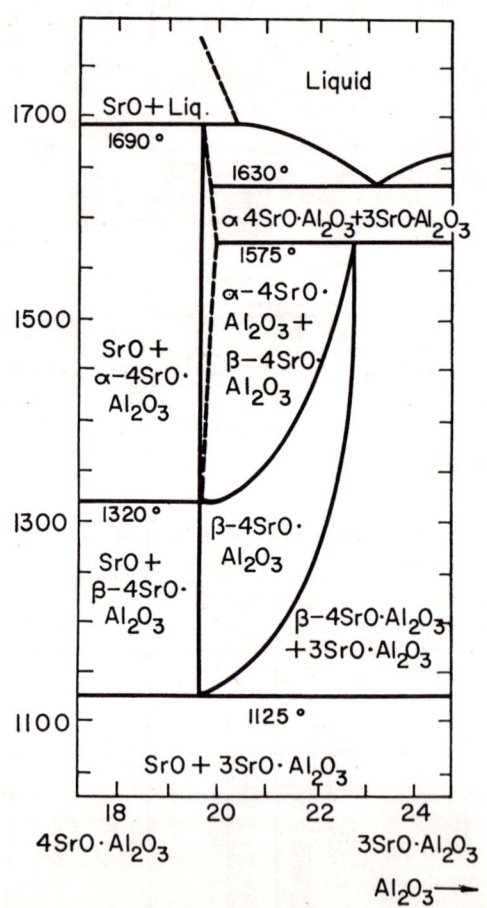

Fig. 295.—System 3SrO·Al$_2$O$_3$–4SrO·Al$_2$O$_3$.

Franco Massazza, *Chim. Ind.* (*Milan*), **41**, 112 (1959).

SrO–SiO$_2$

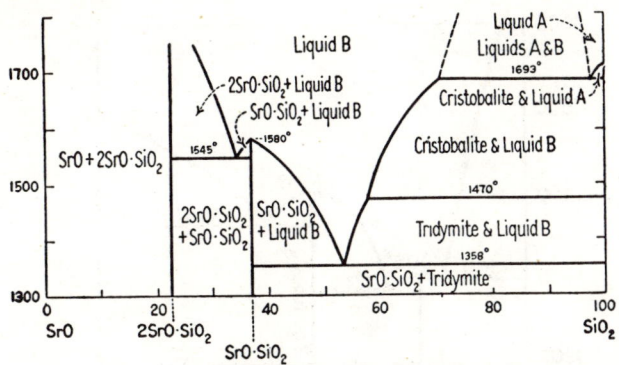

FIG. 296.—System SrO–SiO$_2$.

P. Eskola, *Am. J. Sci.*, 5th Ser., **4**, 336 (1922); modified by J. W. Greig, *ibid.*, 5th Ser., **13**, 19 (1927); see also F. C. Kracek, *J. Am. Chem. Soc.*, **52** [4] 1440 (1930).

SrO–TiO$_2$

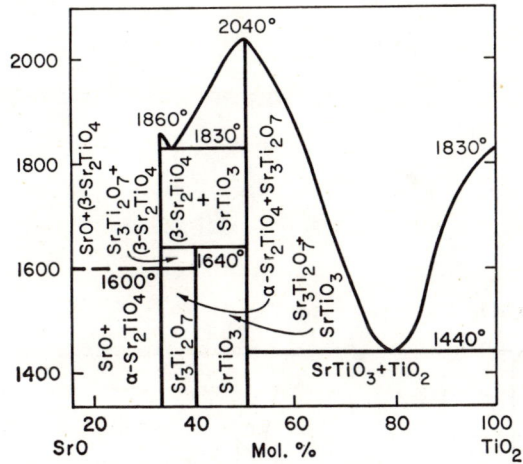

FIG. 297.—System SrO–TiO$_2$.

Miroslawa Dryś and Włodzimierz Trzebiatowski, *Roczniki Chem.*, **31**, 492 (1957).

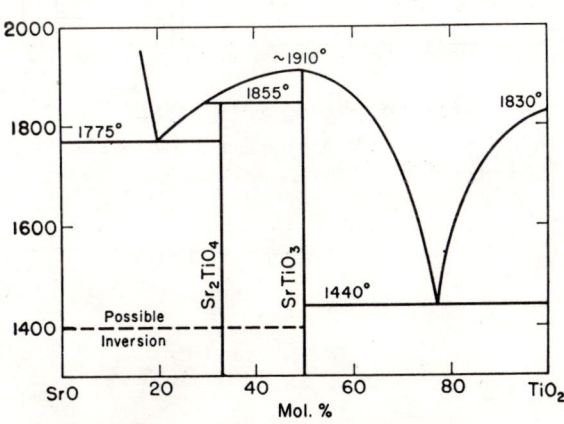

FIG. 298.—System SrO–TiO$_2$; tentative.

Rustum Roy; private communication, 1957.

ZnO–Al$_2$O$_3$

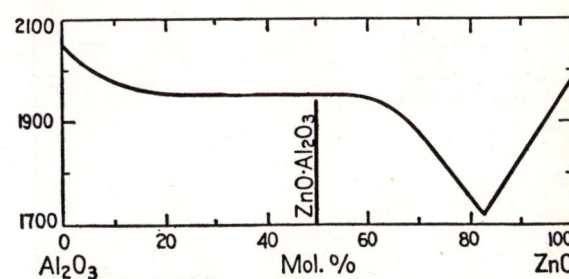

FIG. 299.—Liquidus curve of system ZnO–Al$_2$O$_3$.

E. N. Bunting, *Bur. Standards J. Research*, **8** [2] 280 (1932); R. P. 413.

ZnO–B$_2$O$_3$

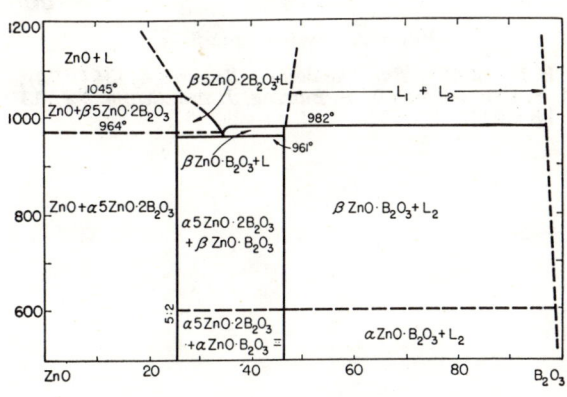

FIG. 300.—System ZnO–B$_2$O$_3$.

D. E. Harrison and F. A. Hummel, *J. Electrochem. Soc.*, **103** [9] 496 (1956); see also, "Structure of Zinc Metaborate, Zn$_4$O(BO$_2$)$_6$," P. Smith, S. Garcia-Blanco, and L. Revoir, *Anales Real Soc. Espan. Fis. Quim.* (Madrid) Ser. A (Nov.-Dec.), 263–268 (1961).

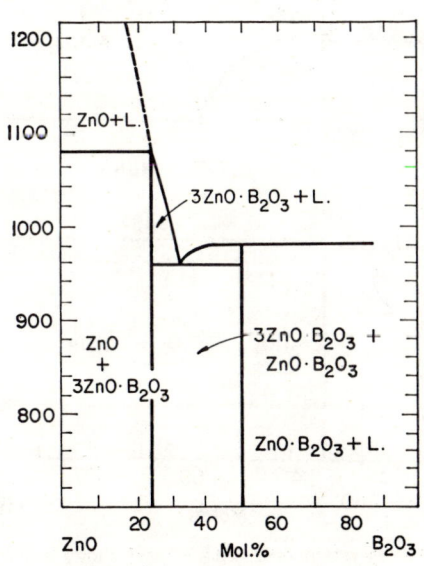

FIG. 301.—System ZnO–B$_2$O$_3$.

Yu. S. Leonov, *Zhur. Neorg. Khim.*, **3**, 1246 (1958).

ZnO–SiO$_2$

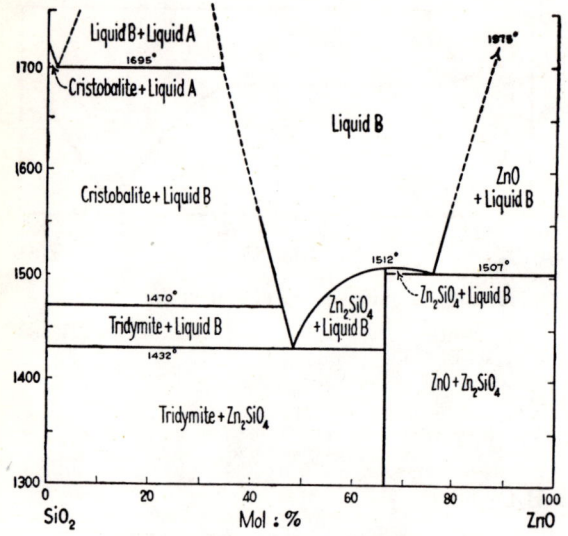

Fig. 302.—System ZnO–SiO$_2$.

E. N. Bunting, *Bur. Standards J. Research*, **4**, 134 (1930); R. P. 136; see also E. N. Bunting, *J. Am. Ceram. Soc.*, **13** [1] 8 (1930).

ZnO–Nb$_2$O$_5$

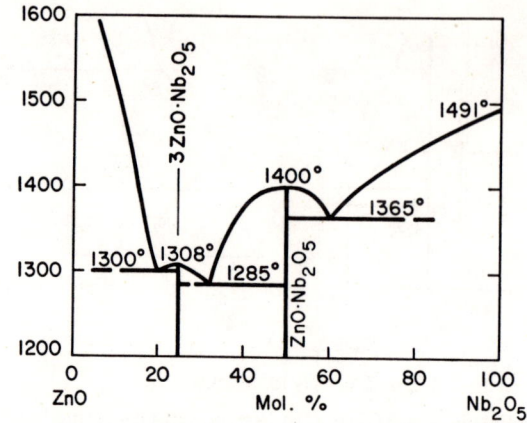

Fig. 304.—System ZnO–Nb$_2$O$_5$; suggested.

A. J. Pollard, *J. Am. Ceram. Soc.*, **44** [12] 630 (1961).

ZnO–TiO$_2$

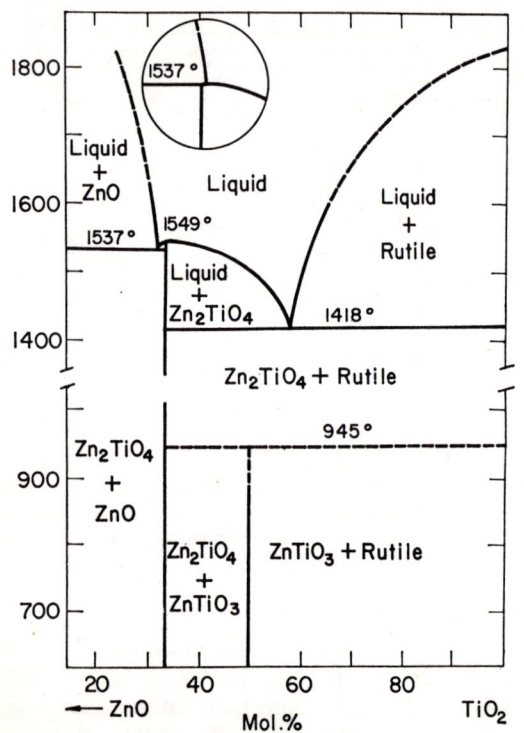

Fig. 303.—System ZnO–TiO$_2$. Inset shows alternative incongruent melting of Zn$_2$TiO$_4$.

F. H. Dulin and D. E. Rase, *J. Am. Ceram. Soc.*, **43** [3] 130 (1960).

ZnO–P$_2$O$_5$

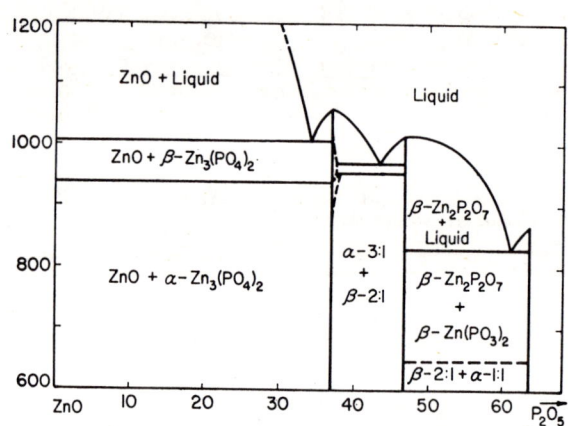

Fig. 305.—System ZnO–Zn(PO$_3$)$_2$.

F. L. Katnack and F. A. Hummel, *J. Electrochem. Soc.*, **105** [3] 132 (1958).

$ZnO-P_2O_5$ (concl.)

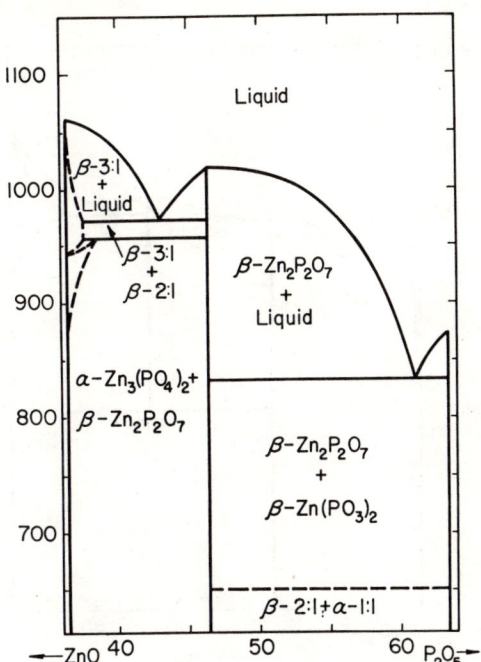

FIG. 306.—System $ZnO-Zn(PO_3)_2$; enlarged section from Fig. 305.

F. L. Katnack and F. A. Hummel, *J. Electrochem. Soc.*, 105 [3] 132 (1958).

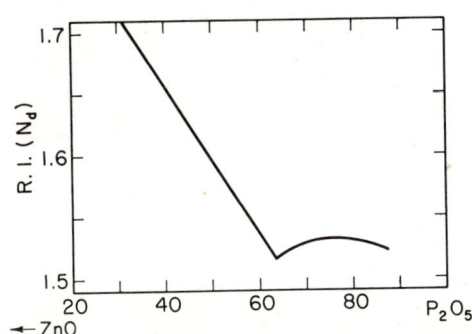

FIG. 307.—System $ZnO-Zn(PO_3)_2$; refractive indices.

F. L. Katnack and F. A. Hummel, *J. Electrochem. Soc.*, 105 [3] 128 (1958).

$Al_2O_3-B_2O_3$

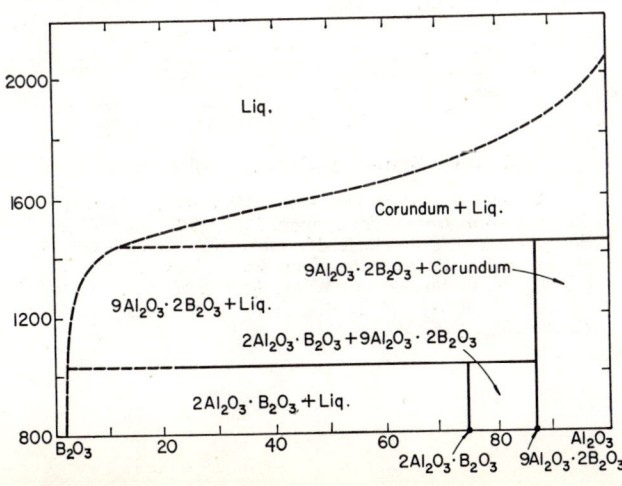

FIG. 308.—System $Al_2O_3-B_2O_3$; tentative.

K. H. Kim and F. A. Hummel, private communication, Dec. 20, 1961.

$Al_2O_3-Cr_2O_3$

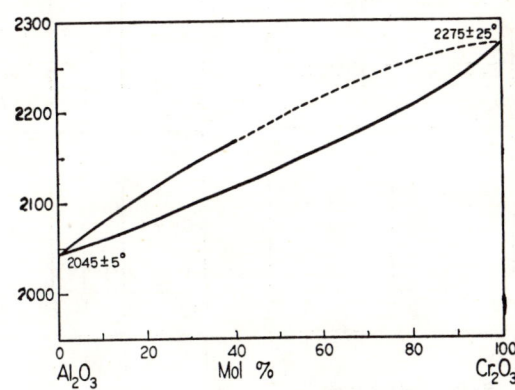

FIG. 309.—System α-Al_2O_3-Cr_2O_3.

E. N. Bunting, *Bur. Standards J. Research*, 6 [6] 948 (1931); R. P. 317.

$Al_2O_3-Ga_2O_3$

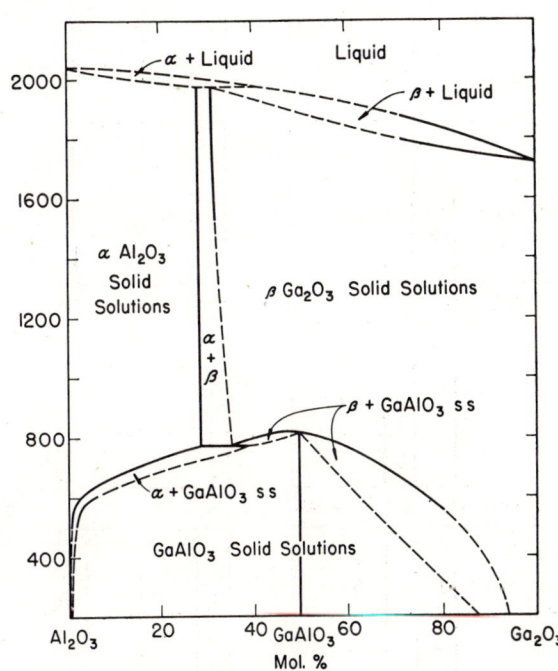

FIG. 310.—System $Al_2O_3-Ga_2O_3$.

V. G. Hill, Rustum Roy, and E. F. Osborn, *J. Am. Ceram. Soc.*, 35 [6] 136 (1952).

Al₂O₃–R₂O₃

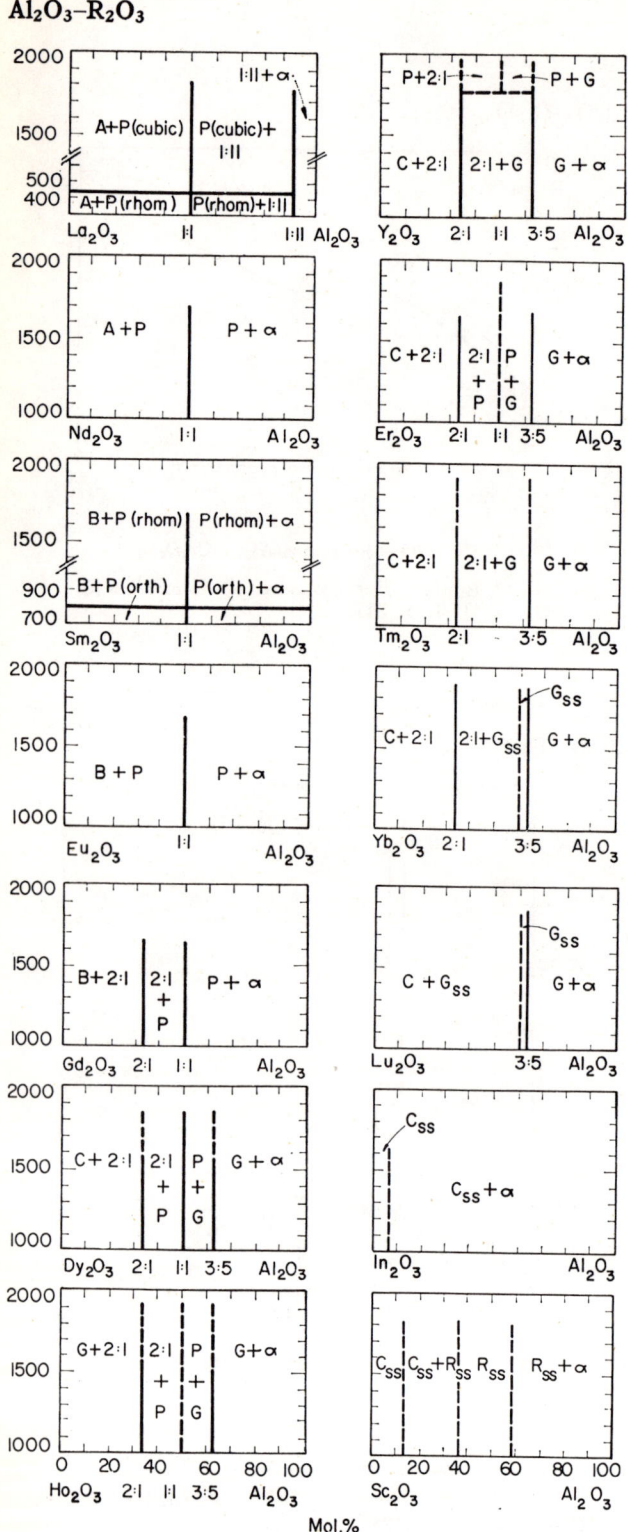

Fig. 312.—System Al_2O_3–R_2O_3; predicted subsolidus. Structure types: A, A-type rare earth oxide; B, B-type rare earth oxide; C, C-type rare earth oxide; G, garnet; 1:11, beta alumina; P, perovskite; R, unknown, rhombohedral symmetry; α, corundum.

S. J. Schneider, R. S. Roth, and J. L. Waring, *J. Research Natl. Bur. Standards*, **65A** [4] 364 (1961).

Al₂O₃–Y₂O₃

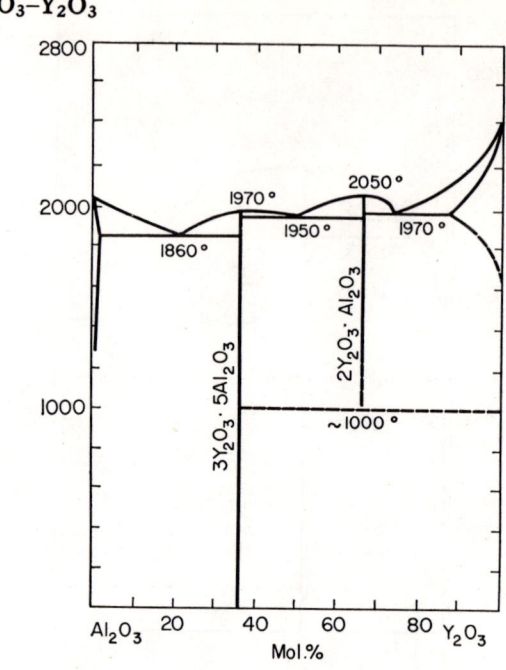

Fig. 311.—System Al_2O_3–Y_2O_3.

L. E. Olds and H. E. Otto, private communication, Dec. 27, 1961. Fig. 312 indicates additional 1:1 compound; see also, I. Warshaw and Rustum Roy, *J. Am. Ceram. Soc.*, **42** [9] 435 (1959).

Al₂O₃–SiO₂

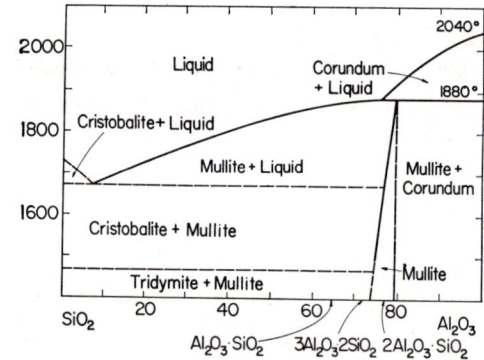

Fig. 313.—System Al_2O_3–SiO_2; redetermined.

J. W. Welch, *Nature*, **186** [4724] 546 (1960); also *Trans. Intern. Ceram. Congr., 7th London*, 1960, **1961**, pp. 197–206. See also: G. Trömel, K.-H. Obst, K. Konopicky, H. Bauer, and I. Patzak, *Ber. deut. keram. Ges.*, **34** [12] 401 (1957); E. C. Shears and W. A. Archibald, *Iron & Steel*, **27** [26] 61 (1954); N. L. Bowen and J. W. Greig, *J. Am. Ceram. Soc.*, **7** [4] 242 (1924).

Al₂O₃–SiO₂ (concl.)

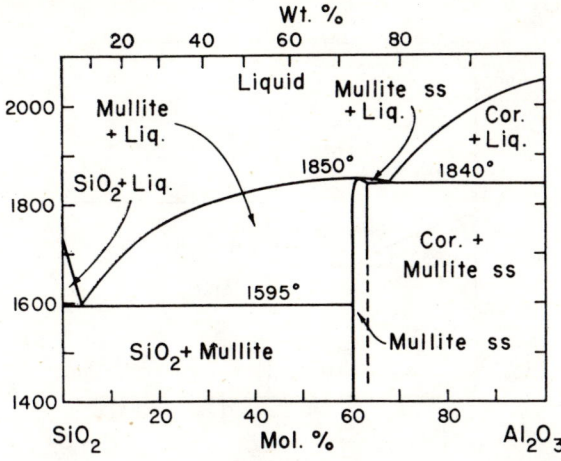

FIG. 314.—System Al₂O₃–SiO₂; revised. Cor = corundum.

Shigeo Aramaki and Rustum Roy, *J. Am. Ceram. Soc.*, **42** [12] 644 (1959); *ibid.*, **45** [5] 239 (1962). See also N. A. Toropov and F. Ya. Galakhov, *Izvest. Akad. Nauk S.S.S.R. Otdel Khim. Nauk*, **9** (1958).

Al₂O₃–SnO₂

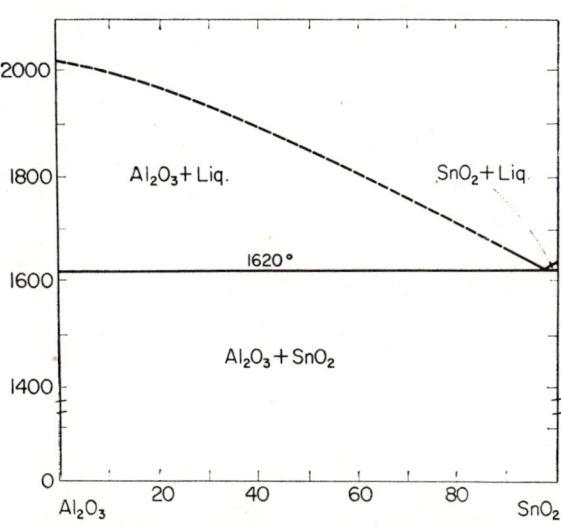

FIG. 315.—System Al₂O₃–SnO₂.

V. J. Barczak and R. H. Insley, *J. Am. Ceram. Soc.*, **45** [3] 144 (1962).

Al₂O₃–TiO₂

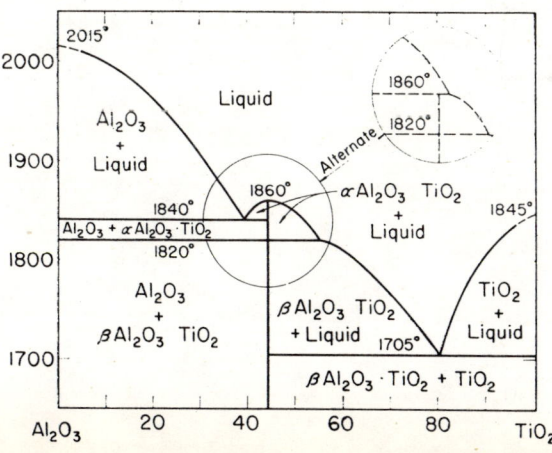

Al₂O₃–UO₂

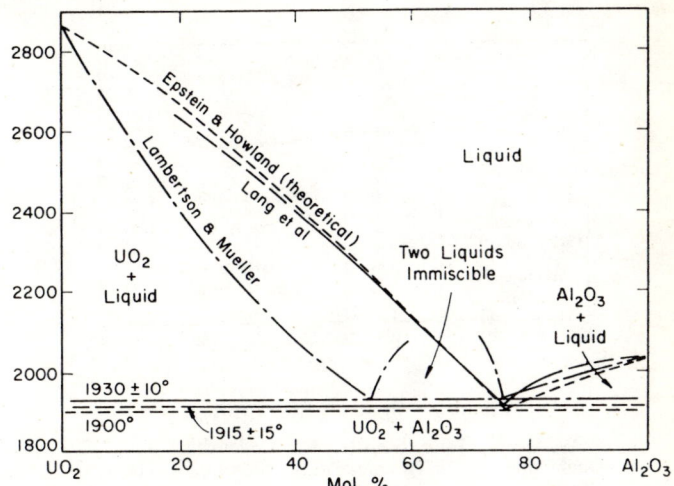

FIG. 317.—System Al₂O₃–UO₂.

S. M. Lang, F. P. Knudsen, C. L. Fillmore, and R. S. Roth, *Natl. Bur. Standards Circ.*, No. 568, p. 14 (Feb. 20, 1956).

Al₂O₃–P₂O₅

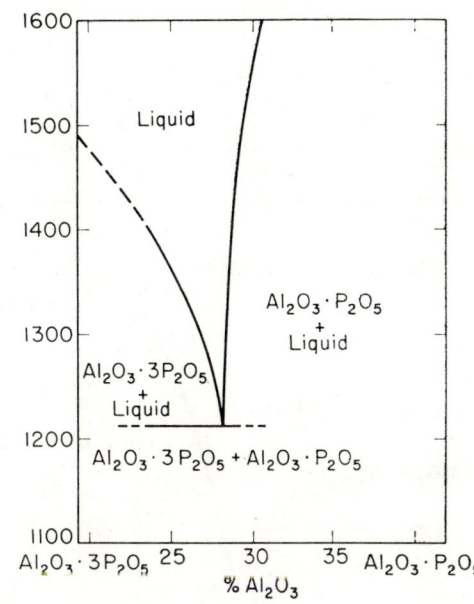

FIG. 318.—System Al₂O₃·P₂O₅–Al₂O₃·3P₂O₅.

P. E. Stone, E. P. Egan, Jr., and J. R. Lehr, *J. Am. Ceram. Soc.*, **39** [3] 92 (1956).

← FIG. 316.—System Al₂O₃–TiO₂; revised.

S. M. Lang, C. L. Fillmore, and L. H. Maxwell, *J. Research Natl. Bur. Standards*, **48** [4] 301 (1952); RP 2316.

Al_2O_3–Ta_2O_5

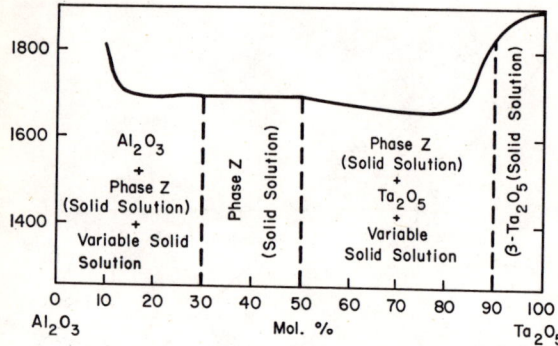

FIG. 319.—System Al_2O_3–Ta_2O_5.

B. W. King, John Schultz, E. A. Durbin, and W. H. Duckworth, U. S. Atomic Energy Comm., BMI-1106, 13 (1956).

Al_2O_3–V_2O_5

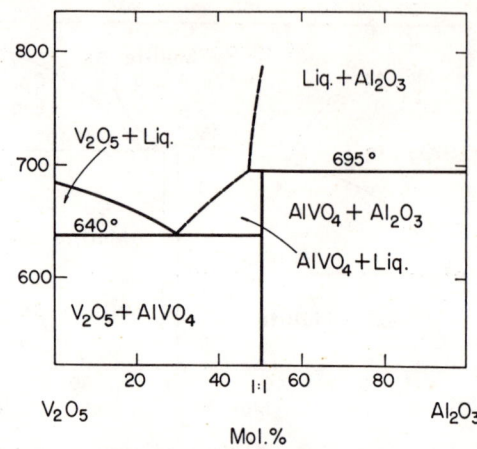

FIG. 320.—System Al_2O_3–V_2O_5.

Aurelio Burdese, *Ann. chim.* (Rome), **47** [7, 8] 804 (1957).

B_2O_3–La_2O_3

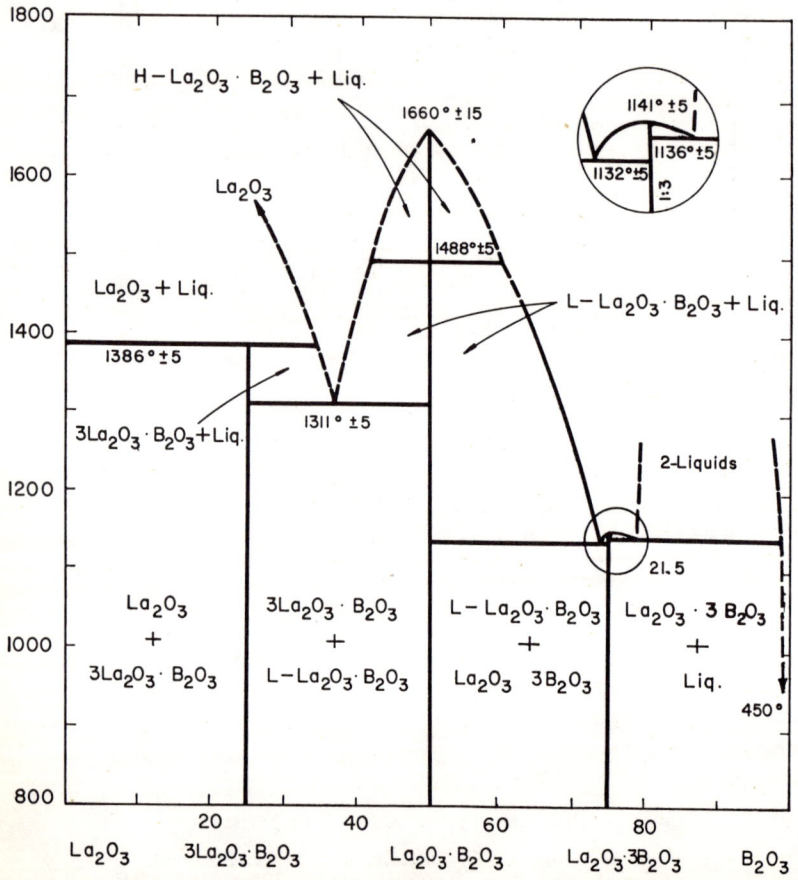

FIG. 321.—System La_2O_3–B_2O_3.

E. M. Levin, C. R. Robbins, and J. L. Waring, *J. Am. Ceram. Soc.*, **44** [2] 89 (1961).

B_2O_3–ThO_2

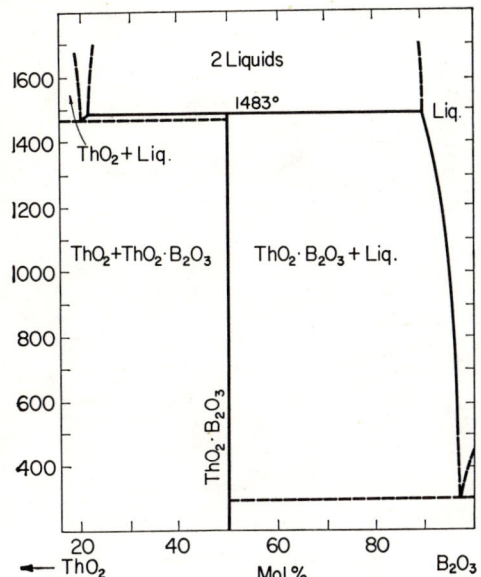

FIG. 322.—System B_2O_3–ThO_2; tentative.

D. E. Rase, *Aeronautical Res. Lab.* Contract No. AF33(616)-6545, p. 7 (1960). See also, D. E. Rase and G. Lane, *J. Am. Ceram. Soc.*, **47** [1] 49 (1964).

Bi_2O_3–Nb_2O_5

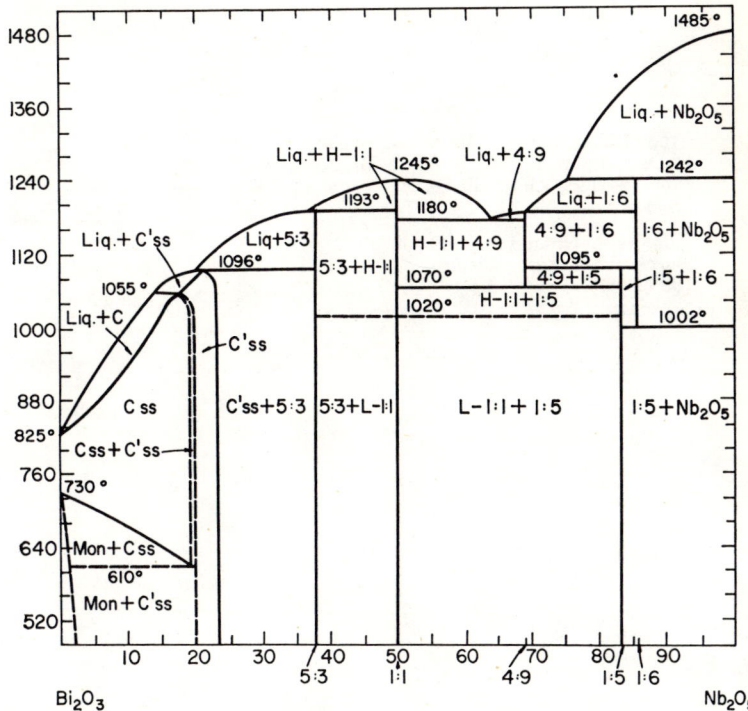

FIG. 324.—System Bi_2O_3–Nb_2O_5. Compounds designated in mole ratios of components. C = cubic Bi_2O_3, Mon = monoclinic Bi_2O_3.

R. S. Roth and J. Waring, *J. Research Natl. Bur. Standards*, **66A** [6] 461 (1962).

Bi_2O_3–B_2O_3

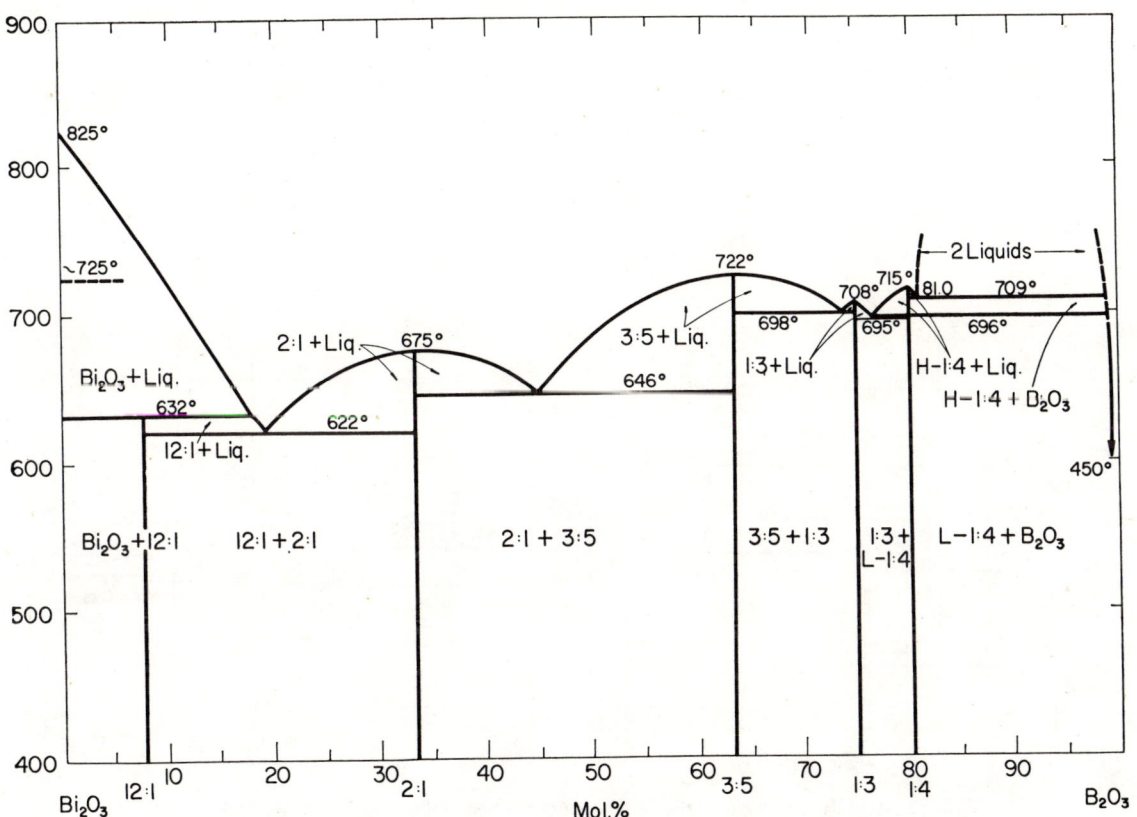

FIG. 323.—System Bi_2O_3–B_2O_3. Compounds designated as mole ratios of components.

E. M. Levin and Clyde McDaniel, *J. Am. Ceram. Soc.*, **45** [8] 356 (1962).

Bi$_2$O$_3$–R$_2$O

Fig. 325.—System Bi$_2$O$_3$–R$_2$O near Bi$_2$O$_3$ component. C = cubic Bi$_2$O$_3$, Mon = monoclinic Bi$_2$O$_3$, ? = unknown.

E. M. Levin and R. S. Roth, *J. Research Natl. Bur. Standards*, **68A** [2] 198 (1964).

Bi$_2$O$_3$–RO

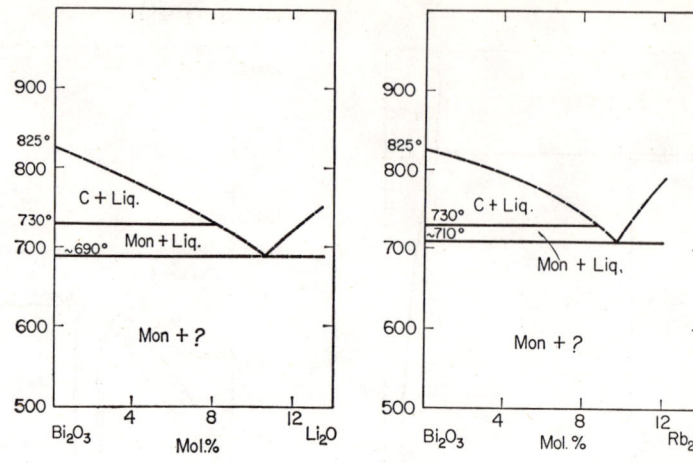

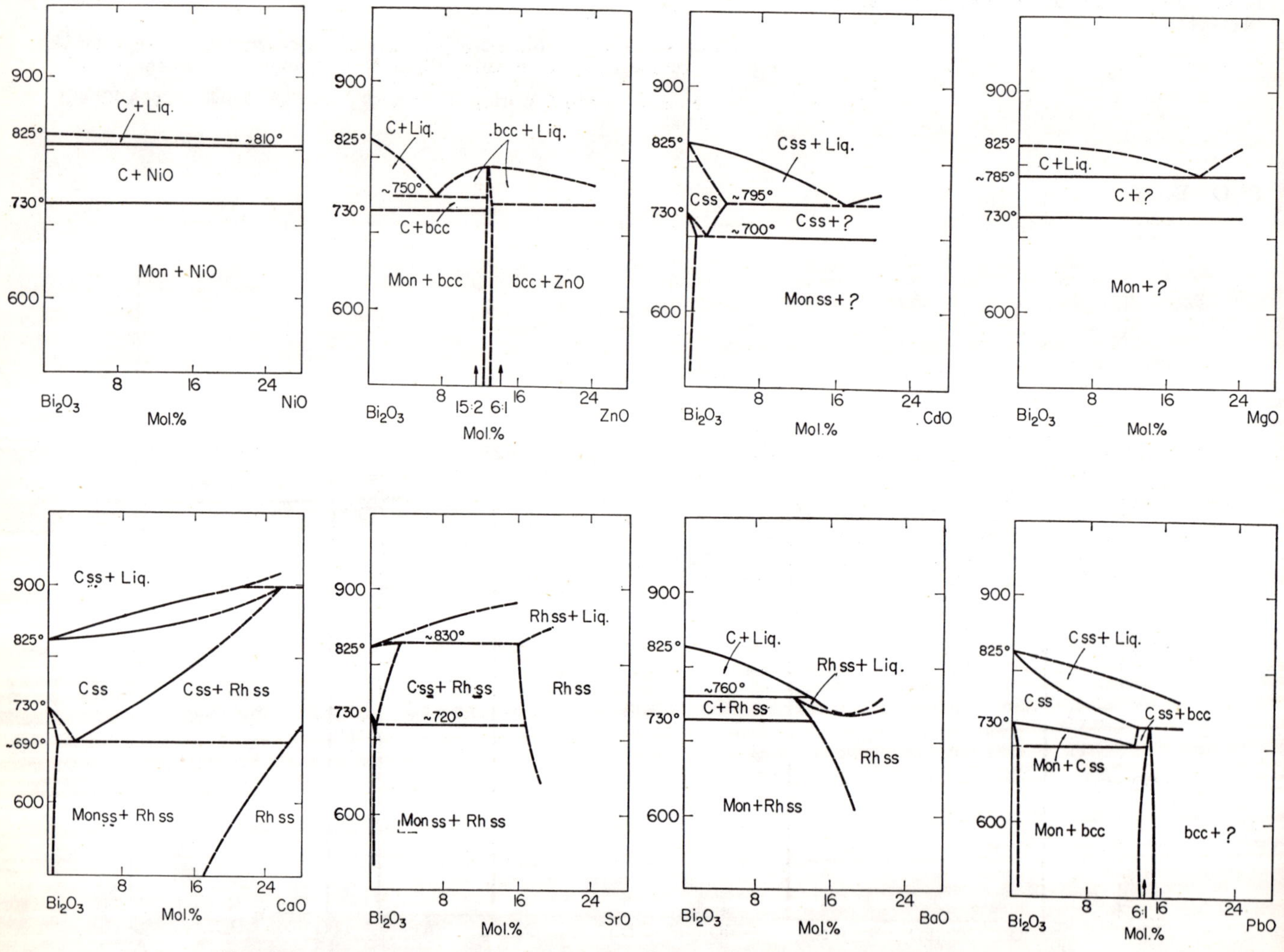

Fig. 326.—System Bi$_2$O$_3$–RO near Bi$_2$O$_3$ component. bcc = body centered cubic phase, C = cubic Bi$_2$O$_3$, Mon = monoclinic Bi$_2$O$_3$, Rh = phase of rhombohedral symmetry, ss = solid solution, ? = unknown.

E. M. Levin and R. S. Roth, *J. Research Natl. Bur. Standards*, **68A** [2] 199 (1964).

Bi$_2$O$_3$–R$_2$O$_3$

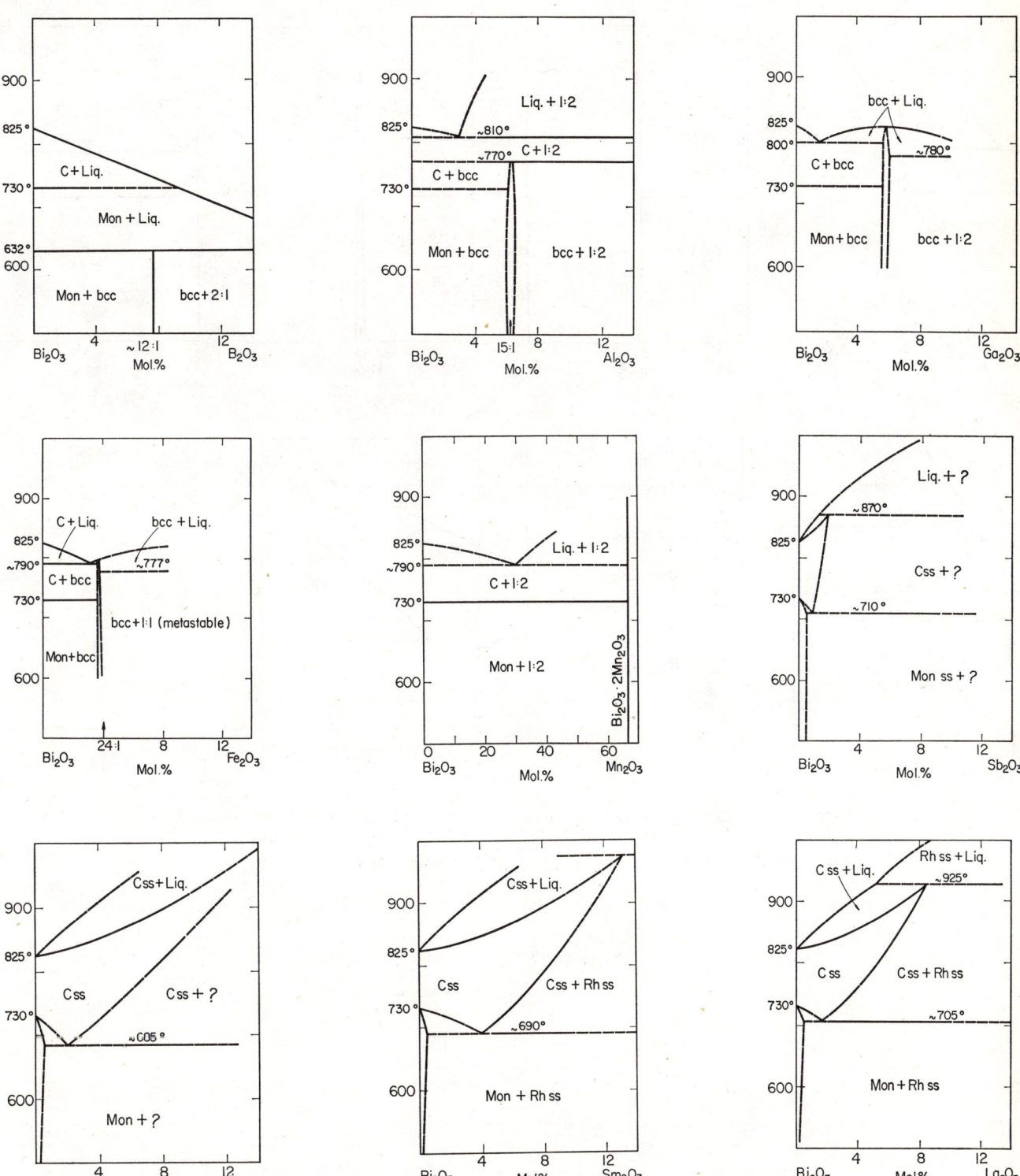

FIG. 327.—System Bi$_2$O$_3$–R$_2$O$_3$ near Bi$_2$O$_3$ component. bcc = body centered cubic phase, C = cubic Bi$_2$O$_3$, Mon = monoclinic Bi$_2$O$_3$, Rh = phase of rhombohedral symmetry, ss = solid solution, ? = unknown.

E. M. Levin and R. S. Roth, *J. Research Natl. Bur. Standards*, **68A** [2] 200 (1964).

Bi$_2$O$_3$–RO$_2$

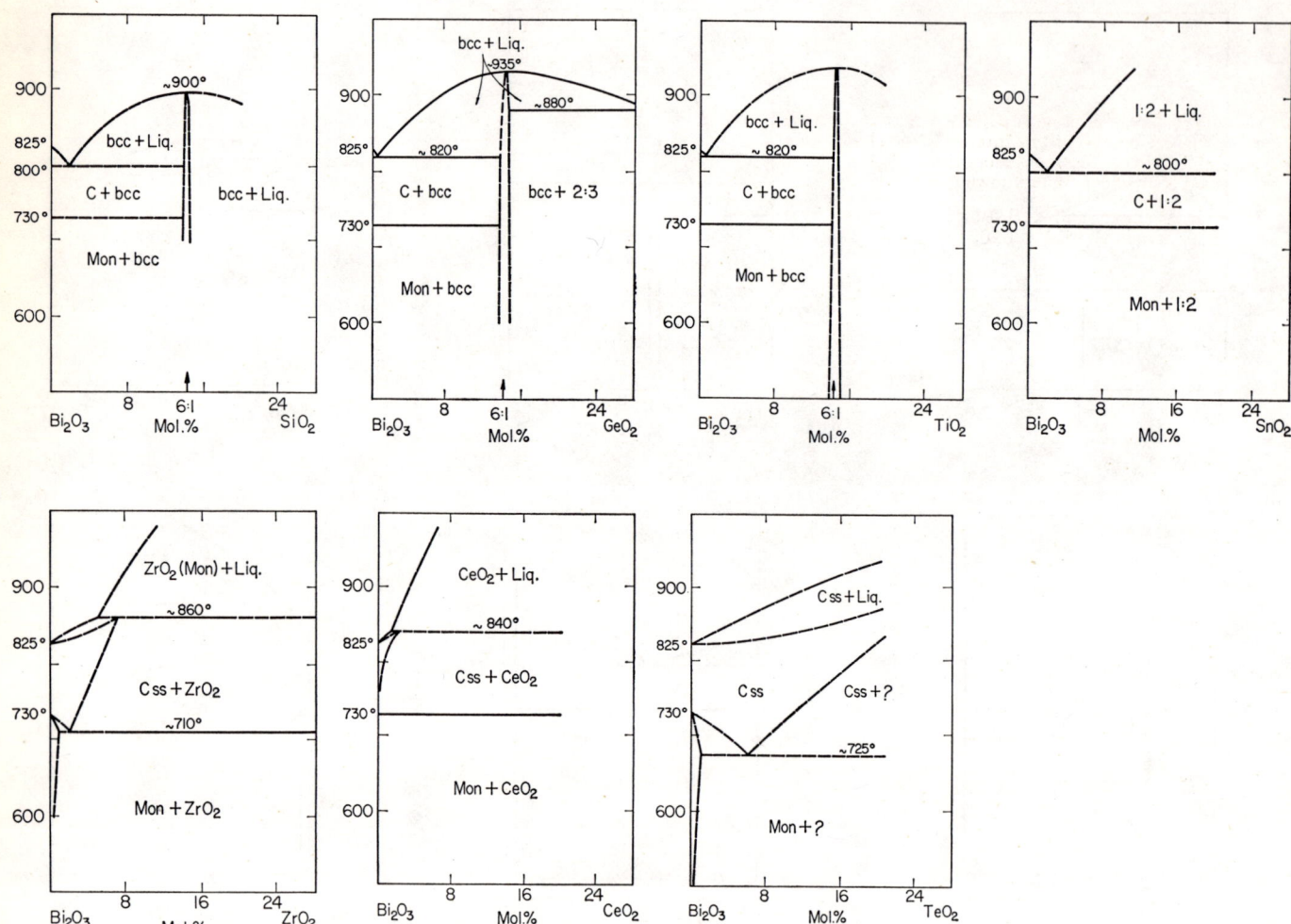

FIG. 328.—System Bi$_2$O$_3$–RO$_2$ near Bi$_2$O$_3$ component. bcc = body centered cubic phase, C = cubic Bi$_2$O$_3$, Mon = monoclinic Bi$_2$O$_3$, ss = solid solution, ? = unknown.

E. M. Levin and R. S. Roth, *J. Research Natl. Bur. Standards*, **68A** [2] 201 (1964).

Bi_2O_3–R_2O_5

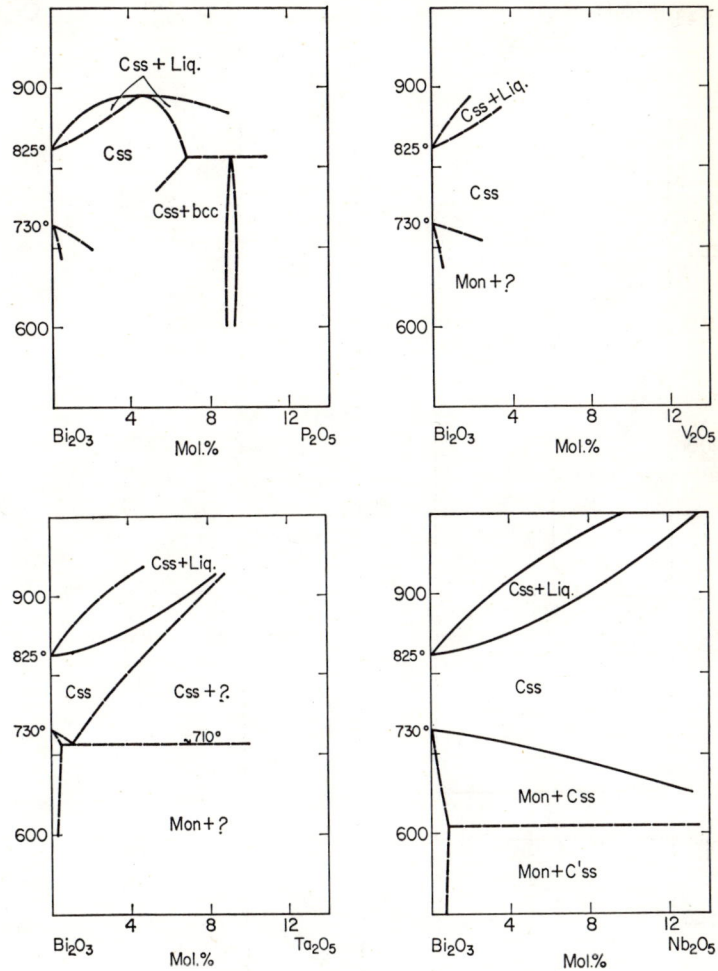

FIG. 329.—System Bi_2O_3–R_2O_5 near Bi_2O_3 component. bcc = body centered cubic phase, C = cubic Bi_2O_3, Mon = monoclinic Bi_2O_3, ss = solid solution, ? = unknown.

E. M. Levin and R. S. Roth, *J. Research Natl. Bur. Standards*, **68A** [2] 202 (1964).

Bi_2O_3–RO_3

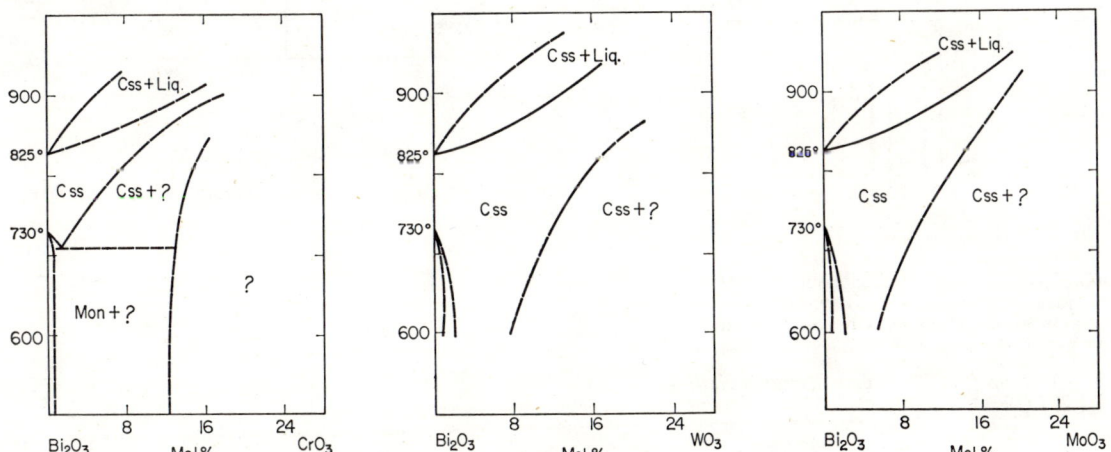

FIG. 330.—System Bi_2O_3–RO_3 near Bi_2O_3 component. Css = cubic Bi_2O_3 solid solution, Mon = monoclinic Bi_2O_3, ? = unknown.

E. M. Levin and R. S. Roth, *J. Research Natl. Bur. Standards*, **68A** [2] 202 (1964).

Cr_2O_3–R_2O_3

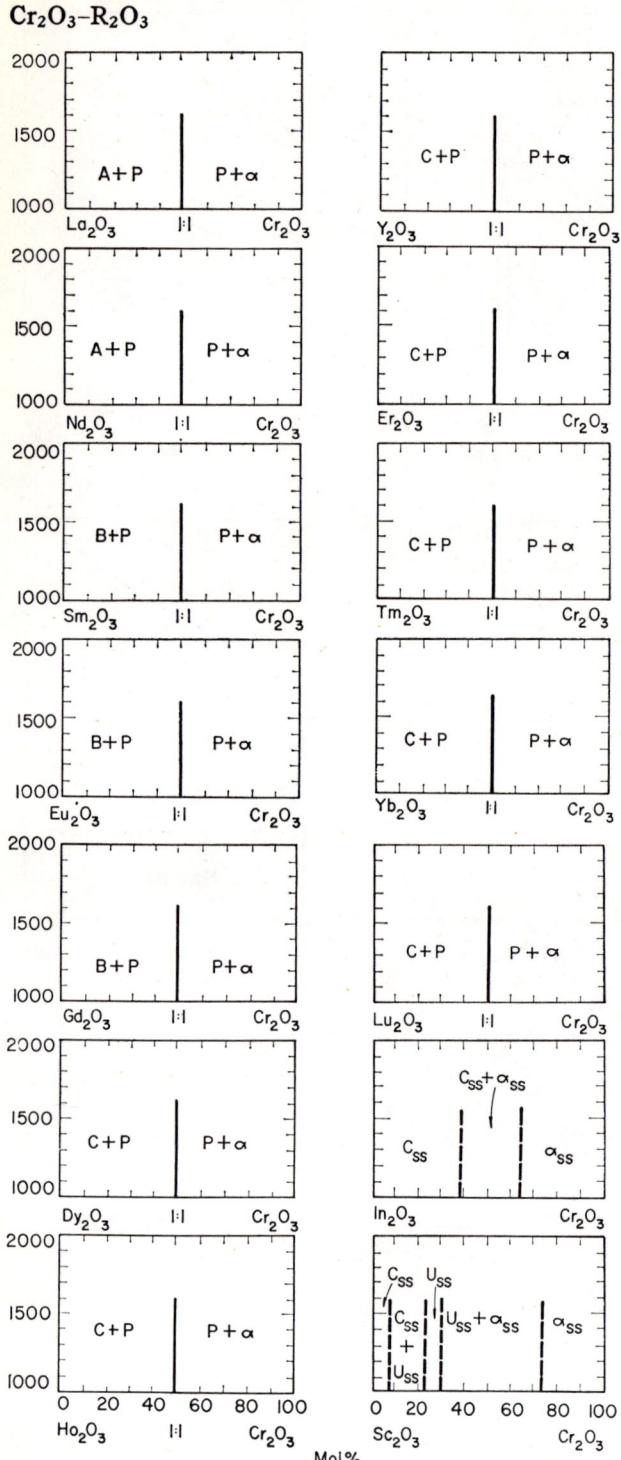

FIG. 331.—System Cr_2O_3–R_2O_3; predicted subsolidus. Structure types: A, A-type rare earth oxide; B, B-type rare earth oxide; C, C-type rare earth oxide; P, perovskite; α, corundum; U, unknown, similar to kappa alumina.

S. J. Schneider, R. S. Roth, and J. L. Waring, *J. Research Natl. Bur. Standards*, **65A** [4] 367 (1961).

Cr_2O_3–SiO_2

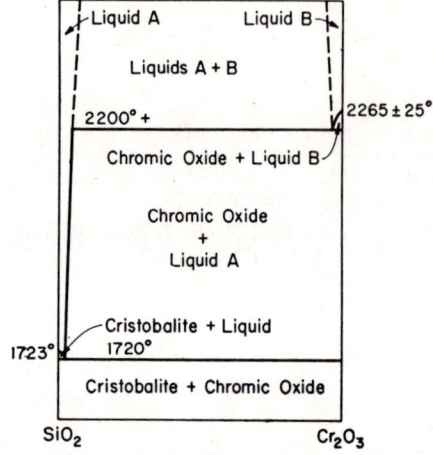

FIG. 332.—System Cr_2O_3–SiO_2; probable.

From data of E. N. Bunting, *J. Research, Natl. Bur. Standards* **5** [2] 325–27 (1930), RP 203; and *ibid.*, **6** [6] 947–49 (1931), RP 317. M. L. Keith, *J. Am. Ceram. Soc.*, **37** [10] 490 (1954).

Cr_2O_3–V_2O_5

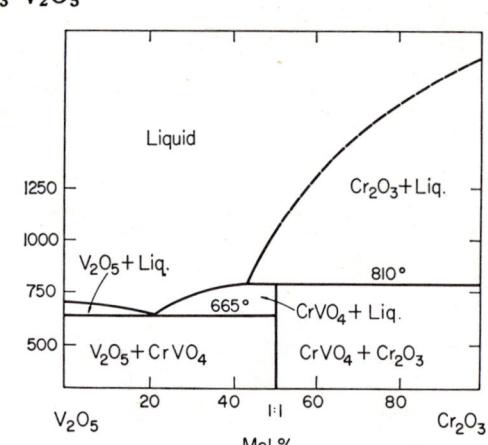

FIG. 333.—System Cr_2O_3–V_2O_5.

Aurelio Burdese, *Ann. chim. (Rome)*, **47** [7, 8] 801 (1957).

Eu$_2$O$_3$–In$_2$O$_3$

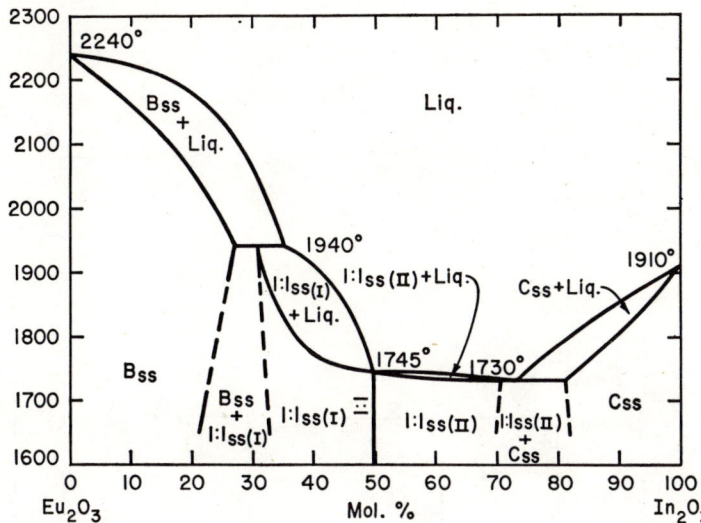

FIG. 334.—System Eu$_2$O$_3$–In$_2$O$_3$.

B = Eu$_2$O$_3$ with B-type rare earth oxide structure;
C = In$_2$O$_3$ with C-type rare earth oxide structure;
1:1 = Eu$_2$O$_3$·In$_2$O$_3$ compound; ss = solid solution.

S. J. Schneider *J. Research Natl. Bur. Standards*, **65A** [5] 431 (1961).

Eu$_2$O$_3$–Ln$_2$O$_3$

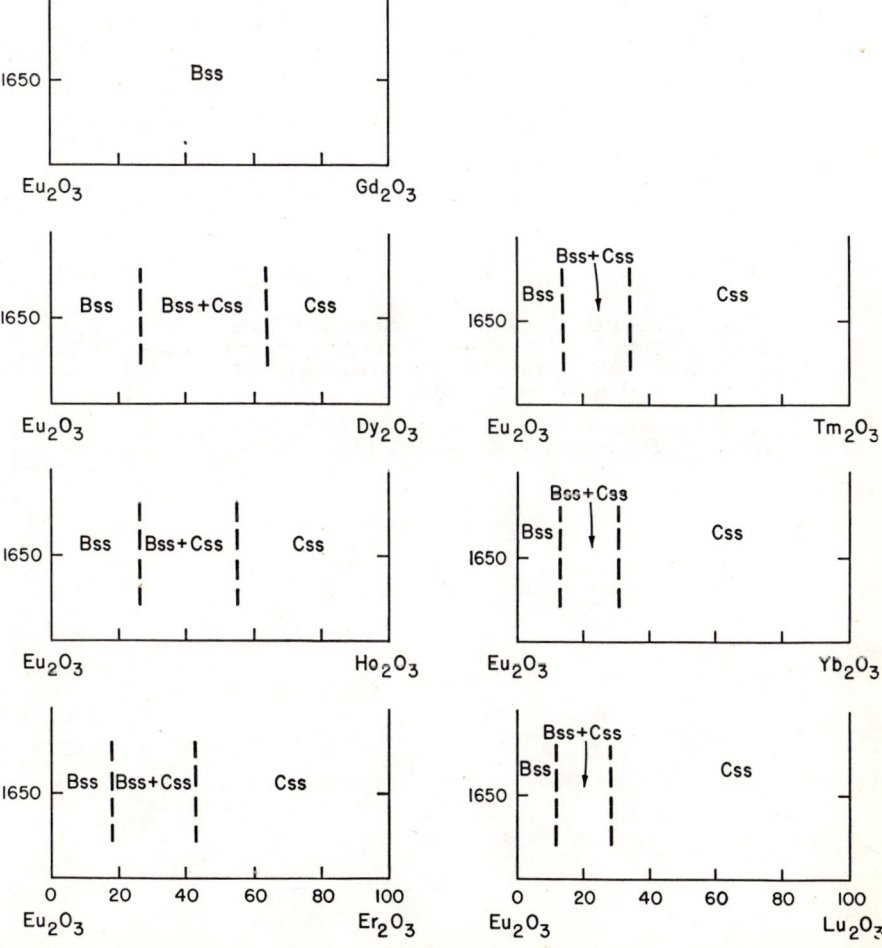

FIG. 335.—System Eu$_2$O$_3$–Ln$_2$O$_3$; predicted subsolidus. B, C refer to rare earth oxide structure types.

S. J. Schneider and R. S. Roth, *J. Research Natl. Bur. Standards*, **64A** [4] 328 (1960).

Fe_2O_3–Gd_2O_3

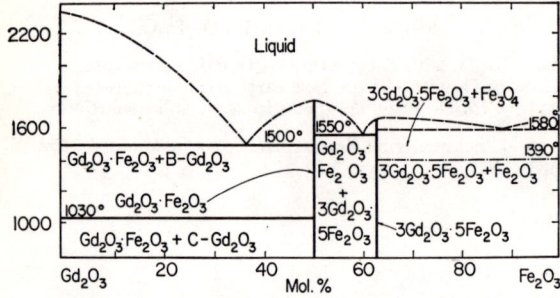

Fig. 336.—System Fe_2O_3–Gd_2O_3. System not binary above the solidus or in the iron-rich portion because of probable FeO content. Dash-dot line at 1390°C represents $Fe_2O_3 \rightarrow Fe_3O_4$ transition.

I. Warshaw and Rustum Roy, *J. Am. Ceram. Soc.*, **42** [9] 437 (1959).

Fe_2O_3–R_2O_3

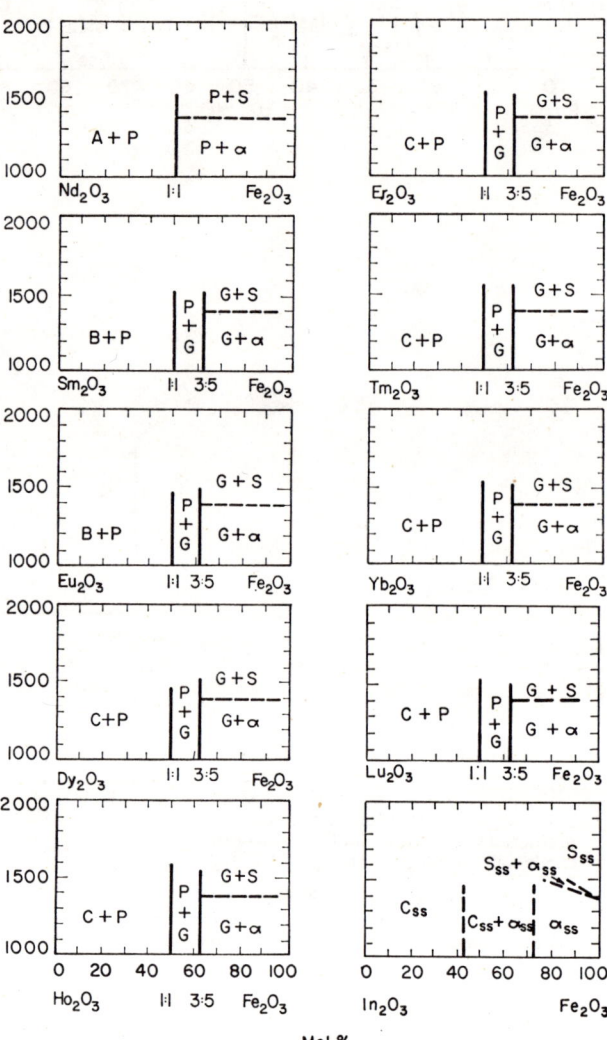

Fig. 337.—System Fe_2O_3–R_2O_3; predicted subsolidus. Structure types: A, A-type rare earth oxide; B, B-type rare earth oxide; C, C-type rare earth oxide; G, garnet; P, perovskite; S, spinel; α, corundum.

S. J. Schneider, R. S. Roth, and J. L. Waring, *J. Research Natl. Bur. Standards*, **65A** [4] 369 (1961).

Fe_2O_3–P_2O_5

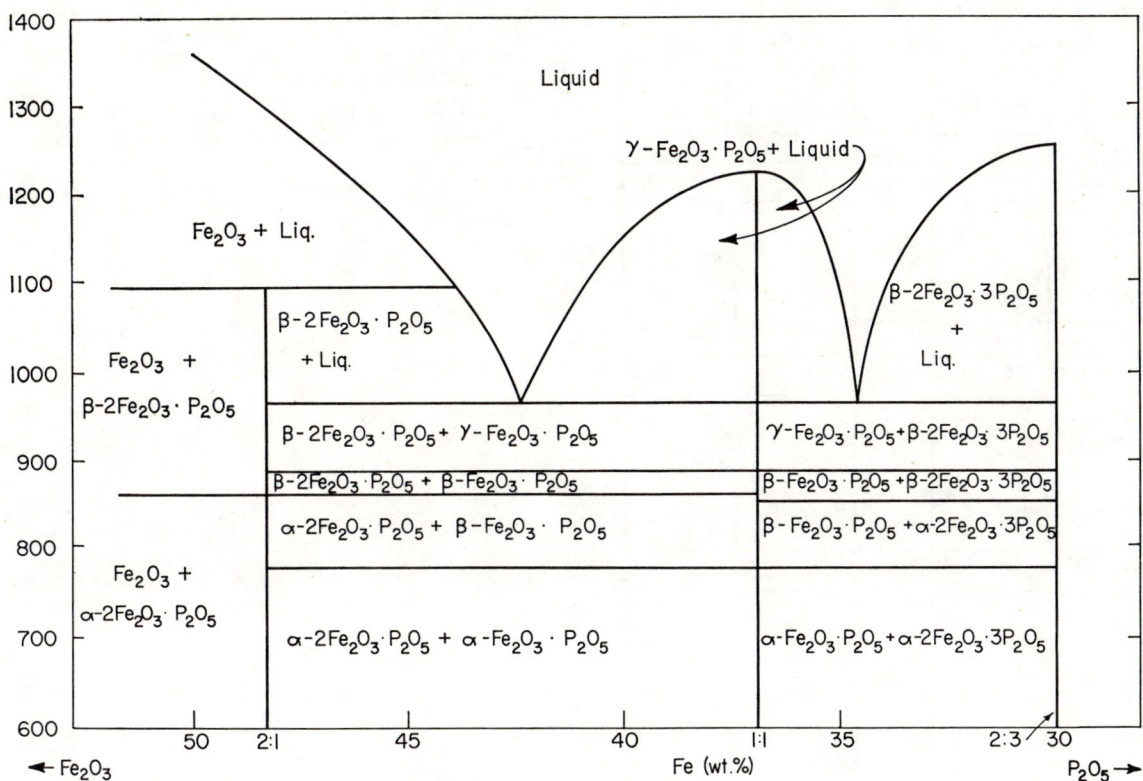

Fig. 338.—System $2Fe_2O_3 \cdot P_2O_5$–$2Fe_2O_3 \cdot 3P_2O_5$.

H. Wentrup, *Arch. Eisenhüettenw.*, **9**, 57 (1935–36).

Ga_2O_3–Sc_2O_3

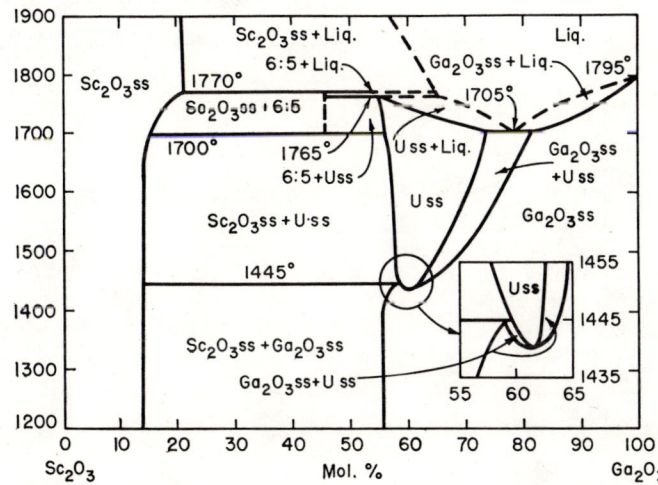

FIG. 339.—System Ga_2O_3–Sc_2O_3. Uss = solid solution phase of unknown structure type.

S. J. Schneider and J. L. Waring, *J. Research Natl. Bur. Standards*, **67A** [1] 21 (1963).

Ga$_2$O$_3$–R$_2$O$_3$

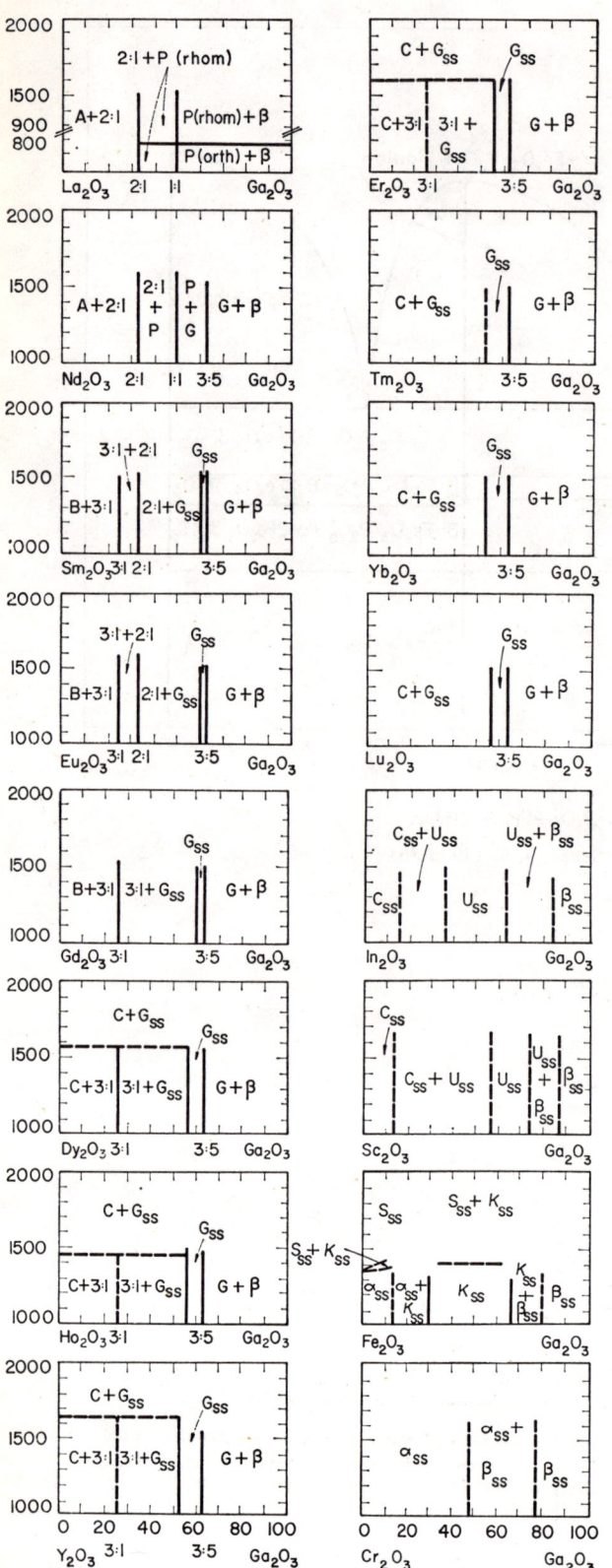

FIG. 340.—System Ga$_2$O$_3$–R$_2$O$_3$; predicted subsolidus. Structure types: A, A-type rare earth oxide; B, B-type rare earth oxide; C, C-type rare earth oxide; G, garnet; P, perovskite; α, corundum; β, beta gallia; K, kappa alumina; U, unknown, similar to kappa alumina.

S. J. Schneider, R. S. Roth, and J. L. Waring, *J. Research Natl. Bur. Standards*, **65A** [4] 365 (1961).

Ga$_2$O$_3$–SiO$_2$

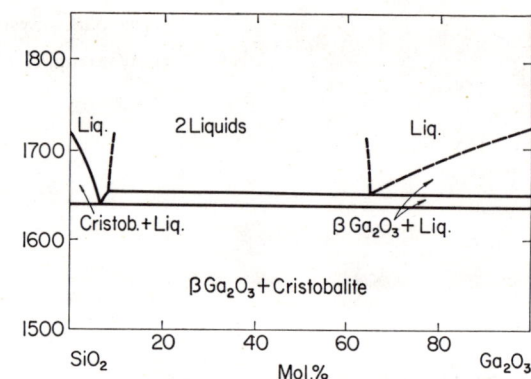

FIG. 341.—System Ga$_2$O$_3$–SiO$_2$.

F. P. Glasser, *J. Phys. Chem.*, **63** [12] 2086 (1959).

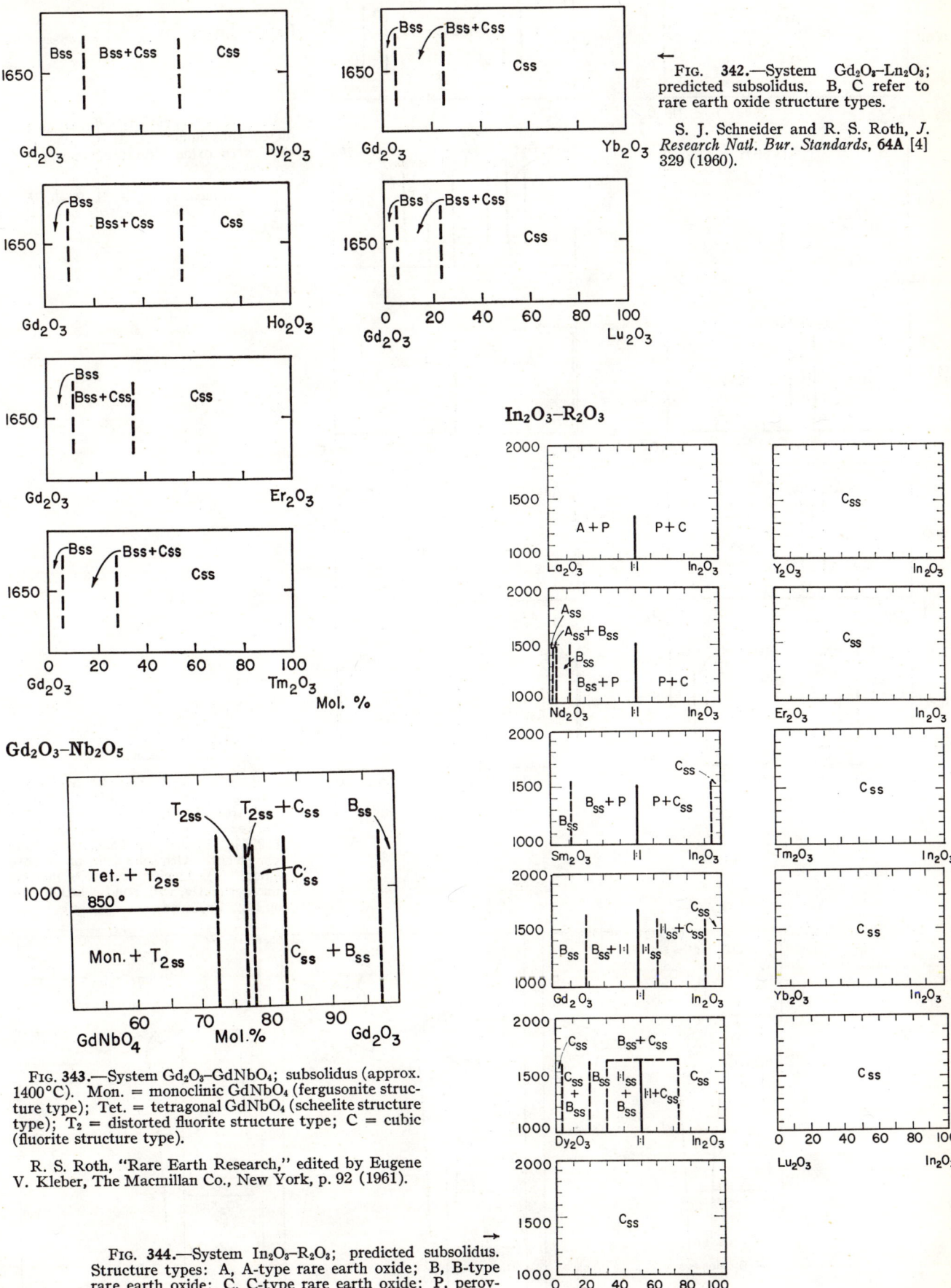

Fig. 342.—System Gd_2O_3–Ln_2O_3; predicted subsolidus. B, C refer to rare earth oxide structure types.

S. J. Schneider and R. S. Roth, J. Research Natl. Bur. Standards, **64A** [4] 329 (1960).

Fig. 343.—System Gd_2O_3–$GdNbO_4$; subsolidus (approx. 1400°C). Mon. = monoclinic $GdNbO_4$ (fergusonite structure type); Tet. = tetragonal $GdNbO_4$ (scheelite structure type); T_2 = distorted fluorite structure type; C = cubic (fluorite structure type).

R. S. Roth, "Rare Earth Research," edited by Eugene V. Kleber, The Macmillan Co., New York, p. 92 (1961).

Fig. 344.—System In_2O_3–R_2O_3; predicted subsolidus. Structure types: A, A-type rare earth oxide; B, B-type rare earth oxide; C, C-type rare earth oxide; P, perovskite.

S. J. Schneider, R. S. Roth, and J. L. Waring, J. Research Natl. Bur. Standards, **65A** [4] 372 (1961).

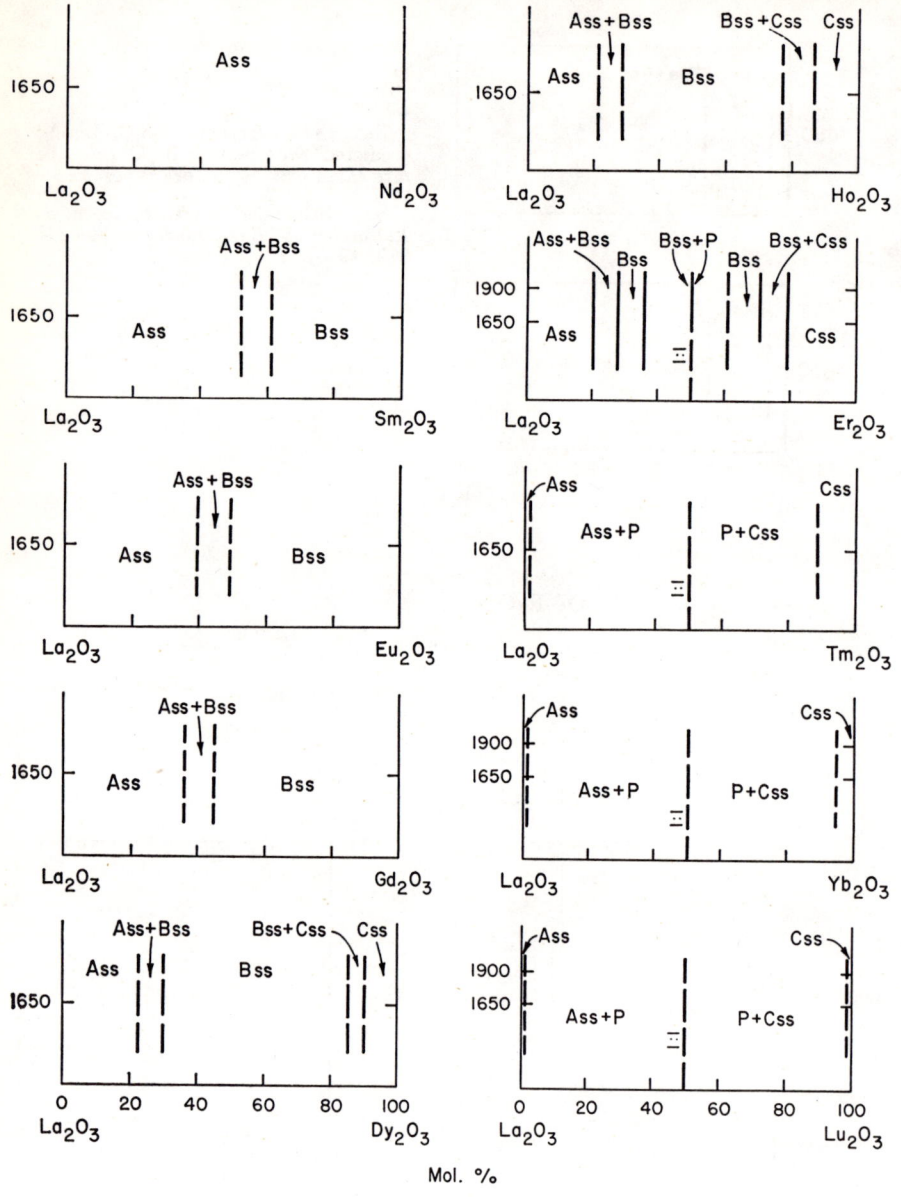

La₂O₃–Ln₂O₃

Fig. 345.—System La$_2$O$_3$–Ln$_2$O$_3$; predicted subsolidus. A, B, C refer to rare earth oxide structure types; P, perovskite.

S. J. Schneider and R. S. Roth, *J. Research Natl. Bur. Standards*, **64A** [4] 325 (1960).

La₂O₃–ZrO₂

Fig. 346.—System La$_2$O$_3$–ZrO$_2$; possible.

R. S. Roth, *J. Research Natl. Bur. Standards*, **56** [1] 23 (1956); RP2643. (a) After F. H. Brown, Jr., and P. Duwez, *J. Am. Ceram. Soc.*, **38** [3] 95 (1955); (b) showing La$_2$Zr$_2$O$_7$ melting congruently with solid solution on both sides; (c) showing La$_2$Zr$_2$O$_7$ melting incongruently, with solid solution only on high La$_2$O$_3$ side.

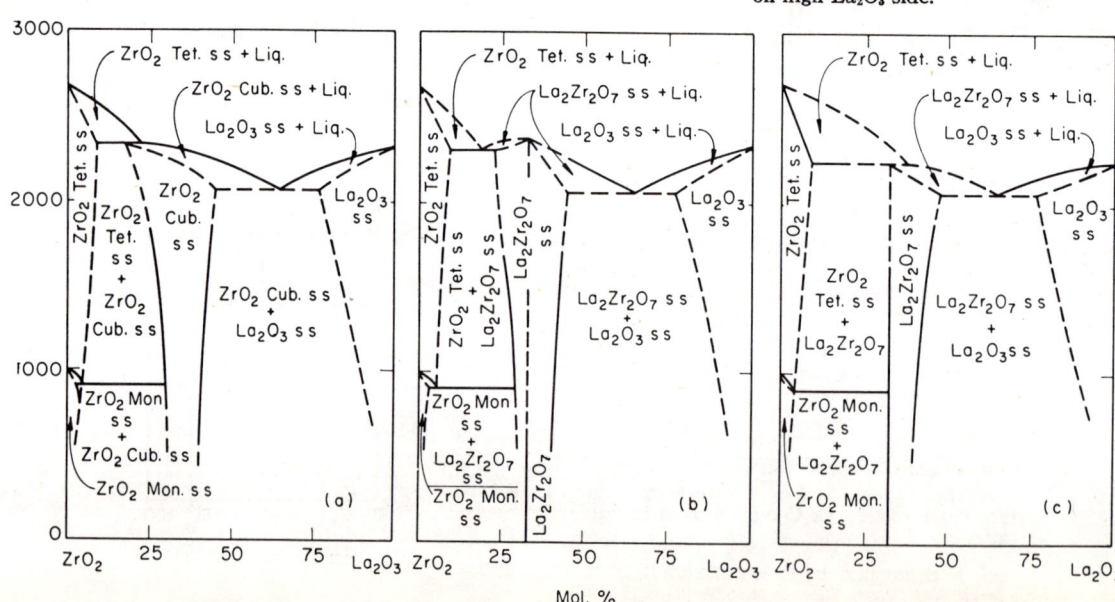

Ln_2O_3–Ln'_2O_3

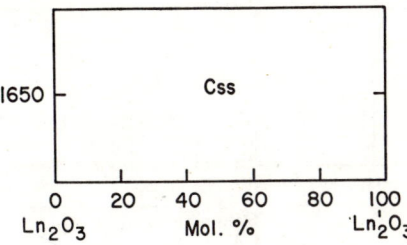

Fig. 347.—System Ln_2O_3–$Ln_2'O_3$; predicted subsolidus for small cations. Ln_2O_3 and $Ln_2'O_3$ refer to Dy_2O_3, Ho_2O_3, Er_2O_3, Tm_2O_3, Yb_2O_3, Lu_2O_3. Css, C-type rare earth oxide solid solutions.

S. J. Schneider and R. S. Roth, *J. Research Natl. Bur. Standards*, **64A** [4] 329 (1960).

Nd_2O_3–Ln_2O_3

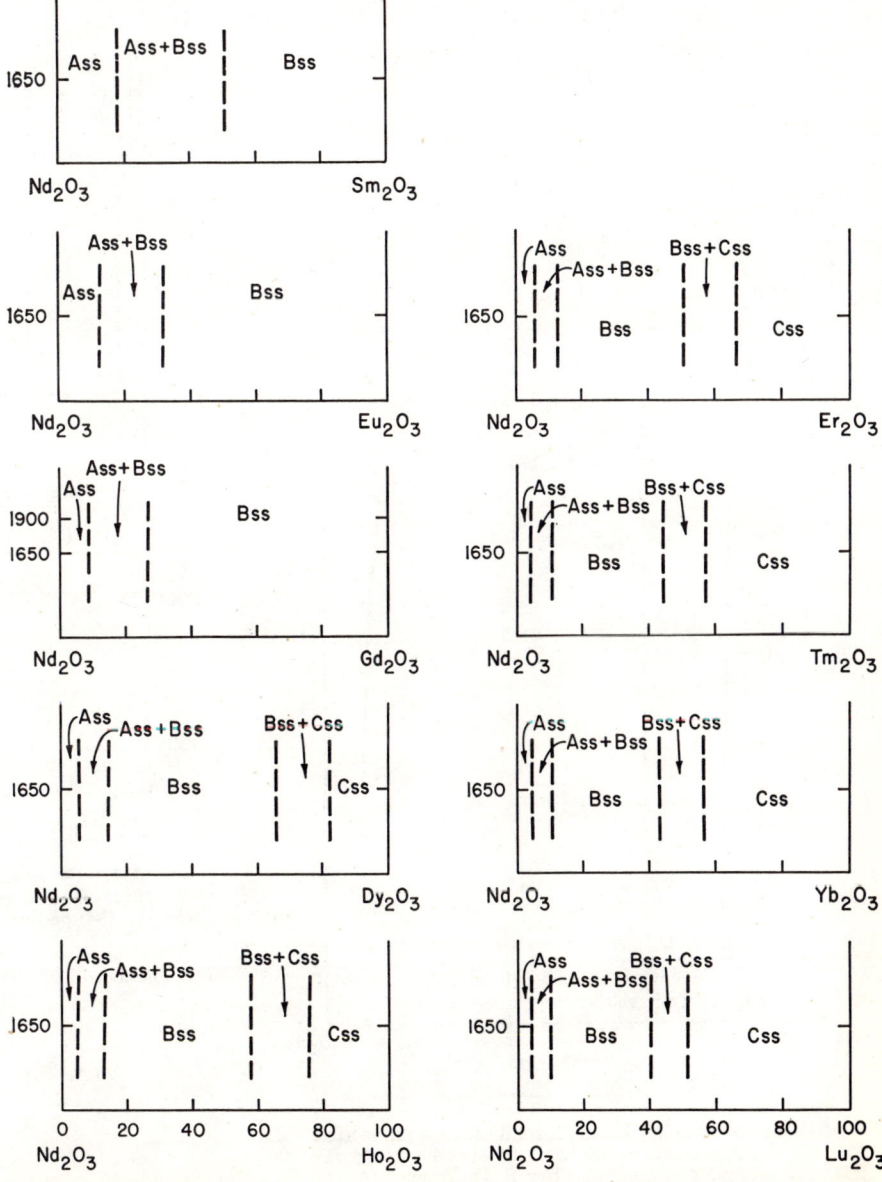

Fig. 348.—System Nd_2O_3–Ln_2O_3; predicted subsolidus. A, B, C refer to rare earth oxide structure types.

S. J. Schneider and R. S. Roth, *J. Research Natl. Bur. Standards*, **64A** [4] 326 (1960).

Nd_2O_3–UO_2

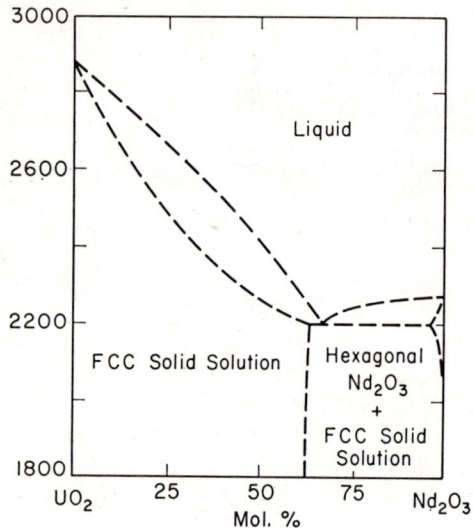

Fig. 349.—System Nd_2O_3–UO_2.

S. M. Lang, F. P. Knudsen, C. L. Fillmore, and R. S. Roth, *Natl. Bur. Standards Circ.*, No. **568**, p. 16 (Feb. 20, 1956).

After W. A. Lambertson and M. H. Mueller, U. S. AEC unclassified report ANL-5312 (Sept. 14, 1954).

Nd_2O_3–ZrO_2

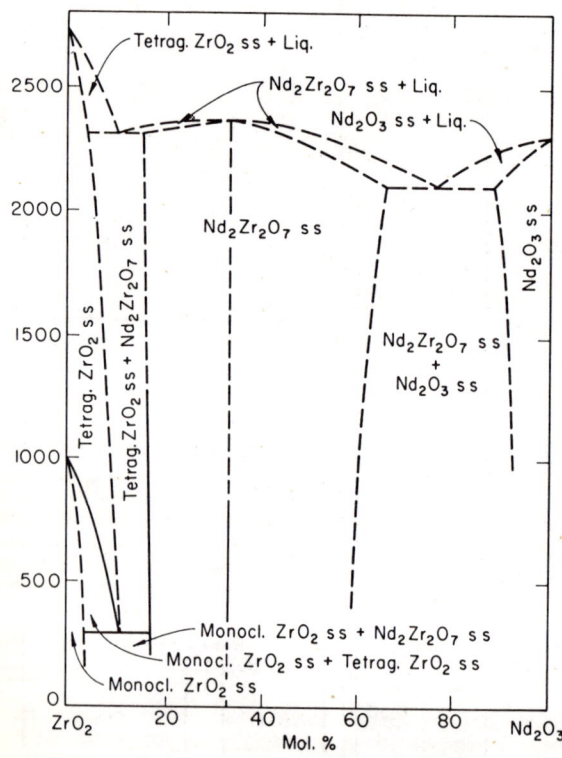

Fig. 350.—System Nd_2O_3–ZrO_2; possible.

Modification showing $Nd_2Zr_2O_7$ solid solution phase after R. S. Roth, *J. Res. Natl. Bur. Std.*, **56** [1] 24 (1956); RP 2643. Remainder of diagram after F. H. Brown, Jr. and Pol Duwez, *J. Am. Ceram. Soc.*, **38** [3] 95 (1955).

Sc_2O_3–R_2O_3

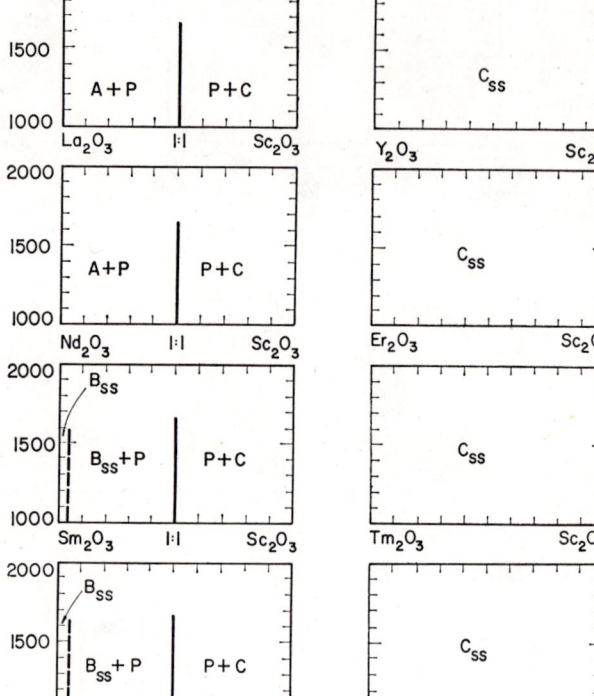

Fig. 351.—System Sc_2O_3–R_2O_3; predicted subsolidus. Structure types: A, A-type rare earth oxide; B, B-type rare earth oxide; C, C-type rare earth oxide; P, perovskite.

S. J. Schneider, R. S. Roth, and J. L. Waring, *J. Research Natl. Bur. Standards*, **65A** [4] 370 (1961).

Sm$_2$O$_3$–Ln$_2$O$_3$

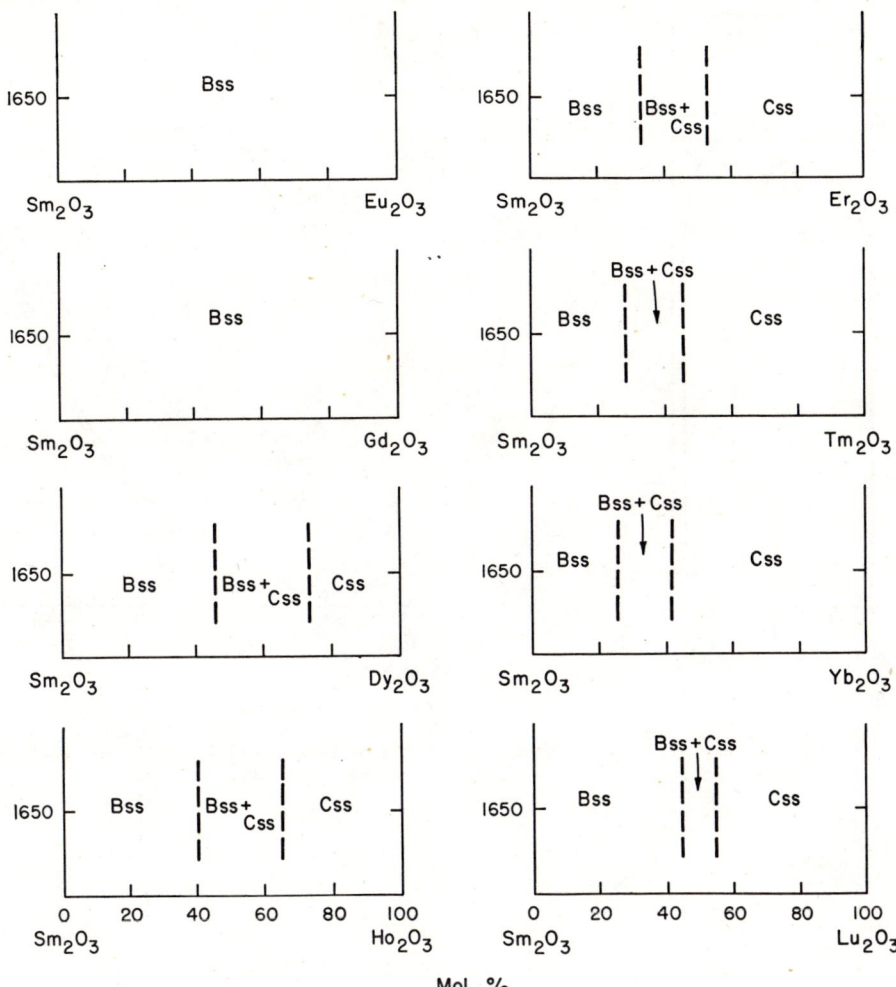

FIG. 352.—System Sm$_2$O$_3$–Ln$_2$O$_3$; predicted subsolidus. B, C refer to rare earth oxide structure types.

S. J. Schneider and R. S. Roth, *J. Research Natl. Bur. Standards*, **64A** [4] 327 (1960).

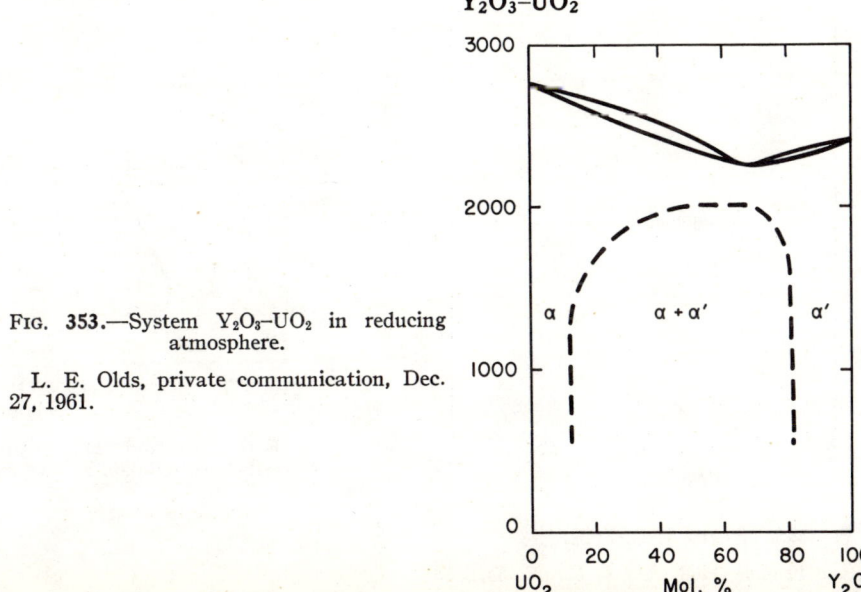

FIG. 353.—System Y$_2$O$_3$–UO$_2$ in reducing atmosphere.

L. E. Olds, private communication, Dec. 27, 1961.

Y_2O_3–ZrO_2

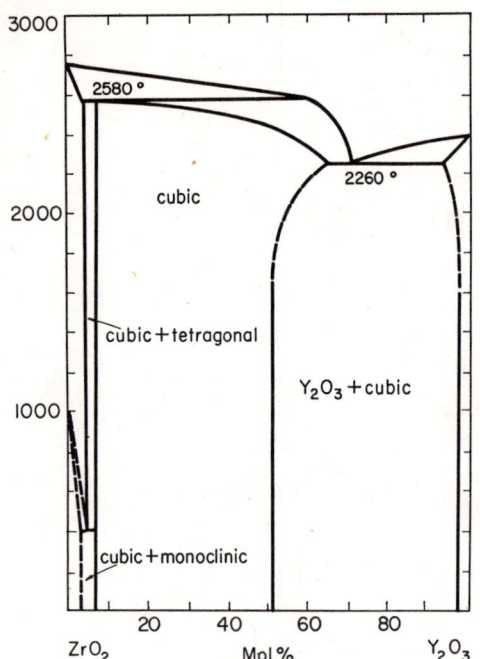

Fig. 354.—System Y_2O_3–ZrO_2.

H. E. Otto, private communication Dec. 27, 1961. See also, P. S. Duwez, F. H. Brown, Jr., and F. Odell, *J. Electrochem. Soc.*, **98**, 360 (1951).

CeO_2–R_2O_3, R_3O_4

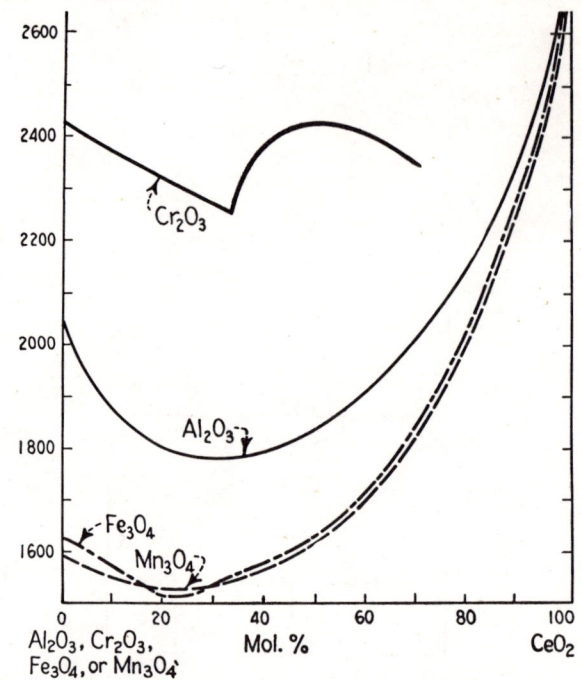

Fig. 356.—Liquidus curves of systems CeO_2–Al_2O_3, CeO_2–Cr_2O_3, CeO_2–Fe_3O_4, CeO_2–Mn_3O_4.

H. von Wartenberg and K. Eckhardt, Part VIII, *Z anorg. u. allgem. Chem.*, **232**, 184 (1937)

CeO_2–ZrO_2

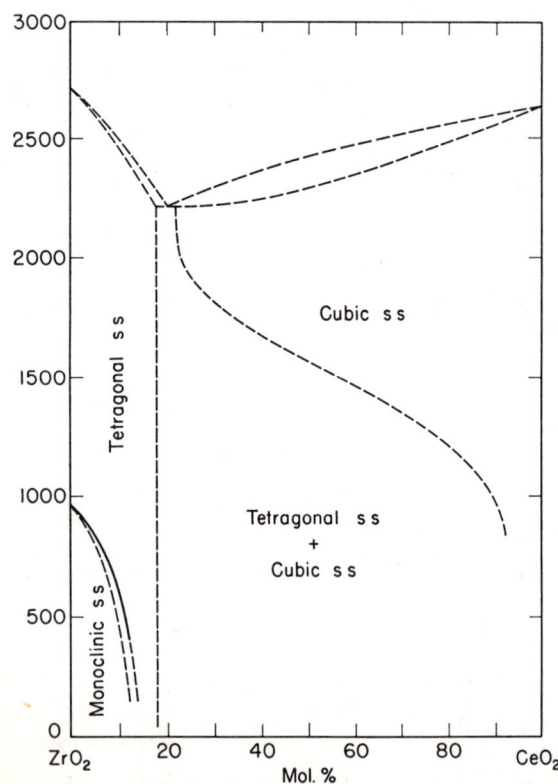

Fig. 355.—System CeO_2–ZrO_2.

Pol Duwez and Francis Odell, *J. Am. Ceram. Soc.*, **33** [9] 280 (1950).

GeO_2–SiO_2

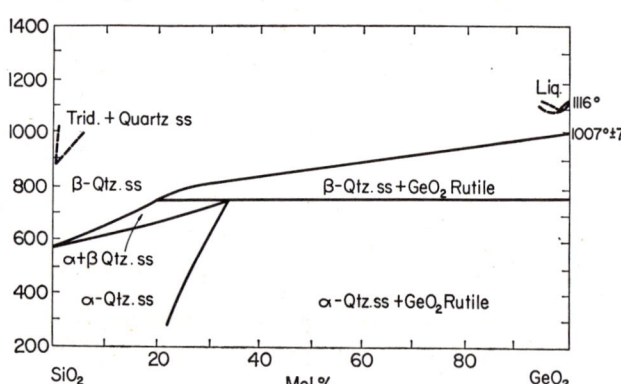

Fig. 357.—System GeO_2–SiO_2. Qtz. = quartz; Trid. = tridymite.

E. C. Shafer and Rustum Roy, *U. S. Army Signal Corps* Contract DA 36-039, SC 63099 (1956).

GeO₂–TiO₂

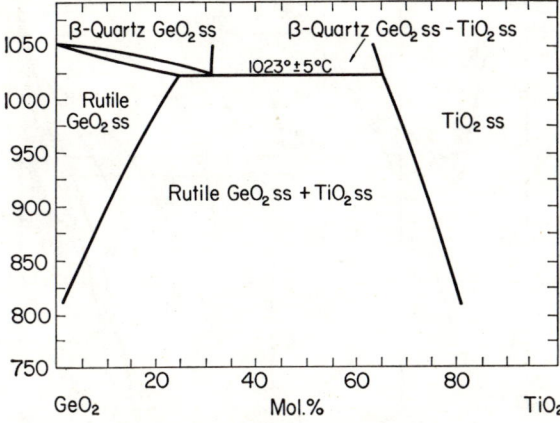

Fig. 358.—System GeO_2–TiO_2.

J. F. Sarver, *Am. J. Sci.*, **259**, 716 (1961).

SiO₂–ThO₂

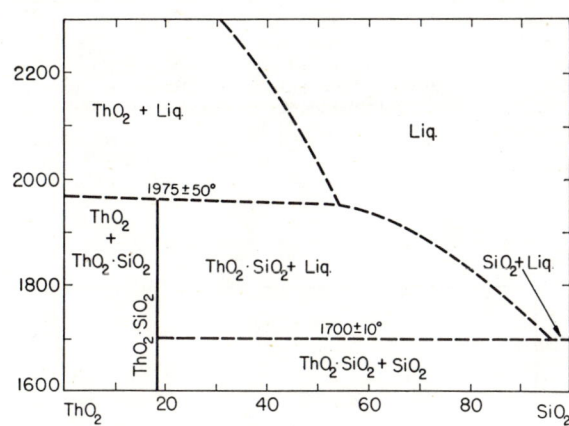

Fig. 359.—System SiO_2–ThO_2; preliminary.

L. A. Harris, *J. Am. Ceram. Soc.*, **42** [3] 74 (1959).

SiO₂–UO₂

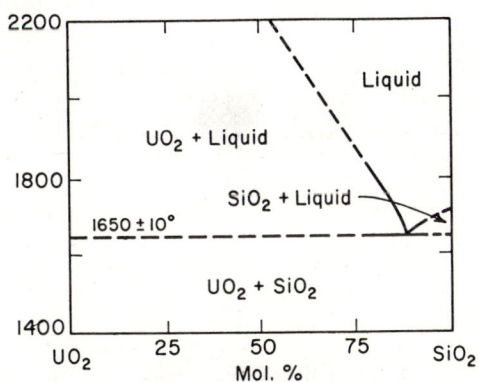

Fig. 360.—System SiO_2–UO_2.

S. M. Lang, F. P. Knudsen, C. L. Fillmore, and R. S. Roth, *Natl. Bur. Standards Circ.*, No. **568**, p. 17 (Feb. 20, 1956).

SiO₂–ZrO₂

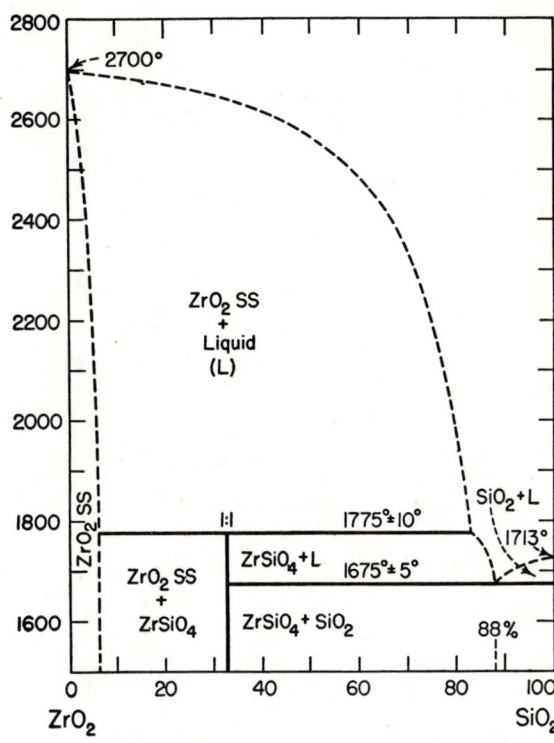

Fig. 361.—System SiO_2–ZrO_2; revised and corrected diagram.

Revised by R. F. Geller and S. M. Lang, National Bureau of Standards, 1959. Decomposition of $ZrSiO_4$ is 1720 ± 20°C according to A. Cocco and N. Schromek, *Ceramica*, (Milan), **12** [8] 45 (1957).

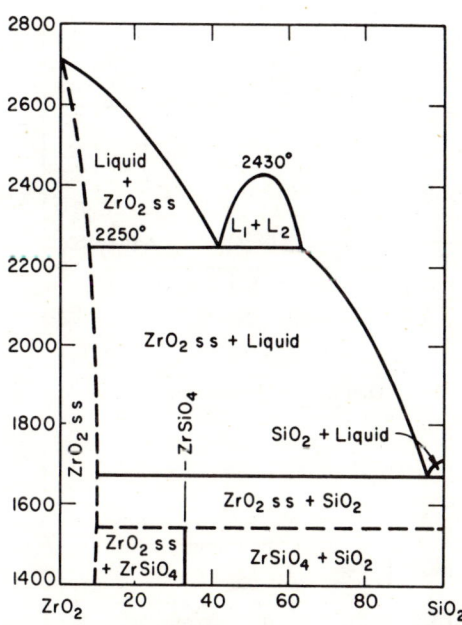

Fig. 362.—System SiO_2–ZrO_2. ss = solid solution.

N. A. Toropov and F. Ya. Galakhov, *Izvest. Akad. Nauk S.S.S.R, Otdel Khim. Nauk*, 1956, 160. See also, C. E. Curtis and H. G. Sowman, *J. Am. Ceram. Soc.*, **36** [6] 198 (1953).

SiO$_2$–Nb$_2$O$_5$

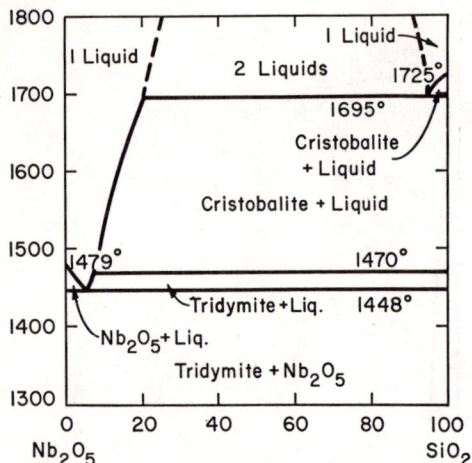

Fig. 363.—System SiO$_2$–Nb$_2$O$_5$.

Mohammed Ibrahim, and N. F. H. Bright, *J. Am. Ceram. Soc.*, **45** [5] 222 (1962).

SiO$_2$–P$_2$O$_5$

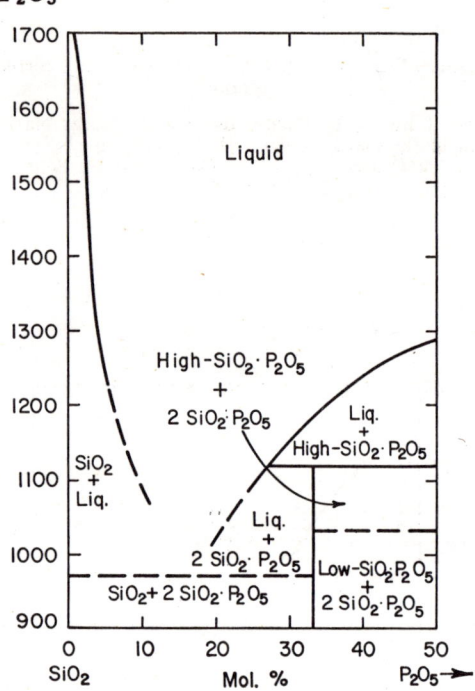

Fig. 364.—System SiO$_2$–SiO$_2$·P$_2$O$_5$.

T. Y. Tien and F. A. Hummel, *J. Am. Ceram Soc.*, **45** [9] 424 (1962).

SiO$_2$–R$_2$O, RO

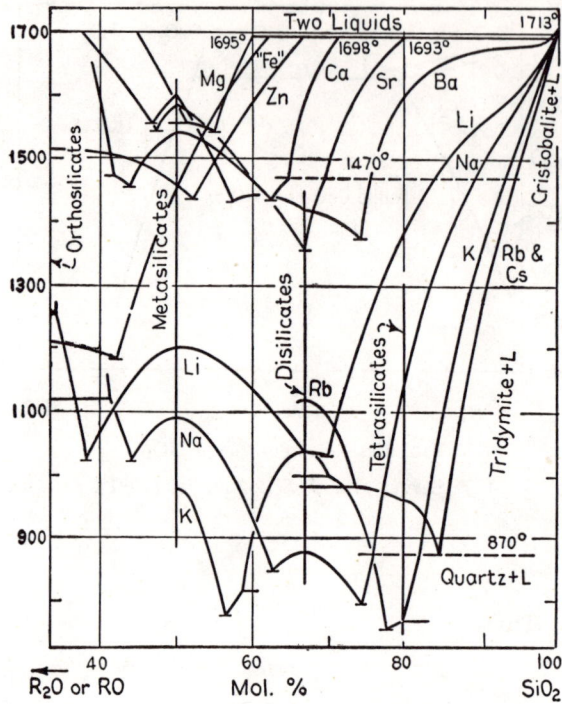

Fig. 365.—Liquidus relations in the alkali and alkali earth-silicate systems.

F. C. Kracek, *Carnegie Inst. Wash., Year Book, Ann Rep. Director Geophys. Lab.*, pp. 61–63 (1932–33); see also Kracek, *J. Am. Chem. Soc.*, **52** [4] 1440 (1930).

ThO$_2$–TiO$_2$

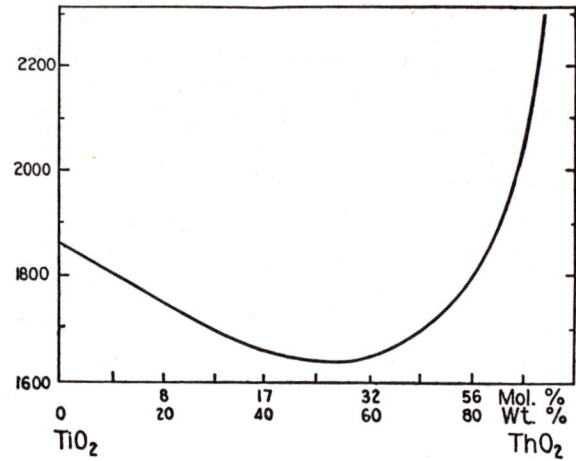

Fig. 367.—Liquidus curve of system ThO$_2$–TiO$_2$.

H. von Wartenberg and K. Eckhardt, Part VIII, *Z. anorg. u. allgem. Chem.*, **232**, 185 (1937).

SnO$_2$–TiO$_2$

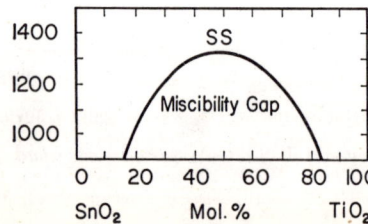

Fig. 366.—System SnO$_2$–TiO$_2$; subsolidus.

N. N. Padurow, *Naturwissenschaften*, **43**, 396 (1956).

ThO₂–ZrO₂

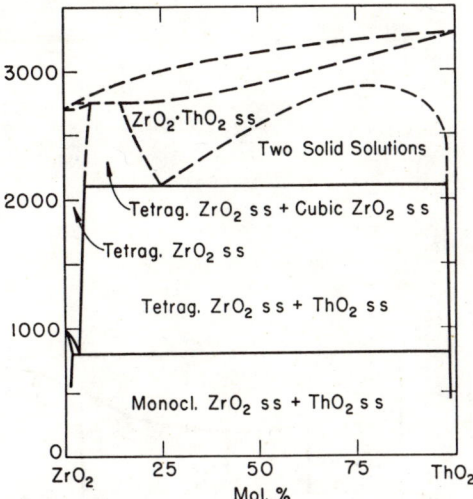

FIG. 368.—System ThO₂–ZrO₂; suggested.

Pol Duwez and Eugene Loh, *J. Am. Ceram. Soc.*, **40** [9] 324 (1957). See also, F. A. Mumpton and Rustum Roy, *J. Am. Ceram. Soc.*, **43** [5] 237 (1960).

TiO₂–ZrO₂

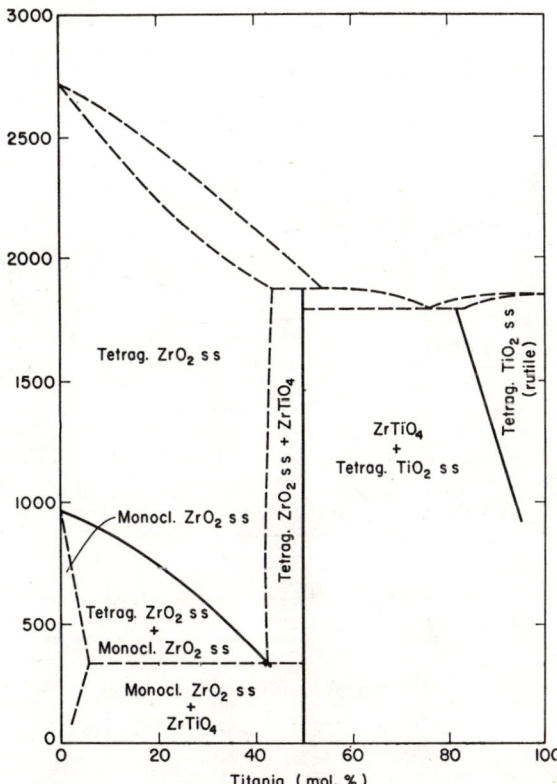

FIG. 369.—System TiO₂–ZrO₂; proposed.

Frank H. Brown, Jr., and Pol Duwez, *J. Am. Ceram. Soc.*, **37** [3] 132 (1954).

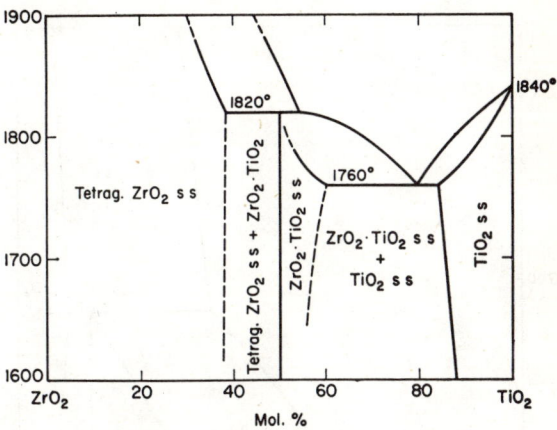

FIG. 370.—System TiO₂–ZrO₂; suggested.

L. W. Coughanour, R. S. Roth, and V. A. DeProsse, *J. Research Natl. Bur. Standards*, **52** [1] 39 (1954); RP 2470.

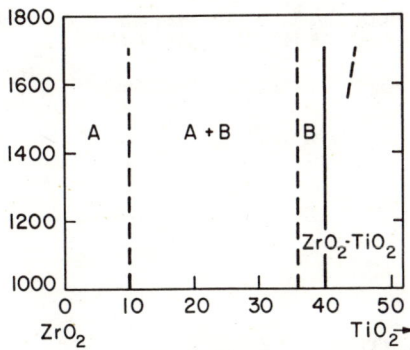

FIG. 371.—System ZrO₂–ZrO₂·TiO₂; subsolidus. A = ZrO₂ss, with tetragonal structure, B = solid solution ending with the composition ZrO₂·TiO₂.

Antonio Cocco, *Ann. chim. (Rome)*, **48**, 598 (1958).

TiO₂–Nb₂O₅

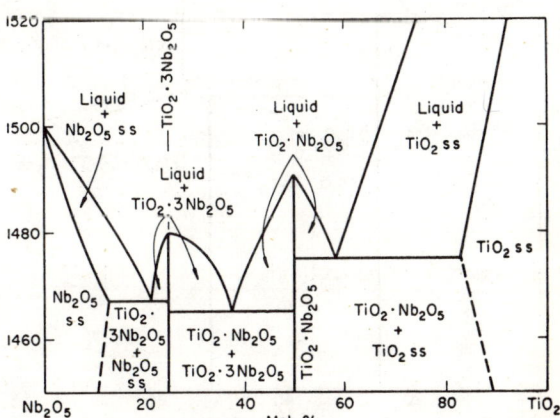

FIG. 372.—System TiO₂–Nb₂O₅. ss = solid solution.

R. S. Roth and L. W. Coughanour, *J. Research. Natl. Bur. Standards*, **55** [4] 211 (1955); RP2621.

ZrO_2–Nb_2O_5

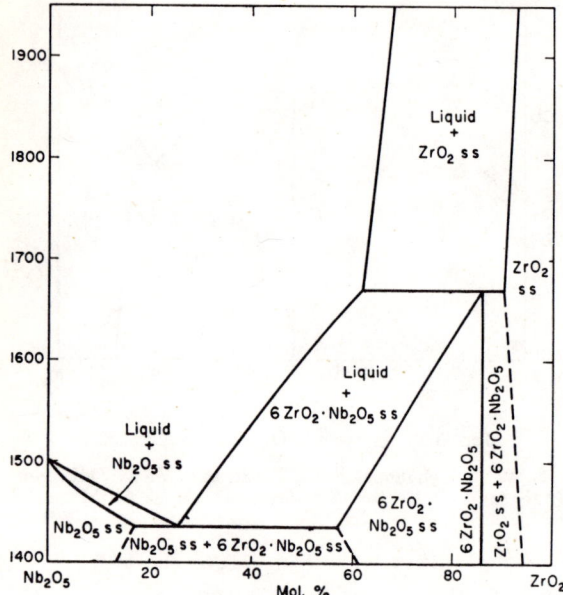

Fig. 373.—System ZrO_2–Nb_2O_5. ss = solid solution.

R. S. Roth and L. W. Coughanour, *J. Research, Natl. Bur. Standards*, **55** [4] 212 (1955); RP2621.

ZrO_2–Ta_2O_5

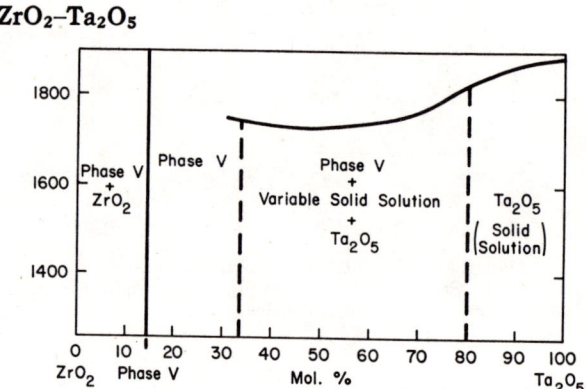

Fig. 374.—System Ta_2O_5–ZrO_2.

B. W. King, John Schultz, E. A. Durbin, and W. H. Duckworth, U. S. Atomic Energy Comm., BMI-1106, 15 (1956).

Nb_2O_5–Ta_2O_5

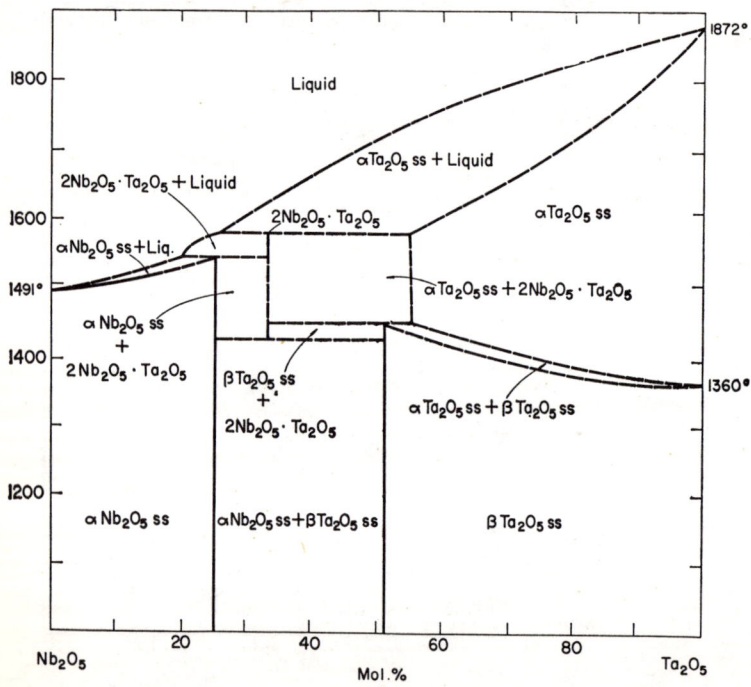

Fig. 375.—System Nb_2O_5–Ta_2O_5.

F. Holtzberg and A. Reisman, *J. Phys. Chem.*, **65**, 1193 (1961).

MoO_3–WO_3

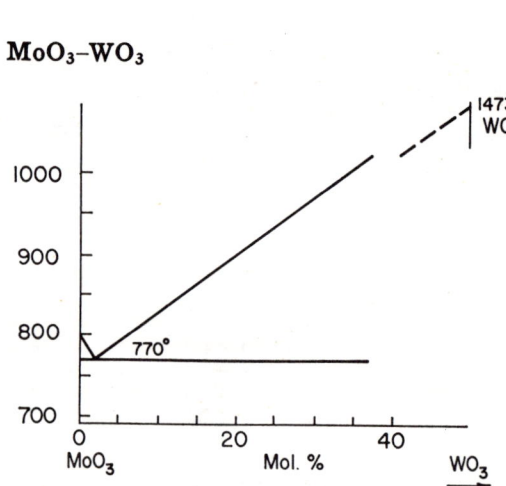

Fig. 376.—System MoO_3–WO_3.

G. D. Rieck, *Rec. Trav. Chim.*, **62**, 429 (1943).

III. Three Oxides

K_2O–Li_2O–SiO_2

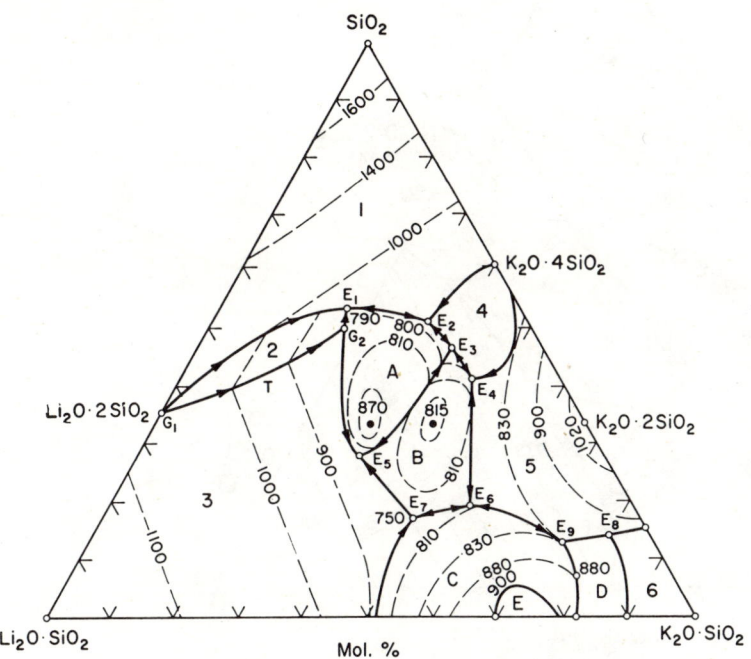

FIG. 377.—System $Li_2O \cdot SiO_2$–$K_2O \cdot SiO_2$–SiO_2.

1. Silica
2. $Li_2O \cdot 2SiO_2$
3. $Li_2O \cdot SiO_2$
4. $K_2O \cdot 4SiO_2$
5. $K_2O \cdot 2SiO_2$
6. $K_2O \cdot SiO_2$

A. $Li_2O \cdot K_2O \cdot 4SiO_2$
B. $Li_2O \cdot 2K_2O \cdot 6SiO_2$
C. $2Li_2O \cdot 5K_2O \cdot 7SiO_2$
D. $Li_2O \cdot 5K_2O \cdot 7SiO_2$
E. $Li_2O \cdot 3K_2O \cdot 4SiO_2$

H. A. Sheybany, *Verres et Réfract.*, **2**, 368 (1948).

K_2O–Li_2O–MoO_3

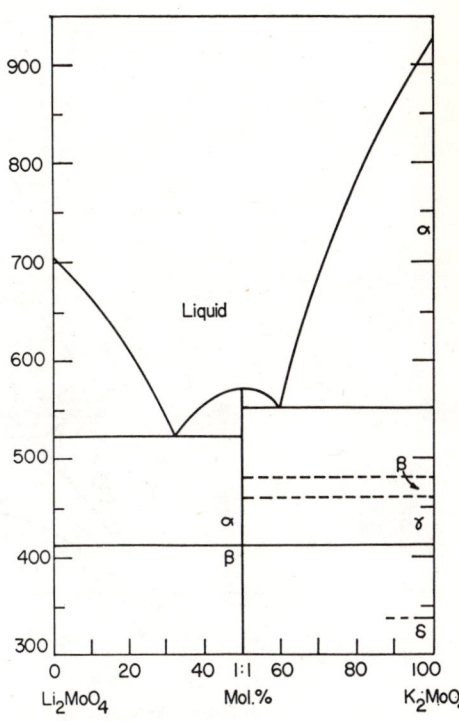

FIG. 379.—System Li_2MoO_4–K_2MoO_4.

F. Hoermann, *Z. anorg. u. allgem. Chem.*, **177**, 182 (1928–29).

K_2O–Li_2O–P_2O_5

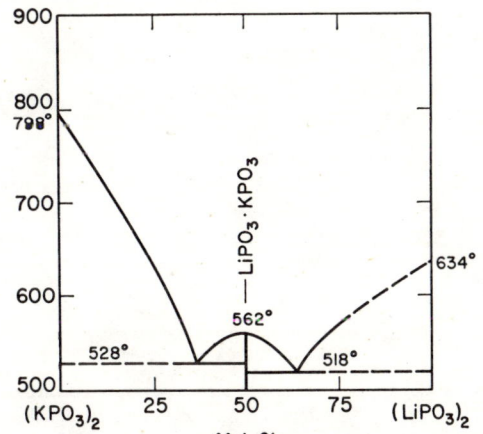

FIG. 378.—System $(KPO_3)_2$–$(LiPO_3)_2$.

A. G. Bergman and M. L. Sholokhovich, *J. Gen. Chem. U.S.S.R.*, **23** [7] 1078 (1953).

K_2O–Na_2O–B_2O_3

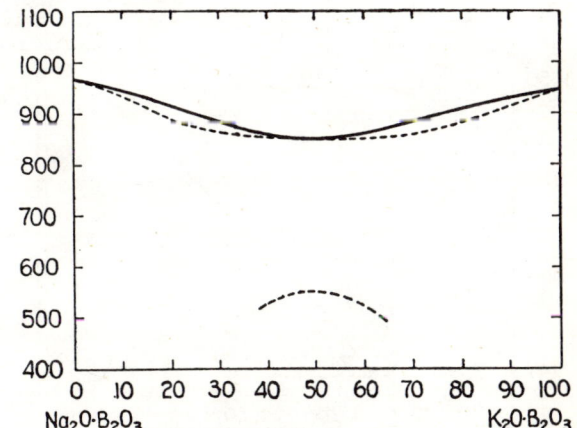

FIG. 380.—System $K_2O \cdot B_2O_3$–$Na_2O \cdot B_2O_3$.

H. S. van Klooster, *Z. anorg. Chem.* **69**, 131 (1910).

K_2O–Na_2O–SiO_2

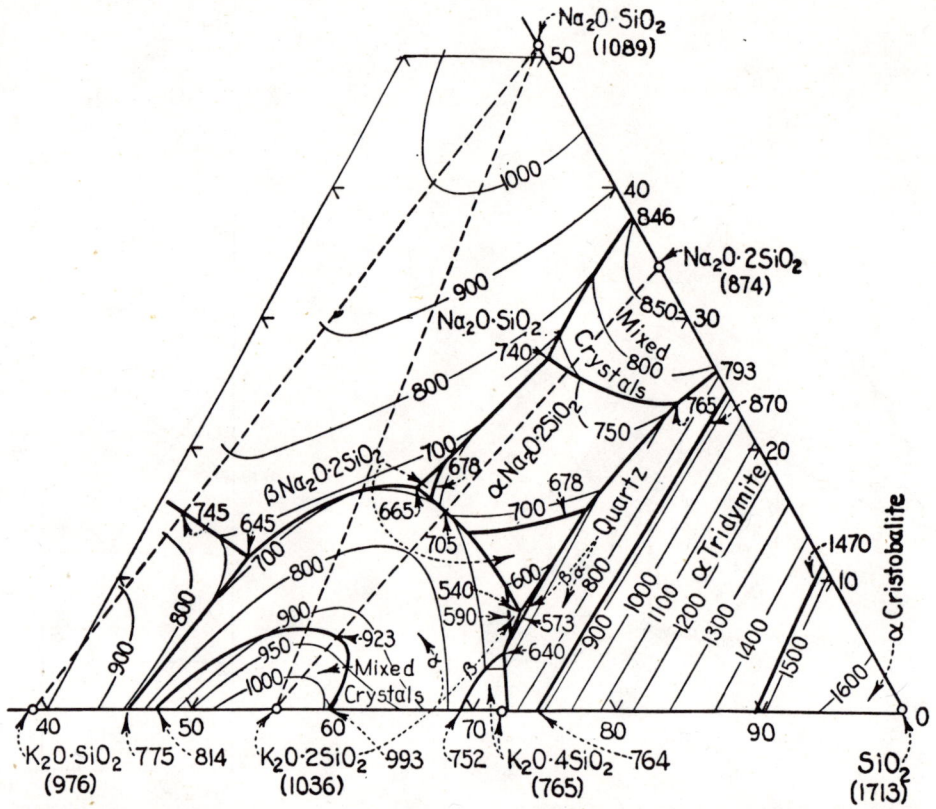

Fig. 381.—System SiO_2–$K_2O \cdot SiO_2$–$Na_2O \cdot SiO_2$.　　F. C. Kracek, *J. Phys. Chem.*, **36**, 2538 (1932).

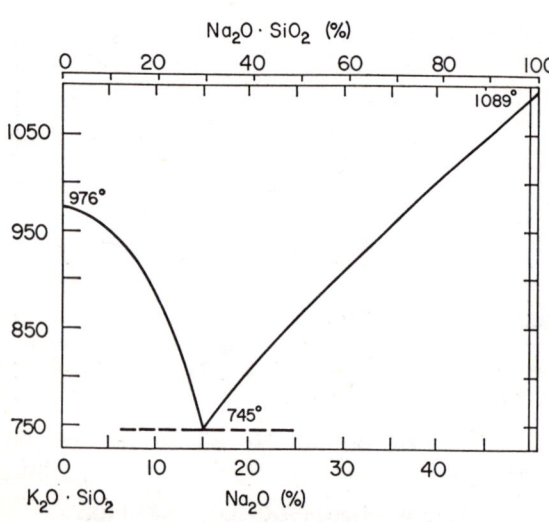

Fig. 382.—System $K_2O \cdot SiO_2$–$Na_2O \cdot SiO_2$.

F. C. Kracek, *J. Phys. Chem.*, **36**, 2540 (1932).

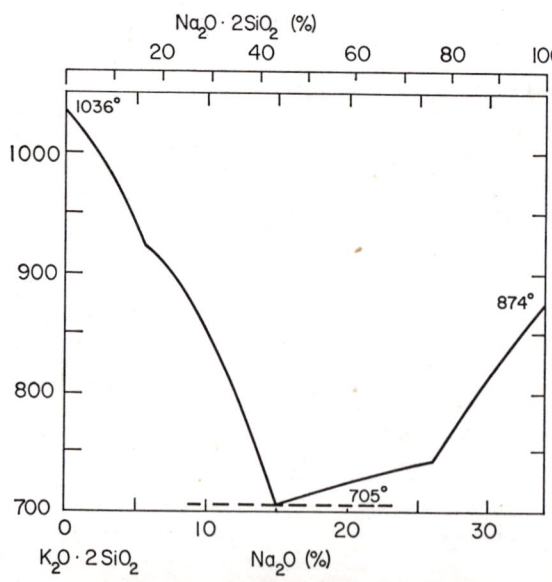

Fig. 383.—System $K_2O \cdot 2SiO_2$–$Na_2O \cdot 2SiO_2$.

F. C. Kracek, *J. Phys. Chem.*, **36**, 2540 (1932).

K_2O–Na_2O–As_2O_5

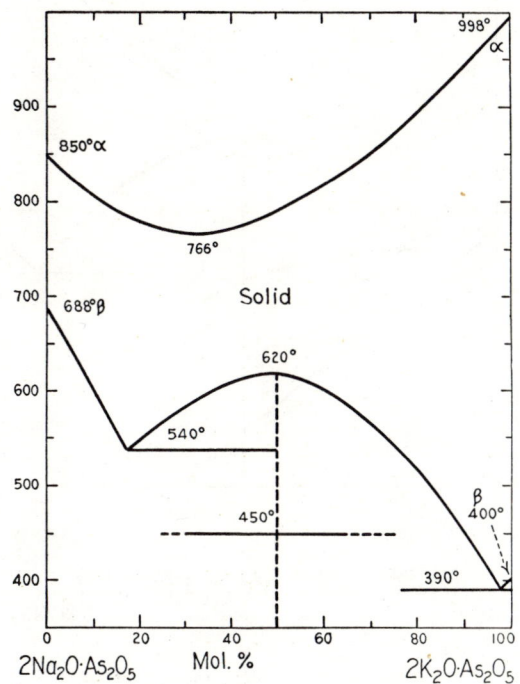

FIG. 384.—System $2K_2O \cdot As_2O_5$–$2Na_2O \cdot As_2O_5$.

M. Amadori, *Atti reale ist. Veneto sci.*, **76** [II] 419 (1917).

K_2O–Na_2O–P_2O_5

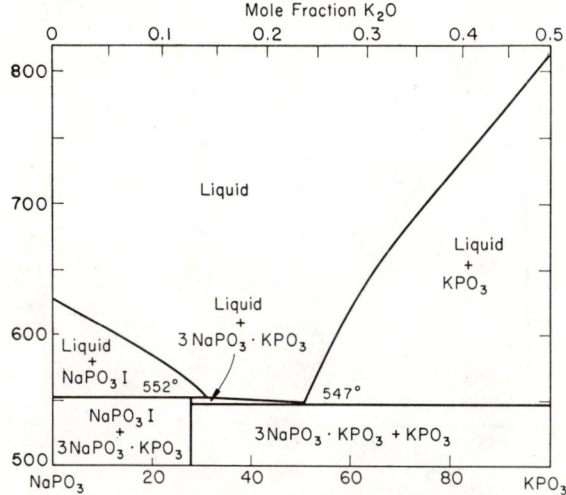

FIG. 385.—System KPO_3–$NaPO_3$.

G. W. Morey, *J. Am. Chem. Soc.*, **76** [18] 4725 (1954).

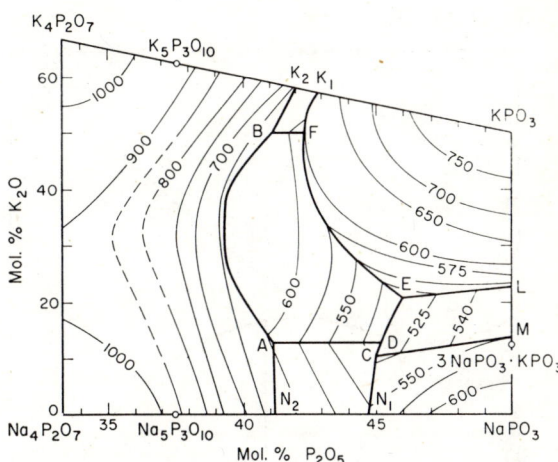

FIG. 386.—System $NaPO_3$–$Na_4P_2O_7$–$K_4P_2O_7$–KPO_3. The field of pyrophosphate solid solutions extends to the boundary N_2ABK_2; that of $Na_5P_3O_{10}$ is the area N_1N_2AD; that of $Na_5P_3O_{10} \cdot K_5P_3O_{10}$, the area ABFED; that of $K_5P_3O_{10}$, the area FBK_2K_1; that of $NaPO_3$, the area $NaPO_3$-N_1CM; that of $3NaPO_3 \cdot KPO_3$, the area MCEL; and that of KPO_3, the area KPO_3-$IEFK_1$.

G. W. Morey, F. R. Boyd, Jr., J. L. England, and W. T. Chen, *J. Am. Chem. Soc.*, **77**, 5003 (1955).

K_2O–Na_2O–CrO_3

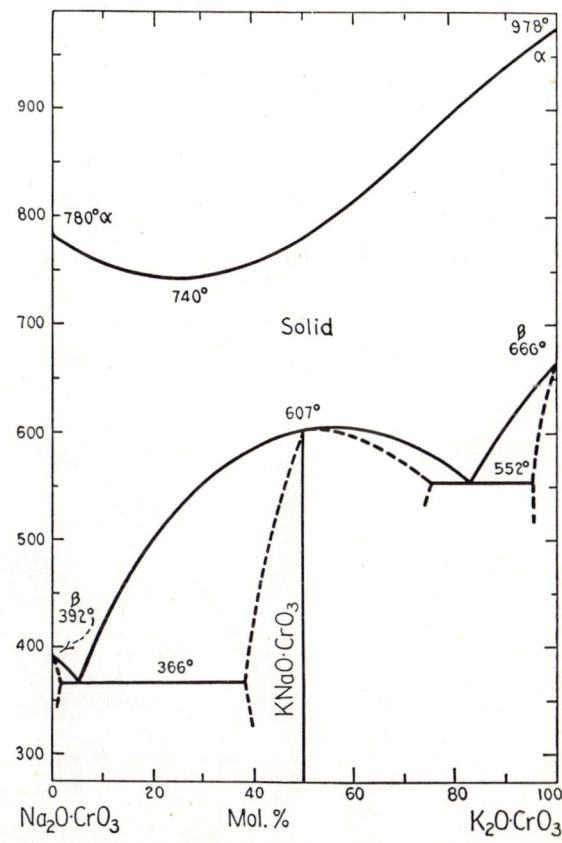

FIG. 387.—System $K_2O \cdot CrO_3$–$Na_2O \cdot CrO_3$.

M. Amadori, *Atti reale ist. Veneto sci.*, **72** [II] 903 (1913).

K₂O–Na₂O–MoO₃

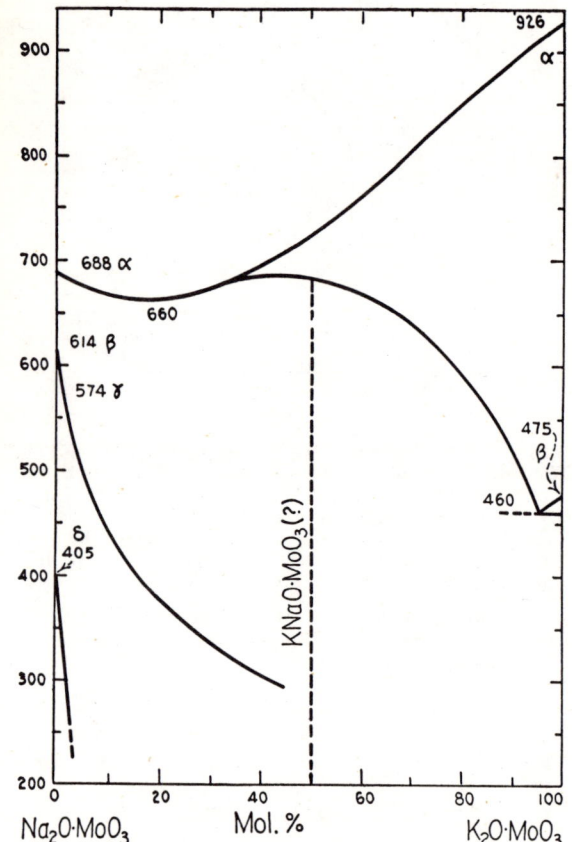

Fig. 388.—System $K_2O \cdot MoO_3$–$Na_2O \cdot MoO_3$.

M. Amadori, *Atti reale ist. Veneto sci.*, **72** [II] 903 (1913).

K₂O–Na₂O–WO₃

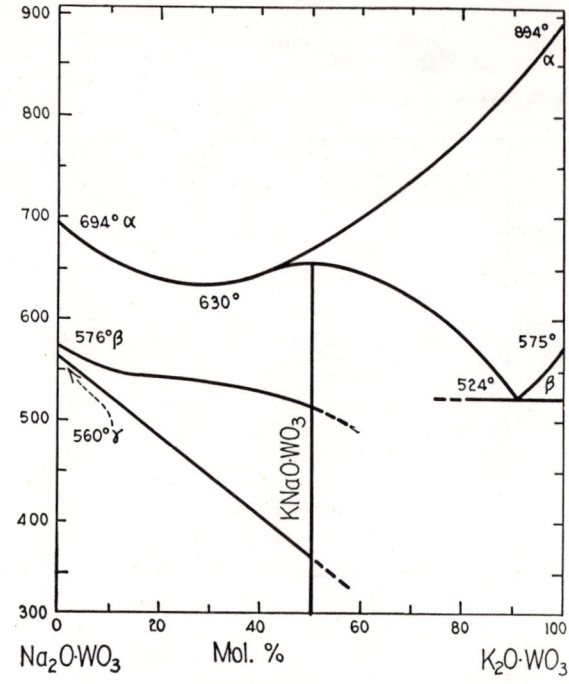

Fig. 389.—System $K_2O \cdot WO_3$–$Na_2O \cdot WO_3$.

M. Amadori, *Atti reale ist. Veneto sci.*, **72** [II] 903 (1913).

K₂O–CaO–Al₂O₃

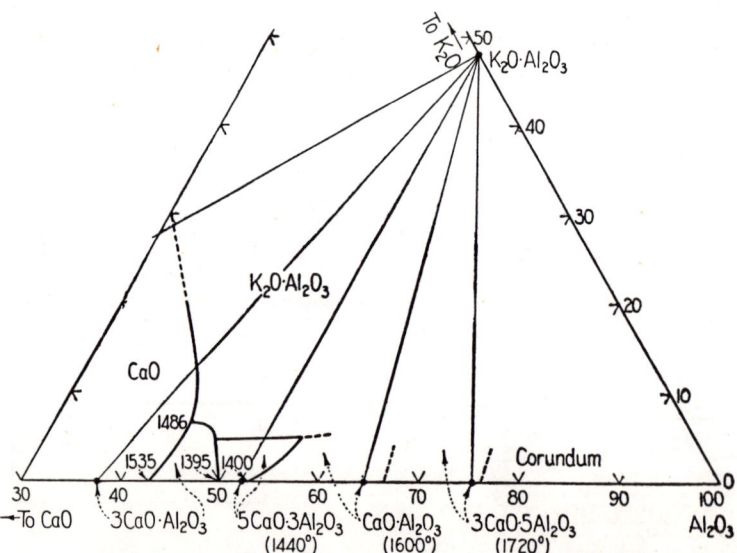

Fig. 390.—System Al_2O_3–$K_2O \cdot Al_2O_3$–$3CaO \cdot Al_2O_3$.

L. T. Brownmiller, *Am. J. Sci.*, 5th Ser., **29**, 268 (1935).

K_2O–CaO–SiO_2

FIG. 391.—The high SiO_2 corner of system K_2O–CaO–SiO_2.

G. W. Morey, F. C. Kracek, and N. L. Bowen, *J. Soc. Glass Technol.*, **14**, 158 (1930).

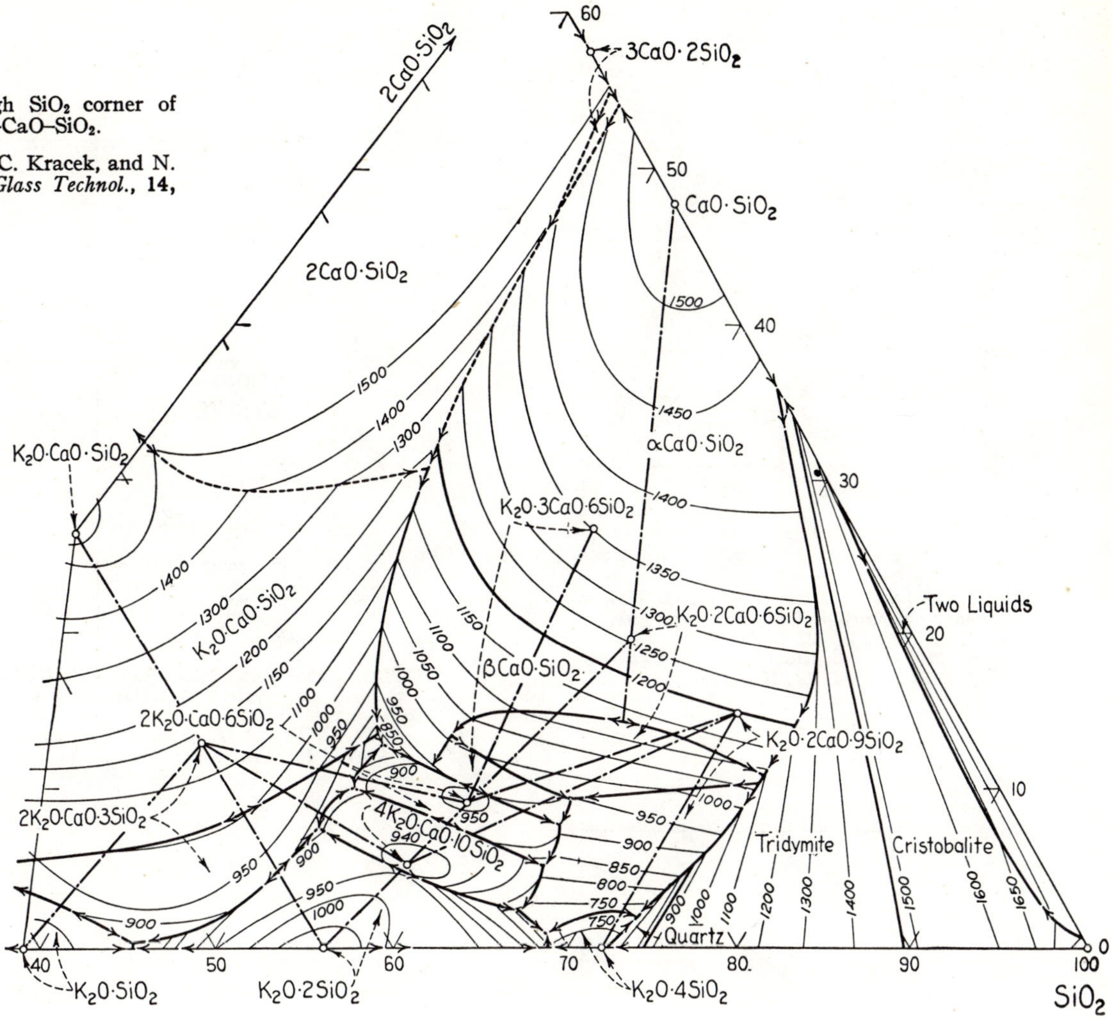

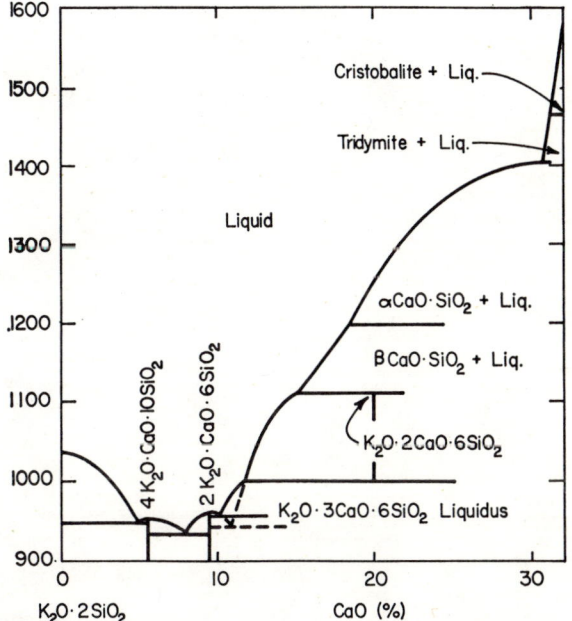

FIG. 392.—Section $K_2O \cdot 2SiO_2$–(70% SiO_2, 30% CaO) through ternary system $K_2O \cdot CaO \cdot SiO_2$, Fig. 391, including the join $K_2O \cdot 2SiO_2$–$K_2O \cdot 2CaO \cdot 6SiO_2$.

G. W. Morey, F. C. Kracek, and N. L. Bowen, *J. Soc. Glass Technol.*, **14**, 162 (1930).

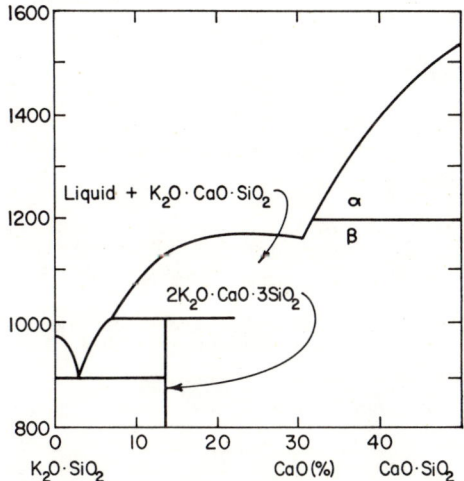

FIG. 393.—Join $K_2O \cdot SiO_2$–$CaO \cdot SiO_2$ of Fig. 391.

G. W. Morey, F. C. Kracek, and N. L. Bowen, *J. Soc. Glass Technol.*, **14**, 160 (1930).

K_2O–CaO–SiO_2 (concl.)

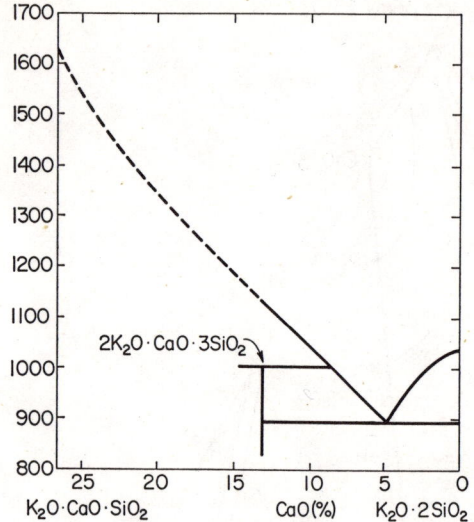

FIG. 394.—Join $K_2O \cdot CaO \cdot SiO_2$–$K_2O \cdot 2SiO_2$ of ternary system K_2O–CaO–SiO_2, Fig. 391.

G. W. Morey, F. C. Kracek, and N. L. Bowen, *J. Soc. Glass Technol.*, **14**, 161 (1930).

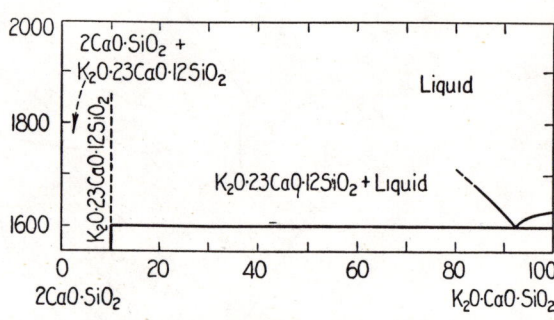

FIG. 396.—Join $2CaO \cdot SiO_2$–$K_2O \cdot CaO \cdot SiO_2$ of ternary system K_2O–CaO–SiO_2, Fig. 391.

W. C. Taylor, *J. Research Nat. Bur. Standards*, **27**, 315 (1941); RP 1421.

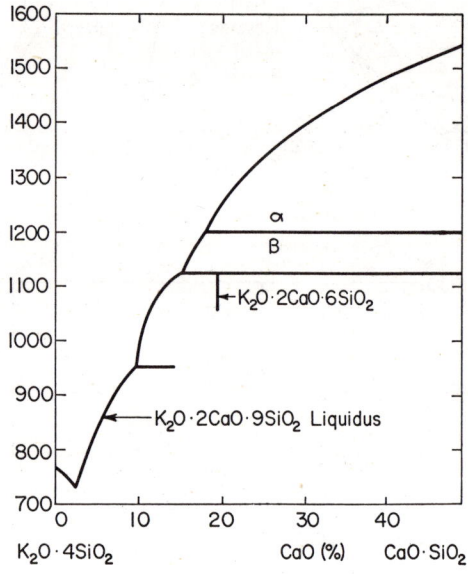

FIG. 395.—Join $K_2O \cdot 4SiO_2$–$CaO \cdot SiO_2$ of ternary system K_2O–CaO–SiO_2, Fig. 391.

G. W. Morey, F. C. Kracek, and N. L. Bowen, *J. Soc. Glass Technol.*, **14**, 163 (1930).

K_2O–FeO–SiO_2

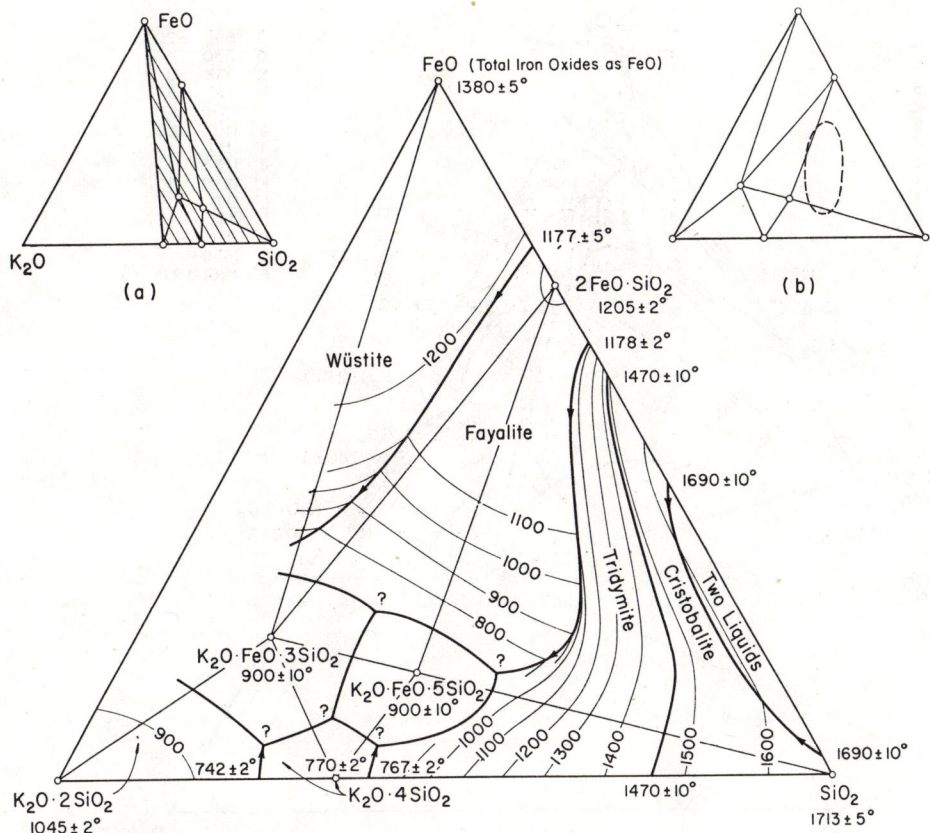

FIG. 397.—System $K_2O \cdot 2SiO_2$–FeO–SiO_2; subject to revision, showing liquidus relationships in equilibrium with metallic iron. Fe_2O_3 is present in all liquids. Figure *a* shows the position of this system (shaded area) in the more general system K_2O–FeO–SiO_2, and figure *b* shows the intersection of the field of low temperature liquid immiscibility in the system leucite–fayalite–silica projected onto this system.

E. W. Roedder, *Am. J. Sci.*, Bowen volume, p. 445 (1952).

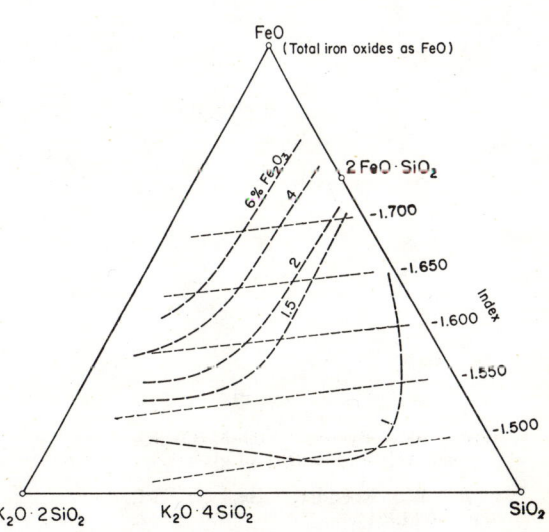

FIG. 398.—System K_2O–FeO–SiO_2; dashed lines indicate lines of equal Fe_2O_3 content in samples quenched from their individual liquidus temperatures; dotted lines refer to isofracts.

E. W. Roedder, *Am. J. Sci.*, Bowen volume p. 443 (1952).

K_2O–MgO–SiO_2

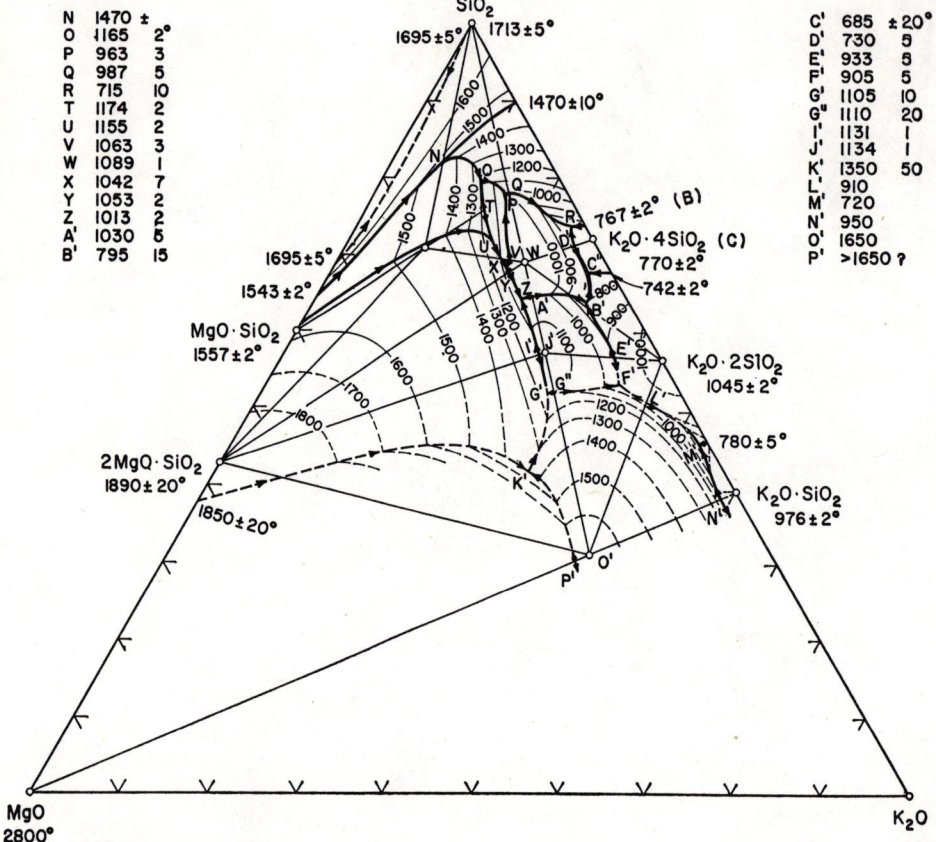

Fig. 399.—System K_2O–MgO–SiO_2. The tridymite field below 867° is metastable, hence binary eutectic point B and ternary eutectic point C are both metastable.

E. W. Roedder, *Am. J. Sci.*, **249**, 97 (1951).

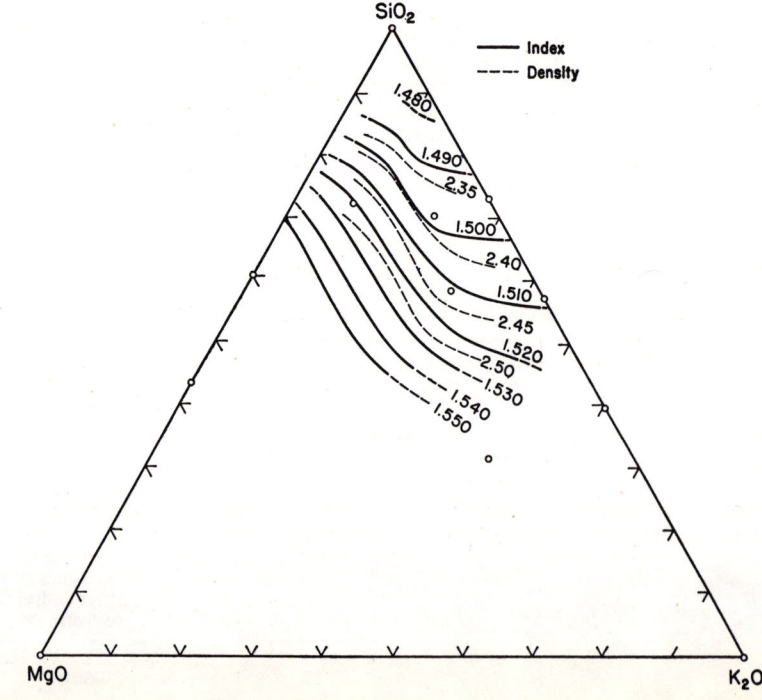

Fig. 400.—System K_2O–MgO–SiO_2; isofracts and density of glasses.

E. W. Roedder, *Am. J. Sci.*, **249**, 95 (1951).

K_2O–MgO–SiO_2 (cont.)

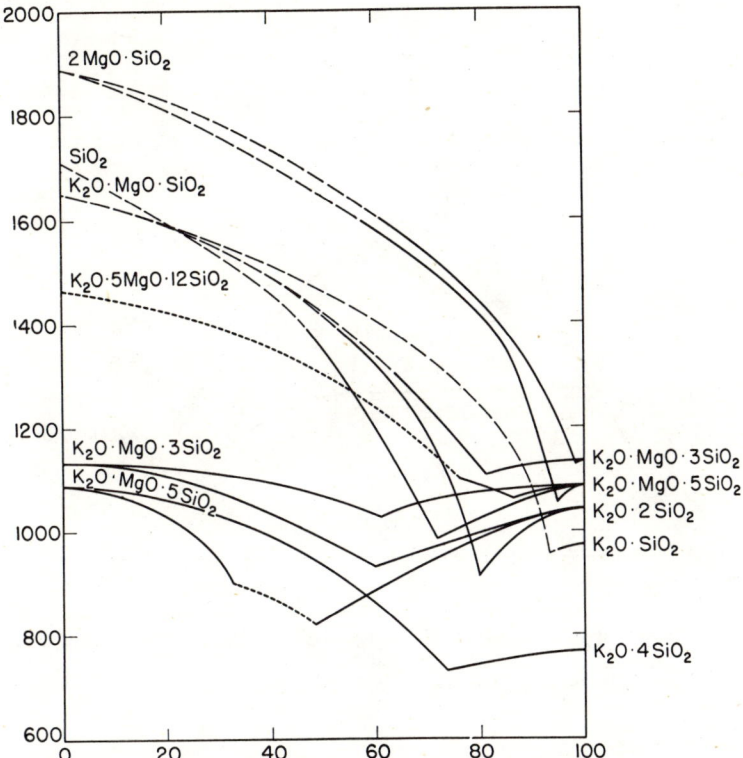

Fig. 401.—System K_2O–MgO–SiO_2; various binary sub-systems; the two dotted lines indicate non-binary (ternary) equilibrium for those portions.

E. W. Roedder, *Am. J. Sci.*, **249**, 97 (1951).

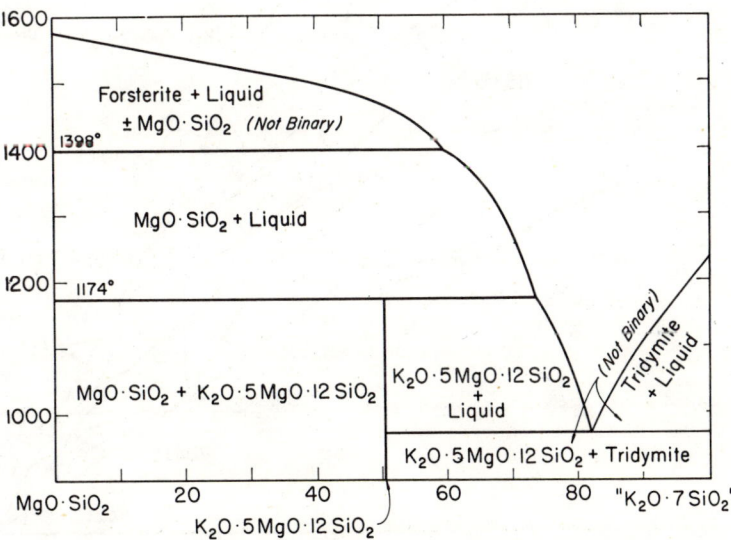

Fig. 402.—System $MgO \cdot SiO_2$–$K_2O \cdot 5\,MgO \cdot 12SiO_2$; partially binary; "$K_2O \cdot 7SiO_2$" not a compound.

E. W. Roedder, *Am. J. Sci.*, **249**, 126 (1951).

K_2O–MgO–SiO_2 (concl.)

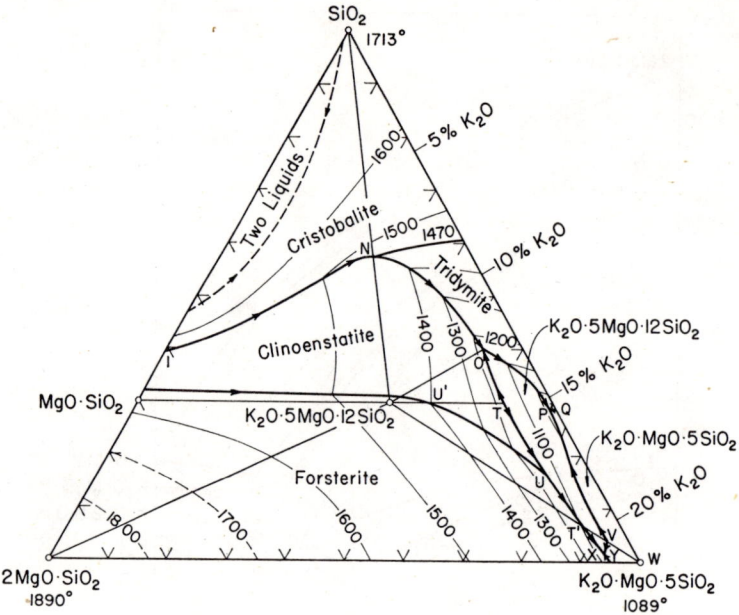

Fig. 403.—System $2MgO \cdot SiO_2$–$K_2O \cdot MgO \cdot 5SiO_2$–$SiO_2$; see Fig. 399 for letter code.
E. W. Roedder, *Am. J. Sci.*, **249**, 127 (1951).

K_2O–PbO–SiO_2

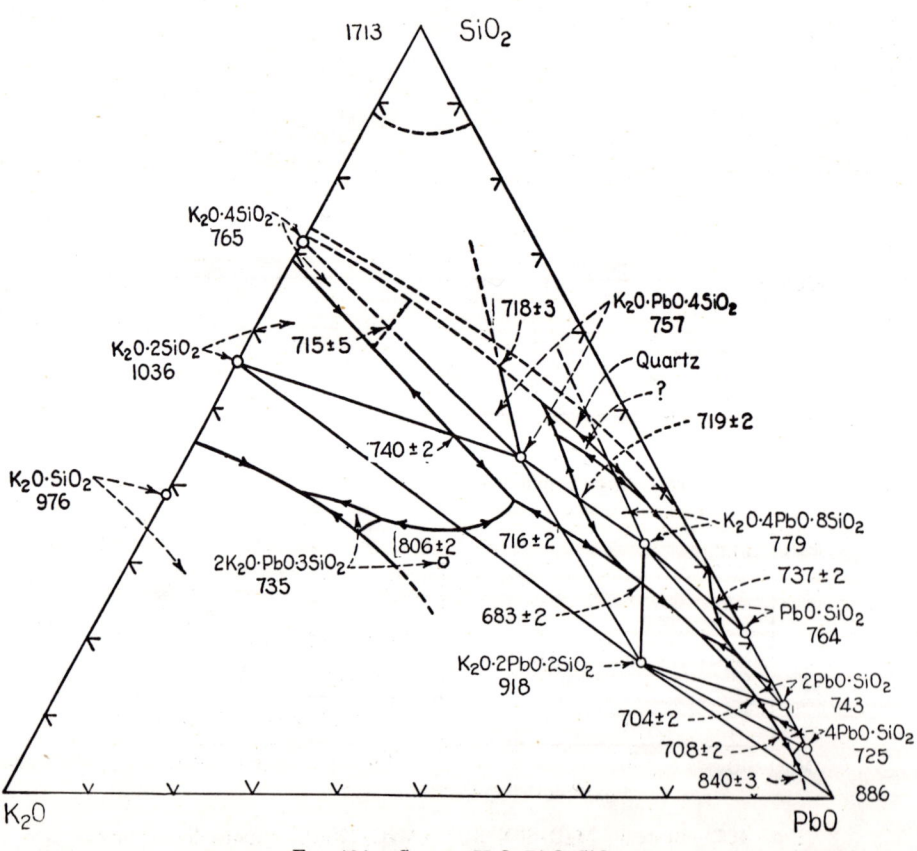

Fig. 404.—System K_2O–PbO–SiO_2.
R. F. Geller and E. N. Bunting, *J. Research Nat. Bur. Standards*, **17**, 283 (1936); RP 911.

K_2O–PbO–SiO_2 (concl.)

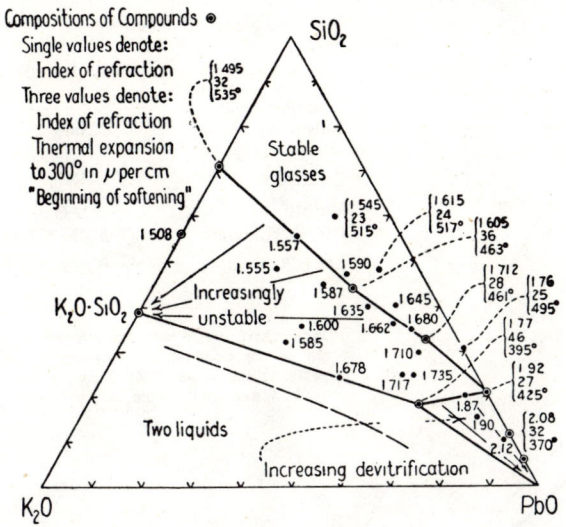

Fig. 405.—Summary of data on glasses of system K_2O–PbO–SiO_2.

R. F. Geller and E. N. Bunting, *J. Research Nat. Bur. Standards*, **17**, 287 (1936); RP 911.

K_2O–ZnO–SiO_2

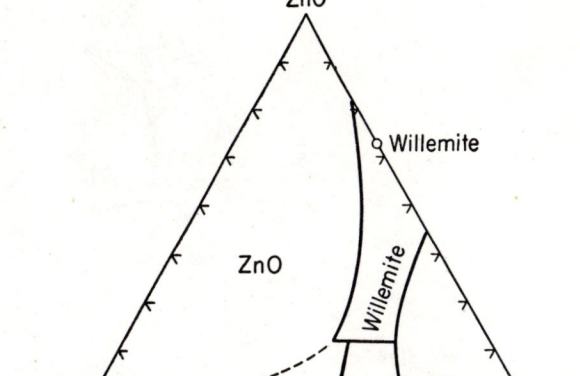

Fig. 406.—System K_2O–ZnO–SiO_2.

E. Ingerson, G. W. Morey, and O. F. Tuttle, *Am. J. Sci.*, **246** [1] 33 (1948).

$K_2O–Al_2O_3–SiO_2$

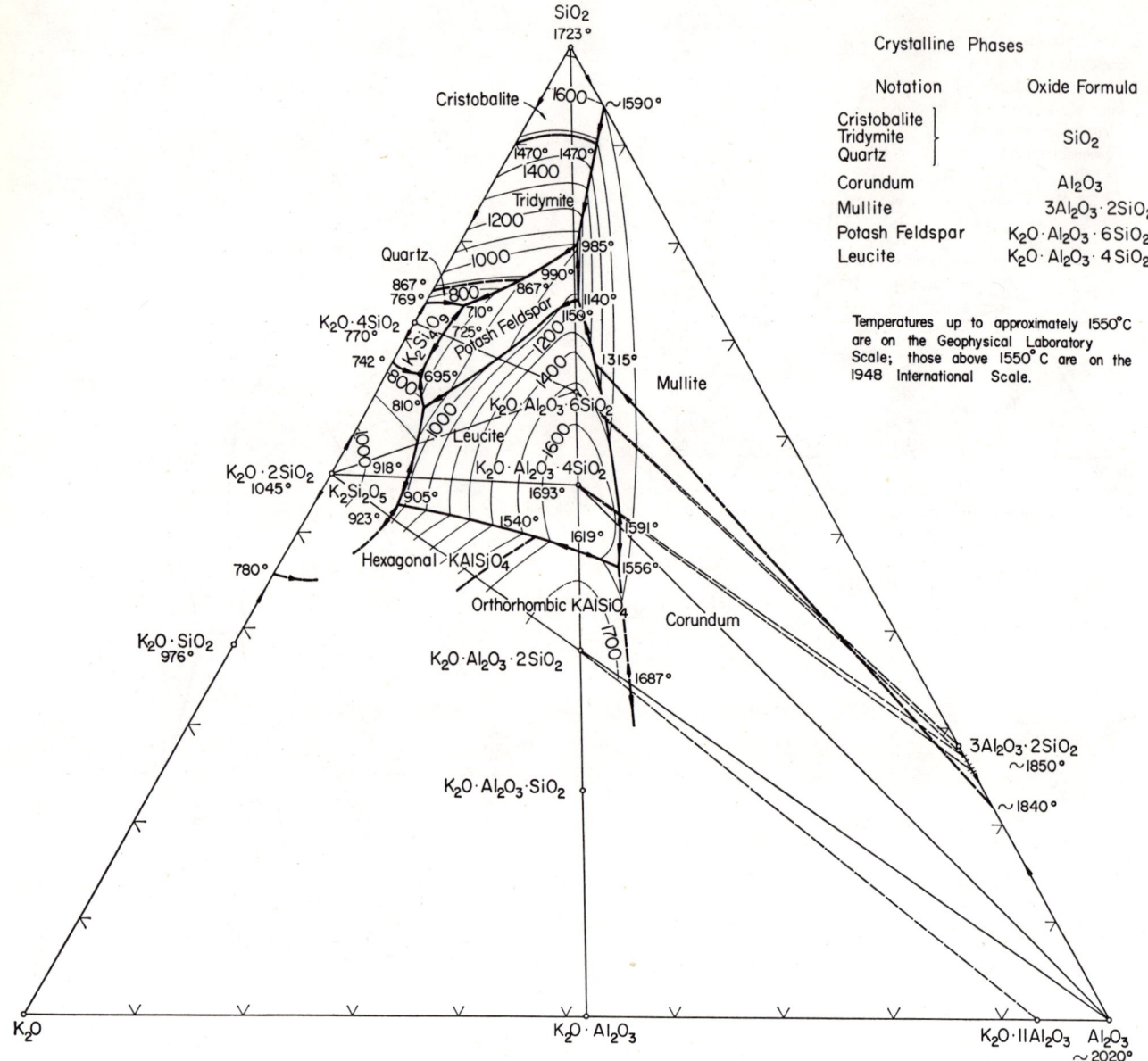

Fig. 407.—System $K_2O–Al_2O_3–SiO_2$; composite.

E. F. Osborn and Arnulf Muan, revised and redrawn "Phase Equilibrium Diagrams of Oxide Systems," Plate 5, Published by the American Ceramic Society and the Edward Orton, Jr., Ceramic Foundation, 1960.

Principal References

F. C. Kracek, N. L. Bowen, and G. W. Morey, *J. Phys. Chem.*, **33**, 1857–79 (1929).
F. C. Kracek, N. L. Bowen, and G. W. Morey, *J. Phys. Chem.*, **41**, 1183–93 (1937).
N. L. Bowen and J. W. Greig, *J. Am. Ceram. Soc.*, **7**, 238–54 (1924); corrections, *ibid.*, 410.
N. A. Toropov and F. Ya. Galakhov, *Vopr. Petrogr. i Mineralog., Akad. Nauk S.S.S.R.*, **2**, 245–55 (1953).
Shigeo Aramaki and Rustum Roy, *Nature*, **184**, 631–32 (1959).
J. F. Schairer and N. L. Bowen, *Am. J. Sci.*, **253**, 681–746 (1955).

$K_2O-Al_2O_3-SiO_2$ (cont.)

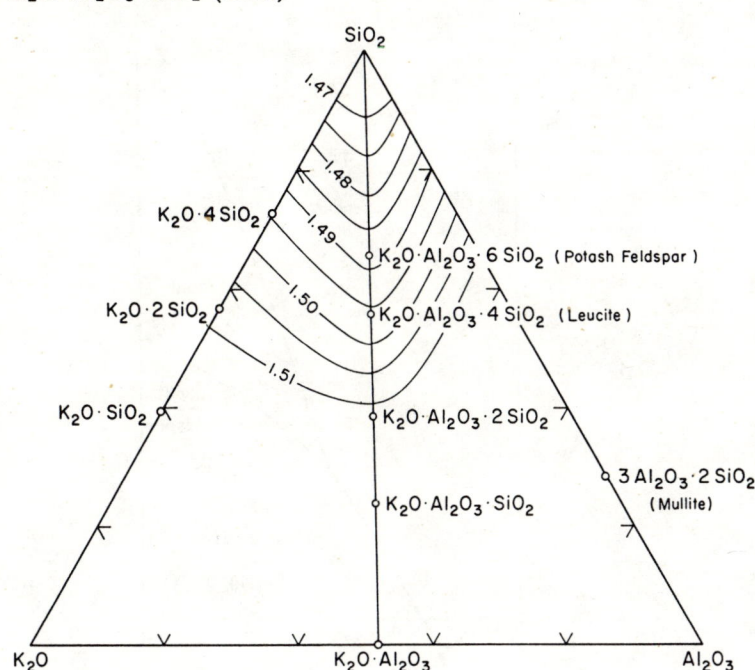

Fig. 408.—System $K_2O-Al_2O_3-SiO_2$; isofracts at 25°C.

J. F. Schairer and N. L. Bowen, *Am. J. Sci.*, **253** [12] 714 (1955).

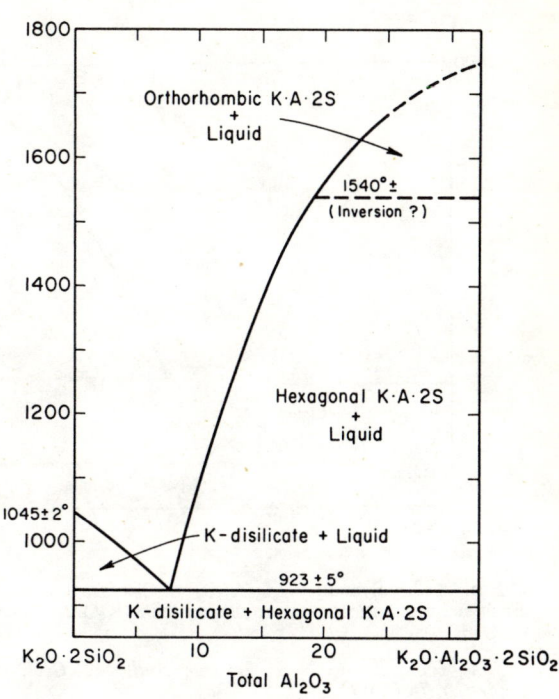

Fig. 410.—System $K_2O \cdot 2SiO_2-K_2O \cdot Al_2O_3 \cdot 2SiO_2$.

J. F. Schairer and N. L. Bowen, *Am. J. Sci.*, **253** [12] 721 (1955).

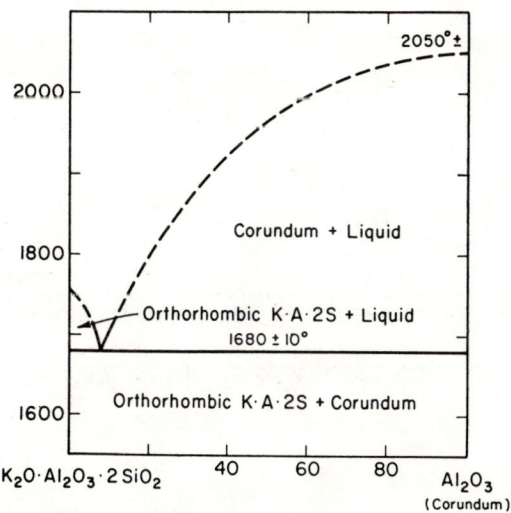

Fig. 409.—System $K_2O \cdot Al_2O_3 \cdot 2SiO_2-Al_2O_3$.

J. F. Schairer and N. L. Bowen, *Am. J. Sci.*, **253** [12] 722 (1955).

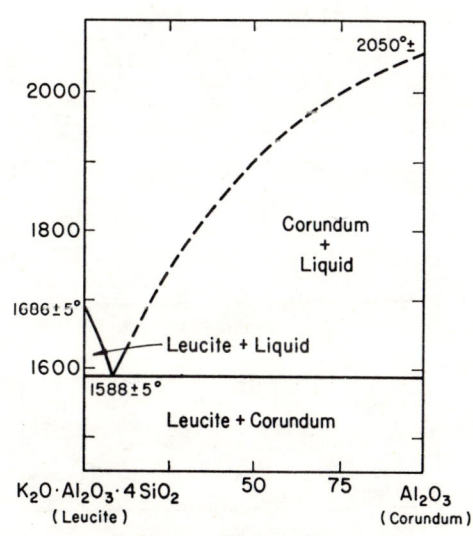

Fig. 411.—System $K_2O \cdot Al_2O_3 \cdot 4SiO_2-Al_2O_3$.

J. F. Schairer and N. L. Bowen, *Am. J. Sci.*, **253** [12] 719 (1955).

K_2O–Al_2O_3–SiO_2 (concl.)

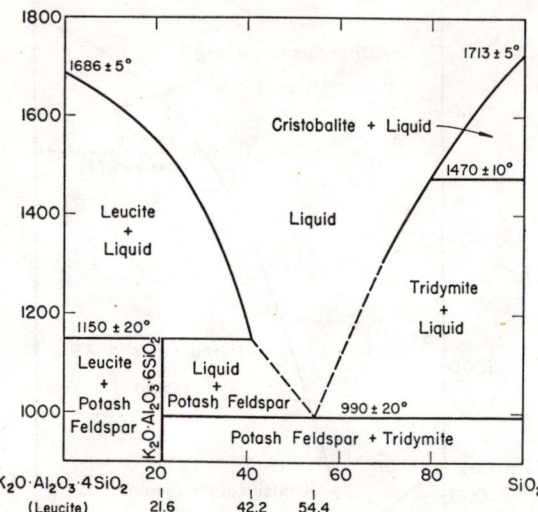

FIG. 412.—System $K_2O \cdot Al_2O_3 \cdot 4SiO_2$ (Leucite)–SiO_2. Broken lines indicate extrapolation from K_2O–Al_2O_3–SiO_2 system.

J. F. Schairer and N. L. Bowen, *Bull. Soc. Géol. Finlande*, **20**, 74 (1947).

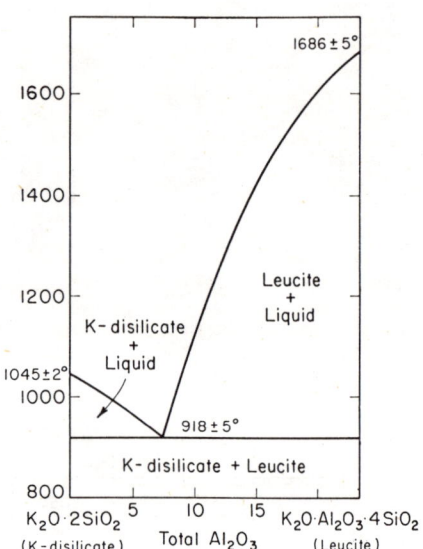

FIG. 413.—System $K_2O \cdot 2SiO_2$–$K_2O \cdot Al_2O_3 \cdot 4SiO_2$.

J. F. Schairer and N. L. Bowen, *Am. J. Sci.*, **253** [12] 720 (1955).

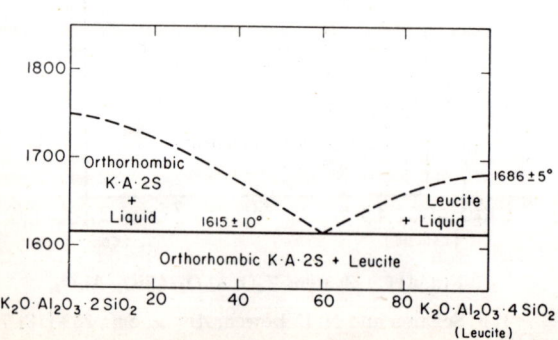

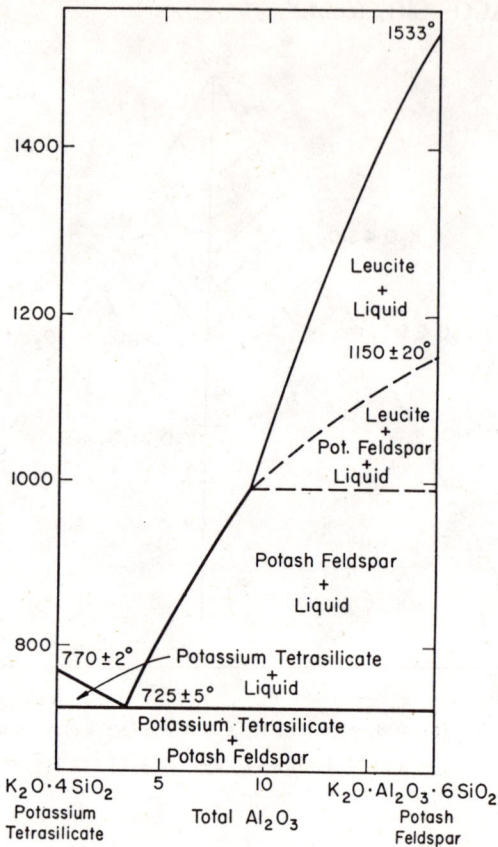

FIG. 415.—System $K_2O \cdot 4SiO_2$–$K_2O \cdot Al_2O_3 \cdot 6SiO_2$.

J. F. Schairer and N. L. Bowen, *Am. J. Sci.*, **253** [12] 723 (1955).

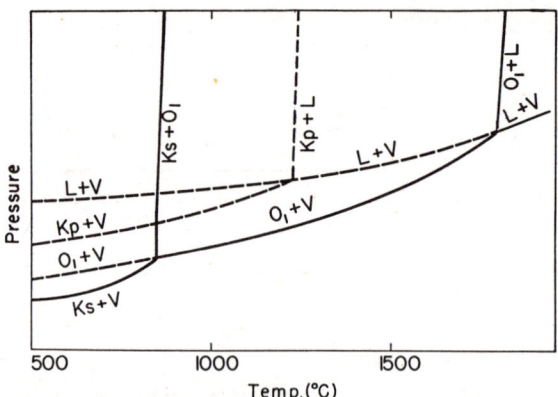

FIG. 416.—System $KAlSiO_4$, p–T; schematic. Dashed lines represent metastable equilibrium. L = liquid, V = vapor, Ks = kalsilite, O_1 = orthorhombic $KAlSiO_4$, Kp = kaliophilite.

O. F. Tuttle and J. V. Smith, *Am. J. Sci.*, **256**, 581 (1958).

FIG. 414.—System $K_2O \cdot Al_2O_3 \cdot 2SiO_2$–$K_2O \cdot Al_2O_3 \cdot 4SiO_2$.

J. F. Schairer and N. L. Bowen, *Am. J. Sci.*, **253** [12] 720 (1955).

$K_2O–B_2O_3–P_2O_5$

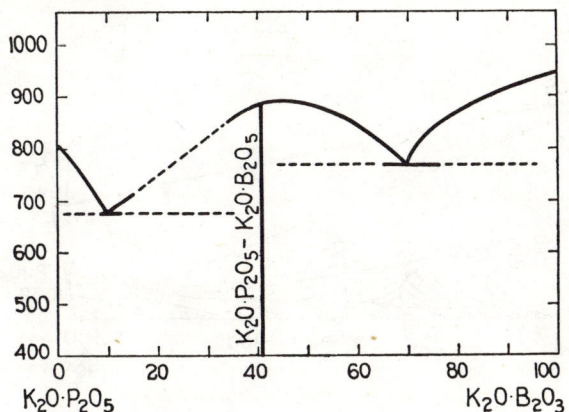

Fig. 417.—System $K_2O \cdot B_2O_3–K_2O \cdot P_2O_5$.

H. S. Van Klooster, *Z. anorg. Chem.* **69**, 124 (1910).

$K_2O–Fe_2O_3–SiO_2$

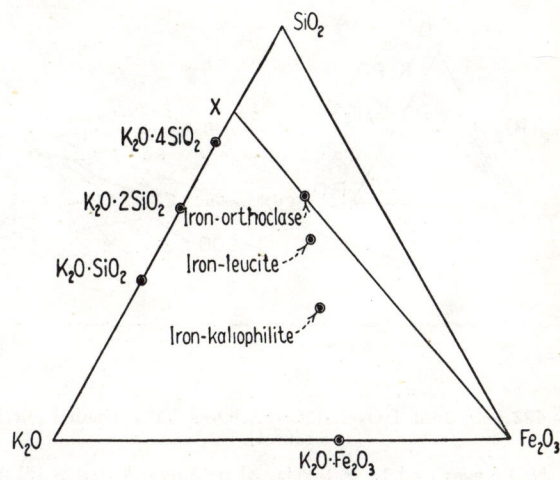

Fig. 419.—Ternary compounds in system $K_2O–Fe_2O_3–SiO_2$.

G. T. Faust, *Am. Mineral,* **21** [12] 752 (1936).

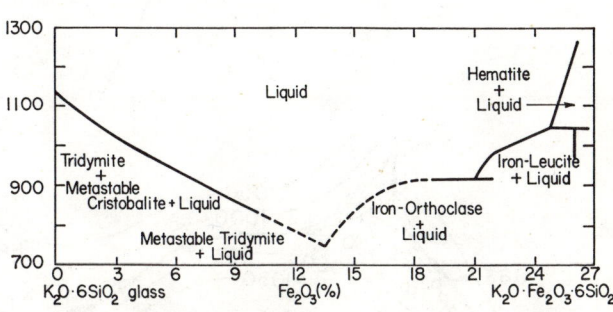

Fig. 420.—System $K_2O \cdot 6SiO_2–K_2O \cdot Fe_2O_3 \cdot 6SiO_2$, pseudobinary. Liquidus temperatures along line X–$K_2O \cdot Fe_2O_3 \cdot 6SiO_2$ of system $K_2O–Fe_2O_3–SiO_2$, Fig. 419.

G. T. Faust, *Am. Mineral.,* **21** [12] 751 (1936).

$K_2O–B_2O_3–CrO_3$

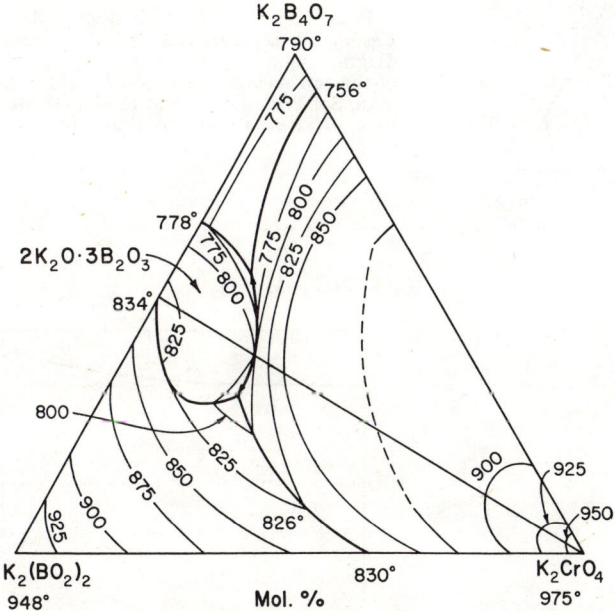

Fig. 418.—System $K_2(BO_2)_2–K_2B_4O_7–K_2CrO_4$.

A. G. Bergman and O. R. Vartbaronov, *Zhur. Neorg. Khim.,* **2** [3] 646 (1957).

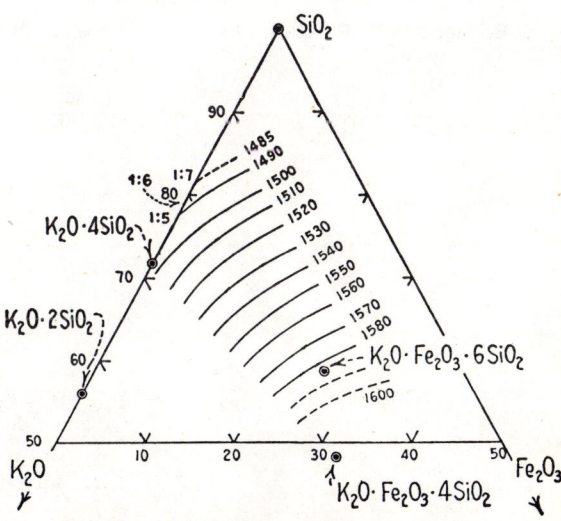

Fig. 421.—Index of refraction of glasses in region of join $K_2O \cdot 4SiO_2–K_2O \cdot Fe_2O_3 \cdot 6SiO_2$ of Fig. 419.

G. T. Faust and A. P. Beck, *J. Am. Ceram. Soc.,* **21**, 322 (1938).

$K_2O–TiO_2–P_2O_5$

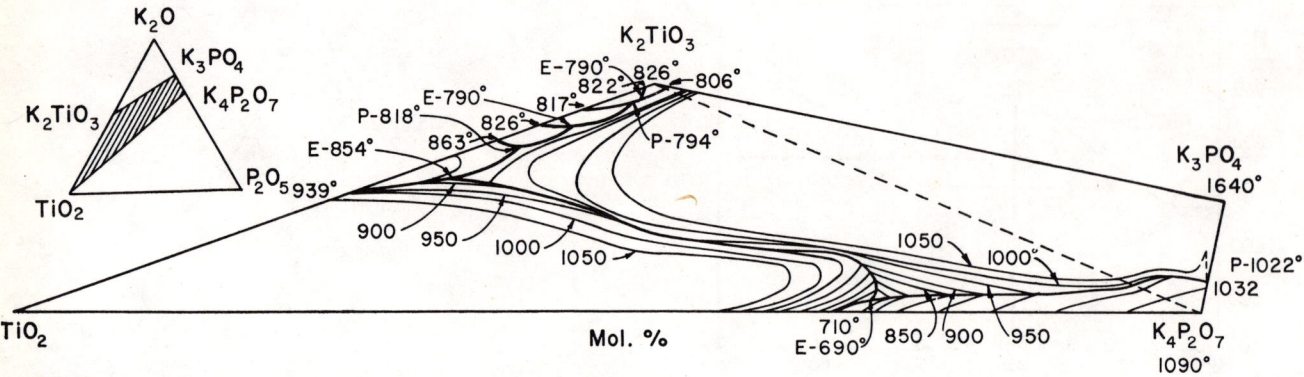

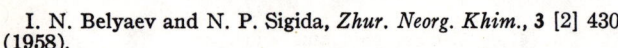

Fig. 422.—System TiO_2–$K_4P_2O_7$–K_3PO_4–K_2TiO_3 (shaded portion of inset).

I. N. Belyaev and N. P. Sigida, *Zhur. Neorg. Khim.*, **3** [2] 430 (1958).

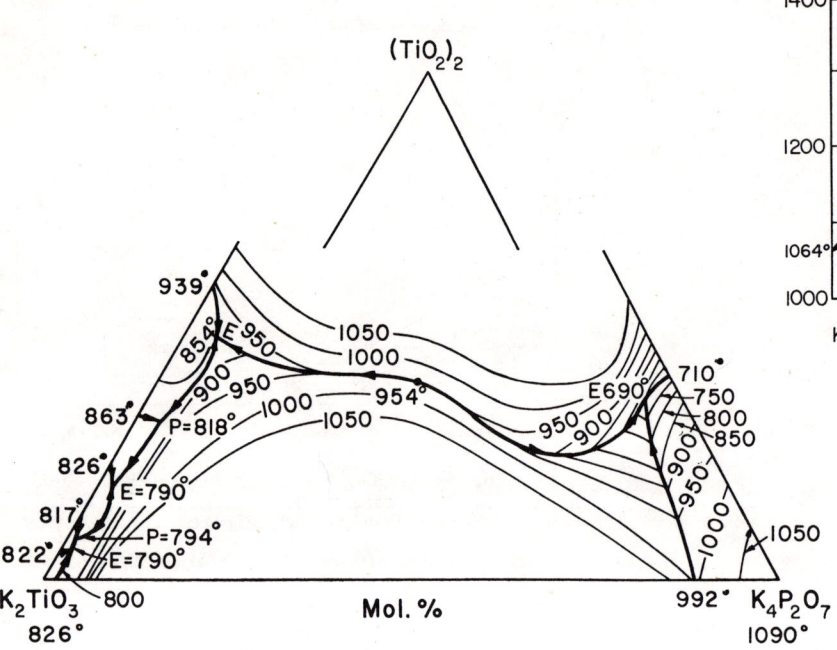

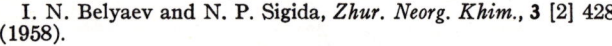

Fig. 423.—System K_2TiO_3–$K_4P_2O_7$–TiO_2.

I. N. Belyaev and N. P. Sigida, *Zhur. Neorg. Khim.*, **3** [2] 428 (1958).

$K_2O–Nb_2O_5–Ta_2O_5$

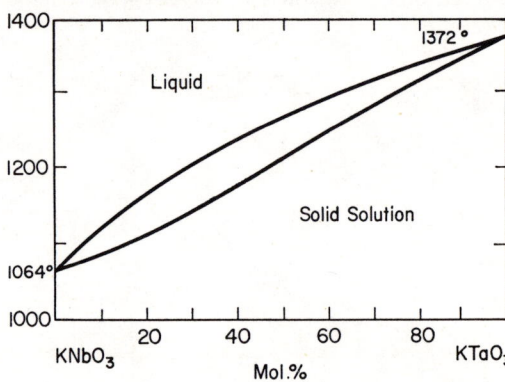

Fig. 424.—System $KNbO_3$–$KTaO_3$.

P. D. Garn and S. S. Flaschen, *Anal. Chem.*, **29** [2] 275 (1957). For similar diagram showing lower melting points of the end members, see also Arnold Reisman, Sol Triebwasser, and Frederic Holtzberg, *J. Am. Chem. Soc.*, **77** [16] 4230 (1955).

$K_2O–CrO_3–MoO_3$

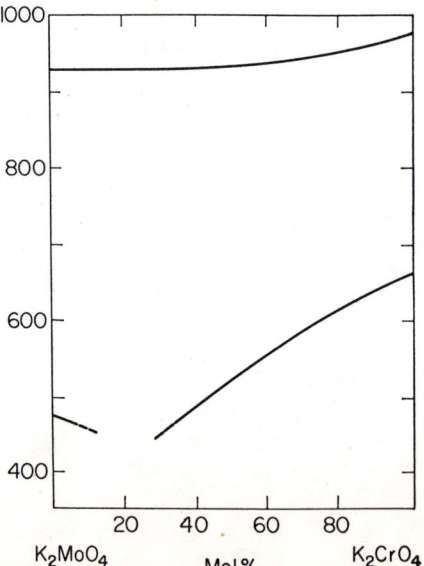

Fig. 425.—System K_2CrO_4–K_2MoO_4. Diagram does not obey phase rule.

M. Amadori, *Atti reale accad. Lincei, Sez. I*, **22**, 457 (1913).

$K_2O-CrO_3-WO_3$

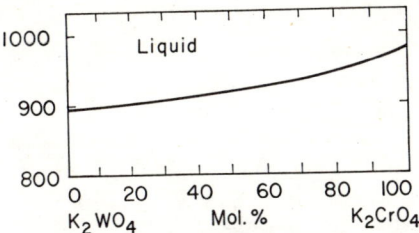

FIG. 426.—System K_2CrO_4–K_2WO_4.

M. Amadori, *Atti reale accad. Lincei, Sez. I*, **22**, 611 (1913).

$K_2O-MoO_3-WO_3$

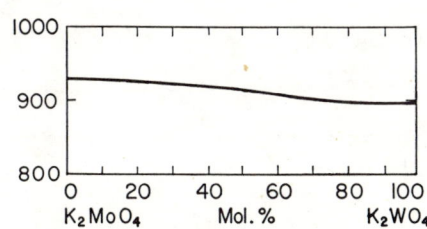

FIG. 427.—System K_2MoO_4–K_2WO_4.

M. Amadori, *Atti reale accad. Lincei, Sez. I*, **22**, 612 (1913).

$Li_2O-Na_2O-B_2O_3$

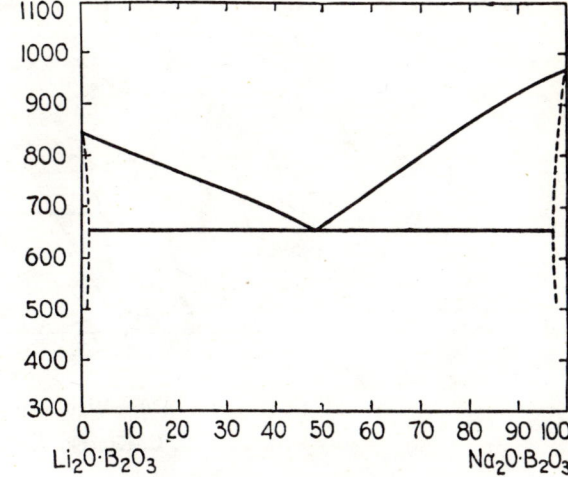

FIG. 429.—System $Li_2O \cdot B_2O_3$–$Na_2O \cdot B_2O_3$.

H. S. van Klooster, *Z. anorg. u. allgem. Chem.*, **69**, 133 (1910).

$K_2O-(CrO_3, SO_3, MoO_3, WO_3)-P_2O_5$

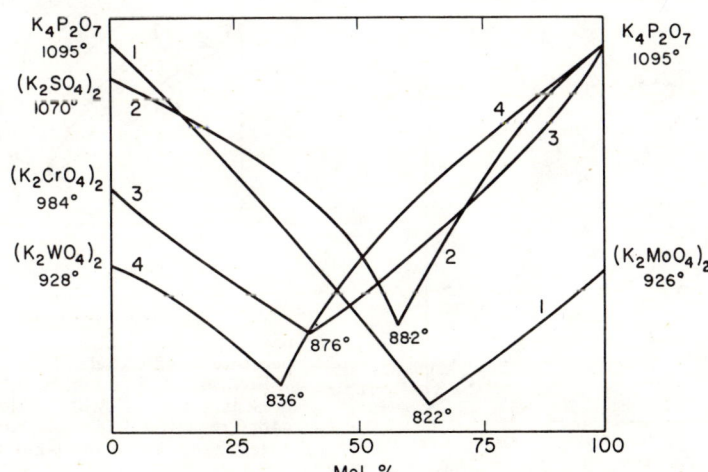

FIG. 428.—Systems: (1) $K_4P_2O_7$–K_2MoO_4; (2) $K_4P_2O_7$–K_2SO_4; (3) $K_4P_2O_7$–K_2CrO_4; (4) $K_4P_2O_7$–K_2WO_4.

A. G. Bergman and M. L. Sholokhovich, *Zhur. Obshcheĭ Khim.*, **24**, 594 (1954).

Li_2O–Na_2O–SiO_2

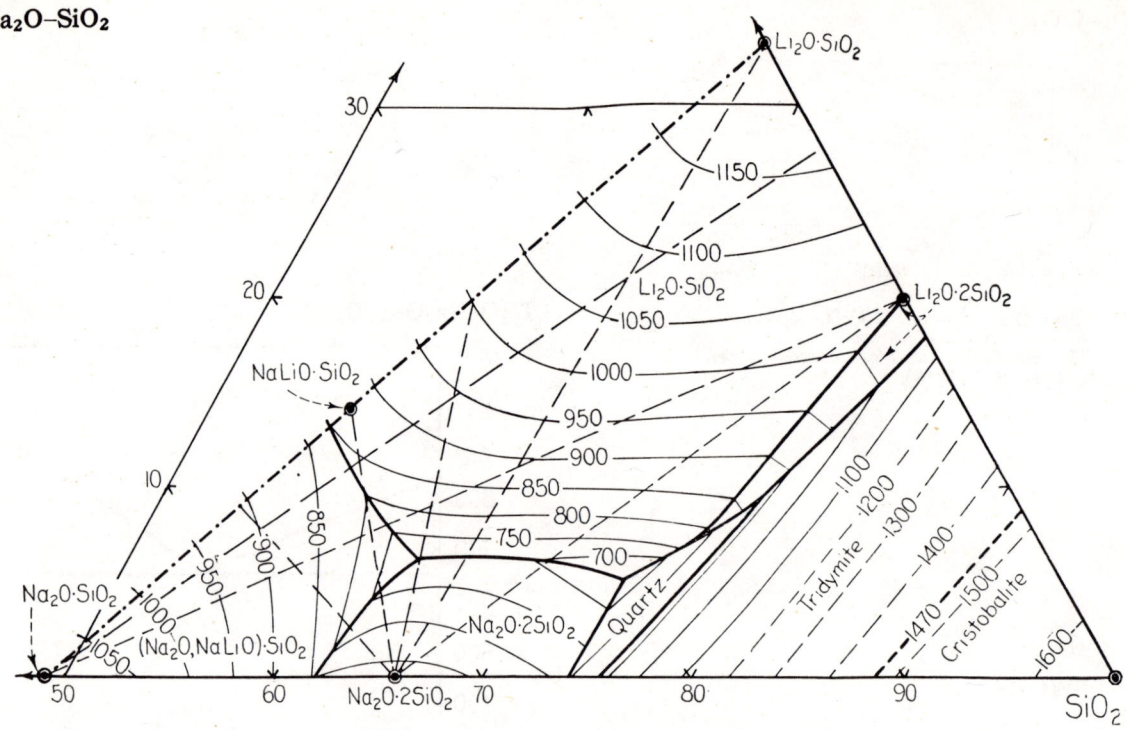

Fig. 430.—High SiO_2 corner of system Li_2O–Na_2O–SiO_2.

F. C. Kracek, *J. Am. Chem. Soc.*, **61**, 2871 (1939).

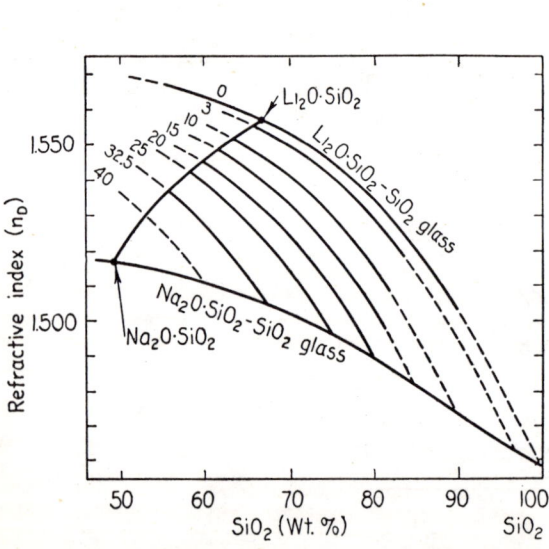

Fig. 431.—Refractive indices of quenched glasses in system SiO_2–$Li_2O\cdot SiO_2$–$Na_2O\cdot SiO_2$; numbers indicate per cent of Na_2O.

F. C. Kracek, *J. Am. Chem. Soc.*, **61**, 2876 (1939).

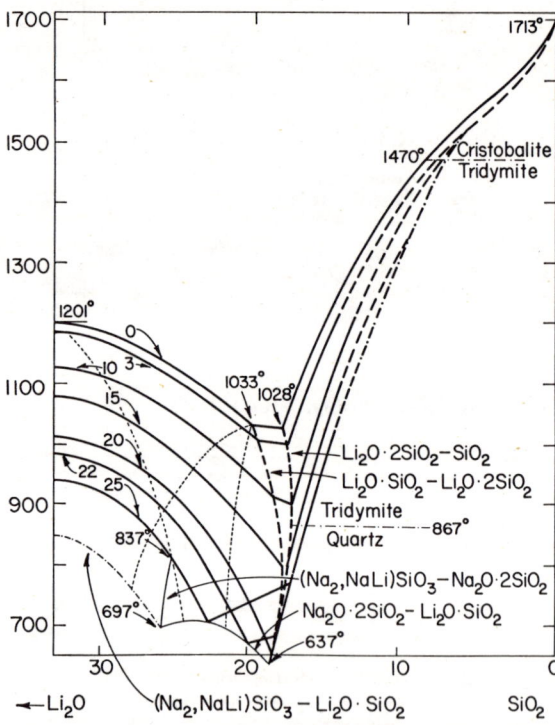

Fig. 432.—Temperature-composition projection of part of system SiO_2–$Li_2O\cdot SiO_2$–$Na_2O\cdot SiO_2$ onto temperature-composition plane of system SiO_2–$Li_2O\cdot SiO_2$.

Figures on liquidus sections indicate per cent Na_2O present; the following table indicates amount of Li_2O added to shift abscissas so that metasilicate composition is at left-hand border of diagram:

Na_2O (%)	Li_2O added (%)	Na_2O (%)	Li_2O added (%)
3	1.96	20	13.08
10	6.54	22	14.39
15	9.81	25	16.35

F. C. Kracek, *J. Am. Chem. Soc.*, **61**, 2673 (1939).

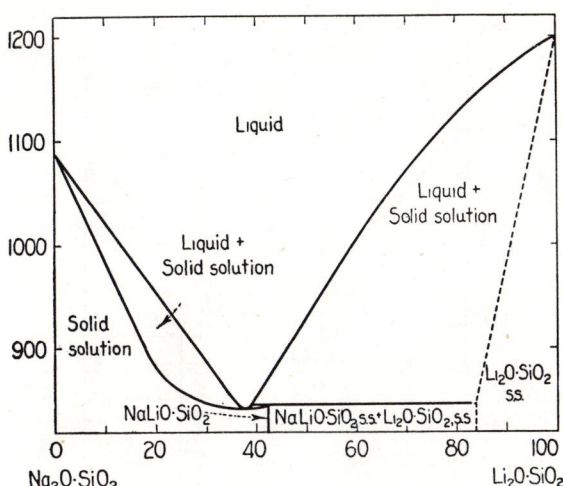

FIG. 433.—System $Li_2O \cdot SiO_2$–$Na_2O \cdot SiO_2$.

F. C. Kracek, *J. Am. Chem. Soc.*, **61**, 2159 (1939).

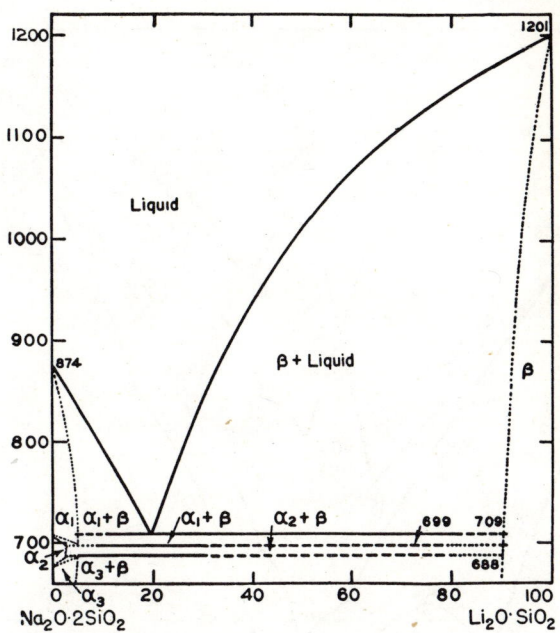

FIG. 435.—System $Li_2O \cdot SiO_2$–$Na_2O \cdot 2SiO_2$; solid solution relations (estimated) indicated by dotted lines.

F. C. Kracek, *J. Am. Chem. Soc.*, **61**, 2874 (1939).

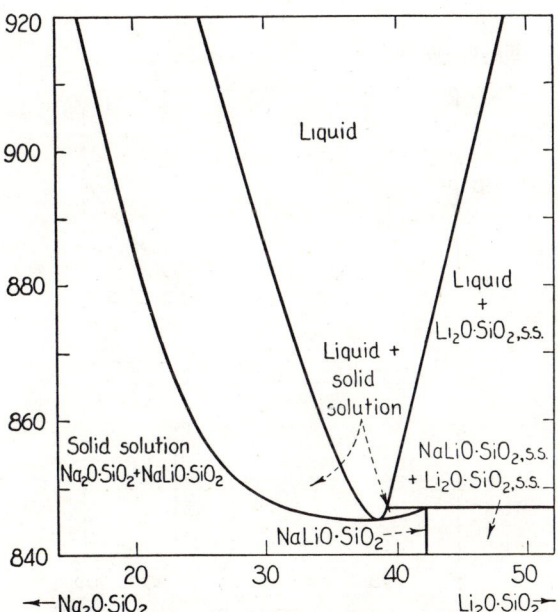

FIG. 434.—Enlargement of portion of system $Li_2O \cdot SiO_2$–$Na_2O \cdot SiO_2$, Fig. 433.

F. C. Kracek, *J. Am. Chem. Soc.*, **61**, 2160 (1939).

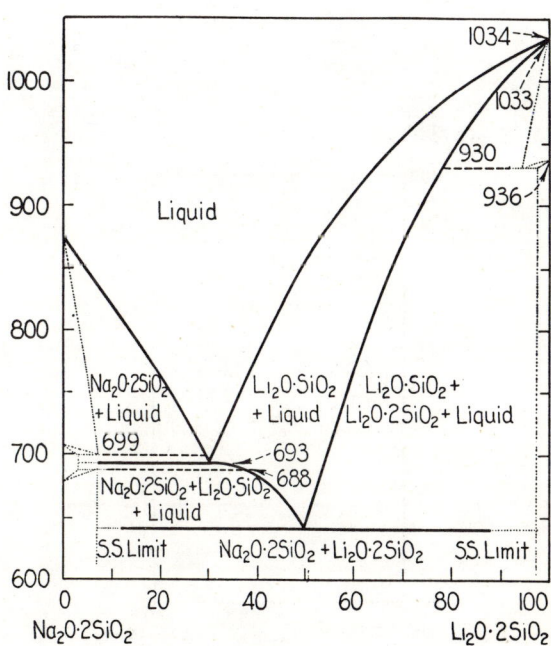

FIG. 436.—System $Li_2O \cdot 2SiO_2$–$Na_2O \cdot 2SiO_2$; solid solution relations (estimated) indicated by dotted lines.

F. C. Kracek, *J. Am. Chem. Soc.*, **61**, 2874 (1939).

$Li_2O-(K_2O, Na_2O, BaO, CaO)-SiO_2$

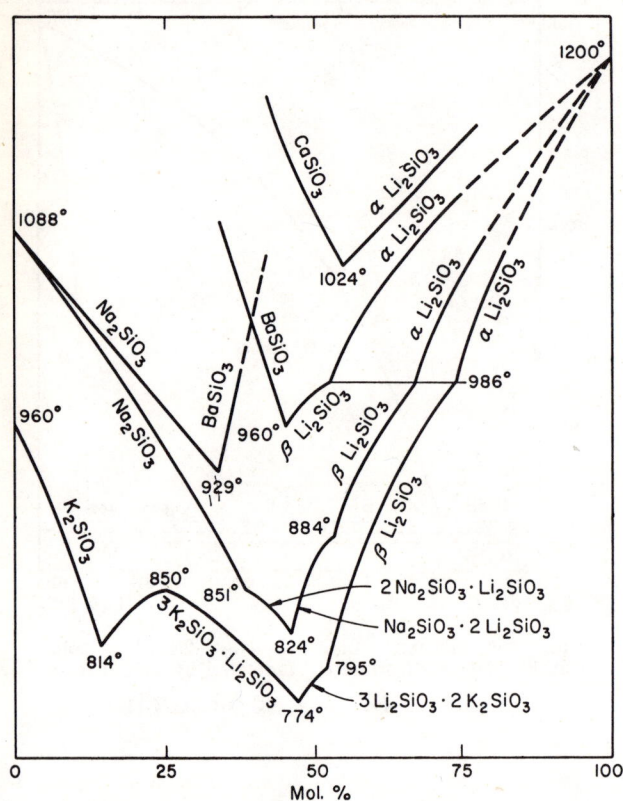

Fig. 437.—Systems $Li_2SiO_3-(R_2O, RO)SiO_2$.

A. G. Bergman, A. K. Nesterova, and N. A. Bychkova, *Doklady Akad. Nauk S.S.S.R.*, **101**, 483 (1955).

$Li_2O-Na_2O-MoO_3$

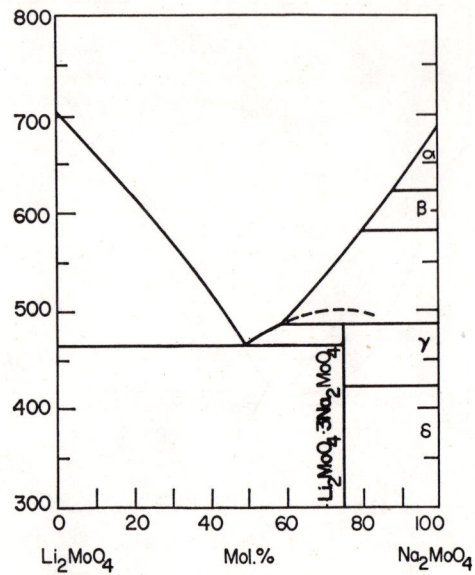

Fig. 439.—System $Li_2MoO_4-Na_2MoO_4$.

F. Hoermann, *Z. anorg. u. allgem. Chem.*, **177**, 175 (1928–29).

$Li_2O-Na_2O-TiO_2$

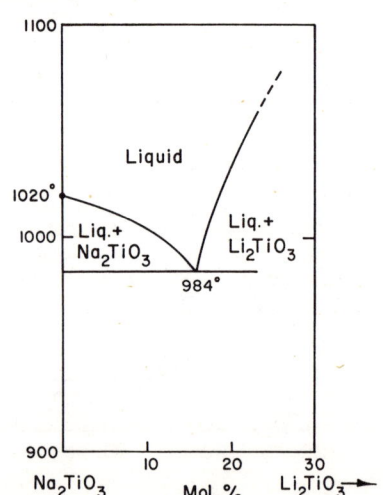

Fig. 438.—System $Li_2TiO_3-Na_2TiO_3$.

I. N. Belyaev and N. P. Sigida, *Zhur. Neorg. Khim.*, **2** [5] 1120 (1957).

$Li_2O-Na_2O-WO_3$

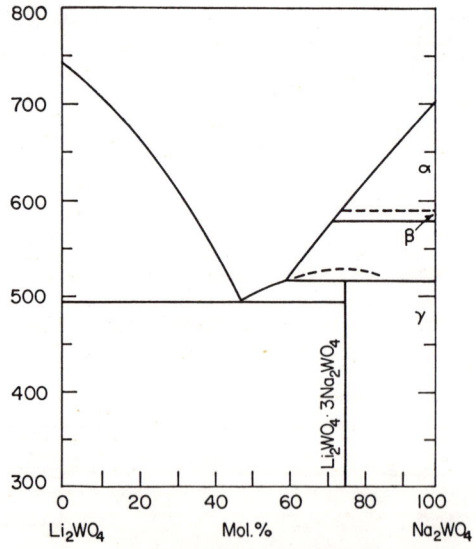

Fig. 440.—System $Li_2WO_4-Na_2WO_4$.

F. Hoermann, *Z. anorg. u. allgem. Chem.*, **177**, 177 (1928–29).

Li_2O–BaO–SiO_2

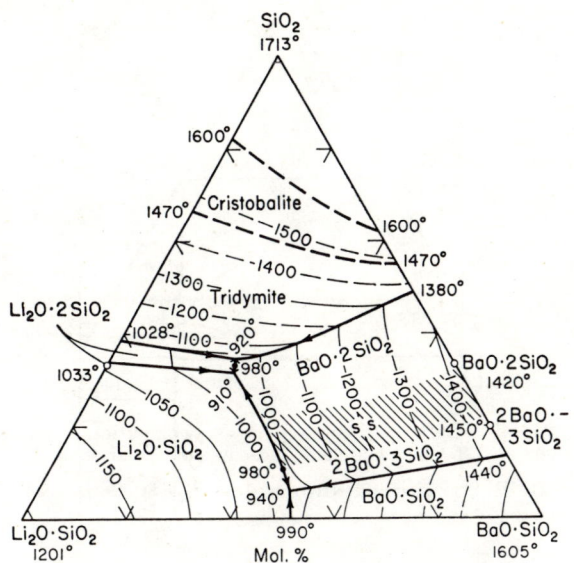

Fig. 441.—System $Li_2O \cdot SiO_2$–$BaO \cdot SiO_2$–SiO_2.

Adolf Dietzel, Helmut Wickert, and Nina Köppen, *Glastech. Ber.*, **27** [5] 150 (1954).

Li_2O–MgO–SiO_2

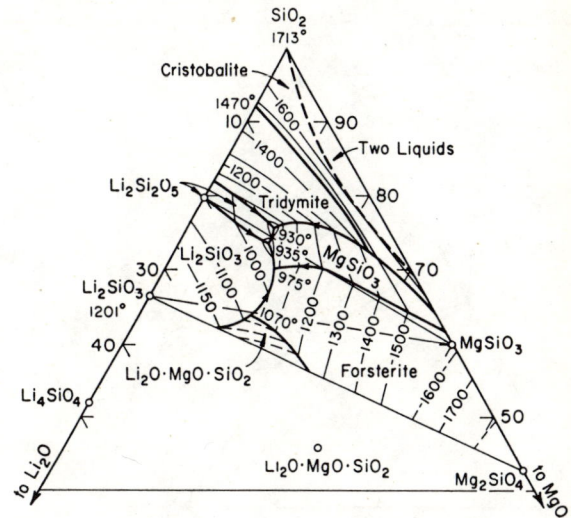

Fig. 443.—System Li_2SiO_3–Mg_2SiO_4–SiO_2.

M. K. Murthy and F. A. Hummel, *J. Am. Ceram. Soc.*, **38** [2] 59 (1955).

Li_2O–CaO–SiO_2

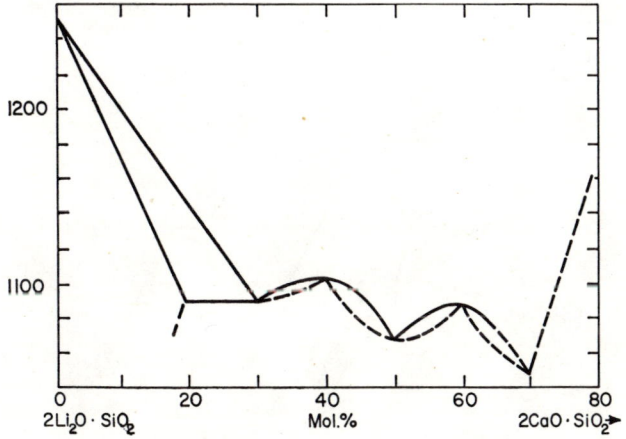

Fig. 442.—System $2Li_2O \cdot SiO_2$–$2CaO \cdot SiO_2$.

R. Schwartz and A. Haacke, *Z. anorg. u. allgem. Chem.*, **115**, 95 (1921).

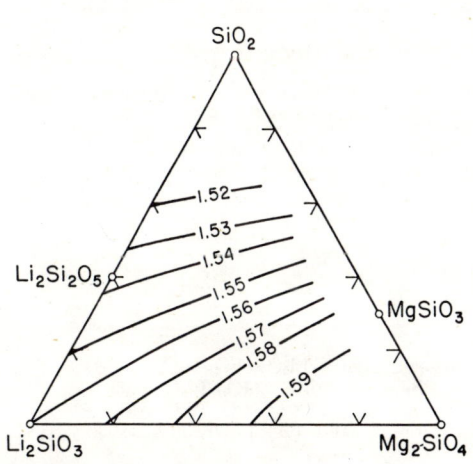

Fig. 444.—System Li_2SiO_3–Mg_2SiO_4–SiO_2, isofracts.

M. K. Murthy and F. A. Hummel, *J. Am. Ceram. Soc.*, **38** [2] 58 (1955).

Li_2O–Al_2O_3–B_2O_3

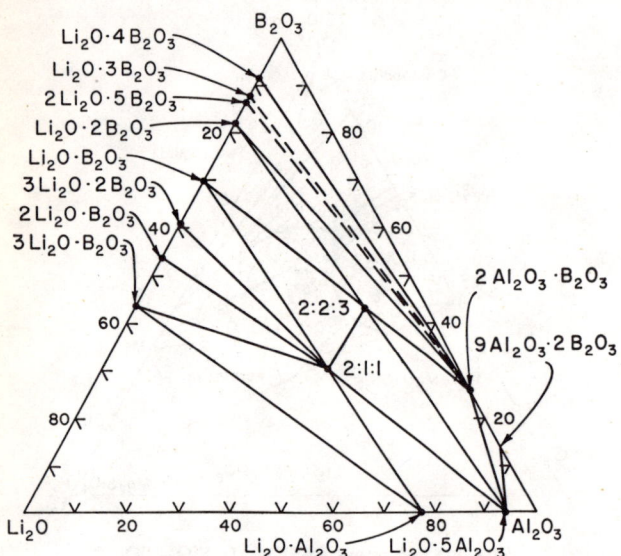

FIG. 445.—System Li_2O–Al_2O_3–B_2O_3; compatibility triangles.

K. H. Kim and F. A. Hummel, *J. Am. Ceram. Soc.*, **45** [10] 488 (1962).

Li_2O–Al_2O_3–Fe_2O_3

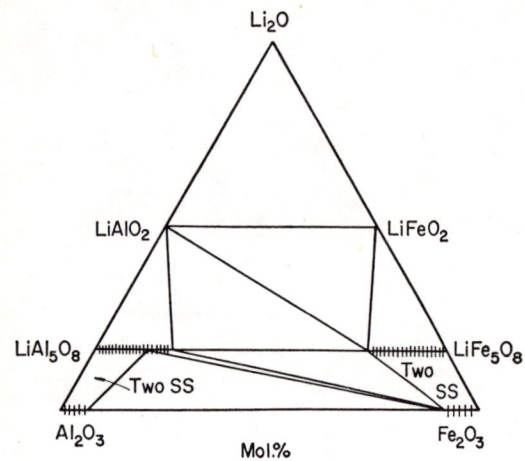

FIG. 446.—System Li_2O–Al_2O_3–Fe_2O_3; compatibility relations at 940°C.

D. W. Strickler and Rustum Roy, *J. Am. Ceram. Soc.*, **44** [5] 229 (1961).

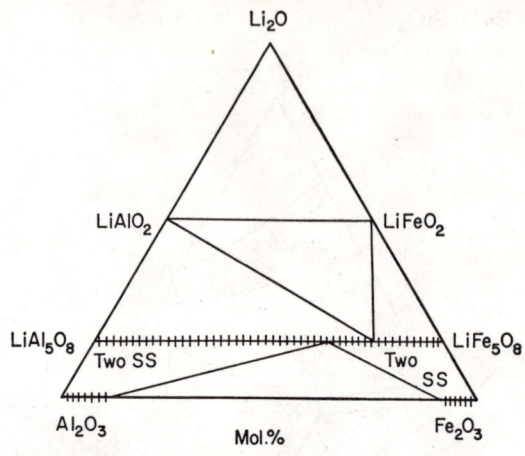

FIG. 447.—System Li_2O–Al_2O_3–Fe_2O_3; compatibility relations at 1250°C. Hatched lines indicate solid solutions.

D. W. Strickler and Rustum Roy, *J. Am. Ceram. Soc.*, **44** [5] 229 (1961).

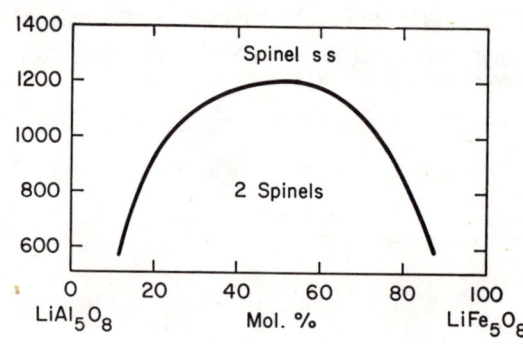

FIG. 448.—System $LiAl_5O_8$–$LiFe_5O_8$.

D. W. Strickler and Rustum Roy, *J. Am. Ceram. Soc.*, **44** [5] 229 (1961).

Li_2O–Al_2O_3–SiO_2

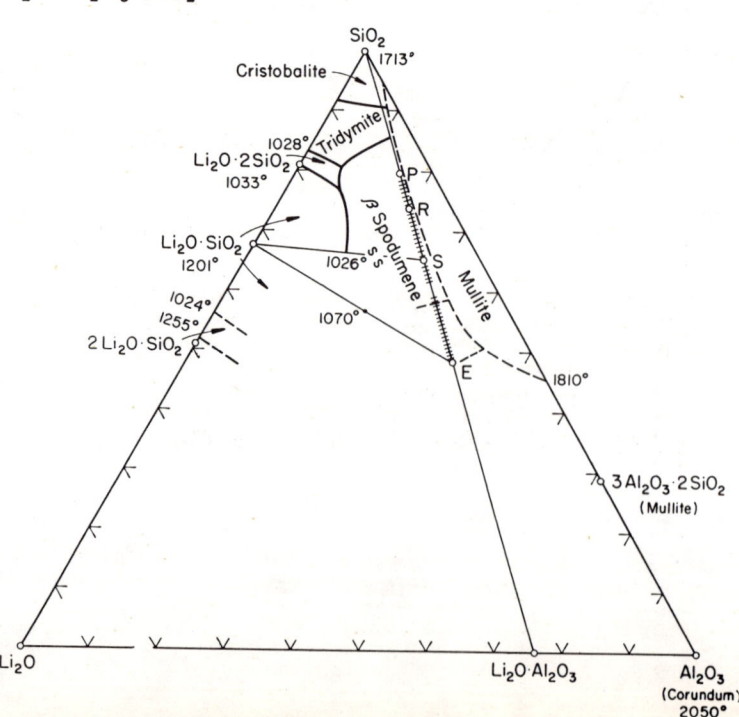

FIG. 449.—System Li_2O–Al_2O_3–SiO_2. (P) $Li_2O \cdot Al_2O_3 \cdot 8SiO_2$, petalite, (R) $Li_2O \cdot Al_2O_3 \cdot 6SiO_2$, "lithium orthoclase," (S) $Li_2O \cdot Al_2O_3 \cdot 4SiO_2$, spodumene, and (E) $Li_2O \cdot Al_2O_3 \cdot 2SiO_2$, eucryptite.

Rustum Roy and E. F. Osborn, *J. Am. Ceram. Soc.*, **71** [6] 2086 (1949); slightly modified by M. Krishna Murthy and F. A. Hummel, *ibid.*, **37** [1] 17 (1954).

$Li_2O-Al_2O_3-SiO_2$ (cont.)

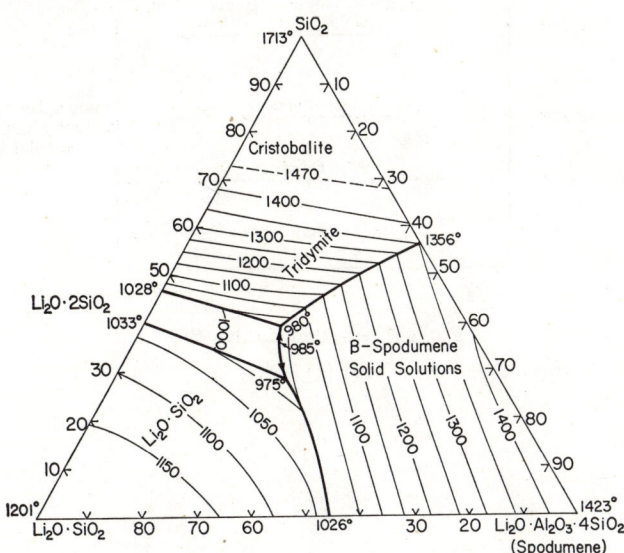

FIG. 450.—System $Li_2O \cdot SiO_2 - Li_2O \cdot Al_2O_3 \cdot 4SiO_2 - SiO_2$.

Rustum Roy and E. F. Osborn, *J. Am. Chem. Soc.*, **71** [6] 2092 (1949).

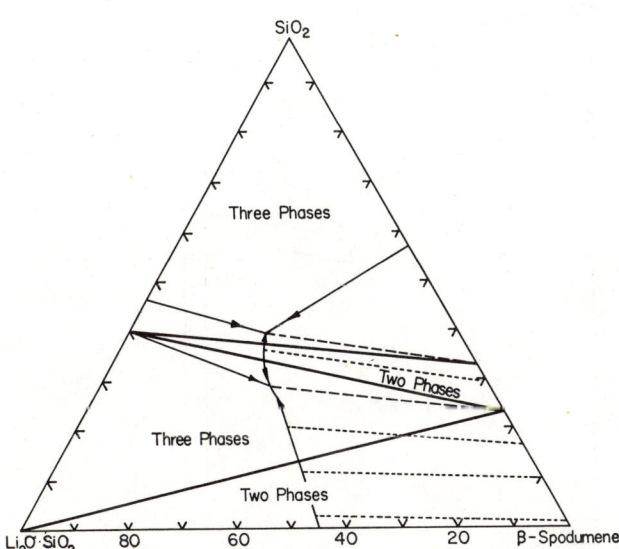

FIG. 451.—System $Li_2O \cdot SiO_2 - Li_2O \cdot Al_2O_3 \cdot 4SiO_2 - SiO_2$. Phase relations just below 975°C. showing three-phase boundaries.

Rustum Roy and E. F. Osborn, *J. Am. Chem. Soc.*, **71** [6] 2095 (1949).

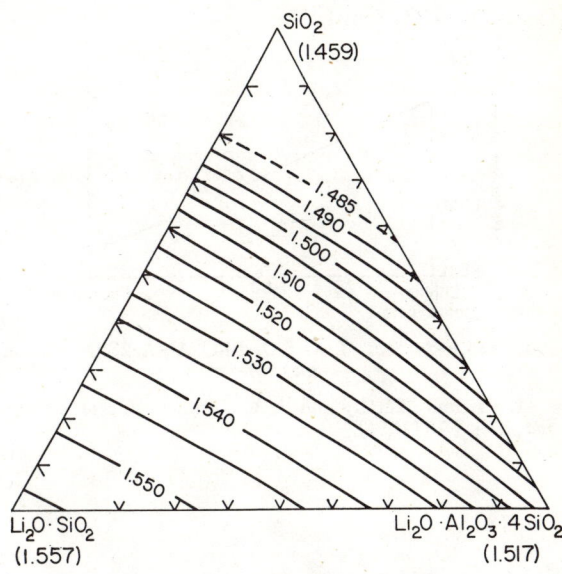

FIG. 452.—System $Li_2O \cdot SiO_2 - Li_2O \cdot Al_2O_3 \cdot 4SiO_2 - SiO_2$; refractive indices of glasses.

Rustum Roy and E. F. Osborn, *J. Am. Chem. Soc.*, **71** [6] 2090 (1949).

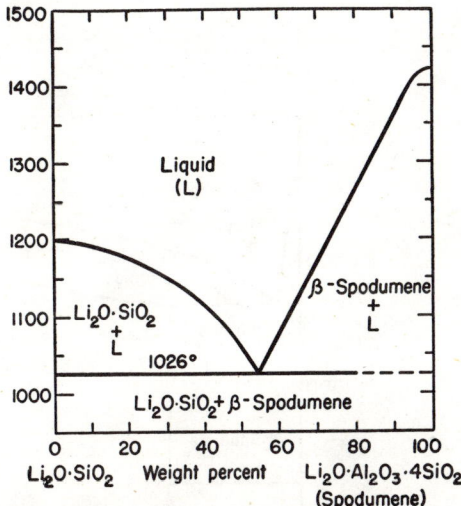

FIG. 453.—System $Li_2O \cdot SiO_2 - Li_2O \cdot Al_2O_3 \cdot 4SiO_2$.

Rustum Roy and E. F. Osborn, *J. Am. Chem. Soc.*, **71** [6] 2091 (1949).

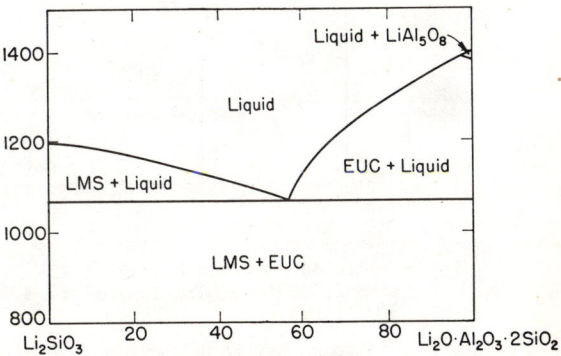

FIG. 454.—System $Li_2O \cdot SiO_2 - Li_2O \cdot Al_2O_3 \cdot 2SiO_2$. LMS—lithium metasilicate (Li_2SiO_3); Euc—eucryptite ($Li_2O \cdot Al_2O_3 \cdot 2SiO_2$).

M. Krishna Murthy and F. A. Hummel, *J. Am. Ceram. Soc.*, **37** [1] 16 (1954).

$Li_2O-Al_2O_3-SiO_2$ (cont.)

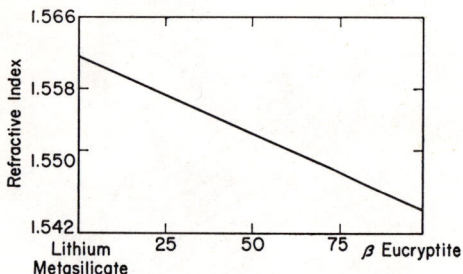

Fig. 455.—System $Li_2O \cdot SiO_2-Li_2O \cdot Al_2O_3 \cdot 2SiO_2$ (n of glasses).

M. Krishna Murthy and F. A. Hummel, *J. Am. Ceram. Soc.*, **37** [1] 15 (1954).

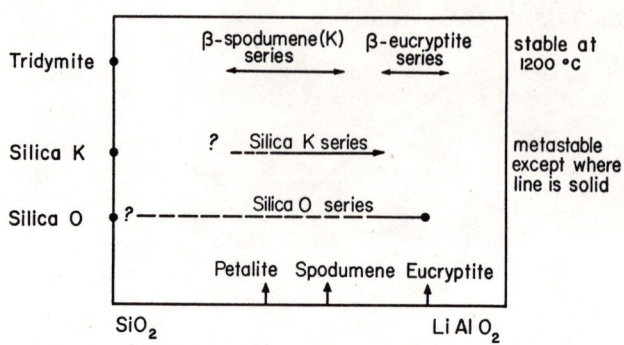

Fig. 457.—System $LiAlO_2-SiO_2$.

Rustum Roy, *Z. Krist.*, **111** [3] 186 (1959).

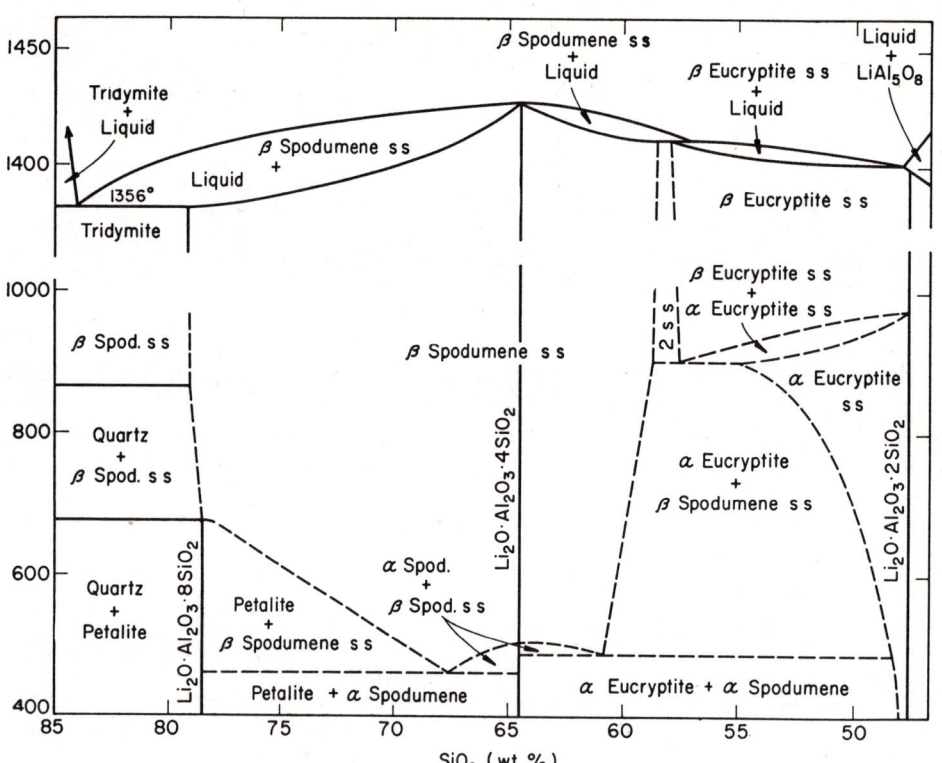

Fig. 456.—System $Li_2O \cdot Al_2O_3 \cdot 2SiO_2$ (eucryptite)–SiO_2; high temperature part of diagram after R. A. Hatch, *Am. Mineral.*, **28**, 471–96 (1943); lower part represents conception at about 10,000 psi.

Rustum Roy, D. M. Roy, and E. F. Osborn, *J. Am. Ceram. Soc.*, **33** [5] 156 (1950).

For subsolidus relationships, see also, H. Saalfeld, *Ber. Deut. Keram. Ges.*, **38** [7] 283 (1961).

Li$_2$O–Al$_2$O$_3$–SiO$_2$ (concl.)

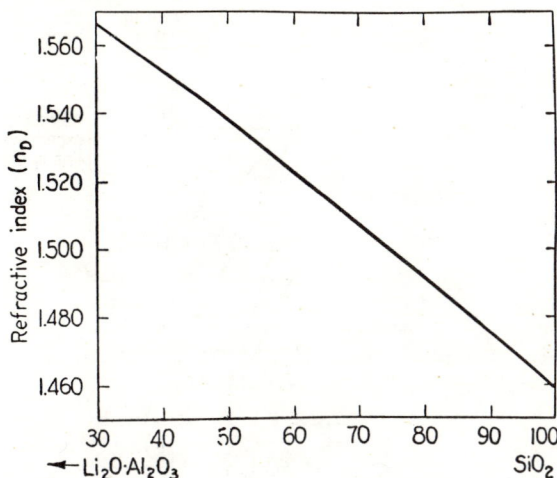

FIG. 458.—Refractive index-composition curves of glasses in system SiO$_2$–Li$_2$O·Al$_2$O$_3$.

R. A. Hatch, *Am. Mineral.*, **28**, 476 (1943).

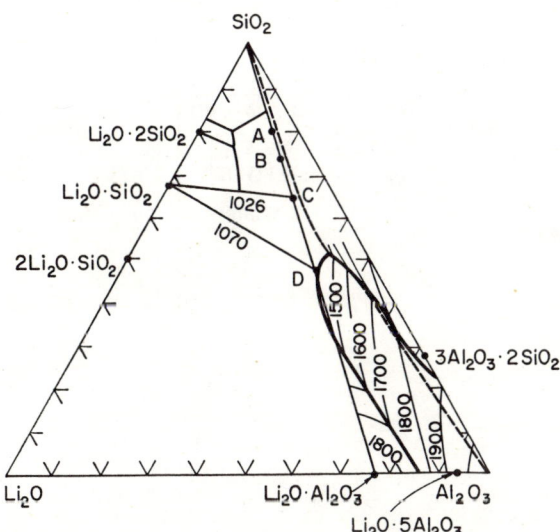

FIG. 459.—System Li$_2$O–Al$_2$O$_3$–SiO$_2$; alumina corner. A = Li$_2$O·Al$_2$O$_3$·8SiO$_2$, B = Li$_2$O·Al$_2$O$_3$·6SiO$_2$, C = Li$_2$O·Al$_2$O$_3$·4SiO$_2$, D = Li$_2$O·Al$_2$O$_3$·2SiO$_2$.

F. Ya. Galakhov, *Izvest. Akad. Nauk S.S.S.R., Otdel. Khim. Nauk*, 579 (1959).

Li$_2$O–Al$_2$O$_3$–TiO$_2$

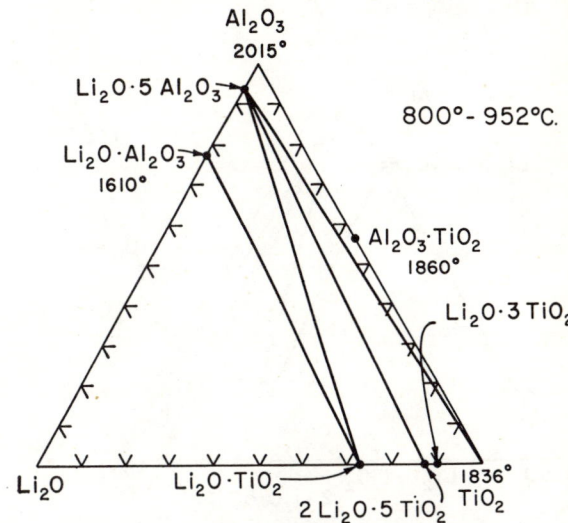

FIG. 460.—System Li$_2$O–Al$_2$O$_3$–TiO$_2$; compatibility relations between 800° and 952°C.

K. H. Kim and F. A. Hummel, *J. Am. Ceram. Soc.*, **43** [12] 612 (1960).

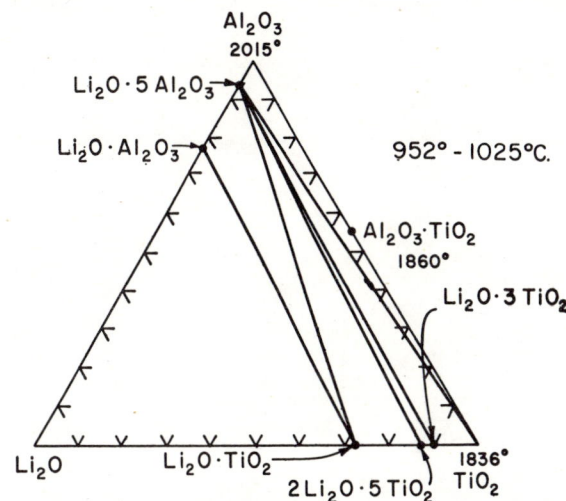

FIG. 461.—System Li$_2$O–Al$_2$O$_3$–TiO$_2$; compatibility relations between 952° and 1025°C.

K. H. Kim and F. A. Hummel, *J. Am. Ceram. Soc.*, **43** [12] 613 (1960).

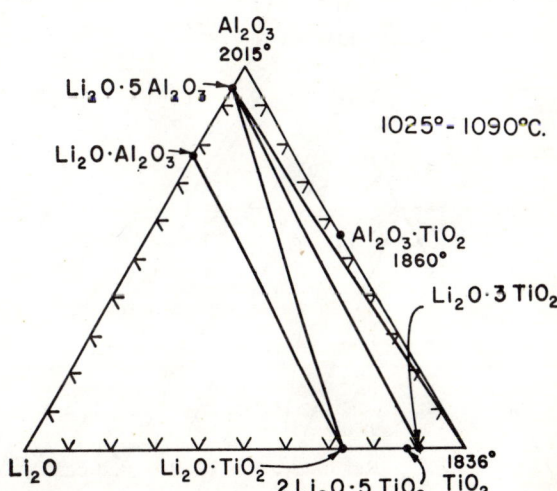

FIG. 462.—System Li$_2$O–Al$_2$O$_3$–TiO$_2$; compatibility relations between 1025° and 1090°C.

K. H. Kim and F. A. Hummel, *J. Am. Ceram. Soc.*, **43** [12] 613 (1960).

Li₂O–Al₂O₃–TiO₂ (concl.)

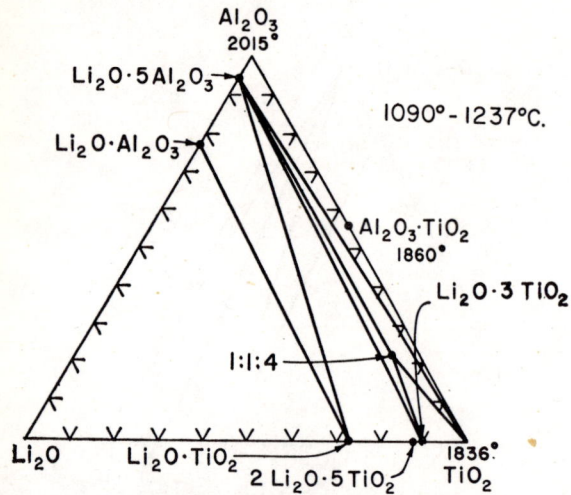

FIG. 463.—System Li₂O–Al₂O₃–TiO₂; compatibility relations between 1090° and 1237°C.

K. H. Kim and F. A. Hummel, *J. Am. Ceram. Soc.*, **43** [12] 613 (1960).

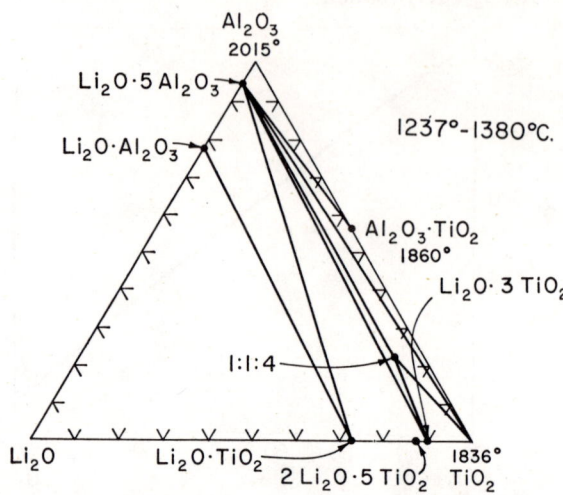

FIG. 464.—System Li₂O–Al₂O₃–TiO₂; compatibility relations between 1237° and 1380°C.

K. H. Kim and F. A. Hummel, *J. Am. Ceram. Soc.*, **43** [12] 613 (1960).

Li₂O–B₂O₃–SiO₂

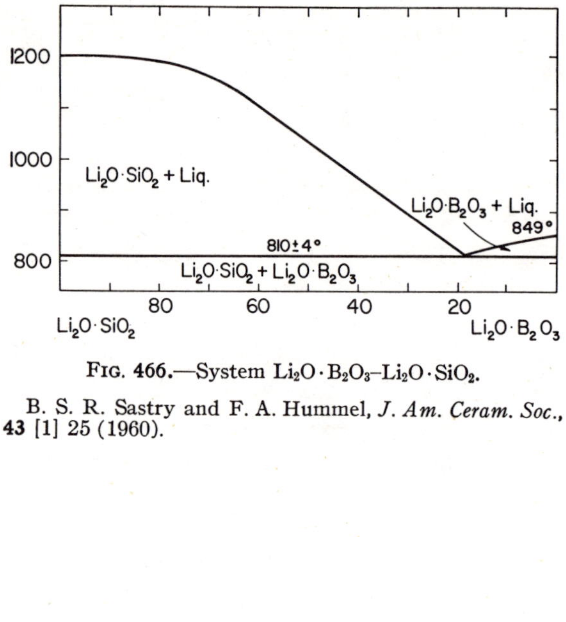

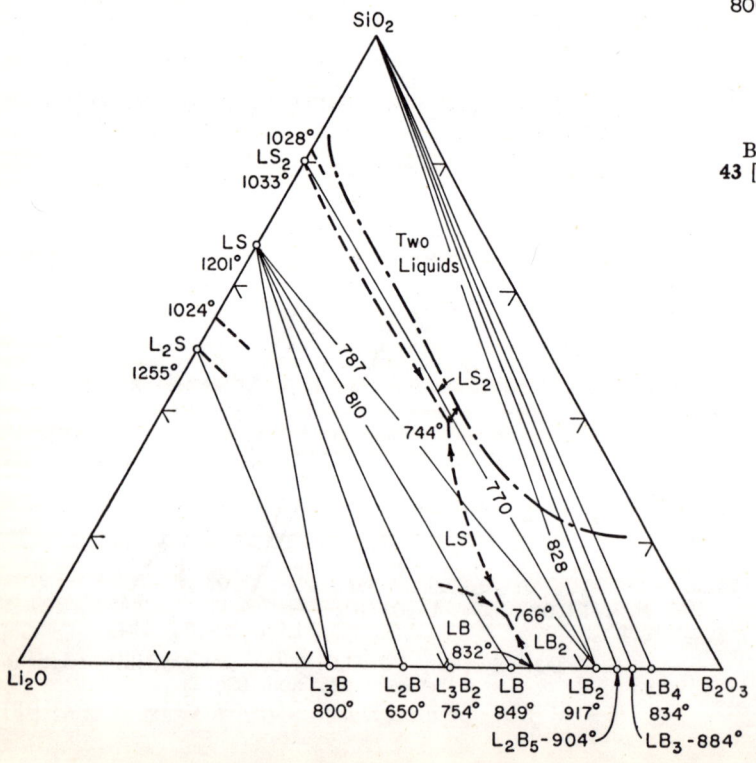

FIG. 466.—System Li₂O·B₂O₃–Li₂O·SiO₂.

B. S. R. Sastry and F. A. Hummel, *J. Am. Ceram. Soc.*, **43** [1] 25 (1960).

FIG. 465.—System Li₂O–B₂O₃–SiO₂, phase relationships. L = Li₂O; B = B₂O₃; S = SiO₂.

B. S. R. Sastry and F. A. Hummel, *J. Am. Ceram. Soc.*, **43** [1] 24 (1960).

$Li_2O-B_2O_3-SiO_2$ (concl.)

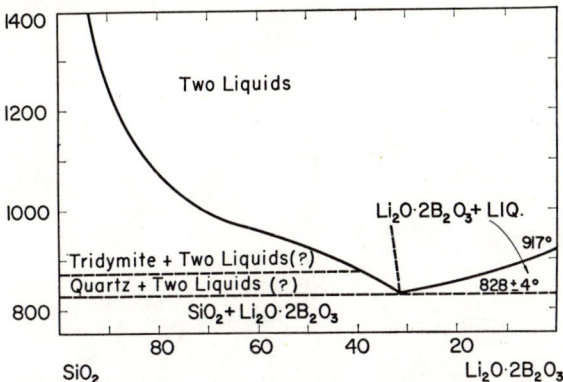

Fig. 467.—System $Li_2O \cdot 2B_2O_3-SiO_2$.

B. S. R. Sastry and F. A. Hummel, *J. Am. Ceram. Soc.*, **43** [1] 28 (1960).

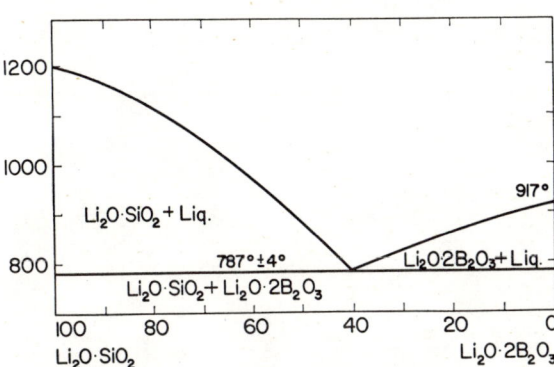

Fig. 468.—System $Li_2O \cdot 2B_2O_3-Li_2O \cdot SiO_2$.

B. S. R. Sastry and F. A. Hummel, *J. Am. Ceram. Soc.*, **43** [1] 27 (1960).

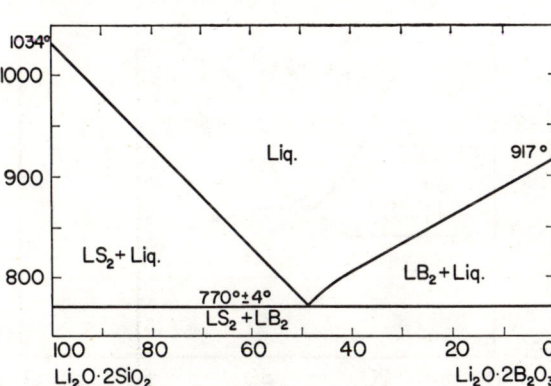

Fig. 469.—System $Li_2O \cdot 2B_2O_3-Li_2O \cdot 2SiO_2$. L = Li_2O, S = SiO_2.

B. S. R. Sastry and F. A. Hummel, *J. Am. Ceram. Soc.*, **43** [1] 27 (1960).

$Li_2O-B_2O_3-P_2O_5$

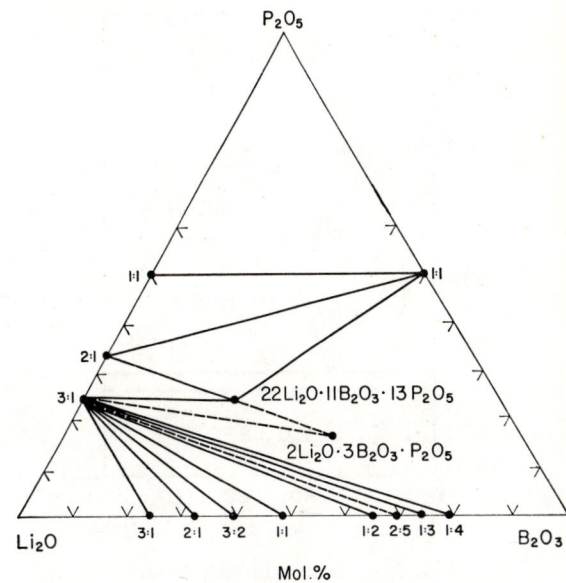

Fig. 470.—System $Li_2O-B_2O_3-P_2O_5$; compatibility relations. Binary compounds designated as mole ratios of components.

T. Y. Tien and F. A. Hummel, *J. Am. Ceram. Soc.*, **44** [8] 393 (1961).

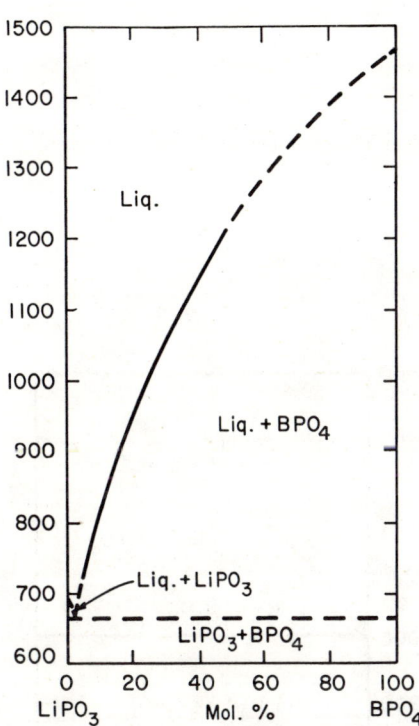

Fig. 471.—System $LiPO_3-BPO_4$.

T. Y. Tien and F. A. Hummel, *J. Am. Ceram. Soc.*, **44** [8] 392 (1961).

$Li_2O-B_2O_3-P_2O_5$ (concl.)

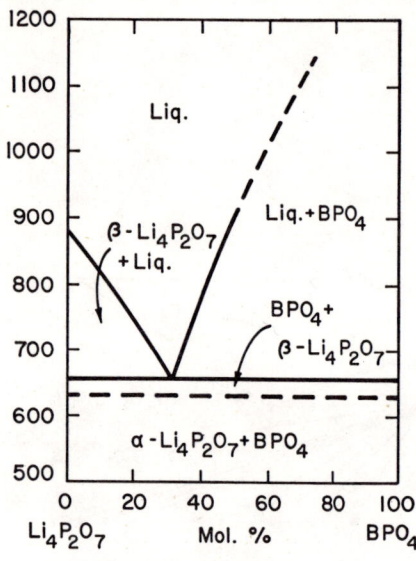

FIG. 472.—System $Li_4P_2O_7$–BPO_4.

T. Y. Tien and F. A. Hummel, *J. Am. Ceram. Soc.*, 44 [8] 393 (1961).

$Li_2O-SiO_2-TiO_2$

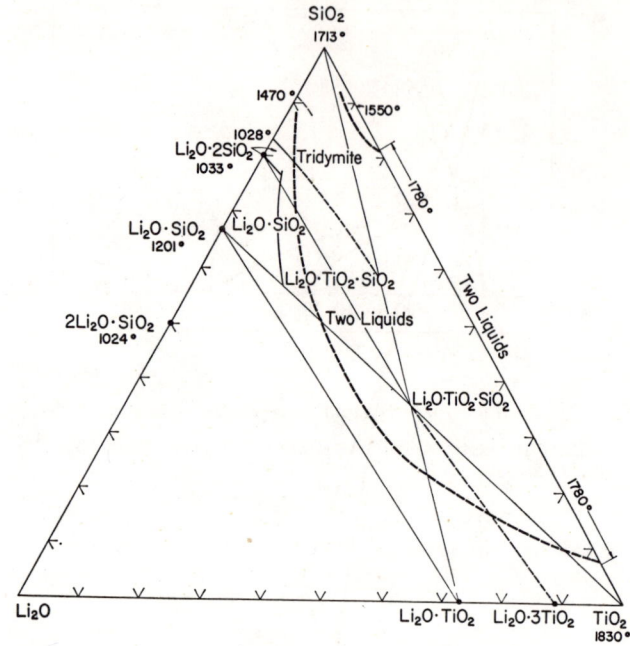

FIG. 474.—System Li_2O–SiO_2–TiO_2.

K. H. Kim and F. A. Hummel, *J. Am. Ceram. Soc.*, 42 [6] 287 (1959).

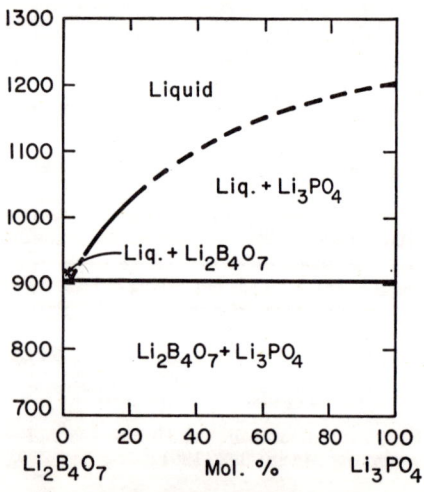

FIG. 473.—System $Li_2B_4O_7$–Li_3PO_4.

T. Y. Tien and F. A. Hummel, *J. Am. Ceram. Soc.*, 44 [8] 393 (1961).

$Li_2O-SiO_2-ZrO_2$

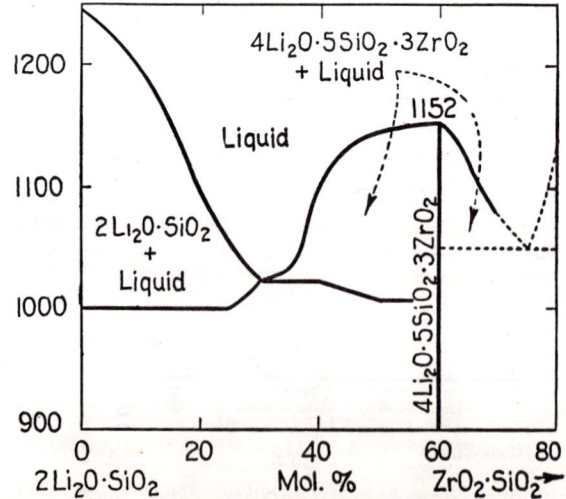

FIG. 475.—System $2Li_2O \cdot SiO_2$–$ZrO_2 \cdot SiO_2$.

R. Schwartz, *Z. anorg. u. allgem. Chem.*, 115, 90 (1921).

$Li_2O-TiO_2-R_2X$

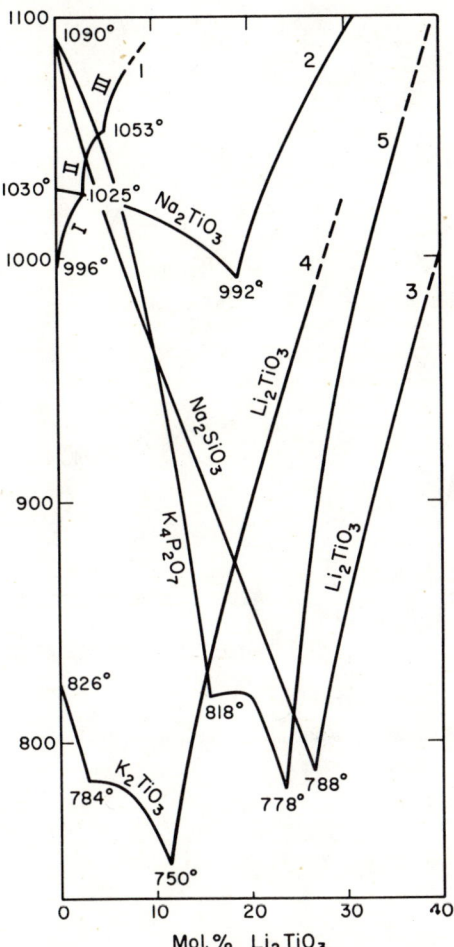

FIG. 476.—Li_2TiO_3–alkali compounds. (1) $Li_4P_2O_7$–$2Li_2TiO_3$. (2) Na_2TiO_3–Li_2TiO_3. (3) Na_2SiO_3–Li_2TiO_3. (4) K_2TiO_3–Li_2TiO_3. (5) $K_4P_2O_7$–$2Li_2TiO_3$.

I. N. Belyaev and N. P. Sigida, *Zhur. Obshcheĭ Khim.*, 26, 1955 (1956).

$Na_2O-BaO-Fe_2O_3$

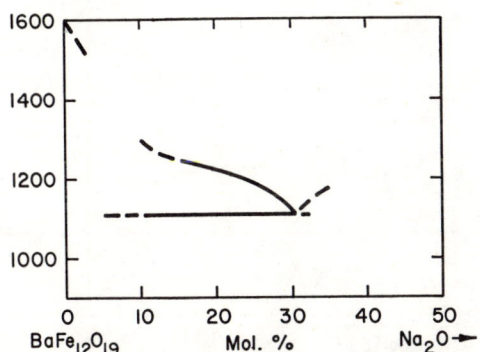

FIG. 477.—System Na_2O–$BaFe_{12}O_{19}$; pseudobinary.

R. J. Gambino and F. Leonhard, *J. Am. Ceram. Soc.*, 44 [5] 222 (1961).

$Na_2O-BaO-SiO_2$

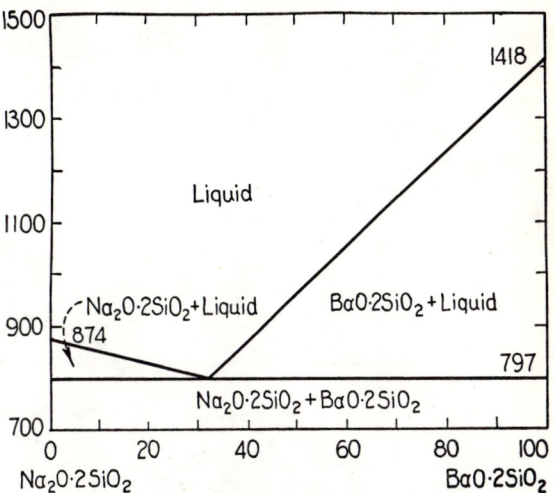

FIG. 478.—System $Na_2O \cdot 2SiO_2$–$BaO \cdot 2SiO_2$.

K. T. Greene and W. R. Morgan, *J. Am. Ceram. Soc.*, 24 [4] 114 (1941).

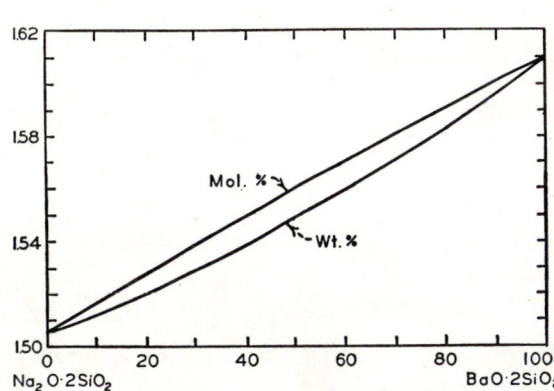

FIG. 479.—Refractive index-composition curves of glasses in system $Na_2O \cdot 2SiO_2$–$BaO \cdot 2SiO_2$.

K. T. Greene and W. R. Morgan, *J. Am. Ceram. Soc.*, 24 [4] 113 (1941).

Na₂O–CaO–Al₂O₃

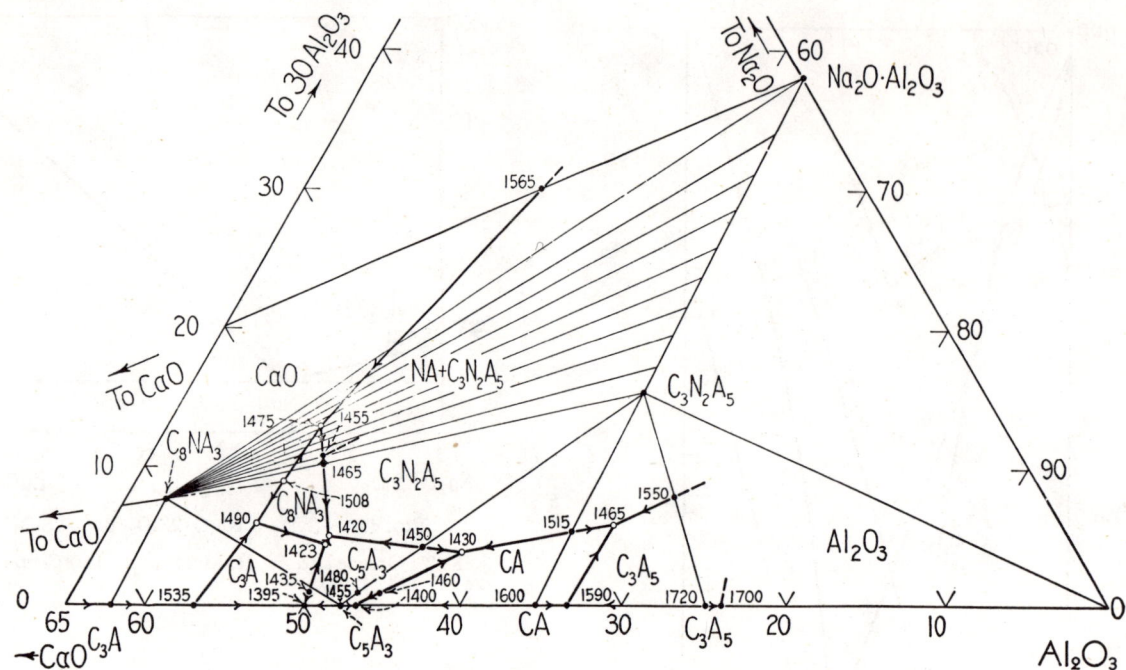

FIG. 480.—High Al₂O₃ corner of system Na₂O–CaO–Al₂O₃.

L. T. Brownmiller and R. H. Bogue, *Bur. Standards J. Research*, 8 [2] 293 (1932); R. P. 414.

Na₂O–CaO–SiO₂

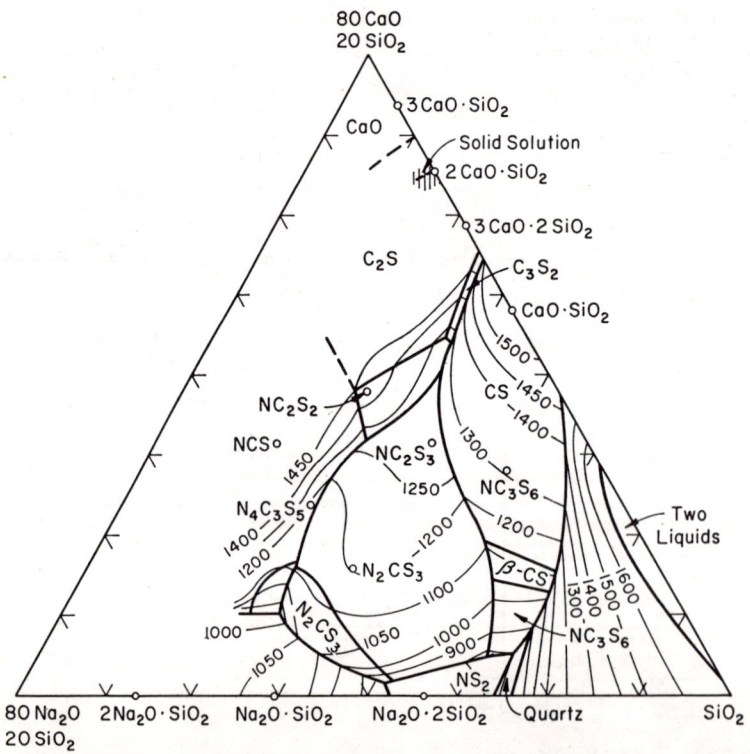

FIG. 481.—System Na₂O–CaO–SiO₂. N = Na₂O; C = CaO; S = SiO₂

The area NS–CS–SiO₂ after Morey and Bowen, Fig 482.
E. R. Segnit, *Am. J. Sci.*, 251 [8] 590 (1953).

Na$_2$O–CaO–SiO$_2$ (cont.)

Compounds	CaO	Na$_2$O	SiO$_2$	Temp.	
SiO$_2$			100.0	1710	M
α CaO·SiO$_2$	48.3		51.7	1540	M
β CaO·SiO$_2$	48.3		51.7	1180	I
Na$_2$O·SiO$_2$		50.8	49.2	1088	M
Na$_2$O·2SiO$_2$		34.1	65.9	874	M
2Na$_2$O·CaO·3SiO$_2$	15.6	34.4	50.0	1141	D
Na$_2$O·3CaO·6SiO$_2$	28.5	10.5	61.0	1047	D
Na$_2$O·2CaO·3SiO$_2$	31.6	17.5	50.9	1284	M

M = Melting Point
D = Decomposition Point
I = Inversion Point

Point		Crystal Phases	CaO	Na$_2$O	SiO$_2$	Temp.
A	△	NS–N$_2$CS$_3$	3.0			1060
B	●	N$_2$CS$_3$–NC$_2$S$_3$	11.5			1141
C	△	NC$_2$S$_3$–αCS	33.0			1280
D	△	αCS–S	37.0			1436
E	⊙	T–Quartz		24.3	75.7	870
F	△	Quartz–NS$_2$		26.4	73.6	790
K	∗	NS$_2$–NS–N$_2$CS$_3$	1.8	37.5	60.7	821
L	○	N$_2$CS$_3$–NC$_2$S$_3$–NS$_2$	2.0	36.6	61.4	827
N	○	NS$_2$–NC$_2$S$_3$–NC$_3$S$_6$	5.2	24.1	70.7	740
O	∗	NC$_3$S$_6$–Q–NS$_2$	5.2	21.3	73.5	725
P	⊙	Q–NC$_3$S$_6$–T	7.0	18.7	74.3	870
Q	○	T–βCS–NC$_3$S$_6$	12.9	13.7	73.4	1035
R	○	NC$_3$S$_6$–NC$_2$S$_3$–βCS	14.5	19.0	66.5	1030
S	○	βCS–NC$_2$S$_3$–αCS	19.5	17.7	62.8	1110
T	●	αCS–S–βCS	15.6	11.4	73.0	1110
I	△	NS–NS$_2$		38.0		840

△ Binary Eutectic
∗ Ternary Eutectic
● Decomposition Point
○ Reaction Point
⊙ Inversion Point

C = CaO N = Na$_2$O
S = SiO$_2$ Q = Quartz
T = Tridymite

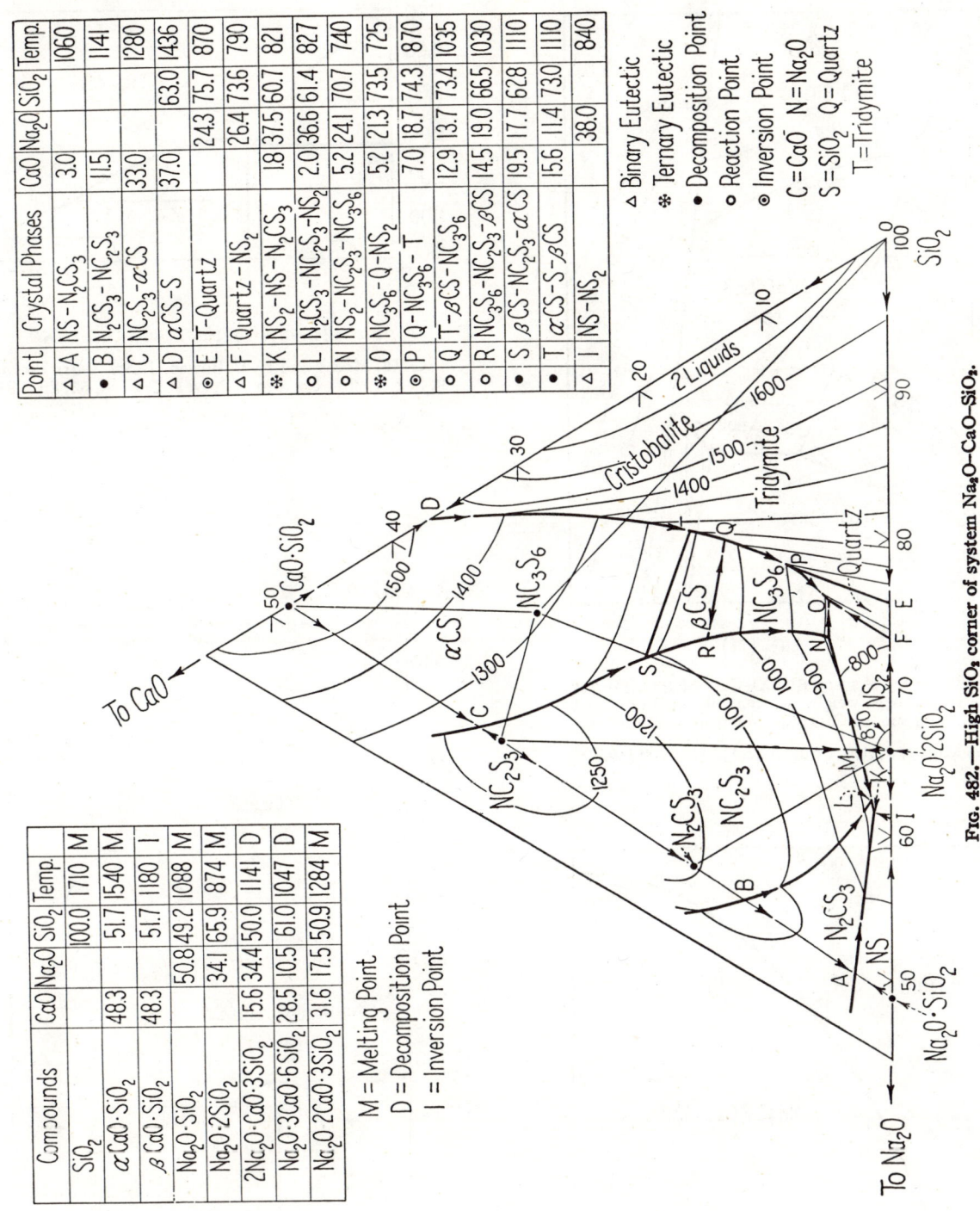

Fig. 482.—High SiO$_2$ corner of system Na$_2$O–CaO–SiO$_2$.

G. W. Morey and N. L. Bowen, *J. Soc. Glass Technol.*, 9, pp. 232, 233 (1925).

Na₂O–CaO–SiO₂ (concl.)

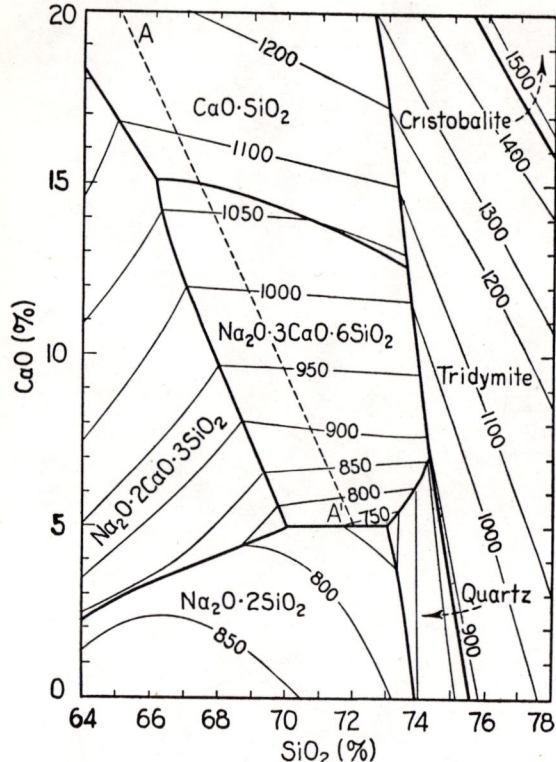

Fig. 483.—Part of system Na₂O–CaO–SiO₂ of interest to glass technology; weight per cent of Na₂O obtained by subtracting sum of CaO plus SiO₂ from 100.

G. W. Morey, *J. Am. Ceram. Soc.*, **13** [10] 700 (1930).

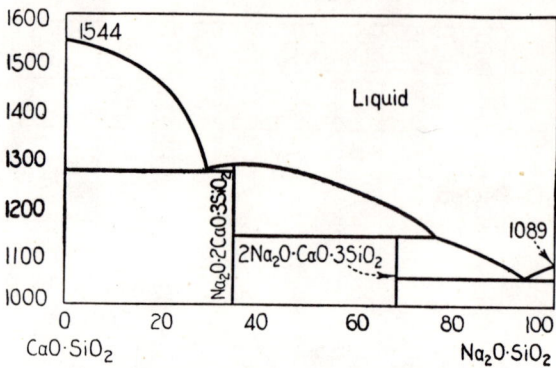

Fig. 484.—System Na₂O·SiO₂–CaO·SiO₂.

G. W. Morey and N. L. Bowen, *J. Soc. Glass Technol.*, **9**, 248 (1925); see also J. Spivak, *J. Geol.*, **52**, 29 (1944).

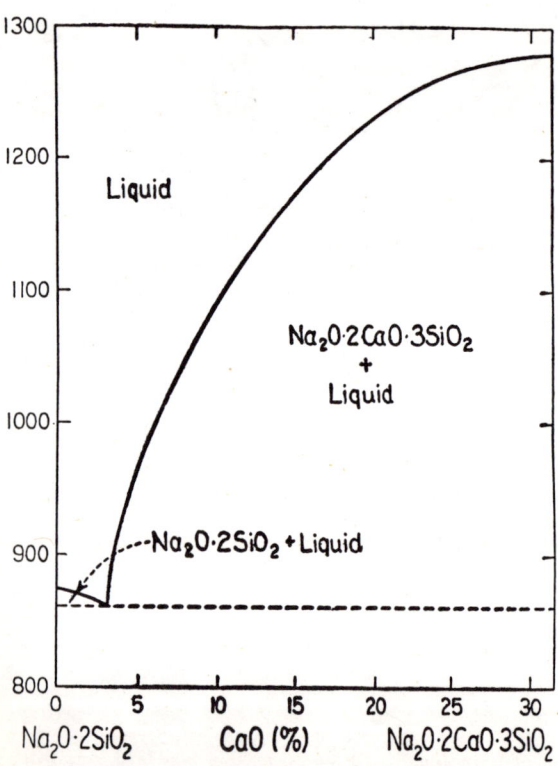

Fig. 485.—System Na₂O·2SiO₂–Na₂O·2CaO·3SiO₂.

Adapted from G. W. Morey and N. L. Bowen *J. Soc. Glass Technol.*, **9**, 249 (1925).

Na₂O–CaO–P₂O₅

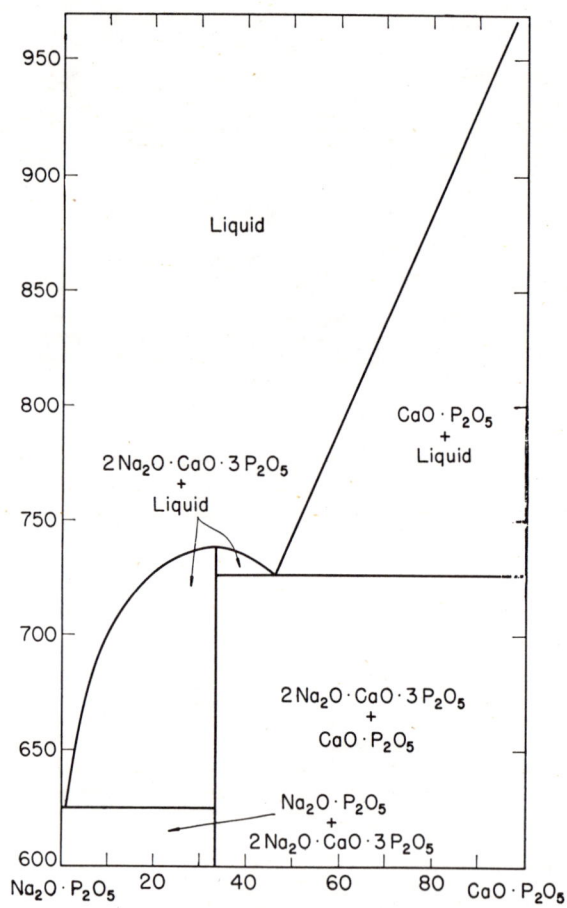

Fig. 486.—System Na₂O·P₂O₅–CaO·P₂O₅.

G. W. Morey, *J. Am. Chem. Soc.*, **74**, 5783 (1952).

Na$_2$O–FeO–SiO$_2$

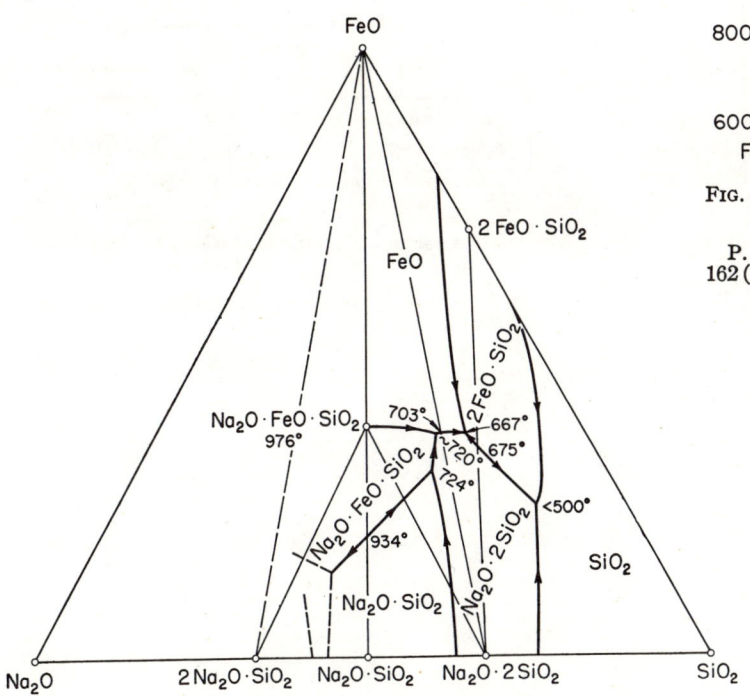

Fig. 487.—System Na$_2$O–FeO–SiO$_2$.

P. T. Carter and M. Ibrahim, *J. Soc. Glass Technol.*, **36**, 156 (1952).

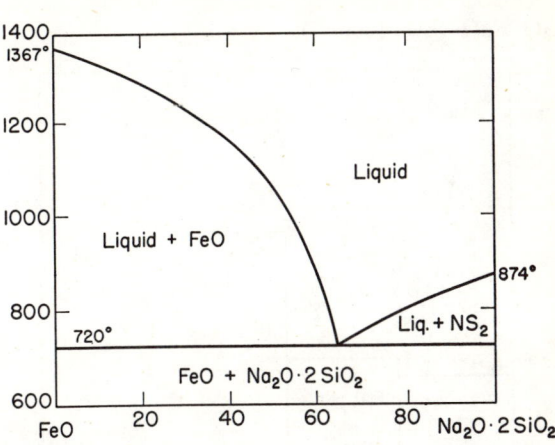

Fig. 489.—System Na$_2$O·2SiO$_2$–FeO (true binary); N = Na$_2$O, S = SiO$_2$.

P. T. Carter and M. Ibrahim, *J. Soc. Glass Technol.*, **36**, 162 (1952).

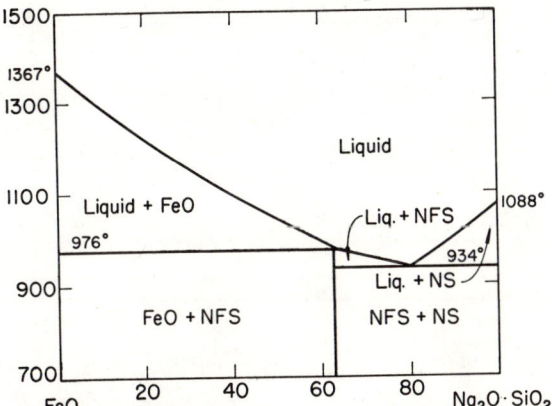

Fig. 488.—System Na$_2$O·SiO$_2$–FeO (true binary); N = Na$_2$O, F = FeO, S = SiO$_2$.

P. T. Carter and M. Ibrahim, *J. Soc. Glass Technol.*, **36**, 161 (1952).

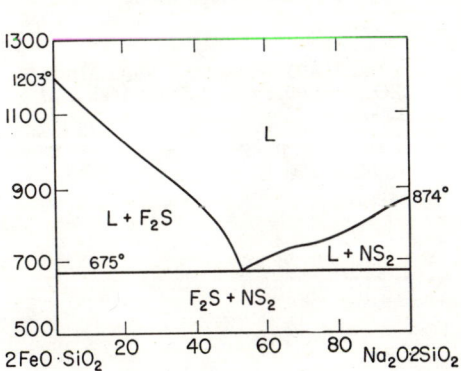

Fig. 490.—System Na$_2$O·2SiO$_2$–2FeO·SiO$_2$ (true binary); F = FeO, N = Na$_2$O, S = SiO$_2$, L = liquid.

P. T. Carter and M. Ibrahim, *J. Soc. Glass Technol.*, **36**, 160 (1952).

Na$_2$O–FeO–SiO$_2$ (concl.)

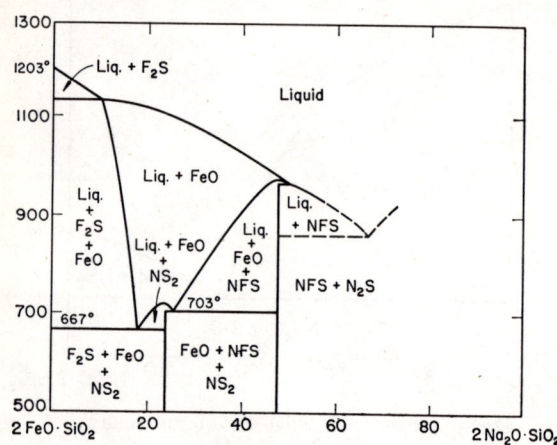

Fig. 491.—System 2Na$_2$O·SiO$_2$–2FeO·SiO$_2$ (not true binary); N = Na$_2$O, F = FeO, S = SiO$_2$.

P. T. Carter and M. Ibrahim, *J. Soc. Glass Technol.*, **36**, 161 (1952).

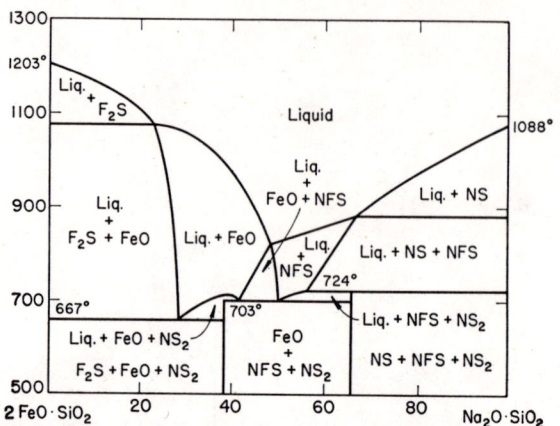

Fig. 492.—System Na$_2$O·SiO$_2$–2FeO·SiO$_2$ (not true binary); N = Na$_2$O, F = FeO, S = SiO$_2$.

P. T. Carter and M. Ibrahim, *J. Soc. Glass Technol.*, **36**, 160 (1952).

Na$_2$O–MnO–SiO$_2$

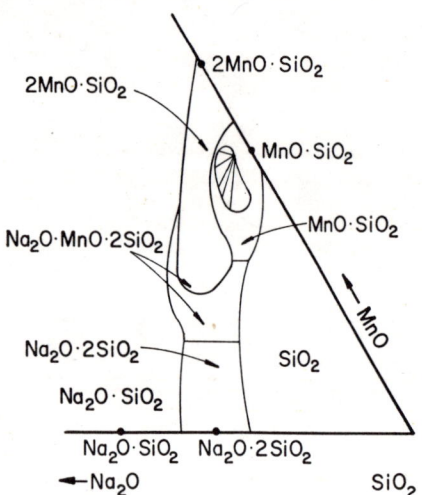

Fig. 493.—System Na$_2$O–MnO–SiO$_2$; boundary of 2-liquid region at 1200°C.

R. Hay, P. T. Carter, and S. K. Kabi, *J. Soc. Glass Technol.*, **40**, 434T (1957).

$Na_2O-PbO-SiO_2$

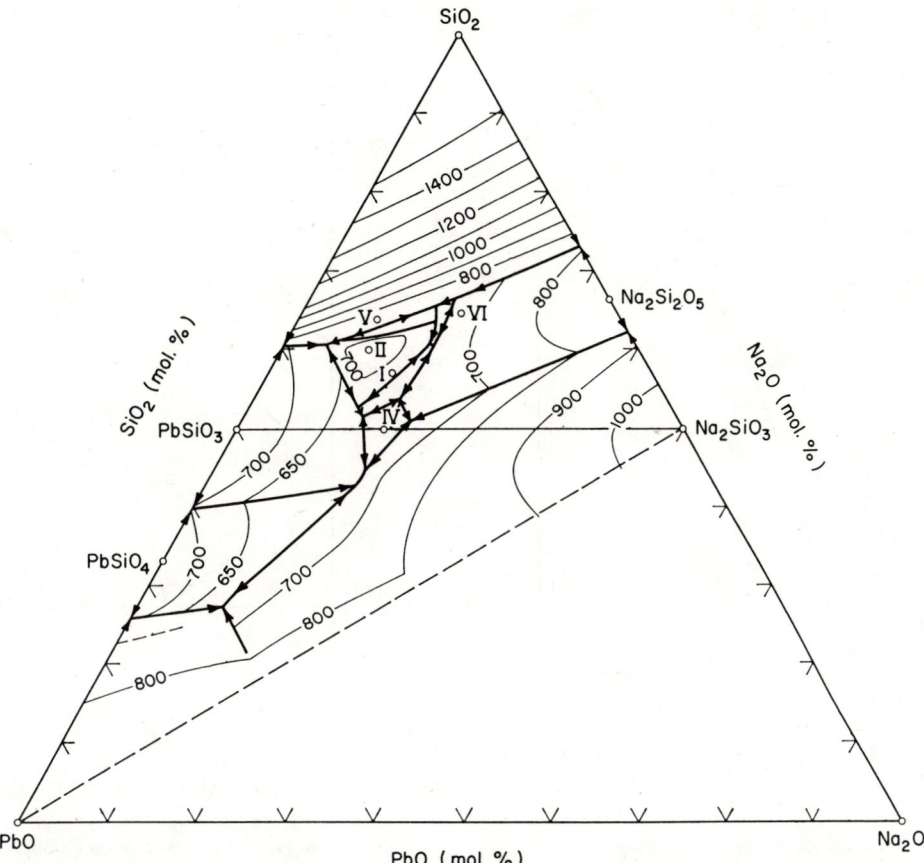

FIG. 494.—System $Na_2SiO_3-PbO-SiO_2$. I—$Na_2O \cdot 2PbO \cdot 4SiO_2$; II—$Na_2O \cdot 3PbO \cdot 6SiO_2$; IV—$Na_2O \cdot 2PbO \cdot 3SiO_2$; V—$Na_2O \cdot 3PbO \cdot 7SiO_2$; VI—$3Na_2O \cdot 3PbO \cdot 11SiO_2$.

On the $PbO-SiO_2$ boundary the compound labeled $PbSiO_4$ should read Pb_2SiO_4.

Modified after K. A. Krakau, E. J. Mukhin, and M. S. Heinrich, *Compt. rend. acad. sci.* (*U.R.S.S.*), **14**, 284 (1937).

I. V. Grebenshchikov, editor; "Physicochemical Properties of the Ternary System Sodium Oxide–Lead Oxide–Silica," p. 26, Moscow and Leningrad (1949).

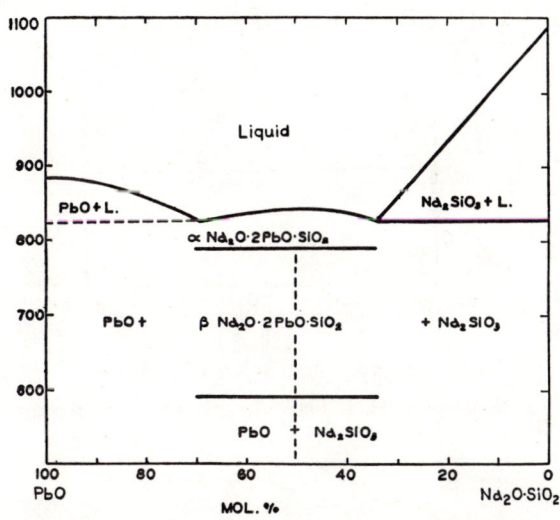

FIG. 495.—Join $PbO-Na_2O \cdot SiO_2$ of system $Na_2O-PbO-SiO_2$.

Abscissa should read Mol.% PbO (based on total moles of oxides).

K. A. Krakau, *Ann. secteur anal. phys.-chim., Inst. chim. gén.* (*U.S.S.R.*), **8**, 331–50 (1936).

FIG. 496. See page 180.

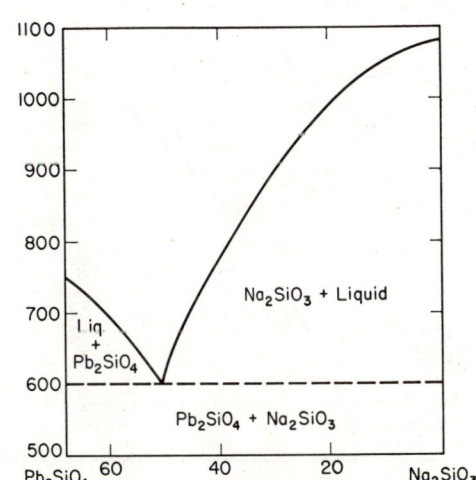

FIG. 497.—System $Na_2SiO_3-Pb_2SiO_4$.

I. V. Grebenshchikov, editor: "Physicochemical Properties of the Ternary System Sodium Oxide—Lead Oxide–Silica," p. 34, Moscow and Leningrad (1949).

Na₂O–PbO–SiO₂ (concl.)

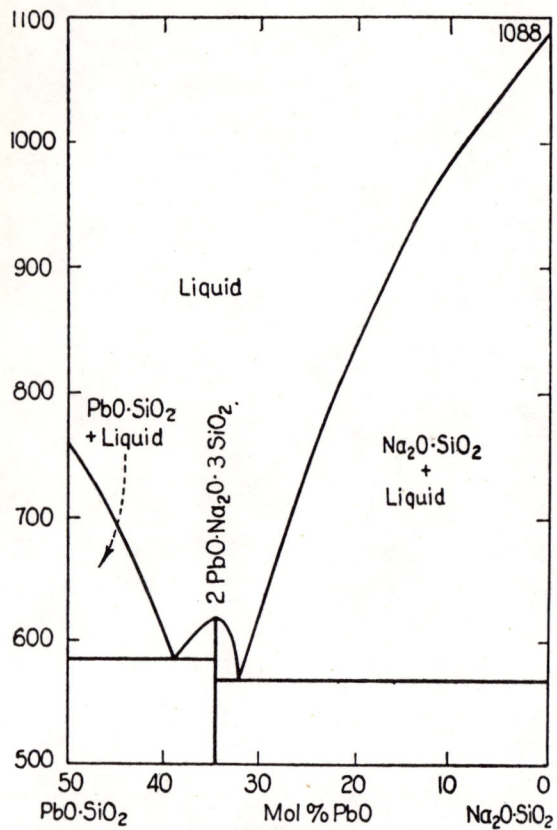

Fig. 496.—System $Na_2O \cdot SiO_2$–$PbO \cdot SiO_2$.

K. A. Krakau, *Ann. secteur anal. phys.-chim., Inst. chim. gén.* (*U.S.S.R.*), **8**, 331–50 (1936).

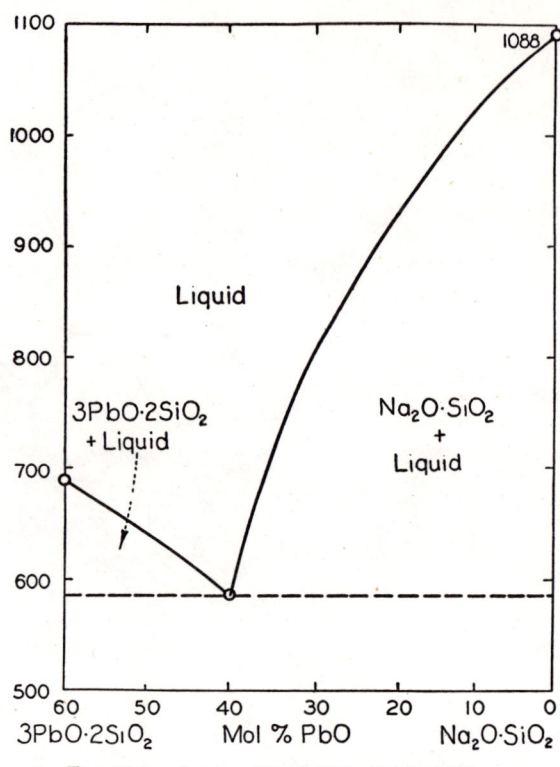

Fig. 499.—System $Na_2O \cdot SiO_2$–$3PbO \cdot 2SiO_2$.

K. A. Krakau, *Ann. secteur anal. phys.-chim., Inst. chim. gén.* (*U.S.S.R.*), **8**, 331–50 (1936).

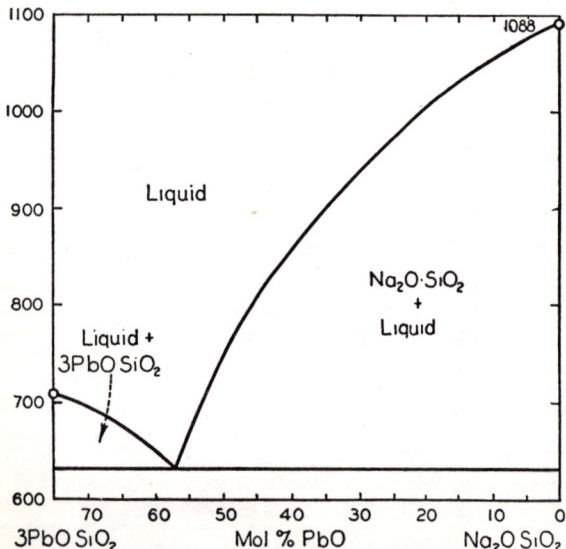

Fig. 498.—System $Na_2O \cdot SiO_2$–$3PbO \cdot SiO_2$

K. A. Krakau, *Ann. secteur anal. phys.-chim. Inst. chim. gén.* (*U.S.S.R.*) **8**, 331–50 (1936).

Na₂O–PbO–WO₃

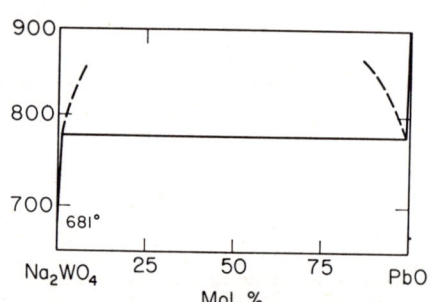

Fig. 500.—System Na_2WO_4–PbO.

I. N. Belyaev, M. L. Sholokhovich, and G. V. Barkova. *Zhur. Obshcheĭ Khim.*, **24**, 215 (1954).

$Na_2O-Al_2O_3-SiO_2$

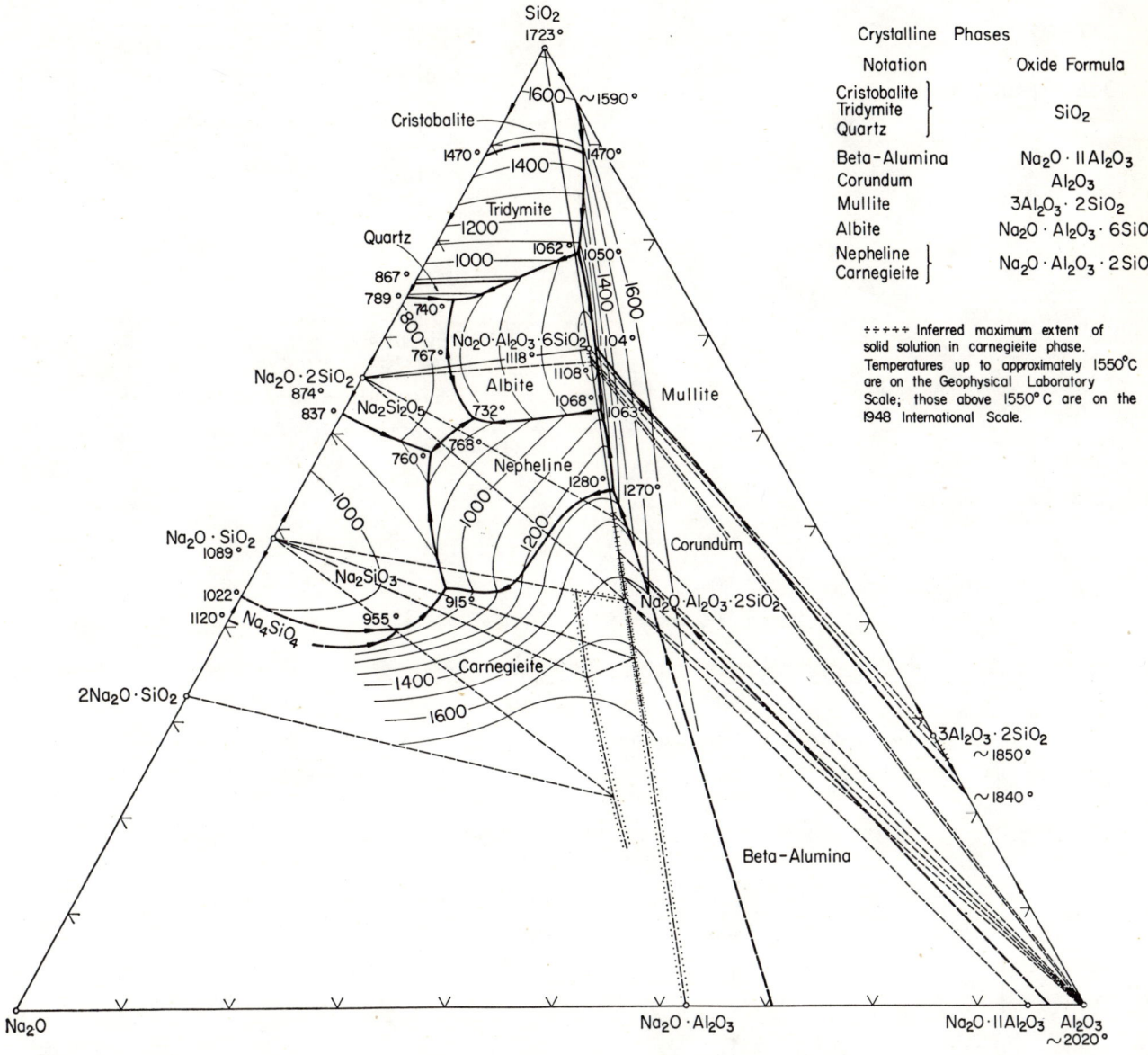

Fig. 501.—System $Na_2O-Al_2O_3-SiO_2$; composite.

E. F. Osborn and Arnulf Muan, revised and redrawn "Phase Equilibrium Diagrams of Oxide Systems," Plate 4, published by the American Ceramic Society and the Edward Orton, Jr., Ceramic Foundation, 1960.

Principal References

G. W. Morey and N. L. Bowen, *J. Phys. Chem.*, **28**, 1167–79 (1924).
F. C. Kracek, *J. Phys. Chem.*, **34**, 1583–98 (1930).
N. L. Bowen and J. W. Greig, *J. Am. Ceram. Soc.*, **7**, 238–54 (1924); corrections, ibid., 410.
N. A. Toropov and F. Ya. Galakhov, *Voprosy Petrogr. i Mineralog., Akad. Nauk S.S.S.R.*, **2**, 245–55 (1953).
Shigeo Aramaki and Rustum Roy, *Nature*, **184**, 631–32 (1959).
J. F. Schairer and N. L. Bowen, *Am. J. Sci.*, **254**, 129–95 (1956).
Liberto De Pablo-Galan and Wilfred R. Foster, *J. Am. Ceram. Soc.*, **42**, 491–98 (1959).

Na₂O–Al₂O₃–SiO₂ (cont.)

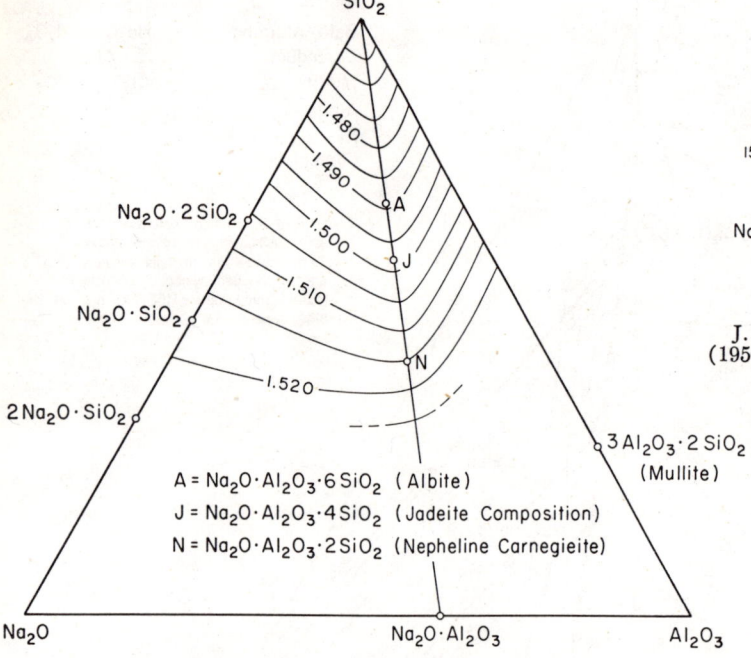

FIG. 502.—System Na₂O–Al₂O₃–SiO₂, isofracts.

J. F. Schairer and N. L. Bowen, *Am. J. Sci.*, **254** [3] 158 (1956).

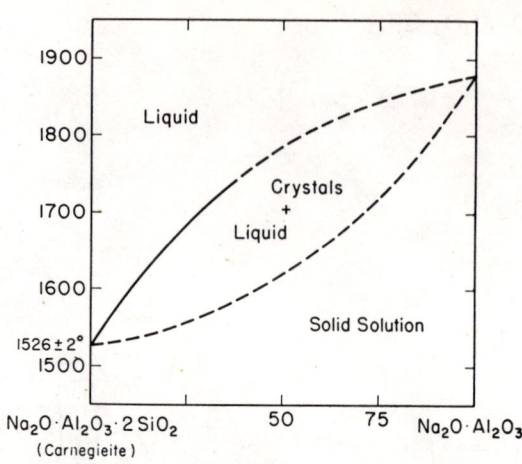

FIG. 504.—System NaAlSiO₄–NaAlO₂.

J. F. Schairer and N. L. Bowen, *Am. J. Sci.*, **254** [3] 167 (1956).

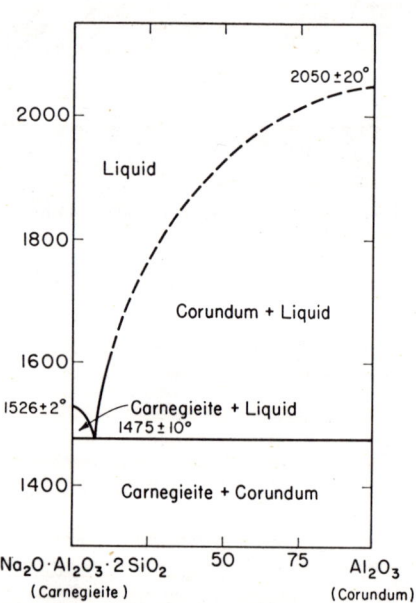

FIG. 503.—System NaAlSiO₄–Al₂O₃.

J. F. Schairer and N. L. Bowen, *Am. J. Sci.*, **254** [3] 166 (1956).

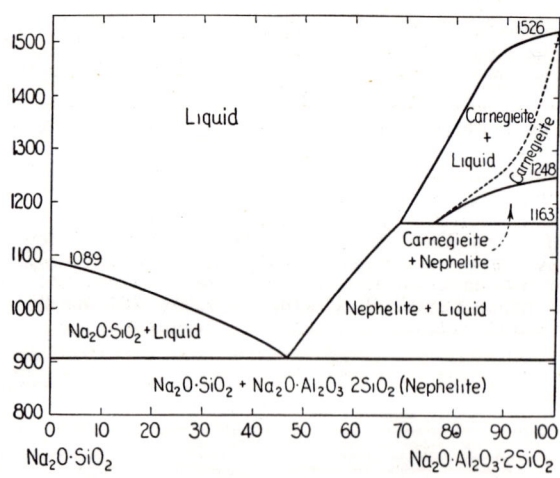

FIG. 505.—System Na₂O·SiO₂–Na₂O·Al₂O₃·2SiO₂.

C. E. Tilley, *Mineralog. u. petrog. Mitt.*, **43**, 411 (1933)

Na$_2$O–Al$_2$O$_3$–SiO$_2$ (cont.)

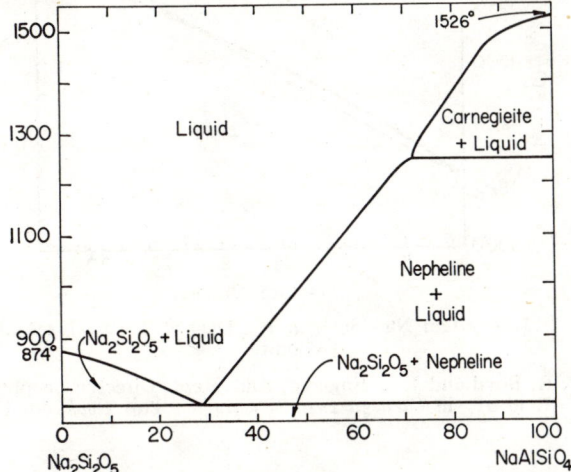

FIG. 506.—System Na$_2$O·2SiO$_2$–Na$_2$O·Al$_2$O$_3$·2SiO$_2$.

C. E. Tilley, *Mineralog. u. petrog. Mitt.*, **43**, 414 (1933).

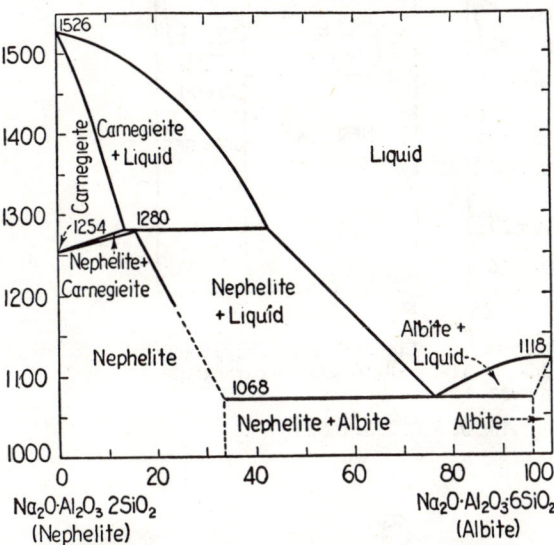

FIG. 507.—System Na$_2$O·Al$_2$O$_3$·2SiO$_2$–Na$_2$O·Al$_2$O$_3$·6SiO$_2$.

J. W. Greig and T. F. W. Barth, *Am. J. Sci.*, 5th Ser., **35A**, 94 (1938).

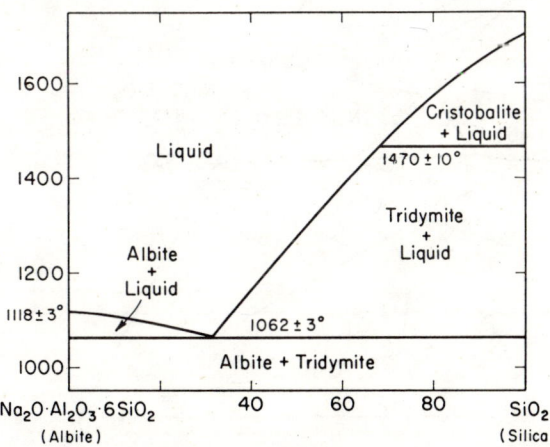

FIG. 508.—System NaAlSi$_3$O$_8$–SiO$_2$.

J. F. Schairer and N. L. Bowen, *Am. J. Sci.*, **254** [3] 161 (1956).

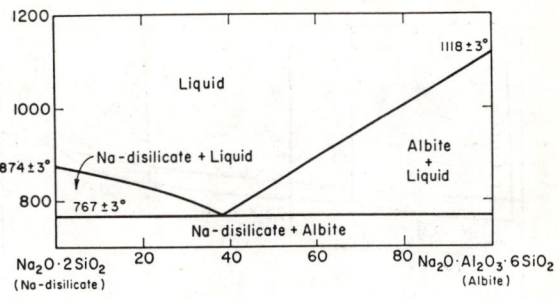

FIG. 509.—System Na$_2$Si$_2$O$_5$–NaAlSi$_3$O$_8$.

J. F. Schairer and N. L. Bowen, *Am. J. Sci.*, **254** [3] 162 (1956).

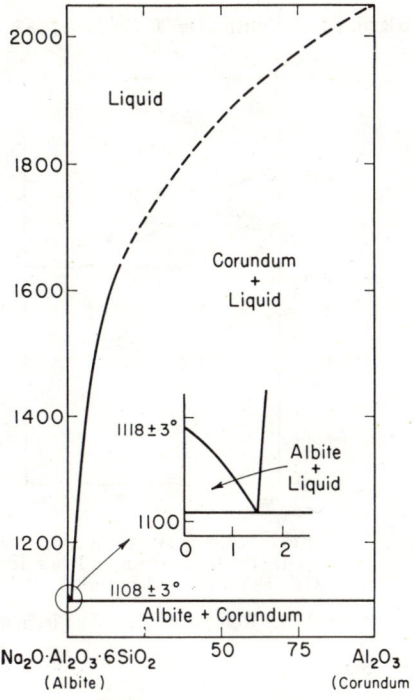

FIG. 510.—System NaAlSi$_3$O$_8$–Al$_2$O$_3$.

J. F. Schairer and N. L. Bowen, *Am. J. Sci.*, **254** [3] 163 (1956).

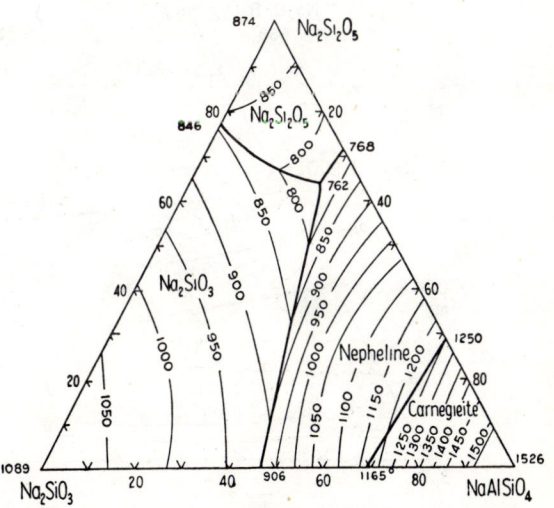

FIG. 511.—System Na$_2$O·SiO$_2$–Na$_2$O·2SiO$_2$–Na$_2$O·Al$_2$O$_3$·2SiO$_2$.

C. E. Tilley, *Mineralog. u. petrog. Mitt.*, **43**, 416 (1933).

Na₂O–Al₂O₃–SiO₂ (concl.)

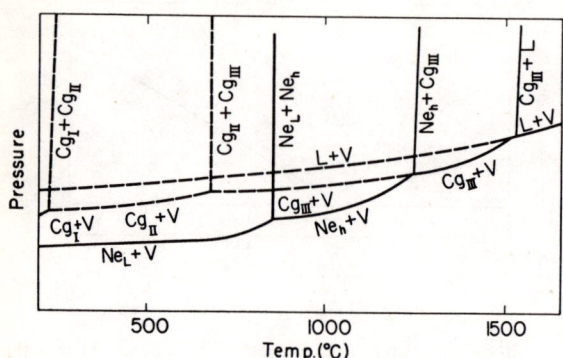

FIG. 512.—System NaAlSiO₄; p-T; schematic. L = liquid, V = vapor, Cg = carnegieite, Ne = nepheline.

O. F. Tuttle and J. V. Smith, *Am. J. Sci.*, **256**, 583 (1958).

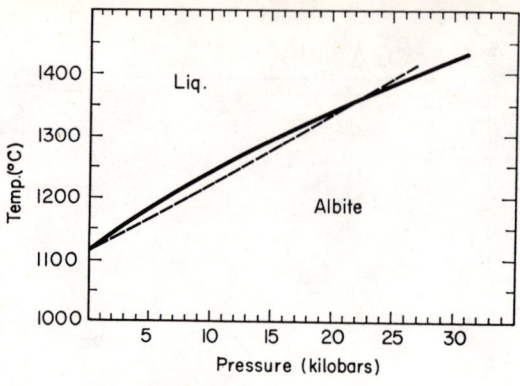

FIG. 513.—System NaAlSi₃O₈; p-T. Dashed line by Birch and LeComte.

F. R. Boyd and J. L. England, Ann. Rept. Director Geophys. Lab. 1960–61; in *Carnegie Inst. Washington Year Book*, **60**, 119 (1961).

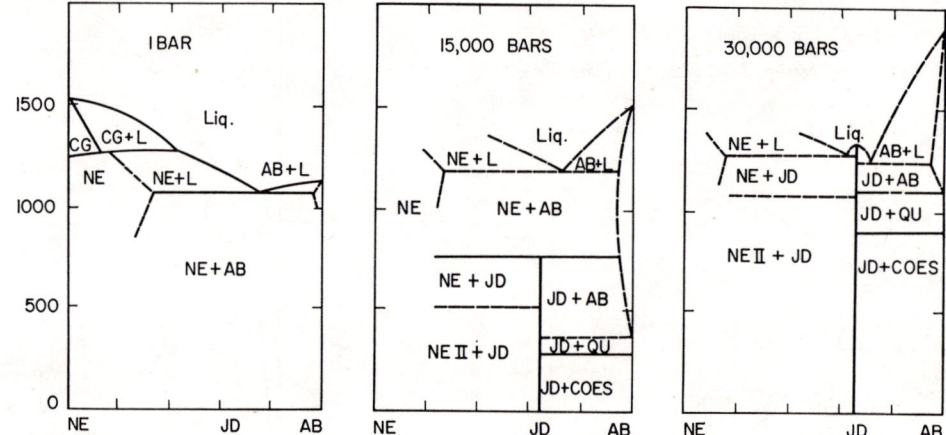

FIG. 514.—System NaAlSiO₄–NaAlSi₃O₆ at several pressures. Liquidus relations largely hypothetical; diagram at 1 bar after Greig and Barth (1938). AB = albite, CG = carnegieite, COES = coesite, JD = jadeite, NE = nepheline, QU = quartz.

E. C. Robertson, Francis Birch, and G. J. F. MacDonald, *Am. J. Sci.*, **255**, 133 (1957).

Na₂O–B₂O₃–SiO₂

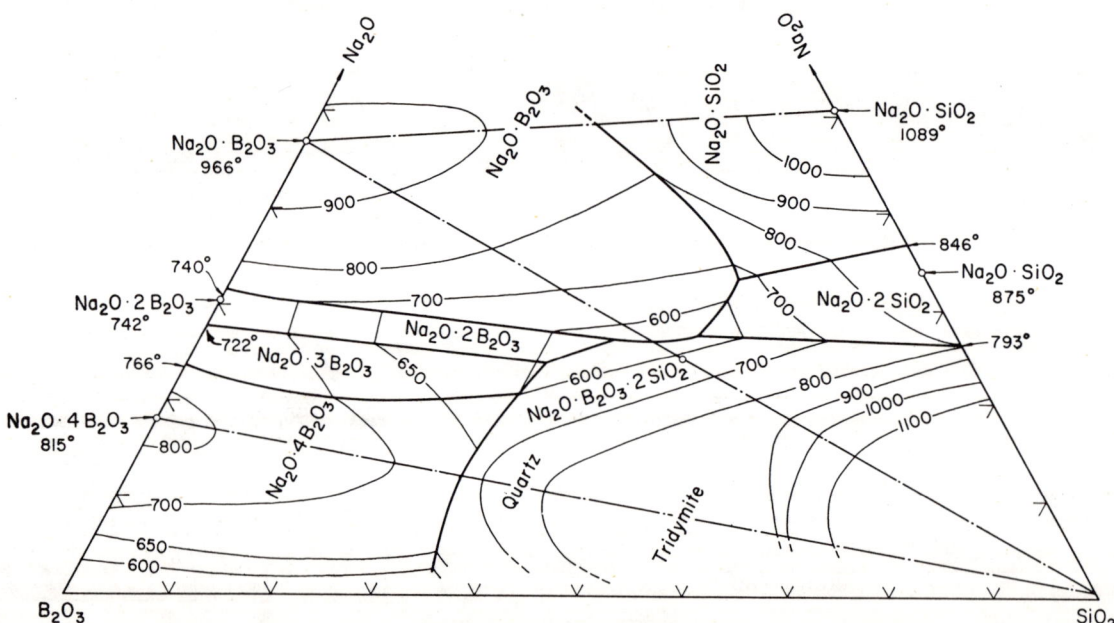

FIG. 515.—System Na₂O–B₂O₃–SiO₂.

G. W. Morey, *J. Soc. Glass Tech.*, **35**, 270 (1951). On the Na₂O–SiO₂ boundary, the lower of the two compounds labeled Na₂O·SiO₂ should read Na₂O·2SiO₂.

Na₂O–B₂O₃–SiO₂ (concl.)

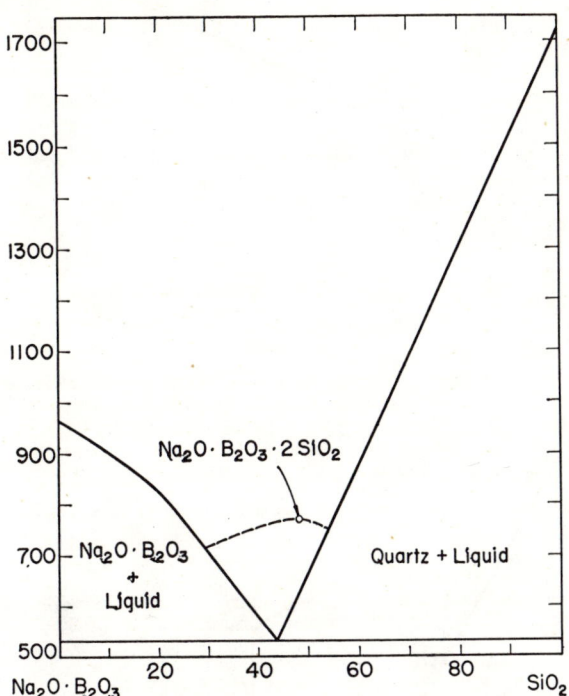

FIG. 516.—System $Na_2O \cdot B_2O_3$–SiO_2; dashed line indicates approximate location of field of $Na_2O \cdot B_2O_3 \cdot 2SiO_2$.

G. W. Morey, *J. Soc. Glass Tech.*, **35**, 270 (1951).

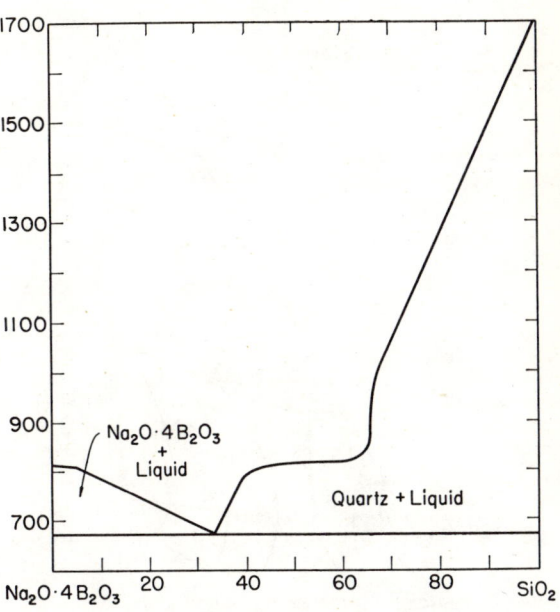

FIG. 518.—System $Na_2O \cdot 4B_2O_3$–SiO_2.

G. W. Morey, *J. Soc. Glass Tech.*, **35**, 270 (1951).

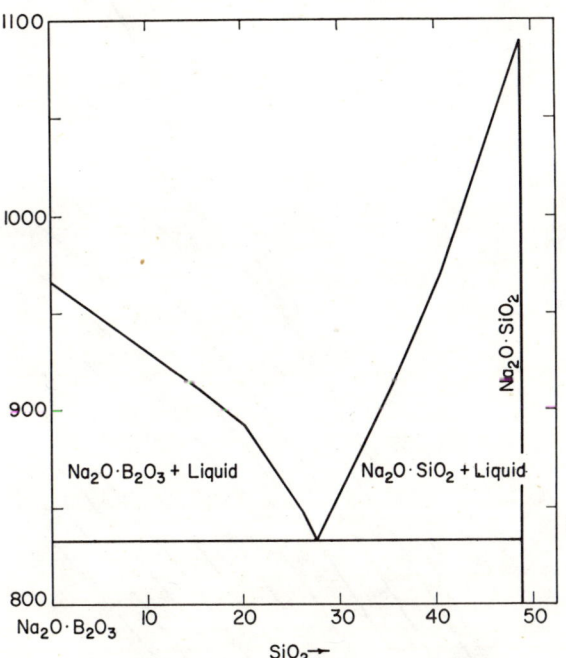

FIG. 517.—System $Na_2O \cdot B_2O_3$–$Na_2O \cdot SiO_2$.

G. W. Morey, *J. Soc. Glass Tech.*, **35**, 270 (1951).

Na₂O–Fe₂O₃–SiO₂

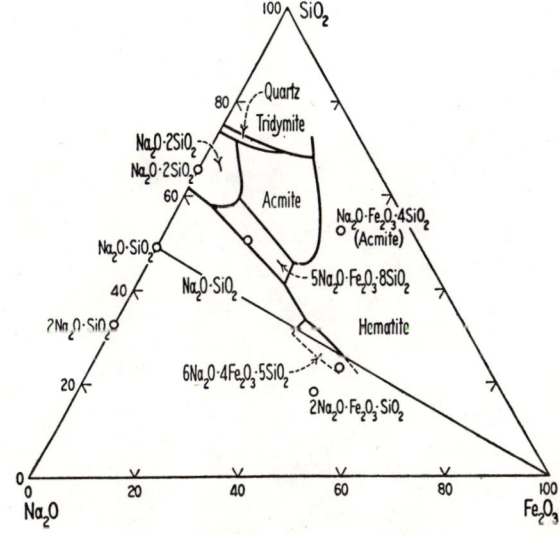

FIG. 519.—System Na_2O–Fe_2O_3–SiO_2.

N. L. Bowen, J. F. Schairer, and H. W. v. Willems, *Am. J. Sci.*, 5th Ser., **20**, 435 (1930).

Na$_2$O–Fe$_2$O$_3$–SiO$_2$ (cont.)

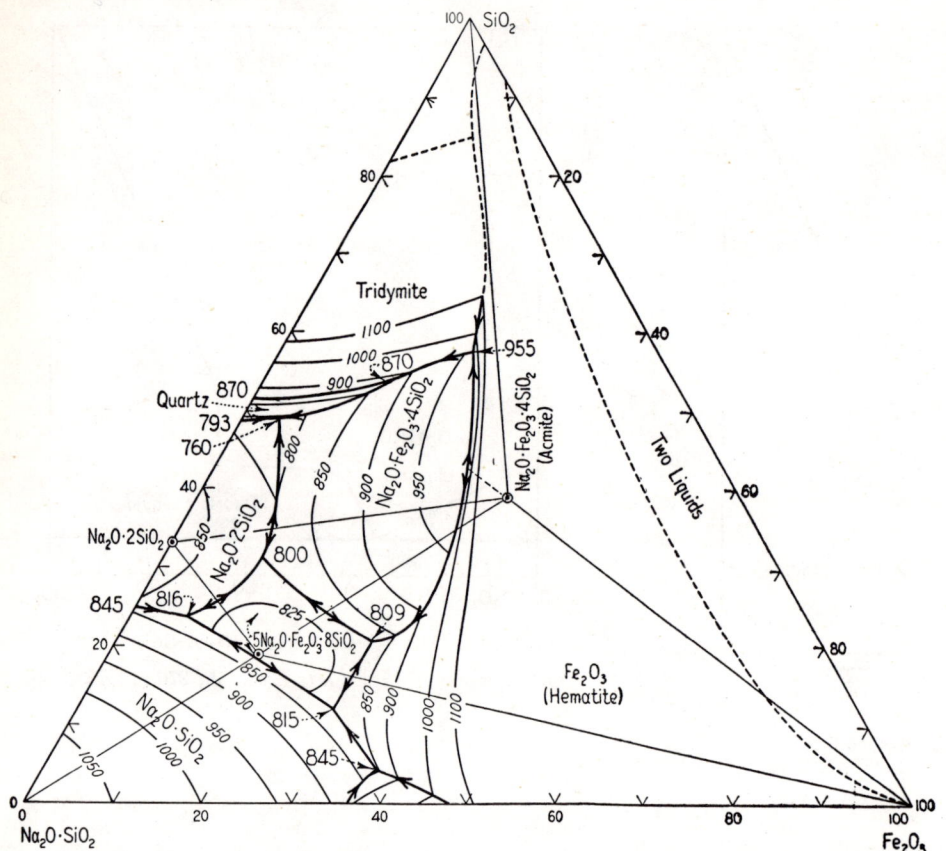

Fig. 520.—System Fe$_2$O$_3$–SiO$_2$–Na$_2$O·SiO$_2$.

N. L. Bowen, J. F. Schairer and H. W. v. Willems *Am. J. Sci.*, 5th Ser., **20**, 419 (1930); two-liquids lines from *ibid.*, p. 425.

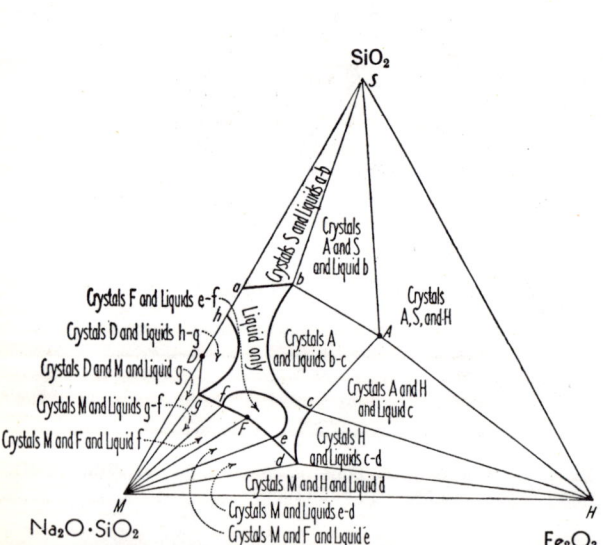

Fig. 521.—Stability relations in system Fe$_2$O$_3$–SiO$_2$–Na$_2$O·SiO$_2$ at 825°C.; A = Na$_2$O·Fe$_2$O$_3$·4SiO$_2$; D = Na$_2$O·2SiO$_2$; F = 5Na$_2$O·Fe$_2$O$_3$·8SiO$_2$; H = Fe$_2$O$_3$; M = Na$_2$O·SiO$_2$; S = SiO$_2$.

N. L. Bowen, J. F. Schairer, and H. W. v. Willems. *Am. J. Sci.*, 5th Ser., **20**, 421 (1930).

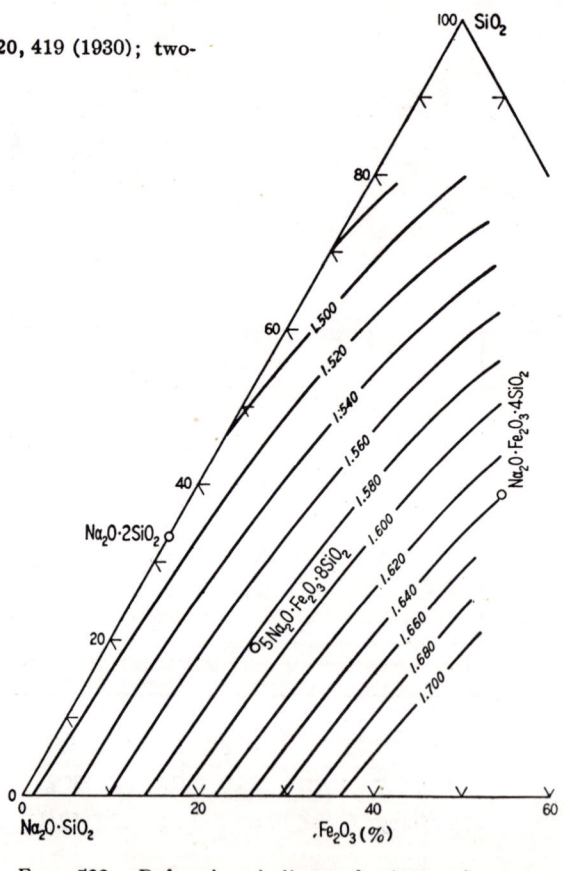

Fig. 522.—Refractive indices of glasses in system Fe$_2$O$_3$–SiO$_2$–Na$_2$O·SiO$_2$.

N. L. Bowen, J. F. Schairer, and H. W. v. Willems, *Am. J. Sci.*, 5th Ser., **20**, 446 (1930).

Na_2O–Fe_2O_3–SiO_2 (cont.)

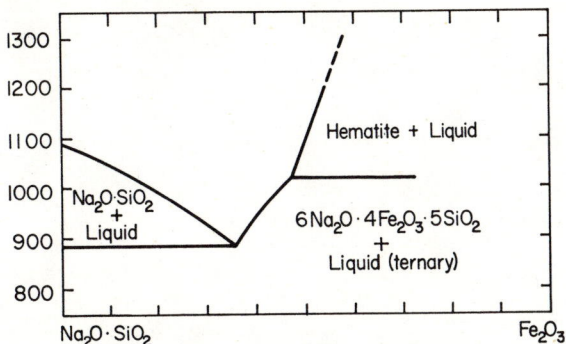

Fig. 523.—Join $Na_2O \cdot SiO_2$–Fe_2O_3, Fig. 519.

N. L. Bowen, J. F. Schairer, and H. W. v. Willems, *Am. J. Sci.* 5th Ser., **20**, 424 (1930).

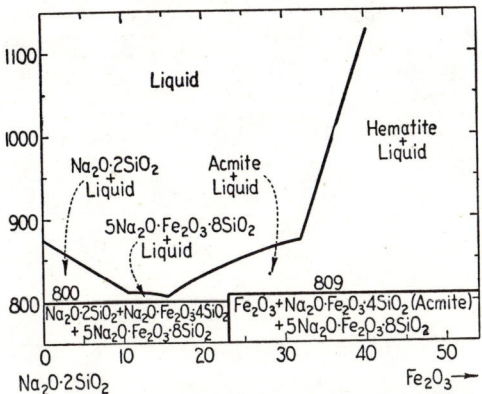

Fig. 524.—Join Fe_2O_3–$Na_2O \cdot 2SiO_2$, Fig. 519. Drawn from data of Bowen, *et al.*, *ibid.* p. 435.

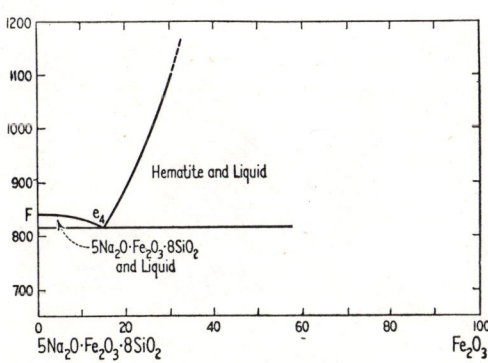

Fig. 525.—System Fe_2O_3–$5Na_2O \cdot Fe_2O_3 \cdot 8SiO_2$, Fig. 519. Bowen, *et al.*, *ibid.*, p. 427.

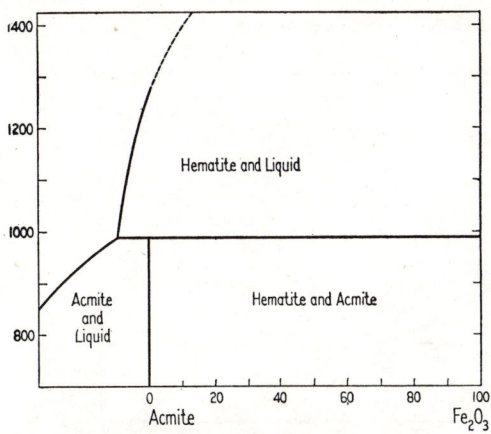

Fig. 526.—Join Fe_2O_3–$Na_2O \cdot Fe_2O_3 \cdot 4SiO_2$ of system Na_2O–Fe_2O_3–SiO_2, Fig. 519.

N. L. Bowen, J. F. Schairer, and H. W. v. Willems, *Am. J. Sci.*, 5th Ser., **20**, 428 (1930).

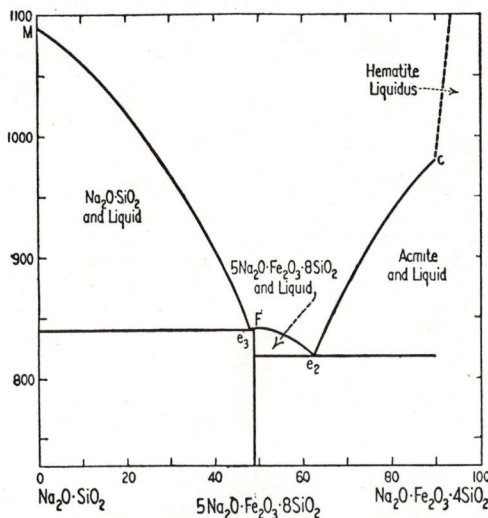

Fig. 527.—Join $Na_2O \cdot SiO_2$–$Na_2O \cdot Fe_2O_3 \cdot 4SiO_2$ of system Na_2O–Fe_2O_3–SiO_2, Fig. 519.

N. L. Bowen, J. F. Schairer, and H. W. v. Willems, *Am. J. Sci.*, 5th Ser., **20**, 429 (1930).

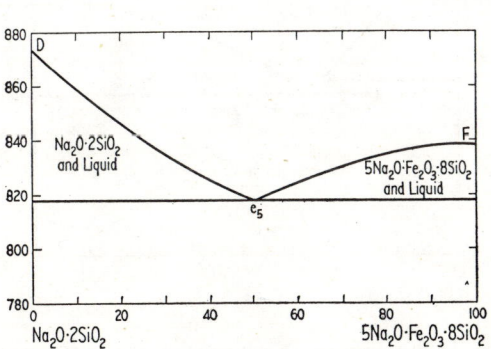

Fig. 528.—System $Na_2O \cdot 2SiO_2$–$5Na_2O \cdot Fe_2O_3 \cdot 8SiO_2$.

Bowen, *et al.*, *ibid.*, p. 428.

$Na_2O-Fe_2O_3-SiO_2$ (concl.)

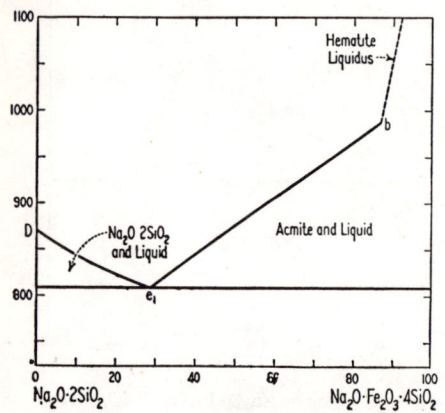

FIG. 529.—Join $Na_2O \cdot 2SiO_2$–$Na_2O \cdot Fe_2O_3 \cdot 4SiO_2$ (acmite) of system Na_2O–Fe_2O_3–SiO_2, Fig. 519.

N. L. Bowen, J. F. Schairer, and H. W. v. Willems. *Am. J. Sci.*, 20, 430 (1930).

$Na_2O-SiO_2-TiO_2$

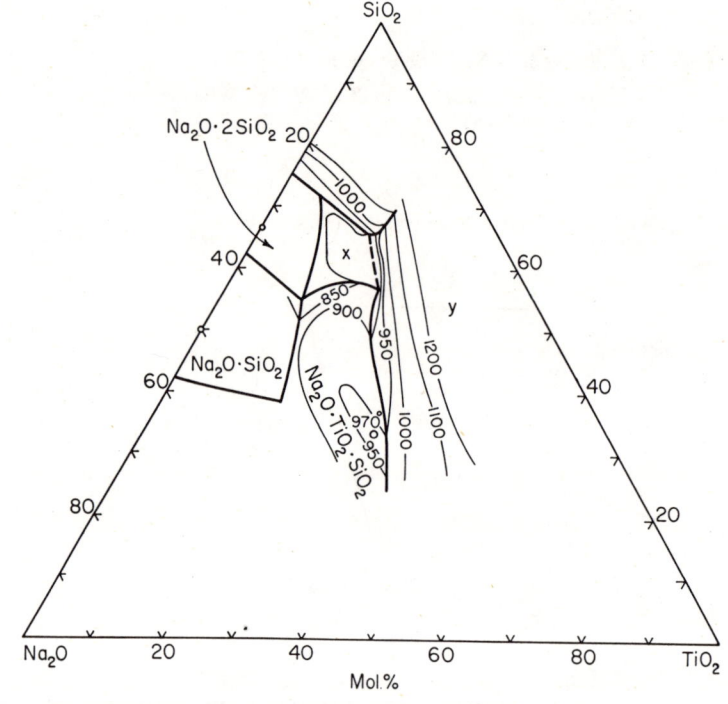

FIG. 531.—System Na_2O–SiO_2–TiO_2; partial. X and Y = unidentified phases

From data by E. H. Hamilton and G. W. Cleek, *J. Research Natl. Bur Standards*, 61 [2] 91 (1958).

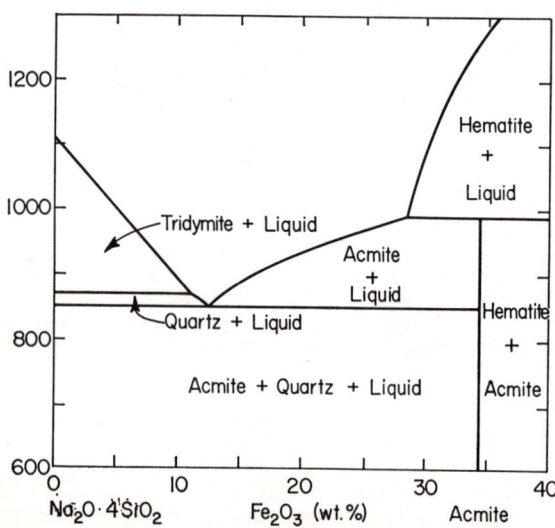

FIG. 530.—Section $Na_2O \cdot 4SiO_2$–$Na_2O \cdot Fe_2O_3 \cdot 4SiO_2$ of system Na_2O–Fe_2O_3–SiO_2.

N. L. Bowen, J. F. Schairer, and H. W. v. Willems, *Am. J. Sci.* 5th Ser., 20, 368 (1930).

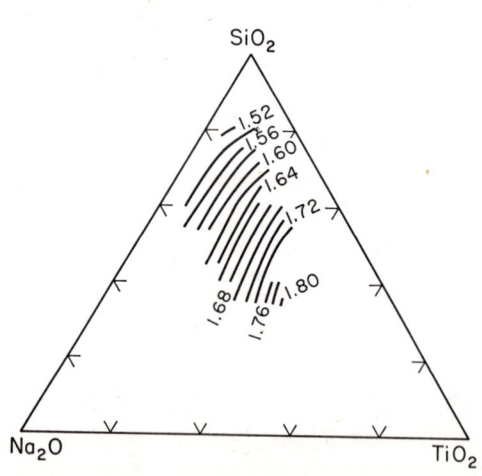

FIG. 532.—System Na_2O–SiO_2–TiO_2, isofracts.

From data by E. H. Hamilton and G. W. Cleek, *J. Research Natl. Bur. Standards*, 61 [2] 90 (1958).

$Na_2O-SiO_2-ZrO_2$

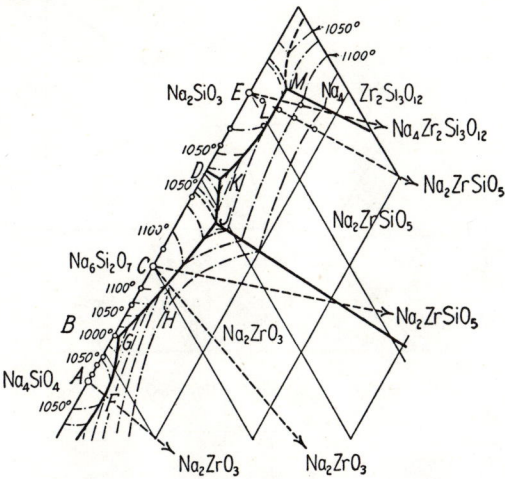

FIG. 533.—Part of system $Na_2O-SiO_2-ZrO_2$.

J. D'Ans and J. Löffler, *Z. anorg. u. allgem. Chem.*, **191**, 22 (1930).

$Na_2O-SiO_2-P_2O_5$

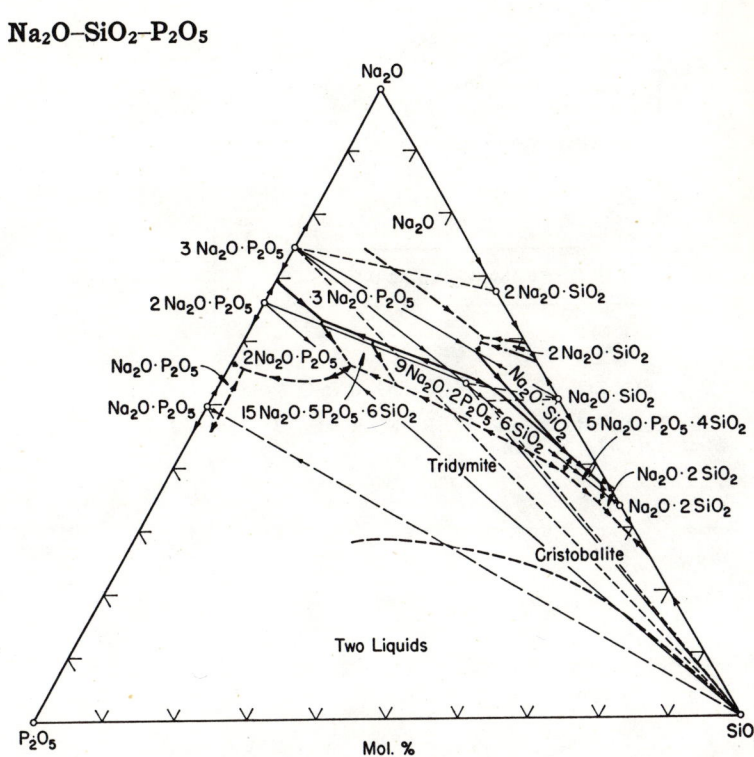

FIG. 535.—System $Na_2O-SiO_2-P_2O_5$.

E. T. Turkdogan and W. R. Maddocks, *J. Iron and Steel Inst.*, **172**, 13 (1952).

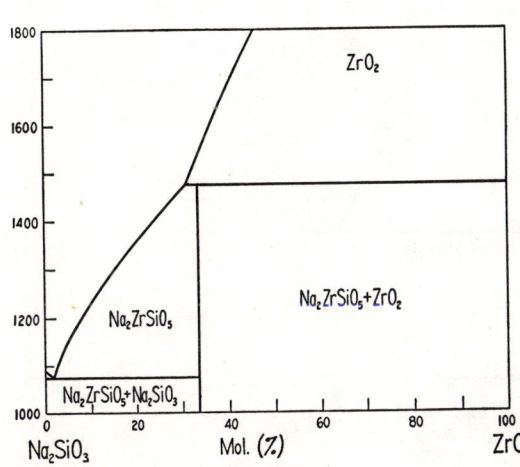

FIG. 534.—System $ZrO_2-Na_2O \cdot SiO_2$. For a particular composition, mol.% $Na_2O \cdot SiO_2$ refers to the sum of the mole percentages of Na_2O and SiO_2.

J. D'Ans and J. Löffler, *Z. anorg. u. allgem. Chem.*, **191**, 19 (1930).

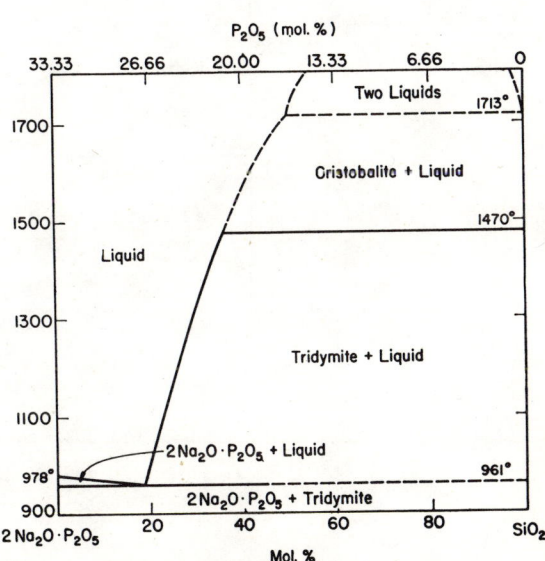

FIG. 536.—System $2Na_2O \cdot P_2O_5-SiO_2$.

E. T. Turkdogan and W. R. Maddocks, *J. Iron and Steel Inst.*, **172**, 11 (1952).

Na_2O–SiO_2–P_2O_5 (cont.)

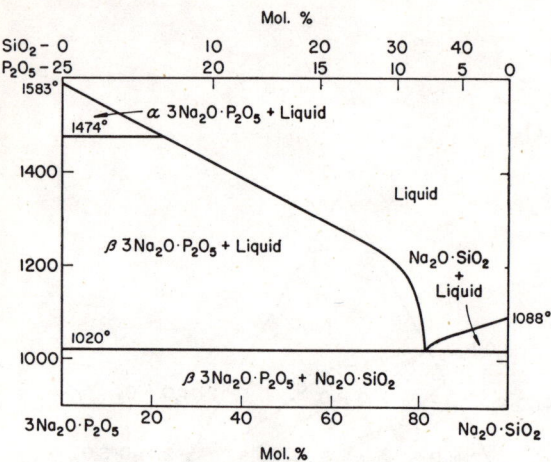

Fig. 537.—System $Na_2O \cdot SiO_2$–$3Na_2O \cdot P_2O_5$.

E. T. Turkdogan and W. R. Maddocks, *J. Iron and Steel Inst.*, **172**, 8 (1952).

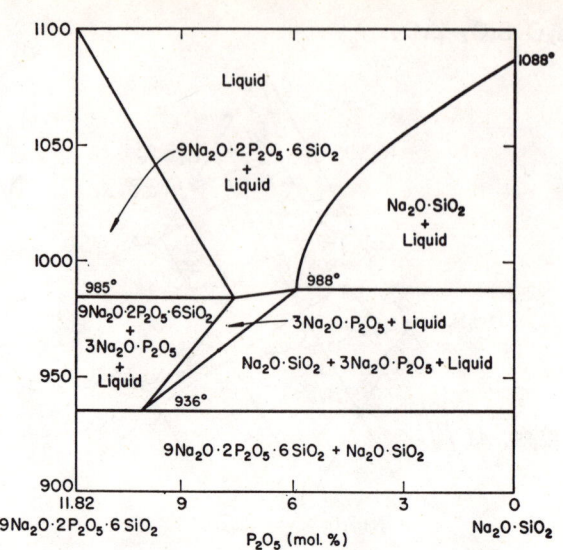

Fig. 539.—System $Na_2O \cdot SiO_2$–$9Na_2O \cdot 2P_2O_5 \cdot 6SiO_2$.

E. T. Turkdogan and W. R. Maddocks, *J. Iron and Steel Inst.*, **172**, 14 (1952).

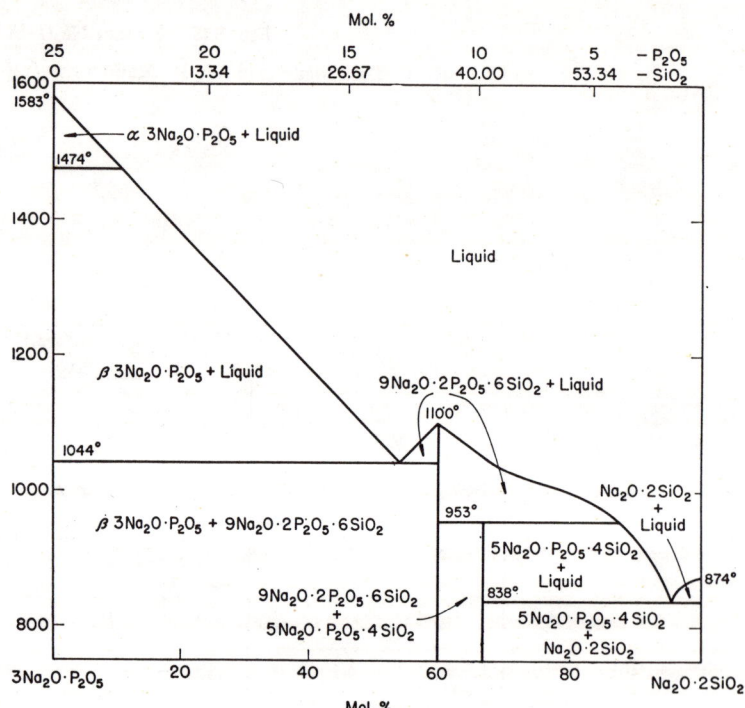

Fig. 538.—System $Na_2O \cdot 2SiO_2$–$3Na_2O \cdot P_2O_5$.

E. T. Turkdogan and W. R. Maddocks, *J. Iron and Steel Inst.*, **172**, 9 (1952).

Na_2O–SiO_2–P_2O_5 (concl.)

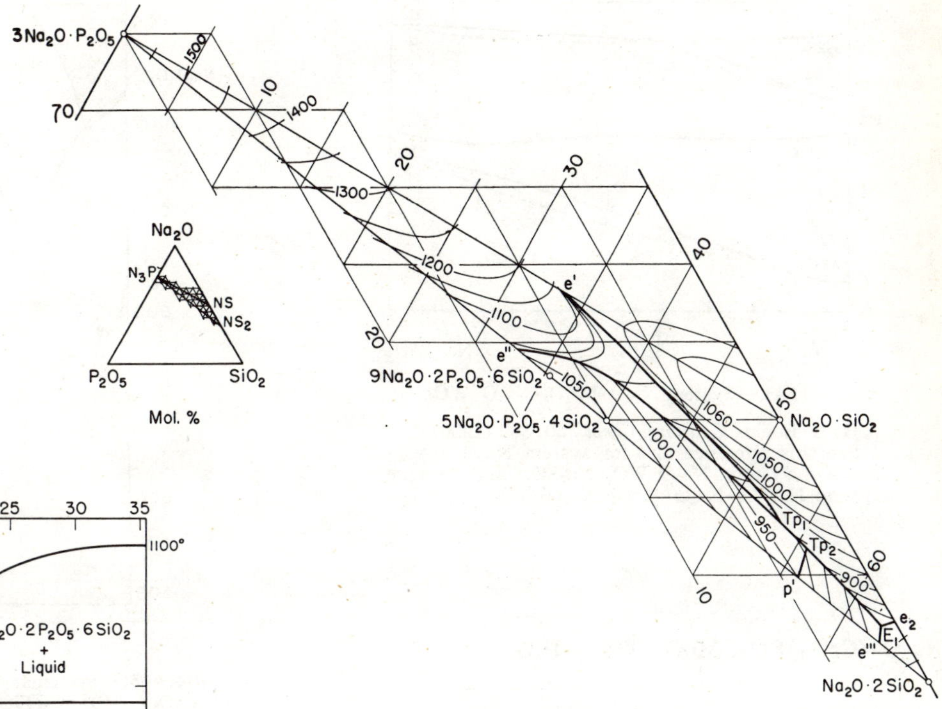

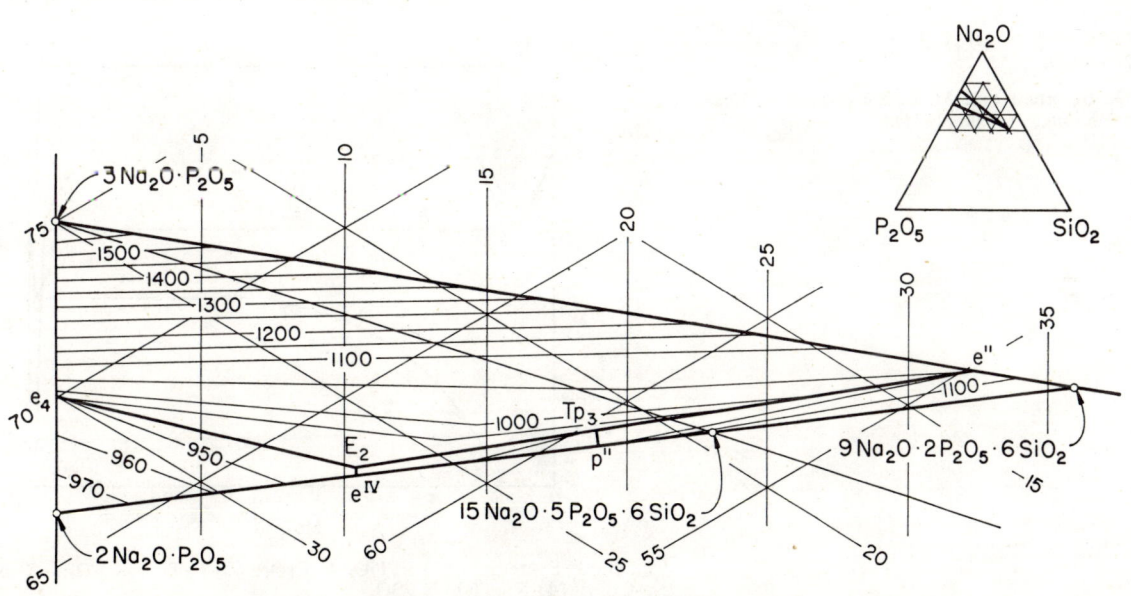

FIG. 541.—System $3Na_2O \cdot P_2O_5$ – $Na_2O \cdot SiO_3$ – $Na_2O \cdot 2SiO_2$.

E. T. Turkdogan and W. R. Maddocks, *J. Iron and Steel Inst.*, **172**, 12 (1952).

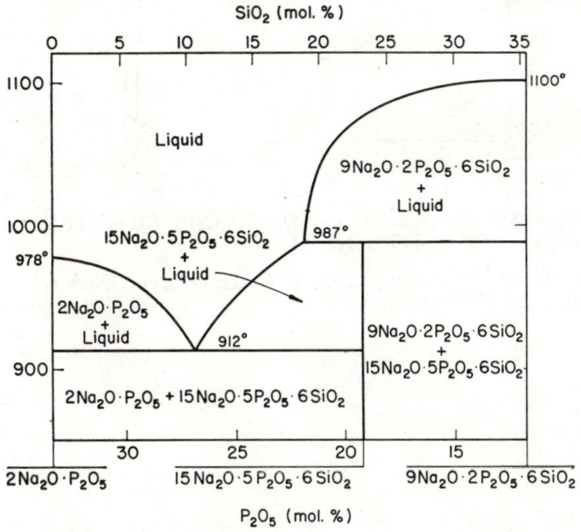

FIG. 540.—System $2Na_2O \cdot P_2O_5$–$9Na_2O \cdot 2P_2O_5 \cdot 6SiO_2$.

E. T. Turkdogan and W. R. Maddocks, *J. Iron and Steel Inst.*, **172**, 10 (1952).

FIG. 542.—System $3Na_2O \cdot P_2O_5$–$2Na_2O \cdot P_2O_5$–$9Na_2O \cdot 2P_2O_5 \cdot 6SiO_2$.

E. T. Turkdogan and W. R. Maddocks, *J. Iron and Steel Inst.*, **172**, 12 (1952).

$Na_2O-MoO_3-WO_3$

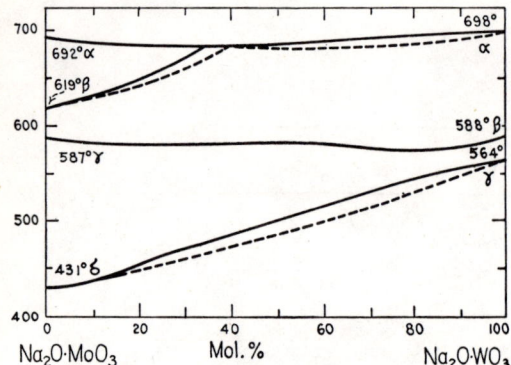

Fig. 543.—System $Na_2O \cdot MoO_3 - Na_2O \cdot WO_3$.

H. E. Boeke, *Z. anorg. Chem.*, **50**, 362 (1906).
See this reference also for system $Na_2O \cdot MoO_3-Na_2SO_4$, system $Na_2O \cdot WO_3-Na_2SO_4$, and ternary system $Na_2SO_4-Na_2O \cdot MoO_3-Na_2O \cdot WO_3$.

$Na_2O-(CrO_3, SO_3, MoO_3, WO_3)-P_2O$

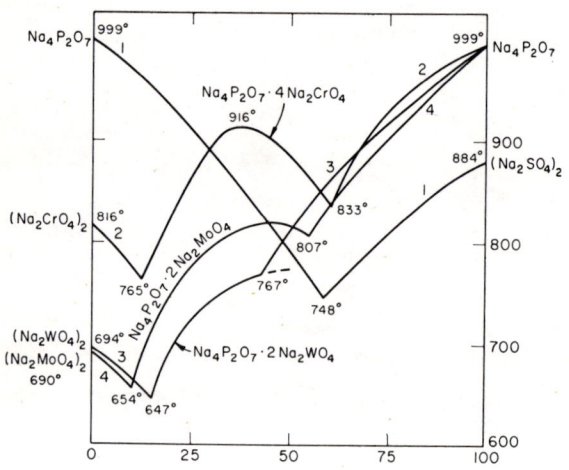

Fig. 544.—Systems: (1) $Na_4P_2O_7-Na_2SO_4$; (2) $Na_4P_2O_7-Na_2CrO_4$; (3) $Na_4P_2O_7-Na_2WO_4$; (4) $Na_4P_2O_7-Na_2MoO_4$.

A. G. Bergman and M. L. Sholokhovich, *Zhur. Obshcheĭ Khim.*, **24**, 596 (1954).

$R_2O-TiO_2-R_2O_5$

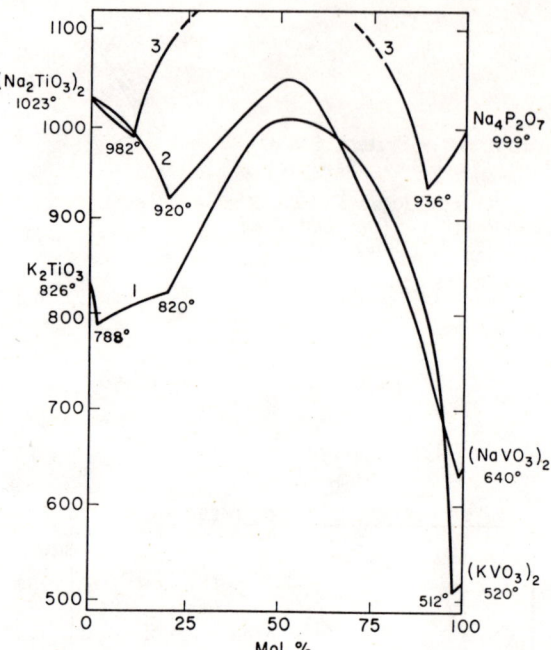

Fig. 545.—Systems: (1) $K_2TiO_3-(KVO_3)_2$; (2) $Na_2TiO_3-(NaVO_3)_2$; (3) $Na_2TiO_3-Na_4P_2O_7$.

M. L. Sholokhovich and G. V. Barkova, *Zhur. Obshcheĭ Khim.*, **26**, 1270 (1956).

$R_2O-TiO_2-RO_3$

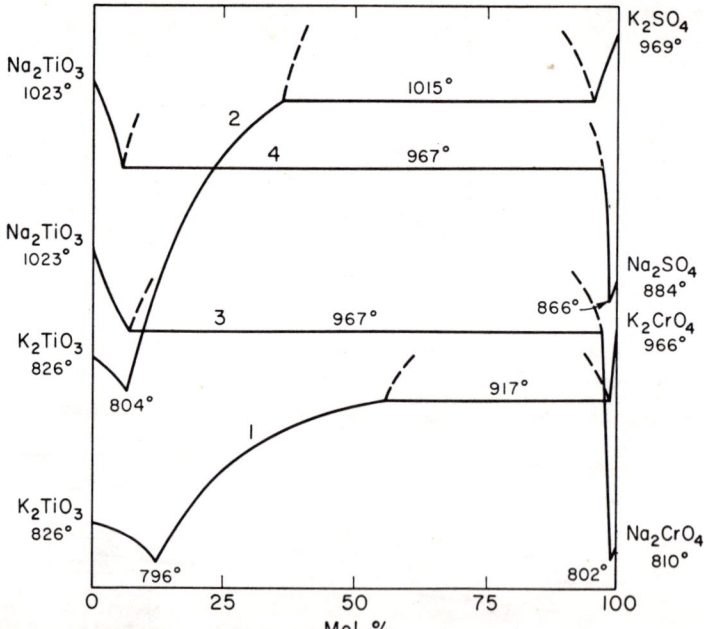

Fig. 546.—Systems: (1) $K_2TiO_3-K_2CrO_4$; (2) $K_2TiO_3-K_2SO_4$; 3) $Na_2TiO_3-Na_2CrO_4$; (4) $Na_2TiO_3-Na_2SO_4$.

M. L. Sholokhovich and G. V. Barkova, *Zhur. Obshcheĭ Khim.*, **26**, 1267 (1956).

R_2O–TiO_2–RO_3 (concl.)

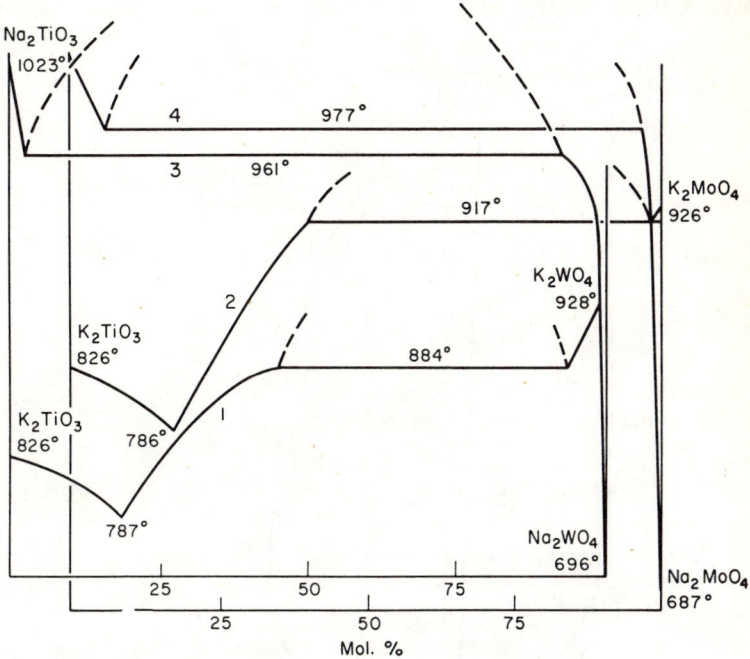

Fig. 547.—Systems: (1) K_2TiO_3–K_2WO_4; (2) K_2TiO_3–K_2MoO_4; (3) Na_2TiO_3–Na_2WO_4; (4) Na_2TiO_3–Na_2MoO_4.

M. L. Sholokhovich and G. V. Barkova, *Zhur. Obshchei Khim.*, **26**, 1268 (1956).

BaO–BeO–La_2O_3

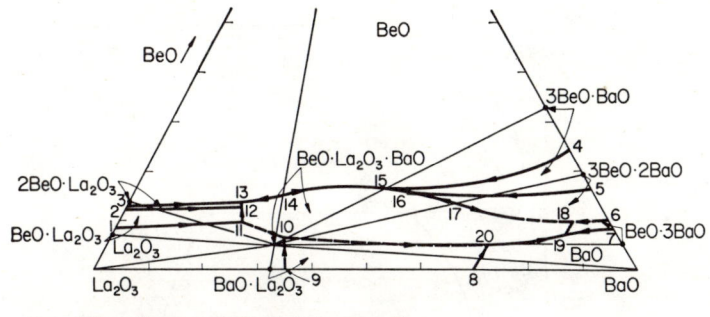

Fig. 548.—System BaO–BeO–La_2O_3. Temperatures of invariant points: 1 = 1514°C., 2 = 1371°C., 3 = 1371°C., 4 = 1430°C., 5 = 1337°C., 6 = 1141°C., 7 = 1284°C., 8 = 1548°C., 9 = 1800°C., 10 = ?, 11 = 1400°C., 12 = 1250° to 1260°C., 13 = 1250° to 1200°C., 14 = 1262°C. (max.), 15 = 1235°C., 16 = 1224°C., 17 = 1250°C. (max.), 18 = 1110° to 1135°C., 19 = 1239°C., 20 = 1284°C.

A. Auriol, G. Hauser, and J. G. Wurm, private communication, Nov. 19, 1961.

BaO–CaO–SiO_2

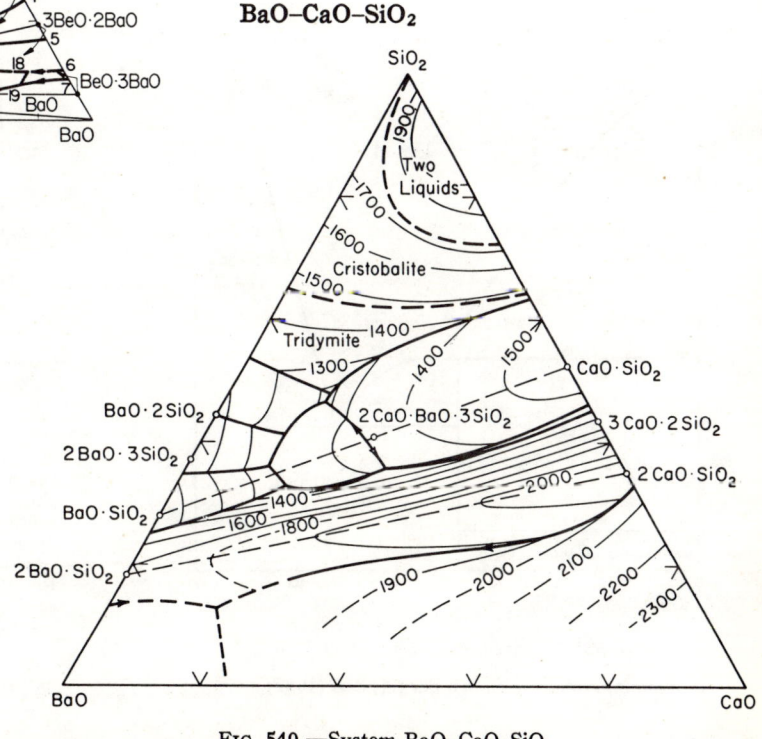

Fig. 549.—System BaO–CaO–SiO_2.

N. A. Toropov, F. Ya. Galakhov, and I. A. Bondar, *Izvest. Akad. Nauk S.S.S.R., Otdel. Khim. Nauk*, 1956 [6] 642.

BaO–CaO–SiO$_2$ (concl.)

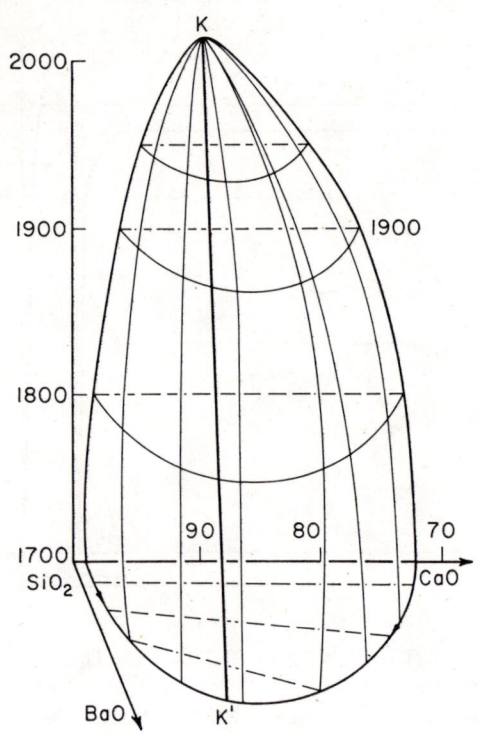

Fig. 550.—System BaO–CaO–SiO$_2$; immiscibility region.

N. A. Toropov, F. Ya. Galakhov, and I. A. Bondar, *Izvest. Akad. Nauk S.S.S.R., Otdel. Khim. Nauk*, **1956** [6] 644; also, I. A. Bondar, *Zhur. Neorg. Khim.*, **1**, 1541 (1956).

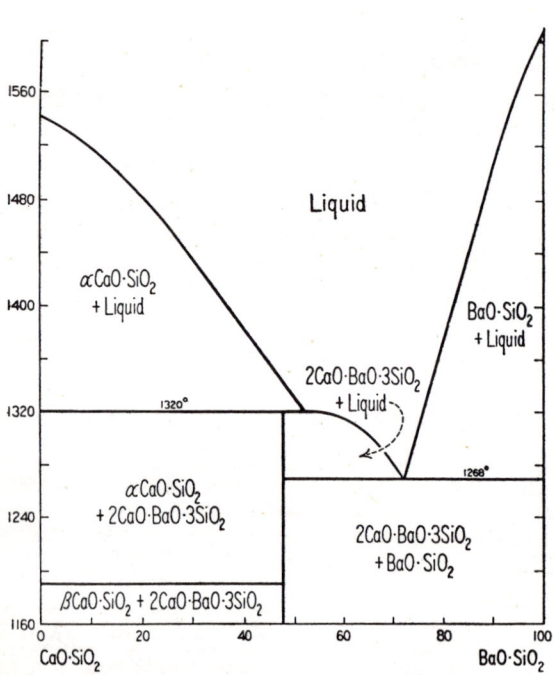

Fig. 551.—System BaO·SiO$_2$–CaO·SiO$_2$.

P. Eskola, *Am. J. Sci.*, 5th Ser., **4**, 358 (1922)

BaO–CaO–TiO$_2$

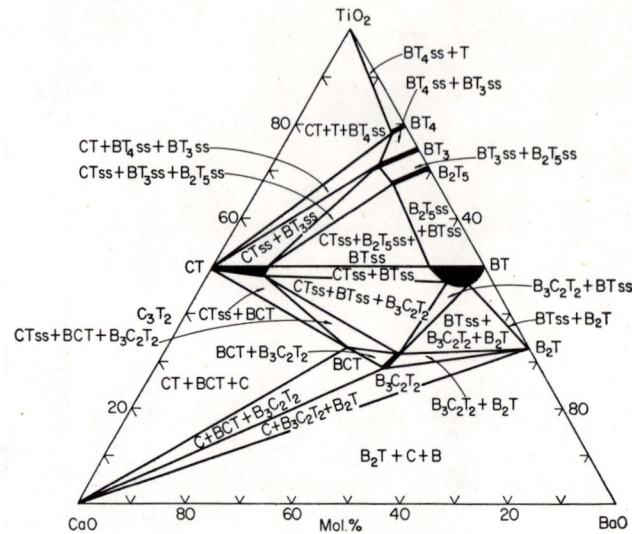

Fig. 552.—System BaO–CaO–TiO$_2$ at 1400°C. B = BaO, C = CaO, T = TiO$_2$.

W. Kwestroo and H. A. M. Paping, *J. Am. Ceram. Soc.*, **42** [6] 295 (1959).

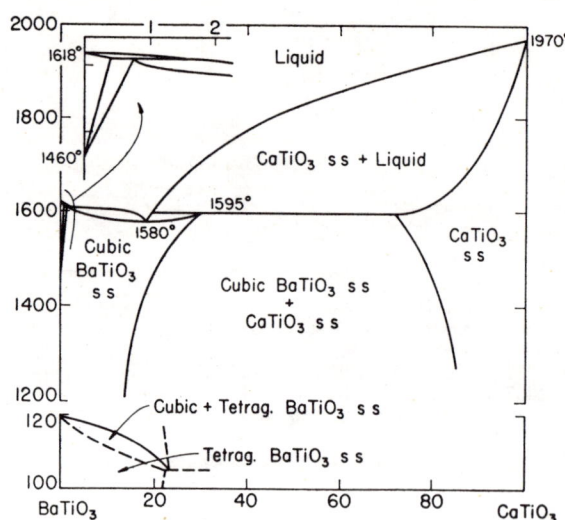

Fig. 553.—System BaTiO$_3$–CaTiO$_3$.

The lower portion represents the effect of CaTiO$_3$ on the tetragonal to cubic inversion of BaTiO$_3$ which takes place in the pure phase at 120°C. The upper insert shows the BaTiO$_3$ rich end in the vicinity of the melting point and the cubic-hexagonal inversion temperature. The fields reading from left to right between 1460°C and 1618°C (not including those which develop liquid) are: hexagonal BaTiO$_3$ solid solutions; hexagonal BaTiO$_3$ solid solutions plus cubic BaTiO$_3$ solid solutions; cubic BaTiO$_3$ solid solutions (indicated on main diagram).

R. C. DeVries and Rustum Roy, *J. Am. Ceram. Soc.*, **38** [4] 145 (1955).

BaO–SrO–TiO$_2$

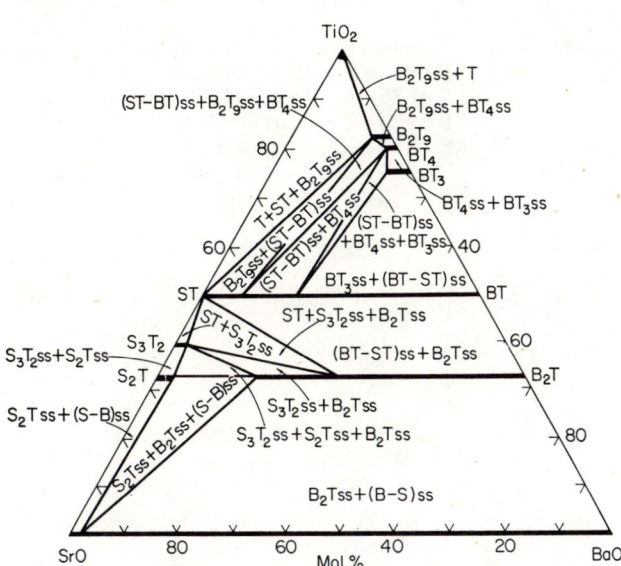

FIG. 554.—System BaO–SrO–TiO$_2$ at 1400°C. Heavy lines represent single-phase solid solutions. B = BaO, S = SrO, T = TiO$_2$.

W. Kwestroo and H. A. M. Paping, *J. Am. Ceram. Soc.*, **42** [6] 295 (1959).

BaO–Al$_2$O$_3$–SiO$_2$

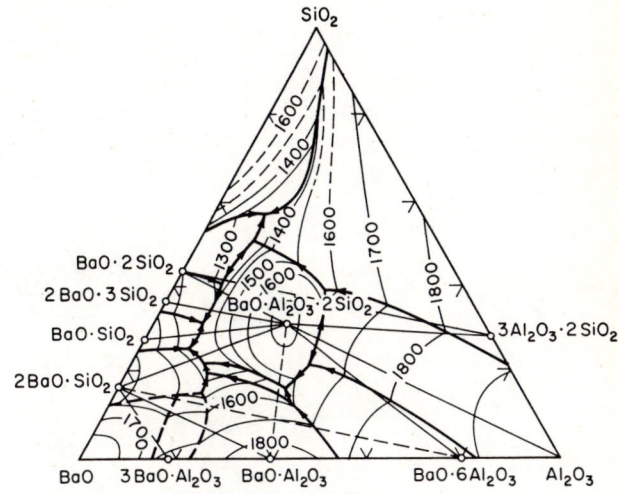

FIG. 556.—System BaO–Al$_2$O$_3$–SiO$_2$.

N. A. Toropov, F. Ya. Galakhov, and I. A. Bondar, *Izvest. Akad. Nauk S.S.S.R., Otdel. Khim. Nauk*, 1954 [5] 756. See also, R. A. Thomas, *J. Am. Ceram. Soc.*, **33** [2] 43 (1950).

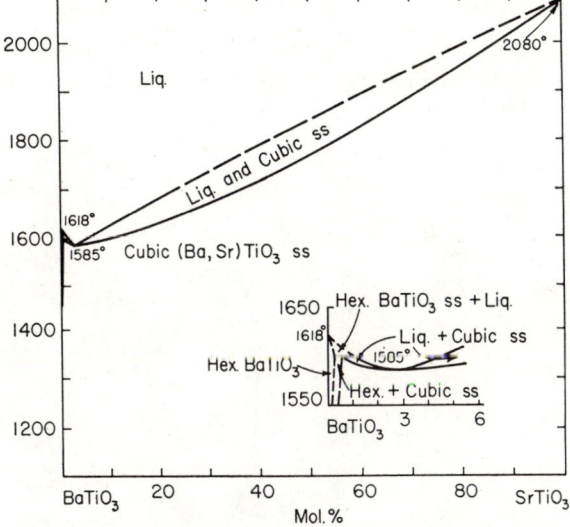

FIG. 555.—System BaTiO$_3$–SrTiO$_3$; proposed.

J. A. Basmajian and R. C. DeVries, *J. Am. Ceram. Soc.*, **40** [11] 374 (1957).

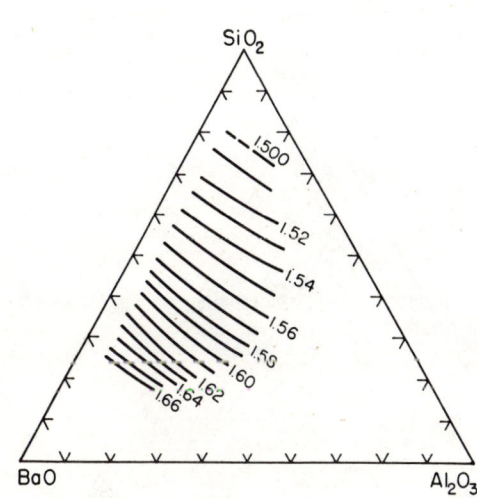

FIG. 557.—System BaO–Al$_2$O$_3$–SiO$_2$; isofracts.

N. A. Toropov, F. Ya. Galakhov, and I. A. Bondar, *Izvest. Akad. Nauk S.S.S.R., Otdel Khim. Nauk*, 1954 [5] 756. See also, R. A. Thomas, *J. Am. Ceram. Soc.*, **33** [2] 44 (1950)

$BaO-B_2O_3-SiO_2$

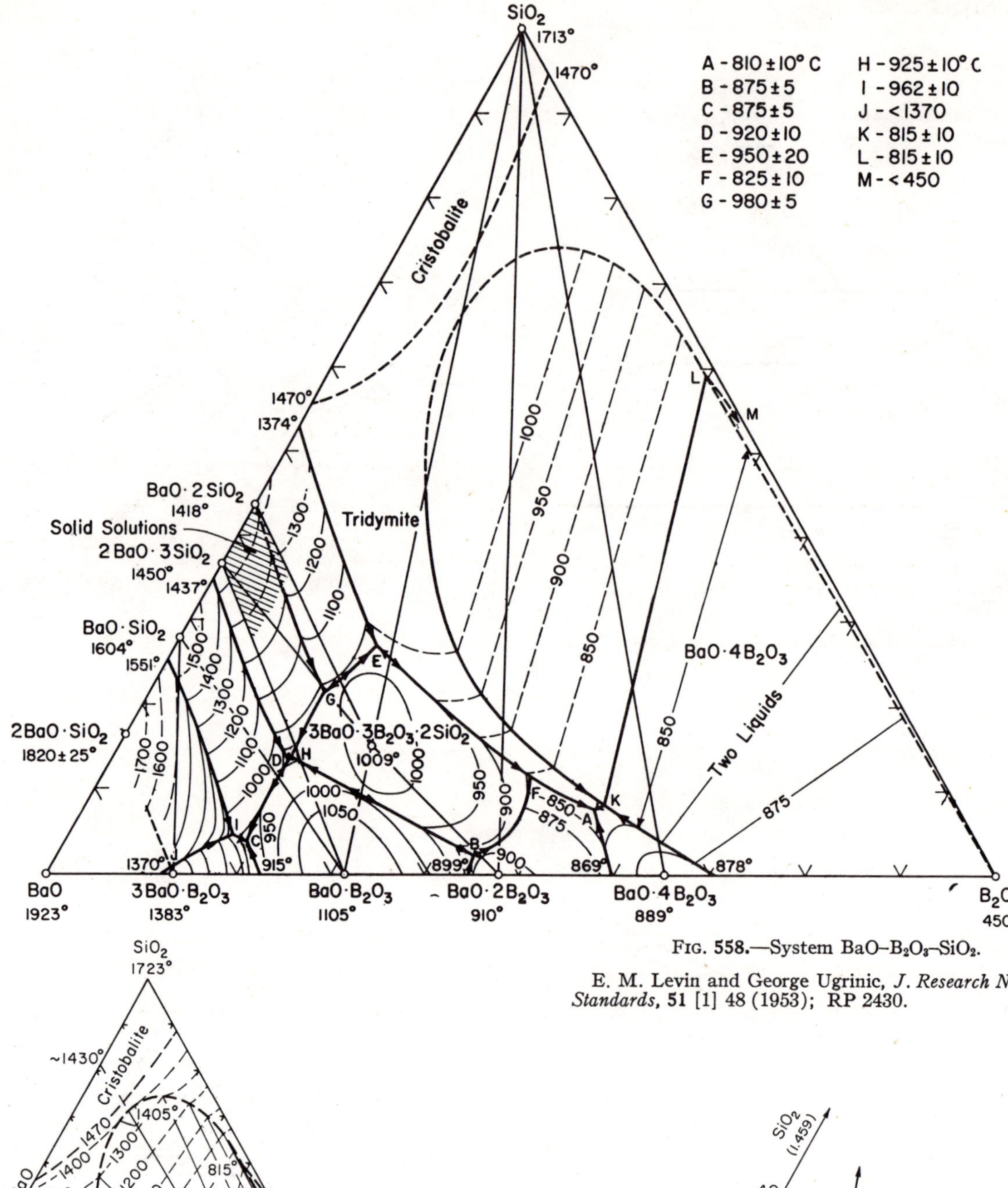

A - 810 ± 10 °C	H - 925 ± 10 °C
B - 875 ± 5	I - 962 ± 10
C - 875 ± 5	J - < 1370
D - 920 ± 10	K - 815 ± 10
E - 950 ± 20	L - 815 ± 10
F - 825 ± 10	M - < 450
G - 980 ± 5	

Fig. 558.—System $BaO-B_2O_3-SiO_2$.

E. M. Levin and George Ugrinic, *J. Research Natl. Bur. Standards*, **51** [1] 48 (1953); RP 2430.

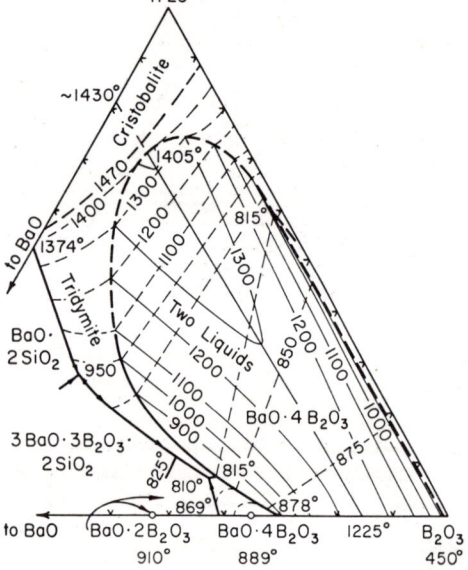

Fig. 559.—System $BaO-B_2O_3-SiO_2$; shape of liquid immiscibility volume. Heavy lines indicate primary-phase boundaries. Isotherms on the surface of the dome are shown as light solid lines. Light dashed straight lines are conodes on the liquidus surface. The critical solution curve is shown as a light solid line approximately parallel to the $B_2O_3-SiO_2$ boundary and extending from 1225° to 1405°C.

E. M. Levin and G. W. Cleek, *J. Am. Ceram. Soc.*, **41** [5] 177 (1958).

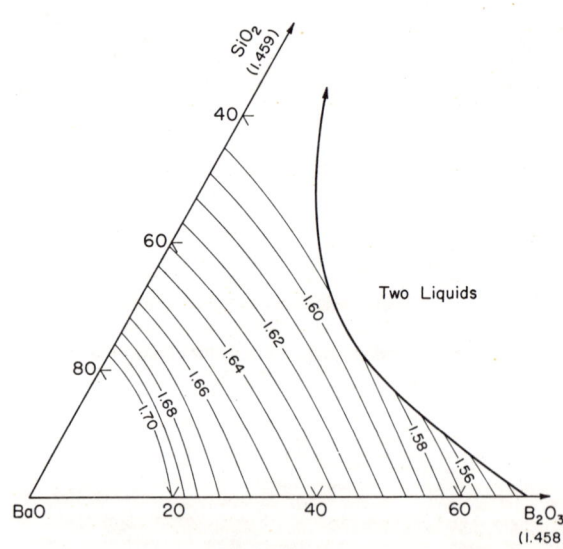

Fig. 560.—System $BaO-B_2O_3-SiO_2$; indices of refraction of glasses.

E. M. Levin and George Ugrinic, *J. Research Natl. Bur. Standards*, **51** [1] 55 (1953); RP 2430.

BaO–B$_2$O$_3$–SiO$_2$ (concl.)

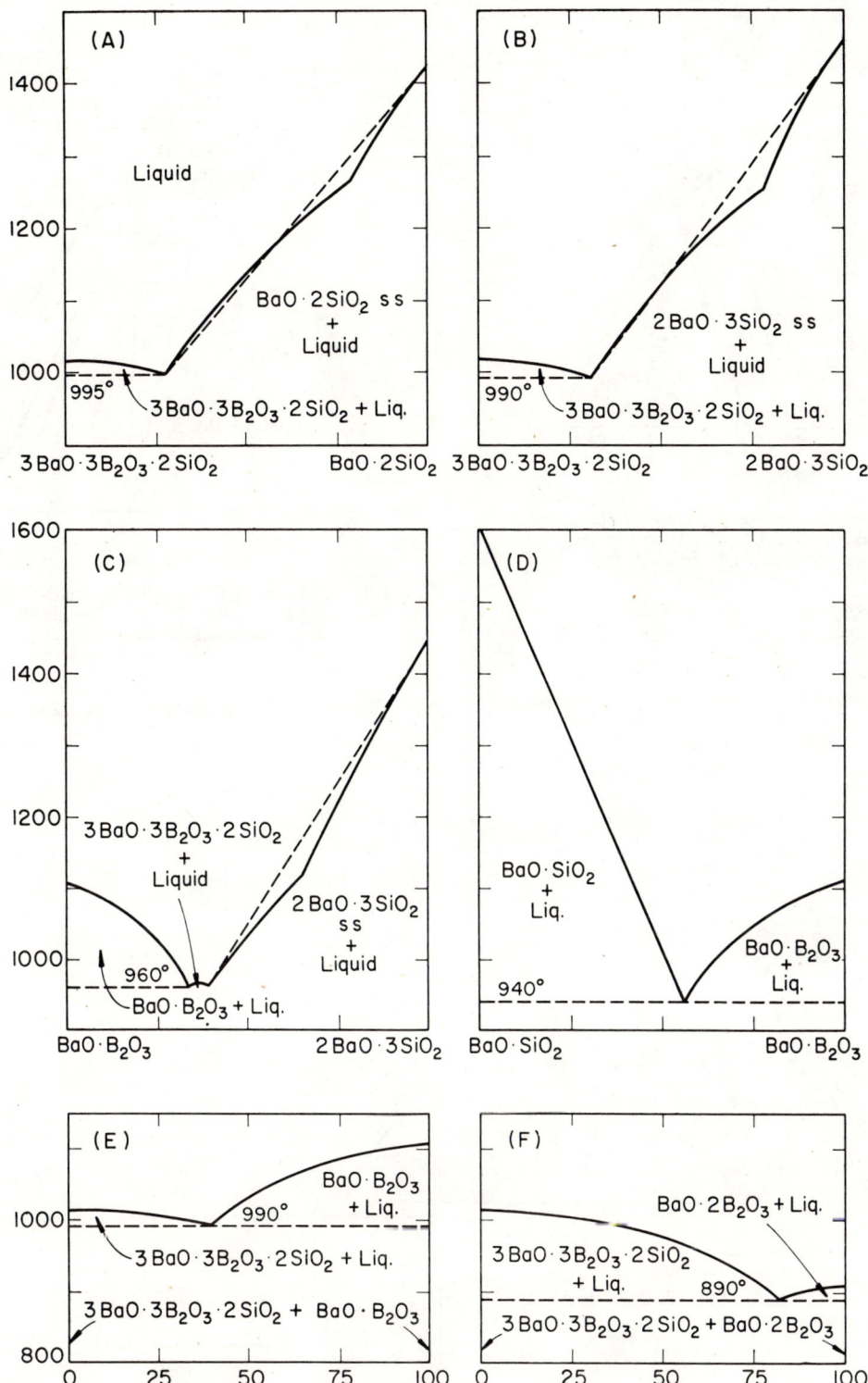

Fig. 561.—BaO–B$_2$O$_3$–SiO$_2$; some binary and pseudobinary systems.

E. M. Levin and George Ugrinic, *J. Research Natl. Bur. Standards*, 51 [1] 50 (1953); RP 2430.

BaO–Fe$_2$O$_3$–Gd$_2$O$_3$

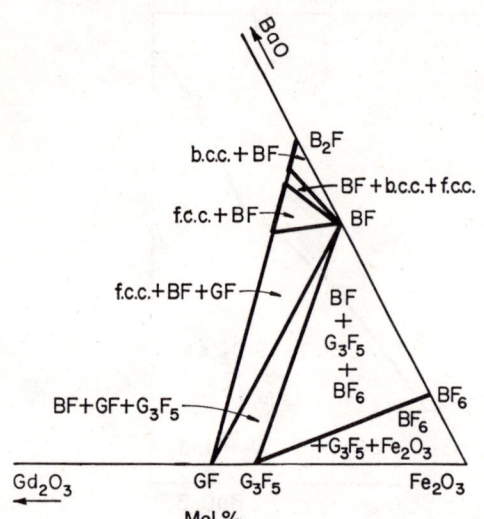

Fig. 562.—System BaO–Fe$_2$O$_3$–Gd$_2$O$_3$; preliminary subsolidus (approx. 1400°C.). b.c.c., body-centered cubic B$_2$Fss; fcc, face-centered cubic B$_2$Fss. B = BaO, F = Fe$_2$O$_3$, G = Gd$_2$O$_3$.

R. S. Roth, "Rare Earth Research," edited by Eugene V. Kleber, The Macmillan Co., New York, p. 92 (1961).

BaO–Gd$_2$O$_3$–Nb$_2$O$_5$

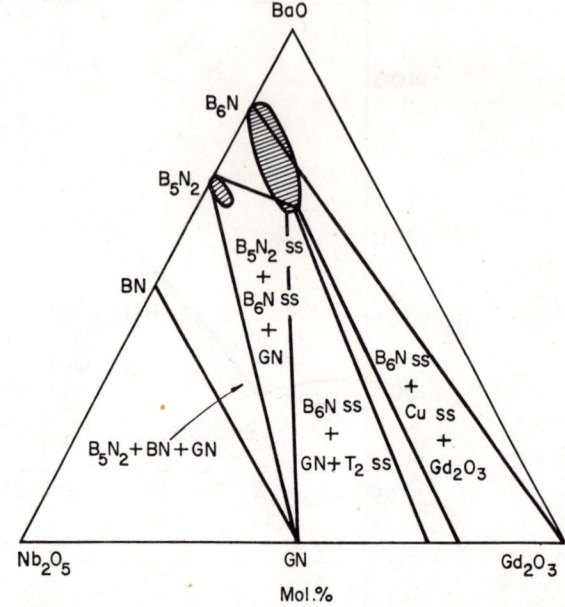

Fig. 563.—System BaO–Gd$_2$O$_3$–Nb$_2$O$_5$; preliminary subsolidus (approx. 1400°C.). C = cubic (fluorite structure type), T$_2$ = distorted fluorite structure type. B = BaO, G = Gd$_2$O$_3$, N = Nb$_2$O$_5$. Shaded area = solid solutions.

R. S. Roth, "Rare Earth Research," edited by Eugene V. Kleber, The Macmillan Co., New York, p. 92 (1961).

BaO–SiO$_2$–TiO$_2$

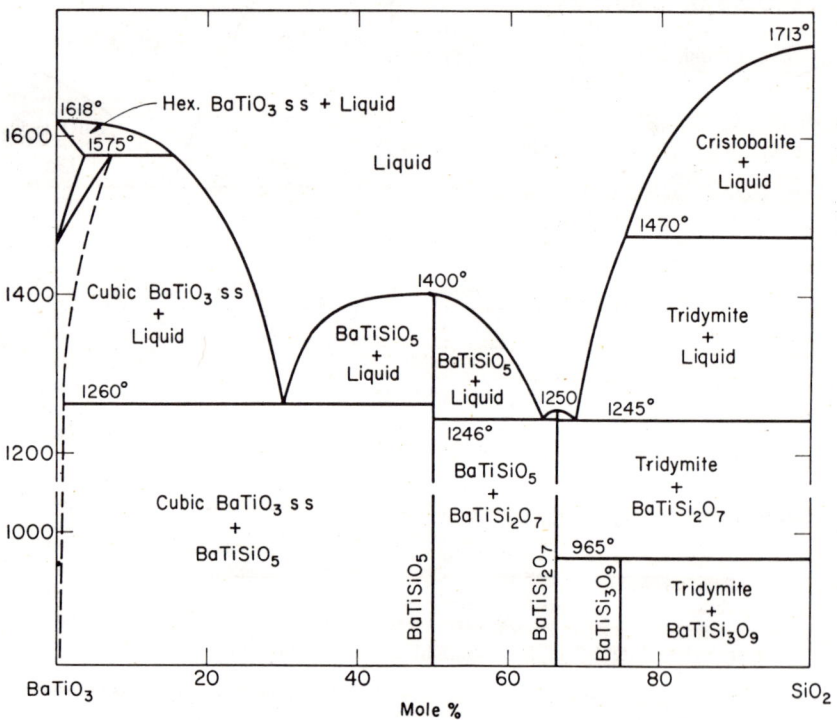

Fig. 564.—System BaTiO$_3$–SiO$_2$.

D. E. Rase and Rustum Roy, *J. Am. Ceram. Soc.*, **38** [11] 393 (1955).

BaO–SnO₂–TiO₂

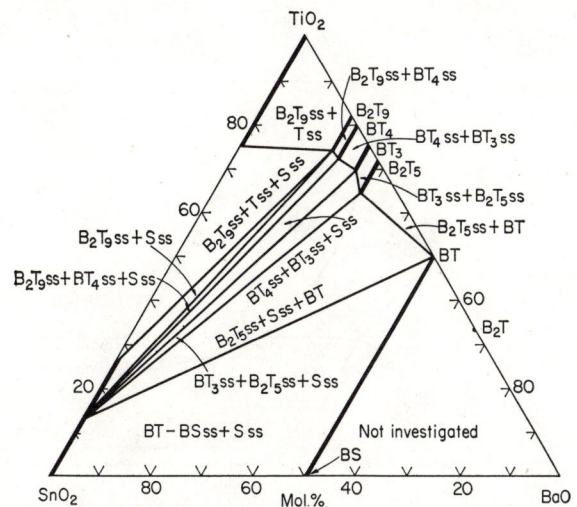

FIG. 565.—System BaO–SnO₂–TiO₂; subsolidus. Heavy lines represent single-phase solid solutions. B = BaO, S = SnO₂, T = TiO₂, ss = solid solution.

G. H. Jonker and W. Kwestroo, *J. Am. Ceram. Soc.*, **41** [10] 392 (1958).

BaO–TiO₂–ZrO₂

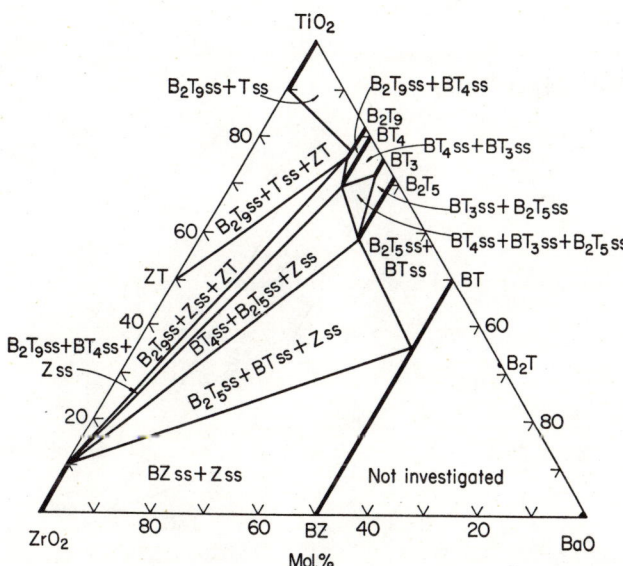

FIG. 566.—System BaO–TiO₂–ZrO₂; subsolidus. Heavy lines represent single-phase solid solutions. B = BaO, T = TiO₂, Z = ZrO₂, ss = solid solution.

G. H. Jonker and W. Kwestroo, *J. Am. Ceram. Soc.*, **41** [10] 392 (1958).

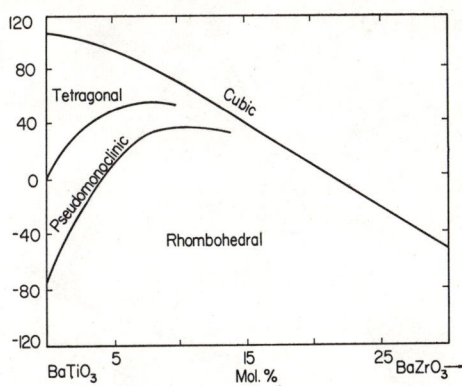

FIG. 567.—System BaTiO₃–BaZrO₃.

T. N. Verbitskaya, G. S. Zhdanov, Yu. N. Venevtsev, and S. P. Solov'ev, *Kristallografiya*, **3**, 189 (1958).

BaO–TiO₂–P₂O₅

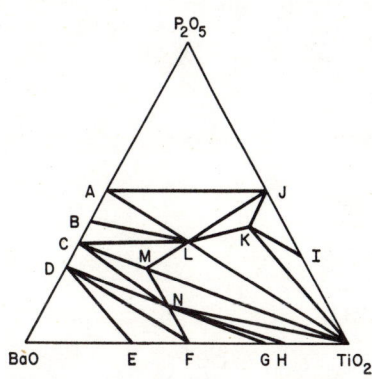

FIG. 568.—System BaO–TiO₂–P₂O₅; compatibility triangles. A = BaO·P₂O₅, B = 3BaO·2P₂O₅, C = 2BaO·P₂O₅, D = 3BaO·P₂O₅, E = 2BaO·TiO₂, F = BaO·TiO₂, G = BaO·3TiO₂, H = BaO·4TiO₂, I = 5TiO₂·2P₂O₅, J = TiO₂·P₂O₅, K = BaO·-4TiO₂·3P₂O₅, L = BaO·TiO₂·P₂O₅, M = 2BaO·TiO₂·P₂O₅, N = 4BaO·3TiO₂·P₂O₅.

D. E. Harrison, *J. Electrochem. Soc.*, **107**, 218 (1960).

BeO–CaO–SrO

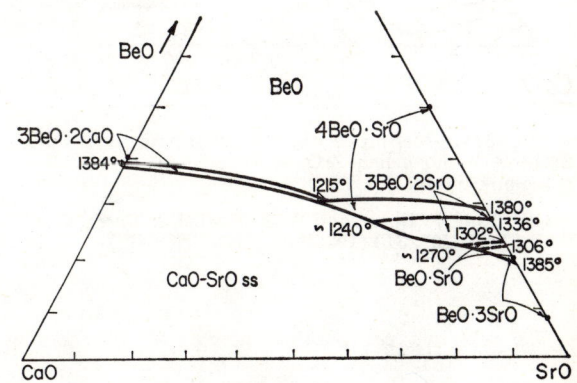

FIG. 569.—System BeO–CaO–SrO.

A. Auriol, G. Hauser, and J. G. Wurm, private communication, Nov. 19, 1961.

BeO–CaO–La$_2$O$_3$

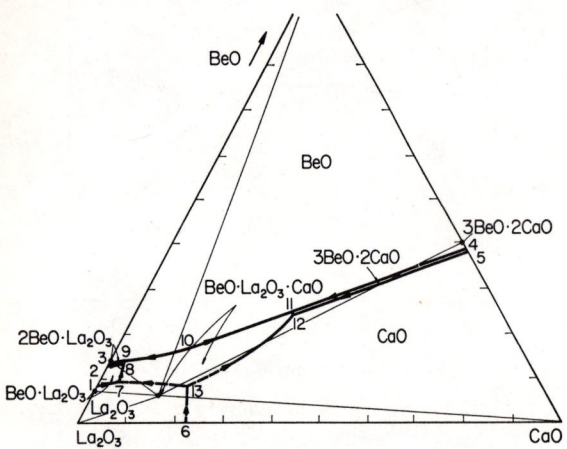

FIG. 570.—System BeO–CaO–La$_2$O$_3$. Temperatures of invariant points: 1 = 1514°C., 2 = 1371°C., 3 = 1371°C., 4 = 1384°C., 5 = 1384°C., 6 = ~1800°C., 7 = 1480°C., 8 = 1328°C., 9 = 1327°C., 10 = 1368°C. (max.), 11 = 1297°C., 12 = 1297°C., 13 = ~1700°C.

A. Auriol, G. Hauser, and J. G. Wurm, private communication, Nov. 19, 1961.

BeO–MgO–Al$_2$O$_3$

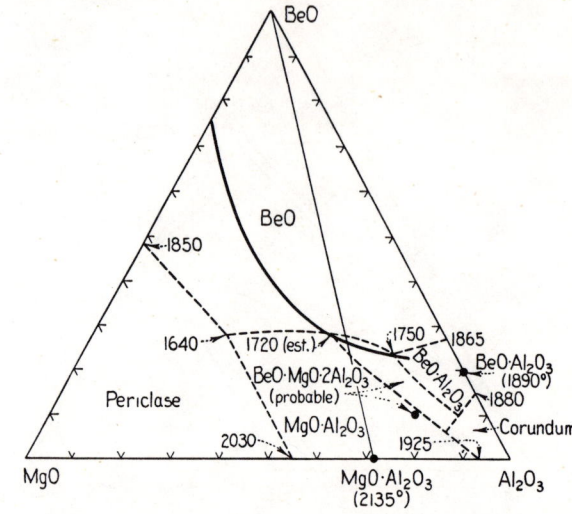

FIG. 572.—System BeO–MgO–Al$_2$O$_3$.

R. F. Geller, P. J. Yavorsky, B. L. Steierman, and A. S. Creamer, *J. Research Natl. Bur. Standards*, **36** [3] 289 (1946); RP 1703.

BeO–CaO–ZrO$_2$

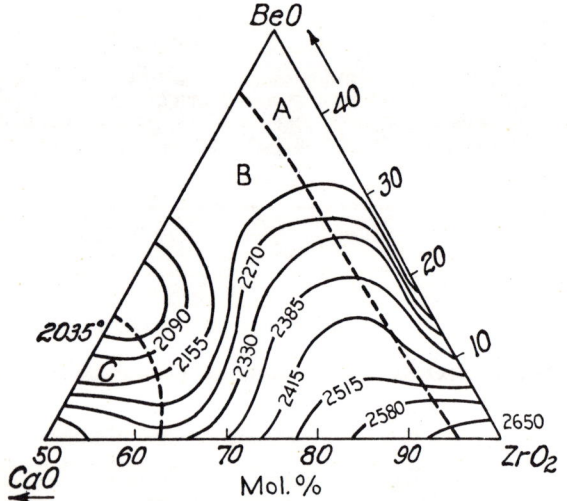

FIG. 571.—Melting isotherms of system BeO–CaO–ZrO$_2$; A = monoclinic ZrO$_2$, B = cubic mix crystal, C = cubic mix crystal + CaZrO$_3$.

Otto Ruff, F. Ebert, and W. Loerpabel, *Z. anorg. u. all-. Chem.*, **207**, 310 (1932).

BeO–MgO–ThO$_2$

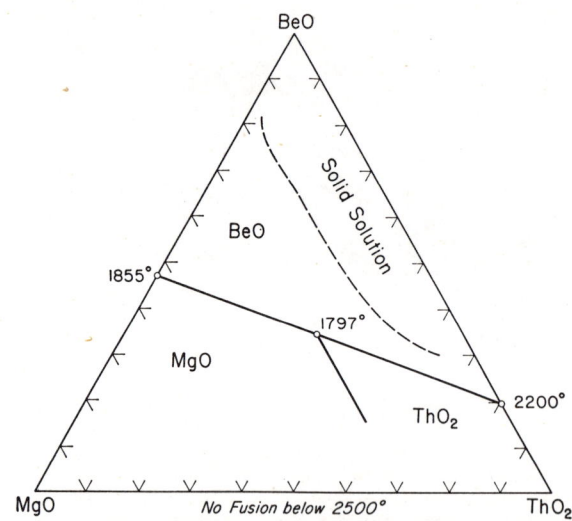

FIG. 573.—System BeO–MgO–ThO$_2$.

S. M. Lang, L. H. Maxwell, and R. F. Geller, *J. Research Natl. Bur. Standards*, **43**, 443 (1949); RP 2034.

BeO–MgO–ZrO₂

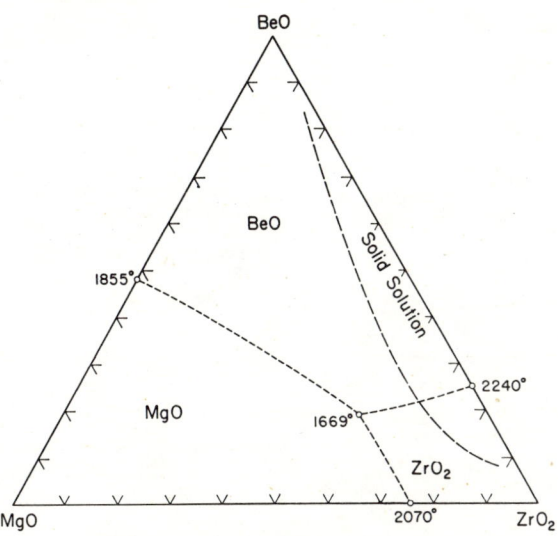

Fig. 574.—System BeO–MgO–ZrO₂.

S. M. Lang, L. H. Maxwell, and R. F. Geller, *J. Research Natl. Bur. Standards,* **43,** 438 (1949); RP 2034.

BeO–Al₂O₃–SiO₂

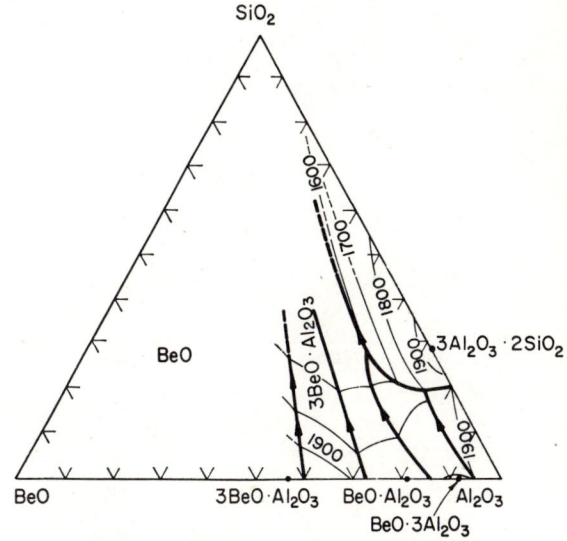

Fig. 576.—System BeO–Al₂O₃–SiO₂; alumina corner.

F. Ya. Galakhov, *Izvest. Akad. Nauk S.S.S.R., Otdel. Khim. Nauk,* 1036 (1957).

BeO–SrO–La₂O₃

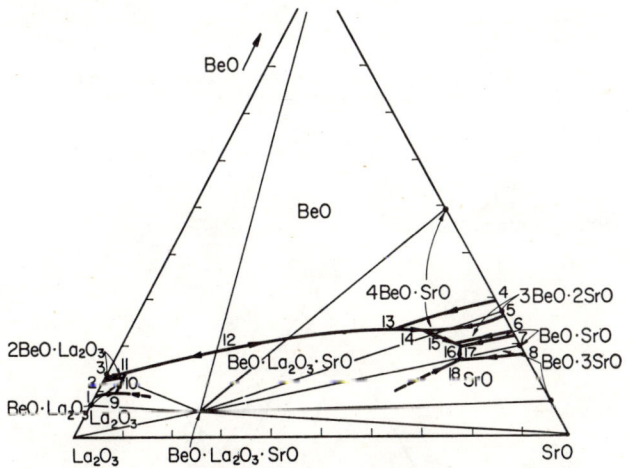

Fig. 575.—System BeO–SrO–La₂O₃. Temperatures of invariant points: 1 = 1514°C., 2 = 1371°C., 3 = 1371°C., 4 = 1380°C., 5 = 1336°C., 6 = 1302°C., 7 = 1306°C., 8 = 1385°C., 9 = 1485°C., 10 = 1349°C., 11 = 1345°C., 12 = 1410°C. (max.), 13 = 1308°C., 14 = 1288°C., 15 = 1297°C. (max.), 16 = 1264°C., 17 = 1271°C. (max.), 18 = 1331°C.

A. Auriol, G. Hauser, and J. G. Wurm, private communication, Nov. 19, 1961.

BeO–Al₂O₃–ThO₂

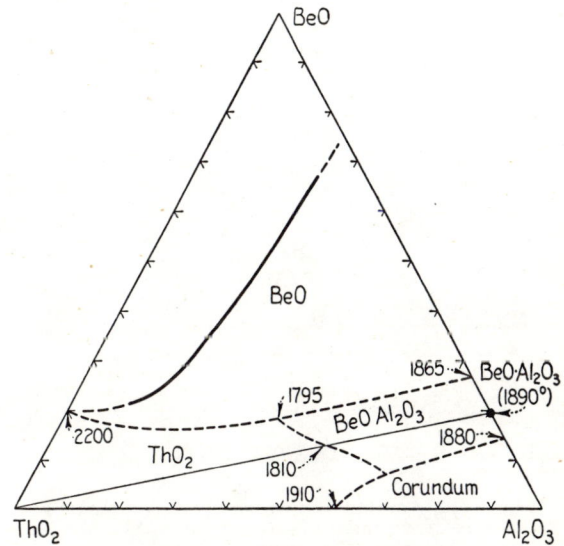

Fig. 577.—System BeO–Al₂O₃–ThO₂.

R. F. Geller, P. J. Yavorsky, B. L. Steierman, and A. S. Creamer, *J. Research Natl. Bur. Standards,* **36** [3] 298 (1946); RP 1703.

BeO–Al₂O₃–TiO₂

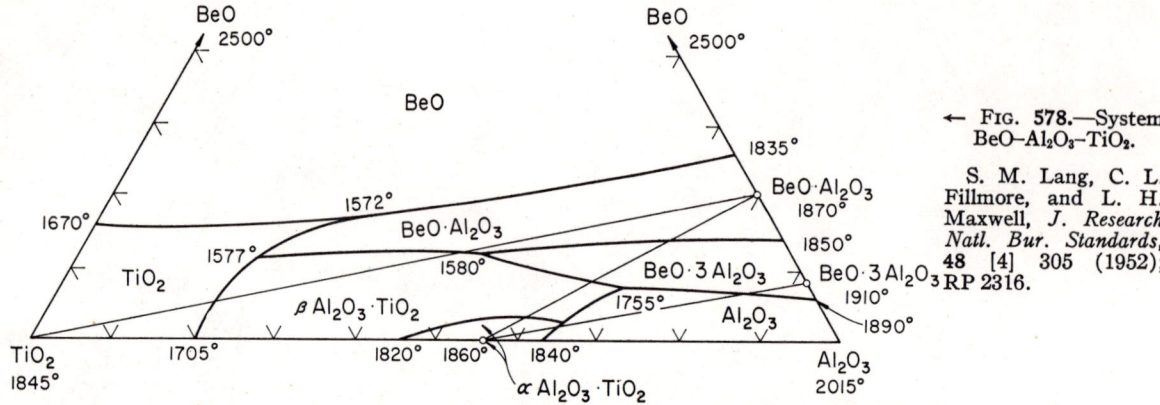

Fig. 578.—System BeO–Al₂O₃–TiO₂.

S. M. Lang, C. L. Fillmore, and L. H. Maxwell, *J. Research Natl. Bur. Standards*, **48** [4] 305 (1952); RP 2316.

BeO–Al₂O₃–ZrO₂

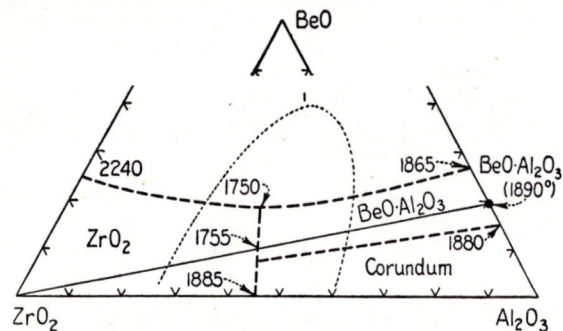

Fig. 579.—System BeO–Al₂O₃–ZrO₂.

R. F. Geller, P. J. Yavorsky, B. L. Steierman, and A. S. Creamer, *J. Research Natl. Bur. Standards*, **36** [3] 303 (1946); RP 1703.

BeO–CeO₂–ZrO₂

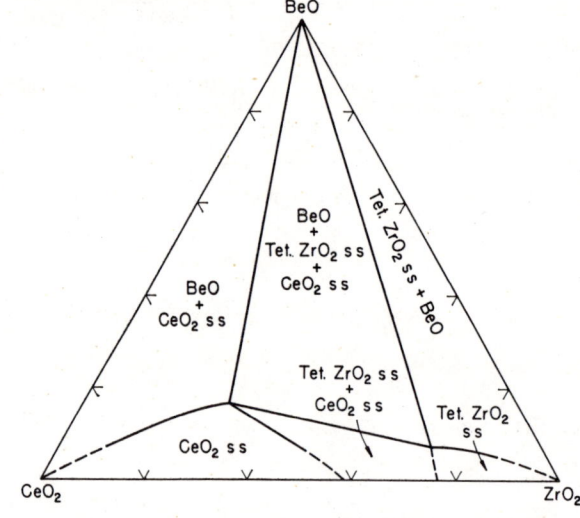

Fig. 581.—System BeO–CeO₂–ZrO₂; isothermal equilibrium at 1750°C.

S. M. Lang, R. S. Roth, and C. L. Fillmore, *J. Research Natl. Bureau Standards*, **53** [4] 206 (1954); RP 2534.

BeO–Cr₂O₃–ZrO₂

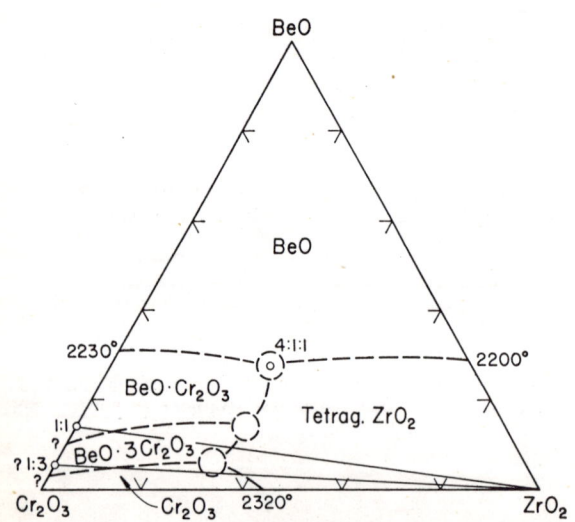

Fig. 580.—System BeO–Cr₂O₃–ZrO₂; postulated.

S. M. Lang, R. S. Roth, and C. L. Fillmore, *J. Research Natl. Bur. Standards*, **53** [4] 208 (1954); RP 2534.

BeO–CeO₂–ZrO₂

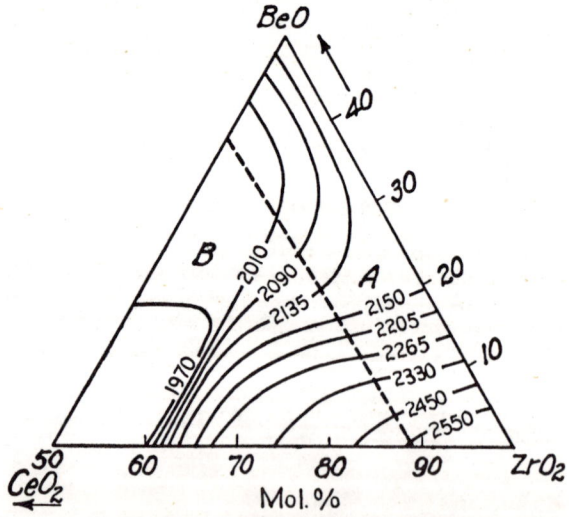

Fig. 582.—Melting isotherms of system BeO–CeO₂–ZrO₂; A = monoclinic ZrO₂, B = cubic mix crystal.

Otto Ruff, F. Ebert, and W. Loerpabel, *Z. anorg. u. allgem. Chem.*, **207**, 310 (1932).

BeO–TiO₂–ZrO₂

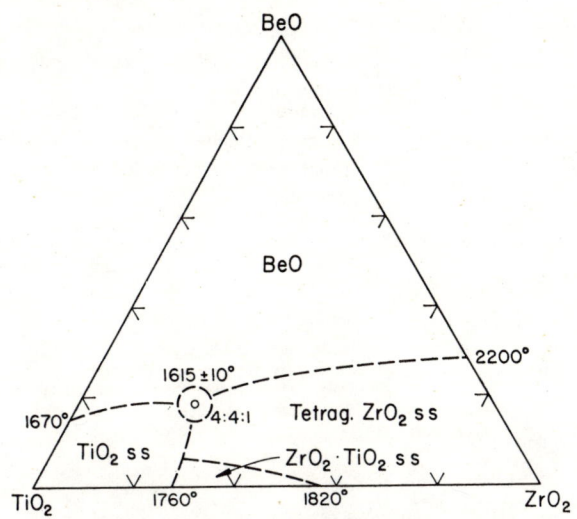

Fig. 583.—System BeO–TiO₂–ZrO₂; liquidus surface.

S. M. Lang, R. S. Roth, and C. L. Fillmore, *J. Research Natl. Bureau Standards,* **53** [4] 203 (1954); RP 2534.

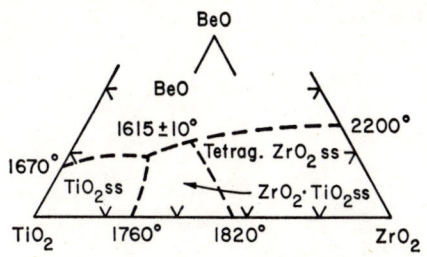

Fig. 584.—System BeO–TiO₂–ZrO₂.

Antonio Cocco, *Ann. Chim.* (*Rome*), **48,** 607 (1958).

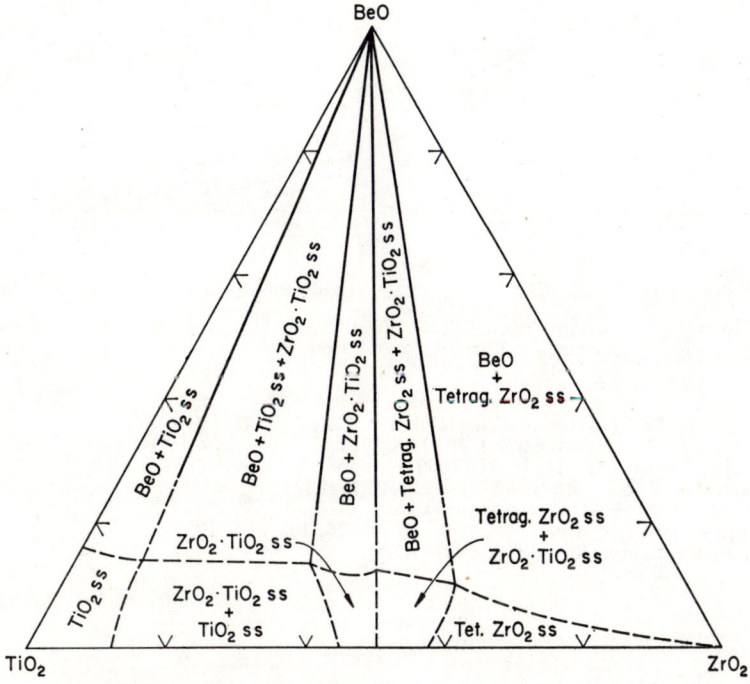

Fig. 585.—System BeO–TiO₂–ZrO₂; isothermal equilibrium at 1550°C.

S. M. Lang, R. S. Roth, and C. L. Fillmore, *J. Research Natl. Bureau Standards,* **53** [4] 203 (1954); RP 2534.

CaO–FeO–SiO₂

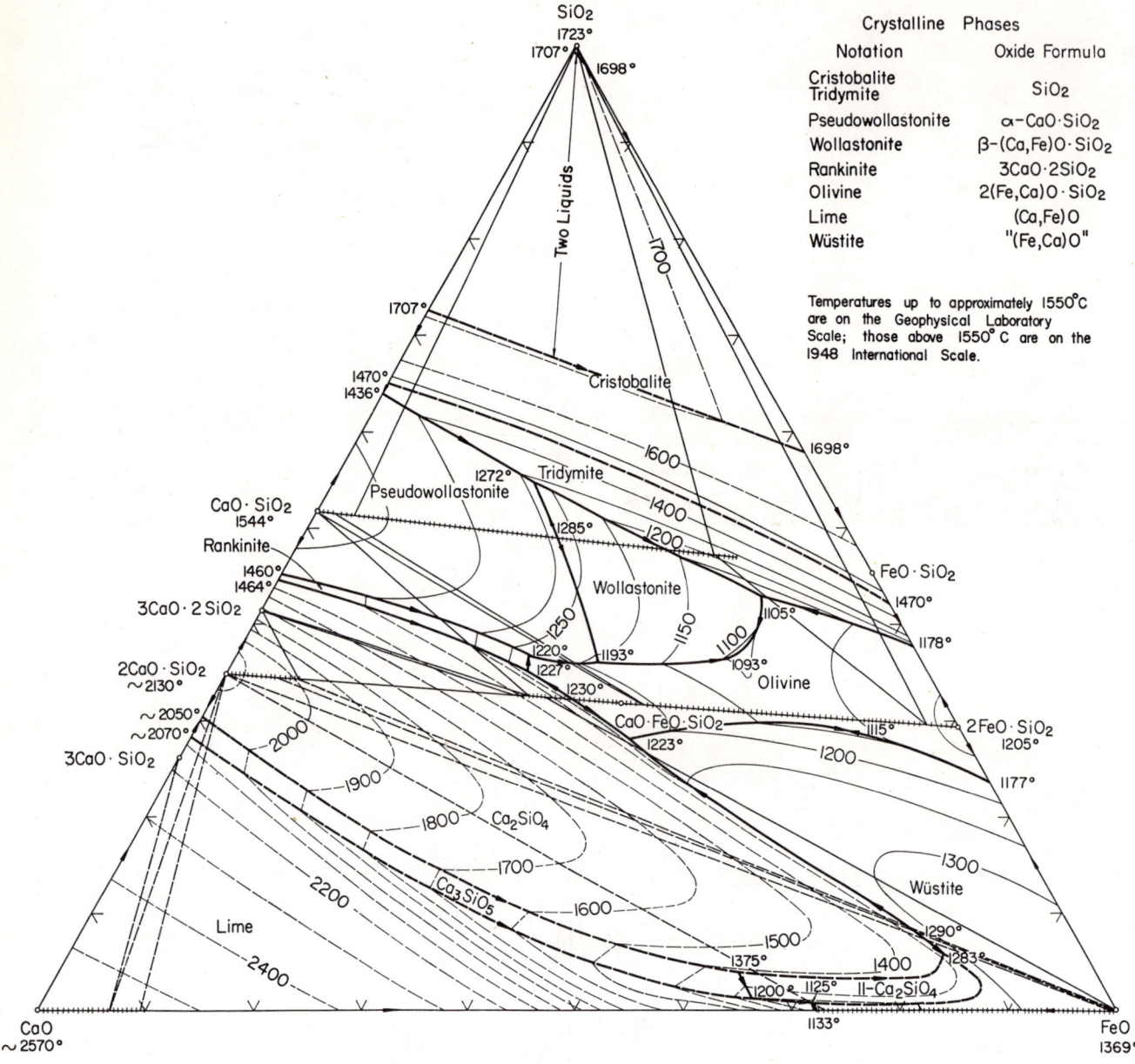

Fig. 586.—System CaO–"FeO"–SiO₂; composite. (Oxide Phases in Equilibrium with Metallic Iron)

E. F. Osborn and Arnulf Muan, revised and redrawn "Phase Equilibrium Diagrams of Oxide Systems," Plate 7, published by the American Ceramic Society and the Edward Orton, Jr., Ceramic Foundation, 1960.

Principal References

A. L. Day, E. S. Shepherd and F. E. Wright, *Am. J. Sci.* (4th Series), **22**, 265–302 (1906).
G. A. Rankin and F. E. Wright, *Am. J. Sci.* (4th Series), **39**, 1–79 (1915).
J. H. Welch and W. Gutt, *J. Am. Ceram. Soc.*, **42**, 11–15 (1959).
N. L. Bowen and J. F. Schairer, *Am. J. Sci.* (5th Series), **24**, 177–213 (1932).
J. W. Greig, *Am. J. Sci.* (5th Series), **13**, 1–44; 133–54 (1927).
N. L. Bowen, J. F. Schairer and E. Posnjak, *Am. J. Sci.* (5th Series), **26**, 193–284 (1933).
W. C. Allen and R. B. Snow, *J. Am. Ceram. Soc.*, **38**, 264–80 (1955).

CaO–FeO–SiO₂ (cont.)

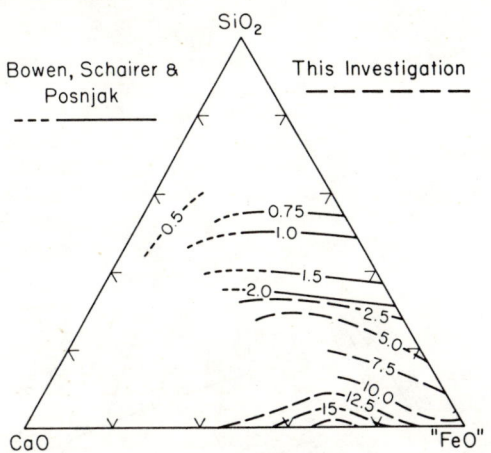

FIG. 587.—System CaO–FeO–SiO₂, showing lines of average Fe₂O₃ content of melts on the liquidus surface

Curves with Fe₂O₃ content below 2.5% after N. L. Bowen, J. F. Schairer, and E. Posnjak, *Am. J. Sci.*, [5th Series], **26** [153] 203 (1933); W. C. Allen and R. B. Snow, *J. Am. Ceram. Soc.*, **38** [8] 271 (1955).

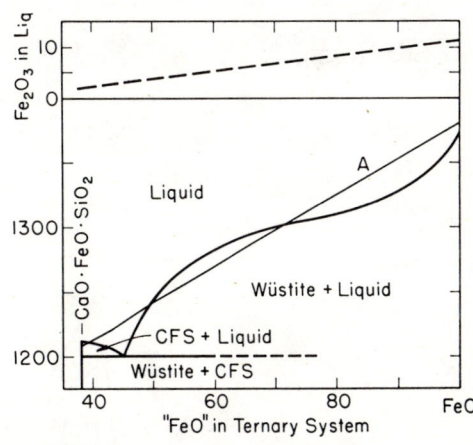

FIG. 590.—System CaO·FeO·SiO₂–FeO.   C = CaO; F = FeO; S = SiO₂.

Curve A after J. F. Schairer, *J. Am. Ceram. Soc.*, **25** [10] 241 (1942).
W. C. Allen and R. B. Snow, *J. Am. Ceram. Soc.*, **38** [8] 267 (1955).

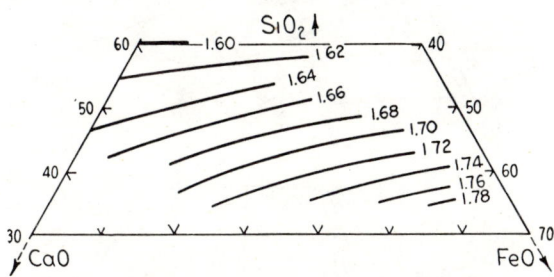

FIG. 588.—Refractive indices of glasses in the system CaO–FeO–SiO₂.

N. L. Bowen, J. F. Schairer, and E. Posnjak, *Am. J. Sci.*, 5th Ser., **26**, 264 (1933).

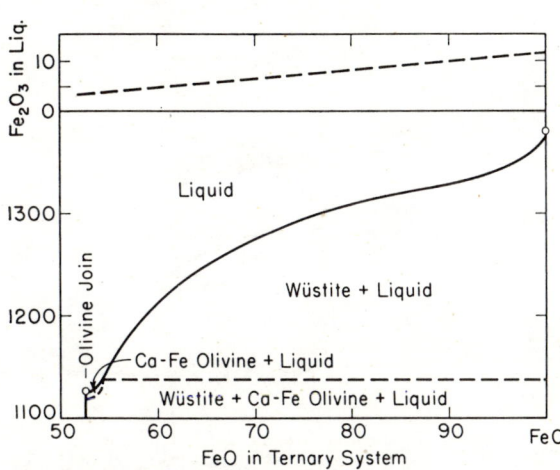

FIG. 591.—System CaO–FeO–SiO₂. Section 16% CaO, 53% FeO, 31% SiO₂ to FeO.

Circles—N. L. Bowen and J. F. Schairer, *Am. J. Sci.*, [5th Series], **24** [141], 177 (1932); W. C. Allen and R. B. Snow, *J. Am. Ceram. Soc.*, **38** [8] 267 (1955).

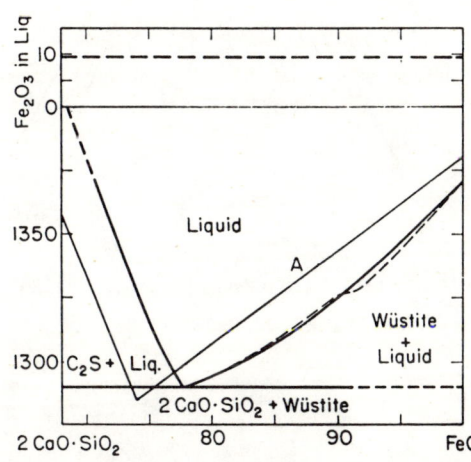

FIG. 589.—System 2CaO·SiO₂–FeO.   C = CaO; S = SiO₂.

Curve A after Ricker and Osborn as published by Arnulf Muan and E. F. Osborn in *Yearbook Am. Iron Steel Inst.*, p. 336 (1951).
W. C. Allen and R. B. Snow, *J. Am. Ceram. Soc.*, **38** [8] 266 (1955).

FIG. 592.   See page 208.

FIG. 593.   See page 208

CaO–FeO–SiO₂ (cont.)

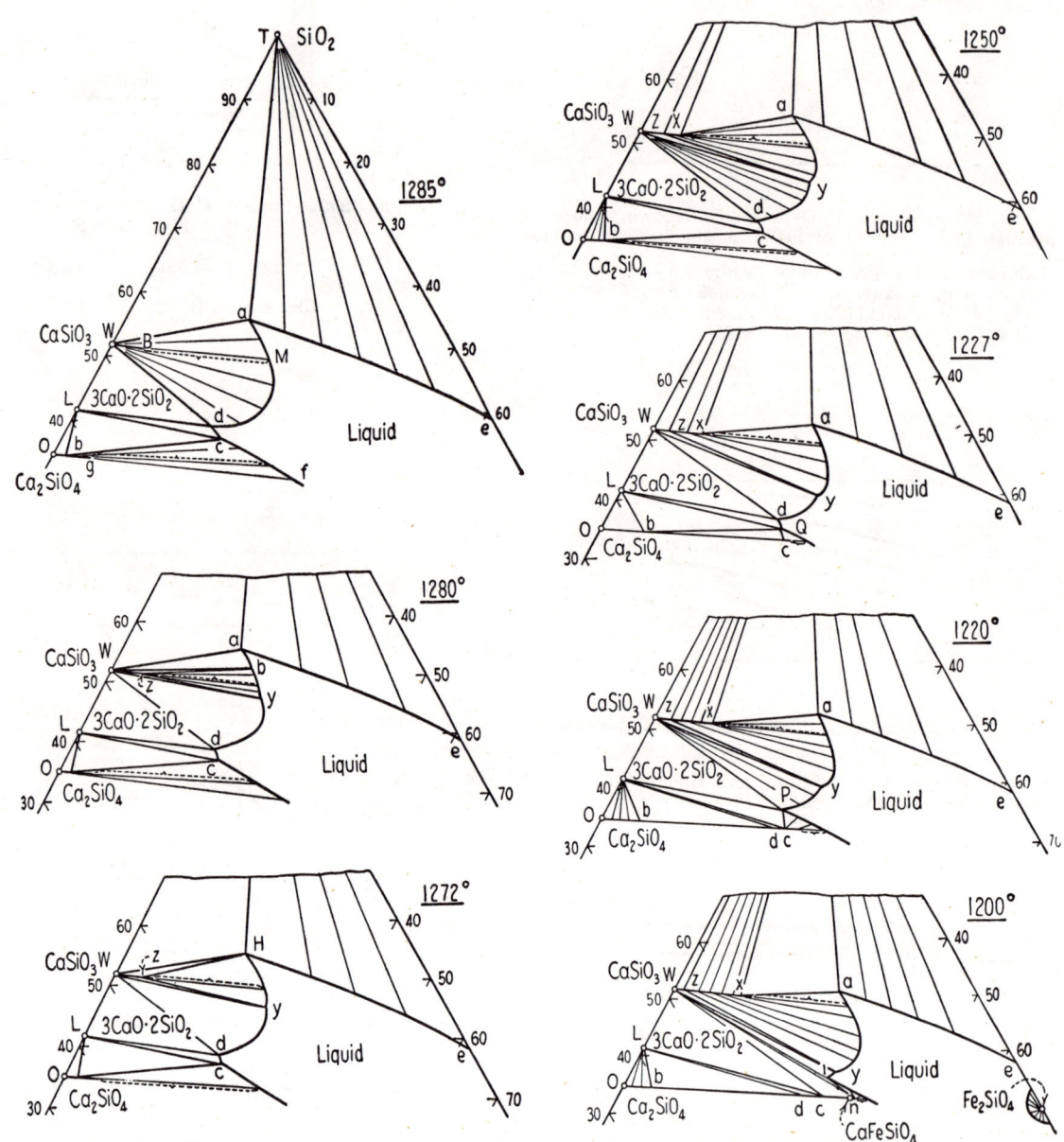

Fig. 594.—Phase relations in the system CaO–FeO–SiO₂ at 1285°, 1280°, 1272°.
N. L. Bowen, J. F. Schairer, and E. Posnjak, *Am. J. Sci.*, 5th Ser., 26, 222–32 (1933).

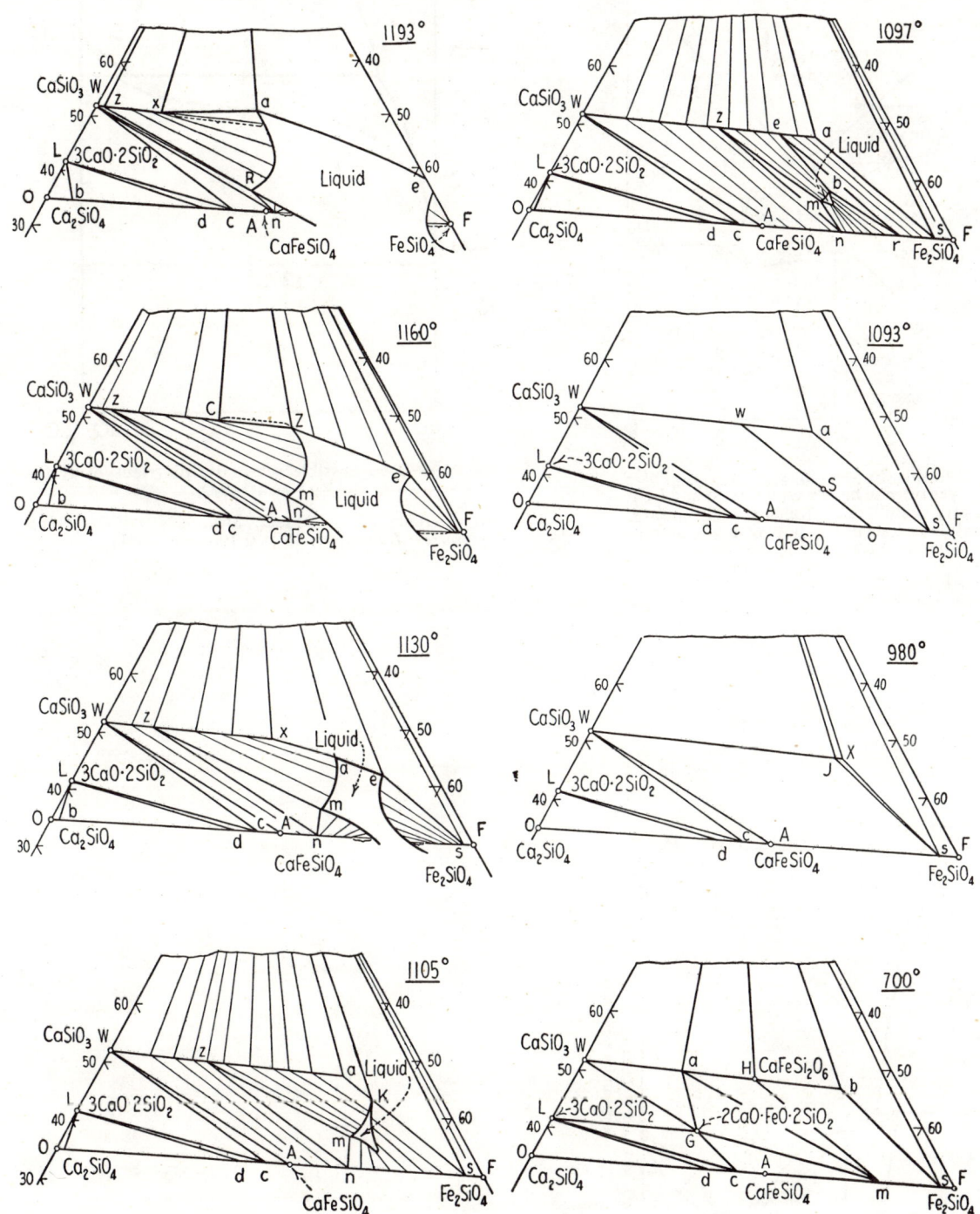

1250°, 1227°, 1220°, 1200°, 1193°, 1160°, 1130°, 1105°, 1097°, 1093°, 980°, and 700°C.

CaO–FeO–SiO₂ (concl.)

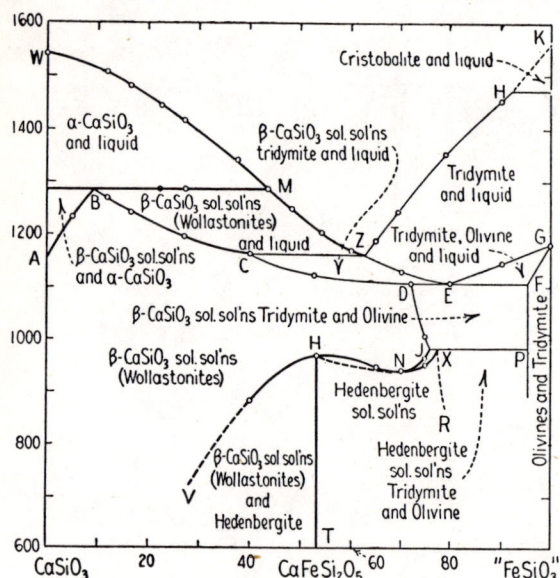

FIG. 592.—Metasilicate join in system CaO–FeO–SiO₂; heavy curves refer to binary equilibrium and light curves to ternary equilibrium (binary and ternary, respectively, only when small amounts of Fe₂O₃ present in liquids are treated as FeO.

N. L. Bowen, J. F. Schairer, and E. Posnjak, *Am. J. Sci.*, 5th Ser., **26**, 213 (1933).

CaO–FeO–P₂O₅

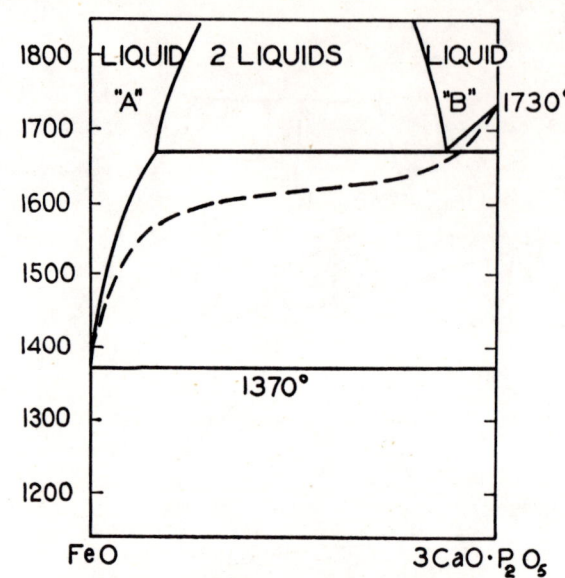

FIG. 595.—Two possible schematic diagrams of system FeO–3CaO·P₂O₅.

W. Oelsen and H. Maetz, *Mitt. Kaiser-Wilhelm-Inst. Eisenforsch. Düsseldorf*, **23** [12] 195–245 (1941).

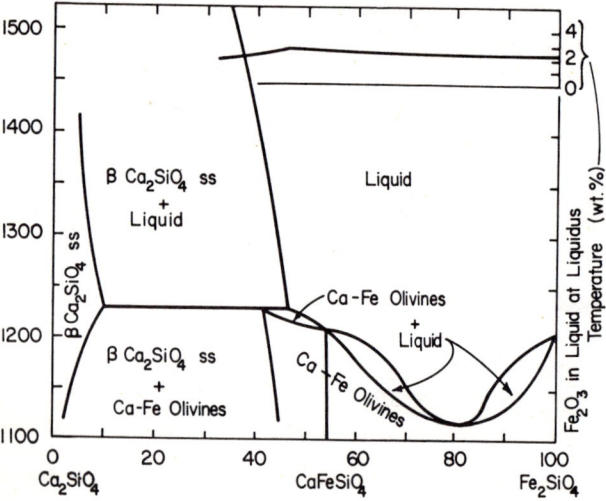

FIG. 593.—System 2CaO·SiO₂–2FeO·SiO₂.

N. L. Bowen, J. F. Schairer, and E. Posnjak, *Am. J. Sci.* 5th Ser., **25**, 281 (1933).

CaO–MgO–Al$_2$O$_3$

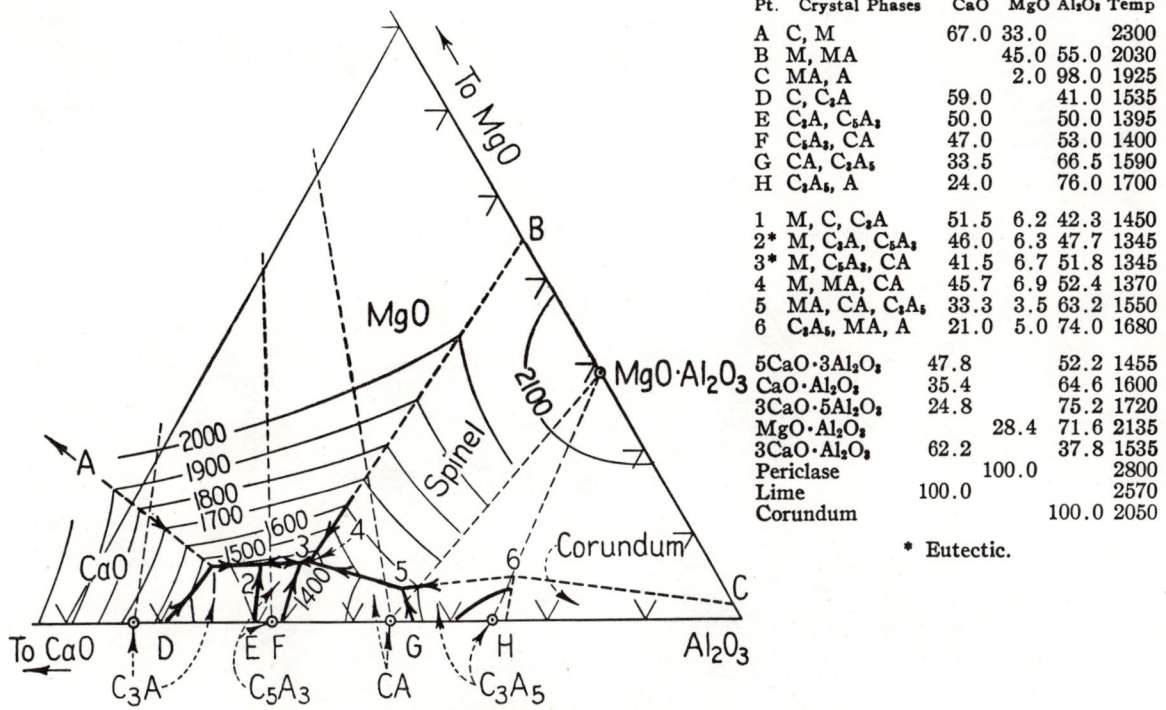

Pt.	Crystal Phases	CaO	MgO	Al$_2$O$_3$	Temp
A	C, M	67.0	33.0		2300
B	M, MA		45.0	55.0	2030
C	MA, A		2.0	98.0	1925
D	C, C$_3$A	59.0		41.0	1535
E	C$_3$A, C$_5$A$_3$	50.0		50.0	1395
F	C$_5$A$_3$, CA	47.0		53.0	1400
G	CA, C$_3$A$_5$	33.5		66.5	1590
H	C$_3$A$_5$, A	24.0		76.0	1700
1	M, C, C$_3$A	51.5	6.2	42.3	1450
2*	M, C$_3$A, C$_5$A$_3$	46.0	6.3	47.7	1345
3*	M, C$_5$A$_3$, CA	41.5	6.7	51.8	1345
4	M, MA, CA	45.7	6.9	52.4	1370
5	MA, CA, C$_3$A$_5$	33.3	3.5	63.2	1550
6	C$_3$A$_5$, MA, A	21.0	5.0	74.0	1680
5CaO·3Al$_2$O$_3$		47.8		52.2	1455
CaO·Al$_2$O$_3$		35.4		64.6	1600
3CaO·5Al$_2$O$_3$		24.8		75.2	1720
MgO·Al$_2$O$_3$			28.4	71.6	2135
3CaO·Al$_2$O$_3$		62.2		37.8	1535
Periclase			100.0		2800
Lime		100.0			2570
Corundum				100.0	2050

* Eutectic.

FIG. 596.—System CaO–MgO–Al$_2$O$_3$; C = CaO, M = MgO, A = Al$_2$O$_3$.

G. A. Rankin and H. E. Merwin, *Z. anorg. u. allgem. Chem.*, **96**, 309 (1916); *J. Am. Chem. Soc.*, **38**, 568 (1916). 1500°C. isotherm should intersect CaO–Al$_2$O$_3$ boundary at 42.8% CaO instead of 41%, according to R. B. Sosman, private communication, April 1, 1958.

CaO–MgO–Cr$_2$O$_3$

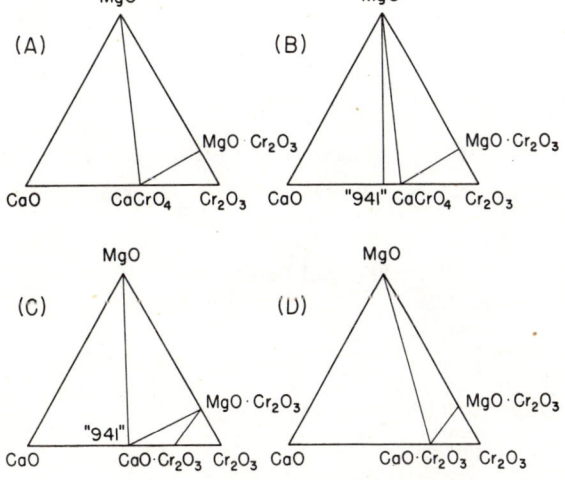

FIG. 597.—System CaO–MgO–Cr$_2$O$_3$.

Phase distribution in the "ternary" system in air (a) up to 800°, (b) 800–1000°, (c) over 1000°C. and (d) in vacuo; (a), (b), (c) not true ternary systems because of oxygen reactions to form CaCrO$_4$ and 9CaO·4CrO$_3$·Cr$_2$O$_3$ ("941"); (d) approaches true ternary system.

W. F. Ford and W. J. Rees, *Trans. Brit. Ceram. Soc.*, **48**, 306 (1949).

Fig. 598

CaO–MgO–SiO₂

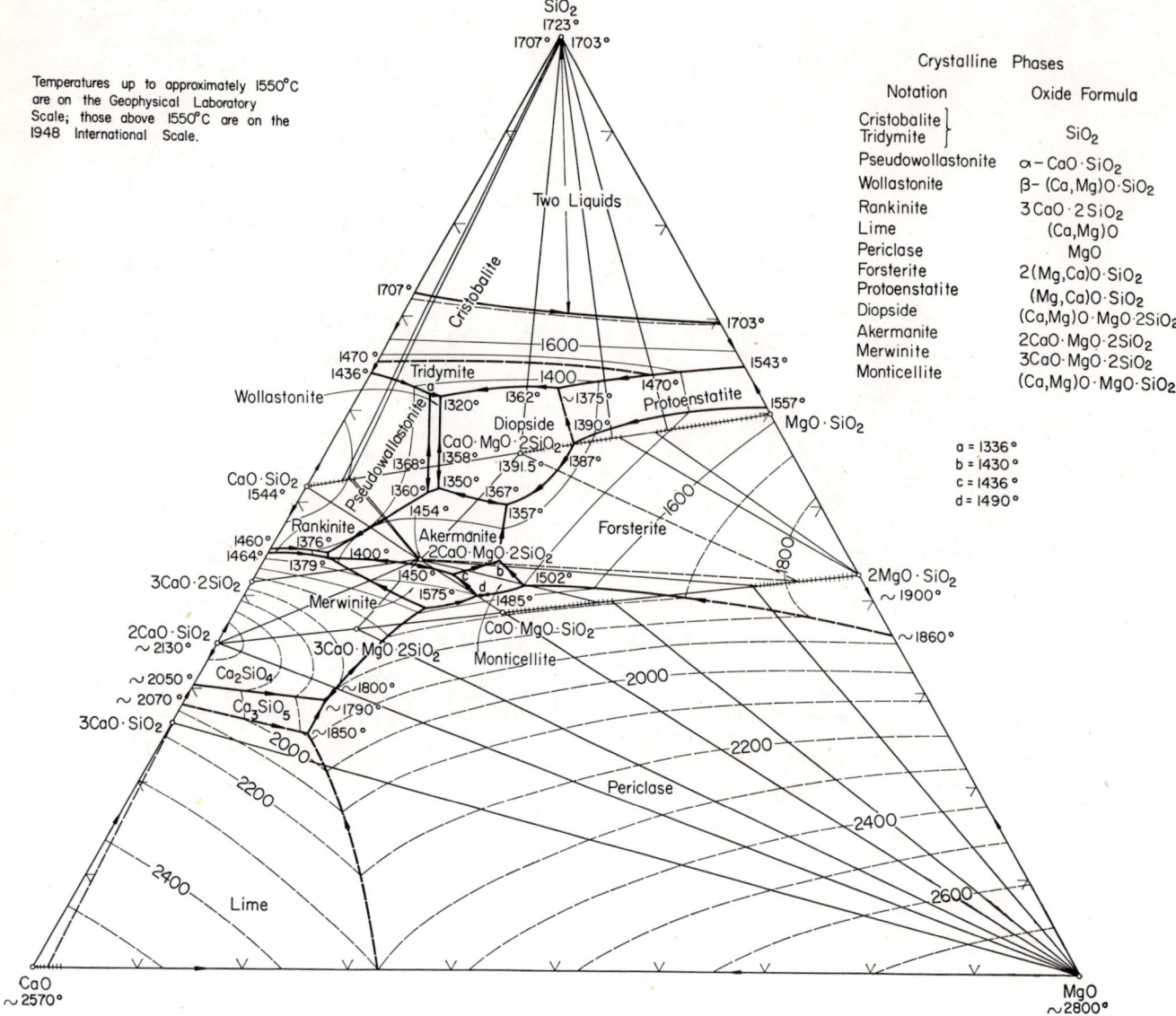

Fig. 598.—System CaO–MgO–SiO₂; composite.

E. F. Osborn and Arnulf Muan, revised and redrawn "Phase Equilibrium Diagrams of Oxide Systems," Plate 2, published by the American Ceramic Society and the Edward Orton, Jr., Ceramic Foundation, 1960.

Principal References

A. L. Day, E. S. Shepherd and F. E. Wright, *Am. J. Sci.* (4th Series), **22**, 265–302 (1906)
J. H. Welch and W. Gutt, *J. Am. Ceram. Soc.*, **42**, 11–15 (1959).
F. Trojer and K. Konopicky, *Radex Rundschau*, **4**, 161–62 (1949).
N. L. Bowen and Olaf Andersen, *Am. J. Sci.* (4th Series), **37**, 487–500 (1914).
N. L. Bowen, *Am. J. Sci.* (4th Series), **38**, 207–64 (1914).
J. B. Ferguson and H. E. Merwin, *Am. J. Sci.* (4th Series), **48**, 81–123 (1919).
J. W. Greig, *Am. J. Sci.* (5th Series), **13**, 1–44; 133–54 (1927).
J. F. Schairer and N. L. Bowen, *Am. J. Sci.*, **240**, 725–42 (1942).
E. F. Osborn, *Am. J. Sci.*, **240**, 751–88 (1942).
E. F. Osborn, *J. Am. Ceram. Soc.*, **26**, 321–32 (1943).
R. W. Ricker and E. F. Osborn, *J. Am. Ceram. Soc.*, **37**, 133–39 (1954).
F. R. Boyd and J. F. Schairer, *Carnegie Inst. Wash. Year Book*, **56**, 223–25 (1957).

CaO–MgO–SiO$_2$ (cont.)

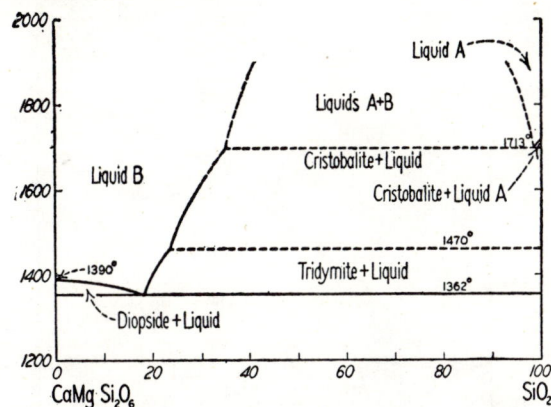

FIG. 599.—System SiO$_2$–CaO·MgO·2SiO$_2$(diopside).

N. L. Bowen, *Am. J. Sci.*, 4th Ser., **38**, 211 (1914).

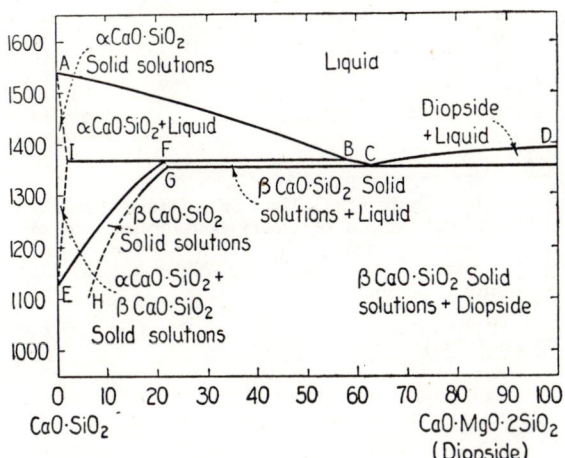

FIG. 600.—System CaO·SiO$_2$–CaO·MgO·2SiO$_2$.

J. F. Schairer and N. L. Bowen, *Am. J. Sci.*, **240**, 730 (1942).

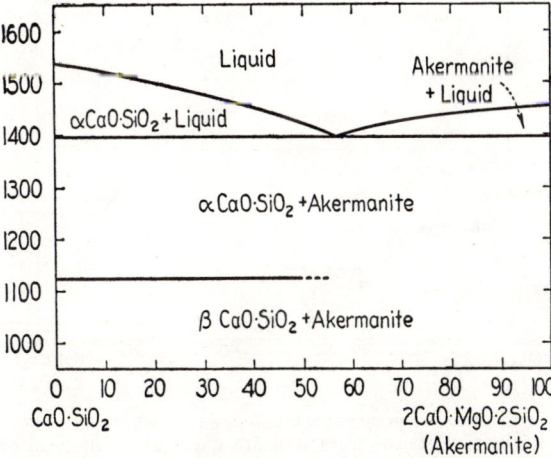

FIG. 601.—System CaO·SiO$_2$–2CaO·MgO·2SiO$_2$; not binary at all temperatures because of instability of akermanite below about 1325°C.

J. F. Schairer and N. L. Bowen, *Am. J. Sci.*, **240**, 741 (1942).

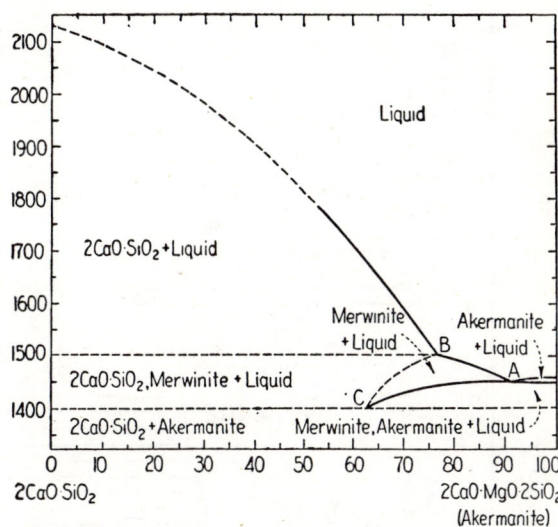

FIG. 602.—Join 2CaO·SiO$_2$–2CaO·MgO·2SiO$_2$., Fig. 598

E. F. Osborn, *J. Am. Ceram. Soc.*, **26** [10] 328 (1943).

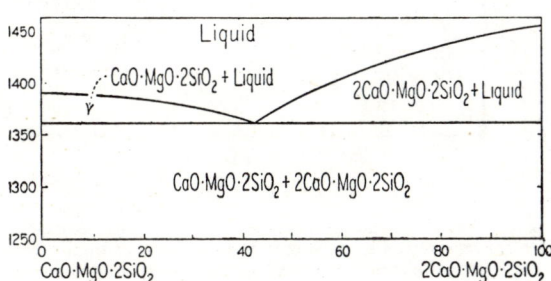

FIG. 603.—System CaO·MgO·2SiO$_2$–2CaO·MgO·2SiO$_2$.

J. B. Ferguson and H. E. Merwin, *Am. J. Sci.*, 4th Ser., **48**, 118 (1919).

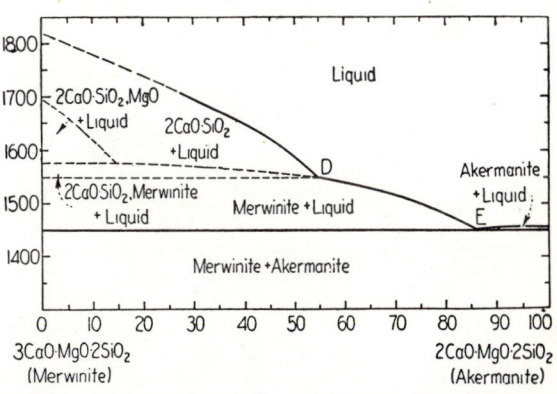

FIG. 604.—Join 2CaO·MgO·2SiO$_2$(akermanite)–3CaO·· MgO·2SiO$_2$(merwinite) in system CaO–MgO–SiO$_2$.

E. F. Osborn, *J. Am. Ceram. Soc.*, **26** [10] 329 (1943).

CaO–MgO–SiO$_2$ (cont.)

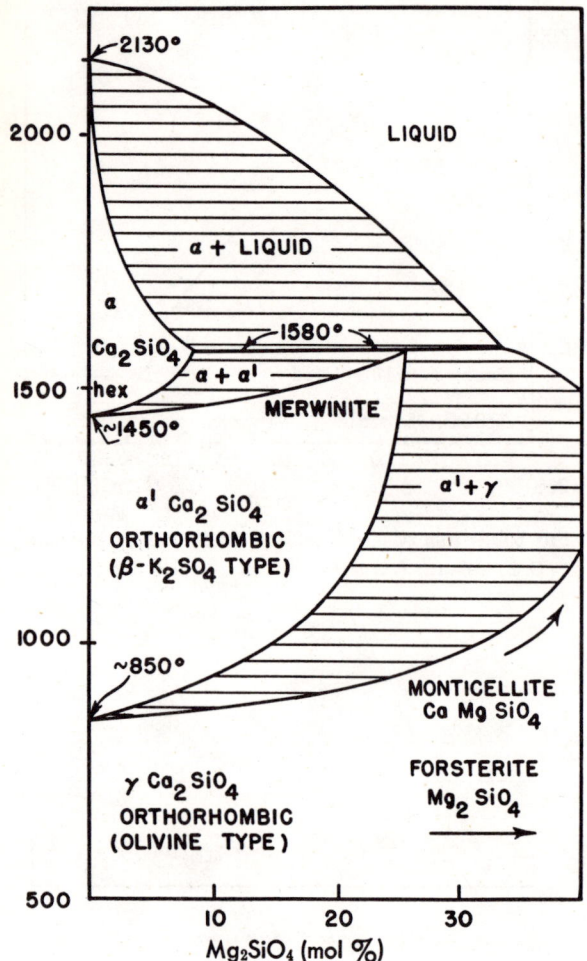

Fig. 605.—Pseudobinary system Ca$_2$SiO$_4$–Mg$_2$SiO$_4$. The curves are to be considered hypothetical or qualitative except for the temperatures of transition and melting points specifically indicated.

M. A. Bredig, *J. Am. Ceram. Soc.*, **33** [6] 190 (1950).

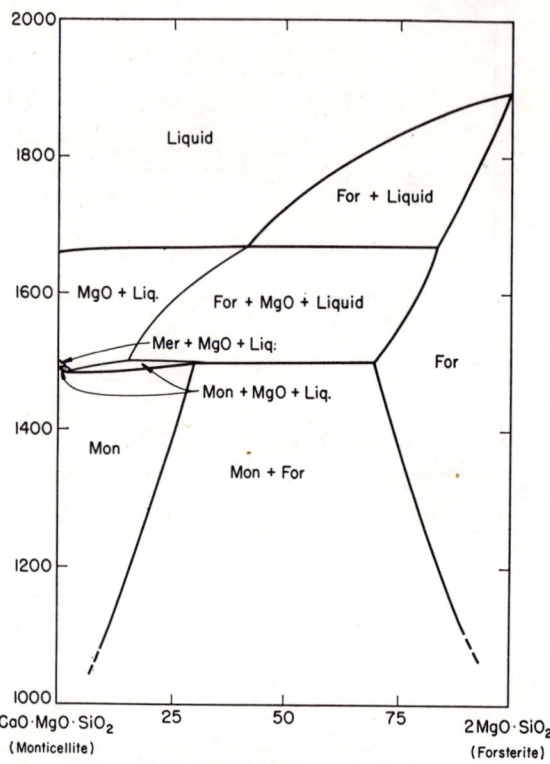

Fig. 606.—System CaO·MgO·SiO$_2$–2MgO·SiO$_2$. Heavy curves bound regions of binary equilibrium; in other regions, ternary relationships obtain. For—forsterite solid solutions; Mon—monticellite solid solutions; Mer—merwinite; Liq—liquid.

R. W. Ricker and E. F. Osborn, *J. Am. Ceram. Soc.*, **37** [3] 136 (1954).

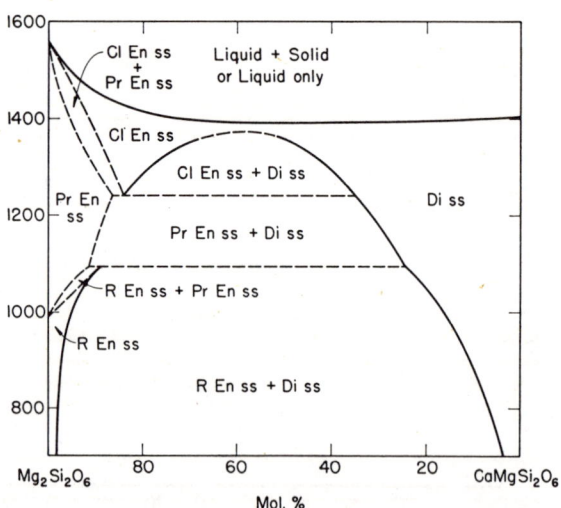

Fig. 607.—System CaMgSi$_2$O$_6$–Mg$_2$Si$_2$O$_6$; subsolidus relations. Dashed lines indicate uncertainty in location of the curves. Liquidus curve adapted from Bowen, *Am. J. Sci.*, 4th Ser. **38**, 207–264 (1914). Di = diopside, R En = rhombic enstatite, Pr En = protoenstatite, Cl En = clinoenstatite, ss = solid solutions.

Leon Atlas, *J. Geol.*, **60**, 146 (1952).

CaO–MgO–SiO$_2$ (concl.)

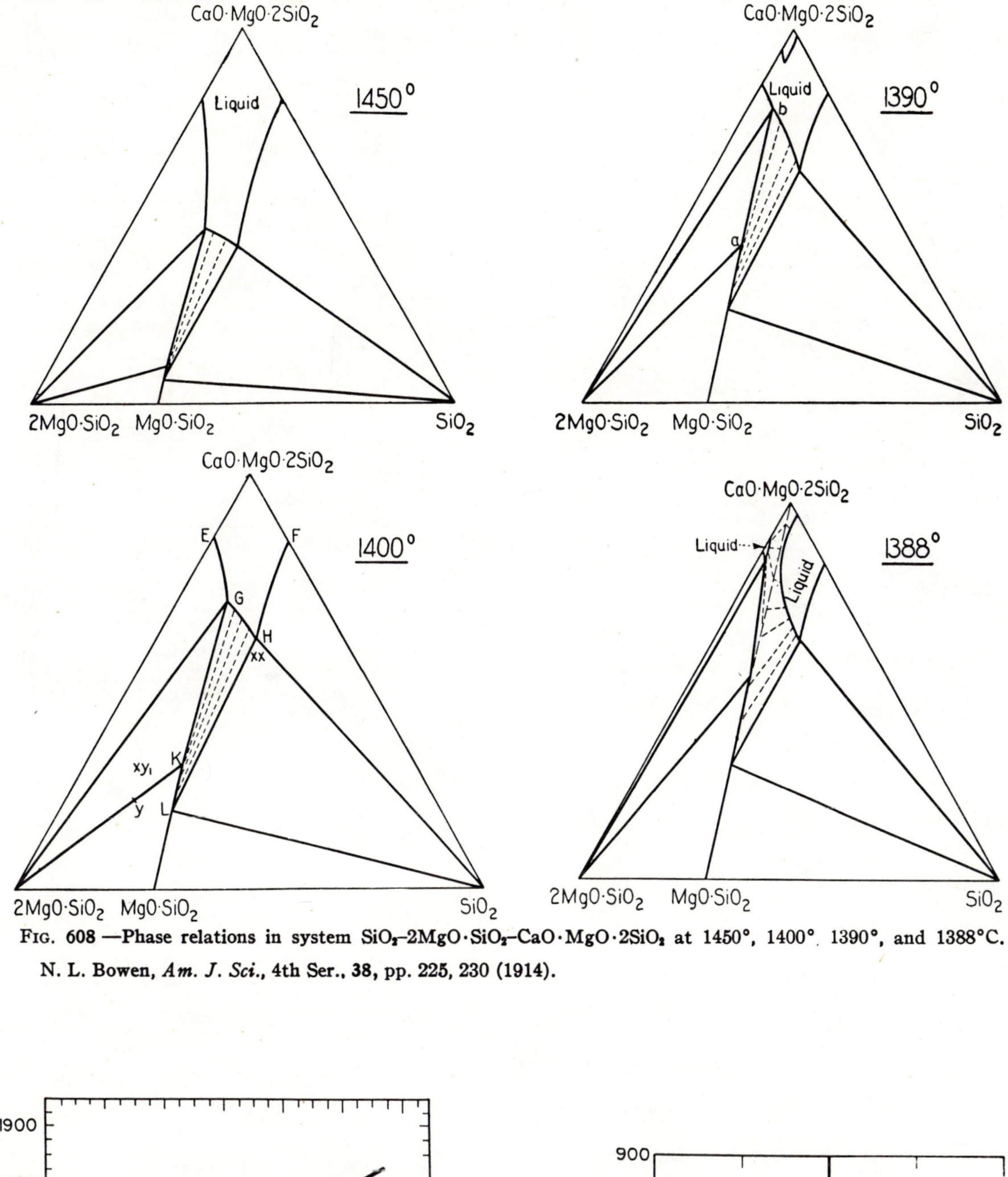

Fig. 608.—Phase relations in system SiO$_2$–2MgO·SiO$_2$–CaO·MgO·2SiO$_2$ at 1450°, 1400°, 1390°, and 1388°C.

N. L. Bowen, *Am. J. Sci.*, 4th Ser., **38**, pp. 225, 230 (1914).

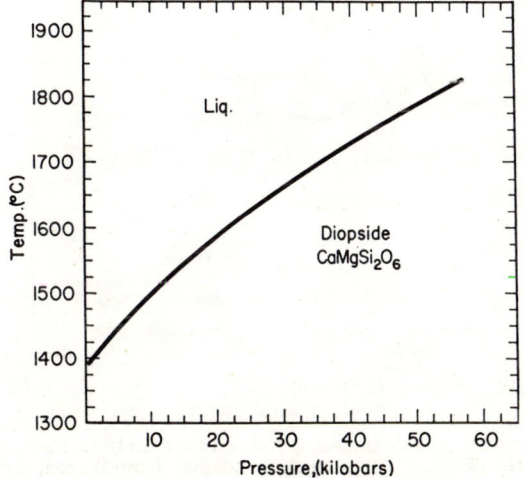

Fig. 609.—System CaMgSi$_2$O$_6$; p-T.

F. R. Boyd and J. L. England, Ann. Rept. Director Geophys. Lab. 1960–61; in *Carnegie Inst. Washington Year Book*, **60**, 117 (1961).

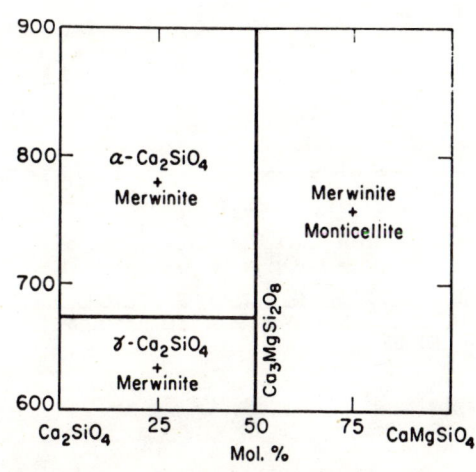

Fig. 610.—System Ca$_2$SiO$_4$–CaMgSiO$_4$; subsolidus, at 2000 p.s.i. H$_2$O pressure.

D. M. Roy, *Mineral. Mag.*, **31**, 191 (1956).

CaO–MgO–SnO$_2$

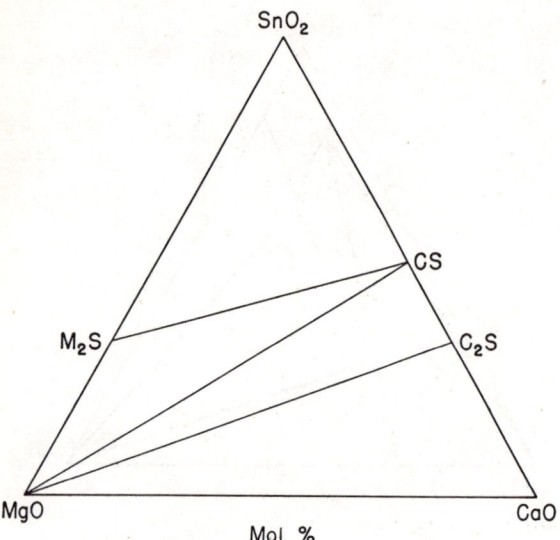

FIG. 611.—System CaO–MgO–SnO$_2$; composition triangles; M = MgO; S = SnO$_2$.

L. W. Coughanour, R. S. Roth, S. Marzullo, and F. E. Sennett, *J. Research Natl. Bur. Standards*, **54** [3] 153 (1955); RP 2576.

CaO–MgO–P$_2$O$_5$

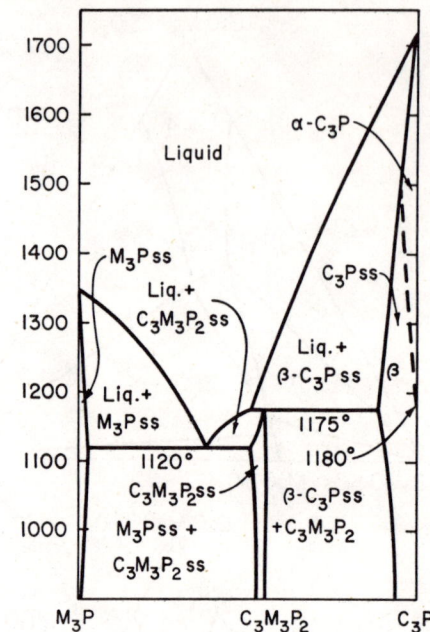

FIG. 613.—System Ca$_3$(PO$_4$)$_2$–Mg$_3$(PO$_4$)$_2$. C = CaO, M = MgO, P = P$_2$O$_5$. Inversion of C$_3$P ss does not follow phase rule.

Jumpei Ando, *Bull. Chem. Soc. Japan*, **31**, 202 (1958). See also, Włodzimierz Bobrownicki and Kazimierz Sławski, *Roczniki Chem.*, **33**, 251 (1959).

CaO–MgO–TiO$_2$

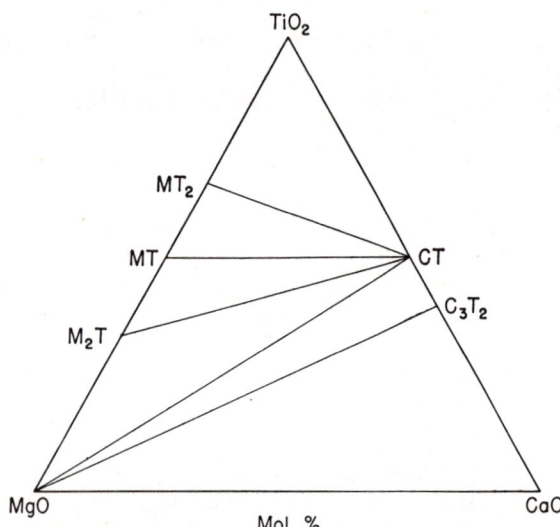

FIG. 612.—System CaO–MgO–TiO$_2$; composition triangles; C = CaO; M = MgO; T = TiO$_2$.

L. W. Coughanour, R. S. Roth, S. Marzullo, and F. E. Sennett, *J. Research Nat. Bur. Standards*, **54** [3] 153 (1955); RP 2576.

CaO–MnO–SiO$_2$

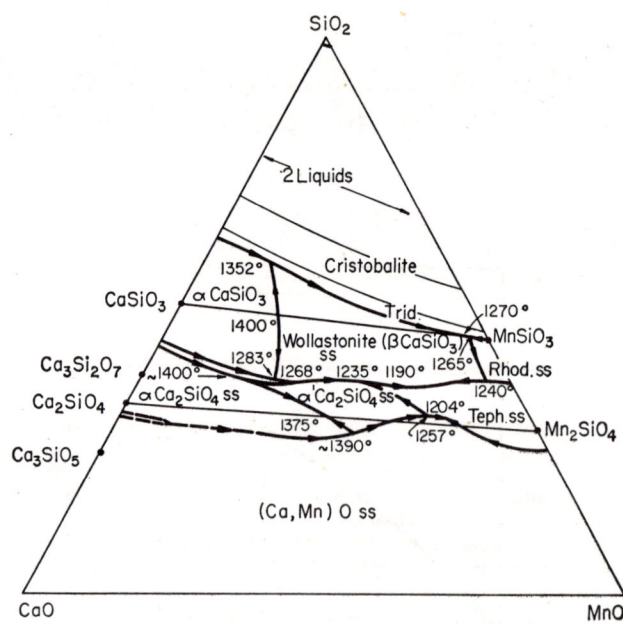

FIG. 614.—System CaO–MnO–SiO$_2$; primary phase fields and selected liquidus temperatures.

F. P. Glasser, *J. Am. Ceram. Soc.*, **45** [5] 245 (1962).

CaO–MnO–SiO$_2$ (concl.)

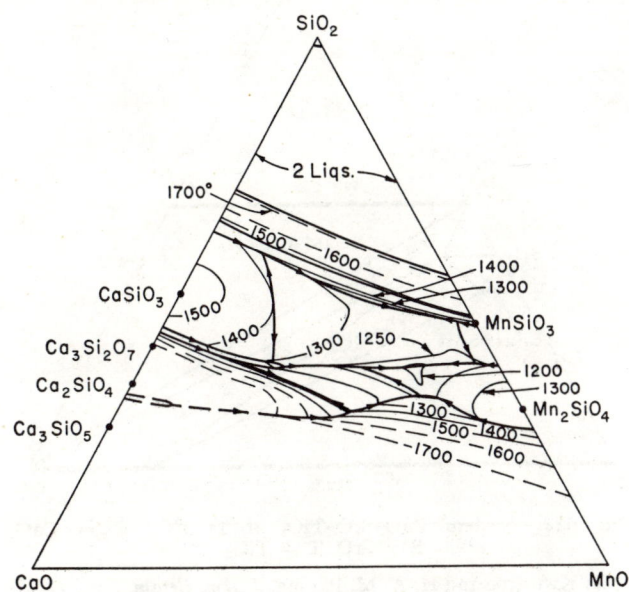

Fig. 615.—System CaO–MnO–SiO$_2$; liquidus.

F. P. Glasser, *J. Am. Ceram. Soc.*, **45** [5] 245 (1962).

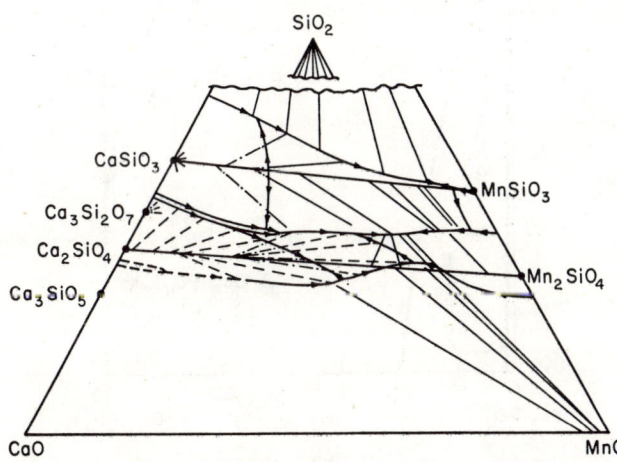

Fig. 616.—System CaO–MnO–SiO$_2$; tie lines. Tie lines closely approximate portions of three-phase triangles.

F. P. Glasser, *J. Am. Ceram. Soc.*, **45** [5] 246 (1962).

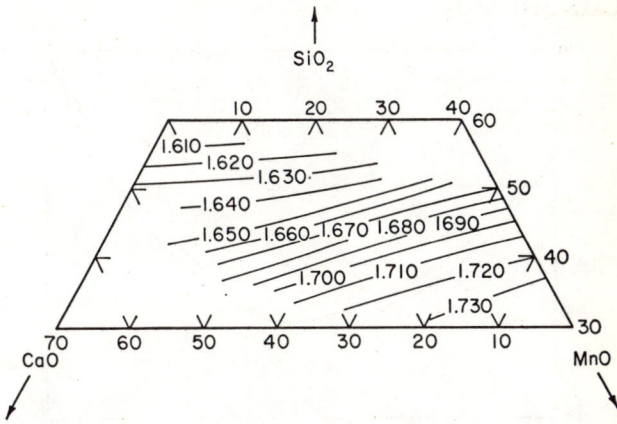

Fig. 617.—System CaO–MnO–SiO$_2$; isofracts of quenched liquids.

F. P. Glasser, *J. Am. Ceram. Soc.*, **45** [5] 246 (1962).

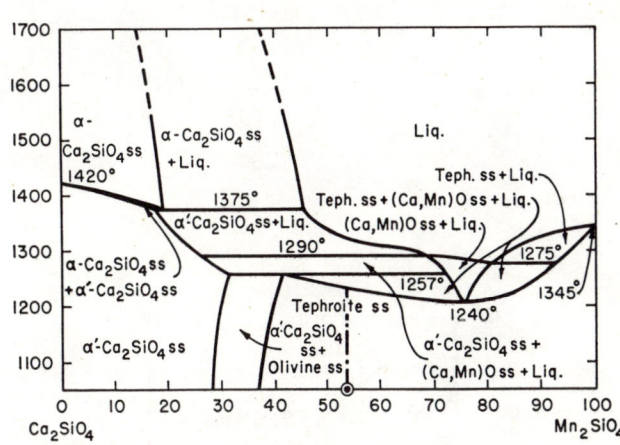

Fig. 618.—System Ca$_2$SiO$_4$–Mn$_2$SiO$_4$ above 1100°C.; pseudobinary. Dash-dot vertical line, glaucochroite composition.

F. P. Glasser, *Am. J. Sci.*, **259**, 51 (1961).

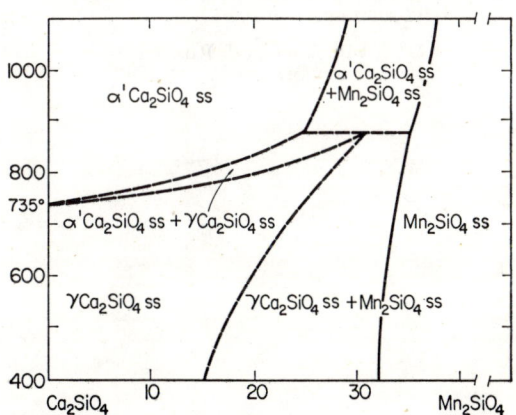

Fig. 619.—System Ca$_2$SiO$_4$–Mn$_2$SiO$_4$; subsolidus.

F. P. Glasser, *Am. J. Sci.*, **259**, 57 (1961).

CaO–SrO–SiO$_2$

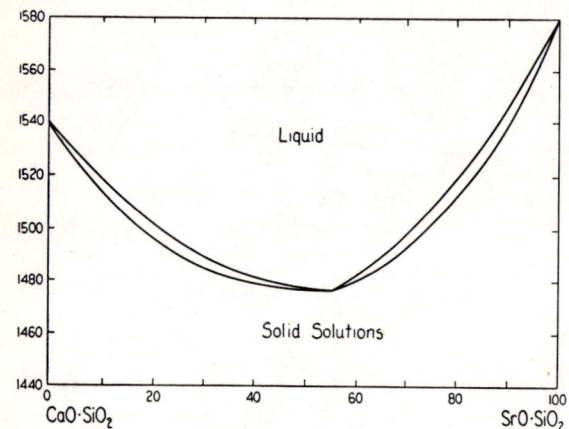

Fig. 620.—System CaO·SiO$_2$–SrO·SiO$_2$.

P. Eskola, *Am. J. Sci.*, 5th Ser., **4**, 353 (1922).

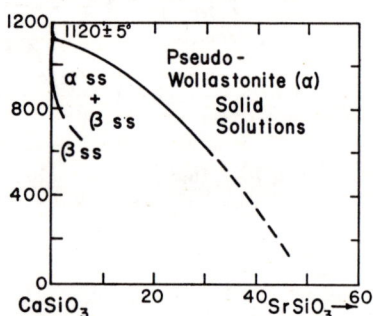

Fig. 621.—System CaSiO$_3$–SrSiO$_3$ showing alpha- and beta-wollastonite solid solutions.

D. A. Buckner and Rustum Roy, *J. Am. Ceram. Soc.*, **43** [1] 53 (1960).

CaO–SrO–TiO$_2$

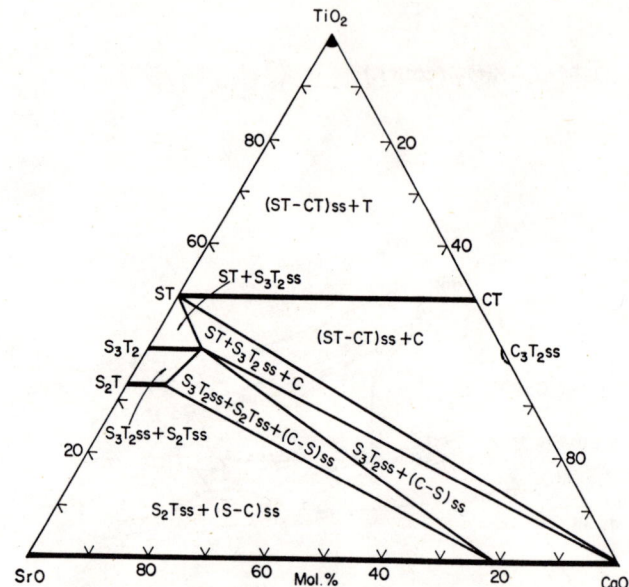

Fig. 622.—System CaO–SrO–TiO$_2$ at 1400°C. C = CaO, S = SrO, T = TiO$_2$.

W. Kwestroo and H. A. M. Paping, *J. Am. Ceram. Soc.*, **42** [6] 295 (1959).

CaO–SrO–P$_2$O$_5$

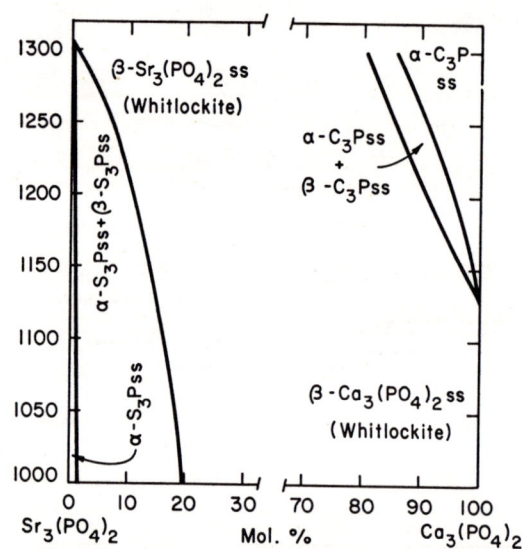

Fig. 623.—System Ca$_3$(PO$_4$)$_2$–Sr$_3$(PO$_4$)$_2$. C$_3$P = Ca$_3$(PO$_4$)$_2$, S$_3$P = Sr$_3$(PO$_4$)$_2$.

J. F. Sarver, F. A. Hummel, and M. V. Hoffman, *J. Electrochem. Soc.*, **108** [12] 1105 (1961).

CaO–ZnO–SiO₂

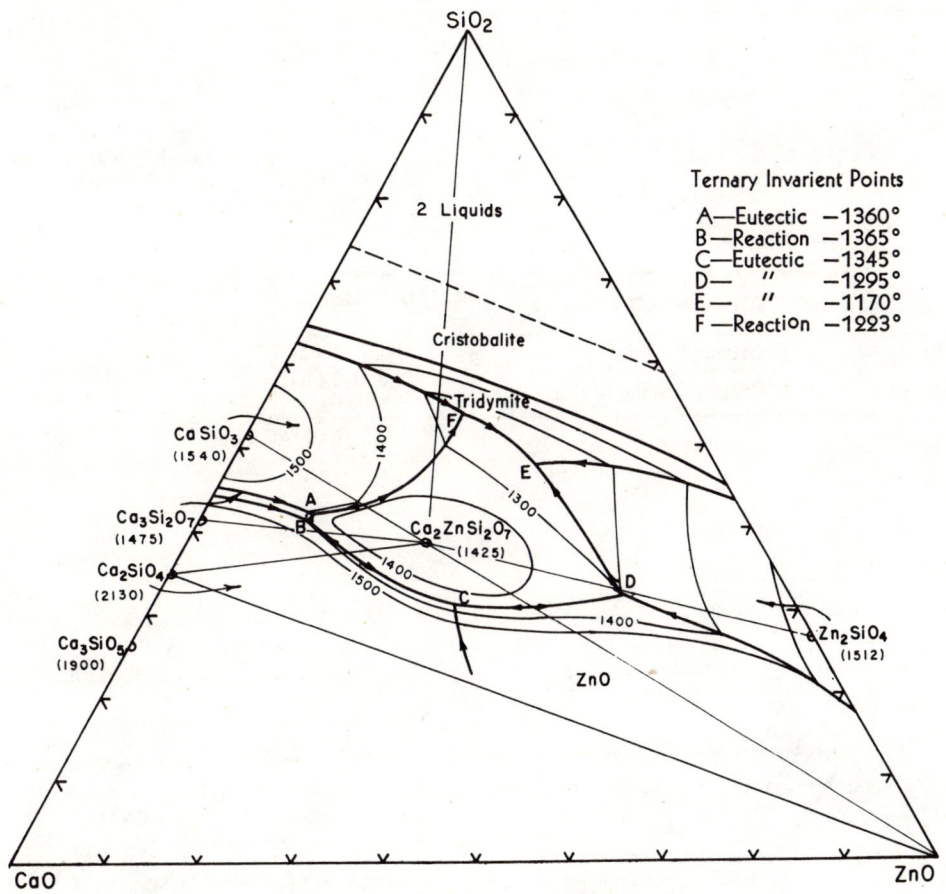

Fig. 624.—System CaO–ZnO–SiO₂.
E. R. Segnit, *J. Am. Ceram. Soc.*, **37** [6] 274 (1954).

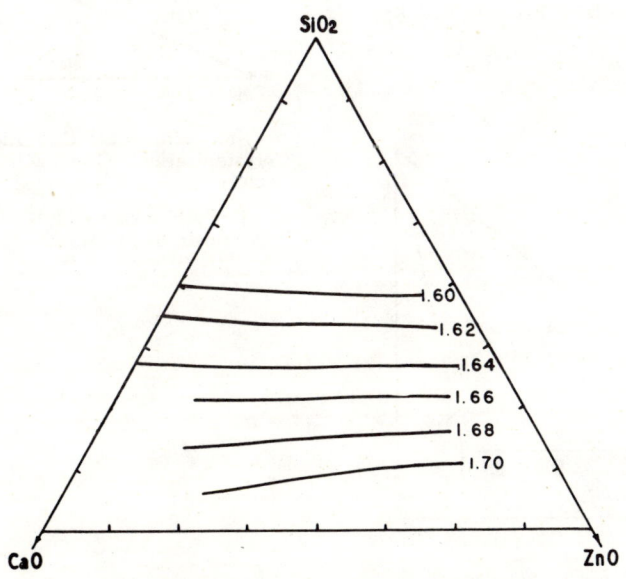

Fig. 625.—System CaO–ZnO–SiO₂; isofracts.
E. R. Segnit, *J. Am. Ceram. Soc.*, **37** [6] 275 (1954).

$CaO–Al_2O_3–Cr_2O_3$

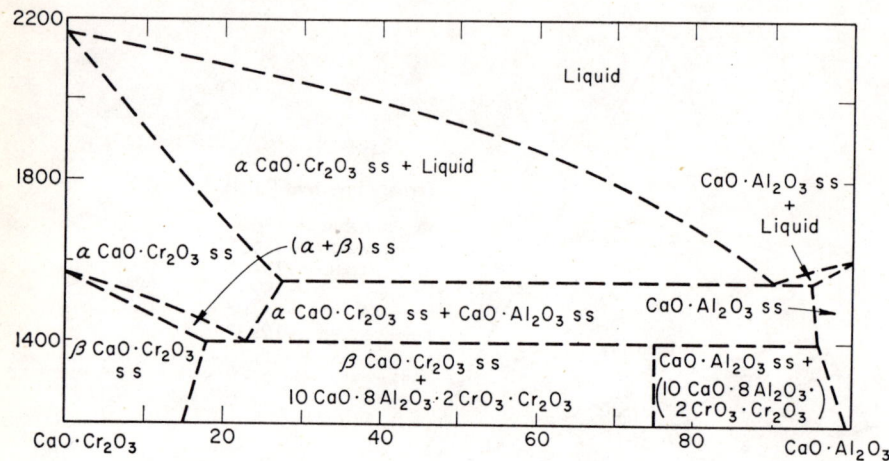

Fig. 626.—System $CaO \cdot Al_2O_3$–$CaO \cdot Cr_2O_3$; tentative.

W. F. Ford and W. J. Rees, Trans. Brit. Ceram. Soc., 57 [5] 237 (1958).

$CaO–Al_2O_3–Fe_2O_3$

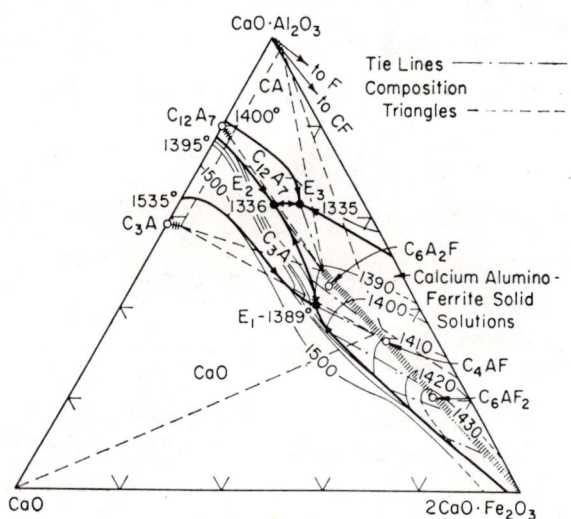

Fig. 627.—System CaO–$CaO \cdot Al_2O_3$–$2CaO \cdot Fe_2O_3$, pseudoternary. $C = CaO$; $A = Al_2O_3$; $F = Fe_2O_3$.

T. F. Newkirk and R. D. Thwaite, J. Research Natl. Bur. Standards, 61 [4] 241 (1958).

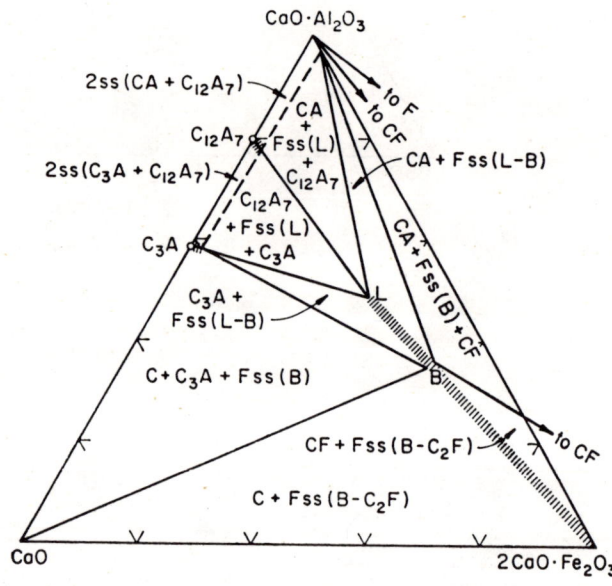

Fig. 628.—CaO–$CaO \cdot Al_2O_3$–$2CaO \cdot Fe_2O_3$; final products of crystallization. $C = CaO$; $A = Al_2O_3$; $F = Fe_2O_3$; ss = solid solution.

T. F. Newkirk and R. D. Thwaite, J. Research Natl. Bur. Standards, 61 [4] 242 (1958).

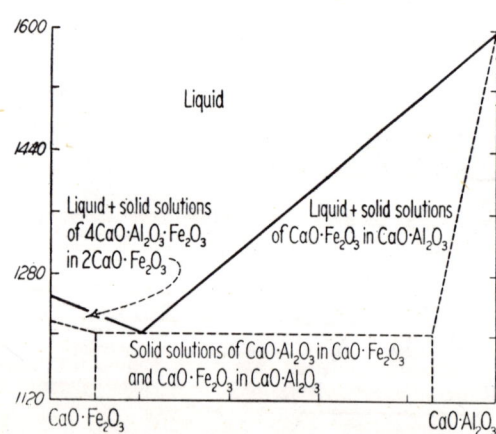

Fig. 629.—System $CaO \cdot Al_2O_3$–$CaO \cdot Fe_2O_3$.

W. C. Hansen, L. T. Brownmiller, and R. H. Bogue, J. Am. Chem. Soc., 50 [2] 404 (1928).

Three Oxides

CaO–Al₂O₃–SiO₂

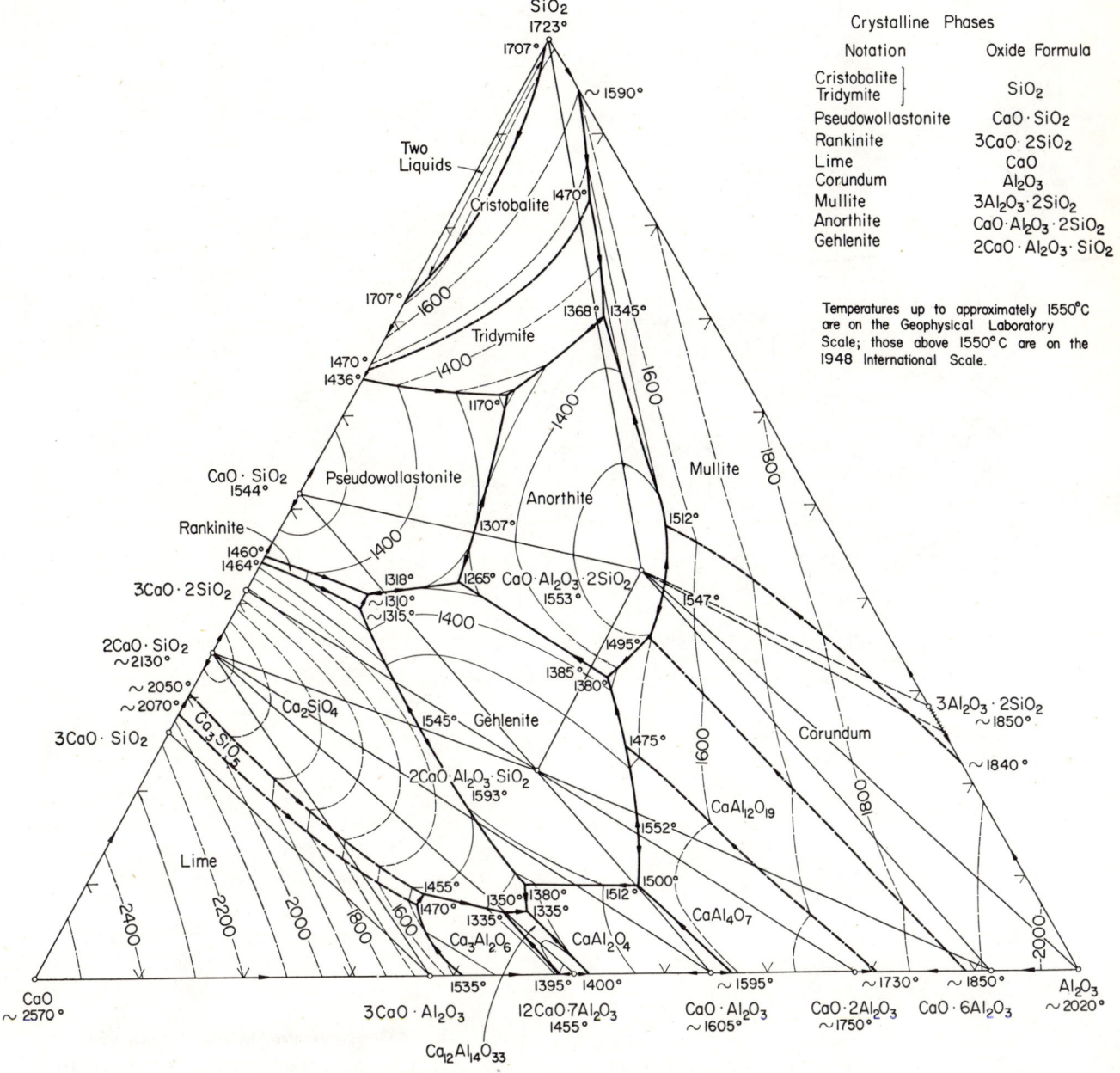

FIG. 630.—System CaO–Al₂O₃–SiO₂; composite.

E. F. Osborn and Arnulf Muan, revised and redrawn "Phase Equilibrium Diagrams of Oxide Systems," Plate 1, published by the American Ceramic Society and the Edward Orton, Jr., Ceramic Foundation, 1960.

Principal References

A. L. Day, E. S. Shepherd, and F. E. Wright, *Am. J. Sci.* (4th series), **22**, 265–302 (1906).
J. H. Welch and W. Gutt, *J. Am. Ceram. Soc.*, **42**, 11–15 (1959).
N. L. Bowen and J. W. Greig, *J. Am. Ceram. Soc.*, **7**, 238–54 (1924); corrections, *ibid.*, 410.
N. A. Toropov and F. Ya. Galakhov, *Voprosy Petrograf. i Mineral.*, Akad. Nauk S.S.S.R., **2**, 245–55 (1953).
G. A. Rankin and F. E. Wright, *Am. J. Sci.* (4th Series), **39**, 1–79 (1915).
J. W. Greig, *Am. J. Sci.* (5th Series), **13**, 1–44; 133–54 (1927).
N. E. Filonenko and I. V. Iavrov, *Zhur. Prik. Khim.*, **23**, 1040–46 (1950); *J. Appl. Chem.* (U.S.S.R.), **23**, 1105–12 (1950) (English translation).
Shigeo Aramaki and Rustum Roy, *J. Am. Ceram. Soc.*, **42**, 644–45 (1959).

CaO–Al$_2$O$_3$–SiO$_2$ (cont.)

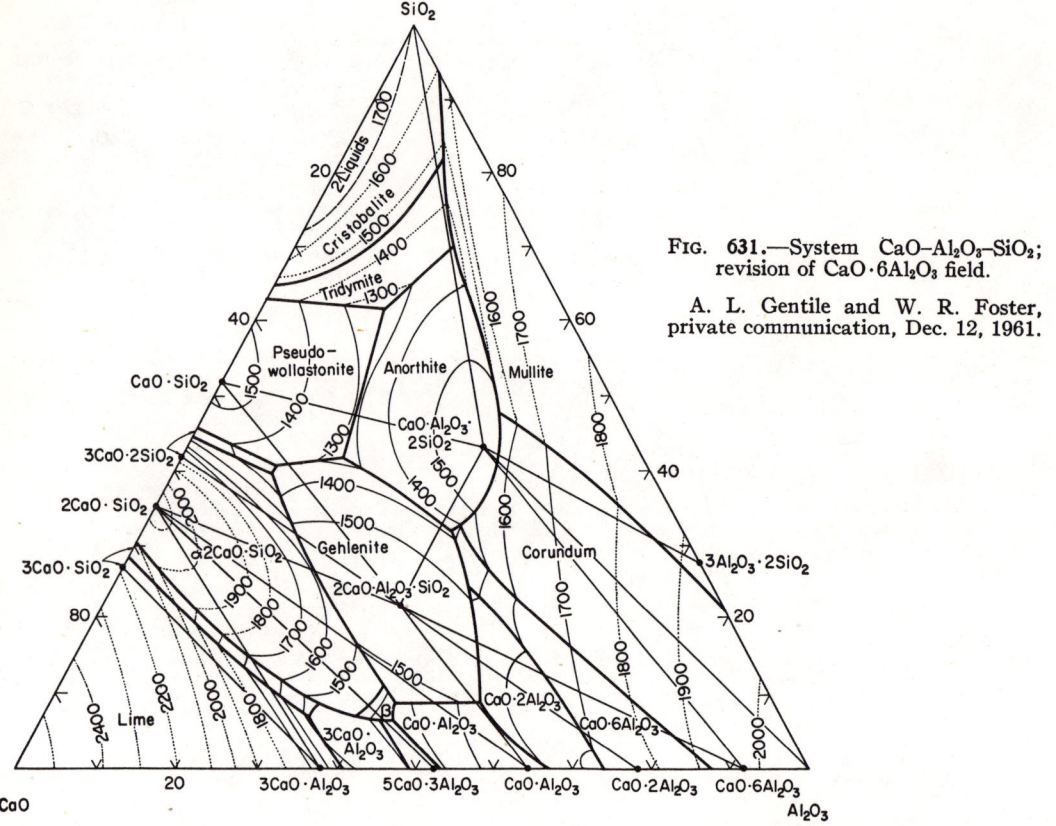

Fig. 631.—System CaO–Al$_2$O$_3$–SiO$_2$; revision of CaO·6Al$_2$O$_3$ field.

A. L. Gentile and W. R. Foster, private communication, Dec. 12, 1961.

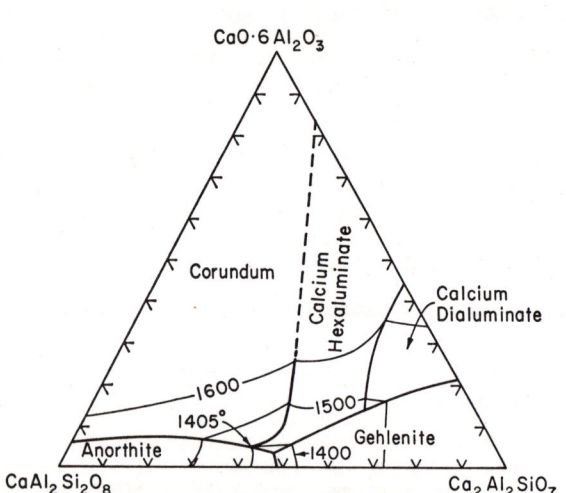

Fig. 632.—System CaO·6Al$_2$O$_3$ (calcium hexaluminate)–CaAl$_2$Si$_2$O$_8$ (anorthite)–Ca$_2$Al$_2$SiO$_7$ (gehlenite); revised.

A. L. Gentile and W. R. Foster, private communication, Dec. 12, 1961.

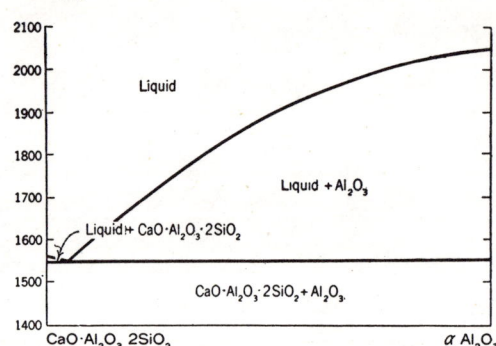

Fig. 633.—System α-Al$_2$O$_3$–CaO·Al$_2$O$_3$·2SiO$_2$.

G. A. Rankin and F. E. Wright, *Am. J. Sci.*, 4th Ser., **39**, 47 (1915).

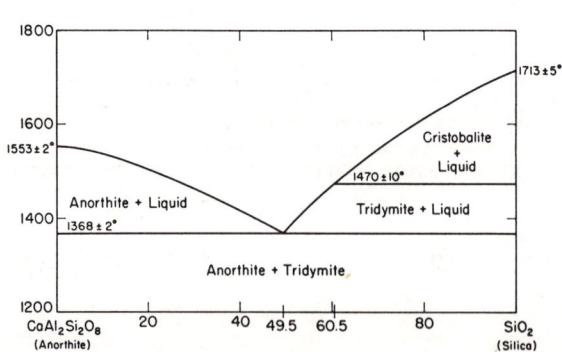

Fig. 634.—System CaAl$_2$Si$_2$O$_8$ (anorthite)—SiO$_2$ (silica).

J. F. Schairer and N. L. Bowen, *Bull. Soc. Geol. de Finlande*, **20**, 71 (1947).

$CaO-Al_2O_3-SiO_2$ (concl.)

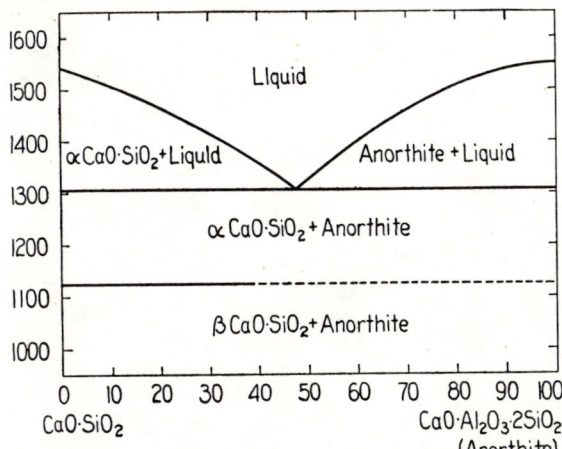

Fig. 635.—System $CaO \cdot SiO_2-CaO \cdot Al_2O_3 \cdot 2SiO_2$.

E. F. Osborn, *Am. J. Sci.*, **240**, 755 (1942).

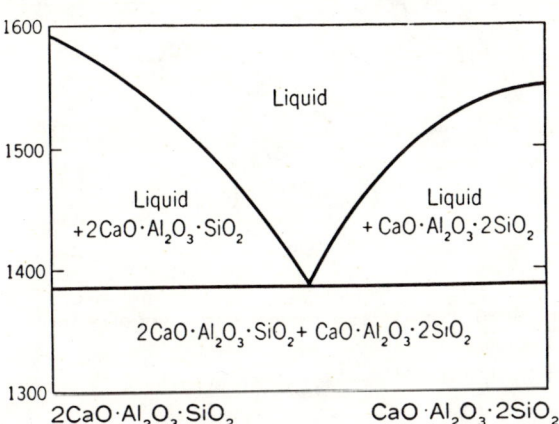

Fig. 636.—System $2CaO \cdot Al_2O_3 \cdot SiO_2-CaO \cdot Al_2O_3 \cdot 2SiO_2$.

G. A. Rankin and F. E. Wright, *Am. J. Sci.*, 4th Ser., **39**, 49 (1915).

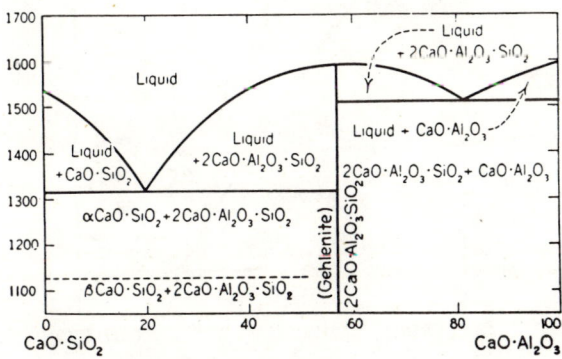

Fig. 637.—System $CaO \cdot SiO_2-CaO \cdot Al_2O_3$.

G. A. Rankin and F. E. Wright, *Am. J. Sci.*, 4th Ser., **39**, 50 (1915); modified by E. F. Osborn and J. F. Schairer, *ibid.*, **239**, 721 (1941).

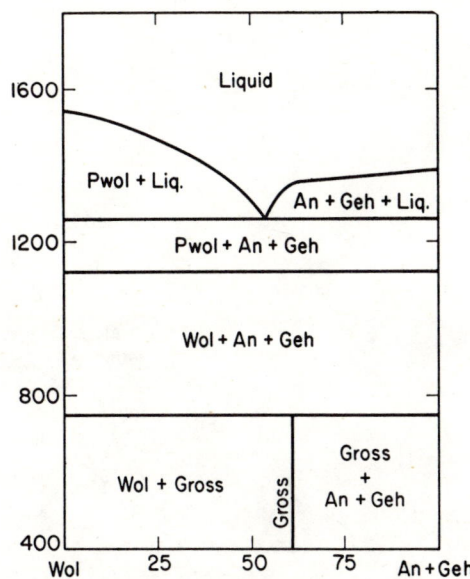

Fig. 638.—Pseudo-binary, (wollastonite)-(anorthite and gehlenite eutectic); postulated. Wol—wollastonite, $CaSiO_3$; Pwol—pseudowollastonite, $CaSiO_3$; An—anorthite, $CaAl_2Si_2O_8$; Geh—gehlenite, $Ca_2Al_2SiO_7$; Gross—grossularite, $Ca_3Al_2(SiO_4)_3$.

H. S. Yoder, Jr., *J. Geol.*, **58** [3] 225 (1950).

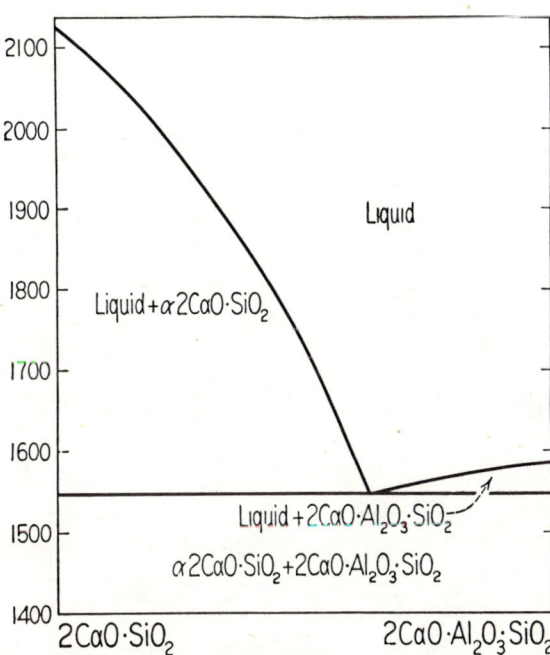

Fig. 639.—System $2CaO \cdot SiO_2-2CaO \cdot Al_2O_3 \cdot SiO_2$.

G. A. Rankin and F. E. Wright, *Am. J. Sci.*, 4th Ser. **39**, 47 (1915).

$CaO-Al_2O_3-P_2O_5$

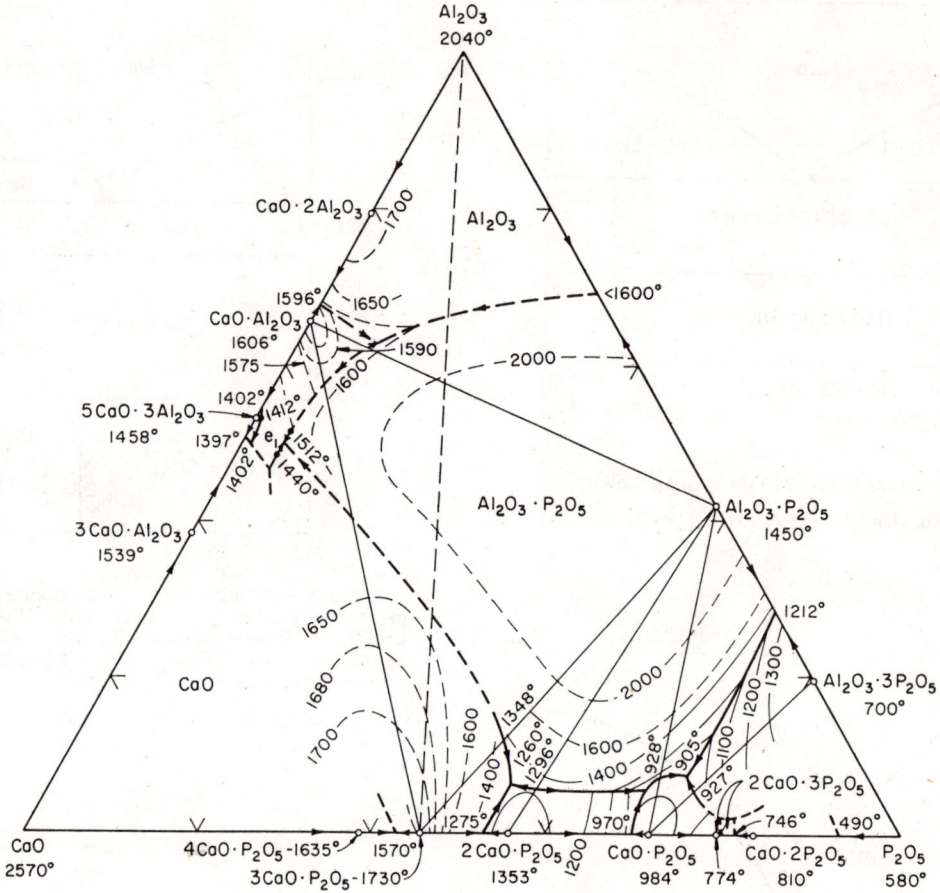

Fig. 640.—System $CaO-Al_2O_3-P_2O_5$. The dashed join $Al_2O_3-3CaO \cdot P_2O_5$ should represent a true binary system according to the data of P. D. S. St. Pierre, *J. Am. Ceram. Soc.*, **39** [10] 361 (1956). Also, eutectic e_1 does not obey the phase rule.

P. E. Stone, E. P. Egan, Jr., and J. R. Lehr, *J. Am. Ceram. Soc.*, **39** [3] 96 (1956).

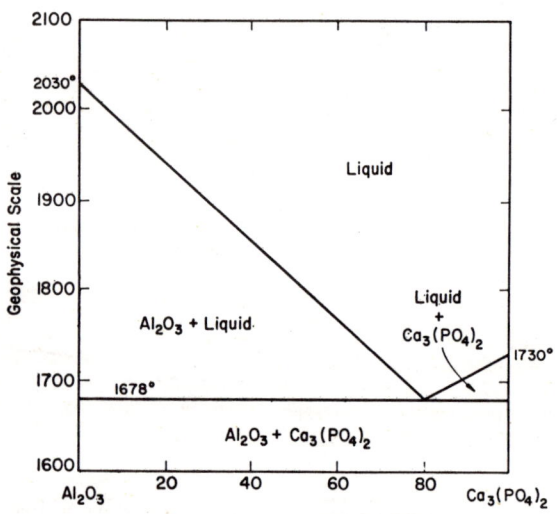

Fig. 641.—System $Al_2O_3-Ca_3(PO_4)_2$.

P. D. S. St. Pierre, Canada Dept. of Mines and Technical Surveys, Mines Branch, Technical Paper No. 2, p. 55 (1953); *J. Am. Ceram. Soc.*, **37** [6] 251 (1954).

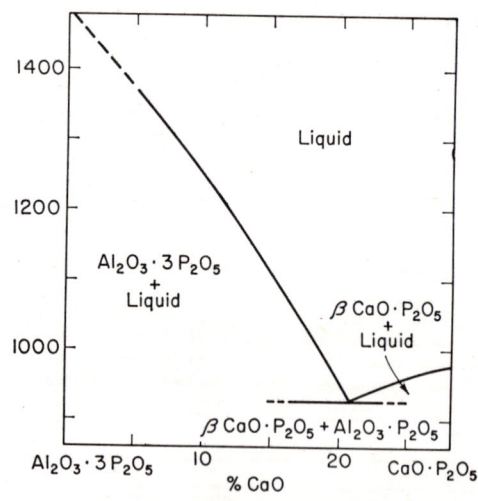

Fig. 642.—System $CaO \cdot P_2O_5-Al_2O_3 \cdot 3P_2O_5$.

P. E. Stone, E. P. Egan, Jr., and J. R. Lehr, *J. Am. Ceram. Soc.*, **39** [3] 92 (1956).

CaO–Al$_2$O$_3$–P$_2$O$_5$ (concl.)

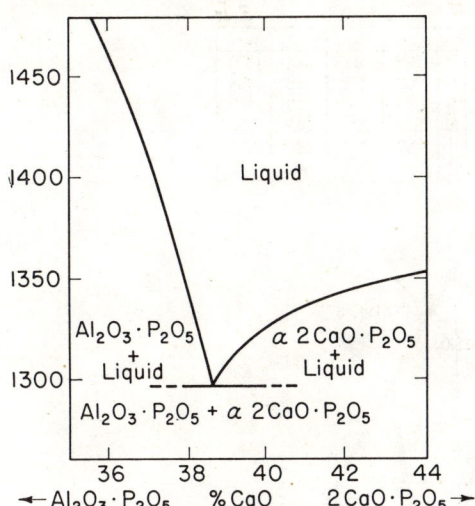

FIG. 643.—System 2CaO·P$_2$O$_5$–Al$_2$O$_3$·P$_2$O$_5$.

P. E. Stone, E. P. Egan, Jr., and J. R. Lehr, *J. Am. Ceram. Soc.*, **39** [3] 94 (1956).

CaO–(FeO + MnO + Fe$_2$O$_3$)–P$_2$O$_5$

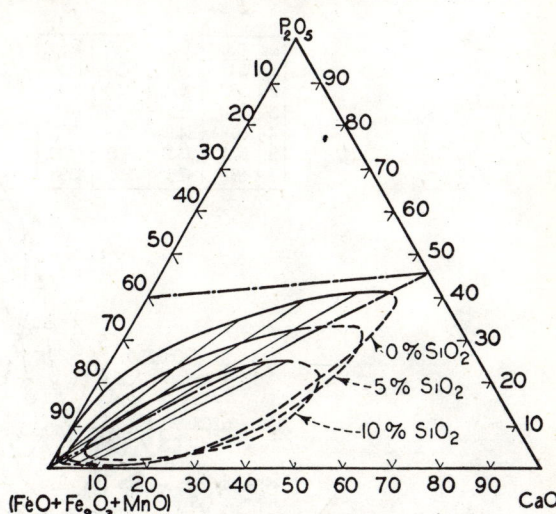

FIG. 645.—System CaO–(FeO + Fe$_2$O$_3$ + MnO)–P$_2$O$_5$; boundaries of solid miscibility.

W. Oelsen and H. Maetz, *Mitt. Kaiser-Wilhelm-Inst. Eisenforsch. Düsseldorf*, **23** [12] 195–245 (1941).

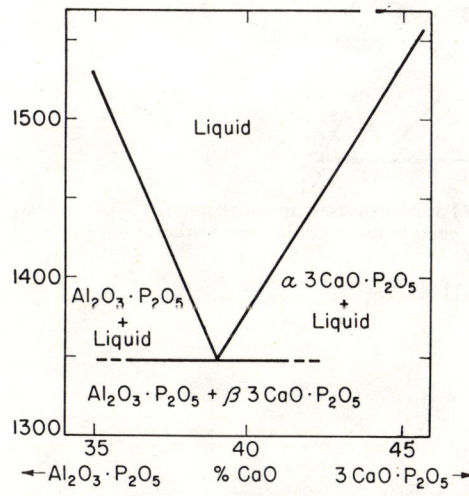

FIG. 644.—System 3CaO·P$_2$O$_5$–Al$_2$O$_3$·P$_2$O$_5$.

P. E. Stone, E. P. Egan, Jr., and J. R. Lehr, *J. Am. Ceram. Soc.*, **39** [3] 94 (1956).

CaO–B$_2$O$_3$–SiO$_2$

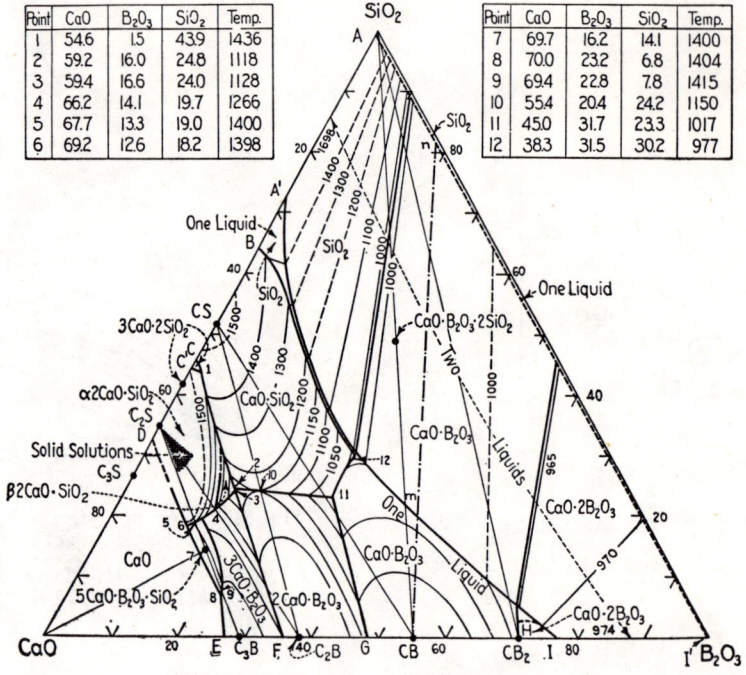

FIG. 646.—System CaO–B$_2$O$_3$–SiO$_2$.

E. P. Flint and L. S. Wells, *J. Research Nat. Bur. Standards*, **17** [5] 745 (1936); R. P. 941.

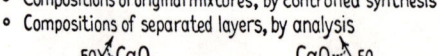
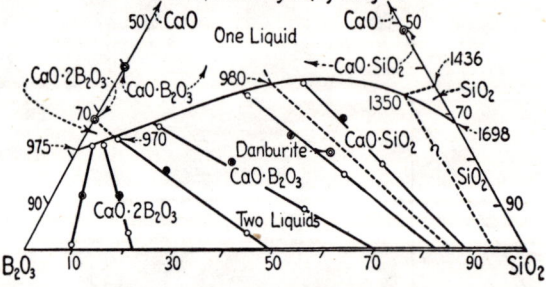

FIG. 647.—System CaO–B$_2$O$_3$–SiO$_2$; compositions of mixtures used in determining the tie lines and the two layers obtained for analysis; each liquid contains some other phase, but separation is much better in CaO-rich layer.

G. W. Morey and Earl Ingerson, *Am. Mineral.*, **22** [1] 38 (1937).

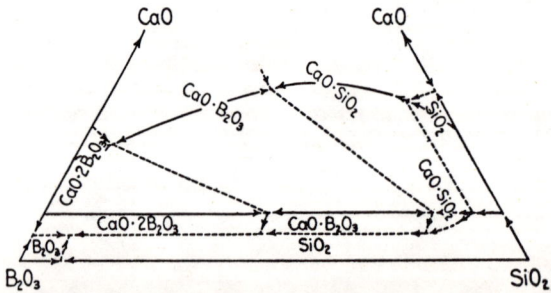

FIG. 648.—Diagrammatic representation of the phase equilibrium relationships in the ternary system CaO–B$_2$O$_3$–SiO$_2$ adjacent to the region of immiscibility.

G. W. Morey and Earl Ingerson, *Am. Mineral.*, **22** [1] 45 (1937).

CaO–B$_2$O$_3$–SiO$_2$ (concl.)

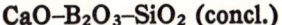

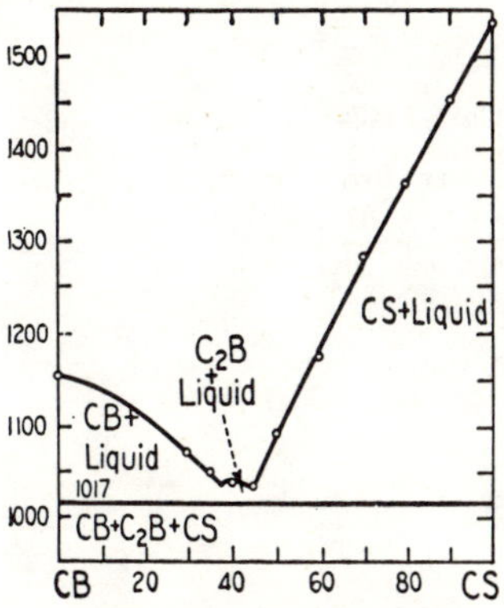

Join CaO·B$_2$O$_3$–CaO·–SiO$_2$.

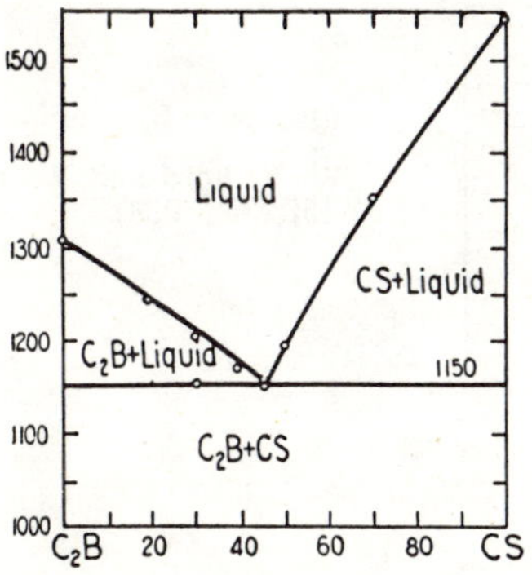

System 2CaO·B$_2$O$_3$–CaO·SiO$_2$.

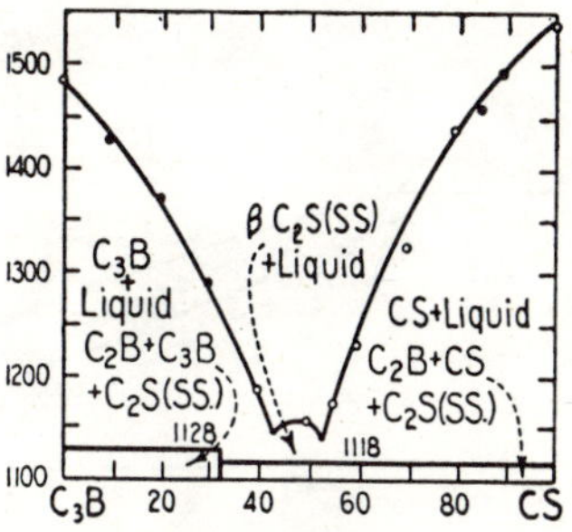

Join 3CaO·B$_2$O$_3$–CaO·SiO$_2$.

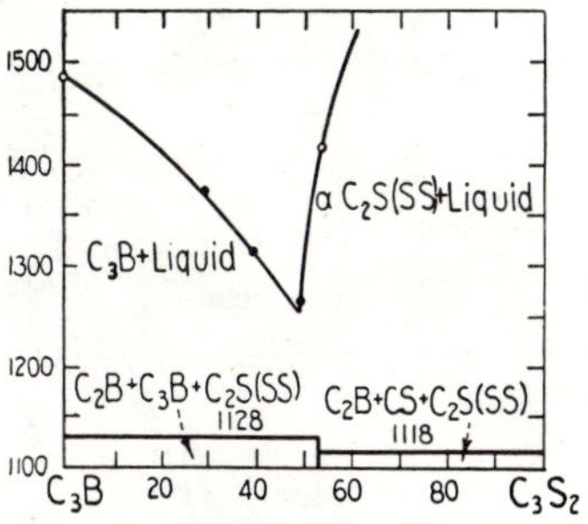

Join 3CaO·B$_2$O$_3$–3CaO·2SiO$_2$.

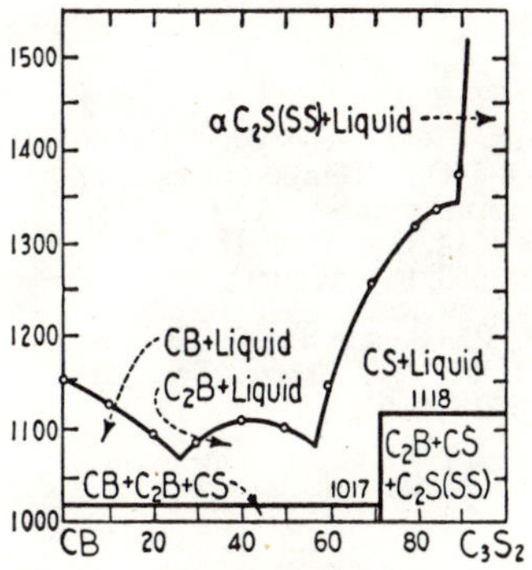

Join CaO·B$_2$O$_3$–3CaO·2SiO$_2$.

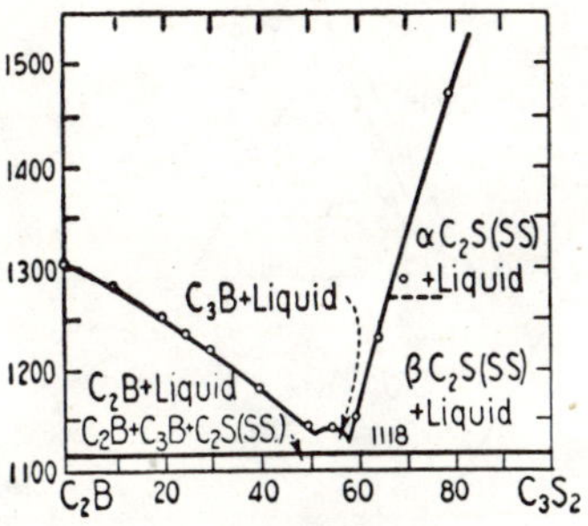

Join 2CaO·B$_2$O$_3$–3CaO·2SiO$_2$.

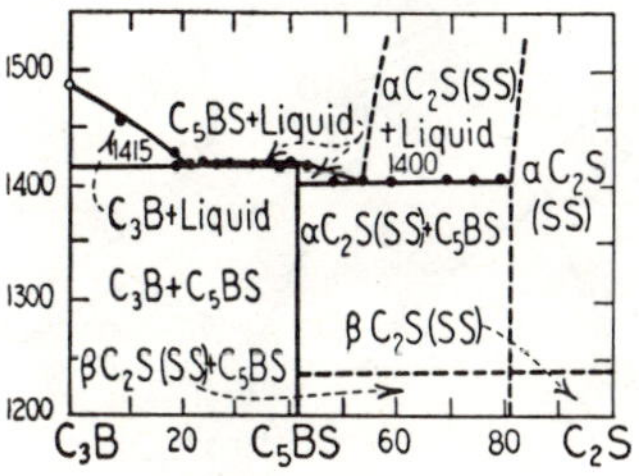

System 3CaO·B$_2$O$_3$–2CaO·SiO$_2$.

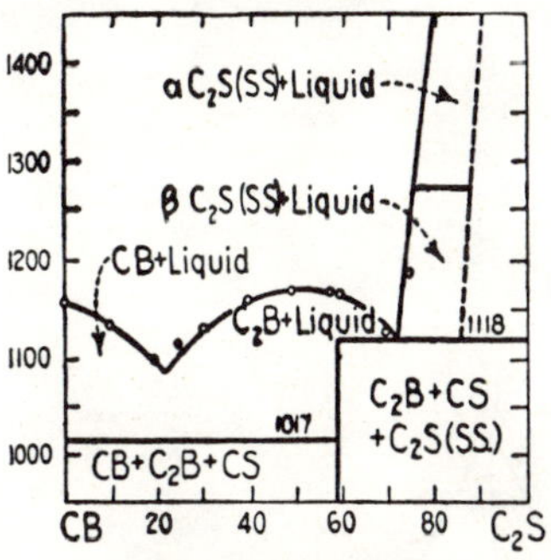

Join CaO·B$_2$O$_3$–2CaO·SiO$_2$.

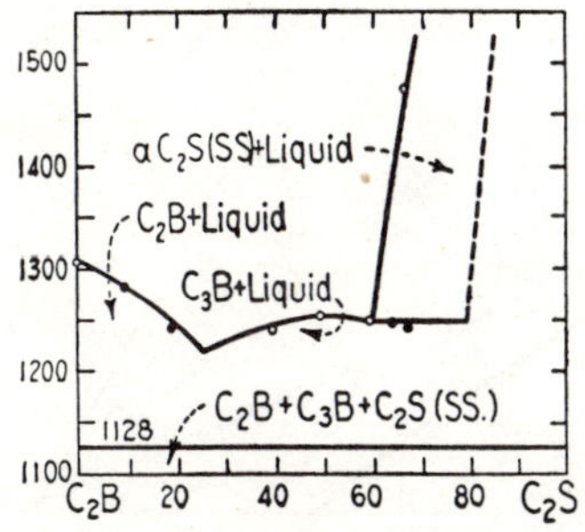

Join 2CaO·B$_2$O$_3$–2CaO·SiO$_2$.

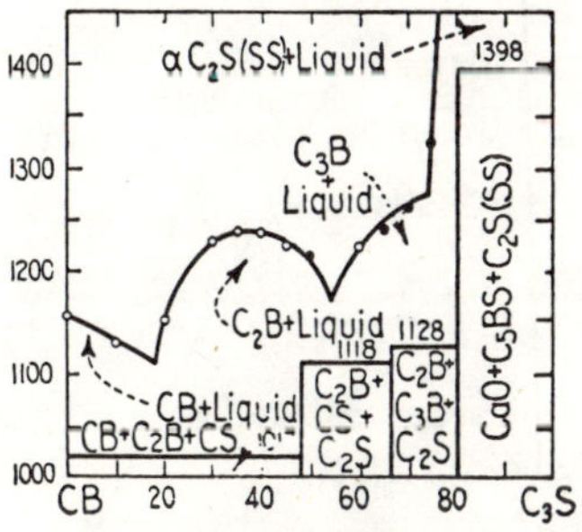

Join CaO·B$_2$O$_3$–3CaO·SiO$_2$.

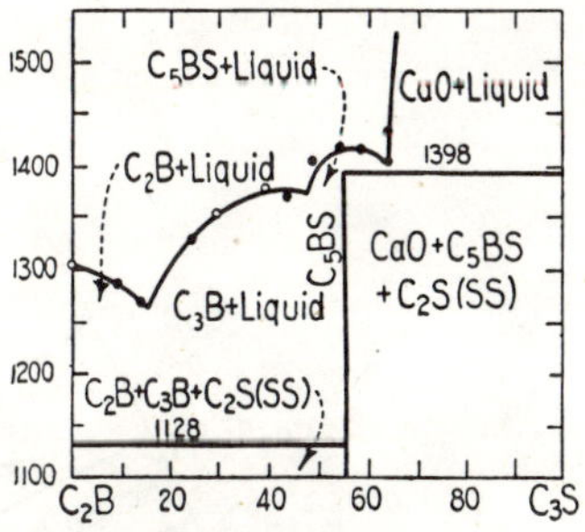

Join 2CaO·B$_2$O$_3$–3CaO·SiO$_2$.

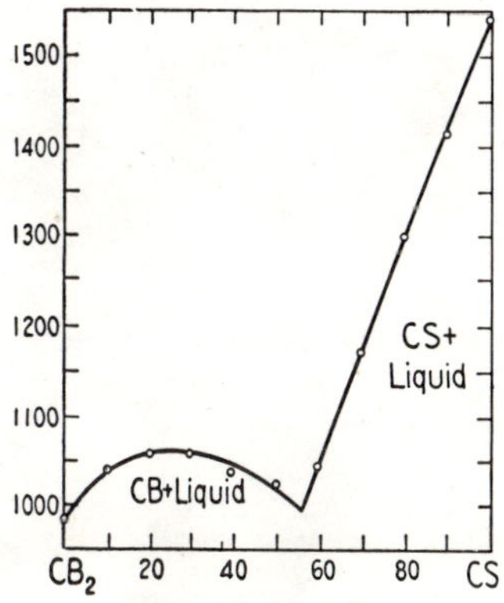

Join CaO·2B$_2$O$_3$–CaO·SiO$_2$.

FIG. 649.—CaO–B$_2$O$_3$–SiO$_2$; some binary and pseudobinary systems. C = CaO, B = B$_2$O$_3$, S = SiO$_2$. All figures adapted from E. P. Flint and L. S. Wells, *J. Research Natl. Bur. Standards*, **17** [5] 734 (1936).

CaO–Cr$_2$O$_3$–Fe$_2$O$_3$

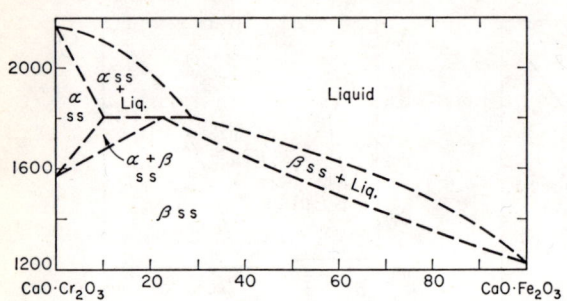

FIG. 650.—System CaO·Cr$_2$O$_3$–CaO·Fe$_2$O$_3$; tentative.

W. F. Ford and W. J. Rees, *Trans. Brit. Ceram. Soc.*, **57** [5] 235 (1958).

CaO–Cr$_2$O$_3$–SiO$_2$

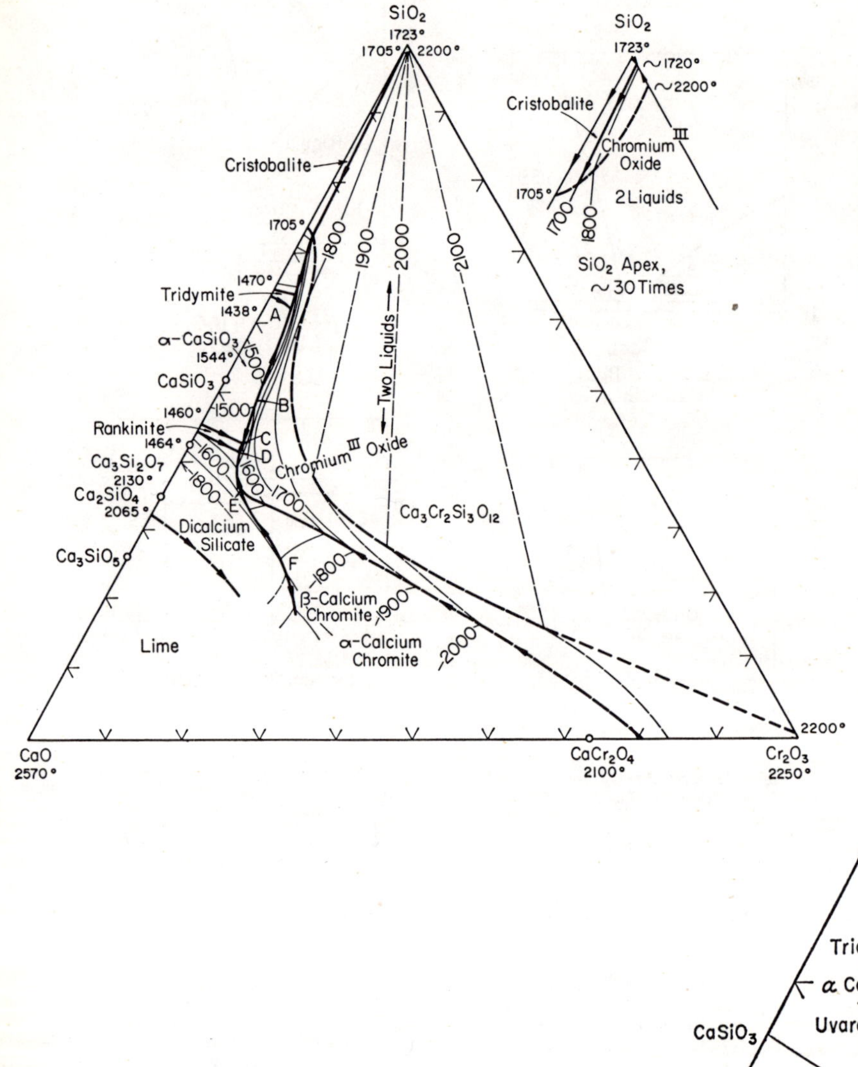

FIG. 651.—System CaO–Cr$_2$O$_3$–SiO$_2$. Temperatures at invariant points: A = 1418, B = 1514, C = 1407, D = 1430, E = 1556, F = 1720°C.

F. P. Glasser and E. F. Osborn, *J. Am. Ceram. Soc.*, **41** [9] 362 (1958).

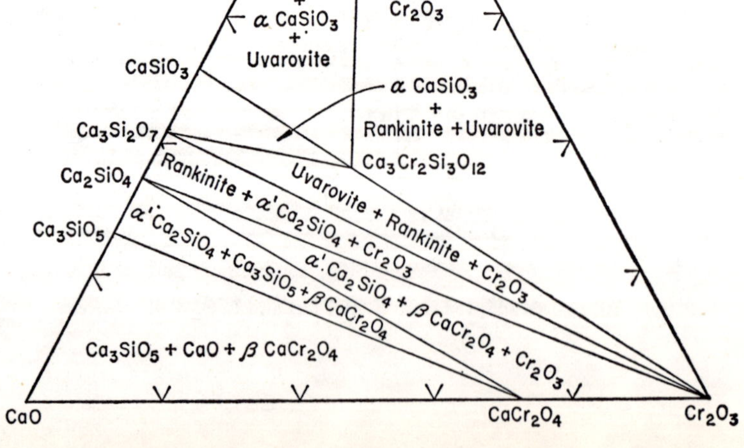

FIG. 652.—System CaO–Cr$_2$O$_3$–SiO$_2$ at 1350°C. T = tridymite, U = uvarovite, R = rankinite.

F. P. Glasser and E. F. Osborn, *J. Am. Ceram. Soc.*, **41** [9] 365 (1958).

CaO–Cr$_2$O$_3$–SiO$_2$ (concl.)

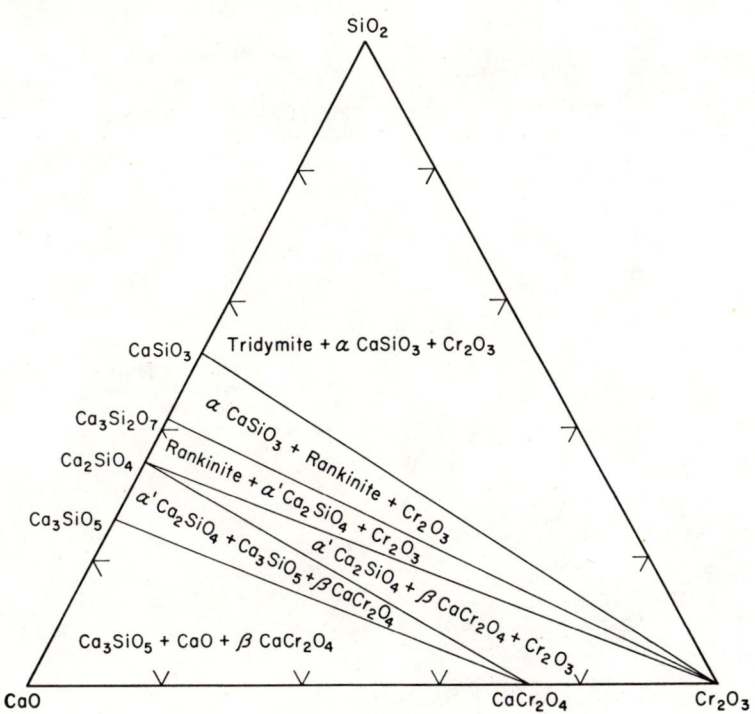

Fig. 653.—System CaO–Cr$_2$O$_3$–SiO$_2$ at 1400°C.

F. P. Glasser and E. F. Osborn, *J. Am. Ceram. Soc.*, **41** [9] 365 (1958).

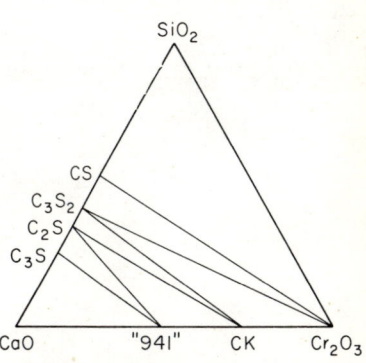

Fig. 655.—System CaO–Cr$_2$O$_3$–SiO$_2$; composition triangles. C = CaO, S = SiO$_2$, K = Cr$_2$O$_3$, "941" = pseudobinary compound 9CaO·4CrO$_3$·Cr$_2$O$_3$

W. F. Ford and W. J. Rees, *Trans. Brit. Ceram. Soc.*, **57** [5] 239 (1958).

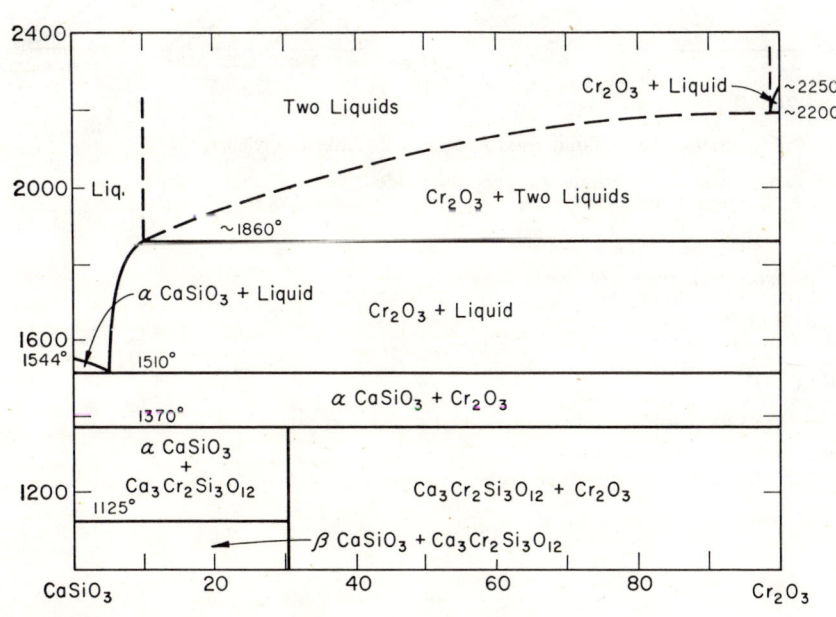

Fig. 654.—System CaSiO$_3$–Cr$_2$O$_3$. Composition of conjugate immiscible liquids cannot be represented on this join.

F. P. Glasser and E. F. Osborn
J. Am. Ceram. Soc., **41** [9] 364 (1958)

Fig. 656

CaO–Fe₂O₃–SiO₂

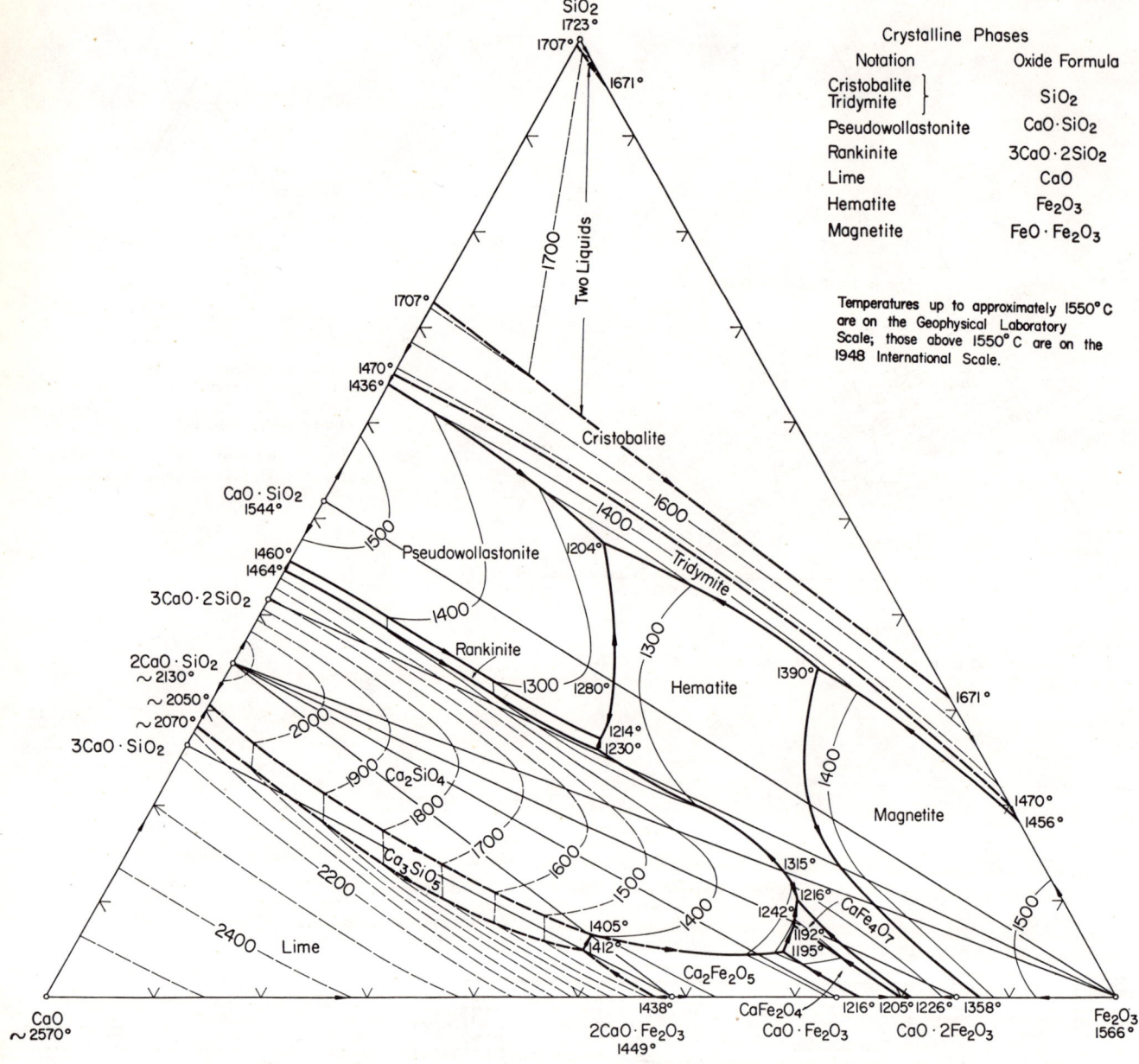

FIG. 656.—System CaO–"Fe₂O₃"–SiO₂; composite. (Condensed Phases in Equilibrium with Air).

E. F. Osborn and Arnulf Muan, revised and redrawn "Phase Equilibrium Diagrams of Oxide Systems," Plate 10, published by the American Ceramic Society and the Edward Orton, Jr., Ceramic Foundation, 1960.

Principal References

A. L. Day, E. S. Shepherd and F. E. Wright, *Am. J. Sci.* (4th Series), **22**, 265–302 (1906).
G. A. Rankin and F. E. Wright, *Am. J. Sci.* (4th Series), **39**, 1–79 (1915).
J. W. Greig, *Am. J. Sci.* (5th Series), **13**, 1–44; 133–54 (1927).
J. H. Welch and W. Gutt, *J. Am. Ceram. Soc.*, **42**, 11–15 (1959).
Bert Phillips and Arnulf Muan, *J. Am. Ceram. Soc.*, **41**, 445–54 (1958).
L. S. Darken, *J. Am. Ceram. Soc.*, **70**, 2046–53 (1948).
J. W. Greig, *Am. J. Sci.* (5th Series), **14**, 473–84 (1927).
M. D. Burdick, *J. Research Natl. Bur. Standards*, **25**, 475–88 (1940).
Bert Phillips and Arnulf Muan, *J. Am. Ceram. Soc.*, **42**, 413–23 (1959).

CaO–Fe$_2$O$_3$–SiO$_2$ (concl.)

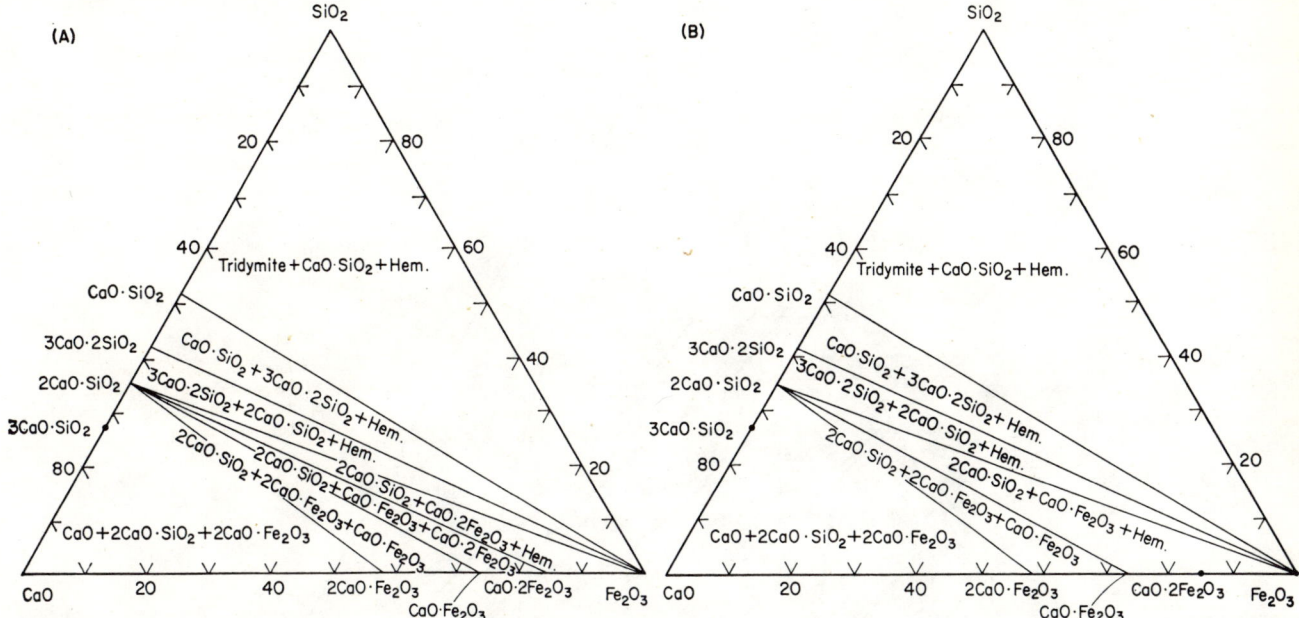

FIG. 657.—System CaO–Fe$_2$O$_3$–SiO$_2$ in air at subsolidus temperatures. (A) for temp. slightly above 1155°C., (B) for temp. slightly below 1155°C. Hem. = hematite.

Bert Phillips and Arnulf Muan, *J. Am. Ceram. Soc.*, **42** [9] 422 (1959).

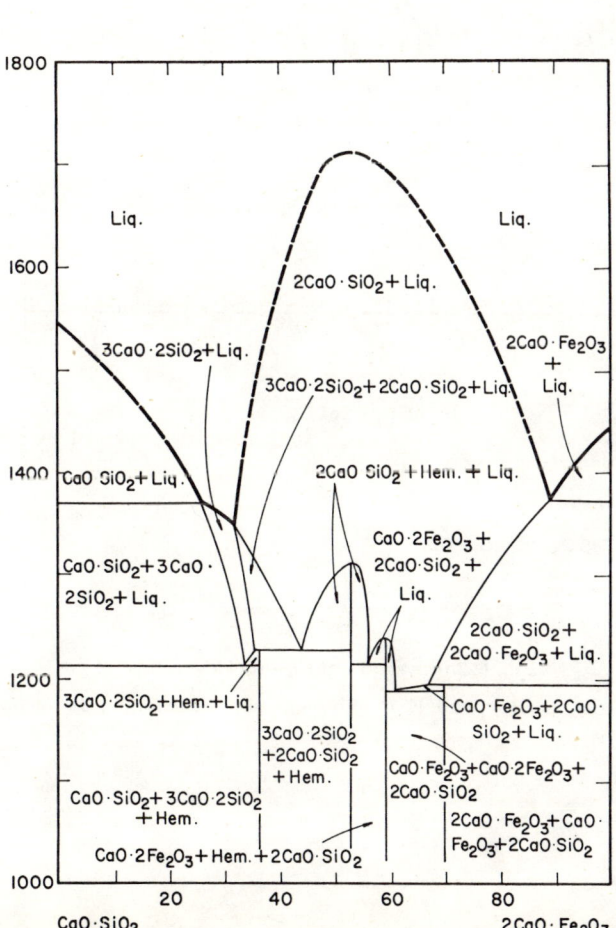

FIG. 658.—CaO–Fe$_2$O$_3$–SiO$_2$ in air; pseudobinary system CaO·SiO$_2$–2CaO·Fe$_2$O$_3$. Small amounts of Fe^{2+} present in the liquid phase. Hem. = hematite.

Bert Phillips and Arnulf Muan, *J. Am. Ceram. Soc.* **42** [9] 421 (1959).

CaO–La$_2$O$_3$–ZrO$_2$

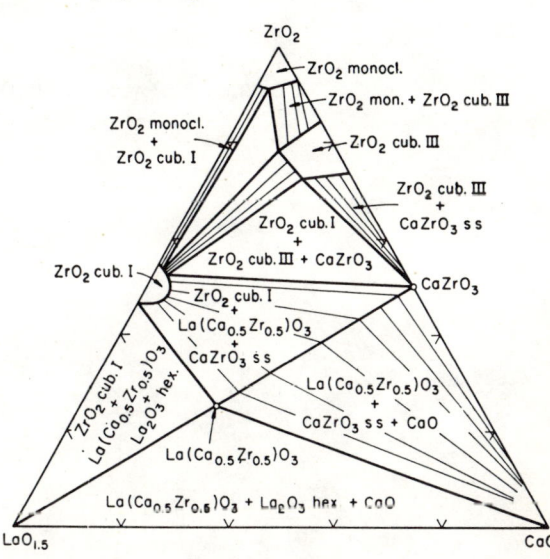

FIG. 659.—System CaO–LaO$_{1.5}$–ZrO$_2$, at approx. 1400°C. ss = solid solutions.

Albrecht Rabenau, *Z. anorg. u. allgem. Chem.*, **288**, 225 (1956).

CaO–SiO₂–TiO₂

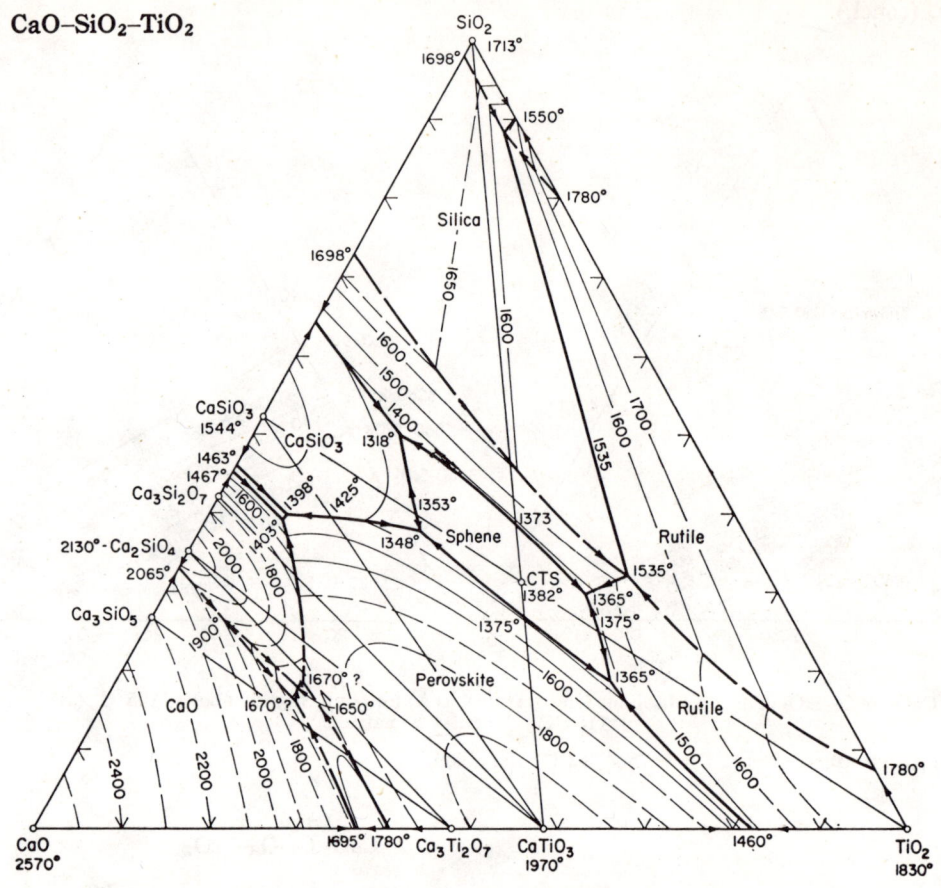

FIG. 660.—System CaO–SiO₂–TiO₂.
— — — two liquid boundary.
R. C. DeVries, R. Roy, and E. F. Osborn, *J. Am. Ceram. Soc.*, **38** [5] 161 (1955).

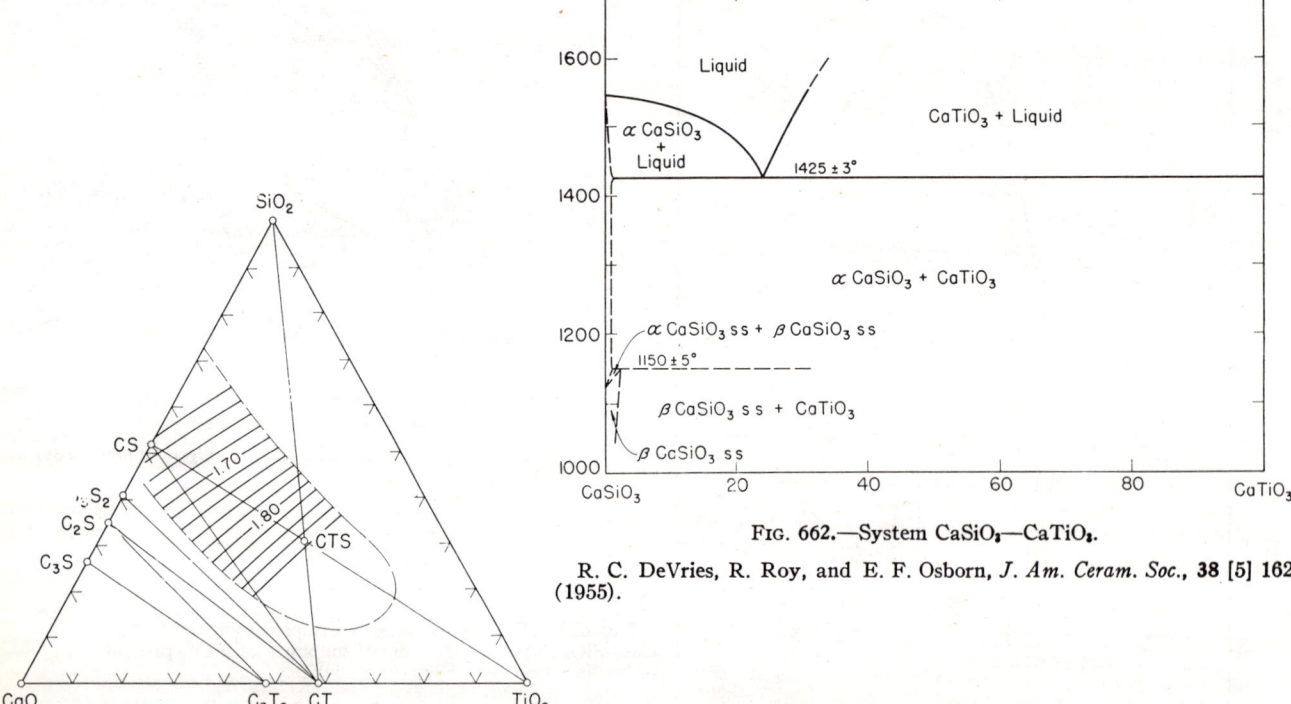

FIG. 661.—System CaO–SiO₂–TiO₂ (n of glasses). C = CaO, S = SiO₂, T = TiO₂.

R. C. DeVries, R. Roy, and E. F. Osborn, *J. Am. Ceram. Soc.*, **38** [5] 165 (1955).

FIG. 662.—System CaSiO₃–CaTiO₃.

R. C. DeVries, R. Roy, and E. F. Osborn, *J. Am. Ceram. Soc.*, **38** [5] 162 (1955).

CaO–SiO$_2$–TiO$_2$ (concl.)

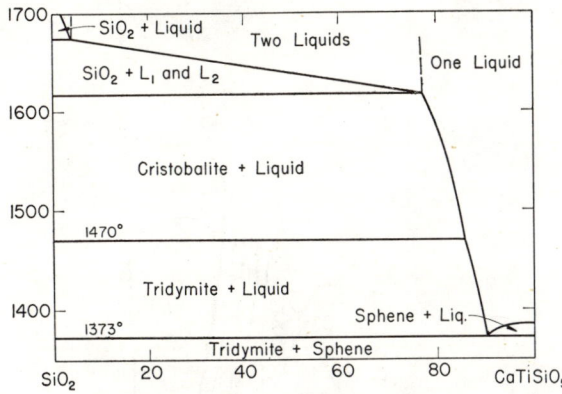

Fig. 663.—System CaO·SiO$_2$·TiO$_2$–SiO$_2$.

R. C. DeVries, R. Roy, and E. F. Osborn, *J. Am. Ceram. Soc.*, **38** [5] 162 (1955).

CaO–SiO$_2$–P$_2$O$_5$

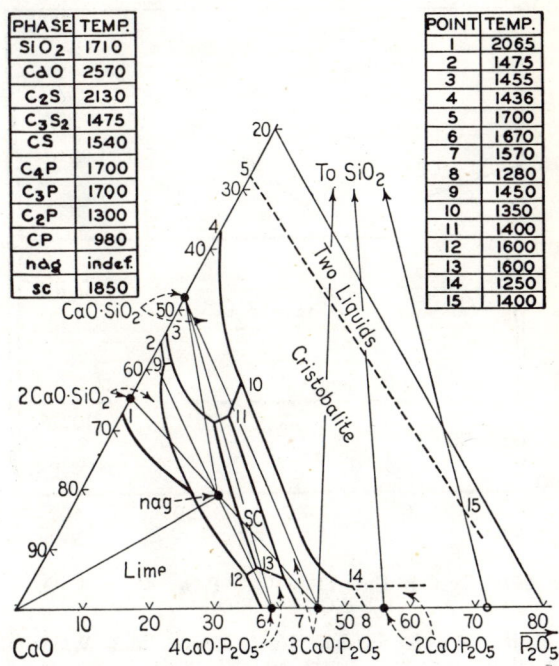

Fig. 665.—System CaO–SiO$_2$–P$_2$O$_5$; temperatures are approximate.

R. L. Barrett and W. J. McCaughey, *Am. Mineral.*, **27**, 687 (1942).

CaO–SiO$_2$–ZrO$_2$

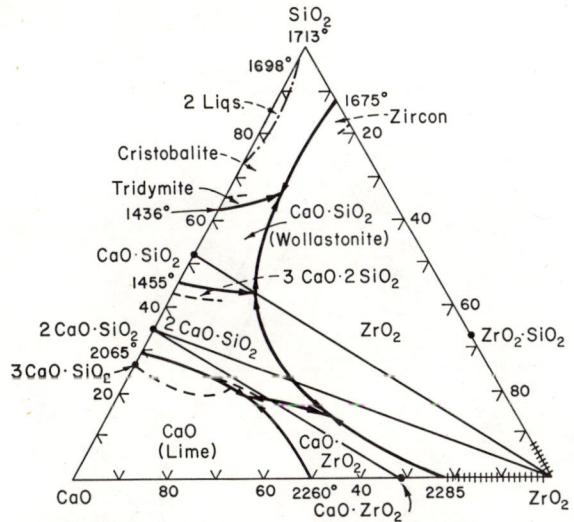

Fig. 664.—System CaO–SiO$_2$–ZrO$_2$; melting.

Kuniharu Matsumoto, Tsuneo Sawamoto, and Shigeaki Koide, *Asahi Garasu Kenkyu Hokoku*, **4** [2] 8 (1954).

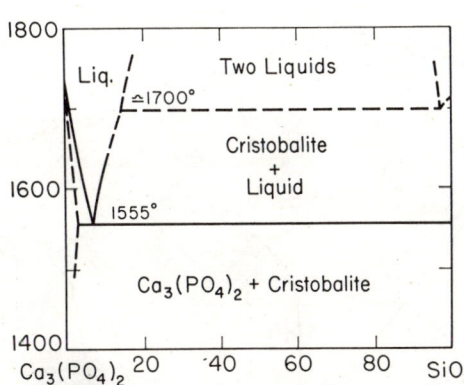

Fig. 666.—System Ca$_3$(PO$_4$)$_2$–SiO$_2$.

P. D. S. St. Pierre, *J. Am. Ceram. Soc.*, **39** [4] 148 (1956).

See also, G. Trömel, H.-J. Harkort, and W. Hotop, *Z. anorg. Chem.*, **256** [5–6] 200 (1948); and Jadwiga Wojciechowska, Józef Berak, and Włodzimierz Trzebiatowski, *Roczniki Chem.*, **30**, 749 (1956).

$CaO-SiO_2-P_2O_5$ (cont.)

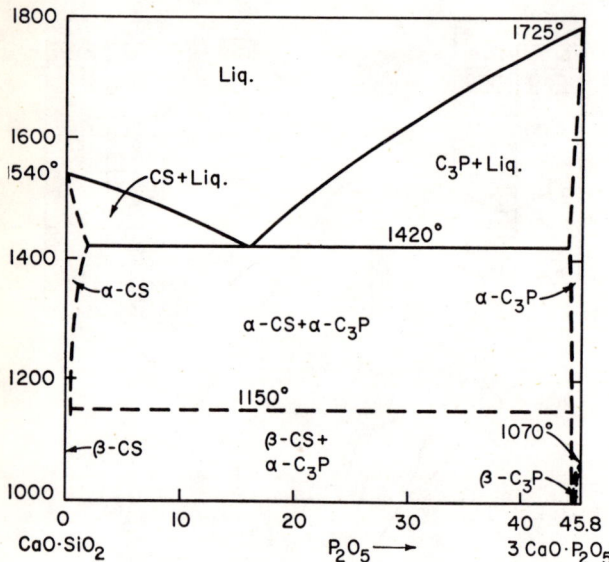

Fig. 667.—System $CaO \cdot SiO_2-3CaO \cdot P_2O_5$. C = CaO, P = P_2O_5, S = SiO_2.

Jadwiga Wojciechowska, Józef Berak, and Włodzimierz Trzebiatowski, *Roczniki Chem.*, **30**, 750 (1956).

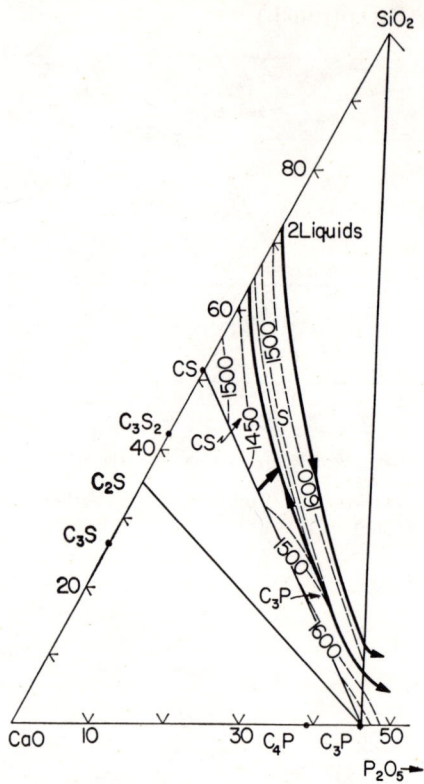

Fig. 669.—System $CaO-SiO_2-3CaO \cdot P_2O_5$. C = CaO, P = P_2O_5, S = SiO_2.

Jadwiga Wojciechowska, Józef Berak, and Włodzimierz Trzebiatowski, *Roczniki Chem.*, **30**, 751 (1956).

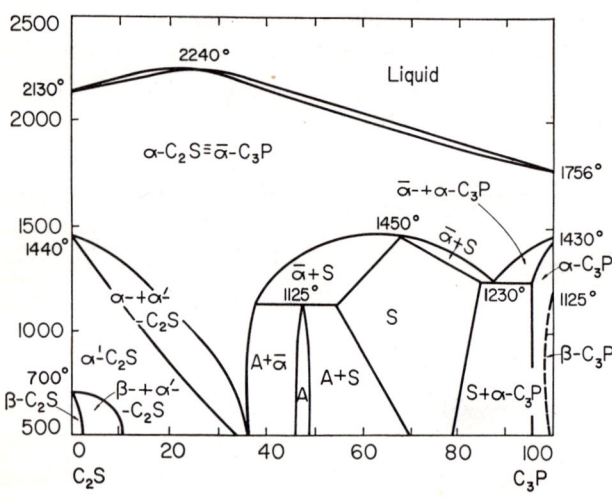

Fig. 668.—System $2CaO \cdot SiO_2-3CaO \cdot P_2O_5$. C = CaO, S = SiO_2, P = P_2O_5, A = new phase, S = silicocarnotite.

R. W. Nurse, J. H. Welch, and W. Gutt, *J. Chem. Soc.*, **1959**, p. 1080. See also, Józef Berak and Jadwiga Wojciechowska, *Roczniki Chem.*, **30**, 758 (1956).

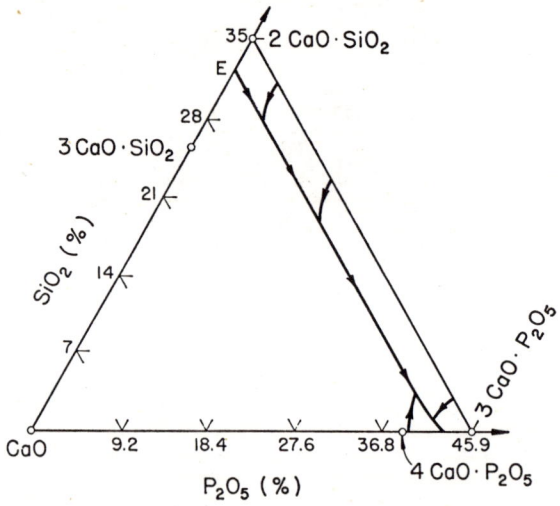

Fig. 670.—System $CaO-2CaO \cdot SiO_2-3CaO \cdot P_2O_5$; at lime corner

G. Trömel, H.-J. Harkort, and W. Hotop. *Z. anorg. Chem.*, **256** [5–6] 269 (1948).

CaO–SiO$_2$–P$_2$O$_5$ (concl.)

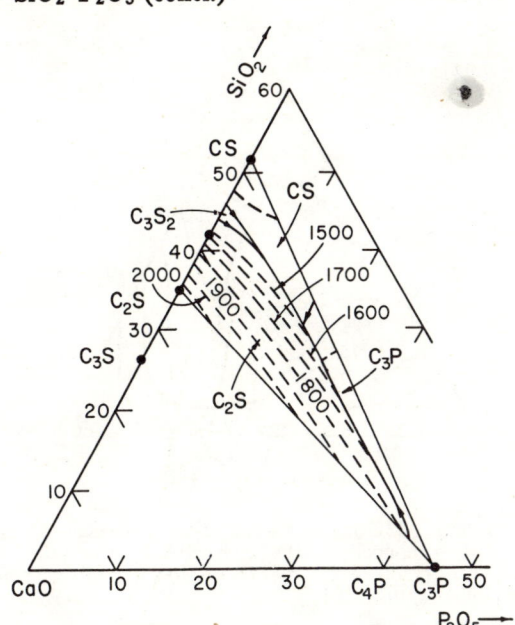

FIG. 671.—System CaO·SiO$_2$–2CaO·SiO$_2$–3CaO·P$_2$O$_5$. C = CaO, S = SiO$_2$, P = P$_2$O$_5$.

Józef Berak and Jadwiga Wojciechowska, *Roczniki Chem.*, **30**, 759 (1956).

CaO–SnO$_2$–TiO$_2$

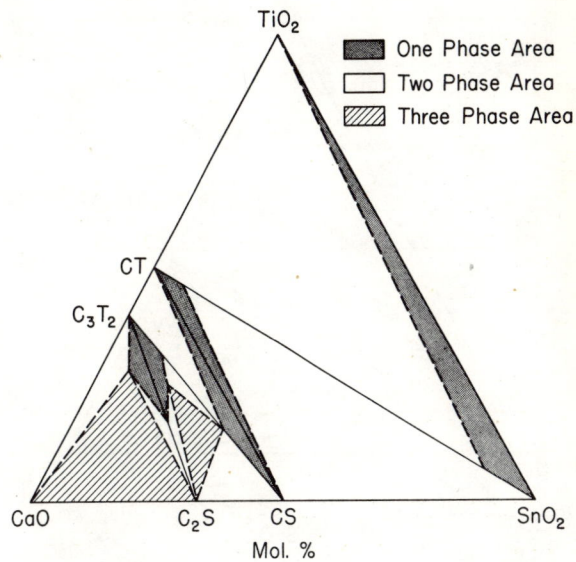

FIG. 673.—System CaO–SnO$_2$–TiO$_2$; compatibility relations at about 1450°C.; C = CaO; S = SnO$_2$; T = TiO$_2$.

L. W. Coughanour, R. S. Roth, S. Marzullo, and F. E. Sennett, *J. Research Natl. Bureau Standards* **54** [3] 155 (1955); RP 2576.

CaO–SiO$_2$–R$_x$O$_y$

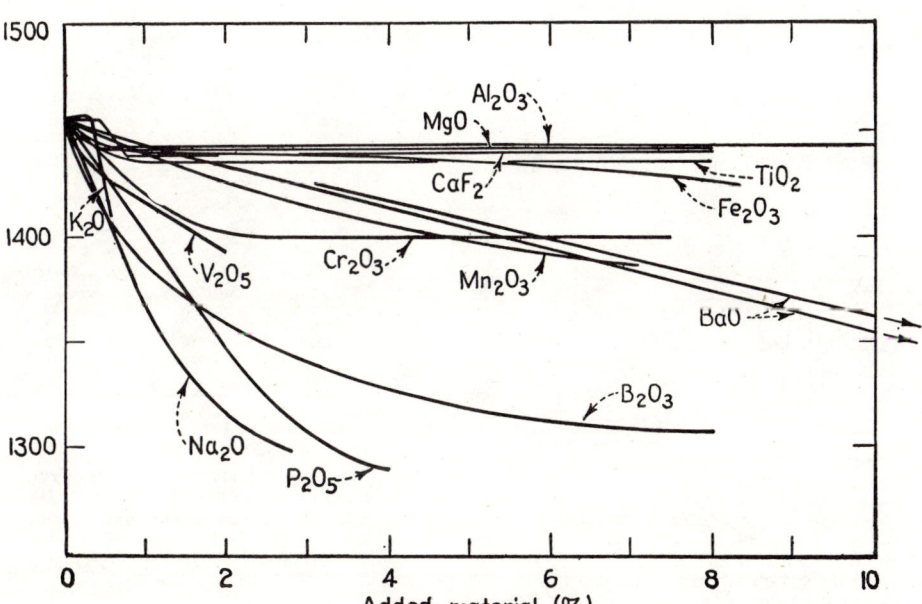

FIG. 672.—Effect on the α-2CaO·SiO$_2$ → β-2CaO·SiO$_2$ inversion temperature caused by addition of third components.
E. S. Newman and L. S. Wells, *J. Research Nat. Bur. Standards*, **36** [2] pp. 151, 152, 153 (1946); R. P. 1696.

CaO–SnO$_2$–TiO$_2$ (concl.)

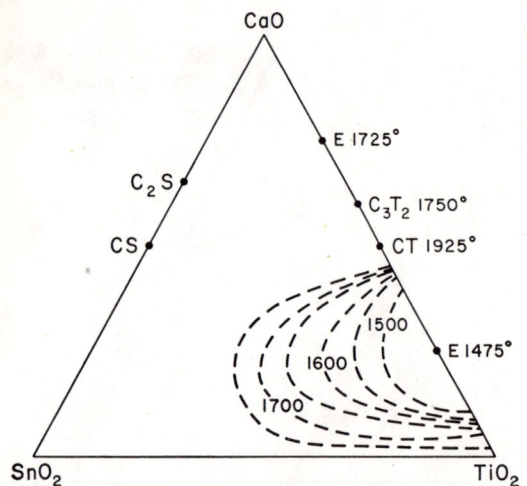

FIG. 674.—System CaO–SnO$_2$–TiO$_2$; softening temps. C = CaO, S = SnO$_2$, T = TiO$_2$.

A. Dietzel and W. Poch, *Radex-Rundschau*, 1960, No. 1, p. 58.

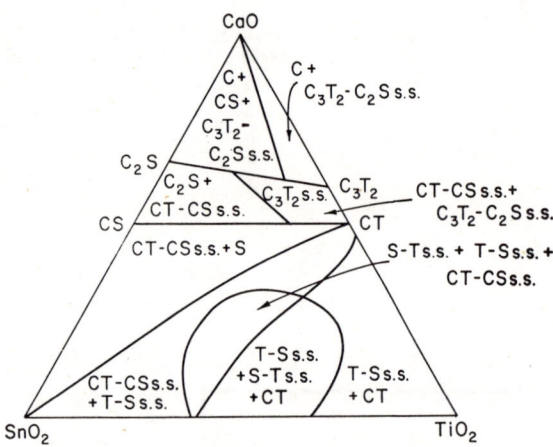

FIG. 675.—System CaO–SnO$_2$–TiO$_2$; phases at 1300°C. C = CaO, S = SnO$_2$, T = TiO$_2$.

A. Dietzel and W. Poch, *Radex-Rundschau*, 1960, No. 1, p. 59.

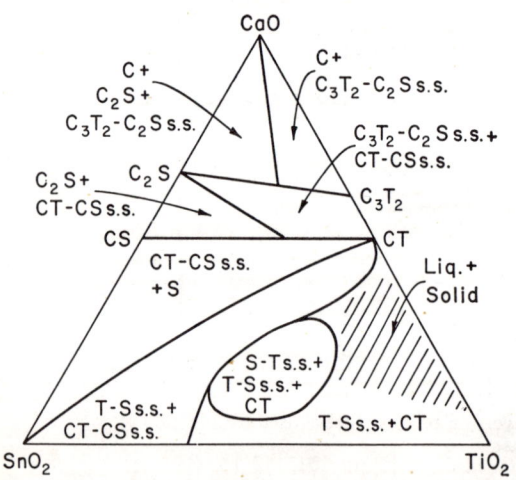

FIG. 676.—System CaO–SnO$_2$–TiO$_2$; phases at 1600°C. C = CaO, S = SnO$_2$, T = TiO$_2$.

A. Dietzel and W. Poch, *Radex-Rundschau*, 1960, No. 1, p. 59.

CaO–ThO$_2$–ZrO$_2$

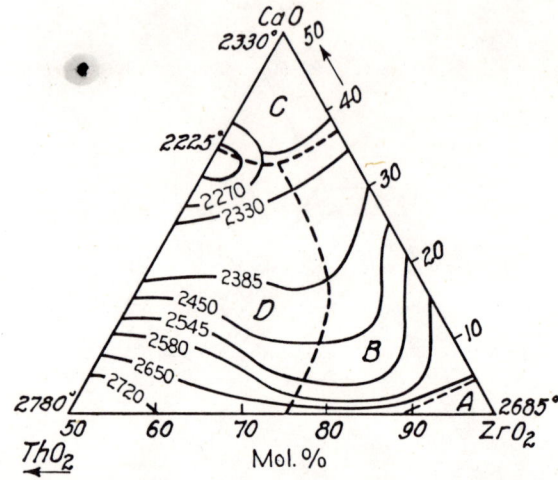

FIG. 677.—Melting isotherms of system CaO–ThO$_2$–ZrO$_2$; A = monoclinic ZrO$_2$, B = cubic mix crystal, C = cubic mix crystal + CaZrO$_3$, D = two cubic mix crystals (miscibility gap).

Otto Ruff, F. Ebert, and W. Loerpabel, *Z. anorg. u. allgem. Chem.*, **207**, 309 (1932).

CaO–TiO$_2$–ZrO$_2$

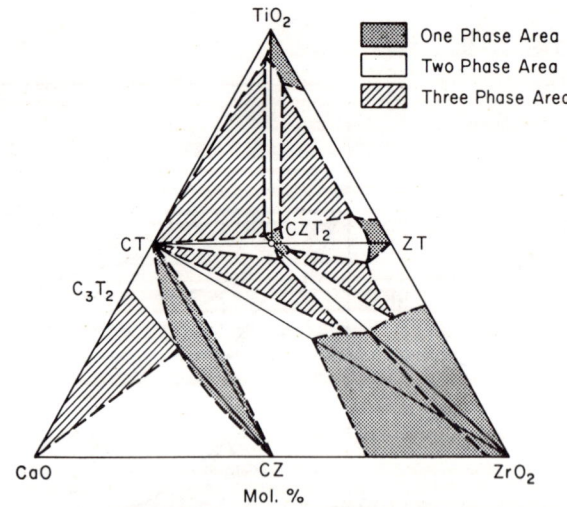

FIG. 678.—System CaO–TiO$_2$–ZrO$_2$; estimated solid solution in range 1450° to 1550°C. Dashed line near ZrO$_2$ apex indicates the approximate location of a two-phase area, which theoretically would separate a cubic ZrO$_2$ field, on the left, from a tetragonal ZrO$_2$ field, on the right. C = CaO; T = TiO$_2$; Z = ZrO$_2$.

L. W. Coughanour, R. S. Roth, S. Marzullo, and F. E. Sennett, *J. Research Natl. Bur. Standards*, **54** [4] 195 (1955); RP2580.

CdO–ZnO–B₂O₃

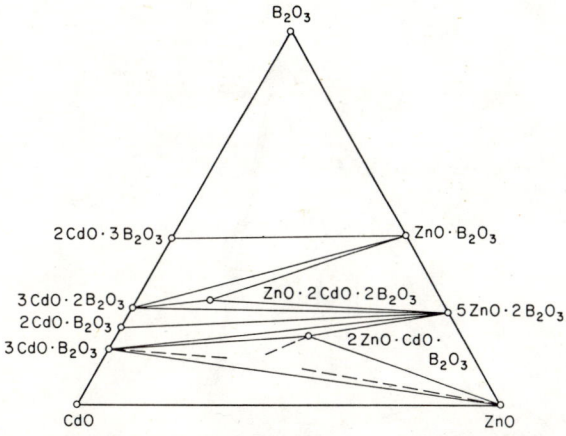

FIG. 679.—System CdO–ZnO–B₂O₃; compatibility triangles.

D. E. Harrison and F. A. Hummel, *J. Electrochem. Soc.*, **106** [1] 25 (1959).

FeO–MgO–Al₂O₃

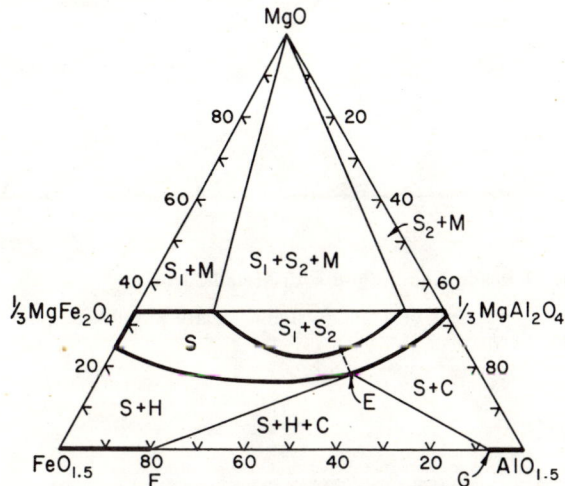

FIG. 680.—System FeO–MgO–Al₂O₃ at 1250°C. S = broad spinel band, S_1 = $MgFe_2O_4$ ss, S_2 = $MgAl_2O_4$ ss, H = hematite ss, C = corundum ss, M = MgO. Thick lines represent one-phase areas. Compositions in the region left of the broken line are ferrimagnetic at room temperature.

W. Kwestroo, *J. Inorg. Nucl. Chem.*, **9**, 67 (1959).

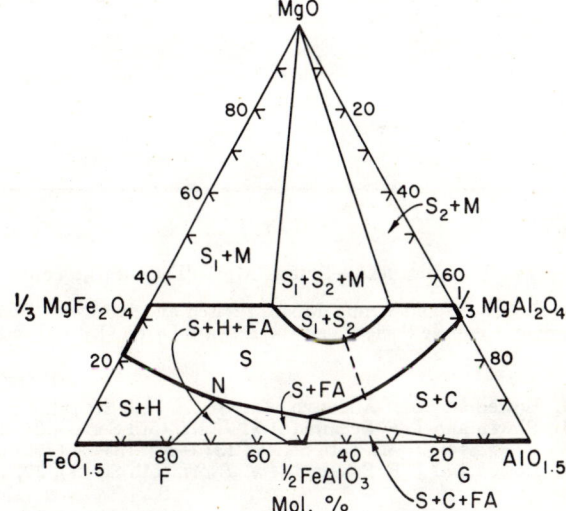

FIG. 681.—System FeO–MgO–Al₂O₃ at 1400°C. S = broad spinel band, S_1 = $MgFe_2O_4$ ss, S_2 = $MgAl_2O_4$ ss, H = hematite ss, C = corundum ss, M = MgO, FA = $FeAlO_3$ ss. Thick lines represent one-phase areas. Compositions in the region left of the broken line are ferrimagnetic at room temperature.

W. Kwestroo, *J. Inorg. Nucl. Chem.*, **9**, 67 (1959).

FeO–MgO–SiO$_2$

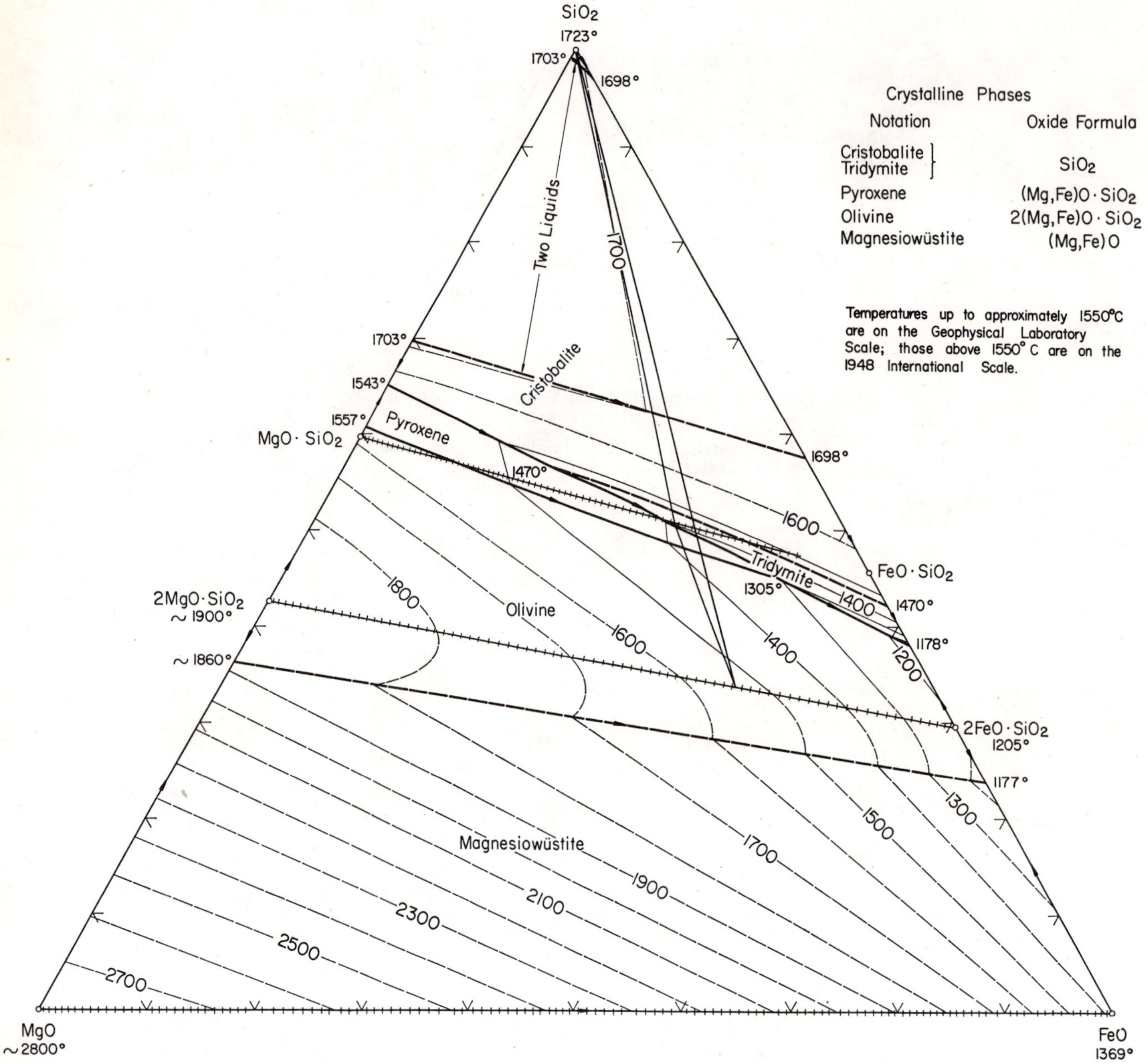

FIG. 682.—System MgO–"FeO"–SiO$_2$; composite. (Oxide Phases in Equilibrium with Metallic Iron).

E. F. Osborn and Arnulf Muan, revised and redrawn "Phase Equilibrium Diagrams of Oxide Systems," Plate 8, published by the American Ceramic Society and the Edward Orton, Jr., Ceramic Foundation, 1960.

Principal References

N. L. Bowen and Olaf Andersen, *Am. J. Sci.* (4th Series), **37**, 487–500 (1914).
N. L. Bowen and J. F. Schairer, *Am. J. Sci.* (5th Series), **24**, 177–213 (1932).
J. W. Greig, *Am. J. Sci.* (5th Series), **13**, 1–44; 133–54 (1927).
N. L. Bowen and J. F. Schairer, *Am. J. Sci.* (5th Series), **29**, 151–217 (1935).

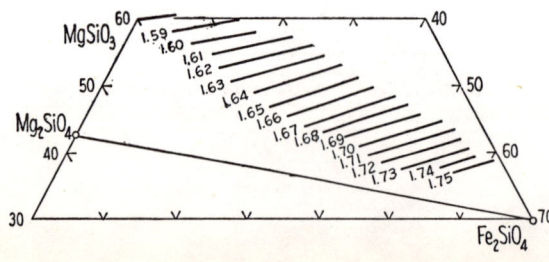

FIG. 683.—Refractive indices of glasses in system FeO–MgO–SiO$_2$.

N. L. Bowen and J. F. Schairer, *Am. J. Sci.*, 5th Ser., **29**, 200 (1935).

FeO–MgO–SiO₂ (cont.)

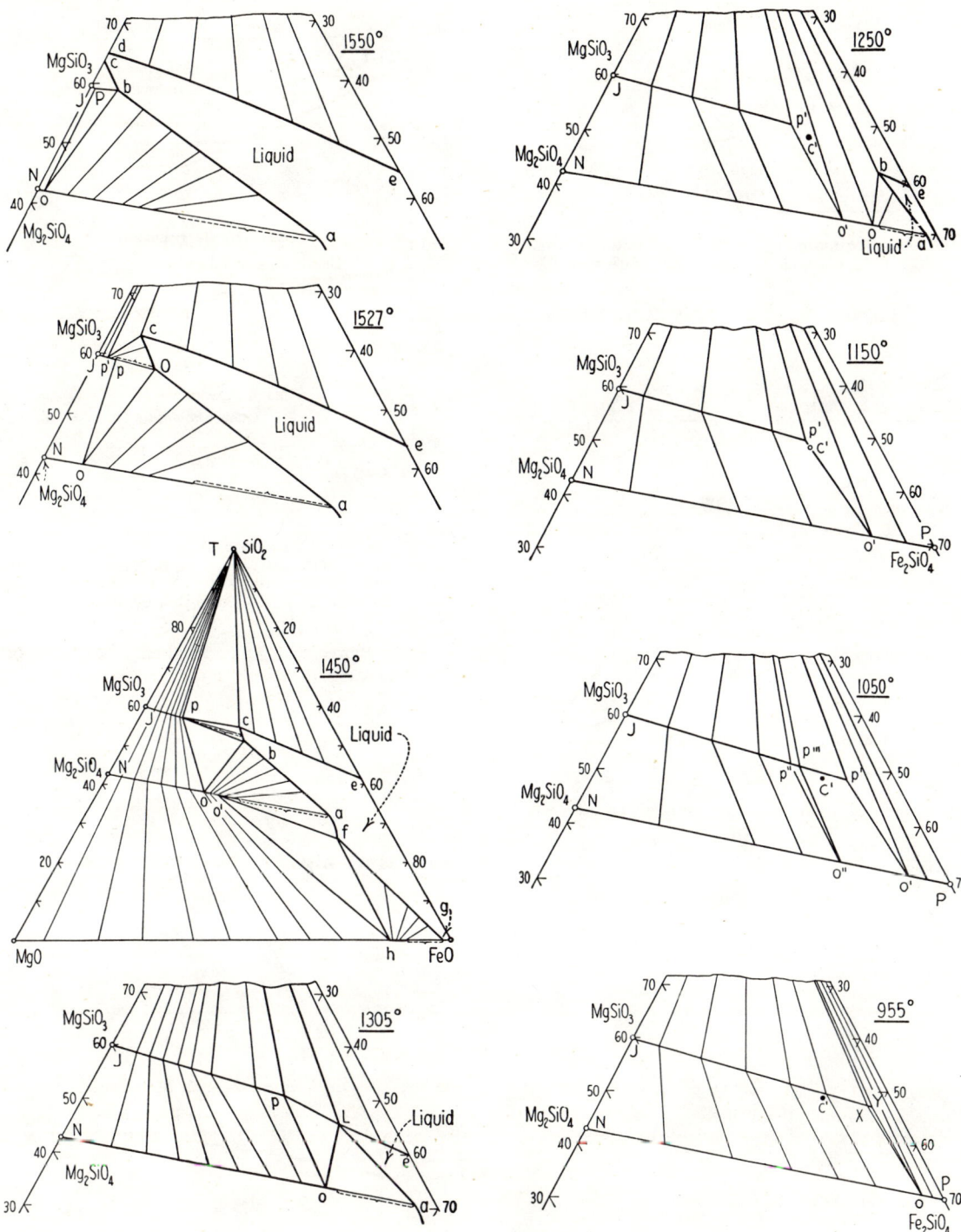

FIG. 684.—Phase relations at 1550°, 1527°, 1450°, 1305°, 1250°, 1150°, 1050°, and 955°C. in system FeO–MgO–SiO₂.

N. L. Bowen and J. F. Schairer, *Am. J. Sci.*, 5th Ser., 29, 174–80 (1935)

FeO–MgO–SiO$_2$ (concl.)

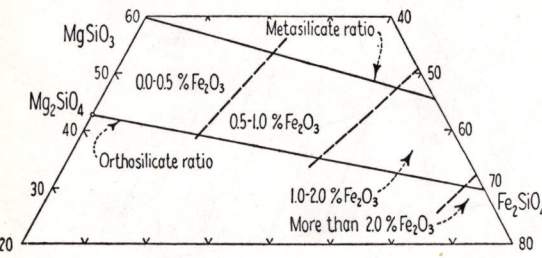

FIG. 685.—Approximate Fe$_2$O$_3$ content of liquid mixtures of MgO, SiO$_2$, and iron oxide in equilibrium with metallic iron.

N. L. Bowen and J. F. Schairer, *Am. J. Sci.* 5th Ser., **29**, 158 (1935).

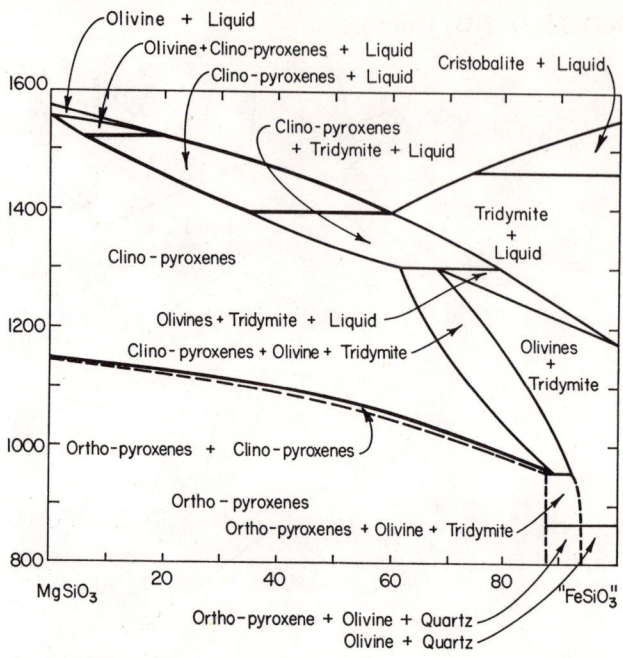

FIG. 686.—Metasilicate join of system FeO–MgO–SiO$_2$. *Heavy curves* refer to binary equilibrium; *light curves*, ternary equilibrium (binary and ternary, respectively, only when the small amounts of Fe$_2$O$_3$ present in liquids are treated as FeO (see Fig. 685)).

N. L. Bowen and J. F. Schairer, *Am. J. Sci.* 5th. Ser., **29**, 164 (1935).

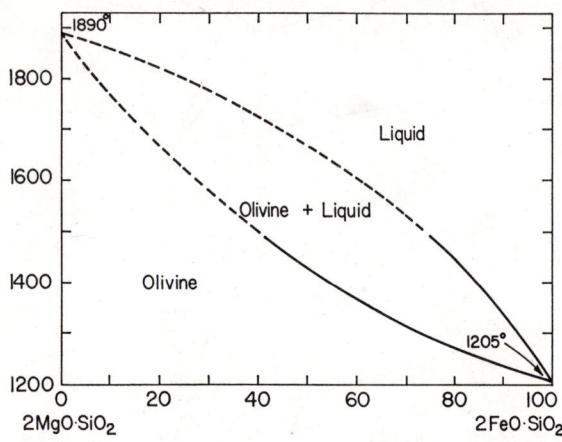

FIG. 687.—System 2FeO·SiO$_2$–2MgO·SiO$_2$.

N. L. Bowen and J. F. Schairer, *Am. J. Sci.* 5th Ser. **29**, 163 (1935).

FeO–MnO–Al$_2$O$_3$

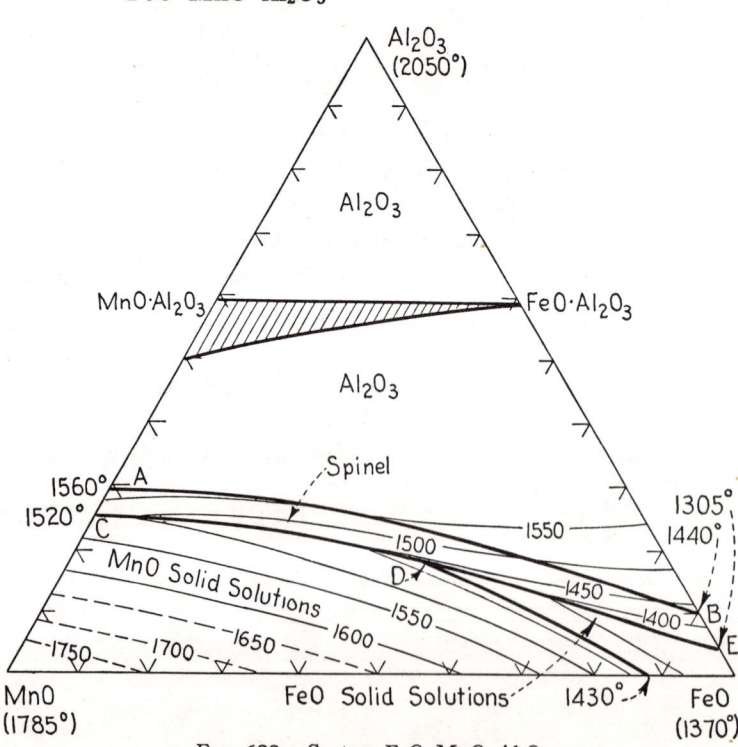

FIG. 688.—System FeO–MnO–Al$_2$O$_3$.

R. Hay, R. McIntosh, A. B. Rait, and James White, *J. West Scot. Iron Steel Inst.*, **44**, 85 (1936–37).

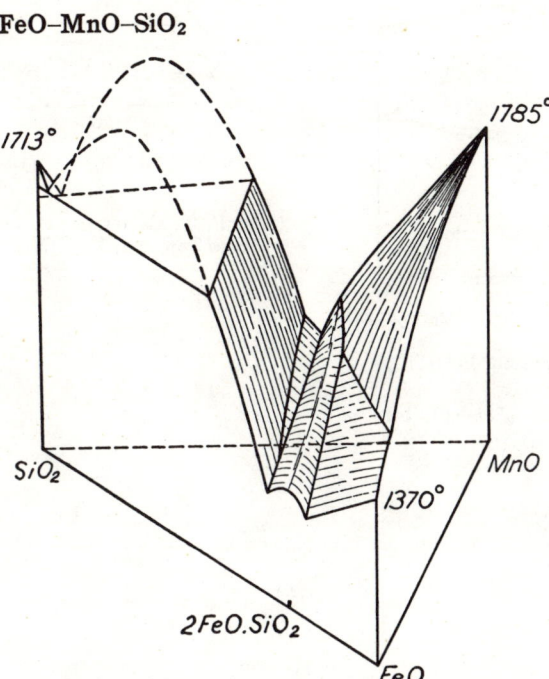

Fig. 689.—Perspective of solid of system FeO–MnO–SiO₂.

R. Hay, James White, and A. B. McIntosh, *J. Wes Scot. Iron Steel Inst.*, **42**, 99 (1934–35).

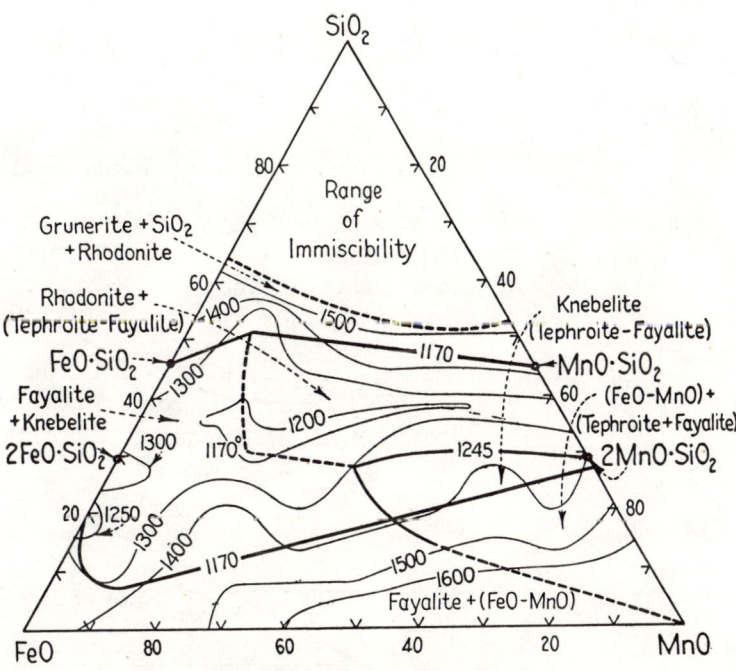

Fig. 690.—System FeO–MnO–SiO₂.

W. R. Maddocks, *Iron Steel Inst. (London) Carnegie Schol. Mem.*, **24**, 64 (1935).

FeO–MnO–SiO₂ (concl.)

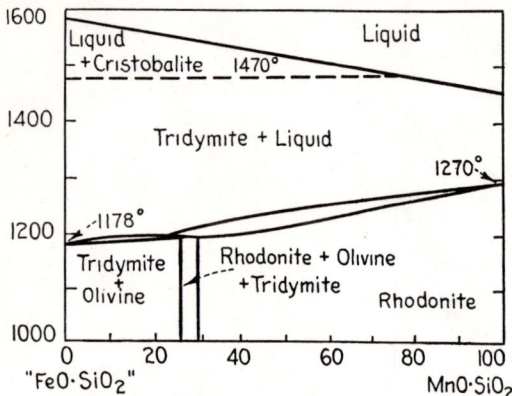

Fig. 691.—Tentative diagram of system FeO·SiO₂–MnO·SiO₂.

James White, *J. Iron Steel Inst. (London),* **148,** No. II, 586 (1943).

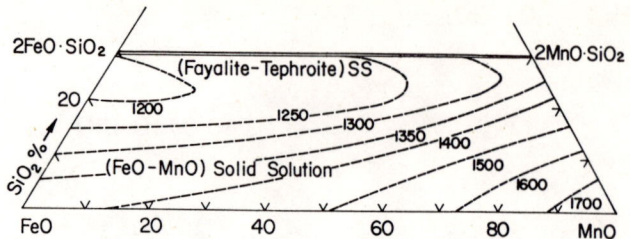

Fig. 694.—System FeO–2FeO·SiO₂–2MnO·SiO₂–MnO; liquidus surface.

P. T. Carter, A. B. Murad, and R. Hay, *J. West Scot. Iron Steel Inst.,* **60,** 128 (1952–53).

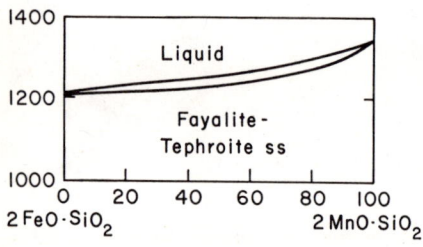

Fig. 692.—System 2FeO·SiO₂–2MnO·SiO₂.

P. T. Carter, A. B. Murad, and R. Hay, *J. West Scot. Iron Steel Inst.,* **60,** 126 (1952–53).

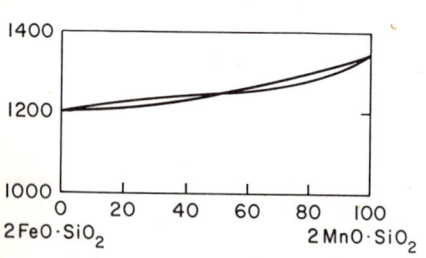

Fig. 693.—System 2FeO·SiO₂–2MnO·SiO₂; possible alternative.

P. T. Carter, A. B. Murad, and R. Hay, *J. West. Scot. Iron Steel Inst.,* **60,** 127 (1952–53).

FeO–MnO–TiO₂

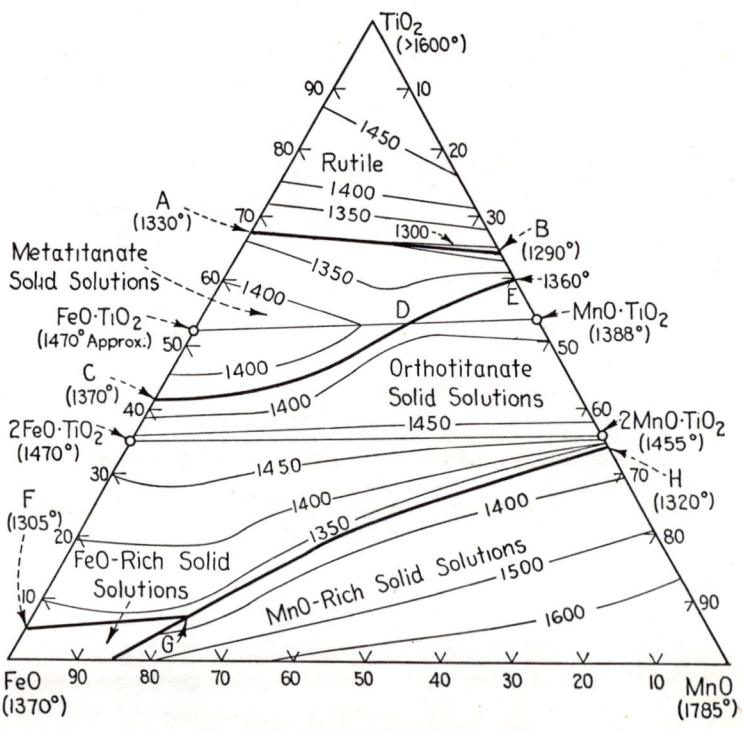

Fig. 695.—System FeO–MnO–TiO₂.

J. Grieve and James White, *J. Roy. Tech. Coll. (Glasgow),* **4,** 665 (1940).

FeO–Al$_2$O$_3$–SiO$_2$

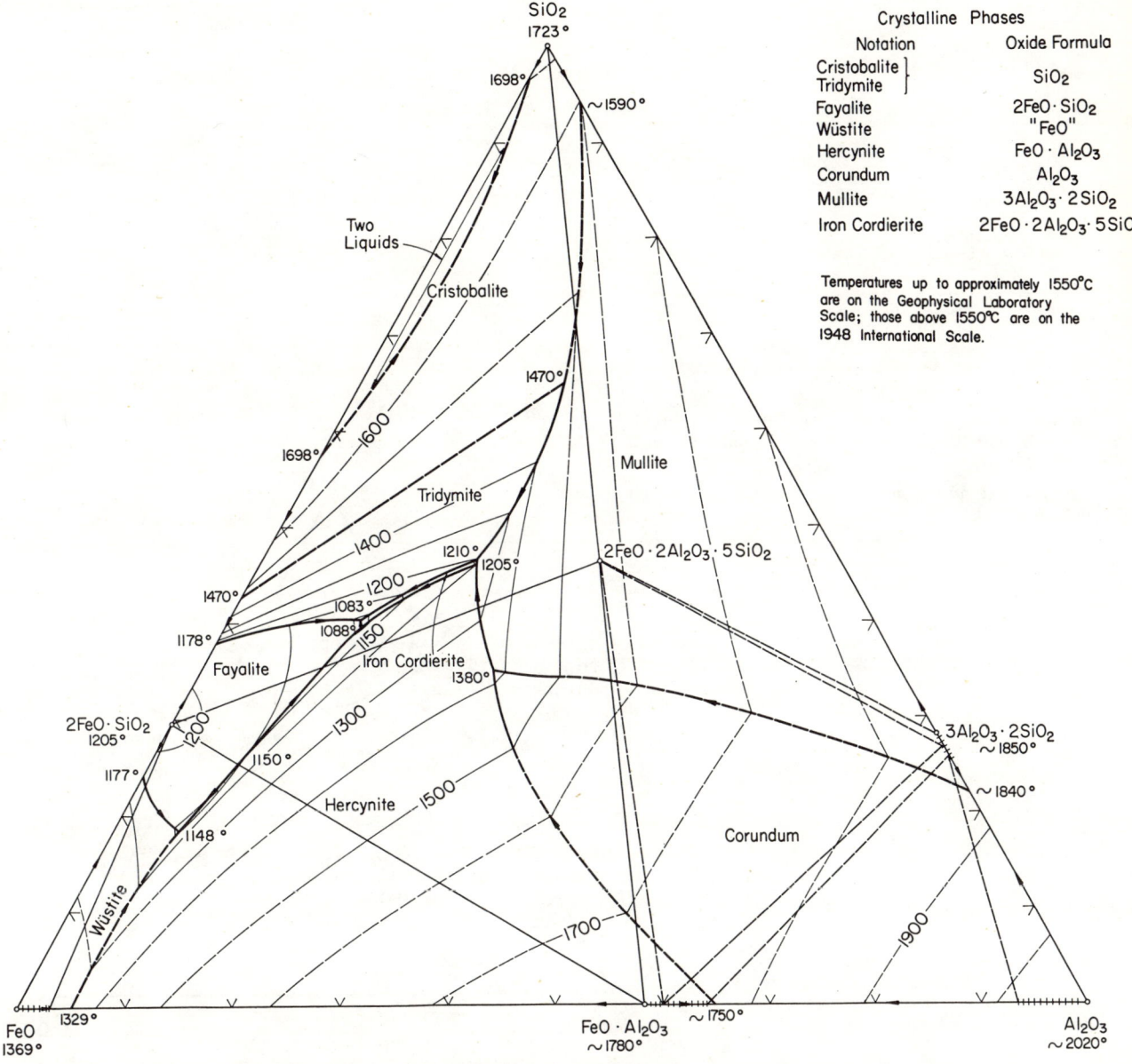

FIG. 696.—System "FeO"–Al$_2$O$_3$–SiO$_2$; composite. (Oxide Phases in Equilibrium with Metallic Iron).

E. F. Osborn and Arnulf Muan, revised and redrawn "Phase Equilibrium Diagrams of Oxide Systems," Plate 9, published by the American Ceramic Society and the Edward Orton, Jr., Ceramic Foundation, 1960.

Principal References

N. L. Bowen and J. F. Schairer, *Am. J. Sci.* (5th Series), **24**, 177–213 (1932).
Willy Oelsen and Gerhard Heynert, *Arch. Eisenhüttenw.*, **26**, 567–75 (1955).
W. A. Fischer and Alfred Hoffmann, *Arch. Eisenhüttenw.*, **27**, 343–46 (1956).
N. L. Bowen and J. W. Greig, *J. Am. Ceram. Soc.*, **7**, 238–54 (1924); corrections, ibid, 410.
N. A. Toropov and F. Ya. Galakhov, *Vopr. Petrogr. i Mineralog., Akad. Nauk S.S.S.R.*, **2**, 245–55 (1953).
Shigeo Aramaki and Rustum Roy, *Nature*, **184**, 631–32 (1959).
J. W. Greig, *Am. J. Sci.* (5th Series), **14**, 473–84 (1927).
J. F. Schairer and Kenzo Yagi, *Am. J. Sci.*, Bowen Volume, Part 2, 471–512 (1952).

FeO–Al₂O₃–SiO₂ (concl.)

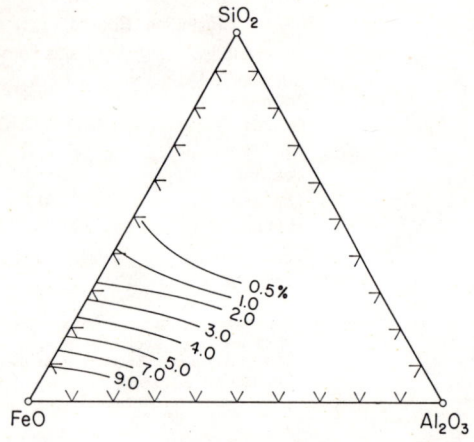

FIG. 697.—System FeO–Al$_2$O$_3$–SiO$_2$; showing approximate Fe$_2$O$_3$ content (%) of liquids in equilibrium with metallic iron.

J. F. Schairer and Kenzo Yagi, *Am. J. Sci.*, Bowen volume, p. 486 (1952).

MgO–MnO–SiO₂

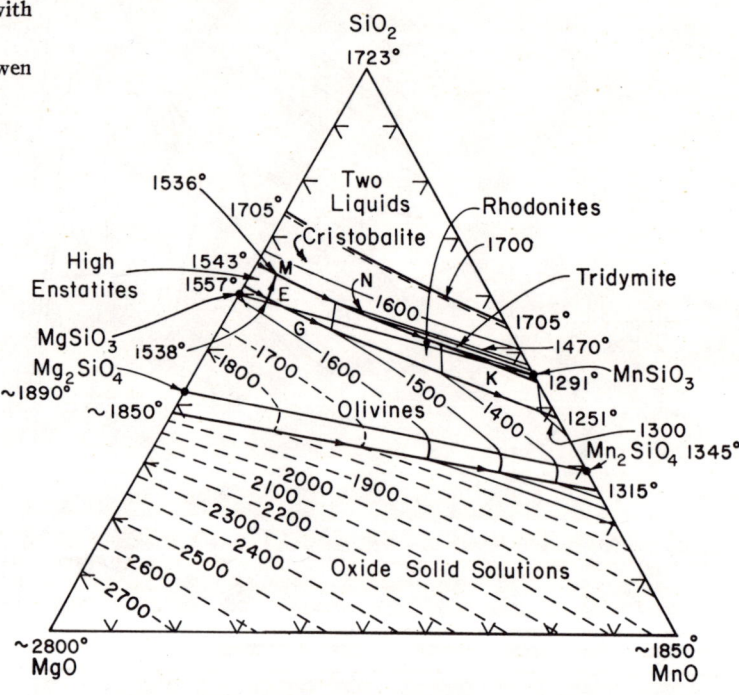

FIG. 699.—System MgO–MnO–SiO$_2$. M,E = ternary peritectics; N = ternary invariant point between tridymite, cristobalite, and rhodonite solid solution. Section of the metasilicate join G to K is binary.

F. P. Glasser and E. F. Osborn, *J. Am. Ceram. Soc.*, **43** [3] 136 (1960).

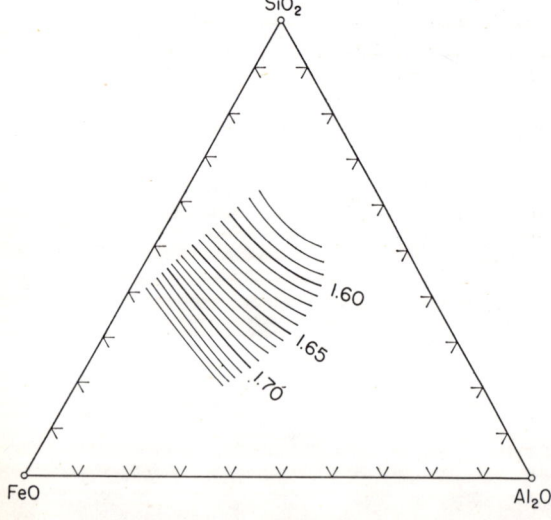

FIG. 698.—System FeO–Al$_2$O$_3$–SiO$_2$; isofracts at 25°C.

J. F. Schairer and Kenzo Yagi, *Am. J. Sci.*, Bowen volume, p. 487 (1952).

MgO–MnO–SiO₂ (concl.)

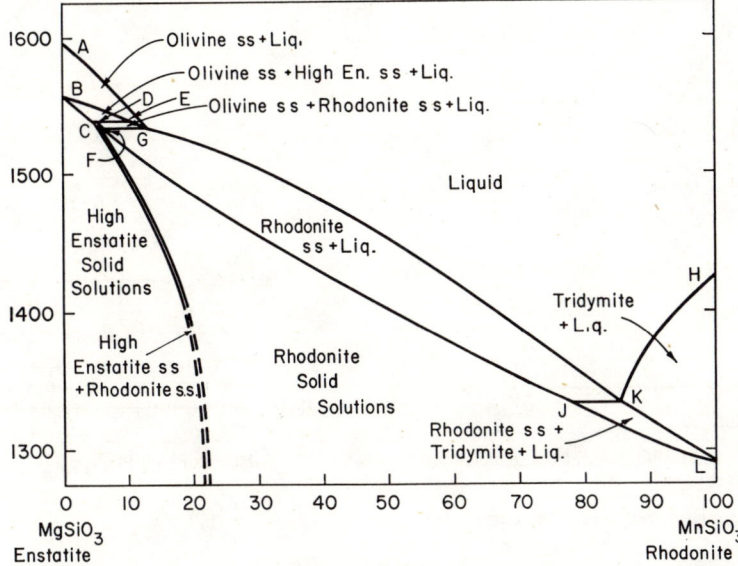

Fig. 700.—System MgSiO₃–MnSiO₃. Ternary behavior in three-phase regions. A = 1595°C.; B = 1557°C.; C, D, E = 1538°C.; F, G = 1533°C.; H = 1425°C.; J, K = 1333°C.; L = 1291°C.

F. P. Glasser and E. F. Osborn, *J. Am. Ceram. Soc.*, **43** [3] 134 (1960).

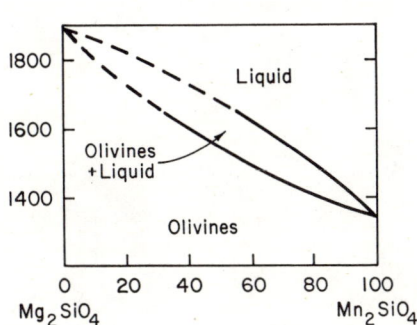

Fig. 701.—System Mg₂SiO₄–Mn₂SiO₄.

F. P. Glasser and E. F. Osborn, *J. Am. Ceram. Soc.*, **43** [3] 135 (1960).

MgO–SrO–P₂O₅

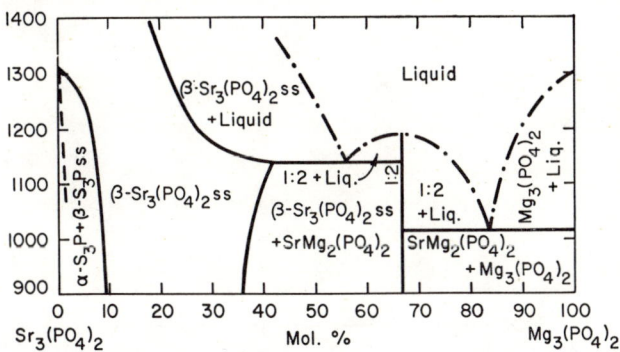

Fig. 703.—System Mg₃(PO₄)₂–Sr₃(PO₄)₂.

J. F. Sarver, F. A. Hummel, and M. V. Hoffman, *J. Electrochem. Soc.*, **108** [12] 1106 (1961).

MgO–NiO–SiO₂

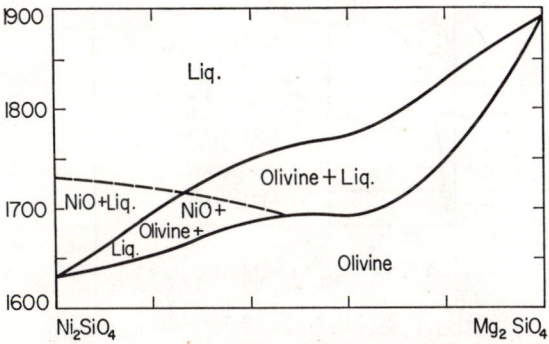

Fig. 702.—System Mg₂SiO₄–Ni₂SiO₄; approximate pseudobinary.

A. E. Ringwood, *Geochim. et Cosmochim. Acta*, **10**, 298 (1956).

MgO–ZnO–GeO₂

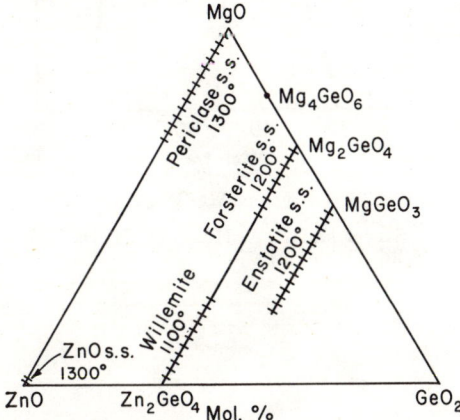

Fig. 704.—System MgO–ZnO–GeO₂; solid solutions (hatched lines).

J. F. Sarver and F. A. Hummel, personal communication, Nov., 1961.

MgO–ZnO–SiO$_2$

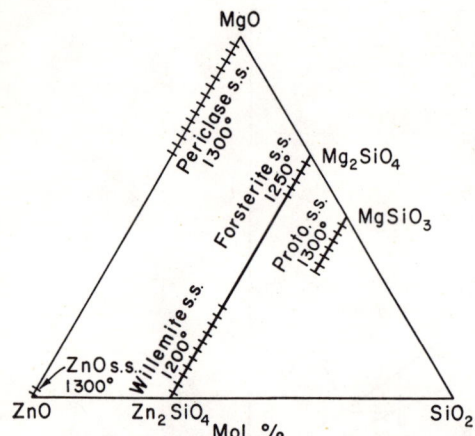

FIG. 705.—System MgO–ZnO–SiO$_2$; solid solutions (hatched lines).

J. F. Sarver and F. A. Hummel, personal communication, Nov., 1961.

MgO–ZnO–P$_2$O$_5$

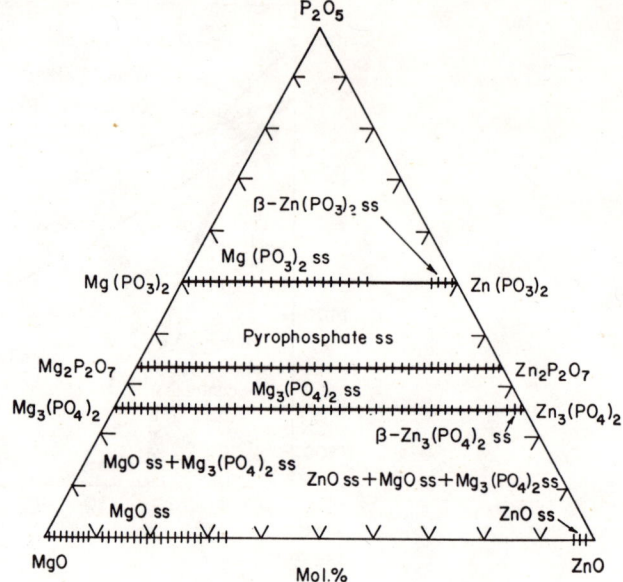

FIG. 707.—System MgO–ZnO–P$_2$O$_5$; subsolidus solid solutions.

J. F. Sarver, F. L. Katnack, and F. A. Hummel, *J. Electrochem. Soc.*, **106** [11] 962 (1959).

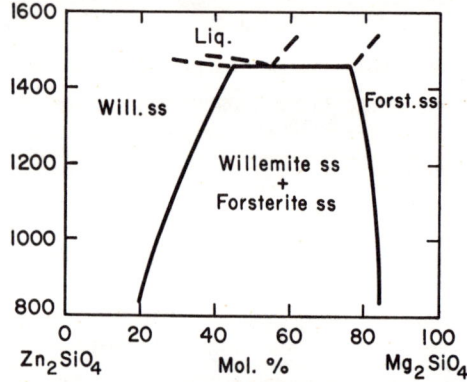

FIG. 706.—System Mg$_2$SiO$_4$–Zn$_2$SiO$_4$.

J. F. Sarver and F. A. Hummel, *J. Am. Ceram. Soc.*, **45** [6] 304 (1962).

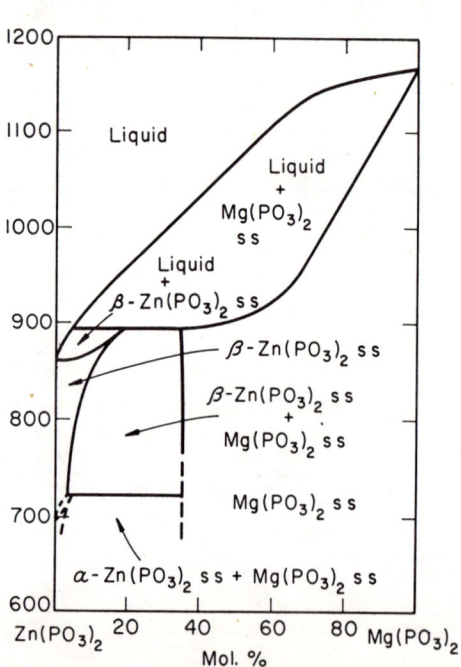

FIG. 708.—System Mg(PO$_3$)$_2$–Zn(PO$_3$)$_2$.

J. F. Sarver and F. A. Hummel, *J. Electrochem. Soc.*, **106**, 500 (1959).

MgO–ZnO–P₂O₅ (concl.)

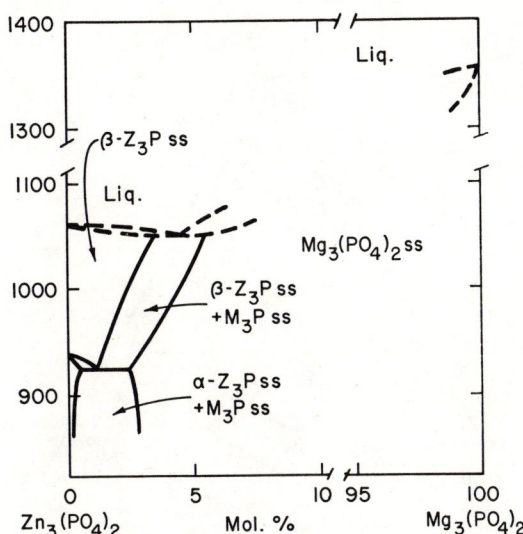

Fig. 709.—System $Mg_3(PO_4)_2$–$Zn_3(PO_4)_2$. M = MgO, P = P_2O_5, Z = ZnO.

J. F. Sarver, F. L. Katnack, and F. A. Hummel, *J. Electrochem. Soc.*, 106 [11] 961 (1959).

MgO–Al₂O₃–Cr₂O₃

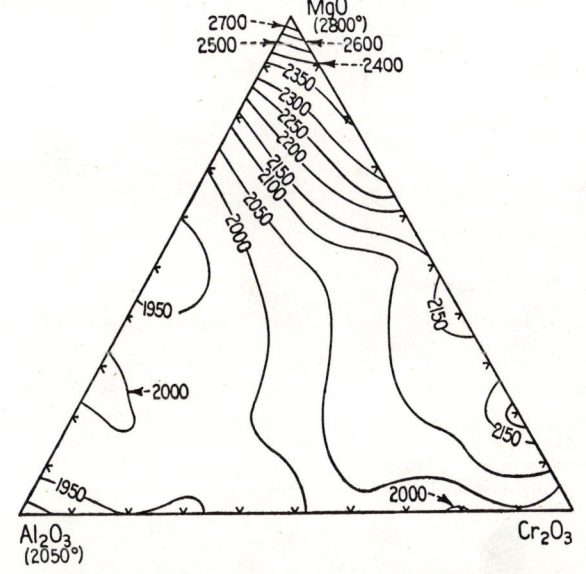

Fig. 710.—Melting isotherms of system MgO–Al₂O₃–Cr₂O₃.

W. T. Wilde and W. J. Rees, *Trans. Brit. Ceram. Soc.*, 42, 127 (1943).

MgO–Al₂O₃–Fe₂O₃

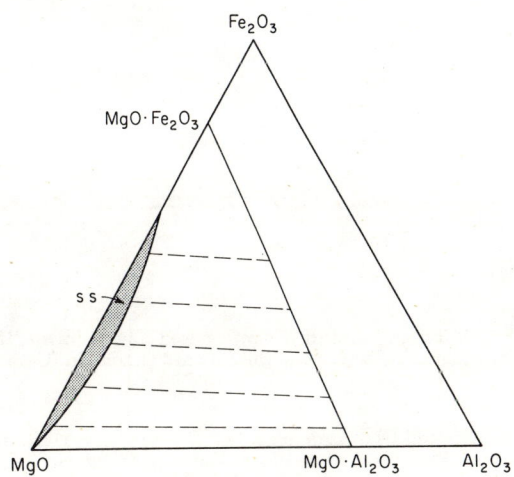

Fig. 711.—System MgO–Al₂O₃–Fe₂O₃; probable at 1600°C.

A. Cocco, *Radex-Rundschau*, 1958, No. 6, p. 288.

MgO–Al$_2$O$_3$–SiO$_2$

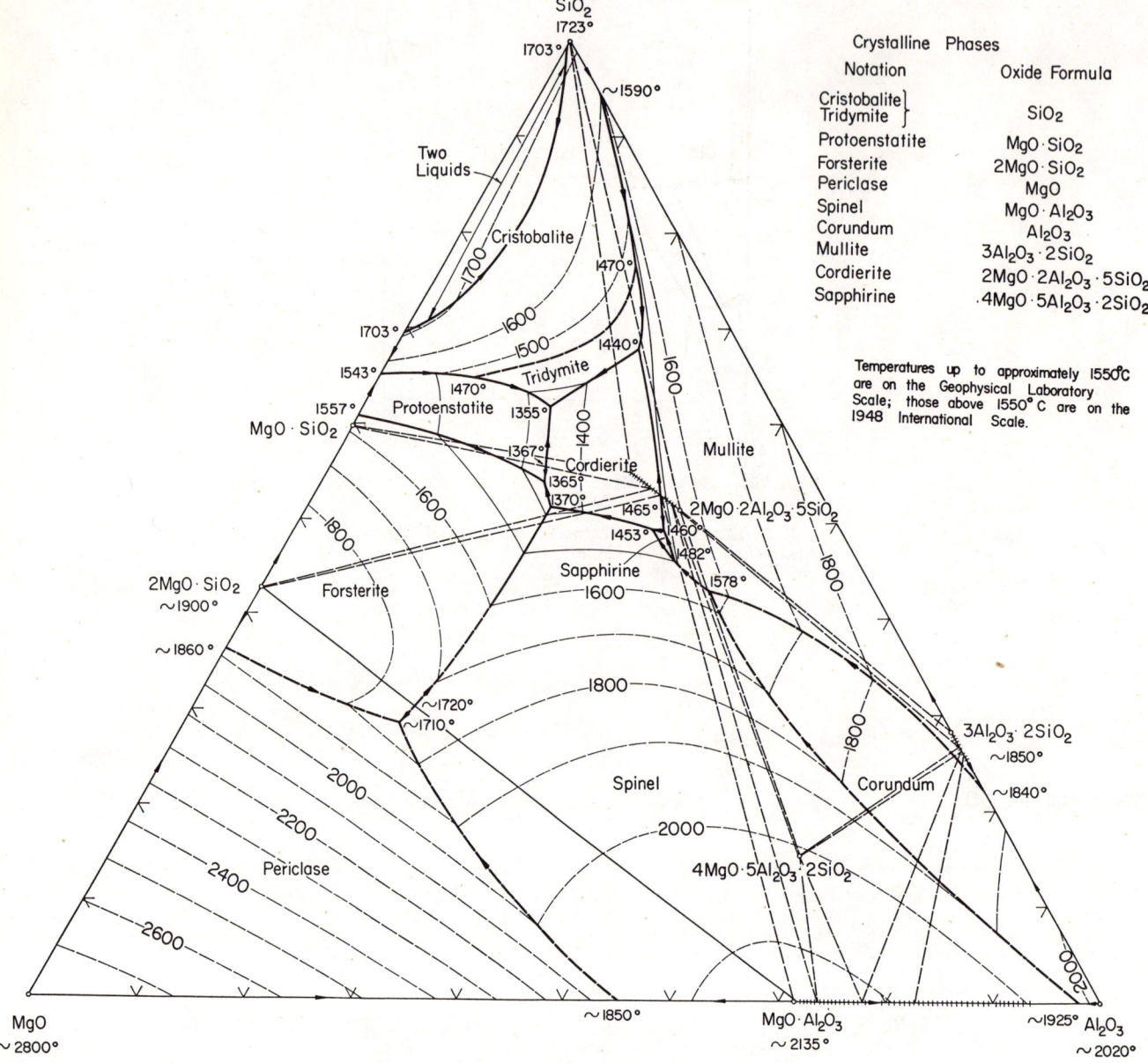

FIG. 712.—System MgO–Al$_2$O$_3$–SiO$_2$; composite.

E. F. Osborn and Arnulf Muan, revised and redrawn "Phase Equilibrium Diagrams of Oxide Systems," Plate 3, published by the American Ceramic Society and the Edward Orton, Jr., Ceramic Foundation, 1960.

Principal References

N. L. Bowen and Olaf Andersen, *Am. J. Sci.* (4th Series), **37**, 487–500 (1914).
D. M. Roy, Rustum Roy and E. F. Osborn, *Am. J. Sci.*, **251**, 337–61 (1953).
N. L. Bowen and J. W. Greig, *J. Am. Ceram. Soc.*, **7**, 238–54 (1924); corrections, ibid., 410.
N. A. Toropov and F. Ya. Galakhov, *Voprosy Petrogr. i Mineralog. Akad. Nauk S.S.S.R.*, **2**, 245–55 (1953).
G. A. Rankin and H. E. Merwin, *Am. J. Sci.* (4th Series), **45**, 301–25 (1918).
J. W. Greig, *Am. J. Sci.* (5th Series), **13**, 1–44; 133–54 (1927).
W. R. Foster, *J. Am. Ceram. Soc.*, **33**, 73–84 (1950).
M. L. Keith and J. F. Schairer, *J. Geol.*, **60**, 181–86 (1952).
Shigeo Aramaki and Rustum Roy, *J. Am. Ceram. Soc.*, **42**, 644–45 (1959).
W. Schreyer and J. F. Schairer, *J. Petrology*, **2**, 324–406 (1961).

$MgO-Al_2O_3-TiO_2$

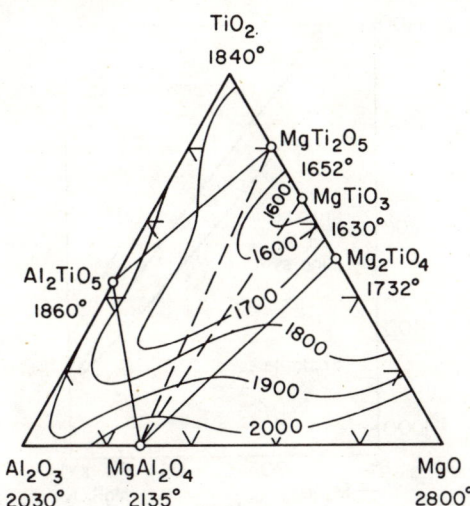

Fig. 713.—System $MgO-Al_2O_3-TiO_2$; melting.

A. S. Berezhnoĭ and N. V. Gul'ko, *Ukrain. Khim. Zhur.*, **21** [2] 162 (1955).

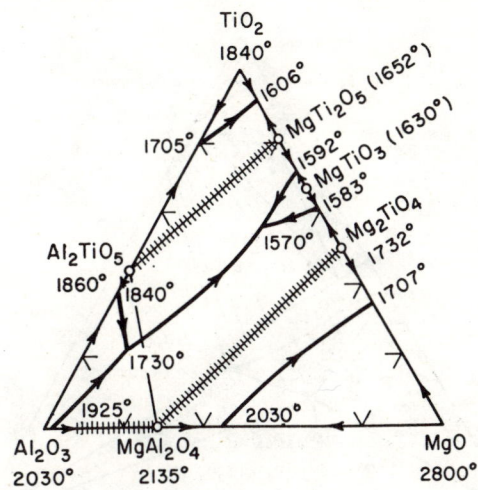

Fig. 714.—System $MgO-Al_2O_3-TiO_2$; primary phases. Cross-hatched lines = solid solutions.

A. S. Berezhnoĭ and N. V. Gul'ko, *Ukrain. Khim. Zhur.*, **21** [2] 162 (1955).

$MgO-Cr_2O_3-SiO_2$

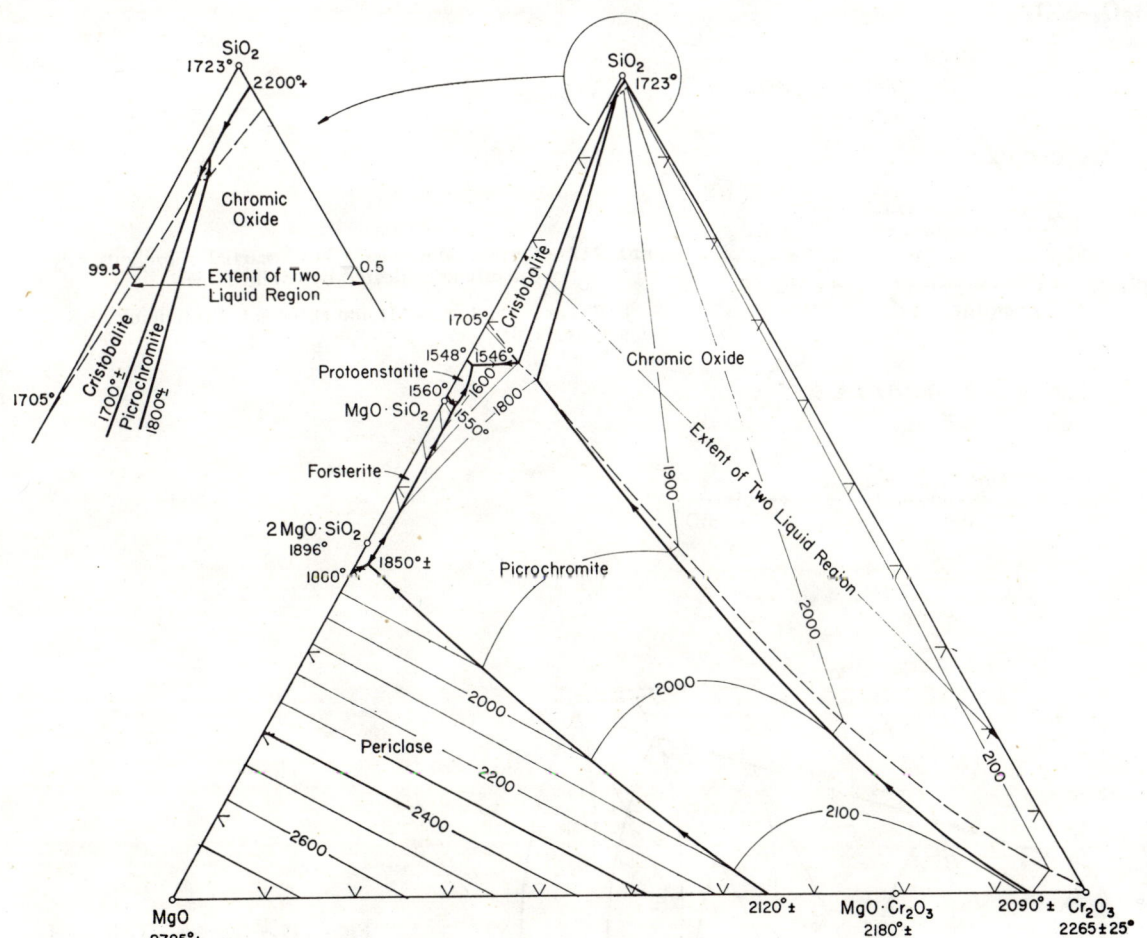

Fig. 715.—System $MgO-Cr_2O_3-SiO_2$.

M. L. Keith, *J. Am. Ceram. Soc.*, **37** [10] 491 (1954).
For compatibility triangles, see also, W. F. Ford and W. J. Rees, *Trans. Brit. Ceram. Soc.*, **57** [5] 238 (1958).
The value 2200°+ in the inset refers to the intersection of the dashed line with the boundary system; the eutectic should read 1720°.

MgO–La$_2$O$_3$–ZrO$_2$

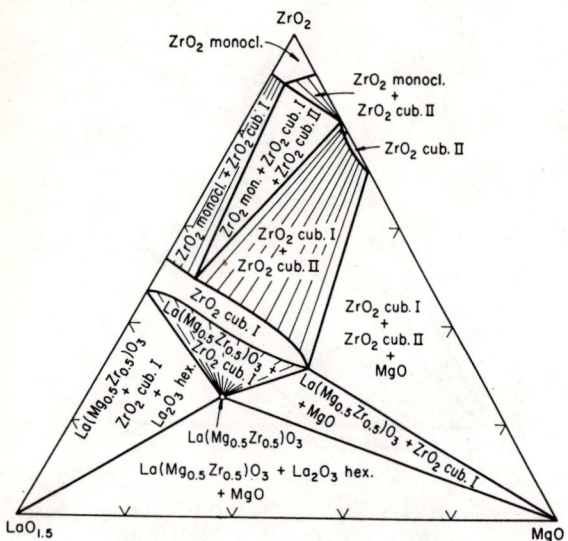

FIG. 716.—System MgO–LaO$_{1.5}$–ZrO$_2$; at approx. 1400°C.

Albrecht Rabenau, *Z. anorg. u. allgem. Chem.*, **288**, 224 (1956).

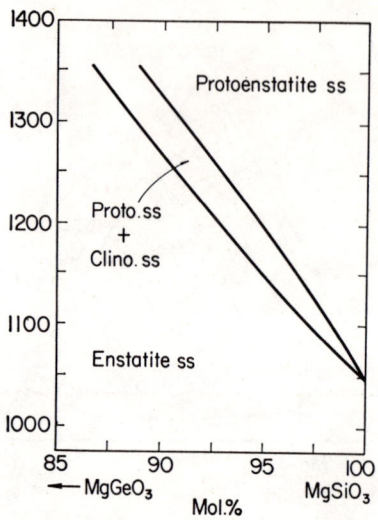

FIG. 718.—System MgGeO$_3$–MgSiO$_3$; partial subsolidus. Clino. = clinoenstatite; Proto. = protoenstatite.

J. F. Sarver and F. A. Hummel, personal communication, Nov., 1961.

MgO–GeO$_2$–SiO$_2$

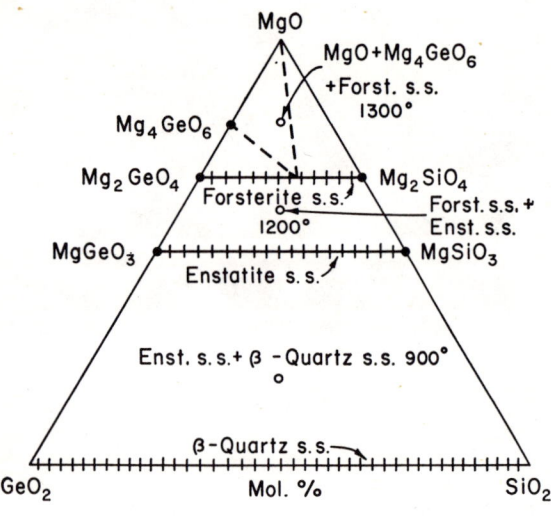

FIG. 717.—System MgO–GeO$_2$–SiO$_2$; partial subsolidus. Solid solutions indicated by hatched lines.

J. F. Sarver and F. A. Hummel, personal communication, Nov., 1961.

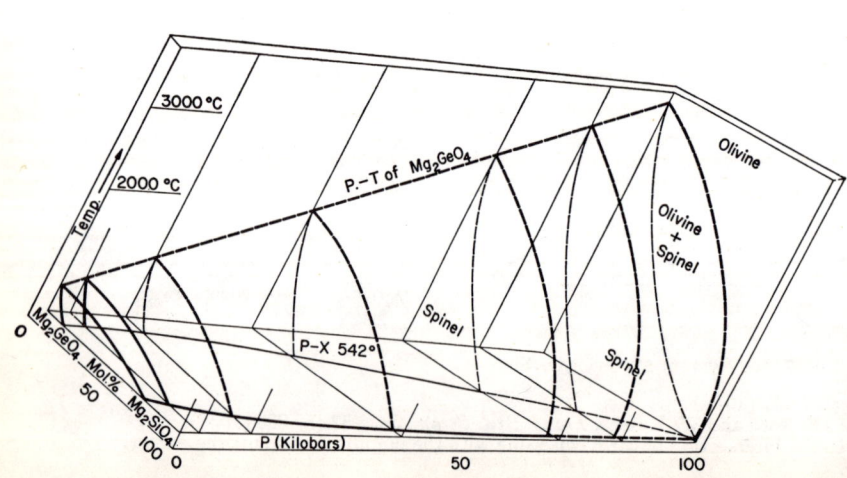

FIG. 719.—System Mg$_2$GeO$_4$–Mg$_2$SiO$_4$; pressure-temperature-composition perspective. Isothermal section at 542°C. (uniaxial pressure) provides basis of construction.

Frank Dachille and Rustum Roy, *Am. J. Sci.*, **258**, 236 (1960).

MgO–GeO$_2$–TiO$_2$

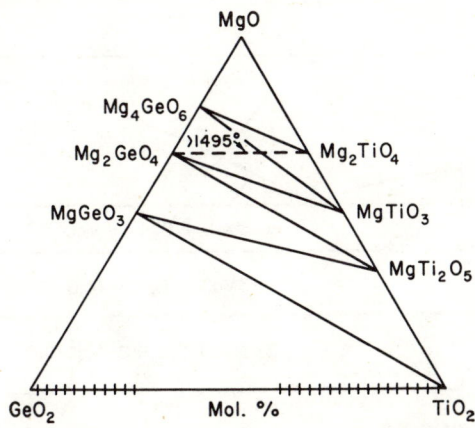

Fig. 720.—System MgO–GeO$_2$–TiO$_2$; compatibility relations. Solid solutions indicated by hatched lines.

J. F. Sarver and F. A. Hummel, personal communication, Nov., 1961.

MgO–HfO$_2$–ThO$_2$

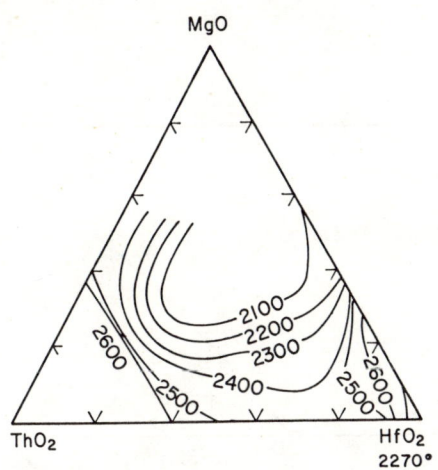

Fig. 721.—System MgO–HfO$_2$–ThO$_2$; apparent melting.

S. D. Mark, Jr., *J. Am. Ceram. Soc.* **42** [4] 208 (1959).

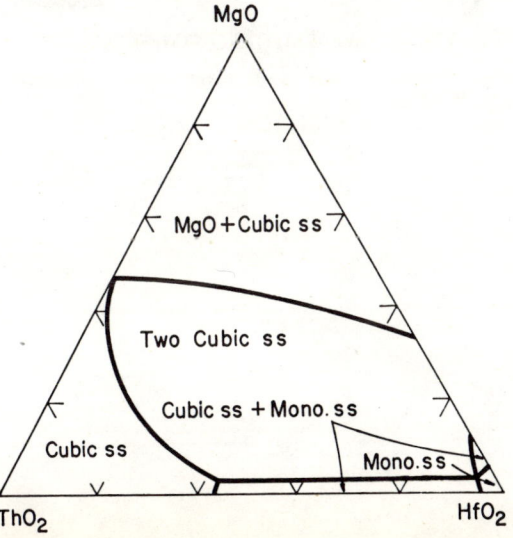

Fig. 722.—System MgO–HfO$_2$–ThO$_2$ at 1600°C.

S. D. Mark, Jr., *J. Am. Ceram. Soc.*, **42** [4] 208 (1959).

MgO–SiO$_2$–TiO$_2$

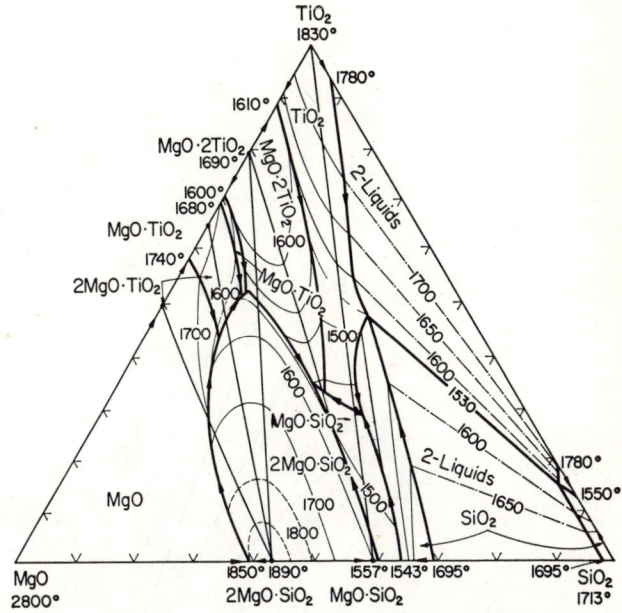

Fig. 723.—System MgO–SiO$_2$–TiO$_2$; proposed revision.

Franco Massazza and Efisia Sirchia, *Chim. Ind.* (*Milan*), **40**, 466 (1958).

MgO–SiO$_2$–ZrO$_2$

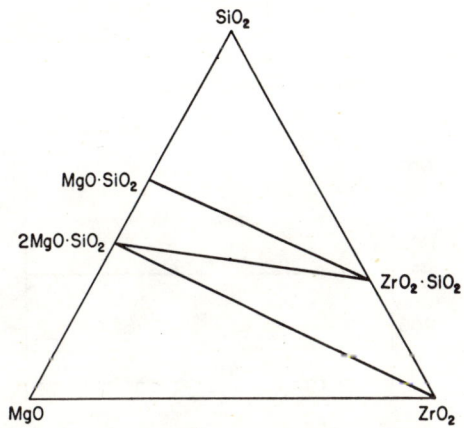

Fig. 724.—System MgO–SiO$_2$–ZrO$_2$; compatibility triangles.

Wilfrid R. Foster, *J. Am. Ceram. Soc.*, **34** [10] 303 (1951).

$MgO–SiO_2–P_2O_5$

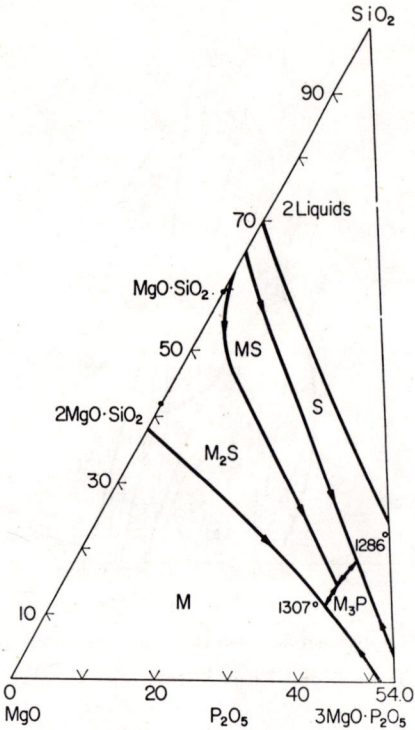

FIG. 725.—System $MgO–3MgO \cdot P_2O_5–SiO_2$.
M = MgO, P = P_2O_5, S = SiO_2.

Jadwiga Wojciechowska and Józef Berak, *Roczniki Chem.*, **33**, 27 (1959).

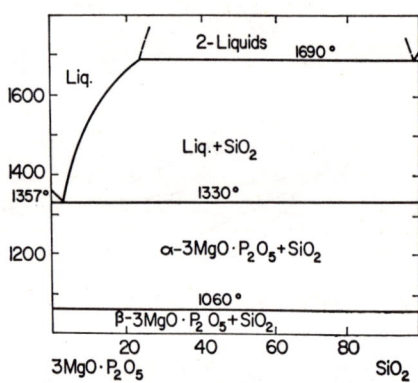

FIG. 726.—System $3MgO \cdot P_2O_5–SiO_2$.

Jadwiga Wojciechowska and Józef Berak, *Roczniki Chem.*, **33**, 24 (1959).

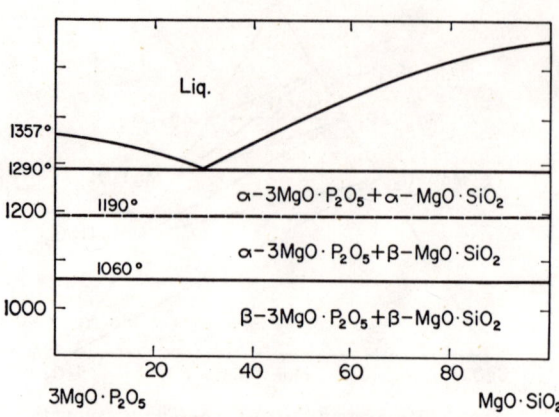

FIG. 727.—System $MgO \cdot SiO_2–3MgO \cdot P_2O_5$.

Jadwiga Wojciechowska and Józef Berak, *Roczniki Chem.*, **33**, 25 (1959).

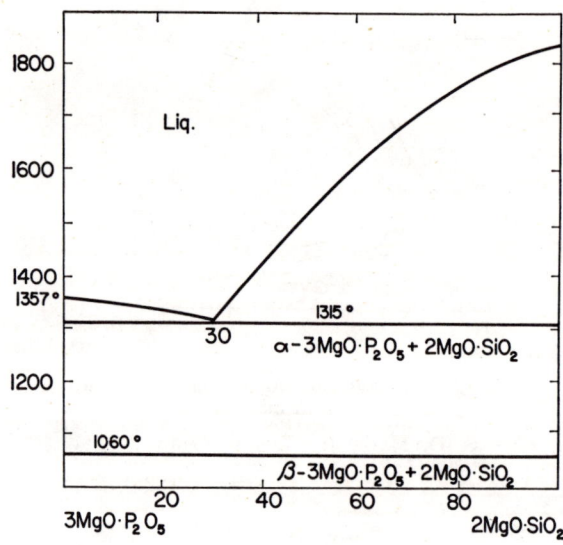

FIG. 728.—System $2MgO \cdot SiO_2–3MgO \cdot P_2O_5$.

Jadwiga Wojciechowska and Józef Berak, *Roczniki Chem.*, **33**, 26 (1959).

MgO–SnO$_2$–TiO$_2$

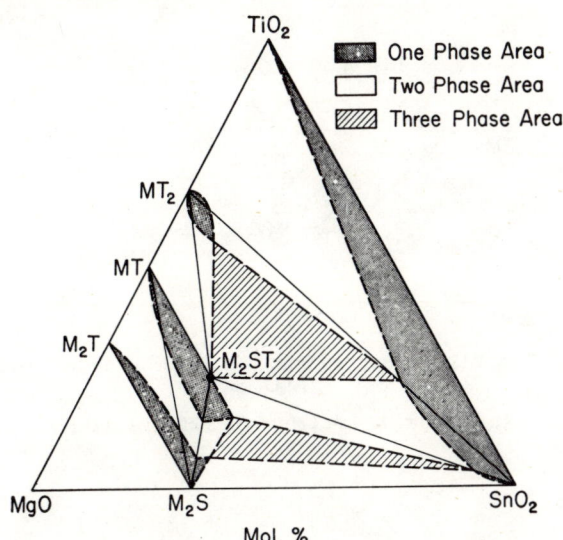

FIG. 729.—System MgO–SnO$_2$–TiO$_2$; compatibility relations at about 1550°C.; M = MgO; S = SnO$_2$; T = TiO$_2$.

L. W. Coughanour, R. S. Roth, S. Marzullo, and F. E. Sennett, *J. Research Natl. Bureau Standards*, **54** [3] 155 (1955); RP 2576.

MgO–TiO$_2$–ZrO$_2$

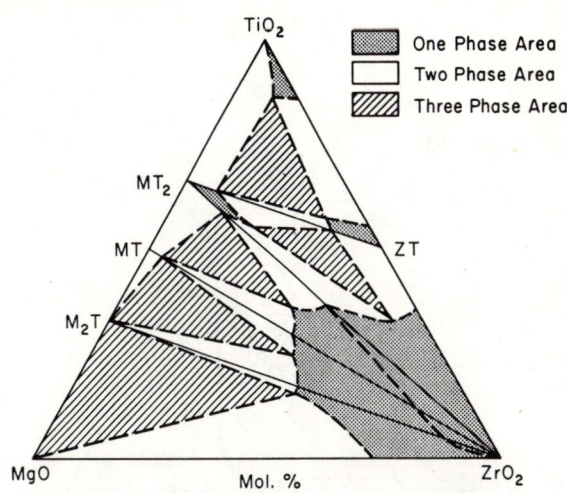

FIG. 731.—System MgO–TiO$_2$–ZrO$_2$; estimated solid solution in range 1400° to 1750°C. Dashed line near ZrO$_2$ apex indicates the approximate location of a two-phase area, which theoretically would separate a cubic ZrO$_2$ field, on the left, from a tetragonal ZrO$_2$ field, on the right. M = MgO; T = TiO$_2$; Z = ZrO$_2$.

L. W. Coughanour, R. S. Roth, S. Marzullo and F. E. Sennett, *J. Research Natl. Bur. Standards*, **54** [4] 193 (1955); RP 2580.

MgO–ThO$_2$–ZrO$_2$

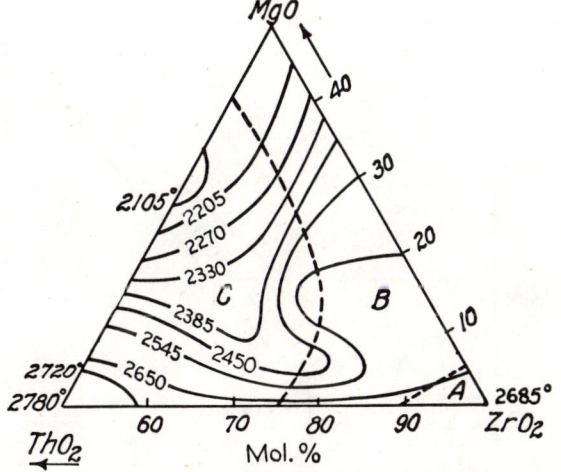

FIG. 730.—Melting isotherms of system MgO–ThO$_2$–ZrO$_2$; A = monoclinic ZrO$_2$, B = cubic mix crystal, C = two cubic mix crystals.

Otto Ruff, F. Ebert, and W. Loerpabel, *Z. anorg. u. all. gem. Chem.*, **207**, 309 (1932).

MnO–ZnO–P$_2$O$_5$

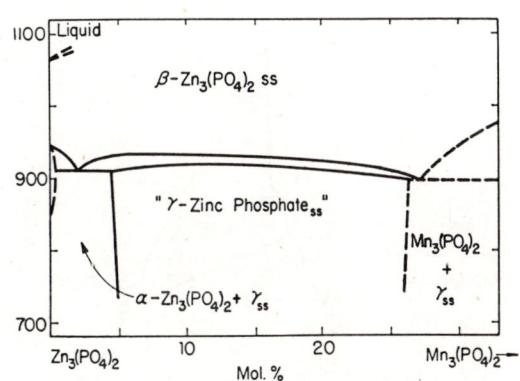

FIG. 732.—System Mn$_3$(PO$_4$)$_2$–Zn$_3$(PO$_4$)$_2$.

F. A. Hummel, *J. Electrochem. Soc.*, **105**, 530 (1958).

MnO–Al$_2$O$_3$–SiO$_2$

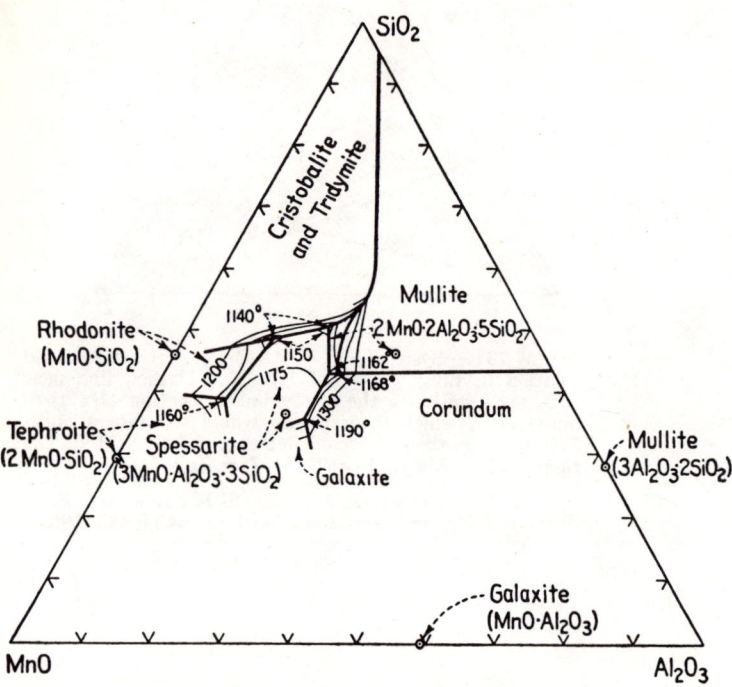

FIG. 733.—Isotherms in part of system MnO–Al$_2$O$_3$–SiO$_2$.

R. B. Snow, *J. Am. Ceram. Soc.*, 26 [1] 15 (1943).

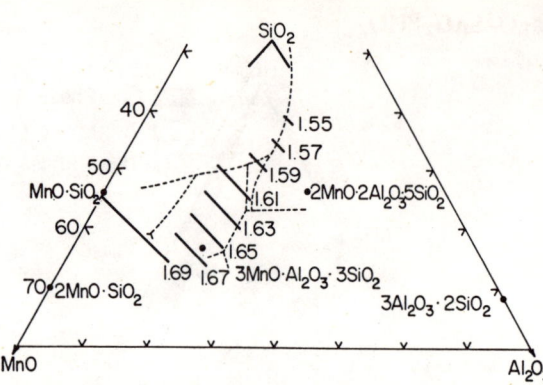

FIG. 735.—Refractive indices of glasses in system MnO–Al$_2$O$_3$–SiO$_2$.

R. B. Snow, *J. Am. Ceram. Soc.*, 26 [1] 15 (1943).

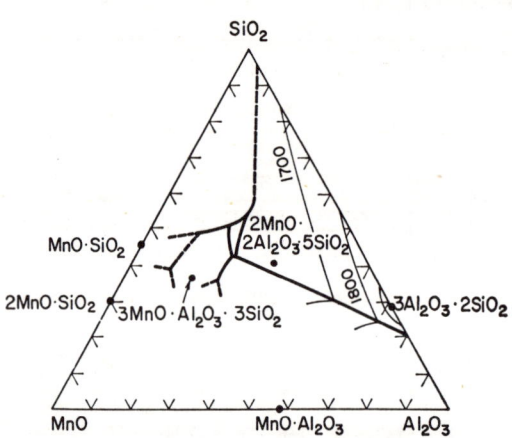

FIG. 734.—System MnO–Al$_2$O$_3$–SiO$_2$; alumina corner.

F. Ya. Galakhov, *Izvest. Akad. Nauk S.S.S.R., Otdel. Khim. Nauk*, No. 5, 530 (1957).

MnO–SiO$_2$–TiO$_2$

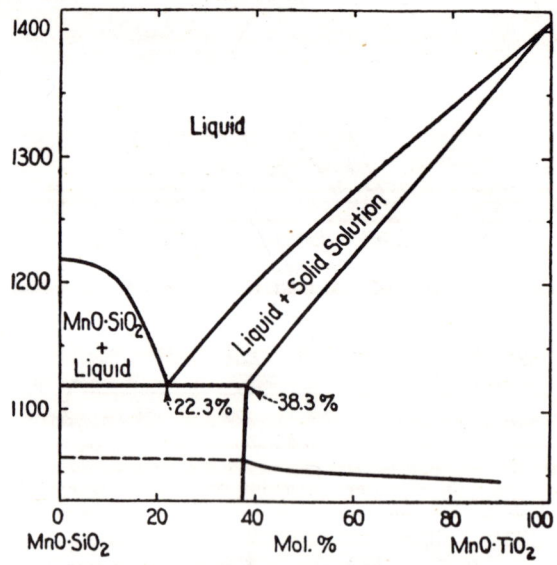

FIG. 736.—System MnO·SiO$_2$–MnO·TiO$_2$.

S. Smolensky, *Z. anorg. Chem.*, 73, 299 (1912).

PbO–Al$_2$O$_3$–SiO$_2$

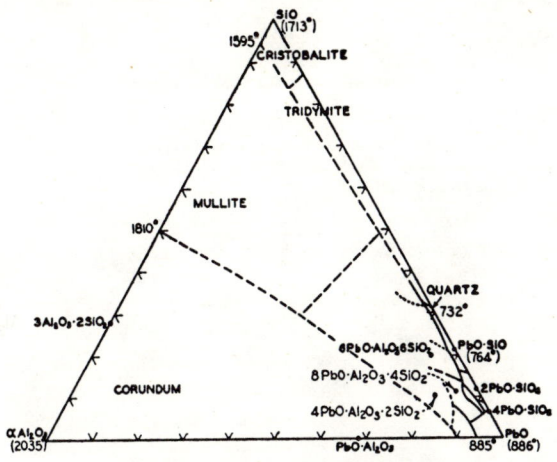

FIG. 737.—System PbO–Al$_2$O$_3$–SiO$_2$.

R. F. Geller and E. N. Bunting, *J. Research Natl. Bur. Standards*, **31**, 261 (1943); R P 1564.

SiO at top vertex of diagram should read SiO$_2$.

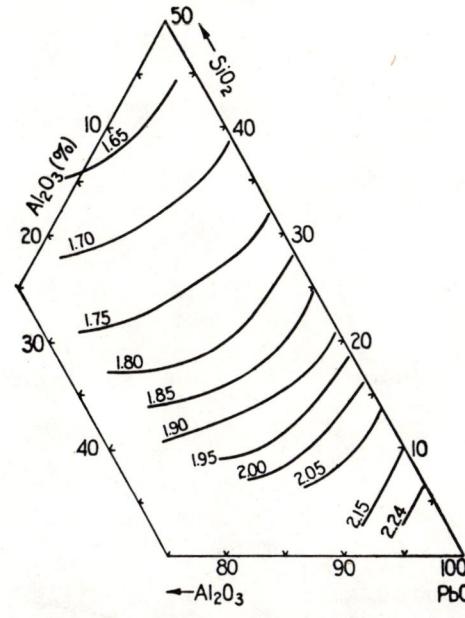

FIG. 739.—Indices of refraction of glasses in system PbO–Al$_2$O$_3$–SiO$_2$.

R. F. Geller and E. N. Bunting, *J. Research Natl. Bur. Standards*, **31**, 268 (1943); R P 1564.

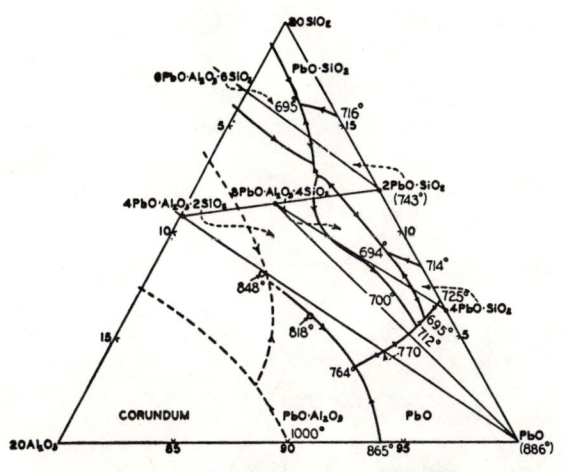

FIG. 738.—System PbO–Al$_2$O$_3$–SiO$_2$; high-PbO portion.

R. F. Geller and E. N. Bunting, *J. Research Natl. Bur. Standards* **31**, 264 (1943); R P 1564.

PbO–B$_2$O$_3$–SiO$_2$

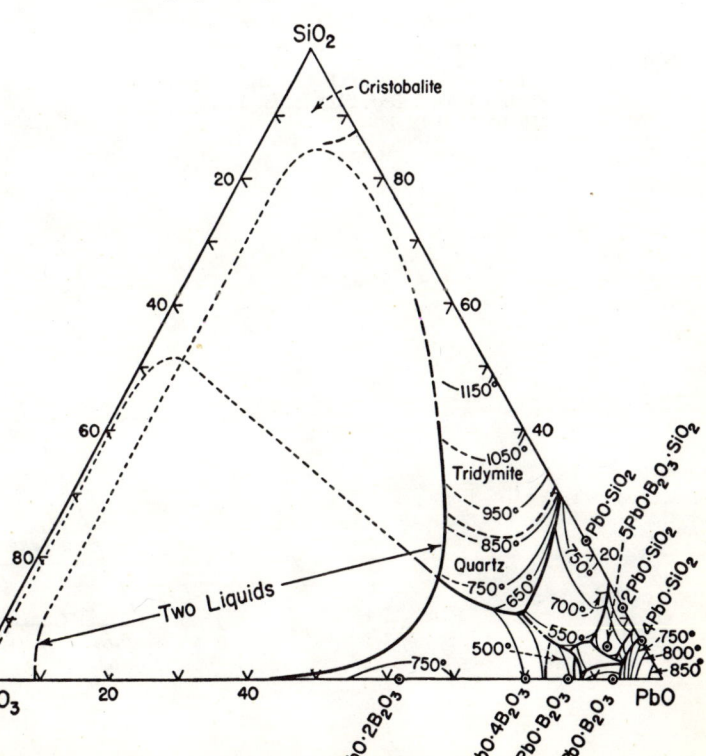

FIG. 740.—System PbO–B$_2$O$_3$–SiO$_2$.

R. F. Geller and E. N. Bunting, *J. Research Natl. Bur. Standards*, **23** [8] 279 (1939); RP 123.

PbO–B$_2$O$_3$–SiO$_2$ (concl.)

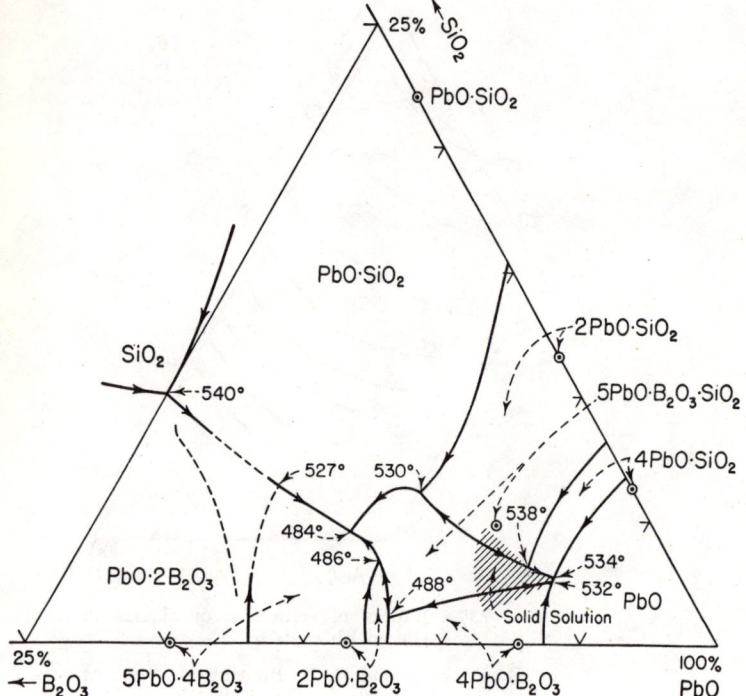

Fig. 741.—System PbO–B$_2$O$_3$–SiO$_2$ above 75% PbO.

R. F. Geller and E. N. Bunting, J. Research Natl. Bur. Standards, **23** [8] 281 (1939); RP 1231.

Fig. 742.—System PbO–B$_2$O$_3$–SiO$_2$; average refractive indices of glasses.

R. F. Geller and E. N. Bunting, J. Research Natl. Bur. Standards, **23** [8] 279 (1939); RP 1231.

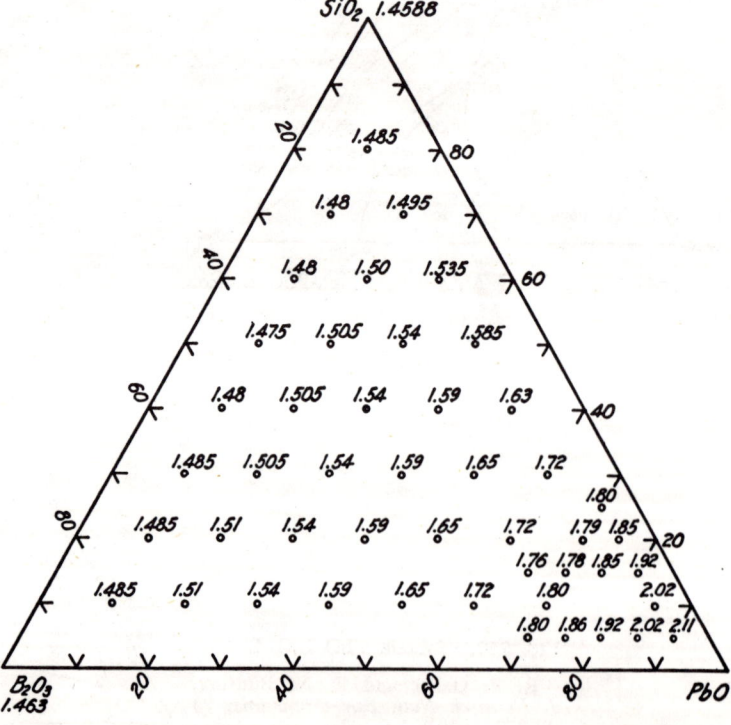

PbO–B₂O₃–TiO₂

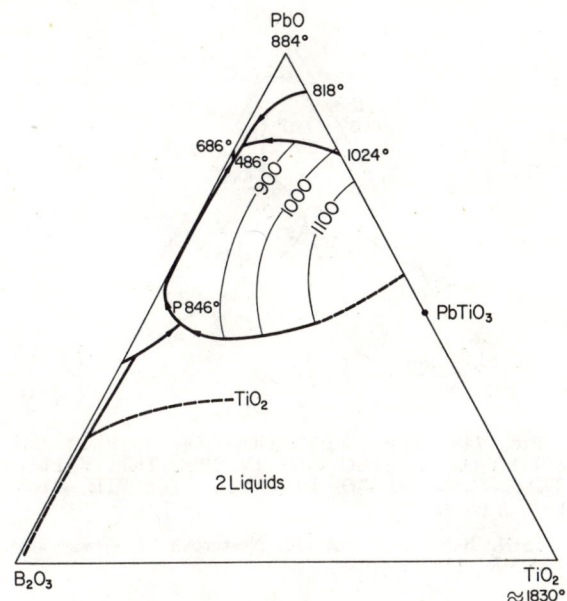

Fig. 743.—System PbO–B₂O₃–TiO₂.

M. L. Sholokhovich, *Zhur. Neorg. Khim.*, **3**, 1215 (1958).

PbO–Bi₂O₃–MoO₃

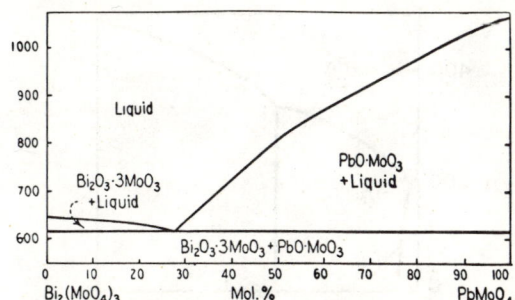

Fig. 744.—System PbO·MoO₃–Bi₂O₃·3MoO₃.

F. Zambonini, *Gazz. chim. ital.*, **50** [II] 141 (1920).

PbO–Bi₂O₃–WO₃

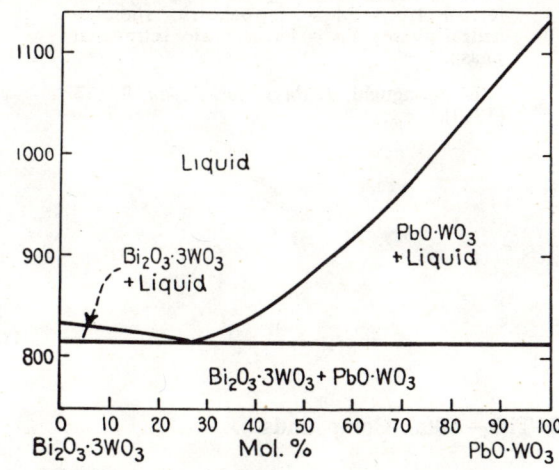

Fig. 745.—System PbO·WO₃–Bi₂O₃·3WO₃.

F. Zambonini, *Gazz. chim. ital.*, **50** [II] 139 (1920).

PbO–SiO₂–P₂O₅

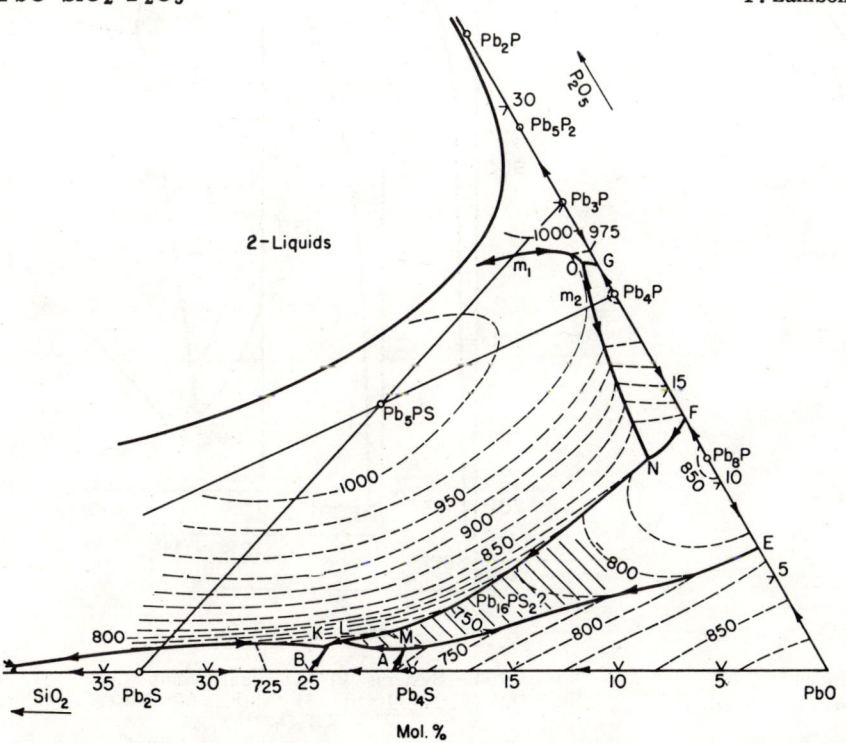

Fig. 746.—System PbO–SiO₂–P₂O₅; lead-rich corner. Pb = PbO; P = P₂O₅; S = SiO₂. Binary and ternary invariant points, °C.: A = 725; B = 710; E = 815; F = 838; G = 954; K = 695; L = 700; M = 710; N = 828; O = 948; m_1 = 985°C.; m_2 = 973°C.

H. H. Paetsch and Adolf Dietzel, *Glastech. Ber.*, **29** [9] 350 (1956).

PbO–TiO$_2$–ZrO$_2$

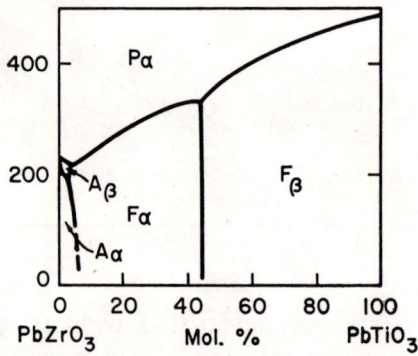

Fig. 747.—System PbTiO$_3$–PbZrO$_3$. P = paraelectric, cubic phase; A$_\alpha$ = antiferroelectric, orthorhombic phase; A$_\beta$ = antiferroelectric; F$_\alpha$ = ferroelectric, rhombohedral phase; F$_\beta$ = ferroelectric, tetragonal phase.

E. Sawaguchi, *J. Phys. Soc. Japan*, **8**, 615 (1953).

PbO–TiO$_2$–V$_2$O$_5$

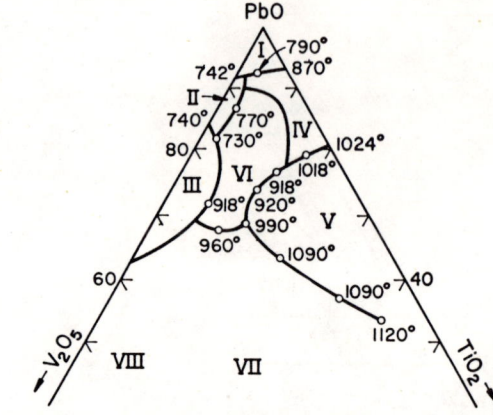

Fig. 748.—System PbO–TiO$_2$–V$_2$O$_5$. I, PbO; II. 8PbO·V$_2$O$_5$; III, 3PbO·V$_2$O$_5$; IV, 2PbO·TiO$_2$; V, PbO·TiO$_2$; VI, 10PbO·V$_2$O$_5$·TiO$_2$; VII, TiO$_2$; VIII, glassy-looking phase.

I. N. Belyaev and A. K. Nesterova, *J. Gen. Chem. U.S.S.R.*, **22**, 402 (1952).

PbO–TiO$_2$–Alkali Compounds

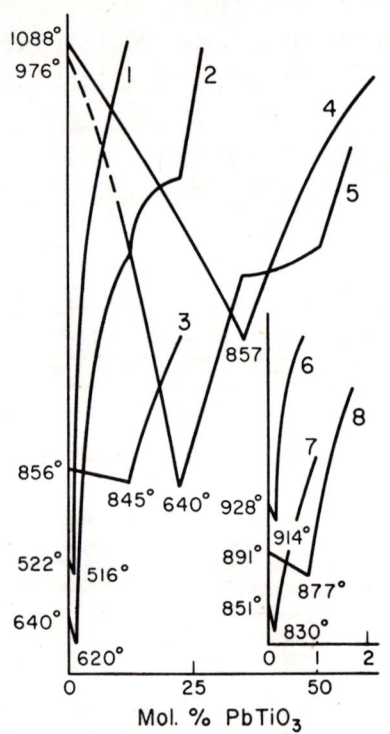

Fig. 749.—Systems PbTiO$_3$–alkali compound: (1) (KVO$_3$)$_2$; (2) (NaVO$_3$)$_2$; (3) K$_2$F$_2$; (4) Na$_2$SiO$_3$; (5) K$_2$SiO$_3$; (6) K$_2$MoO$_4$; (7) Na$_2$CO$_3$; (8) K$_2$CO$_3$.

I. N. Belyaev, M. L. Sholokhovich, and G. V. Barkova, *Zhur. Obshcheĭ Khim.*, **24**, 212 (1954).

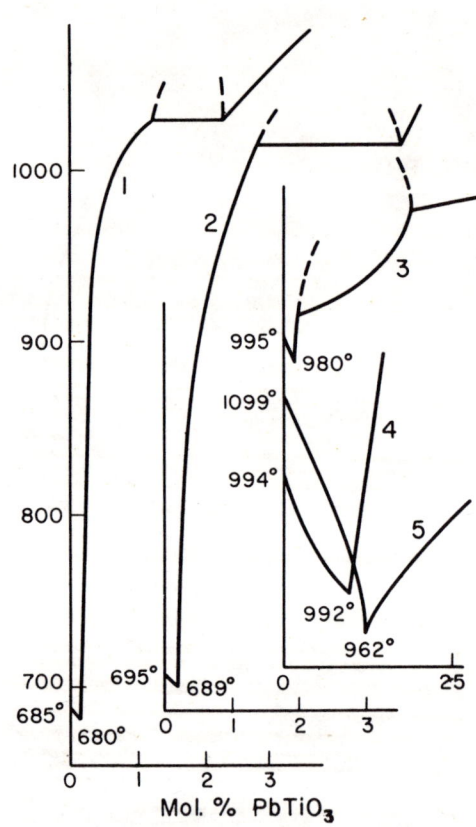

Fig. 750.—Systems PbTiO$_3$–alkali compound: (1) Na$_2$WO$_4$; (2) Na$_2$MoO$_4$; (3) Na$_2$F$_2$; (4) Na$_4$P$_2$O$_7$; (5) K$_4$P$_2$O$_7$.

I. N. Belyaev, M. L. Sholokhovich, and G. V. Barkova, *Zhur. Obshcheĭ Khim.*, **24**, 212 (1954).

PbO–CrO$_3$–MoO$_3$

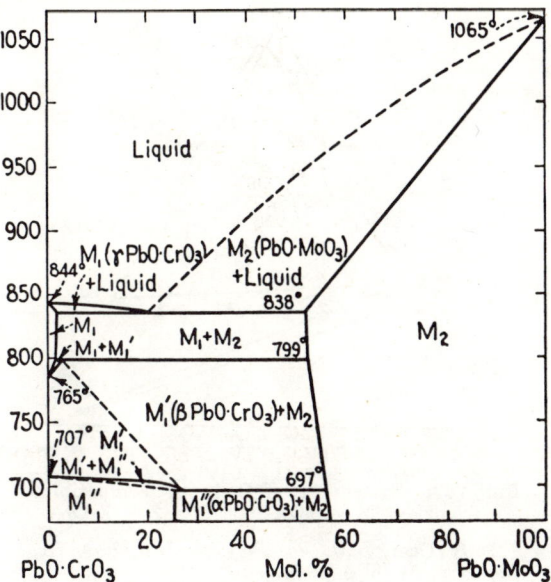

Fig. 751.—System PbO·CrO$_3$–PbO·MoO$_3$.

F. M. Jaeger and H. C. Germs, *Z. anorg. u. allgem. Chem.*, **119**, 169 (1921).

PbO–CrO$_3$–WO$_3$

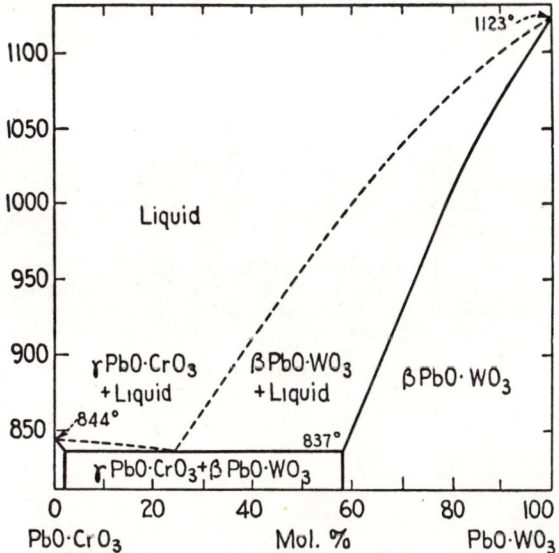

Fig. 752.—System PbO·CrO$_3$–PbO·WO$_3$.

F. M. Jaeger and H. C. Germs, *Z. anorg. u. allgem. Chem.*, **119**, 171 (1921).

PbO–MoO$_3$–WO$_3$

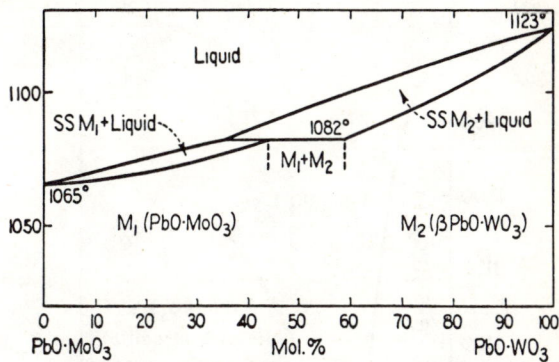

Fig. 753.—System PbO·MoO$_3$–PbO·WO$_3$.

F. M. Jaeger and H. C. Germs, *Z. anorg. u. allgem. Chem.*, **119**, 172 (1921).

SrO–ZnO–P$_2$O$_5$

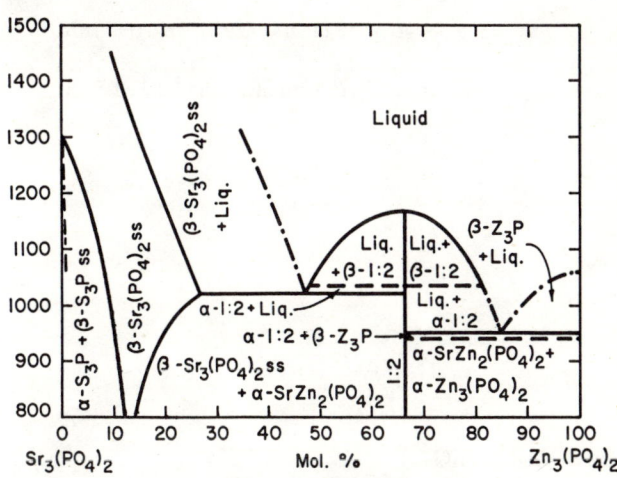

Fig. 754.—System Sr$_3$(PO$_4$)$_2$–Zn$_3$(PO$_4$)$_2$. S$_3$P = Sr$_3$(PO$_4$)$_2$; Z$_3$P = Zn$_3$(PO$_4$)$_2$.

J. F. Sarver, F. A. Hummel, and M. V. Hoffman, *J. Electrochem. Soc.*, **108** [12] 1106 (1961).

SrO–Al$_2$O$_3$–SiO$_2$

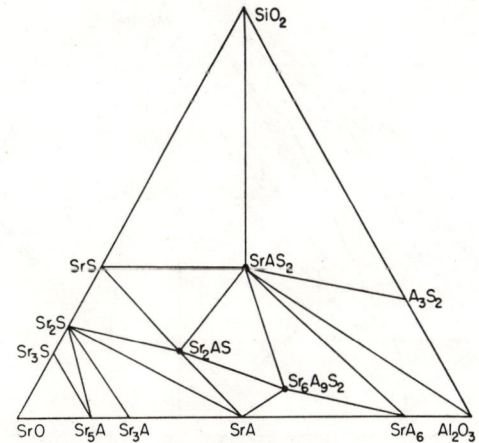

Fig. 755.—System SrO–Al$_2$O$_3$–SiO$_2$, subsolidus compatibility relationships at 1350°C. Sr = SrO; A = Al$_2$O$_3$; S = SiO$_2$.

P. S. Dear, *Bull. Virginia Polytechnic Inst.*, **50** [11] 8 (1957).

SrO–Al$_2$O$_3$–P$_2$O$_5$

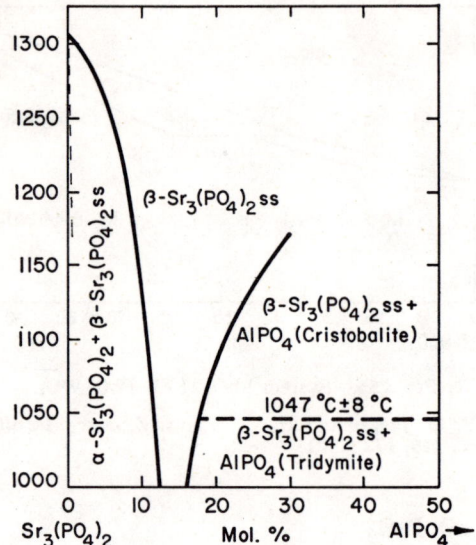

Fig. 756.—System Sr$_3$(PO$_4$)$_2$–AlPO$_4$; partial sub-solidus.

J. F. Sarver and F. A. Hummel, unpublished data.

SrO–SiO$_2$–ZrO$_2$

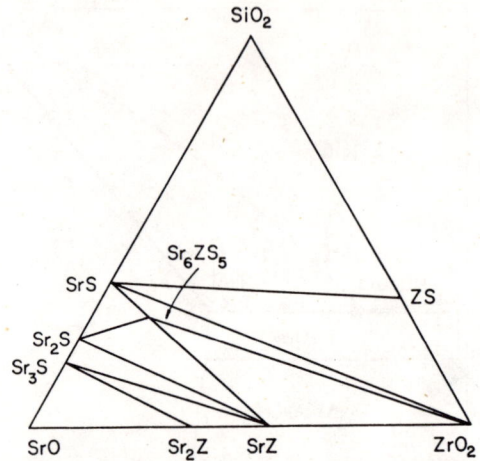

Fig. 757.—System SrO–SiO$_2$–ZrO$_2$; compatibility triangles. S = SiO$_2$, Sr = SrO, Z = ZrO$_2$.

P. S. Dear, *Bull. Va. Polytech. Inst. Eng. Exp. Sta. Ser.*, **51** [8] 6 (1958).

ZnO–B$_2$O$_3$–SiO$_2$

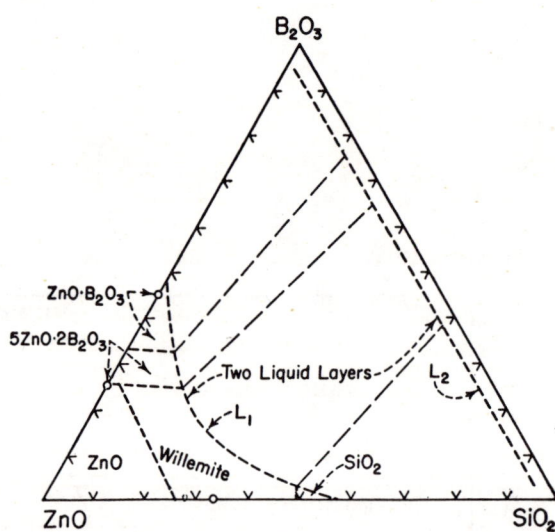

Fig. 759.—System ZnO–B$_2$O$_3$–SiO$_2$. The curve representing the composition of the lighter liquid layer, L_2, is displaced from its proper position; L_2 probably contains less than 1% ZnO.

E. Ingerson, G. W. Morey, and O. F. Tuttle, *Am. J. Sci.*, **246** [1] 39 (1948).

ZnO–Al$_2$O$_3$–SiO$_2$

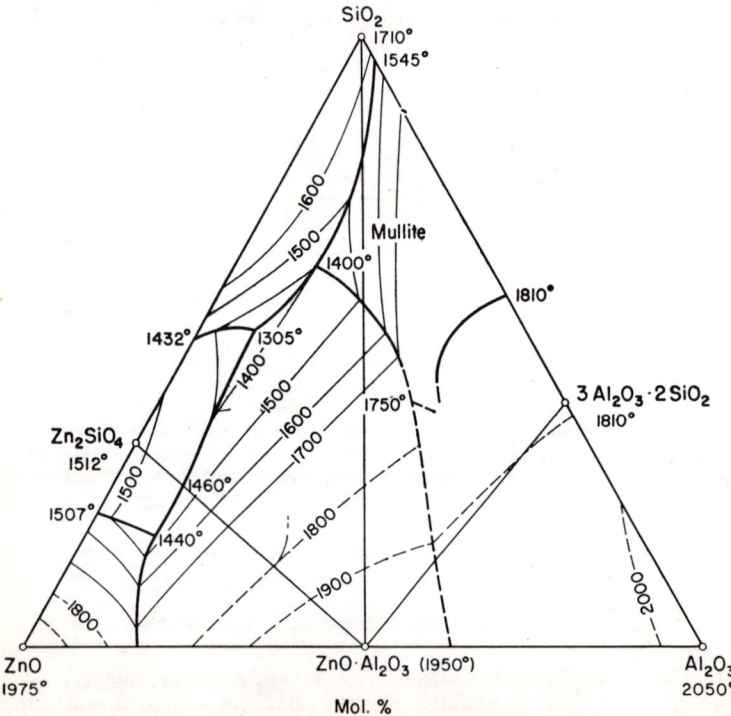

Fig. 758.—System ZnO–Al$_2$O$_3$–SiO$_2$.

E. N. Bunting, *J. Research Natl. Bur. Standards*, **8** [2] 286 (1932); RP 413.

Fe_3O_4–Cr_2O_3–SiO_2

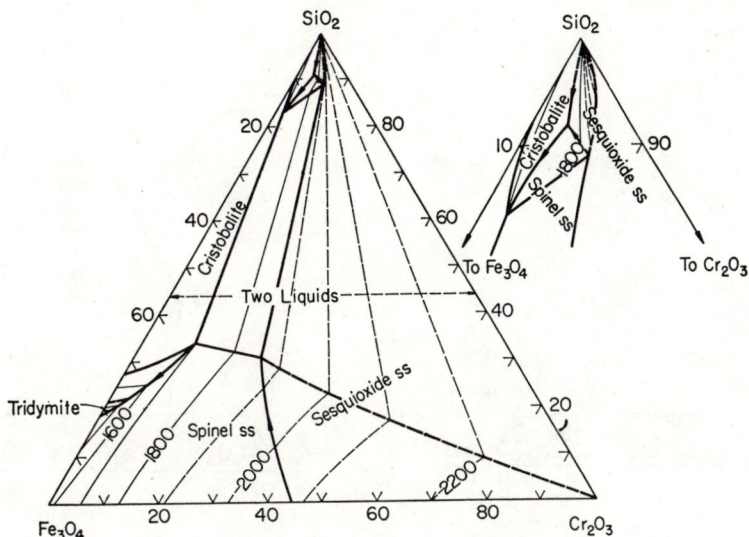

FIG. 760.—System Fe_3O_4–Cr_2O_3–SiO_2 in air.

Arnulf Muan and Shigeyuki Sōmiya, *J. Am. Ceram. Soc.*, **43** [10] 533 (1960).

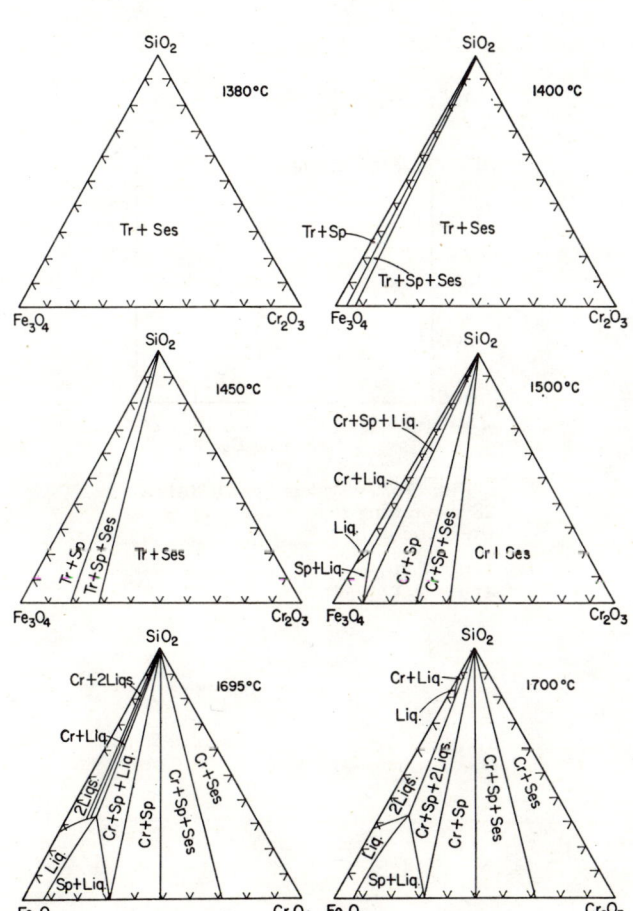

FIG. 761.—System Fe_3O_4–Cr_2O_3–SiO_2; isothermal sections (1380° to 1700°C.), in air. Cr. = cristobalite; Ses. = sesquioxide; Sp. = spinel; Tr. = tridymite.

Arnulf Muan and Shigeyuki Sōmiya, *J. Am. Ceram. Soc.*, **43** [10] 538 (1960).

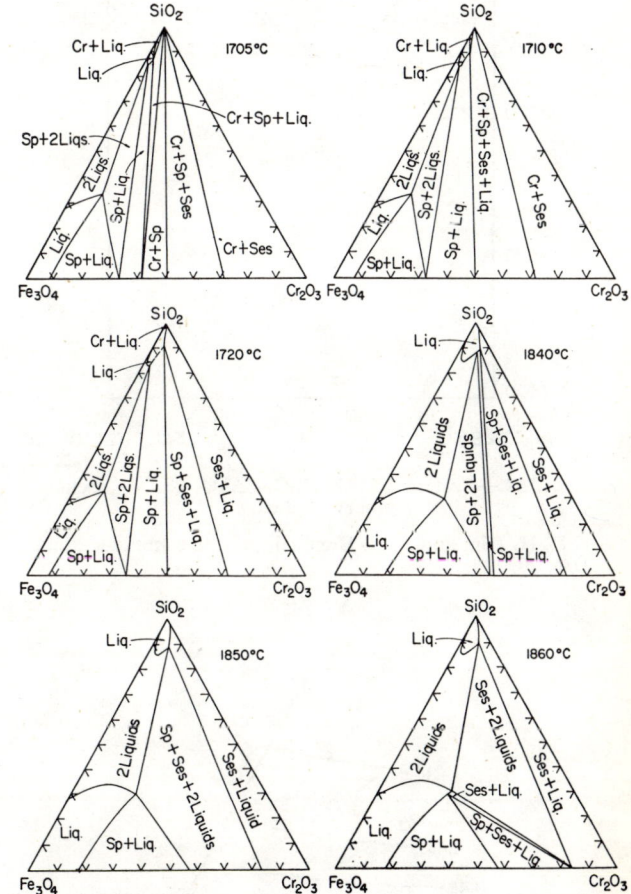

FIG. 762.—System Fe_3O_4–Cr_2O_3–SiO_2; isothermal sections (1705° to 1860°C.), in air. Cr. = cristobalite; Ses. = sesquioxide; Sp. = spinel.

Arnulf Muan and Shigeyuki Sōmiya, *J. Am. Ceram. Soc.*, **43** [10] 539 (1960).

Al_2O_3–B_2O_3–SiO_2

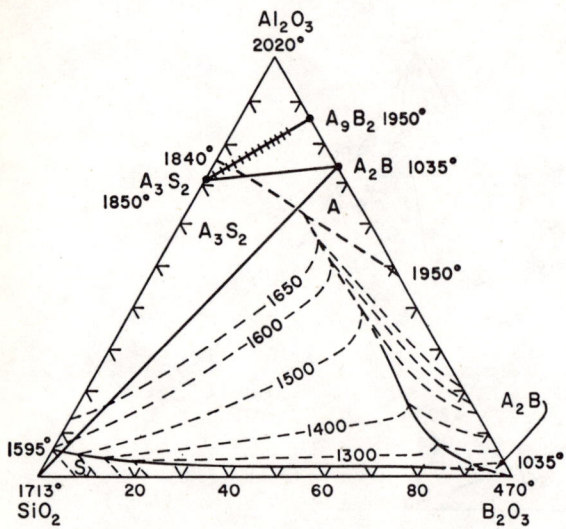

FIG. 763.—System Al_2O_3–B_2O_3–SiO_2; preliminary. A = Al_2O_3, B = B_2O_3, S = SiO_2.

P. J. Gielisse and W. R. Foster, *Quart. Progr. Rept.*, 931-8, The Ohio State Univ. Res. Foundation, p. 6, Oct. (1961).

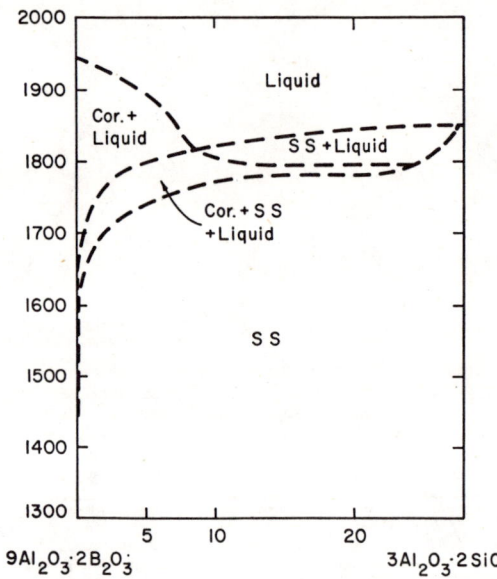

FIG. 764.—System $9Al_2O_3 \cdot 2B_2O_3$–$3Al_2O_3 \cdot 2SiO_2$; pseudobinary, tentative.

K. H. Kim and F. A. Hummel, private communication, Dec. 20, 1961.

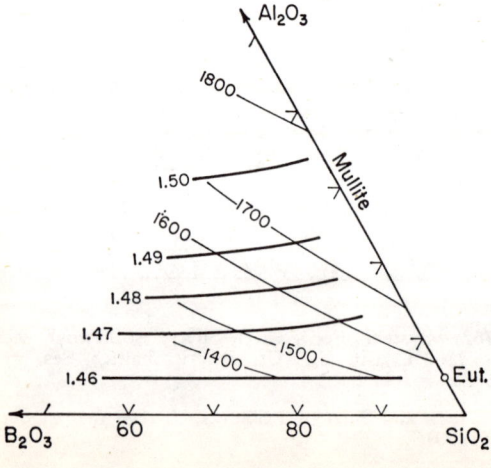

FIG. 765.—System Al_2O_3–B_2O_3–SiO_2. Heavy lines = isofracts at 20°C; light lines = liquidus isotherms.

Adolf Dietzel and Horst Scholze, *Glastech. Ber.*, **28** [2] 48 (1955).

Al_2O_3–Fe_2O_3–SiO_2

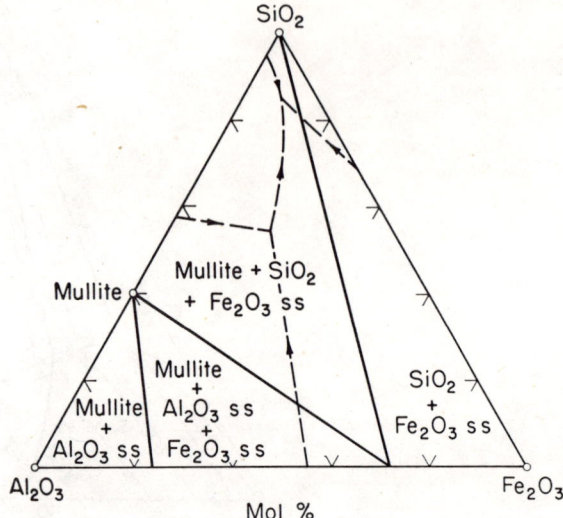

FIG. 766.—System Al_2O_3–Fe_2O_3–SiO_2; equilibrium phases at 1000°C. The small two-phase field mullite + Fe_2O_3 solid solution is not shown.

H. Nowotny and R. Funk, *Radex-Rundschau*, No. 8 337 (1951).

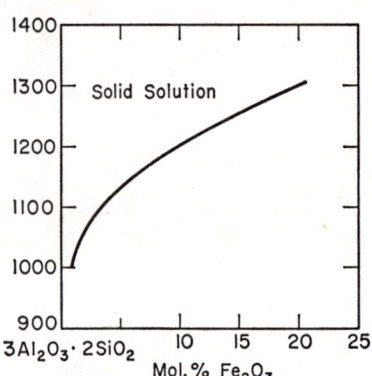

FIG. 767.—System Fe_2O_3–$3Al_2O_3 \cdot 2SiO_2$ showing solid solution of hematite in mullite.

W. E. Brownell, *J. Am. Ceram. Soc.*, **41** [6] 228 (1958).

Al_2O_3–SiO_2–TiO_2

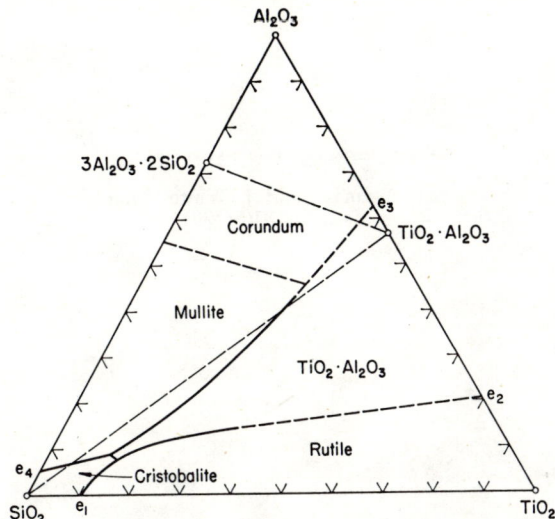

FIG. 768.—Al_2O_3–SiO_2–TiO_2; primary fields.

Y. M. Agamawi and J. White, *Trans. Brit. Ceram. Soc.*, **51** 312 (1951–52).

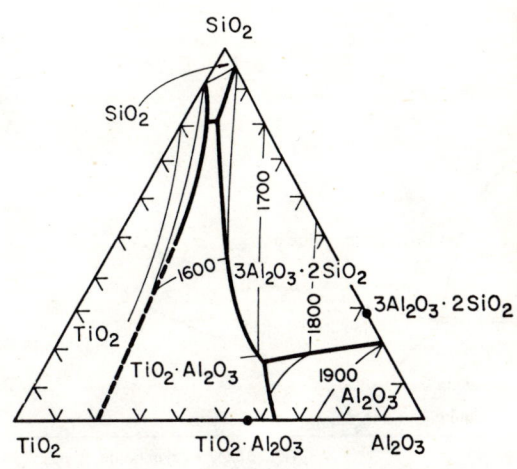

FIG. 769.—System Al_2O_3–SiO_2–TiO_2; alumina corner.

F. Ya. Galakhov, *Izvest. Akad. Nauk S.S.S.R., Otdel. Khim. Nauk*, 533 (1958).

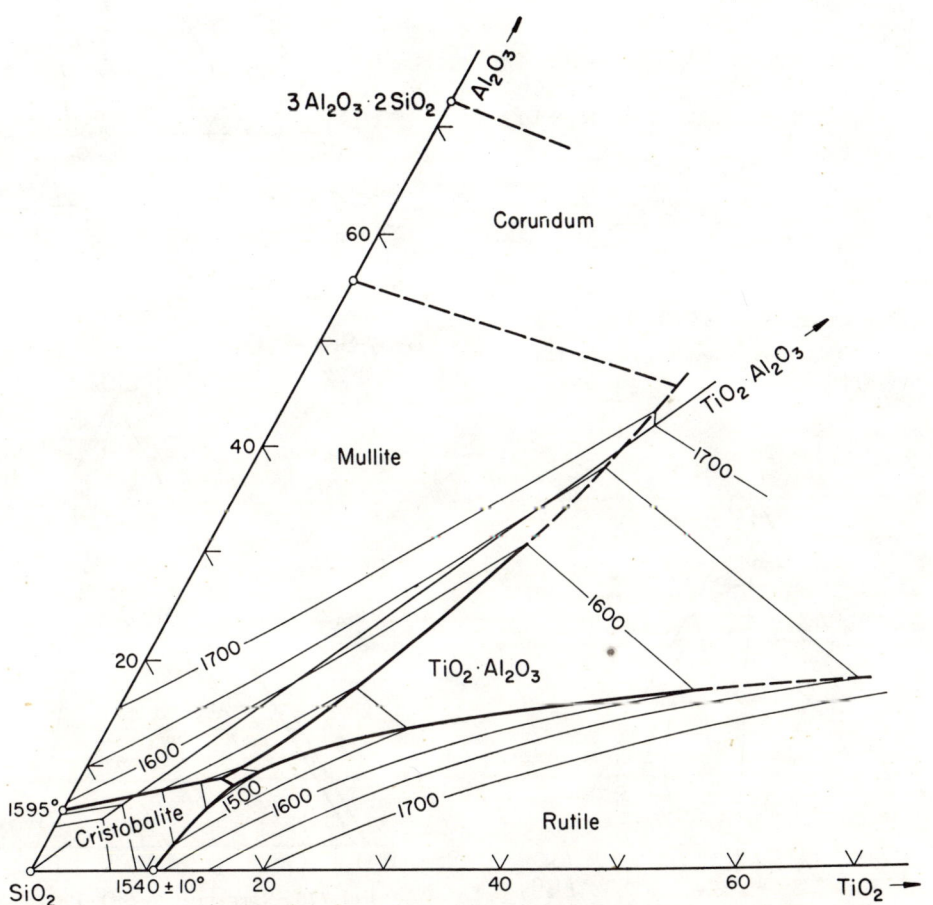

FIG. 770.—System Al_2O_3–SiO_2–TiO_2; high silica region.

Y. M. Agamawi and J. White, *Trans. Brit. Ceram. Soc.*, **51**, 319 (1951–52).

Al_2O_3–SiO_2–TiO_2 (concl.)

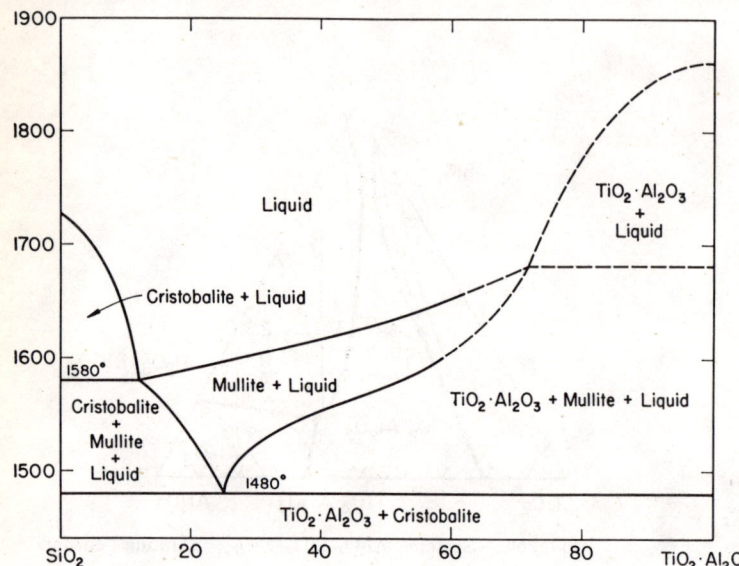

Fig. 771.—System SiO_2–Al_2O_3·TiO_2.

Y. M. Agamawi and J. White, *Trans. Brit. Ceram. Soc.*, **51**, 310 (1951–52).

Al_2O_3–SiO_2–ZrO_2

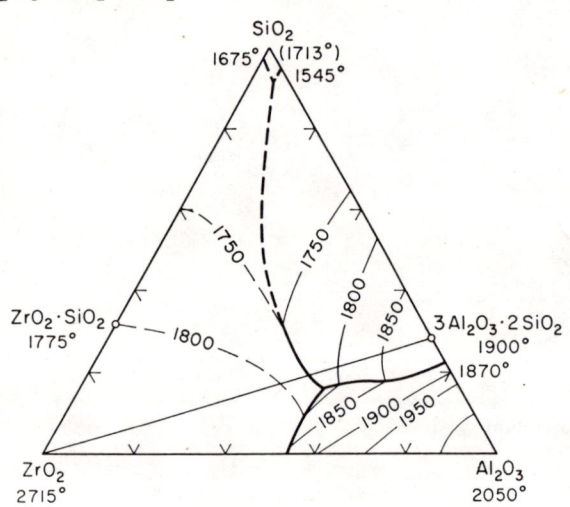

Fig. 772.—System Al_2O_3–SiO_2–ZrO_2.

P. P. Budnikov and A. A. Litvakovskiĭ, *Doklady Akad. Nauk S.S.S.R.*, **106**, 268 (1956).

Al_2O_3–TiO_2–ZrO_2

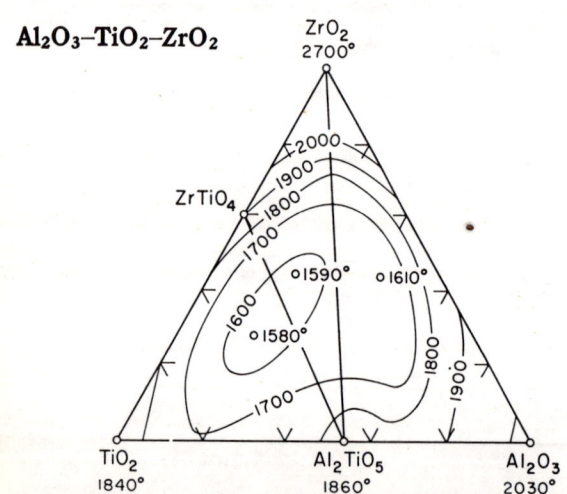

Fig. 773.—System Al_2O_3–TiO_2–ZrO_2; melting isotherms.

A. S. Berezhnoĭ and N. V. Gul'ko, *Dopovidi Akad. Nauk Ukr. R.S.R.*, **1955** [1] 78.

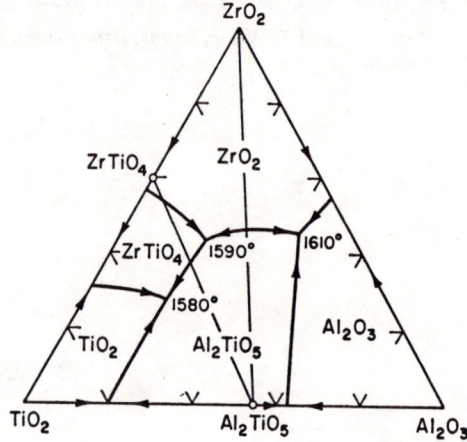

Fig. 774.—System Al_2O_3–TiO_2–ZrO_2; primary phases.

A. S. Berezhnoĭ and N. V. Gul'ko, *Dopovidi Akad. Nauk Ukr. R.S.R.*, **1955** [1] 78.

B_2O_3–SiO_2–P_2O_5

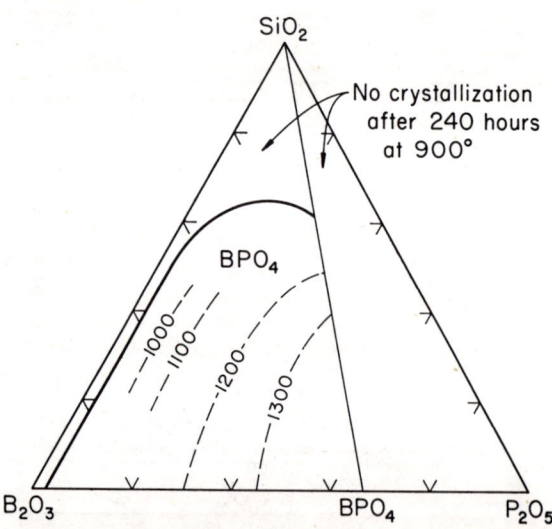

Fig. 775.—System B_2O_3–SiO_2–P_2O_5. Not true ternary at atmospheric pressure and temperature below 1400°C. because of retained water.

W. J. Englert and F. A. Hummel, *J. Soc. Glass Technol.*, **39**, 126T (1955).

B_2O_3–SiO_2–P_2O_5 (concl.)

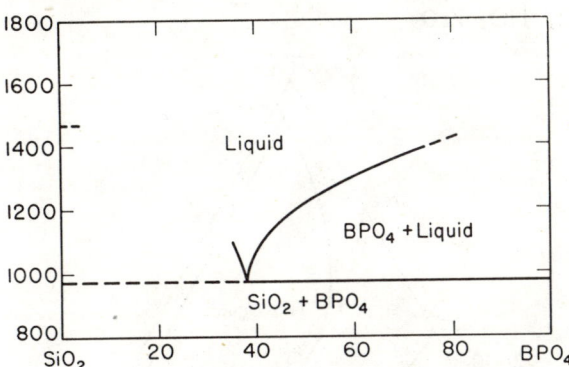

Fig. 776.—System BPO_4–SiO_2.

W. F. Horn and F. A. Hummel, *J. Soc. Glass Technol.*, **39**, 118T (1955).

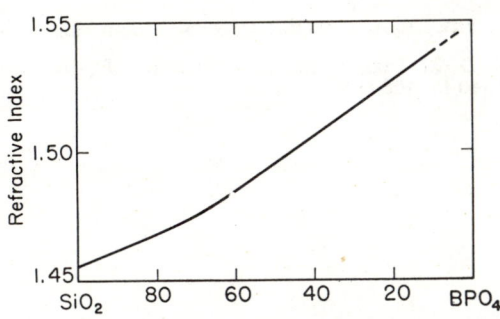

Fig. 777.—System BPO_4–SiO_2, isofracts.

W. F. Horn and F. A. Hummel, *J. Soc. Glass Technol.*, **39**, 117T (1955).

Cr_2O_3–Fe_2O_3–TiO_2

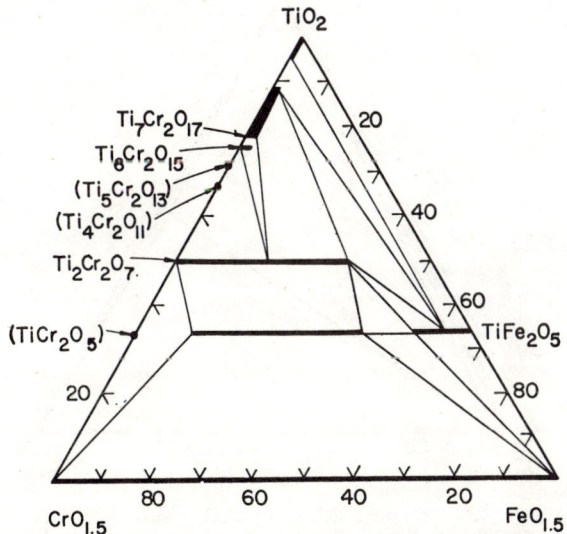

Fig. 778.—System Cr_2O_3–Fe_2O_3–TiO_2 at 1300°C. Formulas in brackets refer to compounds not stable at 1300°C.

W. Kwestroo and A. Roos, *J. Inorg. Nucl. Chem.*, **13**, 325 (1960).

Fe_2O_3–Gd_2O_3–Nb_2O_5

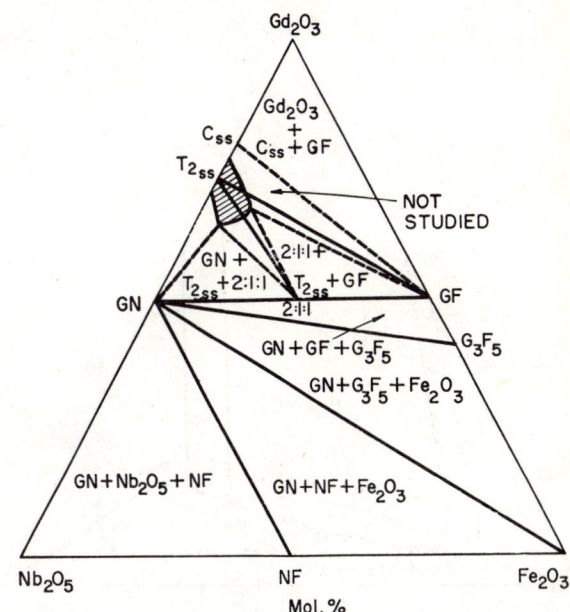

Fig. 779.—System Fe_2O_3–Gd_2O_3–Nb_2O_5; preliminary subsolidus (approx. 1400°C.). C = cubic (fluorite structure type), T_2 = distorted fluorite structure type, F = Fe_2O_3, G = Gd_2O_3, N = Nb_2O_5. Shaded area = solid solutions.

R. S. Roth, "Rare Earth Research," edited by Eugene V. Kleber, The Macmillan Co., New York, p. 92 (1961).

Ga_2O_3–SiO_2–P_2O_5

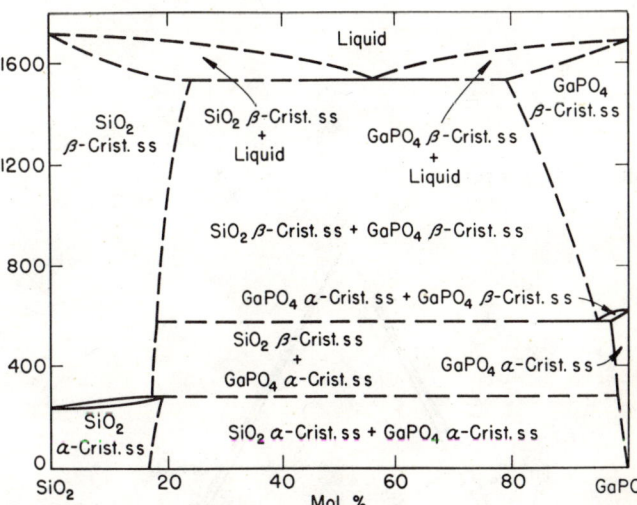

Fig. 780.—System $GaPO_4$–SiO_2; partly metastable. Solid lines = experimentally determined portions. Dashed lines = probable relationships.

E. C. Shafer and Rustum Roy, *J. Am. Ceram. Soc.*, **39** [10] 336 (1956).

La_2O_3–Lu_2O_3–Sm_2O_3

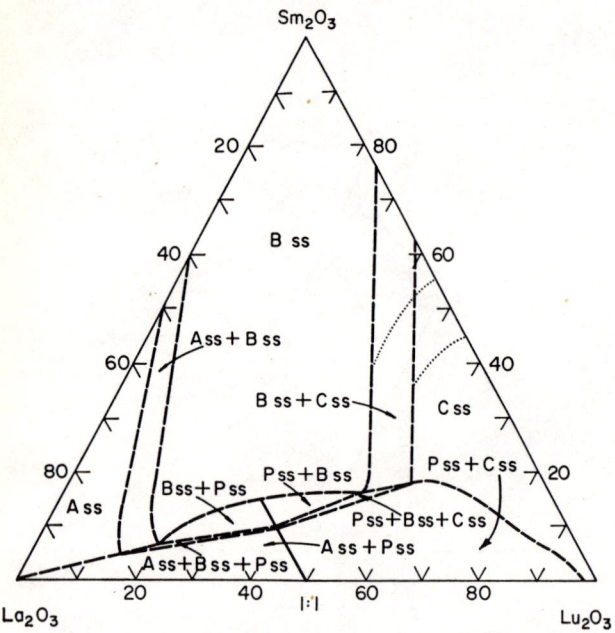

Fig. 781.—System La_2O_3–Lu_2O_3–Sm_2O_3; predicted subsolidus at approx. 1650°. A, B, C refer to rare earth oxide structure types; P = perovskite.

S. J. Schneider and R. S. Roth, *J. Research Natl. Bur. Standards*, **64A** [4] 330 (1960).

SiO_2–TiO_2–ZrO_2

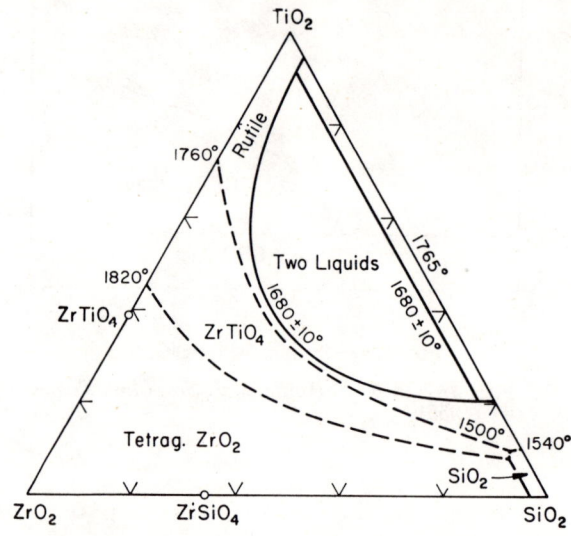

Fig. 783.—System SiO_2–TiO_2–ZrO_2; immiscibility area

G. D. McTaggart and A. I. Andrews, *J. Am. Ceram. Soc.*, **40** [5] 168 (1957).

SiO_2–ThO_2–ZrO_2

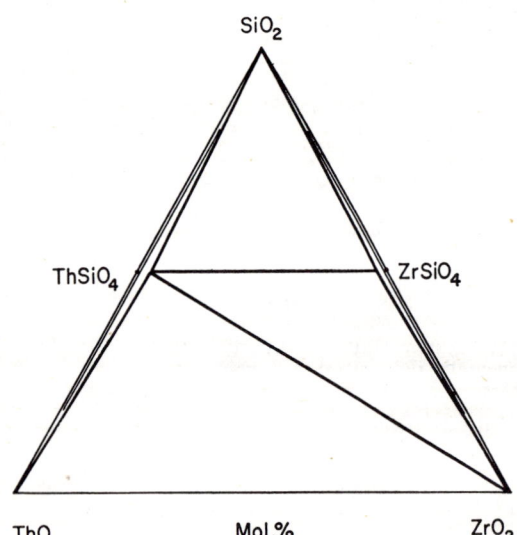

Fig. 782.—System SiO_2–ThO_2–ZrO_2; subsolidus (300° to 1400°C.). The solubilities of $ThSiO_4$ in zircon and of $ZrSiO_4$ in thorite are 4 ± 2 and 6 ± 2 mol.%, respectively.

F. A. Mumpton and R. Roy, *Geochim. et Cosmochim. Acta*, **21**, 230 (1961).

SiO_2–UO_2–ZrO_2

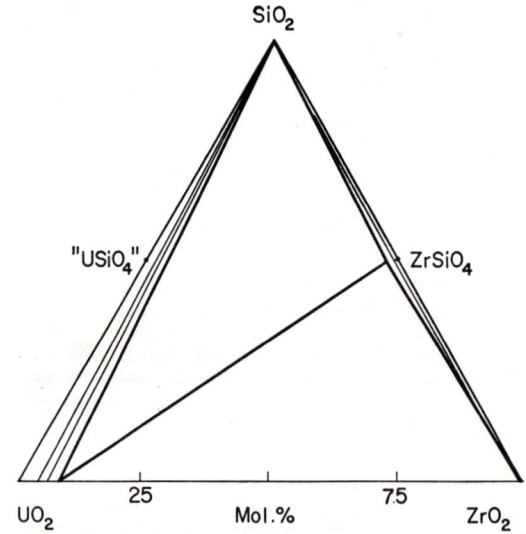

Fig. 784.—System SiO_2–UO_2–ZrO_2; subsolidus (300° to 1350°C.). Solubility of "$USiO_4$" in zircon is about 4 ± 2 mol.%.

F. A. Mumpton and R. Roy, *Geochim. et Cosmochim. Acta*, **21**, 234 (1961).

IV. Four Oxides

K_2O–Li_2O–B_2O_3–WO_3

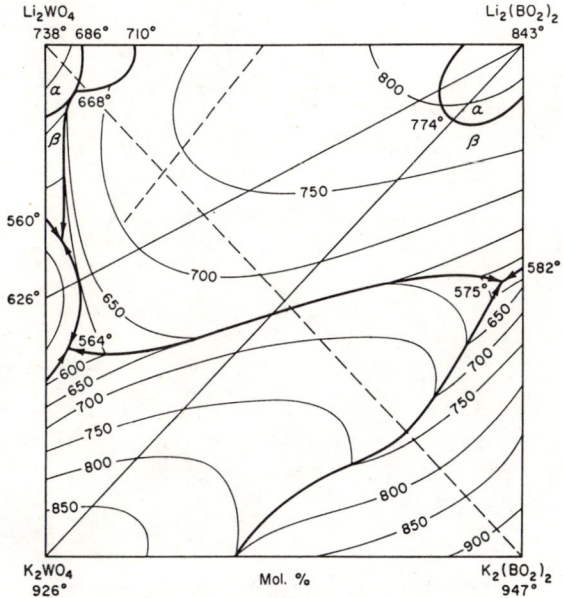

Fig. 785.—System KBO_2–K_2WO_4–$LiBO_2$–Li_2WO_4; irreversible reciprocal.

A. G. Bergman, A. I. Kislova, and V. I. Posypaĭko, *Zhur. Obshcheĭ Khim.*, **25**, 2051 (1955).

K_2O–Na_2O–Al_2O_3–SiO_2

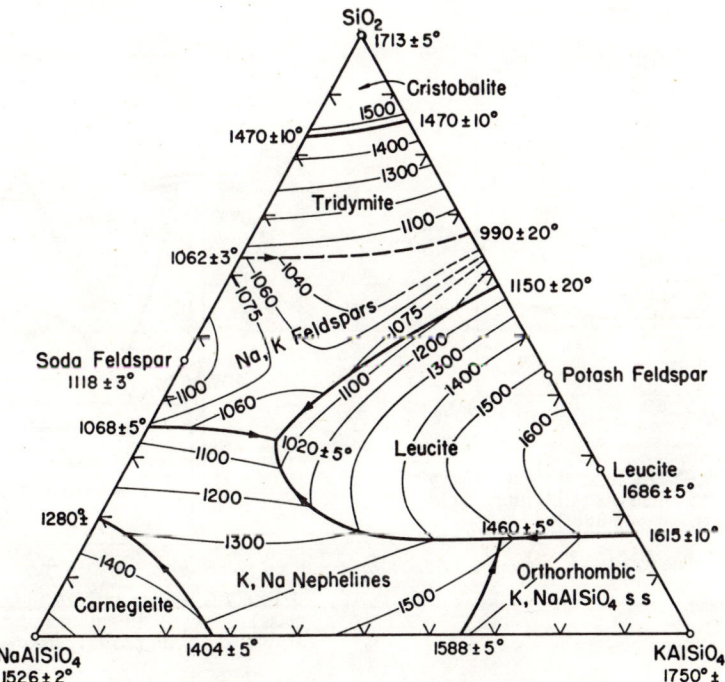

Fig. 786.—System $NaAlSiO_4$–$KAlSiO_4$–SiO_2; revised.

J. F. Schairer, *J. Geol.*, **58**, No. 5, 514 (1950).

K$_2$O–Na$_2$O–Al$_2$O$_3$–SiO$_2$ (concl.)

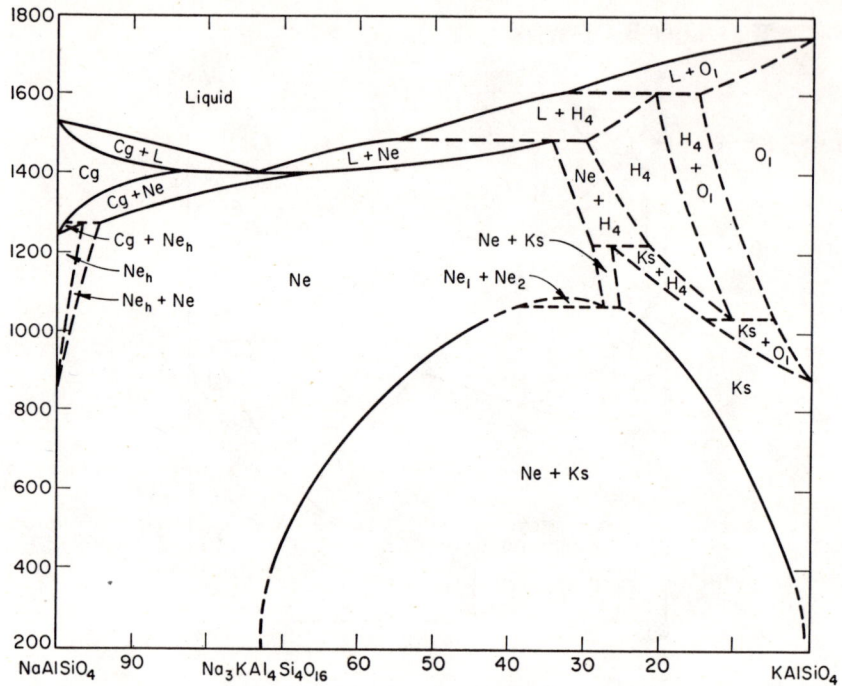

FIG. 787.—System NaAlSiO$_4$–KAlSiO$_4$. Dashed lines are inferred phase boundaries. Cg = carnegieite, L = liquid, Ne$_h$ = high-temperature nepheline, Ne = low-temperature nepheline, O$_1$ = orthorhombic KAlSiO$_4$, Ks = kalsilite, H$_4$ = tetrakalsilite.

O. F. Tuttle and J. V. Smith, *Am. J. Sci.*, **256**, 573 (1958).

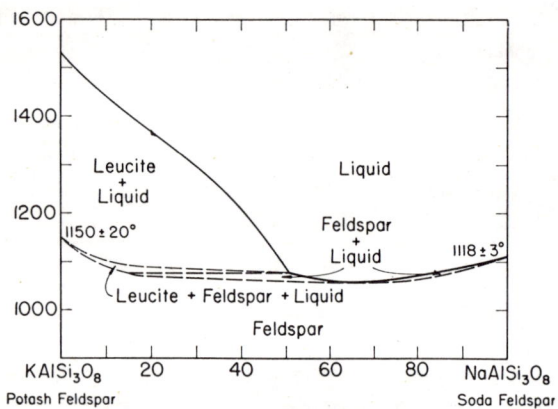

FIG. 788.—System KAlSi$_3$O$_8$–NaAlSi$_3$O$_8$; revised. Heavy solid or dashed lines refer to binary equilibrium; light solid or dashed lines refer to ternary equilibrium.

J. F. Schairer, *J. Geol.*, **58**, No. 5, 515 (1950).

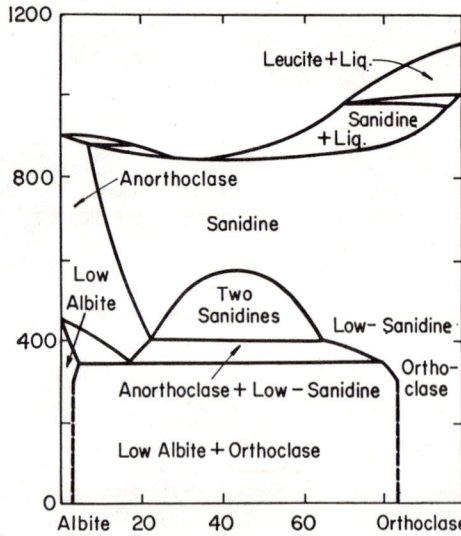

FIG. 789.—System NaAlSi$_3$O$_8$ (albite)–KAlSi$_3$O$_8$ (orthoclase). Proposed equilibrium diagram for alkali feldspars as deduced from phase relations observed in natural rocks.

J. V. Smith, *Intern. Geol. Congr., 21st., Copenhagen 1960*, p. 190, in *Rept. Session, Norden*.

K₂O–Na₂O–B₂O₃–CrO₃

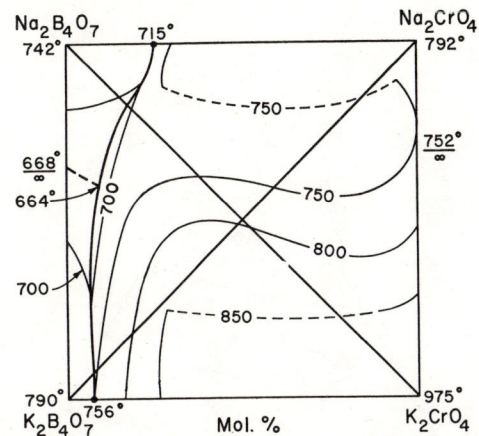

Fig. 791.—System K₂B₄O₇–K₂CrO₄–Na₂B₄O₇–Na₂CrO₄

A. G. Bergman and O. R. Vartbaronov, *Zhur. Neorg. Khim.*, **2** [3] 653 (1957).

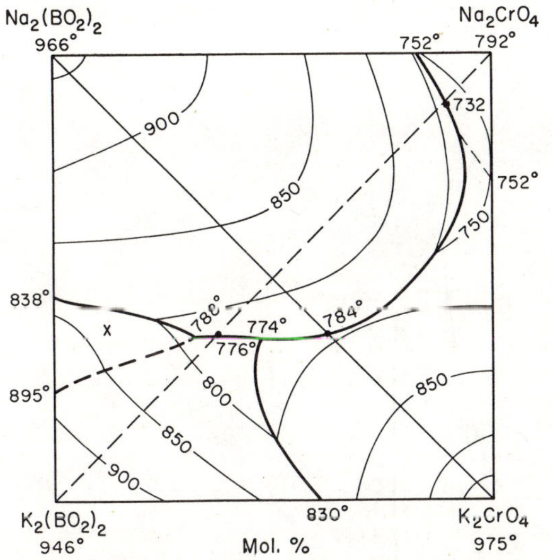

K₂O–Na₂O–P₂O₅–MoO₃

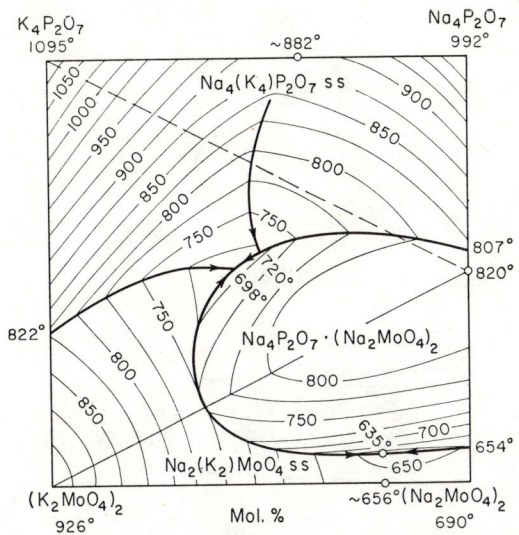

Fig. 792.—System K₄P₂O₇–K₂MoO₄–Na₄P₂O₇–Na₂MoO₄; reciprocal.

I. N. Belyaev and M. L. Sholokhovich, *Zhur. Obshchei Khim.*, **23**, 1271 (1953).

Fig. 790.—System K₂(BO₂)₂–K₂CrO₄–Na₂(BO₂)₂–Na₂CrO₄.

A. G. Bergman and O. R. Vartbaronov, *Zhur. Neorg. Khim.*, **2** [11] 2646 (1957).

K₂O–Na₂O–MoO₃–WO₃

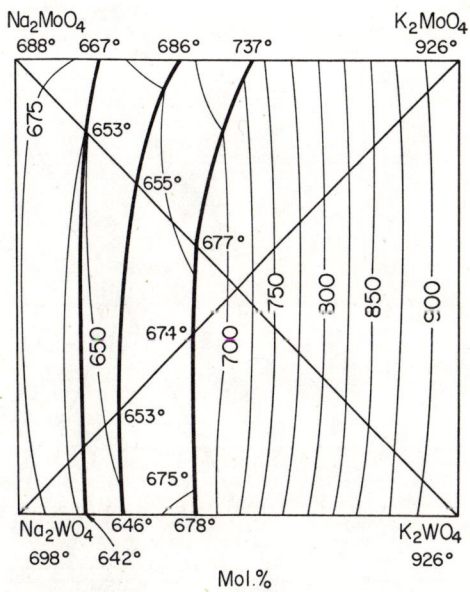

Fig. 793.—System K₂MoO₄–K₂WO₄–Na₂MoO₄–Na₂WO₄.

Z. A. Mateĭko and G. A. Bukhalova, *Zhur. Neorg. Khim.*, **2**, 202 (1957).

K_2O–CaO–Al_2O_3–Fe_2O_3

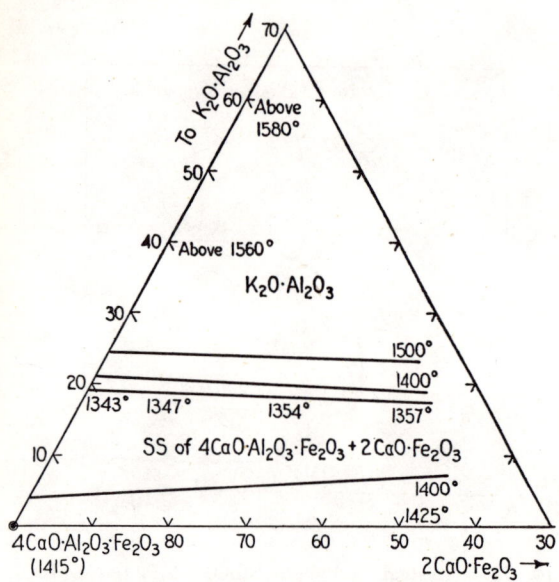

Fig. 794.—System $K_2O \cdot Al_2O_3$–$2CaO \cdot Fe_2O_3$–$4CaO \cdot Al_2O_3 \cdot Fe_2O_3$.

W. C. Taylor, *J. Research Natl. Bur. Standards*, **21**, 323 (1938); R P 1131.

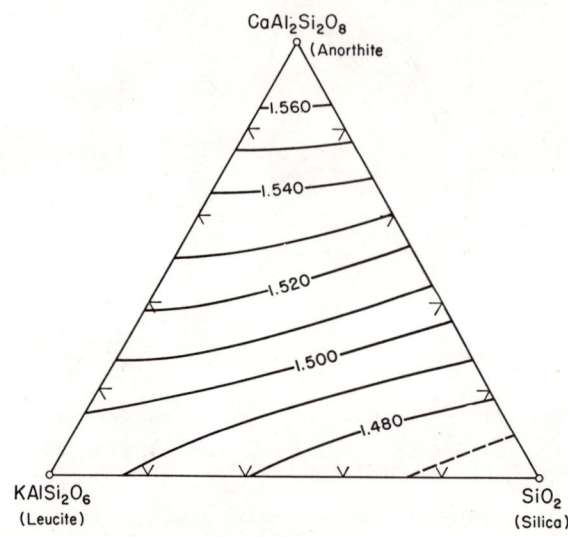

Fig. 796.—System $KAlSi_2O_6$–$CaAl_2Si_2O_8$–SiO_2 (n of glasses at 25°C.).

J. F. Schairer and N. L. Bowen, *Bull. Soc. Geol. de Finlande*, **20**, 81 (1947).

K_2O–CaO–Al_2O_3–SiO_2

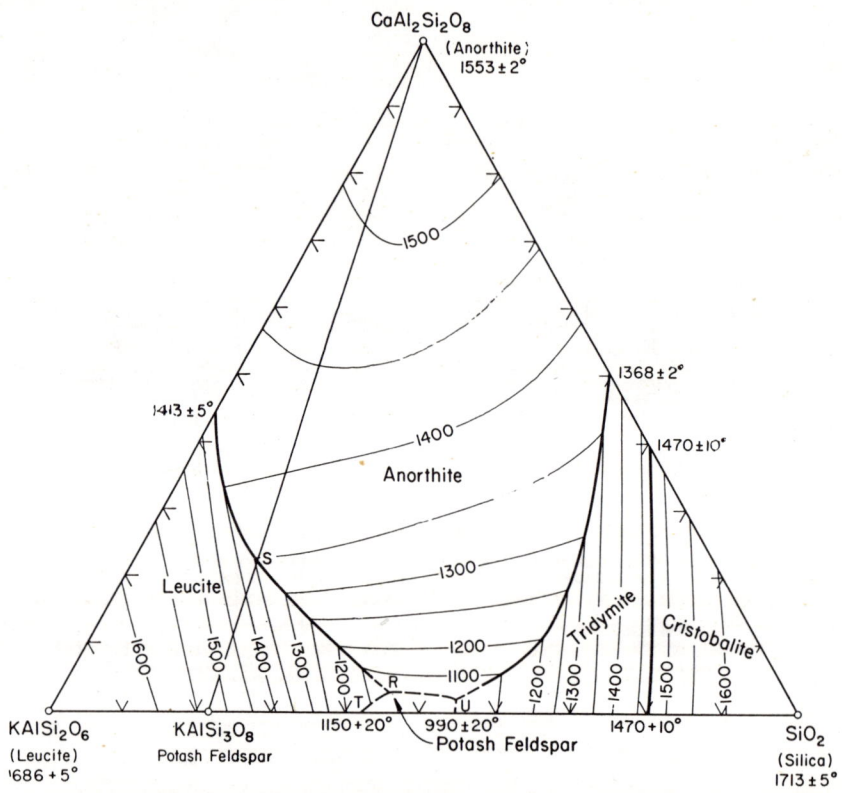

Fig. 795.—System $KAlSi_2O_6$ (leucite)–$CaAl_2Si_2O_8$ (anorthite)–SiO_2.

J. F. Schairer and N. L. Bowen, *Bull. Soc. Geol. de Finlande*, **20**, 75 (1947).

K_2O–CaO–Al_2O_3–SiO_2 (concl.)

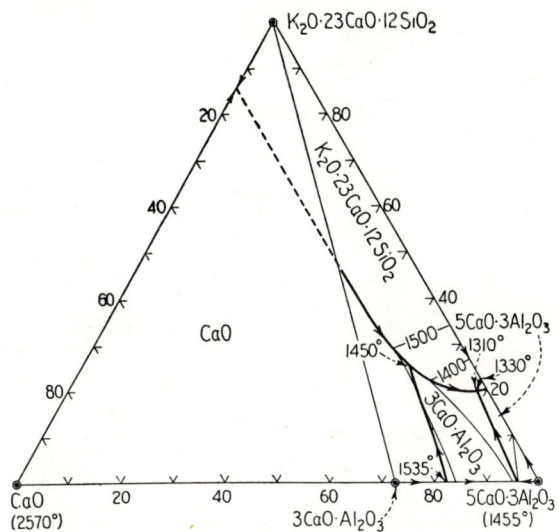

Fig. 797.—System CaO–$5CaO \cdot 3Al_2O_3$–$K_2O \cdot 23CaO \cdot 12SiO_2$.

W. C. Taylor, *J. Research Natl. Bur. Standards*, **29**, 439 (1942); R P 1512.

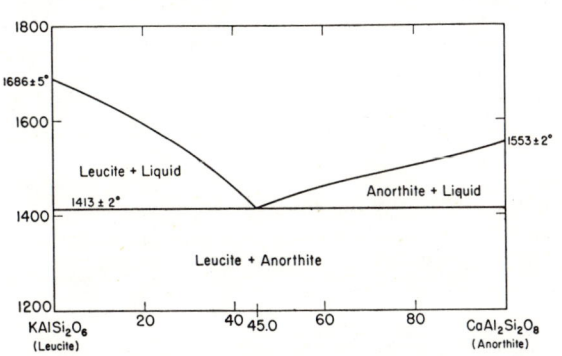

Fig. 798.—System $KAlSi_2O_6$ (leucite)–$CaAl_2Si_2O_8$ (anorthite).

J. F. Schairer and N. L. Bowen, *Bull. Soc. Geol. de Finlande*, **20**, 72 (1947).

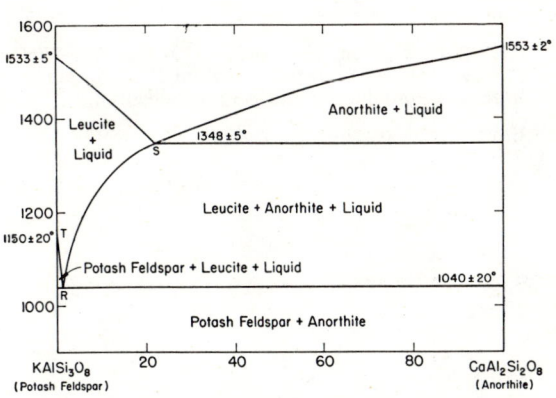

Fig. 799.—System $KAlSi_3O_8$ (potash feldspar)–$CaAl_2Si_2O_8$ (anorthite); heavy curves refer to binary equilibrium; light curves to ternary equilibrium.

J. F. Schairer and N. L. Bowen, *Bull. Soc. Geol. de Finlande*, **20**, 80 (1947).

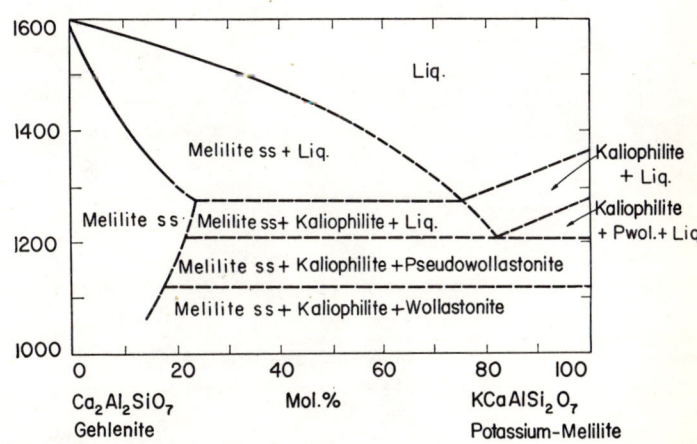

Fig. 800.—System $KCaAlSi_2O_7$ (potassium-melilite)–$Ca_2Al_2SiO_7$ (gehlenite); pseudobinary. pwol. = pseudowollastonite, ss = solid solution.

R. W. Nurse and H. G. Midgley, *J. Iron Steel Inst. (London)*, **174**, 130 (1953).

$K_2O-FeO-Al_2O_3-SiO_2$

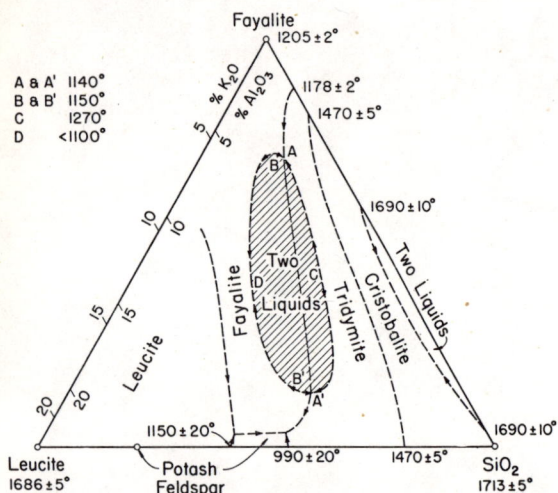

FIG. 801.—System $K_2O \cdot Al_2O_3 \cdot 4SiO_2$ (leucite)–$2FeO \cdot SiO_2$ (fayalite)–SiO_2; all melts contain Fe_2O_3 in amounts representing equilibrium with pure metallic iron.

Edwin Roedder, *Am. Mineralogist*, **36**, 283 (1951).

$K_2O-MgO-Al_2O_3-SiO_2$

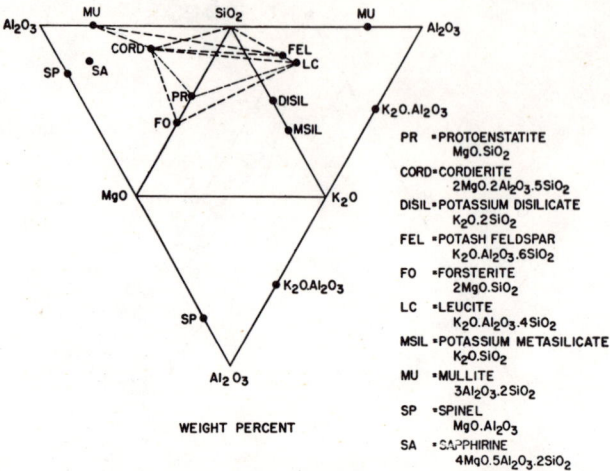

FIG. 802.—System $K_2O-MgO-Al_2O_3-SiO_2$; position of five joins shown in Figs. 803–807. Three of the faces of the tetrahedron have been laid flat in the plane of the base.

J. F. Schairer, *J. Am. Ceram. Soc.*, **37** [11] 505 (1954)

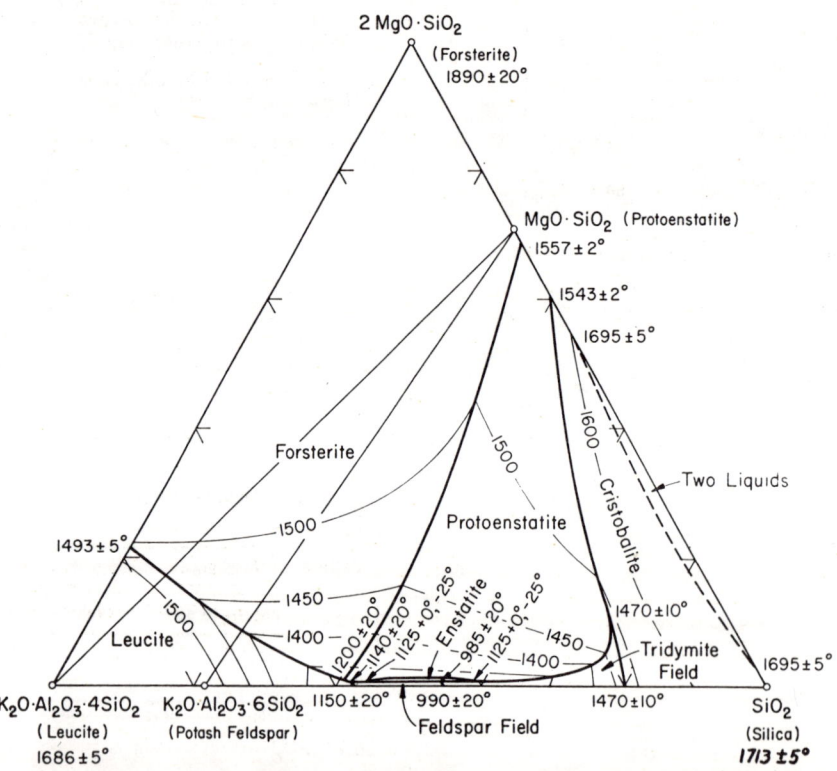

FIG. 803.—System $K_2O \cdot Al_2O_3 \cdot 4SiO_2 - 2MgO \cdot SiO_2 - SiO_2$. System is ternary. The fields of enstatite, and potash feldspar are so narrow and so near the leucite-silica side line that they could not be shown on the scale of this figure without some exaggeration.

J. F. Schairer, *J. Am. Ceram. Soc.*, **37** [11] 506 (1954).

K_2O–MgO–Al_2O_3–SiO_2 (cont.)

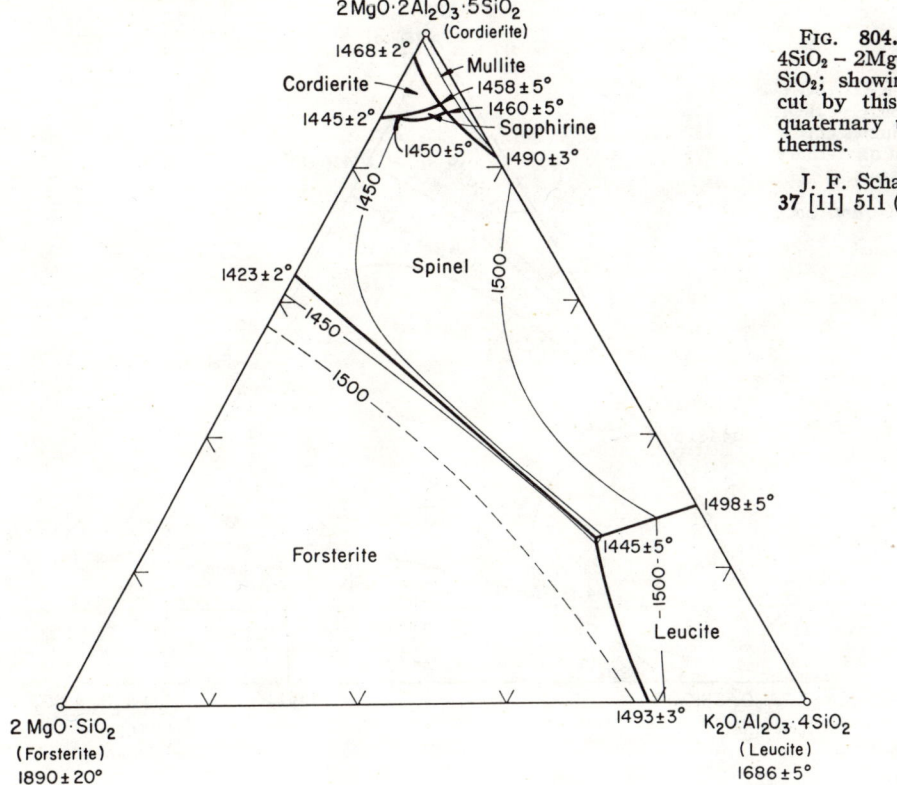

Fig. 804. — System $K_2O \cdot Al_2O_3 \cdot 4SiO_2 - 2MgO \cdot 2Al_2O_3 \cdot 5SiO_2 - 2MgO \cdot SiO_2$; showing primary phase volumes cut by this join, piercing points of quaternary univariant lines, and isotherms.

J. F. Schairer, *J. Am. Ceram. Soc.*, **37** [11] 511 (1954).

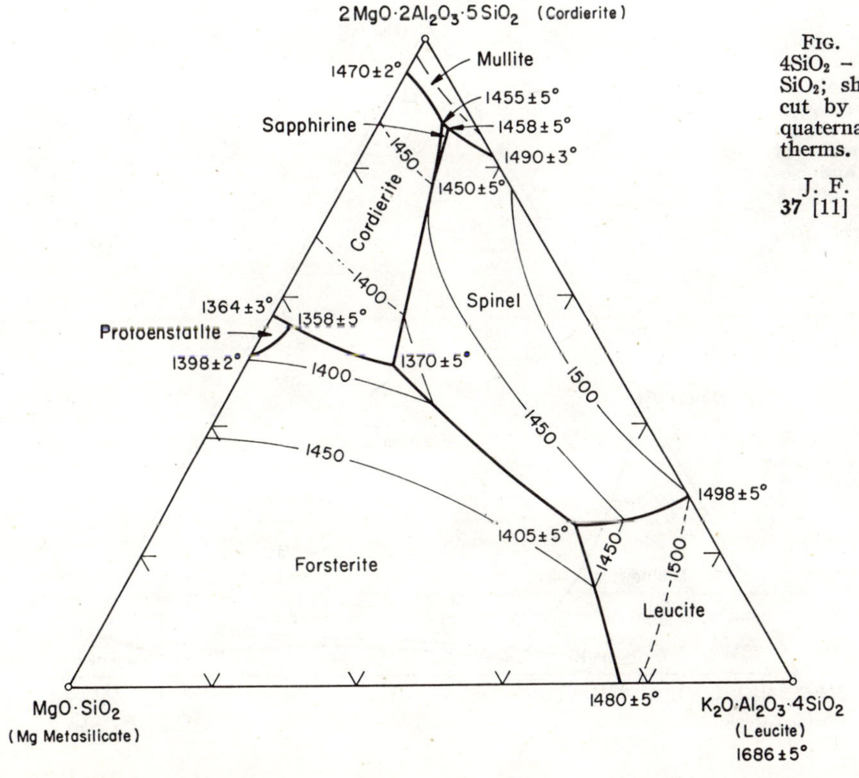

Fig. 805. — System $K_2O \cdot Al_2O_3 \cdot 4SiO_2 - 2MgO \cdot 2Al_2O_3 \cdot 5SiO_2 - MgO \cdot SiO_2$; showing primary phase volumes cut by this join, piercing points of quaternary univariant lines, and isotherms.

J. F. Schairer, *J. Am. Ceram. Soc.*, **37** [11] 518 (1954).

K_2O–MgO–Al_2O_3–SiO_2 (cont.)

Fig. 806. — System $K_2O \cdot Al_2O_3 \cdot 4SiO_2$ – $2MgO \cdot 2Al_2O_3 \cdot 5SiO_2$ – SiO_2; showing primary phase volumes cut by this join, piercing points of quaternary univariant lines, and isotherms.

J. F. Schairer, *J. Am. Ceram. Soc.*, **37** [11] 515 (1954).

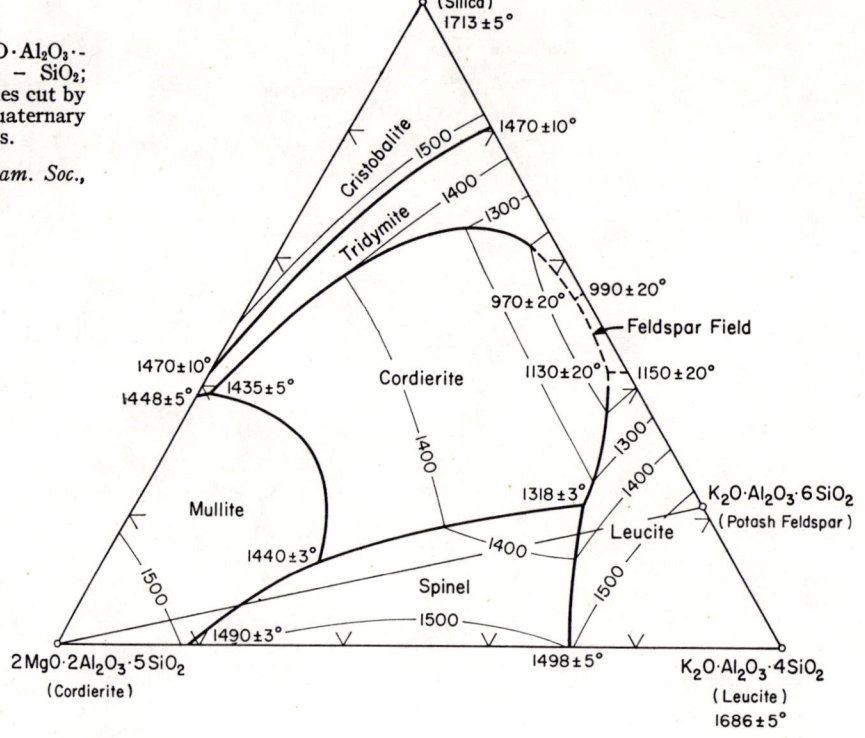

Fig. 807. — System $K_2O \cdot Al_2O_3 \cdot 6SiO_2$ – $2MgO \cdot 2Al_2O_3 \cdot 5SiO_2$ – $3Al_2O_3 \cdot 2SiO_2$; showing primary phase volumes cut by this join, piercing points of quaternary univariant lines, and isotherms.

J. F. Schairer, *J. Am. Ceram. Soc.*, **37** [11] 522 (1954).

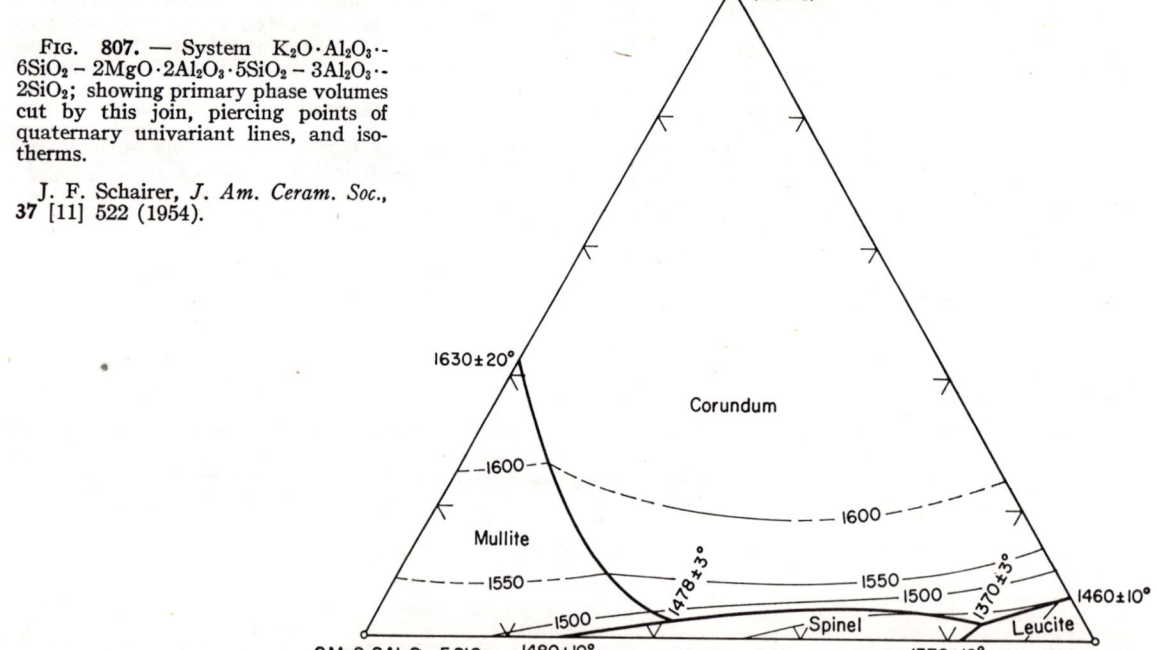

$K_2O-MgO-Al_2O_3-SiO_2$ (cont.)

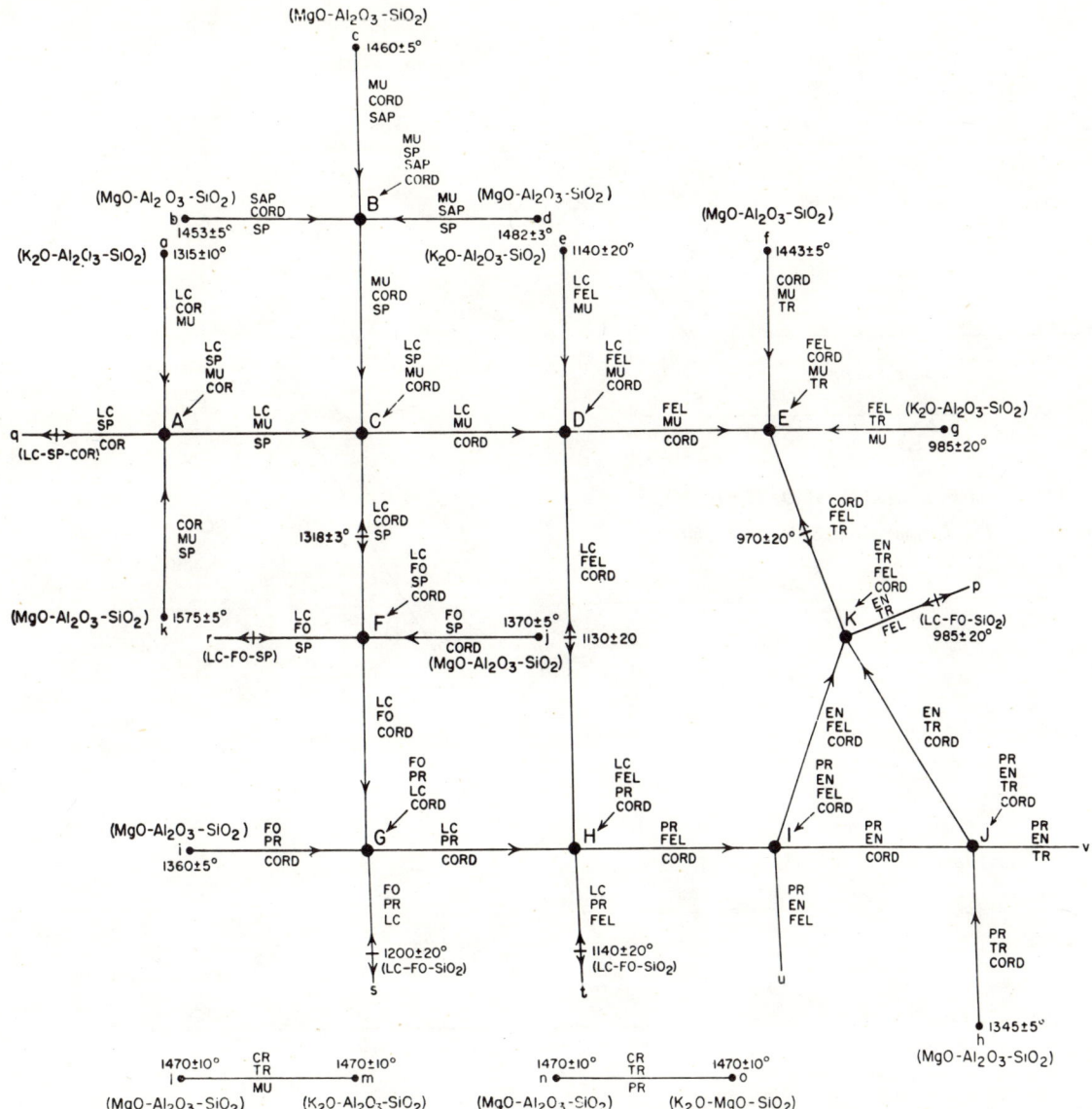

FIG. 808.—System $K_2O-MgO-Al_2O_3-SiO_2$.

Diagram showing univariant lines and their relation to ternary invariant points (small black dots and letters *a* through *o*) in limiting systems and to quaternary invariant points (large black dots and capital letters). These lines and points do not lie in a plane. Only their relations to one another are shown in this diagram, which is not intended to depict their angular spatial relations. The lengths of the lines and the position of a temperature maximum on a line are arbitrary and without significance. Arrows indicate the direction of falling temperature. Abbreviations for crystalline solid phases along the lines and at the points: LC = leucite, COR = corundum, MU = mullite, SP = spinel, CORD = cordierite, SAP = sapphirine, FEL = potash feldspar, TR = tridymite, CR = cristobalite, FO = forsterite, PR = protoenstatite, EN = enstatite.

Temperatures of quaternary invariant points: A = ~1300°, B = 1449 ± 5°, C = ~1290°, D = 1120 ± 20°, E = 960 ± 20° (eutectic), F = near 1318 ± 3°, G = near 1200 ± 20°, H = near 1130 ± 20°, I = ~1112 ± 13° (inversion pt.), J = ~1112 ± 13° (inversion pt.), K = 960 ± 20° (eutectic).

J. F. Schairer, *J. Am. Ceram. Soc.*, **37** [11] 526 (1954).

K_2O–MgO–Al_2O_3–SiO_2 (concl.)

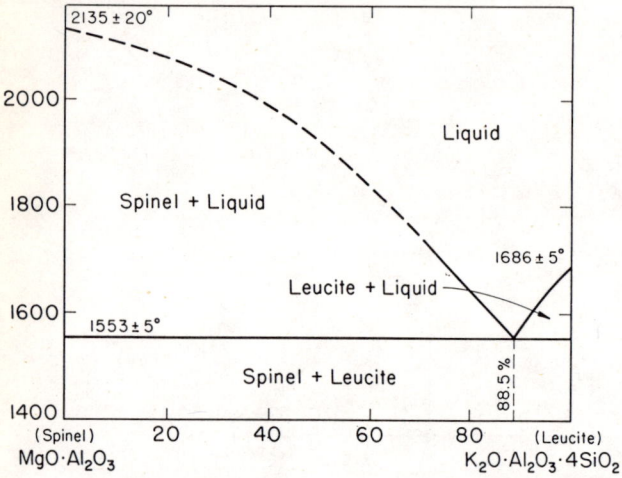

Fig. 809.—System $K_2O \cdot Al_2O_3 \cdot 4SiO_2$–$MgO \cdot Al_2O_3$.

J. F. Schairer, *J. Am. Ceram. Soc.*, **38** [5] 154 (1955).

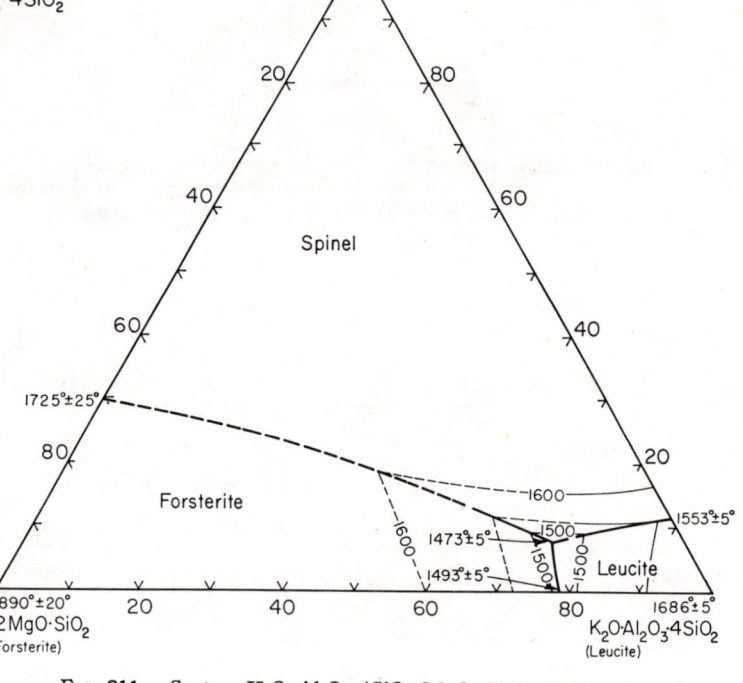

Fig. 811.—System $K_2O \cdot Al_2O_3 \cdot 4SiO_2$–$MgO \cdot Al_2O_3$–$2MgO \cdot SiO_2$.

J. F. Schairer, *J. Am. Ceram. Soc.*, **38** [5] 157 (1955).

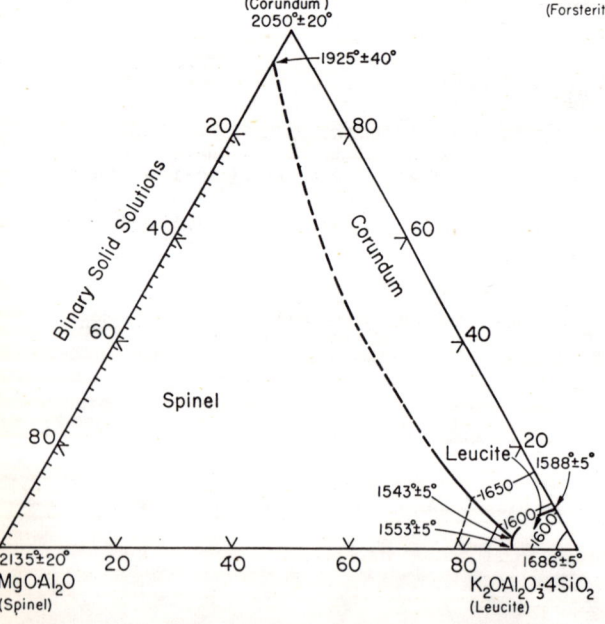

Fig. 810.—System $K_2O \cdot Al_2O_3 \cdot 4SiO_2$–$MgO \cdot Al_2O_3$–$Al_2O_3$.

J. F. Schairer, *J. Am. Ceram. Soc.*, **38** [5] 155 (1955).
On diagram, $MgO \cdot Al_2O$ should read $MgO \cdot Al_2O_3$.

K_2O–PbO–TiO_2–MoO_3

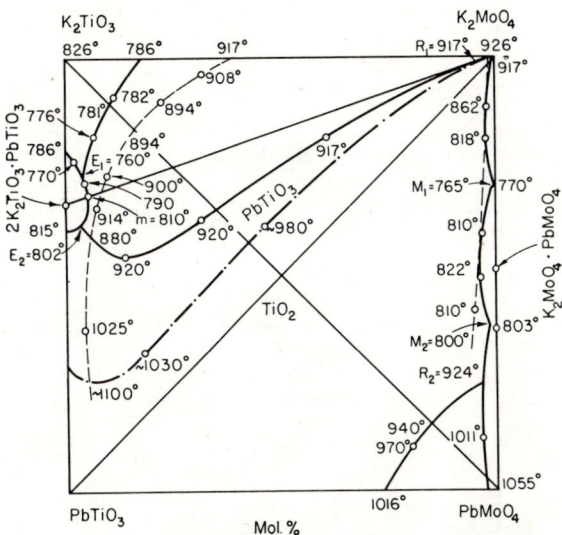

FIG. 812.—System K_2TiO_3–K_2MoO_4–$PbTiO_3$–$PbMoO_4$.

I. N. Belyaev, M. L. Sholokhovich, and G. V. Barkova, *Zhur. Neorg. Khim.*, **1** [5] 1032 (1956).

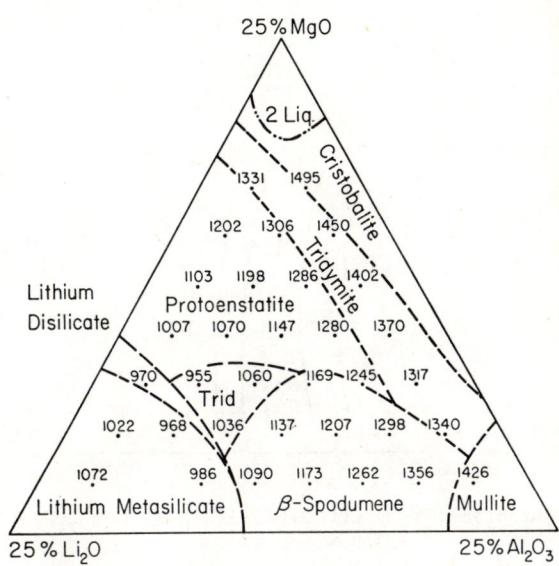

FIG. 814.—System Li_2O–MgO–Al_2O_3–SiO_2, for 75% SiO_2 plane.

T. I. Prokopowicz and F. A. Hummel, *J. Am. Ceram. Soc.*, **39** [8] 270 (1956).

Li_2O–MgO–Al_2O_3–SiO_2

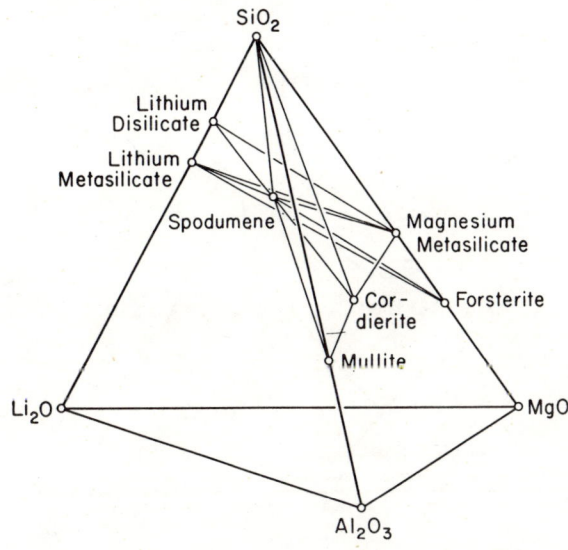

FIG. 813.—System Li_2O–MgO–Al_2O_3–SiO_2, showing compatibility tetrahedra.

T. I. Prokopowicz and F. A. Hummel, *J. Am. Ceram. Soc.*, **39** [8] 274 (1956).

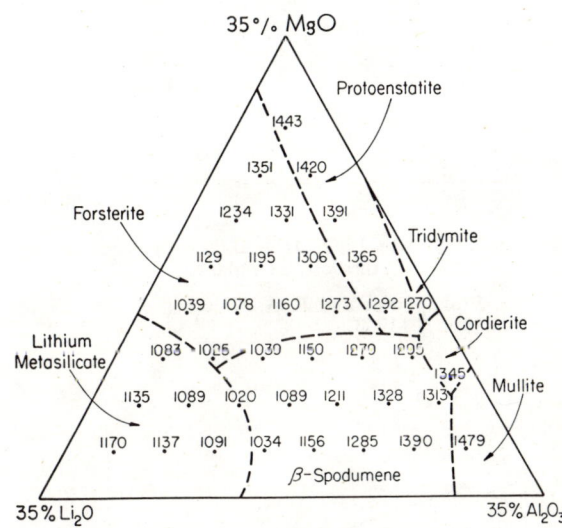

FIG. 815.—System Li_2O–MgO–Al_2O_3–SiO_2, for 65% SiO_2 plane.

T. I. Prokopowicz and F. A. Hummel, *J. Am. Ceram. Soc.*, **39** [8] 271 (1956).

Li₂O–MgO–Al₂O₃–SiO₂ (concl.)

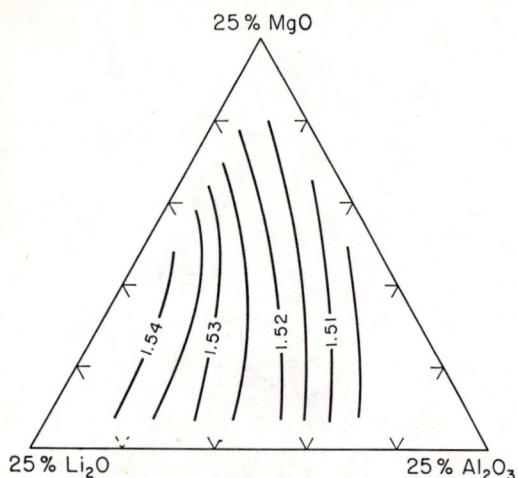

Fig. 816.—System Li₂O–MgO–Al₂O₃–SiO₂; isofracts for the 75% SiO₂ plane.

T. I. Prokopowicz and F. A. Hummel, *J. Am. Ceram. Soc.*, **39** [8] 271 (1956).

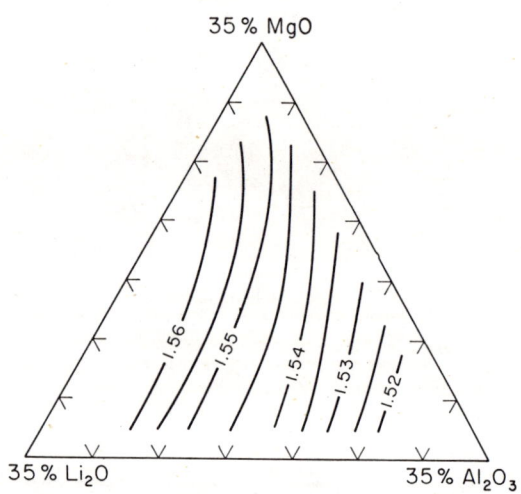

Fig. 817.—System Li₂O–MgO–Al₂O₃–SiO₂; isofracts for the 65% SiO₂ plane.

T. I. Prokopowicz and F. A. Hummel, *J. Am. Ceram. Soc.*, **39** [8] 271 (1956).

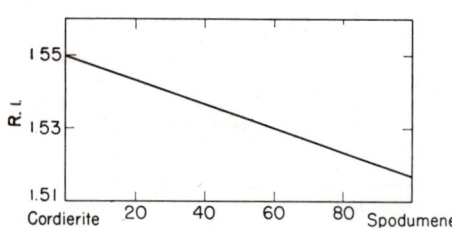

Fig. 819.—System 2MgO·2Al₂O₃·5SiO₂–Li₂O·Al₂O₃·4SiO₂ (n of glasses).

M. D. Karkhanavala and F. A. Hummel, *J. Am. Ceram. Soc.*, **36** [12] 394 (1953).

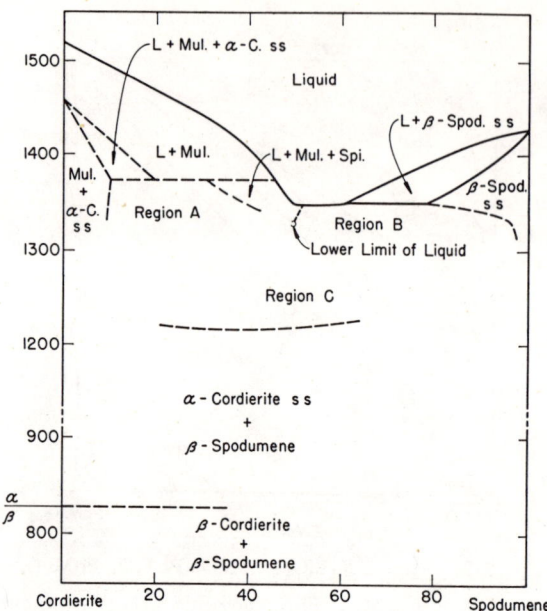

Fig. 818.—System 2MgO·2Al₂O₃·5SiO₂ (cordierite)–Li₂O·Al₂O₃·4SiO₂ (spodumene); non-binary join.

M. D. Karkhanavala and F. A. Hummel, *J. Am. Ceram. Soc.*, **36** [12] 394 (1953).

Li₂O–Al₂O₃–B₂O₃–SiO₂

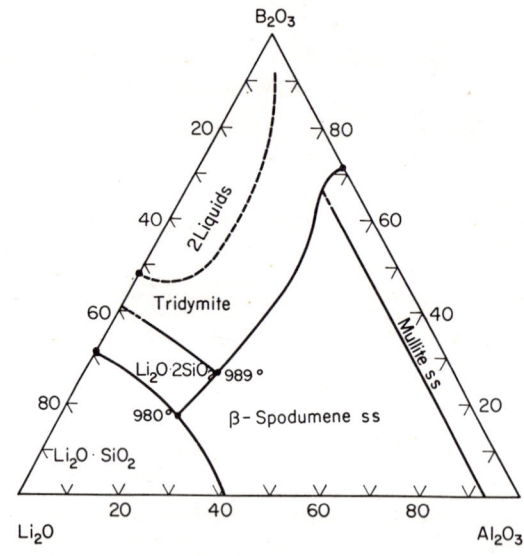

Fig. 820.—System Li₂O–Al₂O₃–B₂O₃–SiO₂ at 70 wt.% SiO₂.

K. H. Kim and F. A. Hummel, private communication, Dec. 20, 1961.

Li$_2$O–Al$_2$O$_3$–B$_2$O$_3$–SiO$_2$ (concl.)

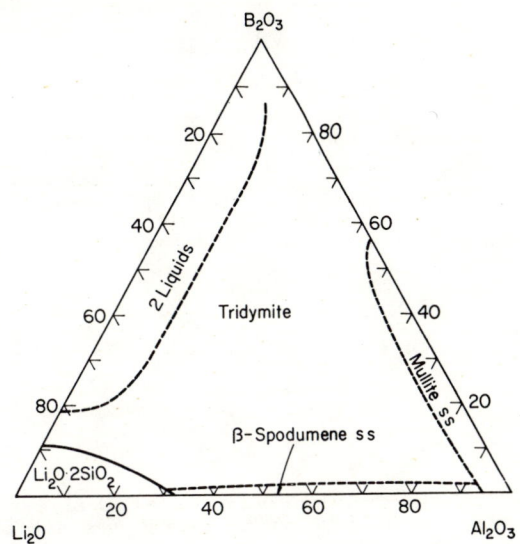

Fig. 821.—System Li$_2$O–Al$_2$O$_3$–B$_2$O$_3$–SiO$_2$ at 80 wt.% SiO$_2$.

K. H. Kim and F. A. Hummel, private communication, Dec. 20, 1961.

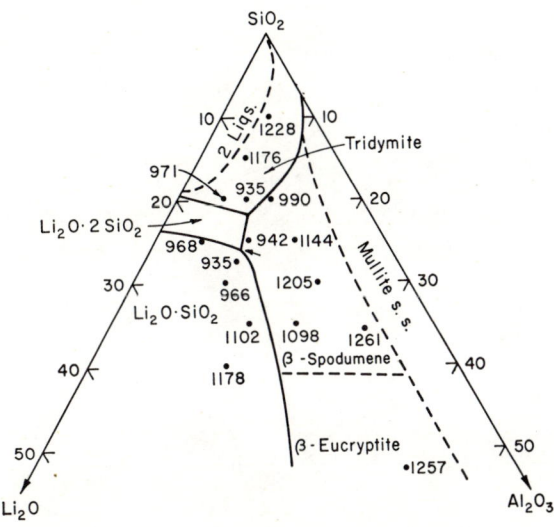

Fig. 822.—System Li$_2$O–Al$_2$O$_3$–B$_2$O$_3$–SiO$_2$ at 10 wt.% B$_2$O$_3$.

K. H. Kim and F. A. Hummel, private communication Dec. 20, 1961.

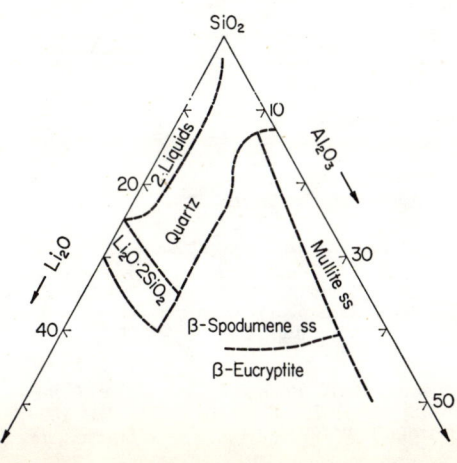

Fig. 823.—System Li$_2$O–Al$_2$O$_3$–B$_2$O$_3$–SiO$_2$ at 30 wt.% B$_2$O$_3$; primary fields.

K. H. Kim and F. A. Hummel, private communication, Dec. 20, 1961.

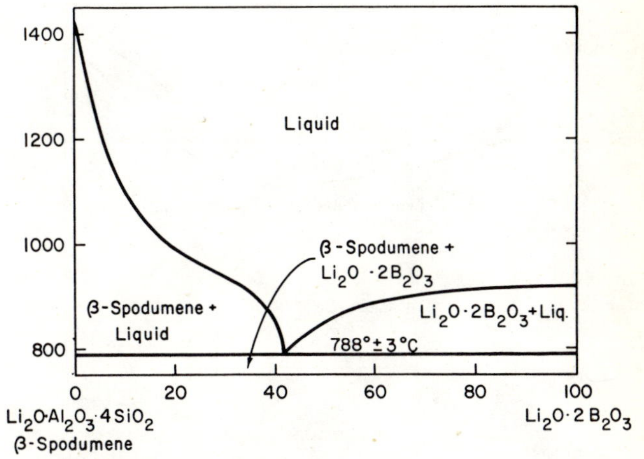

Fig. 824.—System Li$_2$O·2B$_2$O$_3$–Li$_2$O·Al$_2$O$_3$·4SiO$_2$.

K. H. Kim and F. A. Hummel, private communication, Dec. 20, 1961.

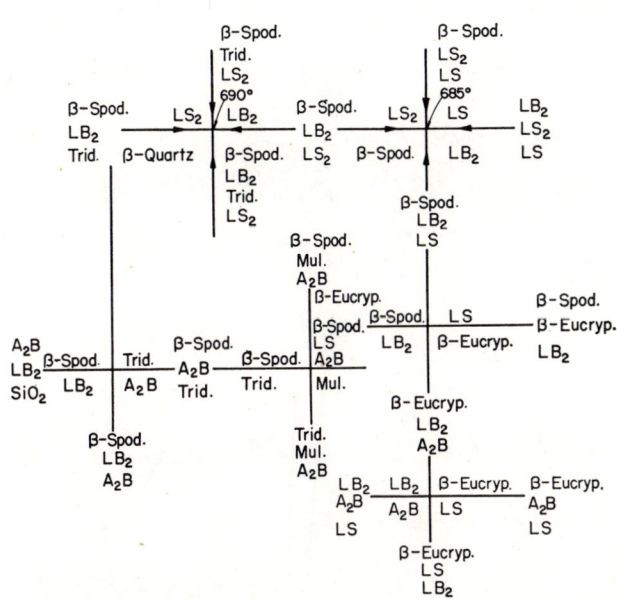

Fig. 825.—System Li$_2$O–Al$_2$O$_3$–B$_2$O$_3$–SiO$_2$, schematic relationships of univariant lines to ternary and quaternary invariant points. Eucryp. = eucryptite; Mul. = mullite; Spod. = spodumene; Trid. = tridymite. A = Al$_2$O$_3$, B = B$_2$O$_3$, L = Li$_2$O, S = SiO$_2$.

K. H. Kim and F. A. Hummel, private communication, Dec. 20, 1961.

$Na_2O-BaO-TiO_2-Nb_2O_5$

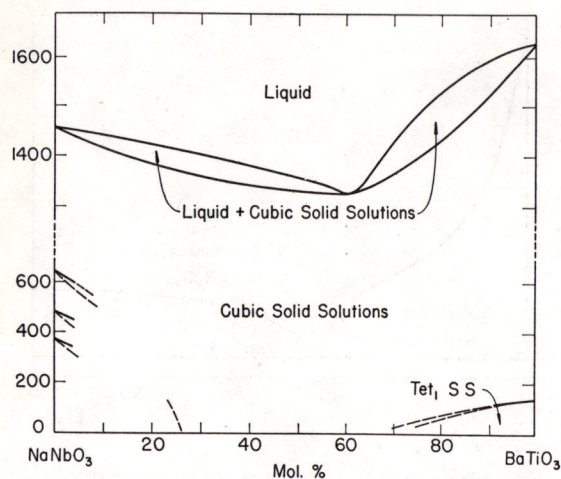

Fig. 826.—System $NaNbO_3-BaTiO_3$.

The Pennsylvania State College, School of Mineral Industries; Eighth Quarterly Progress Report, April 1 to June 30, p. 16 (1953).

$Na_2O-CaO-Al_2O_3-SiO_2$

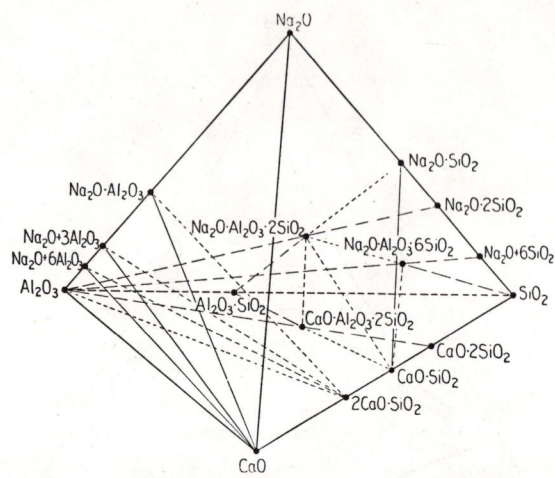

Fig. 828.—Tetrahedron of system $Na_2O-CaO-Al_2O_3-SiO_2$ showing location of the systems given in Figs. 829–851.

K. T. Greene and R. H. Bogue, *J. Research Natl. Bur. Standards*, **36** [2] 188 (1946); RP 1699.

$Na_2O-CaO-Al_2O_3-Fe_2O_3$

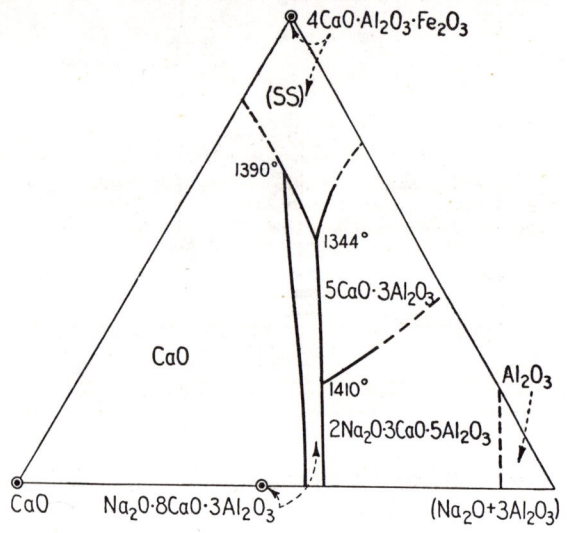

Fig. 827.—System $CaO-4CaO \cdot Al_2O_3 \cdot Fe_2O_3-(Na_2O + 3Al_2O_3)$.

W. R. Eubank and R. H. Bogue, *J. Research Natl. Bur. Standards*, **40**, 227 (1948); Also, Portland Cement Assoc. Fellowship Paper No. 50.

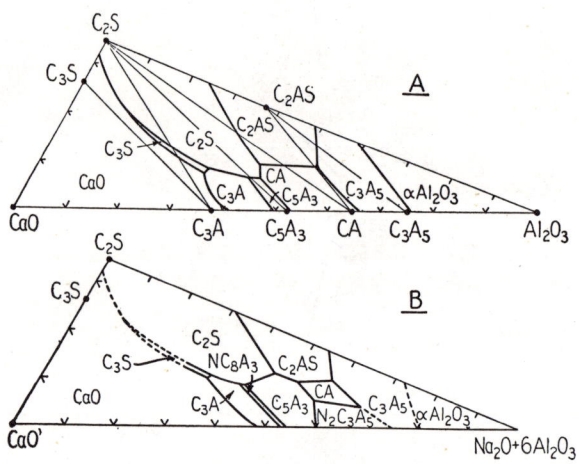

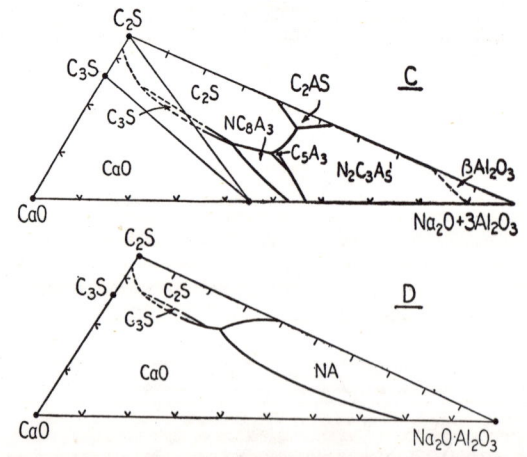

Fig. 829.—Composition planes through system $Na_2O-CaO-Al_2O_3-SiO_2$ shown in Fig. 828; A = Al_2O_3; C = CaO; N = Na_2O; S = SiO_2.

K. T. Greene and R. H. Bogue. *J. Research Natl. Bur. Standards*, **36** [2] pp. 197, 199, 200 (1946); RP 1699; part A taken from J. W. Greig, *Am. J. Sci.* 5th Ser., **13**, (1927).

Na$_2$O–CaO–Al$_2$O$_3$–SiO$_2$ (cont.)

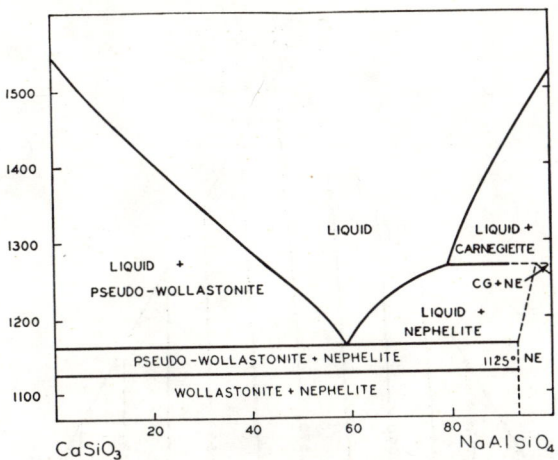

Fig. 830.—System CaO·SiO$_2$–Na$_2$O·Al$_2$O$_3$·2SiO$_2$.

W. R. Foster, *J. Geol.*, **50**, 160 (1942).

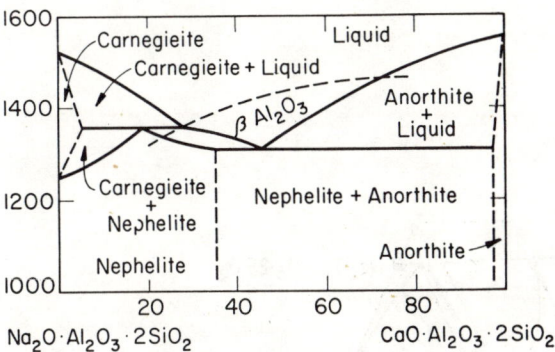

Fig. 832.—System Na$_2$O·Al$_2$O$_3$·2SiO$_2$–CaO·Al$_2$O$_3$·2SiO$_2$.

W. K. Gummer, *J. Geol.*, **51**, 508 (1943); modified after N. L. Bowen, *Am. J. Sci.*, 4th Ser., **33**, 551 (1912).

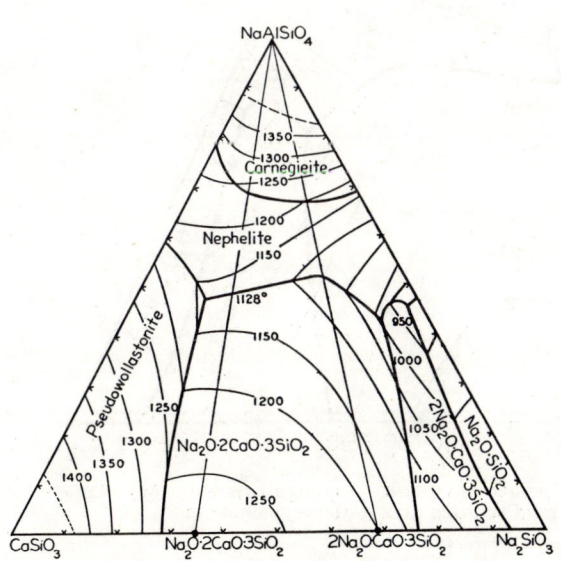

Fig. 834.—System Na$_2$O·SiO$_2$–CaO·SiO$_2$–Na$_2$O·Al$_2$O$_3$·2SiO$_2$.

Joseph Spivak, *J. Geol.*, **52**, 33 (1944).

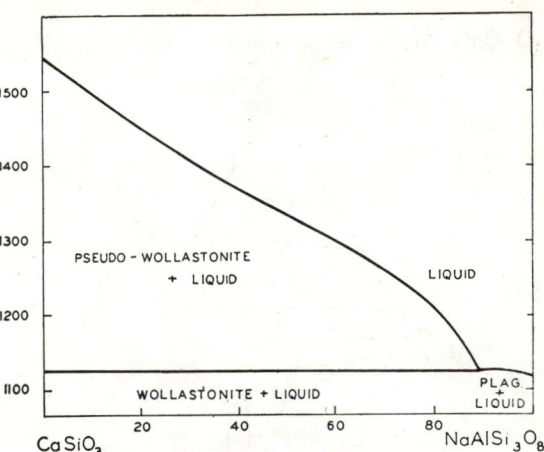

Fig. 831.—System CaO·SiO$_2$–Na$_2$O·Al$_2$O$_3$·6SiO$_2$.

W. R. Foster, *J. Geol.*, **50**, 162 (1942).

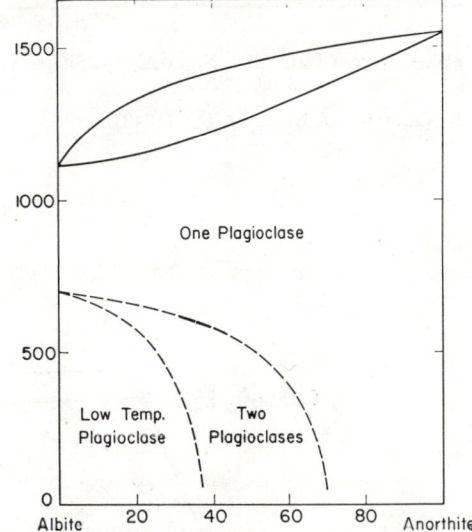

Fig. 833.—System NaAlSi$_3$O$_8$ (albite)–CaO·Al$_2$O$_3$·2SiO$_2$ (anorthite).

O. F. Tuttle and N. L. Bowen, *J. Geol.*, **58**, No. 5, 582 (1950).

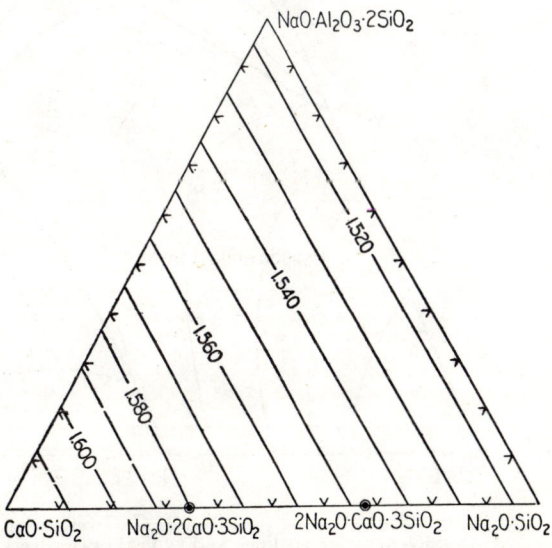

Fig. 835.—Refractive indices for glasses in system Na$_2$O·SiO$_2$–CaO·SiO$_2$–Na$_2$O·Al$_2$O$_3$·2SiO$_2$.

Joseph Spivak, *J. Geol.*, **52**, 26 (1944).

Note: NaO·Al$_2$O$_3$·2SiO$_2$, at top vertex of diagram, should read Na$_2$O·Al$_2$O$_3$·2SiO$_2$

$Na_2O–CaO–Al_2O_3–SiO_2$ (cont.)

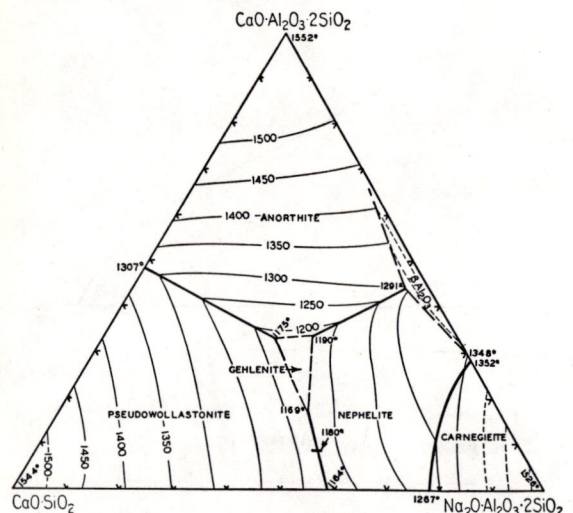

Fig. 836.—System $CaO \cdot SiO_2$–$Na_2O \cdot Al_2O_3 \cdot 2SiO_2$–$CaO \cdot Al_2O_3 \cdot 2SiO_2$.

W. K. Gummer, *J. Geol.*, **51**, 515 (1943).

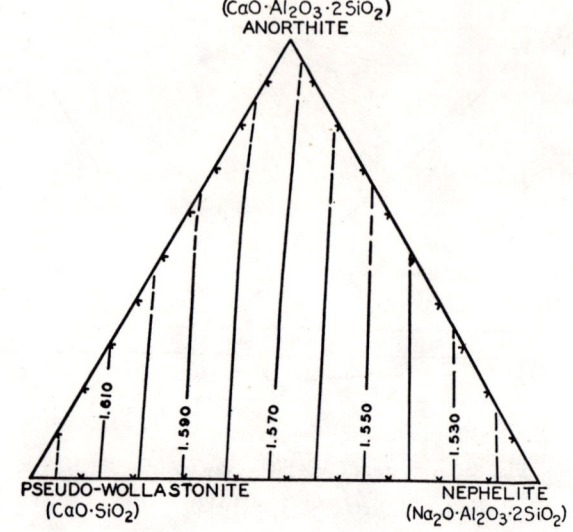

Fig. 837.—Refractive indices for glasses in system $CaO \cdot SiO_2$–$Na_2O \cdot Al_2O_3 \cdot 2SiO_2$–$CaO \cdot Al_2O_3 \cdot 2SiO_2$.

W. K. Gummer, *J. Geol.*, **51**, 510 (1943).

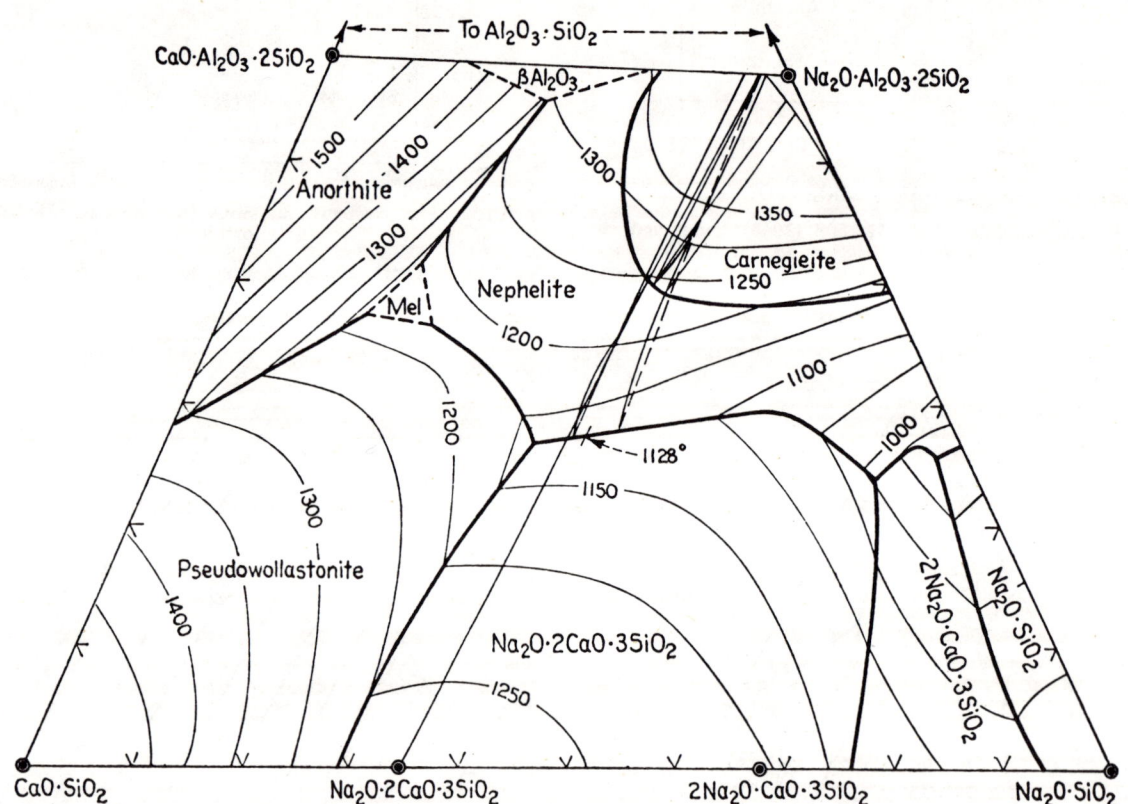

Fig. 838.—Section $Na_2O \cdot SiO_2$–$CaO \cdot SiO_2$–$Na_2O \cdot Al_2O_3 \cdot 2SiO_2$–$CaO \cdot Al_2O_3 \cdot 2SiO_2$ through system Na_2O–CaO–Al_2O_3–SiO_2 showing tie lines and typical crystallization paths in solid solution area; Mel = melilite.

Joseph Spivak, *J. Geol.*, **52**, 45 (1944).

Na_2O–CaO–Al_2O_3–SiO_2 (cont.)

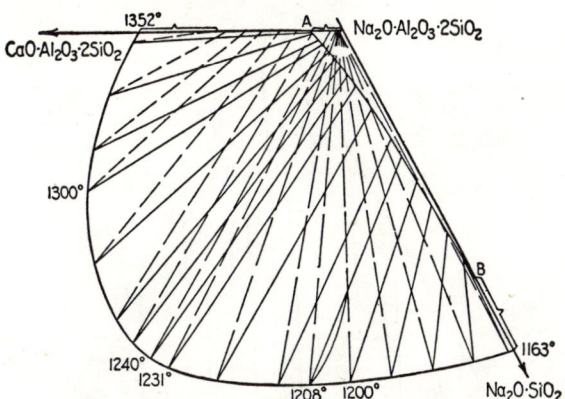

Fig. 839.—Inversion temperatures and three-phase boundaries of nepheline-carnegieite solid solutions; area of ternary solid solutions: A–B–$Na_2O \cdot Al_2O_3 \cdot 2SiO_2$.

Joseph Spivak, *J. Geol.*, **52**, 39 (1944).

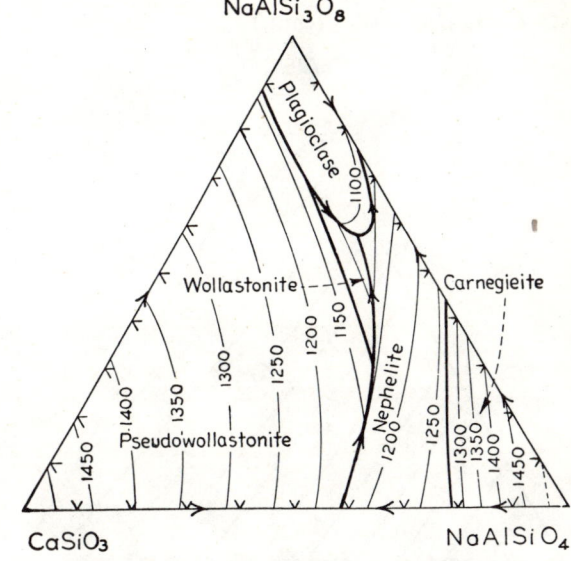

Fig. 841.—System $CaO \cdot SiO_2$–$Na_2O \cdot Al_2O_3 \cdot 2SiO_2$–$Na_2O \cdot Al_2O_3 \cdot 6SiO_2$.

W. R. Foster, *J. Geol.*, **50**, 165 (1942).

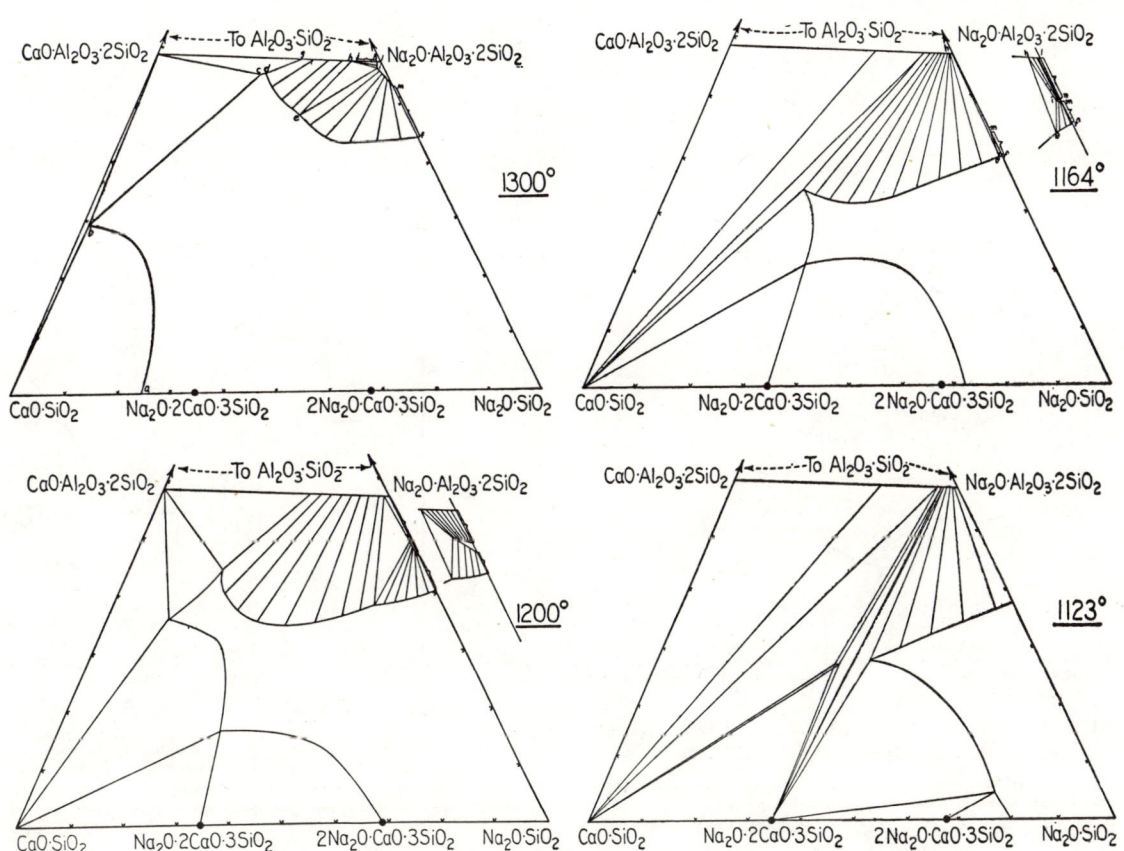

Fig. 840.—Phase relations at 1300°, 1200°, 1164°, and 1123°C. in quaternary section $Na_2O \cdot SiO_2$–$CaO \cdot SiO_2$–$Na_2O \cdot Al_2O_3 \cdot 2SiO_2$–$CaO \cdot Al_2O_3 \cdot 2SiO_2$.

Joseph Spivak, *J. Geol.*, **52**, pp. 40–43 (1944).

Na₂O–CaO–Al₂O₃–SiO₂ (cont.)

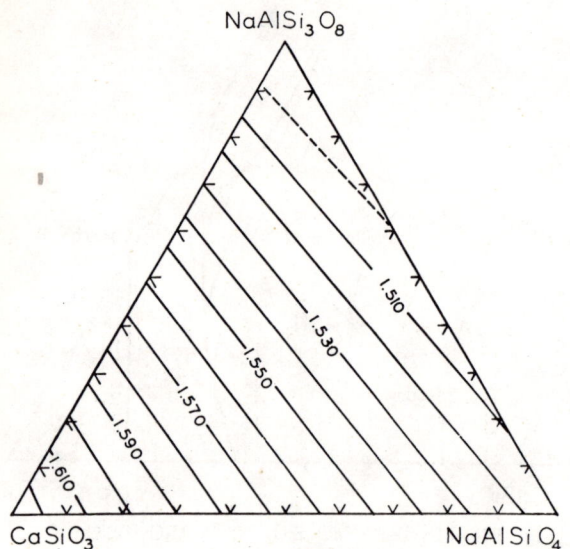

Fig. 842.—Refractive index-composition diagram for glasses in system CaO·SiO₂–Na₂O·Al₂O₃·2SiO₂–Na₂O·Al₂O₃·6SiO₂.

W. R. Foster, *J. Geol.*, **50**, 154 (1942).

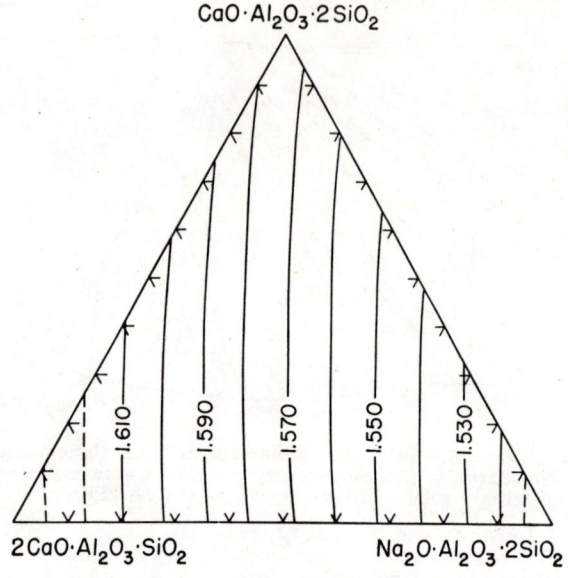

Fig. 844.—System Na₂O·Al₂O₃·2SiO₂–CaO·Al₂O₃·2SiO₂–2CaO·Al₂O₃·SiO₂, refractive indices of glasses.
J. R. Goldsmith, *J. Geol.*, **55** [5] 388 (1947).

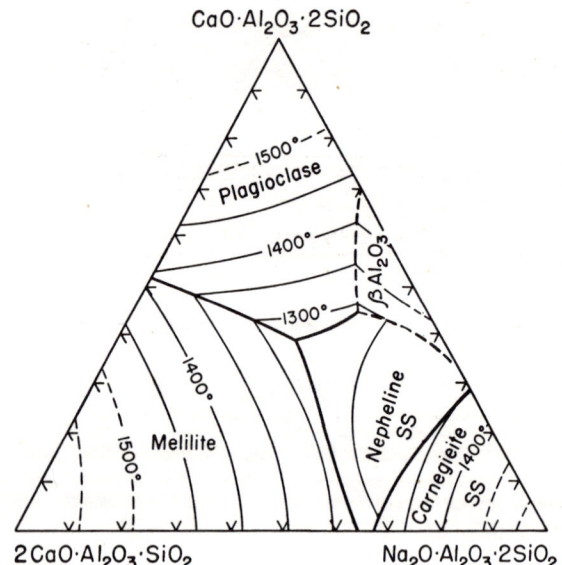

Fig. 843.—System Na₂O·Al₂O₃·2SiO₂–CaO·Al₂O₃·2SiO₂–2CaO·Al₂O₃·SiO₂.
J. R. Goldsmith, *J. Geol.*, **55** [5] 389 (1947).

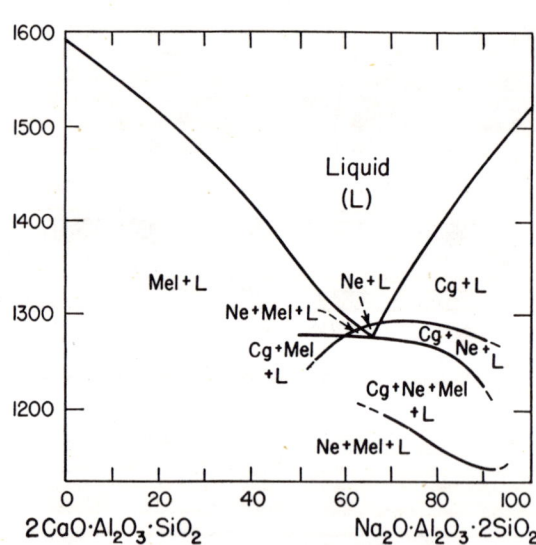

Fig. 845.—System Na₂O·Al₂O₃·2SiO₂–2CaO·Al₂O₃·SiO₂, showing nonbinary nature (MEL = melilite, NE = nepheline CG = carnegieite).
R. G. Smalley, *J. Geol.*, **55** [1] 31 (1947).

Na_2O–CaO–Al_2O_3–SiO_2 (cont.)

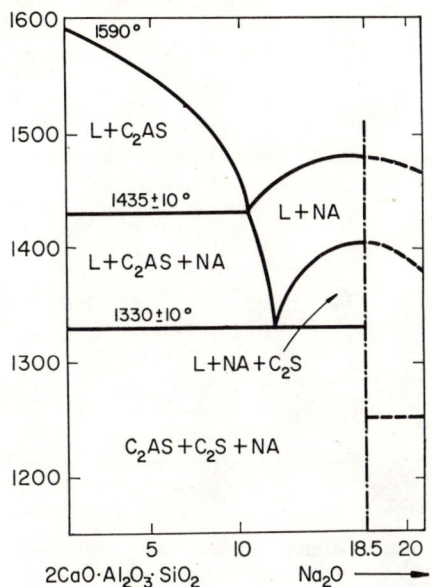

Fig. 846.—System Na_2O–$2CaO \cdot Al_2O_3 \cdot SiO_2$. A = Al_2O_3, C = CaO, N = Na_2O, S = SiO_2.

G. N. Kozhevnikov and P. S. Kusakin, *Izvest. Sibirsk. Otdel. Akad. Nauk S.S.S.R.*, **7**, 18 (1958).

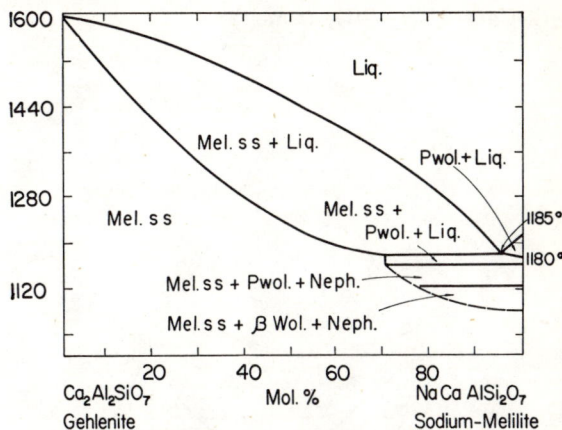

Fig. 847.—System $NaCaAlSi_2O_7$ (sodium-melilite)–$Ca_2Al_2SiO_7$ (gehlenite). Mel. = melilite, Neph. = nepheline, Pwol. = pseudowollastonite, βwol. = low-temp. wollastonite, ss = solid solution.

R. W. Nurse and H. G. Midgley, *J. Iron Steel Inst. (London)*, **174**, 128 (1953).

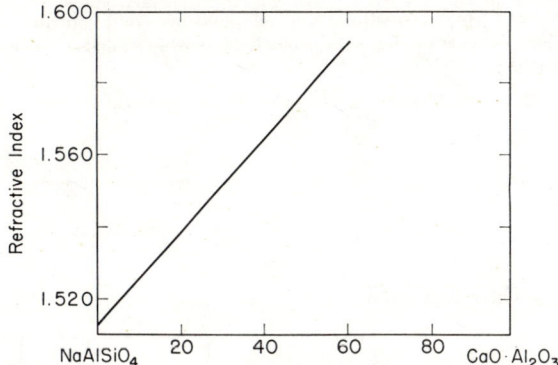

Fig. 849.—System $NaAlSiO_4$–$CaO \cdot Al_2O_3$; refractive indices of the glasses.

Julian R Goldsmith, *Am. Mineral.*, **34**, 478 (1949).

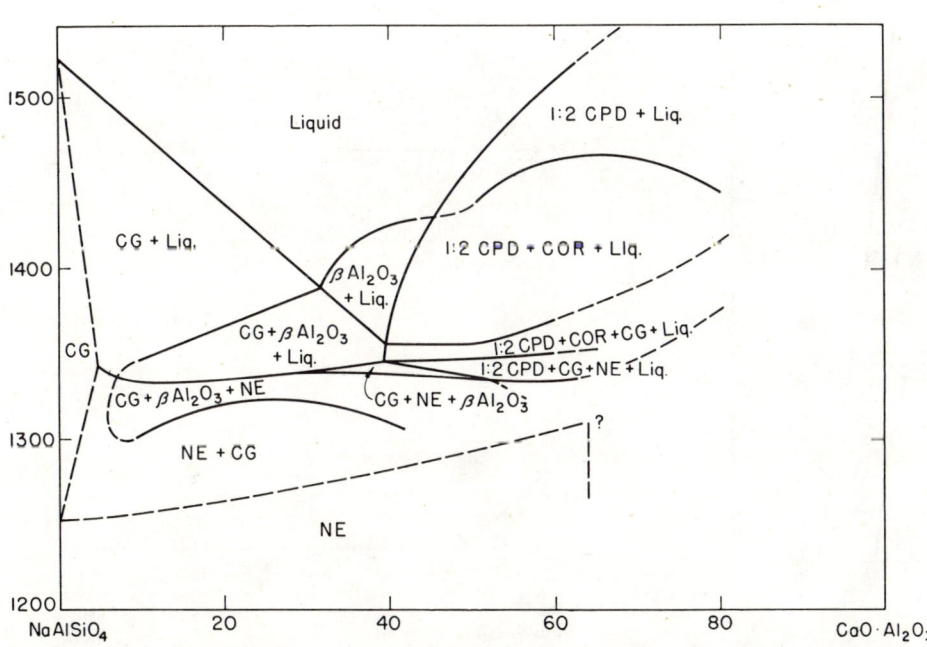

Fig. 848.—System $NaAlSiO_4$–$CaO \cdot Al_2O_3$; CG = carnegieite solid solution, Ne = nepheline solid solution, 1:2 cpd = $CaO \cdot 2Al_2O_3$, cor = corundum.

Julian R. Goldsmith, *Am. Mineral.*, **34**, 477 (1949).

Na$_2$O–CaO–Al$_2$O$_3$–SiO$_2$ (concl.)

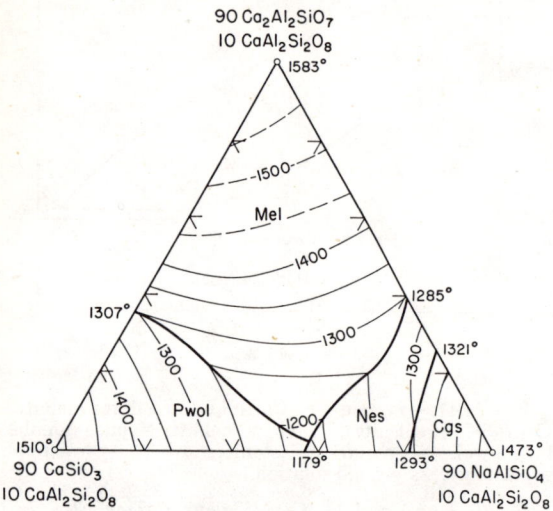

FIG. 850.—Equilibrium diagram for the 10% CaAl$_2$Si$_2$O$_8$ plane in the system CaSiO$_3$–Ca$_2$Al$_2$SiO$_7$–NaAlSiO$_4$–CaAl$_2$Si$_2$O$_8$. Mel = melilite, Nes = nepheline solid solutions, CgS = carnegieite solid solutions, Pwol = pseudowollastonite.

Hatten S. Yoder, Jr., *J. Geol.*, **60**, 591 (1952).

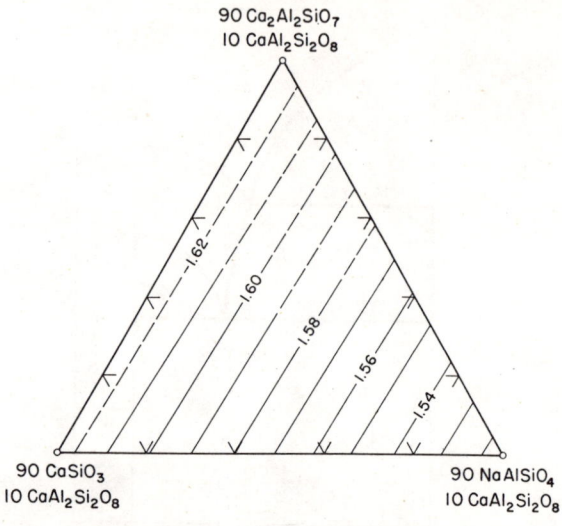

FIG. 851.—Isofract diagram for glasses in the 10% CaAl$_2$Si$_2$O$_8$ plane in system CaSiO$_3$–Ca$_2$Al$_2$SiO$_7$–NaAlSiO$_4$–CaAl$_2$Si$_2$O$_8$.

Hatten S. Yoder, Jr., *J. Geol.*, **60**, 592 (1952).

Na$_2$O–FeO–Al$_2$O$_3$–SiO$_2$

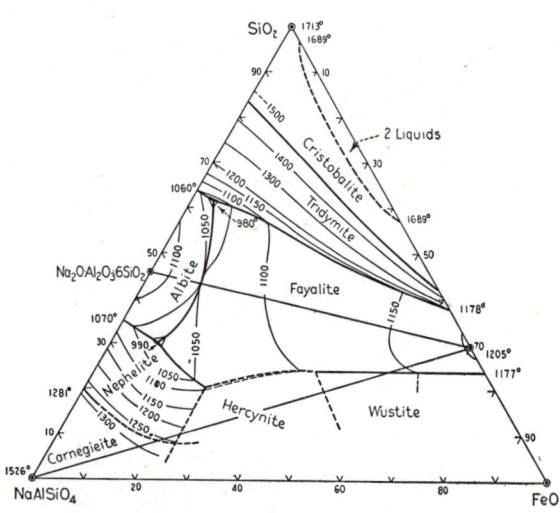

Fig. 853.—System FeO–SiO$_2$–Na$_2$O·Al$_2$O$_3$·2SiO$_2$; hercynite equilibria are not ternary.

N. L. Bowen, *Am. J. Sci.*, **33**, 9 (1937).

Na$_2$O–CaO–B$_2$O$_3$–SiO$_2$

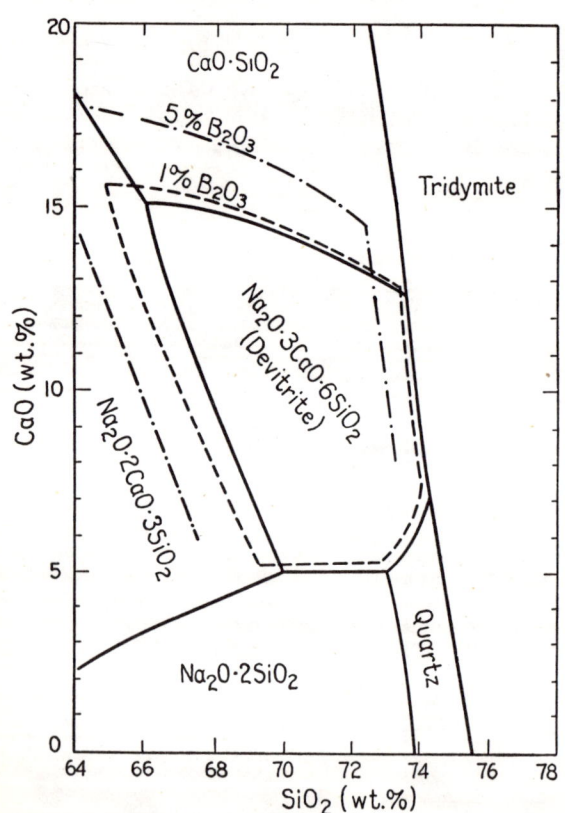

Fig. 852.—Effects of additions of 1% and 5% of B$_2$O$_3$ on devitrite (Na$_2$O·3CaO·6SiO$_2$); see system Na$_2$O–CaO–SiO$_2$, Fig. 483.

G. W. Morey. *J. Am. Ceram. Soc.*, **15** [9] 470 (1932).

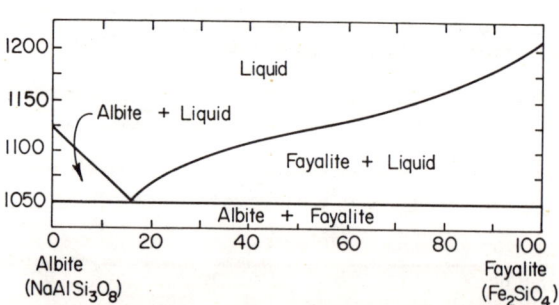

FIG. 854.—System 2FeO·SiO$_2$–Na$_2$O·Al$_2$O$_3$·6SiO$_2$.

N. L. Bowen and J. F. Schairer, *Proc. Natl. Acad. Sci., U.S.*, **22** [6] 349 (1936).

$Na_2O–MgO–Al_2O_3–SiO_2$

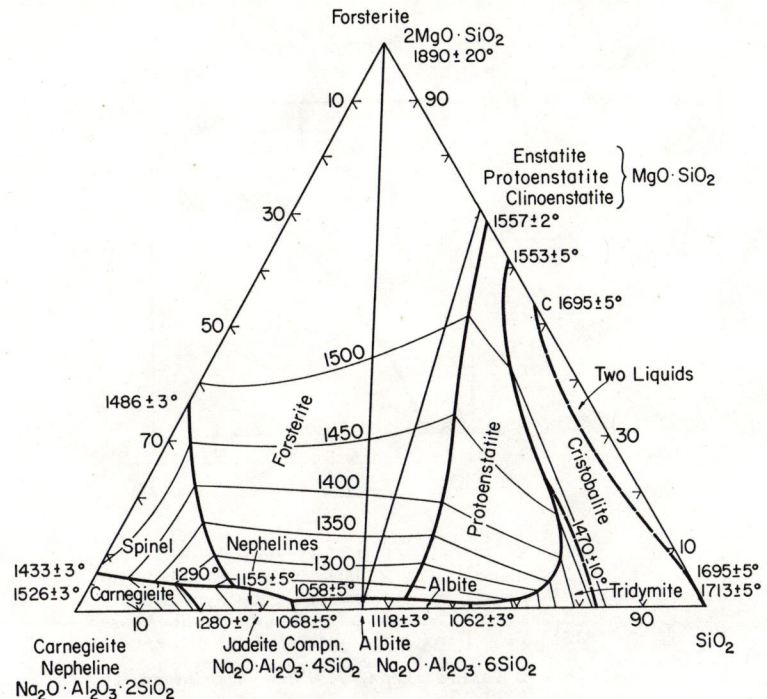

Fig. 855.—System $Na_2O \cdot Al_2O_3 \cdot 2SiO_2 – 2MgO \cdot SiO_2 – SiO_2$.

J. F. Schairer and H. S. Yoder, Jr., Ann. Rept. Director Geophys. Lab. 1960–61; in *Carnegie Inst. Washington Year Book*, **60**, 142 (1961).

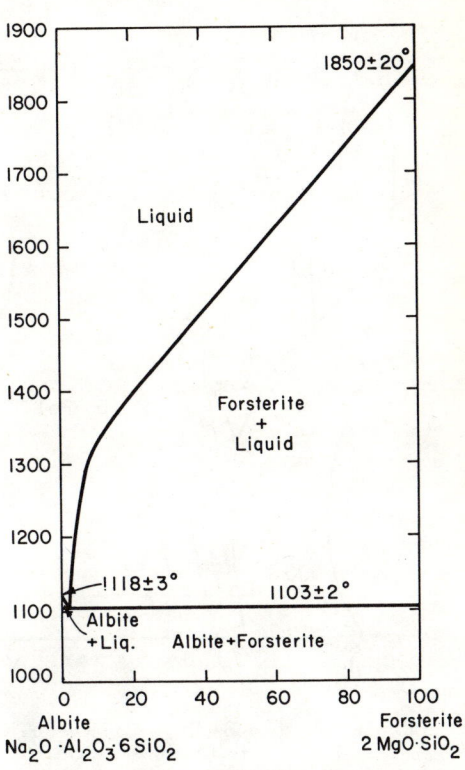

Fig. 856.—System $Na_2O \cdot Al_2O_3 \cdot 6SiO_2 – 2MgO \cdot SiO_2$.

J. F. Schairer and H. S. Yoder, Jr., Ann. Rept. Director Geophys. Lab. 1960–61; in *Carnegie Inst. Washington Year Book*, **60**, 143 (1961).

$BaO–CaO–MgO–SiO_2$

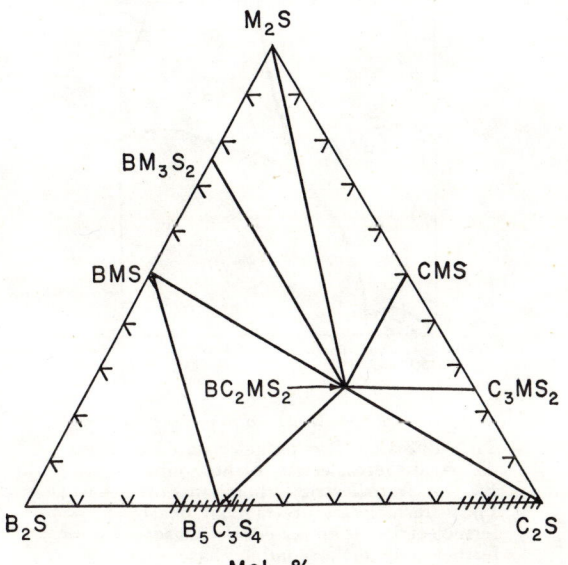

Fig. 857.—System $2BaO \cdot SiO_2 – 2CaO \cdot SiO_2 – 2MgO \cdot SiO_2$; compatibility triangles. B = BaO, C = CaO, M = MgO, S = SiO_2.

Franciszak Nadachowski and Miroslaw Grylicki, *Silikat Tech.*, **10**, 79 (1959).

$BaO–CaO–SrO–TiO_2$

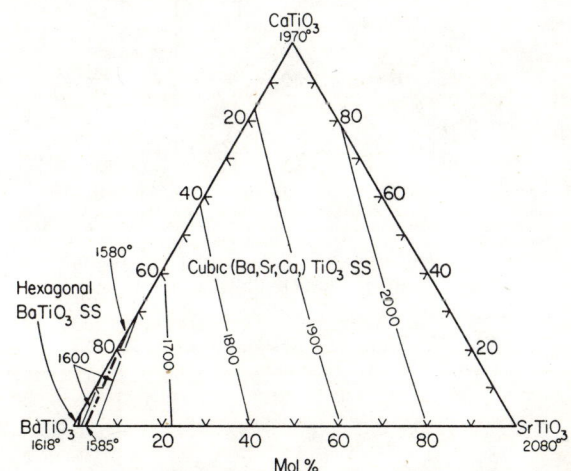

Fig. 858.—System $BaTiO_3–CaTiO_3–SrTiO_3$; not impossible liquidus surface.

J. A. Basmajian and R. C. DeVries, *J. Am. Ceram. Soc.*, **40** [11] 376 (1957).

BaO–PbO–B₂O₃–TiO₂

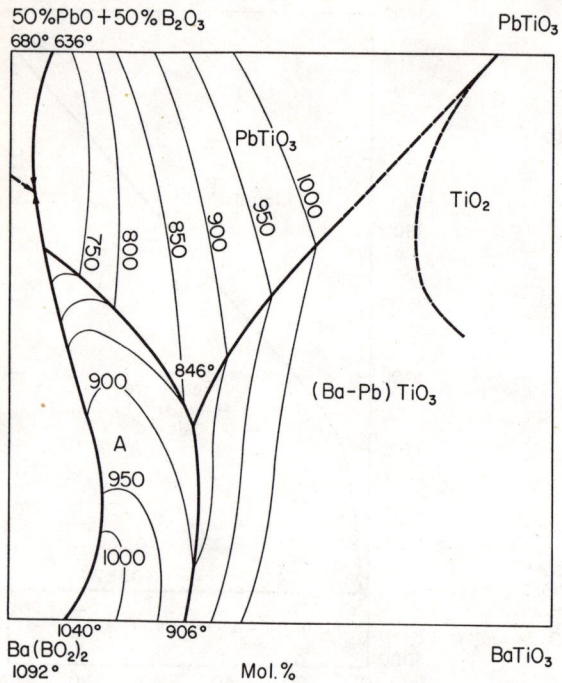

FIG. 859.—System Ba(BO₂)₂–BaTiO₃–PbTiO₃–(50% PbO + 50% B₂O₃); a cut.

M. L. Sholokhovich and V. I. Varicheva, *Izvest. Akad. Nauk S.S.S.R., Ser. Fiz.*, **22**, 1450 (1958).

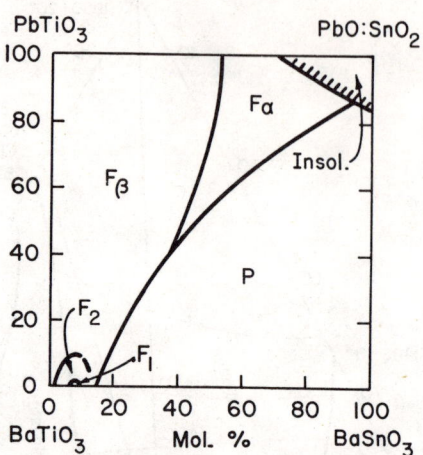

FIG. 861.—System BaSnO₃–BaTiO₃–PbSnO₃–PbTiO₃. P = paraelectric, cubic phase; F_α = ferroelectric, rhombohedral phase; F_β = ferroelectric, tetragonal phase; F_1 = ferroelectric, rhombohedral phase; F_2 = ferroelectric, orthorhombic phase.

Takuro Ikeda, *J. Phys. Soc. Japan*, **14**, 1292 (1959).

BaO–PbO–SnO₂–TiO₂

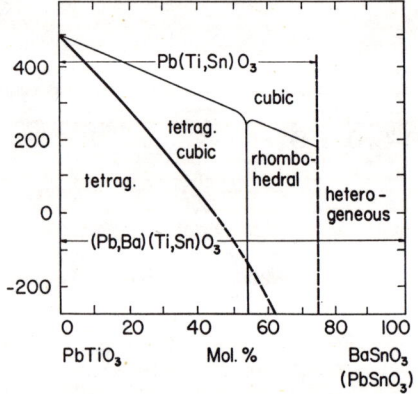

FIG. 860.—System BaSnO₃–PbTiO₃, heavy lines; system PbSnO₃–PbTiO₃, light lines.

Yu. N. Venevtsev, A. G. Kapyshev, and Yu. V. Shumov, *Kristallografiya*, **2**, 227 (1957).

BaO–PbO–TiO₂–ZrO₂

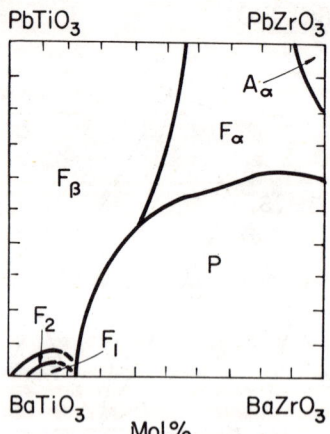

FIG. 862.—System BaTiO₃–BaZrO₃–PbTiO₃–PbZrO₃. P = paraelectric, cubic phase; A_α = antiferroelectric, orthorhombic phase; F_α = ferroelectric, rhombohedral phase; F_β = ferroelectric, tetragonal phase; F_1 = ferroelectric, rhombohedral phase; F_2 = ferroelectric, orthorhombic phase.

Takuro Ikeda, *J. Phys. Soc. Japan*, **14**, 173 (1959).

$BaO-SrO-R_2O_3-TiO_2$

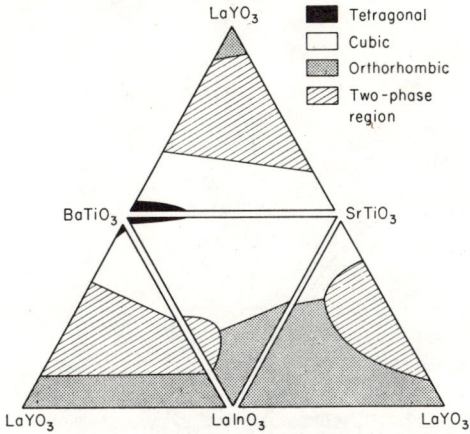

Fig. 863.—System $BaTiO_3-SrTiO_3-LaInO_3-LaYO_3$.

N. N. Padurow and C. Schusterius, *Ber. deut. keram. Ges.*, **33** [9] 291 (1956).

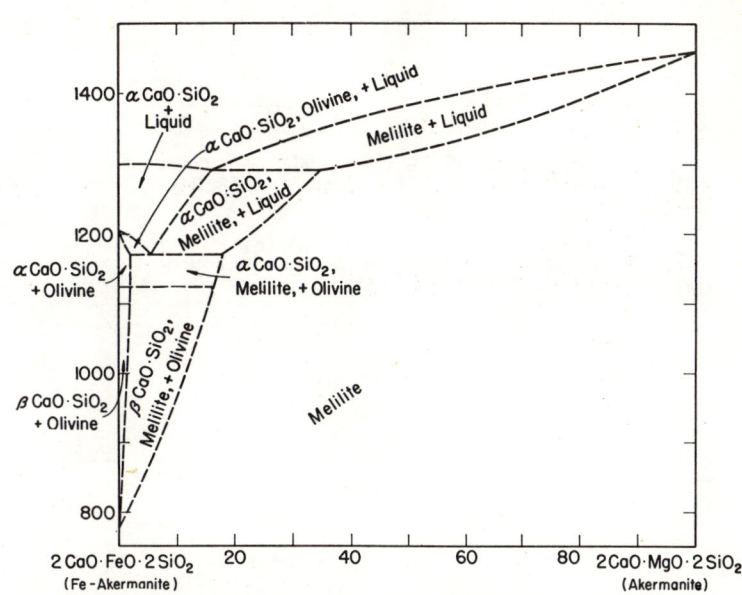

Fig. 865.—System $2CaO \cdot FeO \cdot 2SiO_2$ (Fe-Akermanite)–$2CaO \cdot MgO \cdot 2SiO_2$ (Akermanite); heavy dashed lines refer to binary equilibrium, and light dashed lines enclose areas where one or more phases cannot be expressed in terms of the melilite components; some Fe_2O_3 is present in the liquids.

J. F. Schairer and E. F. Osborn, *J. Am. Ceram. Soc.*, **33** [5] 167 (1950).

$CaO-FeO-MgO-SiO_2$

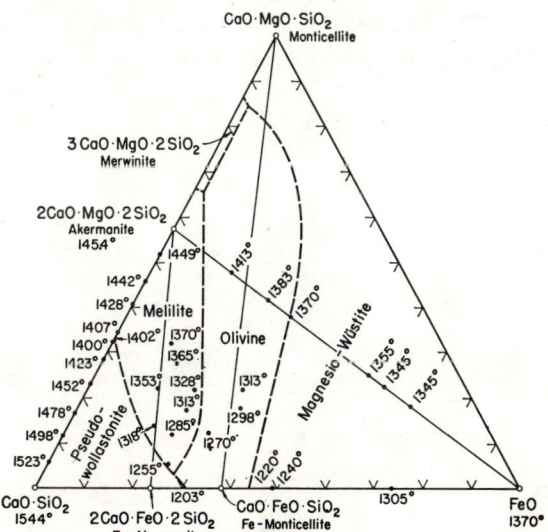

Fig. 864.—System $CaO \cdot SiO_2-CaO \cdot MgO \cdot SiO_2$ (monticellite)–FeO.

J. F. Schairer and E. F. Osborn, *J. Am. Ceram. Soc.*, **33** [5] 164 (1950).

$CaO-FeO-Al_2O_3-SiO_2$

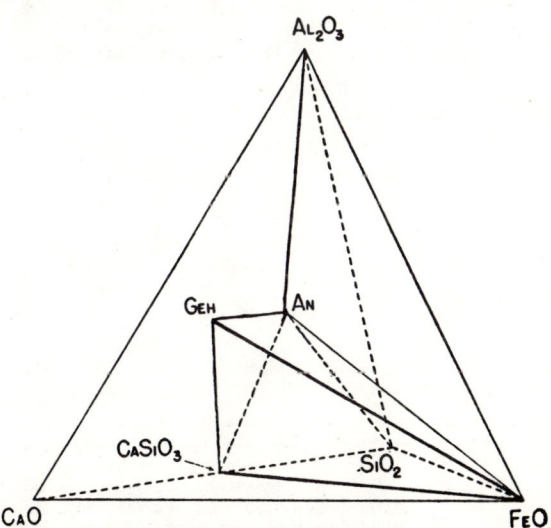

Fig. 866.—System $CaO-FeO-Al_2O_3-SiO_2$; diagram showing position of joins $FeO-Al_2O_3-CAS_2$, $FeO-SiO_2-CAS_2$, $FeO-CS-CAS_2$, $FeO-CS-C_2AS$, and $FeO-CAS_2-C_2AS$; An = anorthite ($CaO \cdot Al_2O_3 \cdot 2SiO_2$); Geh = gehlenite ($2CaO \cdot Al_2O_3 \cdot SiO_2$); see Figs. 868–872.

J. F. Schairer, *J. Am. Ceram. Soc.*, **25** [10] 248 (June, 1942).

CaO–FeO–Al$_2$O$_3$–SiO$_2$ (cont.)

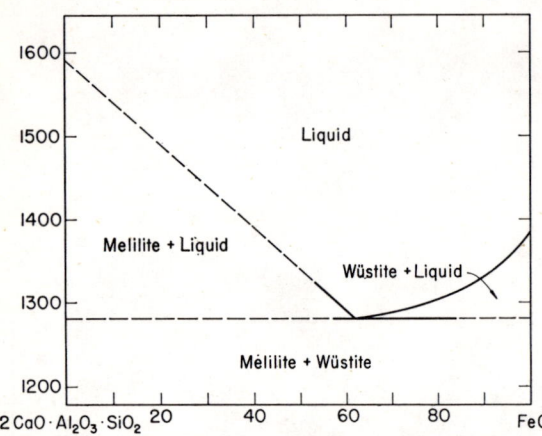

FIG. 867.—System 2CaO·Al$_2$O$_3$·SiO$_2$–FeO. Combined with data of J. F. Schairer, *J. Am. Ceram. Soc.*, **25**, 241–274 (1942).

A. Muan and E. F. Osborn, preprint of paper presented before General Meeting of American Iron and Steel Institute, at New York, May 23–24 (1951).

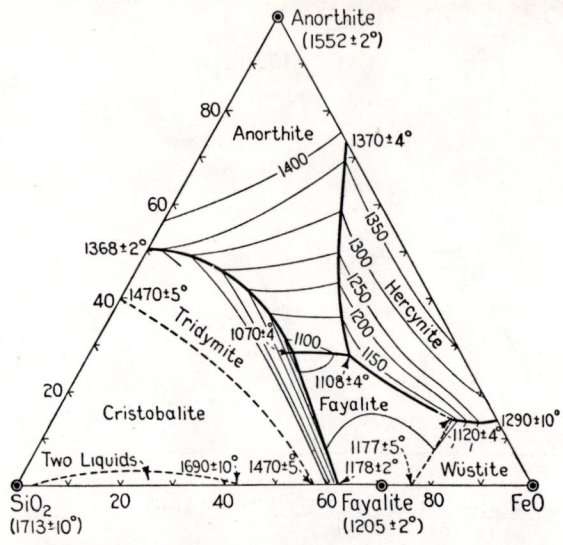

Fig. 869.—Join-plane FeO–SiO$_2$–CaO·Al$_2$O$_3$·2SiO$_2$.

Modified from J. F. Schairer, *J. Am. Ceram. Soc.*, **25** [10] 252 (June, 1942).

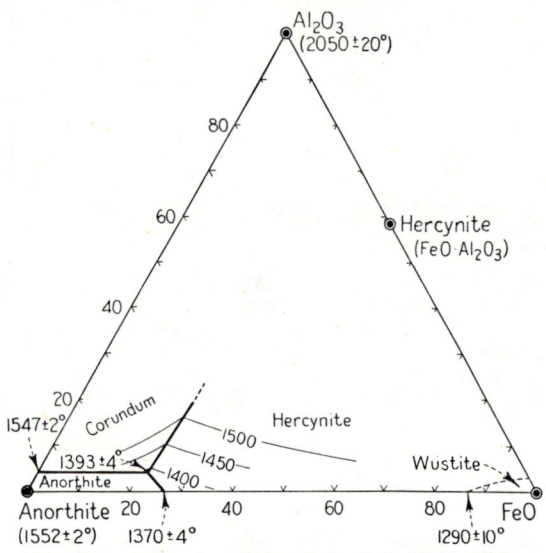

Fig. 868.—Join-plane FeO–Al$_2$O$_3$–CaO·Al$_2$O$_3$·2SiO$_2$.

Modified from J. F. Schairer, *J. Am. Ceram. Soc.*, **25** [10] 254 (June, 1942).

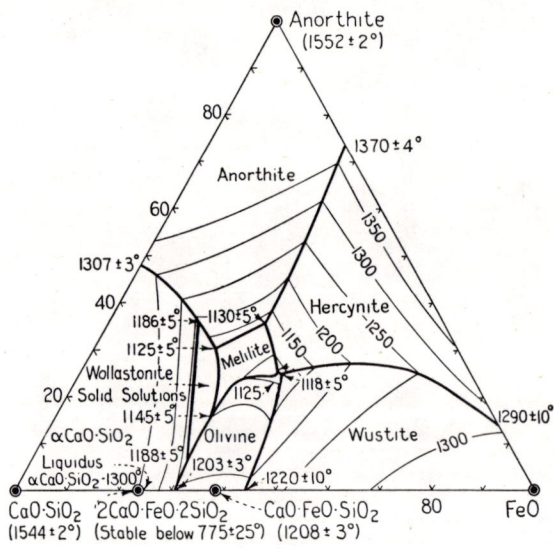

Fig. 870.—Join-plane FeO–CaO·SiO$_2$–CaO·Al$_2$O$_3$·2SiO$_2$.

Modified from J. F. Schairer, *J. Am. Ceram. Soc.*, **25** [10] 256 (June, 1942).

FIG. 871.—System CaO·SiO$_2$–2CaO·Al$_2$O$_3$·SiO$_2$–FeO. After Schairer, *J. Am. Ceram. Soc.*, **25**, 241–274 (1942), revised with respect to the position of the wüstite-melilite boundary curve, and with lines added to indicate liquidus temperature (light solid lines) and Fe$_2$O$_3$ content at the liquidus temperature (light dashed lines).

A. Muan and E. F. Osborn, preprint of paper presented before General Meeting of American Iron and Steel Institute, at New York, May 23–24 (1951).

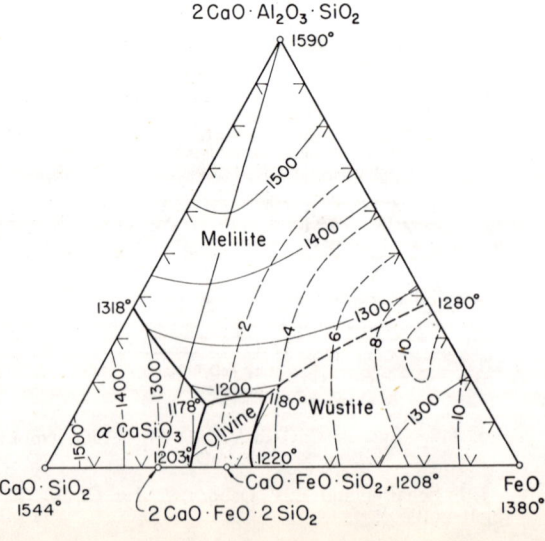

$CaO-FeO-Al_2O_3-SiO_2$ (cont.)

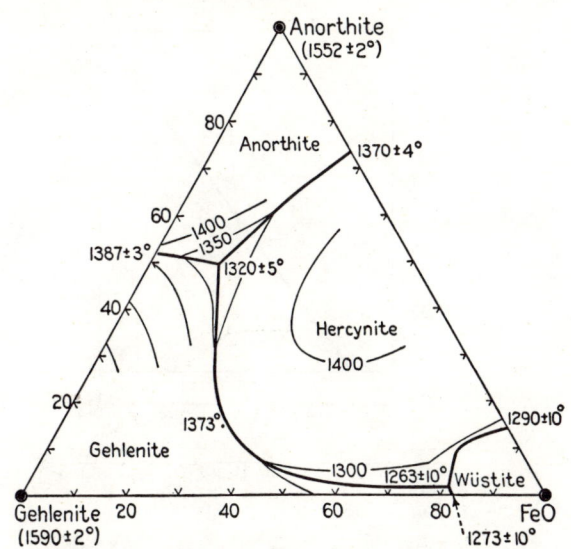

Fig. 872.—Join-plane $FeO-2CaO \cdot Al_2O_3 \cdot SiO_2-CaO \cdot Al_2O_3 \cdot 2SiO_2$.

Modified from J. F. Schairer, *J. Am. Ceram. Soc.*, **25** [10] 260 (June, 1942).

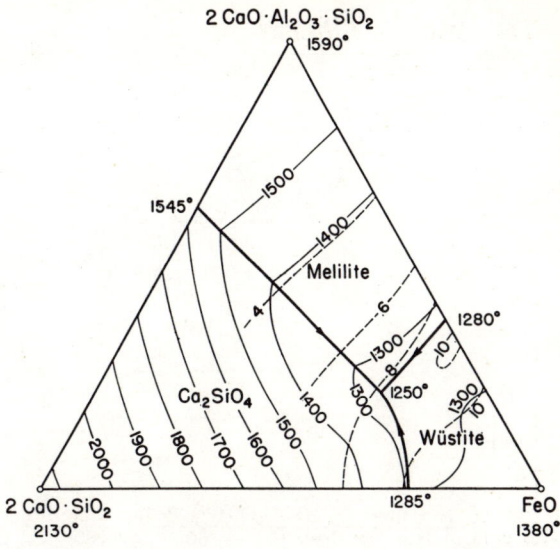

Fig. 874.—System $2CaO \cdot SiO_2-2CaO \cdot Al_2O_3 \cdot SiO_2-FeO$; light dashed lines pass approximately through points of equal Fe_2O_3 content at liquidus temperatures. Ca_2SiO_4 is treated as a single phase owing to the uncertainty of phase relations in the system Ca_2SiO_4.

A. Muan and E. F. Osborn, preprint of paper presented before General Meeting of American Iron and Steel Institute, at New York, May 23–24 (1951).

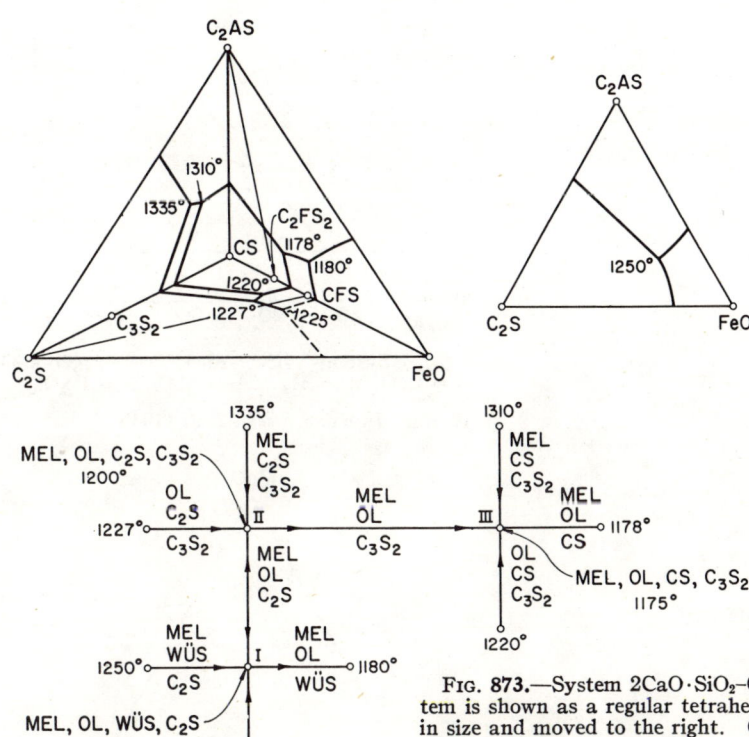

Fig. 873.—System $2CaO \cdot SiO_2-CaO \cdot SiO_2-2CaO \cdot Al_2O_3 \cdot SiO_2-FeO$; the system is shown as a regular tetrahedron with the front face removed, reduced in size and moved to the right. Compounds, boundary curves on the faces, and two binary joins are shown. The diagram in the lower part of the figure indicates schematically the relationships of univariant lines and invariant points existing in the tetrahedron. The three crystalline phases in equilibrium with liquid along a line, the four crystalline phases coexisting with liquid at a point, and the temperature of invariant points are shown. $C_2S = Ca_2SiO_4$, $C_3S_2 = Ca_3Si_2O_7$, $CS = \alpha\text{-}CaSiO_3$, $C_2FS_2 = Ca_2FeSi_2O_7$, $CFS = CaFeSiO_4$, $C_2AS = Ca_2Al_2SiO_7$, Mel = Melilite $(Ca_2(Al, Fe)(Al, Si)SiO_7)$, Ol = Olivine $(Ca(Ca, Fe)SiO_4)$, Wüs = Wüstite.

A. Muan and E. F. Osborn, preprint of paper presented before General Meeting of American Iron and Steel Institute, at New York, May 23–24 (1951).

CaO–FeO–MgO–SiO₂ (concl.)

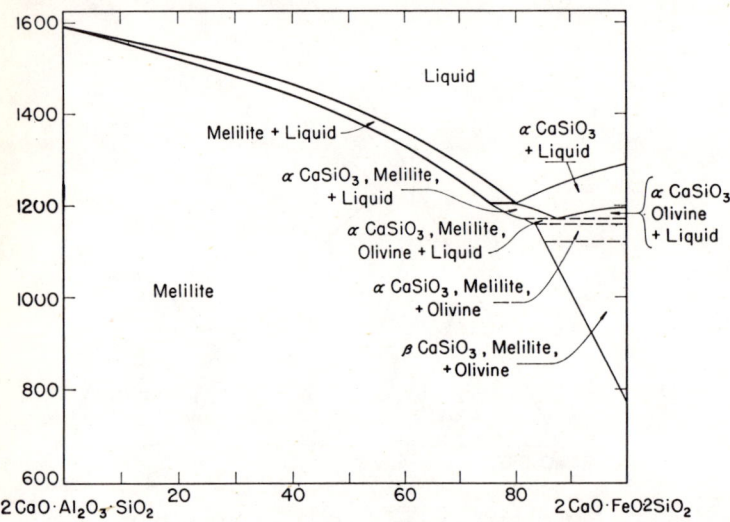

FIG. 875.—System $2CaO \cdot Al_2O_3 \cdot SiO_2$–$2CaO \cdot FeO \cdot 2SiO_2$; heavy lines refer to binary, and light lines to quaternary relations.

A. Muan and E. F. Osborn, preprint of paper presented before General Meeting of American Iron and Steel Institute, at New York, May 23–24, (1951).

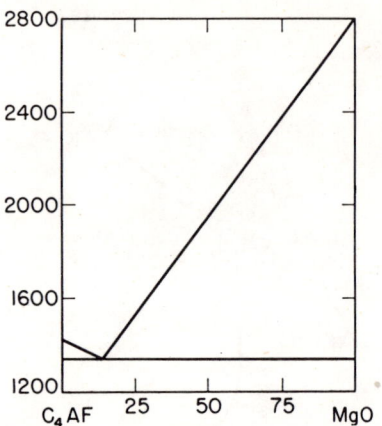

FIG. 877.—System $4CaO \cdot Al_2O_3 \cdot Fe_2O_3$–MgO; C = CaO, A = Al_2O_3, F = Fe_2O_3.

J. R. Rait, *Iron and Steel*, **22**, 90 (1949).

CaO–MgO–Al₂O₃–Fe₂O₃

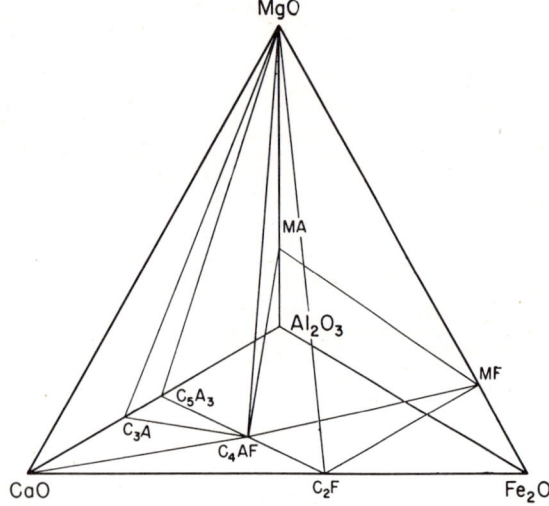

FIG. 876.—Phase distribution in the system CaO–MgO–Al_2O_3–Fe_2O_3; C = CaO, M = MgO, A = Al_2O_3, F = Fe_2O_3.

J. R. Rait, *Iron and Steel*, **22**, 89 (1949).

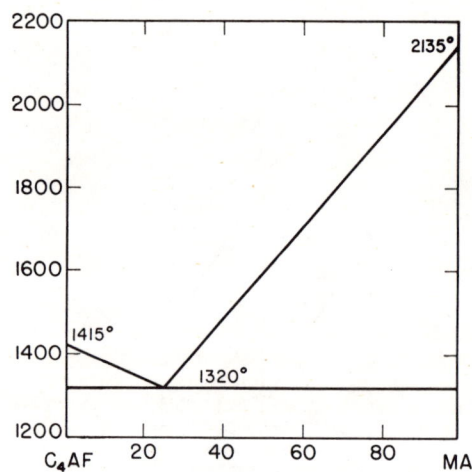

FIG. 878.—System $4CaO \cdot Al_2O_3 \cdot Fe_2O_3$–$MgO \cdot Al_2O_3$; C = CaO, M = MgO, A = Al_2O_3, F = Fe_2O_3.

J. R. Rait, *Iron and Steel*, **22**, 91 (1949).

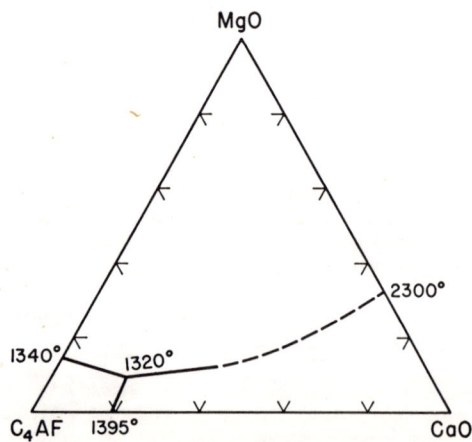

FIG. 879.—System CaO–MgO–$4CaO \cdot Al_2O_3 \cdot Fe_2O_3$; C = CaO, A = Al_2O_3, F = Fe_2O_3.

J. R. Rait, *Iron and Steel*, **22**, 90 (1949).

$CaO-MgO-Al_2O_3-SiO_2$

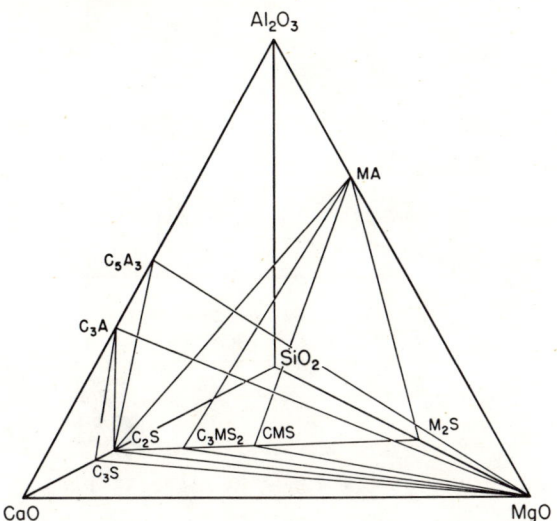

FIG. 880.—Phase distribution in the system $CaO-MgO-Al_2O_3-SiO_2$; C = CaO, M = MgO, A = Al_2O_3, S = SiO_2.

J. R. Rait, *Iron & Steel*, **22**, 18 (1949).

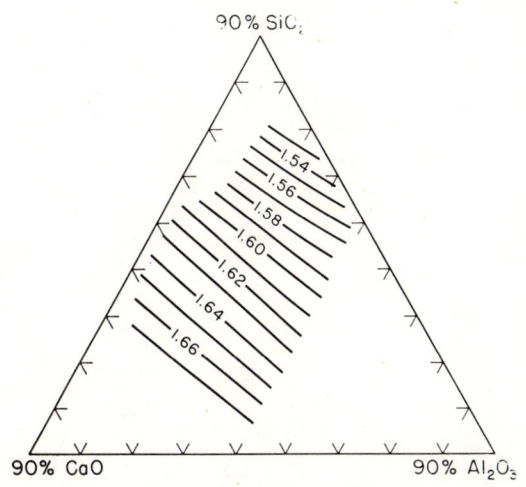

FIG. 883.—System $CaO-MgO-Al_2O_3-SiO_2$; isofracts for 10% MgO plane.

A. T. Prince, *J. Am. Ceram. Soc.*, **37** [9] 407 (1954).

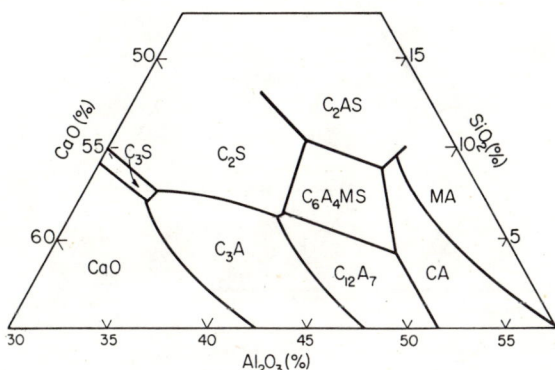

FIG. 881.—System $CaO-MgO-Al_2O_3-SiO_2$, at 5% MgO plane. C = CaO; M = MgO; A = Al_2O_3; S = SiO_2.

F. M. Lea and C. H. Desch, The Chemistry of Cement and Concrete, 2d ed., p. 71. Edward Arnold & Co., London, 1956.

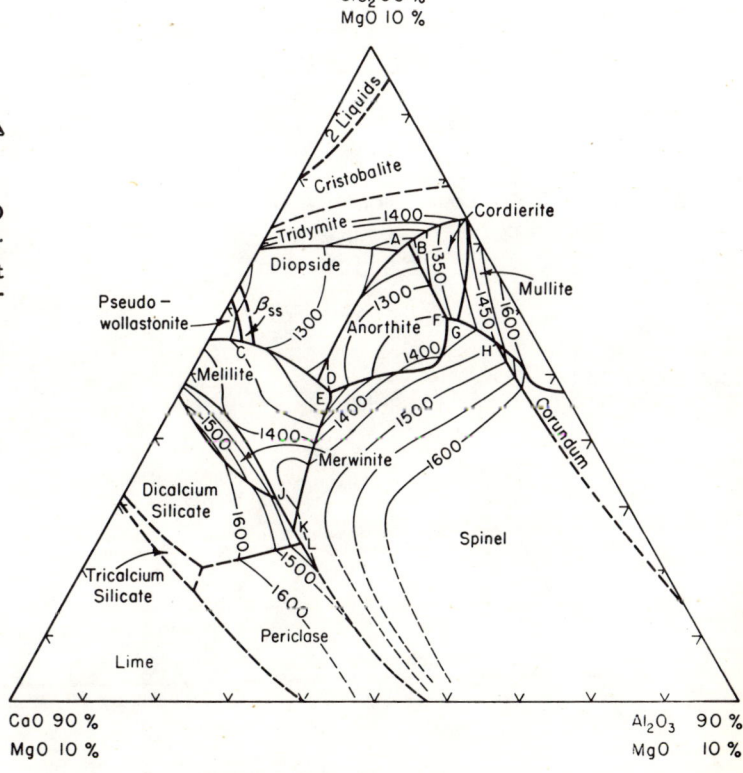

FIG. 882.—System $CaO-MgO-Al_2O_3-SiO_2$, for 10% MgO plane. Temperatures of quaternary piercing points; ±5°C.: A = 1230; B = 1245; C = 1330; D = 1235; E = 1250; F = 1345; G = 1370; H = 1485; J = 1425; K = 1410; L = 1410.

A. T. Prince, *J. Am. Ceram. Soc.*, **37** [9] 406 (1954).

CaO–MgO–Al₂O₃–SiO₂ (cont.)

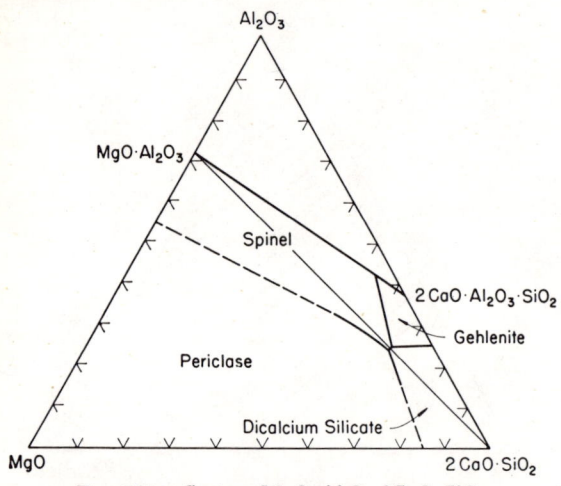

Fig. 884.—System MgO–Al₂O₃–2CaO·SiO₂.

A. T. Prince, *J. Am. Ceram. Soc.*, **34** [2] 50 (1951).

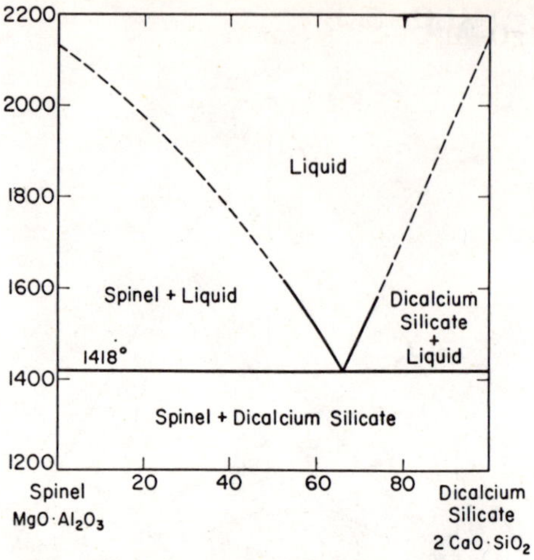

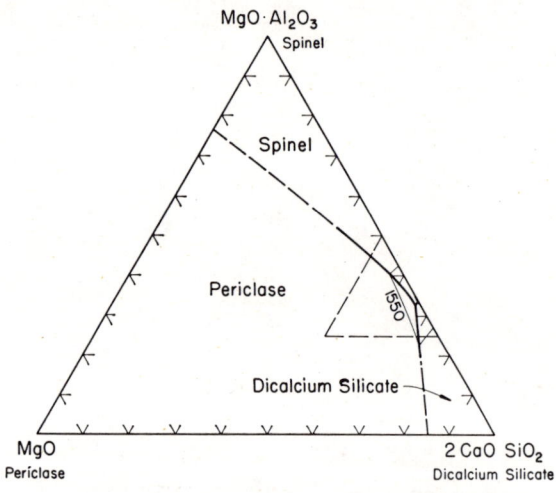

Fig. 885.—System MgO–2CaO·SiO₂–MgO·Al₂O₃.

A. T. Prince, *J. Am. Ceram. Soc.*, **34** [2] 47 (1951).

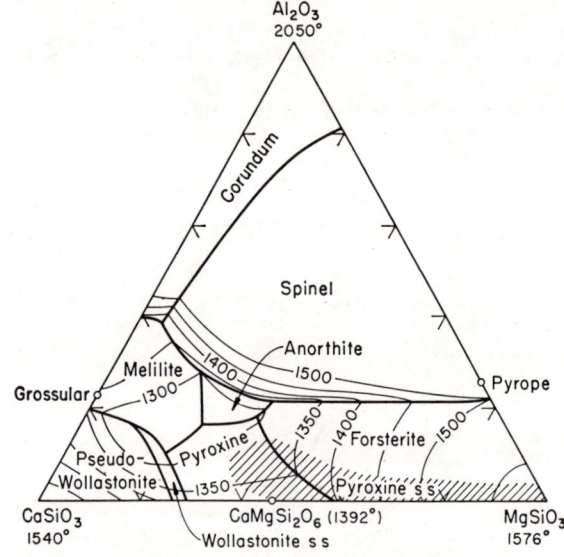

Fig. 887 — System MgO·Al₂O₃–2CaO·SiO₂.

A. T. Prince, *J. Am. Ceram. Soc.*, **34** [2] 46 (1951).

Fig. 888.—Join CaSiO₃–MgSiO₃–Al₂O₃.

E. R. Segnit, *Mineral. Mag.*, **31**, 257 (1956).

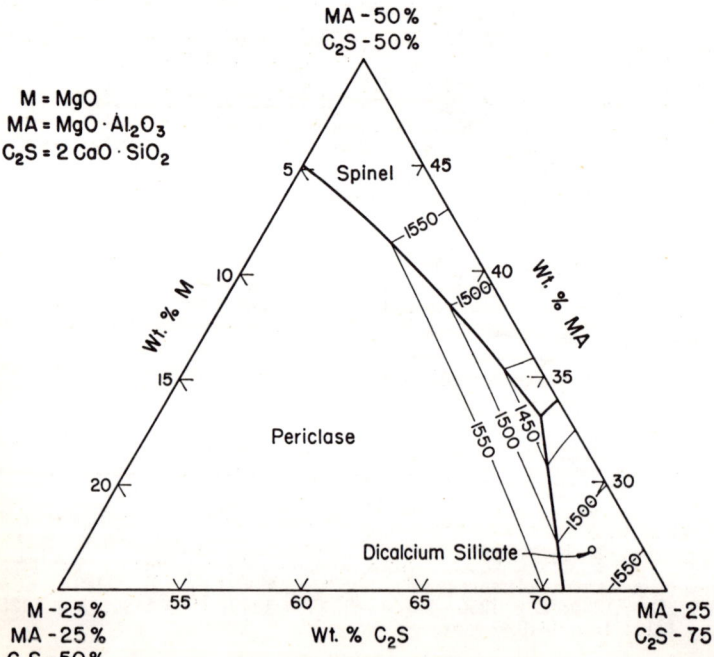

Fig. 886.—System MgO–2CaO·SiO₂–MgO·Al₂O₃; enlarged portion.

A. T. Prince, *J. Am. Ceram. Soc.* **34** [2] 47 (1951).

CaO–MgO–Al$_2$O$_3$–SiO$_2$ (cont.)

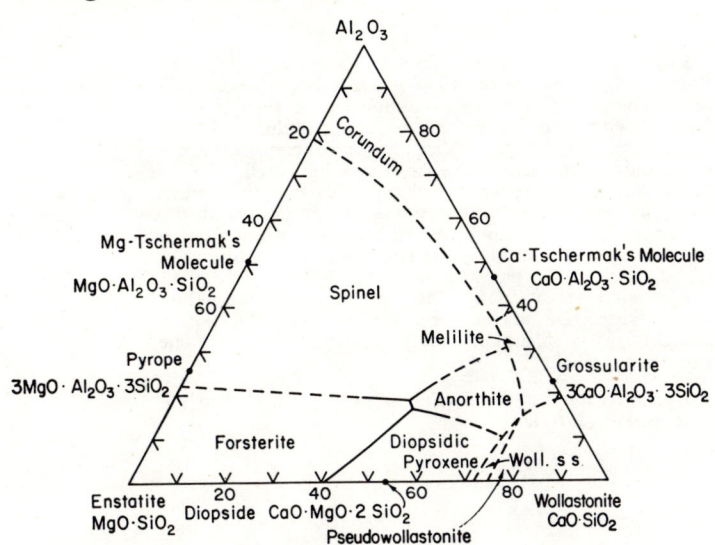

FIG. 889.—System CaO·SiO$_2$–MgO·SiO$_2$–Al$_2$O$_3$; join.

Kai Hytönen and J. F. Schairer, Ann. Rept. Director Geophys. Lab. 1960–61; in *Carnegie Inst. Washington Year Book*, **60**, 134 (1961).

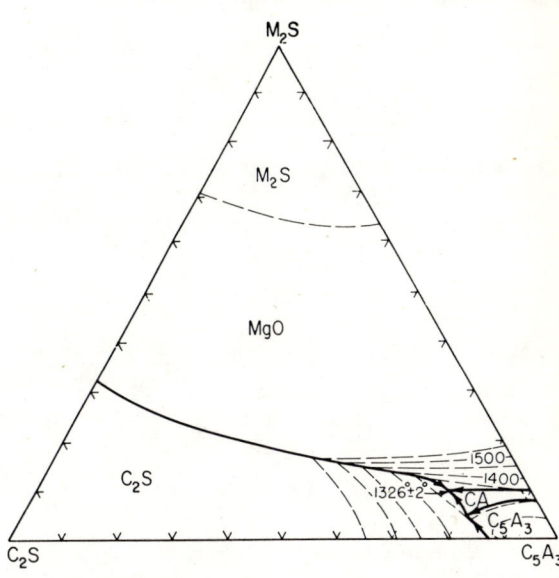

FIG. 890.—System 2CaO·SiO$_2$–2MgO·SiO$_2$–5CaO·3Al$_2$O$_3$; C = CaO; M = MgO; A = Al$_2$O$_3$; S = SiO$_2$.

E. R. Segnit and J. H. Weymouth, *Trans. Brit. Ceram. Soc.*, **56** [5] 254 (1957).

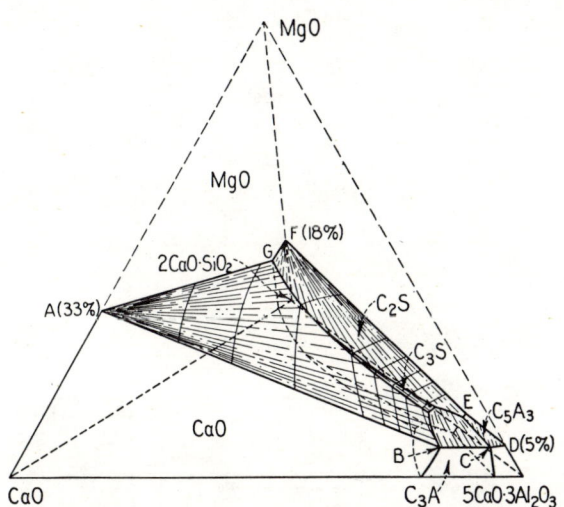

Fig. 891.—System CaO–MgO–2CaO·SiO$_2$–5CaO·3Al$_2$O$_3$; the surface intersecting the sides of the tetrahedron at A–B–C–D–E–F–G indicates the lower level of the primary-phase volume of MgO; C$_3$S = 3CaO·SiO$_2$; C$_2$S = 2CaO·SiO$_2$; C$_3$A = 3CaO·Al$_2$O$_3$; C$_5$A$_3$ = 5CaO·3Al$_2$O$_3$.

H. F. McMurdie and Herbert Insley, *J. Research Natl. Bur. Standards*, **16** [5] 471 (1936); R P 884.

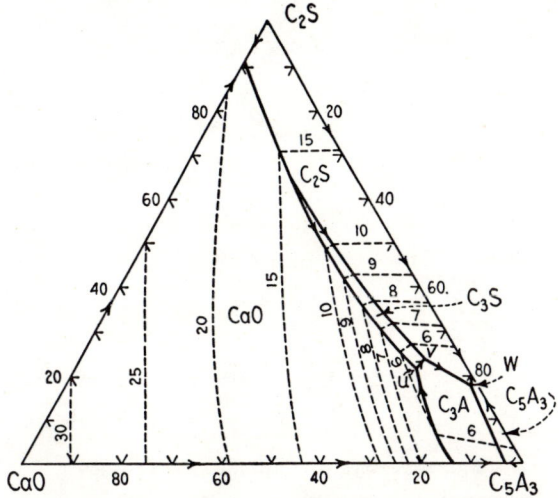

Fig. 892.—Projection of surface A–B–C–D–E–F–G of Fig. 891 onto the base of the tetrahedron with the eye at apex MgO; contours represent level of the surface with respect to MgO content; C = CaO; A = Al$_2$O$_3$; S = SiO$_2$.

H. F. McMurdie and Herbert Insley, *J. Research Natl. Bur. Standards*, **16** [5] 472 (1936); RP 884.

FIG. 893.—Plane at 5% MgO through the CaO–Al$_2$O$_3$–SiO$_2$–MgO System. A = Al$_2$O$_3$, C = CaO, M = MgO, S = SiO$_2$.

F. M. Lea and C. H. Desch, "The Chemistry of Cement and Concrete," 2nd ed., p. 71, Edward Arnold Ltd., London, 1956. Based on work of H. F. McMurdie and Herbert Insley, *J. Research Natl. Bur. Standards*, **16** [5] 473 (1936) and T. W. Parker, *Sym.*, London, 485 (1952).

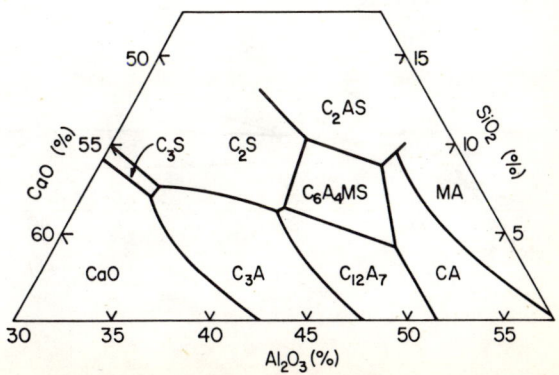

CaO–MgO–Al$_2$O$_3$–SiO$_2$ (cont.)

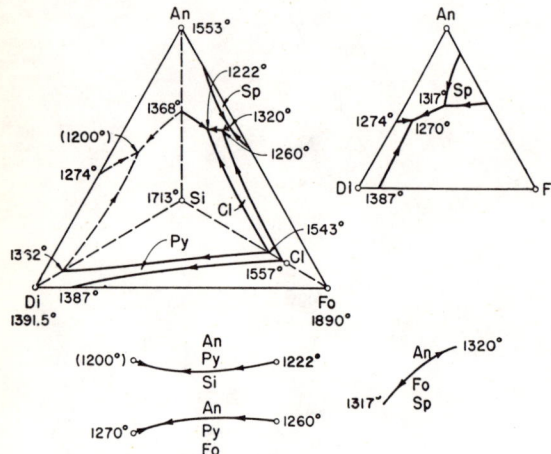

Fig. 894. — System CaO·MgO·2SiO$_2$ (diopside)–2MgO·SiO$_2$ (forsterite)–CaO·Al$_2$O$_3$·2SiO$_2$ (anorthite)–SiO$_2$ (silica). The system is shown as a regular tetrahedron with the front face removed, reduced in size and shown at the right. The position of compounds and of boundary curves on the faces are indicated. In the lower part of the figure, the three univariant lines which pass through the tetrahedron are shown schematically, with the three crystalline phases indicated which exist in equilibrium with liquids along the line. The temperature of 1200°, shown in parentheses, is the approximate metastable eutectic of cristobalite, diopside, and anorthite, after Grieg. Di = diopside, Fo = forsterite, An = anorthite, Si = silica, Cl = clinoenstatite, Py = pyroxene, and Sp = spinel.

E. F. Osborn and D. B. Tait, *Am. J. Sci.*, Bowen volume, p. 426 (1952).

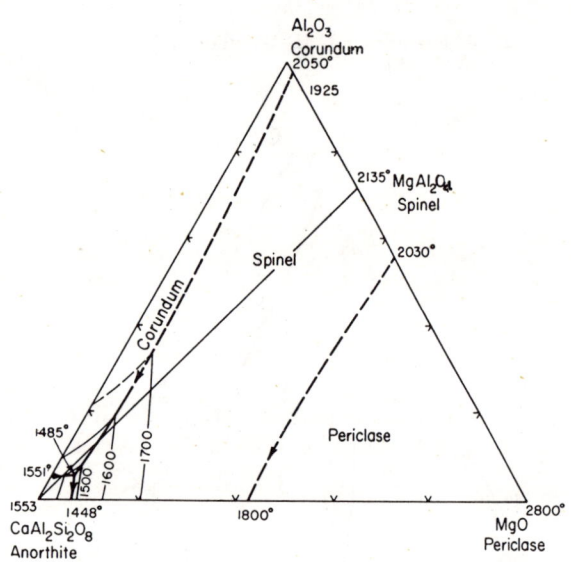

Fig. 895.—System CaAl$_2$Si$_2$O$_8$–MgO–Al$_2$O$_3$.

R. C. DeVries and E. F. Osborn, *J. Am. Ceram. Soc.*, **40** [1] 9 (1957).

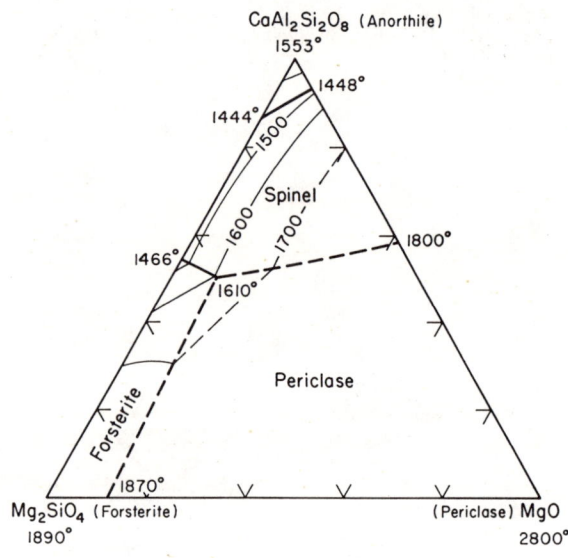

Fig. 897.—System CaAl$_2$Si$_2$O$_8$–MgO–Mg$_2$SiO$_4$.

R. C. DeVries and E. F. Osborn, *J. Am. Ceram. Soc.*, **40** [1] 10 (1957)

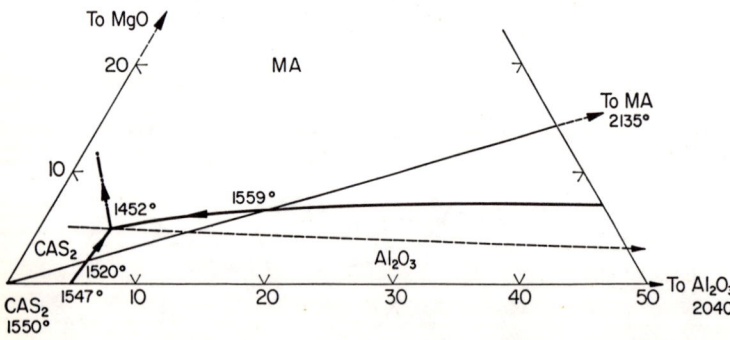

Fig. 896.—System CaO·Al$_2$O$_3$·2SiO$_2$–MgO–Al$_2$O$_3$; portion of the plane. A = Al$_2$O$_3$, C = CaO, M = MgO, S = SiO$_2$.

J. H. Welch, *J. Iron Steel Inst.*, (*London*), **183** [3] 280 (1956).

CaO–MgO–Al$_2$O$_3$–SiO$_2$ (cont.)

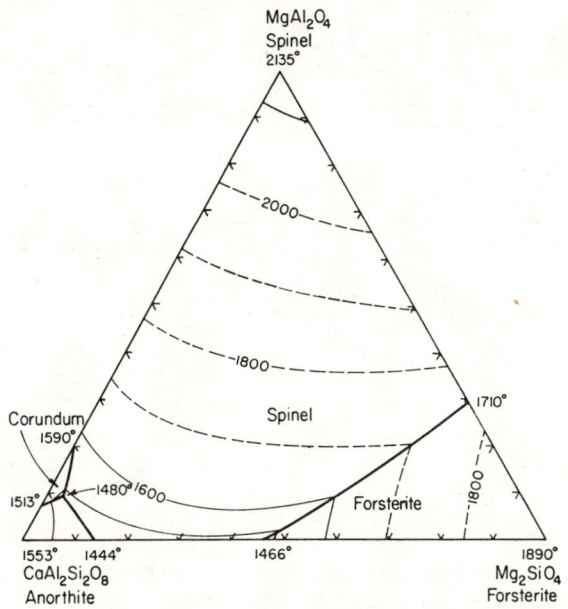

Fig. 898.—System CaAl$_2$Si$_2$O$_8$–MgAl$_2$O$_4$–Mg$_2$SiO$_4$.

R. C. DeVries and E. F. Osborn, *J. Am. Ceram. Soc.*, **40** [1] 9 (1957).

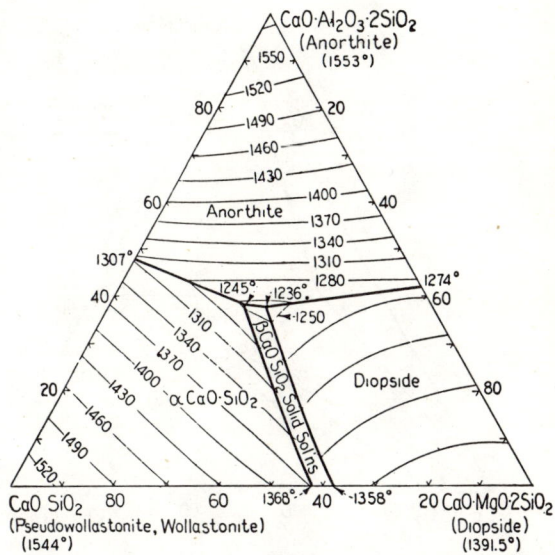

Fig. 900.—System CaO·SiO$_2$–CaO·MgO·2SiO$_2$–CaO·Al$_2$O$_3$·2SiO$_2$.

E. F. Osborn, *Am. J. Sci.*, **240**, 761 (1942).

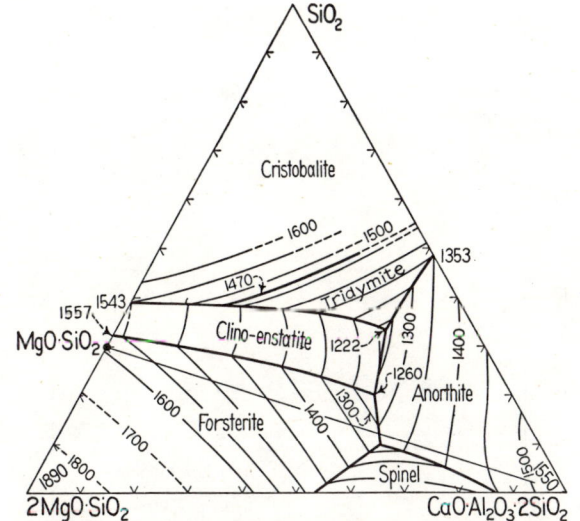

Fig. 899.—System SiO$_2$–2MgO·SiO$_2$–CaO·Al$_2$O$_3$·2SiO$_2$.

Olaf Andersen, *Am. J. Sci.*, 4th Ser., **39**, 440 (1915).

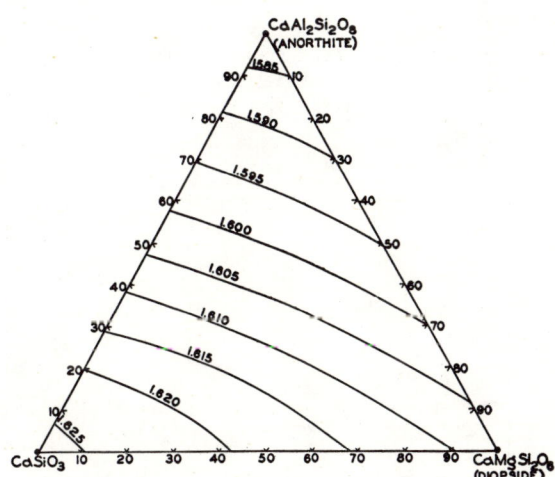

Fig. 901.—Refractive indices of glasses in system CaO·SiO$_2$–CaO·MgO·2SiO$_2$–CaO·Al$_2$O$_3$·2SiO$_2$.

E. F. Osborn, *Am. J. Sci.*, **240**, 786 (1942).

CaO–MgO–Al$_2$O$_3$–SiO$_2$ (cont.)

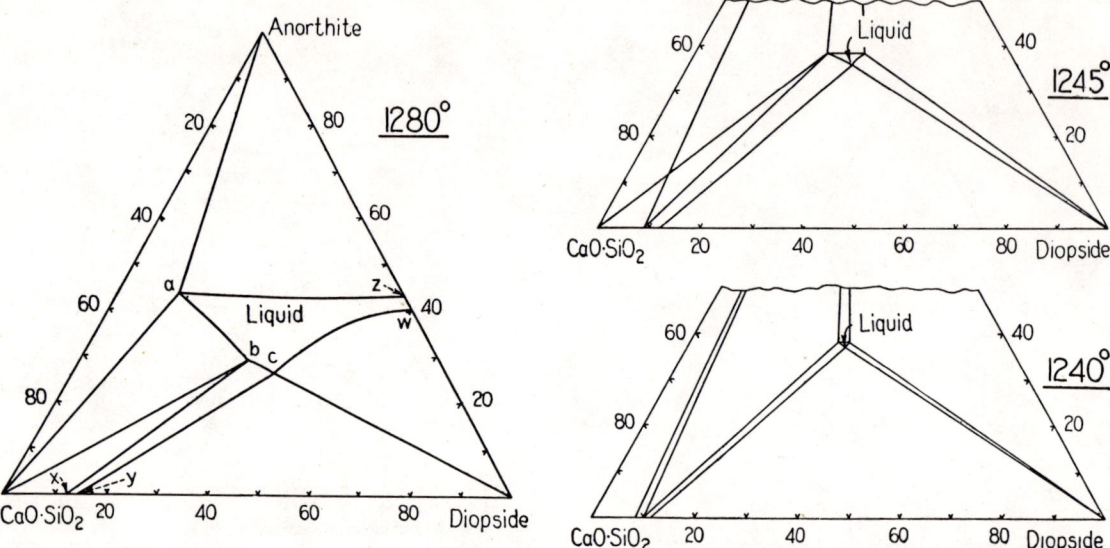

Fig. 902.—Phase relations at 1280°, 1245°, and 1240°C. in system CaO·SiO$_2$–CaO·MgO·2SiO$_2$–CaO·Al$_2$O$_3$·2SiO$_2$.

E. F. Osborn, *Am. J. Sci.*, **240**, pp. 768, 769, 770 (1942).

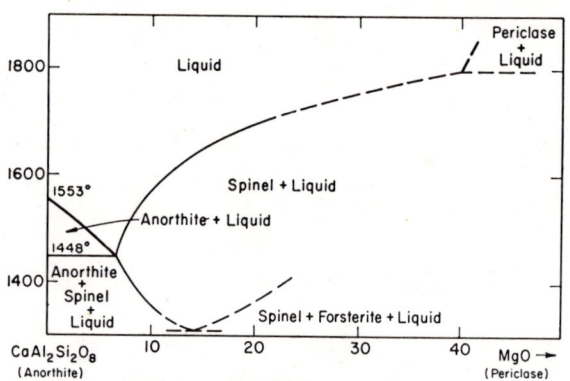

FIG. 903.—System CaAl$_2$Si$_2$O$_8$–MgO. Heavy lines indicate binary relationship.

R. C. DeVries and E. F. Osborn, *J. Am. Ceram. Soc.*, **40** [1] 8 (1957).

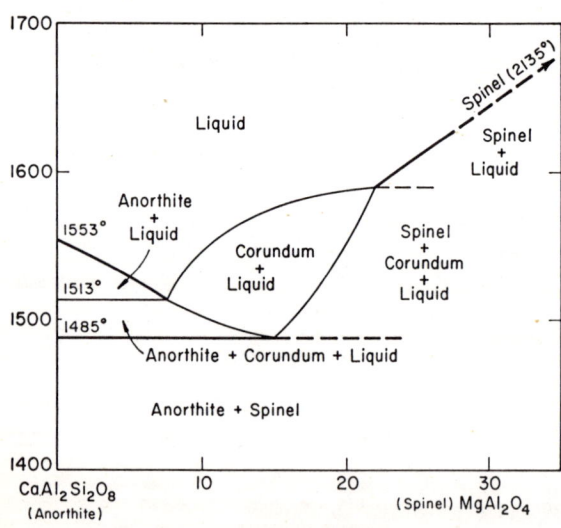

FIG. 904.—System CaAl$_2$Si$_2$O$_8$–MgAl$_2$O$_4$. Heavy lines indicate binary relationships.

R. C. DeVries and E. F. Osborn, *J. Am. Ceram. Soc.*, **40** [1] 8 (1957).

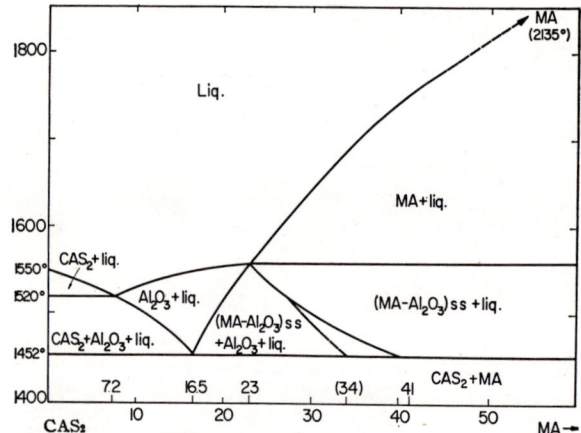

FIG. 905.—System CaO·Al$_2$O$_3$·2SiO$_2$ (anorthite)–MgO·Al$_2$O$_3$ (spinel). A = Al$_2$O$_3$, C = CaO, M = MgO, S = SiO$_2$.

J. H. Welch, *J. Iron Steel Inst.* (*London*), **183** [3] 279 (1956).

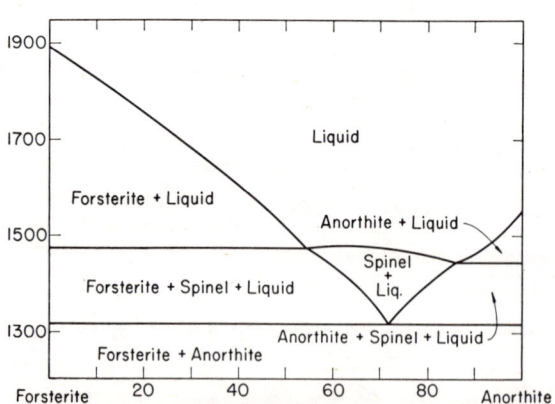

FIG. 906.—System 2MgO·SiO$_2$–CaO·Al$_2$O$_3$·2SiO$_2$; modification of Andersen's diagram (Fig. 899) to show subliquidus relationships.

E. F. Osborn and D. B. Tait, *Am. J. Sci.*, Bowen volume, p. 418 (1952).

CaO–MgO–Al₂O₃–SiO₂ (cont.)

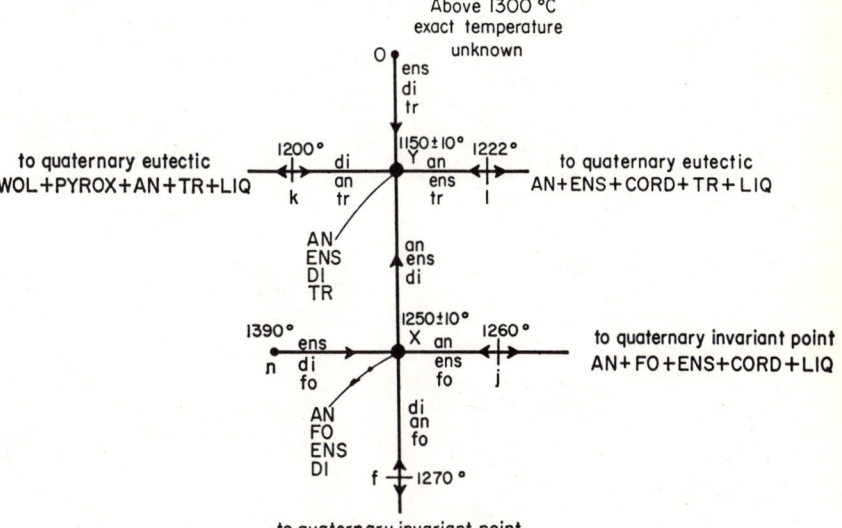

Fig. 908.—System enstatite-anorthite-diopside-silica; schematic relationships of ternary invariant points (small circles) to univariant lines and to quaternary invariant points (large circles, X and Y). An = anorthite; Cord = cordierite; Di = diopside; Ens = enstatite; Fo = forsterite; Mel = melilite; Pyrox = pyroxene; Tr = tridymite; Wol = wollastonite.

Kai Hytönen and J. F. Schairer, Ann. Rept. Director Geophys. Lab. 1960–61; in *Carnegie Inst. Washington Year Book*, **60**, 140 (1961).

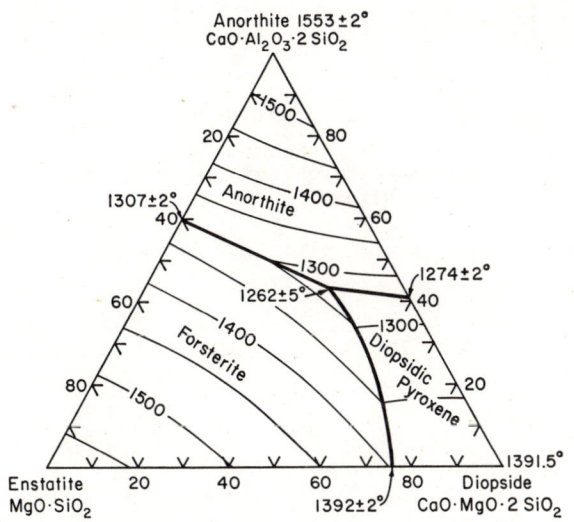

Fig. 907.—System MgO·SiO₂–CaO·MgO·2SiO₂–CaO·Al₂O₃·2SiO₂; join.

Kai Hytönen and J. F. Schairer, Ann. Rept. Director Geophys. Lab. 1960–61; in *Carnegie Inst. Washington Year Book*, **60**, 126 (1961).

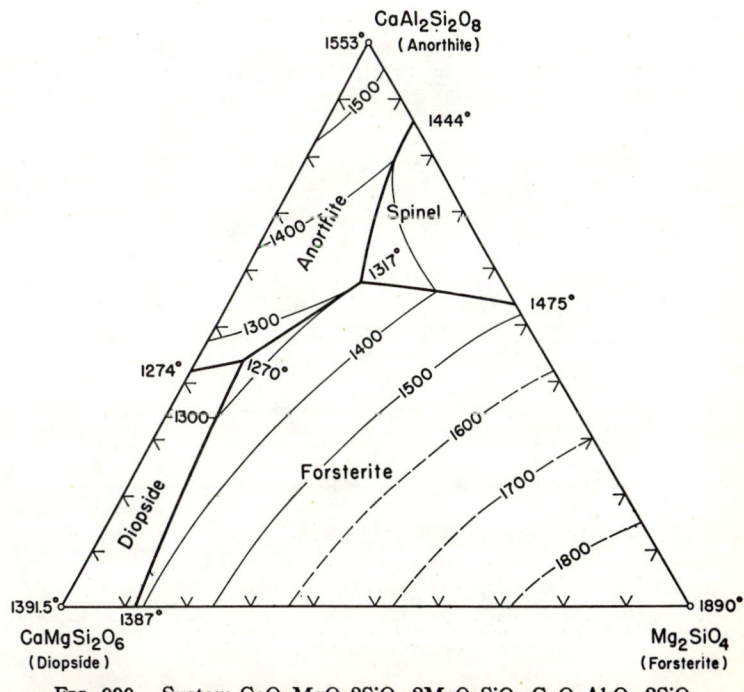

Fig. 909.—System CaO·MgO·2SiO₂–2MgO·SiO₂–CaO·Al₂O₃·2SiO₂.
E. F. Osborn and D. B. Tait, *Am. J. Sci.*, Bowen volume, p. 419 (1952).

CaO–MgO–Al$_2$O$_3$–SiO$_2$ (cont.)

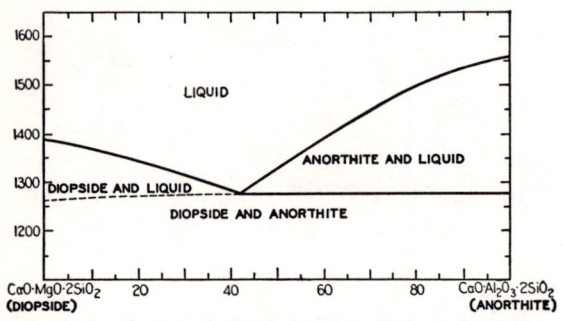

Fig. 910.—System CaO·MgO·2SiO$_2$–CaO·Al$_2$O$_3$·2SiO$_2$.

E. F. Osborn, *Am. J. Sci.*, **240**, 758 (1942).

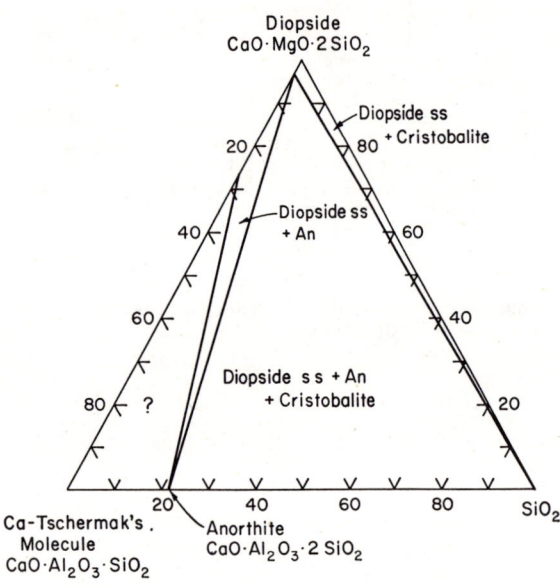

Fig. 911.—System CaO·MgO·2SiO$_2$–CaO·Al$_2$O$_3$·SiO$_2$–SiO$_2$; isothermal section for 1135°C.

Kai Hytönen and J. F. Schairer, Ann. Rept. Director Geophys. Lab. 1960–61; in *Carnegie Inst. Washington Year Book*, **60**, 137 (1961).

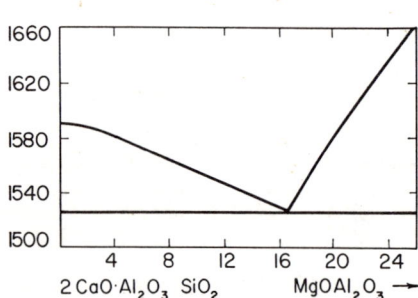

Fig. 912.—System 2CaO·Al$_2$O$_3$·SiO$_2$–MgO·Al$_2$O$_3$.

R. W. Nurse and N. Stutterheim, *J. Iron Steel Inst. (London)*, **165**, 138 (1950).

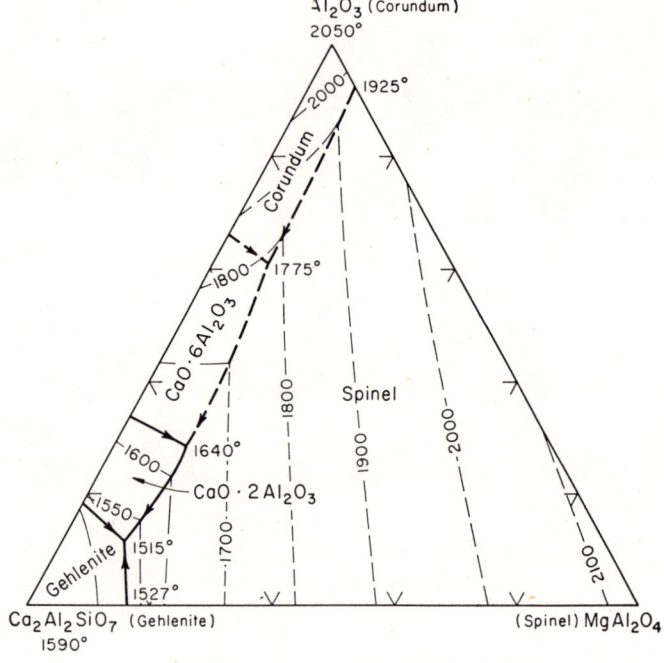

Fig. 913.—System Ca$_2$Al$_2$SiO$_7$–MgAl$_2$O$_4$–Al$_2$O$_3$.

R. C. DeVries and E. F. Osborn, *J. Am. Ceram. Soc.*, **40** [1] 13 (1957).

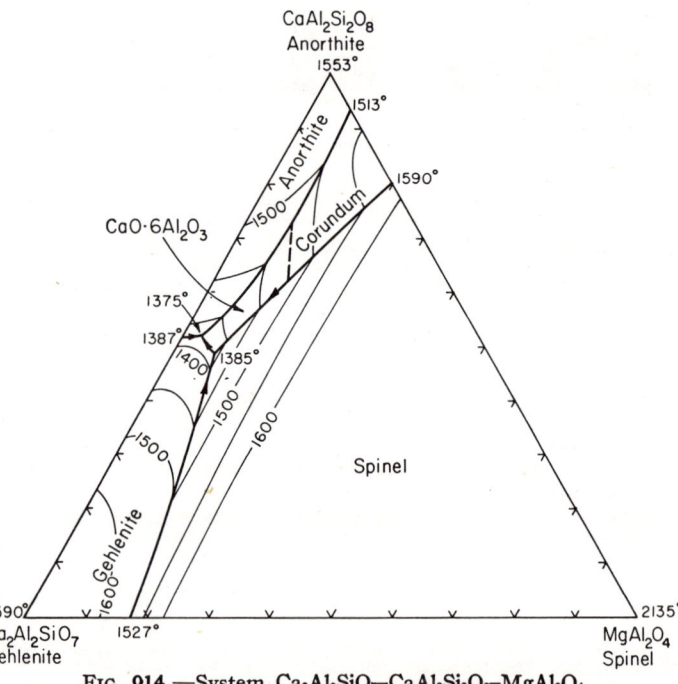

Fig. 914.—System Ca$_2$Al$_2$SiO$_7$–CaAl$_2$Si$_2$O$_8$–MgAl$_2$O$_4$.

R. C. DeVries and E. F. Osborn, *J. Am. Ceram. Soc.*, **40** [1] 10 (1957).

CaO–MgO–Al₂O₃–SiO₂ (cont.)

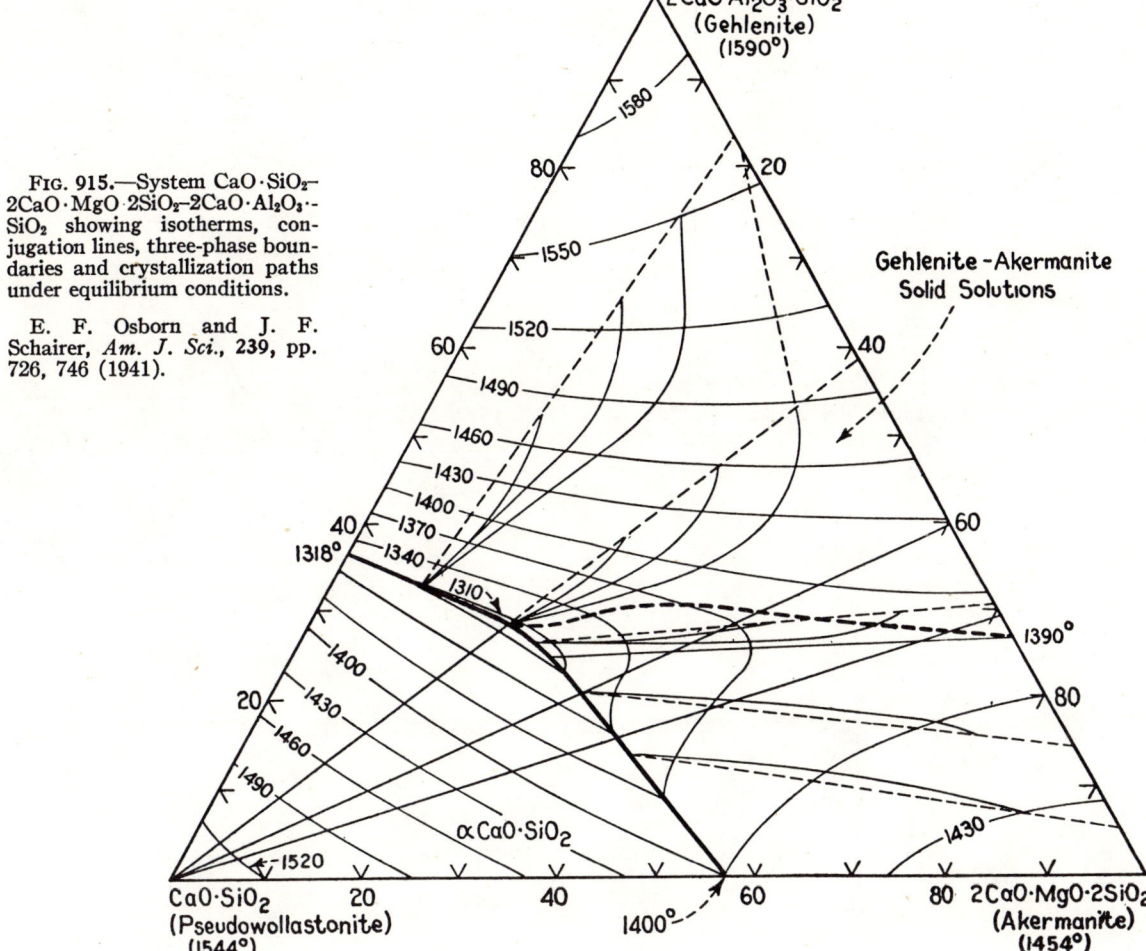

FIG. 915.—System CaO·SiO₂–2CaO·MgO·2SiO₂–2CaO·Al₂O₃·SiO₂ showing isotherms, conjugation lines, three-phase boundaries and crystallization paths under equilibrium conditions.

E. F. Osborn and J. F. Schairer, *Am. J. Sci.*, **239**, pp. 726, 746 (1941).

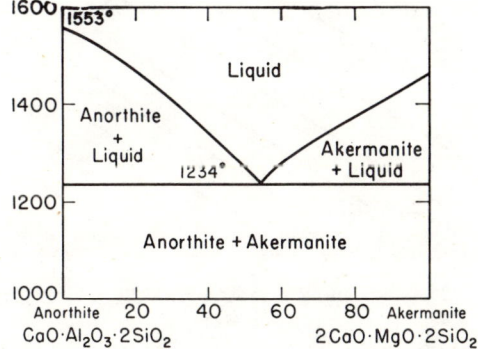

FIG. 916.—System 2CaO·MgO·2SiO₂–CaO·Al₂O₃·2SiO₂.

E. C. De Wys and W. R. Foster, *J. Am. Ceram. Soc.*, **39** [11] 375 (1956).

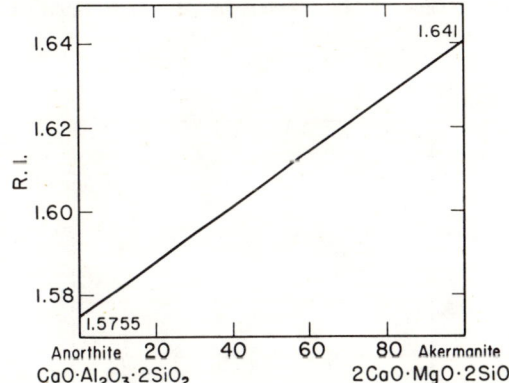

FIG. 917.—System 2CaO·MgO·2SiO₂–CaO·Al₂O₃·2SiO₂; *n* of glasses.

E. C. De Wys and W. R. Foster, *J. Am. Ceram. Soc.*, **39** [11] 374 (1956).

CaO–MgO–Al$_2$O$_3$–SiO$_2$ (cont.)

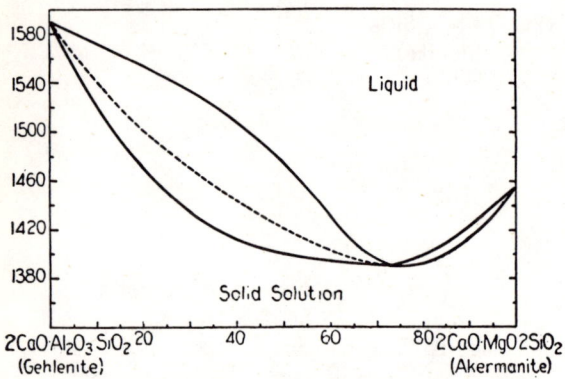

Fig. 918.—System 2CaO·MgO·2SiO$_2$–2CaO·Al$_2$O$_3$·SiO$_2$.

E. F. Osborn and J. F. Schairer, *Am. J. Sci.*, **239**, 724 (1941).

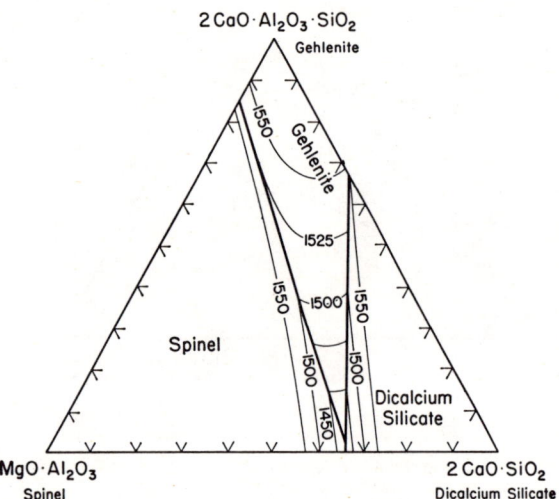

Fig. 919.—System MgO·Al$_2$O$_3$–2CaO·SiO$_2$–2CaO·Al$_2$O$_3$·SiO$_2$.

A. T. Prince, *J. Am. Ceram. Soc.*, **34** [2] 50 (1951).

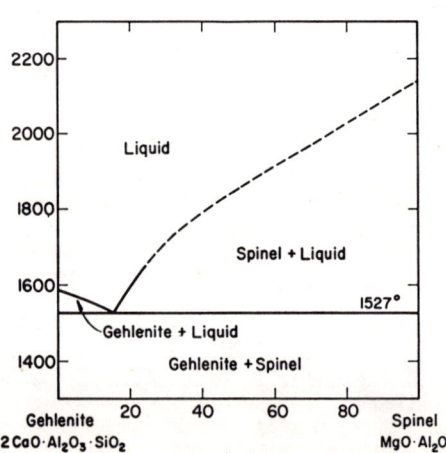

Fig. 920.—System 2CaO·Al$_2$O$_3$·SiO$_2$–MgO·Al$_2$O$_3$.

A. T. Prince, *J. Am. Ceram. Soc.*, **34** [2] 46 (1951).

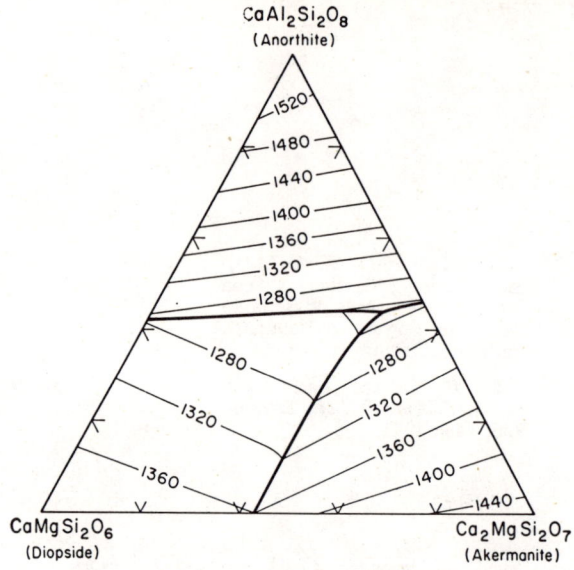

Fig. 921.—System CaMgSi$_2$O$_6$ (diopside)–CaAl$_2$Si$_2$O$_8$ (anorthite)–Ca$_2$MgSi$_2$O$_7$ (akermanite).

E. C. De Wys and W. R. Foster, *Mineral. Mag.*, **31** [240] 741 (1958).

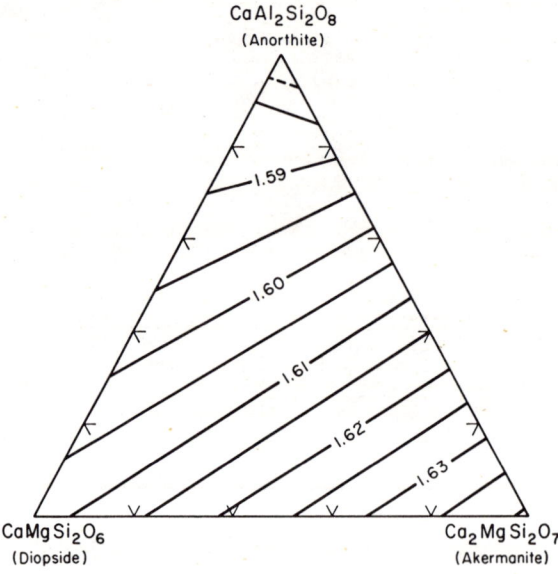

Fig. 922.—System CaMgSi$_2$O$_6$–CaAl$_2$Si$_2$O$_8$–Ca$_2$MgSi$_2$O$_7$, isofracts.

E. C. De Wys and W. R. Foster, *Mineral. Mag.*, **31** [240] 739 (1958).

CaO–MgO–Al₂O₃–SiO₂ (concl.)

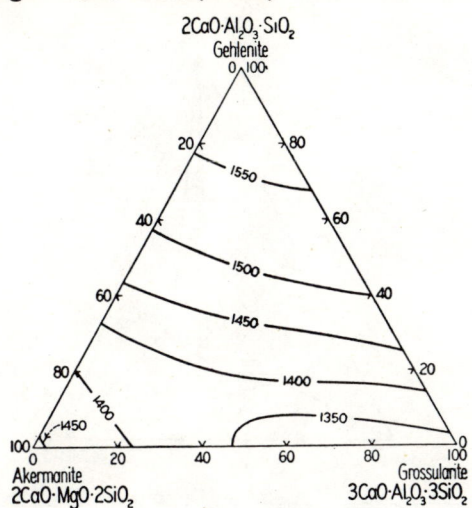

Fig. 923.—Melting isotherms in system $2CaO \cdot Al_2O_3 \cdot SiO_2 - 2CaO \cdot MgO \cdot 2SiO_2 - 3CaO \cdot Al_2O_3 \cdot 3SiO_2$ (Gehlenite–akermanite–grossularite).

A. F. Buddington, *Am. J. Sci.*, 5th Ser., **3**, 48 (1922).

CaO–MgO–Cr₂O₃–SiO₂

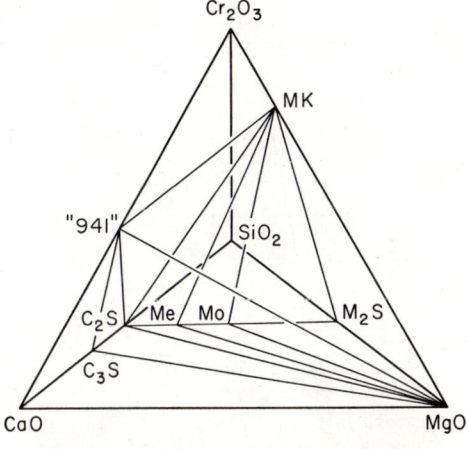

Fig. 924.—System CaO–MgO–Cr₂O₃–SiO₂; phase fields, section 1. C = CaO, M = MgO, S = SiO₂, K = Cr₂O₃, "941" = pseudobinary compound $9CaO \cdot 4CrO_3 \cdot Cr_2O_3$, Me = merwinite, Mo = monticellite.

W. F. Ford and W. J. Rees, *Trans. Brit. Ceram. Soc.*, **57** [5] 239 (1958).

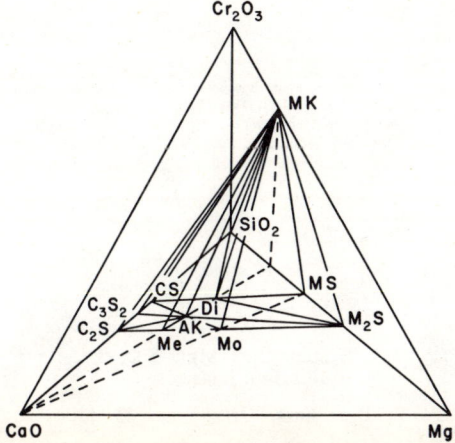

Fig. 925.—System CaO–MgO–Cr₂O₃–SiO₂; phase fields, section 2. C = CaO, M = MgO, S = SiO₂, K = Cr₂O₃, Ak = akermanite, Di = diopside, Me = merwinite, Mo = monticellite.

W. F. Ford and W. J. Rees, *Trans. Brit. Ceram. Soc.*, **57** [5] 240 (1958).

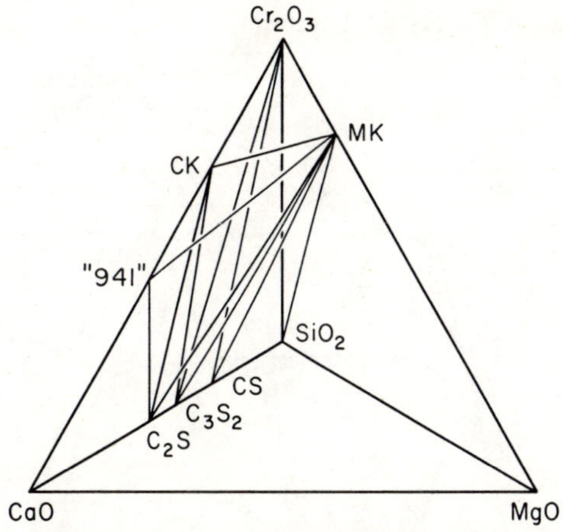

Fig. 926.—System CaO–MgO–Cr₂O₃–SiO₂; phase fields, section 3. C = CaO, M = MgO, S = SiO₂, K = Cr₂O₃, "941" = pseudobinary compound $9CaO \cdot 4CrO_3 \cdot Cr_2O_3$.

W. F. Ford and W. J. Rees, *Trans. Brit. Ceram. Soc.*, **57** [5] 240 (1958).

CaO–MgO–Fe₂O₃–SiO₂

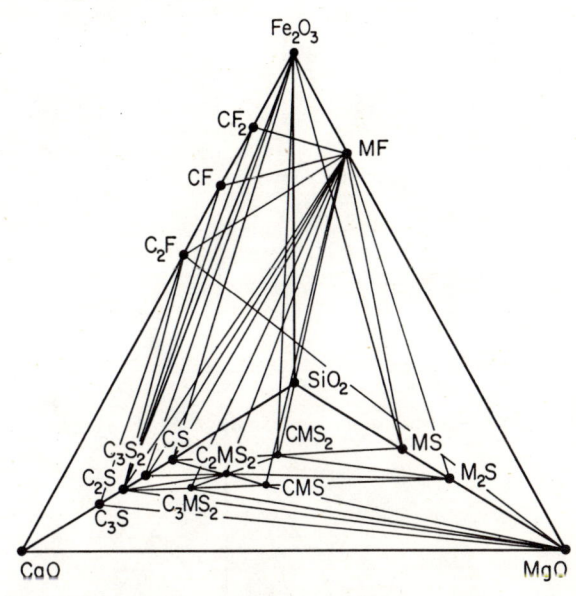

Fig. 927.—System CaO–MgO–Fe₂O₃–SiO₂, primary tetrahedra. C = CaO; M = MgO; F = Fe₂O₃; S = SiO₂.

A. S. Berezhnoĭ, *Voprosy Petrograf. i Mineral.*, Akad. Nauk S.S.S.R., **2**, 295 (1953). See also, J. R. Rait, *Iron and Steel*, **22**, 15 (1949).

CaO–MgO–Fe$_2$O$_3$–SiO$_2$ (concl.)

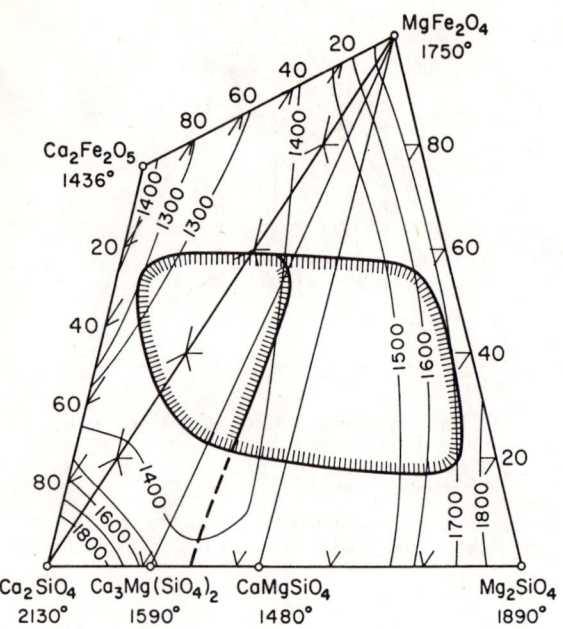

FIG. 928.—System Ca$_2$Fe$_2$O$_5$–Ca$_2$SiO$_4$–MgFe$_2$O$_4$–Mg$_2$SiO$_4$, melting isotherms.

A. S. Berezhnoĭ, *Voprosy Petrograf. i Mineral.*, Akad. Nauk S.S.S.R., **2**, 297 (1953).

CaO–MgO–CO$_2$–SiO$_2$

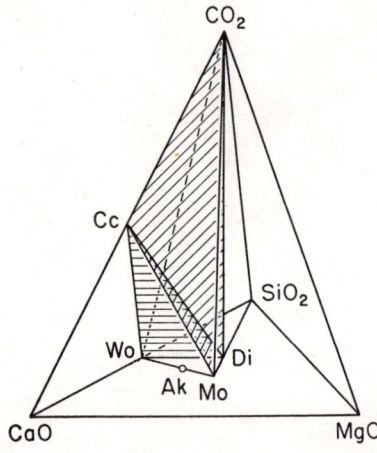

FIG. 929.—CaO–MgO–SiO$_2$–CO$_2$ tetrahedron, showing CO$_2$ deficient region below the Cc-Di-Wo and the Cc-Wo-Mo planes (shaded). Cc = calcite; Wo = wollastonite; Ak = akermanite; Di = diopside; Mo = monticellite.

R. I. Harker and O. F. Tuttle, *Am. J. Sci.*, **254**, 476 (1956).

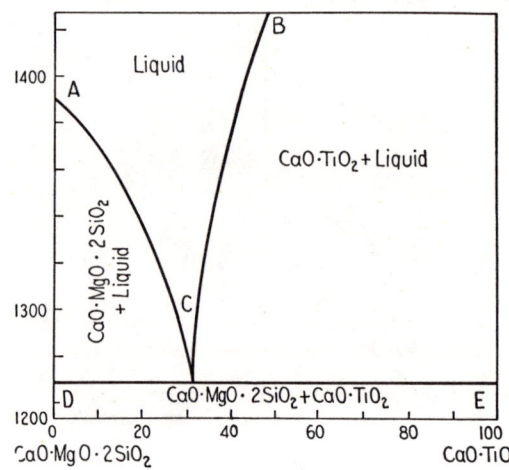

Fig. 931.—System CaO·TiO$_2$–CaO·MgO·2SiO$_2$.

Usaburô Nisioka, *Science Repts., Tôhoku Imp. Univ.*, Ser. 1, **24**, 715 (1935–36).

CaO–MgO–SiO$_2$–TiO$_2$

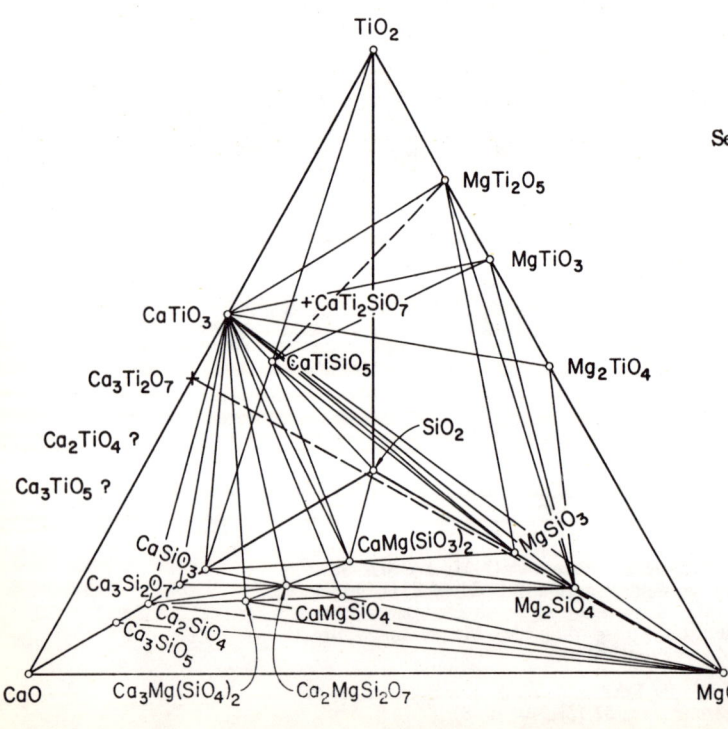

FIG. 930.—System CaO–MgO–SiO$_2$–TiO$_2$; equilibrium phases.

A. S. Berezhnoi, *Ogneupory*, **15** [10] 453 (1950)

CaO–MgO–SiO₂–TiO₂ (concl.)

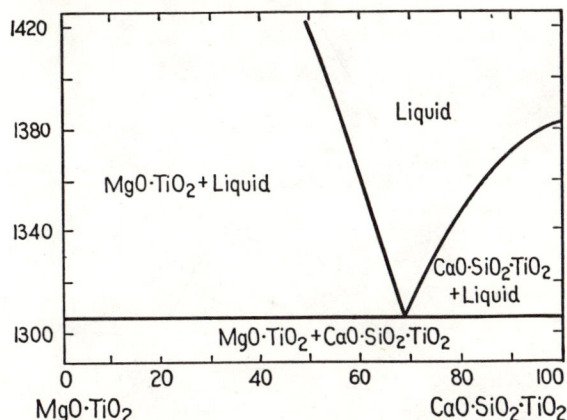

Fig. 932.—System MgO·TiO₂–CaO·SiO₂·TiO₂.

Kiezo Iwasé and Usaburô Nisioka, *Science Repts., Tôhoku Imp. Univ.*, Ser. 1, **26**, 593 (1937–38).

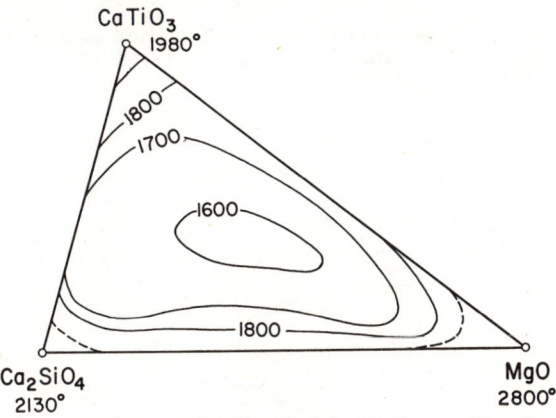

Fig. 933.—System CaTiO₃–Ca₂SiO₄–MgO; melting isotherms.

A. S. Berezhnoi, *Ogneupory*, **15**, No. 10, 453 (1950).

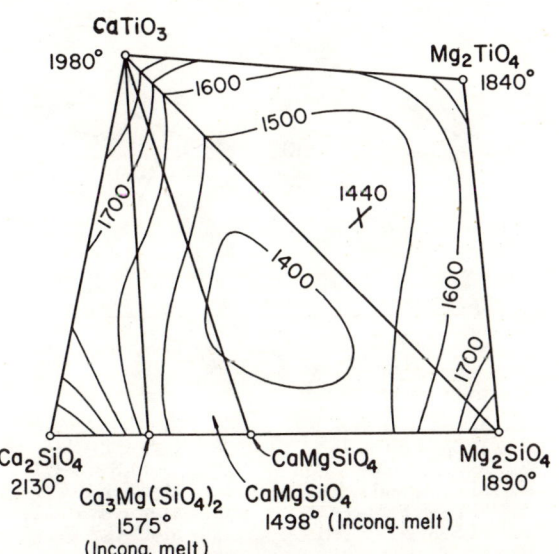

Fig. 934.—System CaTiO₃–Ca₂SiO₄–Mg₂SiO₄ and CaTiO₃–Mg₂SiO₄–Mg₂TiO₄; melting isotherms.

A. S. Berezhnoi, *Ogneupory*, **15**, No. 10, 453 (1950).

CaO–MgO–SiO₂–P₂O₅

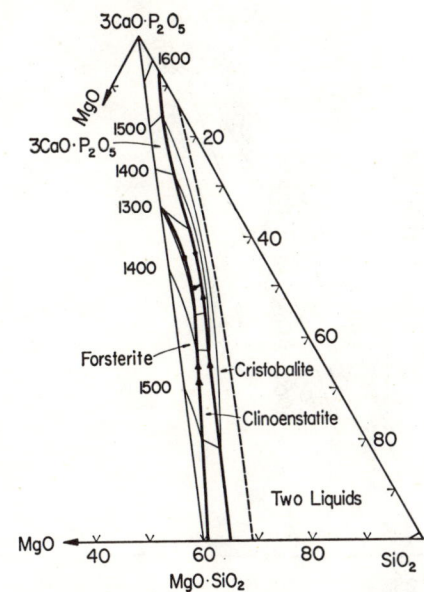

Fig. 935.—System 3CaO·P₂O₅–MgO·SiO₂–SiO₂.

Toshiyuki Sata, *Bull. Chem. Soc. Japan*, **31**, 412 (1958).

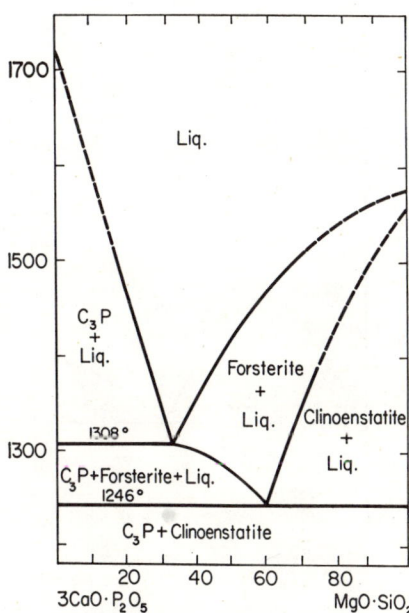

Fig. 936.—System 3CaO·P₂O₅–MgO·SiO₂. C = CaO, P = P₂O₅.

Toshiyuki Sata, *Bull. Chem. Soc. Japan*, **31**, 410 (1958).

$CaO-MgO-SiO_2-P_2O_5$ (concl.)

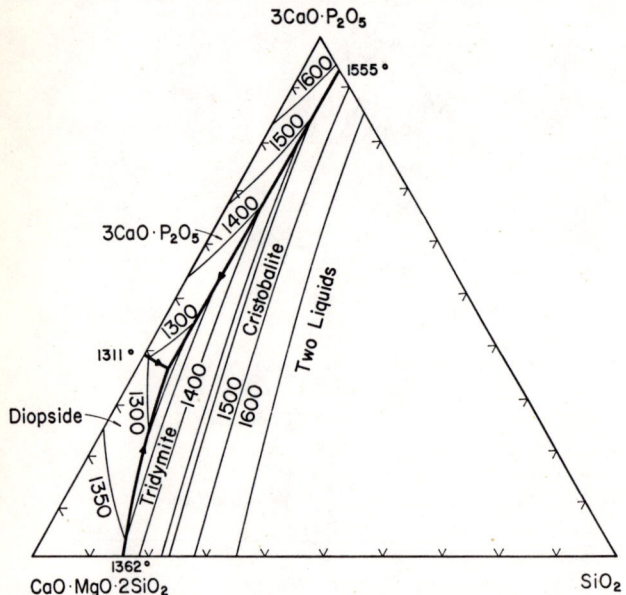

Fig. 937.—System $CaO \cdot MgO \cdot 2SiO_2 - 3CaO \cdot P_2O_5 - SiO_2$.

Toshiyuki Sata, *Bull. Chem. Soc. Japan*, **32**, 106 (1959).

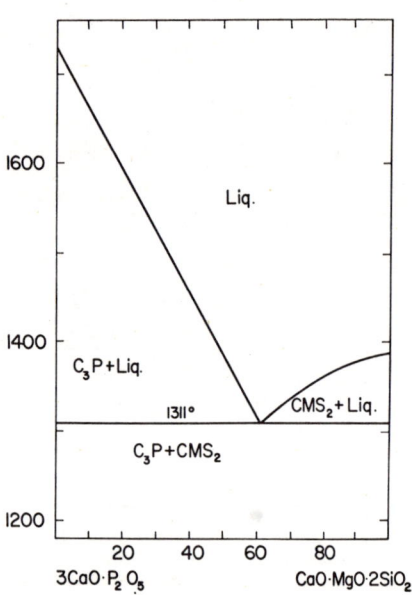

Fig. 938.—System $CaO \cdot MgO \cdot 2SiO_2 - 3CaO \cdot P_2O_5$. C = CaO, M = MgO, P = P_2O_5, S = SiO_2.

Toshiyuki Sata, *Bull. Chem. Soc. Japan*, **32**, 105 (1959).

$CaO-MgO-SnO_2-TiO_2$

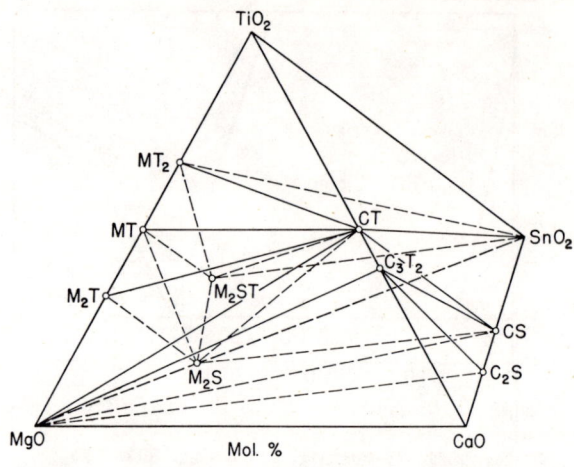

Fig. 939.—System $CaO-MgO-SnO_2-TiO_2$; compatibility relations; C = CaO; M = MgO; S = SnO_2; T = TiO_2. Extensive solid solutions exist within the system, but no attempt has been made to determine the limits or to indicate solid solution fields in the diagram. See ternary boundary systems for indicated solid solutions.

L. W. Coughanour, R. S. Roth, S. Marzullo, and F. E. Sennett, *J. Research Natl. Bureau Standards* **54** [3] 156 (1955); RP 2576.

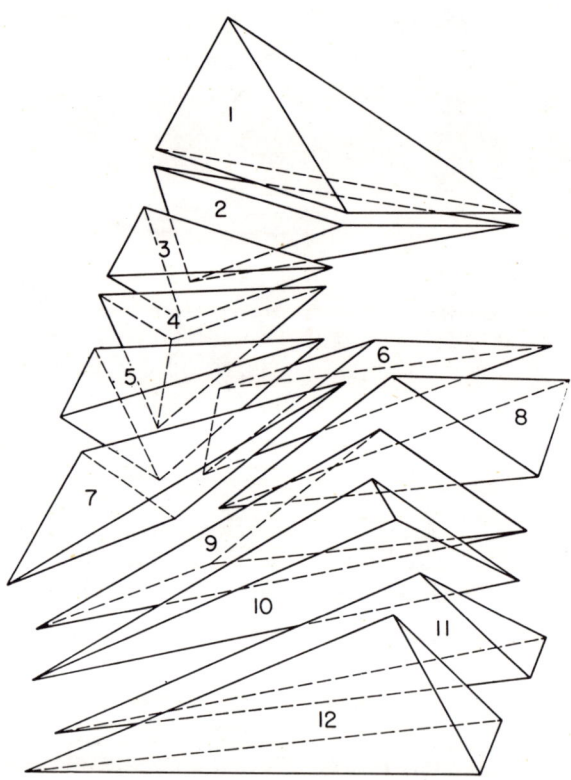

Fig. 940.—System $CaO-MgO-SnO_2-TiO_2$; compatibility tetrahedra exploded (see Fig. 939).

Sub-tetrahedra vertices:

1. $MT_2-CT-T-S$
2. $MT_2-M_2ST-CT-S$
3. MT_2-MT-M_2ST-CT
4. $MT-M_2S-M_2ST-CT$
5. $MT-M_2T-M_2S-CT$
6. $M_2S-M_2ST-CT-S$
7. $M-M_2T-M_2S-CT$
8. $M_2S-CT-CS-S$
9. $M-M_2S-CS-CT$
10. $M-CS-CT-C_3T_2$
11. $M-C_2S-CS-C_3T_2$
12. $M-C-C_2S-C_3T_2$

C = CaO; M = MgO; S = SnO_2; T = TiO_2.

L. W. Coughanour, R. S. Roth, S. Marzullo, and F. E. Sennett, *J. Research Natl. Bureau Standards* **54** [3] 156 (1955); RP 2576.

CaO–MnO–SiO$_2$–TiO$_2$

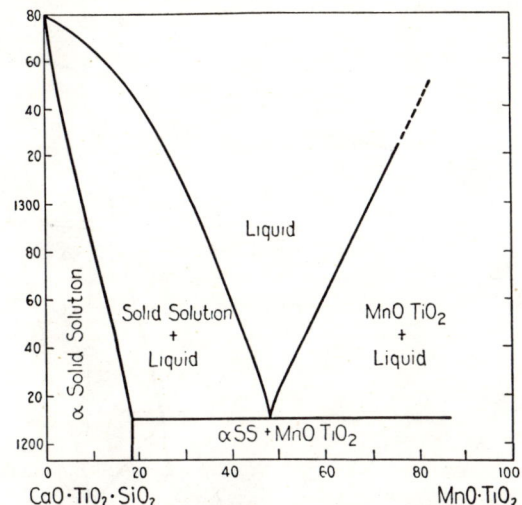

Fig. 941.—System MnO·TiO$_2$–CaO·SiO$_2$·TiO$_2$.
Kiezo Iwasé and Usaburô Nisioka, *Science Repts. Tôhoku Imp. Univ.*, Ser. 1, **25**, 505 (1936–37).

CaO–MnO–SiO$_2$–WO$_3$

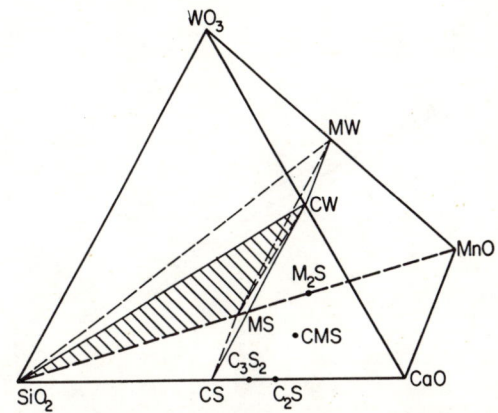

Fig. 942.—System CaO–MnO–SiO$_2$–WO$_3$, compatibility tetrahedra in high SiO$_2$-WO$_3$ region, at 1100°C. C = CaO; M = MnO; S = SiO$_2$; W = WO$_3$.

D. E. Harrison and F. A. Hummel, *J. Electrochem. Soc.*, **105** [1] 36 (1958).

CaO–Al$_2$O$_3$–Fe$_2$O$_3$–SiO$_2$

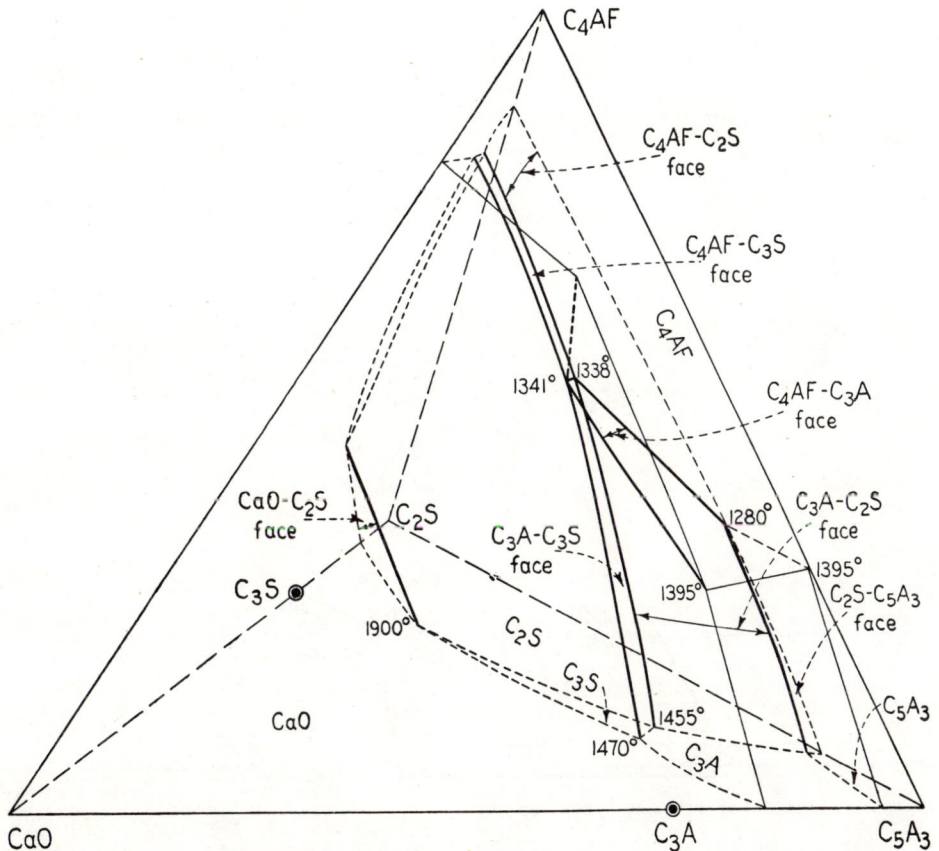

Fig. 943.—System CaO–5CaO·3Al$_2$O$_3$–2CaO·SiO$_2$–4CaO·Al$_2$O$_3$·Fe$_2$O$_3$; perspective diagram, showing primary-phase volumes; C = CaO; A = Al$_2$O$_3$; F = Fe$_2$O$_3$; S = SiO$_2$; see Fig. 944 for side systems.

F. M. Lea and T. W. Parker *Trans. Roy. Soc. (London)*, Ser. A, **234A** [731] 16 (1934).

CaO–Al$_2$O$_3$–Fe$_2$O$_3$–SiO$_2$ (cont.)

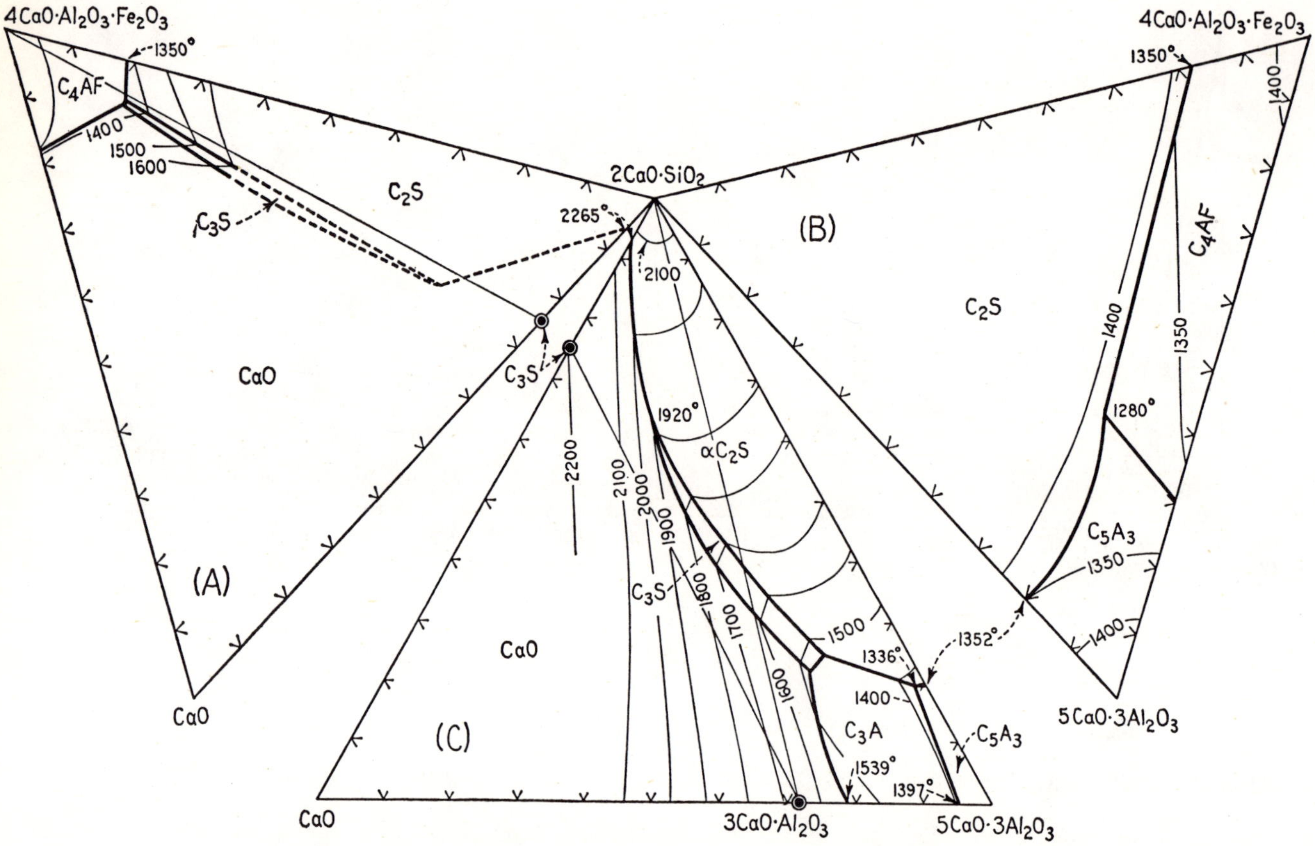

Fig. 944.—The base (system CaO–5CaO·3Al$_2$O$_3$–2CaO·SiO$_2$), left rear side (system CaO–2CaO·SiO$_2$–4CaO·Al$_2$O$_3$·Fe$_2$O$_3$), and right rear side (system 5CaO·3Al$_2$O$_3$–2CaO·SiO$_2$–4CaO·Al$_2$O$_3$·Fe$_2$O$_3$) of the tetrahedron of Fig. 943.

Systems of (A) and (B) from F. M. Lea and T. W. Parker, *Trans. Roy. Soc.* (*London*), Ser. A, **234A**, [731] 16 (1934); (C) adapted from J. W. Greig, *Am. J. Sci.*, 5th Ser., **13**, 41 (1927).

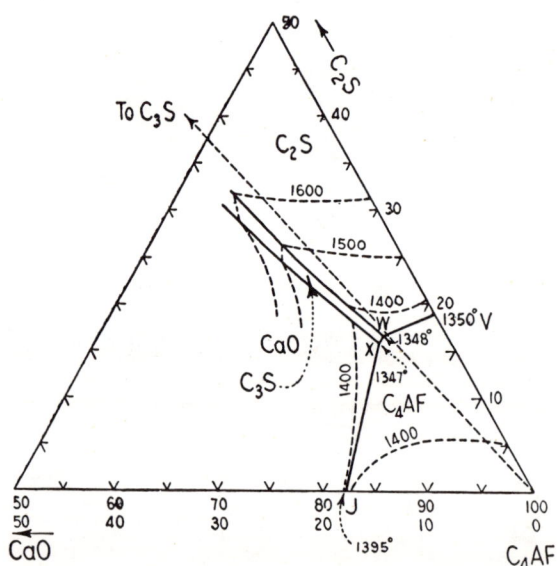

Fig. 945.—Detail of the high 4CaO·Al$_2$O$_3$·Fe$_2$O$_3$ portion of system CaO–2CaO·SiO$_2$–4CaO·Al$_2$O$_3$·Fe$_2$O$_3$ shown in Fig. 944(A); C = CaO; A = Al$_2$O$_3$; F = Fe$_2$O$_3$; S = SiO$_2$.

F. M. Lea and T. W. Parker, *Trans. Roy. Soc.* (*London*), Ser. A, **234A** [731] 16 (1934).

CaO–Al$_2$O$_3$–Fe$_2$O$_3$–SiO$_2$ (cont.)

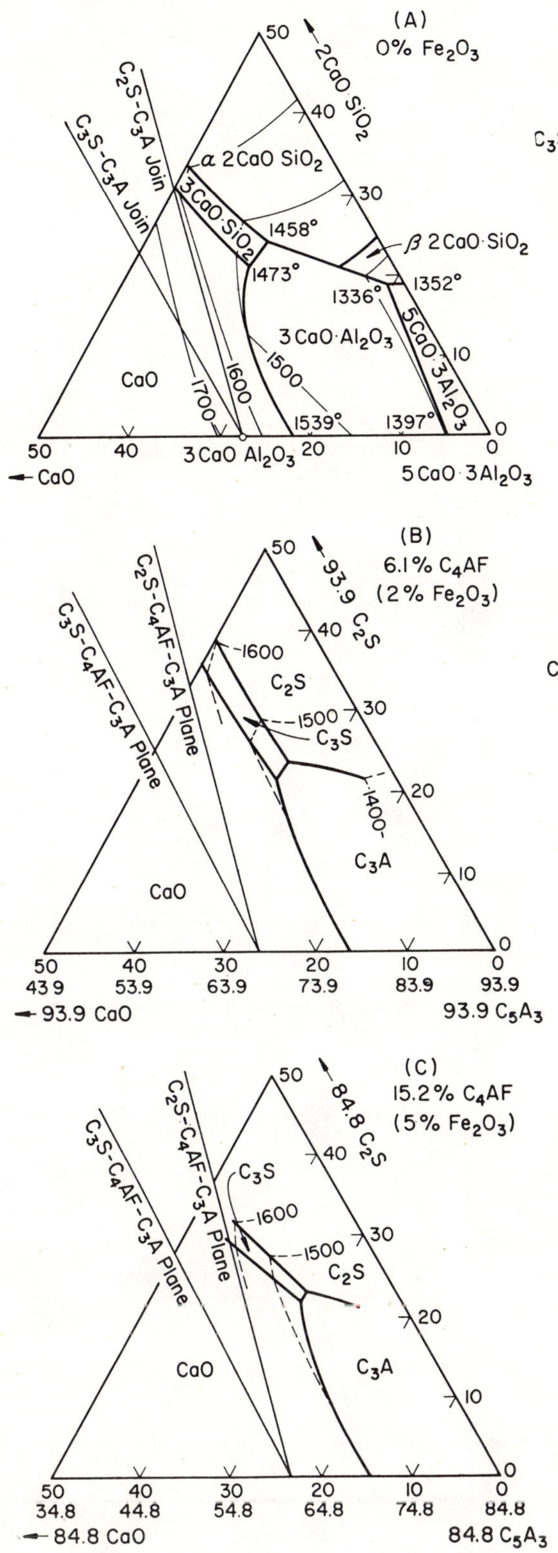

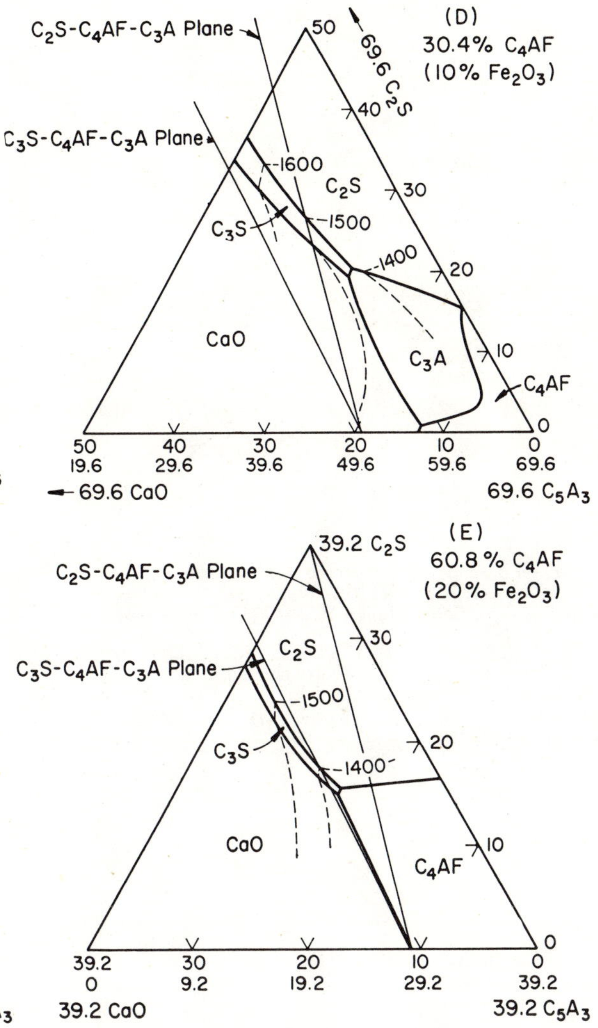

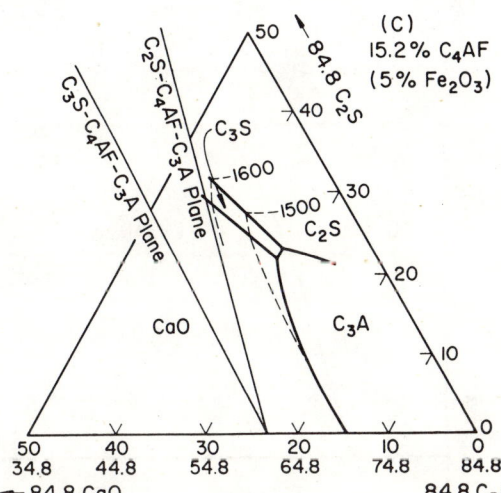

FIG. 946.—Planes through quaternary system CaO–5CaO·3Al$_2$O$_3$–2CaO·SiO$_2$–4CaO·Al$_2$O$_3$·Fe$_2$O$_3$ parallel to the base levels indicated (0, 2, 5, 10, and 20%, respectively).
Position of the trace of the C$_3$A–C$_3$S–C$_4$AF plane in (B), (C), (D), and (E) corrected to make an angle of 62°53' to the base, according to F. M. Lea, Private communication, Dec., 1953.

F. M. Lea and T. W. Parker, *Trans. Roy. Soc. (London)* A, **234** [731] 16 (1934); Part (A) adapted from J. W. Grieg, *Am. J. Sci.*, 5th Ser., **13**, 41 (1927).

Note: Use these diagrams in conjunction with diagrams of Figs. 943 and 944.

CaO–Al$_2$O$_3$–Fe$_2$O$_3$–SiO$_2$ (cont.)

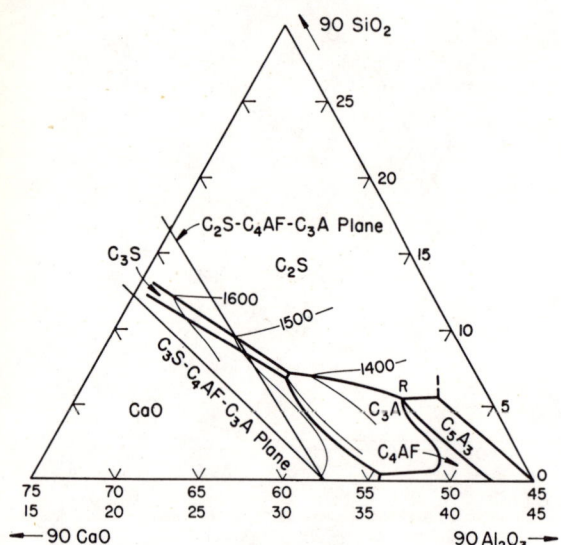

Fig. 947.—Plane passing through tetrahedron Fig. 943 at 30.4% 4CaO·Al$_2$O$_3$·Fe$_2$O$_3$ (10% Fe$_2$O$_3$) parallel to base; see Fig. 946 (D) for detail of high 4CaO·Al$_2$O$_3$·Fe$_2$O$_3$ area.

Portion of the trace of the C$_3$A–C$_3$S–C$_4$AF plane corrected to make an angle of 43° to the base, according to F. M. Lea, Private communication, Dec. 1953.

F. M. Lea and T. W. Parker, *Trans. Roy. Soc. (London)*, A, **234** [731] 16 (1934).

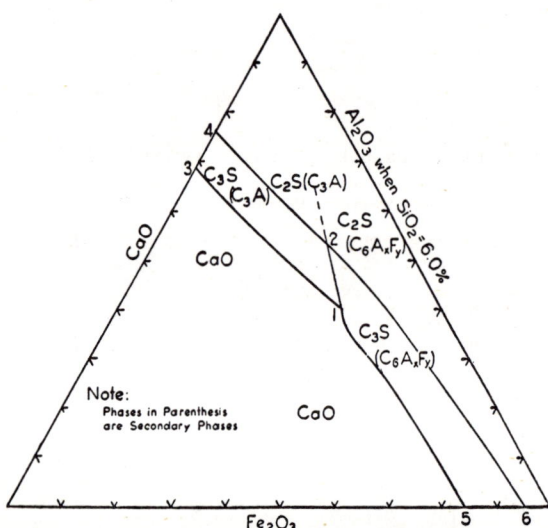

Fig. 949.—Portion of system CaO–5CaO·3Al$_2$O$_3$–2CaO·Fe$_2$O$_3$–2CaO·SiO$_2$ showing location of invariant points.

M. A. Swayze, *Am. J. Sci.*, **244**, 15 (1946).

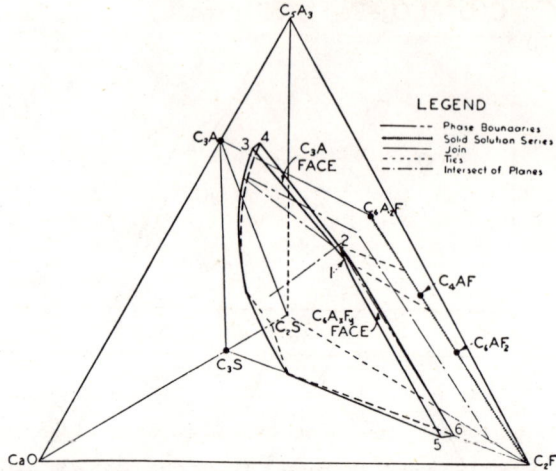

Fig. 948.—Location of the phase-volume for 3CaO·SiO$_2$ in system CaO–5CaO·3Al$_2$O$_3$–2CaO·Fe$_2$O$_3$–2CaO·SiO$_2$.

M. A. Swayze, *Am. J. Sci.*, **244**, 18 (1946).

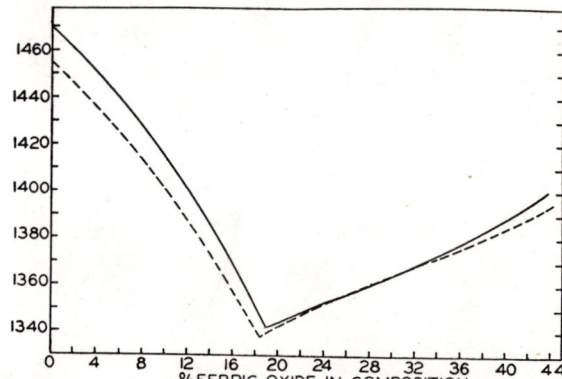

Fig. 950.—System CaO–2CaO·SiO$_2$–5CaO·3Al$_2$O$_3$–2CaO·Fe$_2$O$_3$; liquidus temperatures along C$_3$S phase boundaries; solid line = CaO–C$_3$S–C$_3$A:C$_2$F boundary curve (line 3–1–5 of Figs. 948 and 949); dashed line = C$_3$S–C$_2$S–C$_3$A:C$_2$F boundary curve (line 4–2–6 of Figs 948 and 949).

M. A. Swayze, *Am. J. Sci.*, **244**, 19 (1946).

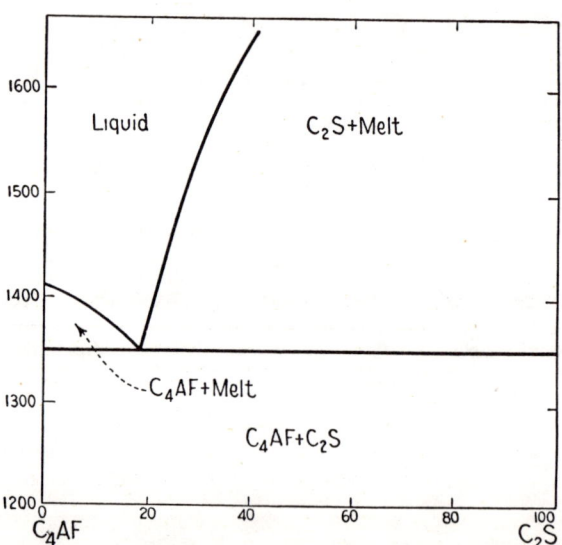

Fig. 951.—System 2CaO·SiO$_2$–4CaO·Al$_2$O$_3$·Fe$_2$O$_3$; C = CaO; A = Al$_2$O$_3$; F = Fe$_2$O$_3$; S = SiO$_2$.

F. M. Lea and T. W. Parker, *Trans. Roy. Soc. (London)*, Ser. A, **234A** [731] 13 (1934).

CaO–Al₂O₃–Fe₂O₃–SiO₂ (concl.)

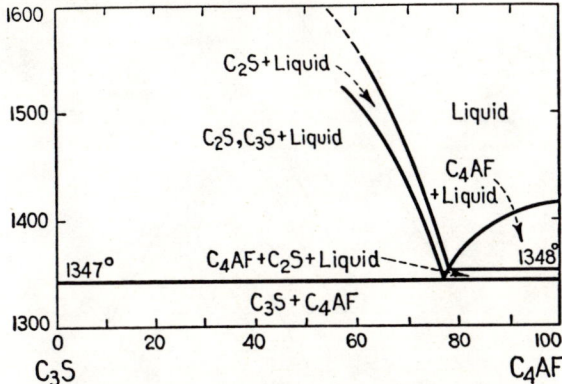

Fig. 952.—System $3CaO \cdot SiO_2$–$4CaO \cdot Al_2O_3 \cdot Fe_2O_3$.

F. M. Lea and T. W. Parker, *Trans. Roy. Soc. (London)*, Ser. A, **234A** [731] 16 (1934).

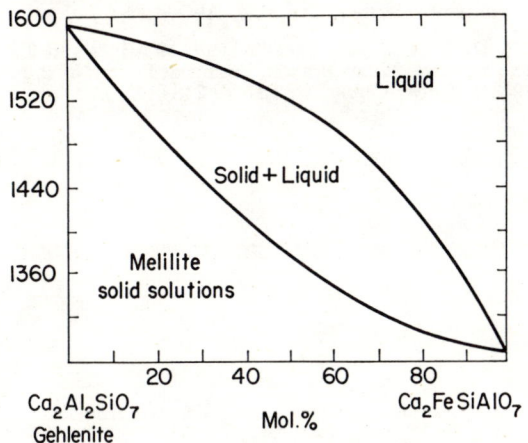

Fig. 953.—System $Ca_2Al_2SiO_7$ (gehlenite)–$Ca_2FeSiAlO_7$.

R. W. Nurse and H. G. Midgley, *J. Iron Steel Inst. (London)*, **174**, 124 (1953).

CaO–Al₂O₃–SiO₂–TiO₂

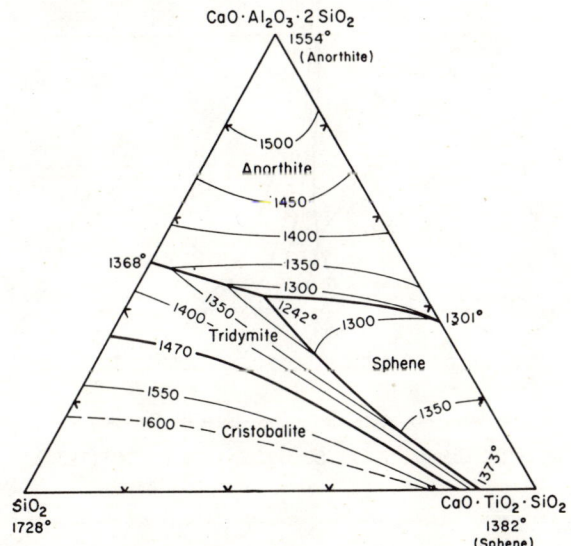

Fig. 954.—System $CaO \cdot Al_2O_3 \cdot 2SiO_2$–$CaO \cdot SiO_2 \cdot TiO_2$–$SiO_2$.

Y. M. Agamawi and J. White, *Trans. Brit. Ceram. Soc.*, **53** [1] 23 (1954).

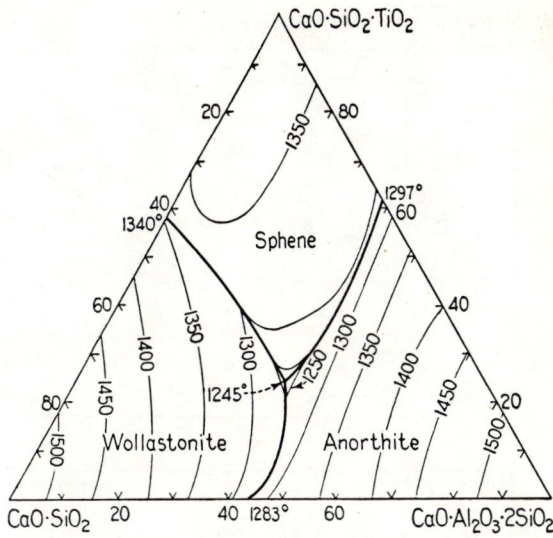

Fig. 955.—System $CaO \cdot SiO_2$–$CaO \cdot Al_2O_3 \cdot 2SiO_2$–$CaO \cdot SiO_2 \cdot TiO_2$.

Keizo Iwasé and Usaburô Nisioka, *Science Repts. Tôhoku Imp. Univ.*, Ser. 1, Honda Anniv. Vol., pp. 1–10 (1936).

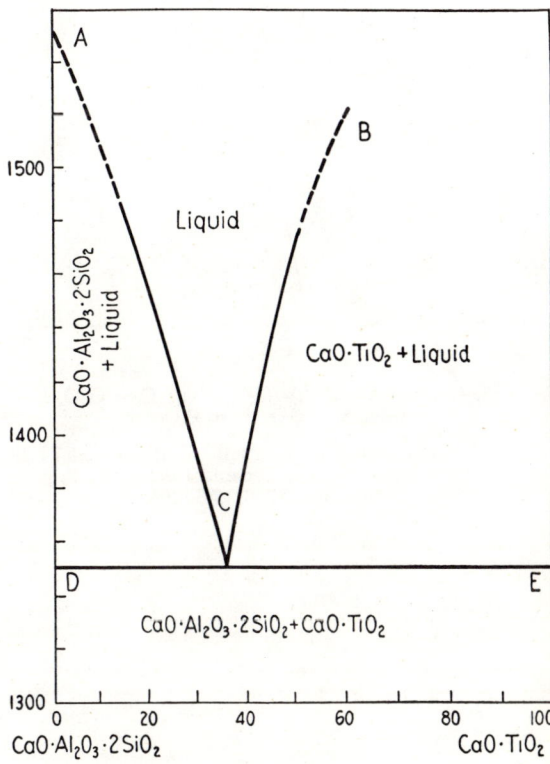

Fig. 956.—System $CaO \cdot TiO_2$–$CaO \cdot Al_2O_3 \cdot 2SiO_2$.

Usaburô Nisioka, *Science Repts., Tôhoku Imp. Univ.*, Ser. 1, **24**, 708 (1935–36).

CaO–Al$_2$O$_3$–SiO$_2$–TiO$_2$ (concl.)

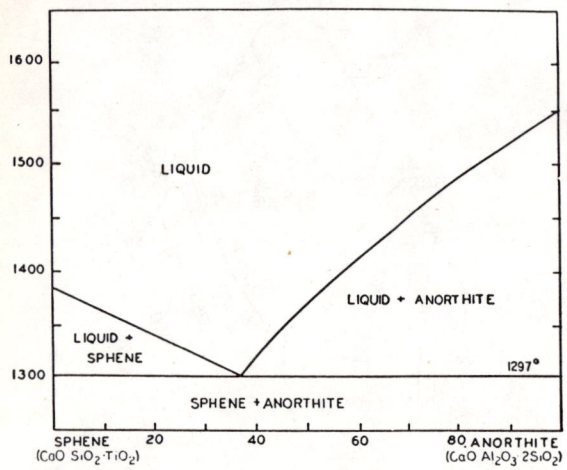

Fig. 957.—System CaO·Al$_2$O$_3$·2SiO$_2$–CaO·SiO$_2$·TiO$_2$.

A. T. Prince, *J. Geol.*, **51**, 4 (1943).

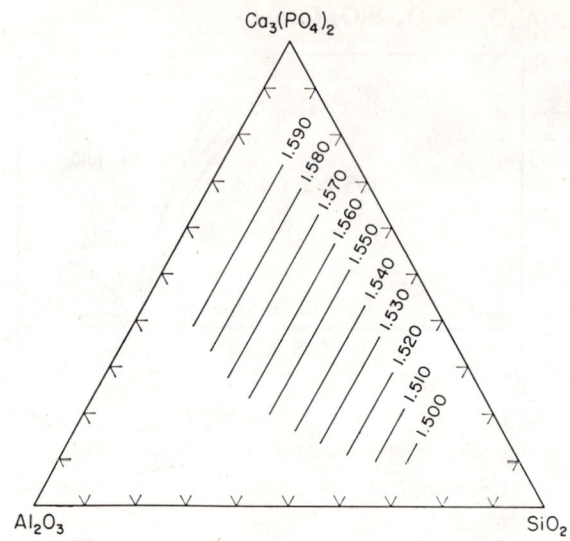

Fig. 959.—System Ca$_3$(PO$_4$)$_2$–Al$_2$O$_3$–SiO$_2$; isofracts.

P. D. S. St. Pierre, Canada Dept. of Mines and Technical Surveys, Mines Branch, Technical Paper No. 2, p. 23 (1953); *J. Am. Ceram. Soc.*, **37** [6] 248 (1954).

CaO–Al$_2$O$_3$–SiO$_2$–P$_2$O$_5$

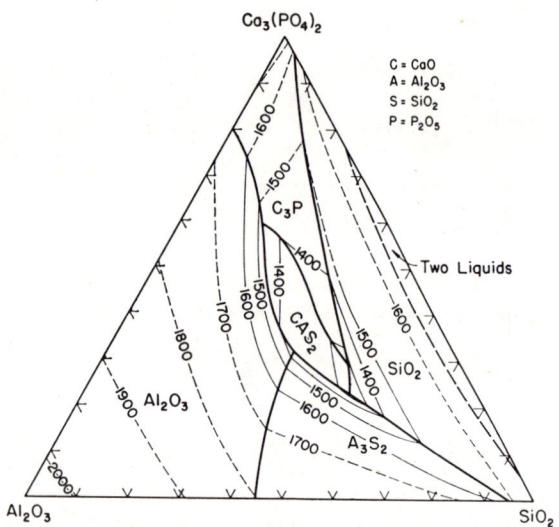

Fig. 958.—System Ca$_3$(PO$_4$)$_2$–Al$_2$O$_3$–SiO$_2$; C = CaO, A = Al$_2$O$_3$, S = SiO$_2$, P = P$_2$O$_5$.

P. D. S. St. Pierre, Canada Dept. of Mines and Technical Surveys, Mines Branch, Technical Paper No. 2, p. 69 (1953); *J. Am. Ceram. Soc.*, **37** [6] 253 (1954).

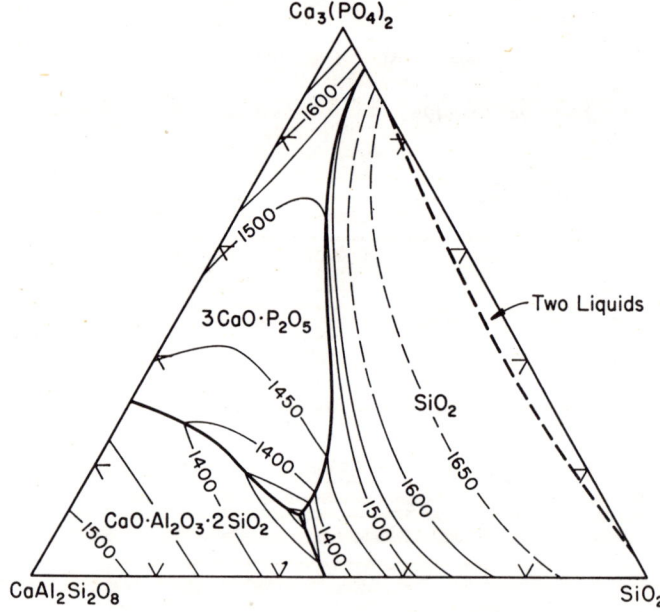

Fig. 960.—System Ca$_3$(PO$_4$)$_2$–CaAl$_2$Si$_2$O$_8$–SiO$_2$.

P. D. S. St. Pierre, *J. Am. Ceram. Soc.*, **39** [4] 149 (1956).

$CaO-Al_2O_3-SiO_2-P_2O_5$ (concl.)

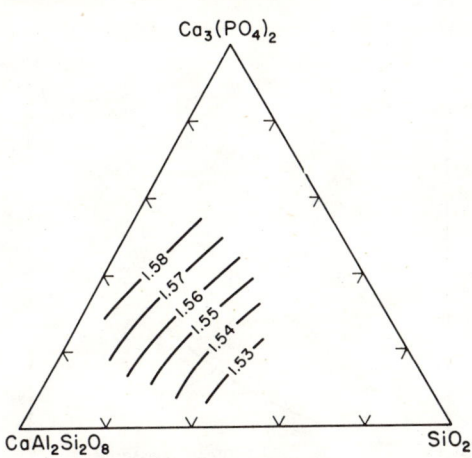

Fig. 961.—System $Ca_3(PO_4)_2$–$CaAl_2Si_2O_8$–SiO_2; isofracts.

P. D. S. St. Pierre, *J. Am. Ceram. Soc.*, **39** [4] 150 (1956).

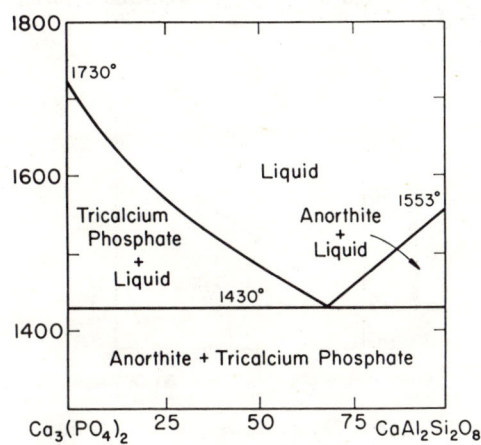

Fig. 962.—System $Ca_3(PO_4)_2$–$CaAl_2Si_2O_8$.

P. D. S. St. Pierre, *J. Am. Ceram. Soc.*, **39** [4] 149 (1956).

$FeO-MgO-Al_2O_3-SiO_2$

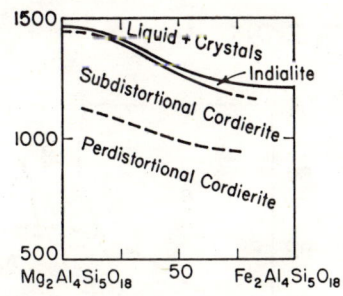

Fig. 963.—System $Mg_2Al_4Si_5O_{18}$–$Fe_2Al_4Si_5O_{18}$. "Indialite" and "subdistortional cordierite" represent the high modifications while "perdistortional cordierite" represents both the high and the low modifications.

Akiho Miyashiro, *Am. J. Sci.*, **255**, 58 (1957).

$MgO-NiO-GeO-SiO_2$

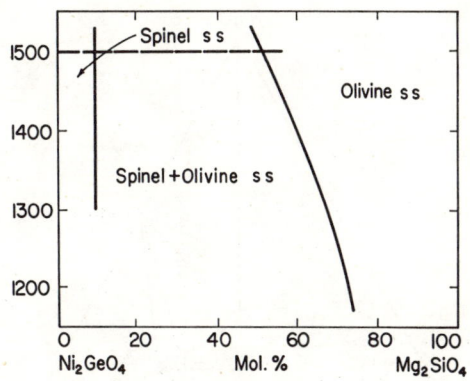

Fig. 964.—System Mg_2SiO_4–Ni_2GeO_4; pseudobinary subsolidus.

A. E. Ringwood, *Geochim. et Cosmochim. Acta*, **13**, 307 (1958).

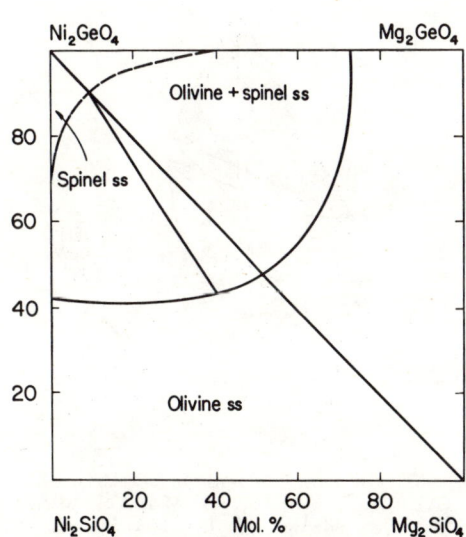

Fig. 965.—System Mg_2GeO_4–Mg_2SiO_4–Ni_2GeO_4–Ni_2SiO_4; subsolidus at 1500°C.

A. E. Ringwood, *Geochim. et Cosmochim. Acta*, **13**, 310 (1958).

MgO–Al$_2$O$_3$–SiO$_2$–ZrO$_2$

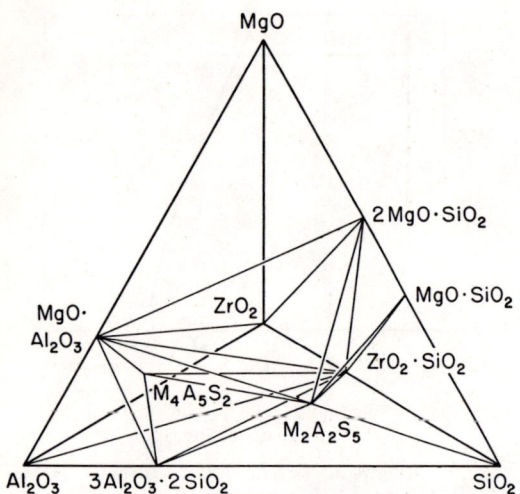

FIG. 966.—System MgO–Al$_2$O$_3$–SiO$_2$–ZrO$_2$, solid-state relationships. M = MgO; A = Al$_2$O$_3$; S = SiO$_2$; Z = ZrO$_2$.

P. G. Herold and W. J. Smothers, *J. Am. Ceram. Soc.* **37** [8] 353 (1954).

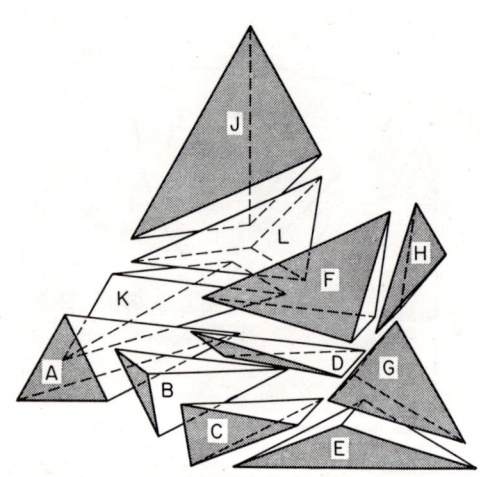

FIG. 967.—System MgO–Al$_2$O$_3$–SiO$_2$–ZrO$_2$, exploded view. (A) A, A$_3$S$_2$, MA, ZS; (B) MA, A$_3$S$_2$, M$_4$A$_5$S$_2$, ZS; (C) A$_3$S$_2$, M$_4$A$_5$S$_2$, M$_2$A$_2$S$_5$, ZS; (D) MA, M$_4$A$_5$S$_2$, M$_2$A$_2$S$_5$, ZS; (E) A$_3$S$_2$, S, M$_2$A$_2$S$_5$, ZS; (F) MA, M$_2$A$_2$S$_5$, M$_2$S, ZS; (G) M$_2$A$_2$S$_5$, S, MS, ZS; (H) MS, M$_2$A$_2$S$_5$, M$_2$S, ZS; (J) MA, Z, M, M$_2$S; (K) A, ZS, Z, MA; (L) MA, Z, M$_2$S, ZS. M = MgO; A = Al$_2$O$_3$; S = SiO$_2$; Z = ZrO$_2$.

P. G. Herold and W. J. Smothers, *J. Am. Ceram. Soc.*, **73** [8] 353 (1954).

MnO–Al$_2$O$_3$–Y$_2$O$_3$–SiO$_2$

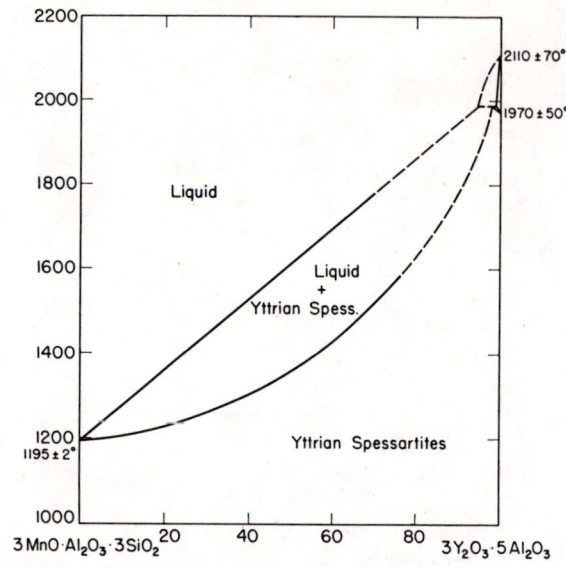

FIG. 968.—System 3MnO·Al$_2$O$_3$·3SiO$_2$–3Y$_2$O$_3$·5Al$_2$O$_3$.

H. S. Yoder and M. L. Keith, *American Mineral.*, **36**, 524 (1951).

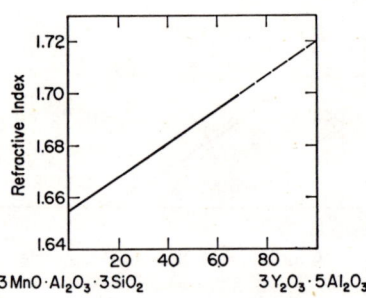

FIG. 969.—System 3MnO·Al$_2$O$_3$·3SiO$_2$–3Y$_2$O$_3$·5Al$_2$O$_3$; index of refraction of glasses.

H. S. Yoder and M. L. Keith, *American Mineral.*, **36**, 525 (1951).

PbO–SrO–TiO$_2$–ZrO$_2$

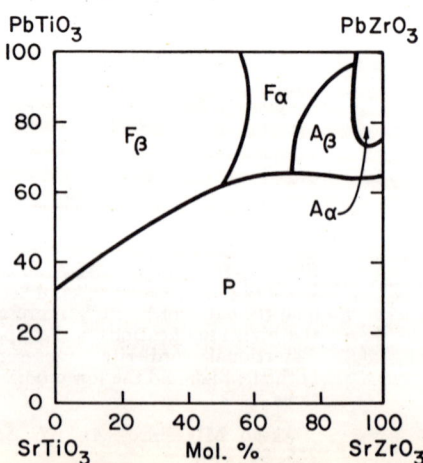

FIG. 970.—System PbTiO$_3$–PbZrO$_3$–SrTiO$_3$–SrZrO$_3$. P = paraelectric, cubic phase; A$_\alpha$ = antiferroelectric, orthorhombic phase; A$_\beta$ = antiferroelectric; F$_\alpha$ = ferroelectric, rhombohedral phase; F$_\beta$ = ferroelectric, tetragonal phase.

Takuro Ikeda, *J. Phys. Soc. Japan*, **14**, 1290 (1959).

V. Systems Containing Five Oxides

K₂O–Na₂O–CaO–Al₂O₃–SiO₂

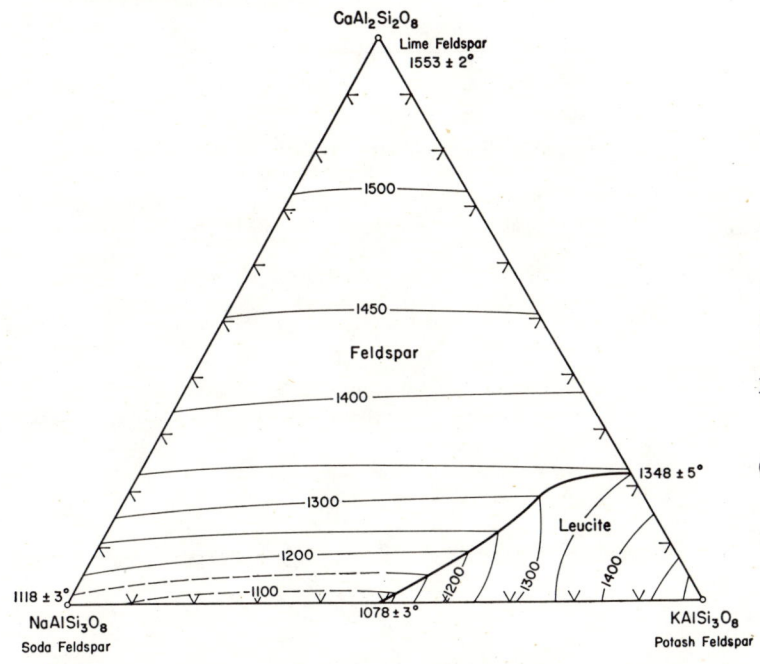

FIG. 971.—System NaAlSi₃O₈ (soda feldspar)–KAlSi₃O₈ (potash feldspar)–CaAl₂Si₂O₈ (lime feldspar).

R. R. Franco and J. F. Schairer, *J. Geol.*, **59**, 264 (1951).

K₂O–CaO–MgO–Al₂O₃–SiO₂

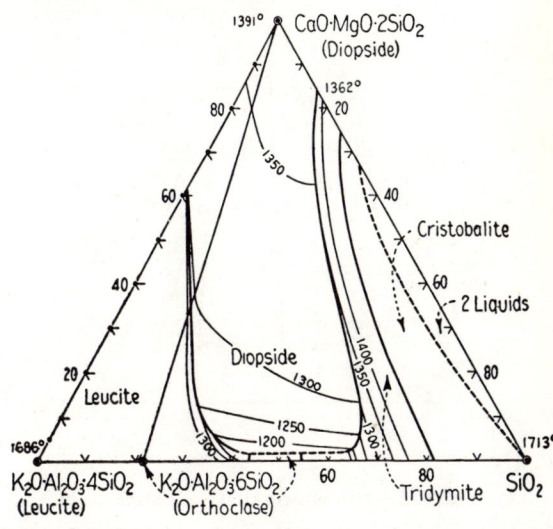

Fig. 973.—System SiO₂–K₂O·Al₂O₃·4SiO₂–CaO·MgO·2SiO₂ (silica-leucite-diopside).

J. F. Schairer and N. L. Bowen, *Am. J. Sci.*, **33**, 8 (1937).

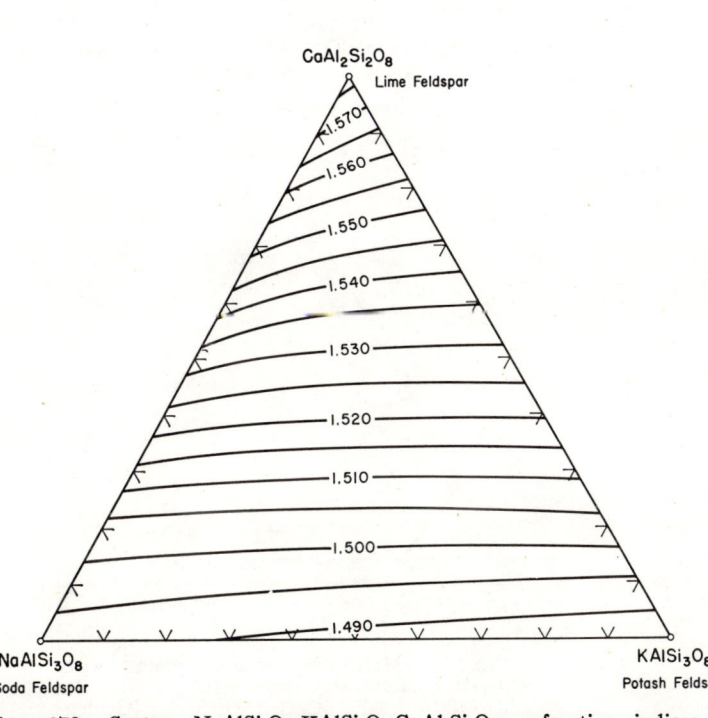

FIG. 972.—System NaAlSi₃O₈–KAlSi₃O₈–CaAl₂Si₂O₈; refractive indices of glasses at 25°C.

R. R. Franco and J. F. Schairer, *J. Geol.*, **59**, 266 (1951).

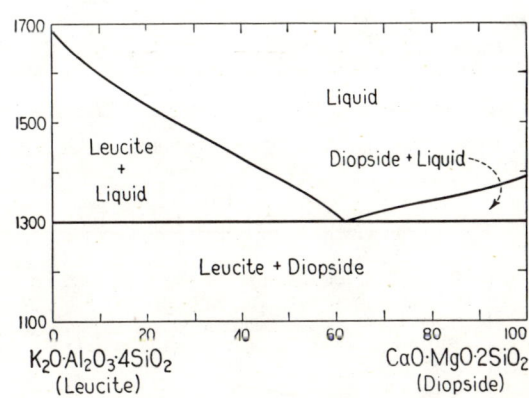

Fig. 974.—System K₂O·Al₂O₃·4SiO₂ (leucite)–CaO·MgO·2SiO₂ (diopside).

N. L. Bowen and J. F. Schairer, *Am. J. Sci.*, 5th Ser., **18**, 304 (1929).

$Na_2O-CaO-MgO-Al_2O_3-SiO_2$

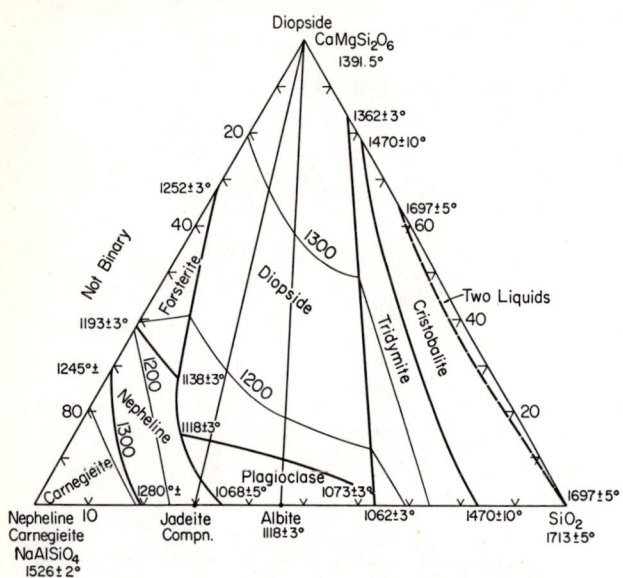

FIG. 975.—System $NaAlSiO_4-CaMgSi_2O_6-SiO_2$.

J. F. Schairer and H. S. Yoder, Jr., *Am. J. Sci.*, **258-A**, 278 (1960).

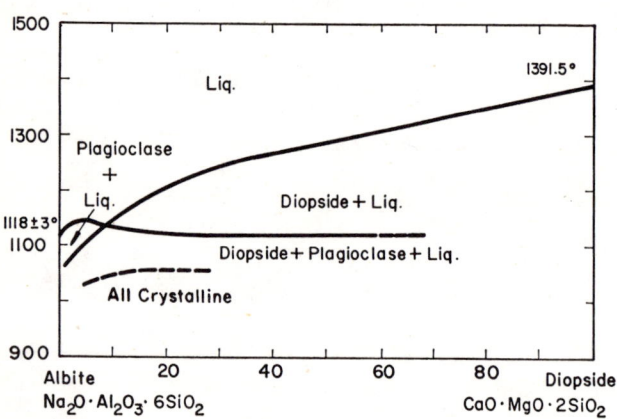

FIG. 976.—System $Na_2O \cdot Al_2O_3 \cdot 6SiO_2-CaO \cdot MgO \cdot 2SiO_2$.

J. F. Schairer and H. S. Yoder, Jr., *Am. J. Sci.*, **258-A**, 279, (1960).

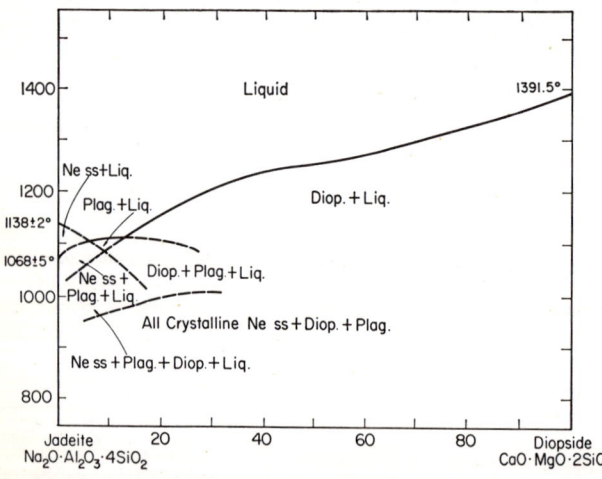

FIG. 977.—System $Na_2O \cdot Al_2O_3 \cdot 4SiO_2-CaO \cdot MgO \cdot 2SiO_2$. Diop. = diopside; Ne ss = nepheline ($NaAlSiO_4$) solid solution; Plag. = plagioclase.

J. F. Schairer and H. S. Yoder, Jr., *Am. J. Sci.*, **258-A**, 280 (1960).

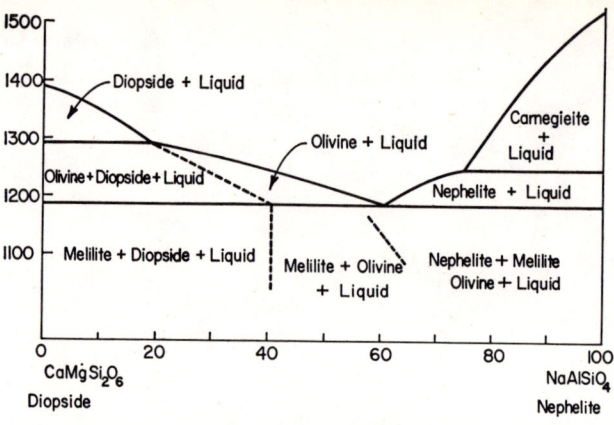

FIG. 978.—Pseudobinary system $Na_2O \cdot Al_2O_3 \cdot 2SiO_2-CaO \cdot MgO \cdot 2SiO_2$.

N. L. Bowen, Evolution of Igneous Rocks, Princeton, 1928, p. 261; see also *Am. J. Sci.*, 5th Ser., **3**, 18 (1922).

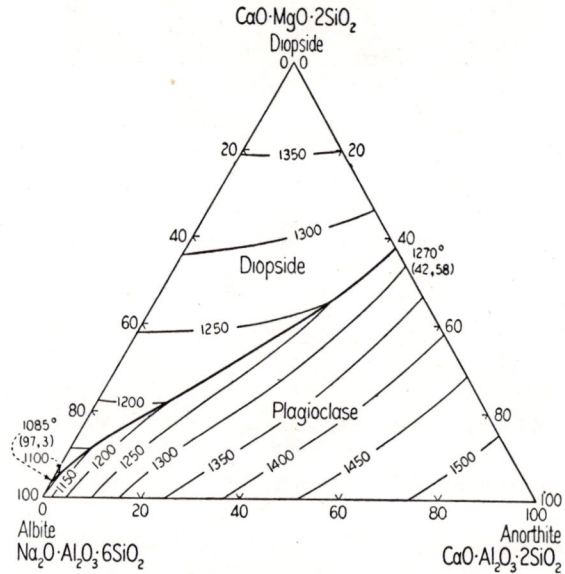

Fig. 979.—System $Na_2O \cdot Al_2O_3 \cdot 6SiO_2-CaO \cdot MgO \cdot 2SiO_2-CaO \cdot Al_2O_3 \cdot 2SiO_2$ (albite-anorthite-diopside).

N. L. Bowen, *Am. J. Sci.*, 4th Ser., **40**, 167 (1915).

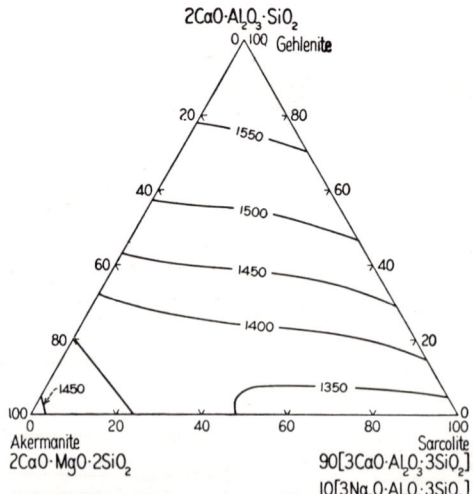

FIG. 980.—Melting isotherms of system $2CaO \cdot Al_2O_3 \cdot SiO_2-2CaO \cdot MgO \cdot 2SiO_2-(10\% \ 3Na_2O \cdot Al_2O_3 \cdot 3SiO_2 + 90\% \ 3CaO \cdot Al_2O_3 \cdot 3SiO_2)$ (Gehlenite-akermanite-sarcolite).

A. F. Buddington, *Am. J. Sci.*, 5th Ser., **3**, 57 (1922).

Na_2O–CaO–MgO–Al_2O_3–SiO_2 (concl.)

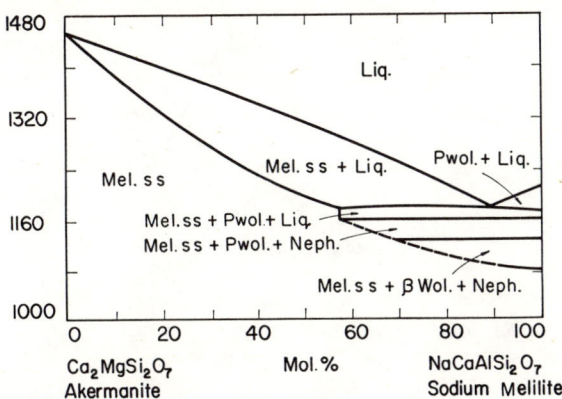

Fig. 981.—System $NaCaAlSi_2O_7$ (sodium-melilite)–$Ca_2MgSi_2O_7$ (akermanite). Mel. = melilite, Neph. = nepheline, Pwol. = pseudowollastonite, βWol. = low-temp. wollastonite, ss = solid solution.

R. W. Nurse and H. G. Midgley, *J. Iron Steel Inst. (London)*, **174**, 129 (1953).

Na_2O–CaO–Al_2O_3–SiO_2–TiO_2

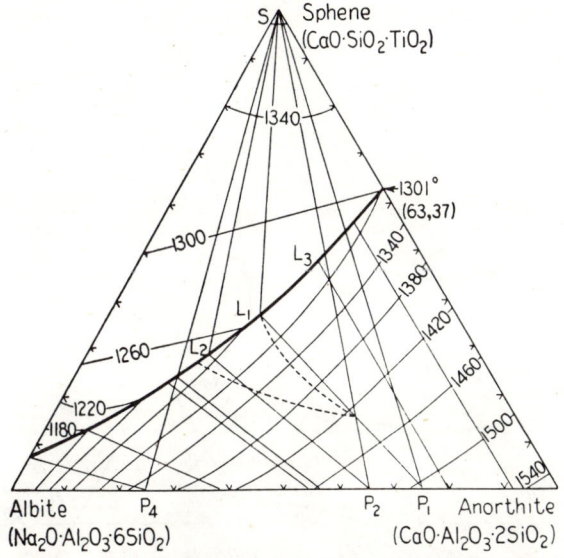

Fig. 982.—System $Na_2O \cdot Al_2O_3 \cdot 6SiO_2$–$CaO \cdot Al_2O_3 \cdot 2SiO_2$–$CaO \cdot SiO_2 \cdot TiO_2$; boundary curve and three-phase boundaries in plagioclase field, with two complete three-phase triangles, SL_1P_1 and SL_2P_2; composition of liquid phase on lower end of boundary line follows boundary outside of triangle.

A. T. Prince, *J. Geol.*, **51**, 8 (1943).

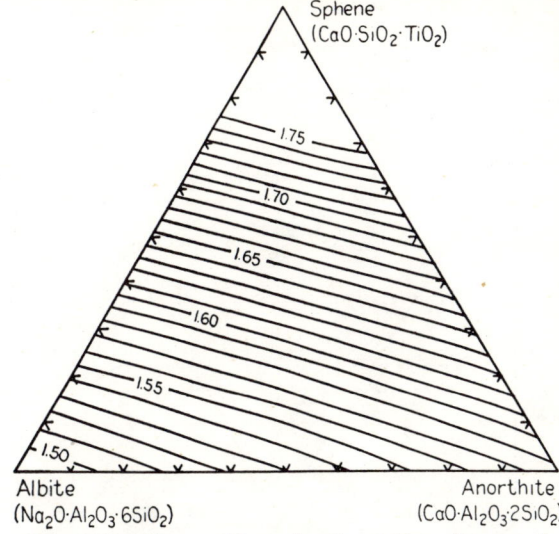

Fig. 983.—Composition-refractive index diagram for glasses in system $Na_2O \cdot Al_2O_3 \cdot 6SiO_2$–$CaO \cdot Al_2O_3 \cdot 2SiO_2$–$CaO \cdot SiO_2 \cdot TiO_2$.

A. T. Prince, *J. Geol.*, **51**, 12 (1943).

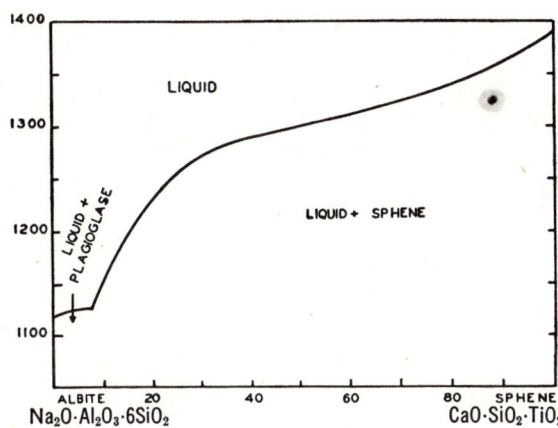

Fig. 984.—System $Na_2O \cdot Al_2O_3 \cdot 6SiO_2$–$CaO \cdot SiO_2 \cdot TiO_2$.

A. T. Prince, *J. Geol.*, **51**, 3 (1943).

Na_2O–CaO–Al_2O_3–SiO_2–P_2O_5

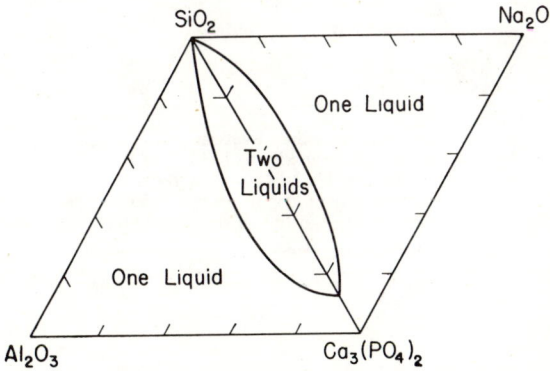

Fig. 985.—System Na_2O–$Ca_3(PO_4)_2$–Al_2O_3–SiO_2, non-miscible liquid equilibrium.

B. N. Melent'ev and Ya. I. Ol'shanskiĭ, *Doklady Akad. Nauk S.S.S.R.*, **86**, 1126 (1952).

CaO–MgO–MnO–SiO₂–TiO₂

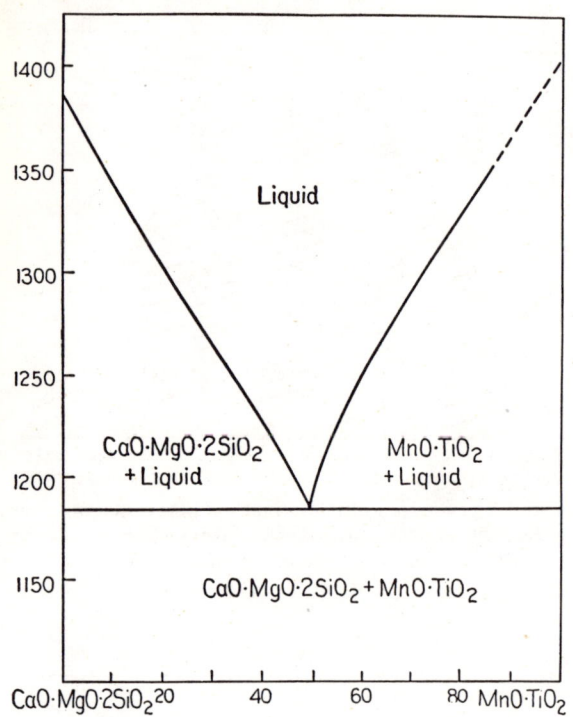

Fig. 986.—System MnO·TiO₂–CaO·MgO·2SiO₂.

Kiezo Iwasé and Usaburô Nisioka, *Science Repts., Tôhoku Imp. Univ.*, Ser. 1, **26**, 596 (1937–38).

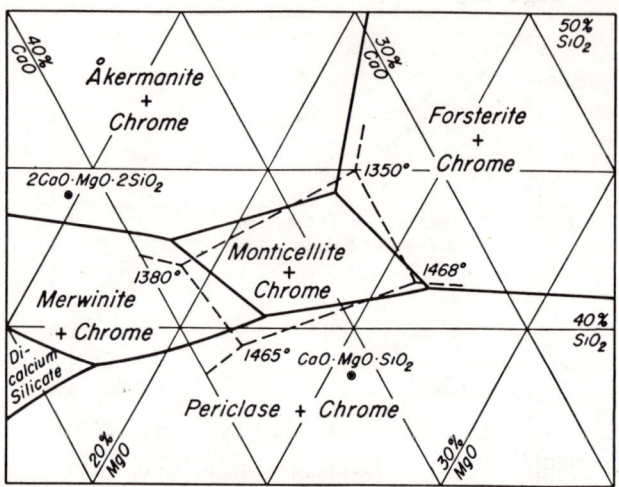

Fig. 988.—Part of the system CaO–MgO–Al₂O₃–Cr₂O₃–SiO₂ on the 7% R₂O₃ plane (50MgO·Al₂O₃, 50MgO·Cr₂O₃) parallel to the CaO–MgO–SiO₂ base.
Solid line shows projected boundary curve in the CaO–MgO–SiO₂ system; dashed line shows projected boundary curve in the CaO–MgO–Al₂O₃–Cr₂O₃–SiO₂ system.

T. F. Berry, W. C. Allen, and R. B. Snow, *J. Am. Ceram. Soc.*, **33** [4] 126 (1950).

CaO–MgO–Al₂O₃–Cr₂O₃–SiO₂

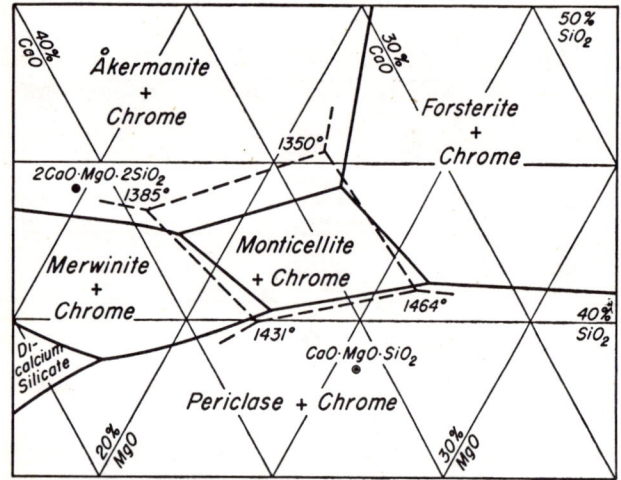

Fig. 987.—Part of the system CaO–MgO–Al₂O₃–Cr₂O₃–SiO₂ on the 7% R₂O₃ plane (75MgO·Al₂O₃, 25MgO·Cr₂O₃) parallel to the CaO–MgO–SiO₂ base.
Solid line shows projected boundary curve in the CaO–MgO–SiO₂ system; dashed line shows projected boundary curve in the CaO–MgO–Al₂O₃–Cr₂O₃–SiO₂ system.

T. F. Berry, W. C. Allen, and R. B. Snow, *J. Am. Ceram. Soc.*, **33** [4] 126 (1950).

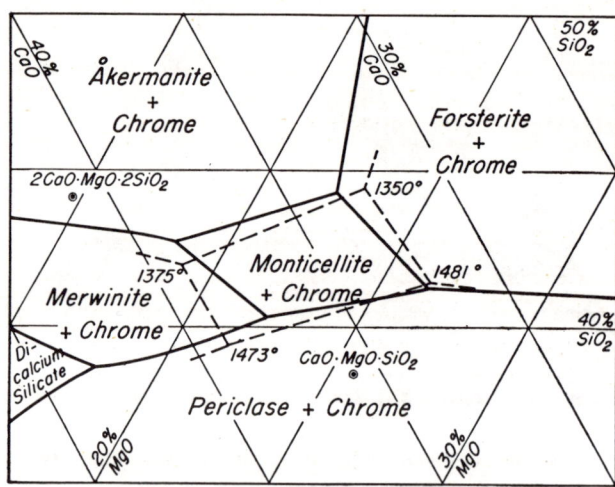

Fig. 989.—Part of the system CaO–MgO–Al₂O₃–Cr₂O₃–SiO₂ on the 7% R₂O₃ plane (25MgO·Al₂O₃, 75MgO·Cr₂O₃) parallel to the CaO–MgO–SiO₂ base.
Solid line shows projected boundary curve in the CaO–MgO–SiO₂ system; dashed line shows projected boundary curve in the CaO–MgO–Al₂O₃–Cr₂O₃–SiO₂ system.

T. F. Berry, W. C. Allen, and R. B. Snow, *J. Am. Ceram. Soc.*, **33** [4] 126 (1950).

CaO–MgO–Al$_2$O$_3$–Cr$_2$O$_3$–SiO$_2$ (concl.)

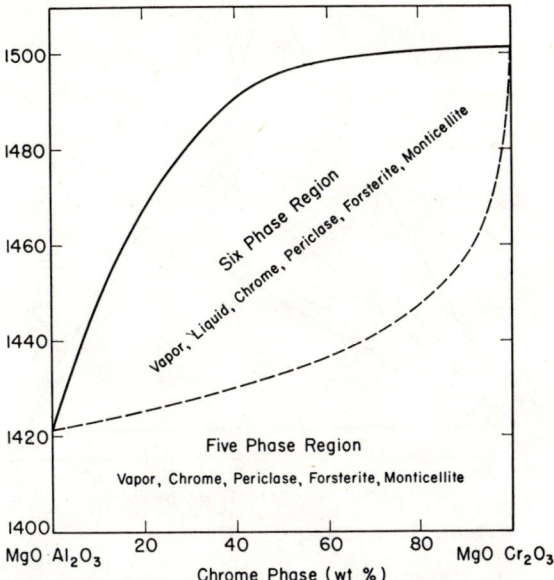

Fig. 990.—Pseudobinary diagram showing the relationship between temperature and the weight proportion of MgO·Al$_2$O$_3$ and MgO·Cr$_2$O$_3$ in the chrome phase for the sextuple equilibrium points in the system CaO–MgO–Al$_2$O$_3$–Cr$_2$O$_3$–SiO$_2$ for the solid phases chrome, periclase, monticellite, and forsterite.

T. F. Berry, W. C. Allen, and R. B. Snow, J. Am. Ceram. Soc., 33 [4] 127 (1950).

CaO–MgO–Al$_2$O$_3$–Fe$_2$O$_3$–SiO$_2$

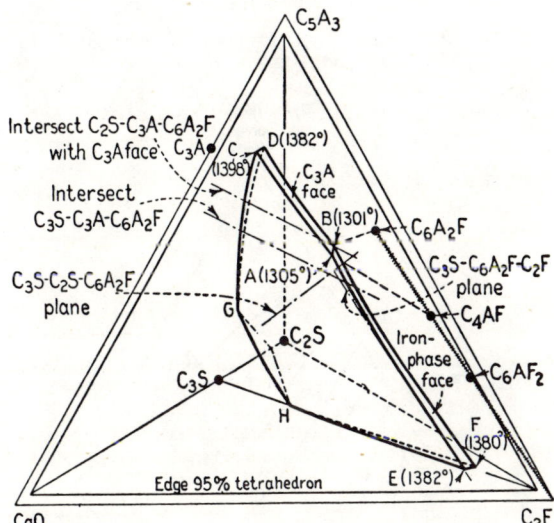

Fig. 991.—System CaO–5CaO·3Al$_2$O$_3$–2CaO·Fe$_2$O$_3$–2CaO·SiO$_2$ modified by 5% MgO; see Fig. 948 for system without MgO.

M. A. Swayze, Am. J. Sci., 244, 70 (1946).

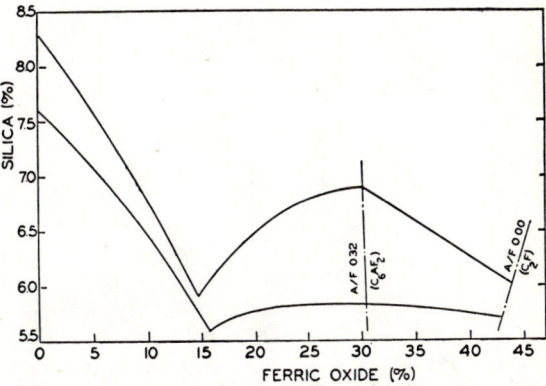

Fig. 992.—Silica content of liquids modified by 5% MgO in system CaO–5CaO·3Al$_2$O$_3$–2CaO·Fe$_2$O$_3$–2CaO·SiO$_2$; upper curve, CaO–C$_3$S–C$_3$A:C$_2$F; lower curve, C$_3$S–C$_2$S–C$_3$A:C$_2$F.

M. A. Swayze, Am. J. Sci., 244, 73 (1946).

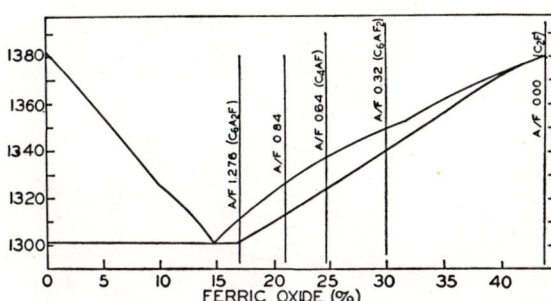

Fig. 993.—Melting temperatures along C$_3$S–C$_2$S–C$_3$A:C$_2$F boundary curve modified by 5% MgO in system CaO–5CaO·3Al$_2$O$_3$–2CaO·Fe$_2$O$_3$–2CaO·SiO$_2$.

M. A. Swayze, Am. J. Sci., 244, 71 (1946).

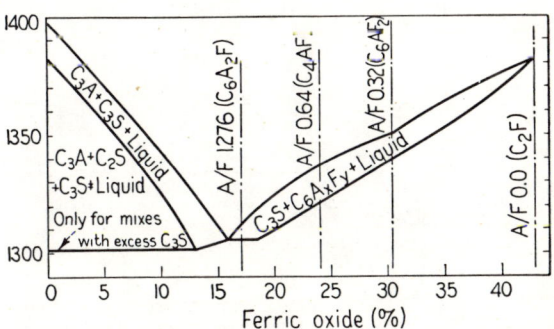

Fig. 994.—Melting temperatures along CaO–C$_3$S–C$_3$A:C$_2$F boundary curve modified by 5% MgO in system CaO–5CaO·3Al$_2$O$_3$–2CaO·Fe$_2$O$_3$–2CaO·SiO$_2$.

M. A. Swayze, Am. J. Sci., 244, 70 (1946).

CaO–MgO–Al$_2$O$_3$–Fe$_2$O$_3$–SiO$_2$ (cont.)

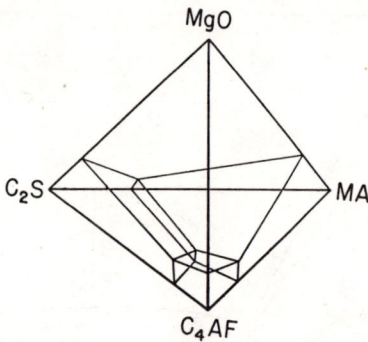

Fig. 995.—System MgO–2CaO·SiO$_2$–4CaO·Al$_2$O$_3$·Fe$_2$O$_3$–MgO·Al$_2$O$_3$; C = CaO, M = MgO, A = Al$_2$O$_3$, F = Fe$_2$O$_3$, S = SiO$_2$.

J. R. Rait, *Iron and Steel*, **22**, 188 (1949).

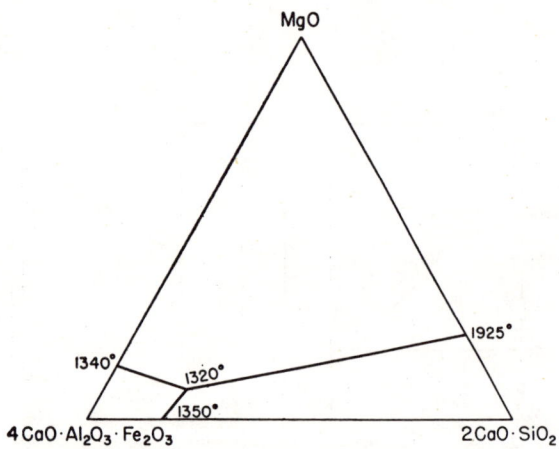

Fig. 996.—System 4CaO·Al$_2$O$_3$·Fe$_2$O$_3$–2CaO·SiO$_2$–MgO

J. R. Rait, *Iron and Steel*, **22**, 186 (1949).

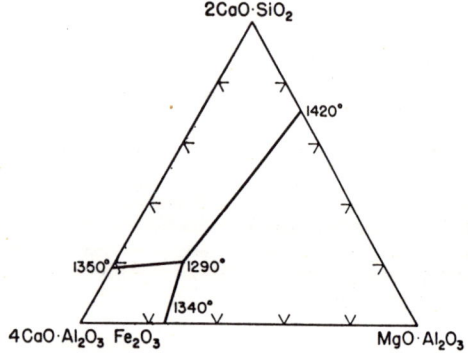

Fig. 997.—System 4CaO·Al$_2$O$_3$·Fe$_2$O$_3$–MgO·Al$_2$O$_3$–2CaO·SiO$_2$.

J. R. Rait, *Iron and Steel*, **22**, 91 (1949).

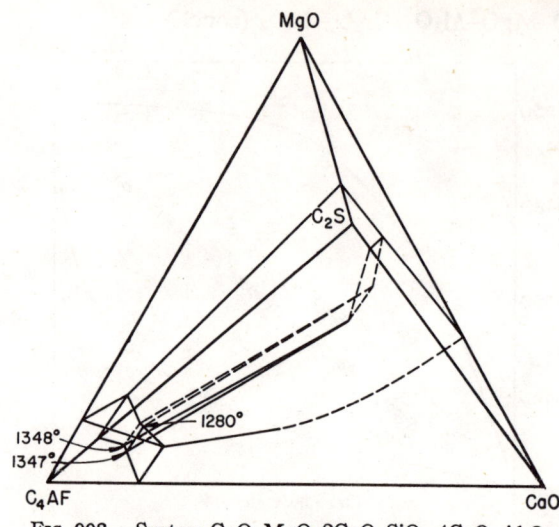

Fig. 998.—System CaO–MgO–2CaO·SiO$_2$–4CaO·Al$_2$O$_3$·Fe$_2$O$_3$; C = CaO, A = Al$_2$O$_3$, F = Fe$_2$O$_3$, S = SiO$_2$.

J. R. Rait, *Iron and Steel*, **22**, 187 (1949).

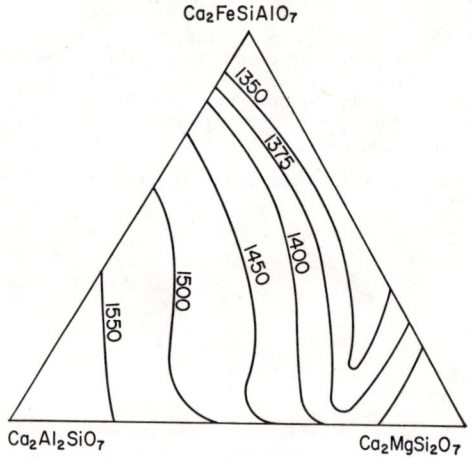

Fig. 999.—System Ca$_2$FeSiAlO$_7$–Ca$_2$MgSi$_2$O$_7$–Ca$_2$Al$_2$SiO$_7$; liquidus surface.

R. W. Nurse and H. G. Midgley, *J. Iron Steel Ins.* (London), **174**, 125 (1953).

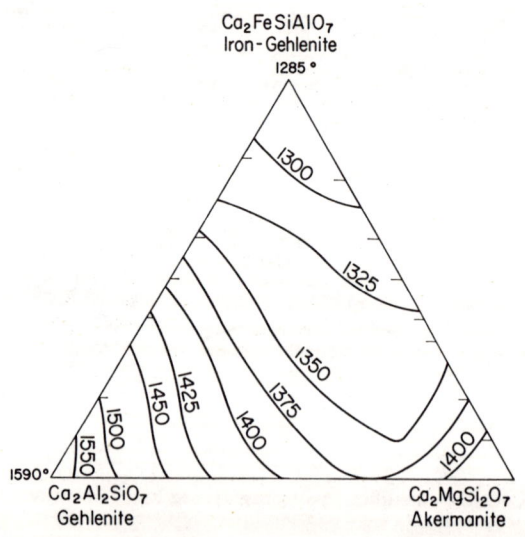

Fig. 1000.—System Ca$_2$FeSiAlO$_7$–CaMgSi$_2$O$_7$–Ca$_2$Al$_2$SiO$_7$; solidus surface.

R. W. Nurse and H. G. Midgley, *J. Iron Steel Inst.* (London), **174**, 125 (1953).

CaO–MgO–Al$_2$O$_3$–Fe$_2$O$_3$–SiO$_2$ (concl.)

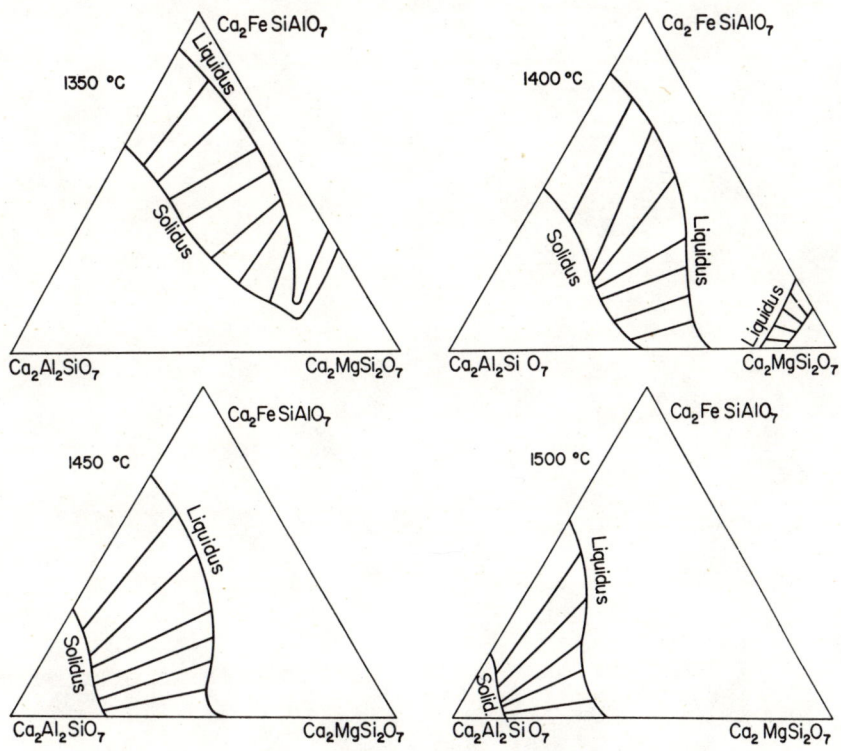

Fig. 1001.—System Ca$_2$FeSiAlO$_7$–Ca$_2$MgSi$_2$O$_7$–Ca$_2$Al$_2$SiO$_7$; isothermal planes.

R. W. Nurse and H. G. Midgley, *J. Iron Steel Inst.* (*London*), **174**, 126 (1953).

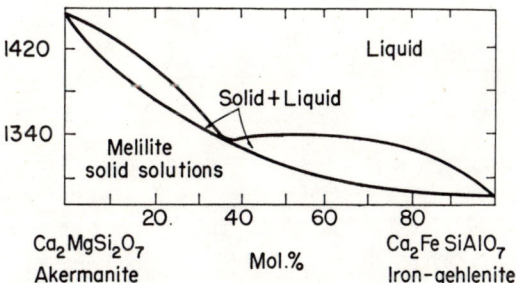

Fig. 1002.—System Ca$_2$MgSi$_2$O$_7$ (akermanite)–Ca$_2$FeSiAlO$_7$ (iron-gehlenite).

R. W. Nurse and H. G. Midgley, *J. Iron Steel Inst.* (*London*), **174**, 124 (1953).

C. SYSTEMS WITH OXYGEN-CONTAINING RADICALS

I. Carbonates Only

(a) One Carbonate

CaCO₃

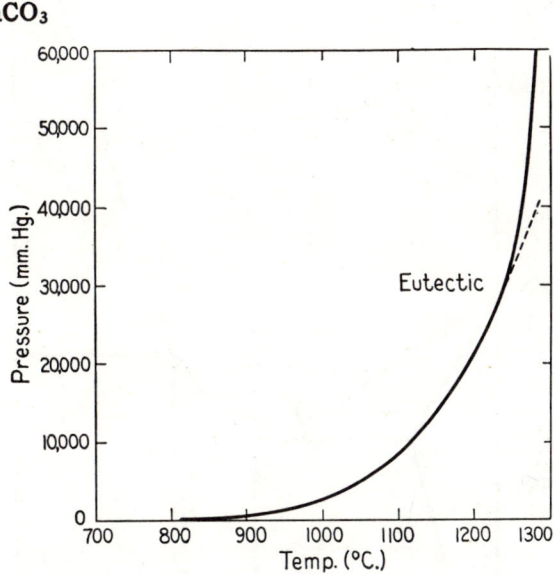

FIG. 1003.—System CaO–CO₂; pressure-temperature curve.

F. H. Smyth and L. H. Adams, *J. Am. Chem. Soc.*, **45**, 1178 (1923).

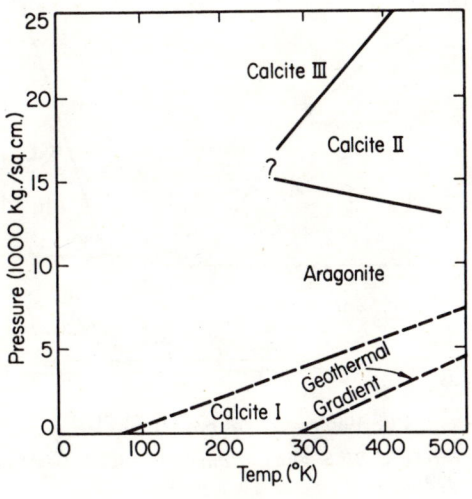

FIG. 1005.—System CaCO₃.

J. C. Jamieson, *J. Chem. Phys.*, **21**, 1389 (1953).

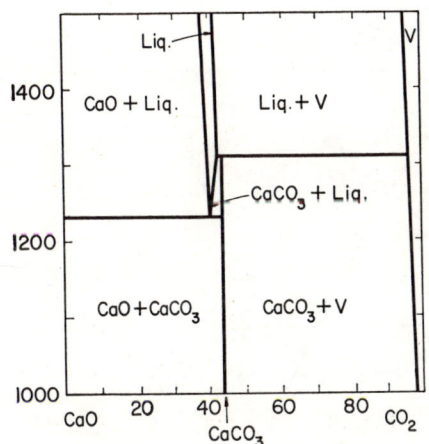

FIG. 1004.—System CaO–CO₂ at 1,000 bars. Vapor, V, is almost pure CO₂.

P. J. Wyllie and O. F. Tuttle, *J. Petrol.* **1** [1] 10 (1960).

MgCO₃

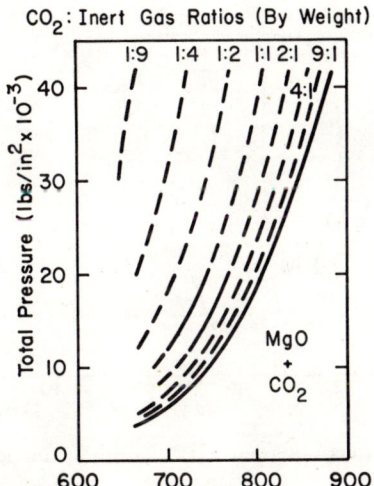

FIG. 1006.—System MgO–CO₂–A. Thermal equilibrium curves for the reaction MgCO₃ = MgO + CO₂, as deduced for varying proportions of CO₂ to inert pressure. The continuous lines are the regions in which experimental results confirm the curves.

R. I. Harker, *Am. J. Sci.*, **256**, 136 (1958).

(b) Two Carbonates

K₂CO₃–Li₂CO₃

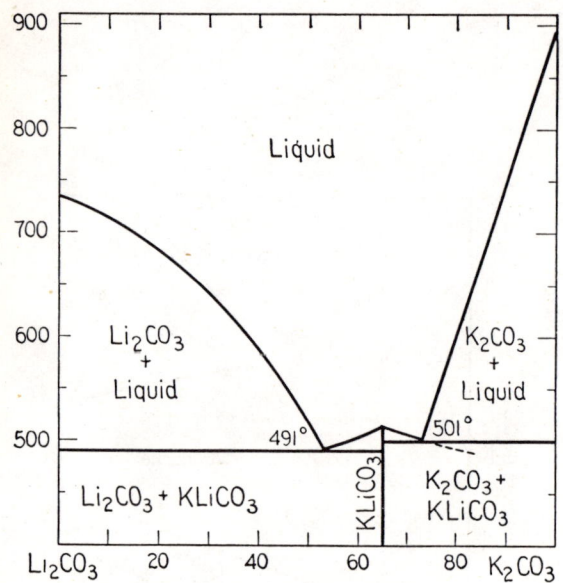

Fig. 1007.—System K_2CO_3–Li_2CO_3.

W. Eitel and W. Skaliks, Z. anorg. u. allgem. Chem., **183**, 263 (1929).

K₂CO₃–Na₂CO₃

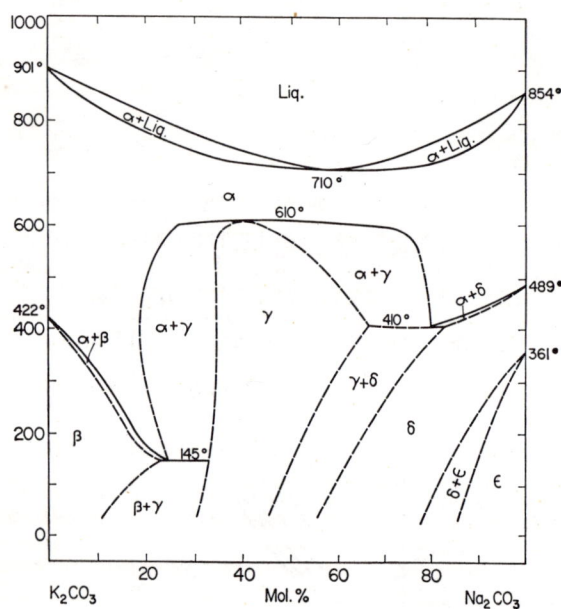

Fig. 1008.—System K_2CO_3–Na_2CO_3; proposed.

Arnold Reisman, J. Am. Chem. Soc., **81**, 810 (1959).

K₂CO₃–CaCO₃

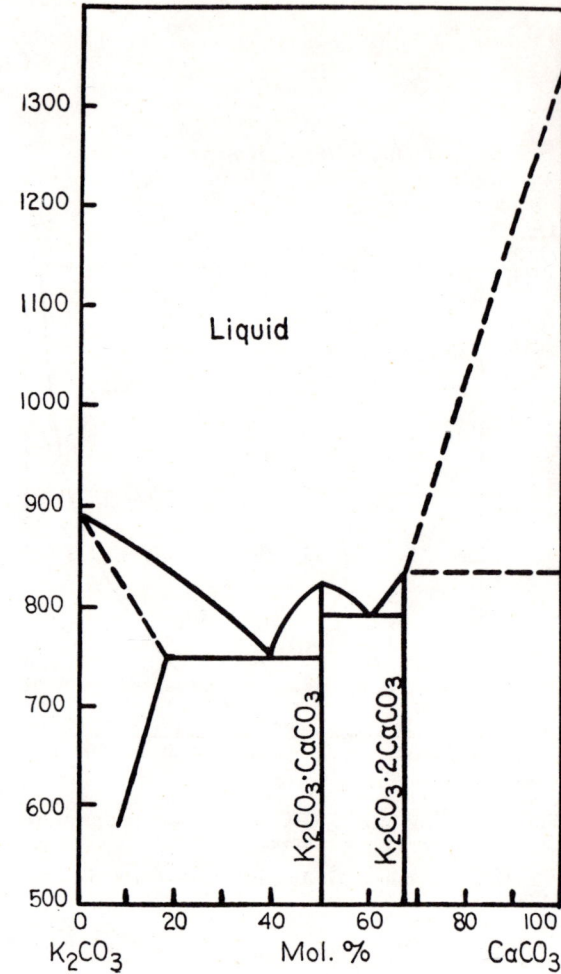

Fig. 1009.—System K_2CO_3–$CaCO_3$; equilibrium diagram determined under CO_2 pressure of 50 atmospheres.

C. Kroger, K. W. Illner, and W. Glaeser, Z. anorg. u. allgem. Chem., **251**, 271 (1943).

Li₂CO₃–Na₂CO₃

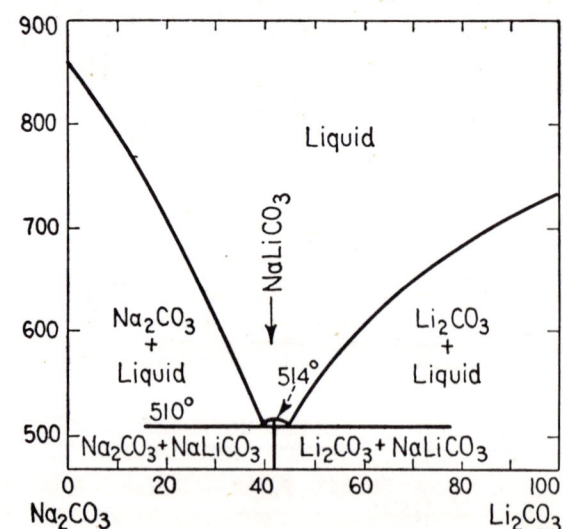

Fig. 1010.—System Li_2CO_3–Na_2CO_3.

W. Eitel and W. Skaliks, Z. anorg. u. allgem. Chem., **183**, 270 (1929).

Li$_2$CO$_3$–CaCO$_3$

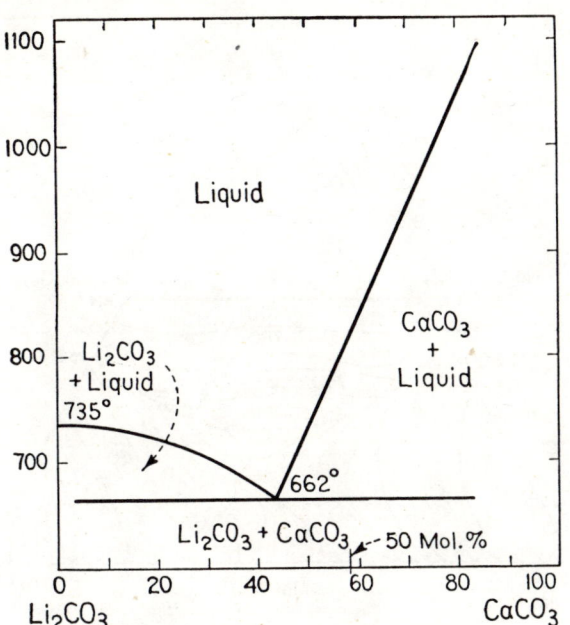

FIG. 1011.—System Li$_2$CO$_3$–CaCO$_3$.

W. Eitel and W. Skaliks, *Z. anorg. u. allgem. Chem.*, **183**, 268 (1929).

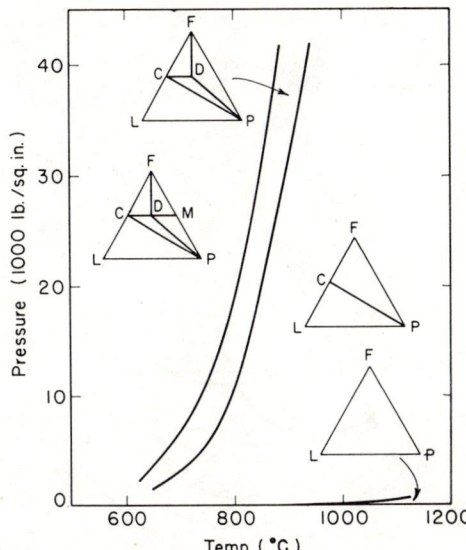

FIG. 1013.—System CaCO$_3$–MgCO$_3$, dissociation. Univariant pressure (CO$_2$)-temp. curves for the dissociation of calcite, dolomite, and magnesite. The triangular diagrams indicate the phases which would be stable in the divariant regions defined by the P-T curves ignoring any solid solution. F = CO$_2$; C = calcite; D = dolomite; M = magnesite; L = lime; P = periclase.

R. I. Harker and O. F. Tuttle, *Am. J. Sci.*, **253**, 221 1955).

CaCO$_3$–MgCO$_3$

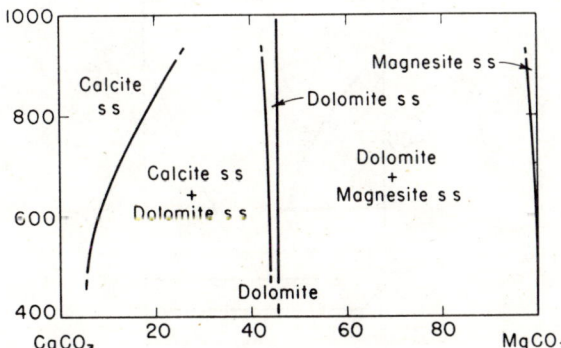

FIG. 1012.—System CaCO$_3$–MgCO$_3$, limits of solid solution (ss).

R. I. Harker and O. F. Tuttle, *Am. J. Sci.*, **253**, 279 (1955).

CaCO$_3$–MnCO$_3$

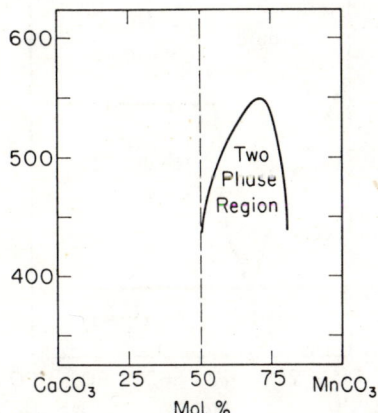

FIG. 1014.—System CaCO$_3$–MnCO$_3$, subsolidus.

J. R. Goldsmith and D. L. Graft, *Geochim. et Cosmochim. Acta*, **11**, 321 (1957).

(c) Three Carbonates

K_2CO_3–Li_2CO_3–Na_2CO_3

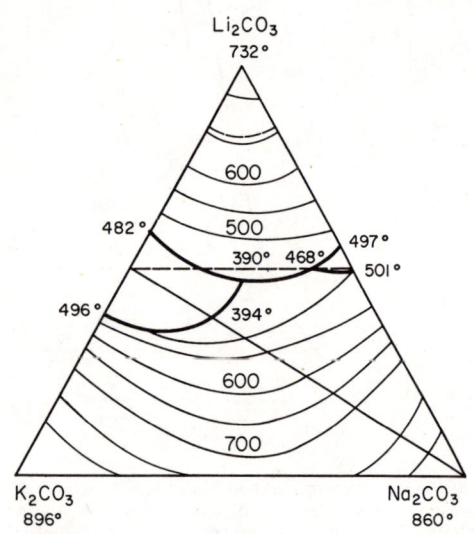

FIG. 1015.—System K_2CO_3–Li_2CO_3–Na_2CO_3.

L. F. Volkova, *Izvest. Sibirsk. Otdel. Akad. Nauk S.S.S.R.*, **1958**, No. 7, p. 34.

K_2CO_3–Na_2CO_3–$CaCO_3$

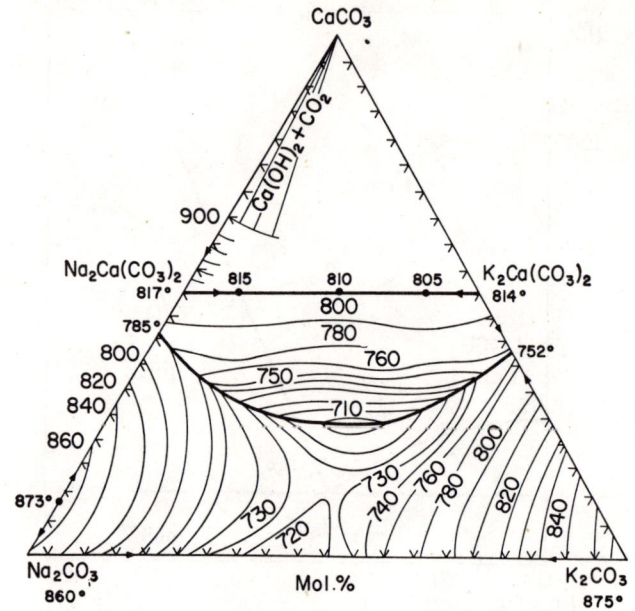

FIG. 1016.—System K_2CO_3–Na_2CO_3–$CaCO_3$ at 1 atm. in CO excess.

Paul Niggli, *Z. anorg. u. allgem. Chem.*, **106**, 131 (1919).

II. Perchlorates Only

$LiClO_4$–NH_4ClO_4

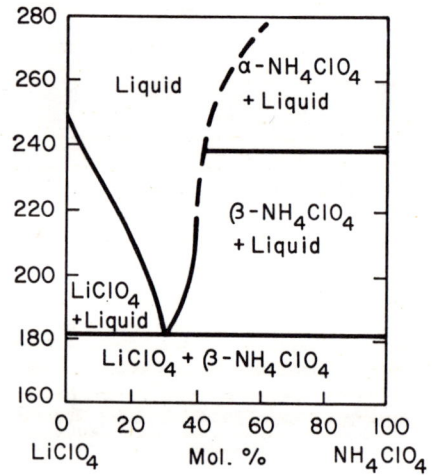

FIG. 1017.—System $LiClO_4$–NH_4ClO_4.

M. M. Markowitz and R. F. Harris, *J. Phys. Chem.*, **63**, 1520 (1959).

$NaClO_4$–$Ba(ClO_4)_2$

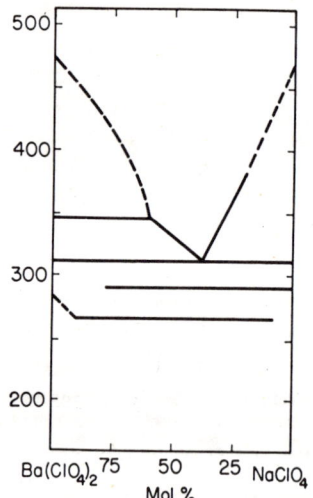

FIG. 1018.—System $NaClO_4$–$Ba(ClO_4)_2$.

A. A. Zinov'ev, L. I. Chudinova, and L. P. Smolina, *Zhur. Neorg. Khim.*, **1**, 1854 (1956).

III. Hydroxides

(a) Two Substances

(K, Na)OH–(K, Na)NO₂

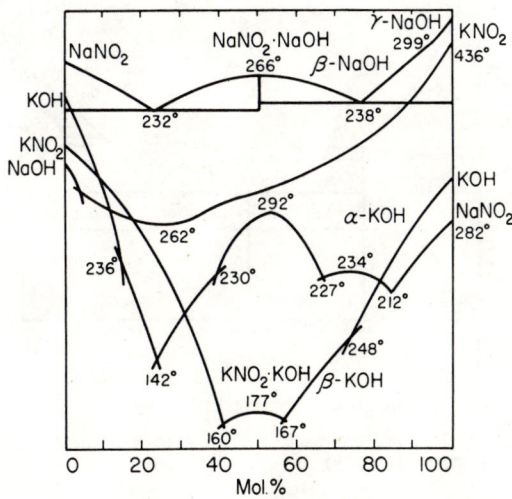

FIG. 1019.—Systems: KOH–KNO₂, KOH–NaNO₂, NaOH–KNO₂, NaOH–NaNO₂.

A. G. Bergman and N. A. Reshetnikov, *Izvest. Sektora Fiz.-Khim. Anal., Inst. Obshchei Neorg. Khim., Akad. Nauk S.S.S.R.*, **25**, 213 (1954).

(K, Na)OH–(K, Na)NO₃

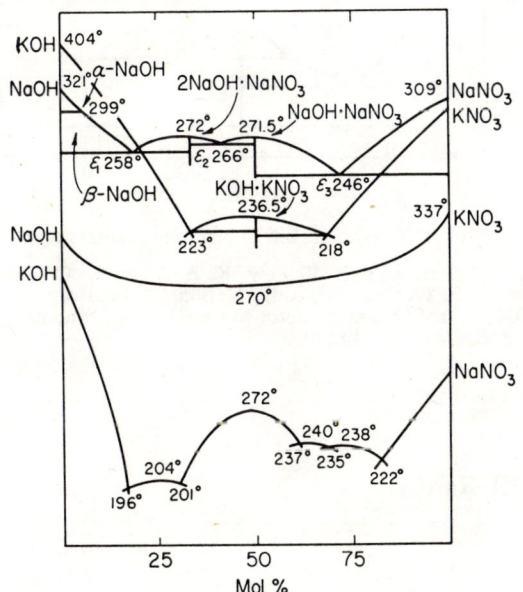

FIG. 1020.—Systems: KOH–KNO₃, KOH–NaNO₃, NaOH–KNO₃, NaOH–NaNO₃.

A. G. Bergman and N. A. Reshetnikov, *Izvest. Sektora Fiz.-Khim. Anal., Inst. Obshchei Neorg. Khim., Akad. Nauk S.S.S.R.*, **25**, 212 (1954).

KOH–KNO₃

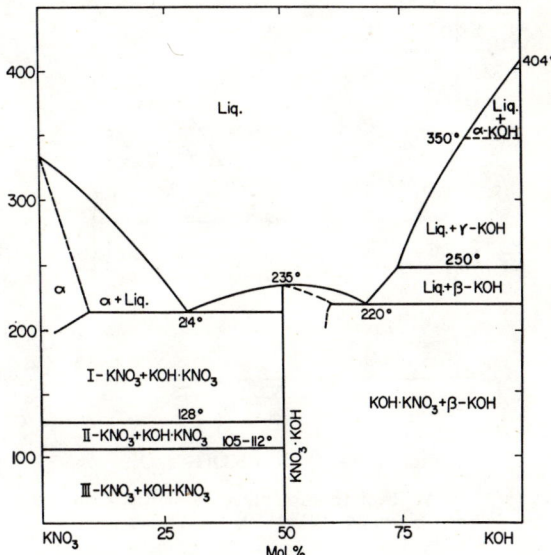

FIG. 1021. System KNO₃–KOH.

N. A. Reshetnikov and N. I. Vilutis, *Zhur. Neorg. Khim.*, **3**, 371 (1958).

KOH–LiOH

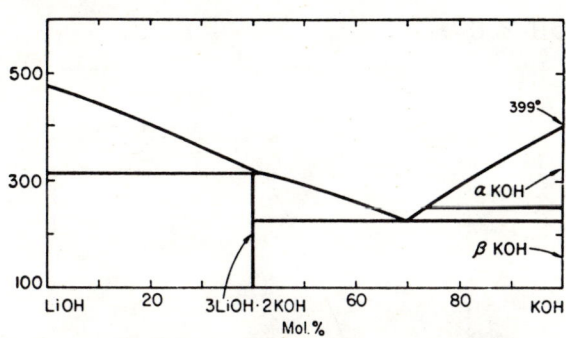

FIG. 1022. System KOH–LiOH; preliminary.

C. J. Barton, J. P. Blakely, L. M. Bratcher, and W. R. Grimes, Oak Ridge National Laboratory, Phase Diagrams of Nuclear Reactor Materials, R. E. Thoma, ed., ORNL-2548, p. 141 (1959).

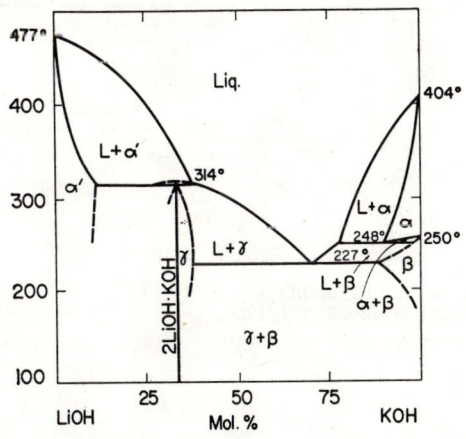

FIG. 1023.—System KOH–LiOH.

N. A. Reshetnikov and N. I. Vilutis, *Zhur. Neorg. Khim.*, **4**, 124 (1959).

KOH–NaOH

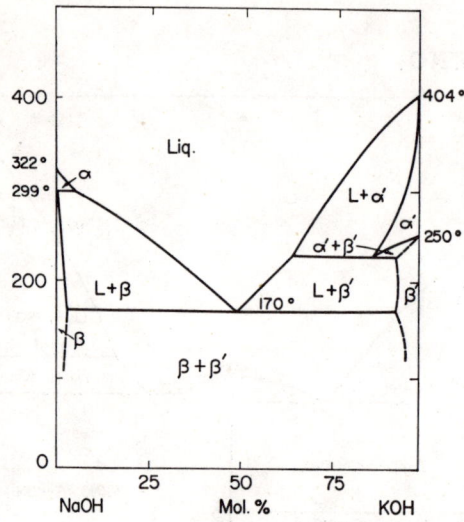

Fig. 1024.—System KOH–NaOH.

N. A. Reshetnikov and N. I. Vilutis, *Zhur. Neorg. Khim.*, **4**, 124 (1959).

KOH–K₂CrO₄

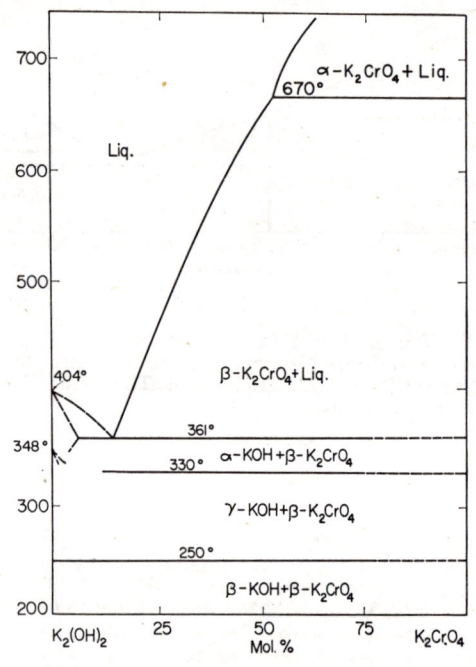

Fig. 1025.—System KOH–K$_2$CrO$_4$.

N. A. Reshetnikov and N. I. Vilutis, *Zhur. Neorg. Khim.*, **3**, 374 (1958).

LiOH–NaOH

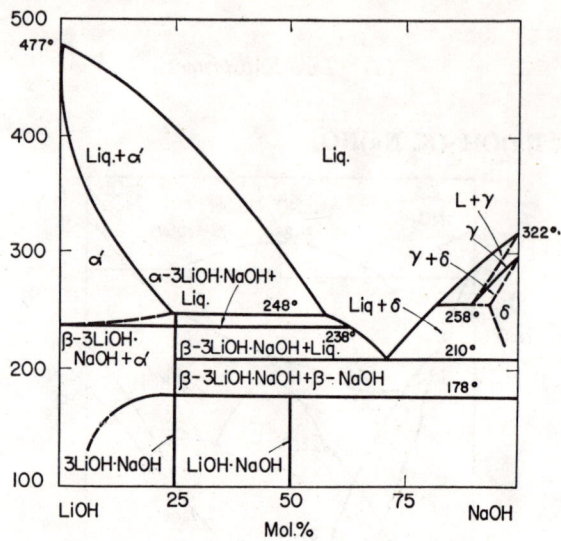

Fig. 1026.—System LiOH–NaOH.

N. A. Reshetnikov and G. M. Unzhakov, *Zhur. Neorg. Khim.*, **3**, 1435 (1958).

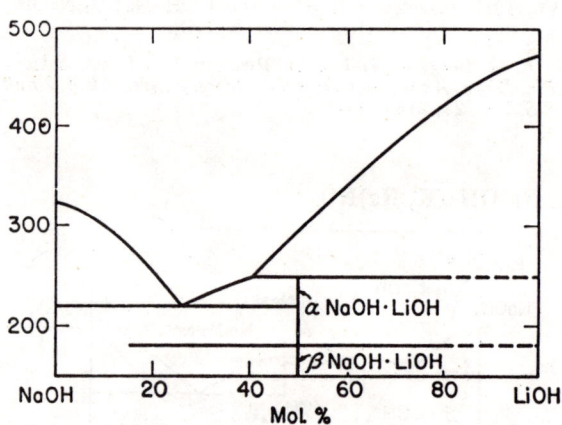

Fig. 1027. System LiOH–NaOH; preliminary.

C. J. Barton, J. P. Blakely, K. A. Allen, W. C. Davis, and B. S. Weaver, Oak Ridge National Laboratory, Phase Diagrams of Nuclear Reactor Materials, R. E. Thoma, ed., ORNL-2548, p. 140 (1959).

NaOH–NaNO₂

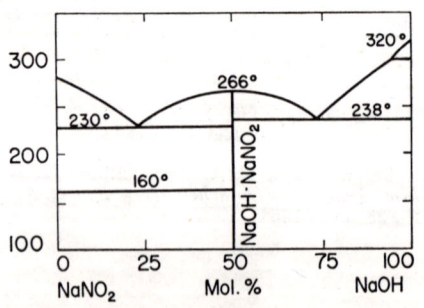

Fig. 1028.—System NaNO$_2$–NaOH.

N. A. Reshetnikov and N. I. Vilutis, *Zhur. Neorg. Khim.*, **3**, 372 (1958).

NaOH–KNO₃

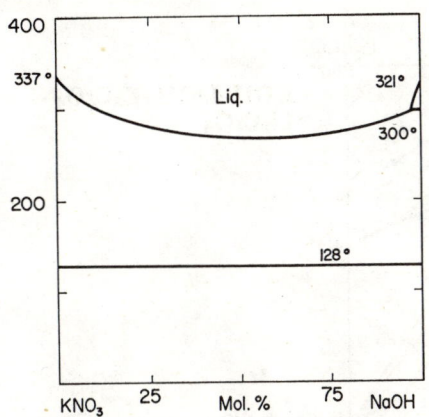

Fig. 1029.—System KNO₃–NaOH.

N. A. Reshetnikov and N. I. Vilutis, *Zhur. Neorg. Khim.*, **3**, 375 (1958).

Ba(OH)₂–Sr(OH)₂

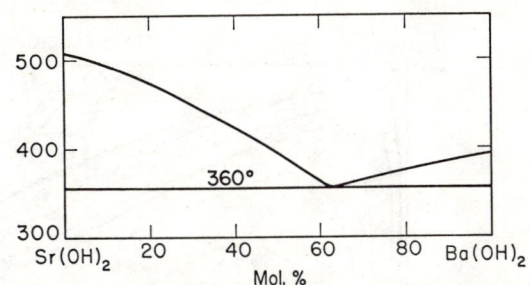

Fig. 1030.—System Ba(OH)₂–Sr(OH)₂; tentative.

K. A. Allen, W. C. Davis, B. S. Weaver, Oak Ridge National Laboratory, "Phase Diagrams of Nuclear Reactor Materials," R. E. Thoma, ed., ORNL-2548, p. 144 (1959).

(b) Three Substances

KOH–LiOH–NaOH

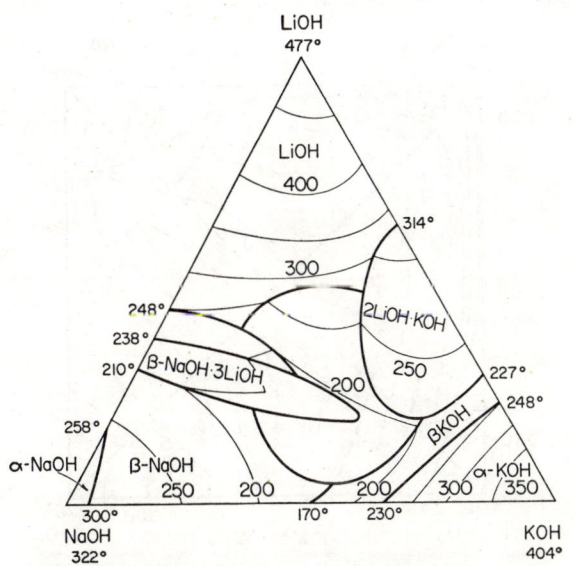

Fig. 1031.—System KOH–LiOH–NaOH.

N. A. Reshetnikov and N. I. Vilutis, *Zhur. Neorg. Khim.*, **4**, 128 (1959).

NaOH–NaCl–Na₂SO₄

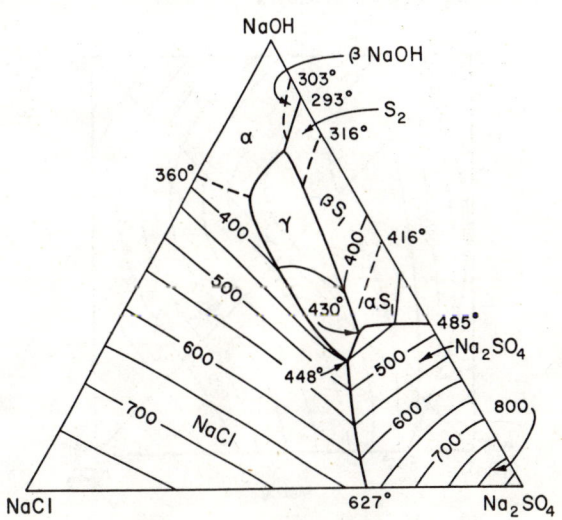

Fig. 1032.—System NaCl–NaOH–Na₂SO₄.

M. I. Ravich and V. M. Elenevskaya, *Zhur. Neorg. Khim.*, **2**, 1136 (1957).

(c) Four Substances

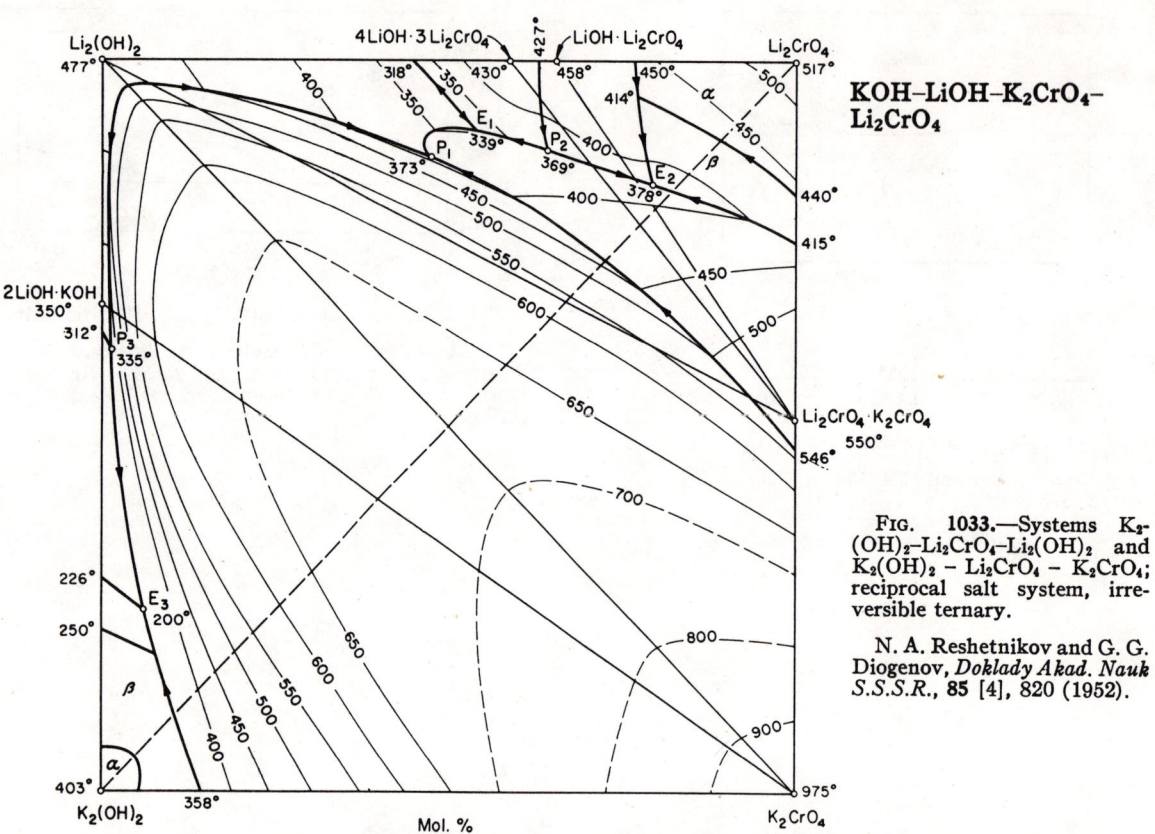

Fig. 1033.—Systems $K_2(OH)_2$–Li_2CrO_4–$Li_2(OH)_2$ and $K_2(OH)_2$ – Li_2CrO_4 – K_2CrO_4; reciprocal salt system, irreversible ternary.

N. A. Reshetnikov and G. G. Diogenov, *Doklady Akad. Nauk S.S.S.R.*, **85** [4], 820 (1952).

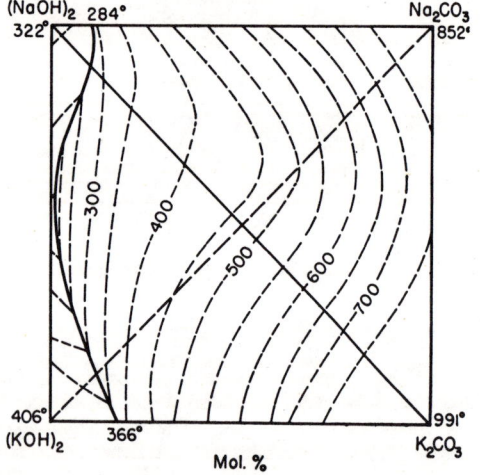

Fig. 1034.—System $(KOH)_2$–K_2CO_3–$(NaOH)_2$–Na_2CO_3.

V. A. Khitrov, *Izvest. Sektora Fiz.-Khim. Anal., Inst. Obshcheĭ Neorg. Khim., Akad. Nauk S.S.S.R.*, **25**, 243 (1954).

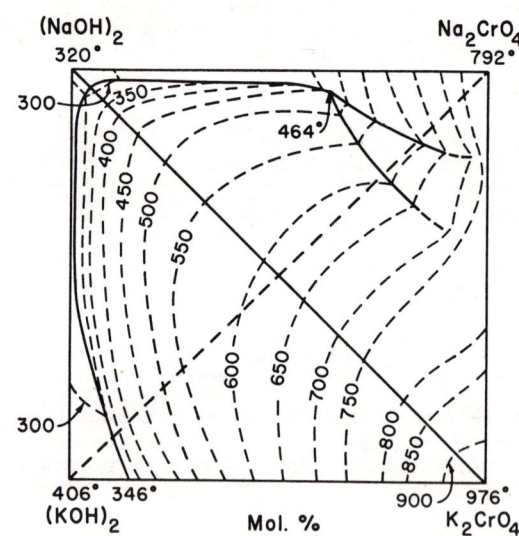

Fig. 1035.—System KOH–K_2CrO_4–NaOH–Na_2CrO_4.

A. G. Bergman and V. A. Khitrov, *Izvest. Sektora Fiz.-Khim. Anal., Inst. Obshcheĭ Neorg. Khim., Akad. Nauk S.S.S.R.*, **24**, 210 (1954).

KOH–NaOH–K₂SO₄–Na₂SO₄

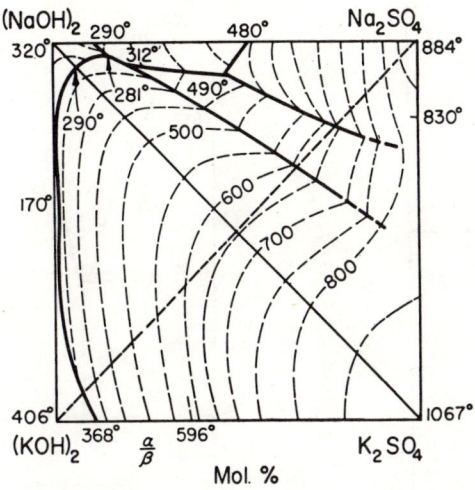

Fig. 1036.—System (KOH)₂–K₂SO₄–(NaOH)₂–Na₂SO₄.

A. G. Bergman and V. A. Khitrov, *Izvest. Sektora Fiz.-Khim. Anal., Inst. Obshcheĭ Neorg. Khim., Akad. Nauk S.S.S.R.*, **21**, 207 (1952).

LiOH–NaOH–Li₂CrO₄–Na₂CrO₄

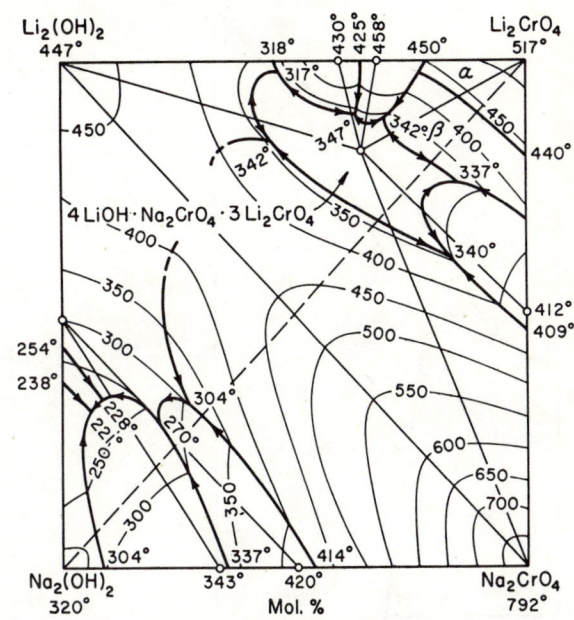

Fig. 1038.—System Li₂(OH)₂–Li₂CrO₄–Na₂(OH)₂–Na₂CrO₄, reciprocal.

N. A. Reshetnikov and G. G. Diogenov, *Sbornik Stateĭ Obshcheĭ Khim., Akad. Nauk S.S.S.R.*, **1**, 123 (1953).

LiOH–NaOH–LiNO₃–NaNO₃

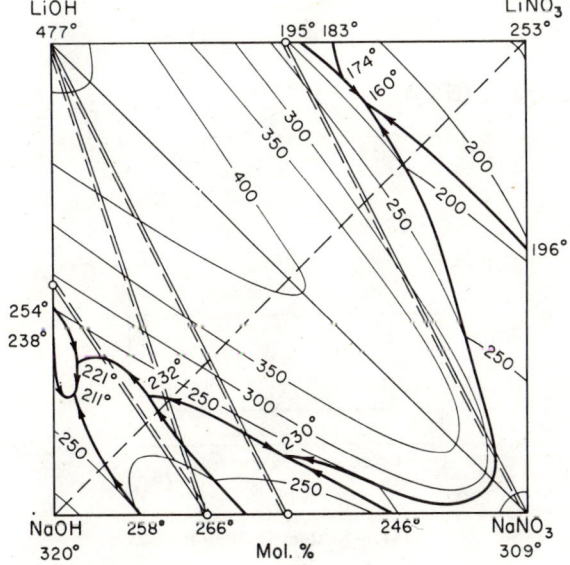

Fig. 1037.—System LiOH–LiNO₃–NaOH–NaNO₃.

G. G. Diogenov, *Doklady Akad. Nauk S.S.S.R.*, **89**, 305 (1953).

IV. Nitrates Only

(a) *Two Nitrates*

AgNO₃–KNO₃

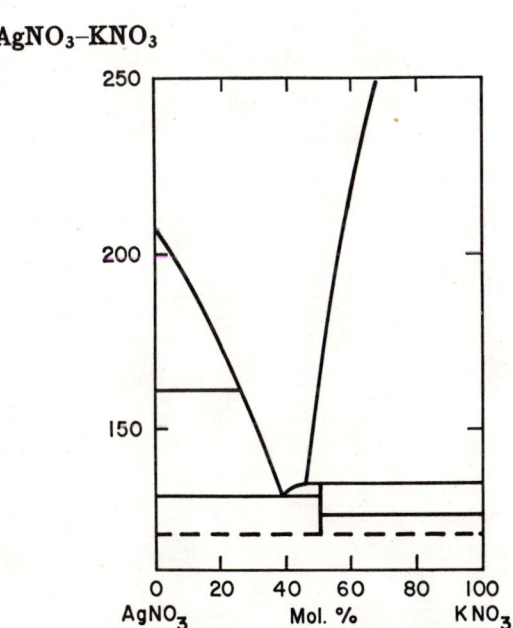

Fig. 1039.—System AgNO₃–KNO₃.

A. Ussow, *Z. anorg. Chem.*, **38**, 423 (1904).

AgNO₃–NaNO₃

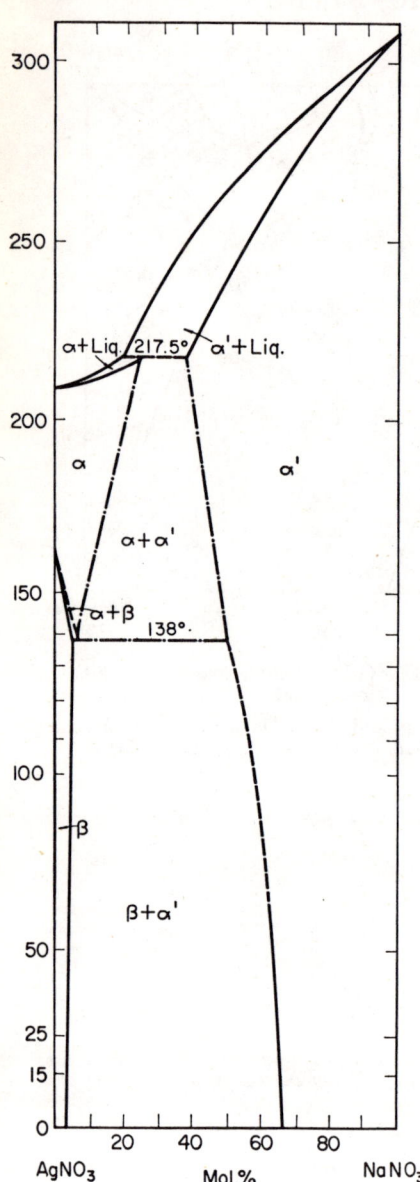

Fig. 1040.—System AgNO₃–NaNO₃.

D. H. Hissink, *Z. physik. Chem.*, **32**, 545 (1900).

AgNO₃–NH₄NO₃

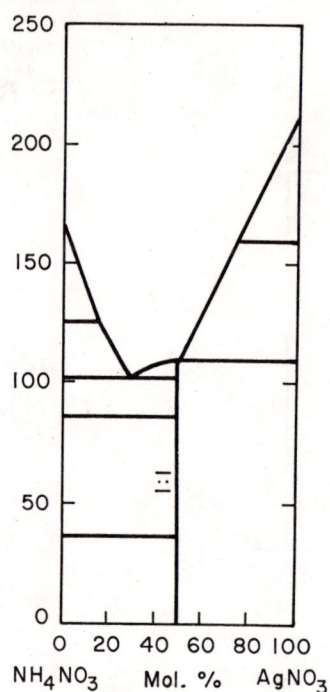

Fig. 1041.—System AgNO₃–NH₄NO₃.

Jan Zawidzki, *Z. physik. Chem.*, **47**, 725 (1904).

AgNO₃–TlNO₃

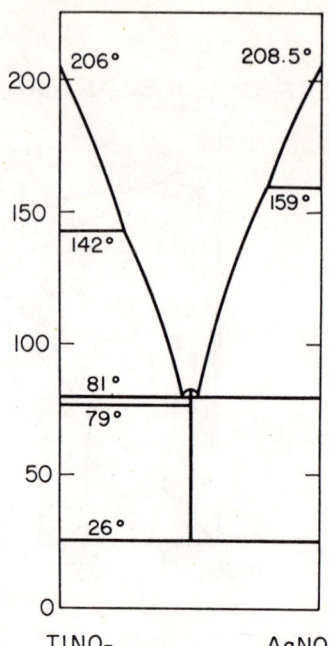

Fig. 1042.—System AgNO₃–TlNO₃.

C. van Eyk, *Z. physik. Chem.*, **51**, 726 (1905).

AgNO₃–Ca(NO₃)₂

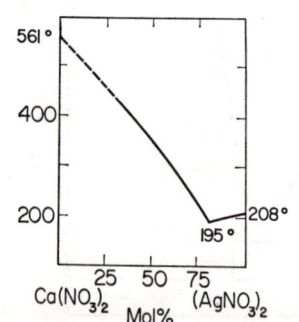

Fig. 1043.—System AgNO₃–Ca(NO₃)₂.

P. I. Protsenko and Z. I. Belova, *Zhur. Neorg. Khim.*, **2**, 2618 (1957).

CsNO₃–Ba(NO₃)₂

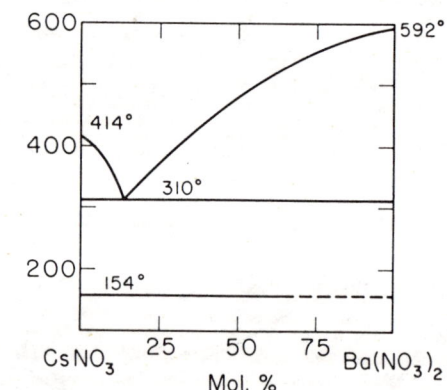

Fig. 1044.—System CsNO₃–Ba(NO₃)₂.

V. E. Plyushchev, I. B. Markina, and L. P. Shklover, *Doklady Akad. Nauk S.S.S.R.*, **108** [4] 646 (1956).

CsNO$_3$–Ca(NO$_3$)$_2$

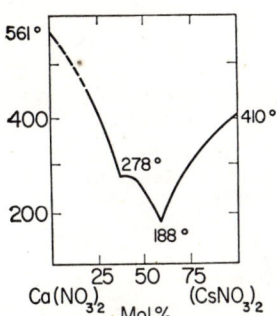

FIG. 1045.—System CsNO$_3$–Ca(NO$_3$)$_2$.

P. I. Protsenko and Z. I. Belova, *Zhur. Neorg. Khim.*, **2**, 2618 (1957).

CsNO$_3$–Sr(NO$_3$)$_2$

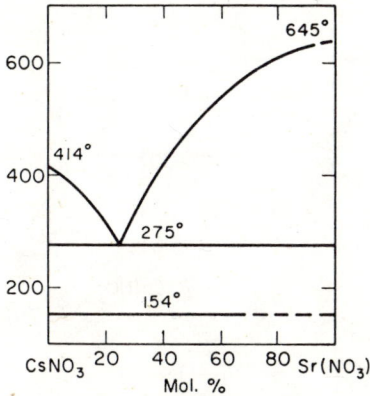

FIG. 1046.—System CsNO$_3$–Sr(NO$_3$)$_2$.

V. E. Plyushchev, I. B. Markina, and L. P. Shklover, *Zhur. Neorg. Khim.*, **1** [7] 1614 (1956).

KNO$_3$–NaNO$_3$

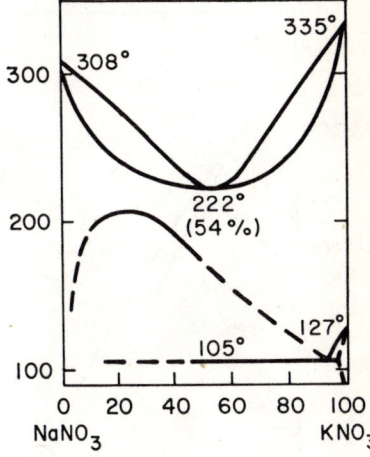

FIG. 1047.—System KNO$_3$–NaNO$_3$.

Adelheid Kofler, *Monatsh. Chem.*, **86**, 646 (1955).

KNO$_3$–TlNO$_3$

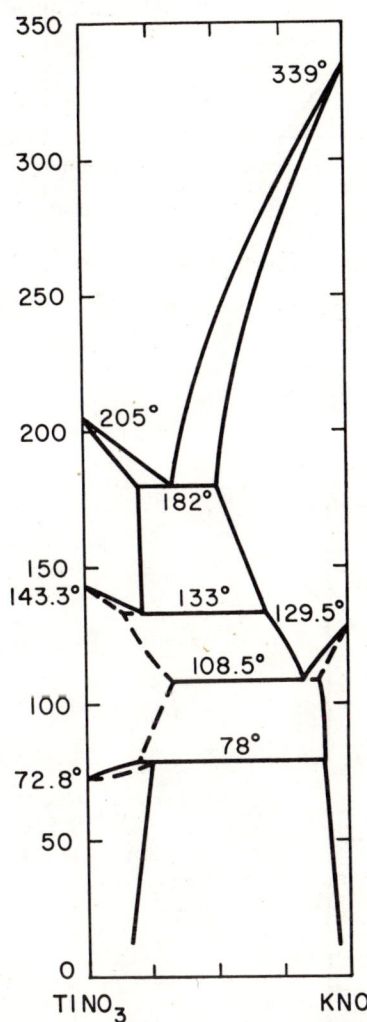

FIG. 1048.—System KNO$_3$–TlNO$_3$.

C. van Eyk, *Z. physik. Chem.*, **51**, 725 (1905).

KNO$_3$–Ba(NO$_3$)$_2$

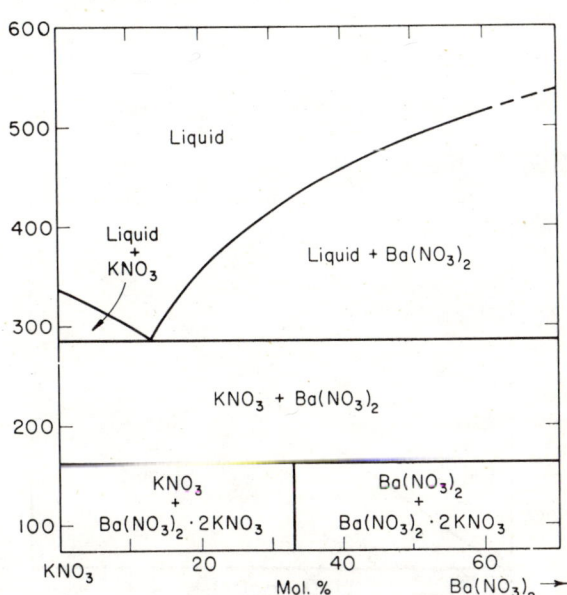

FIG. 1049.—System KNO$_3$–Ba(NO$_3$)$_2$.

Meyer Markowitz, J. E. Ricci, and P. F. Winternitz, *J. Am. Chem. Soc.*, **77**, 3484 (1955).

(KNO₃, NaNO₃)–Ca(NO₃)₂

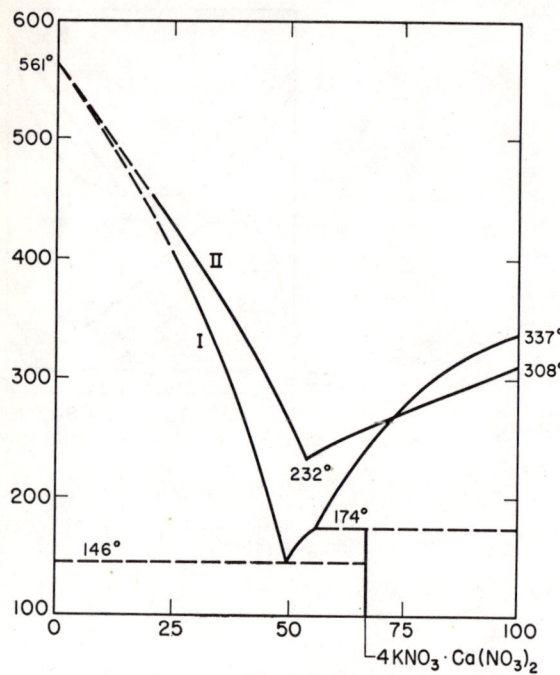

FIG. 1050.—I. System Ca(NO₃)₂–(KNO₃)₂; II. System Ca(NO₃)₂–(NaNO₃)₂; liquidus diagrams.

P. I. Protsenko and A. G. Bergman, *J. Gen. Chem. S.S.S.R.*, **20**, 1367 (1950).

LiNO₃–Ba(NO₃)₂

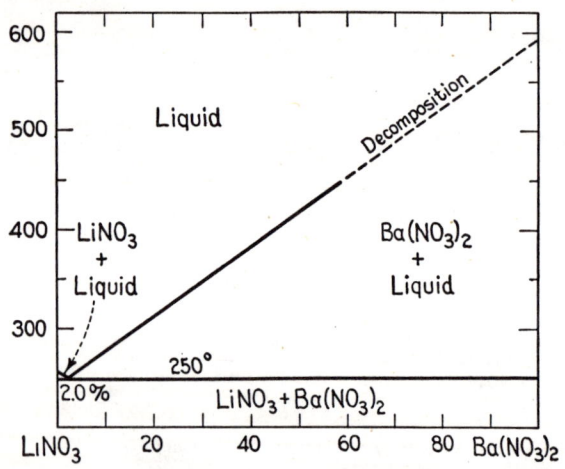

FIG. 1051.—System LiNO₃–Ba(NO₃)₂.

Drawn by Metalloy Corporation from data of W. D. Harkins and G. L. Clark, *J. Am. Chem. Soc.*, **37**, 1816 (1915).

LiNO₃–Ca(NO₃)₂

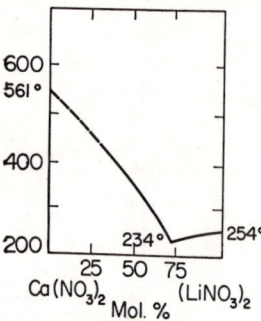

FIG. 1052.—System LiNO₃–Ca(NO₃)₂.

P. I. Protsenko and Z. I. Belova, *Zhur. Neorg. Khim.*, **2**, 2618 (1957).

NaNO₃–NH₄NO₃

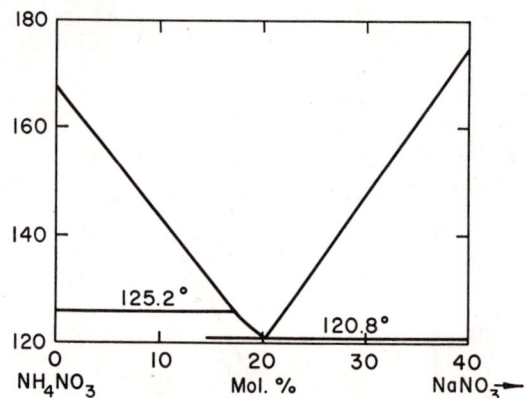

FIG. 1053.—System NaNO₃–NH₄NO₃.

R. G. Early and T. M. Lowry, *J. Chem. Soc.*, **121**, 965 (1922).

NaNO₃–TlNO₃

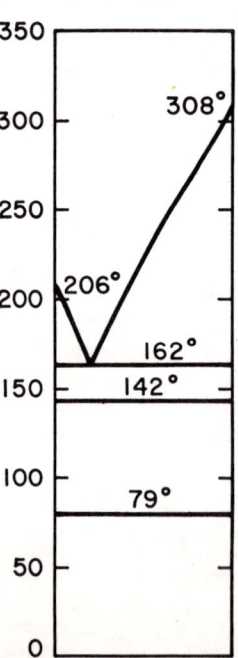

FIG. 1054.—System NaNO₃–TlNO₃.

C. van Eyk, *Z. physik. Chem.*, **51**, 731 (1905).

NH_4NO_3–$Pb(NO_3)_2$

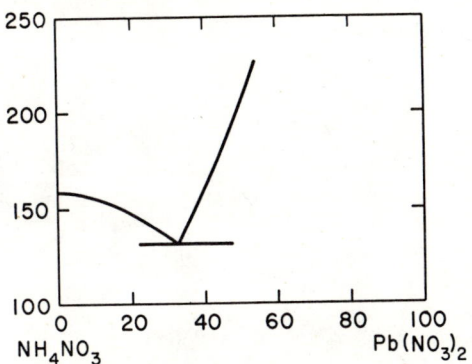

FIG. 1055.—System NH_4NO_3–$Pb(NO_3)_2$.

M. B. Bogitch, *Compt. rend.*, **161**, 791 (1915).

$RbNO_3$–$Ba(NO_3)_2$

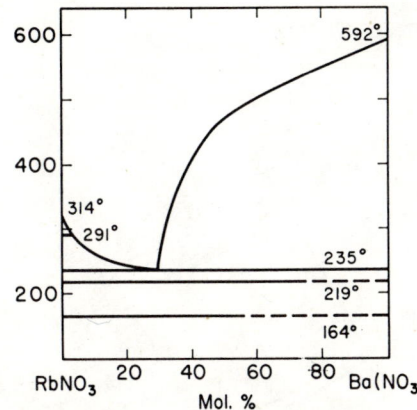

FIG. 1056.—System $RbNO_3$–$Ba(NO_3)_2$.

V. E. Plyushchev, I. B. Markina, and L. P. Shklover, *Doklady Akad. Nauk S.S.S.R.*, **108** [4] 646 (1956).

Note: The component labeled $Ba(NO_3)$ should read $Ba(NO_3)_2$.

$RbNO_3$–$Ca(NO_3)_2$

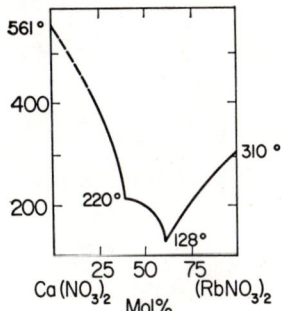

FIG. 1057.—System $RbNO_3$–$Ca(NO_3)_2$.

P. I. Protsenko and Z. I. Belova, *Zhur. Neorg. Khim.*, **2**, 2618 (1957).

$RbNO_3$–$Sr(NO_3)_2$

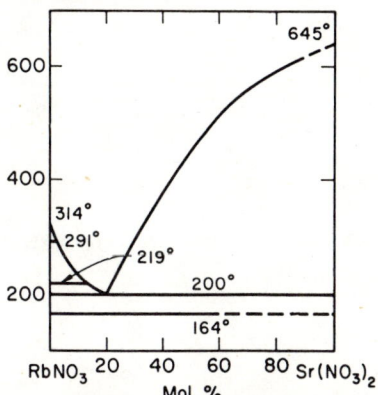

FIG. 1058.—System $RbNO_3$–$Sr(NO_3)_2$.

V. E. Plyushchev, I. B. Markina, and L. P. Shklover, *Zhur. Neorg. Khim.*, **1** [7] 1614 (1956).

$TlNO_3$–$Ca(NO_3)_2$

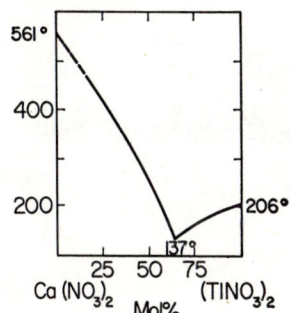

FIG. 1059.—System $TlNO_3$–$Ca(NO_3)_2$.

P. I. Protsenko and Z. I. Belova, *Zhur. Neorg. Khim.*, **2**, 2619 (1957).

$Ba(NO_3)_2$–$Ca(NO_3)_2$

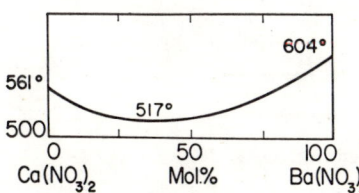

FIG. 1060.—System $Ba(NO_3)_2$–$Ca(NO_3)_2$.

P. I. Protsenko and Z. I. Belova, *Zhur. Neorg. Khim.*, **2**, 2619 (1957).

$Ba(NO_3)_2$–$Sr(NO_3)_2$

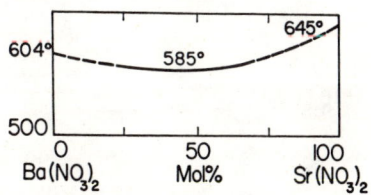

FIG. 1061.—System $Ba(NO_3)_2$–$Sr(NO_3)_2$.

P. I. Protsenko and Z. I. Belova, *Zhur. Neorg. Khim.*, **2**, 2619 (1957).

$Ca(NO_3)_2$–$Sr(NO_3)_2$

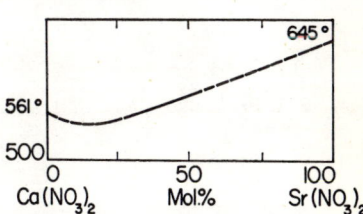

FIG. 1062.—System $Ca(NO_3)_2$–$Sr(NO_3)_2$.

P. I. Protsenko and Z. I. Belova, *Zhur. Neorg. Khim.*, **2**, 2619 (1957).

(b) *Three Nitrates*

AgNO₃–KNO₃–TlNO₃

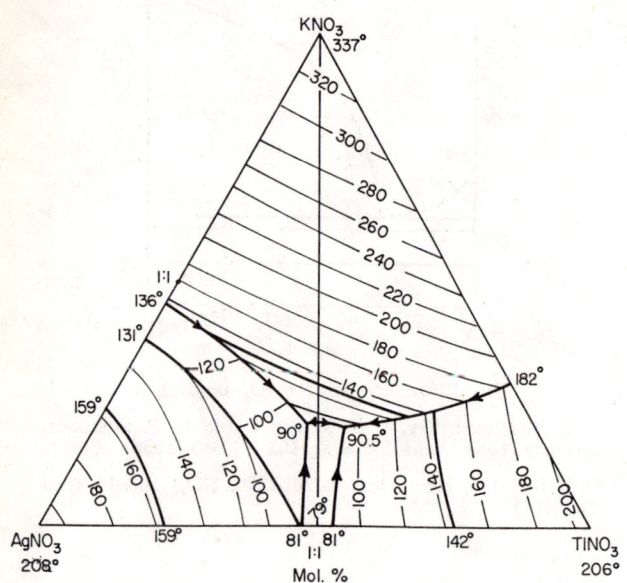

Fig. 1063.—System AgNO₃–KNO₃–TlNO₃.
V. A. Palkin, *Doklady Akad. Nauk S.S.S.R.*, **66**, 72 (1949).

AgNO₃–LiNO₃–RbNO₃

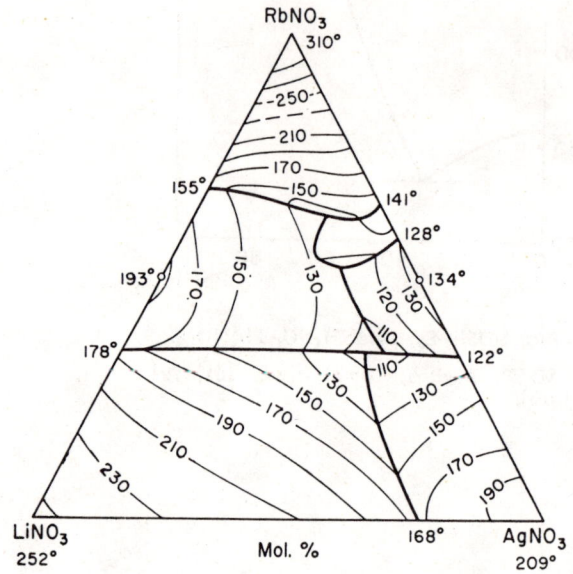

Fig. 1065.—System AgNO₃–LiNO₃–RbNO₃.
P. I. Protsenko and L. M. Kiparenko, *Zhur. Obshcheĭ Khim.*, **25**, 450 (1955).

AgNO₃–KNO₃–Cd(NO₃)₂

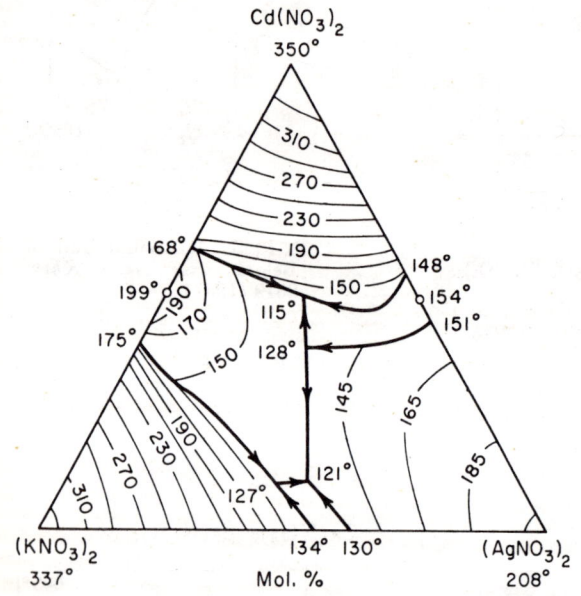

Fig. 1064.—System (AgNO₃)₂–(KNO₃)₂–Cd(NO₃)₂.
P. I. Protsenko, *Zhur. Obshcheĭ Khim.*, **23**, 1616 (1953)

AgNO₃–NaNO₃–TlNO₃

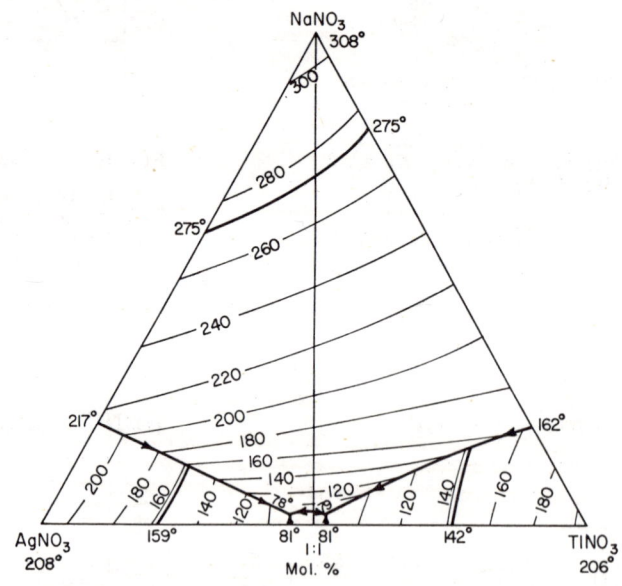

Fig. 1066.—System AgNO₃–NaNO₃–TlNO₃.
V. A. Palkin, *Doklady Akad. Nauk S.S.S.R.*, **66**, 651 (1949).

CsNO$_3$–RbNO$_3$–Ca(NO$_3$)$_2$

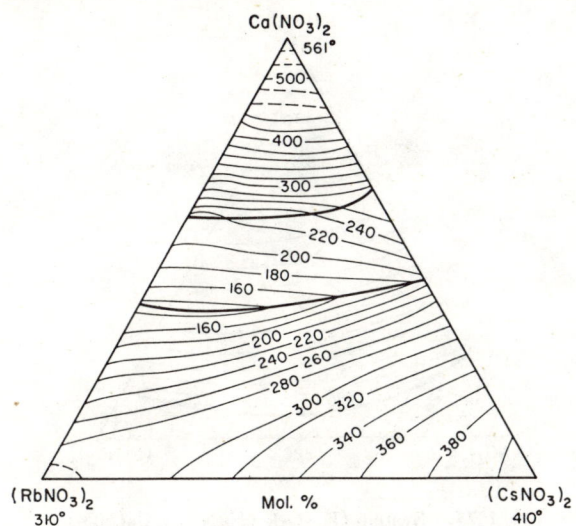

Fig. 1067.—System CsNO$_3$–RbNO$_3$–Ca(NO$_3$)$_2$.

P. I. Protsenko and Z. I. Belova, *Zhur. Obshcheĭ Khim.*, **25**, 248 (1955).

KNO$_3$–LiNO$_3$–NaNO$_3$

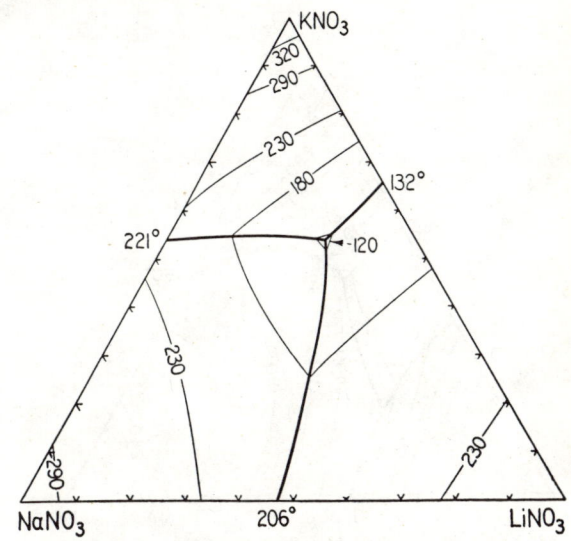

Fig. 1069.—System KNO$_3$–LiNO$_3$–NaNO$_3$.

H. R. Carveth, *J. Phys. Chem.*, **2**, 216 (1898).

CsNO$_3$–TlNO$_3$–Cd(NO$_3$)$_2$

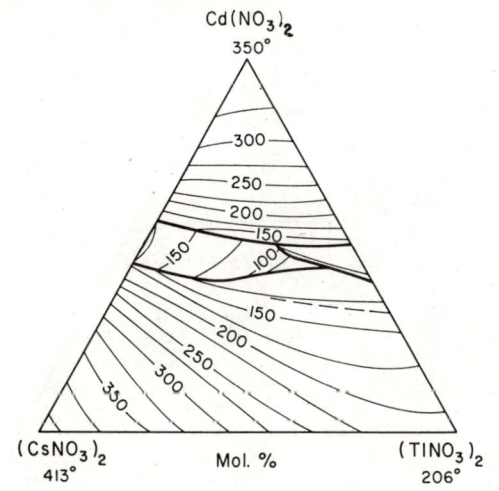

Fig. 1068.—System CsNO$_3$–TlNO$_3$–Cd(NO$_3$)$_2$.

P. I. Protsenko and V. V. Rubleva, *Zhur. Obshcheĭ Khim.*, **25**, 242 (1955).

KNO$_3$–LiNO$_3$–Ca(NO$_3$)$_2$

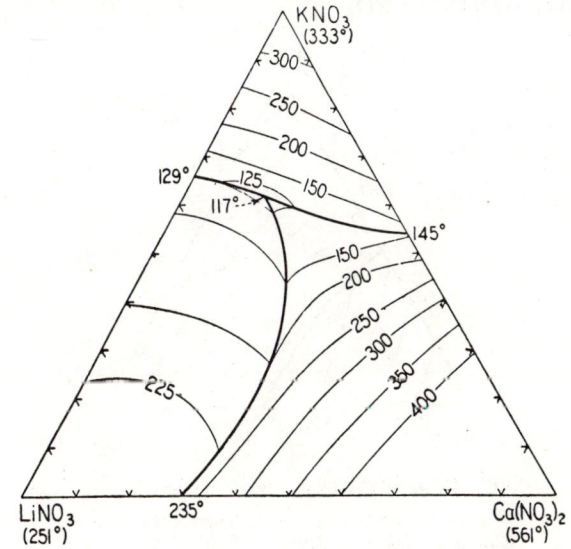

Fig. 1070.—System KNO$_3$–LiNO$_3$–Ca(NO$_3$)$_2$.

A. Lehrman, et al., *J. Am. Chem. Soc.*, **59**, 179 (1937).

KNO₃–LiNO₃–Cd(NO₃)₂

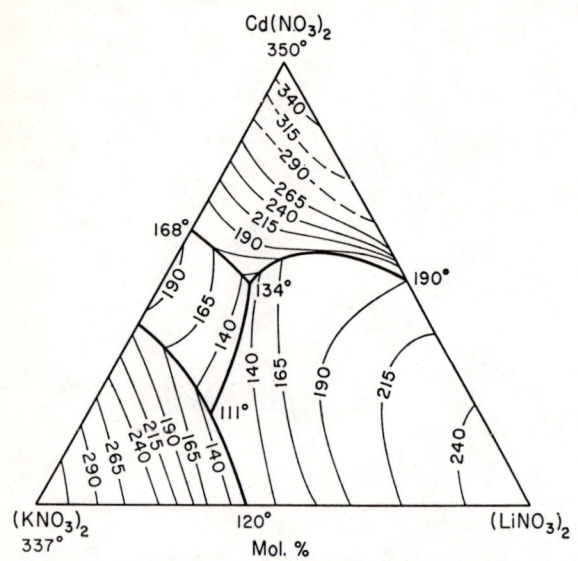

Fig. 1071.—System KNO₃–LiNO₃–Cd(NO₃)₂.

P. I. Protsenko, *Zhur. Obshcheĭ Khim.*, **22**, 1317 (1952).

KNO₃–NaNO₃–Ba(NO₃)₂

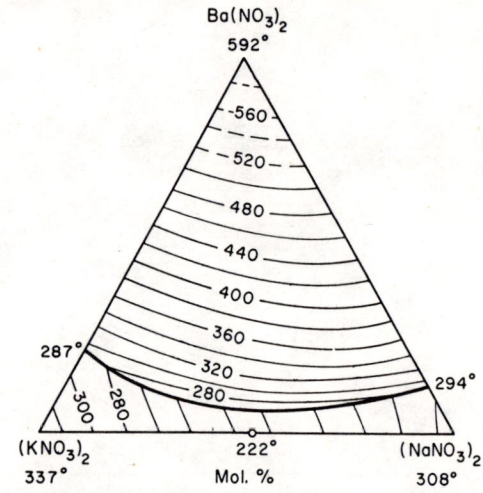

Fig. 1073.—System (KNO₃)₂–(NaNO₃)₂–Ba(NO₃)₂.

P. I. Protsenko and A. G. Bergman, *Zhur. Obshcheĭ Khim.*, **21**, 1583 (1951).

KNO₃–NaNO₃–RbNO₃

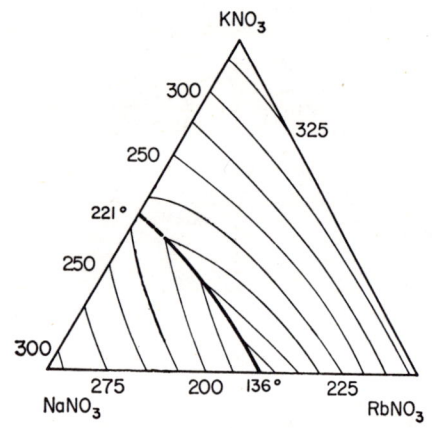

Fig. 1072.—System KNO₃–NaNO₃–RbNO₃.

S. D. Gromakov and A. I. Kostromin, *Uch. Zap. Kazansk. Gos. Univ.*, **115**, 95 (1955).

KNO₃–NaNO₃–Ca(NO₃)₂

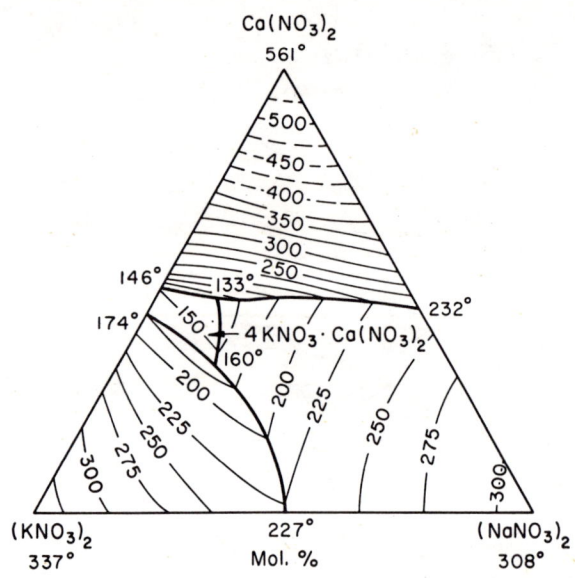

Fig. 1074.—System (KNO₃)₂–(NaNO₃)₂–Ca(NO₃)₂.

A. G. Bergman, I. S. Rassonskaya, and N. E. Shmidt, *Izvest. Sektora Fiz.-Khim. Anal., Inst. Obshcheĭ Neorg. Khim., Akad. Nauk S.S.S.R.*, **26**, 156 (1955).

KNO₃–RbNO₃–Cd(NO₃)₂

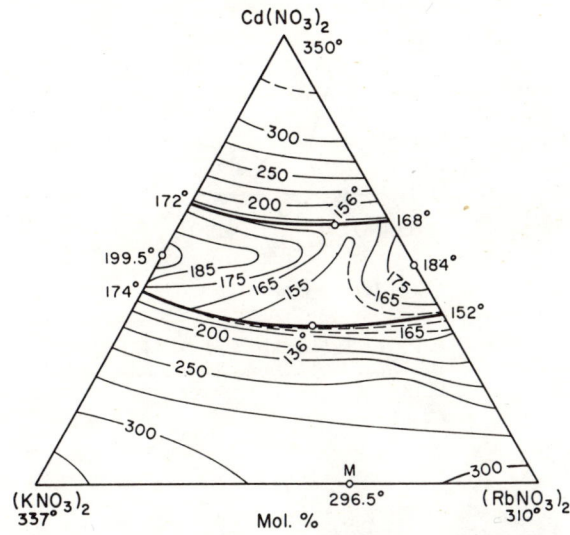

FIG. 1075.—System KNO₃–RbNO₃–Cd(NO₃)₂.

P. I. Protsenko and N. P. Popovskaya, *Zhur. Obshcheĭ Khim.*, **23**, 1249 (1953).

V. Nitrites Only

KNO₂–NaNO₂

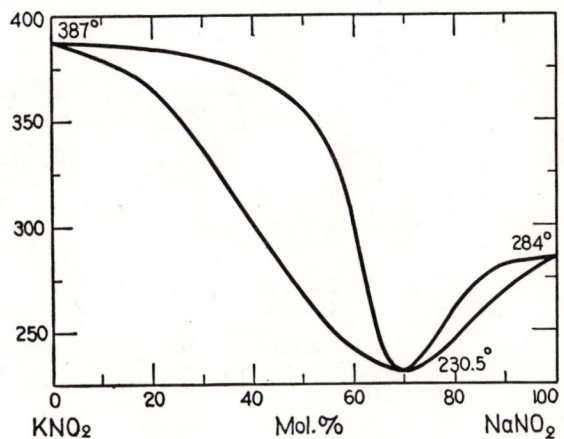

FIG. 1077.—System KNO₂–NaNO₂.

Von J. Ettinger, *Z. anorg. u. allgem. Chem.*, **206**, 262 (1932).

LiNO₃–NaNO₃–Cd(NO₃)₂

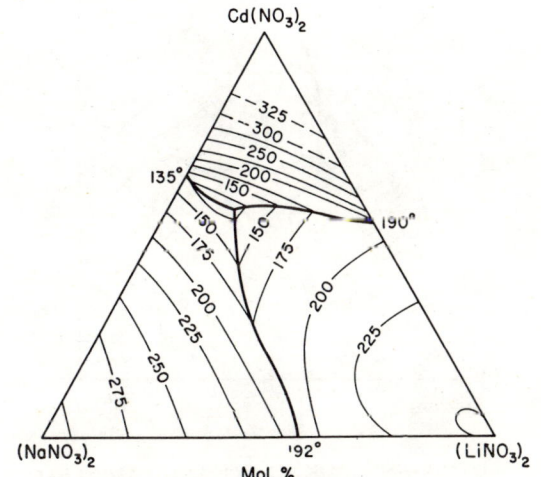

FIG. 1076.—System LiNO₃–NaNO₃–Cd(NO₃)₂.

P. I. Protsenko, *Zhur. Obshcheĭ Khim.*, **22**, 1311 (1952).

VI. Nitrites–Nitrates, Mixed

(a) Two Substances

KNO₂–KNO₃

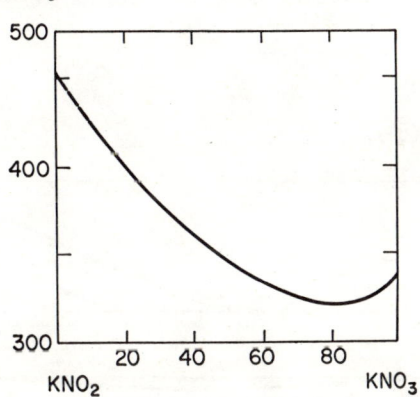

FIG. 1078.—System KNO₂–KNO₃.

D. Meneghini, *Gazz. chim. ital.*, **42 II**, 474 (1912).

KNO$_2$–Ca(NO$_3$)$_2$

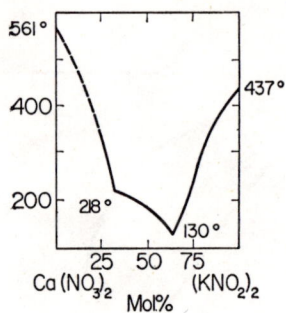

Fig. 1079.—System KNO$_2$–Ca(NO$_3$)$_2$.

P. I. Protsenko and Z. I. Belova, *Zhur. Neorg. Khim.*, **2**, 2619 (1957).

NaNO$_2$–KNO$_3$

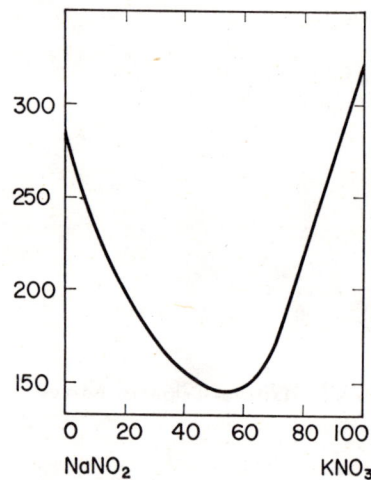

Fig. 1080.—System NaNO$_2$–KNO$_3$.

Kaoru Sakai, *Bull. Chem. Soc. Japan*, **27**, 463 (1954).

NaNO$_2$–NaNO$_3$

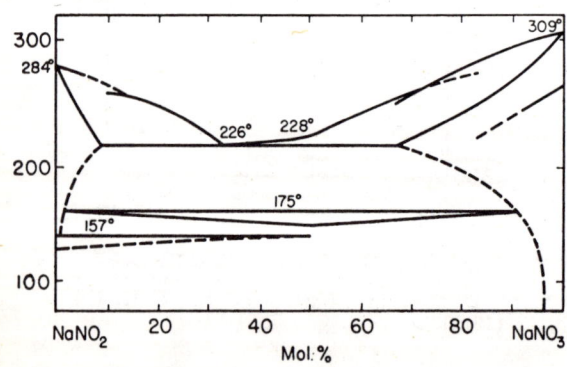

Fig. 1081.—System NaNO$_2$–NaNO$_3$.

A. G. Bergman, S. I. Berul, and I. N. Nikonova, *Izvest. Sektora Fiz.-Khim. Anal., Inst. Obshcheĭ Neorg. Khim., Akad. Nauk S.S.S.R.*, **23**, 183 (1953).

NaNO$_2$–Ca(NO$_3$)$_2$

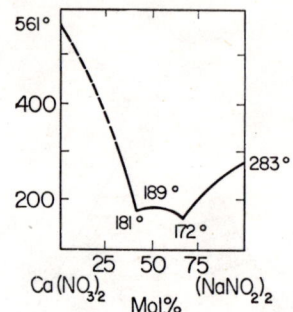

Fig. 1082.—System NaNO$_2$–Ca(NO$_3$)$_2$.

P. I. Protsenko and Z. I. Belova, *Zhur. Neorg. Khim.*, **2**, 2619 (1957).

(b) Four Substances

KNO$_2$–NaNO$_2$–KNO$_3$–NaNO$_3$

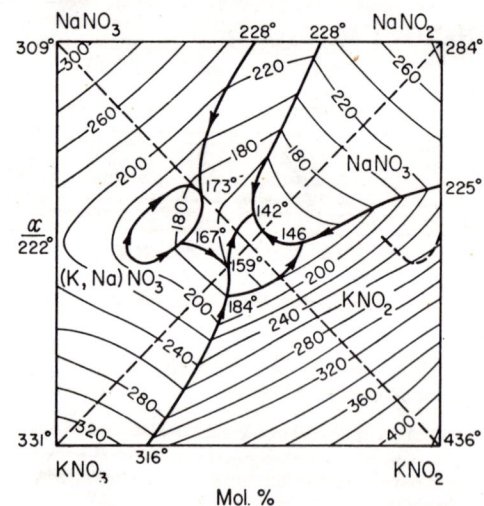

Fig. 1083.—System KNO$_2$–NaNO$_2$–KNO$_3$–NaNO$_3$.

S. I. Berul and A. G. Bergman, *Izvest. Sektora Fiz.-Khim. Anal., Inst. Obshcheĭ Neorg. Khim., Akad. Nauk S.S.S.R.*, **25**, 233 (1954).

KNO₂–Ba(NO₂)₂–KNO₃–Ba(NO₃)₂

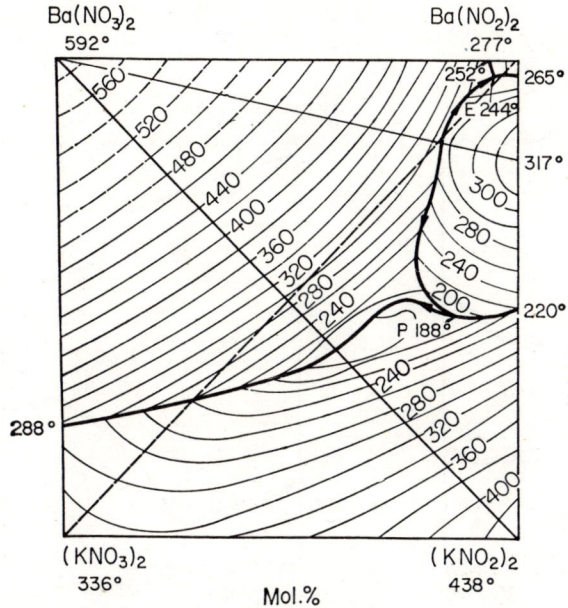

FIG. 1084.—System KNO₂–KNO₃–Ba(NO₂)₂–Ba(NO₃)₂.

P. I. Protsenko and A. Ya. Malakhova, *Zhur. Neorg. Khim.*, **2**, 2152 (1957).

LiNO₃–Li₂CO₃

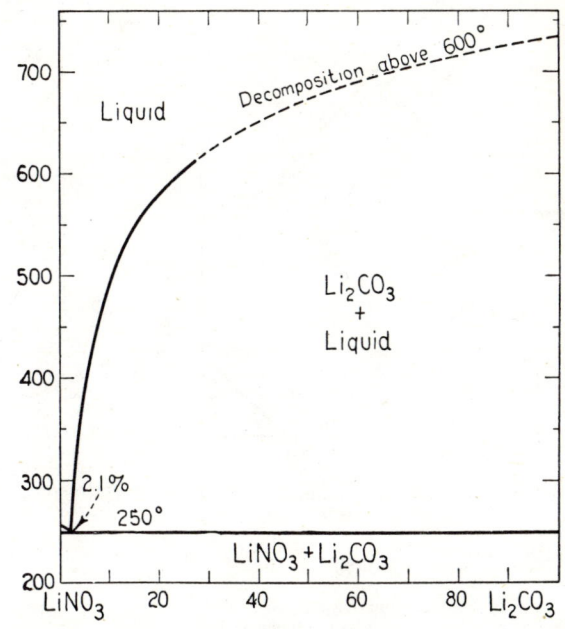

FIG. 1086.—System LiNO₃–Li₂CO₃

M. Amadori, *Atti reale accad. Lincei, Sez. II*, **22** 332 (1913).

VII. Nitrates-Carbonates, Mixed

KNO₃–K₂CO₃

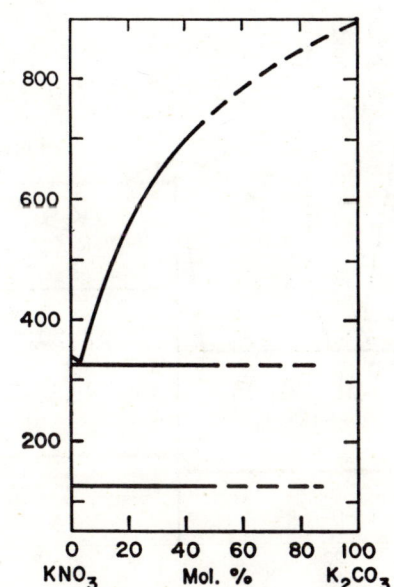

FIG. 1085.—System KNO₃–K₂CO₃.

M. Amadori, *Atti reale accad. Lincei, Sez. II*, **22**, 337 (1913).

NaNO₃–Na₂CO₃

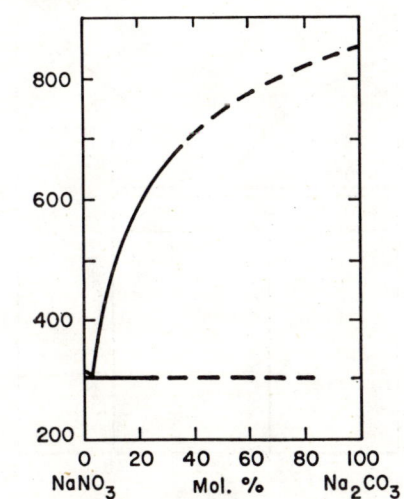

FIG. 1087.—System NaNO₃–Na₂CO₃.

M. Amadori, *Atti reale accad. Lincei, Sez. II*, **22**, 336 (1913).

VIII. Sulfates Only

(a) Two Sulfates

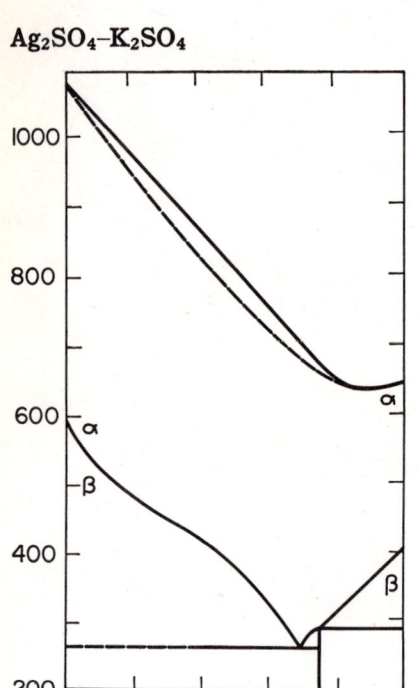

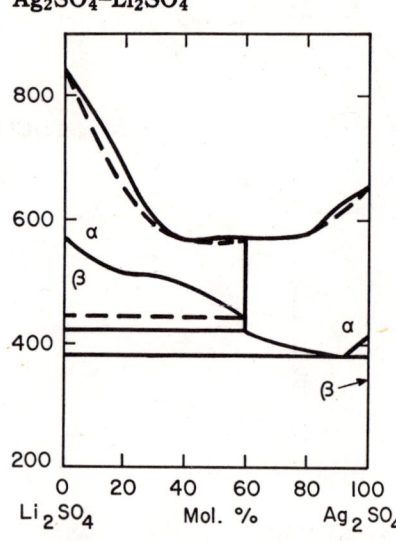

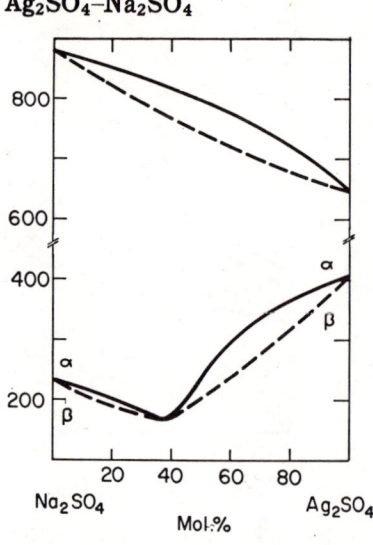

Fig. 1088.—System Ag_2SO_4–K_2SO_4. Diagram does not obey phase rule.

Richard Nacken, *Neues Jahrb Mineral., Geol.*, **24**, 65 (1907).

Fig. 1089.—System Ag_2SO_4–Li_2SO_4. Diagram does not obey phase rule.

Richard Nacken, *Neues Jahrb. Mineral., Geol.*, **24**, 47 (1907).

Fig. 1090.—System Ag_2SO_4–Na_2SO_4.

Richard Nacken, *Neues Jahrb. Mineral., Geol.*, **24**, 63 (1907).

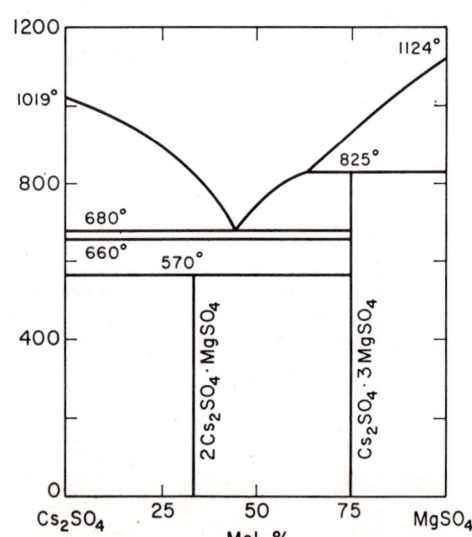

Fig. 1091.—System Cs_2SO_4–$MgSO_4$.

V. E. Plyushchev and N. F. Markovskaya, *Doklady Akad. Nauk S.S.S.R.*, **95**, 555 (1954).

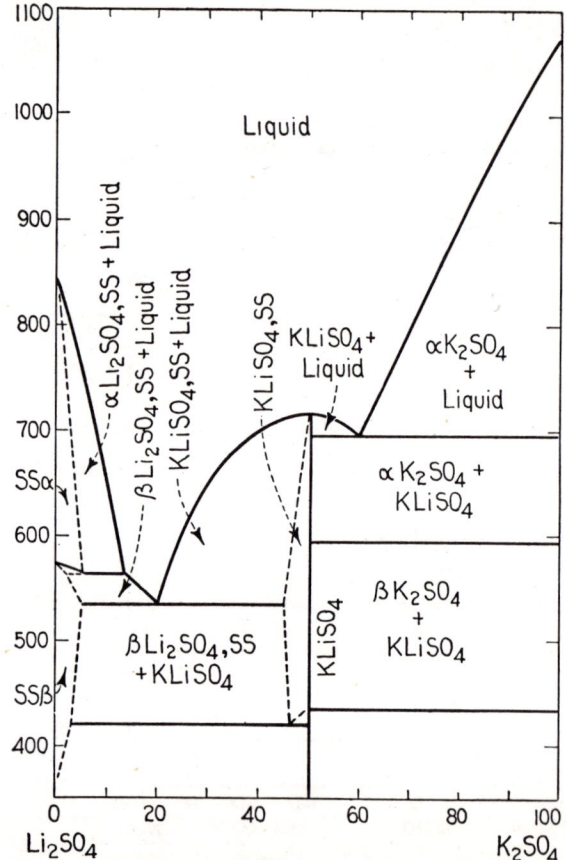

Fig. 1092.—System Li_2SO_4–K_2SO_4.

A. A. Nacken, *Neues Jahrb. Mineral. Geol., Beilage Bd.* **24A**, 43 (1910).

K_2SO_4–Na_2SO_4

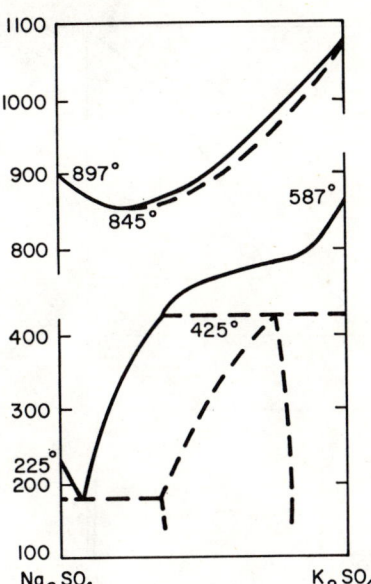

Fig. 1093.—System K_2SO_4–Na_2SO_4.
Diagram does not obey phase rule.

Ernst Jänecke, *Z. physik. Chem.*, **64**, 346 (1908). See also Richard Nacken, *Neues Jahrb. Mineral., Geol.*, **24**, 55 (1907).

K_2SO_4–$BaSO_4$

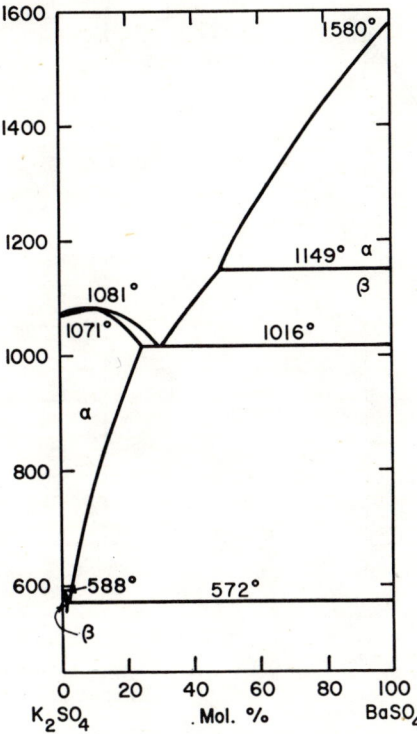

Fig. 1094.—System K_2SO_4–$BaSO_4$.

Werner Grahmann, *Z. anorg. Chem.*, **81**, 289 (1913).

K_2SO_4–$BeSO_4$

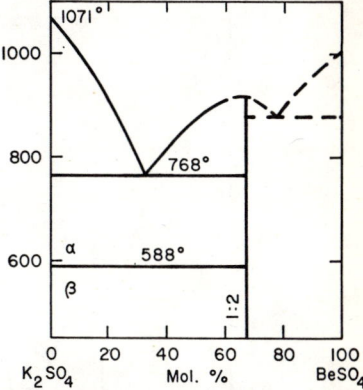

Fig. 1095.—System K_2SO_4–$BeSO_4$.

Werner Grahmann, *Z. anorg. Chem.*, **81**, 266 (1913).

K_2SO_4–$CaSO_4$

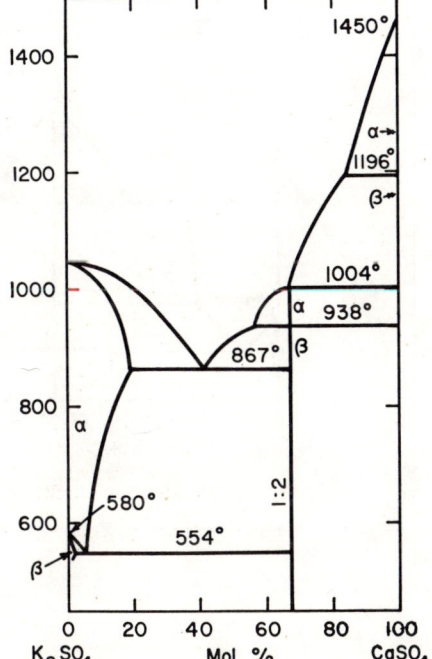

Fig. 1096.—System K_2SO_4–$CaSO_4$.

Werner Grahmann, *Z. anorg. Chem.*, **81**, 272 (1913).

K_2SO_4–$CdSO_4$

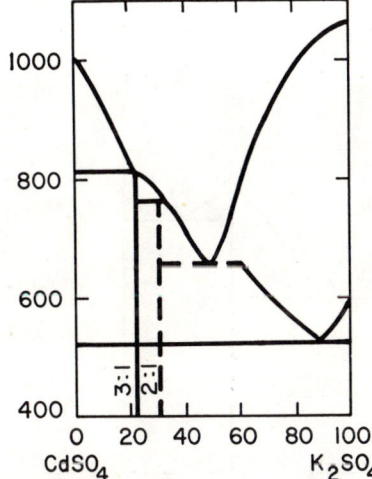

Fig. 1097.—System K_2SO_4–$CdSO_4$.
Diagram does not obey phase rule.

C. Calcagni and D. Marotta, *Gazz. chim. ital.* **44 I**, 495 (1914).

K₂SO₄–CoSO₄

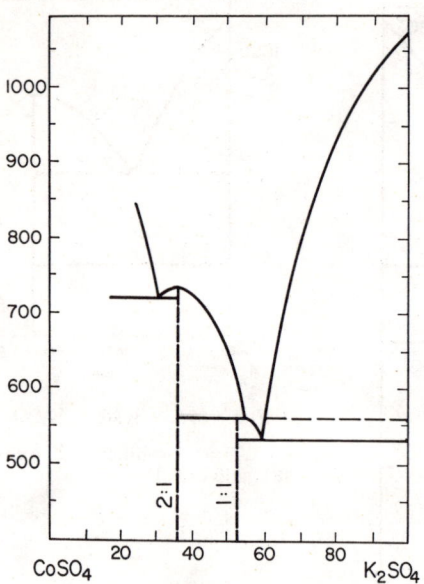

Fig. 1098.—System K₂SO₄–CoSO₄.

G. Calcagni and D. Marotta, *Gazz. chim. ital.*, **43 II**, 388 (1913).

K₂SO₄–CuSO₄

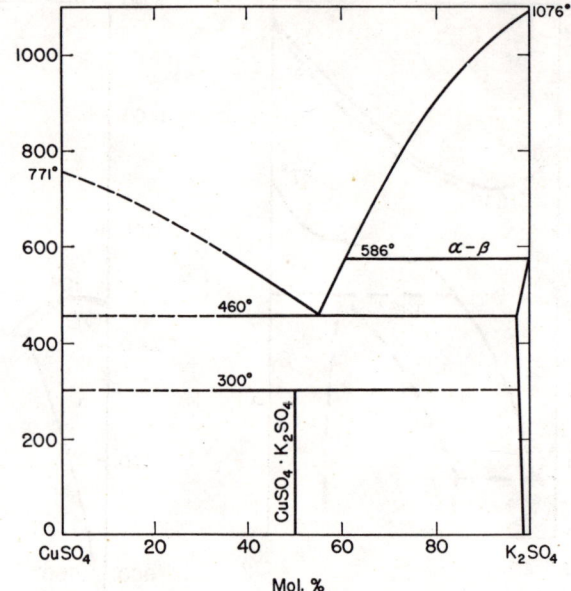

Fig. 1099.—System K₂SO₄–CuSO₄.

Angelo Bellanca and Marcello Carapezza, *Periodico Mineral. (Rome)*, **20**, 280 (1951).

K₂SO₄–MgSO₄

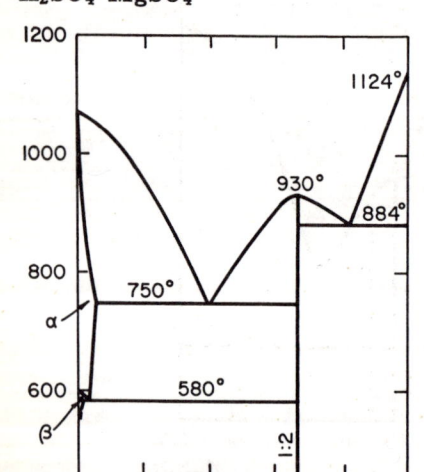

Fig. 1100.—System K₂SO₄–MgSO₄.

Werner Grahmann, *Z. anorg. Chem.*, **81**, 270 (1913).

K₂SO₄–MnSO₄

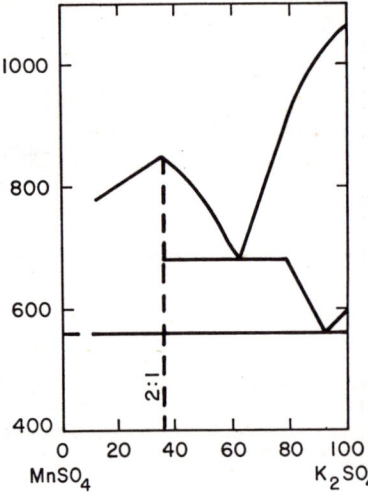

Fig. 1101.—System K₂SO₄–MnSO₄.

G. Calcagni and D. Marotta, *Gazz. chim. ital.*, **45 II**, 375 (1919).

K₂SO₄–PbSO₄

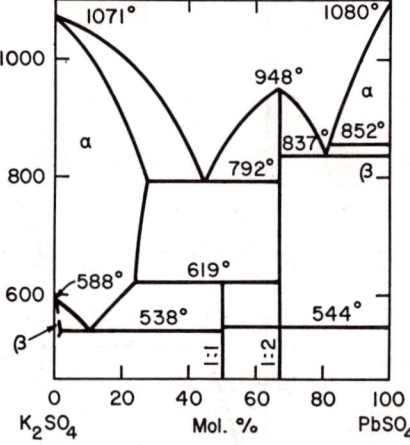

Fig. 1102.—System K₂SO₄–PbSO₄.

Werner Grahmann, *Z. anorg. Chem.*, **81**, 295 (1913).

K₂SO₄–SrSO₄

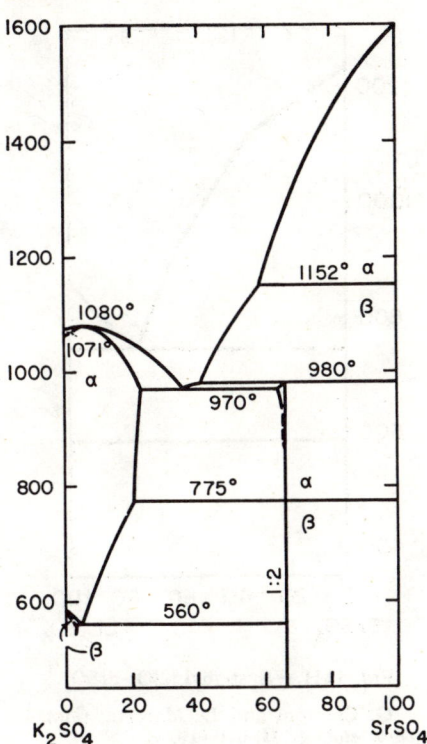

Fig. 1103.—System Li₂SO₄–SrSO₄.

Werner Grahmann, *Z. anorg. Chem.*, **81**, 277 (1913).

Li₂SO₄–Na₂SO₄

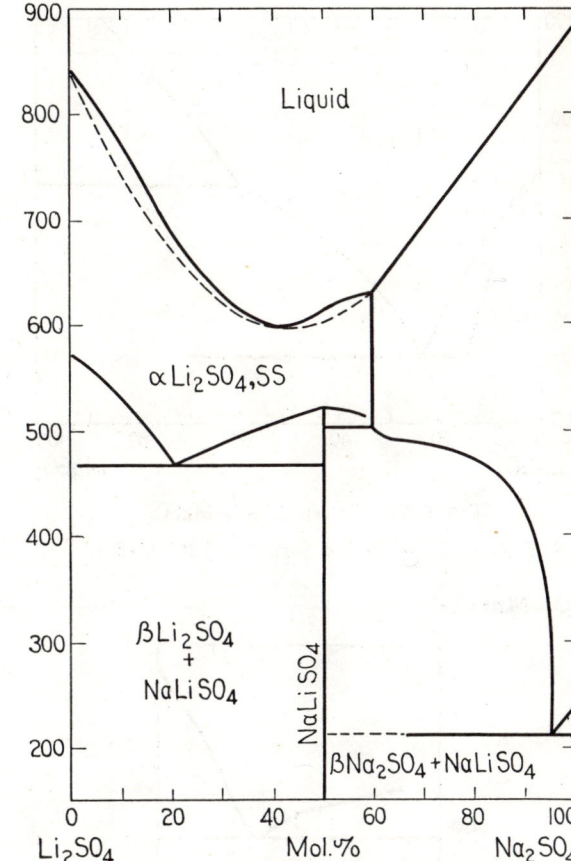

Fig. 1104.—System Li₂SO₄–Na₂SO₄.

A. A. Nacken, *Neues Jahrb. Mineral. Geol., Beilage Bd.*, **24A**, 34 (1910).

Li₂SO₄–CaSO₄

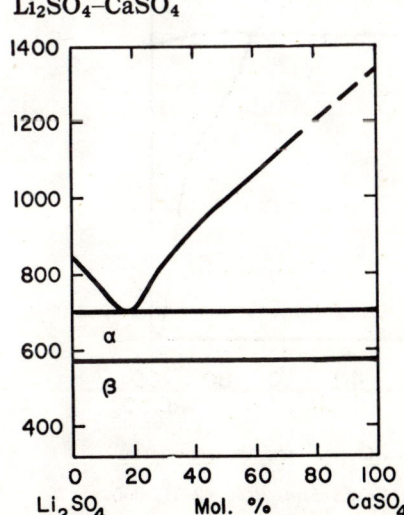

Fig. 1105.—System Li₂SO₄–CaSO₄.

Hans Müller, *Neues Jahrb. Mineral., Geol.*, **30**, 45 (1910).

Li₂SO₄–CdSO₄

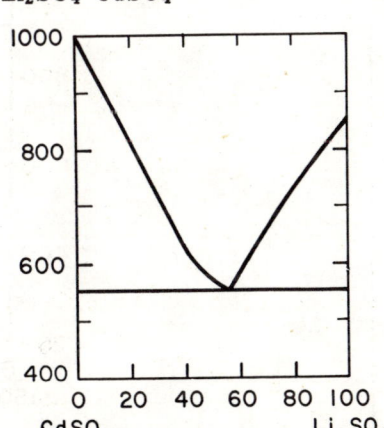

Fig. 1106.—System Li₂SO₄–CdSO₄.

G. Calcagni and D. Marotta, *Atti reale accad. Lincei, Sez. II*, **22**, 375 (1913).

Li₂SO₄–CoSO₄

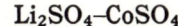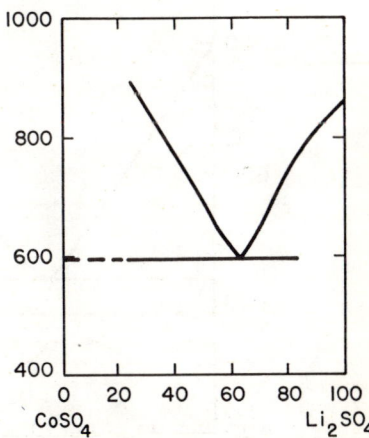

Fig. 1107.—System Li₂SO₄–CoSO₄.

G. Calcagni and D. Marotta, *Gazz. chim. ital.*, **43 II**, 382 (1913).

Li₂SO₄–MgSO₄

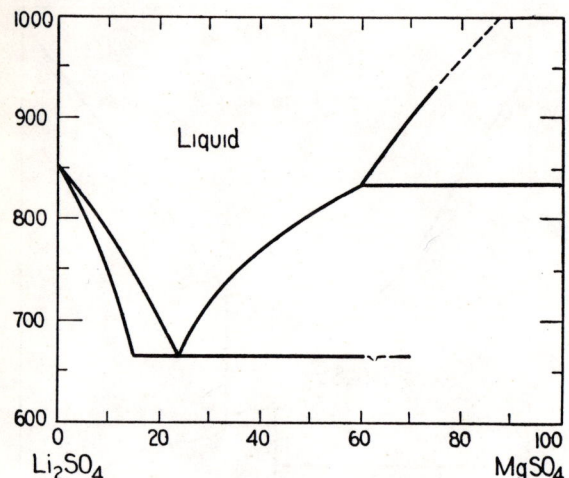

Fig. 1108.—System Li$_2$SO$_4$–MgSO$_4$.

R. F. Rea, *J. Am. Ceram. Soc.*, **21** [3] 100 (1938).

Li₂SO₄–MnSO₄

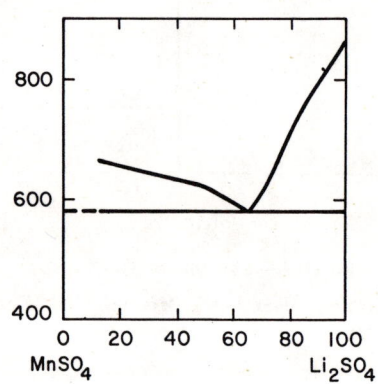

Fig. 1109.—System Li$_2$SO$_4$–MnSO$_4$.

G. Calcagni and D. Marotta, *Gazz. chim. ital.*, **45 II**, 371 (1915).

Li₂SO₄–PbSO₄

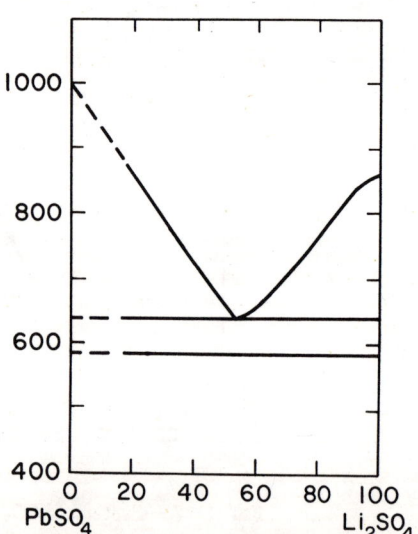

Fig. 1110.—System Li$_2$SO$_4$–PbSO$_4$.

G. Calcagni and D. Marotta, *Gazz. chim. ital.*, **42 II**, 679 (1912).

Li₂SO₄–SrSO₄

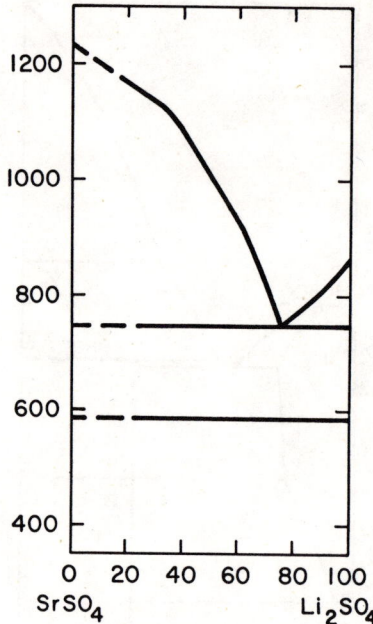

Fig. 1111.—System Li$_2$SO$_4$–SrSO$_4$.

G. Calcagni and D. Marotta, *Gazz. chim. ital.*, **42 II**, 670 (1912).

Na₂SO₄–BaSO₄

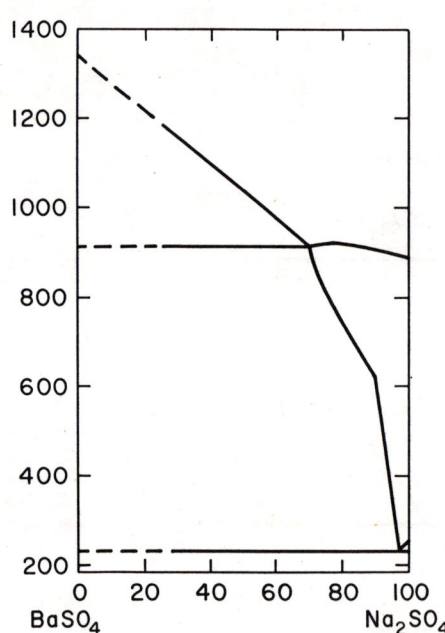

Fig. 1112.—System Na$_2$SO$_4$–BaSO$_4$.

G. Calcagni, *Gazz. chim. ital.*, **42 II**, 655 (1912).

Na$_2$SO$_4$–CaSO$_4$

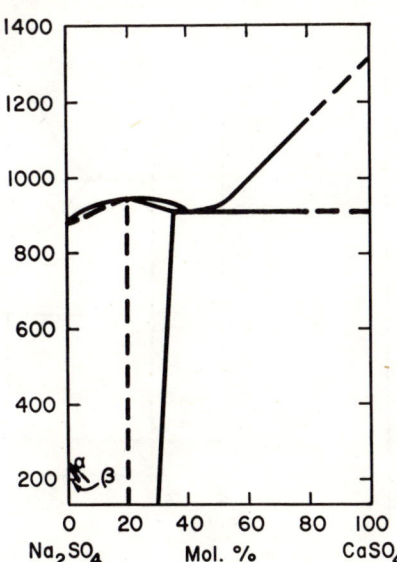

FIG. 1113.—System Na$_2$SO$_4$–CaSO$_4$.

Hans Müller, *Neues Jahrb. Mineral., Geol.*, **30,** 36 (1910).

Na$_2$SO$_4$–CoSO$_4$

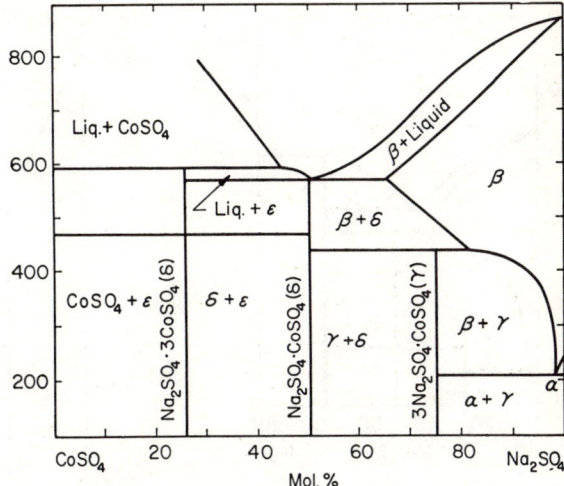

FIG. 1115.—System Na$_2$SO$_4$–CoSO$_4$.

K. A. Bol'shakov and P. I. Fedorov, *Zhur. Obshcheĭ Khim.*, **26,** 348 (1956).

Na$_2$SO$_4$–CdSO$_4$

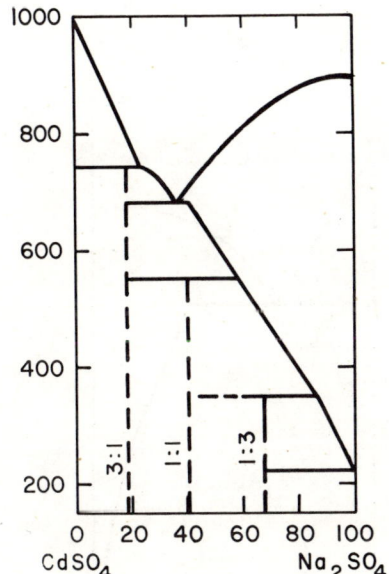

FIG. 1114.—System Na$_2$SO$_4$–CdSO$_4$.

G. Calcagni and D. Marotta, *Atti reale accad. Lincei, Sez. II*, **22,** 377 (1913).

Na$_2$SO$_4$–CuSO$_4$

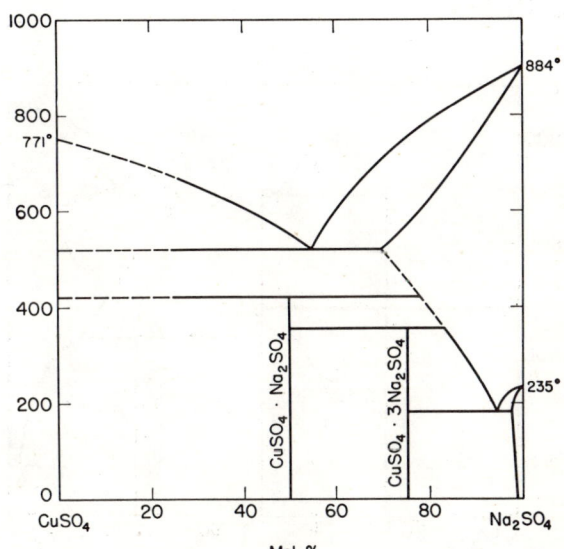

FIG. 1116.—System Na$_2$SO$_4$–CuSO$_4$.

Angelo Bellanca and Marcello Carapezza, *Periodico Mineral. (Rome)* **20,** 278 (1951).

Na₂SO₄–MgSO₄

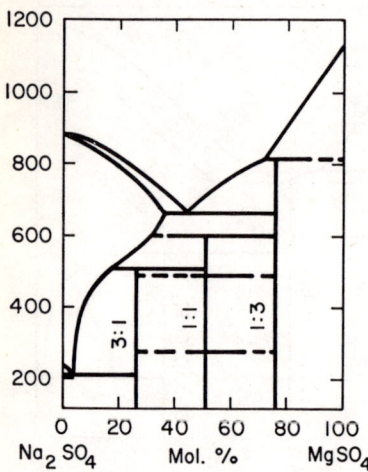

Fig. 1117.—System Na₂SO₄–MgSO₄.

A. S. Ginsberg, *Z. anorg. Chem.*, **61**, 126 (1909).

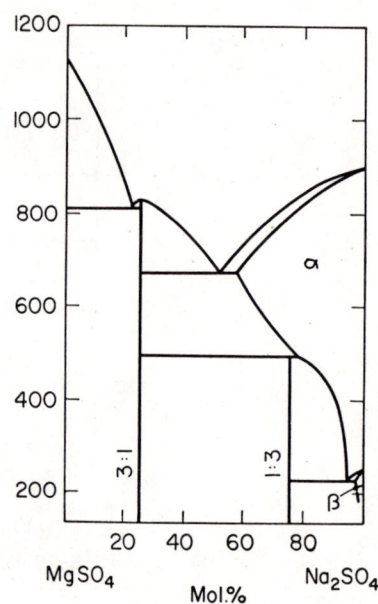

Fig. 1118.—System Na₂SO₄–MgSO₄.

A. S. Ginsberg, *Z. anorg. Chem.*, **61**, 130 (1909).

Na₂SO₄–MnSO₄

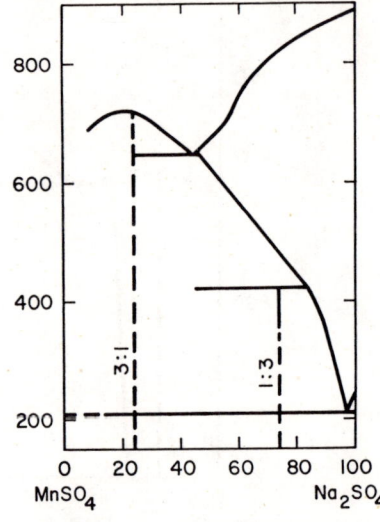

Fig. 1119.—System Na₂SO₄–MnSO₄.

G. Calcagni and D. Marotta, *Gazz. chim. ital.*, **45 II**, 373 (1915).

Na₂SO₄–NiSO₄

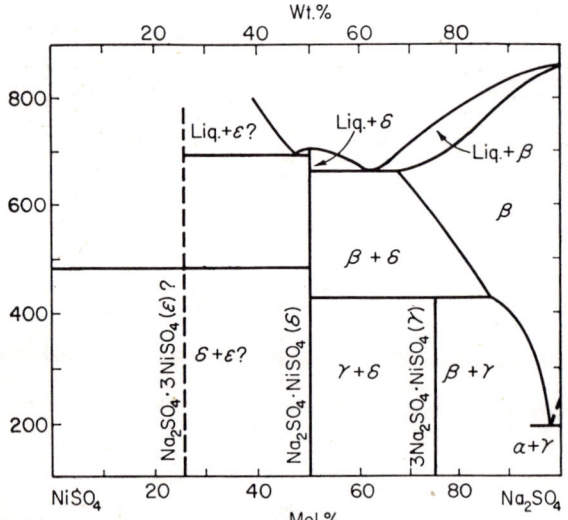

Fig. 1120.—System Na₂SO₄–NiSO₄.

K. A. Bol'shakov and P. I. Fedorov, *Zhur. Obshchei Khim.*, **26**, 349 (1956).

Na₂SO₄–PbSO₄

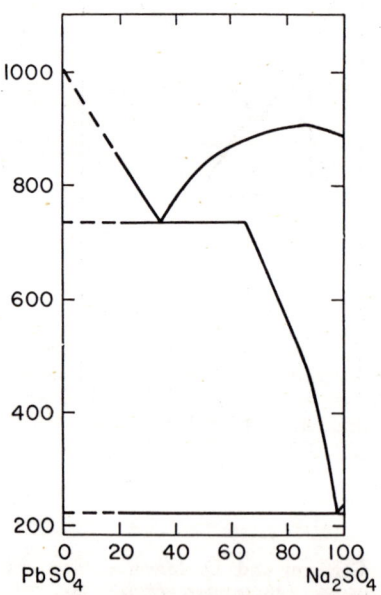

Fig. 1121.—System Na₂SO₄–PbSO₄.

G. Calcagni and D. Marotta, *Gazz. chim. ital.*, **42 II**, 681 (1912).

Na₂(SO₄, WO₄)–Pb(SO₄, WO₄)

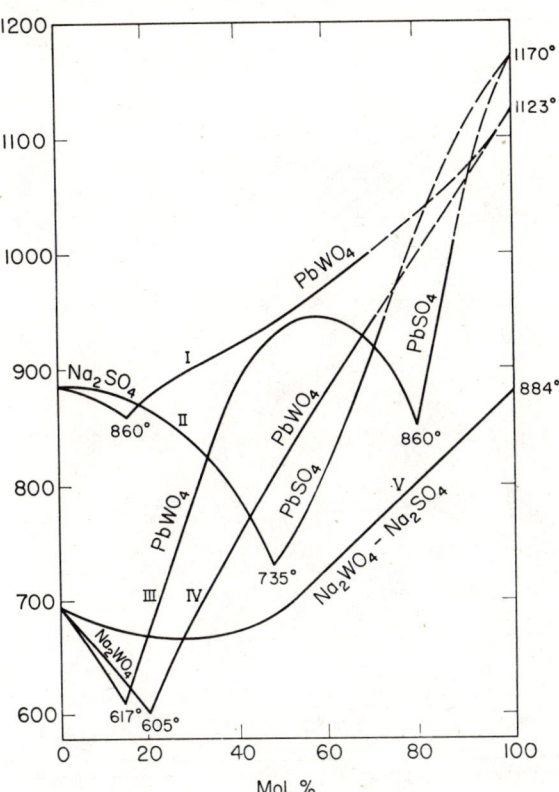

Fig. 1122.—Na₂O(SO₃, WO₃)–PbO(SO₃, WO₃); binary systems. I, Na₂SO₄–PbWO₄; II, Na₂SO₄–PbSO₄; III, Na₂WO₄–PbSO₄; IV, Na₂WO₄–PbWO₄; V, Na₂WO₄–Na₂SO₄.

I. N. Belyaev, *J. Gen. Chem. U.S.S.R.*, **22**, 1746 (1952).

Na₂SO₄–ZnSO₄

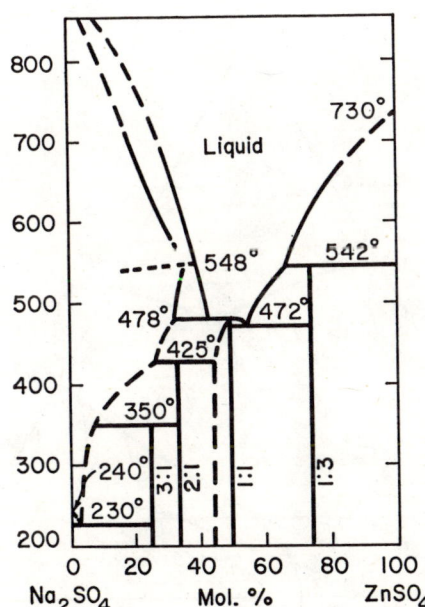

Fig. 1123.—System Na₂SO₄–ZnSO₄.

N. N. Evseeva, *Izvest. Sektora Fiz.-Khim. Anal., Inst. Obshchei Neorg. Khim., Akad. Nauk S.S.S.R.*, **22**, 165 (1953).

Rb₂SO₄–CaSO₄

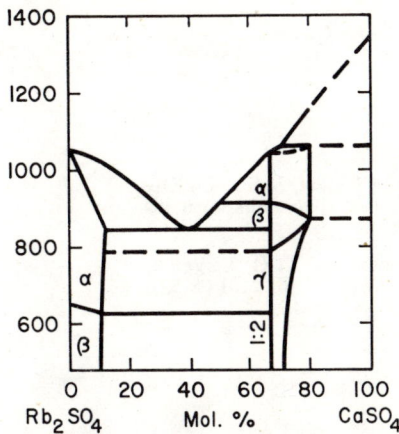

Fig. 1124.—System Rb₂SO₄–CaSO₄. Solid solution regions do not obey phase rule.

Hans Müller, *Neues Jahrb. Mineral., Geol.*, **30**, 22 (1910).

Rb₂SO₄–MgSO₄

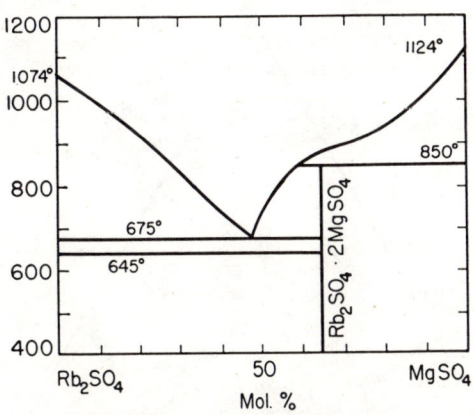

Fig. 1125.—System Rb₂SO₄–MgSO₄.

V. E. Plyushchev and N. F. Markovskaya, *Zhur. Obshchei Khim.*, **24**, 1302 (1954).

(b) *Three Sulfates*

K₂SO₄–Li₂SO₄–Na₂SO₄

Fig. 1126.—System K₂SO₄–Li₂SO₄–Na₂SO₄. [I]–[VIII] are unknown interior phases.

E. K. Akopov and A. G. Bergman, *Zhur. Neorg. Khim.*, **4** [5] 1151 (1959).

K₂SO₄–Li₂SO₄–Tl₂SO₄

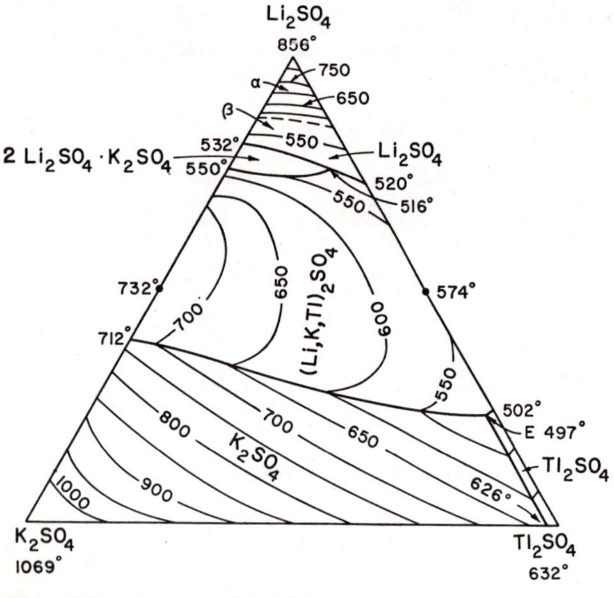

Fig. 1127.—System K₂SO₄–Li₂SO₄–Tl₂SO₄.

E. K. Akopov and A. G. Bergman, *Zhur. Neorg. Khim.*, **2** [1] 199 (1957).

K₂SO₄–Li₂SO₄–CdSO₄

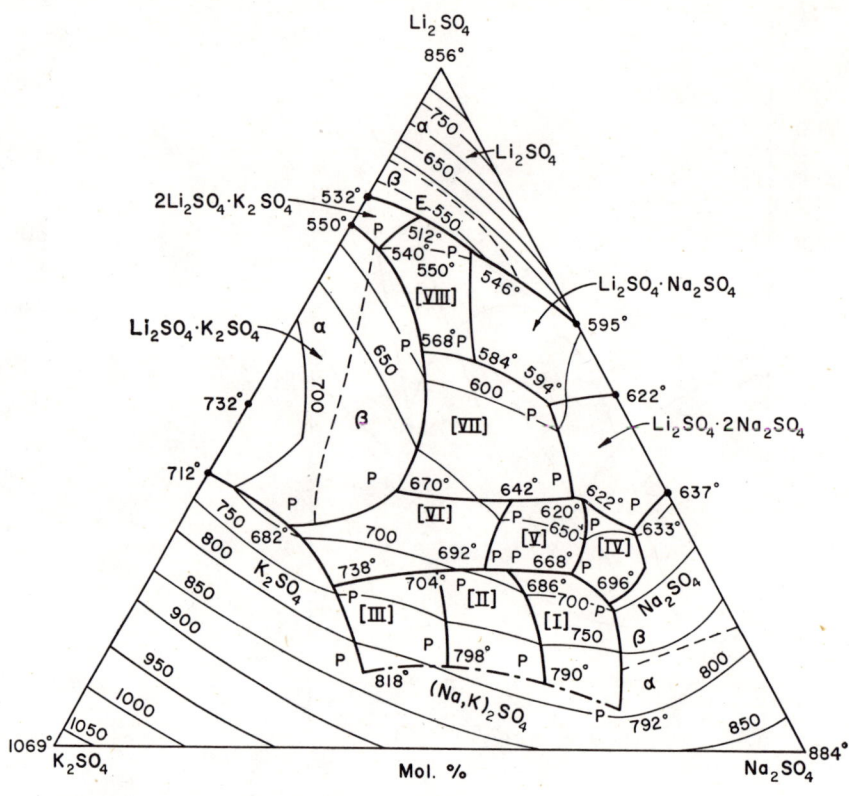

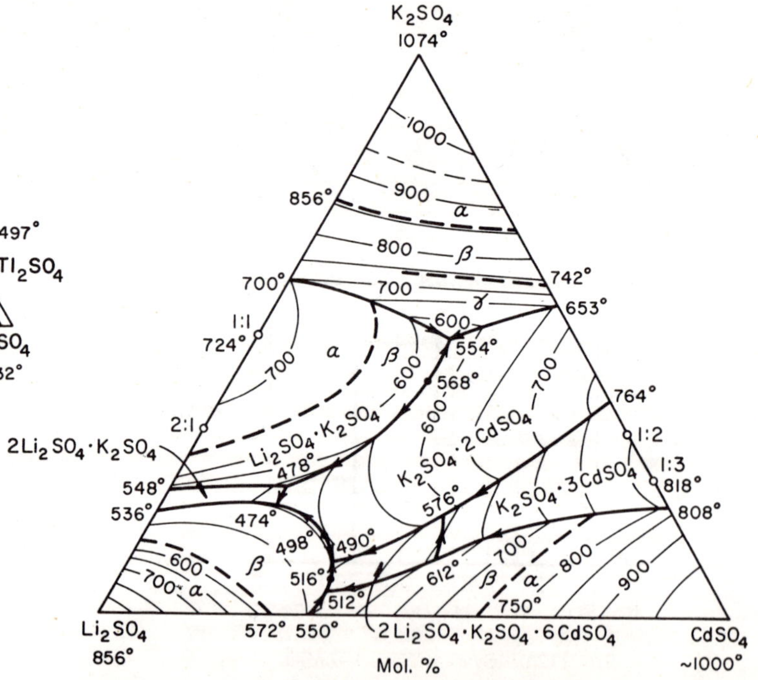

Fig. 1128.—System K₂SO₄–Li₂SO₄–CdSO₄.

A. G. Bergman and E. L. Bakumskaya, *Zhur. Neorg. Khim.*, **1**, 2091 (1956).

K_2SO_4–Na_2SO_4–$CdSO_4$

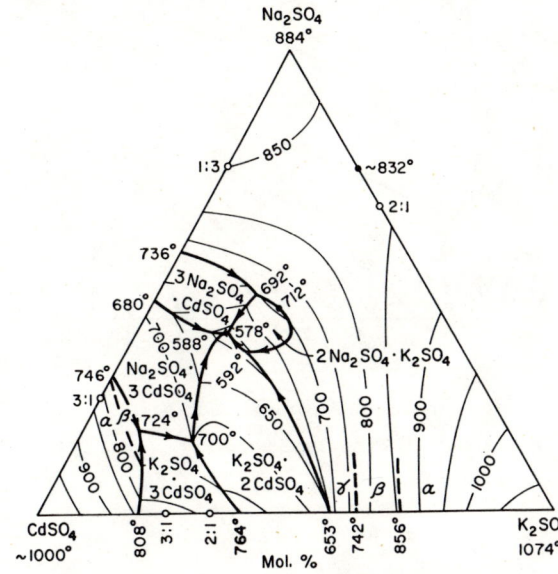

Fig. 1129.—System K_2SO_4–Na_2SO_4–$CdSO_4$.

E. L. Bakumskaya and A. G. Bergman, *Zhur. Neorg. Khim.*, **1** [5] 1039 (1956).

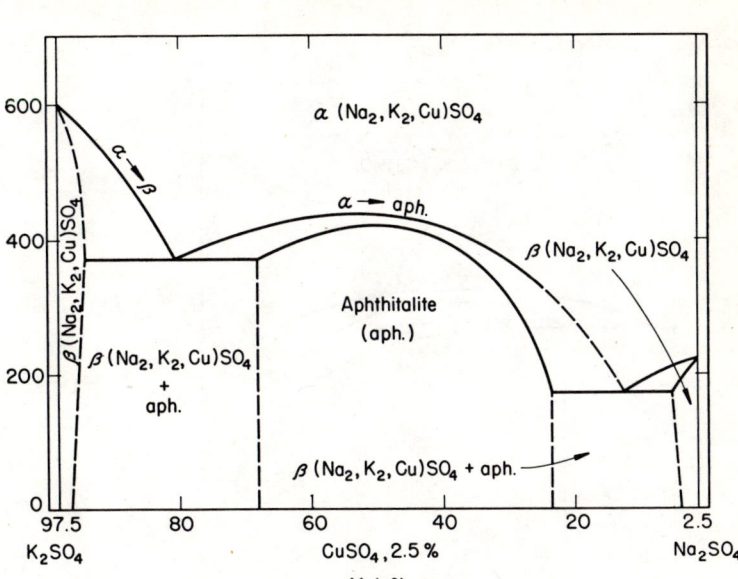

Fig. 1131.—System K_2SO_4–Na_2SO_4–2.5% $CuSO_4$; subsolidus.

Angelo Bellanca and Marcello Carapezza, *Periodico Mineral. (Rome)*, **20**, 291 (1951).

K_2SO_4–Na_2SO_4–$CuSO_4$

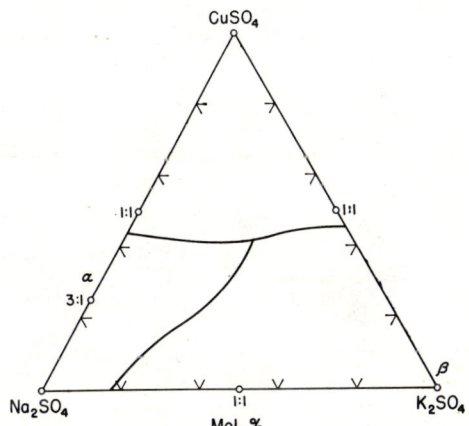

Fig. 1130.—System K_2SO_4–Na_2SO_4–$CuSO_4$.

Angelo Bellanca and Marcello Carapezza, *Periodico Mineral. (Rome)*, **20**, 282 (1951).

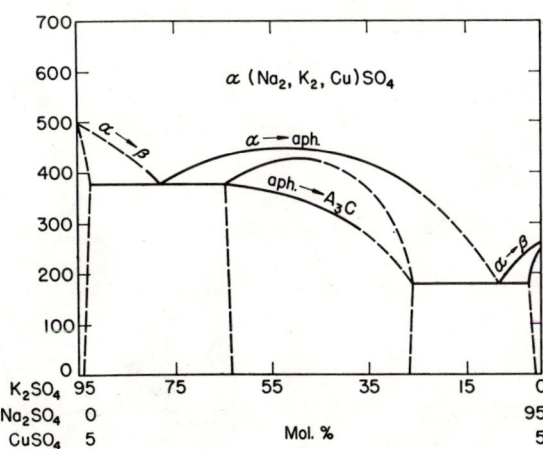

Fig. 1132.—System K_2SO_4–Na_2SO_4–5% $CuSO_4$; subsolidus. $A_3C = 3Na_2SO_4 \cdot CuSO_4$, aph. = aphthitalite.

Angelo Bellanca and Marcello Carapezza, *Periodico Mineral. (Rome)*, **20**, 293 (1951).

K_2SO_4–Na_2SO_4–$CuSO_4$ (concl.)

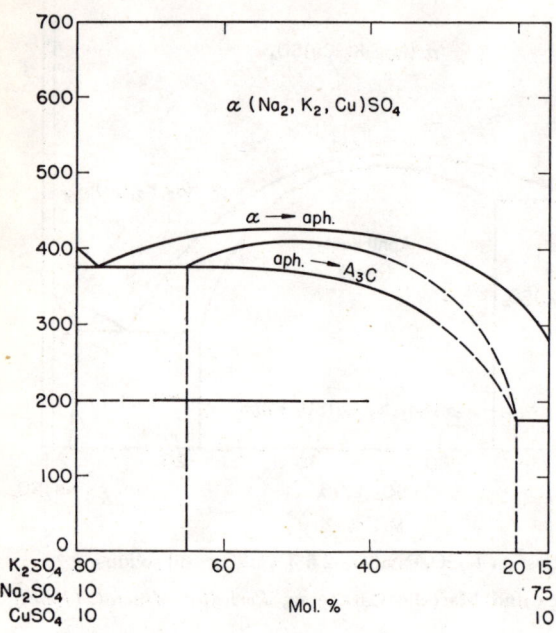

Fig. 1133.—System K_2SO_4–Na_2SO_4–10% $CuSO_4$; subsolidus; A_3C = $3Na_2SO_4 \cdot CuSO_4$, aph. = aphthitalite.

Angelo Bellanca and Marcello Carapezza, *Periodico Mineral.* (Rome), **20**, 294 (1951).

Li_2SO_4–Na_2SO_4–$CdSO_4$

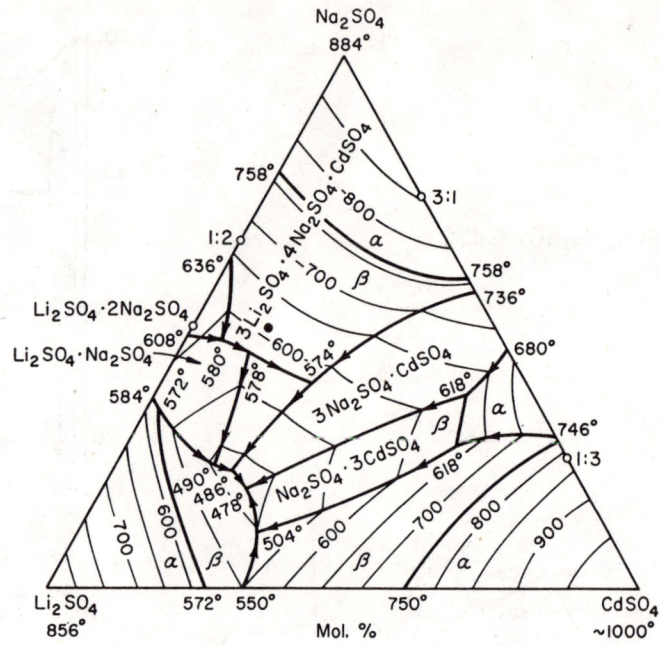

Fig. 1135.—System Li_2SO_4–Na_2SO_4–$CdSO_4$.

E. L. Bakumskaya and A. G. Bergman, *Zhur. Neorg. Khim.*, **1**, 1636 (1956).

K_2SO_4–Na_2SO_4–$ZnSO_4$

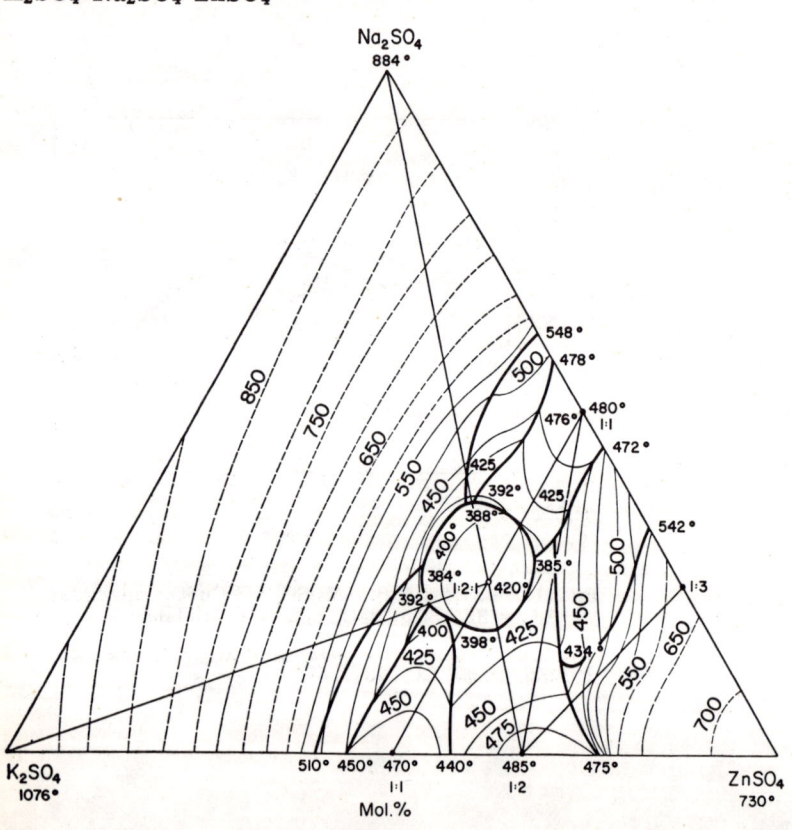

Fig. 1134.—System K_2SO_4–Na_2SO_4–$ZnSO_4$. 1:2:1 = $K_2SO_4 \cdot 2ZnSO_4 \cdot Na_2SO_4$.

N. V. Khakhlova and N. S. Dombrovskaya, *Zhur. Neorg. Khim.*, **4**, 924 (1959).

Na_2SO_4–$CoSO_4$–$NiSO_4$

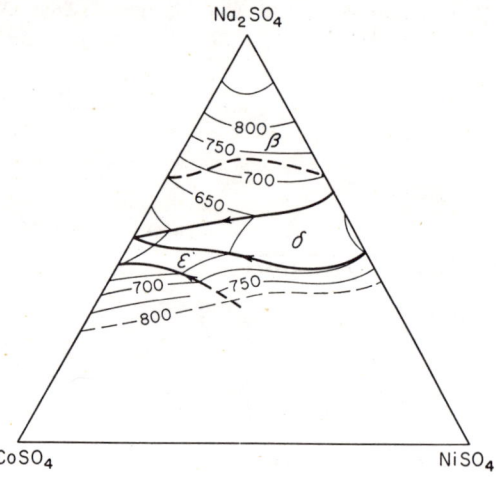

Fig. 1136.—System Na_2SO_4–$CoSO_4$–$NiSO_4$.

K. A. Bol'shakov and P. I. Fedorov, *Zhur. Neorg. Khim.*, **3** [8] 1899 (1958).

IX. Sulfates with Metal Oxides

(a) Two Substances

K$_2$SO$_4$–V$_2$O$_5$

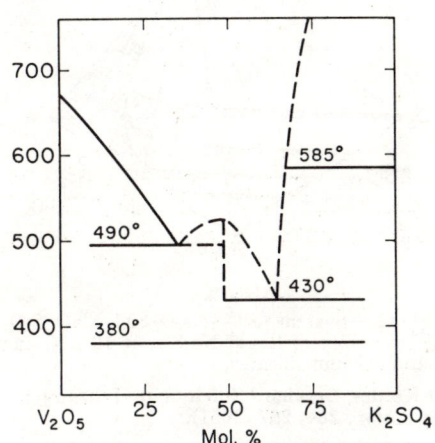

FIG. 1137.—System K$_2$SO$_4$–V$_2$O$_5$.

G. K. Boreskov, V. V. Illarionov, and R. P. Ozerov, *Zhur. Obshchei Khim.*, **24**, 28 (1954).

PbSO$_4$–PbO

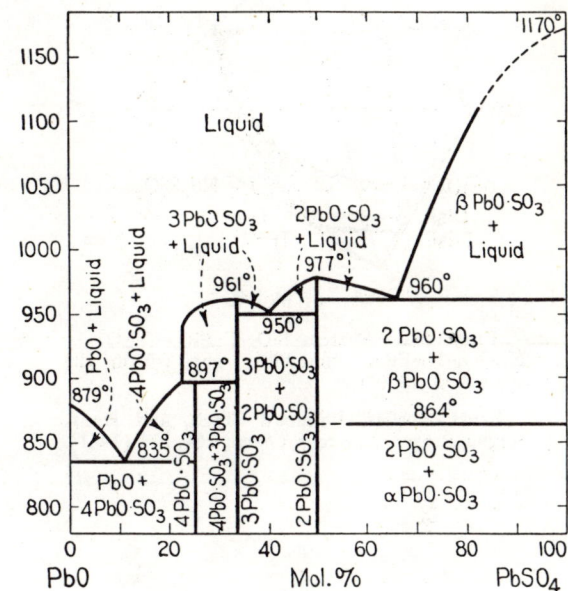

FIG. 1139.—System PbO–PbSO$_4$.

F. M. Jaeger and H. C. Germs, *Z. anorg. u. allgem. Chem.*, **119**, 152 (1921).

Na$_2$SO$_4$–V$_2$O$_5$

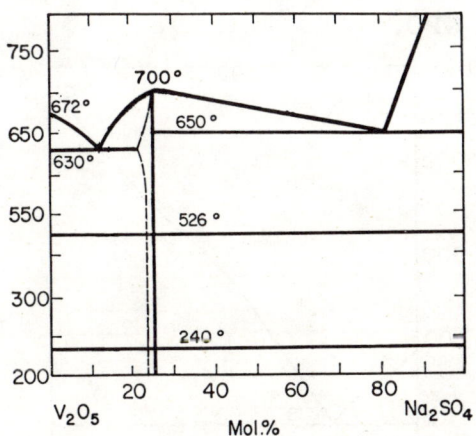

FIG. 1138.—System Na$_2$SO$_4$–V$_2$O$_5$.

V. V. Illarionov, R. P. Ozerov, and E. V. Kil'disheva, *Zhur. Neorg. Khim.*, **2**, 884 (1957).

(b) Three Substances

Na$_2$O–V$_2$O$_5$–SO$_3$

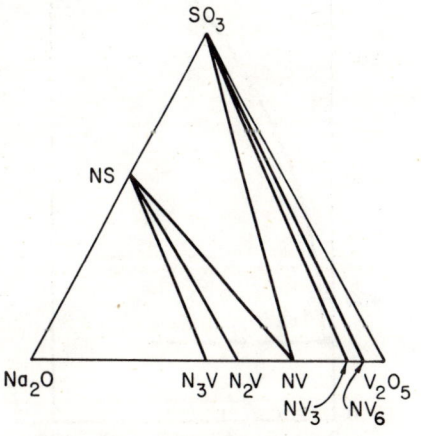

FIG. 1140.—System Na$_2$O–SO$_3$–V$_2$O$_5$; compatibility triangles. N = Na$_2$O, S = SO$_3$, V = V$_2$O$_5$. NV$_3$ and NV$_6$ are oxygen deficient.

W. R. Foster, M. H. Leipold, and T. S. Shevlin, *Corrosion*, **12** [11] 543T (1956).

(c) Four Substances

Na₂O–CaO–SiO₂–SO₃

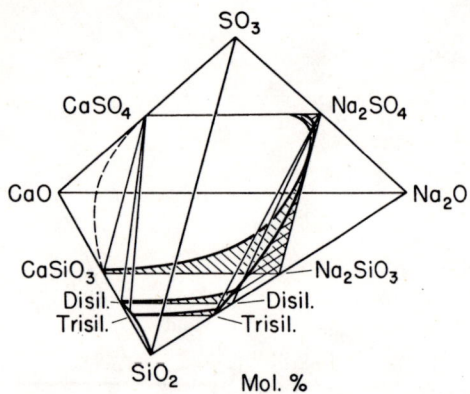

Fig. 1141.—System SiO₂–CaSiO₃–CaSO₄–Na₂SO₄–Na₂SiO₃; miscibility gap within the system at 1200°C.

Ernst Kordes, Bernhard Zöfelt, and Hansjürgen Pröger, *Z. anorg. Chem.*, **264**, 271 (1951).

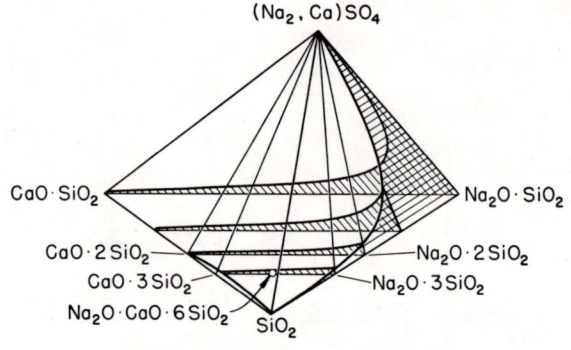

Fig. 1142.—System SiO₂–Na₂O·SiO₂–CaO·SiO₂–(Na₂,Ca)SO₄; solubility of liquid Na₂SO₄ at 1200°C. in melted sodium and calcium silicates.

Ernst Kordes, Bernhard Zöfelt, and Hansjürgen Pröger, *Z. anorg. Chem.*, **264**, 267 (1951).

X. Sulfates with Other Oxygen-Containing Radicals

(a) Two Substances

K₂SO₄–KNO₃

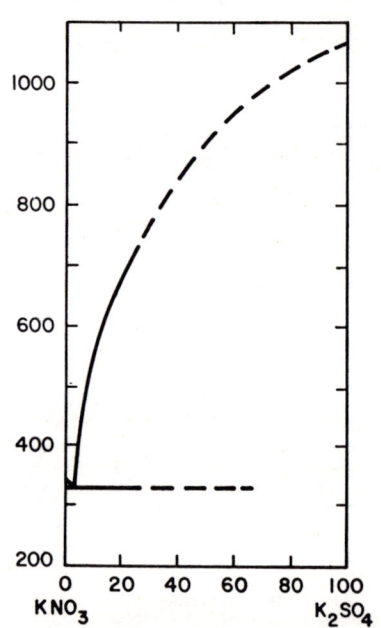

Fig. 1143.—System KNO₃–K₂SO₄.

M. Amadori, *Atti reale accad. Lincei*, Sez. II, **22**, 334 (1913).

K₂SO₄–KPO₃

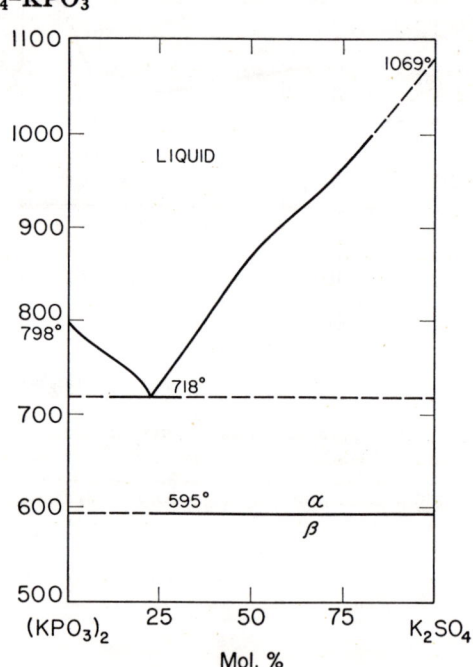

Fig. 1144.—System KPO₃–K₂SO₄.

A. G. Bergman and M. L. Sholokhovich, *J. Gen. Chem. U.S.S.R.*, **23** [7] 1077 (1953).

$(K_2, Li_2)SO_4-(K, Li)PO_3$

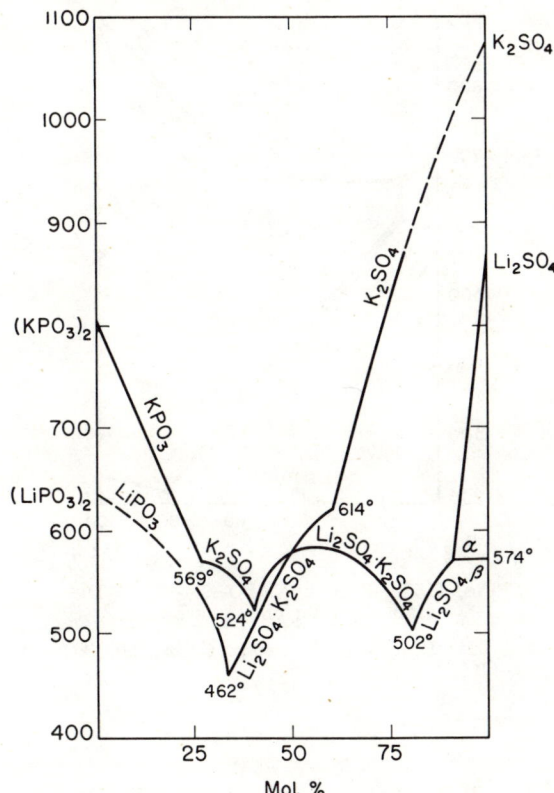

Fig. 1145—Systems $(KPO_3)_2-Li_2SO_4$ and $(LiPO_3)_2-K_2SO_4$.

A. G. Bergman and M. L. Sholokhovich, *J. Gen. Chem. U.S.S.R.*, **23** [7] 1079 (1953).

$K_2SO_4-K_2MoO_4$

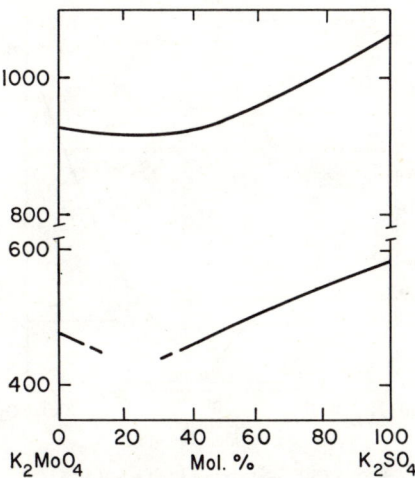

Fig. 1147.—System $K_2MoO_4-K_2SO_4$. Diagram does not obey phase rule.

M. Amadori, *Atti reale accad. Lincei, Sez. I*, **22**, 458 (1913).

$K_2SO_4-K_2WO_4$

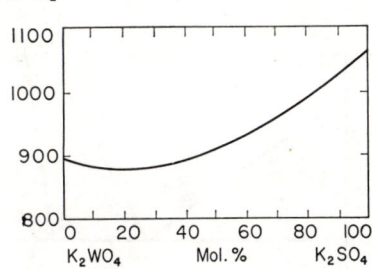

Fig. 1148.—System $K_2SO_4-K_2WO_4$.

M. Amadori, *Atti reale accad. Lincei., Sez. I*, **22**, 610 (1913).

$K_2SO_4-K_2CrO_4$

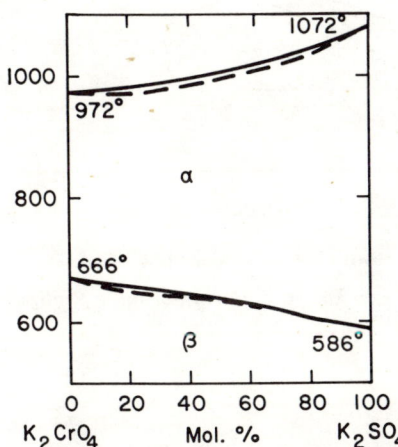

Fig. 1146.—System $K_2CrO_4-K_2SO_4$.

E. Groschuff, *Z. anorg. Chem.*, **58**, 107 (1908).

$Li_2SO_4-LiNO_3$

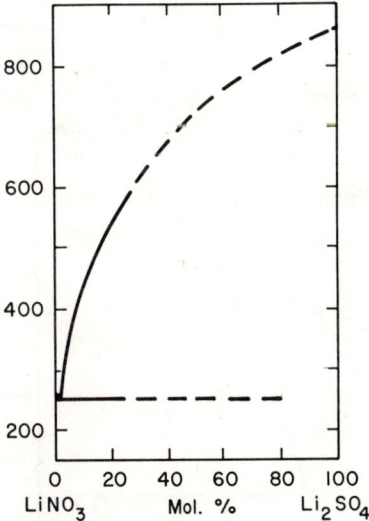

Fig. 1149.—System $LiNO_3-Li_2SO_4$.

M. Amadori, *Atti reale accad. Lincei, Sez. II*, **22**, 333 (1913).

Li$_2$SO$_4$–LiPO$_3$

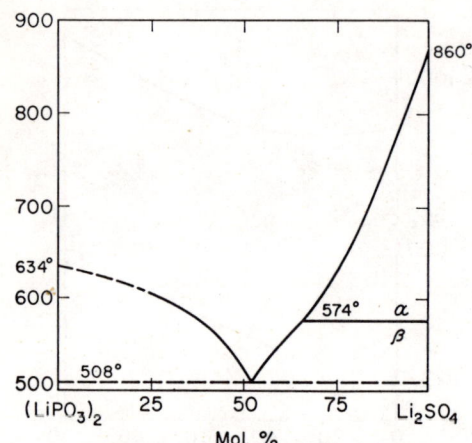

Fig. 1150.—System (LiPO$_3$)$_2$–Li$_2$SO$_4$.

A. G. Bergman and M. L. Sholokhovich, *J. Gen. Chem. U.S.S.R.*, **23** [7] 1078 (1953).

Li$_2$SO$_4$–Li$_2$CO$_3$

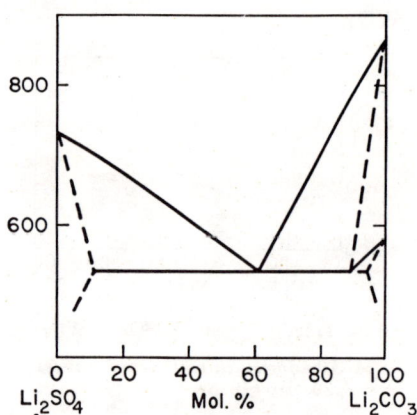

Fig. 1151.—System Li$_2$SO$_4$–Li$_2$CO$_3$.

M. Amadori, *Atti reale accad. Lincei, Sez. II*, **21**, 68 (1912).

Na$_2$SO$_4$–NaNO$_3$

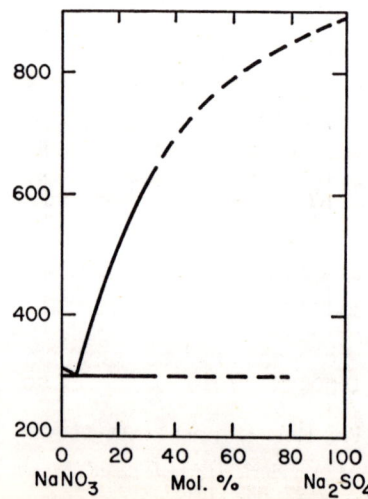

Fig. 1152.— System NaNO$_3$–Na$_2$SO$_4$.

M. Amadori, *Atti reale accad. Lincei, Sez. II*, **22**, 334 (1913).

Na$_2$SO$_4$–Na$_2$WO$_4$

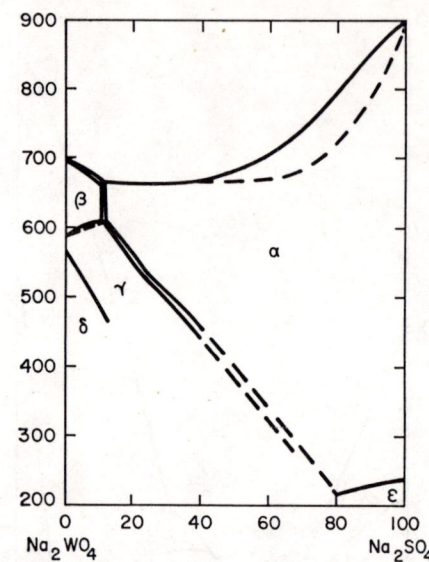

Fig. 1153.—System Na$_2$SO$_4$–Na$_2$WO$_4$. Diagram does not obey phase rule.

H. E. Boeke, *Z. anorg. Chem.*, **50**, 364 (1907).

PbSO$_4$–PbCrO$_4$

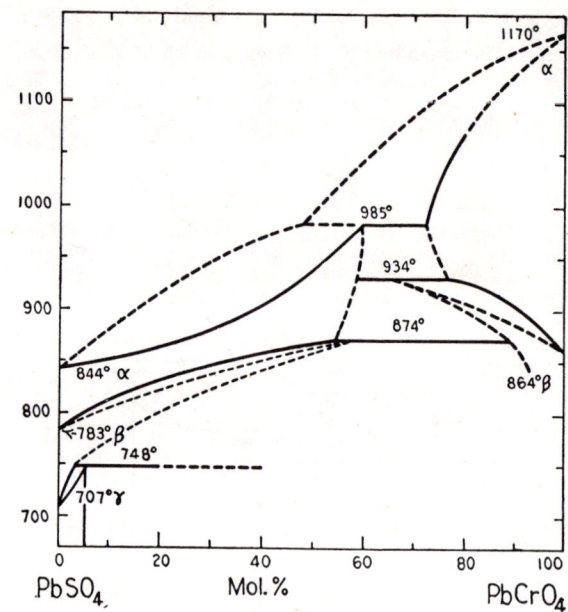

Fig. 1154.—System PbSO$_4$–PbCrO$_4$.

F. H. Jaeger and H. C. Germs, *Z. anorg. u. allgem. Chem.*, **119**, 164 (1921).

$PbSO_4$–$PbMoO_4$

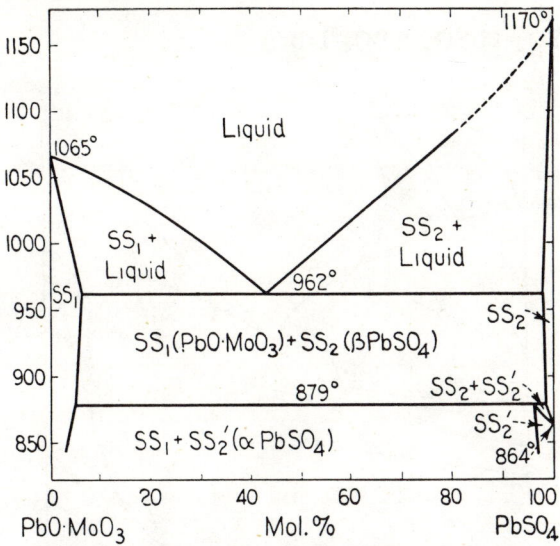

Fig. 1155.—System $PbSO_4$–$PbO·MoO_3$.

F. M. Jaeger and H. C. Germs, *Z. anorg. u. allgem. Chem.*, **119**, 165 (1921).

$PbSO_4$–$PbWO_4$

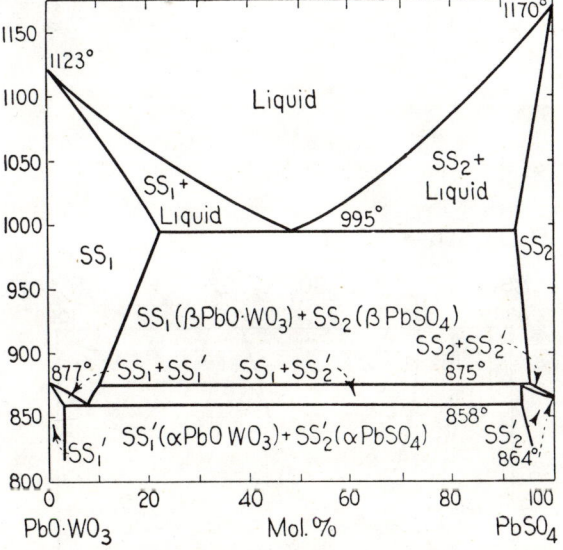

Fig. 1156.—System $PbSO_4$–$PbO·WO_3$.

F. M. Jaeger and H. C. Germs, *Z. anorg. u. allgem Chem.*, **119**, 167 (1921).

(b) Three Substances

K_2SO_4–KNO_3–K_2CrO_4

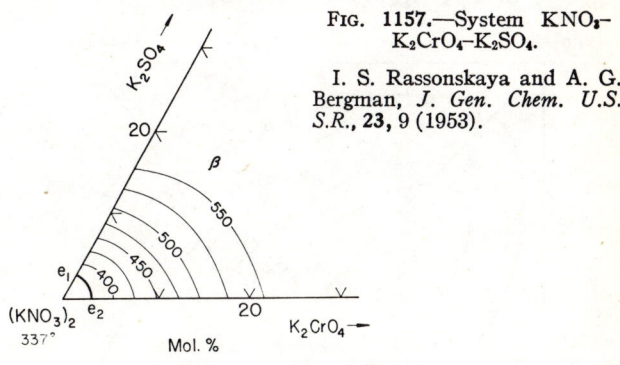

Fig. 1157.—System KNO_3–K_2CrO_4–K_2SO_4.

I. S. Rassonskaya and A. G. Bergman, *J. Gen. Chem. U.S.S.R.*, **23**, 9 (1953).

Li_2SO_4–$LiBO_2$–Li_2WO_4

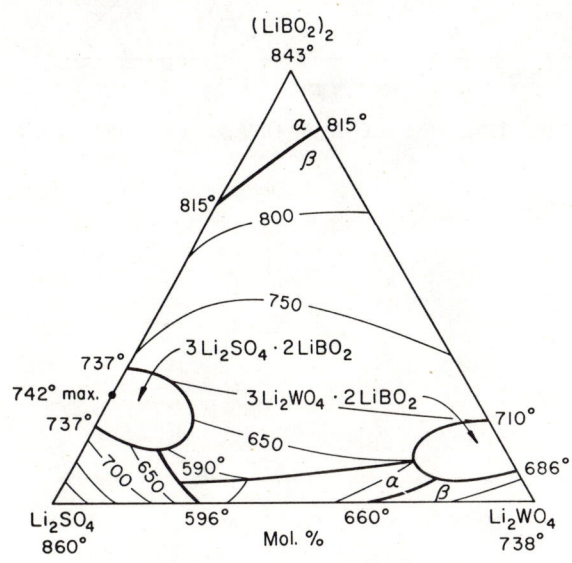

Fig. 1158.—System $(LiBO_2)_2$–Li_2SO_4–Li_2WO_4.

V. I. Posypaĭko, A. I. Kislova, and A. G. Bergman, *Zhur. Neorg. Khim.*, **1**, 814 (1956).

Na_2SO_4–$NaNO_3$–Na_2CrO_4

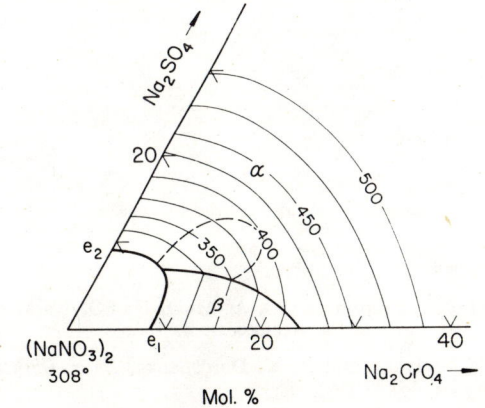

Fig. 1159.—System $NaNO_3$–Na_2CrO_4–Na_2SO_4.

I. S. Rassonskaya and A. G. Bergman, *J. Gen. Chem. U.S.S.R.*, **23**, 13 (1953).

(c) Four Substances

Ag_2SO_4–Li_2SO_4–Ag_2MoO_4–Li_2MoO_4

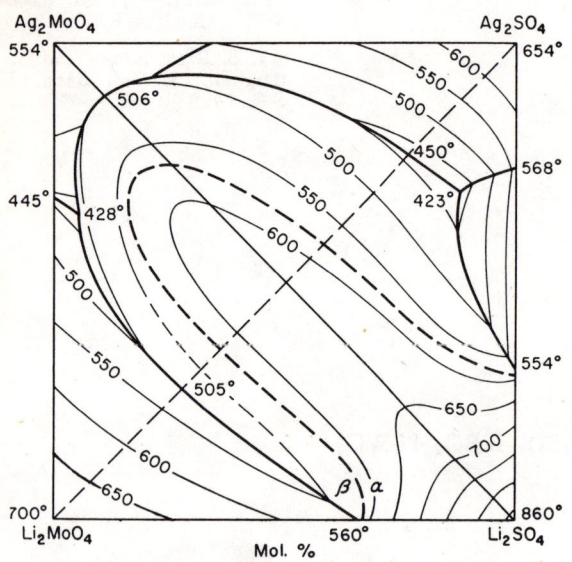

FIG. 1160.—System Ag_2MoO_4–Ag_2SO_4–Li_2MoO_4–Li_2SO_4, reciprocal.

I. N. Belyaev and S. S. Doroshenko, *Zhur. Obshcheĭ Khim.*, **26**, 1820 (1956).

K_2SO_4–Li_2SO_4–KBO_2–$LiBO_2$

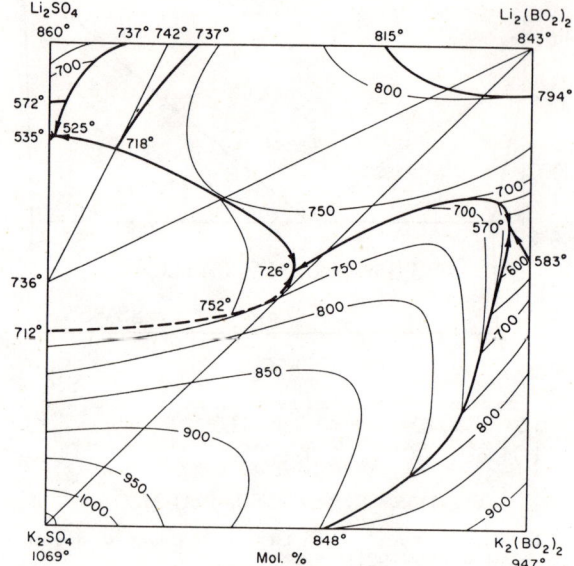

FIG. 1162.—System KBO_2–K_2SO_4–$LiBO_2$–Li_2SO_4, reciprocal.

A. G. Bergman, A. I. Kislova, and V. I. Posypaĭko, *Zhur. Obshcheĭ Khim.*, **25**, 1897 (1955).

Ag_2SO_4–Na_2SO_4–Ag_2MoO_4–Na_2MoO_4

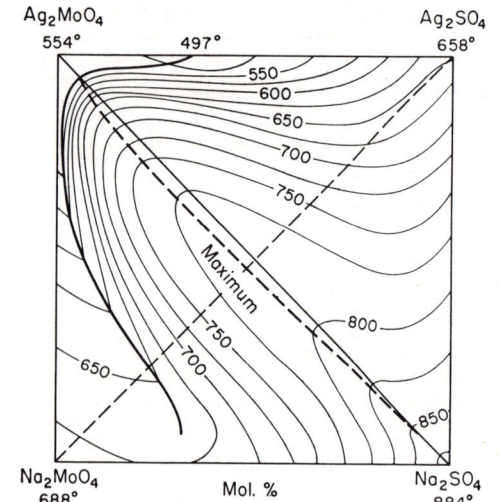

FIG. 1161.—System Ag_2SO_4–Ag_2MoO_4–Na_2SO_4–Na_2MoO_4, reciprocal.

I. N. Belyaev and A. K. Doroshenko, *Zhur. Obshcheĭ Khim.*, **24**, 431 (1954).

K_2SO_4–Li_2SO_4–KNO_3–$LiNO_3$

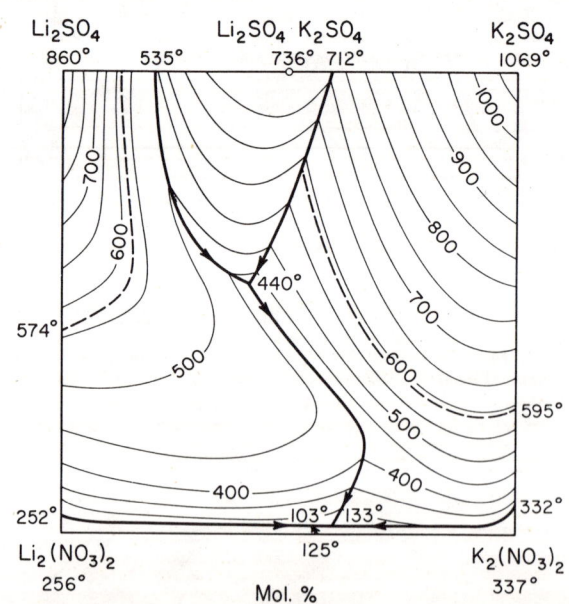

FIG. 1163.—System $K_2(NO_3)_2$–K_2SO_4–$Li_2(NO_3)_2$–Li_2SO_4; reciprocal salt pairs, ternary systems.

G. A. Bukhalova, M. L. Sholokhovich, and A. G. Bergman, *Doklady Akad. Nauk S.S.S.R.*, **71** [2] 288 (1950).

K_2SO_4–Li_2SO_4–KPO_3–$LiPO_3$

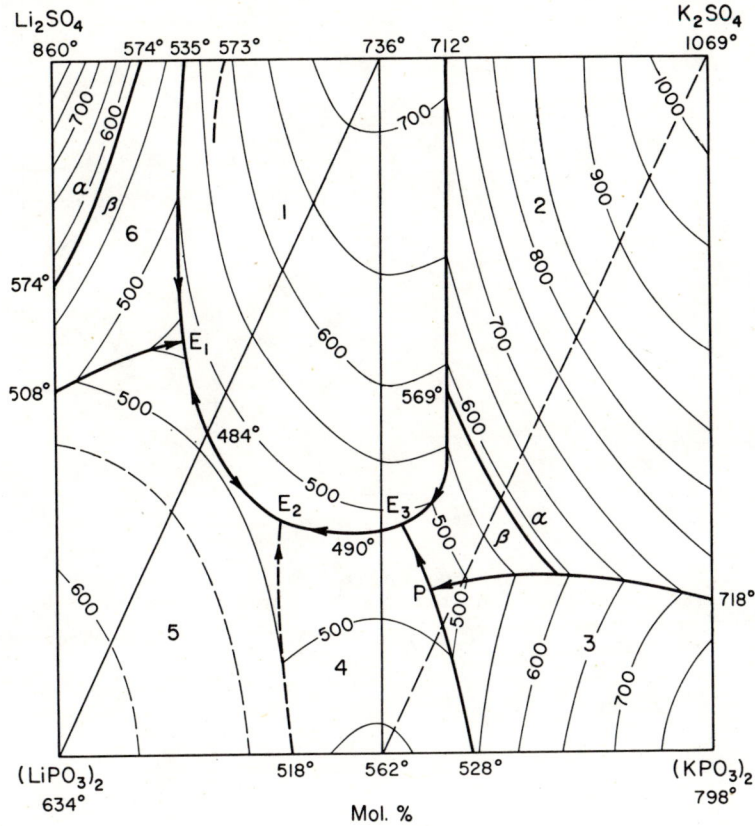

Fig. 1164.—System $(LiPO_3)_2$–$(KPO_3)_2$–K_2SO_4–Li_2SO_4; reciprocal salt pairs, ternary systems, non-diagonal fields.

1—$Li_2SO_4 \cdot K_2SO_4$; 2—K_2SO_4; 3—KPO_3; 4—$LiPO_3 \cdot KPO_3$; 5—$LiPO_3$; 6—Li_2SO_4; E_1—440°; E_2—460°; E_3—453°; P—474°.

A. G. Bergman and M. L. Sholokhovich, *J. Gen. Chem. U.S.S.R.*, **23** [7] 1082 (1953).

K_2SO_4–Li_2SO_4–K_2MoO_4–Li_2MoO_4

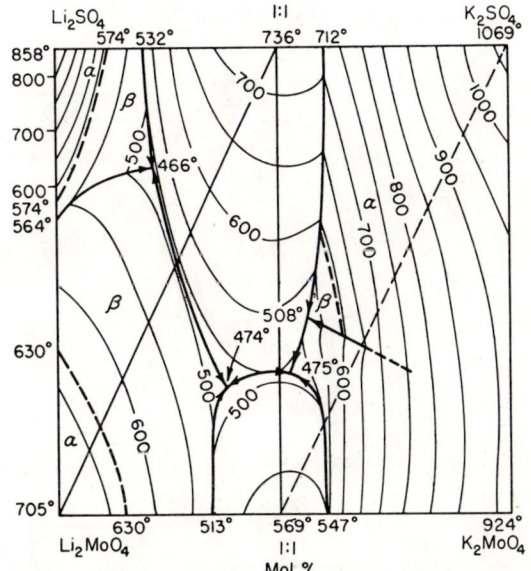

Fig. 1165.—System K_2SO_4–K_2MoO_4–Li_2SO_4–Li_2MoO_4.

A. G. Bergman, A. I. Kislova, and E. I. Korobka, *Zhur. Obshchei Khim.*, **24**, 1133 (1954).

K_2SO_4–Li_2SO_4–K_2WO_4–Li_2WO_4

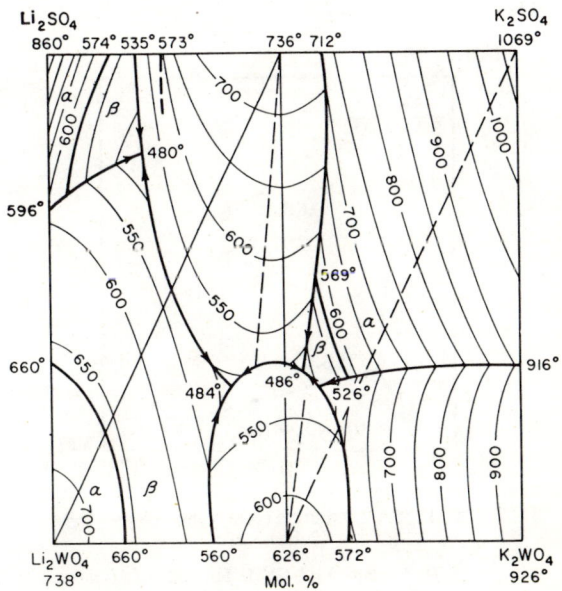

Fig. 1166.—System K_2SO_4–K_2WO_4–Li_2SO_4–Li_2WO_4; reciprocal.

A. G. Bergman and A. I. Kislova, *Zhur. Obshchei Khim.*, **25**, 860 (1955).

K₂SO₄–Na₂SO₄–K₂CO₃–Na₂CO₃

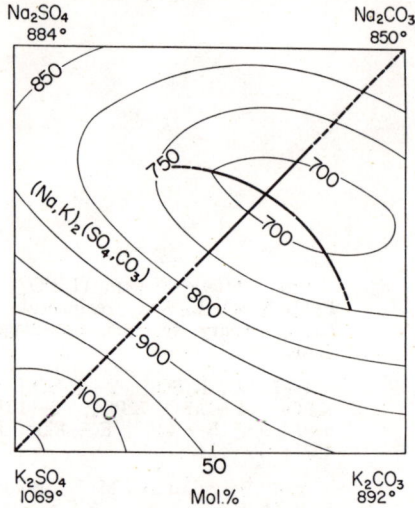

FIG. 1167.—System K₂CO₃–K₂SO₄–Na₂CO₃–Na₂SO₄.

A. K. Sementsova, K. A. Evdokimova, and A. G. Bergman, *Zhur. Neorg. Khim.*, **4**, 146 (1958).

K₂SO₄–PbSO₄–K₂WO₄–PbWO₄

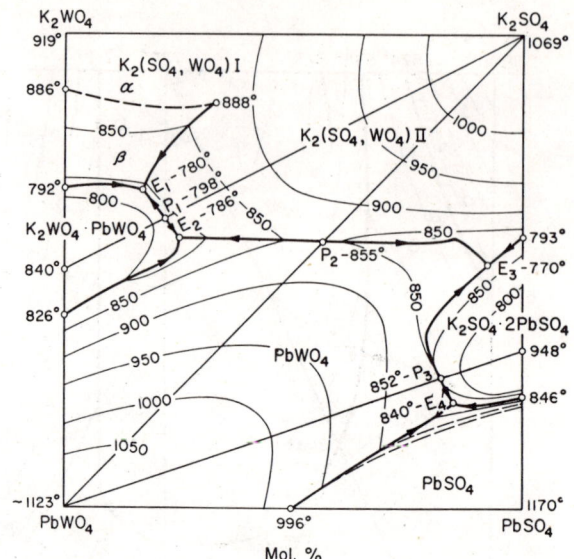

FIG. 1169.—PbWO₄–K₂SO₄–K₂WO₄ and PbWO₄–K₂SO₄–PbSO₄; reciprocal salt systems, irreversible ternary.

I. N. Belyaev and A. K. Nesterova, *Doklady Akad. Nauk S.S.S.R.*, **86** [5] 949 (1952).

K₂SO₄–(NH₄)₂SO₄–KNO₃–NH₄NO₃

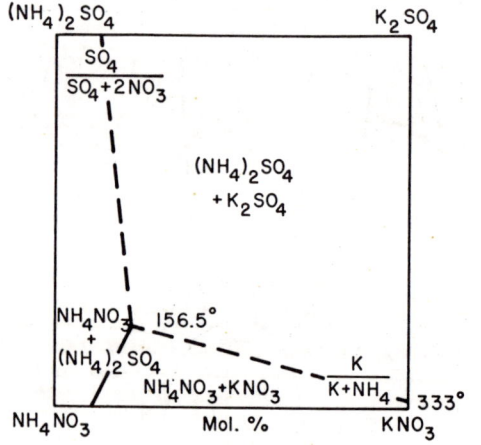

FIG. 1168.—System KNO₃–K₂SO₄–NH₄NO₃–(NH₄)₂SO₄.

E. P. Perman and W. J. Howells, *J. Chem. Soc.*, **123**, 2132 (1923).

Li₂SO₄–Na₂SO₄–Li₂MoO₄–Na₂MoO₄

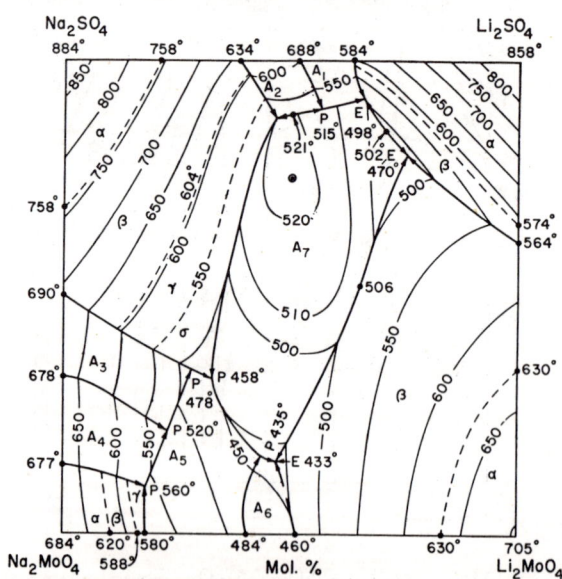

FIG. 1170.—System Li₂SO₄–Li₂MoO₄–Na₂SO₄–Na₂MoO₄. A₁ to A₇ designate primary phase fields.

A. G. Bergman and E. I. Korobka, *Zhur. Neorg. Khim.*, **4** [1] 110 (1959).

Li_2SO_4–$PbSO_4$–Li_2WO_4–$PbWO_4$

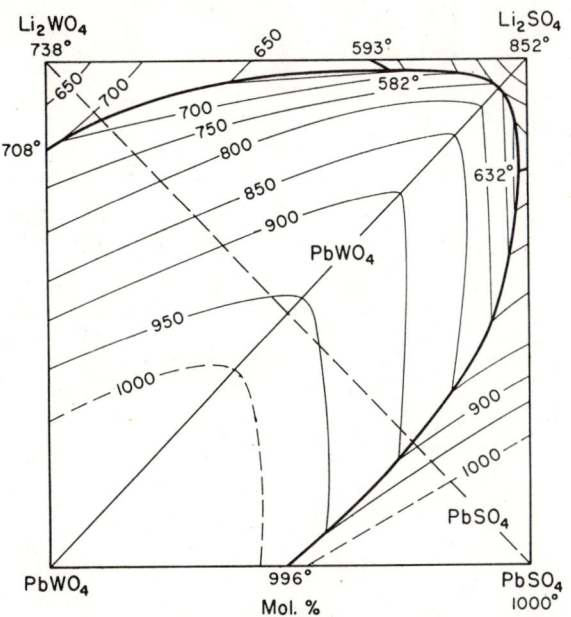

FIG. 1171.—System Li_2SO_4–$PbSO_4$–Li_2WO_4–$PbWO_4$; irreversible reciprocal.

I. N. Belyaev, *Zhur. Obshchei Khim.*, **25**, 233 (1955).

Na_2SO_4–$PbSO_4$–Na_2MoO_4–$PbMoO_4$

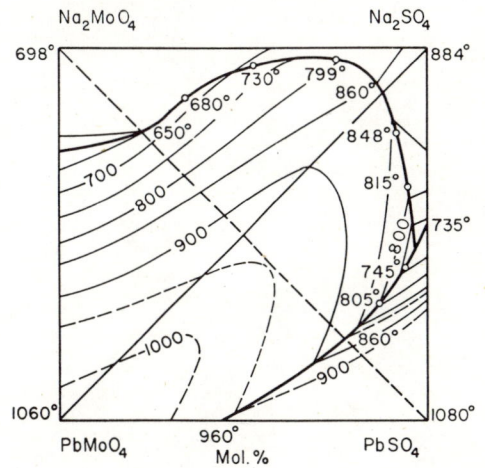

FIG. 1173.—System Na_2MoO_4–Na_2SO_4–$PbMoO_4$–$PbSO_4$.

I. N. Belyaev, *Zhur. Obshchei Khim.*, **22**, 1323 (1952).

Na_2SO_4–$(NH_4)_2SO_4$–$NaNO_3$–NH_4NO_3

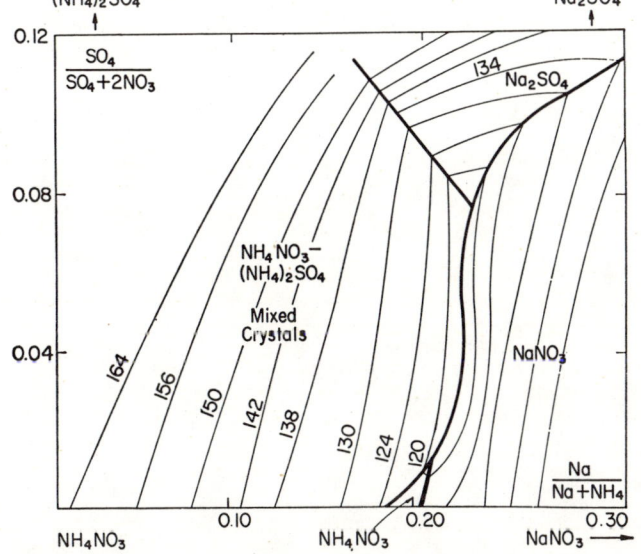

FIG. 1172.—System $NaNO_3$–Na_2SO_4–NH_4NO_3–$(NH_4)_2SO_4$.

E. P. Perman and W. R. Harrison, *J. Chem. Soc.*, **125**, 365 (1924).

Na_2SO_4–$PbSO_4$–Na_2WO_4–$PbWO_4$

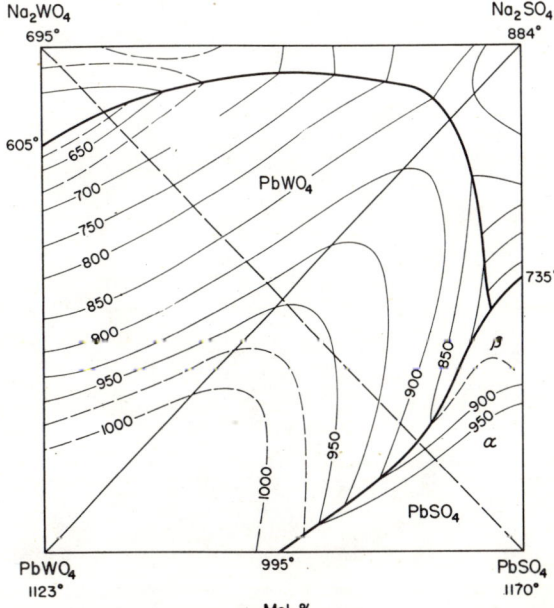

FIG. 1174.—System Na_2SO_4–Na_2WO_4–$PbSO_4$–$PbWO_4$; reciprocal salt pairs, ternary systems.

I. N. Belyaev, *J. Gen. Chem. U.S.S.R.*, **22**, 1754 (1952).

D. SYSTEMS CONTAINING HALIDES ONLY

I. Bromides Only

(a) Two Bromides

AgBr–KBr

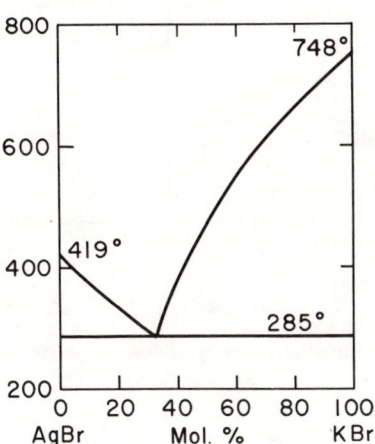

Fig. 1175.—System AgBr–KBr.

S. F. Zhemchuzhnui, *Z. anorg. u. allgem. Chem.*, **153**, 56 (1926).

AgBr–LiBr

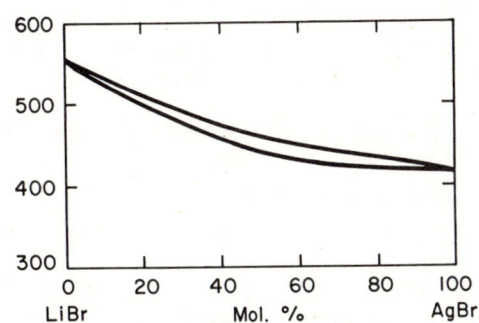

Fig. 1176.—System AgBr–LiBr.

C. Sandonnini and G. Scarpa, *Atti reale accad. Lincei Sez. II*, **22**, 519 (1913).

AgBr–NaBr

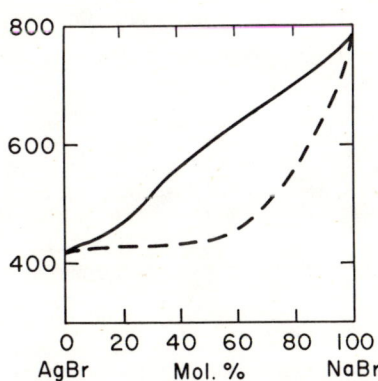

Fig. 1177.—System AgBr–NaBr.

S. F. Zhemchuzhnui, *Z. anorg. u. allgem. Chem.*, **153**, 52 (1926).

AgBr–RbBr

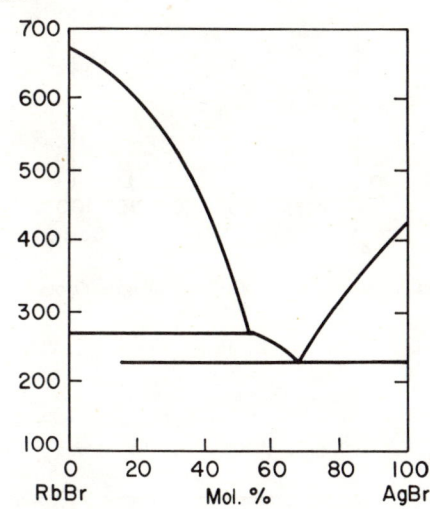

Fig. 1178.—System AgBr–RbBr.
Carlo Sandonnini, *Atti reale accad. Lincei, Sez. I*, **21**, 198 (1912).

AgBr–CdBr$_2$

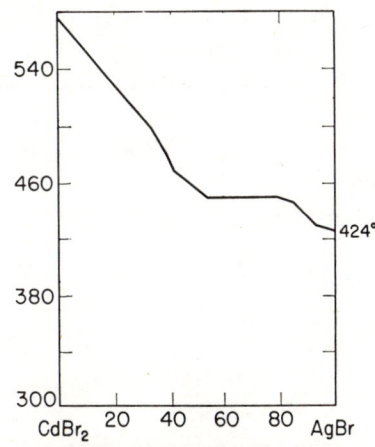

Fig. 1179.—System AgBr–CdBr$_2$.
G. H. Zakharchenko, *Zhur. Obshcheĭ Khim.*, **21**, 455 (1951).

AgBr–PbBr$_2$

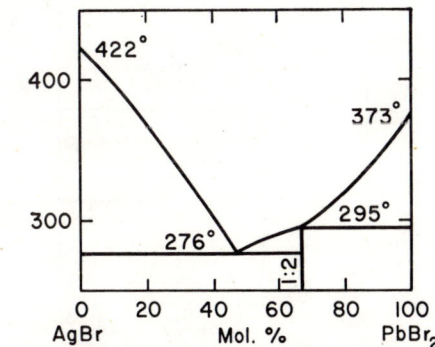

Fig. 1180.—System AgBr–PbBr$_2$.
C. Tubandt and Sophie Eggert, *Z. anorg. u. allgem. Chem.*, **110**, 222 (1920).

CuBr–CdBr₂

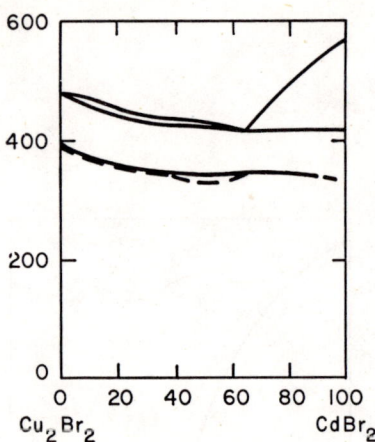

Fig. 1181.— System CuBr–CdBr₂. Diagram does not obey phase rule.

Gottfried Herrmann, *Z. anorg. Chem.*, **71**, 293 (1911).

KBr–LiBr

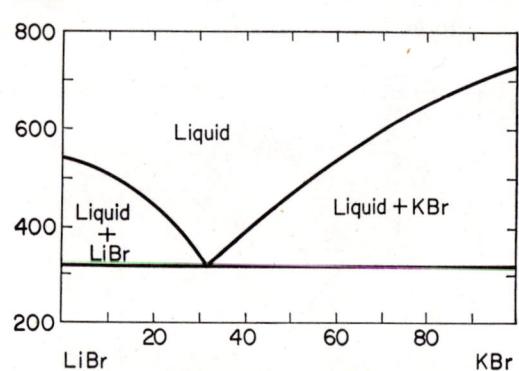

Fig. 1182.—System KBr–LiBr.

P. D. Garn, *Bell Lab. Record*, **33** [12] 452 (1955). See also Georg Kellner, *Z. anorg. u. allgem. Chem.*, **99**, 146 (1917).

KBr–BaBr₂

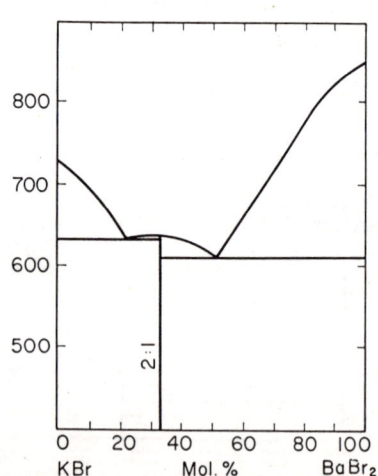

Fig. 1183.—System KBr–BaBr₂.

Georg Kellner, *Z. anorg. u. allgem. Chem.*, **99**, 175 (1917).

KBr–CaBr₂

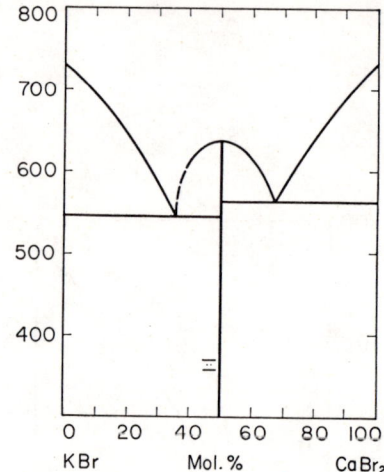

Fig. 1184.—System KBr–CaBr₂.

Georg Kellner, *Z. anorg. u. allgem. Chem.*, **99**, 169 (1917).

KBr–CdBr₂

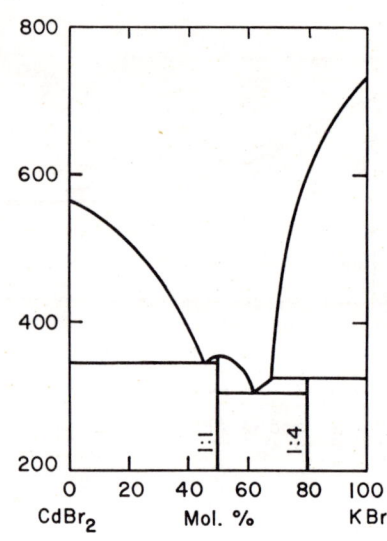

Fig. 1185.—System KBr–CdBr₂.

Hermann Brand, *Neues Jahrb. Mineral., Geol.*, **35**, 11 (1913I).

KBr–HgBr₂

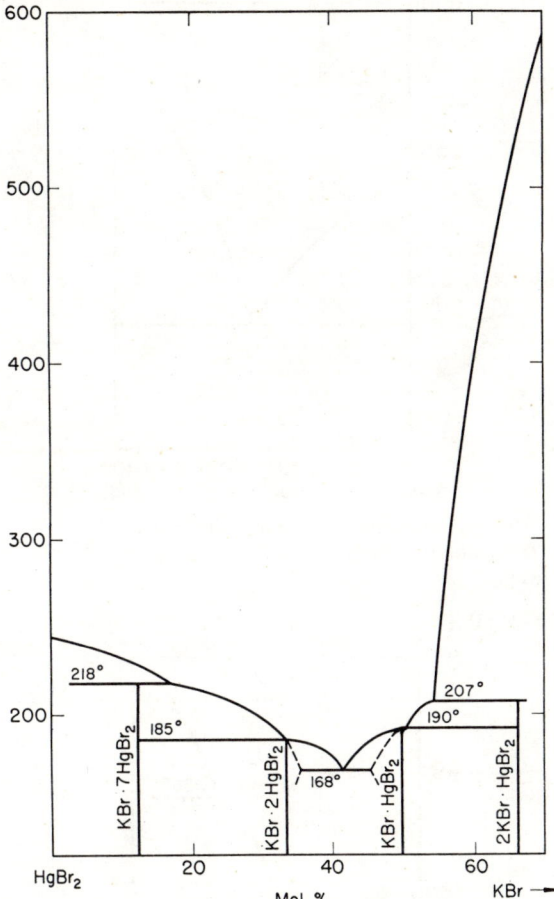

Fig. 1186.—System KBr–HgBr₂.

I. N. Belyaev and K. E. Mironov, *Zhur. Obshcheĭ Khim.*, **22**, 1491 (1952).

KBr–MgBr₂

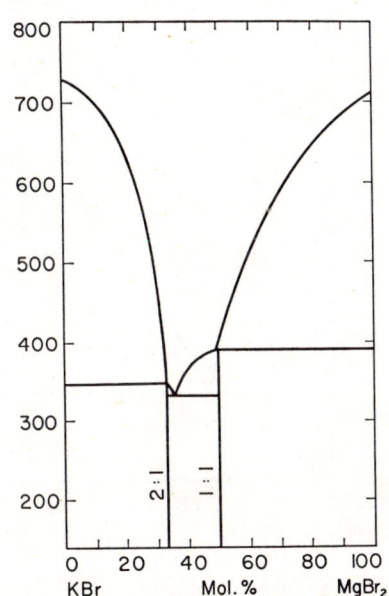

Fig. 1187.—System KBr–MgBr₂.

Georg Kellner, *Z. anorg. u. allgem. Chem.*, **99**, 166 (1917).

KBr–SrBr₂

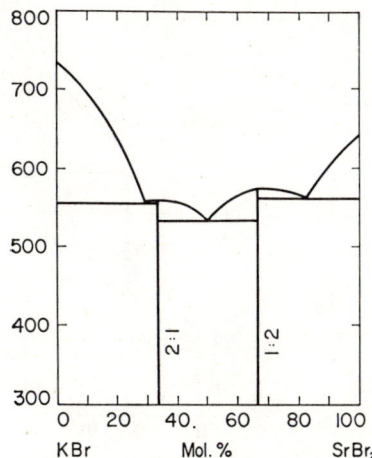

Fig. 1188.—System KBr–SrBr₂.

Georg Kellner, *Z. anorg. u. allgem. Chem.*, **99**, 172 (1917).

LiBr–NaBr

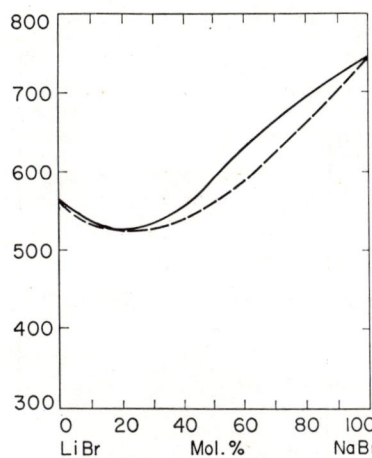

Fig. 1189.—System LiBr–NaBr.

Georg Kellner, *Z. anorg. u. allgem. Chem.*, **99**, 144 (1917).

LiBr–BaBr₂

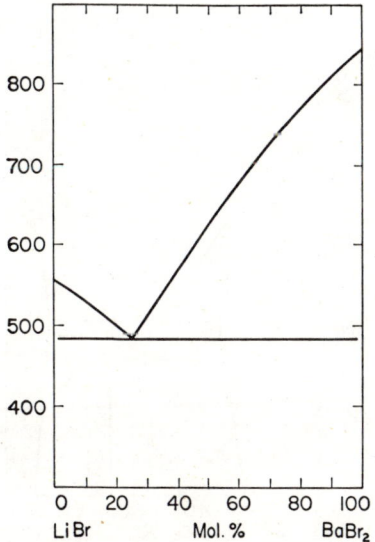

Fig. 1190.—System LiBr–BaBr₂.

Georg Kellner, *Z. anorg. u. allgem. Chem.*, **99**, 156 (1917).

LiBr–SrBr₂

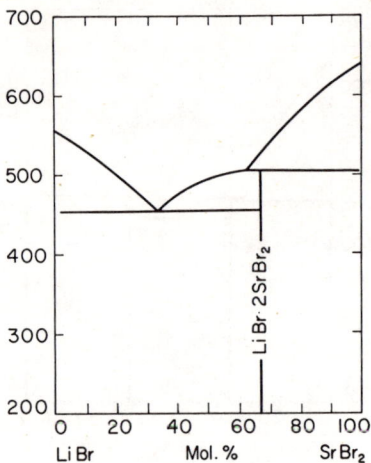

Fig. 1191.—System LiBr–SrBr₂.

Georg Kellner, *Z. anorg. u. allgem. Chem.*, **99**, 154 (1917).

NaBr–BaBr₂

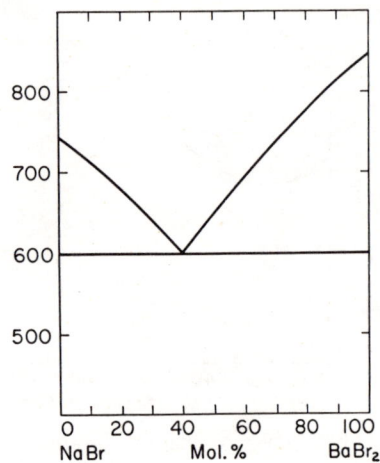

Fig. 1192.—System NaBr–BaBr₂.

Georg Kellner, *Z. anorg. u. allgem. Chem.*, **99**, 164 (1917).

NaBr–CaBr₂

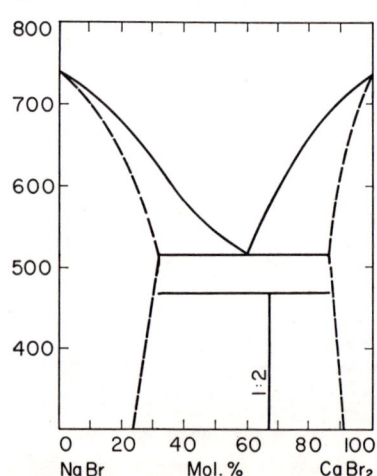

Fig. 1193.—System NaBr–CaBr₂.

Georg Kellner, *Z. anorg. u. allgem. Chem.*, **99**, 161 (1917).

NaBr–CdBr₂

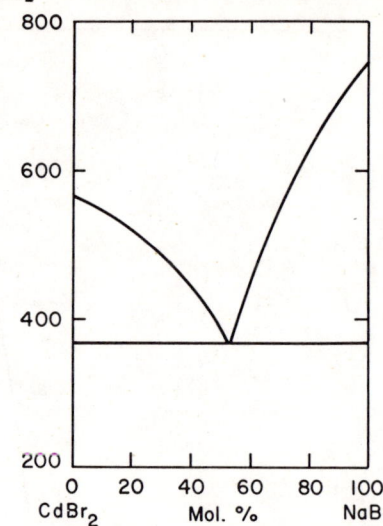

Fig. 1194.—System NaBr–CdBr₂.

Hermann Brand, *Neues Jahrb. Mineral., Geol.*, **35**, 15 (1913I).

NaBr–HgBr₂

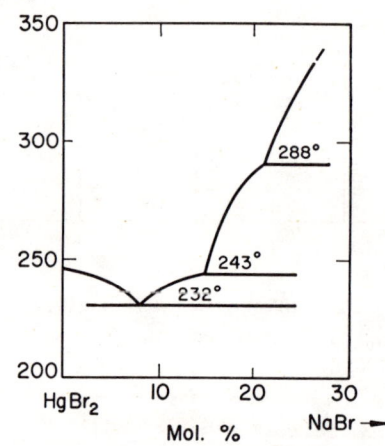

Fig. 1195.—System NaBr–HgBr₂.

I. N. Belyaev and K. E. Mironov, *Zhur. Obshcheĭ Khim.*, **22**, 1491 (1952).

NaBr–MgBr₂

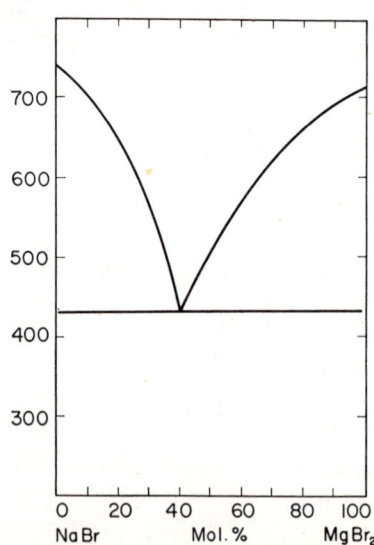

Fig. 1196.—System NaBr–MgBr₂.

Georg Kellner, *Z. anorg. u. allgem. Chem.*, **99**, 158 (1917).

NaBr–SrBr$_2$

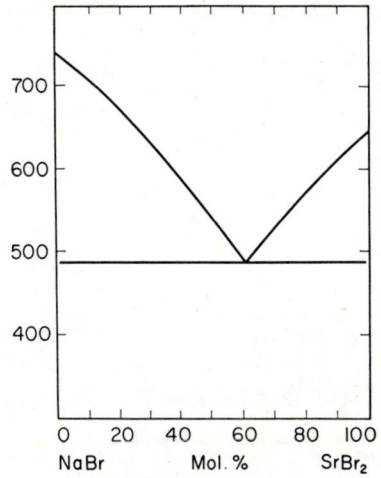

Fig. 1197.—System NaBr–SrBr$_2$.

Georg Kellner, *Z. anorg. u. allgem. Chem.*, **99**, 163 (1917).

NH$_4$Br–HgBr$_2$

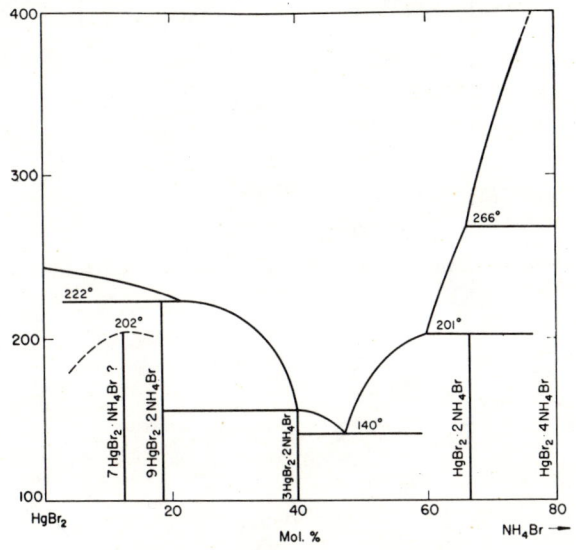

Fig. 1198.—System NH$_4$Br–HgBr$_2$.

I. N. Belyaev and K. E. Mironov, *Zhur. Obshcheĭ Khim.*, **22**, 1492 (1952).

CdBr$_2$–PbBr$_2$

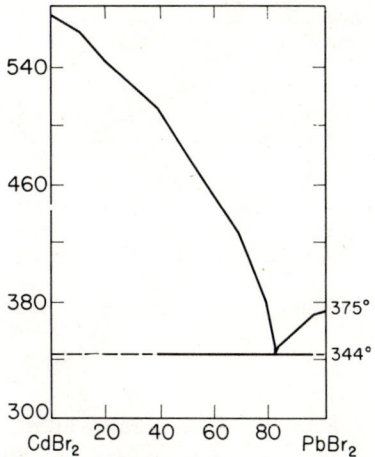

Fig. 1199.—System CdBr$_2$–PbBr$_2$.

G. H. Zakharchenko, *Zhur. Obshcheĭ Khim.*, **21**, 454 (1951).

CdBr$_2$–ZnBr$_2$

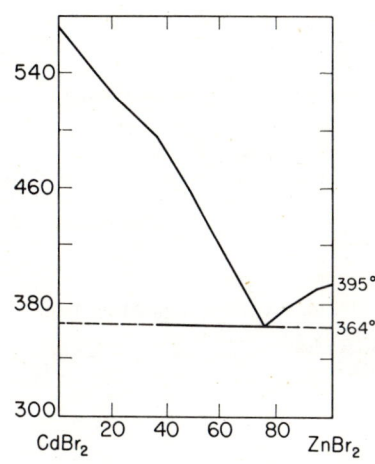

Fig. 1200.—System CdBr$_2$–ZnBr$_2$.

G. H. Zakharchenko, *Zhur. Obshcheĭ Khim.*, **21**, 454 (1951).

HgBr$_2$–PbBr$_2$

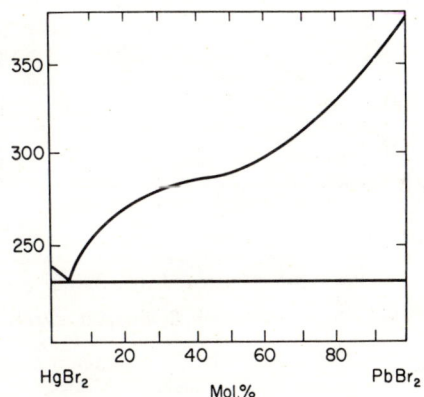

Fig. 1201.—System HgBr$_2$–PbBr$_2$.

Carlo Sandonnini, *Gazz. chim. ital.*, **44 I**, 369 (1914).

PbBr₂–BiBr₃

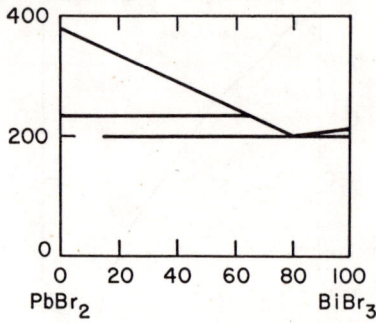

Fig. 1202.—System PbBr₂–BiBr₃.

Gottfried Herrmann, *Z. anorg. Chem.*, **71**, 288 (1911).

AlBr₃–BBr₃

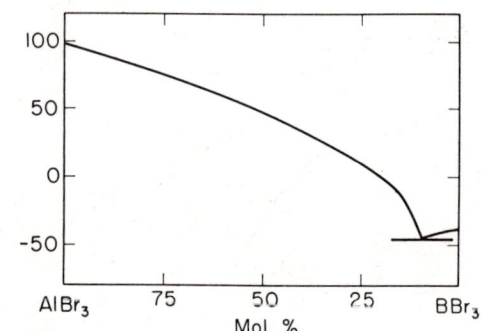

Fig. 1203.—System AlBr₃–BBr₃.

R. F. Adamsky and C. M. Wheeler, Jr., *J. Phys. Chem.*, **58**, 225 (1954).

AsBr₃–SbBr₃

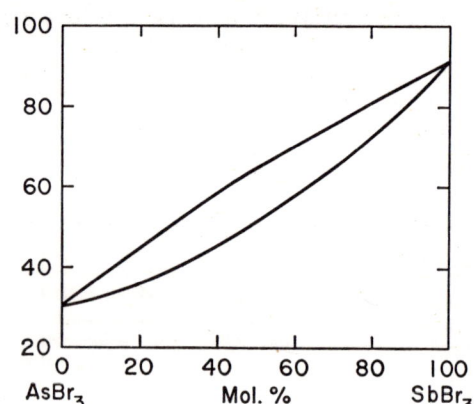

Fig. 1204.—System AsBr₃–SbBr₃.

N. A. Puschin and S. Löwy, *Z. anorg. u. allgem. Chem.*, **150**, 168 (1926).

BBr₃–SnBr₄

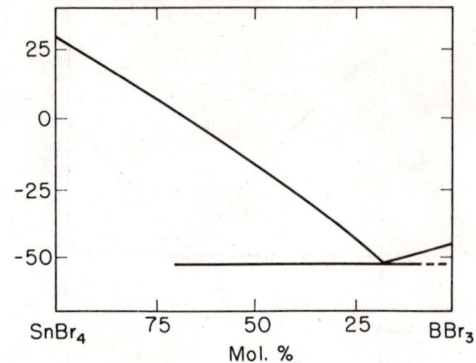

Fig. 1205.—System BBr₃–SnBr₄.

R. F. Adamsky and C. M. Wheeler, Jr., *J. Phys. Chem.*, **58**, 225 (1954).

BBr₃–AsBr₅

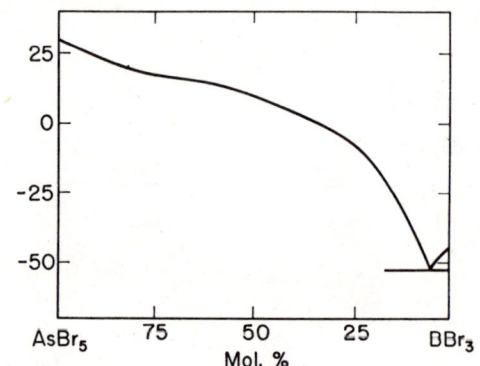

Fig. 1206.—System BBr₃–AsBr₅.

R. F. Adamsky and C. M. Wheeler, Jr., *J. Phys. Chem.*, **58**, 226 (1954).

(b) *Three Bromides*

KBr–NaBr–CdBr₂

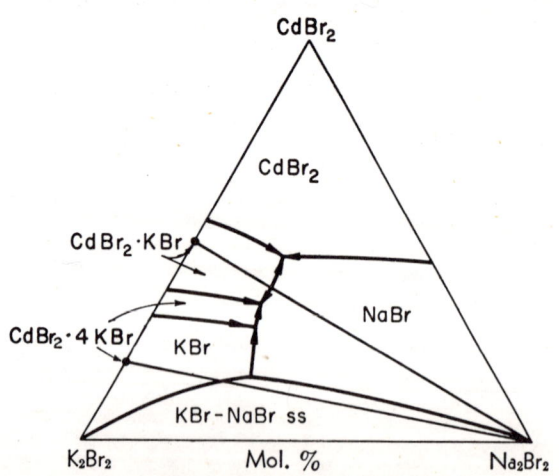

Fig. 1207.—System KBr–NaBr–CdBr₂.

Hermann Brand, *Neues Jahrb. Mineral., Geol.*, **35**, 17 (1913I).

KBr–NaBr–CdBr₂–(concl.)

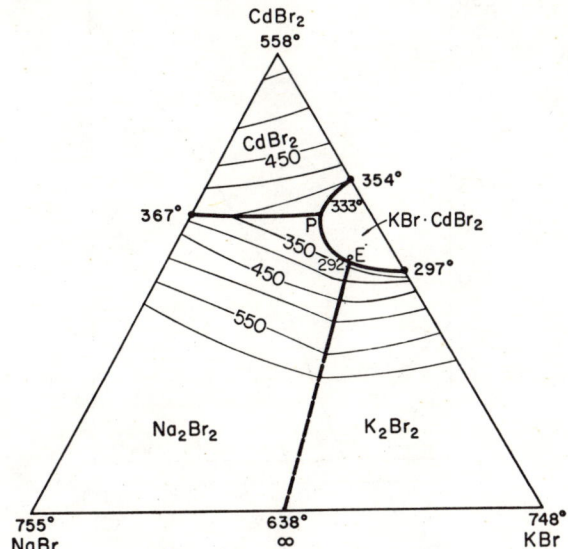

FIG. 1208.—System KBr–NaBr–CdBr₂.

I. I. Il'yasov, V. N. Mirsoyapov, and Yu. V. Korotkov, *Zhur. Neorg. Khim.*, **4**, 911 (1959).

NaBr–CdBr₂–PbBr₂

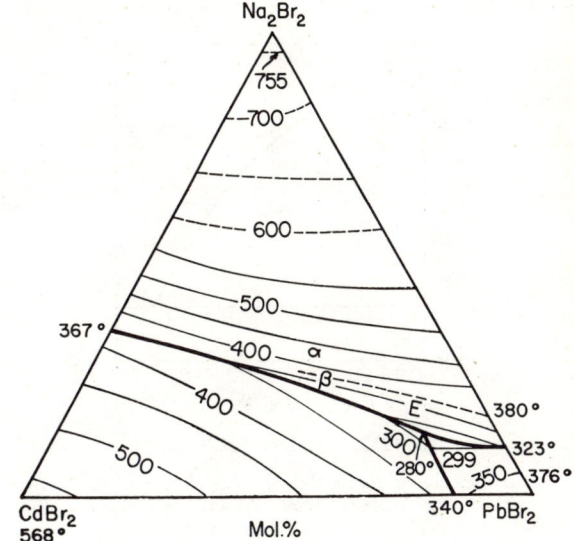

FIG. 1209.—System NaBr–CdBr₂–PbBr₂.

I. I. Il'yasov, G. G. Shchemeleva, and A. G. Bergman, *Zhur. Neorg. Khim.*, **4**, 908 (1959).

II. Chlorides Only

(a) One Chloride

RCl

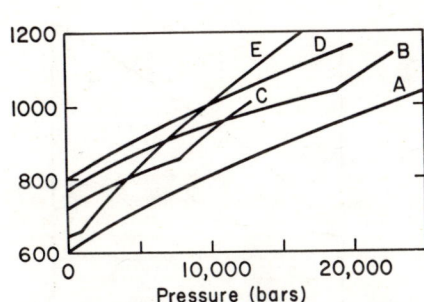

FIG. 1210.—System RCl; melting curves. A = LiCl, B = KCl, C = RbCl, D = NaCl, E = CsCl.

S. P. Clark, Jr., *J. Chem. Phys.*, **31** [6] 1527 (1959).

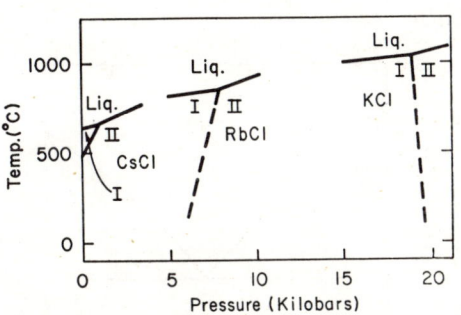

FIG. 1211.—System RCl, near the triple points.

S. P. Clark, Jr., *J. Chem. Phys.*, **31** [6] 1530 (1959).

(b) *Two Chlorides*

AgCl–BiCl

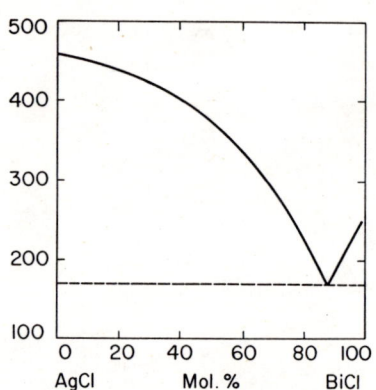

FIG. 1212.—System AgCl–BiCl.

M. A. Sokolova, *Izvest. Sektora Fiz. Khim. Anal., Inst. Obshchei Neorg. Khim., Akad. Nauk S.S.S.R.*, **21**, 163 (1952).

AgCl–CsCl

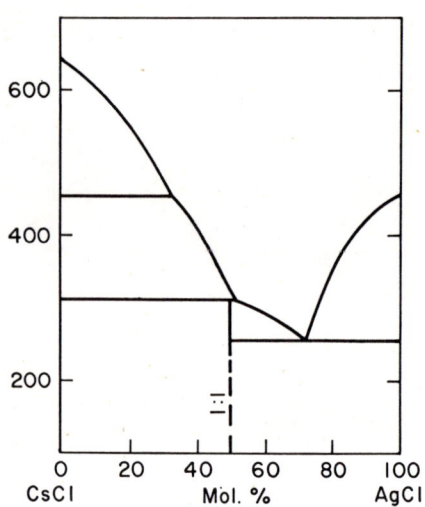

FIG. 1213.—System AgCl–CsCl.

M. Amadori, *Atti reale accad. Lincei, Sez. II*, **21**, 81 (1912).

AgCl–CuCl

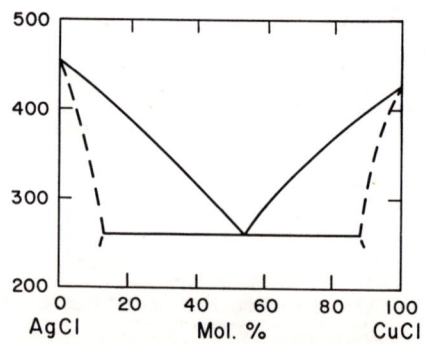

FIG. 1214.—System AgCl–CuCl.

Carlo Sandonnini, *Gazz. chim. ital.*, **44 I**, 325 (1914).

AgCl–KCl

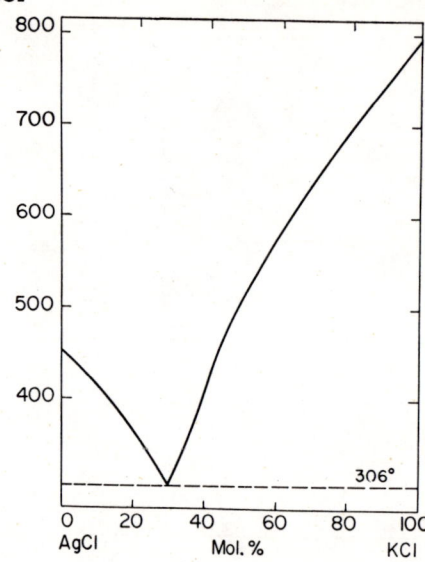

FIG. 1215.—System AgCl–KCl.

S. Zhemchuzhnui, *Z. anorg. Chem.*, **57**, 276 (1908).

AgCl–LiCl

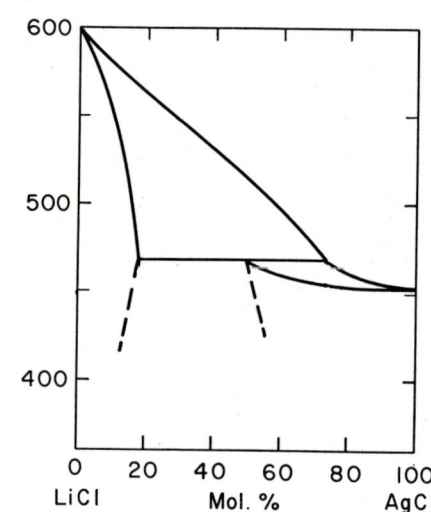

FIG. 1216.—System AgCl–LiCl.

Carlo Sandonnini, *Gazz. chim. ital.*, **44 I**, 299 (1914).

AgCl–NaCl

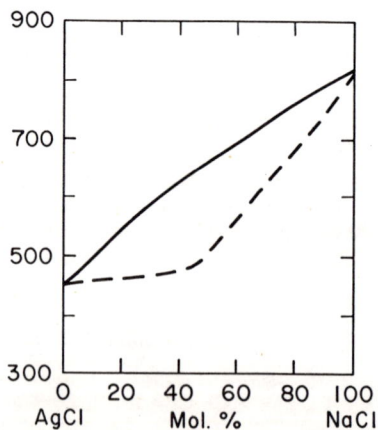

FIG. 1217.—System AgCl–NaCl.

S. Zhemchuzhnui, *Z. anorg. u. allgem. Chem.*, **153**, 53 (1926).

AgCl–RbCl

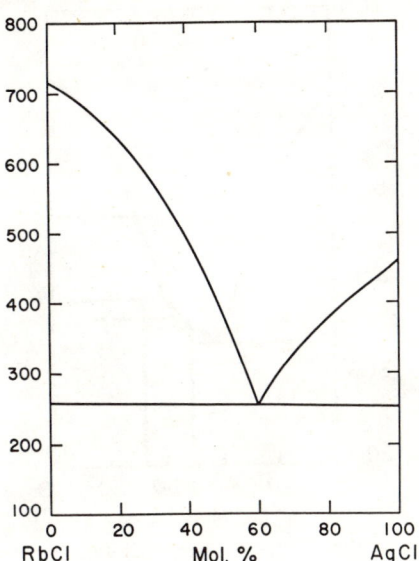

FIG. 1218.—System AgCl–RbCl.

Carlo Sandonnini, *Gazz. chim. ital.*, **44 I**, 315 (1914).

AgCl–TlCl

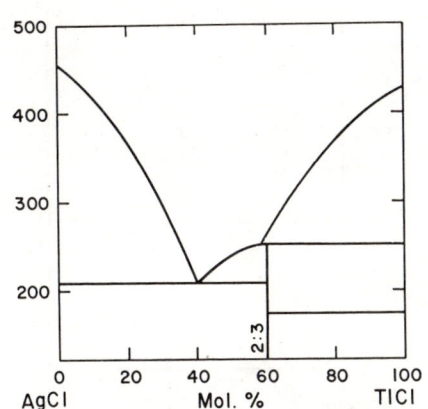

FIG. 1219.—System AgCl–TlCl.

Carlo Sandonnini, *Gazz. chim. ital.*, **44 I**, 328 (1914).

AgCl–BeCl₂

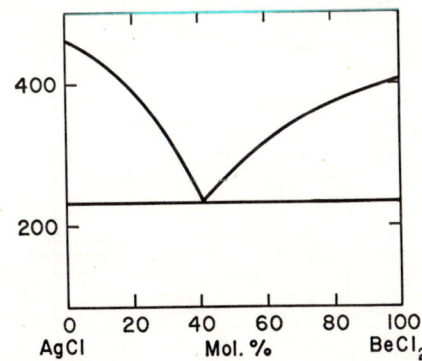

FIG. 1220.—System AgCl–BeCl₂.

M. J. M. Schmidt, *Bull. soc. chim., France*, **39**, 1695 (1926).

AgCl–CaCl₂

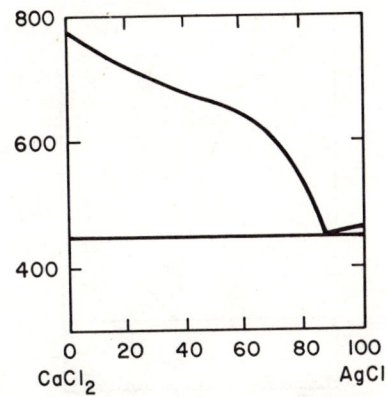

FIG. 1221.—System AgCl–CaCl₂.

Otto Menge, *Z. anorg. Chem.*, **72**, 204 (1911).

AgCl–MgCl₂

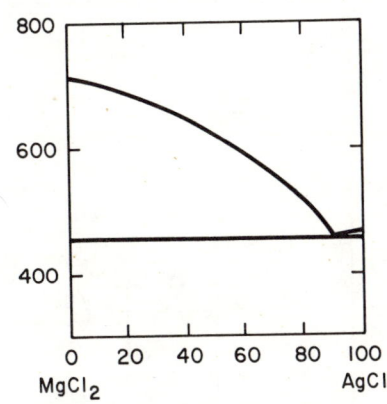

FIG. 1222.—System AgCl–MgCl₂.

Otto Menge, *Z. anorg. Chem.*, **72**, 184 (1911).

AgCl–PbCl₂

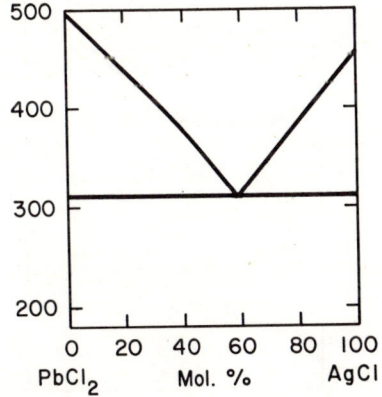

FIG. 1223.—System AgCl–PbCl₂.

Kurt Treis, *Neues Jahrb. Mineral., Geol.*, **37**, 775 (1914).

AgCl–ZnCl$_2$

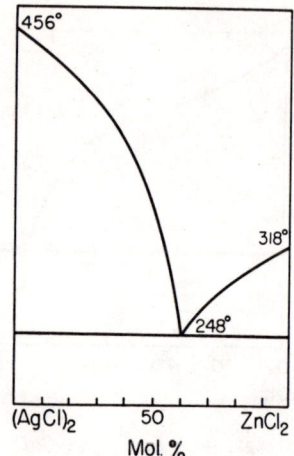

Fig. 1224.—System AgCl–ZnCl$_2$.

A. P. Palkin and N. A. Shchirova, *Zhur. Neorg. Khim.*, 1, 2156 (1956).

CsCl–CuCl

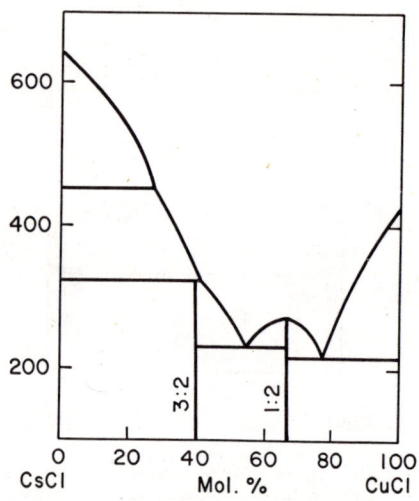

Fig. 1225.—System CsCl–CuCl.

M. Amadori, *Atti reale accad. Lincei, Sez. II*, 21, 79 (1912).

CsCl–KCl

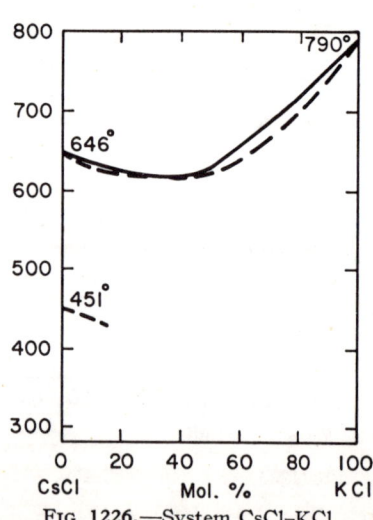

Fig. 1226.—System CsCl–KCl.

S. Zhemchuzhnui and F. Rambach, *Z. anorg. Chem.*, 65, 419 (1910).

CsCl–LiCl

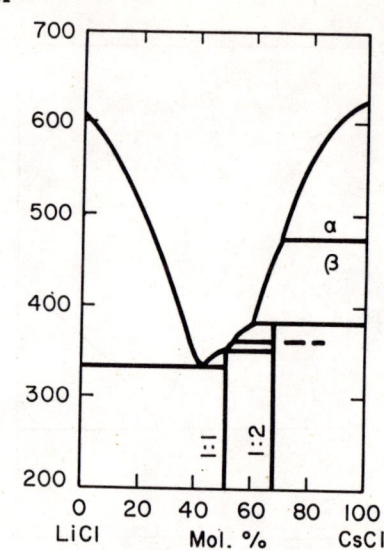

Fig. 1227.—System CsCl–LiCl.

E. Korreng, *Z. anorg. Chem.*, 91, 200 (1915).

CsCl–NaCl

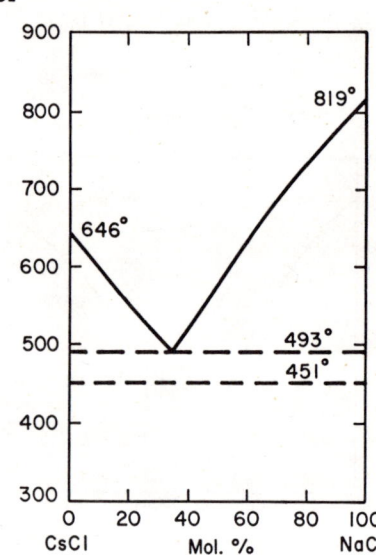

Fig. 1228.—System CsCl–NaCl.

S. Zhemchuzhnui and F. Rambach, *Z. anorg. Chem.*, 65, 421 (1910).

CsCl–RbCl

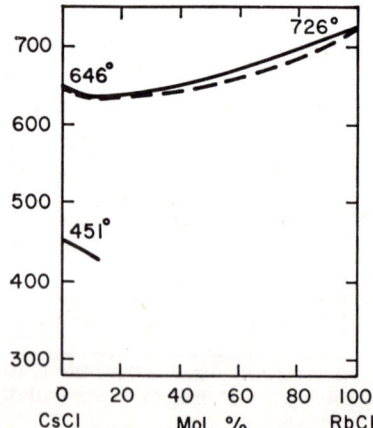

Fig. 1229.—System CsCl–RbCl.

S. Zhemchuzhnui and F. Rambach, *Z. anorg. Chem.*, 65, 417 (1910).

CsCl–TlCl

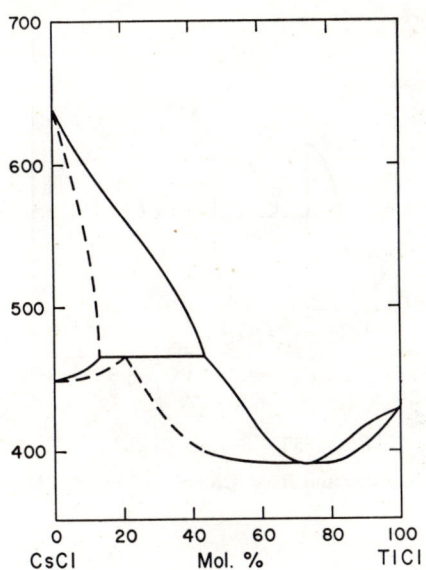

Fig. 1230.—System CsCl–TlCl.

Carlo Sandonnini, *Gazz. chim. ital.*, **44 I**, 323 (1914).

CsCl–MgCl$_2$

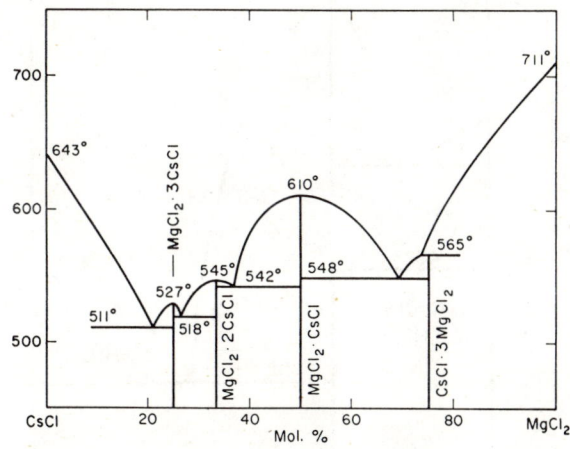

Fig. 1231.—System CsCl–MgCl$_2$.

B. F. Markov and I. D. Panchenko, *Zhur. Obshcheĭ Khim.*, **25**, 2042 (1055).

CsCl–VCl$_2$

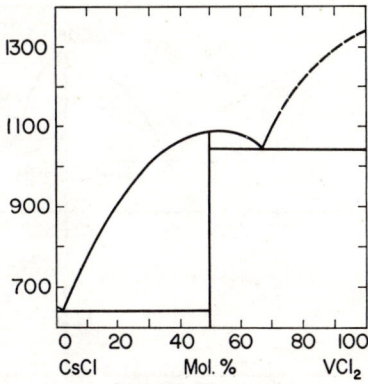

Fig. 1232.—System CsCl–VCl$_2$.

H. J. Seifert and Paul Ehrlich, *Z. anorg. u. allgem. Chem.*, **302**, 287 (1959).

CsCl–ZnCl$_2$

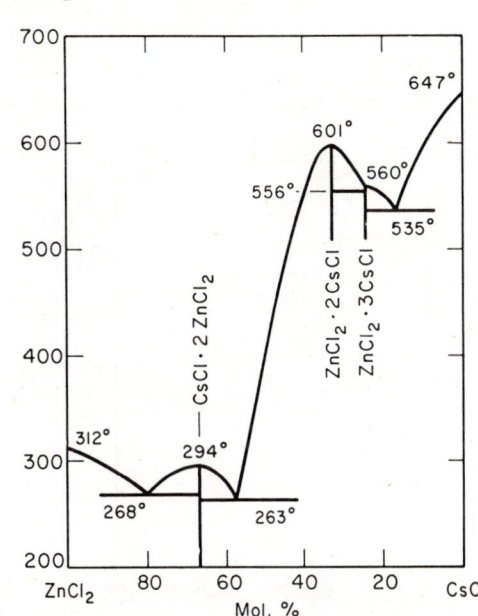

Fig. 1233.—System CsCl–ZnCl$_2$.

B. F. Markov, I. D. Panchenko, and T. G. Kostenko, *Ukrain. Khim. Zhur.*, **22**, 290 (1956) (in Russian).

CsCl–AlCl$_3$

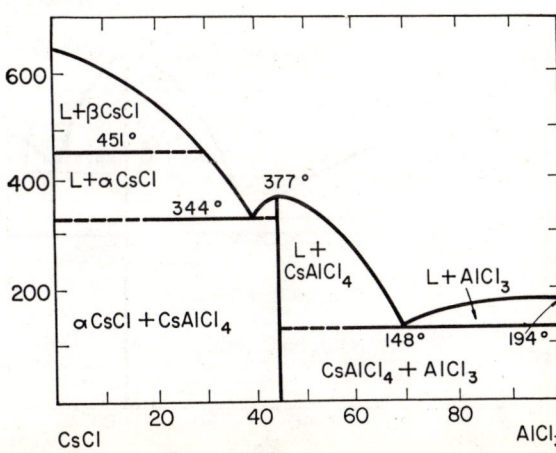

Fig. 1234.—System CsCl–AlCl$_3$.

I. S. Morozov and A. T. Simonich, *Zhur. Neorg. Khim.*, **2**, 1908 (1957).

CsCl–CeCl₃

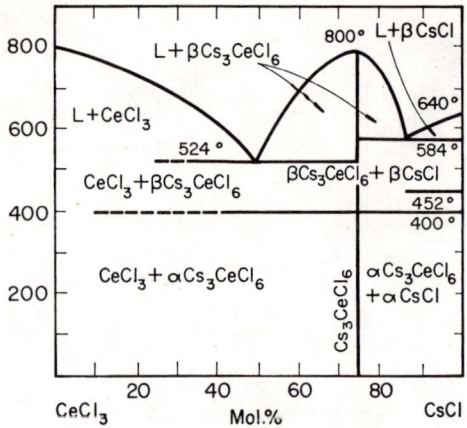

Fig. 1235.—System CsCl–CeCl₃.

In-Chzhu Sun and I. S. Morozov, *Zhur. Neorg. Khim.*, **3**, 1916 (1958).

CsCl–TiCl₃

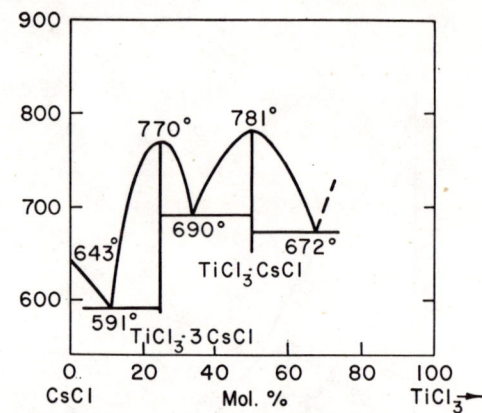

Fig. 1238.—System CsCl–TiCl₃.

B. F. Markov and R. V. Chernov, *Ukrain. Khim. Zhur.*, **25**, 283 (1959).

CsCl–LaCl₃

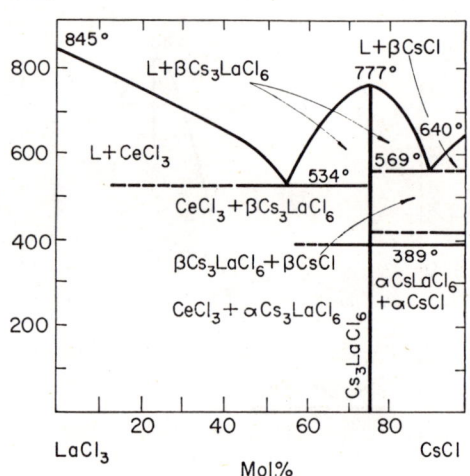

Fig. 1236.—System CsCl–LaCl₃.

In-Chzhu Sun and I. S. Morozov, *Zhur. Neorg. Khim.*, **3**, 1916 (1958).

CsCl–HfCl₄

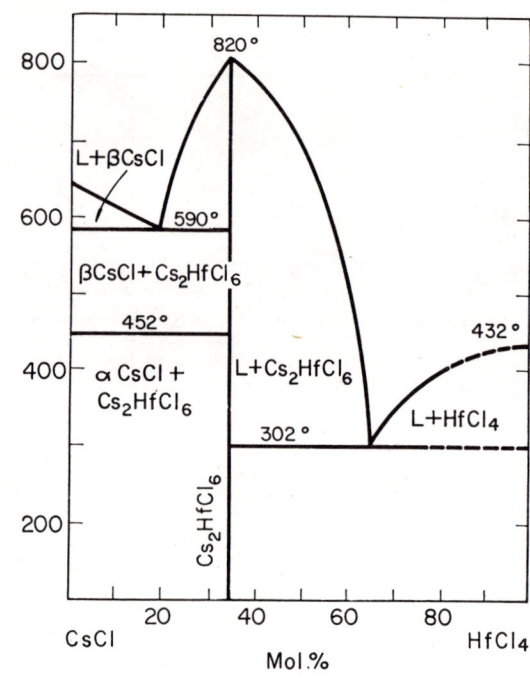

Fig. 1239.—System CsCl–HfCl₄.

I. S. Morozov and In-Chzhu Sun, *Zhur. Neorg. Khim.*, **4**, 681 (1959).

CsCl–NdCl₃

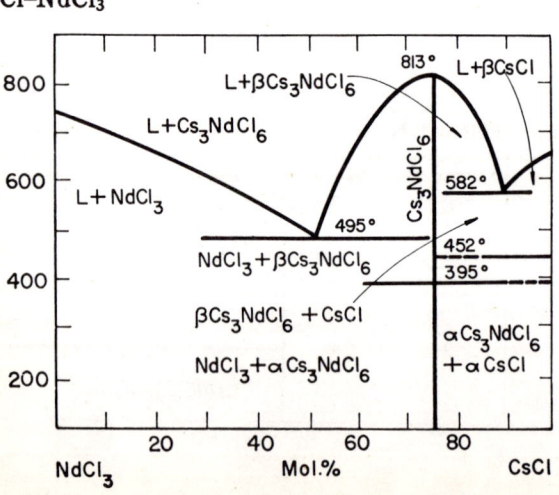

Fig. 1237.—System CsCl–NdCl₃.

In-Chzhu Sun and I. S. Morozov, *Zhur. Neorg. Khim.*, **3**, 1916 (1958).

CsCl–UCl₄

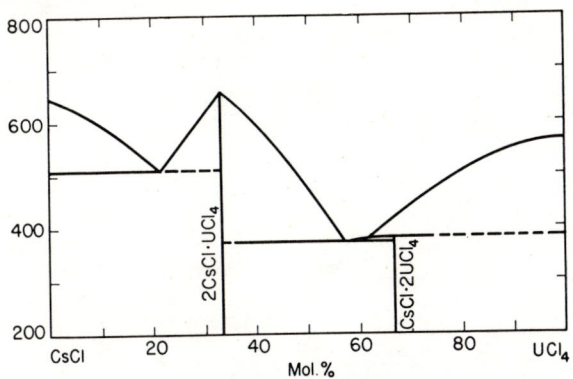

Fig. 1240.—System CsCl–UCl₄, tentative.

C. J. Barton, A. B. Wilkerson, and W. R. Grimes, Oak Ridge National Laboratory, Phase Diagrams of Nuclear Reactor Materials, R. F. Thoma, ed., ORNL-2548, p. 137 (1959).

CsCl–ZrCl₄

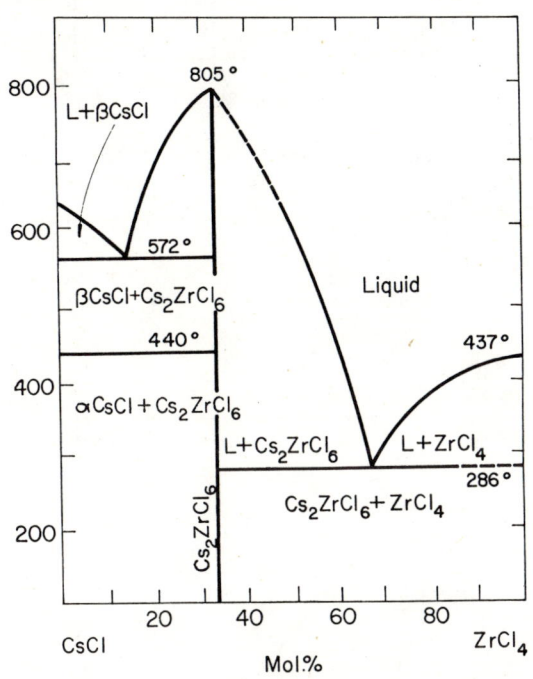

Fig. 1241.—System CsCl–ZrCl₄.

I. S. Morozov and In-Chzhu Sun, *Zhur. Neorg. Khim.*, **4**, 680 (1959).

CsCl–TaCl₅

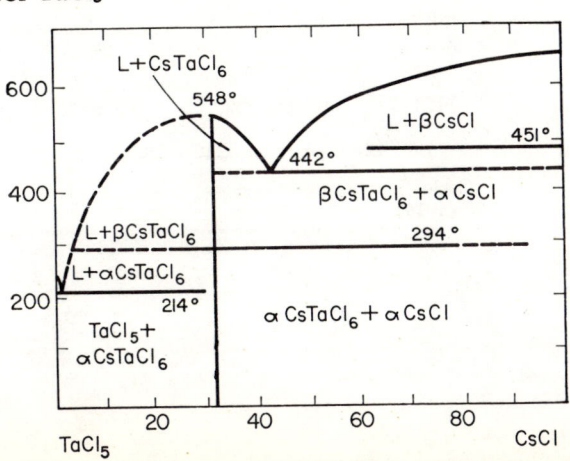

Fig. 1242.—System CsCl–TaCl₅.

I. S. Morozov and A. T. Simonich, *Zhur. Neorg. Khim.*, **2**, 1908 (1957).

CuCl–KCl

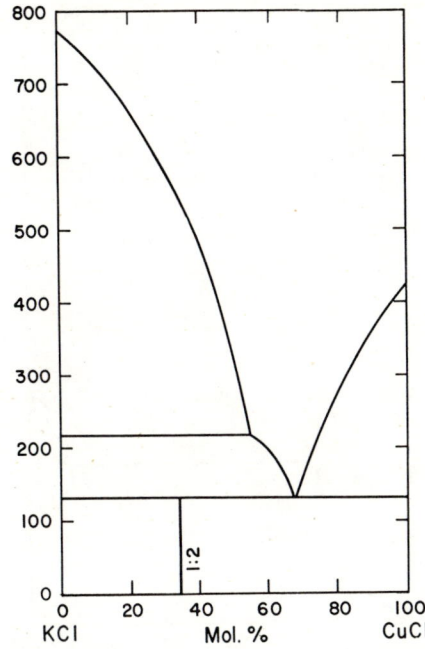

Fig. 1243.—System CuCl–KCl.

Carlo Sandonnini, *Gazz. chim. ital.*, **44 I**, 307 (1914). See also G. Poma and G. Gabbi, *Atti reale accad. Lincei*, **20 I**, 469 (1911).

CuCl–LiCl

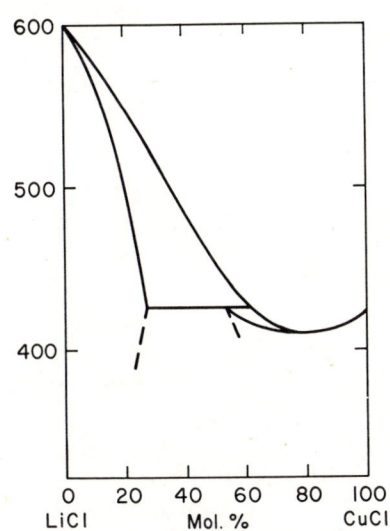

Fig. 1244.—System CuCl–LiCl.

Carlo Sandonnini, *Gazz. chim. ital.*, **44 I**, 297 (1914).

CuCl–NaCl

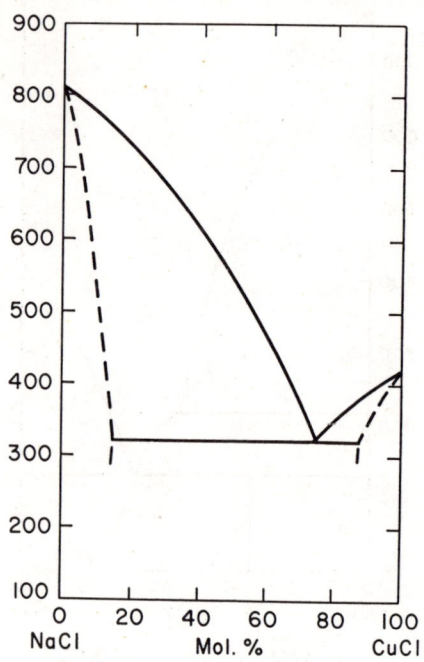

Fig. 1245.—System CuCl–NaCl.

Carlo Sandonnini, *Gazz. chim. ital.* **44 I**, 302 (1914).

CuCl–RbCl

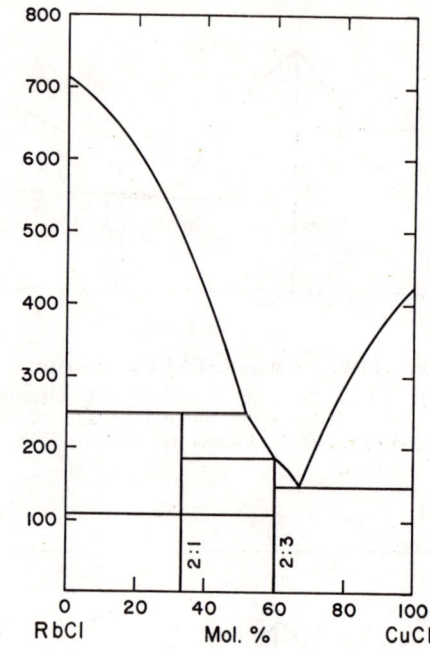

Fig. 1247.—System CuCl–RbCl.

Carlo Sandonnini, *Gazz. chim. ital.*, **44 I**, 311 (1914).

CuCl–NH$_4$Cl

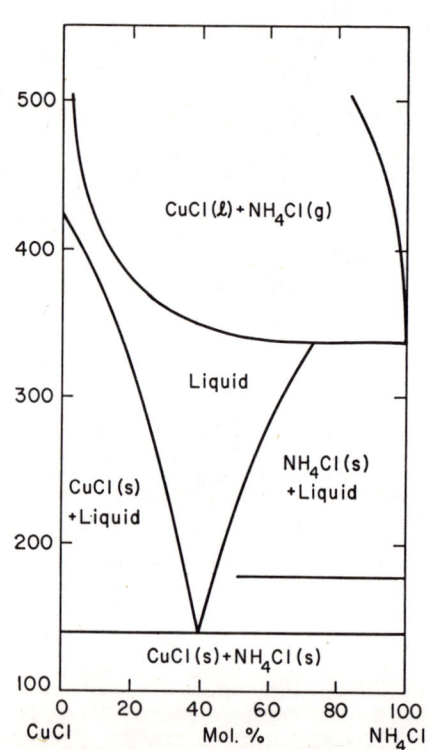

Fig. 1246.—System CuCl–NH$_4$Cl.

Kurt Hachmeister, *Z. anorg. u. allgem. Chem.*, **109**, 165 (1919).

CuCl–TlCl

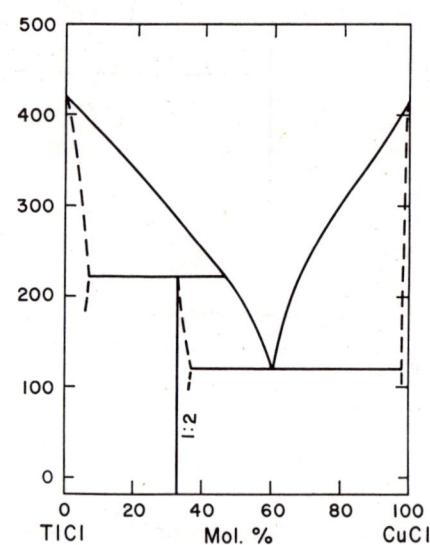

Fig. 1248.—System CuCl–TlCl.

Carlo Sandonnini, *Gazz. chim. ital.*, **44 I**, 327 (1914).

CuCl–CaCl$_2$

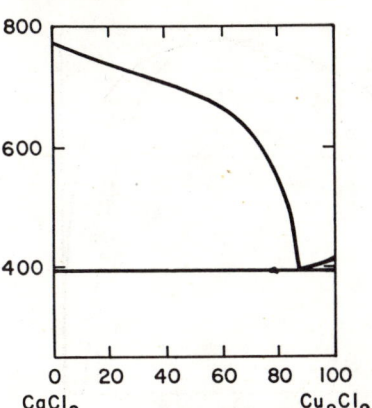

FIG. 1249.—System CaCl$_2$–CuCl.

Otto Menge, *Z. anorg. Chem.*, **72**, 209 (1911).

CuCl–CdCl$_2$

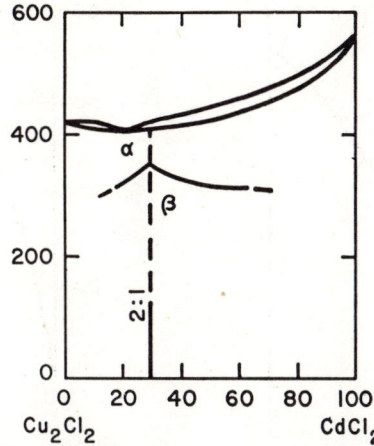

FIG. 1250.—System CuCl–CdCl$_2$.

Gottfried Herrmann, *Z. anorg. Chem.* **71**, 289 (1911).

CuCl–MgCl$_2$

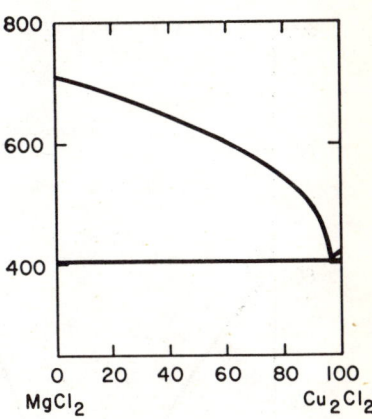

FIG. 1251.—System CuCl–MgCl$_2$.

Otto Menge, *Z. anorg. Chem.*, **72**, 189 (1911).

CuCl–PbCl$_2$

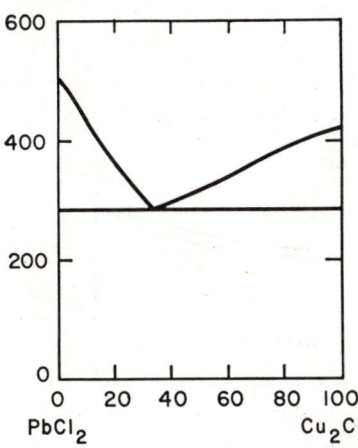

FIG. 1252.—System CuCl–PbCl$_2$.

Gottfried Herrmann, *Z. anorg. Chem.* **71**, 262 (1911).

CuCl–SnCl$_2$

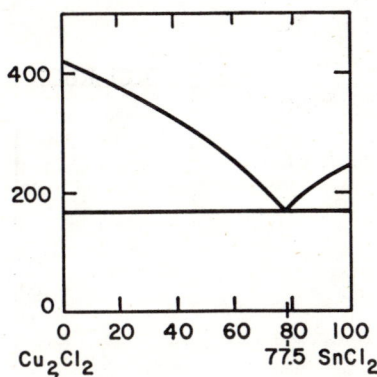

FIG. 1253.—System CuCl–SnCl$_2$.

Gottfried Herrmann, *Z. anorg. Chem.*, **71**, 269 (1911).

CuCl–ZnCl$_2$

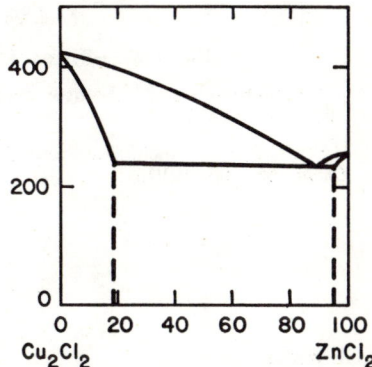

FIG. 1254.—System CuCl–ZnCl$_2$.

Gottfried Herrmann, *Z. anorg. Chem.*, **71**, 270 (1911).

CuCl–BiCl$_3$

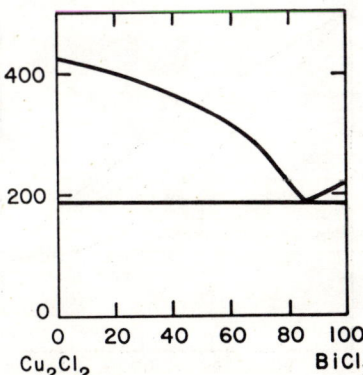

FIG. 1255.—System CuCl–BiCl$_3$.

Gottfried Herrmann, *Z. anorg. Chem.*, **71**, 272 (1911).

CuCl–FeCl$_3$

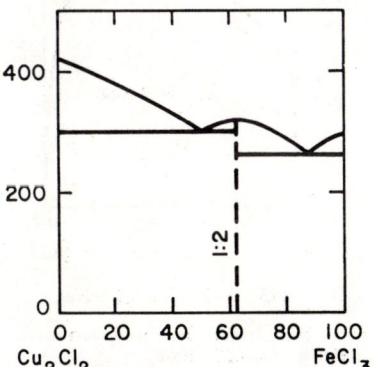

FIG. 1256.—System CuCl–FeCl$_3$.

Gottfried Herrmann, *Z. anorg. Chem.*, **71**, 273 (1911).

KCl–LiCl

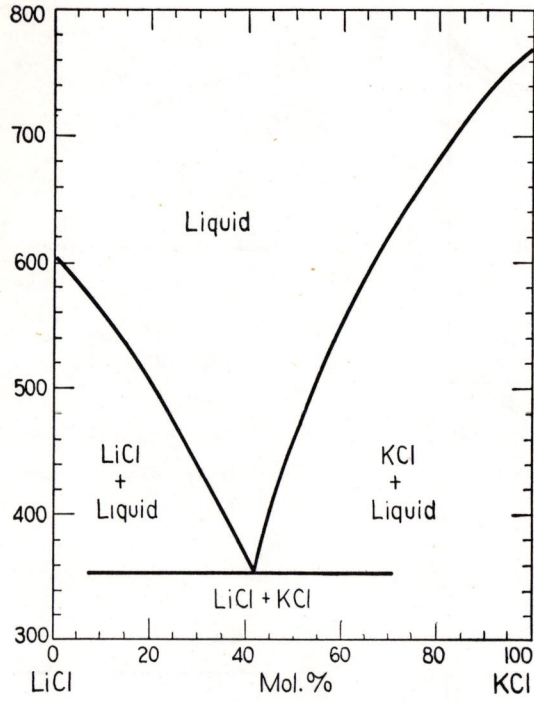

FIG. 1257.—System KCl–LiCl.

E. Elchardus and P. Laffitte, *Bull. soc. chim.*, *France*, **51**, 1572 (1932).

See also S. Zhemchuzhnui and F. Rambach, *Z. anorg Chem.*, **65**, 406 (1910).

KCl–NaCl

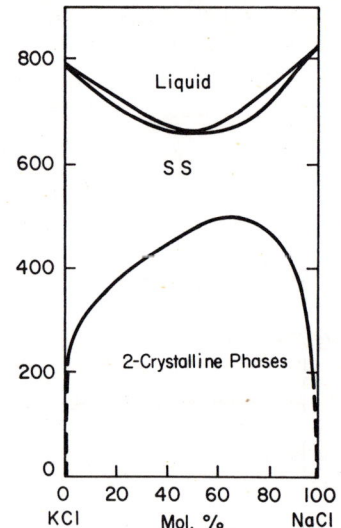

FIG. 1258.—System KCl–NaCl.

E. Scheil and H. Stadelmaier, *Z. Metallk.*, **43**, 227 (1952).

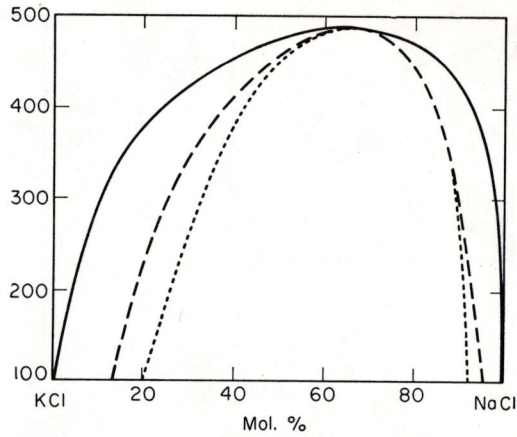

FIG. 1259.—System KCl–NaCl; solubility gap.

Solid line = experimental data; dashed line = spinodal curve; dotted curve = spinodal curve according to Erich Scheil and Hans Stadelmaier, *Z. Metallk.*, **43**, 227 (1952). The limits of metastable equilibrium are defined by the solubility curve and the spinodal.

A. J. H. Bunk and G. W. Tichelaar, *Koninkl. Ned. Akad. Wetenschap. Proc. Ser. B*, **56**, 378 (1953).

KCl–RbCl

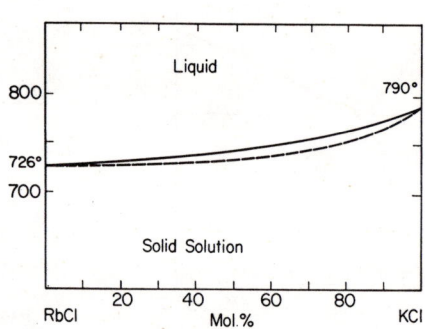

FIG. 1260.—System KCl–RbCl.

S. Zhemchuzhnui and F. Rambach, *Z. anorg. Chem.*, **65**, 412 (1910).

KCl–TlCl

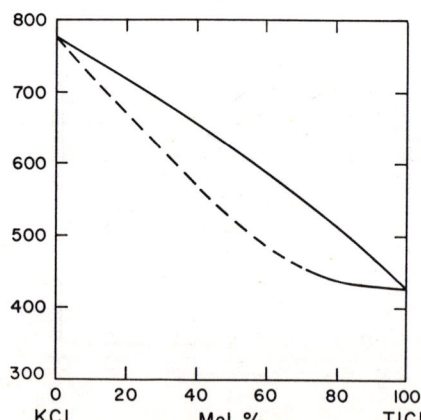

FIG. 1261.—System KCl–TlCl.

Carlo Sandonnini, *Gazz. chim. ital.*, **44 I**, 309 (1914).

KCl–BaCl₂

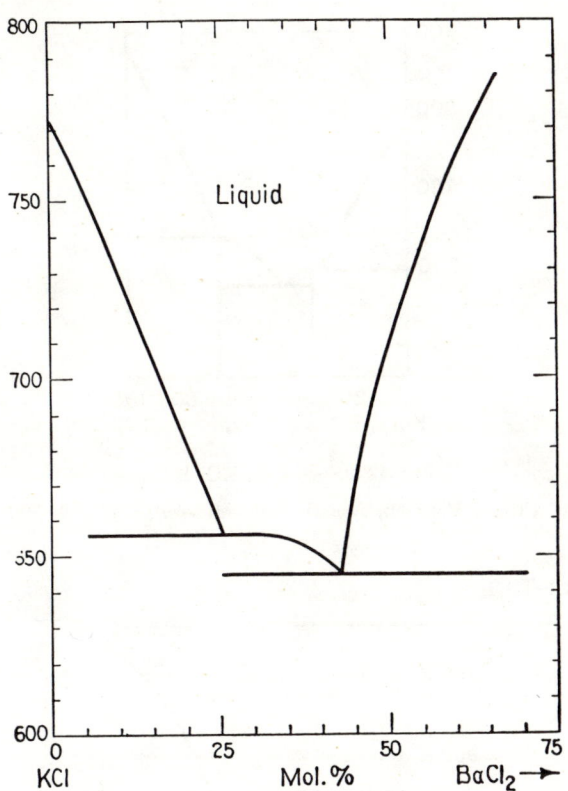

Fig. 1262.—System KCl–BaCl₂.

E. Elchardus and P. Laffitte, *Bull. soc. chim., France*, **51**, 1577 (1932).

KCl–CaCl₂

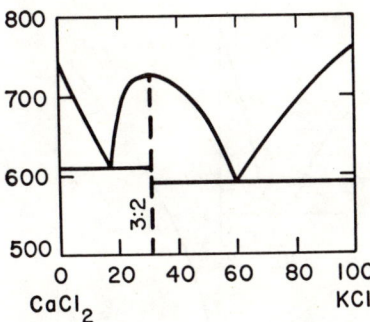

Fig. 1263.—System KCl–CaCl₂.

F. C. A. H. Lantsberry and R. A. Page, *J. Soc. Chem. Ind.*, **39**, 38T (1920).

KCl–CdCl₂

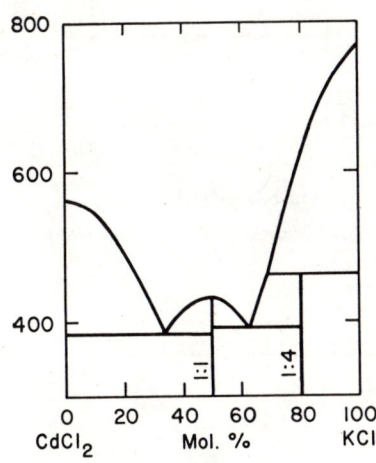

Fig. 1264.—System KCl–CdCl₂.

Hermann Brand, *Neues Jahrb. Mineral., Geol.*, **32**, 630 (1911).

KCl–FeCl₂

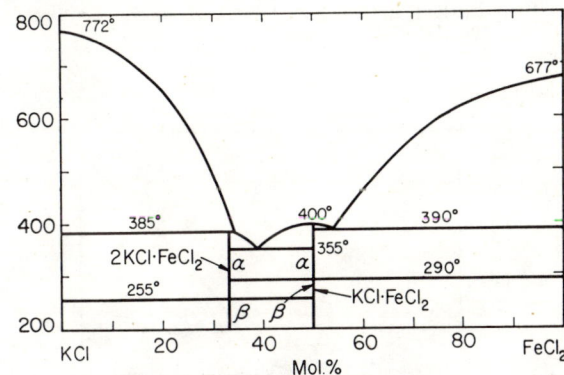

Fig. 1265.—System KCl–FeCl₂.

C. Beusman, Doctors thesis, University of Cincinnati (1957); see also, H. L. Pinch and J. M. Hirshon, *J. Am. Chem. Soc.*, **79**, 6149 (1957).

KCl–HgCl₂

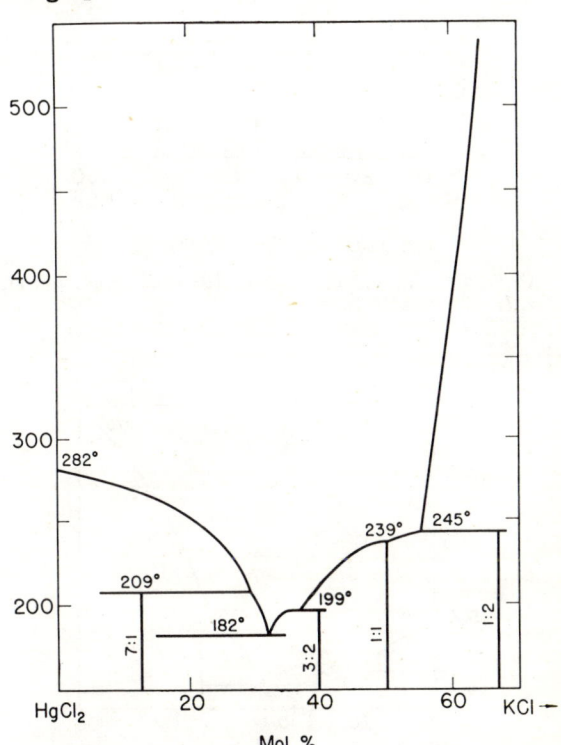

Fig. 1266.—System KCl–HgCl₂.

I. N. Belyaev and K. E. Mironov, *Zhur. Obshcheĭ Khim.*, **22**, 1486 (1952).

KCl–MgCl₂

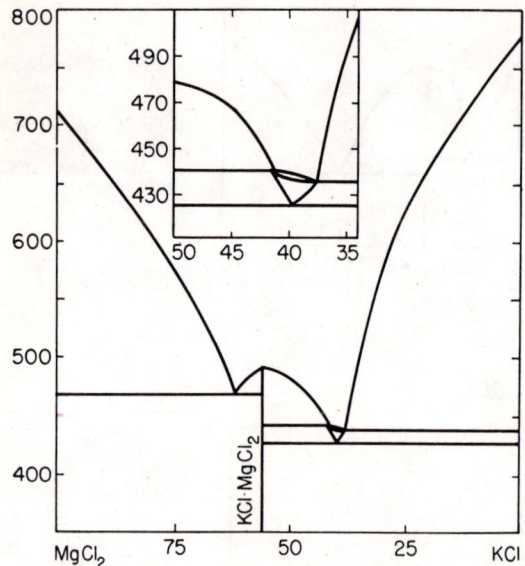

FIG. 1267.—System KCl–MgCl₂.

A. I. Ivanov, *Sbornik Statei Obshchei Khim., Akad. Nauk S.S.S.R.*, **1**, 758 (1953).

KCl–MnCl₂

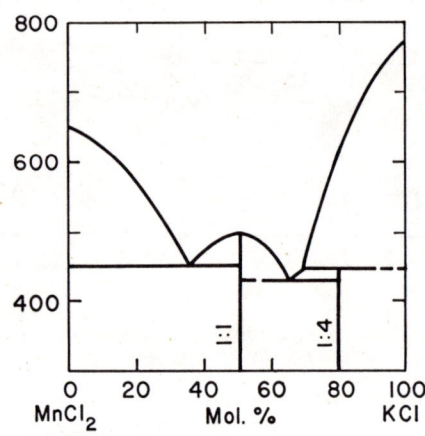

FIG. 1268.—System KCl–MnCl₂.

C. Sandonnini and G. Scarpa, *Atti reale accad. Lincei, Sez. II*, **22**, 167 (1913).

KCl–PbCl₂

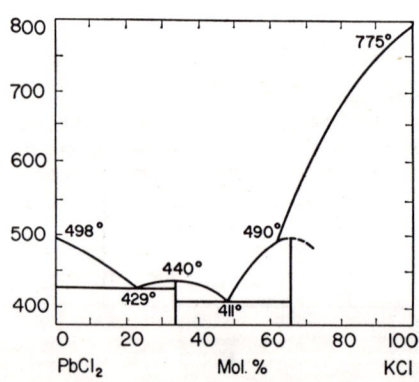

FIG. 1269.—System KCl–PbCl₂.

Ya A. Ugai and V. A. Shatillo, *J. Phys. Chem. U.S.S.R.*, **23** [6] 745 (1949).

KCl–SrCl₂

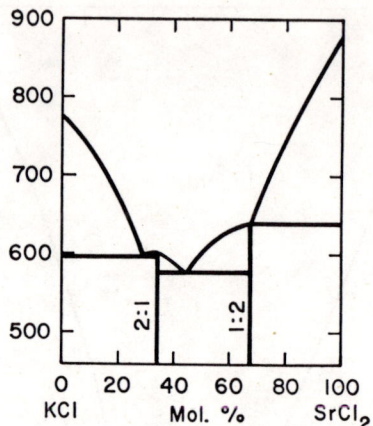

FIG. 1270.—System KCl–SrCl₂.

Erhard Vortisch, *Neues Jahrb. Mineral., Geol.*, **38**, 220 (1914).

KCl–TiCl₂

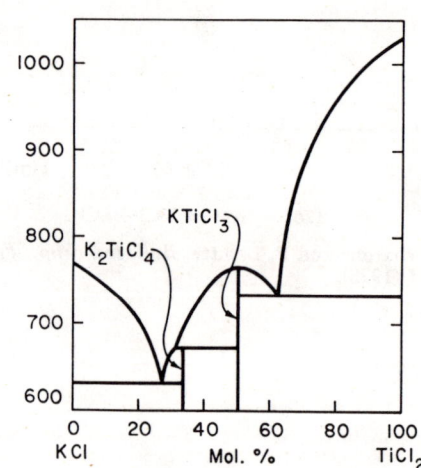

FIG. 1271.—System KCl–TiCl₂.

Paul Ehrlich and Hubert Kühnl, *Z. anorg. u. allgem. Chem.*, **292**, 148 (1957).

KCl–VCl₂

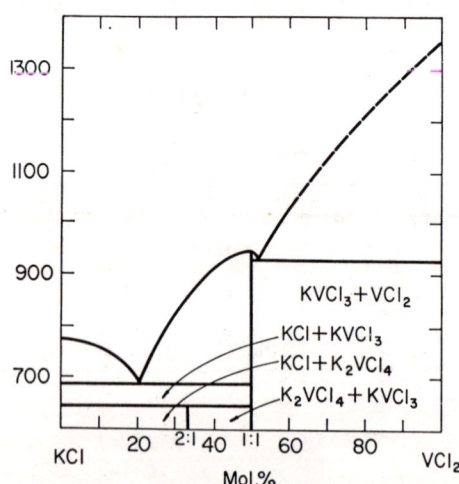

FIG. 1272.—System KCl–VCl₂.

H. J. Seifert and Paul Ehrlich, *Z. anorg. u. allgem. Chem.*, **302**, 286 (1959).

KCl–ZnCl₂

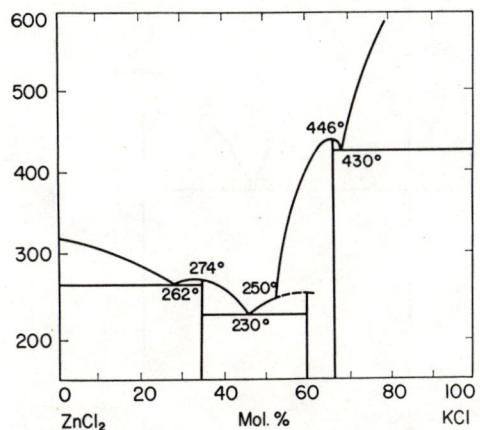

FIG. 1273.—System KCl–ZnCl₂.
Ya A. Ugai and V. A. Shatillo, *J. Phys. Chem. U.S.S.R.*, **23** [6] 745 (1949). See also F. R. Duke and R. A. Fleming, *J. Electrochem. Soc.*, **104**, 253 (1957).

KCl–AlCl₃

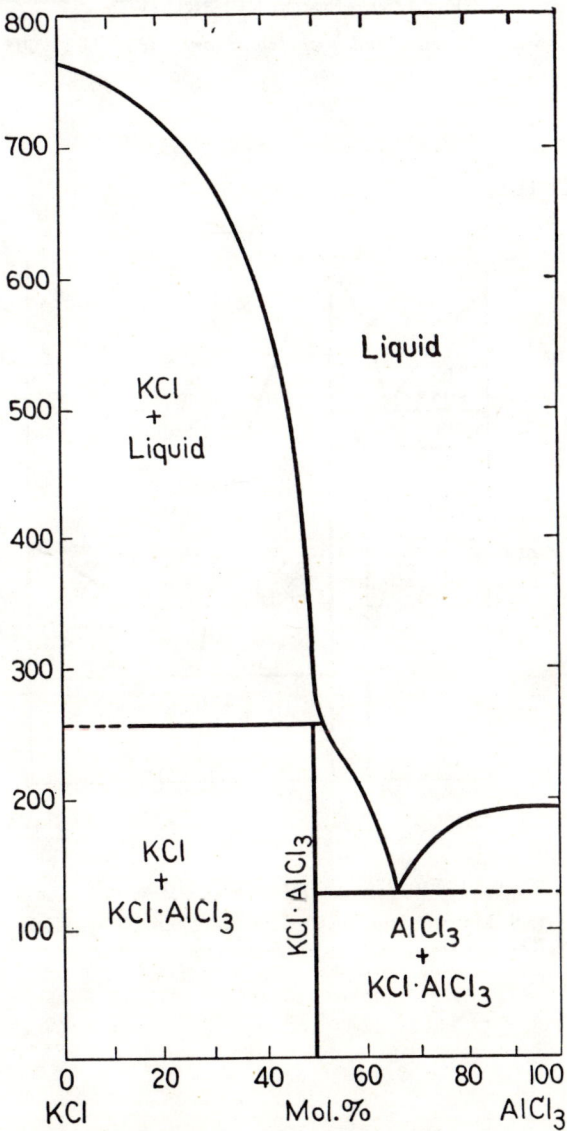

FIG. 1274 System KCl–AlCl₃.
U. I. Shvartsman, *J. Phys. Chem. (U.S.S.R.)*, **14**, 254 (1940).

KCl–CeCl₃

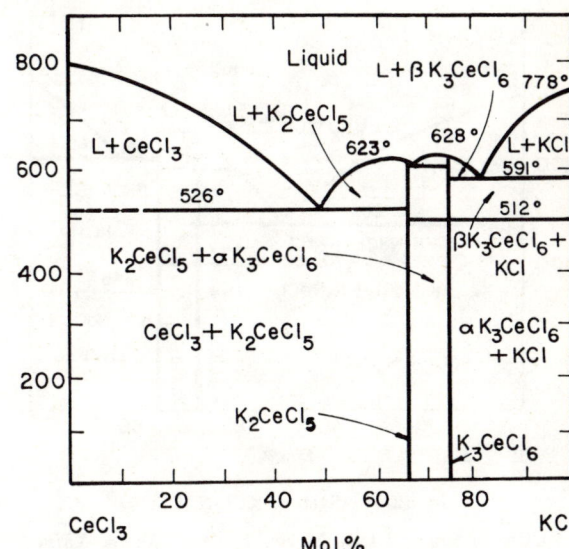

FIG. 1275.—System KCl–CeCl₃.
In-Chzhu Sun and I. S. Morozov, *Zhur. Neorg. Khim.*, **3**, 1916 (1958).

KCl–LaCl₃

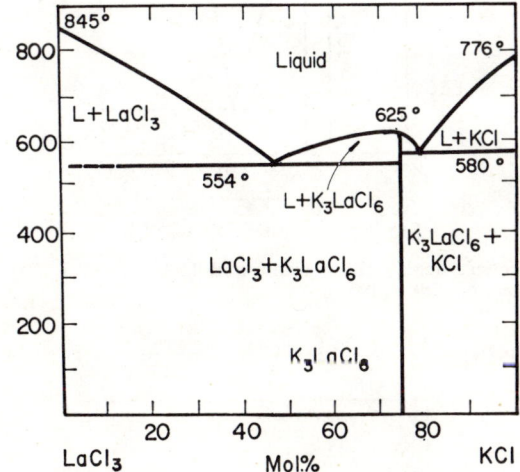

FIG. 1276.—System KCl–LaCl₃.
In-Chzhu Sun and I. S. Morozov, *Zhur. Neorg. Khim.*, **3**, 1916 (1958).

KCl–NdCl₃

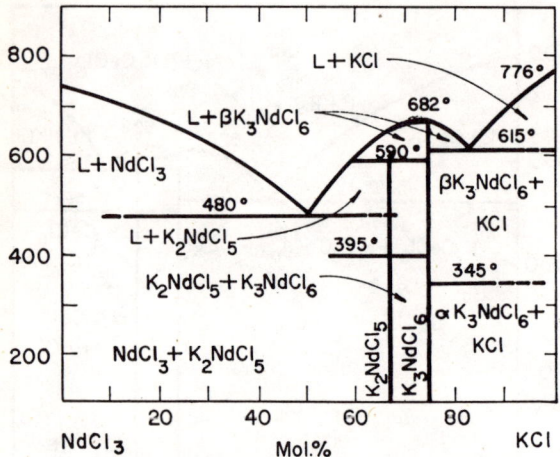

Fig. 1277.—System KCl–NdCl₃.

In-Chzhu Sun and I. S. Morozov, *Zhur. Neorg. Khim.*, **3**, 1916 (1958).

KCl–PuCl₃

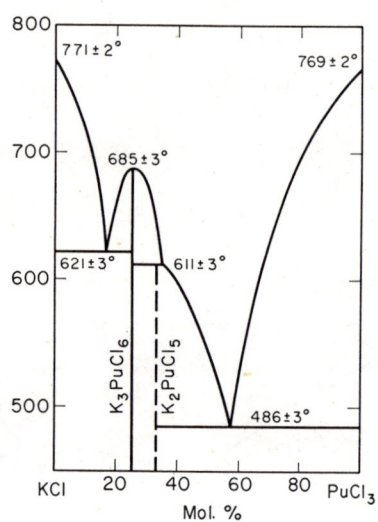

Fig. 1278.—System KCl–PuCl₃.

Robert Benz, Milton Kahn, and J. A. Leary, *J. Phys. Chem.*, **63**, 1984 (1959).

KCl–TiCl₃

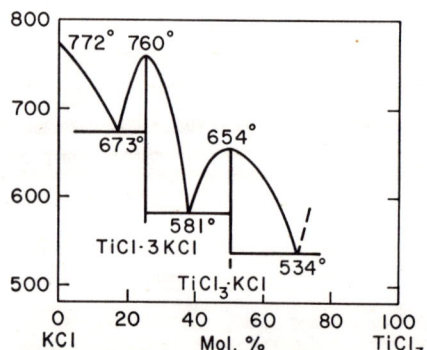

Fig. 1279.—System KCl–TiCl₃.

B. F. Markov and R. V. Chernov, *Ukrain. Khim. Zhur.*, **25**, 281 (1959). See also M. V. Kamenetskiĭ, *Tsvetn. Metal.*, **31** [2] 42 (1958).

KCl–VCl₃

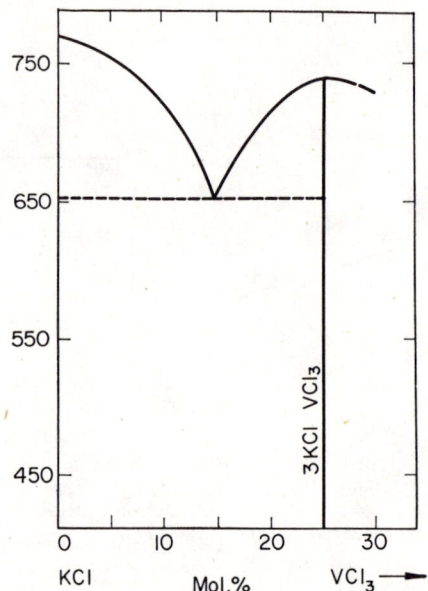

Fig. 1280.—System KCl–3KCl·VCl₃.

Claude Grena, *Bull. soc. chim. France*, **130**, 656 (1960).

KCl–HfCl₄

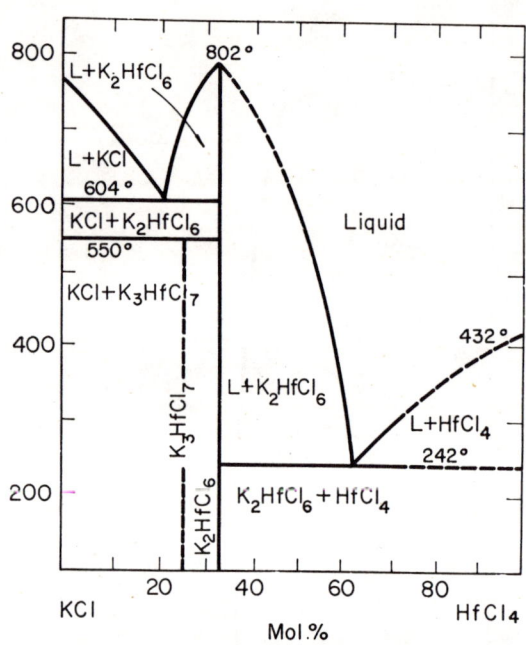

Fig. 1281.—System KCl–HfCl₄.

I. S. Morozov and In-Chzhu Sun, *Zhur. Neorg. Khim.* **4**, 680 (1959).

KCl–UCl$_4$

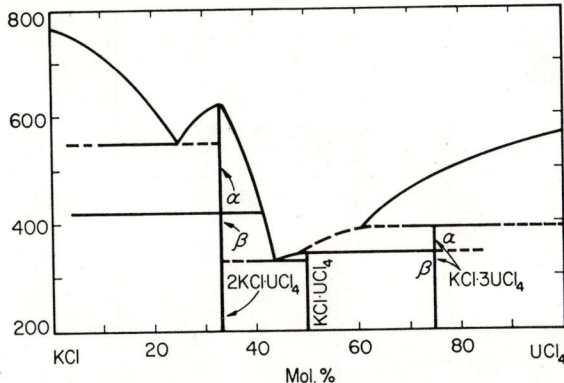

Fig. 1282.—System KCl–UCl$_4$; tentative.

C. J. Barton, A. B. Wilkerson, T. N. McVay, R. J. Sheil, and W. R. Grimes, Oak Ridge National Laboratory, Phase Diagrams of Nuclear Reactor Materials, R. E. Thoma, ed., ORNL-2548 p. 135 (1959).

KCl–ZrCl$_4$

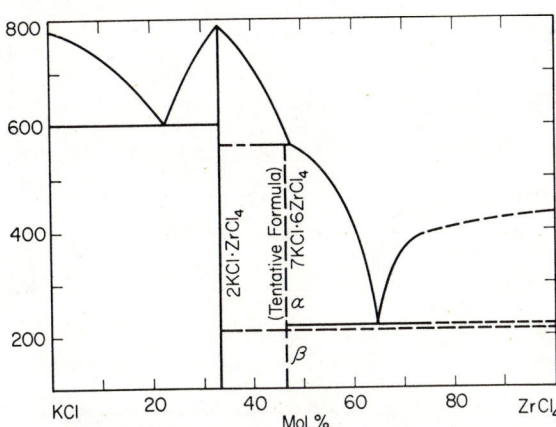

Fig. 1283.—System KCl–ZrCl$_4$; tentative.

C. J. Barton, R. J. Sheil, and W. R. Grimes, Oak Ridge National Laboratory, Phase Diagrams of Nuclear Reactor Materials, R. E. Thoma, ed., ORNL-2548, p. 130 (1959).

See also I. S. Morozov and In-Chzhu Sun, *Zhur. Neorg. Khim.*, **4**, 680 (1959).

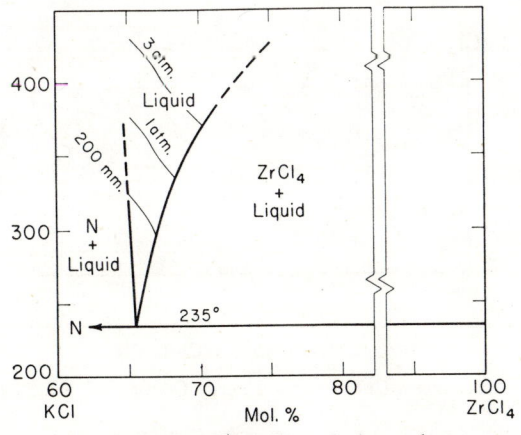

Fig. 1284.—System KCl–ZrCl$_4$. Isobars show v.p. of ZrCl$_4$ in equilibrium with melt.

L. J. Howell, R. C. Sommer, and H. H. Kellogg, *J. Metals*, **9**, *Trans. Am. Inst. Mining Met. Engrs.*, **209**, 197 (1959).

LiCl–NaCl

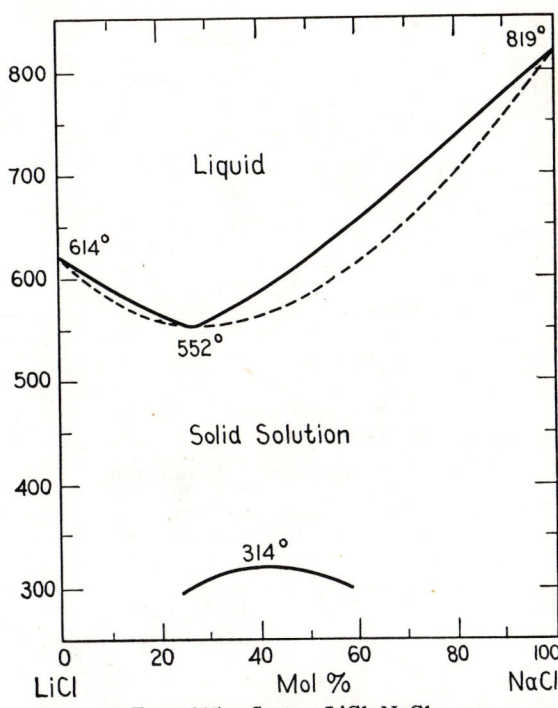

Fig. 1285.—System LiCl–NaCl.

A. Smits, J. Elgersma and H. v. Hardenberg, *Rec. trav. chim.*, **43**, 671 (1924).

See also S. Zhemchuzhnui and F. Rambach, *Z. anorg. Chem.*, **65**, 411 (1910).

LiCl–NH$_4$Cl

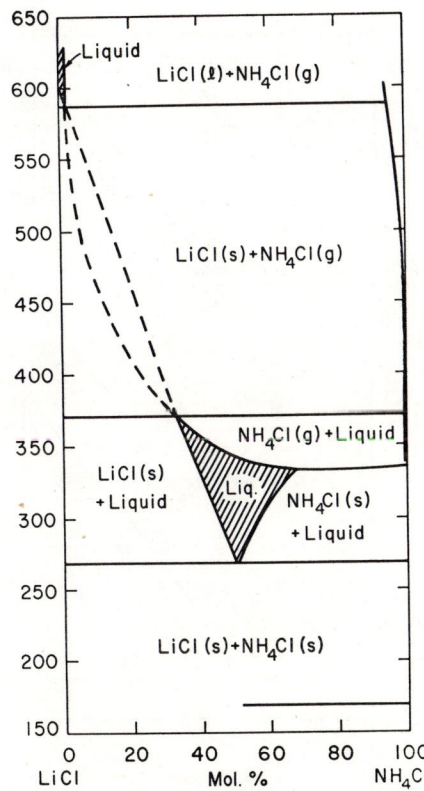

Fig. 1286.—System LiCl–NH$_4$Cl.

Kurt Hachmeister, *Z. anorg. u. allgem. Chem.*, **109**, 167 (1919).

LiCl–RbCl

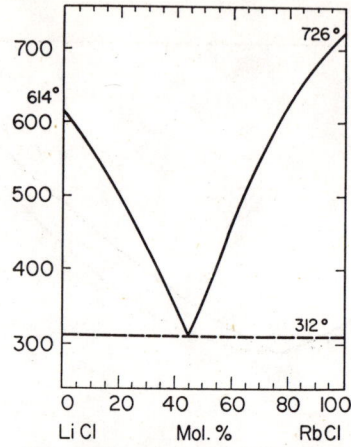

FIG. 1287.—System LiCl–RbCl.

S. Zhemchuzhnui and F. Rambach, *Z. anorg. Chem.*, **65**, 408 (1910). See also T. W. Richards and W. B. Meldrum, *J. Am. Chem. Soc.*, **39**, 1823 (1917).

LiCl–TlCl

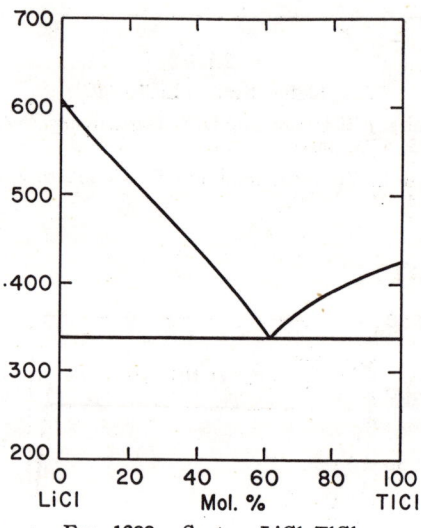

FIG. 1288.—System LiCl–TlCl.

Carlo Sandonnini, *Gazz. chim. ital.* **44 I**, 300 (1914).

LiCl–BaCl$_2$

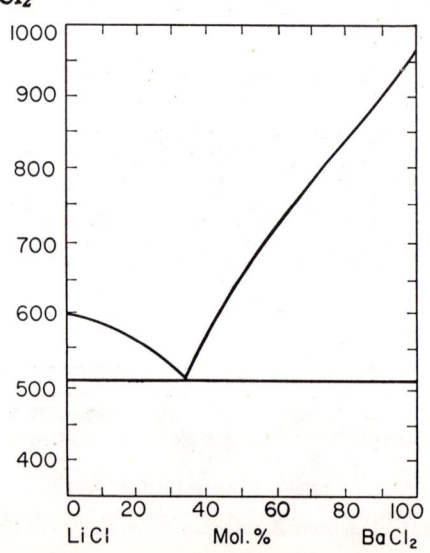

LiCl–BeCl$_2$

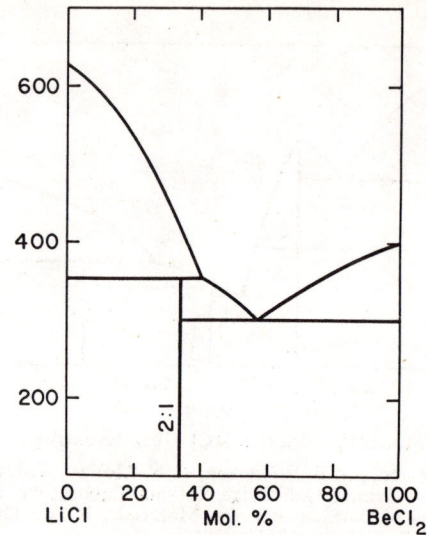

FIG. 1290.—System LiCl–BeCl$_2$.

M. J. M. Schmidt, *Bull. soc. chim., France*, **39**, 1694 (1926).

LiCl–CaCl$_2$

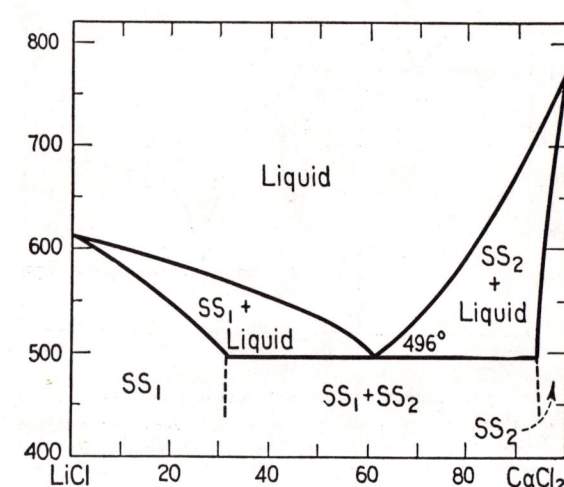

FIG. 1291.—System LiCl–CaCl$_2$.

G. Grube and W. Rüdel, *Z. anorg. u. allgem. Chem.*, **133**, 375 (1924).

LiCl–FeCl$_2$

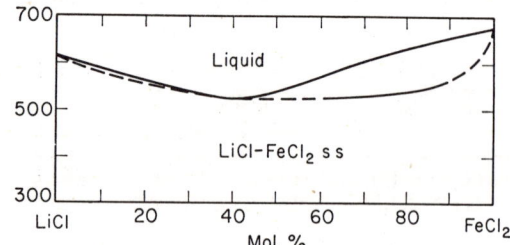

FIG. 1292.—System LiCl–FeCl$_2$.

C. Beusman, U. S. Atomic Energy Comm., ORNL-2323, 20 (1957).

← FIG. 1289.—System LiCl–BaCl$_2$.

Carlo Sandonnini, *Gazz. chim. ital.*, **44 I**, 383 (1914).

LiCl–MnCl$_2$

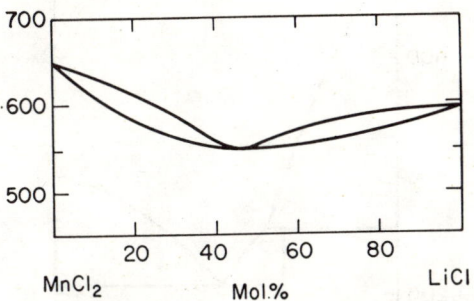

FIG. 1293.—System LiCl–MnCl$_2$.

C. Sandonnini and G. Scarpa, *Atti reale accad. Lincei, Sez. II*, **22**, 165 (1913).

LiCl–PbCl$_2$

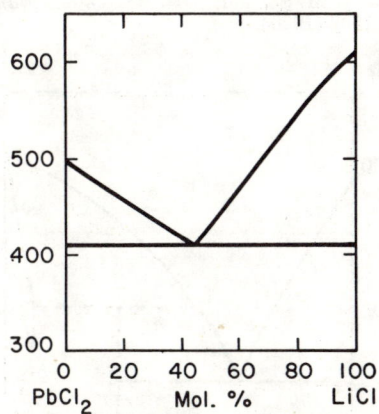

FIG. 1294.—System LiCl–PbCl$_2$.

Kurt Treis, *Neues Jahrb. Mineral., Geol.*, **37**, 774 (1914).

LiCl–SrCl$_2$

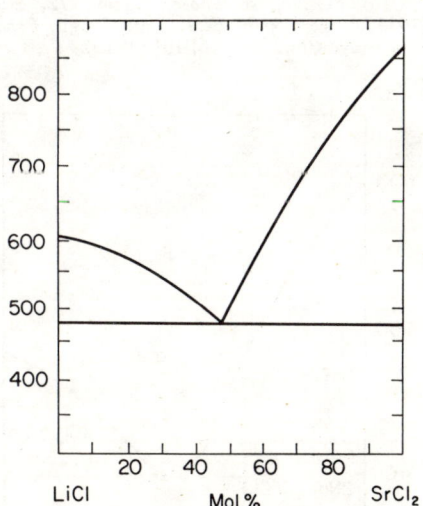

FIG. 1295.—System LiCl–SrCl$_2$.

Carlo Sandonnini, *Gazz. chim. ital.*, **44 I**, 381 (1914).

LiCl–PuCl$_3$

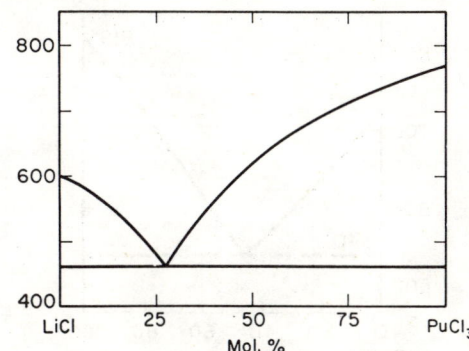

FIG. 1296.—System LiCl–PuCl$_3$.

C. W. Bjorklund, J. G. Reavis, J. A. Leary, and K. A. Walsh, *J. Phys. Chem.*, **63**, 1776 (1959).

LiCl–UCl$_3$

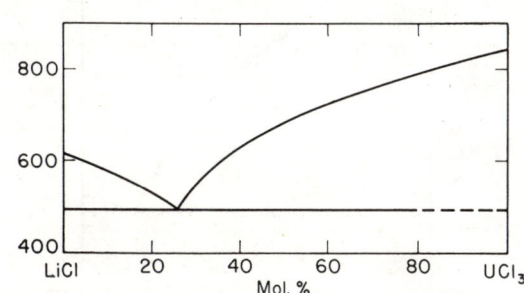

FIG. 1297.—System LiCl–UCl$_3$.

C. J. Barton, A. B. Wilkerson, and W. R. Grimes, Oak Ridge National Laboratory, Phase Diagrams of Nuclear Reactor Materials, R. E. Thoma, ed., ORNL-2548, p. 131 (1959).

LiCl–UCl$_4$

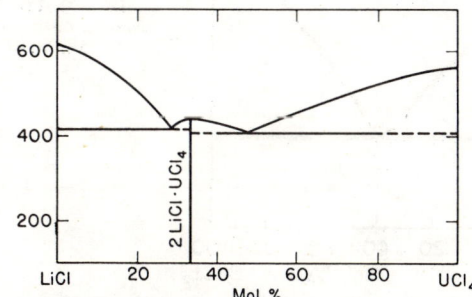

FIG. 1298.—System LiCl–UCl$_4$; tentative.

C. J. Barton, R. J. Sheil, and W. R. Grimes, Oak Ridge National Laboratory, Phase Diagrams of Nuclear Reactor Materials, R. E. Thoma, ed., ORNL-2548, p. 132 (1959).

NaCl–RbCl

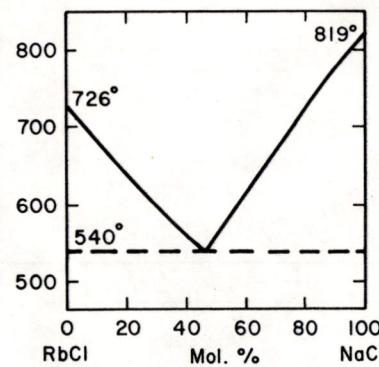

Fig. 1299.—System NaCl–RbCl.

S. Zhemchuzhnui and F. Rambach, *Z. anorg. Chem.*, **65**, 415 (1910).

NaCl–TlCl

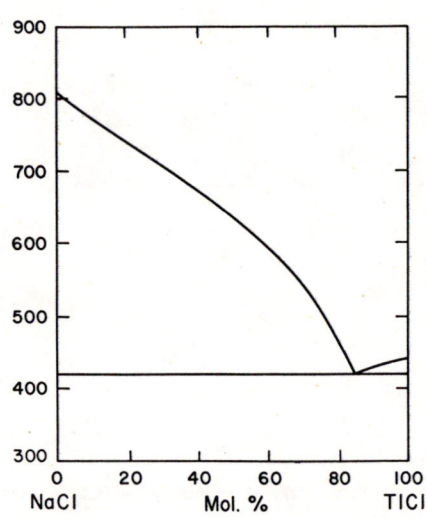

Fig. 1300.—System NaCl–TlCl.

Carlo Sandonnini, *Gazz. chim. ital.*, **44 I**, 305 (1914).

NaCl–BaCl$_2$

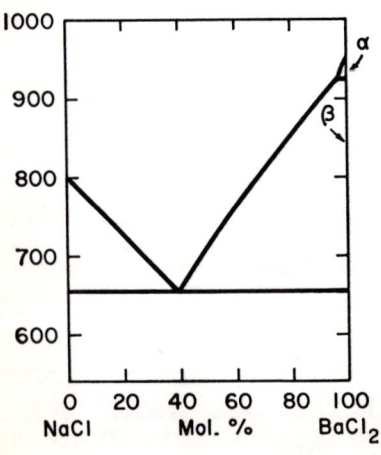

Fig. 1301.—System NaCl–BaCl$_2$.

Erhard Vortisch, *Neues Jahrb. Mineral., Geol.*, **38**, 204 (1914).

NaCl–BeCl$_2$

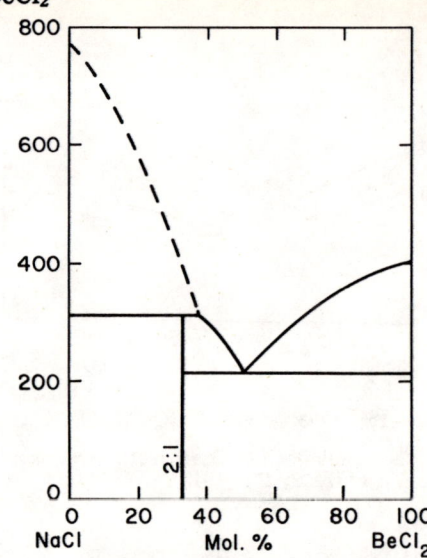

Fig. 1302.—System NaCl–BeCl$_2$.

M. J. M. Schmidt, *Bull. soc. chim., France*, **39**, 1692 (1926).

NaCl–CaCl$_2$

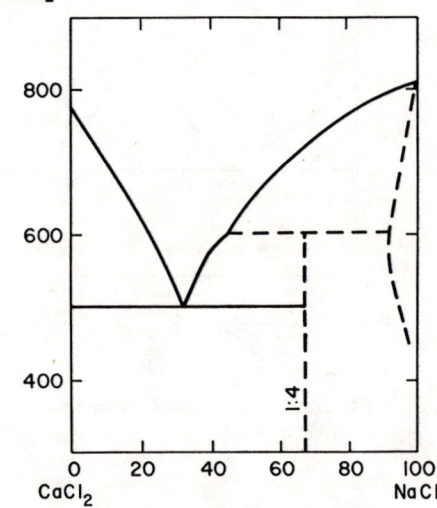

Fig. 1303.—System NaCl–CaCl$_2$.

Otto Menge, *Z. anorg. Chem.*, **72**, 202 (1911). See also F. E. E. Lamplough, *Proc. Cambridge Phil. Soc.*, **16**, 194 (1912).

NaCl–CdCl$_2$

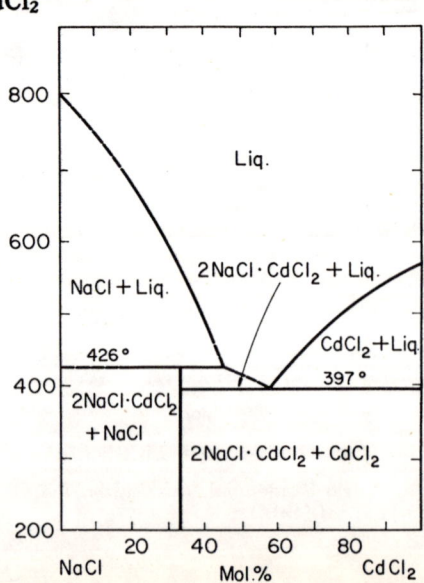

Fig. 1304.—System NaCl–CdCl$_2$.

D. L. Deadmore and J. S. Machin, *J. Am. Ceram. Soc.*, **43** [11] 593 (1960).

NaCl–CoCl$_2$

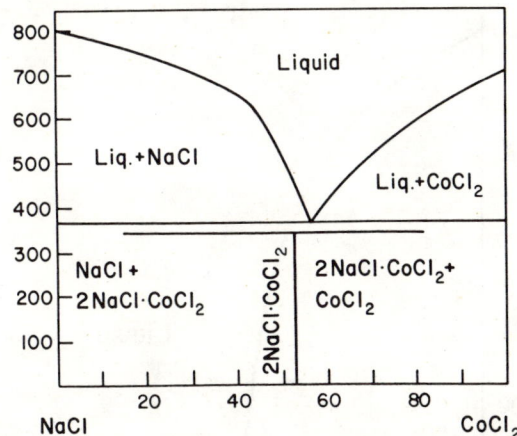

Fig. 1305.—System NaCl–CoCl$_2$.

K. A. Bol'shakov, P. I. Fedorov, and G. D. Agashkina, *Zhur. Neorg. Khim.*, **2** [5] 1116 (1957).

NaCl–HgCl$_2$

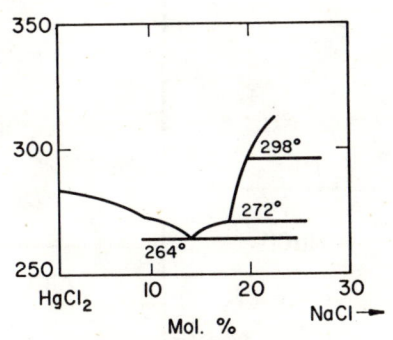

Fig. 1306.—System NaCl–HgCl$_2$.

I. N. Belyaev and K. E. Mironov, *Zhur. Obshcheĭ Khim.* **22**, 1486 (1952).

NaCl–MgCl$_2$

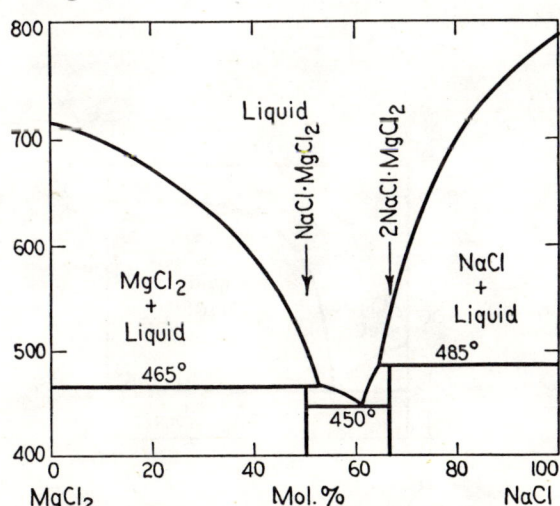

Fig. 1307.—System NaCl–MgCl$_2$.

W. Klemm and P. Weiss, *Z. anorg. u. allgem. Chem.*, **245**, 281 (1940).

NaCl–MnCl$_2$

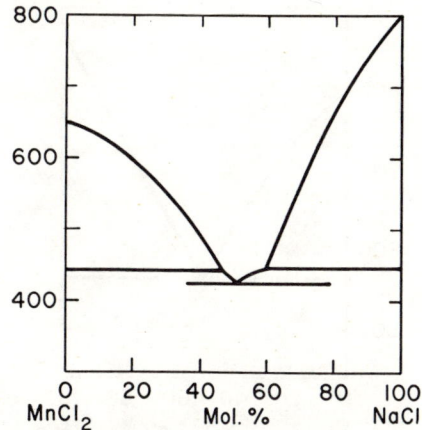

Fig. 1308.—System NaCl–MnCl$_2$.

C. Sandonnini and G. Scarpa, *Atti reale accad. Lincei, Sez. II*, **22**, 166 (1913).

NaCl–NiCl$_2$

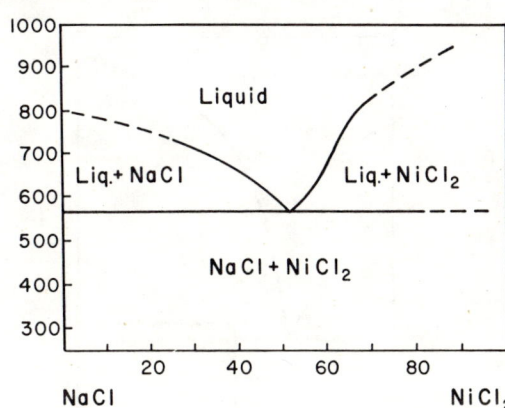

Fig. 1309.—System NaCl–NiCl$_2$.

K. A. Bol'shakov, P. I. Fedorov, and G. D. Agashkina, *Zhur. Neorg. Khim.*, **2** [5] 1117 (1957).

NaCl–PbCl$_2$

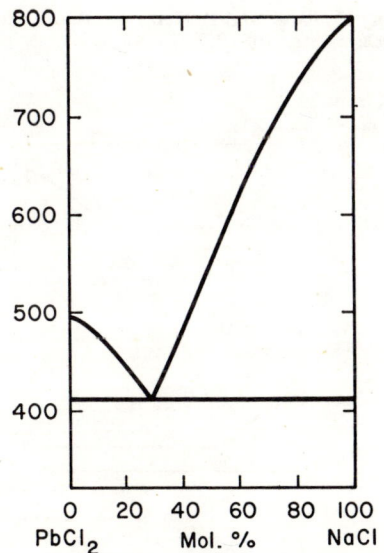

Fig. 1310.—System NaCl–PbCl$_2$.

Kurt Treis, *Neues Jahrb. Mineral., Geol.*, **37**, 772 (1914)

NaCl–SrCl₂

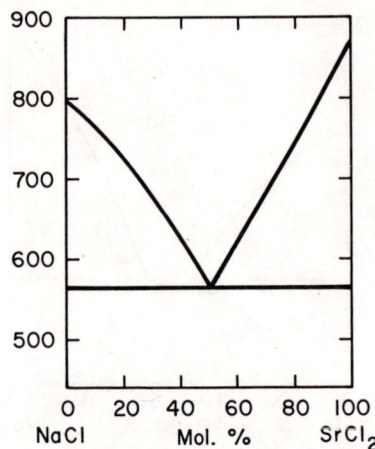

FIG. 1311.—System NaCl–SrCl₂.

Erhard Vortisch, *Neues Jahrb. Mineral., Geol.*, **38**, 202 (1914).

NaCl–TiCl₂

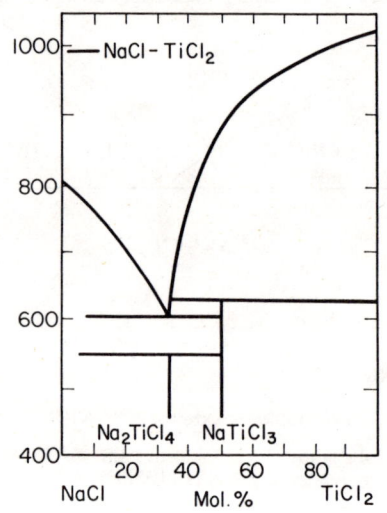

FIG. 1312.—System NaCl–TiCl₂.

K. Komarek and P. Herasymenko, *J. Electrochem. Soc.*, **105**, 217 (1958).

NaCl–VCl₂

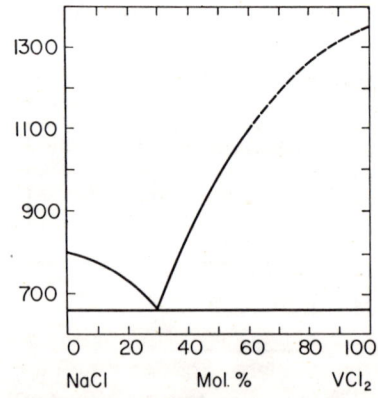

FIG. 1313.—System NaCl–VCl₂.

H. J. Seifert and Paul Ehrlich, *Z. anorg. u. allgem. Chem.*, **302**, 286 (1959).

NaCl–AlCl₃

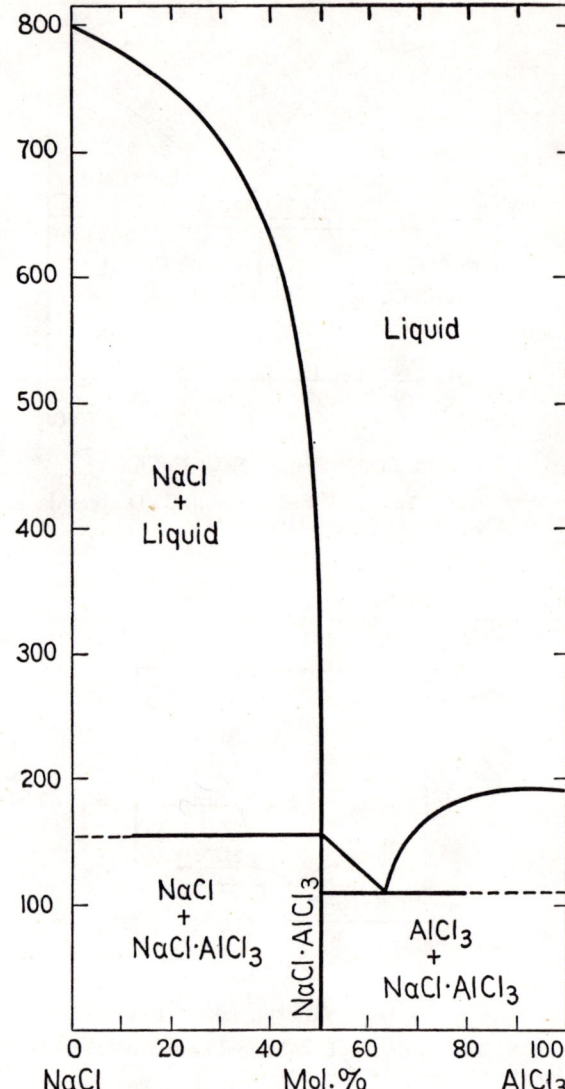

FIG. 1314.—System NaCl–AlCl₃.

U. I. Shvartsman, *J. Phys. Chem. (U.S.S.R.)*, **14**, 254 (1940).

NaCl–FeCl₃

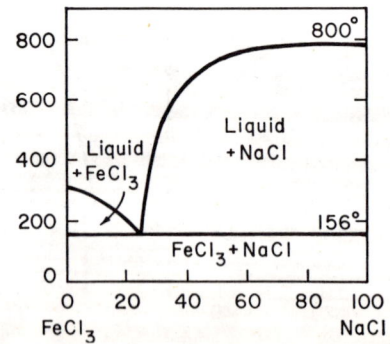

FIG. 1315.—System NaCl–FeCl₃.

I. S. Morozov and D. Ya. Toptygin, *Zhur. Neorg. Khim.*, **2**, 2133 (1957).

NaCl–PuCl$_3$

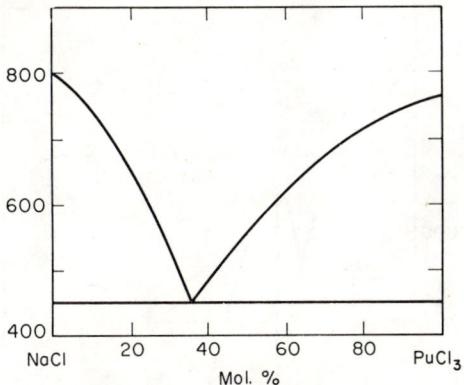

Fig. 1316.—System NaCl–PuCl$_3$.

C. W. Bjorklund, J. G. Reavis, J. A. Leary, and K. A. Walsh, *J. Phys. Chem.*, **63**, 1776 (1959).

NaCl–TiCl$_3$

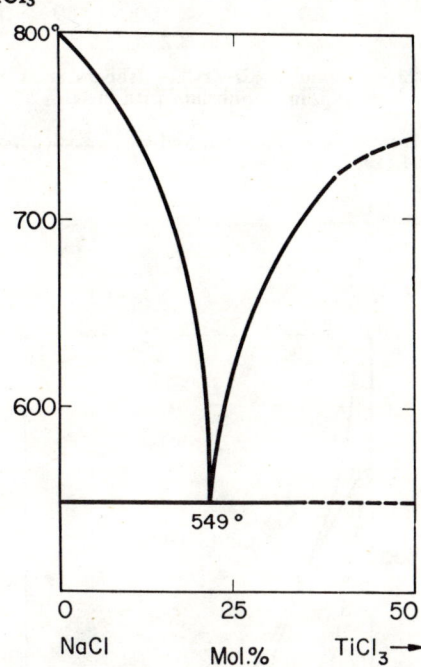

Fig. 1317.—System NaCl–TiCl$_3$.

B. F. Markov and R. V. Chernov, *Ukrain. Khim. Zhur.*, **25**, 280 (1959).

NaCl–UCl$_3$

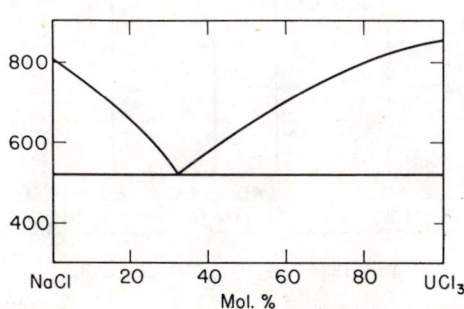

Fig. 1318.—System NaCl–UCl$_3$.

C. A. Kraus, Report No. M-251, July 1, 1943; confirmed at Oak Ridge National Laboratory, Oak Ridge, Tenn.; by C. J. Barton, *et al*. Phase Diagrams of Nuclear Reactor Materials, R. F. Thoma, ed., ORNL-2548, p. 133 (1959).

NaCl–VCl$_3$

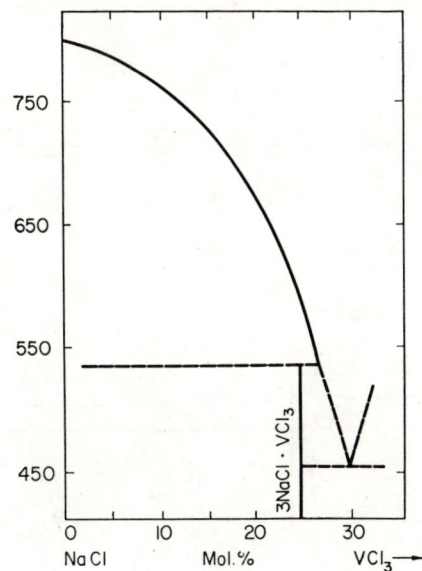

Fig. 1319.—System NaCl–3NaCl·VCl$_3$.

Claude Grena, *Bull. soc. chim.*, France, **130**, 656 (1960).

NaCl–HfCl$_4$

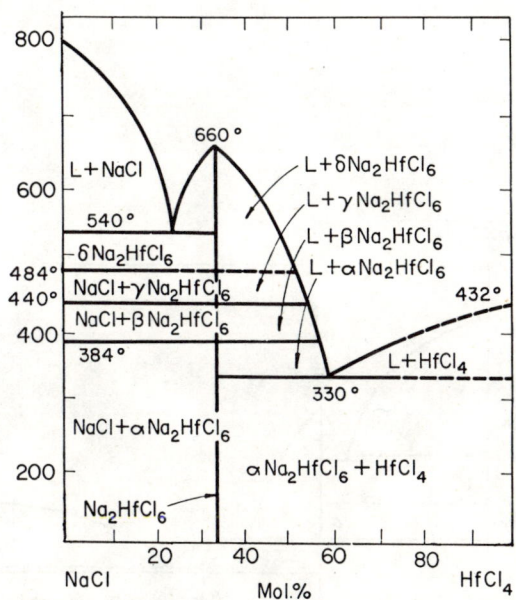

Fig. 1320.—System NaCl–HfCl$_4$.

I. S. Morozov and In-Chzhu Sun, *Zhur. Neorg. Khim.*, **4**, 680 (1959). See also A. S. Roy, L. J. Howell, and H. H. Kellogg, *Trans. AIME*, **212**, 818 (1958).

NaCl–UCl₄

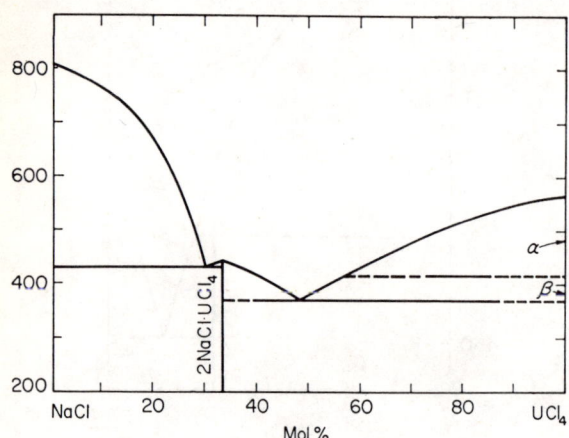

Fig. 1321.—System NaCl–UCl₄; tentative.

C. J. Barton, R. J. Sheil, A. B. Wilkerson, and W. R. Grimes, Oak Ridge National Laboratory, Phase Diagrams of Nuclear Reactor Materials, R. F. Thoma, ed., ORNL-2548, p. 134 (1959).

NaCl–ZrCl₄

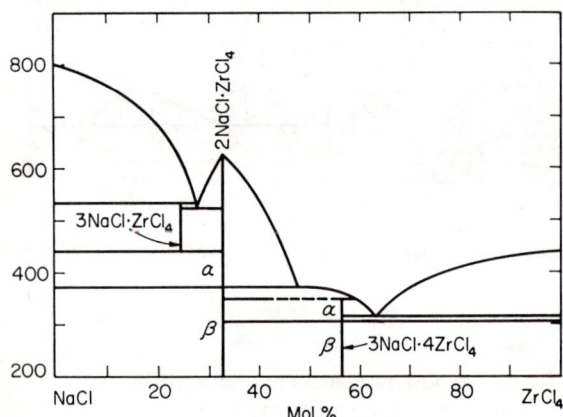

Fig. 1322.—System NaCl–ZrCl₄; tentative.

C. J. Barton, R. J. Sheil, and W. R. Grimes, Oak Ridge National Laboratory, Phase Diagrams of Nuclear Reactor Materials, R. E. Thoma, ed., ORNL-2548, p. 129 (1959).

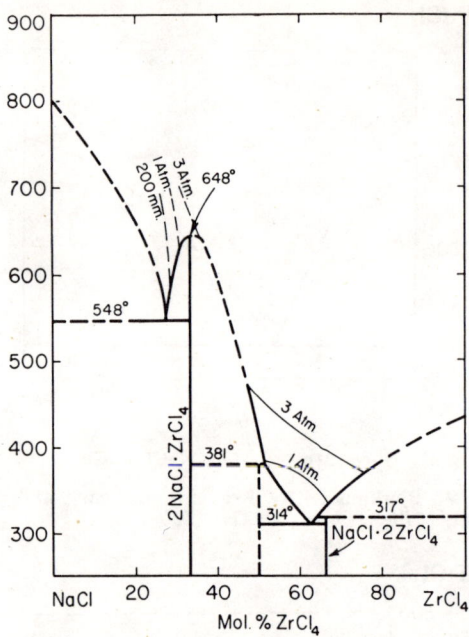

Fig. 1323.—System NaCl–ZrCl₄. Isobars show v.p. of ZrCl₄ in equilibrium with melt.

L. J. Howell, R. C. Sommer, and H. H. Kellogg, *J. Metals*, **9** 194 (1957).

NH₄Cl–CdCl₂

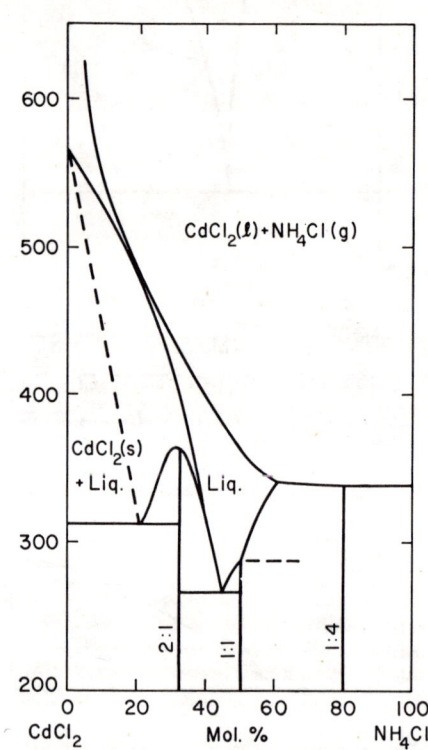

Fig. 1324.—System NH₄Cl–CdCl₂.

Kurt Hachmeister, *Z. anorg. u. allgem. Chem.*, **109**, 176 (1919).

NH_4Cl–$HgCl_2$

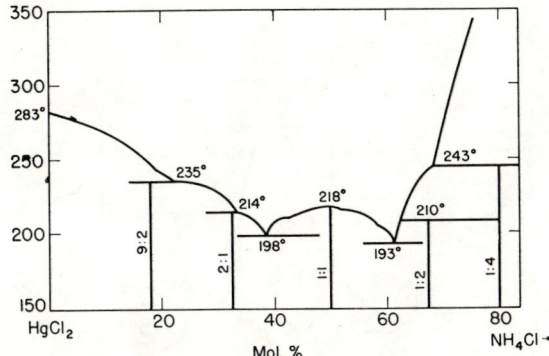

Fig. 1325.—System NH_4Cl–$HgCl_2$.

I. N. Belyaev and K. E. Mironov, *Zhur. Obshcheĭ Khim.*, **22**, 1488 (1952).

NH_4Cl–$ZnCl_2$

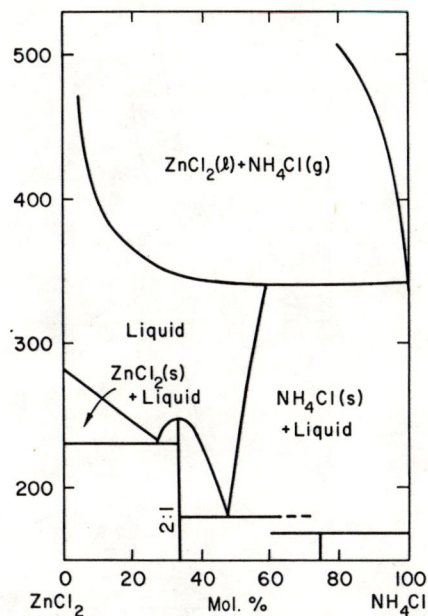

Fig. 1326.—System NH_4Cl–$ZnCl_2$.

Kurt Hachmeister, *Z. anorg. u. allgem. Chem.*, **109**, 173 (1919).

NH_4Cl–$FeCl_3$

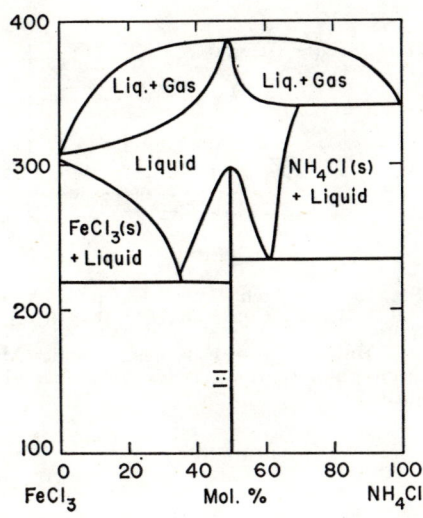

Fig. 1327.—System NH_4Cl–$FeCl_3$.

Kurt Hachmeister, *Z. anorg. u. allgem. Chem.*, **109**, 180 (1919).

RCl–$NbCl_5$

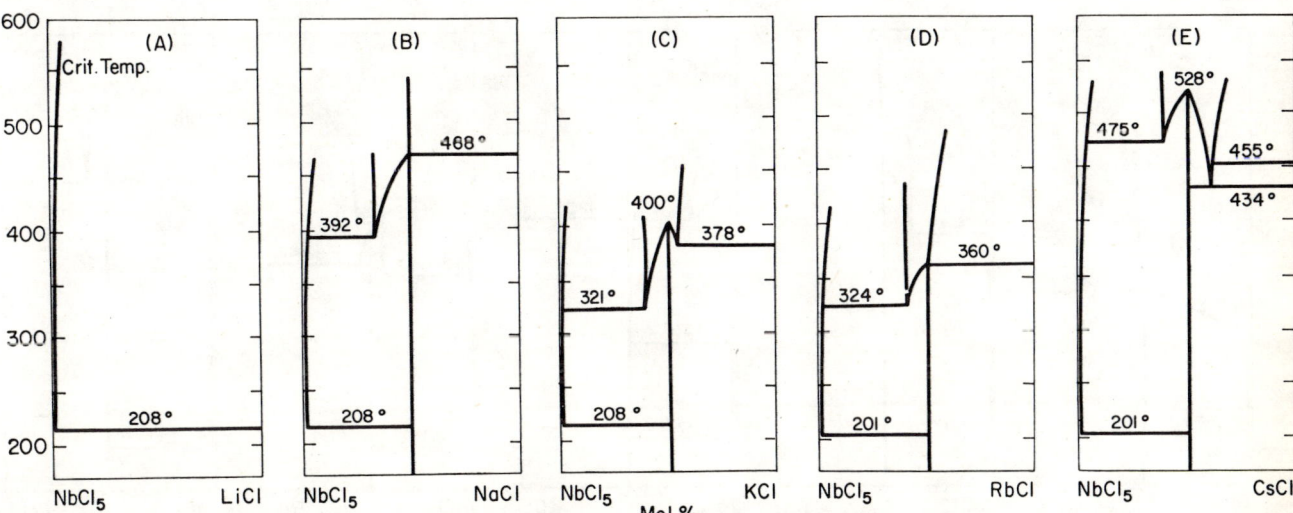

Fig. 1328.—System RCl–$NbCl_5$. A = System LiCl–$NbCl_5$; B = system NaCl–$NbCl_5$; C = system KCl–$NbCl_5$; D = system RbCl–$NbCl_5$; E = system CsCl–$NbCl_5$.

K. Huber, E. Jost, E. Neuenschwander, M. Studer, and B. Roth, *Helv. Chim. Acta*, **41**, 2414 (1958).

RCl–NbCl₅ (concl.)

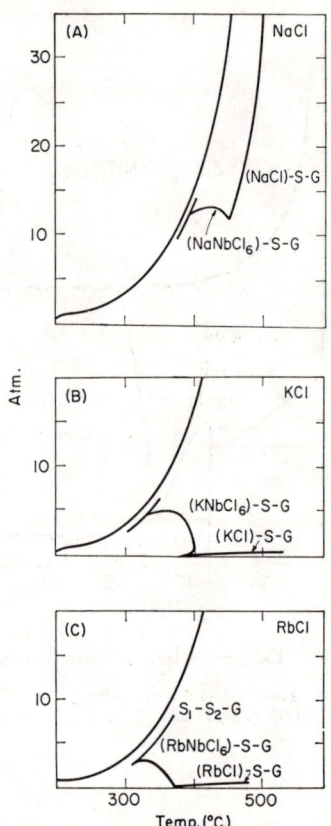

Fig. 1329.—System RCl–NbCl₅; p-T. A = NaCl; B = KCl; C = RbCl.

K. Huber, E. Jost, E. Neuenschwander, M. Studer, and B. Roth, *Helv. Chim. Acta*, **41**, 2419 (1958).

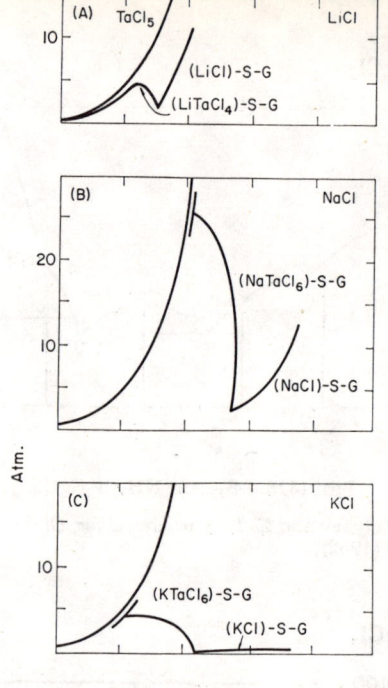

Fig. 1331.—System RCl–TaCl₅; p-T. A = LiCl; B = NaCl; C = KCl; D = RbCl.

K. Huber, E. Jost, E. Neuenschwander, M. Studer, and B. Roth, *Helv. Chim. Acta*, **41**, 2419 (1958).

RCl–TaCl₅

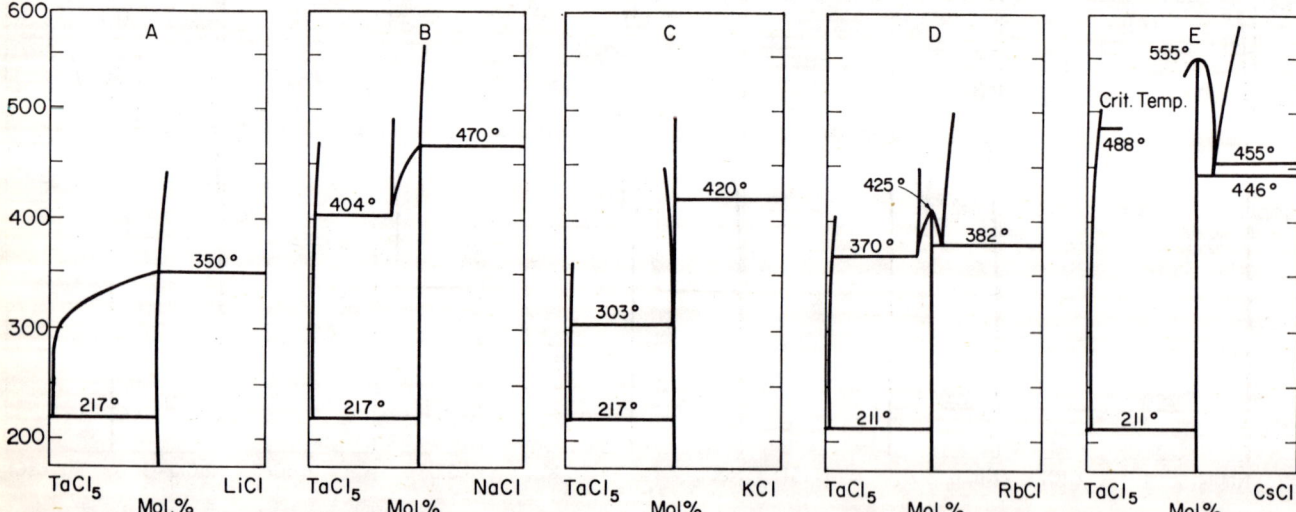

Fig. 1330.—System RCl–TaCl₅. A = System LiCl–TaCl₅; B = system NaCl–TaCl₅; C = system KCl–TaCl₅; D = system RbCl–TaCl₅; E = system CsCl–TaCl₅.

K. Huber, E. Jost, E. Neuenschwander, M. Studer, and B. Roth, *Helv. Chim. Acta*, **41**, 2414 (1958).

RbCl–TlCl

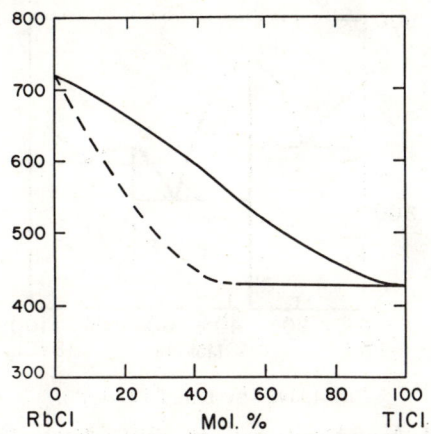

Fig. 1332.—System RbCl–TlCl.

Carlo Sandonnini, *Gazz. chim. ital.*, **44 I**, 317 (1914).

RbCl–PbCl$_2$

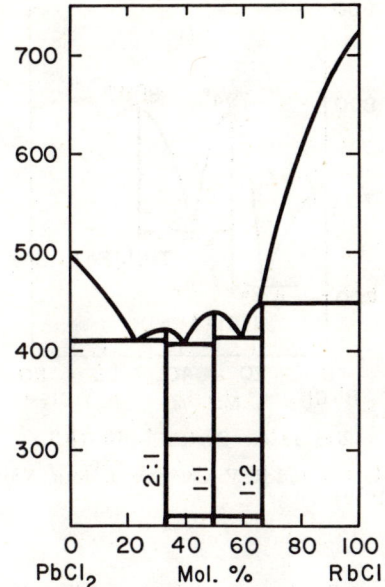

Fig. 1334.—System RbCl–PbCl$_2$.

Kurt Treis, *Neues Jahrb. Mineral., Geol.*, **37**, 784 (1914).

RbCl–MgCl$_2$

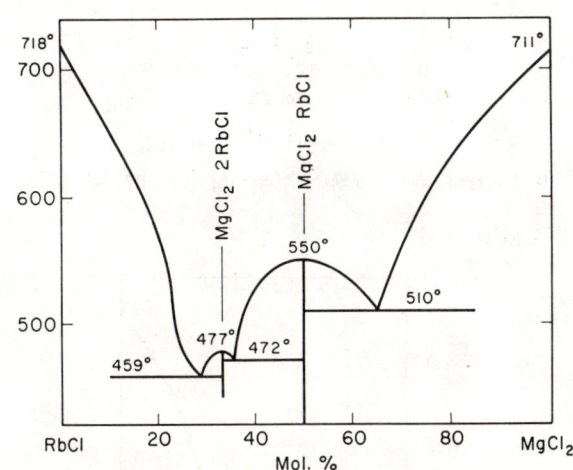

Fig. 1333.—System RbCl–MgCl$_2$.

B. F. Markov and I. D. Panchenko, *Zhur. Obshchei Khim.*, **25**, 2039 (1955).

RbCl–ZnCl$_2$

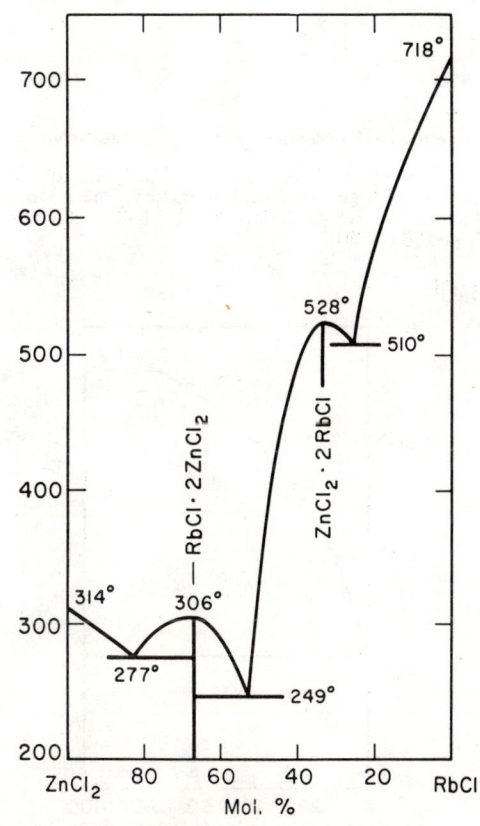

Fig. 1335.—System RbCl–ZnCl$_2$.

B. F. Markov, I. D. Panchenko, and T. G. Kostenko, *Ukrain. Khim. Zhur.*, **22**, 290 (1956) (in Russian).

RbCl–TiCl₃

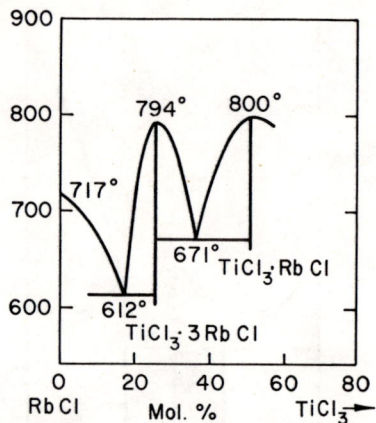

Fig. 1336.—System RbCl–TiCl₃.

B. F. Markov and R. V. Chernov, *Ukrain. Khim. Zhur.*, **25**, 282 (1959).

RbCl–UCl₃

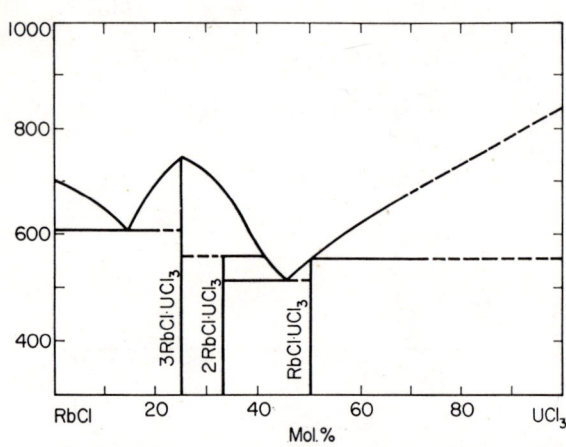

Fig. 1337.—System RbCl–UCl₃; tentative.

C. J. Barton, R. J. Sheil, A. B. Wilkerson, and W. R. Grimes, Oak Ridge, National Laboratory, Phase Diagrams of Nuclear Reactor Materials, R. E. Thoma, ed., ORNL–2548, p. 136 (1959).

TlCl–BaCl₂

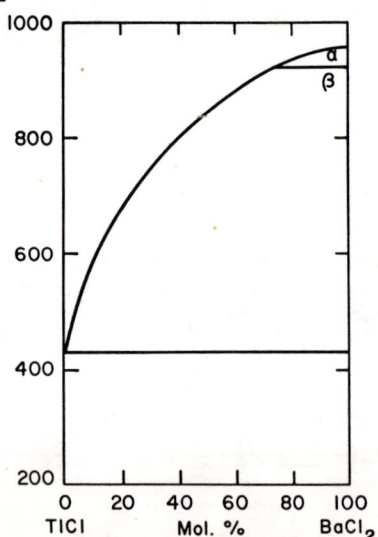

Fig. 1338.—System TlCl–BaCl₂.

E. Korreng, *Neues Jahrb. Mineral., Geol.*, **37**, 103 (1914).

TlCl–BeCl₂

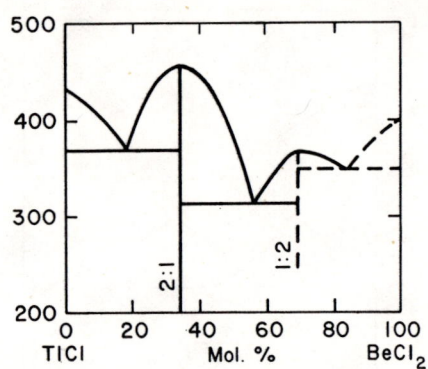

Fig. 1339.—System TlCl–BeCl₂.

M. J. M. Schmidt, *Bull. soc. chim., France*, **39**, 1700 (1926).

TlCl–CaCl₂

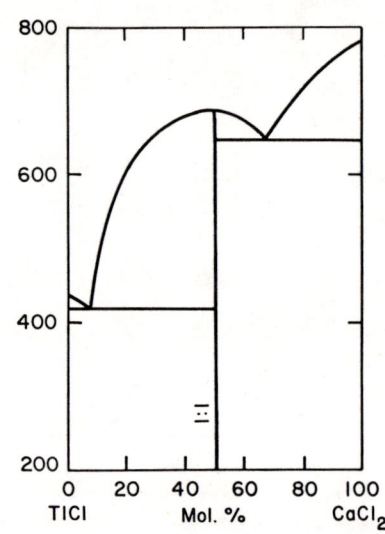

Fig. 1340.—System TlCl–CaCl₂.

E. Korreng, *Neues Jahrb. Mineral., Geol.*, **37**, 95 (1914).

TlCl–CdCl₂

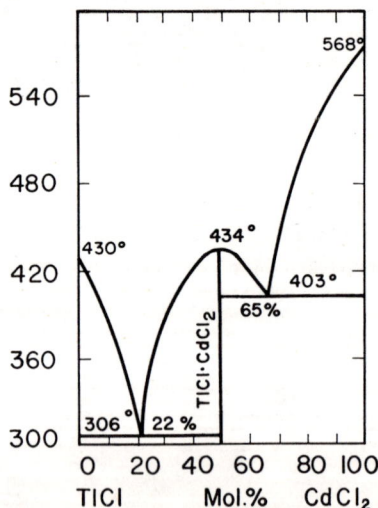

Fig. 1341.—System TlCl–CdCl₂.

I. P. Palyura and A. P. Palkin, *Zhur. Neorg. Khim.*, **4**, 238 (1959).

TlCl–MgCl₂

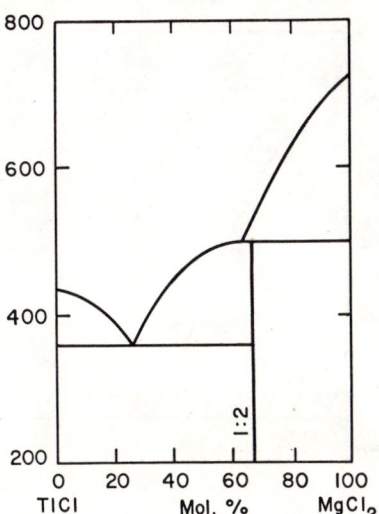

Fig. 1342.—System TlCl–MgCl₂.

E. Korreng, *Neues Jahrb. Mineral., Geol.*, **37**, 92 (1914).

TlCl–PbCl₂

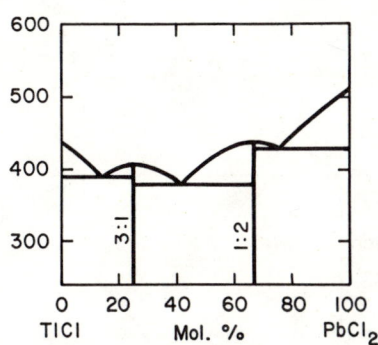

Fig. 1343.—System TlCl–PbCl₂.

E. Korreng, *Neues Jahrb. Mineral., Geol.*, **37**, 117 (1914).

TlCl–SnCl₂

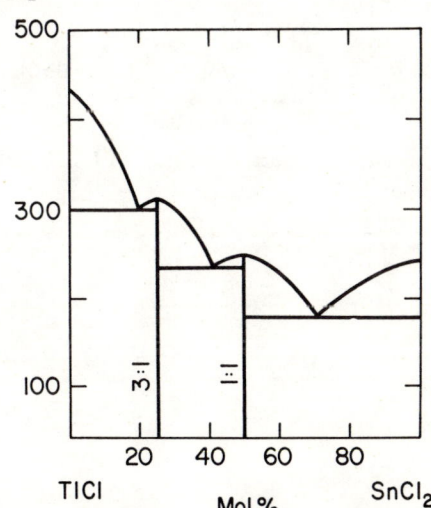

Fig. 1344.—System TlCl–SnCl₂.

E. Korreng, *Neues Jahrb. Mineral., Geol.*, **37**, 114 (1914).

TlCl–SrCl₂

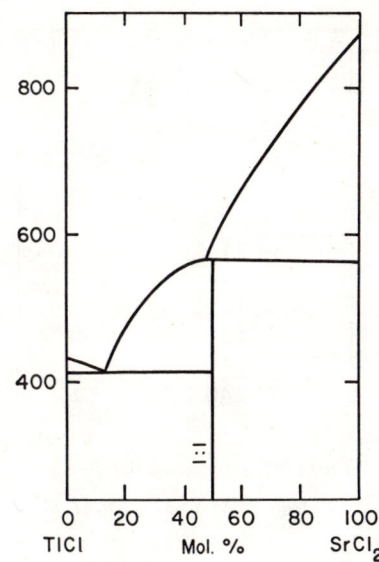

Fig. 1345.—System TlCl–SrCl₂.

E. Korreng, *Neues Jahrb. Mineral., Geol.*, **37**, 98 (1914).

TlCl–ZnCl₂

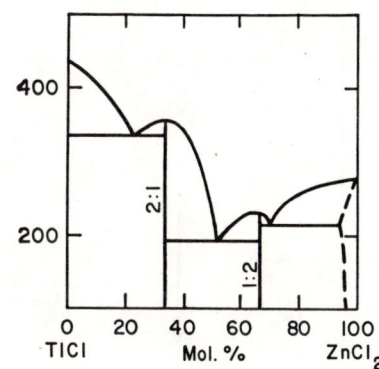

Fig. 1346.—System TlCl–ZnCl₂.

E. Korreng, *Neues Jahrb. Mineral., Geol.*, **37**, 106 (1914).

TlCl–BiCl₃

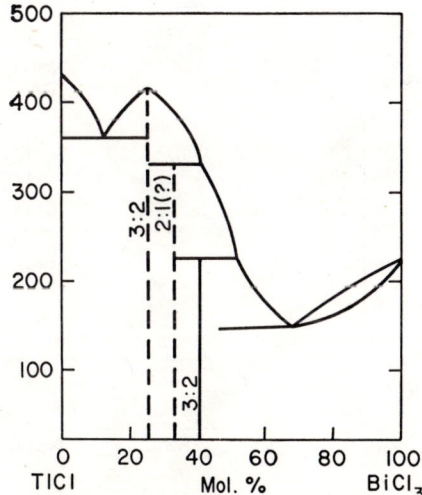

Fig. 1347.—System TlCl–BiCl₃.

G. Scarpa, *Atti reale accad. Lincei, Sez. II*, **21**, 724 (1912).

TlCl–FeCl₃

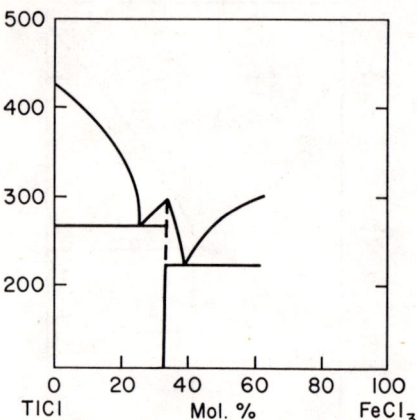

Fig. 1348.—System TlCl–FeCl₃.

G. Scarpa, *Atti reale accad. Lincei, Sez. II*, **21**, 722 (1912).

BaCl₂–BeCl₂

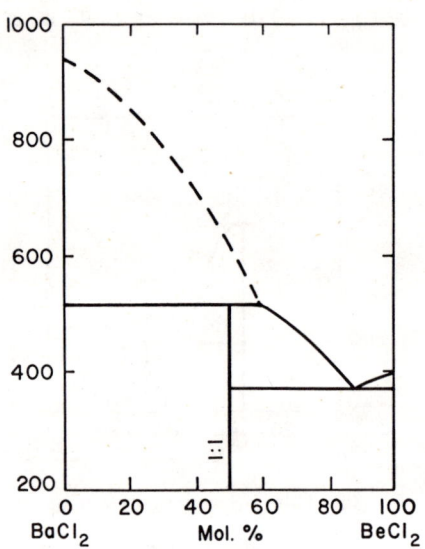

Fig. 1349.—System BaCl₂–BeCl₂.

M. J. M. Schmidt, *Bull. soc. chim., France*, **39**, 1697 (1926).

BaCl₂–CaCl₂

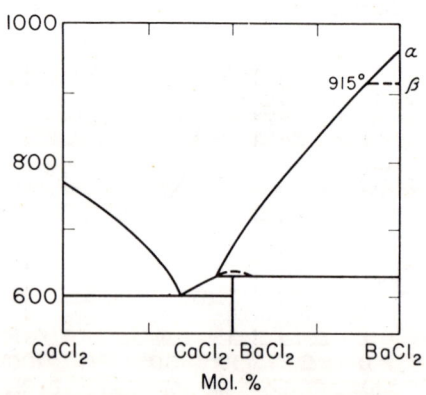

Fig. 1350.—System BaCl₂–CaCl₂.

P. P. Budnikov, P. L. Volodin, and S. G. Tresvyatskiĭ, *Ukrain. Khim. Zhur.*, **22**, 293 (1956).

BaCl₂–CdCl₂

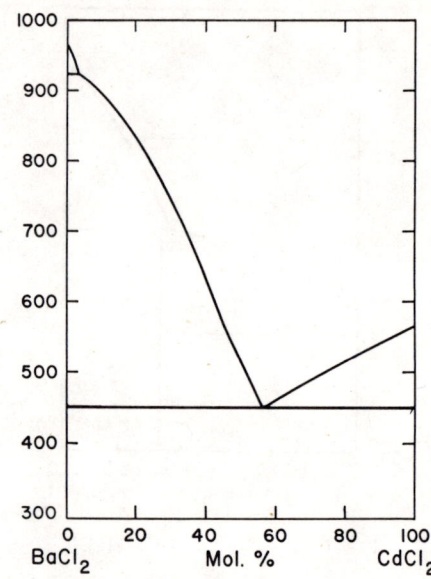

Fig. 1351.—System BaCl₂–CdCl₂.

Carlo Sandonnini, *Gazz. chim. ital.*, **44 I**, 355 (1914).

BaCl₂–MgCl₂

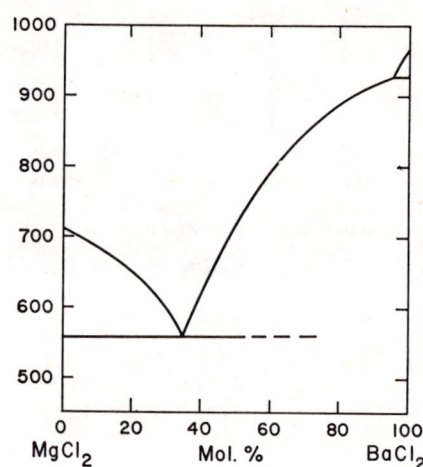

Fig. 1352.—System BaCl₂–MgCl₂.

Carlo Sandonnini, *Gazz. chim. ital.*, **44 I**, 351 (1914).

BaCl₂–MnCl₂

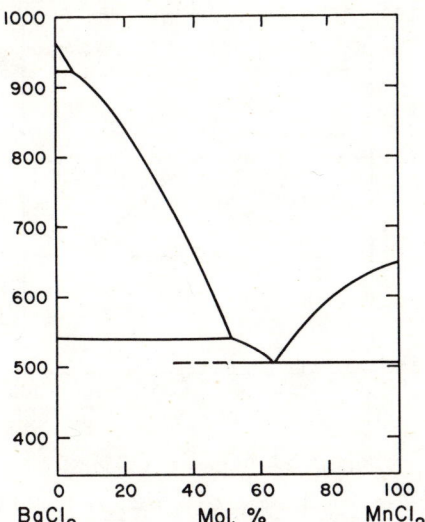

Fig. 1353.—System BaCl₂–MnCl₂.

Carlo Sandonnini, *Gazz. chim. ital.*, **44 I**, 359 (1914).

BaCl₂–PbCl₂

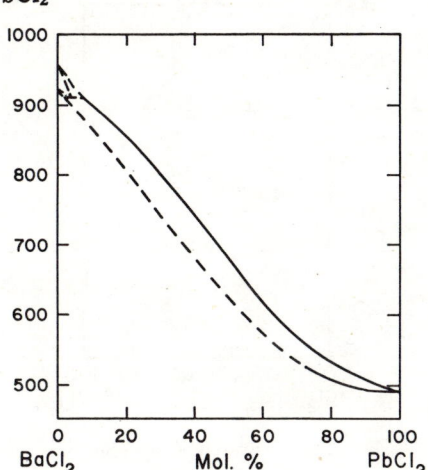

Fig. 1354.—System BaCl₂–PbCl₂.

Carlo Sandonnini, *Gazz. chim. ital.*, **44 I**, 357 (1914).

BaCl₂–SrCl₂

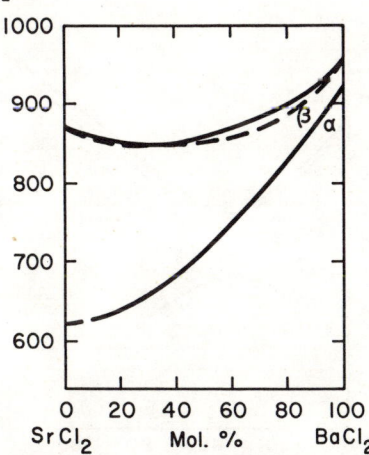

Fig. 1355.—System BaCl₂–SrCl₂.
Diagram does not obey phase rule.

Erhard Vortisch, *Neues Jahrb. Mineral., Geol.*, **38**, 198 (1914).

BaCl₂–ZnCl₂

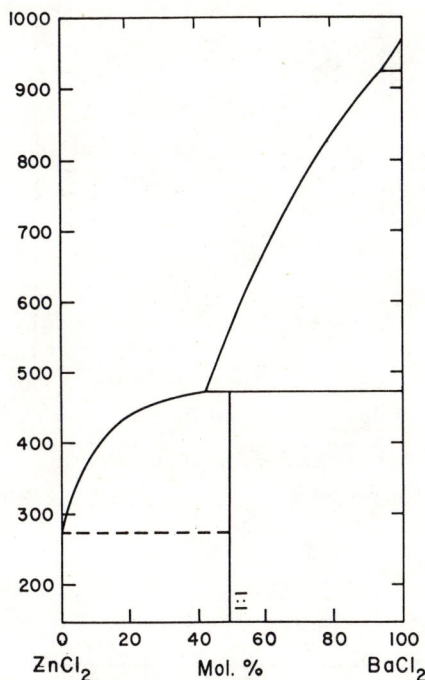

Fig. 1356.—System BaCl₂–ZnCl₂.

Carlo Sandonnini, *Gazz. chim. ital.*, **44 I**, 353 (1914).

BeCl₂–CdCl₂

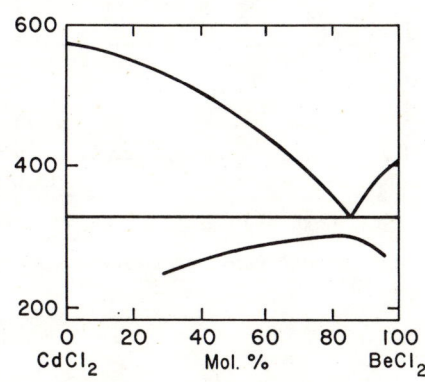

Fig. 1357.—System BeCl₂–CdCl₂.

M. J. M. Schmidt, *Bull. soc. chim., France*, **39**, 1702 (1926).

BeCl₂–PbCl₂

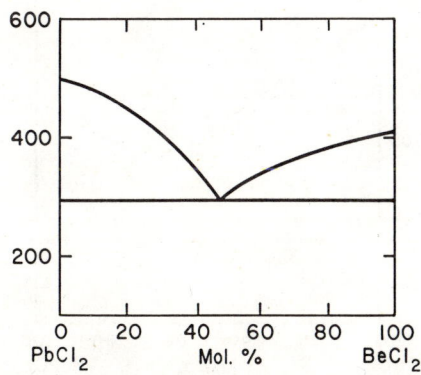

Fig. 1358.—System BeCl₂–PbCl₂.

M. J. M. Schmidt, *Bull. soc. chim., France*, **39**, 1699 (1926).

CaCl$_2$–CdCl$_2$

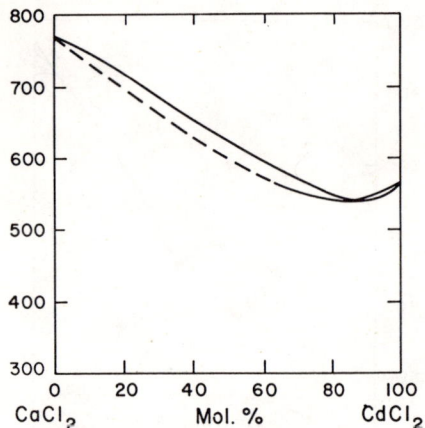

Fig. 1359.—System CaCl$_2$–CdCl$_2$.

Carlo Sandonnini, *Gazz. chim. ital.*, **44 I**, 338 (1914).

CaCl$_2$–MgCl$_2$

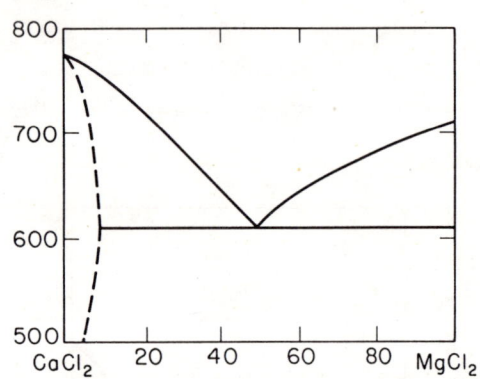

Fig. 1360.—System CaCl$_2$–MgCl$_2$.

A. I. Ivanov, *Sbornik Stateĭ Obshcheĭ Khim., Akad. Nauk S.S.S.R.*, **1**, 755 (1953).

CaCl$_2$–MnCl$_2$

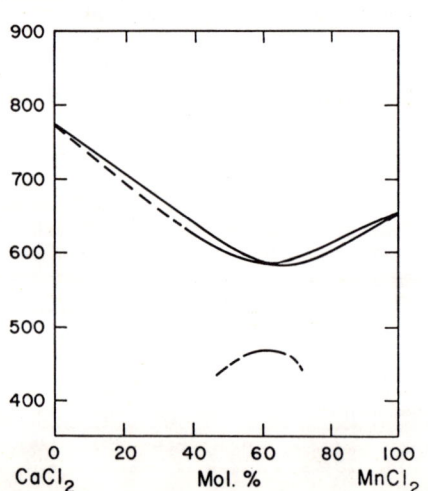

Fig. 1361.—System CaCl$_2$–MnCl$_2$.

Carlo Sandonnini, *Gazz. chim. ital.*, **44 I**, 341 (1914).

CaCl$_2$–PbCl$_2$

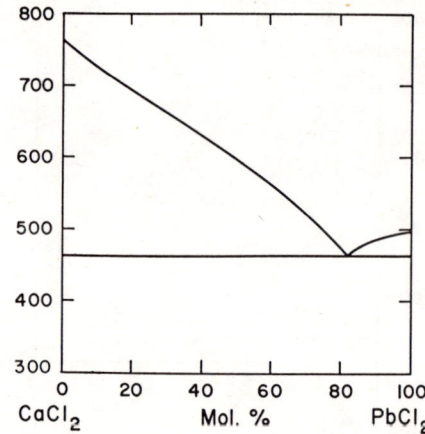

Fig. 1362.—System CaCl$_2$–PbCl$_2$.

Carlo Sandonnini, *Gazz. chim. ital.*, **44 I**, 340 (1914).
See also Otto Menge, *Z. anorg. chem.*, **72**, 206 (1911).

CaCl$_2$–SnCl$_2$

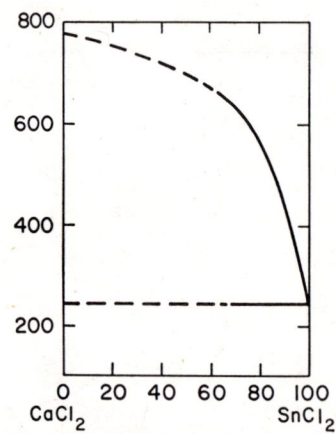

Fig. 1363.—System CaCl$_2$–SnCl$_2$.

Otto Menge, *Z. anorg. Chem.*, **72**, 212 (1911).

CaCl$_2$–SrCl$_2$

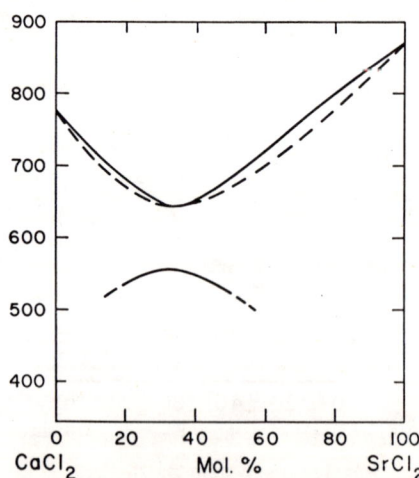

Fig. 1364.—System CaCl$_2$–SrCl$_2$.

Carlo Sandonnini, *Gazz. chim. ital.*, **44 I**, 335 (1914).

CaCl₂–ZnCl₂

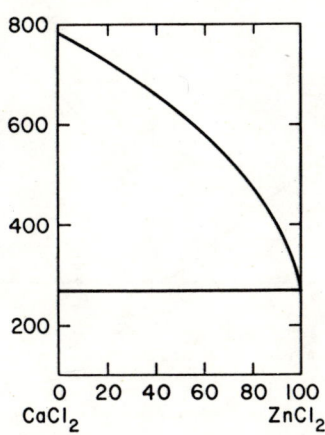

Fig. 1365.—System CaCl₂–ZnCl₂.
Otto Menge, *Z. anorg. Chem.*, **72**, 211 (1911).

CdCl₂–MgCl₂

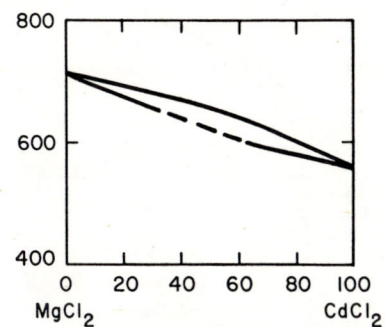

Fig. 1366.—System CdCl₂–MgCl₂.
Otto Menge, *Z. anorg. Chem.*, **72**, 196 (1911).

CdCl₂–MnCl₂

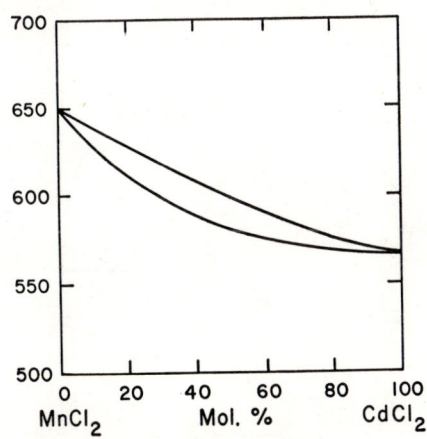

Fig. 1367.—System CdCl₂–MnCl₂.
Carlo Sandonnini, *Gazz. chim. ital.*, **44 I**, 367 (1914).

CdCl₂–PbCl₂

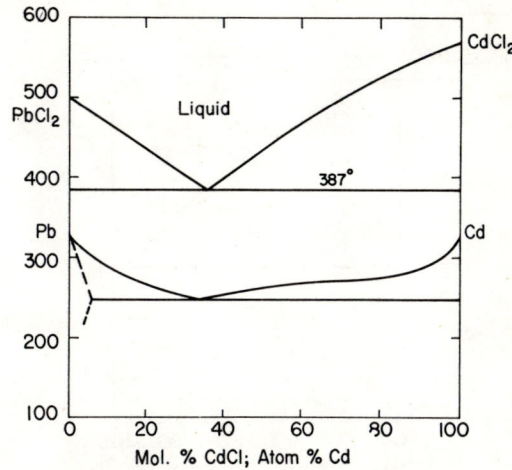

Fig. 1368.—System CdCl₂–PbCl₂; comparison with system Cd–Pb.
F. Körber and W. Oelsen, *Mitt. Kaiser-Wilhelm-Inst. Eisenforsch., Düsseldorf*, **14**, 127 (1932).

CdCl₂–SnCl₂

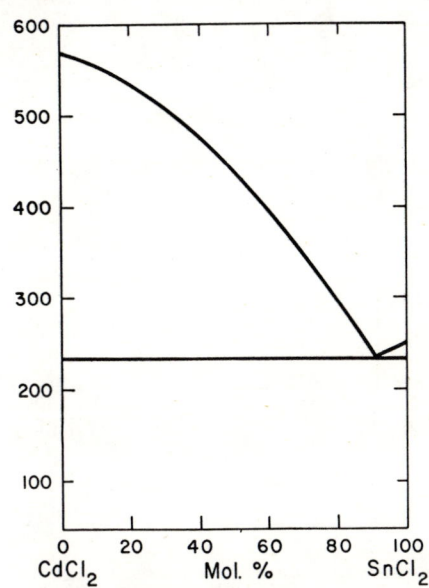

Fig. 1369.—System CdCl₂–SnCl₂.
Carlo Sandonnini, *Gazz. chim. ital.*, **44 I**, 365 (1914).

CdCl₂–SrCl₂

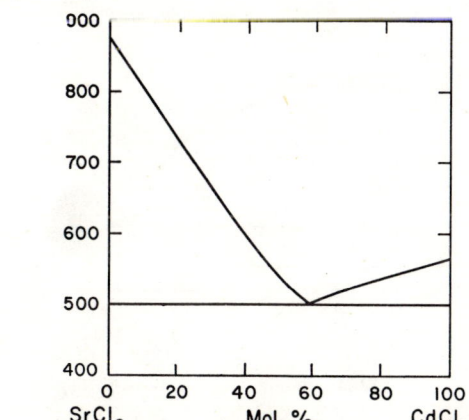

Fig. 1370.—System CdCl₂–SrCl₂.
Carlo Sandonnini, *Gazz. chim. ital.*, **44 I**, 347 (1914).

CdCl$_2$–ZnCl$_2$

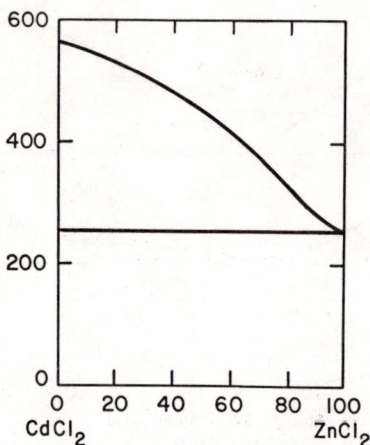

Fig. 1371.—System CdCl$_2$–ZnCl$_2$.
Gottfried Herrmann, *Z. anorg. Chem.*, **71**, 280 (1911).

MgCl$_2$–MnCl$_2$

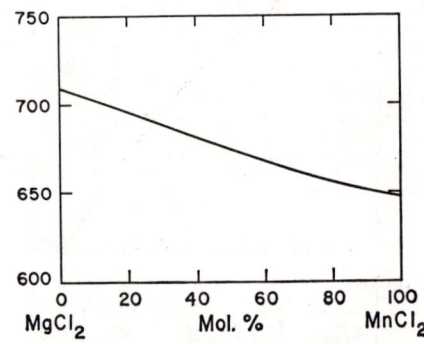

Fig. 1372.—System MgCl$_2$–MnCl$_2$.
Carlo Sandonnini, *Gazz. chim. ital.*, **44 I**, 360 (1914).

MgCl$_2$–PbCl$_2$

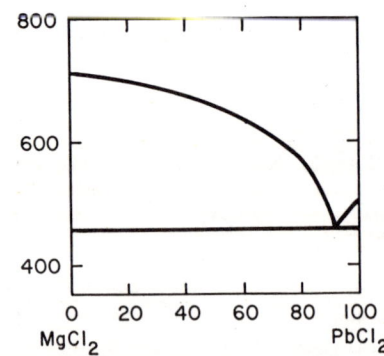

Fig. 1373.—System MgCl$_2$–PbCl$_2$.
Otto Menge, *Z. anorg. Chem.*, **72**, 187 (1911).

MgCl$_2$–SrCl$_2$

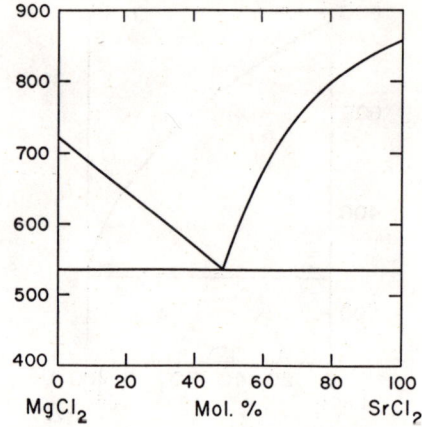

Fig. 1374.—System MgCl$_2$–SrCl$_2$.
Carlo Sandonnini, *Gazz. chim. ital.*, **44 I**, 344 (1914).

MgCl$_2$–TiCl$_2$

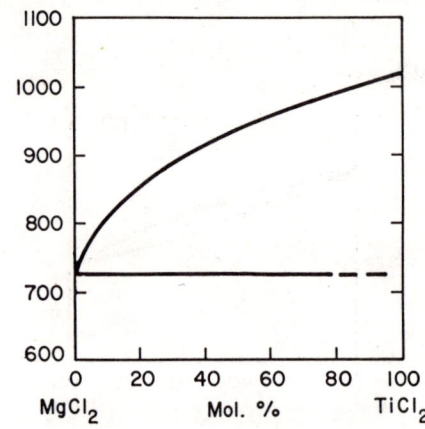

Fig. 1375.—System MgCl$_2$–TiCl$_2$. Solubility of TiCl$_2$ in MgCl$_2$(s) less than 0.35 wt% at 715°C.
K. Komarek and P. Herasymenko, *J. Electrochem. Soc.*, **105**, 214 (1958).

MgCl$_2$–ZnCl$_2$

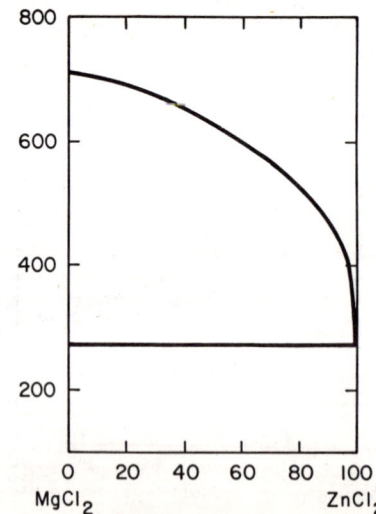

Fig. 1376.—System MgCl$_2$–ZnCl$_2$.
Otto Menge, *Z. anorg. Chem.*, **72**, 191 (1911).

MgCl$_2$–CeCl$_3$

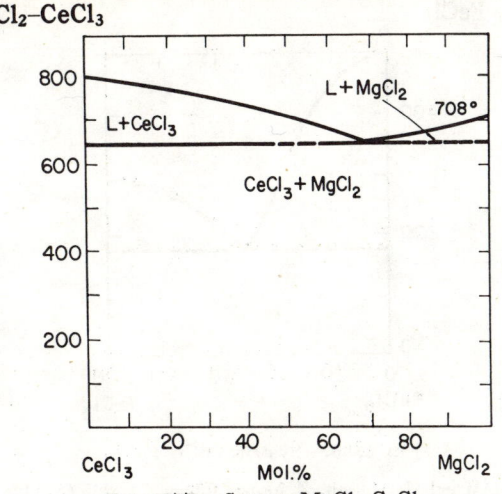

Fig. 1377.—System MgCl$_2$–CeCl$_3$.

In-Chzhu Sun and I. S. Morozov, *Zhur. Neorg. Khim.*, **3**, 1918 (1958).

MnCl$_2$–PbCl$_2$

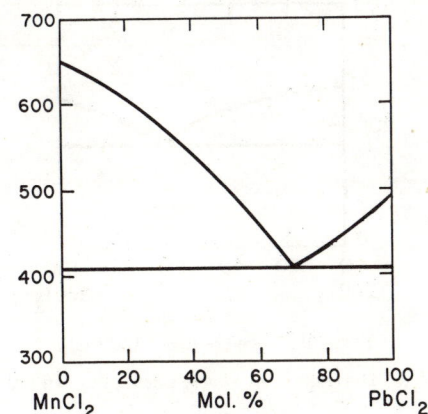

Fig. 1378.—System MnCl$_2$–PbCl$_2$.

Carlo Sandonnini, *Gazz. chim. ital.*, **44 I**, 373 (1914).

MnCl$_2$–SnCl$_2$

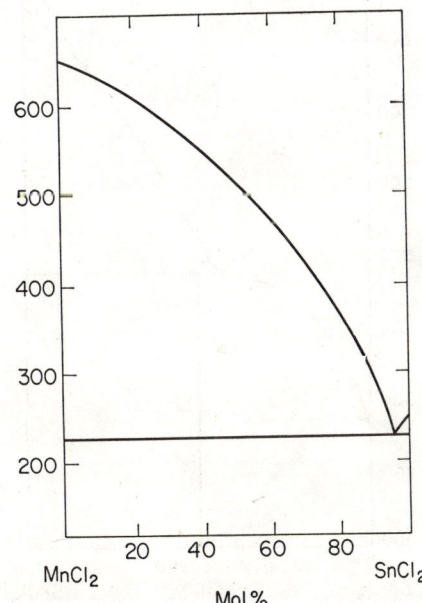

Fig. 1379.—System MnCl$_2$–SnCl$_2$.

Carlo Sandonnini, *Gazz. chim. ital.*, **44 I**, 371 (1914).

MnCl$_2$–SrCl$_2$

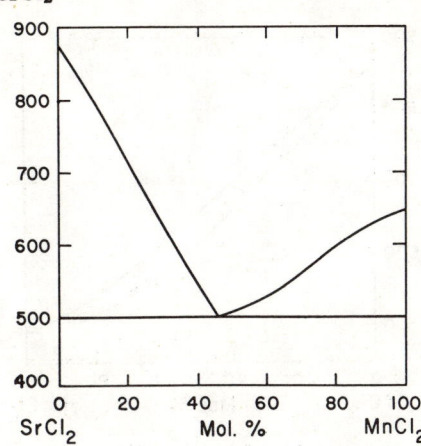

Fig. 1380.—System MnCl$_2$–SrCl$_2$.

Carlo Sandonnini, *Gazz. chim. ital.*, **44 I**, 349 (1914).

MnCl$_2$–ZnCl$_2$

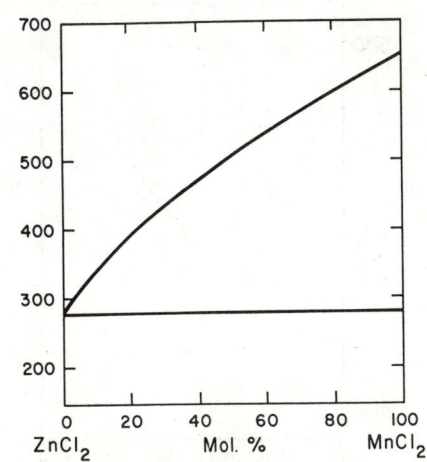

Fig. 1381.—System MnCl$_2$–ZnCl$_2$.

Carlo Sandonnini, *Gazz. chim. ital.*, **44 I**, 362 (1914).

PbCl$_2$–SnCl$_2$

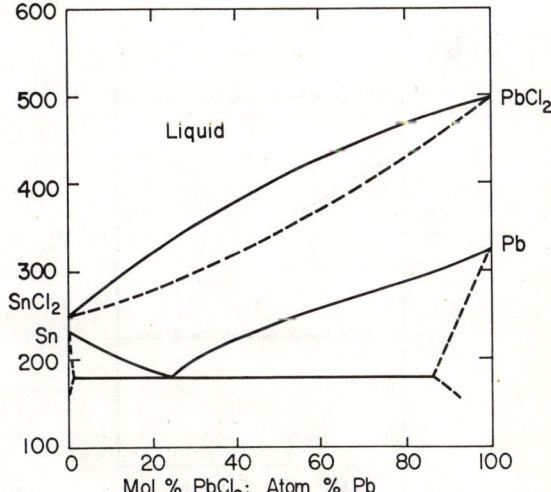

Fig. 1382.—System PbCl$_2$–SnCl$_2$; comparison with system Sn–Pb.

F. Körber and W. Oelsen, *Mitt. Kaiser-Wilhelm-Inst. Eisenforsch., Düsseldorf*, **14**, 127 (1932).

$PbCl_2$–$SrCl_2$

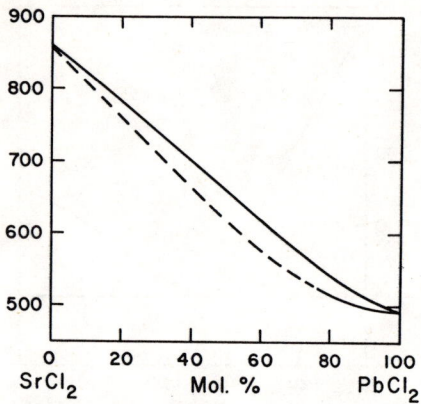

Fig. 1383.—System $PbCl_2$–$SrCl_2$.

Carlo Sandonnini, *Gazz. chim. ital.*, **44** I, 348 (1914).

$PbCl_2$–$ZnCl_2$

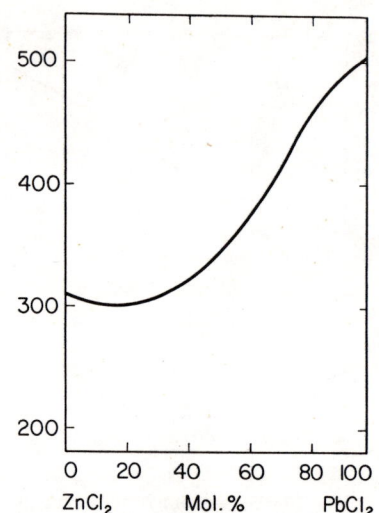

Fig. 1384.—System $PbCl_2$–$ZnCl_2$.

Ya A. Ugai and V. A. Shatillo, *J. Phys. Chem., U.S.S.R.*, **23** [7] 746 (1949). See also Gottfried Herrmann, *Z. anorg. Chem.*, **71**, 282 (1911).

$PbCl_2$–$BiCl_3$

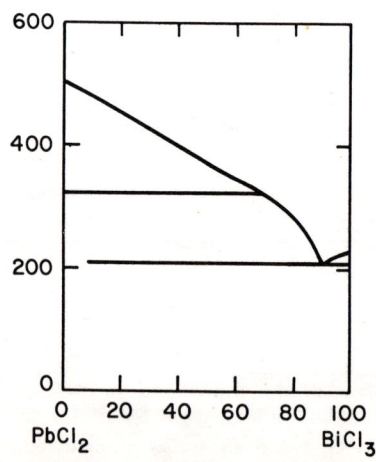

Fig. 1385.—System $PbCl_2$–$BiCl_3$.

Gottfried Herrmann, *Z. anorg. Chem.*, **71**, 285 (1911).

$PbCl_2$–$FeCl_3$

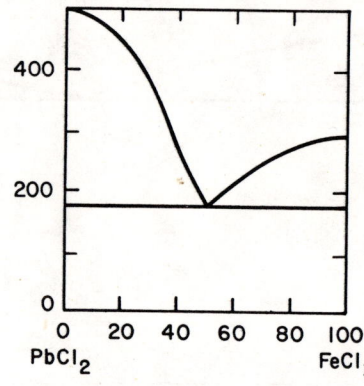

Fig. 1386.—System $PbCl_2$–$FeCl_3$.

Gottfried Herrmann, *Z. anorg. Chem.*, **71**, 265 (1911).

$SnCl_2$–$ZnCl_2$

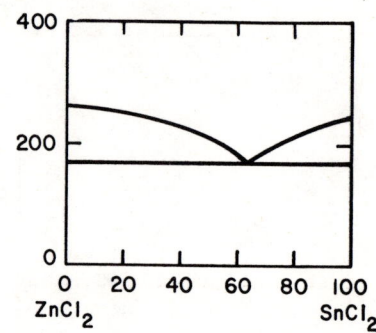

Fig. 1387.—System $SnCl_2$–$ZnCl_2$.

Gottfried Herrmann, *Z. anorg. Chem.*, **71**, 276 (1911).

$SrCl_2$–$ZnCl_2$

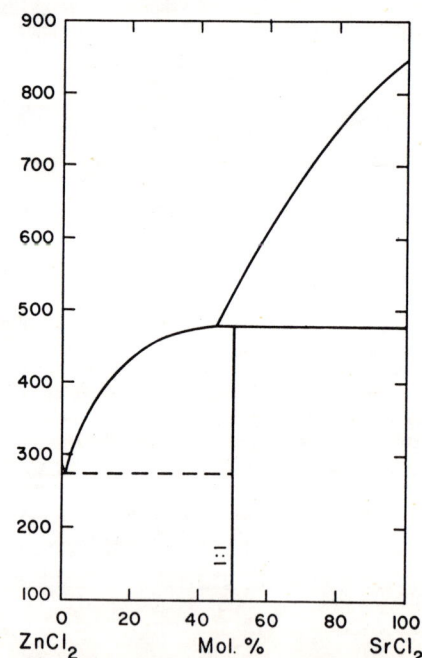

Fig. 1388.—System $SrCl_2$–$ZnCl_2$.

Carlo Sandonnini, *Gazz. chim. ital.*, **44** I, 346 (1914).

ZnCl₂–BiCl₃

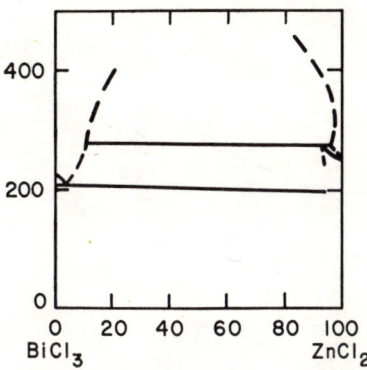

Fig. 1389.—System $ZnCl_2$–$BiCl_3$.

Gottfried Herrmann, Z. anorg. Chem., **71**, 283 (1911).

ZnCl₂–FeCl₃

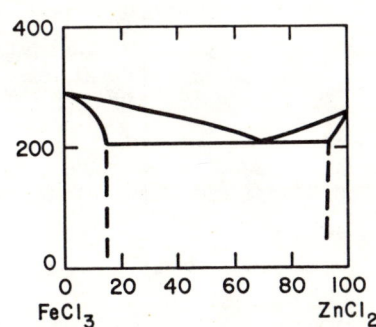

Fig. 1390.—System $ZnCl_2$–$FeCl_3$.

Gottfried Herrmann, Z. anorg. Chem., **71**, 278 (1911).

AlCl₃–SeCl₄

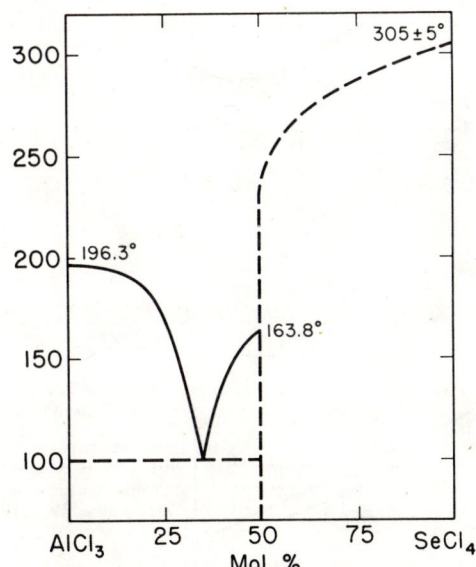

Fig. 1391.—System $AlCl_3$–$SeCl_4$.

H. Houtgraaf, H. J. Rang, and L. Vollbracht, Rec. trav. chim., **72**, 983 (1953).

AlCl₃–TeCl₄

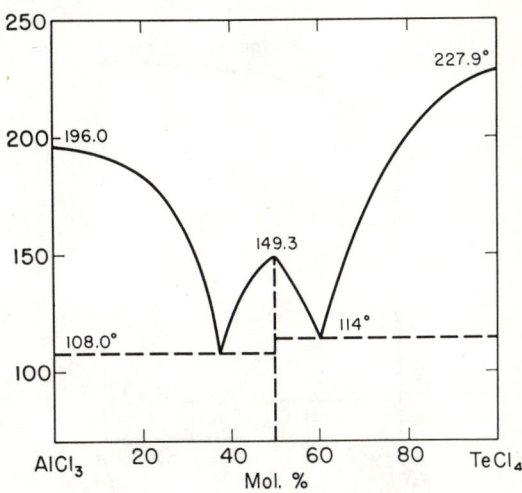

Fig. 1392.—System $AlCl_3$–$TeCl_4$.

H. Houtgraaf, H. J. Rang, and L. Vollbracht, Rec. trav. chim., **72**, 987 (1953).

AlCl₃–TiCl₄

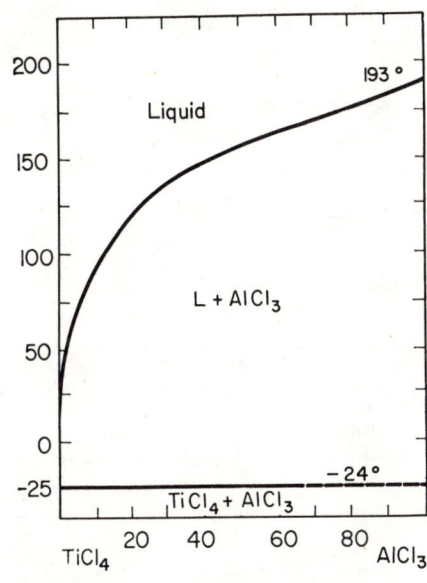

Fig. 1393.—System $AlCl_3$–$TiCl_4$.

I. S. Morozov and D. Ya. Toptygin, Zhur. Neorg. Khim., **2**, 1917 (1957).

BiCl₃–FeCl₃

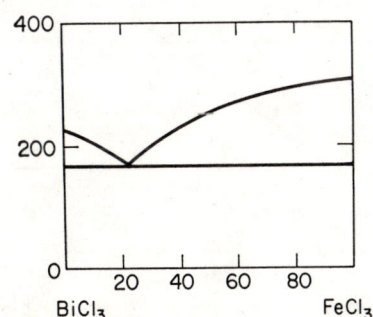

Fig. 1394.—System $BiCl_3$–$FeCl_3$.

Gottfried Herrmann, Z. anorg. Chem., **71**, 277 (1911).

FeCl$_3$–TiCl$_4$

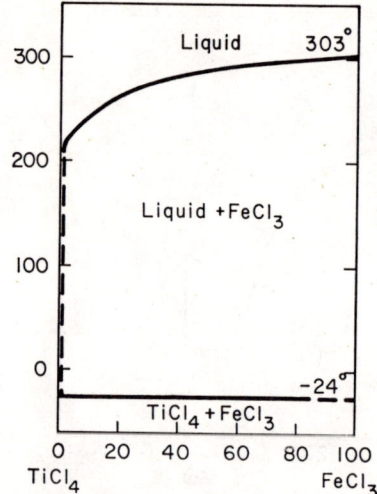

Fig. 1395.—System FeCl$_3$–TiCl$_4$.
I. S. Morozov and D. Ya. Toptygin, *Zhur. Neorg. Khim.*, **2**, 2130 (1957).

CCl$_4$–PbCl$_4$

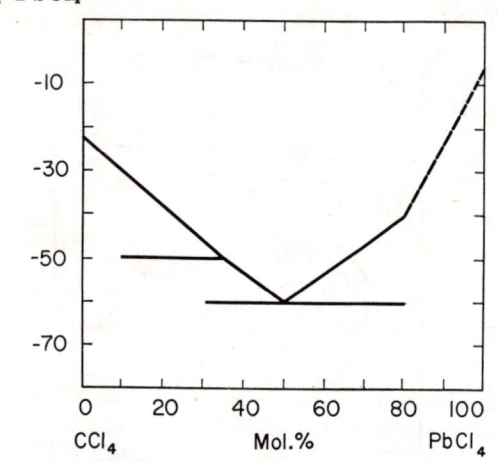

Fig. 1396.—System CCl$_4$–PbCl$_4$.
Horst Sackmann, *Z. Physik. Chem. (Leipzig)*, **207** 243 (1957).

PbCl$_4$–SiCl$_4$

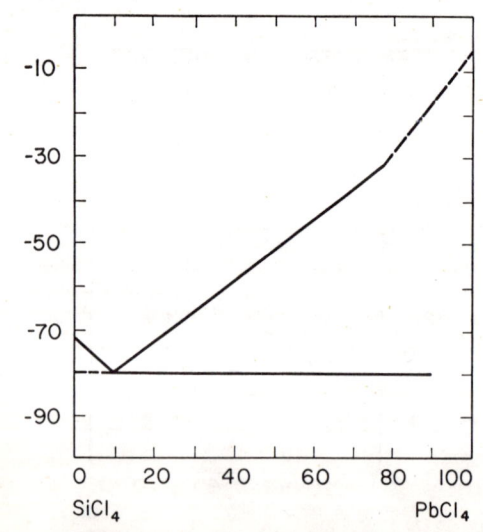

Fig. 1397.—System PbCl$_4$–SiCl$_4$.
Horst Sackmann, *Z. Physik. Chem. (Leipzig)*, **207**, 245 (1957).

PbCl$_4$–SnCl$_4$

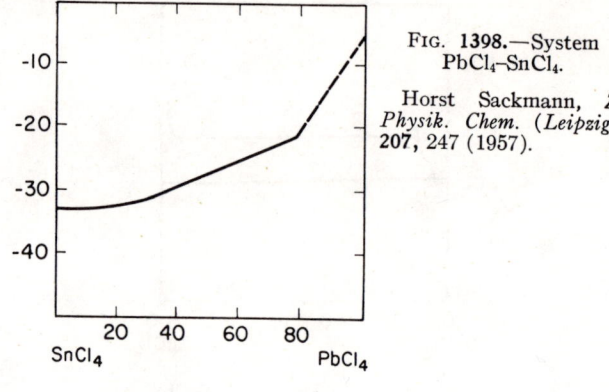

Fig. 1398.—System PbCl$_4$–SnCl$_4$.

Horst Sackmann, *Z. Physik. Chem. (Leipzig)*, **207**, 247 (1957).

PbCl$_4$–TiCl$_4$

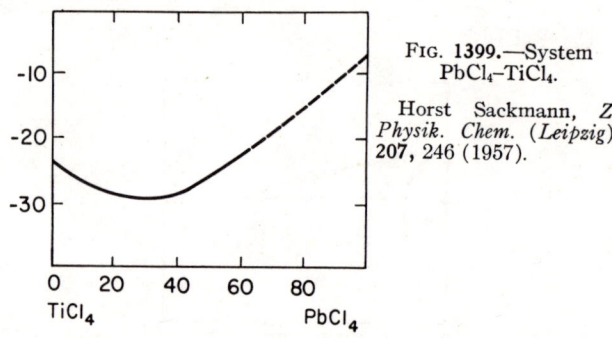

Fig. 1399.—System PbCl$_4$–TiCl$_4$.

Horst Sackmann, *Z. Physik. Chem. (Leipzig)*, **207**, 246 (1957).

SiCl$_4$–SnCl$_4$

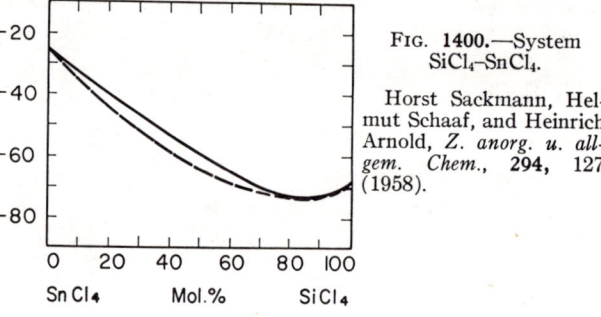

Fig. 1400.—System SiCl$_4$–SnCl$_4$.

Horst Sackmann, Helmut Schaaf, and Heinrich Arnold, *Z. anorg. u. allgem. Chem.*, **294**, 127 (1958).

SiCl$_4$–TiCl$_4$

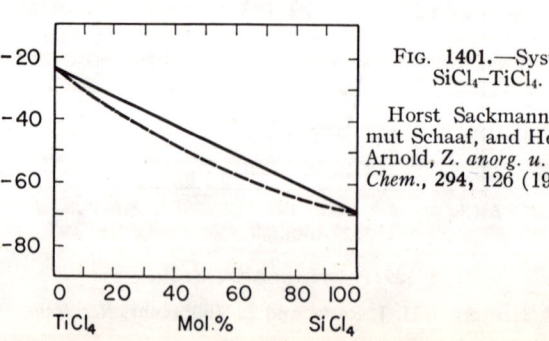

Fig. 1401.—System SiCl$_4$–TiCl$_4$.

Horst Sackmann, Helmut Schaaf, and Heinrich Arnold, *Z. anorg. u. allgem. Chem.*, **294**, 126 (1958).

(c) Three Chlorides

TiCl₄–NbCl₅

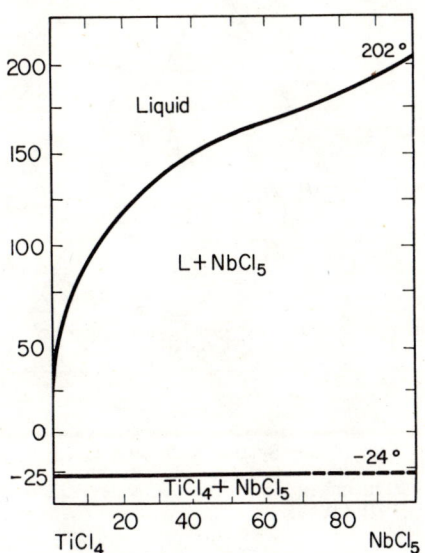

Fig. 1402.—System TiCl₄–NbCl₅.

I. S. Morozov and D. Ya. Toptygin, *Zhur. Neorg. Khim.*, **2**, 1918 (1957).

CsCl–KCl–CaCl₂

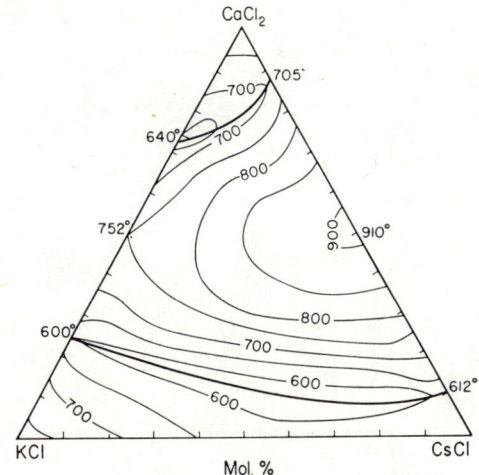

Fig. 1404.—System CsCl–KCl–CaCl₂.

V. E. Plyushchev, L. N. Komissarova, L. V. Meshchaninova, and L. M. Akulkina, *Zhur. Neorg. Khim.*, **1**, 830 (1956).

TiCl₄–TaCl₅

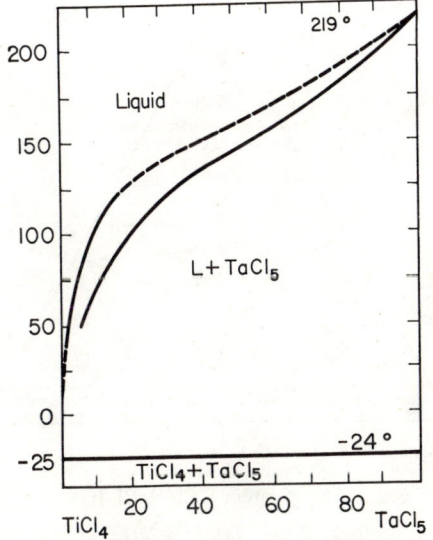

Fig. 1403.—System TiCl₄–TaCl₅.

I. S. Morozov and D. Ya. Toptygin, *Zhur. Neorg. Khim.*, **2**, 1917 (1957).

CsCl–NaCl–CaCl₂

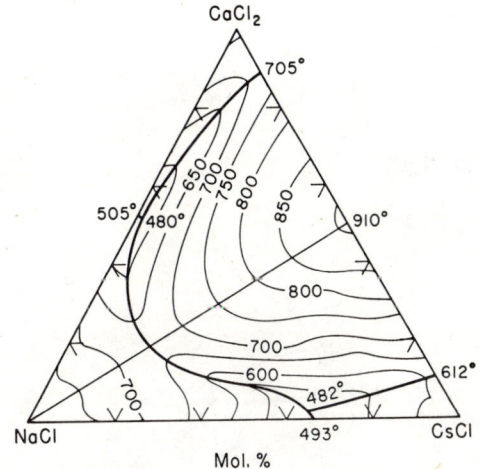

Fig. 1405.—System CsCl–NaCl–CaCl₂.

V. E. Plyushchev, I. V. Shakhno, and S. A. Pozhitkova, *Zhur. Obshchei Khim.*, **25**, 1073 (1955).

CsCl–AlCl₃–TaCl₅

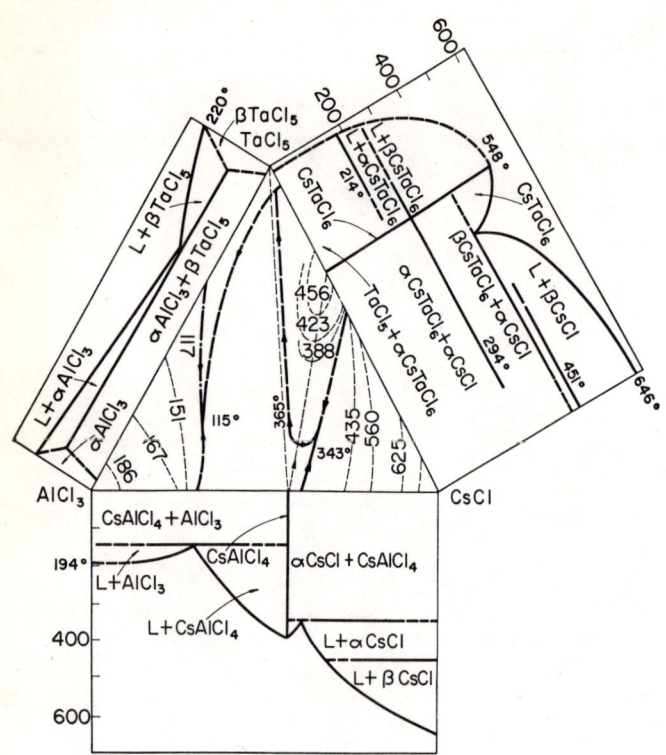

Fig. 1406.—System CsCl–AlCl₃–TaCl₅.

I. S. Morozov and A. T. Simonich, *Zhur. Neorg. Khim.*, **2**, 1909 (1957).

KCl–LiCl–CaCl₂

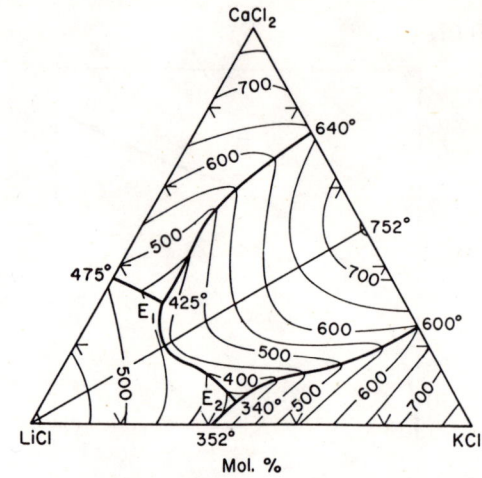

Fig. 1408.—System KCl–LiCl–CaCl₂.

V. E. Plyushchev and F. V. Kovalev, *Zhur. Neorg. Khim.*, **1**, 1016 (1956).

KCl–NaCl–BaCl₂

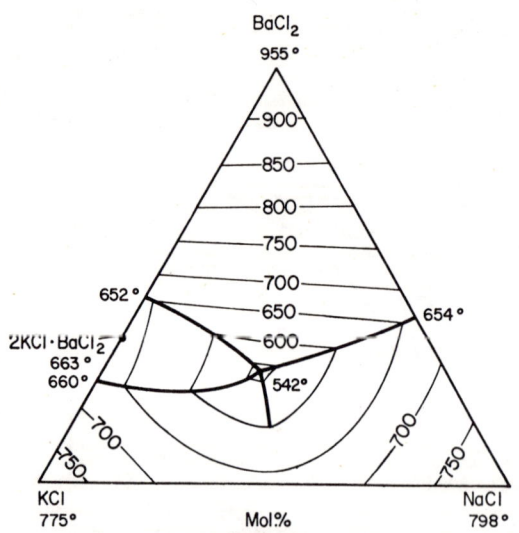

Fig. 1409.—System KCl–NaCl–BaCl₂.

Erhard Vortisch, *Neues Jahrb. Mineral., Geol.*, **38**, 521 (1914).

KCl–LiCl–NaC

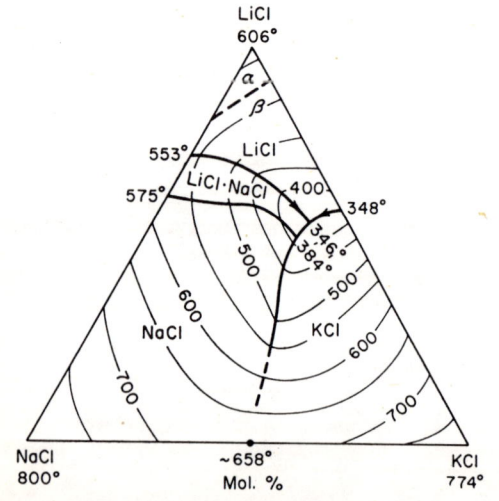

Fig. 1407.—System KCl–LiCl–NaCl.

E. K. Akopov, *Zhur. Neorg. Khim.*, **1**, 1024 (1956).

KCl–NaCl–CaCl$_2$

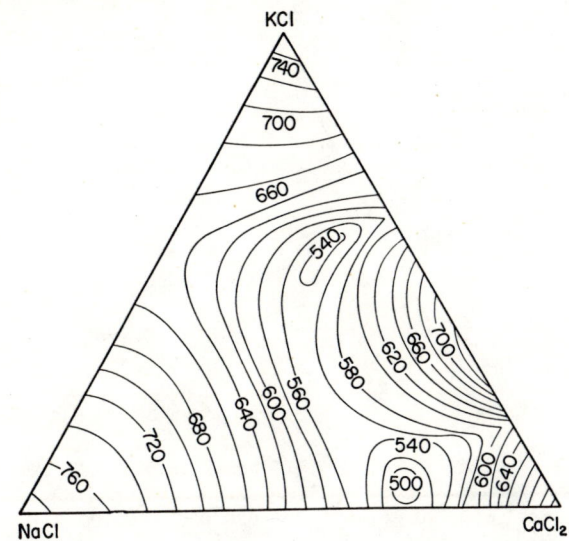

Fig. 1410.—System KCl–NaCl–CaCl$_2$.

F. C. A. H. Lantsberry and R. A. Page, *J. Soc. Chem. Ind.*, **39**, 40T (1920).

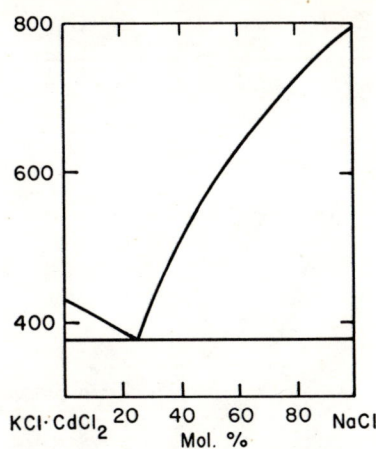

Fig. 1412.—System NaCl–KCl·CdCl$_2$.

Hermann Brand, *Neues Jahrb. Mineral., Geol.*, **32**, 663 (1911).

KCl–NaCl–PbCl$_2$

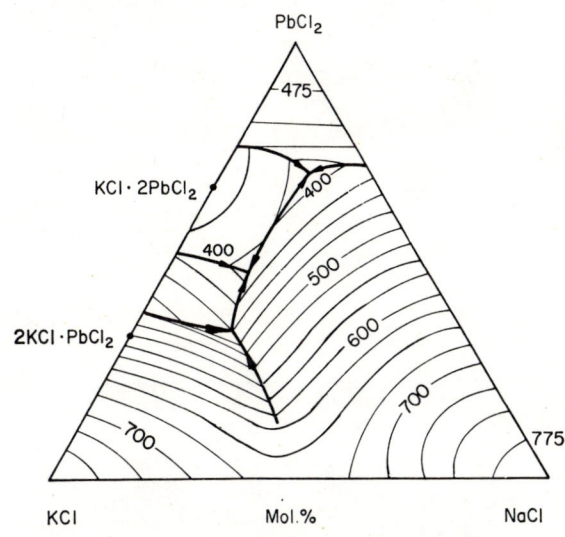

Fig. 1413.—System KCl–NaCl–PbCl$_2$.

Kurt Treis, *Neues Jahrb. Mineral., Geol.*, **37**, 812 (1914)

KCl–NaCl–CdCl$_2$

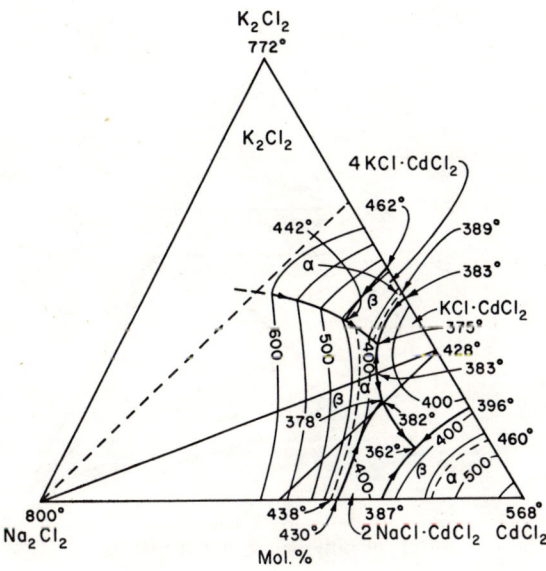

Fig. 1411.—System KCl–NaCl–CdCl$_2$.

I. I. Il'yasov, A. K. Bostandzhiyan, and A. G. Bergman' *Zhur. Neorg. Khim.*, **2**, 177 (1957).

KCl–NaCl–TiCl$_3$

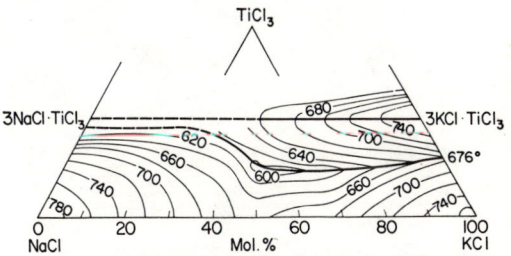

Fig. 1414.—System KCl–NaCl–TiCl$_3$.

M. V. Kamenetskiĭ, *Tsvetn. Metal.*, **31**, [2] 44 (1958).

KCl–NaCl–ZrCl₄

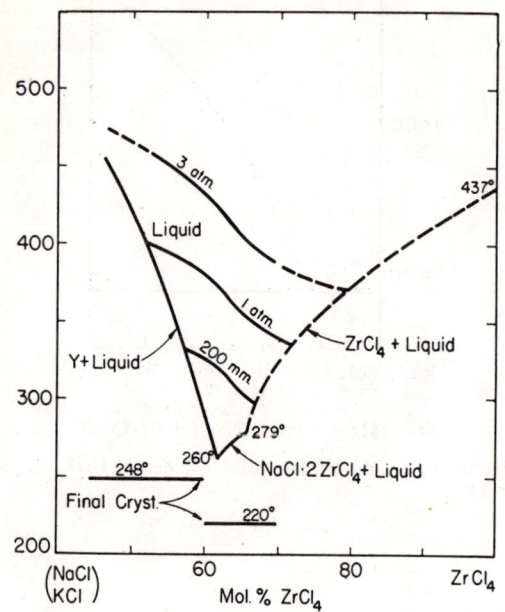

Fig. 1415.—System NaCl·KCl–ZrCl₄. Isobars show v.p. of ZrCl₄ in equilibrium with melt.

L. J. Howell, R. C. Sommer, and H. H. Kellogg, *J. Metals*, **9**, *Trans. Am. Inst. Mining Met. Engrs.*, **209**, 197 (1957).

KCl–BaCl₂–MgCl₂

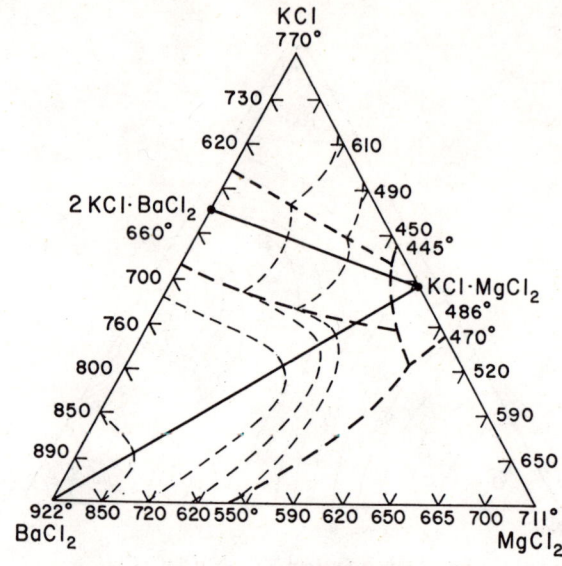

Fig. 1417.—System KCl–BaCl₂–MgCl₂.

C. Matignon and J. Valentin, *Bull. soc. chim.*, France, **33**, 279 (1923).

KCl–RbCl–CaCl₂

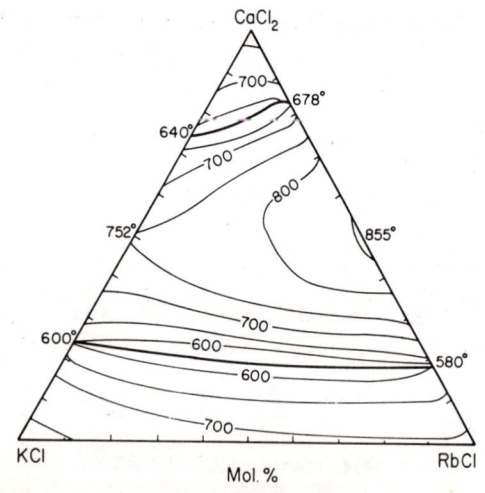

Fig. 1416.—System KCl–RbCl–CaCl₂.

V. E. Plyushchev, L. N. Komissarova, L. V. Meshchaninova, and L. M. Akulkina, *Zhur. Neorg. Khim.*, **1**, 826 (1956).

KCl–BaCl₂–SrCl₂

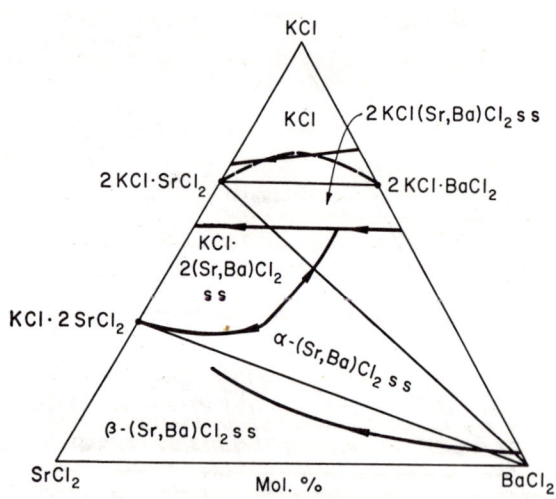

Fig. 1418.—System KCl–BaCl₂–SrCl₂.

Erhard Vortisch, *Neues Jahrb. Mineral., Geol.*, **38**, 260 (1914).

KCl–CaCl$_2$–MgCl$_2$

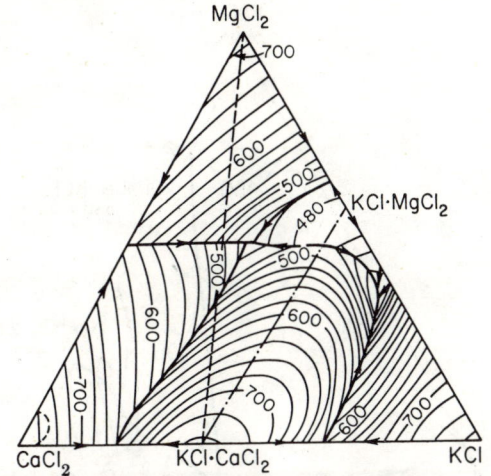

FIG. 1419.—System KCl–CaCl$_2$–MgCl$_2$.

A. I. Ivanov, *Izvest. Sektora Fiz.-Khim. Anal., Inst. Obshchei Neorg. Khim., Akad. Nauk S.S.S.R.*, **23**, 197 (1953).

KCl–CaCl$_2$–PbCl$_2$

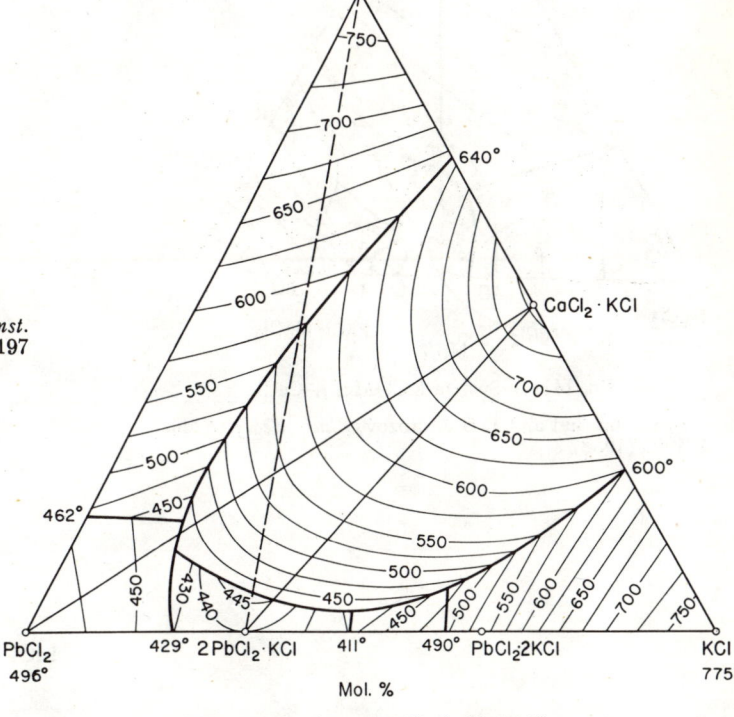

FIG. 1421.—System KCl–CaCl$_2$–PbCl$_2$.

Ya. A. Ugai, *Doklady Akad. Nauk S.S.S.R.*, **70** [4] 653 (1950).

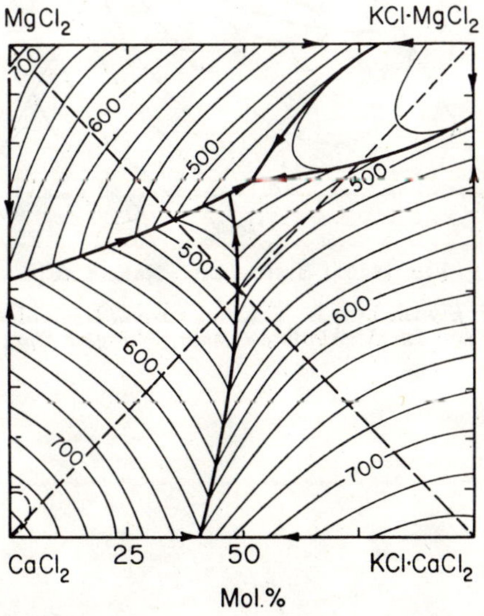

FIG. 1420.—System CaCl$_2$–MgCl$_2$–KCl·MgCl$_2$–KCl·CaCl$_2$.

A. I. Ivanov, *Izvest. Sektora Fiz.-Khim. Anal., Inst. Obshchei Neorg. Khim., Akad. Nauk S.S.S.R.*, **23**, 197 (1953).

KCl–CdCl$_2$–PbCl$_2$

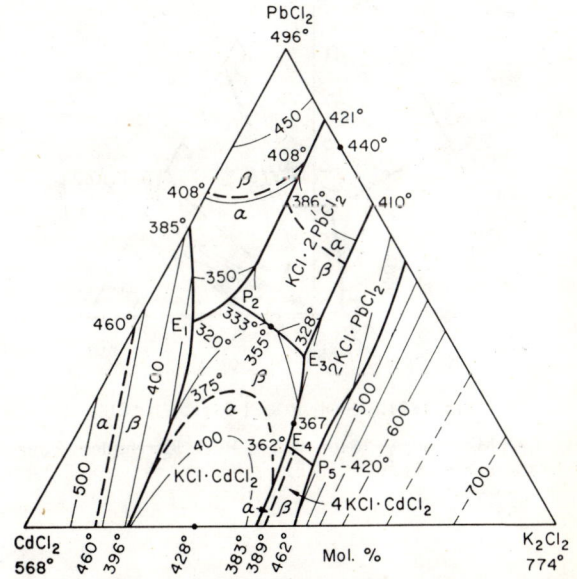

FIG. 1422.—System KCl–CdCl$_2$–PbCl$_2$.

I. I. Il'yasov, A. K. Bostandzhian, and A. G. Bergman, *Zhur. Neorg. Khim.*, **1** [11] 2549 (1956).

KCl–MgCl₂–CeCl₃

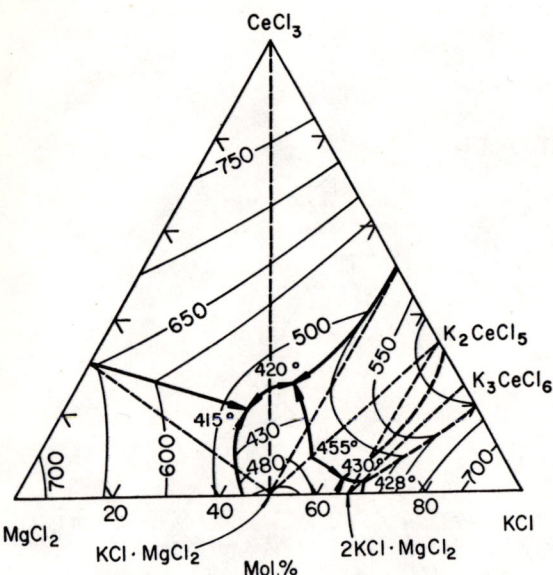

Fig. 1423.—System KCl–MgCl₂–CeCl₃.

In-Chzhu Sun and I. S. Morozov, *Zhur. Neorg. Khim.*, **3**, 1923 (1958).

KCl–PbCl₂–ZnCl₂

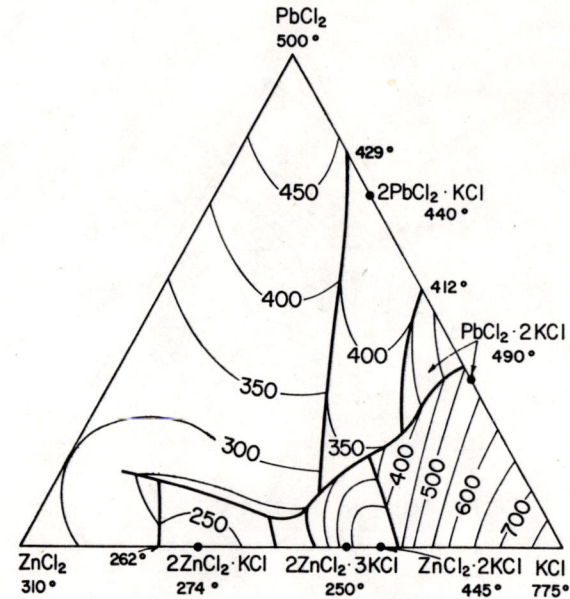

Fig. 1425.—System KCl–PbCl₂–ZnCl₂.

Ya A. Ugai and V. A. Shatillo, *J. Phys. Chem. U.S.S.R.*, **23**, [6] 751 (1949).

KCl–MgCl₂–NdCl₃

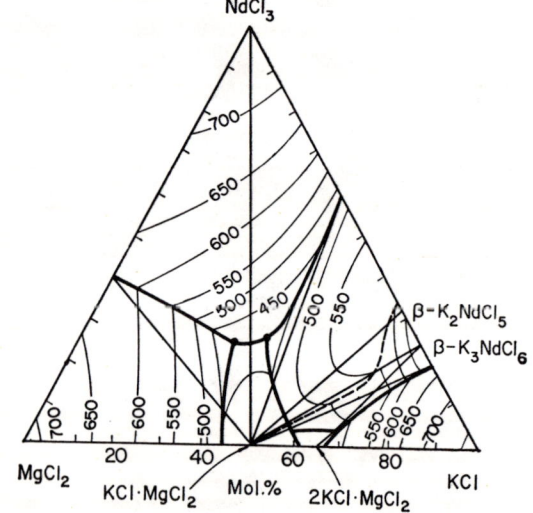

Fig. 1424.—System KCl–MgCl₂–NdCl₃.

I. S. Morozov, V. I. Ionov, and B. G. Korshunov, *Zhur. Neorg. Khim.*, **4**, 1458 (1959).

LiCl–NaCl–CaCl₂

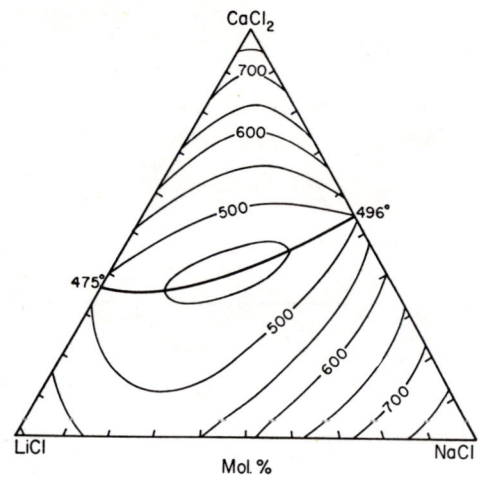

Fig. 1426.—System LiCl–NaCl–CaCl₂.

V. E. Plyushchev, L. N. Komissarova, L. V. Meshchaninova, and L. M. Akulkina, *Zhur. Neorg. Khim.*, **1**, 822 (1956).

NaCl–RbCl–CaCl₂

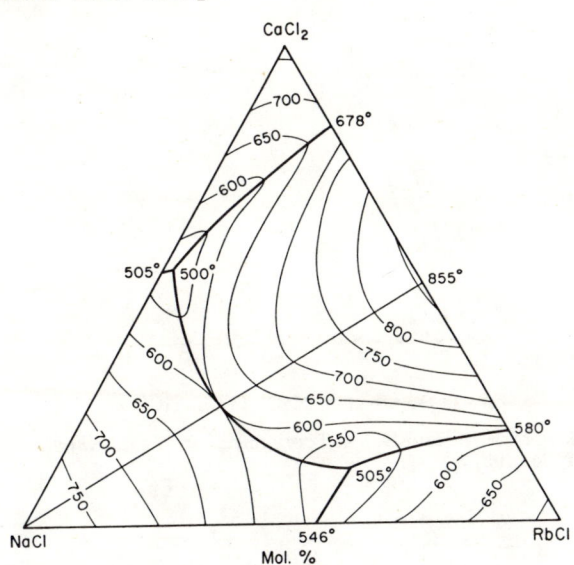

FIG. 1427.—System NaCl–RbCl–CaCl₂.

V. E. Plyushchev, F. V. Kovalev, and I. V. Shakhno, *Zhur. Obshcheĭ Khim.*, **25**, 857 (1955).

NaCl–CaCl₂–LaCl₃

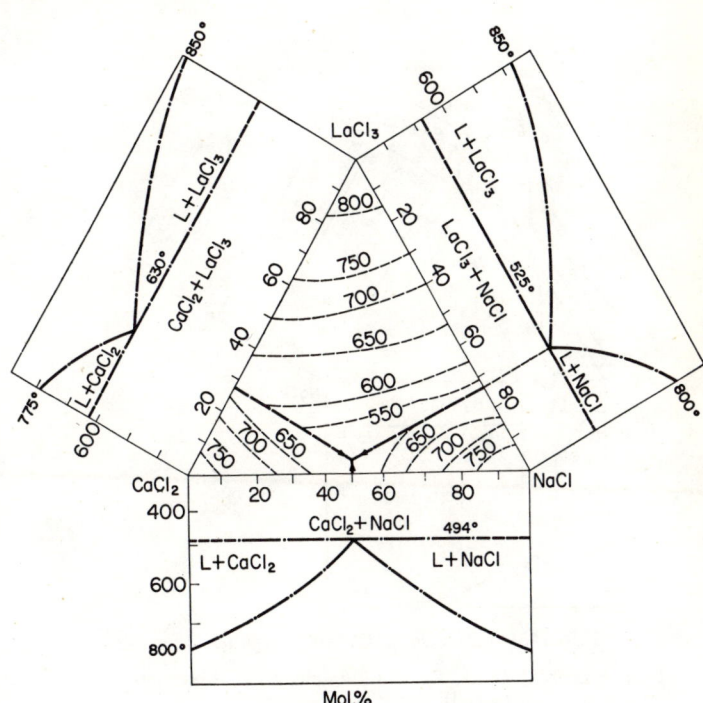

FIG. 1429.—System NaCl–CaCl₂–LaCl₃.

I. S. Morozov, Z. N. Shevtsova, and L. V. Klyukina, *Zhur. Neorg. Khim.*, **2**, 1640 (1957).

NaCl–BaCl₂–SrCl₂

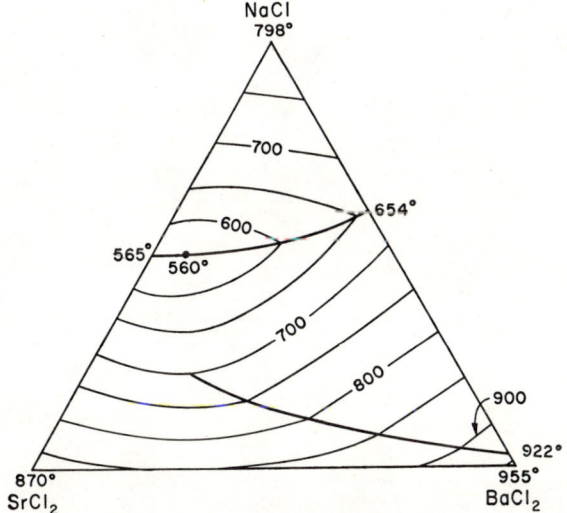

FIG. 1428.—System NaCl–BaCl₂–SrCl₂.

Erhard Vortisch, *Neues Jahrb. Mineral., Geol.*, **38**, 215 (1914).

NaCl–CaCl₂–NdCl₃

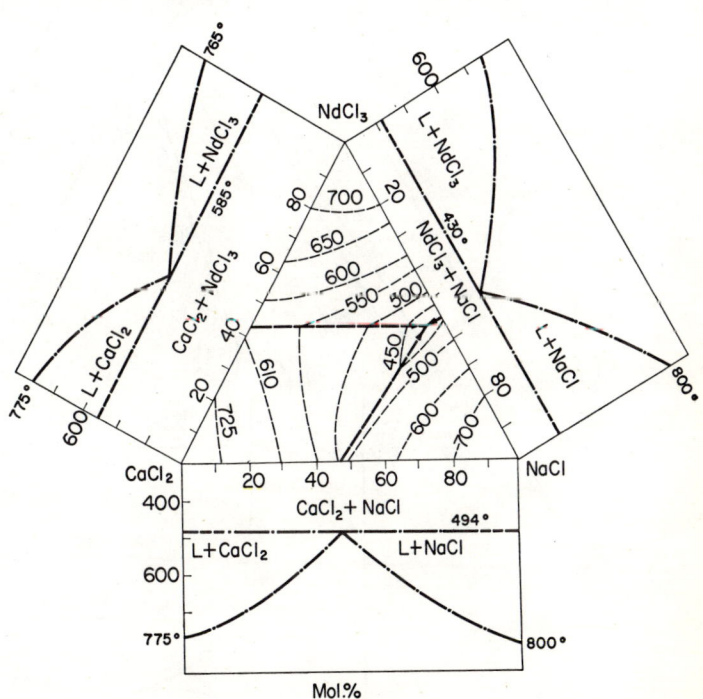

FIG. 1430.—System NaCl–CaCl₂–NdCl₃.

I. S. Morozov, Z. N. Shevtsova, and L. V. Klyukina, *Zhur. Neorg. Khim.*, **2**, 1639 (1957).

NaCl–AlCl$_3$–NbCl$_5$

NaCl–FeCl$_3$–TiCl$_4$

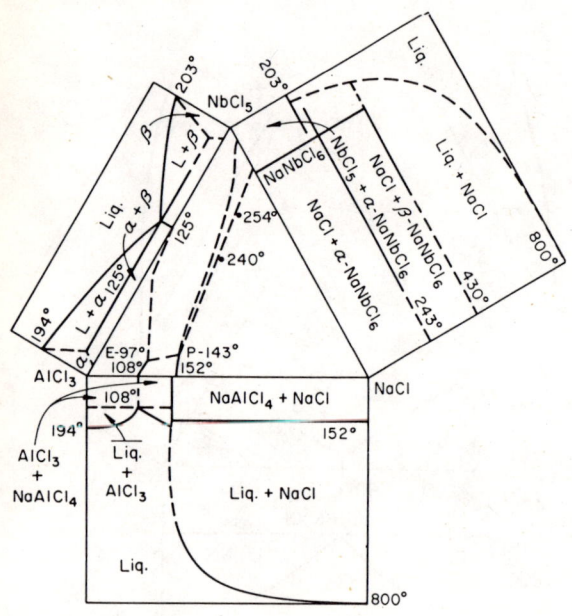

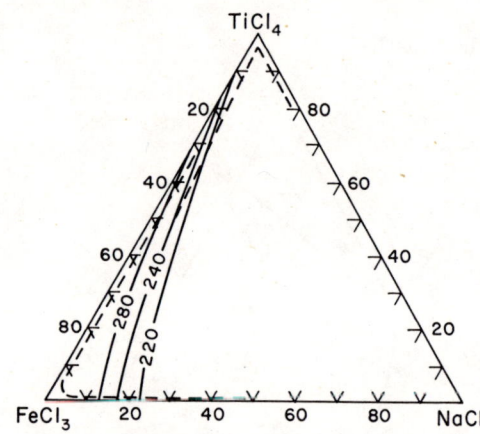

Fig. 1433.—System NaCl–FeCl$_3$–TiCl$_4$.

I. S. Morozov and D. Ya. Toptygin, *Zhur. Neorg. Khim.*, **2**, 2134 (1957).

Fig. 1431.—System NaCl–AlCl$_3$–NbCl$_5$.

I. S. Morozov, B. G. Korshunov, and A. T. Simonich, *Zhur. Neorg. Khim.*, **1**, 1650 (1956).

NaCl–AlCl$_3$–TaCl$_5$

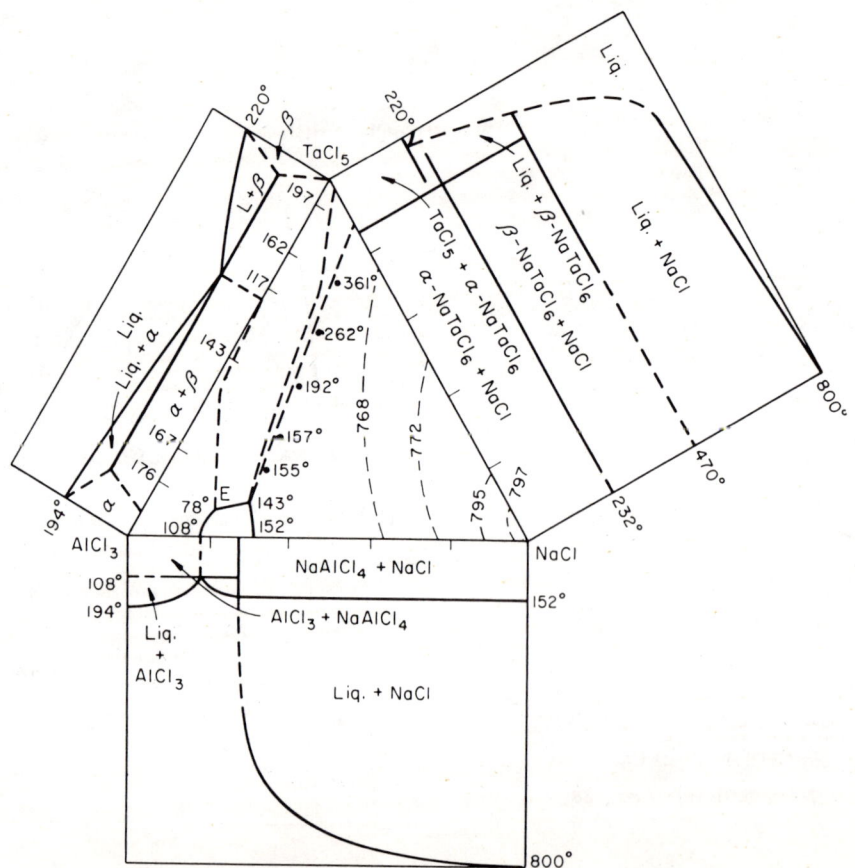

Fig. 1432.—System NaCl–AlCl$_3$–TaCl$_5$.

I. S. Morozov, B. G. Korshunov, and A. T. Simonich, *Zhur. Neorg. Khim.*, **1**, 1650 (1956).

NaCl–ZrCl$_4$–NbCl$_5$

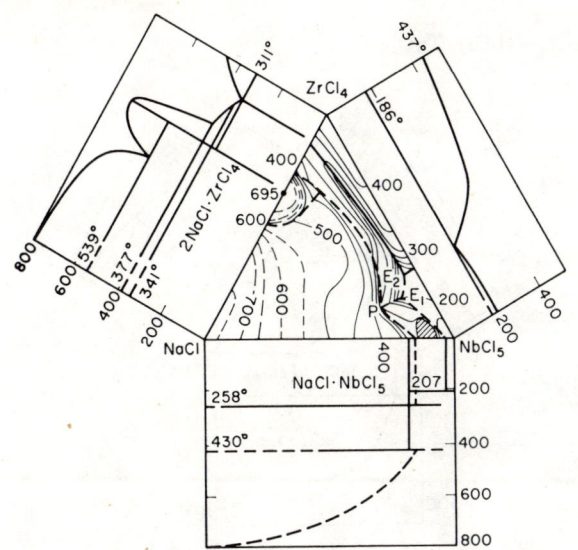

Fig. 1434.—System NaCl–ZrCl$_4$–NbCl$_5$.

I. S. Morozov and B. G. Korshunov, *Zhur. Neorg. Khim.*, 1 [1] 151 (1956).

AlCl$_3$–FeCl$_3$–NbCl$_5$

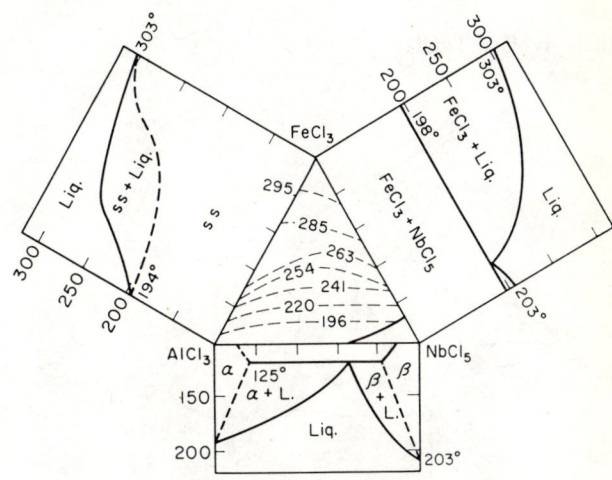

Fig. 1436.—System AlCl$_3$–FeCl$_3$–NbCl$_5$.

I. S. Morozov, *Zhur. Neorg. Khim.*, 1, 2798 (1956).

AlCl$_3$–FeCl$_3$–TiCl$_4$

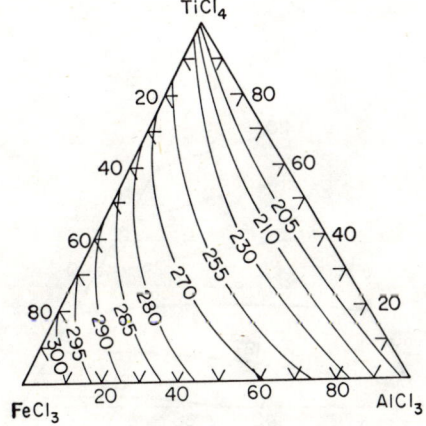

Fig. 1435.—System AlCl$_3$–FeCl$_3$–TiCl$_4$.

I. S. Morozov and D. Ya. Toptygin, *Zhur. Neorg. Khim.*, 2, 2131 (1957).

AlCl$_3$–FeCl$_3$–TaCl$_5$

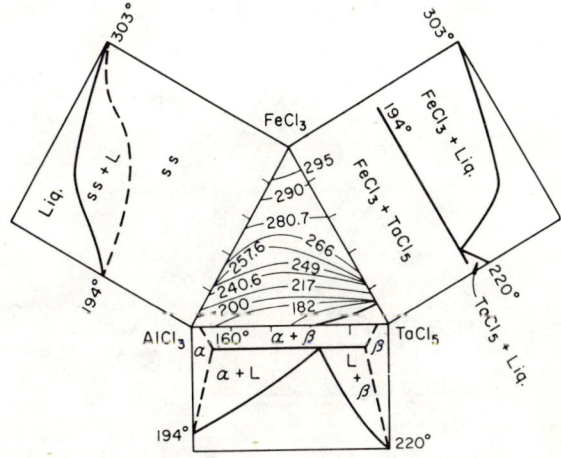

Fig. 1437.—System AlCl$_3$–FeCl$_3$–TaCl$_5$.

I. S. Morozov, *Zhur. Neorg. Khim.*, 1, 2796 (1956).

AlCl₃–NbCl₅–TaCl₅

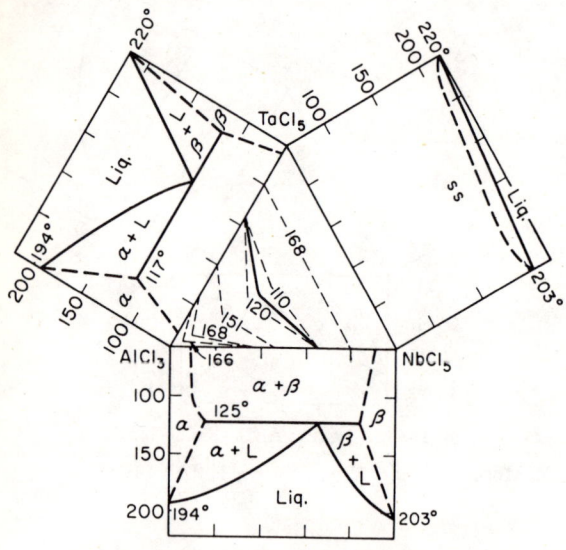

FIG. 1438.—System AlCl₃–NbCl₅–TaCl₅.

I. S. Morozov, *Zhur. Neorg. Khim.*, **1**, 2795 (1956).

FeCl₃–NbCl₅–TaCl₅

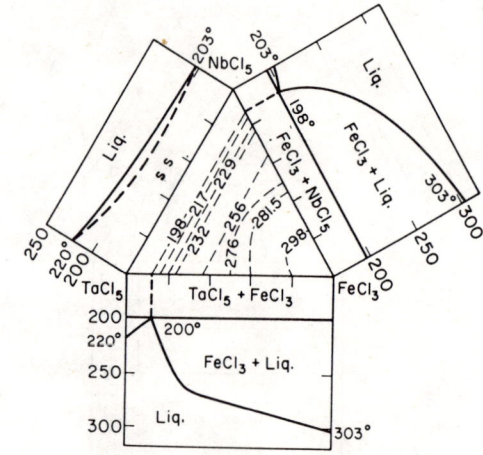

FIG. 1440.—System FeCl₃–NbCl₅–TaCl₅.

I. S. Morozov, *Zhur. Neorg. Khim.*, **1**, 2795 (1956).

FeCl₃–TiCl₄–NbCl₅

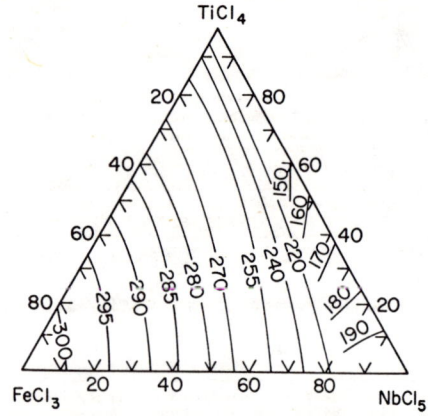

FIG. 1439.—System FeCl₃–TiCl₄–NbCl₅.

I. S. Morozov and D. Ya. Toptygin, *Zhur. Neorg. Khim.*, **2**, 2133 (1957).

TiCl₄–NbCl₅–TaCl₅

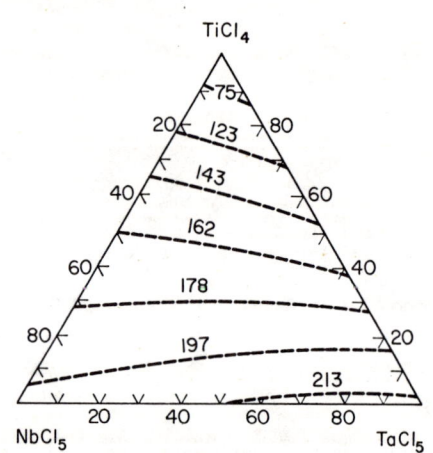

FIG. 1441.—System TiCl₄–NbCl₅–TaCl₅.

I. S. Morozov and D. Ya. Toptygin, *Zhur. Neorg. Khim.*, **2**, 1918 (1957).

III. Fluorides Only

(a) One Fluoride

MnF$_2$

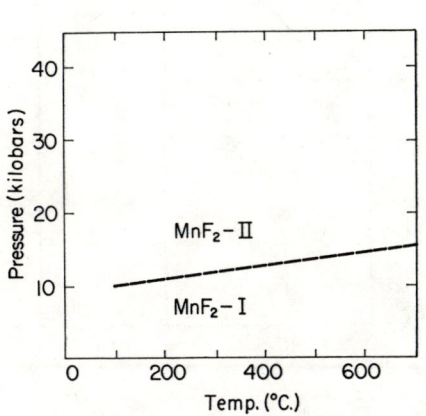

Fig. 1442.—System MnF$_2$; p-T.

L. M. Azzaria and Frank Dachille, *J. Phys. Chem.*, **65**, 889 (1961).

AgF–ZnF$_2$

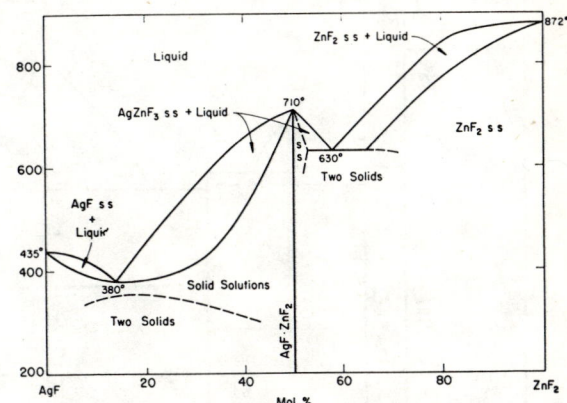

Fig. 1444.—System AgF–ZnF$_2$.

R. C. DeVries and Rustum Roy, *J. Am. Chem. Soc.*, **75**, 2483 (1953).

(b) Two Fluorides

AgF–PbF$_2$

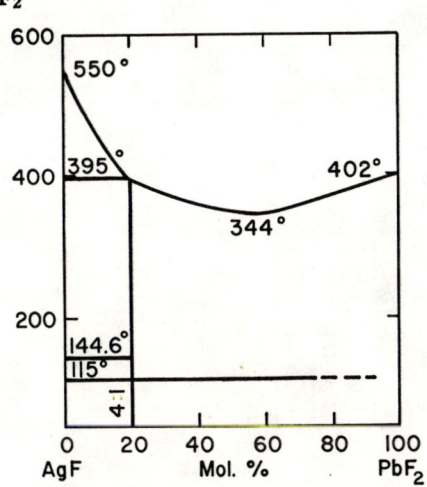

Fig. 1443.—System AgF–PbF$_2$. Diagram does not obey phase rule.

C. Tubandt and Sophie Eggert, *Z. anorg. u. allgem. Chem.*, **110**, 223 (1920).

CsF–LiF

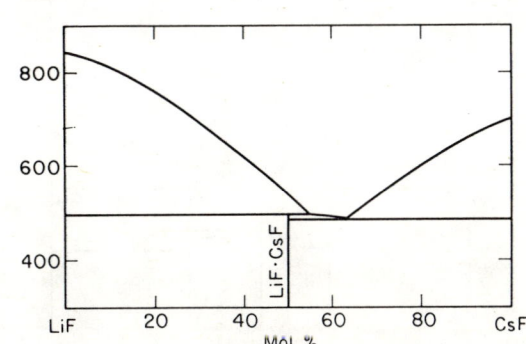

Fig. 1445.—System LiF–CsF.

C. J. Barton, L. M. Bratcher, T. N. McVay, and W. R. Grimes, Oak Ridge National Laboratory, Phase Diagrams of Nuclear Reactor Materials, R. E. Thoma, ed., ORNL-2548, p. 17 (1959).

CsF–BeF$_2$

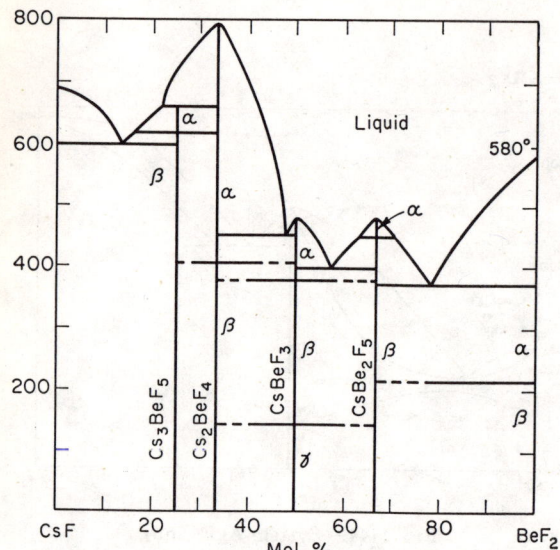

FIG. 1446.—System CsF–BeF$_2$.
α, β, γ refer to polymorphs of the designated compounds.

O. N. Breusov, A. V. Novoselova, and Yu. P. Simanov, *Doklady Akad. Nauk S.S.S.R.*, **118** [5] 936 (1958).

CsF–AlF$_3$

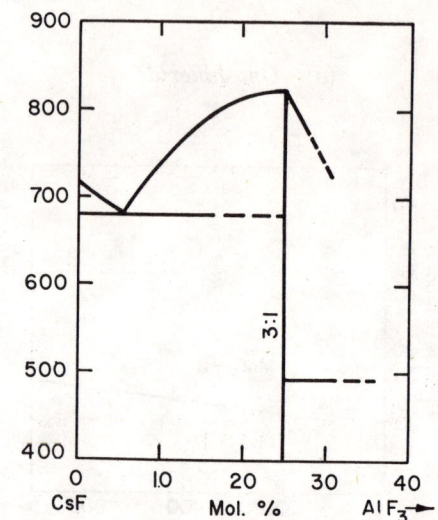

FIG. 1448.—System CsF–AlF$_3$.

N. Puschin and A. Baskow, *Z. anorg. Chem.*, **81**, 357 (1913).

CsF–ZnF$_2$

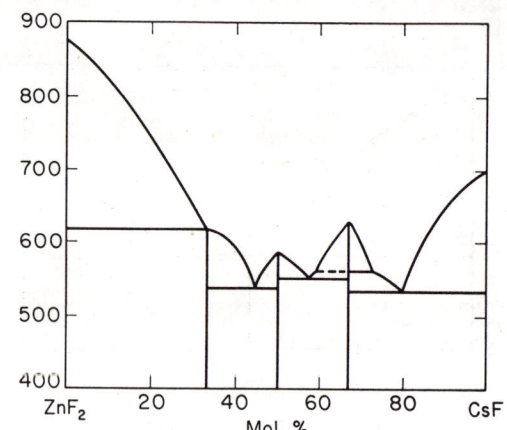

FIG. 1447.—System CsF–ZnF$_2$.

O. Schmitz-Dumont and Horst Bornefeld, *Z. anorg. u. allgem. Chem.*, **287**, 122 (1956).

CsF–UF$_4$

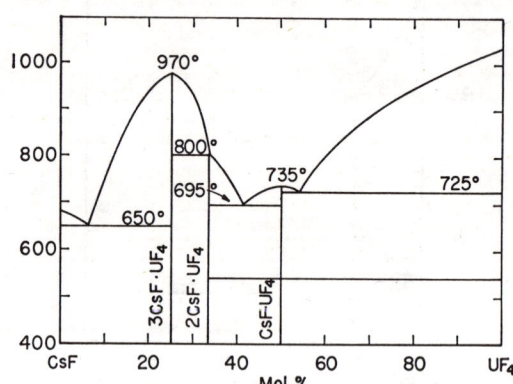

FIG. 1449.—System CsF–UF$_4$; tentative.

C. J. Barton, L. M. Bratcher, J. P. Blakely, G. J. Nessle, and W. R. Grimes, Oak Ridge National Laboratory, Phase Diagrams of Nuclear Reactor Materials, R. E. Thoma, ed., ORNL-2548, p. 92 (1959).

CsF–ZrF$_4$

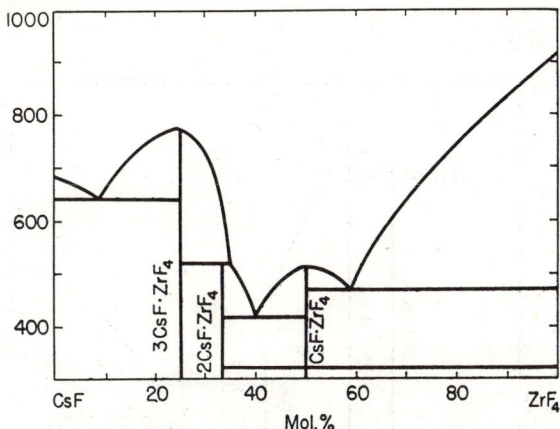

Fig. 1450.—System CsF–ZrF$_4$; tentative.

C. J. Barton, L. M. Bratcher, and W. R. Grimes, Oak Ridge National Laboratory, Phase Diagrams of Nuclear Reactor Materials, R. E. Thoma, ed., ORNL-2548, p. 59 (1959).

KF–LiF

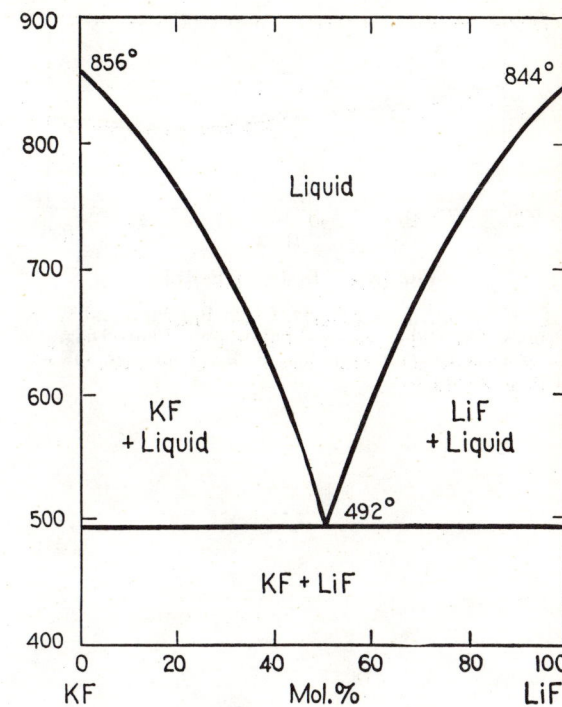

Fig. 1452.—System KF–LiF.

A. G. Bergman and E. P. Dergunov, *Compt. rend. acad. sci., U.R.S.S.*, **31**, 753 (1941).

HF–NaF

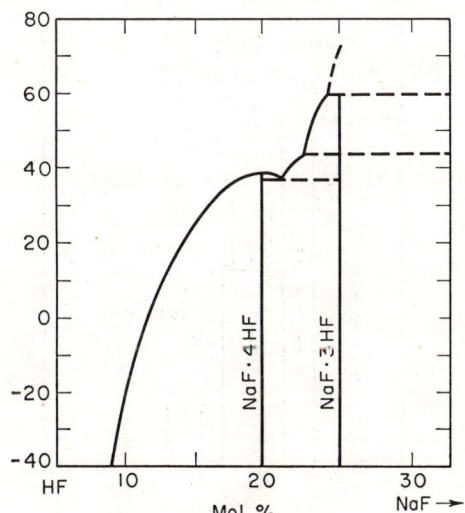

Fig. 1451.—System HF–NaF.

R. L. Adamczak, J. A. Mattern, and Howard Tieckelmann, *J. Phys. Chem.*, **63**, 2064 (1959).

KF–NaF

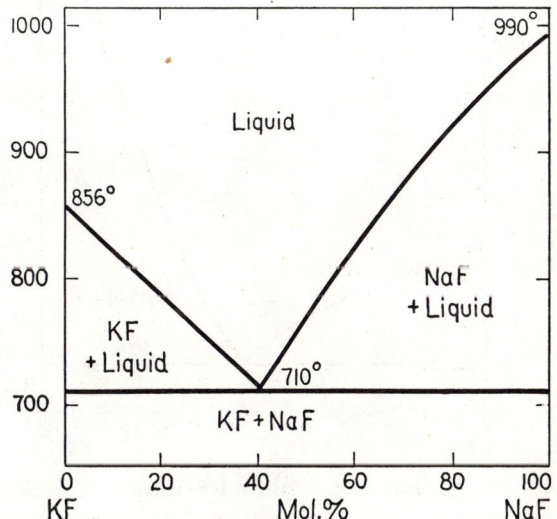

Fig. 1453.—System KF–NaF.

A. G. Bergman and E. P. Dergunov, *Compt. rend. acad. sci., U.R.S.S.*, **31**, 753 (1941).

KF–RbF

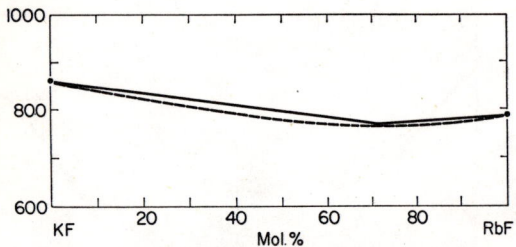

FIG. 1454.—System KF–RbF.

C. J. Barton, J. P. Blakely, L. M. Bratcher, and W. R. Grimes, Oak Ridge National Laboratory, Phase Diagrams for Nuclear Reactor Materials, R. E. Thoma, ed., ORNL-2548, p. 20 (1959).

KF–BeF$_2$

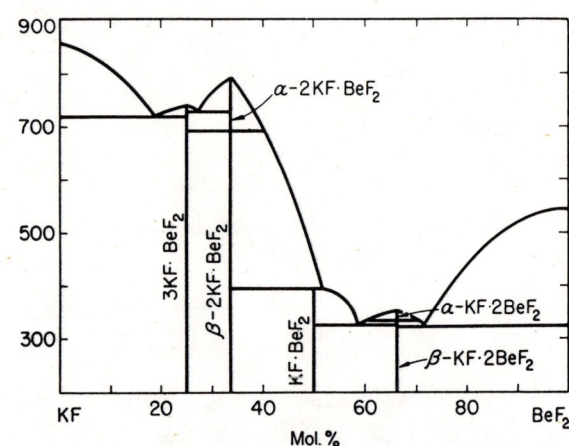

FIG. 1456.—System KF–BeF$_2$.

R. E. Moore, C. J. Barton, L. M. Bratcher, T. N. McVay, G. D. White, R. J. Sheil, W. R. Grimes, R. E. Meadows, and L. A. Harris, Oak Ridge National Laboratory, Phase Diagrams of Nuclear Reactor Materials, R. E. Thoma, ed., ORNL-2548, p. 37 (1959).

KF–BaF$_2$

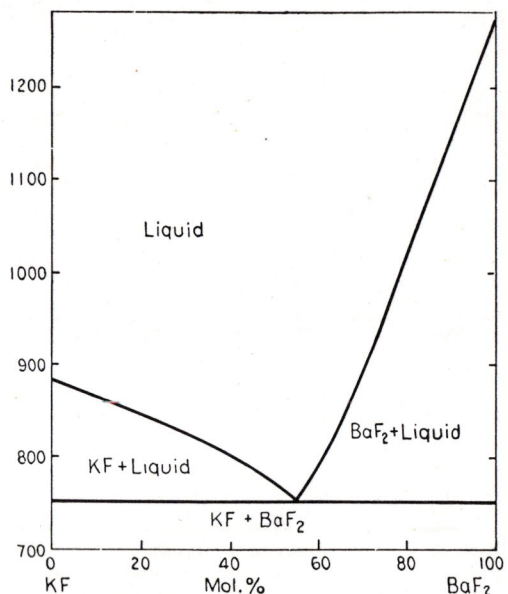

FIG. 1455.—System KF–BaF$_2$.

N. Puschin and A. Baskow, *Z. anorg. Chem.*, **81**, 360 (1913).

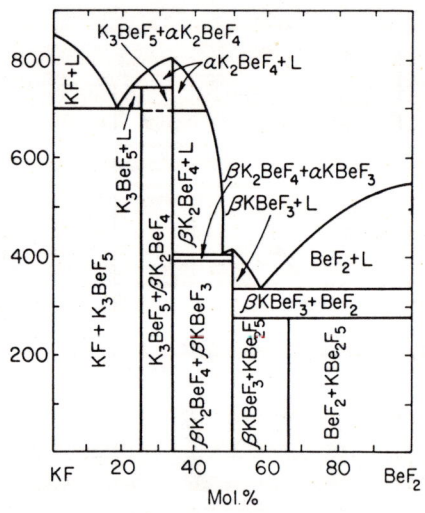

FIG. 1457.—System KF–BeF$_2$.

M. P. Borzenkova, A. V. Novoselova, Yu. P. Simanov, V. I. Chernykh, and E. I. Yarembash, *Zhur. Neorg. Khim.*, **1**, 2074 (1956).

KF–MgF$_2$

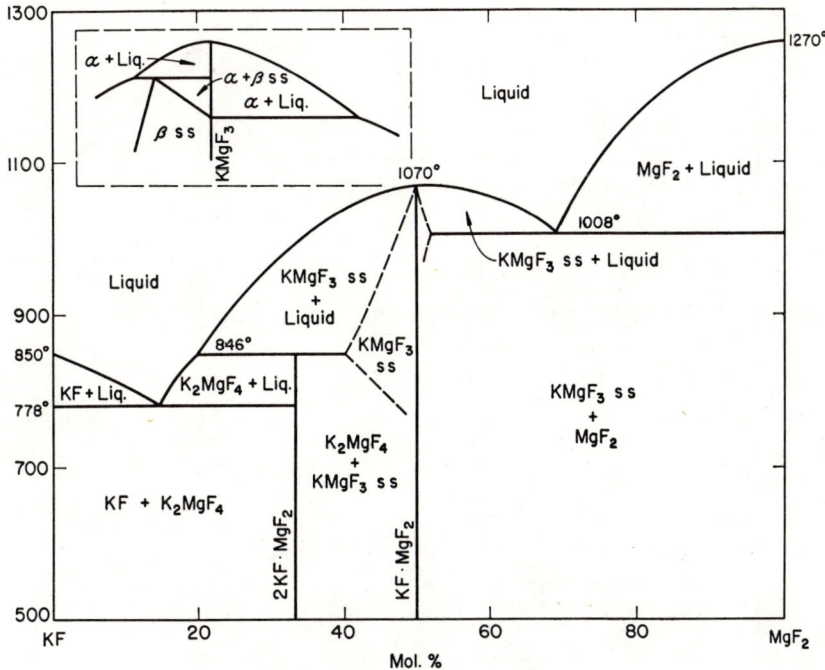

Fig. 1458.—System KF–MgF$_2$. Insert shows possible alternative for high-temperature region near KF·MgF$_2$.

R. C. DeVries and Rustum Roy, *J. Am. Chem. Soc.*, **75**, 2481 (1953).

KF–NiF$_2$

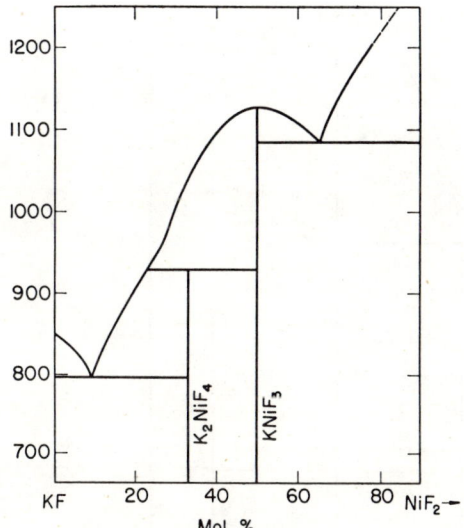

Fig. 1459.—System KF–NiF$_2$.

G. Wagner and D. Balz, *Z. Elektrochem.*, **56**, 576 (1952).

KF–ZnF$_2$

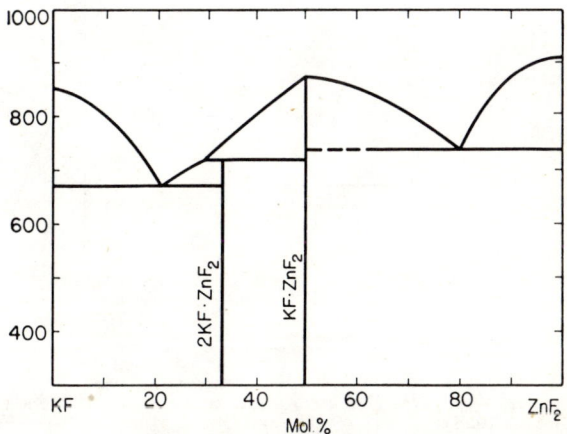

Fig. 1460.—System KF–ZnF$_2$, tentative.

C. J. Barton, L. M. Bratcher, and W. R. Grimes, Oak Ridge National Laboratory, Phase Diagrams of Nuclear Reactor Materials, R. E. Thoma, ed., ORNL-2548, p. 49 (1959).

See also O. Schmitz-Dumont and Horst Bornefeld, *Z. anorg. u. allgem. Chem.*, **287**, 122 (1956).

KF–AlF$_3$

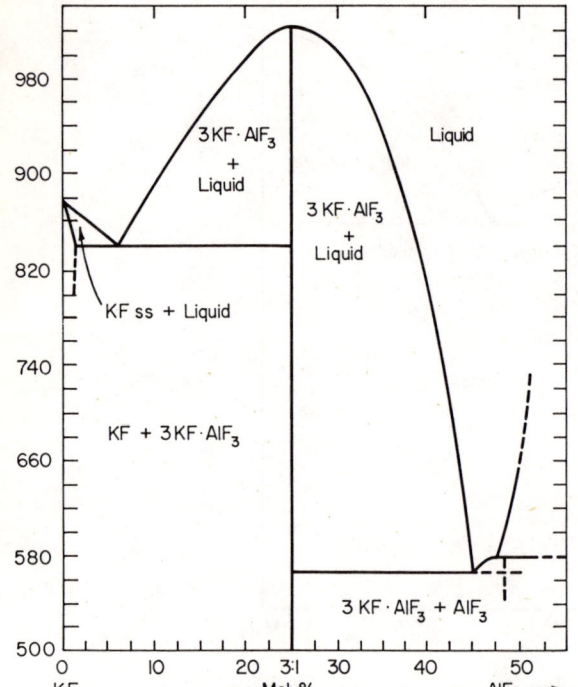

FIG. 1461.—Partial system KF–AlF$_3$.

P. P. Fedotieff and K. Timofeeff, *Z. anorg. u. allgem. Chem.*, **206**, 265 (1932).

KF–CeF$_3$

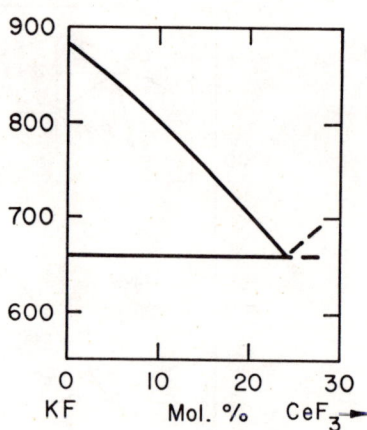

FIG. 1463.—System KF–CeF$_3$.

N. Puschin and A. Baskow, *Z. anorg. Chem.*, **81**, 361 (1913).

KF–BF$_3$

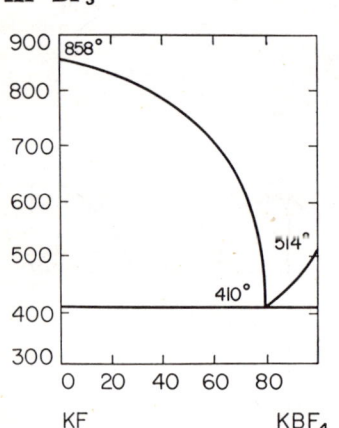

FIG. 1462.—System KF–KBF$_4$.

V. G. Selivanov and V. V. Stender, *Zhur. Neorg. Khim.*, **3** [2] 448 (1958).

KF–ThF$_4$

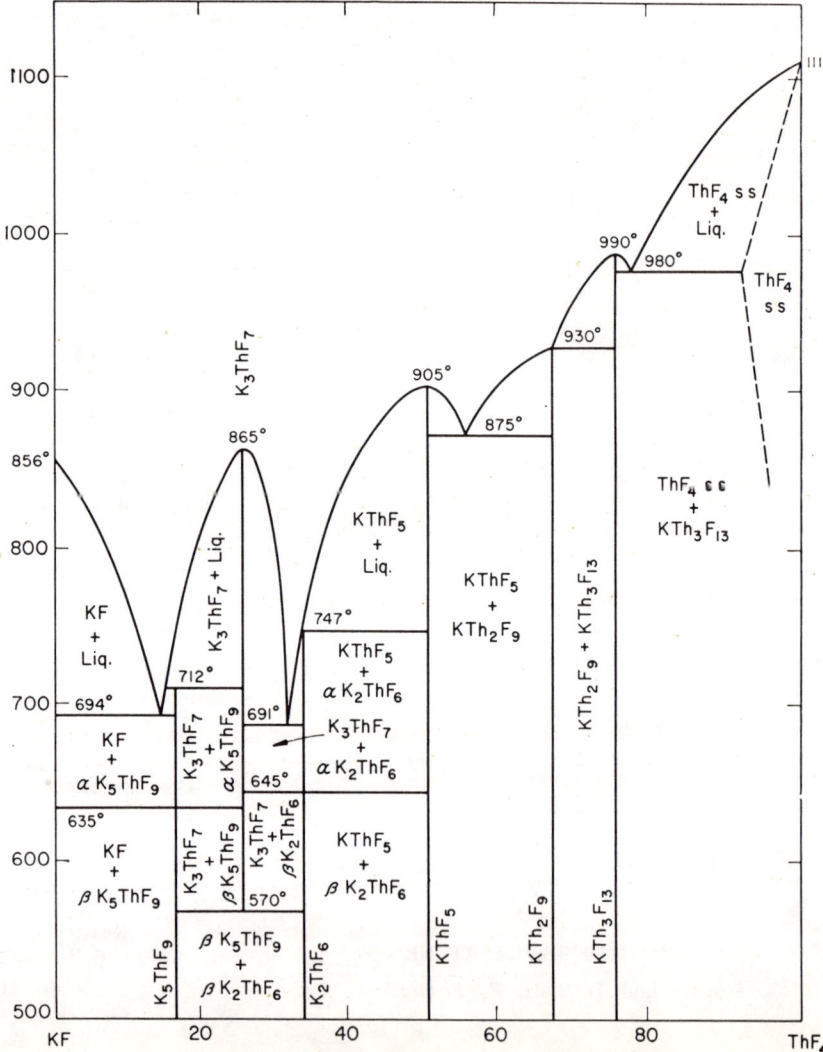

FIG. 1464.—System KF–ThF$_4$.

W. J. Asker, E. R. Segnit, and A. W. Wylie, *J. Chem. Soc.* (*London*), p. 4471, Nov. (1952).

KF–UF$_4$

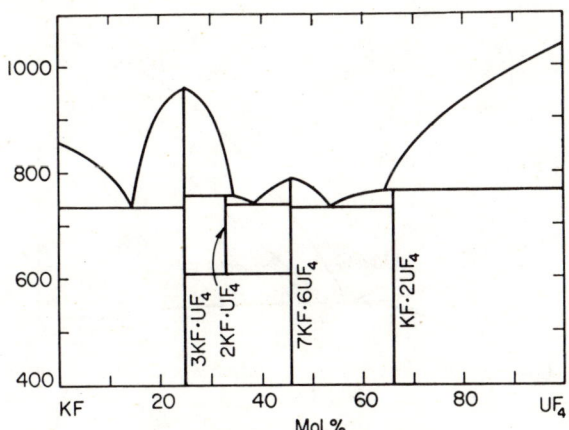

Fig. 1465.—System KF–UF$_4$.

R. E. Thoma, H. Insley, B. S. Landau, H. A. Friedman, and W. R. Grimes, *J. Am. Ceram. Soc.*, **41** [12] 539 (1958).

KF–ZrF$_4$

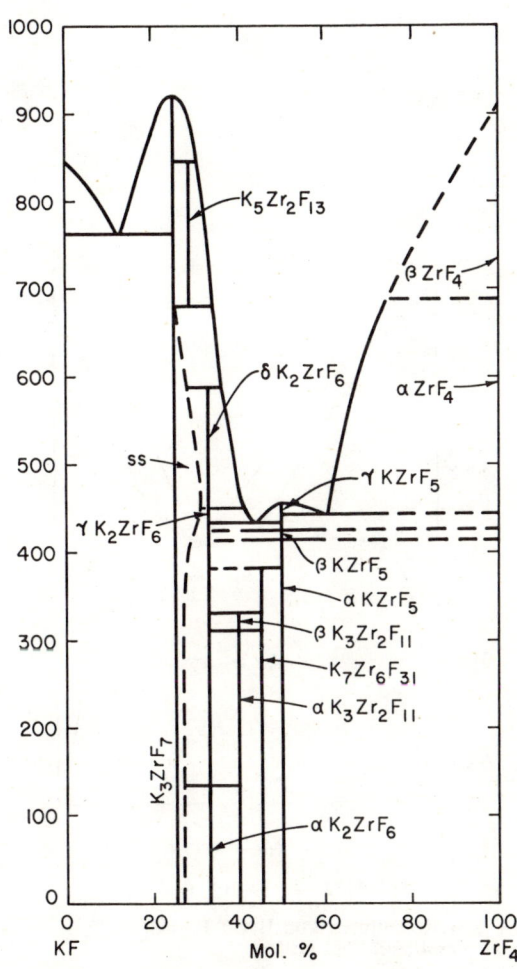

Fig. 1466.—System KF–ZrF$_4$.

A. V. Novoselova, Yu. M. Korenev, and Yu. P. Simanov, *Doklady Akad. Nauk S.S.S.R.*, **139** [4] 893 (1961). See also C. J. Barton, H. Insley, R. E. Metcalf, R. E. Thoma, and W. R. Grimes, Oak Ridge National Laboratory, Phase Diagrams of Nuclear Reactor Materials, R. E. Thoma, ed., ORNL-2548 (1959).

LiF–NaF

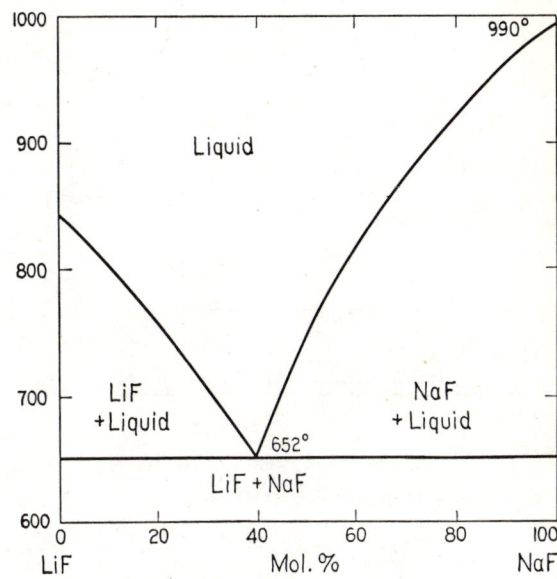

Fig. 1467.—System LiF–NaF.

A. G. Bergman and E. P. Dergunov, *Compt. rend. acad. sci., U.R.S.S.*, **31**, 755 (1941).

LiF–RbF

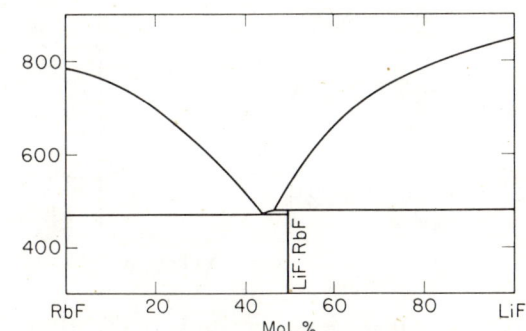

Fig. 1468.—System LiF–RbF.

C. J. Barton, L. M. Bratcher, T. N. McVay, and W. R Grimes, Oak Ridge National Laboratory, Phase Diagrams of Nuclear Reactor Materials, R. E. Thoma, ed., ORNL-2548, p. 16 (1959).

LiF–BeF$_2$

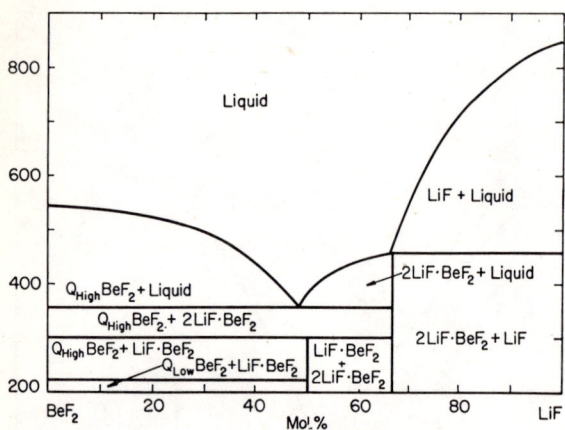

FIG. 1469.—System LiF–BeF$_2$.

L. V. Jones, D. E. Etter, C. R. Hudgens, A. A. Huffman, T. B. Rhinehammer, N. E. Rogers, P. A. Tucker, and L. J. Wittenberg, *J. Am. Ceram. Soc.*, **45** [2] 80 (1962). For diagram showing the additional subsolidus compound LiBe$_2$F$_5$, see D. M. Roy, R. Roy, and E. F. Osborn, *J. Am. Ceram. Soc.*, **37** [7] 302 (1954).

LiF–MgF$_2$

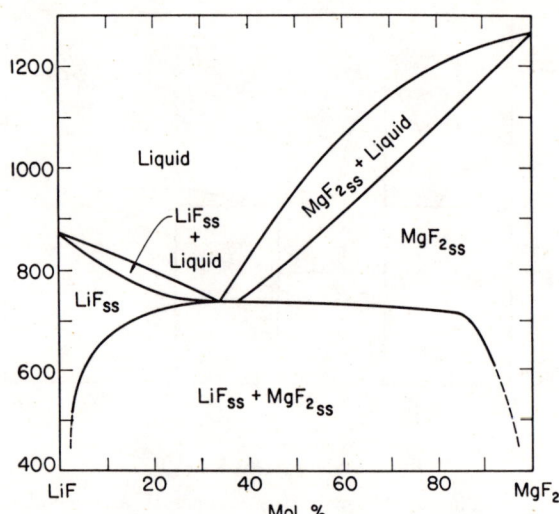

FIG. 1471.—System LiF–MgF$_2$.

W. E. Counts, Rustum Roy, and E. F. Osborn, *J. Am. Ceram. Soc.*, **36** [1] 15 (1953).

LiF–CaF$_2$

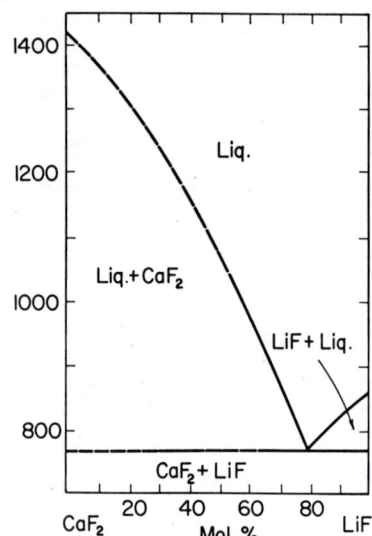

FIG. 1470.—System LiF–CaF$_2$.

W. E. Roake, *J. Electrochem. Soc.*, **104**, 661 (1957).

LiF–ZnF$_2$

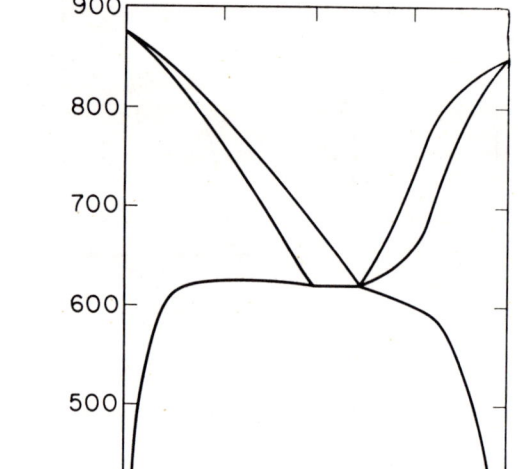

FIG. 1472.—System LiF–ZnF$_2$.

O. Schmitz-Dumont and Horst Bornefeld, *Z. anorg. u. allgem. Chem.*, **287**, 121 (1956).

LiF–AlF₃

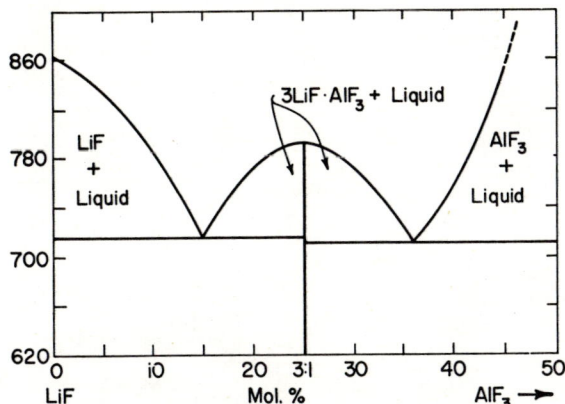

Fig. 1473.—System LiF–AlF₃.

P. P. Fedotieff and K. Timofeeff, *Z. anorg. u. allgem. Chem.*, **206**, 266 (1932).

LiF–CeF₃

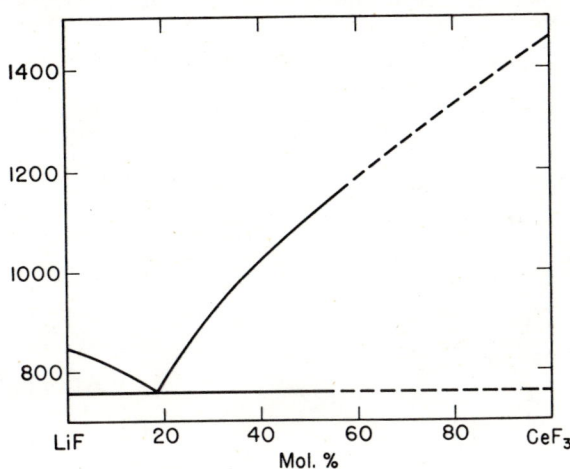

Fig. 1474.—System LiF–CeF₃.

C. J. Barton, L. M. Bratcher, R. J. Sheil, and W. R. Grimes, Oak Ridge National Laboratory, Phase Diagrams of Nuclear Reactor Materials, R. E. Thoma, ed., ORNL-2548, p. 69 (1959).

LiF–PuF₃

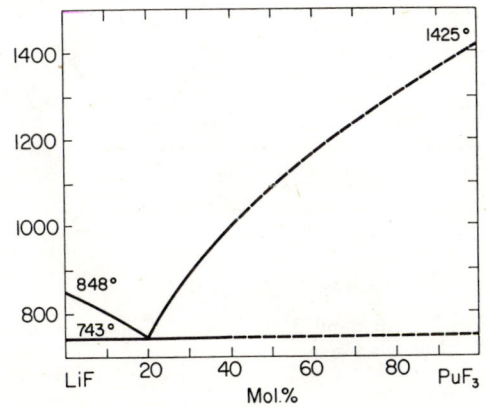

Fig. 1475.—System LiF–PuF₃.

C. J. Barton and R. A. Strehlow, *J. Inorg. Nucl. Chem.*, **18**, 143–147 (1961).

LiF–UF₃

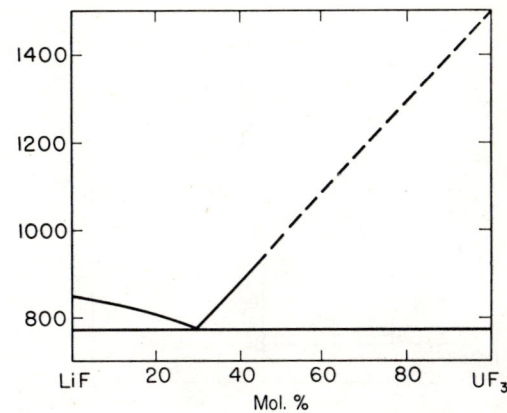

Fig. 1476.—System LiF–UF₃; tentative.

C. J. Barton, V. S. Coleman, L. M. Bratcher, and W. R. Grimes, Oak Ridge National Laboratory, Phase Diagrams of Nuclear Reactor Materials, R. E. Thoma, ed., ORNL-2548 p. 84, (1959).

LiF–YF₃

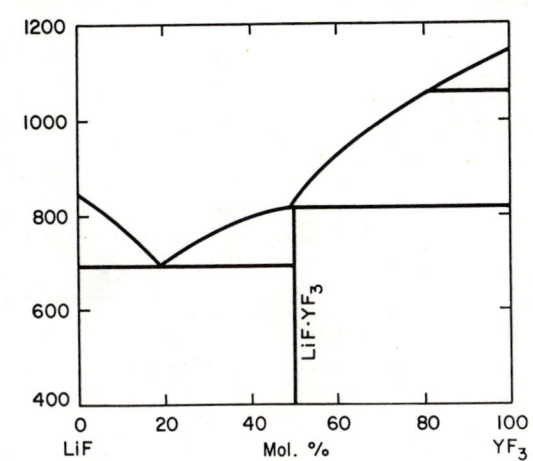

Fig. 1477.—System LiF–YF₃.

R. E. Thoma, C. F. Weaver, H. A. Friedman, H. Insley, L. A. Harris, and H. L. Yakel, Jr., *J. Phys. Chem.*, **65**, 1096–1099 (1961).

LiF–ThF₄

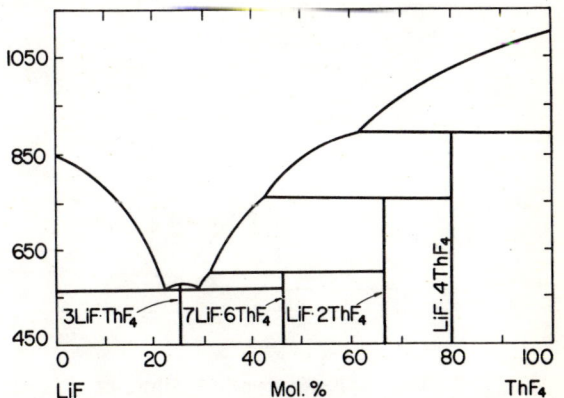

Fig. 1478.—System LiF–ThF₄.

R. E. Thoma, H. Insley, B. S. Landau, H. A. Friedman, and W. R. Grimes, *J. Phys. Chem.*, **63**, 1267 (1959).

LiF–UF$_4$

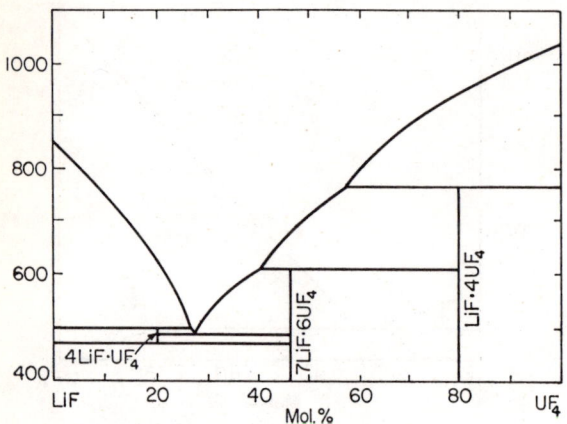

FIG. 1479.—System LiF–UF$_4$.

C. J. Barton, H. A. Friedman, W. R. Grimes, H. Insley, R. E. Moore, and R. E. Thoma, *J. Am. Ceram. Soc.*, **41** [2] 67 (1958).

LiF–ZrF$_4$

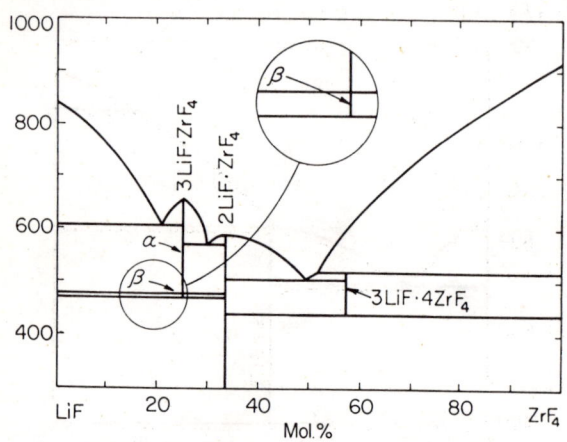

FIG. 1480.—System LiF–ZrF$_4$; preliminary.

R. E. Moore, F. F. Blankenship, W. R. Grimes, H. A. Friedman, C. J. Barton, R. E. Thoma, and H. Insley, Oak Ridge National Laboratory, Phase Diagrams of Nuclear Reactor Materials, R. E. Thoma, ed., ORNL-2548, p. 53 (1959).

NaF–RbF

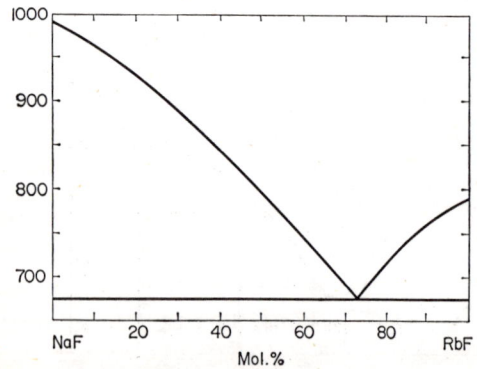

FIG. 1481.—System NaF–RbF.

C. J. Barton, J. P. Blakely, L. M. Bratcher, and W. R. Grimes, Oak Ridge National Laboratory, Phase Diagrams for Nuclear Reactor Materials, R. E. Thoma, ed., ORNL-2548, p. 19 (1959).

NaF–BeF$_2$

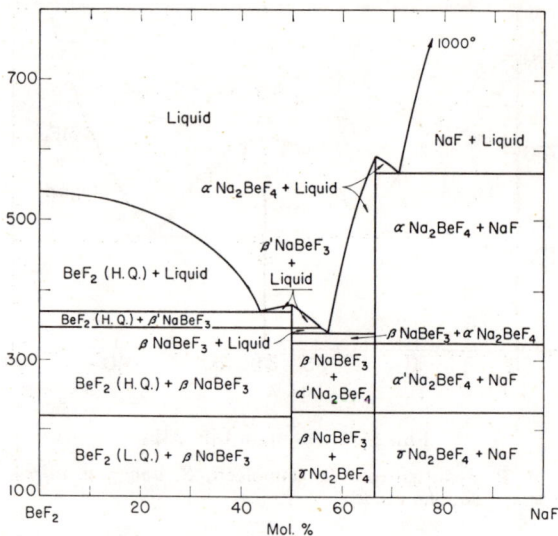

FIG. 1482.—System NaF–BeF$_2$.

Della M. Roy, Rustum Roy, and E. F. Osborn, *J. Am. Ceram. Soc.*, **36** [6] 185 (1953).

See also Erich Thilo and Hansjürgen Schröder, *Z. physik. chem.*, **197**, 41 (1951).

NaF–CaF$_2$

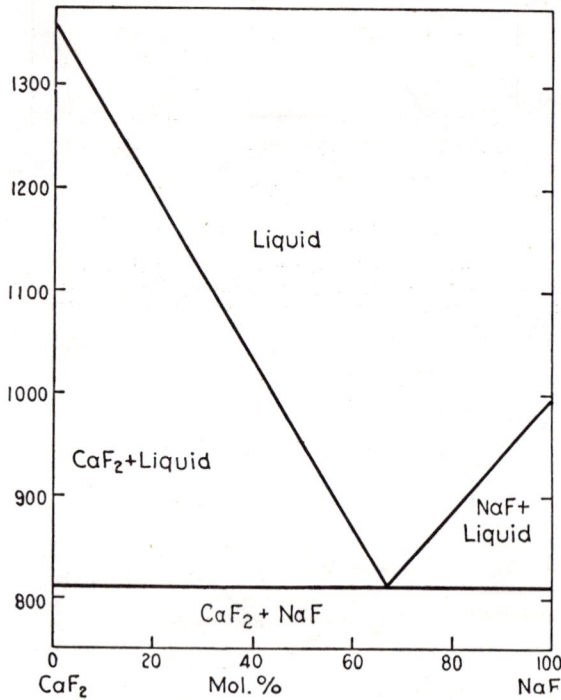

FIG. 1483.—System NaF–CaF$_2$.

P. P. Fedotieff and W. P. Iljinskiĭ, *Z. anorg. u. allgem. Chem.*, **129**, 101 (1923).

NaF–CdF$_2$

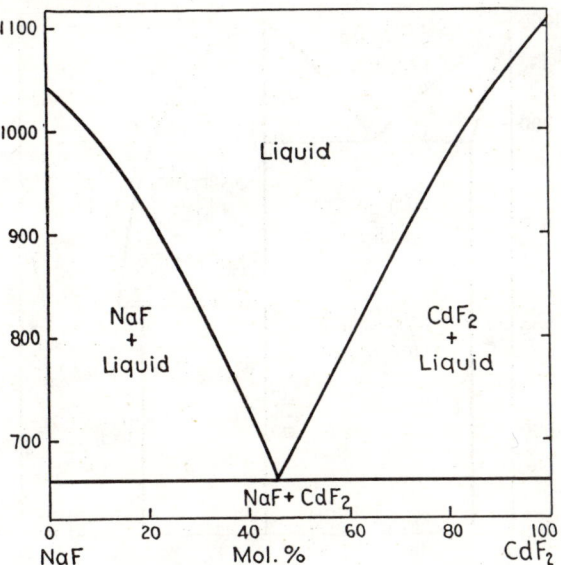

Fig. 1484.—System NaF–CdF$_2$.

N. A. Puschin and A. V. Baskow, *Z. anorg. Chem.*, **81**, 359 (1913).

NaF–MgF$_2$

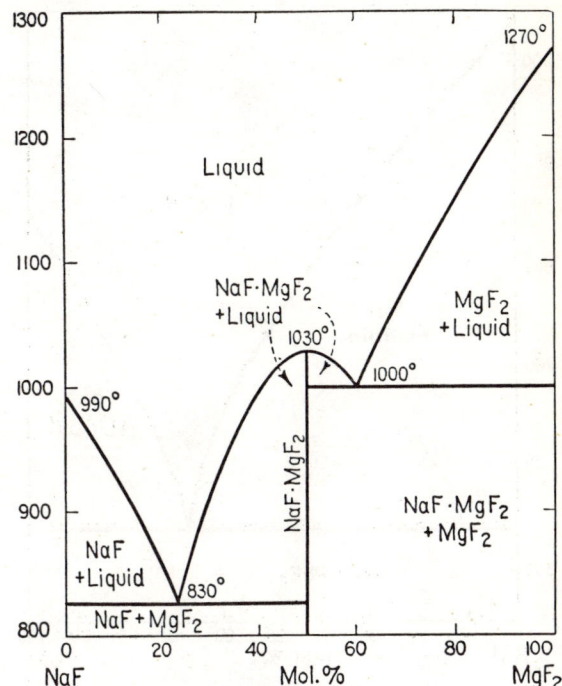

Fig. 1486.—System NaF–MgF$_2$.

A. G. Bergman and E. P. Dergunov, *Compt. rend. acad. sci., U.R.S.S.*, **31**, 755 (1941).

NaF–FeF$_2$

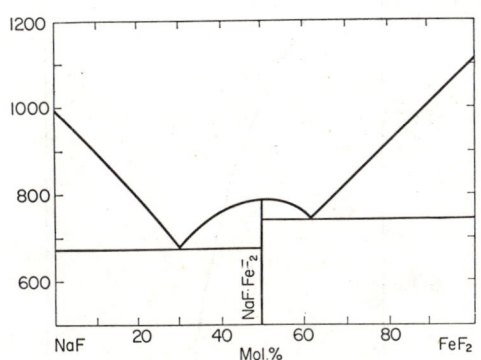

Fig. 1485.—System NaF–FeF$_2$.

R. E. Thoma, H. A. Friedman, B. S. Landau, and W. R. Grimes, Oak Ridge National Laboratory Phase Diagrams of Nuclear Reactor Materials, R. E. Thoma, ed., ORNL 2548, p. 26 (1959).

NaF–NiF$_2$

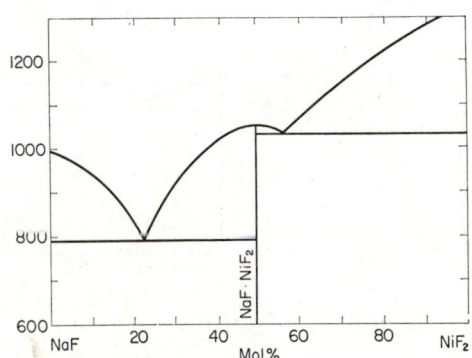

Fig. 1487.—System NaF–NiF$_2$.

R. E. Thoma, H. A. Friedman, B. S. Landau, and W. R. Grimes, Oak Ridge National Laboratory Phase Diagrams for Nuclear Reactor Materials, R. E. Thoma, ed., ORNL-2548, p. 27 (1959).

NaF–PbF$_2$

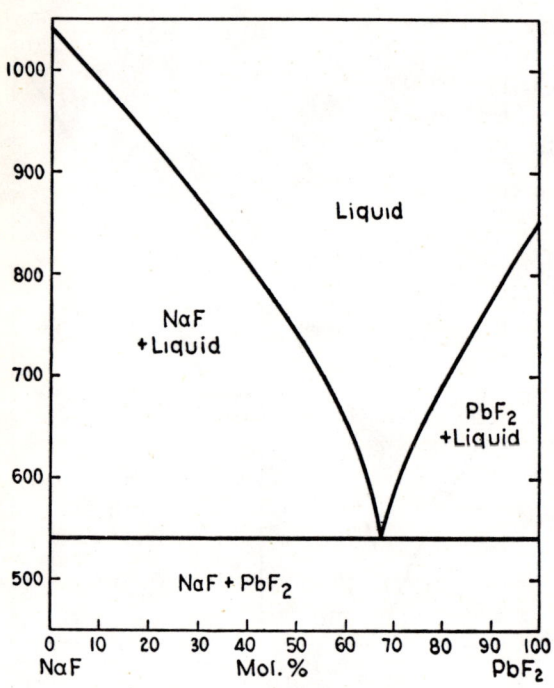

Fig. 1488.—System NaF–PbF$_2$.

N. A. Puschin and A. V. Baskow, *Z. anorg. Chem.*, **81**, 359 (1913).

NaF–AlF$_3$

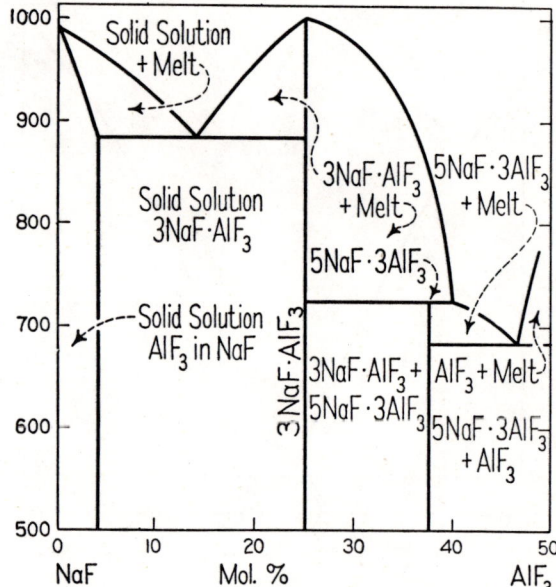

Fig. 1490.—System NaF–AlF$_3$.

P. P. Fedotieff and W. P. Iljinskiĭ, *Z. anorg. Chem.*, **80**, 121 (1913); see also N. A. Puschin and A. V. Baskow, *ibid.*, **81**, 350 (1913).

NaF–ZnF$_2$

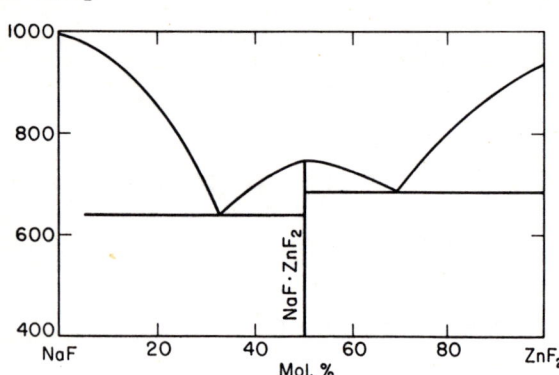

Fig. 1489.—System NaF–ZnF$_2$; tentative.

C. J. Barton, L. M. Bratcher, and W. R. Grimes, Oak Ridge National Laboratory, Phase Diagrams of Nuclear Reactor Materials, R. E. Thoma, ed., ORNL-2548, p. 48 (1959).

See also O. Schmitz-Dumont and Horst Bornefeld, *Z. anorg. u. allgem. Chem.*, **287**, 121 (1956).

NaF–BF$_3$

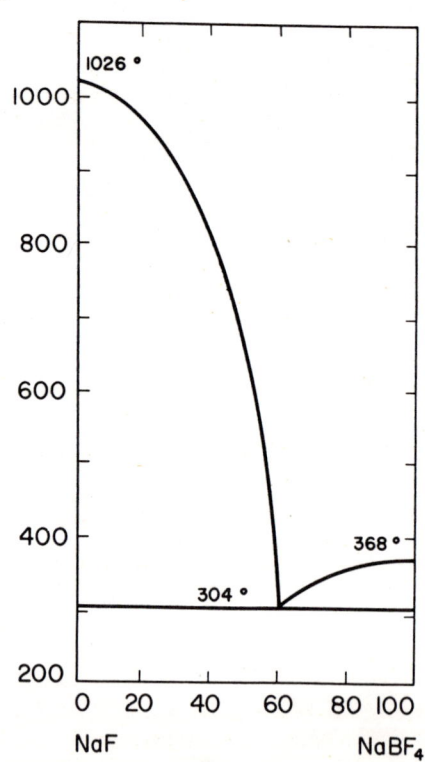

Fig. 1491.—System NaF–NaBF$_4$.

V. G. Selivanov and V. V. Stender, *Zhur. Neorg. Khim.*, **3** [2] 448 (1958).

NaF–CeF₃

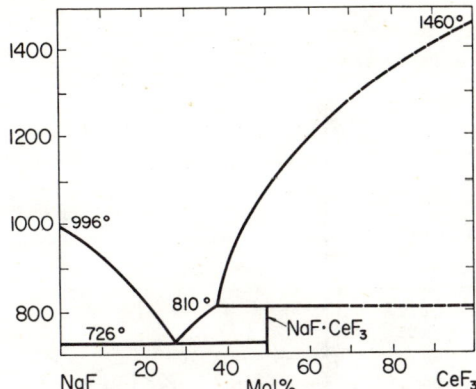

Fig. 1492.—System NaF–CeF₃.

C. J. Barton, J. D. Redman, and R. A. Strehlow, *J. Inorg. Nucl. Chem.*, **20**, 45–54 (1961).

NaF–FeF₃

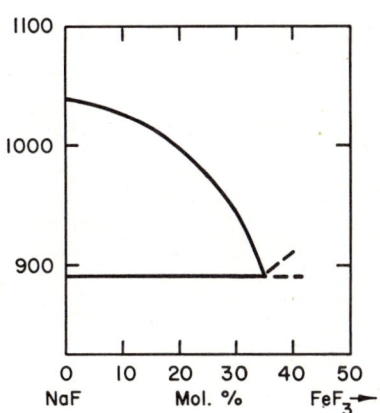

Fig. 1493.—System NaF–FeF₃.

N. Puschin and A. Baskow, *Z. anorg. Chem.*, **81**, 361 (1913).

NaF–PuF₃

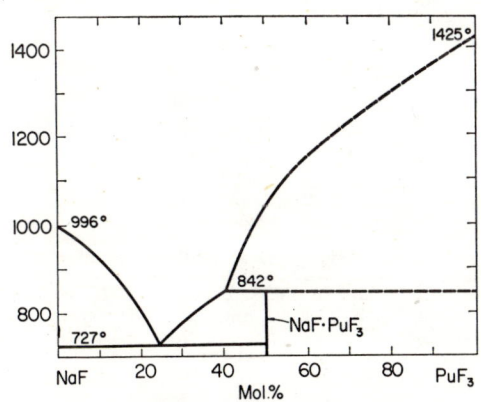

Fig. 1494.—System NaF–PuF₃.

C. J. Barton, J. D. Redman, and R. A. Strehlow, *J. Inorg. Nucl. Chem.*, **20**, 45–54 (1961).

NaF–UF₃

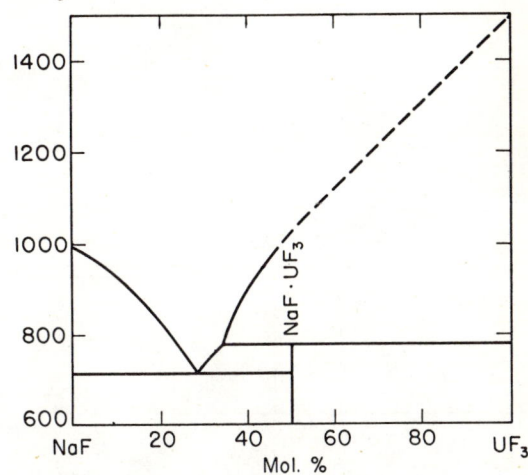

Fig. 1495.—System NaF–UF₃; tentative.

C. J. Barton, V. S. Coleman, T. N. McVay, and W. R. Grimes, Oak Ridge National Laboratory, Phase Diagrams of Nuclear Reactor Materials, R. E. Thoma, ed., ORNL-2548, p. 87 (1959).

NaF–HfF₄

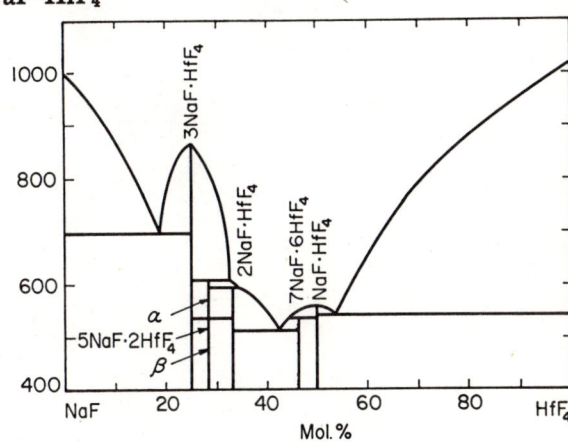

Fig. 1496.—System NaF–HfF₄; preliminary.

R. E. Thoma, C. F. Weaver, T. N. McVay, H. A. Friedman, and W. R. Grimes, Oak Ridge National Laboratory, Phase Diagrams of Nuclear Reactor Materials, R. E. Thoma, ed., ORNL-2548, p. 71 (1959).

NaF–ThF₄

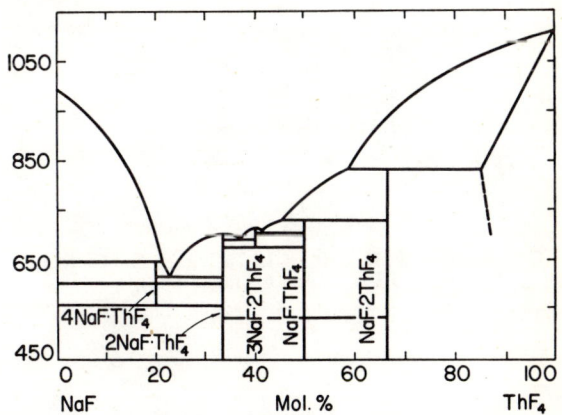

Fig. 1497.—System NaF–ThF₄.

R. E. Thoma, H. Insley, B. S. Landau, H. A. Friedman, and W. R. Grimes, *J. Phys. Chem.*, **63**, 1269 (1959).

NaF–UF₄

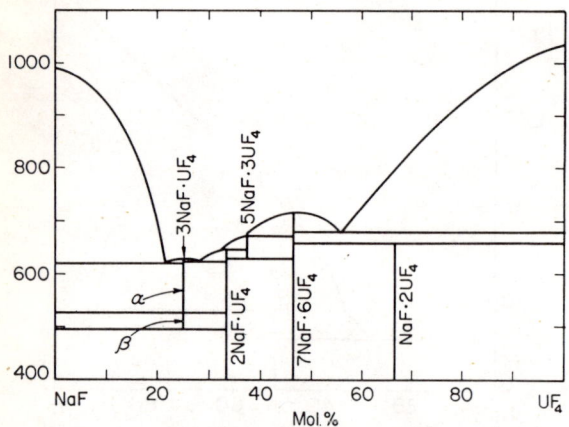

FIG. 1498.—System NaF–UF₄.

C. J. Barton, H. A. Friedman, W. R. Grimes, H. Insley, R. E. Moore, and R. E. Thoma, *J. Am. Ceram. Soc.*, **41** [2] 68 (1958).

RbF–BeF₂

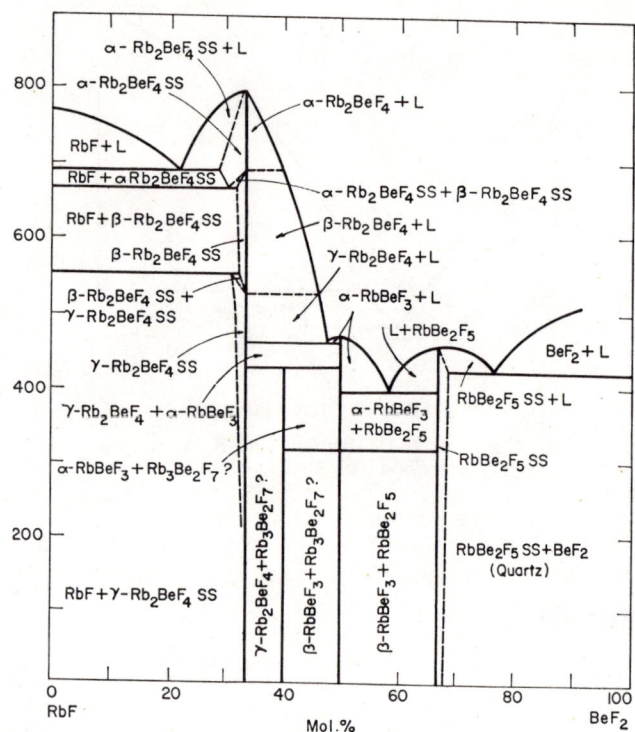

FIG. 1500.—System RbF–BeF₂.

R. G. Grebenshchikov, *Doklady Akad. Nauk S.S.S.R.*, **114**, 317 (1957).

NaF–ZrF₄

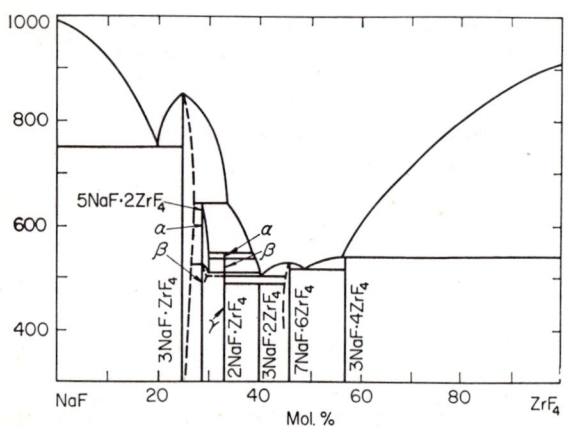

FIG. 1499.—System NaF–ZrF₄.

C. J. Barton, W. R. Grimes, H. Insley, R. E. Moore, and R. E. Thoma, *J. Phys. Chem.*, **62**, 666 (1958).

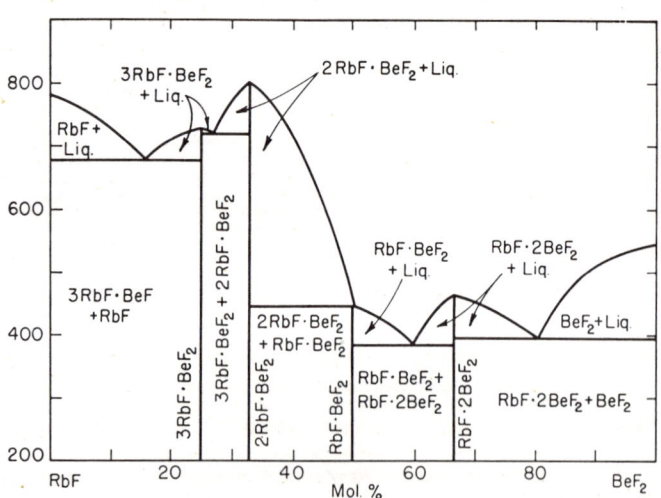

FIG. 1501.—System RbF–BeF₂.

R. E. Moore, C. J. Barton, L. M. Bratcher, T. N. McVay, G. D. White, R. J. Sheil, W. R. Grimes, and R. E. Meadows, Oak Ridge National Laboratory, Phase Diagrams of Nuclear Reactor Materials, R. E. Thoma, ed., ORNL-2548, p. 39 (1959).

RbF–CaF$_2$

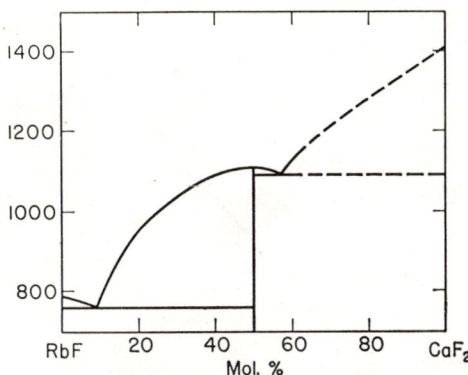

FIG. 1502.—System RbF–CaF$_2$.

C. J. Barton, L. M. Bratcher, R. J. Sheil, and W. R. Grimes, Oak Ridge National Laboratory, Phase Diagrams of Nuclear Reactor Materials, R. E. Thoma, ed., ORNL-2548, p. 28 (1959).

RbF–ZnF$_2$

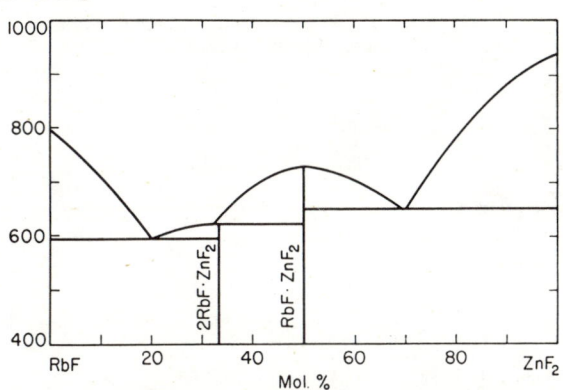

FIG. 1503.—System RbF–ZnF$_2$; tentative.

C. J. Barton, L. M. Bratcher, and W. R. Grimes, Oak Ridge National Laboratory, Phase Diagrams of Nuclear Reactor Materials, R. E. Thoma, ed., ORNL-2548, p. 50 (1959).

See also O. Schmitz-Dumont and Horst Bornefeld, Z. anorg. u. allgem. Chem., 287, 122 (1956).

RbF–AlF$_3$

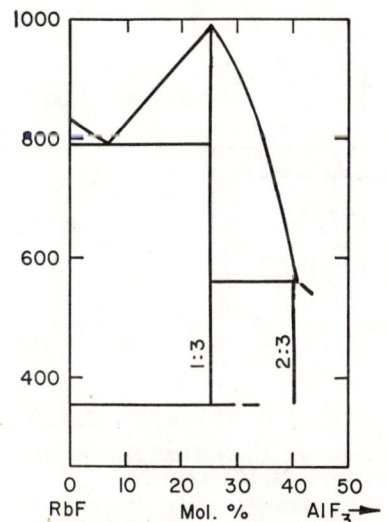

FIG. 1504.—System RbF–AlF$_3$.

N. Puschin and A. Baskow, Z. anorg. Chem., 81, 356 (1913).

RbF–ThF$_4$

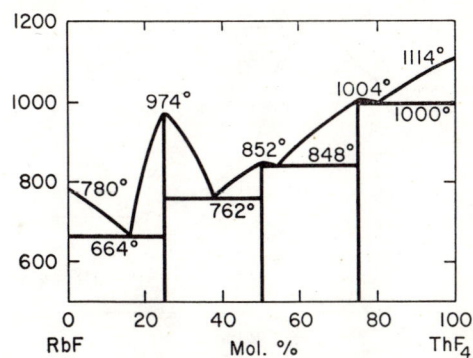

FIG. 1505.—System RbF–ThF$_4$.

E. P. Dergunov and A. G. Bergman, Doklady Akad. Nauk S.S.S.R. 60, 391–4 (1948). See also, Oak Ridge National Laboratory, Phase Diagrams of Nuclear Reactor Materials, R. E. Thoma, ed., ORNL-2548, p. 76 (1959).

RbF–UF$_4$

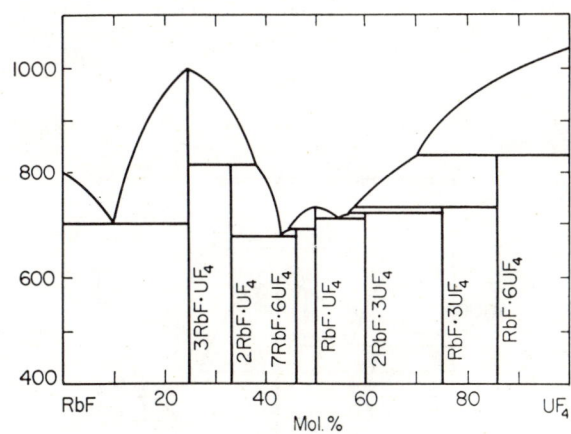

FIG. 1506.—System RbF–UF$_4$.

R. E. Thoma, H. Insley, B. S. Landau, H. A. Friedman, and W. R. Grimes, J. Am. Ceram. Soc., 41 [12] 542 (1958).

RbF–ZrF$_4$

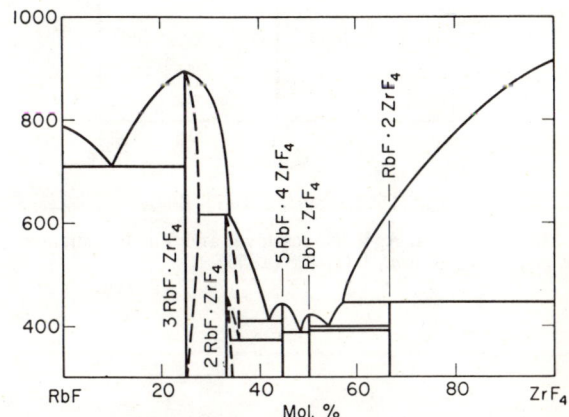

FIG. 1507.—System RbF–ZrF$_4$.

R. E. Moore, R. E. Thoma, C. J. Barton, W. R. Grimes, H. Insley, B. S. Landau, and H. A. Friedman, Oak Ridge National Laboratory, Phase Diagrams of Nuclear Reactor Materials, R. E. Thoma, ed., ORNL-2548, p. 57 (1959).

RbF–ZrF₄(concl.)

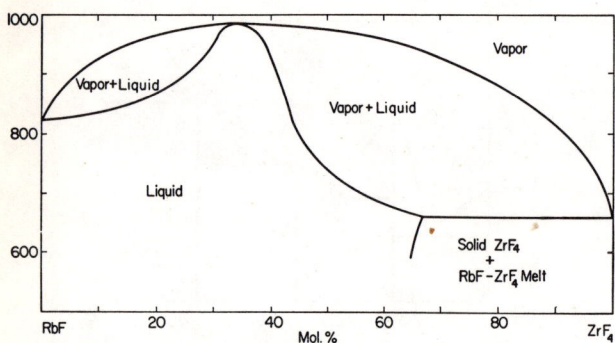

Fig. 1508.—System RbF–ZrF₄ at 1 mm Hg pressure.

K. A. Sense, R. W. Stone, and Robert B. Filbert, Jr., *U.S. At. Energy Comm.*, **BMI–1199**, 17 (1957).

RF–ZrF₄

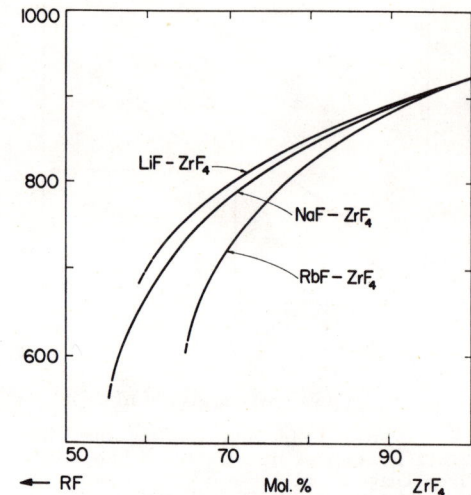

Fig. 1509.—(LiF-, NaF-, RbF-)-ZrF₄; liquidus curves.

K. A. Sense, R. W. Stone, and Robert B. Filbert, Jr., *U.S. At. Energy Comm.*, **BMI–1199**, 31 (1957).

BaF₂–BeF₂

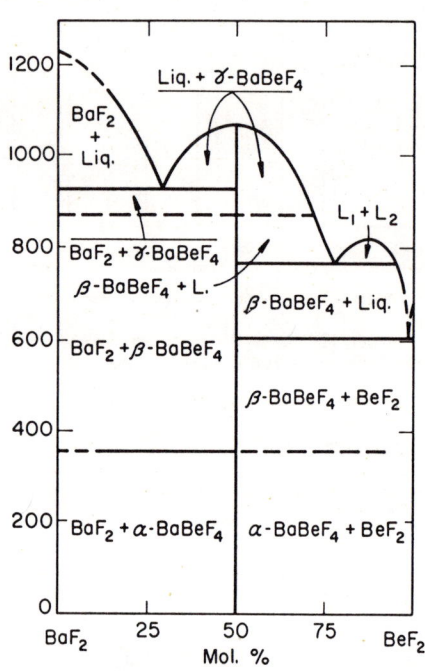

Fig. 1510.—System BaF₂–BeF₂.

D. F. Kirkina, A. V. Novoselova, and Yu. P. Simanov, *Zhur. Neorg. Khim.*, **1** [1] 128 (1956).

BaF₂–MgF₂

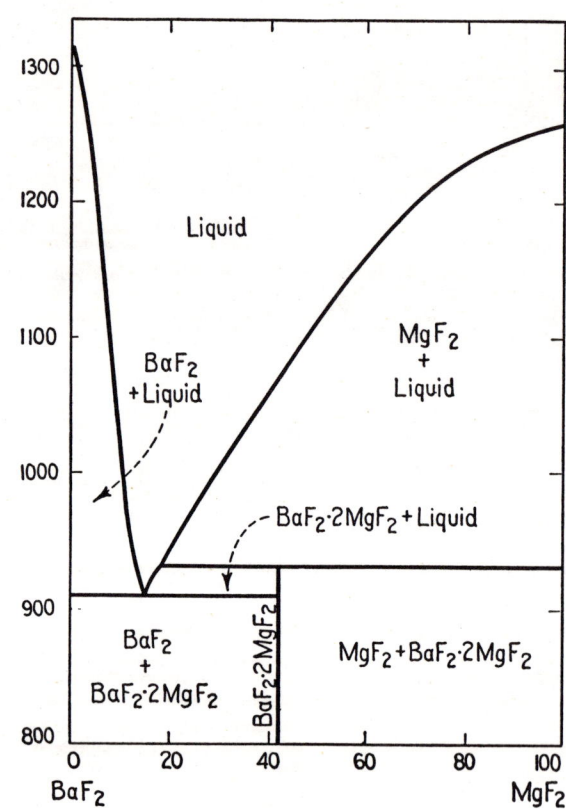

Fig. 1511.—System BaF₂–MgF₂.

M. Okamoto and U. Nisioka, *Science Repts., Tôhoku Imp. Univ.*, Ser. 1, **24**, 142 (1935–36).

BaF_2–UF_3

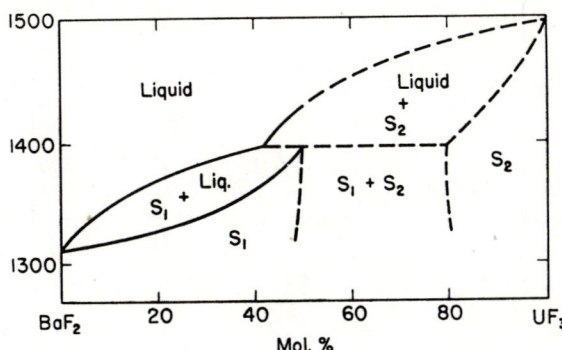

FIG. 1512.—System BaF_2–UF_3, proposed. S_1 = solid solution of UF_3 in BaF_2. S_2 = solid solution of BaF_2 in UF_3.

R. W. M. D'Eye and F. S. Martin, *J. Chem. Soc.*, **1957**, p. 1851.

BeF_2–MgF_2

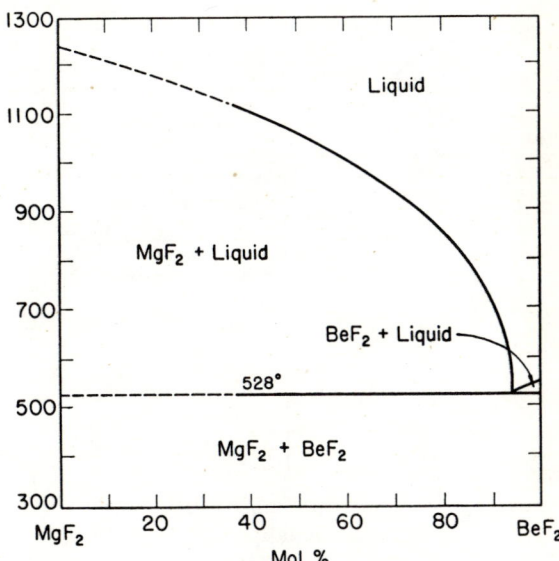

FIG. 1514.—System MgF_2–BeF_2.

W. E. Counts, Rustum Roy, and E. F. Osborn, *J. Am. Ceram. Soc.*, **36** [1] 14 (1953).

BeF_2–CaF_2

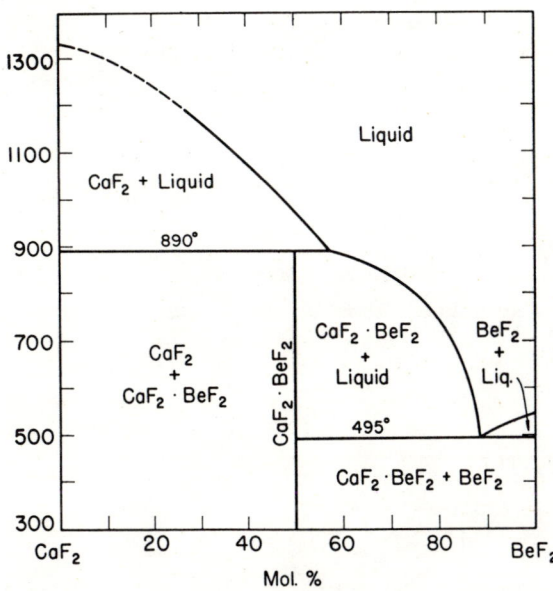

FIG. 1513.—System CaF_2–BeF_2.

W. E. Counts, Rustum Roy, and E. F. Osborn, *J. Am. Ceram. Soc.*, **36** [1] 12 (1953).

BeF_2–PbF_2

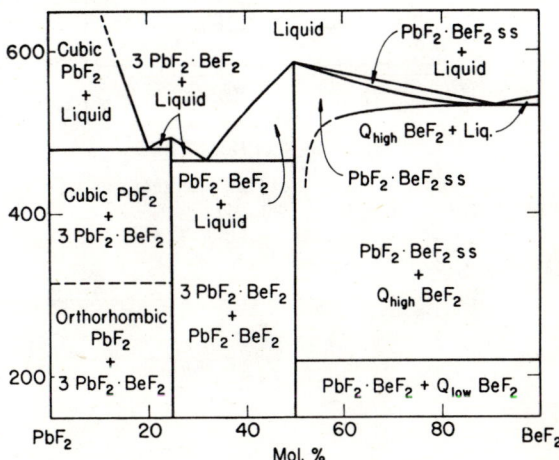

FIG. 1515.—System BeF_2–PbF_2. Q-high = high-quartz form; Q-low = low-quartz form.

D. M. Roy, R. Roy, and E. F. Osborn, *J. Am. Ceram. Soc.*, **37** [7] 303 (1954).

BeF$_2$–SrF$_2$

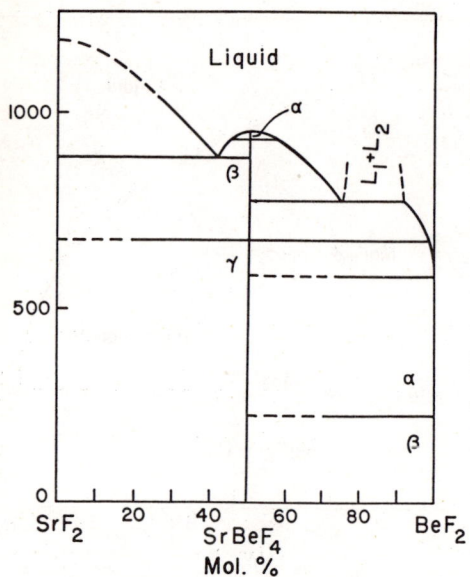

Fig. 1516.—System SrF$_2$–BeF$_2$.

O. N. Breusov, G. Trapp, A. V. Novoselova, and Yu. P. Simanov, *Zhur. Neorg. Khim.*, **4** [3] 672 (1959).

BeF$_2$–ThF$_4$

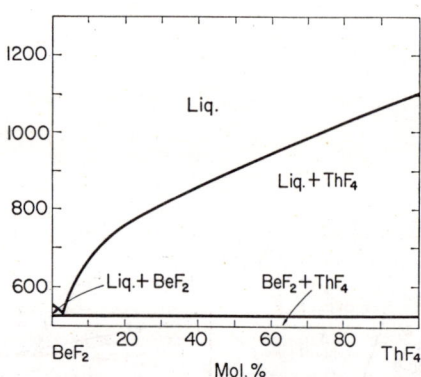

Fig. 1517.—System BeF$_2$–ThF$_4$.

R. E. Thoma, H. Insley, H. A. Friedman, and C. F. Weaver, *J. Phys. Chem.*, **64**, 865–70 (1960).

BeF$_2$–UF$_4$

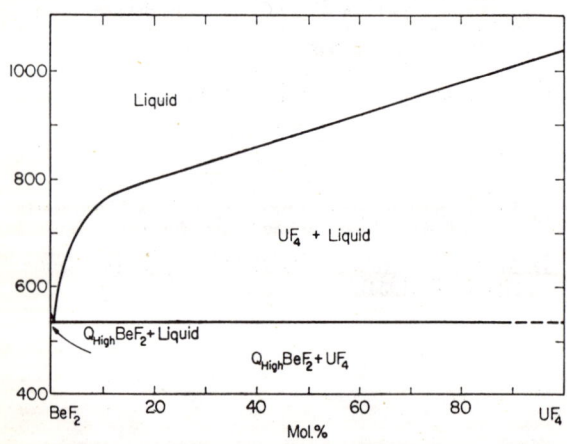

Fig. 1518.—System BeF$_2$–UF$_4$.

L. V. Jones, D. E. Etter, C. R. Hudgens, A. A. Huffman, T. B. Rhinehammer, N. E. Rogers, P. A. Tucker, and L. J. Wittenberg, *J. Am. Ceram. Soc.*, **45** [2] 79 (1962).

CaF$_2$–MgF$_2$

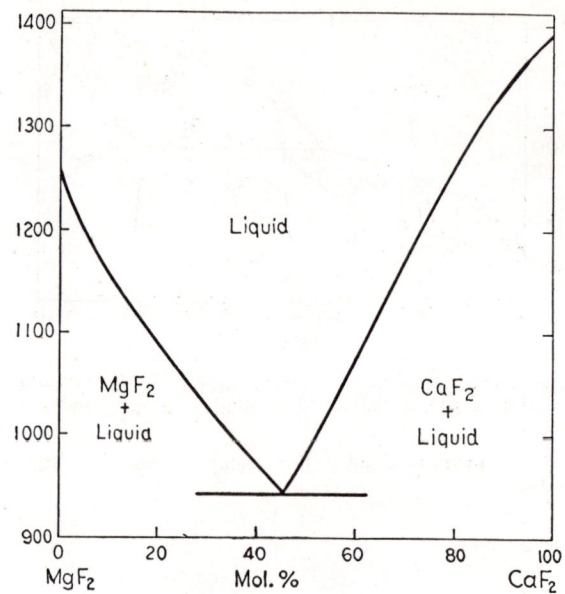

Fig. 1519.—System CaF$_2$–MgF$_2$.

E. Beck, *Metallurgie*, **5**, 504 (1908).

CaF$_2$–YF$_3$

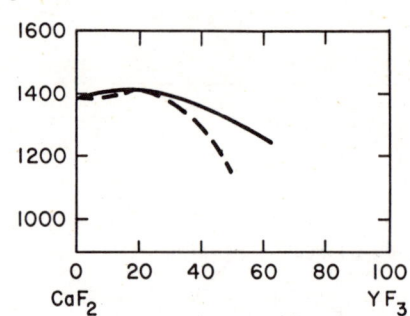

Fig. 1520.—System CaF$_2$–YF$_3$.

Thoroff Vogt, *Neues Jahrb. Mineral., Geol.*, **2**, 12 (1914 II).

MgF$_2$–ThF$_4$

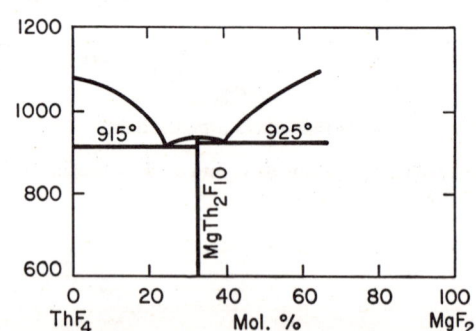

Fig. 1521.—System MgF$_2$–ThF$_4$.

J. O. Blomeke, Oak Ridge National Laboratory, Phase Diagrams of Nuclear Reactor Materials, R. E. Thoma, ed., ORNL-2548, p. 79 (1959).

PbF_2–UF_4

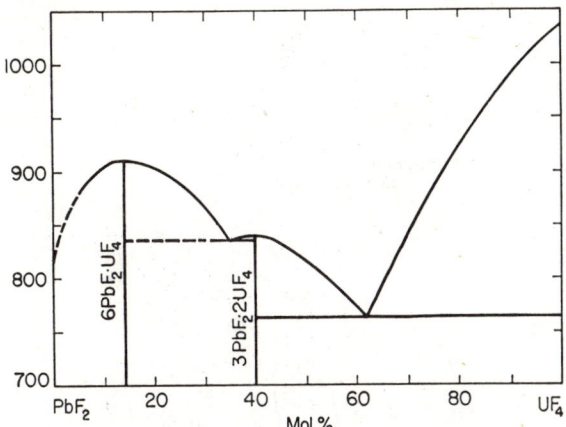

Fig. 1522.—System PbF_2–UF_4; tentative.

C. J. Barton, L. M. Bratcher, J. P. Blakely, G. J. Nessle, and W. R. Grimes, Oak Ridge National Laboratory, Phase Diagrams of Nuclear Reactor Materials, R. E. Thoma, ed., ORNL-2548, p. 95 (1959).

SnF_2–UF_4

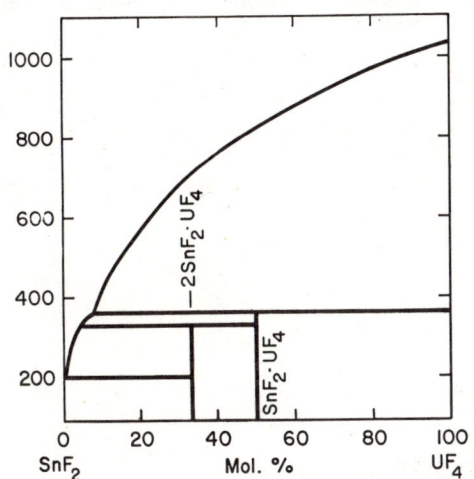

Fig. 1523.—System SnF_2–UF_4.

B. J. Thamer and G. E. Meadows, Oak Ridge National Laboratory, Phase Diagrams of Nuclear Reactor Materials, R. E. Thoma, ed., ORNL-2548, p. 94 (1959).

BF_3–UF_6

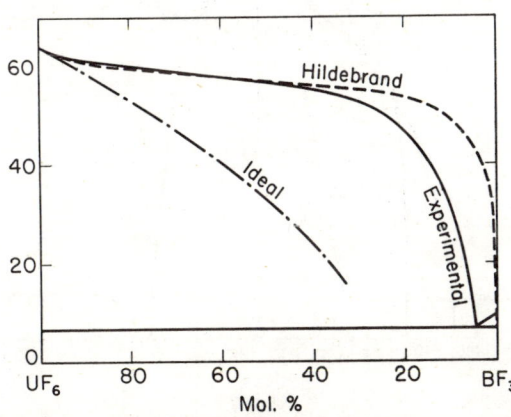

Fig. 1524.—System BF_3–UF_6.

Jack Fischer and R. C. Vogel, J. Am. Chem. Soc., 76, 4831 (1954).

ClF_3–UF_6

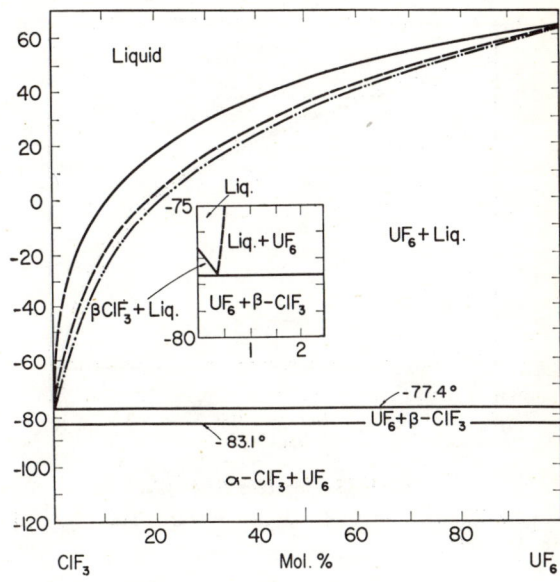

Fig. 1525.—System ClF_3–UF_6. Solid line = experimental, dashed line = theoretical, dash-dot line = ideal.

W. S. Wendolkowski and E. J. Barber, J. Phys. Chem., 62, 751 (1958).

ThF_4–UF_4

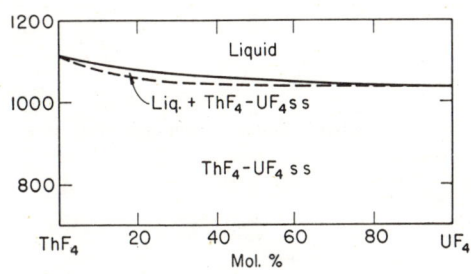

Fig. 1526.—System ThF_4–UF_4.

R. E. Thoma, C. F. Weaver, H. A. Friedman, H. Insley, and W. R. Grimes, Oak Ridge National Laboratory, Phase Diagrams of Nuclear Reactor Materials, R. F. Thoma, ed., ORNL-2548, p. 97 (1959).

UF_4–ZrF_4

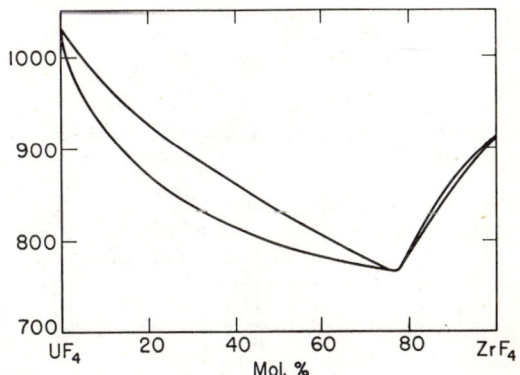

Fig. 1527.—System UF_4–ZrF_4.

C. J. Barton, W. R. Grimes, H. Insley, R. E. Moore, and R. E. Thoma, J. Phys. Chem., 62, 671 (1958).

BF₅–UF₆

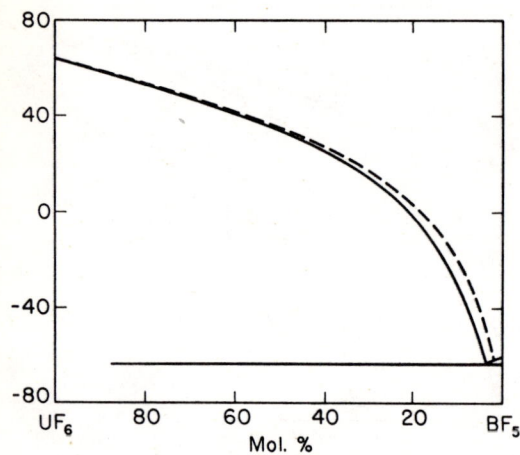

Fig. 1528.—System BF₅–UF₆.

Jack Fischer and R. C. Vogel, *J. Am. Chem. Soc.*, **76**, 4832 (1954).

KF–LiF–RbF

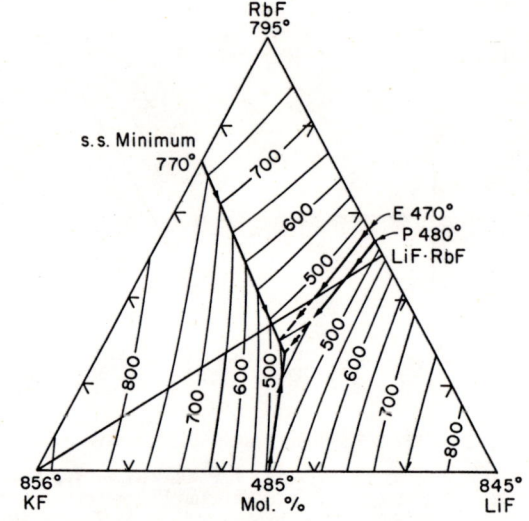

Fig. 1530.—System KF–LiF–RbF.

C. J. Barton, J. P. Blakely, L. M. Bratcher, and W. R. Grimes, Oak Ridge National Laboratory, Phase Diagrams for Nuclear Reactor Materials, R. E. Thoma, ed., ORNL-2548, p. 23 (1959).

(c) *Three Fluorides*

KF–LiF–NaF

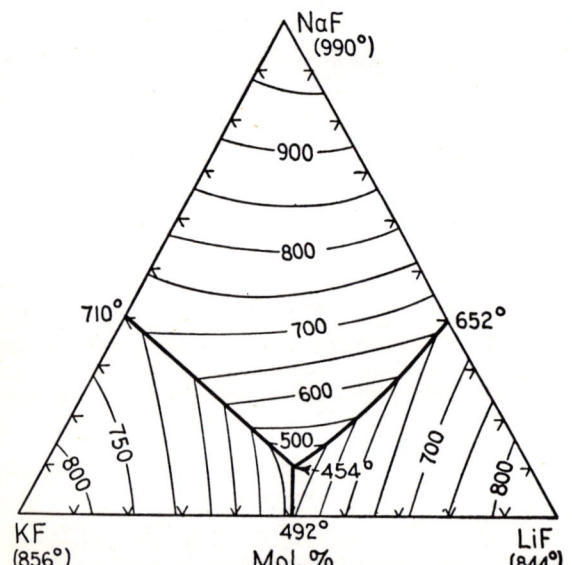

Fig. 1529.—System KF–LiF–NaF.

A. G. Bergman and E. P. Dergunov, *Compt. rend. acad. sci., U.R.S.S.*, **31**, 754 (1941).

KF–LiF–MgF₂

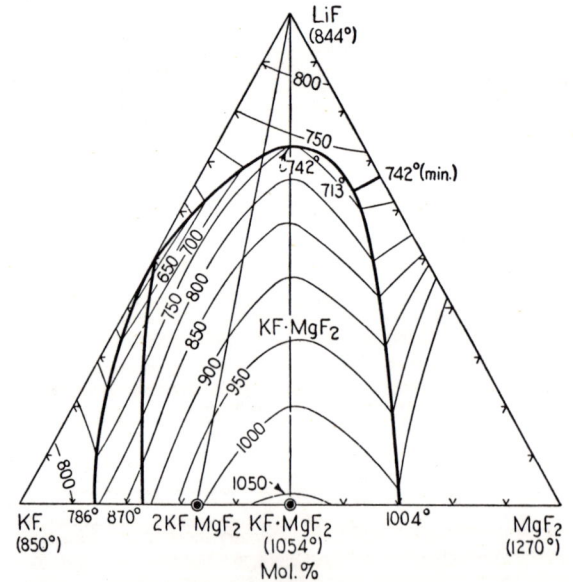

Fig. 1531.—System KF–LiF–MgF₂.

A. G. Bergman and S. P. Parlenko, *Compt. rend. acad. sci., U.R.S.S.*, **31**, 818–19 (1941).

KF–LiF–UF₄

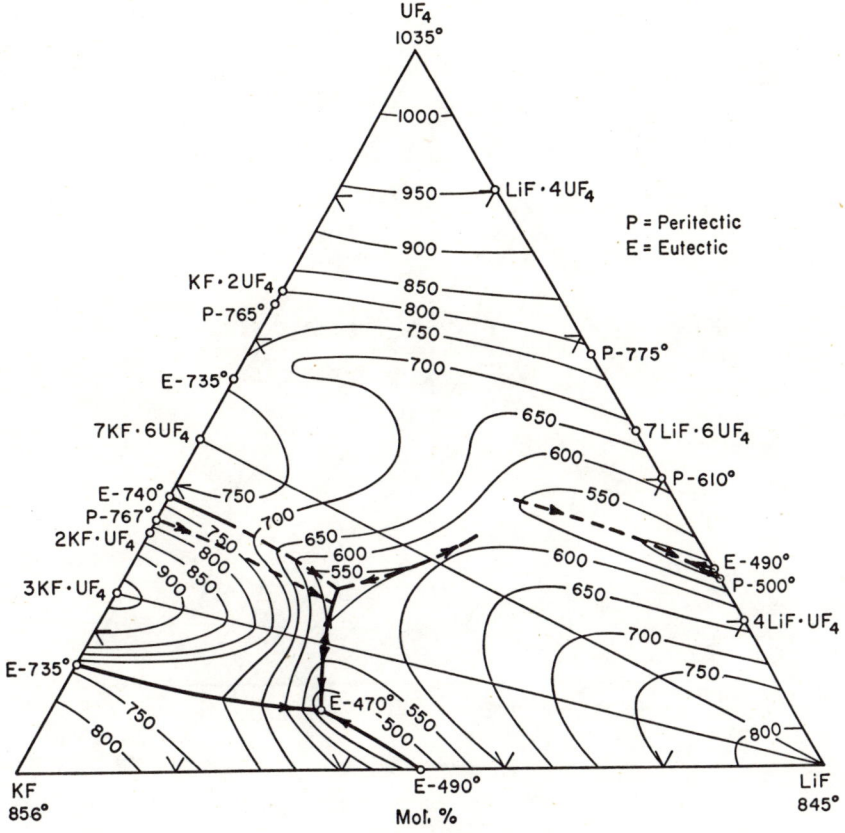

FIG. 1532.—System KF–LiF–UF₄; tentative.

C. J. Barton, J. P. Blakely, L. M. Bratcher, and W. R. Grimes, Oak Ridge National Laboratory, Phase Diagrams of Nuclear Reactor Materials, R. E. Thoma, ed., ORNL-2548, p. 101 (1959).

KF–NaF–RbF

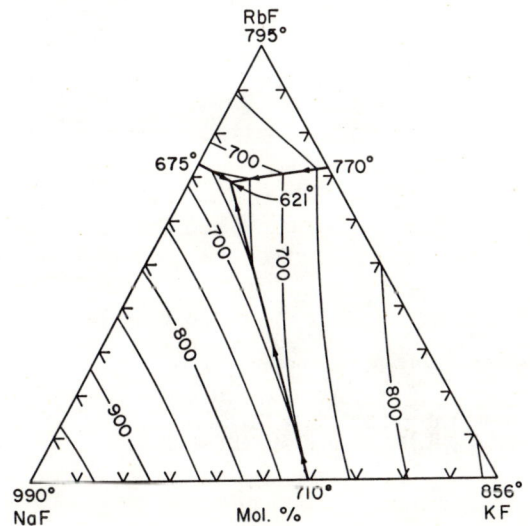

FIG. 1533.—System KF–NaF–RbF.

C. J. Barton, J. P. Blakely, L. M. Bratcher, and W. R. Grimes, Oak Ridge National Laboratory, Phase Diagrams for Nuclear Reactor Materials, R. E. Thoma, ed., ORNL-2548, p. 24 (1959).

KF–NaF–BeF₂

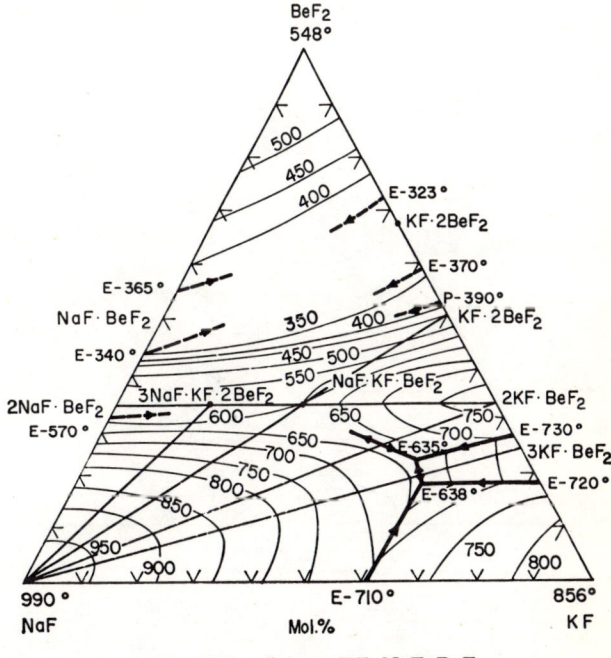

FIG. 1534.—System KF–NaF–BeF₂.

R. E. Moore, C. J. Barton, L. M. Bratcher, T. N. McVay, and W. R. Grimes, Oak Ridge National Laboratory, Phase Diagrams of Nuclear Reactor Materials, R. E. Thoma, ed., ORNL-2548, p. 46 (1959).

KF–NaF–MgF$_2$

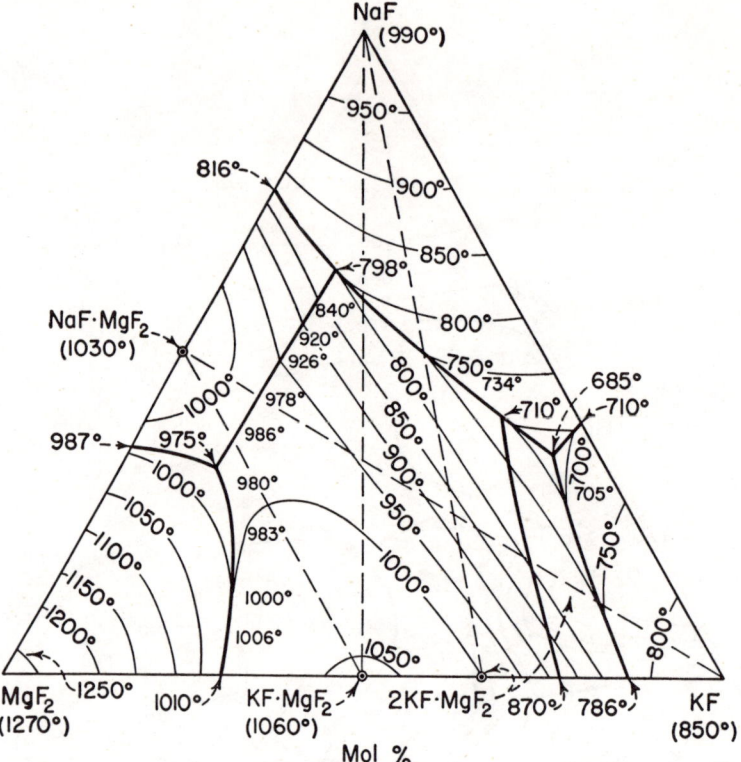

Fig. 1535.—System KF–NaF–MgF$_2$.

A. G. Bergman and E. P. Dergunov, *Compt. rend. acad. sci., U.R.S.S.*, **48** 330 (1945).

The join between KF and NaF·MgF$_2$ should be eliminated.

KF–NaF–AlF$_3$

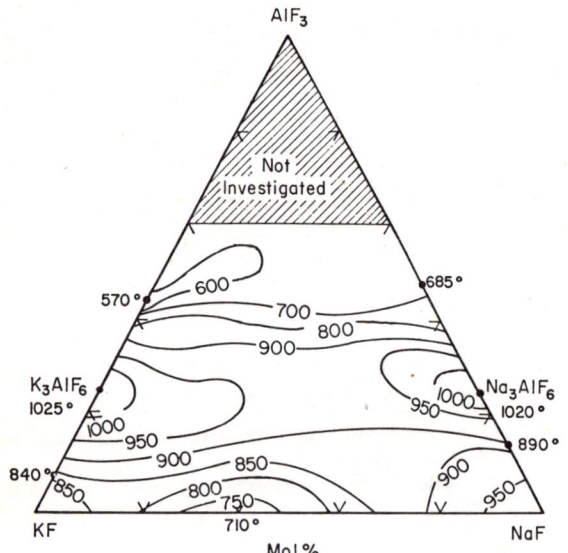

Fig. 1536.—System KF–NaF–AlF$_3$; preliminary.

C. J. Barton, L. M. Bratcher, and W. R. Grimes, Oak Ridge National Laboratory, Phase Diagrams of Nuclear Reactor Materials, R. E. Thoma, ed., ORNL-2548, p. 32 (1959).

KF–NaF–BF$_3$

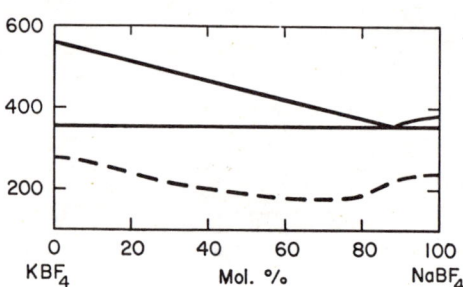

Fig. 1537.—System KBF$_4$–NaBF$_4$. Dotted line obtained from thermal effects representing incompletely understood solid-phase transformations.

R. E. Moore, J. G. Surak, and W. R. Grimes, Oak Ridge National Laboratory, Phase Diagrams of Nuclear Reactor Materials, R. E. Thoma, ed., ORNL-2548, p. 25 (1959).

KF–NaF–UF$_4$

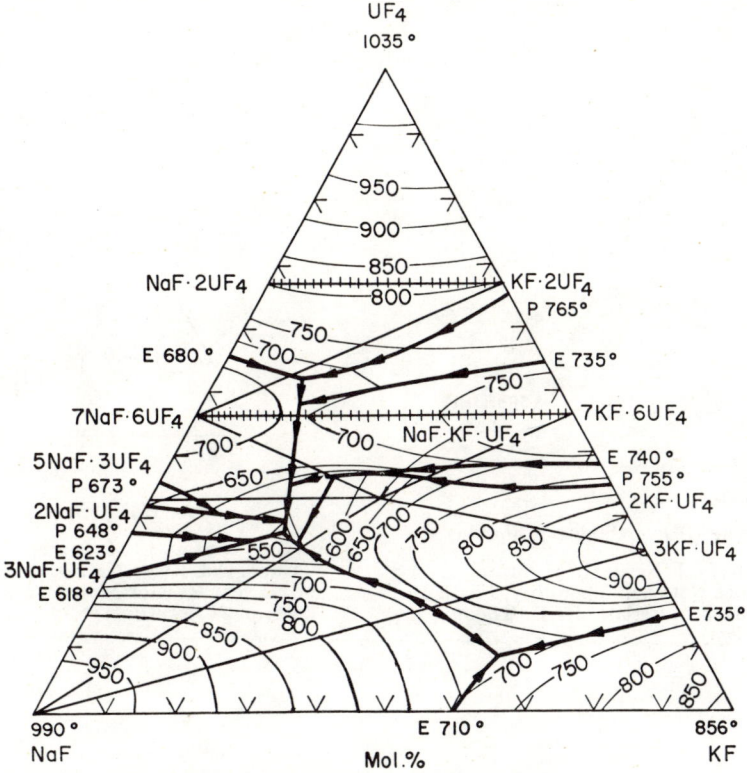

Fig. 1538.—System KF–NaF–UF$_4$; preliminary.

R. E. Thoma, C. J. Barton, J. P. Blakely, R. E. Moore, G. J. Nessle, H. Insley, and H. A. Friedman, Oak Ridge National Laboratory, Phase Diagrams of Nuclear Reactor Materials, R. E. Thoma, ed., ORNL-2548, p. 103 (1959).

KF–NaF–ZrF$_4$

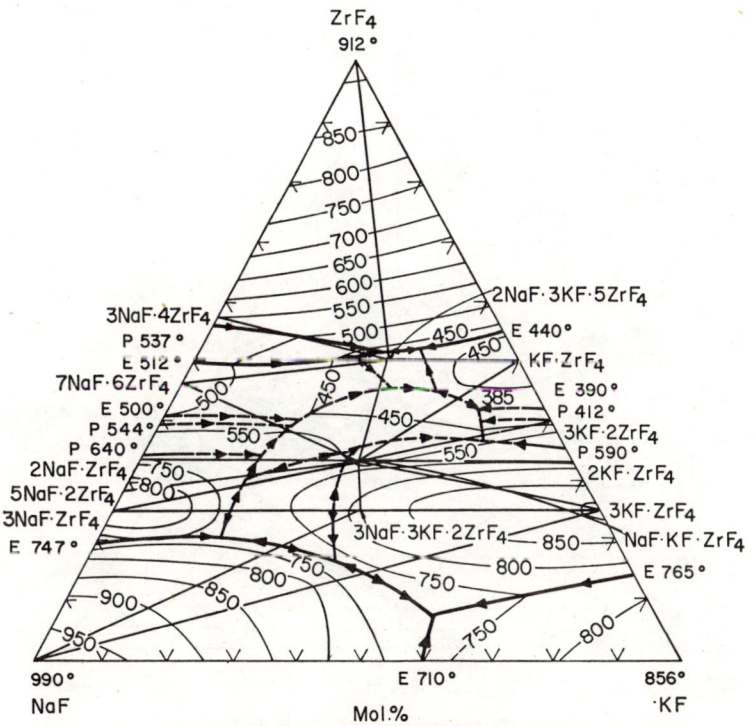

Fig. 1539.—System KF–NaF–ZrF$_4$.

R. E. Thoma, C. J. Barton, H. Insley, H. A. Friedman, and W. R. Grimes, Oak Ridge National Laboratory, Phase Diagrams of Nuclear Reactor Materials, R. E. Thoma, ed., ORNL-2548, p. 63 (1959).

KF–PbF$_2$–UF$_4$

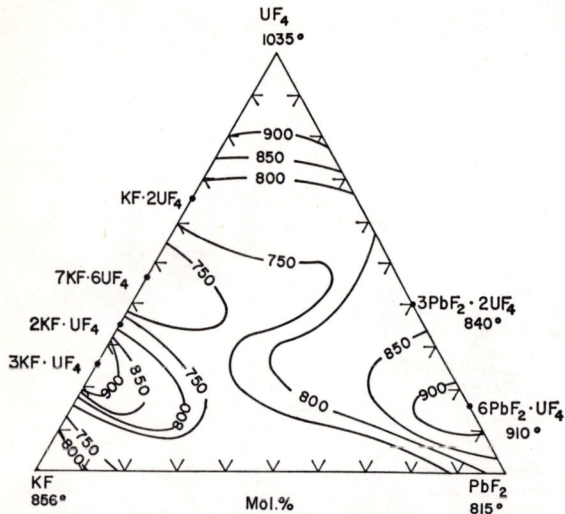

FIG. 1540.—System KF–PbF$_2$–UF$_4$; preliminary, liquidus

C. J. Barton, J. P. Blakely, G. J. Nessle, L. M. Bratcher and W. R. Grimes, Oak Ridge National Laboratory, Phase Diagrams of Nuclear Reactor Materials, R. E. Thoma, ed., ORNL-2548, p. 115 (1959).

LiF–NaF–RbF

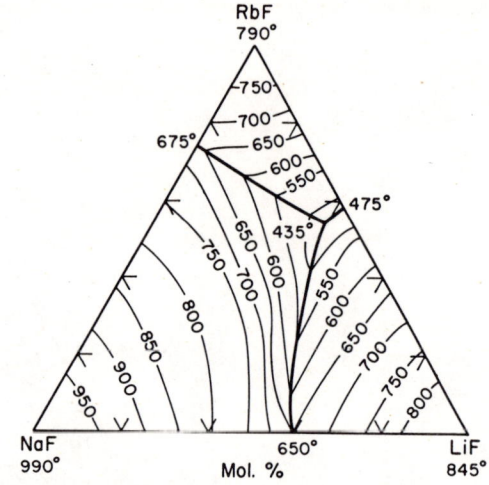

FIG. 1541.—System LiF–NaF–RbF.

C. J. Barton, L. M. Bratcher, J. P. Blakely and W. R. Grimes, Oak Ridge National Laboratory, Phase Diagrams of Nuclear Reactor Materials, R. E. Thoma, ed., ORNL-2548, p. 22 (1959).

LiF–NaF–BeF$_2$

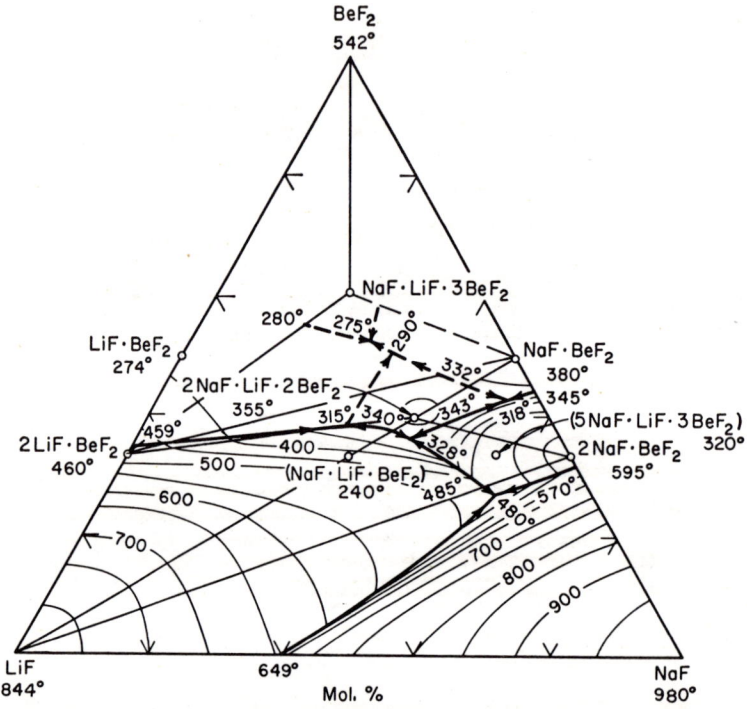

FIG. 1542.—System LiF–NaF–BeF$_2$.

R. E. Moore, C. J. Barton, W. R. Grimes, R. E. Meadows, L. M. Bratcher, G. D. White, T. N. McVay, and L. A. Harris, Oak Ridge National Laboratory, Phase Diagrams of Nuclear Reactor Materials, R. E. Thoma, ed., ORNL-2548, p. 43 (1959).

LiF–NaF–BeF₂(concl.)

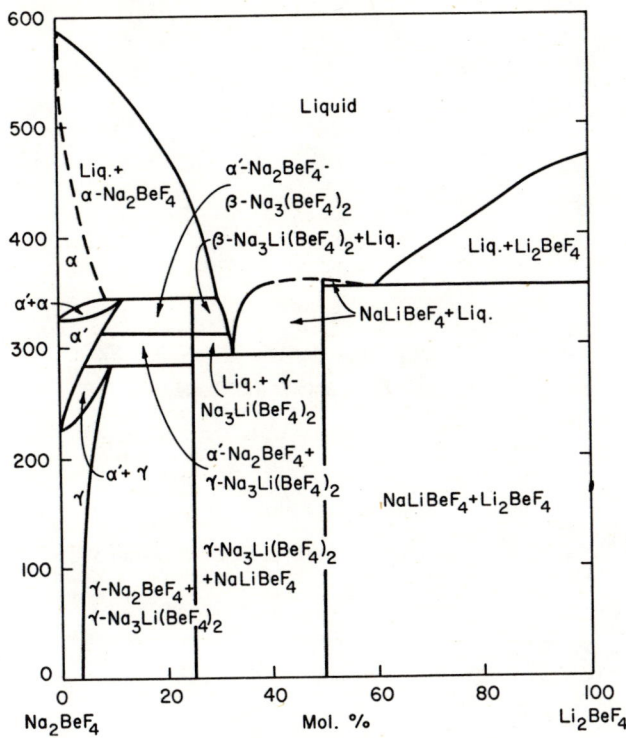

Fig. 1543.—System Li₂BeF₄–Na₂BeF₄.

N. A. Toropov and I. L. Shchetnikova, *Zhur. Neorg. Khim.*, **2**, 1857 (1957).

LiF–NaF–MgF₂

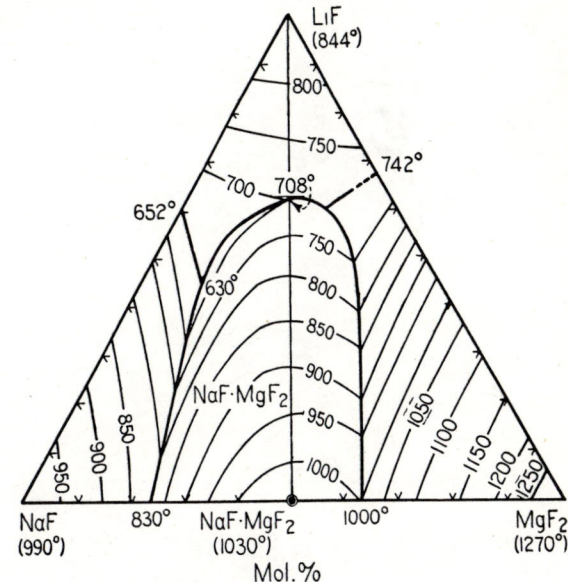

Fig. 1545.—System LiF–NaF–MgF₂.

A. G. Bergman and E. P. Dergunov, *Compt. rend. acad. sci., U.R.S.S.*, **31**, 755 (1941).

LiF–NaF–CaF₂

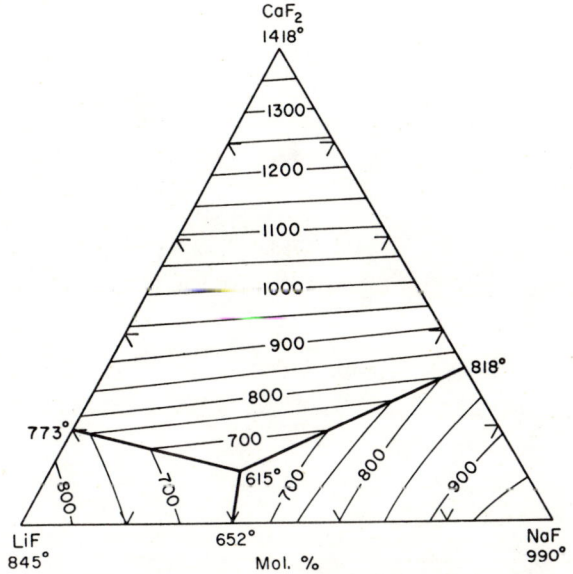

Fig. 1544.—System LiF–NaF–CaF₂.

C. J. Barton, L. M. Bratcher, and W. R. Grimes, Oak Ridge National Laboratory, Phase Diagrams of Nuclear Reactor Materials, R. E. Thoma, ed., ORNL-2548, p. 29 (1959).

See also G. A. Bukhalova, K. Sulaĭmankulov, and A. K. Bostandzhiyan, *Zhur. Neorg. Khim.*, **4**, 1138 (1959).

LiF–NaF–AlF₃

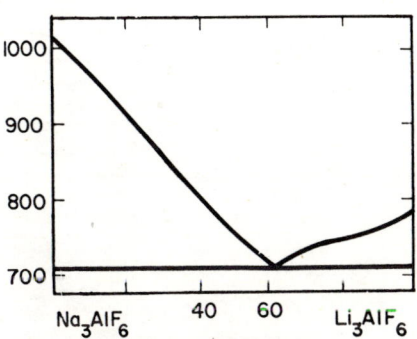

Fig. 1546.—System Li₃AlF₆–Na₃AlF₆.

V. P. Mashovets and V. I. Petrov, *Zhur. Prikl. Khim.*, **30**, 1696 (1957).

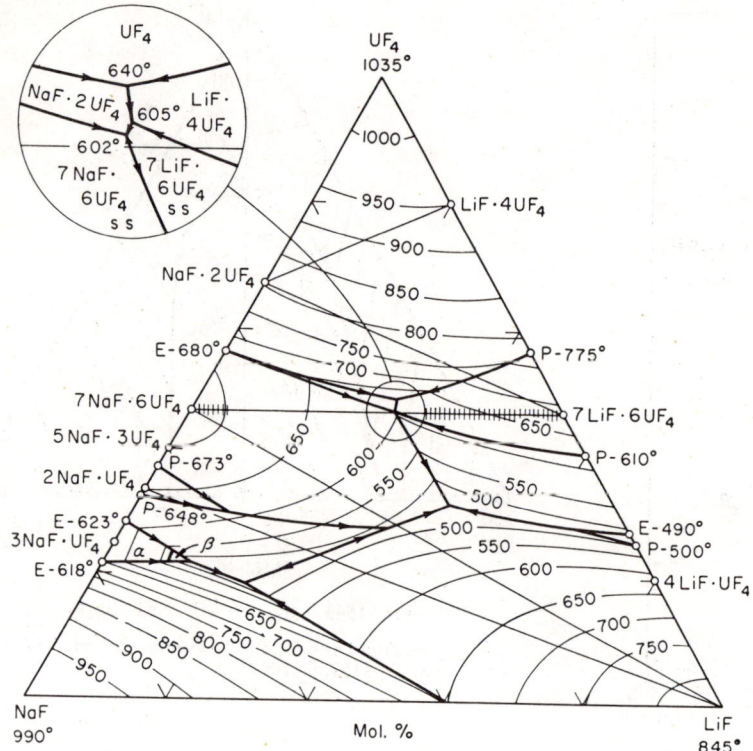

Fig. 1547.—System LiF–NaF–UF$_4$.

R. E. Thoma, H. Insley, B. S. Landau, H. A. Friedman, and W. R. Grimes, *J. Am. Ceram. Soc.*, **42** [1] 22 (1959).

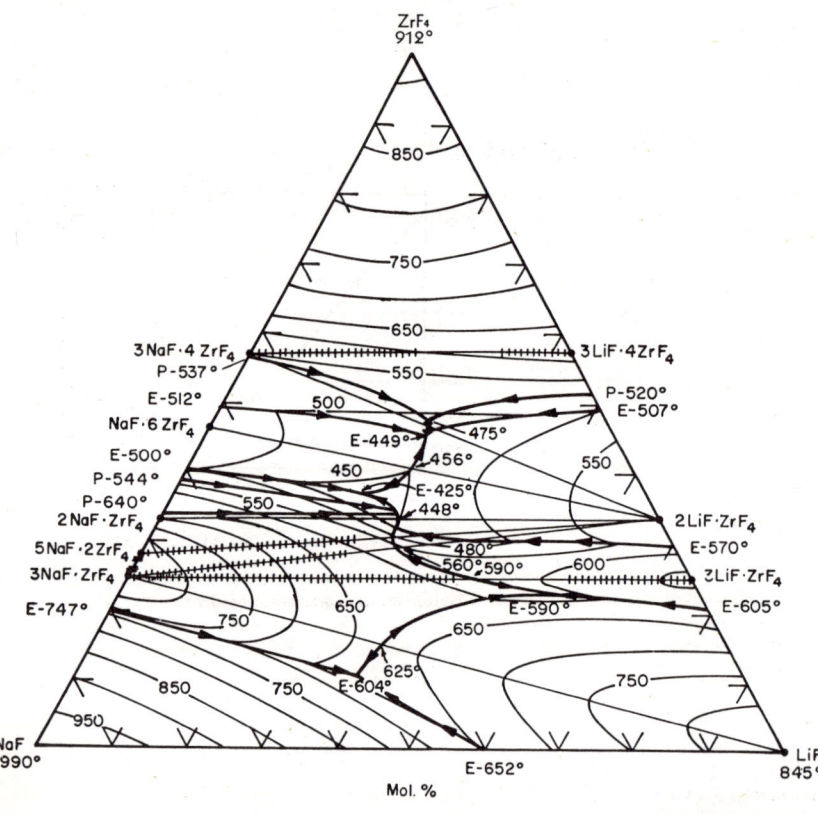

Fig. 1548.—System LiF–NaF–ZrF$_4$.

F. F. Blankenship, H. A. Friedman, R. E. Thoma, and W. R. Grimes, Oak Ridge National Laboratory, Phase Diagrams of Nuclear Reactor Materials, R. E. Thoma, ed., ORNL-2548 p. 61 (1959).

LiF–RbF–BeF$_2$

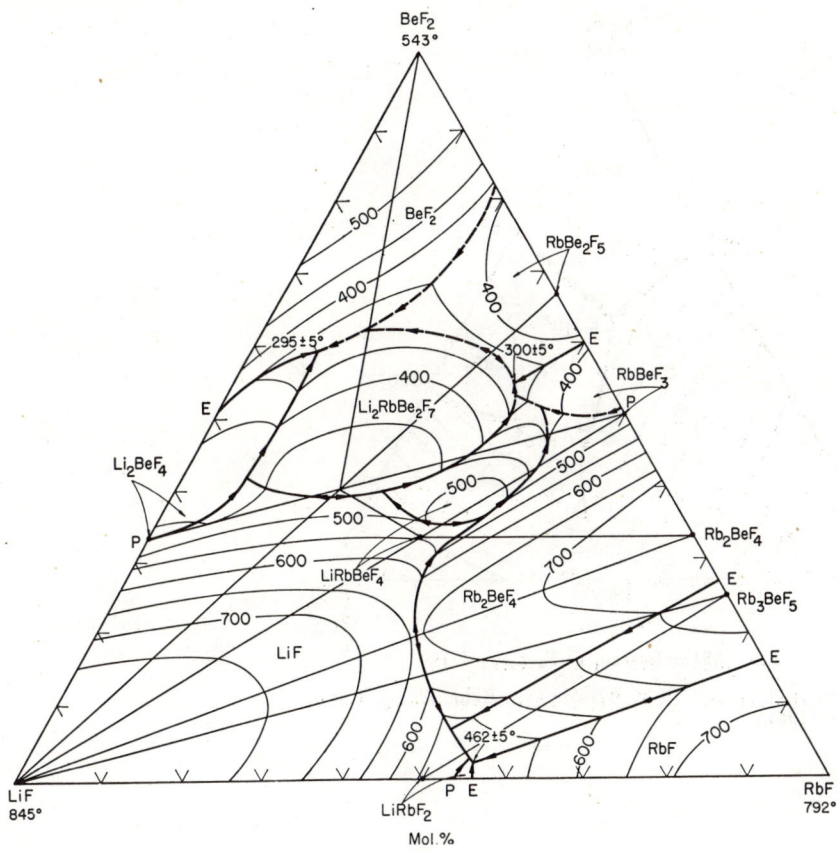

FIG. 1549.—System LiF–RbF–BeF$_2$.

T. B. Rhinehammer, D. E. Etter, C. R. Hudgens, N. E. Rogers and P. A. Tucker, Oak Ridge National Laboratory, Phase Diagrams of Nuclear Reactor Materials, R. E. Thoma, ed., ORNL-2548, p. 44 (1959).

LiF–RbF–UF$_4$

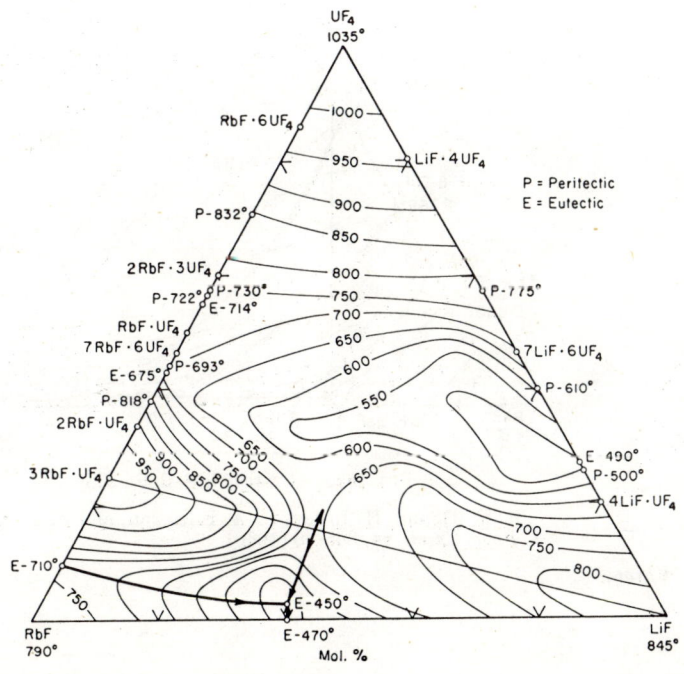

FIG. 1550.—System LiF–RbF–UF$_4$; tentative.

C. J. Barton, J. P. Blakely, L. M. Bratcher, and W. R. Grimes, Oak Ridge National Laboratory, Phase Diagrams of Nuclear Reactor Materials, R. E. Thoma, ed., ORNL-2548, p. 102 (1959).

LiF–BaF$_2$–CaF$_2$

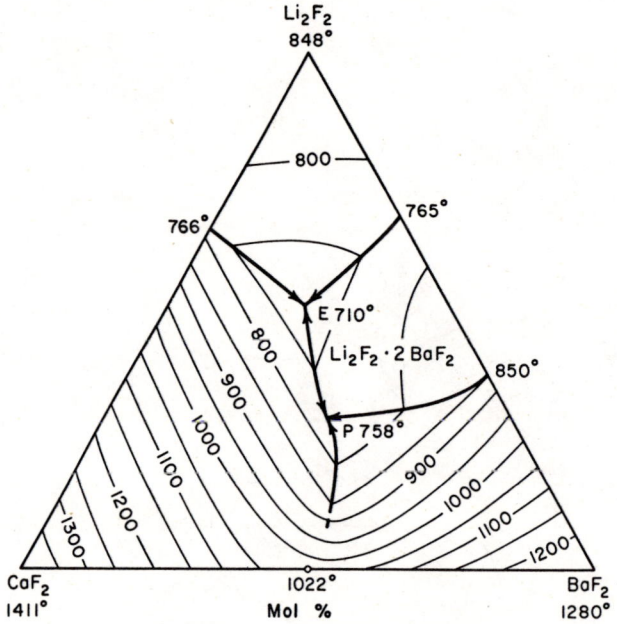

Fig. 1551.—System BaF$_2$–CaF$_2$–LiF.

G. A. Bukhalova and V. T. Berezhnaya, *Zhur. Neorg. Khim.*, **2** [6] 1409 (1957).

LiF–BaF$_2$–MgF$_2$

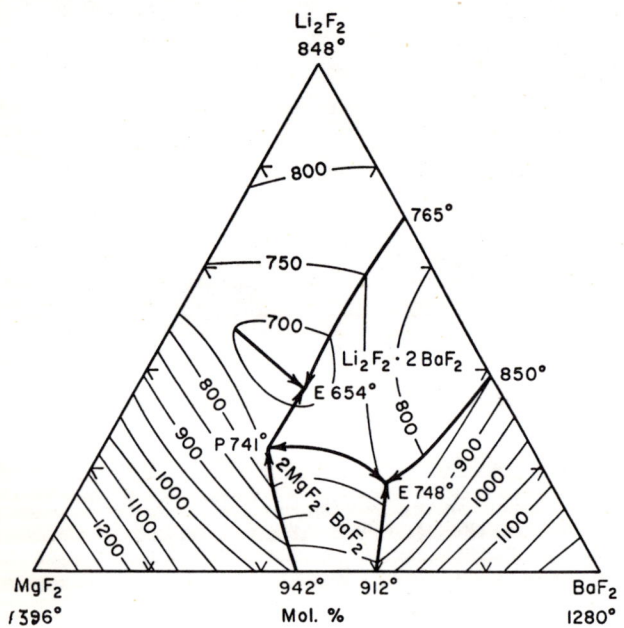

Fig. 1552.—System BaF$_2$–LiF–MgF$_2$.

G. A. Bukhalova and V. T. Berezhnaya, *Zhur. Neorg. Khim.*, **4** [5] 1141 (1959).

LiF–BeF$_2$–ThF$_4$

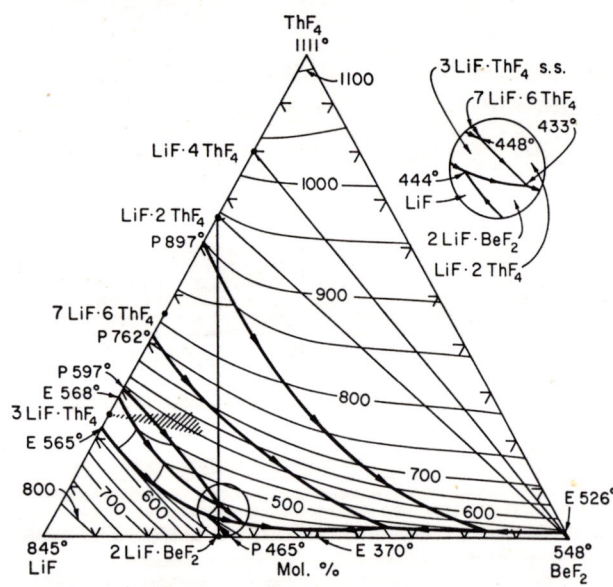

Fig. 1553.—System LiF–BeF$_2$–ThF$_4$.

R. E. Thoma, H. Insley, H. A. Friedman, and C. F. Weaver. *J. Phys. Chem.*, **64**, 865–70 (1960).

LiF–BeF$_2$–UF$_4$

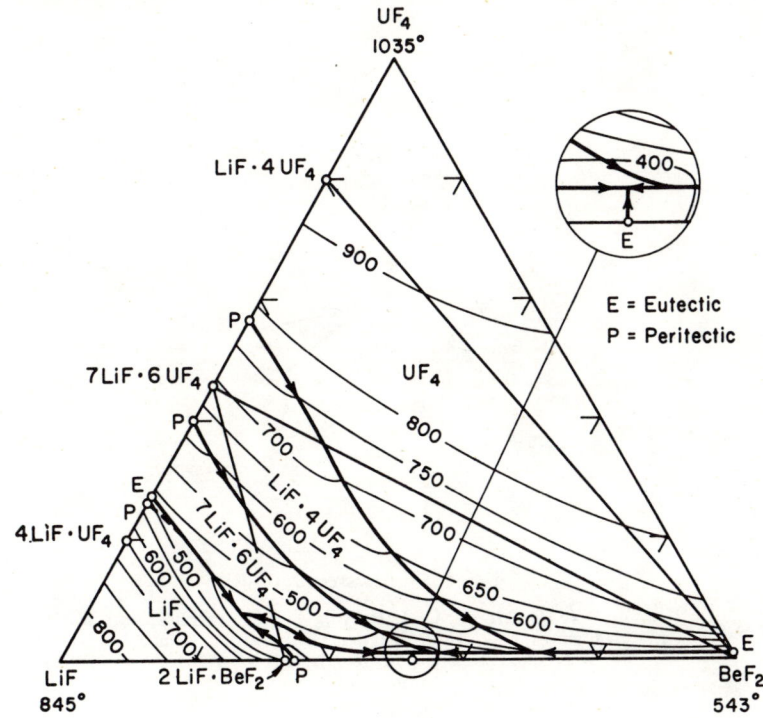

Fig. 1554.—System LiF–BeF$_2$–UF$_4$.

L. V. Jones, D. E. Etter, C. R. Hudgens, A. A. Huffman, T. B. Rhinehammer, N. E. Rogers, P. A. Tucker, and L. J. Wittenberg, *J. Am. Ceram. Soc.*, **45** [2] 81 (1962).

LiF–CaF$_2$–MgF$_2$

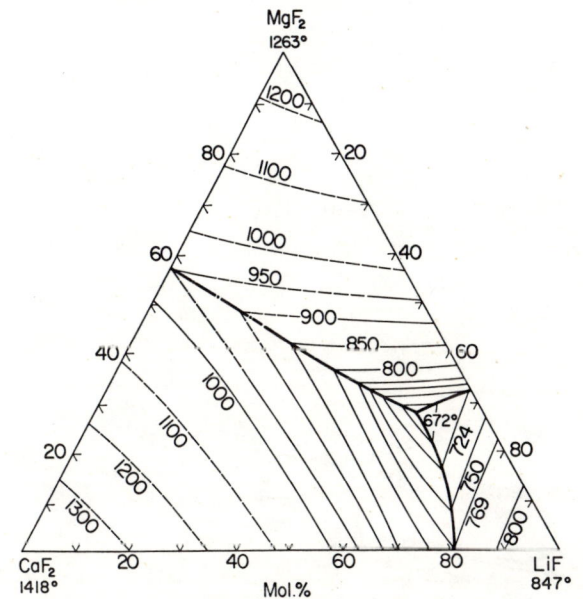

Fig. 1555.—System LiF–CaF$_2$–MgF$_2$ (solid solutions not shown).

W. E. Roake, *J. Electrochem. Soc.*, **104**, 662 (1957). See also V. T. Berezhnaya and G. A. Bukhalova, *Zhur. Neorg. Khim.*, **4**, 903 (1959).

LiF–ThF$_4$–UF$_4$

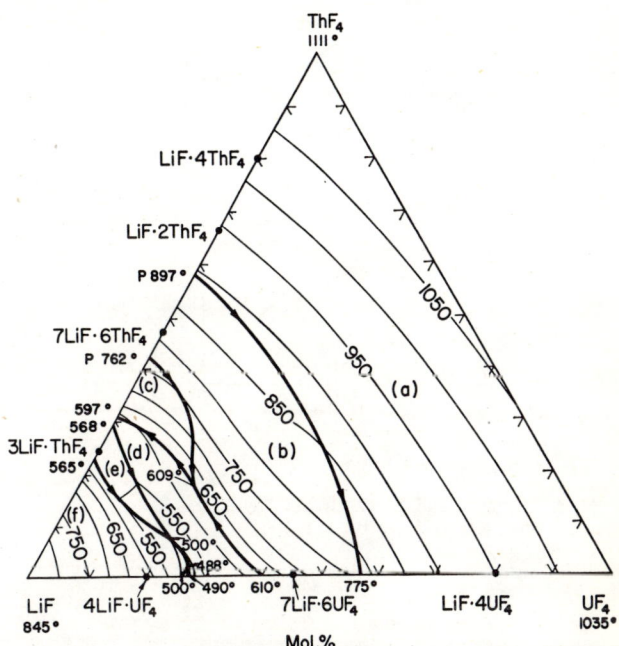

Fig. 1556.—System LiF–ThF$_4$–UF$_4$. (*a*) = UF$_4$–ThF$_4$ ss, (*b*) = LiF·4UF$_4$–LiF·4ThF$_4$ ss, (*c*) = LiF·2Th(U)F$_4$ ss, (*d*) = 7LiF·6UF$_4$–7LiF·6ThF$_4$ ss, (*e*) = 3LiF·Th(U)F$_4$ ss, (*f*) = LiF.

C. F. Weaver, R. E. Thoma, H. Insley, and H. A. Friedman, *J. Am. Ceram. Soc.*, **43** [4] 214 (1960). On the LiF–ThF$_4$ binary, the eutectic temperatures labeled 568° and 565°C. (now correct) are interchanged in the reference cited, according to R. E. Thoma, private communication, 1962.

LiF–ThF₄–UF₄ (concl.)

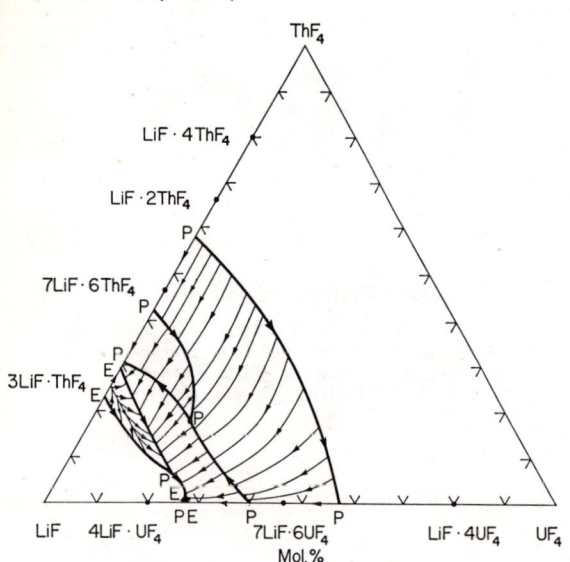

Fig. 1557.—System LiF–ThF$_4$–UF$_4$; fractionation paths.

C. F. Weaver, R. E. Thoma, H. Insley, and H. A. Friedman, *J. Am. Ceram. Soc.*, **43** [4] 216 (1960).

NaF–RbF–BeF₂

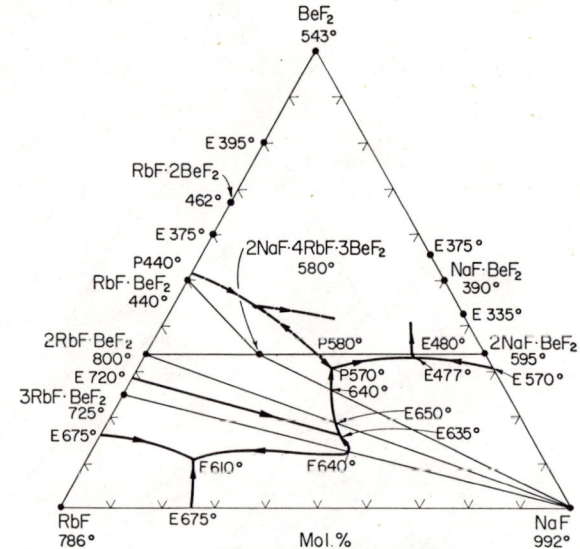

Fig. 1558.—System NaF–RbF–BeF$_2$.

R. E. Moore, C. J. Barton, L. M. Bratcher, and W. R. Grimes, Oak Ridge National Laboratory, Phase Diagrams of Nuclear Reactor Materials, ORNL-2548, p. 47 (1959).

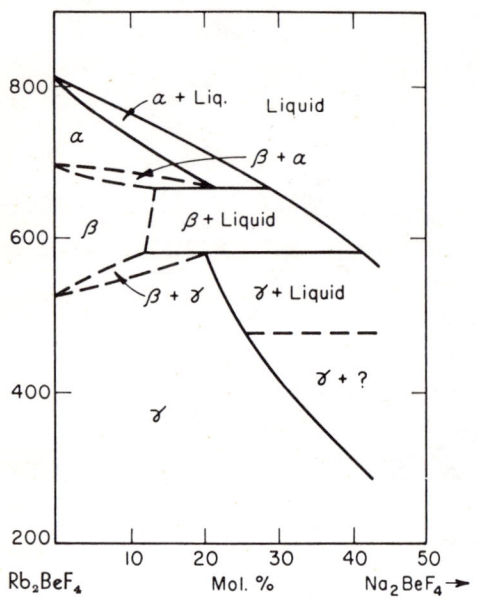

Fig. 1559.—System Na$_2$BeF$_4$–Rb$_2$BeF$_4$.

N. A. Toropov and R. G. Grebenshchikov, *Zhur. Neorg. Khim.*, **1** [7] 1623 (1956).

NaF–RbF–UF₄

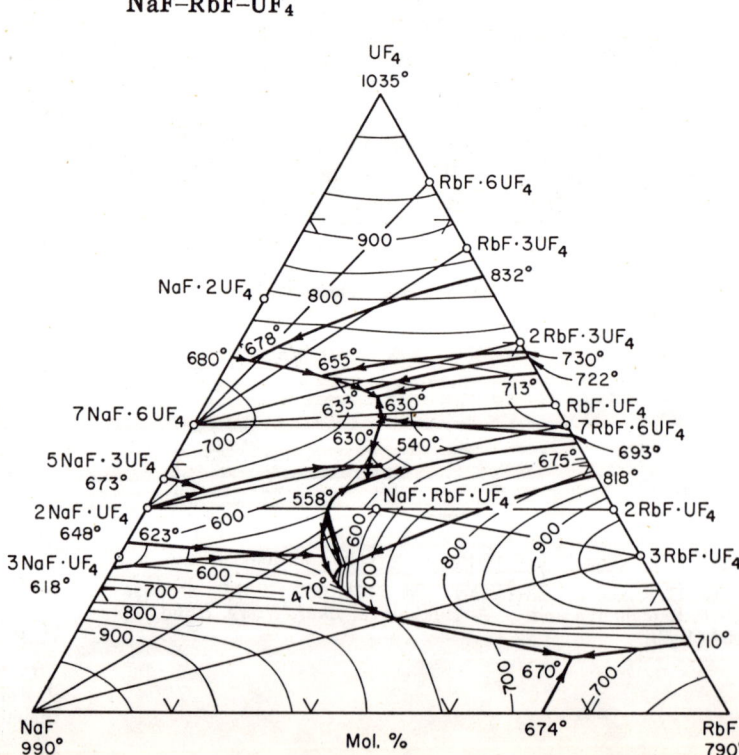

Fig. 1560.—System NaF–RbF–UF$_4$.

R. E. Thoma, H. Insley, H. A. Friedman, and W. R. Grimes, Oak Ridge National Laboratory, Phase Diagrams of Nuclear Reactor Materials, R. E. Thoma, ed., ORNL-2548, p. 105 (1959).

NaF–RbF–ZrF$_4$

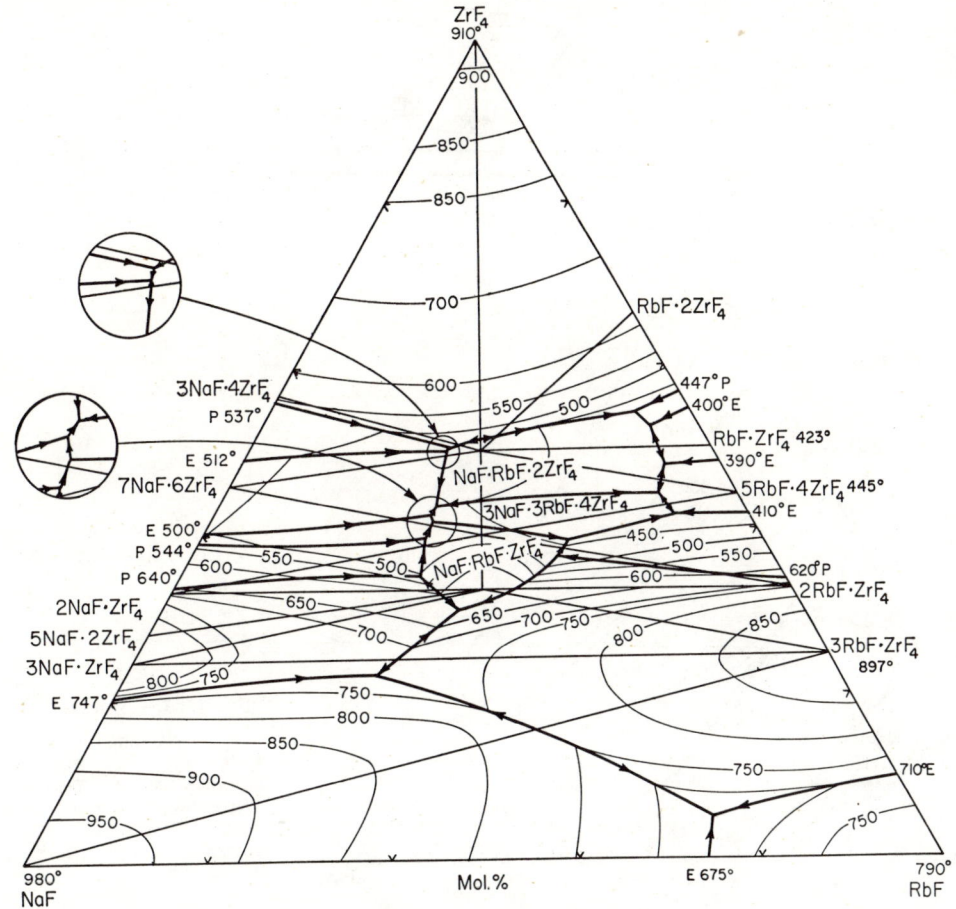

Fig. 1561.—System NaF–RbF–ZrF$_4$.

R. E. Thoma, H. Insley, H. A. Friedman, and W. R. Grimes, Oak Ridge National Laboratory, Phase Diagrams of Nuclear Reactor Materials, R. E. Thoma, ed., ORNL-2548 p. 65 (1959).

NaF–BeF$_2$–ThF$_4$

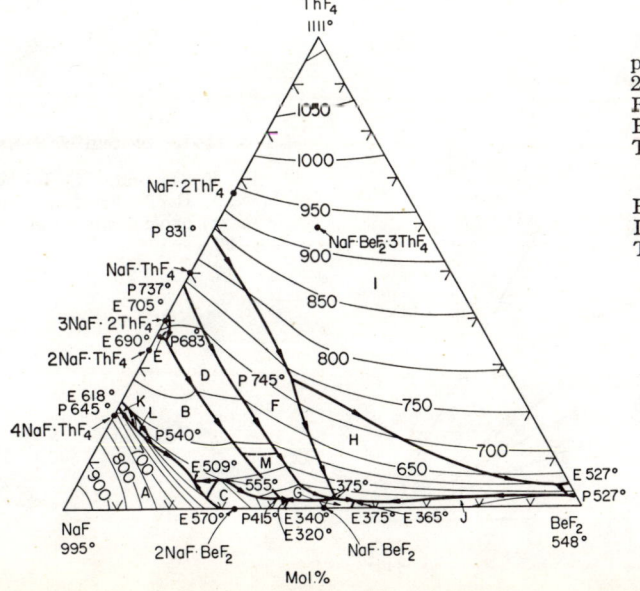

Fig. 1562.—System NaF–BeF$_2$–ThF$_4$. Primary phase fields: A = NaF, B = 2NaF·ThF$_4$, C = 2NaF·BeF$_2$, D = αNaF·ThF$_4$, E = 3NaF·2ThF$_4$, F = NaF·2ThF$_4$, G = NaF·BeF$_2$, H = NaF·BeF$_2$·3ThF$_4$, I = ThF$_4$, J = BeF$_2$, K = α4NaF·ThF$_4$, L = β4NaF·ThF$_4$, M = βNaF·ThF$_4$.

C. F. Weaver, R. E. Thoma, H. Insley, and H. A. Friedman, Oak Ridge National Laboratory, Phase Diagrams of Nuclear Reactor Materials, R. E. Thoma, ed., ORNL-3122, p. 113 (1961).

NaF–BeF₂–UF₄

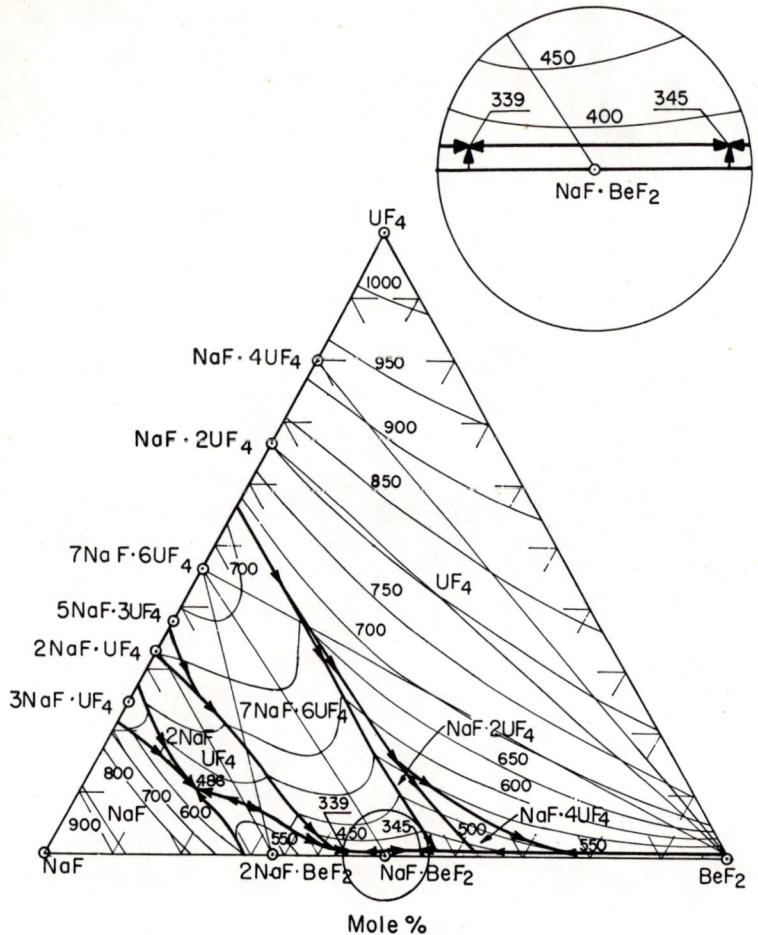

FIG. 1563.—System NaF–BeF₂–UF₄.

J. F. Eichelberger, C. R. Hudgens, L. V. Jones, G. Pish, T. B. Rhinehammer, P. A. Tucker, and L. J. Wittenberg, Mound Laboratory Report MLM-1081. See also same authors, *J. Am. Ceram. Soc.*, **46** [6] 282 (1963).

NaF–BeF₂–ZrF₄

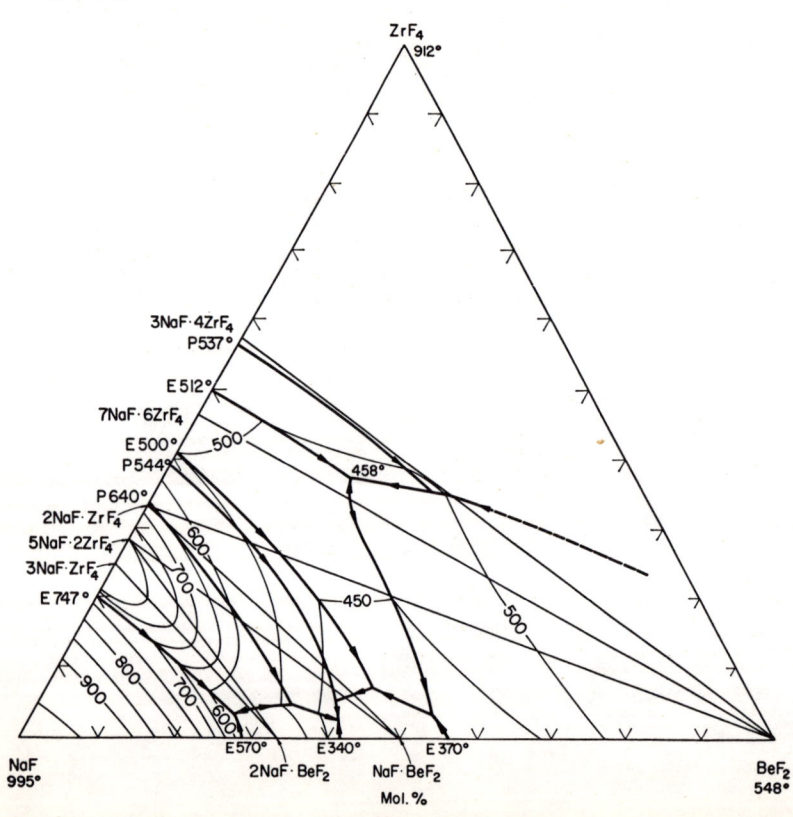

FIG. 1564.—System NaF–BeF₂–ZrF₄.

R. E. Thoma, H. Insley, T. N. McVay, H. A. Friedman, and C. F. Weaver, private communication, Dec. 20, 1961.

NaF–CaF₂–MgF₂

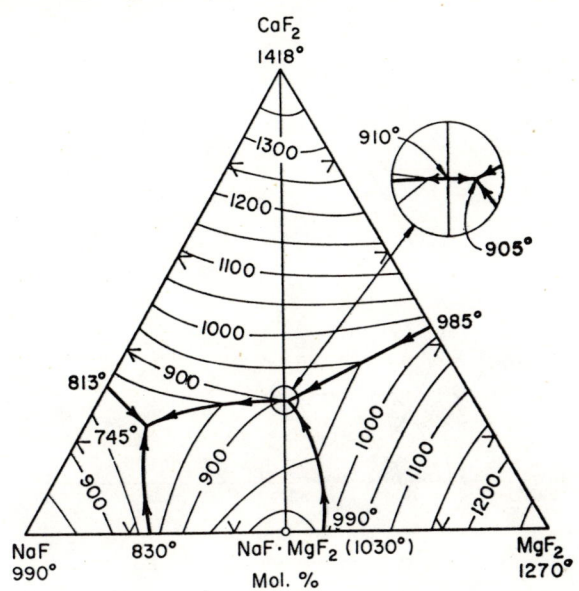

FIG. 1565.—System NaF–CaF₂–MgF₂; preliminary.

C. J. Barton, L. M. Bratcher, J. P. Blakely, and W. R. Grimes, Oak Ridge National Laboratory, Phase Diagrams of Nuclear Reactor Materials, R. E. Thoma, ed., ORNL-2548, p. 30 (1959).

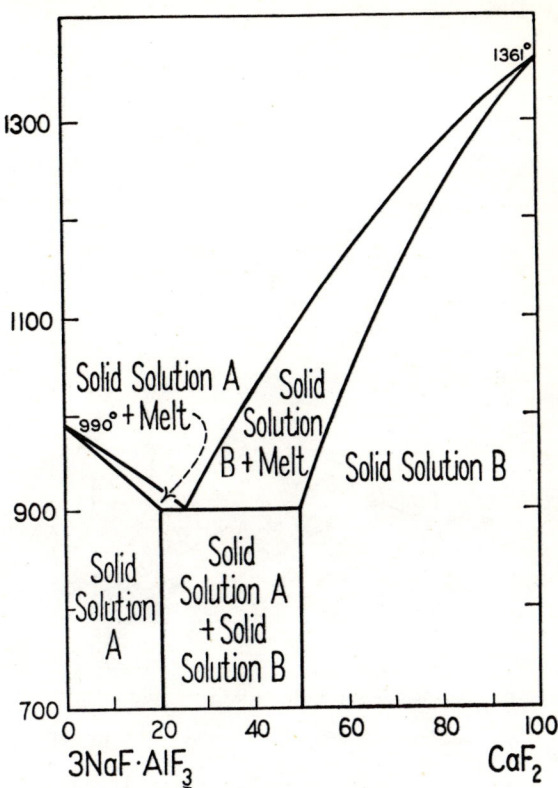

FIG. 1567.—System CaF₂–3NaF·AlF₃.

Von P. Pascal, *Z. Elektrochem.*, **19**, 611 (1913).

NaF–CaF₂–AlF₃

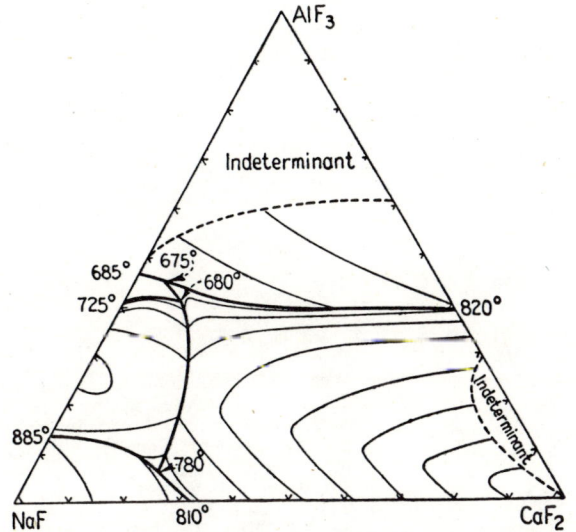

FIG. 1566.—System NaF–CaF₂–AlF₃.

P. P. Fedotieff and W. P. Iljinskiĭ, *Z. anorg. u. allgem. Chem.*, **129**, pp. 106, 107 (1923).

NaF–MgF₂–AlF₃

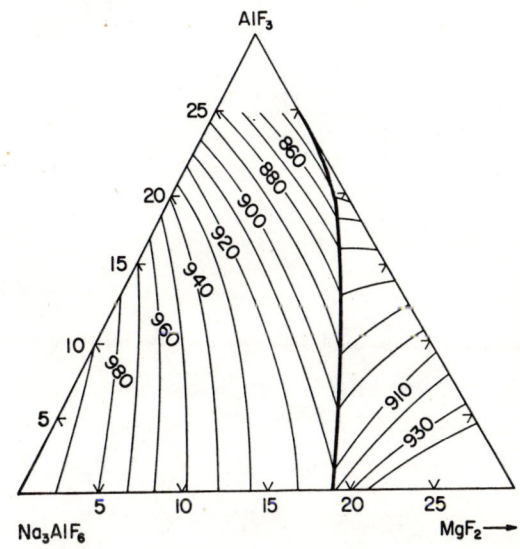

FIG. 1568.—System Na₃AlF₆–MgF₂–AlF₃.

E. Vatslavik and A. I. Belyaev, *Zhur. Neorg. Khim.*, **3**, 1045 (1958).

NaF–PbF$_2$–UF$_4$

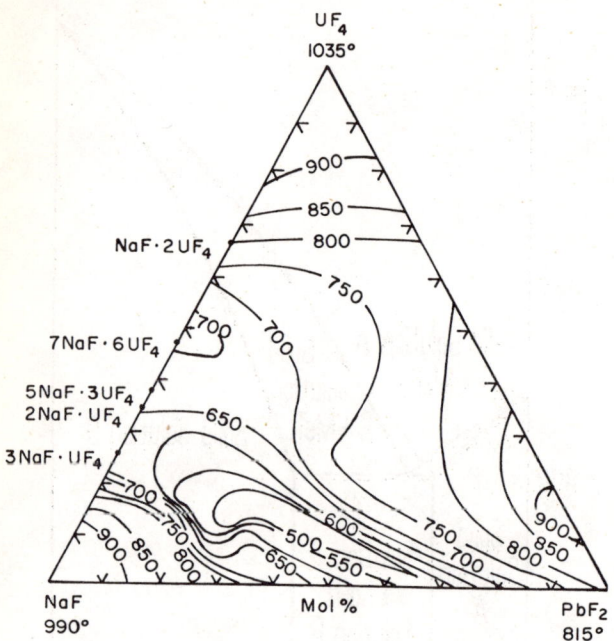

Fig. 1569.—System NaF–PbF$_2$–UF$_4$; preliminary.

C. J. Barton, J. P. Blakely, G. J. Nessle, L. M. Bratcher, and W. R. Grimes, Oak Ridge National Laboratory, Phase Diagrams of Nuclear Reactor Materials, R. E. Thoma, ed., ORNL-2548, p. 113 (1959).

NaF–CeF$_3$–ZrF$_4$

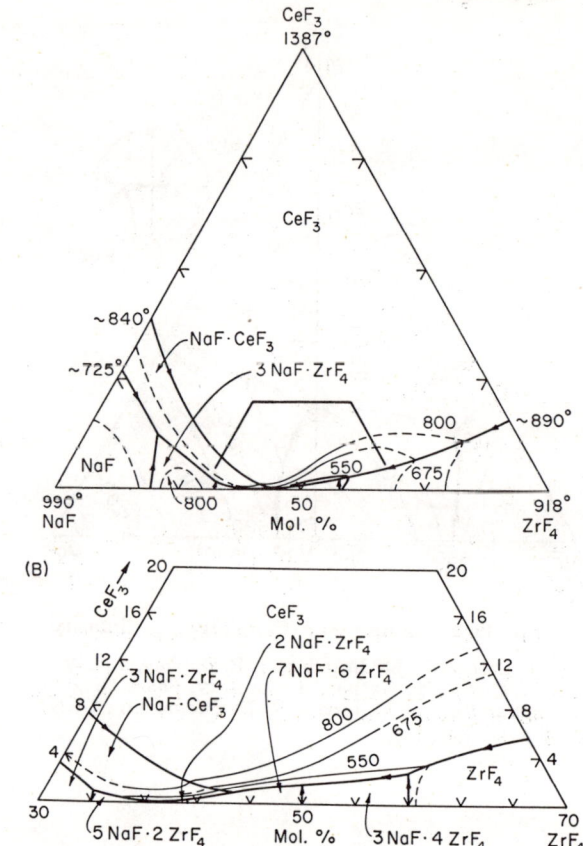

Fig. 1570.—System NaF–CeF$_3$–ZrF$_4$. (B) Enlargement showing probable NaF–ZrF$_4$ primary phase fields.

W. T. Ward, R. A. Strehlow, W. R. Grimes, and G. M. Watson, Oak Ridge National Laboratory, Phase Diagrams of Nuclear Reactor Materials, R. E. Thoma, ed., ORNL-2548, p. 67 (1959).

NaF–ThF$_4$–UF$_4$

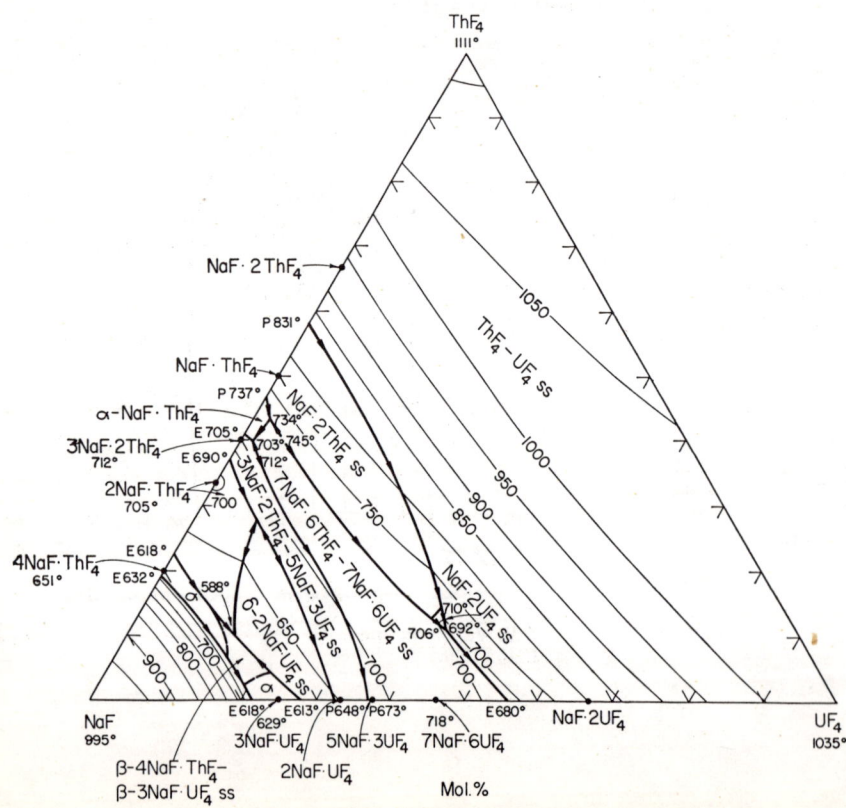

Fig. 1571.—System NaF–ThF$_4$–UF$_4$.

R. E. Thoma, H. Insley, H. A. Friedman, G. M. Hebert, and C. F. Weaver, private communication, Dec. 20, 1961. Presented at the 63rd Annual Meeting of The American Ceramic Society, Toronto, Canada, 1961. For more recent diagram showing minor variations and a field labeled 7NaF·2ThF$_4$–7NaF·2UF$_4$ ss (rather than β4NaF·ThF$_4$–β3NaF·UF$_4$ ss), see same authors, J. Am. Ceram. Soc., 46 [1] 39 (1963).

NaF–ThF$_4$–ZrF$_4$

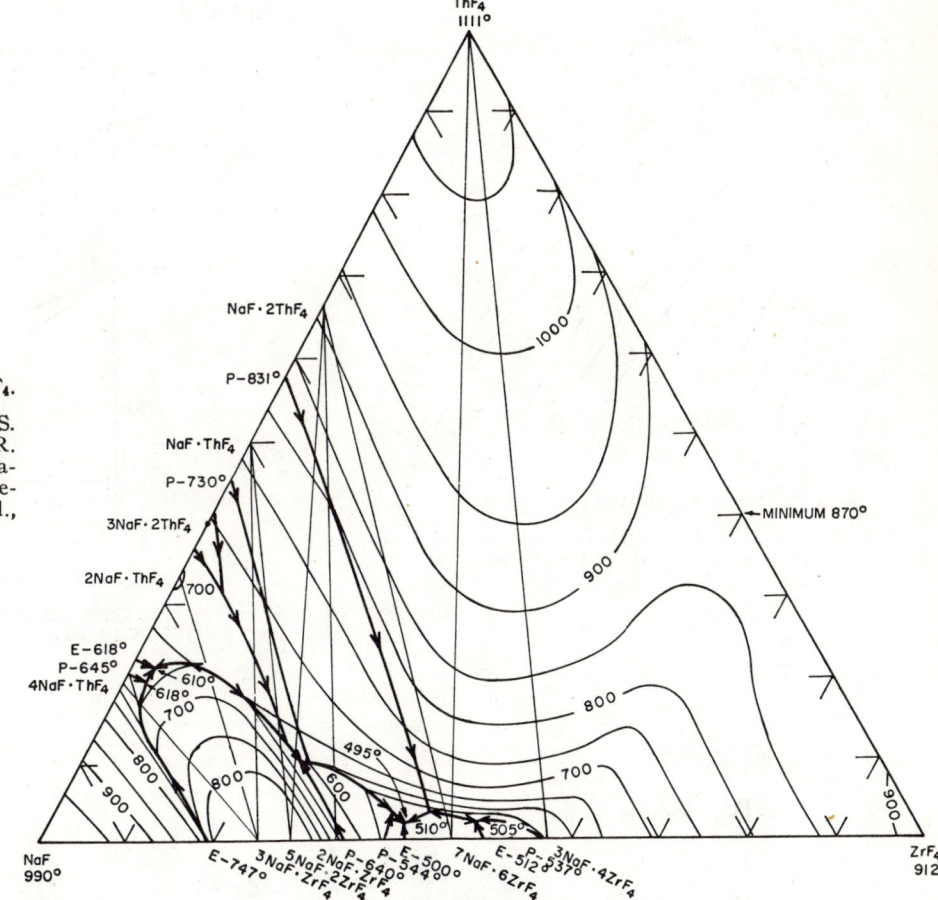

Fig. 1572.—System NaF–ThF$_4$–ZrF$_4$.

R. E. Thoma, H. Insley, B. S. Landau, H. A. Friedman, and W. R. Grimes, Oak Ridge National Laboratory, Phase Diagrams of Nuclear Reactor Materials, R. E. Thoma, ed., ORNL-2548, p. 82 (1959).

NaF–UF$_4$–ZrF$_4$

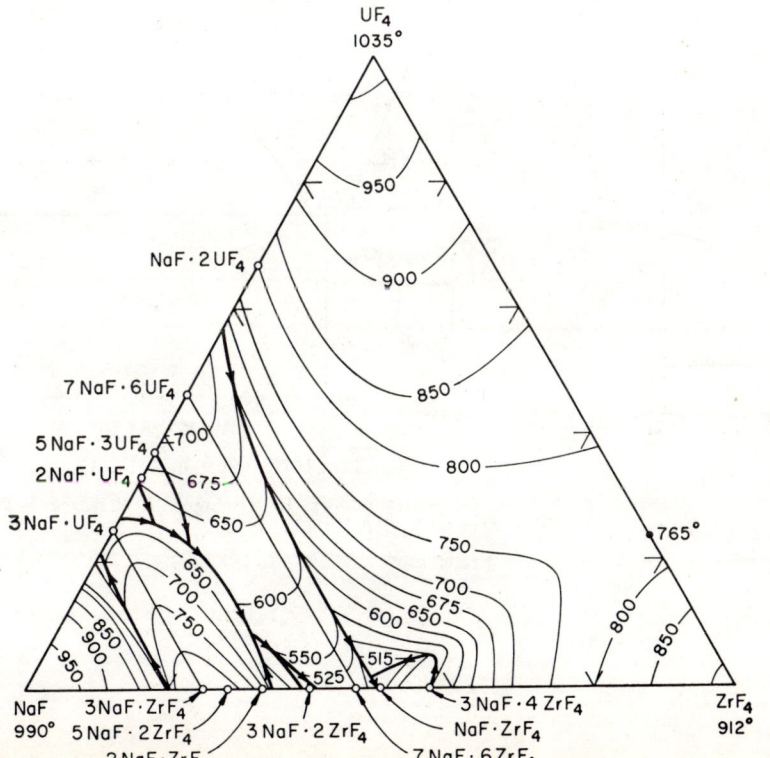

Fig. 1573.—System NaF–UF$_4$–ZrF$_4$.

C. J. Barton, W. R. Grimes, H. Insley, R. E. Moore, and R. E. Thoma, J. Phys. Chem., 62, 671 (1958).

BeF$_2$–ThF$_4$–UF$_4$

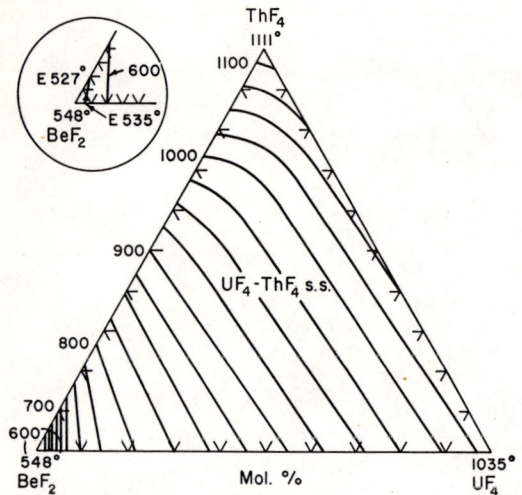

Fig. 1574.—System BeF$_2$–ThF$_4$–UF$_4$.

C. F. Weaver, R. E. Thoma, H. A. Friedman, and G. M. Hebert, *J. Am. Ceram. Soc.*, **44** [3] 146 (1961).

(b) *Two Iodides*

AgI–CuI

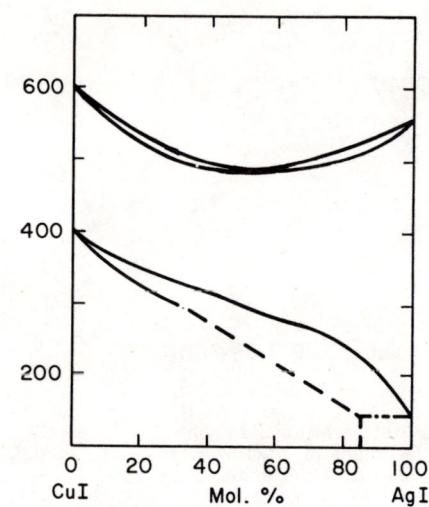

Fig. 1576.—System AgI–CuI.

E. Quercigh, *Atti reale accad. Lincei, Sez. I*, **23**, 831 (1914).

IV. Iodides Only

(a) *One Iodide*

GaI$_x$

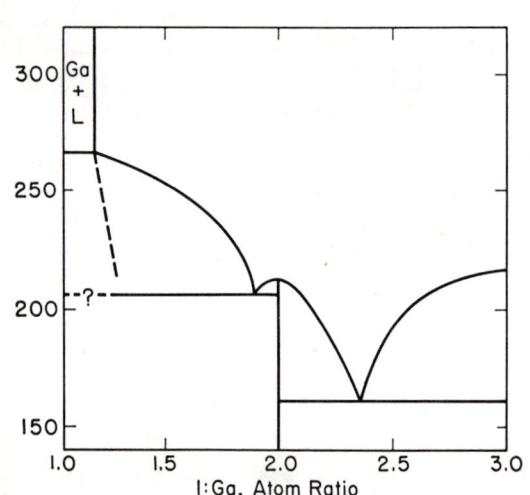

Fig. 1575.—System GaI–GaI$_3$.

J. D. Corbett and R. K. McMullan, *J. Am. Chem. Soc.*, **77**, 4219 (1955).

AgI–KI

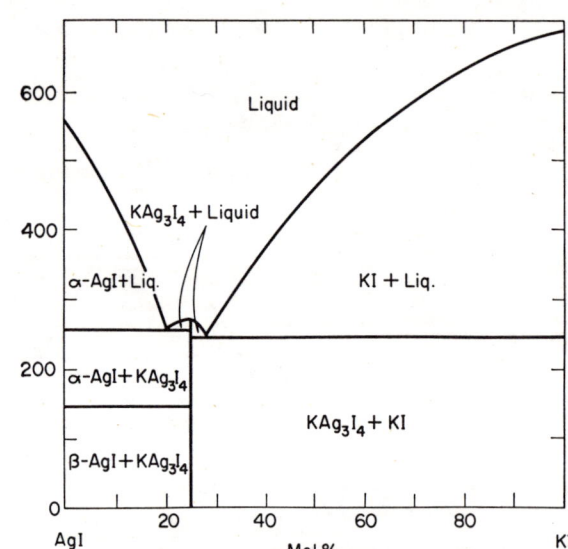

Fig. 1577.—System AgI–KI.

G. Burley and H. E. Kissinger, *J. Research Natl. Bur. Standards*, **64A** [5] 404 (1960).

AgI–LiI

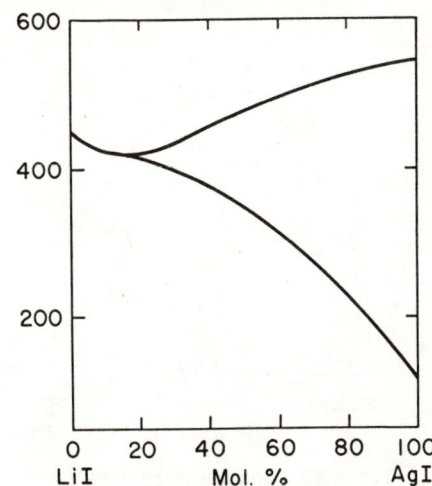

Fig. 1578.—System AgI–LiI. Diagram does not obey phase rule.

Carlo Sandonnini and G. Scarpa, *Atti reale accad. Lincei, Sez. II*, **22**, 521 (1913).

AgI–NaI

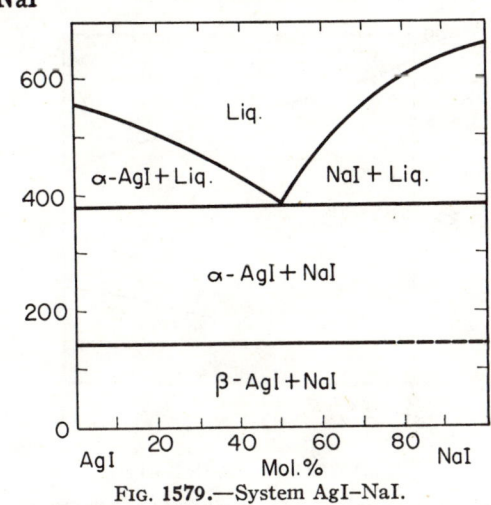

Fig. 1579.—System AgI–NaI.

G. Burley and H. E. Kissinger, *J. Research Natl. Bur. Standards*, **64A** [5] 404 (1960).

AgI–RbI

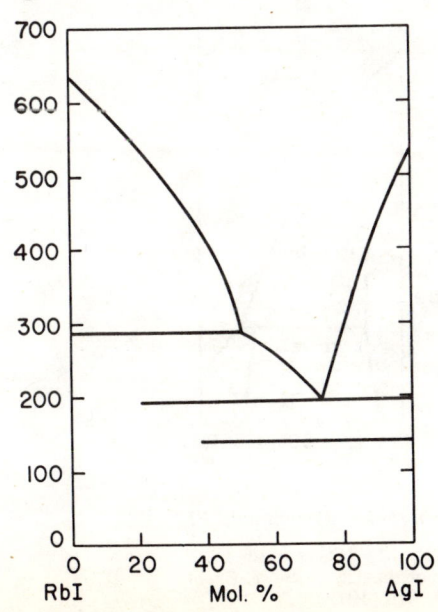

← Fig. 1580.—System AgI–RbI.

Carlo Sandonnini, *Atti reale accad. Lincei, Sez. I*, **21**, 201 (1912).

AgI–HgI$_2$

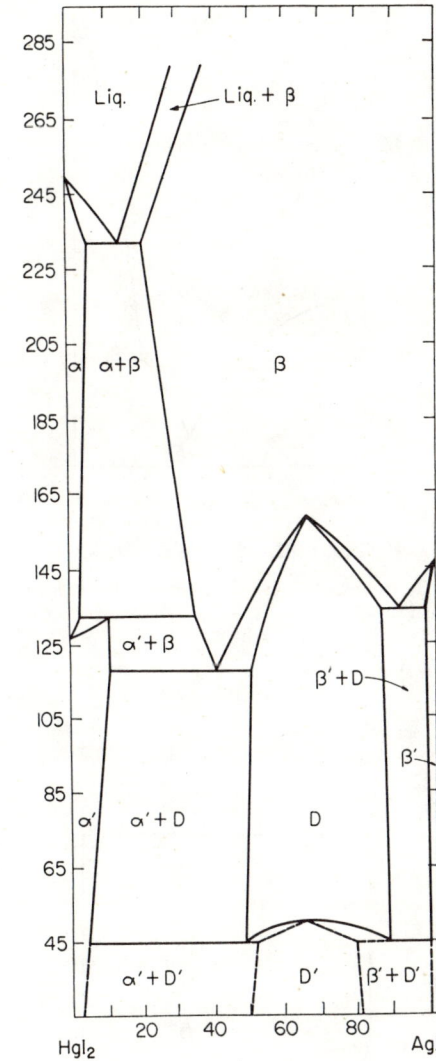

Fig. 1581.—System AgI–HgI$_2$.

Alph Steger, *Z. physik. Chem.*, **43**, 625 (1903).

CuI–CdI$_2$

Fig. 1582.—System CuI–CdI$_2$. →

Gottfried Herrmann, *Z. anorg. Chem.* **71**, 296 (1911).

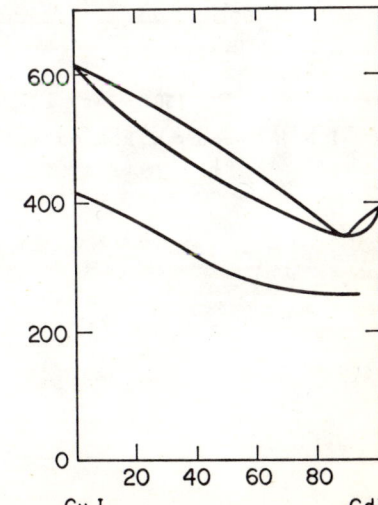

KI–HgI$_2$

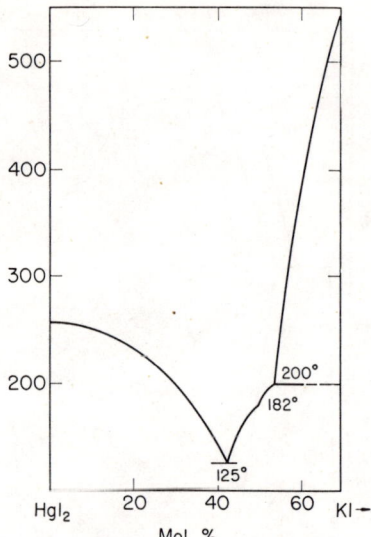

Fig. 1583.—System KI–HgI$_2$.

I. N. Belyaev and K. E. Mironov, *Zhur. Obshcheĭ Khim.* **22**, 1495 (1952).

NaI–HgI$_2$

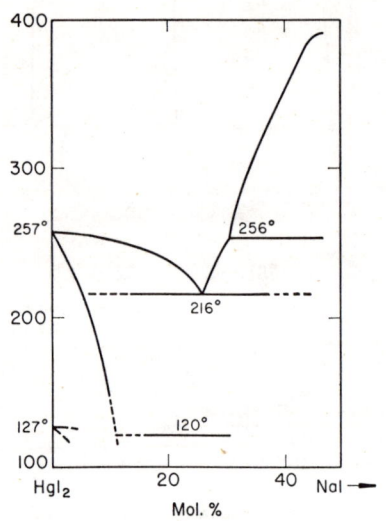

Fig. 1584.—System NaI–HgI$_2$.

I. N. Belyaev and K. E. Mironov, *Zhur. Obshcheĭ Khim.*, **22**, 1495 (1952).

NH$_4$I–HgI$_2$

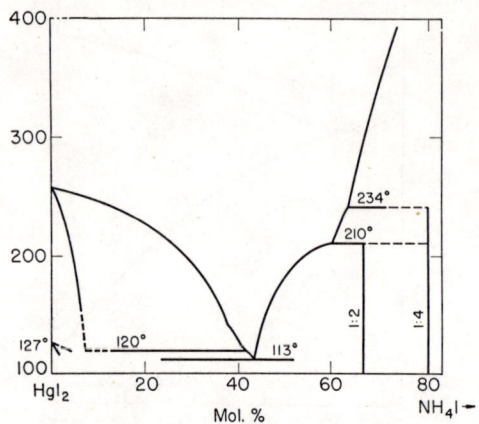

Fig. 1585.—System NH$_4$I–HgI$_2$.

I. N. Belyaev and K. E. Mironov. *Zhur. Obshcheĭ Khim.*, **22**, 1496 (1952)

CdI$_2$–HgI$_2$

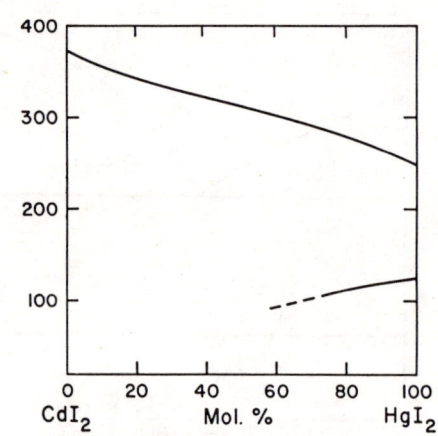

Fig. 1586.—System CdI$_2$–HgI$_2$.

Carlo Sandonnini, *Gazz. chim. ital.*, **44 I**, 363 (1914).

AsI$_3$–PI$_3$

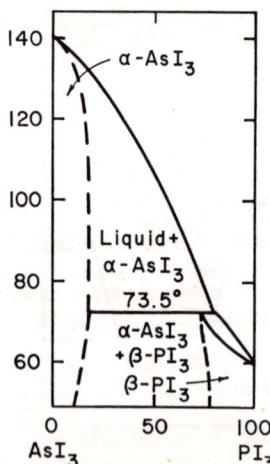

Fig. 1587.—System AsI$_3$–PI$_3$.

F. M. Jaeger and H. J. Doornbosch, *Z. anorg. Chem.*, **75**, 270 (1912).

AsI₃–SbI₃

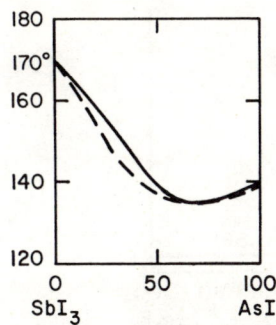

Fig. 1588.—System AsI₃–SbI₃.

F. M. Jaeger and H. J. Doornbosch, *Z. anorg. Chem.*, **75**, 270 (1912).

PI₃–SbI₃

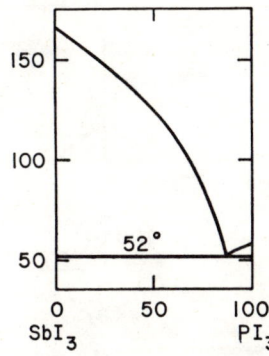

Fig. 1589.—System SbI₃–PI₃.

F. M. Jaeger and H. J. Doornbosch, *Z. anorg. Chem.*, **75**, 270 (1912).

(c) Three Iodides

NaI–CdI₂–PbI₂

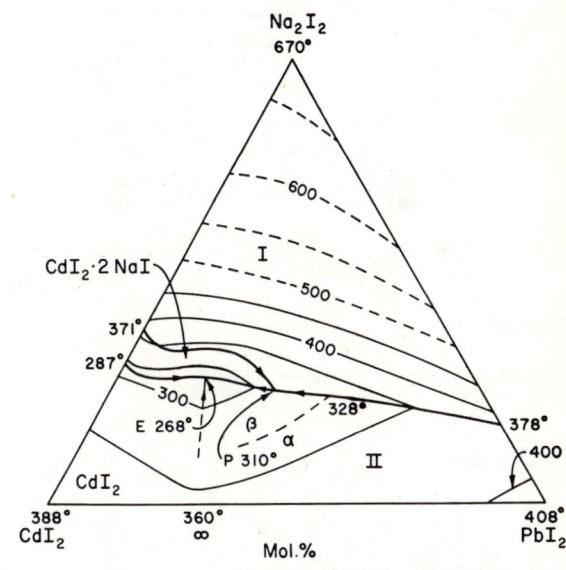

Fig. 1590.—System NaI–CdI₂–PbI₂.

I. I. Il'yasov and A. K. Bostandzhiyan, *Zhur. Neorg. Khim.*, **2**, 169 (1957).

V. Mixed Halides Only

(a) Two Halides

NaX

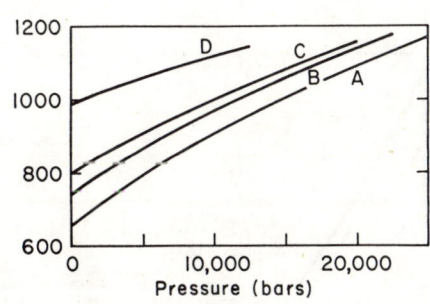

Fig. 1591.—System NaX; melting curves.
A = NaI, B = NaBr, C = NaCl, D = NaF
S. P. Clark, Jr., *J. Chem. Phys.*, **31** [6] 1527 (1959).

Br₂–UF₆

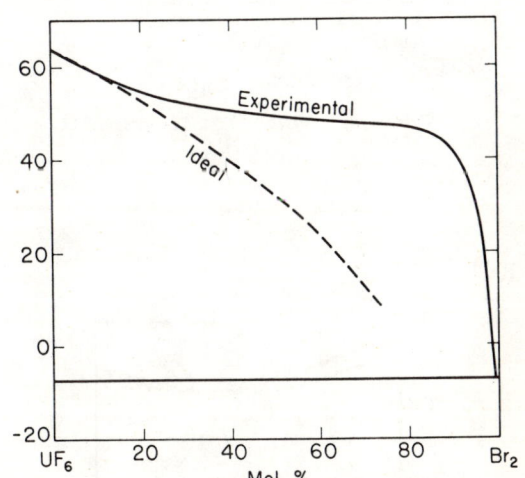

Fig. 1592.—System Br₂–UF₆.

Jack Fischer and R. C. Vogel, *J. Am. Chem. Soc.*, **76**, 4863 (1954).

AgBr–AgCl

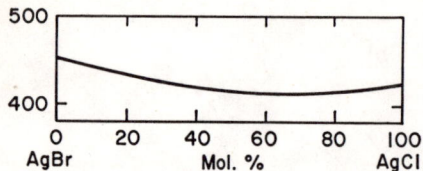

Fig. 1593.—System AgBr–AgCl, liquidus of solid solutions.

K. Mönkemeyer, *Neues Jahrb. Mineral., Geol.*, **22**, 29 (1906).

AgBr–AgI

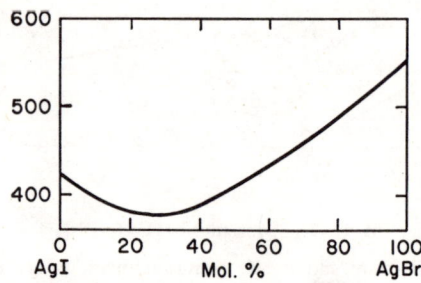

Fig. 1594.—System AgBr–AgI; liquidus of solid solutions.

K. Mönkemeyer, *Neues Jahrb. Mineral., Geol.*, **22**, 30 (1906).

CuBr–CuCl

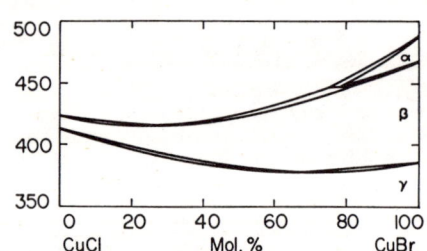

Fig. 1595.—System CuBr–CuCl.

J. Teltow, *Z. physik. Chem.* (*Leipzig*), **211**, 242 (1959).

KBr–KCl

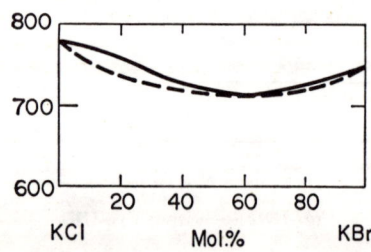

Fig. 1596.—System KBr–KCl.

J. B. Wrzesnewsky, *Z. anorg. Chem.*, **74**, 111 (1912).

KBr–KF

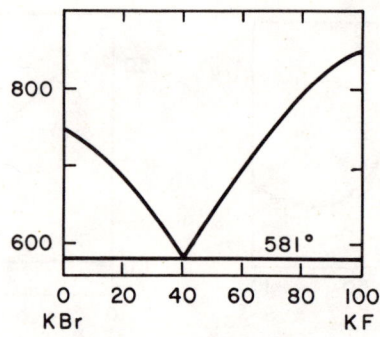

Fig. 1597.—System KBr–KF.

N. S. Kurnakow and J. B. Wrzesnewsky, *Z. anorg. Chem.*, **74**, 90 (1912).

KBr–KI

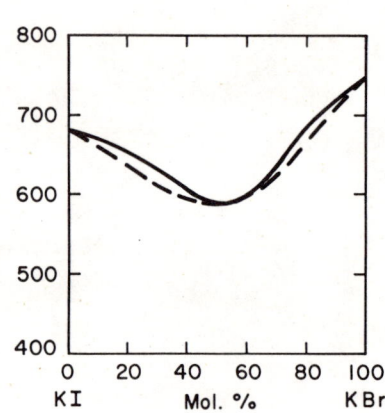

Fig. 1598.—System KBr–KI.

J. B. Wrzesnewsky, *Z. anorg. Chem.*, **74**, 110 (1912).

LiBr–LiCl

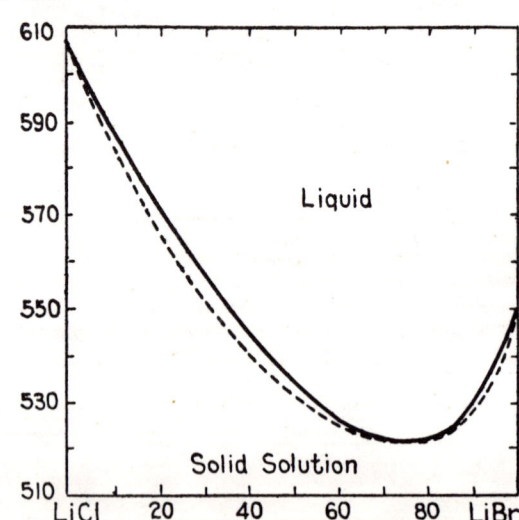

Fig. 1599.—System LiBr–LiCl.

A. A. Botschwar, *Z. anorg. u. allgem. Chem.*, **210**, 163 (1933).

LiBr–LiF

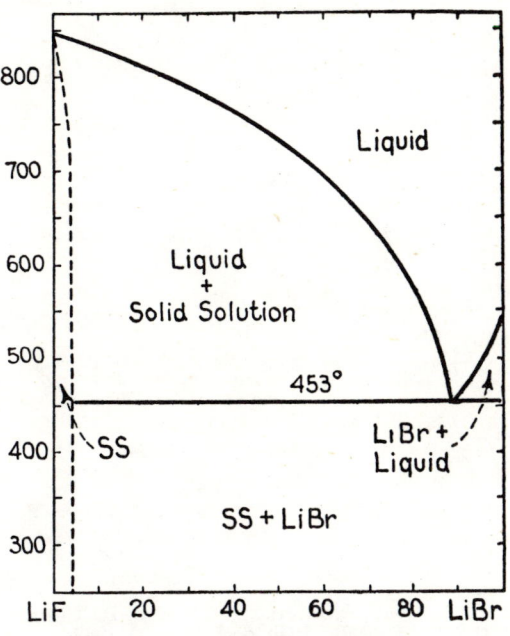

Fig. 1600.—System LiBr–LiF.

A. A. Botschwar, *Z. anorg. u. allgem. Chem.*, **210**, 163 (1933).

TlBr–TlCl

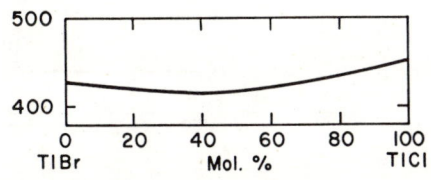

Fig. 1601.—System TlBr–TlCl; liquidus of solid solutions.

K. Mönkemeyer, *Neues Jahrb. Mineral., Geol.*, **22**, 38 (1906).

TlBr–TlI

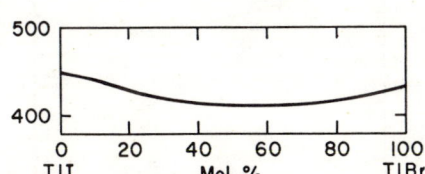

Fig. 1602.—System TlBr–TlI; liquidus of solid solutions.

K. Mönkemeyer, *Neues Jahrb. Mineral., Geol.*, **22**, 39 (1906).

$HgBr_2$–HgI_2

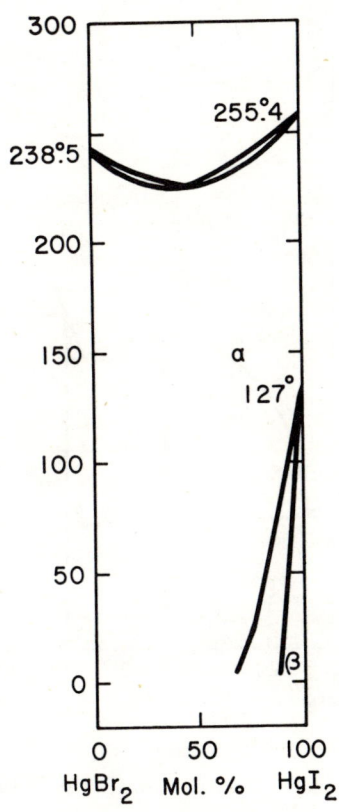

Fig. 1603.—System $HgBr_2$–HgI_2.

W. Reinders, *Z. physik. Chem.*, **32**, 497 (1900).

$PbBr_2$–$PbCl_2$

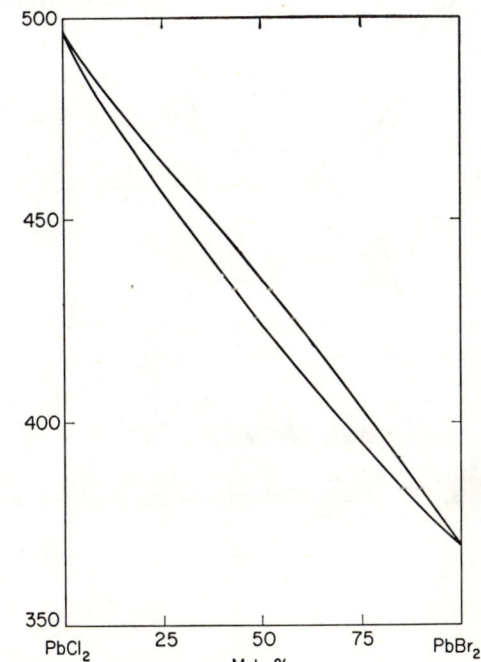

Fig. 1604.—System $PbCl_2$–$PbBr_2$.

G. Calingaert, F. W. Lamb, and F. Meyers, *J. Am. Chem. Soc.*, **71**, 3712 (1949).

PbBr$_2$–PbF$_2$

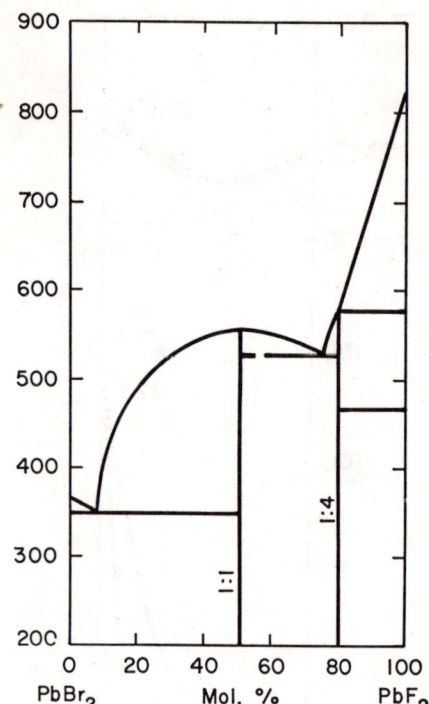

Fig. 1605.—System PbBr$_2$–PbF$_2$.

Carlo Sandonnini, *Gazz. chim. ital.*, 41 II, 151 (1911).

PbBr$_2$–PbI$_2$

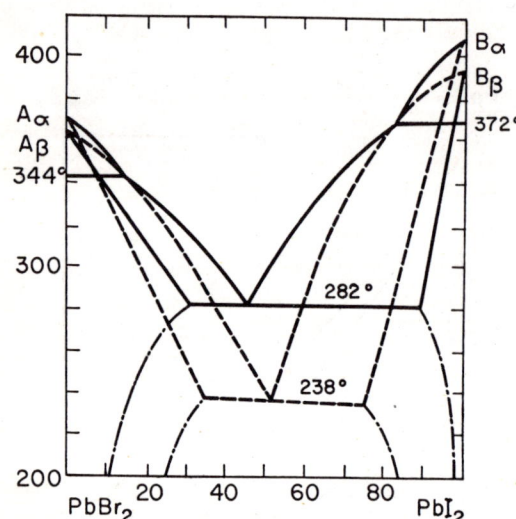

Fig. 1606.—System PbBr$_2$–PbI$_2$.

T. Modestova and T. N. Sumarokova, *Zhur. Neorg. Khim.*, 3, 1656 (1958). Subscripts α and β refer to high and low forms, respectively.

SrBr$_2$–UBr$_3$

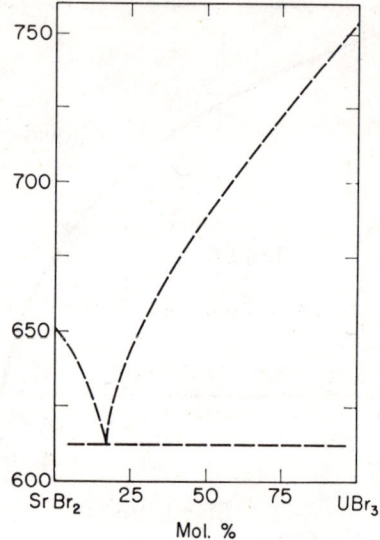

Fig. 1607.—System SrBr$_2$–UBr$_3$.

E. D. Eastman, A. E. Stickland, and C. D. Thurmond, AECD-2312; Sept. 28, 1945, decl. May 20, 1947.

BBr$_3$–SnI$_4$

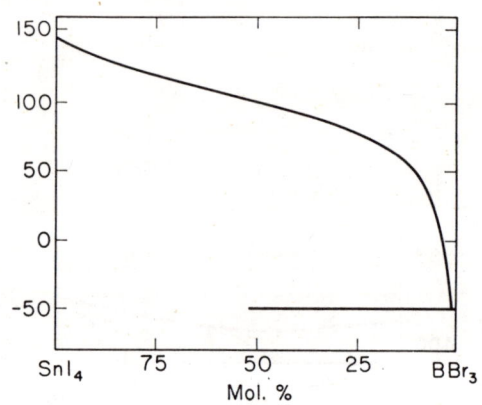

Fig. 1608.—System BBr$_3$–SnI$_4$.

R. F. Adamsky and C. M. Wheeler, Jr., *J. Phys. Chem.*, 58, 225 (1954).

SbBr$_3$–SbCl$_3$

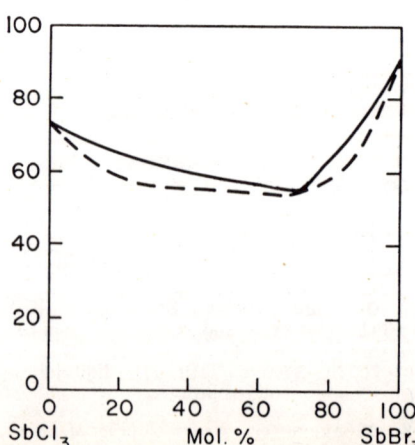

Fig. 1609.—System SbBr$_3$–SbCl$_3$.

G. B. Bernardis, *Atti reale accad. Lincei, Sez. I*, 21, 442 (1912).

SbBr₃–SbI₃

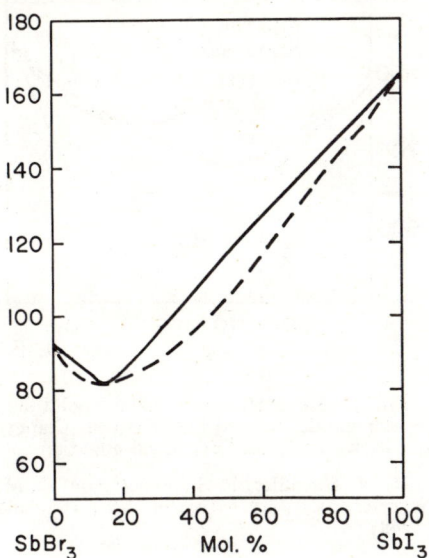

Fig. 1610.—System SbBr₃–SbI₃.

G. B. Bernardis, *Atti reale accad. Lincei*, Sez. I, **21**, 443 (1912).

UBr₃–SrI₂

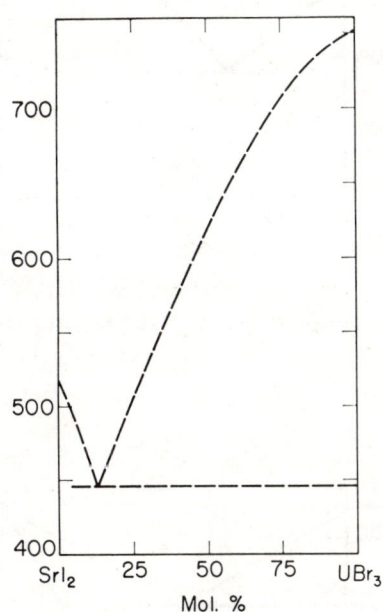

Fig. 1611.—System SrI₂–UBr₃.

E. D. Eastman, A. E. Stickland, and C. D. Thurmond AECD-2312; Sept. 28, 1945, decl. May 20, 1947.

AgCl–AgI

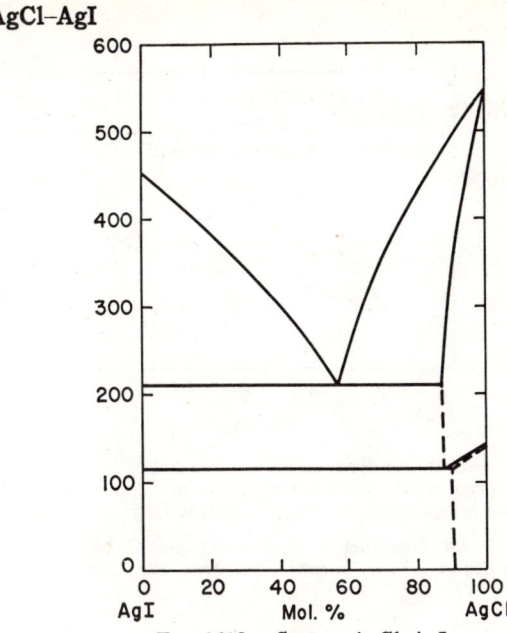

Fig. 1612.—System AgCl–AgI.

K. Mönkemeyer, *Neues Jahrb. Mineral., Geol.*, **22**, 33 (1906).

CuCl–CuI

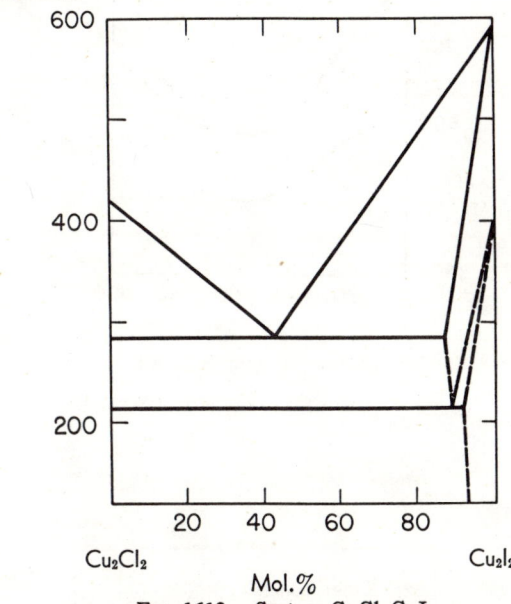

Fig. 1613.—System CuCl–CuI.

K. Mönkemeyer, *Neues Jahrb. Mineral., Geol.*, **22**, 47 (1906).

KCl–KF

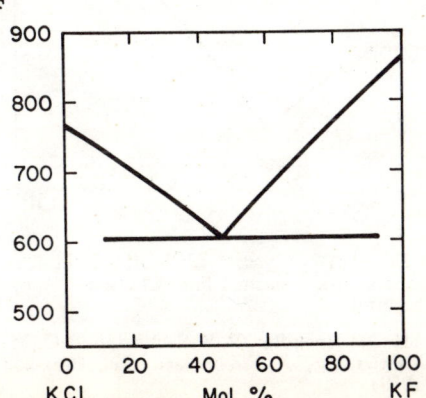

Fig. 1614.—System KCl–KF.

W. Plato, *Z. physik. Chem.*, **58**, 364 (1907).

KCl–LiF

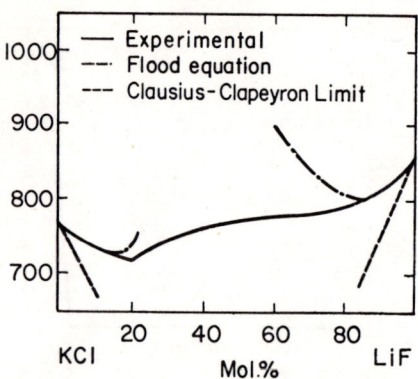

FIG. 1615.—System KCl–LiF. Solid line, experimental; dashed line, Clausius-Clapeyron limit; dash-dot line, Flood equation.

H. M. Haendler, P. S. Sennett, and C. M. Wheeler, Jr., *J. Electrochem. Soc.*, **106**, 266 (1959).

KCl–KI

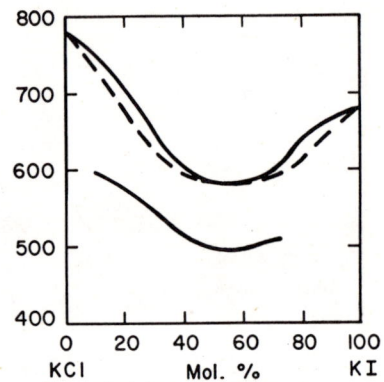

FIG. 1616.—System KCl–KI.

J. B. Wrzesnewsky, *Z. anorg. Chem.*, **74**, 107 (1912).

LiCl–LiF

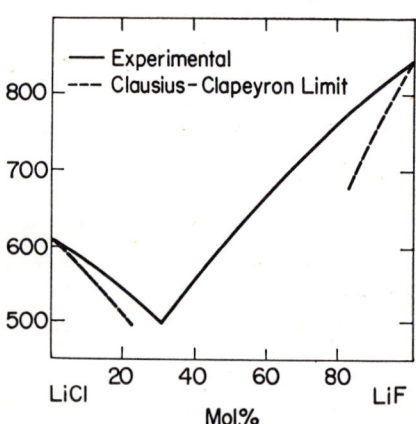

FIG. 1617.—System LiCl–LiF. Solid line, experimental; dashed line, Clausius-Clapeyron limit.

H. M. Haendler, P. S. Sennett, and C. M. Wheeler, Jr., *J. Electrochem. Soc.*, **106**, 265 (1959).

NaCl–LiF

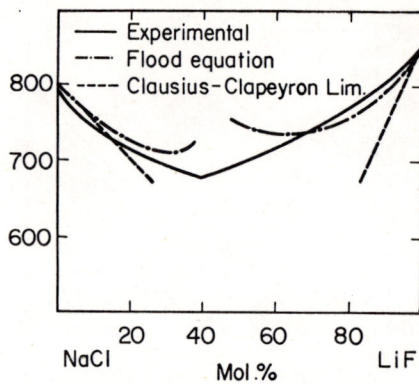

FIG. 1618.—System LiF–NaCl. Solid line, experimental; dashed line, Clausius-Clapeyron limit; dash-dot line, Flood equation.

H. M. Haendler, P. S. Sennett, and C. M. Wheeler, Jr., *J. Electrochem. Soc.*, **106**, 265 (1959).

NaCl–NaF

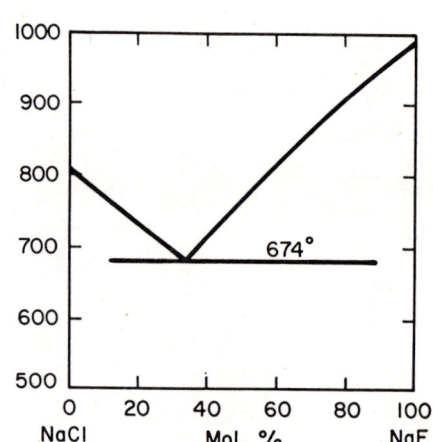

FIG. 1619.—System NaCl–NaF.

W. Plato, *Z. physik. Chem.*, **58**, 364 (1907).

TlCl–TlI

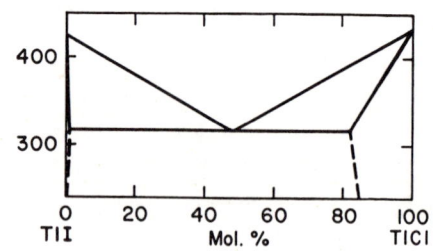

FIG. 1620.—System TlCl–TlI.

K. Mönkemeyer, *Neues Jahrb. Mineral., Geol.*, **22**, 40 (1906).

BaCl₂–BaF₂

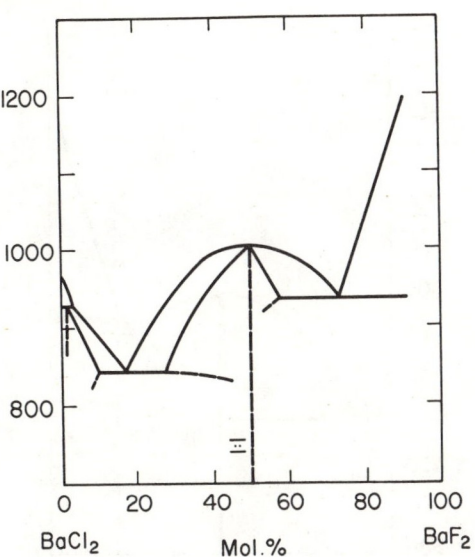

Fig. 1621.—System BaCl₂–BaF₂.

W. Plato, *Z. physik. Chem.*, **58**, 360 (1907).

CaCl₂–CaF₂

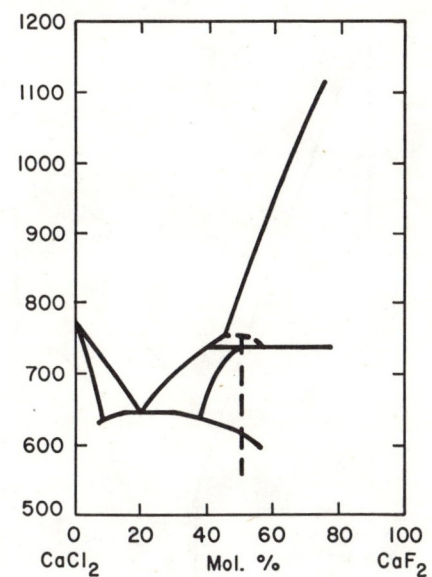

Fig. 1623.—System CaCl₂–CaF₂. Diagram does not obey phase rule.

W. Plato, *Z. physik. Chem.*, **58**, 363 (1907).

BeCl₂–BeF₂

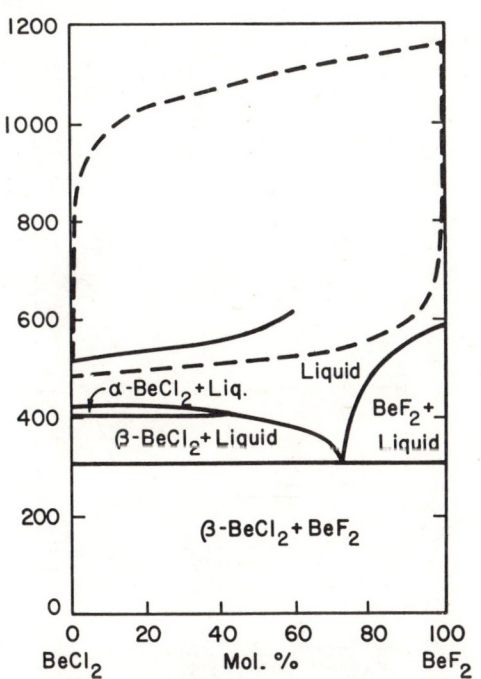

Fig. 1622.—System BeCl₂–BeF₂.

O. N. Kuvyrkin, O. N. Breusov, and A. V. Novoselova, *Nauchn. Dokl. Vysshei Shkoly, Khim. i Khim. Tekhnol.*, 1958, p. 662.

CdCl₂–LiF

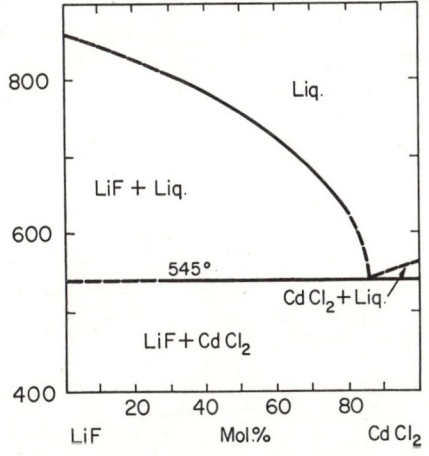

Fig. 1624.—System LiF–CdCl₂.

D. L. Deadmore and J. S. Machin, *J. Am. Ceram. Soc.*, **43** [11] 592 (1960).

CdCl$_2$–NaF

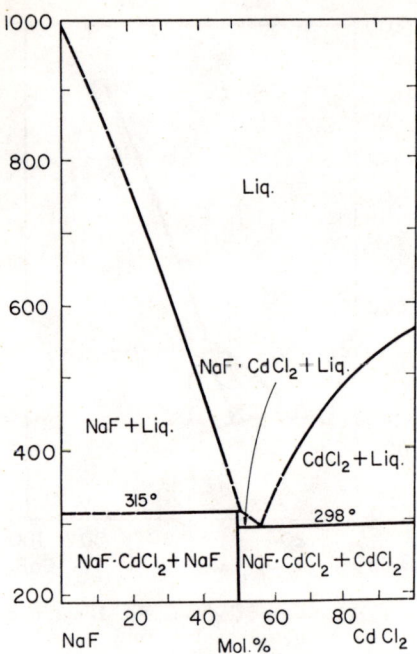

Fig. 1625.—System NaF–CdCl$_2$.

D. L. Deadmore and J. S. Machin, *J. Am. Ceram. Soc.*, **43** [11] 593 (1960).

PbCl$_2$–PbF$_2$

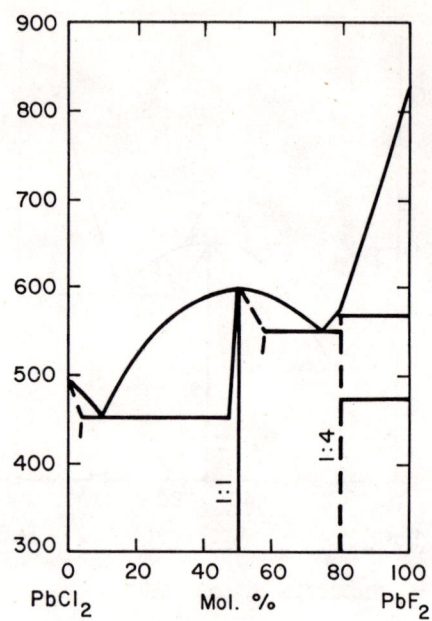

Fig. 1627.—System PbCl$_2$–PbF$_2$.

Carlo Sandonnini, *Gazz. chim. ital.*, **41**·II, 148 (1911).

CdCl$_2$–CdF$_2$

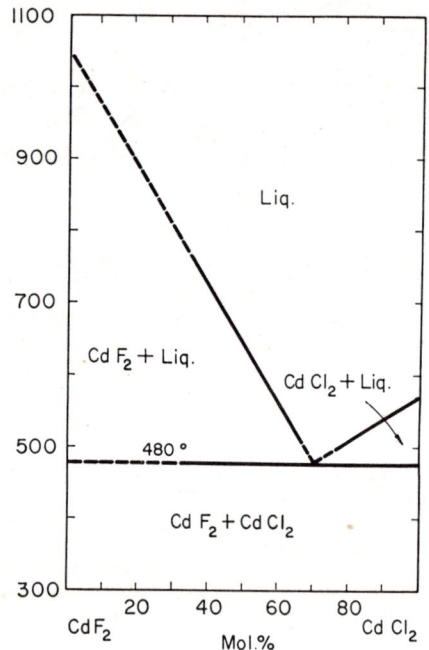

Fig. 1626.—System CdCl$_2$–CdF$_2$.

D. L. Deadmore and J. S. Machin, *J. Am. Ceram. Soc.*, **43** [11] 593 (1960).

PbCl$_2$–PbI$_2$

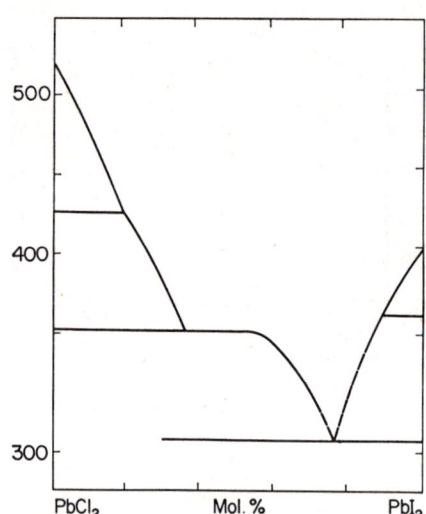

Fig. 1628.—System PbCl$_2$–PbI$_2$.

T. Sumarokova and T. Modestova, *Zhur. Neorg. Khim.*, **1**, 2029 (1956).

SrCl$_2$–SrF$_2$

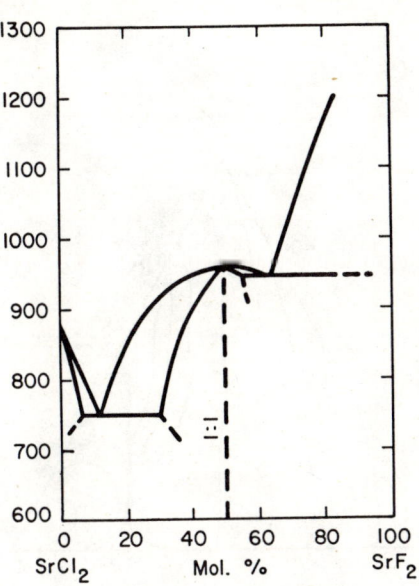

Fig. 1629.—System SrCl$_2$–SrF$_2$.

W. Plato, *Z. physik. Chem.*, **58**, 356 (1907).

CaF$_2$–CaI$_2$

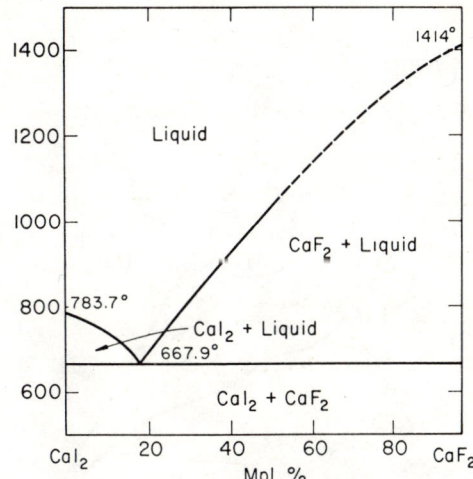

Fig. 1631.—System CaI$_2$–CaF$_2$.

W. J. McCreary, *J. Am. Chem. Soc.*, **77**, 2114 (1955).

SbCl$_3$–SbI$_3$

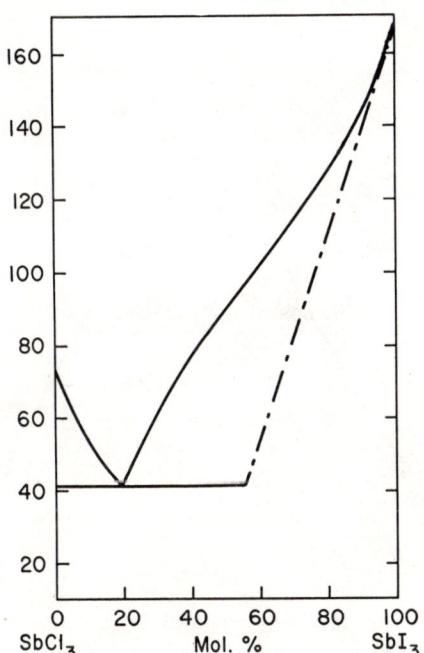

Fig. 1630.—System SbCl$_3$–SbI$_3$.

G. B. Bernardis, *Atti reale Accad. Lincei*, Sez. I, **21**, 444 (1912).

PbF$_2$–PbI$_2$

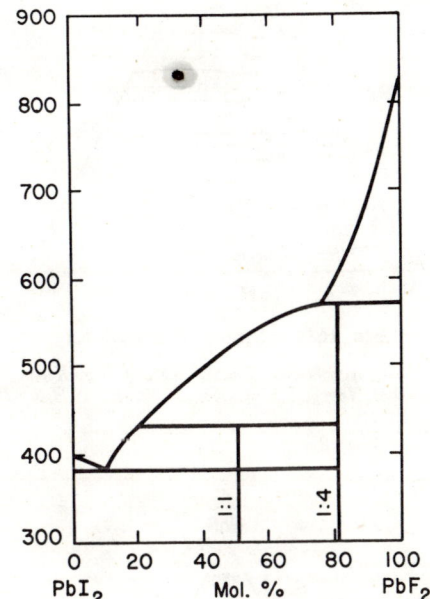

Fig. 1632.—System PbF$_2$–PbI$_2$.

Carlo Sandonnini, *Gazz. chim. ital.*, **41 II**, 154 (1911).

(b) Three Halides

AgBr–AgCl–AgI

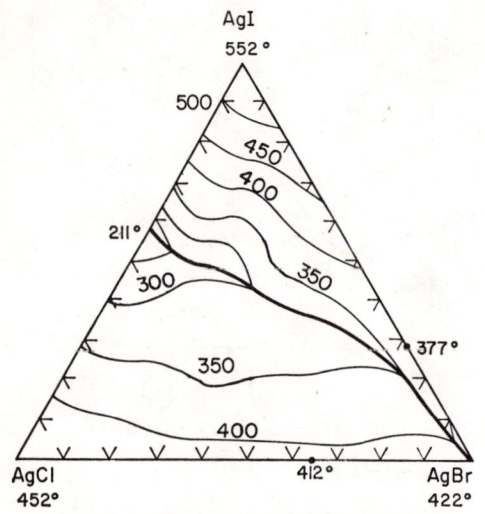

FIG. 1633.—System AgBr–AgCl–AgI.

Fritz Matthes, *Neues Jahrb. Mineral., Geol.*, **31**, 355 (1911).

PbBr$_2$–PbCl$_2$–PbI$_2$

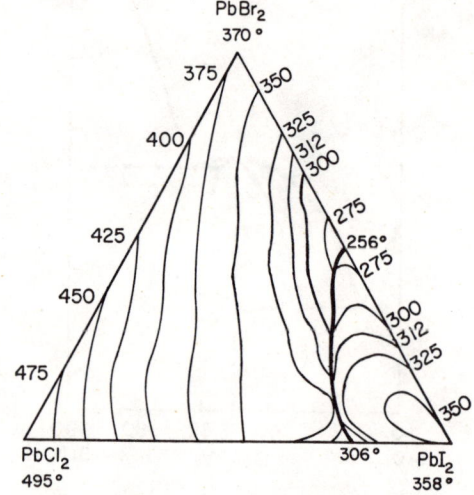

FIG. 1636.—System PbBr$_2$–PbCl$_2$–PbI$_2$.

Fritz Matthes, *Neues Jahrb. Mineral., Geol.*, **31**, 382 (1911).

KBr–KF–KI

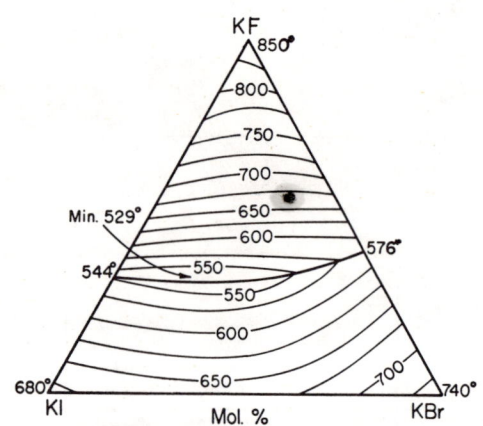

FIG. 1634.—System KBr–KF–KI.

N. S. Dombrovskaya, *Izvest. Sektora Fiz.-Khim. Anal., Inst. Obshcheĭ Neorg. Khim., Akad. Nauk S.S.S.R.*, **20**, 127 (1950).

KCl–BaCl$_2$–CaF$_2$

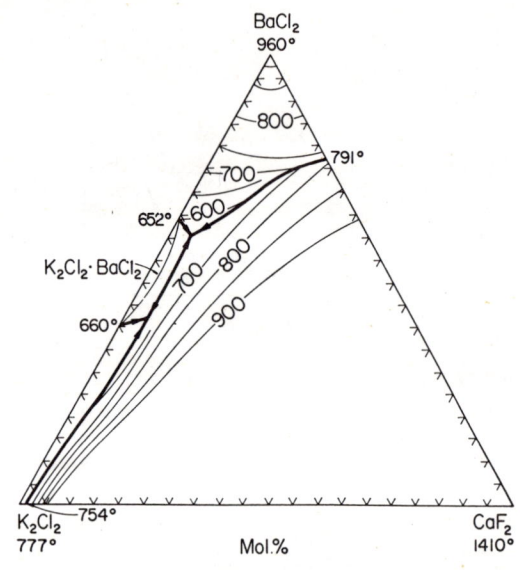

FIG. 1637.—System KCl–BaCl$_2$–CaF$_2$.

N. A. Shul'ga and G. A. Bukhalova, *Zhur. Neorg. Khim.*, **2**, 2141 (1957).

NaBr–NaF–NaI

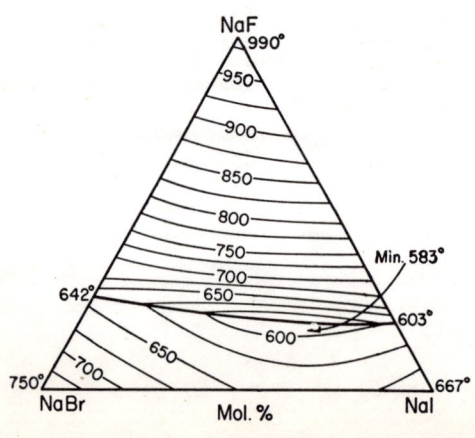

← FIG. 1635.—System NaBr–NaF–NaI.

N. S. Dombrovskaya, *Izvest. Sektora. Fiz.-Khim. Anal., Inst. Obshcheĭ Neorg. Khim., Akad. Nauk S.S.S.R.*, **20**, 127 (1950).

KCl–KF–BF₃

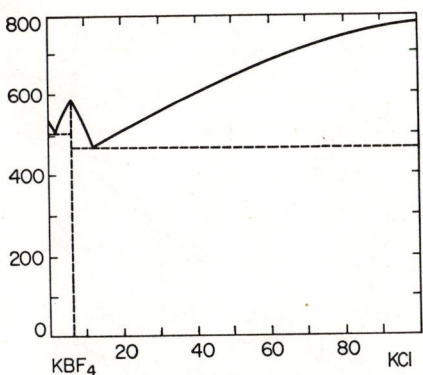

Fig. 1638.—System KBF₄–KCl.

G. V. Samsonov, V. A. Obolonchik, and G. N. Kulichkina, *Khim. Nauka i Promy.*, **4**, 804 (1959).

(c) Four Halides

KBr–NaBr–KF–NaF

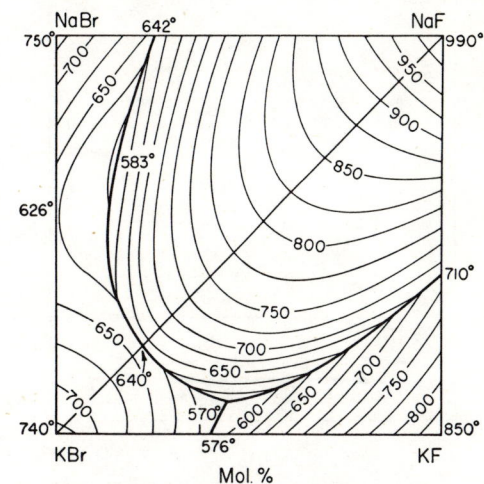

Fig. 1640.—System KBr–KF–NaBr–NaF.

N. S. Dombrovskaya, *Izvest. Sektora Fiz.-Khim. Anal., Inst. Obshcheĭ Neorg. Khim., Akad. Nauk S.S.S.R.*, **20**, 128 (1950).

BaCl₂–NaF–CaF₂

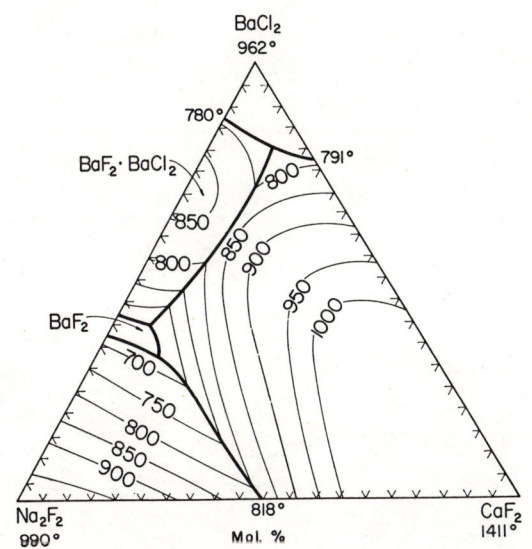

Fig. 1639.—System NaF–BaCl₂–CaF₂.

N. A. Shul'ga and G. A. Bukhalova, *Zhur. Neorg. Khim.*, **2**, 2137 (1957).

KBr–TlBr–KCl–TlCl

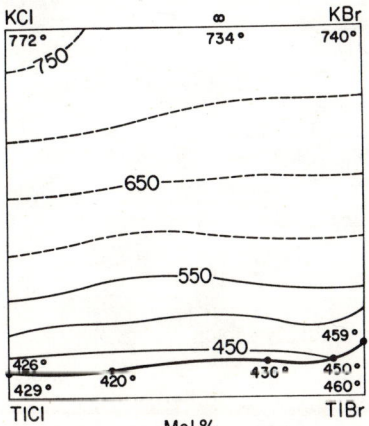

Fig. 1641.—System KBr–KCl–TlBr–TlCl.

I. I. Il'yasov, L. V. Rozhkovskaya, and A. G. Bergman, *Zhur. Neorg. Khim.*, **2**, 1886 (1957).

NaBr–TlBr–NaCl–TlCl

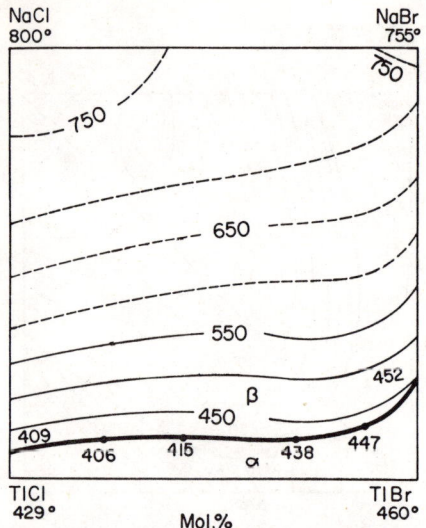

Fig. 1642.—System NaBr–NaCl–TlBr–TlCl.

I. I. Il'yasov, V. M. Fonardzhyan, and A. G. Bergman, *Zhur. Neorg. Khim.*, 2157 (1957).

NaBr–PbBr$_2$–NaCl–PbCl$_2$

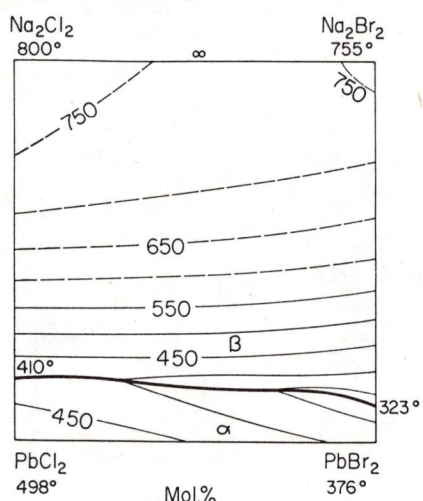

Fig. 1643.—System NaBr–NaCl–PbBr$_2$–PbCl$_2$.

I. I. Il'yasov, G. G. Shchemeleva, and A. G. Bergman, *Zhur. Neorg. Khim.*, **2**, 2171 (1957).

CdBr$_2$–PbBr$_2$–CdCl$_2$–PbCl$_2$

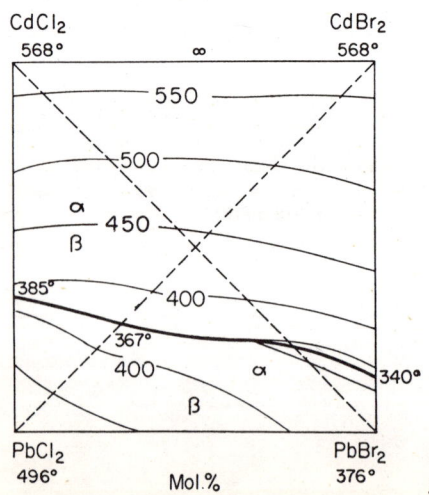

Fig. 1644.—System CdBr$_2$–CdCl$_2$–PbBr$_2$–PbCl$_2$.

I. I. Il'yasov, L. V. Rozhkovskaya, and A. G. Bergman, *Zhur. Neorg. Khim.*, **2**, 2176 (1957).

KCl–BaCl$_2$–KF–BaF$_2$

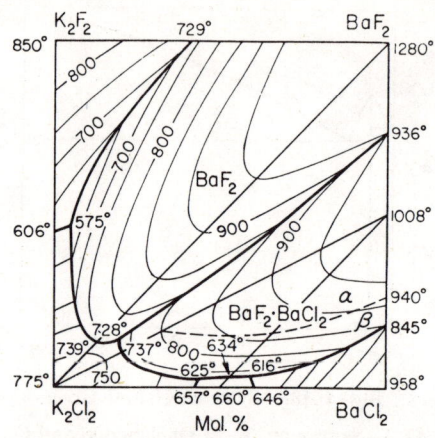

Fig. 1645.—System K$_2$Cl$_2$–K$_2$F$_2$–BaCl$_2$–BaF$_2$.

E. I. Banashek and A. G. Bergman, *Izvest. Sektora Fiz.-Khim. Anal., Inst. Obshcheĭ Neorg. Khim., Akad. Nauk S.S.S.R.*, **20**, 105 (1950).

KCl–PbCl$_2$–KI–PbI$_2$

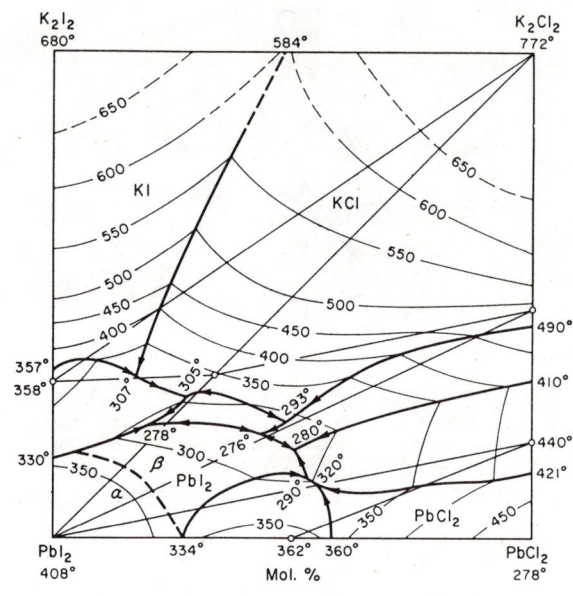

Fig. 1646.—System KCl–KI–PbCl$_2$–PbI$_2$; reciprocal.

I. I. Il'yasov and A. G. Bergman, *Zhur. Obshcheĭ Khim.*, **26**, 987 (1956).

LiCl–BaCl$_2$–LiF–BaF$_2$

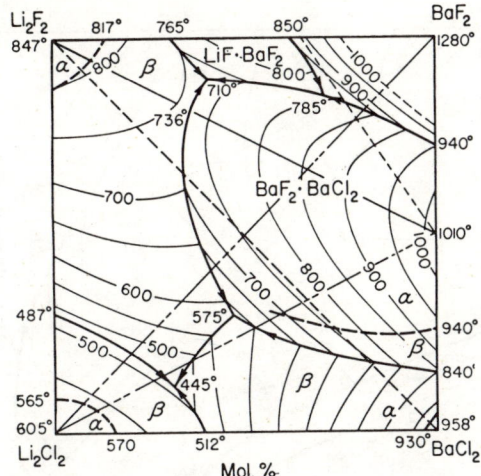

Fig. 1647.—System Li$_2$Cl$_2$–Li$_2$F$_2$–BaCl$_2$–BaF$_2$.

A. G. Bergman and E. I. Banashek, *Izvest. Sektora Fiz.-Khim. Anal., Inst. Obshcheĭ Neorg. Khim., Akad. Nauk S.S.S.R.*, **23**, 202 (1953).

LiCl–SrCl$_2$–LiF–SrF$_2$

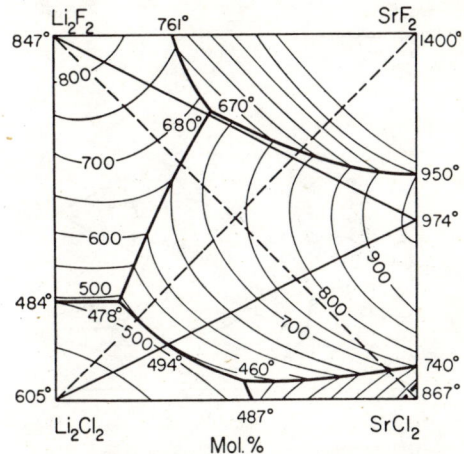

Fig. 1649.—System Li$_2$Cl$_2$–Li$_2$F$_2$–SrCl$_2$–SrF$_2$.

E. I. Banashek and A. G. Bergman, *Izvest. Sektora Fiz.-Khim. Anal., Inst. Obshcheĭ Neorg. Khim., Akad. Nauk S.S.S.R.*, **25**, 252 (1954).

LiCl–CaCl$_2$–LiF–CaF$_2$

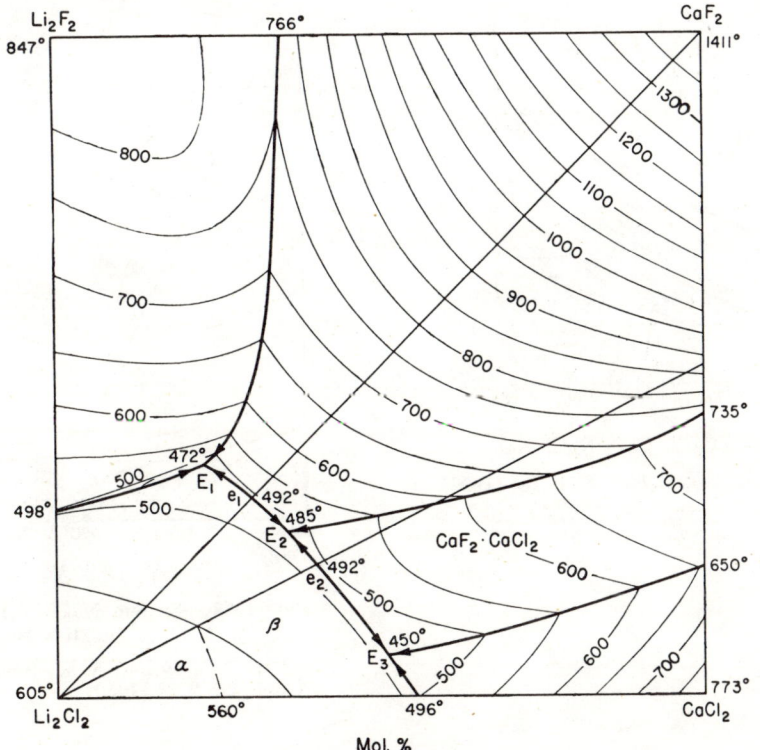

Fig. 1648.—Li$_2$Cl$_2$–CaF$_2$–Li$_2$F$_2$ and Li$_2$Cl$_2$–CaF$_2$–CaCl$_2$; reciprocal salt pairs, ternary systems.

G. A. Bukhalova and A. G. Bergman, *Doklady Akad. Nauk S.S.S.R.*, **66** [1] 69 (1949).

NaCl–BaCl$_2$–NaF–BaF$_2$

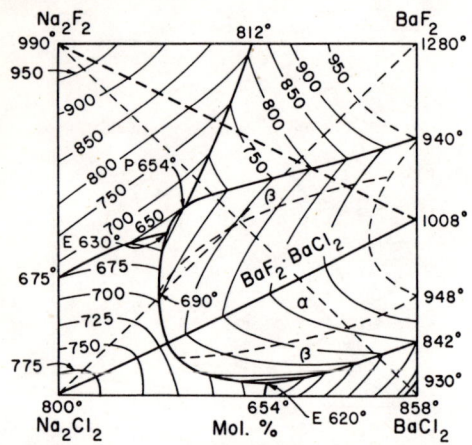

Fig. 1650.—System NaCl–NaF–BaCl$_2$–BaF$_2$.

A. G. Bergman and E. I. Banashek, *Izvest. Sektora Fiz.-Khim. Anal., Inst. Obshcheĭ Neorg. Khim., Akad. Nauk S.S.S.R.*, **22**, 203 (1953).

NaCl–CdCl$_2$–NaI–CdI$_2$

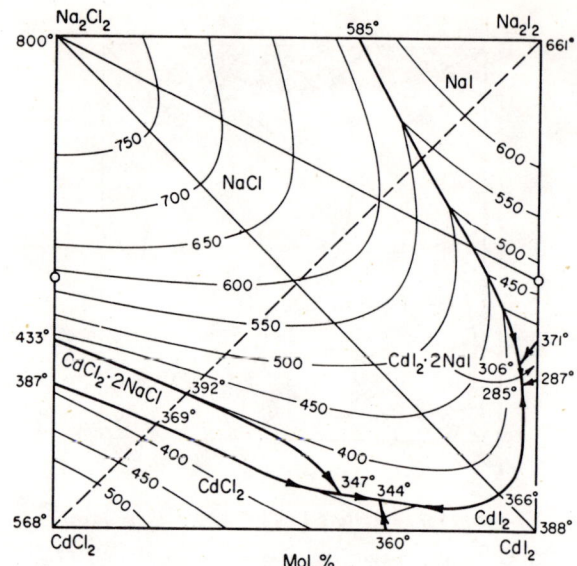

Fig. 1652.—System NaCl–NaI–CdCl$_2$–CdI$_2$; irreversible reciprocal.

I. I. Il'yasov and A. G. Bergman, *Zhur. Obshcheĭ Khim.*, **26**, 1295 (1956).

NaCl–CaCl$_2$–NaF–CaF$_2$

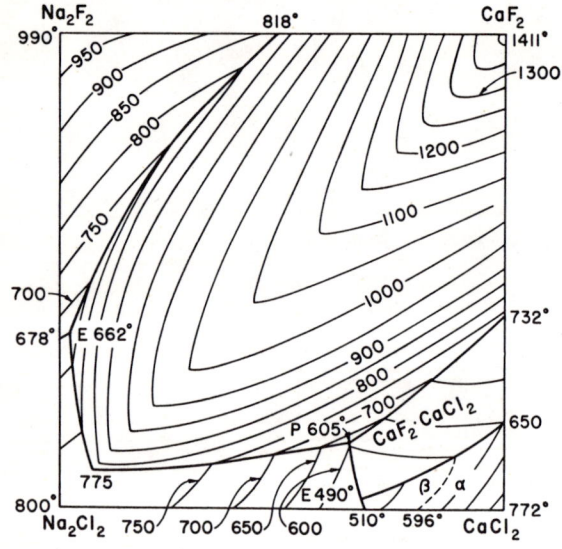

Fig. 1651.—System CaCl$_2$–CaF$_2$–NaCl–NaF.

G. A. Bukhalova, *Zhur. Neorg. Khim.*, **4** [1] 121 (1959).

NaCl–PbCl$_2$–NaI–PbI$_2$

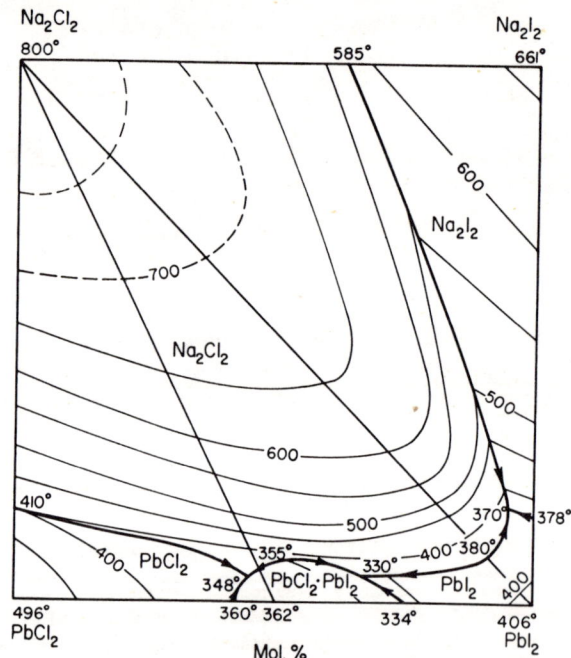

Fig. 1653.—System NaCl–NaI–PbCl$_2$–PbI$_2$; irreversible reciprocal.

I. I. Il'yasov and A. K. Bostandzhian, *Zhur. Obshcheĭ Khim.*, **26**, 2396 (1956).

NaCl–SrCl$_2$–NaF–SrF$_2$

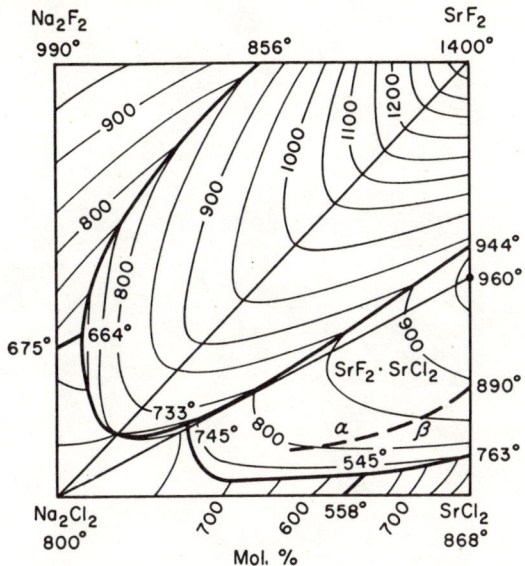

FIG. 1654.—System Na$_2$Cl$_2$–Na$_2$F$_2$–SrCl$_2$–SrF$_2$.

G. A. Bukhalova, *Izvest. Sektora Fiz.-Khim. Anal., Inst. Obshchei Neorg. Khim., Akad. Nauk S.S.S.R.*, **26**, 142 (1955).

BaCl$_2$–CaCl$_2$–BaF$_2$–CaF$_2$

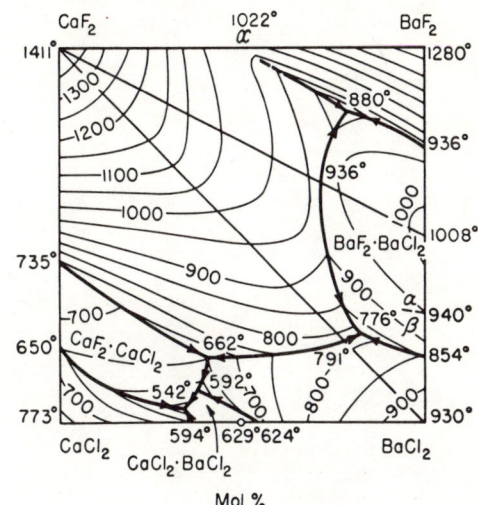

FIG. 1656.—System BaCl$_2$–BaF$_2$–CaCl$_2$–CaF$_2$.

G. A. Bukhalova and A. G. Bergman, *Zhur. Obshchei Khim.*, **21**, 1571 (1951).

RbCl–BaCl$_2$–RbF–BaF$_2$

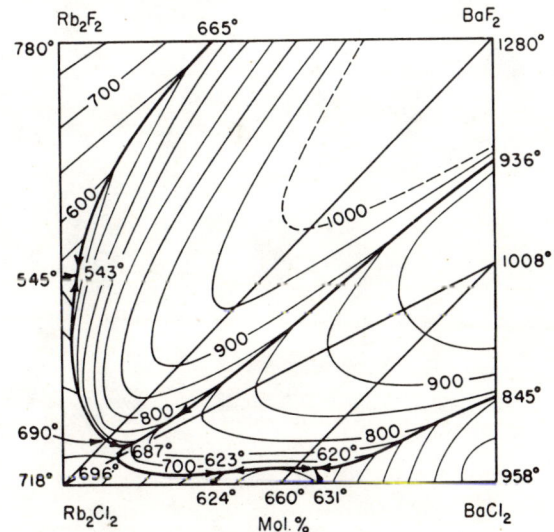

FIG. 1655.—System Rb$_2$Cl$_2$–Rb$_2$F$_2$–BaCl$_2$–BaF$_2$.

E. I. Banashek, *Izvest. Sektora Fiz.-Khim. Anal., Inst. Obshchei Neorg. Khim., Akad. Nauk S.S.S.R.*, **20**, 120 (1950).

CdCl$_2$–PbCl$_2$–CdI$_2$–PbI$_2$

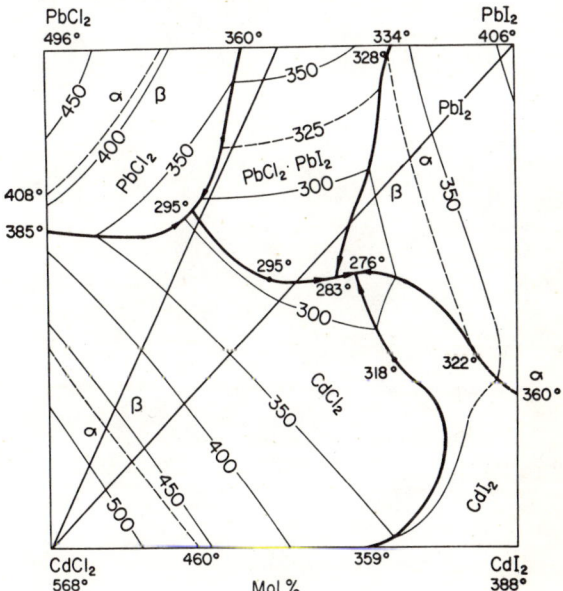

FIG. 1657.—System CdCl$_2$–CdI$_2$–PbCl$_2$–PbI$_2$.

I. I. Il'yasov and A. G. Bergman, *Zhur. Neorg. Khim.*, **2**, 2166 (1957).

E. SYSTEMS CONTAINING HALIDES WITH OTHER SUBSTANCES

I. Halides with Metals

(a) Two Substances

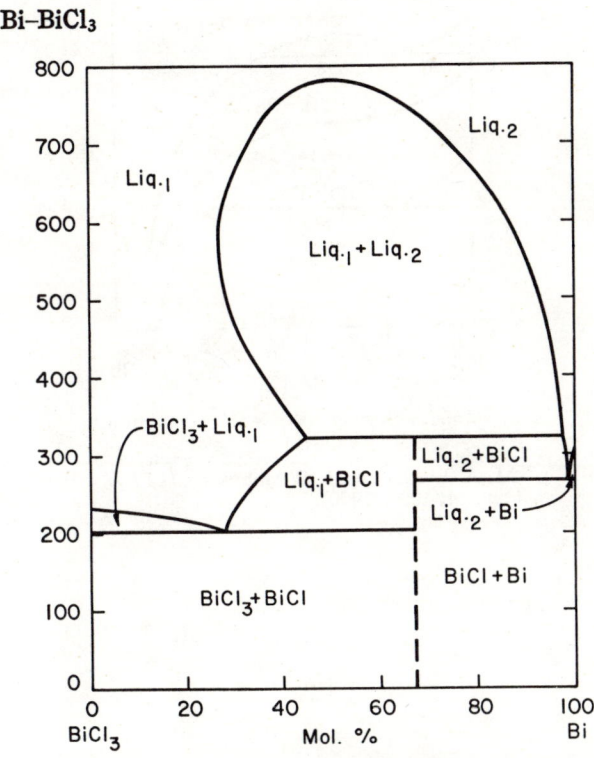

Fig. 1658.—System Bi–BiCl₃.

S. J. Yosim, A. J. Darnell, W. G. Gehman and S. W. Mayer, *J. Phys. Chem.*, 63, 231 (1959).

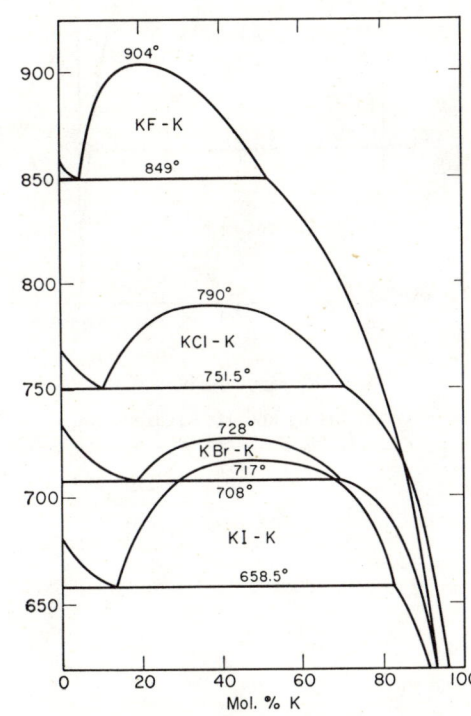

Fig. 1660.—Systems K–KX (X = F, Cl, Br, I).

J. W. Johnson and M. A. Bredig, *J. Phys. Chem.*, 62, 606 (1958).

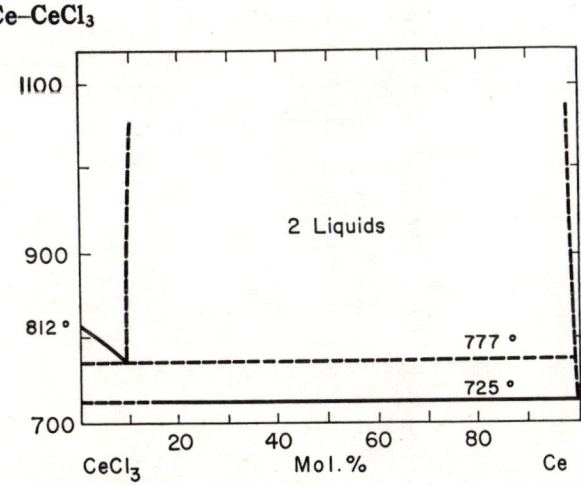

Fig. 1659.—System Ce–CeCl₃.

G. W. Mellors and S. Senderoff, *J. Phys. Chem.*, 63, 1111 (1959).

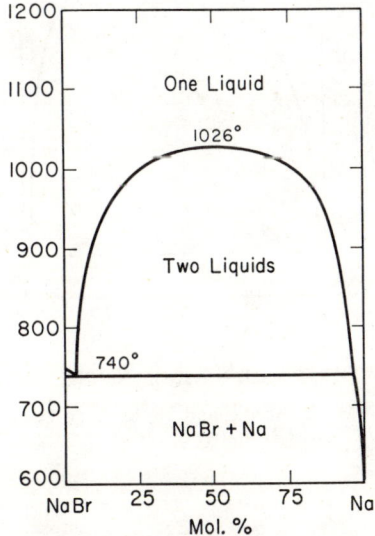

Fig. 1661.—System Na–NaBr.

M. A. Bredig and H. R. Bronstein, *J. Phys. Chem.*, 64, 65 (1960).

Na–NaCl

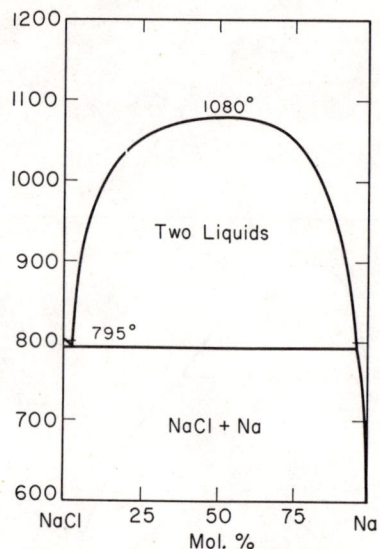

Fig. 1662.—System Na–NaCl.

M. A. Bredig and H. R. Bronstein, *J. Phys. Chem.*, **64**, 65 (1960).

Rb–RbX

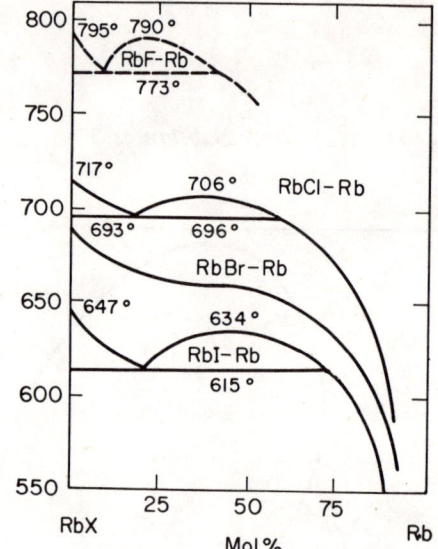

Fig. 1664.—System Rb–RbX (rubidium halides).

M. A. Bredig and J. W. Johnson, *J. Phys. Chem.*, **64**, 1900 (1960).

Na–NaI

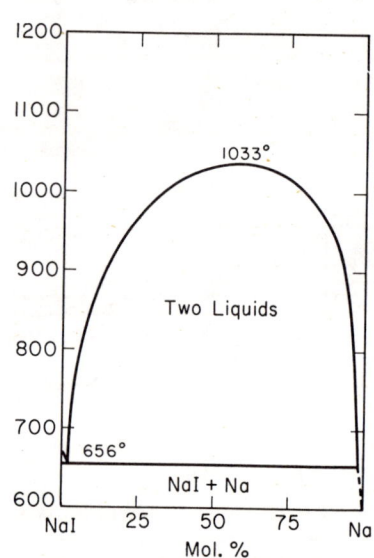

Fig. 1663.—System Na–NaI.

M. A. Bredig and H. R. Bronstein, *J. Phys. Chem.*, **64**, 65 (1960).

R–RF

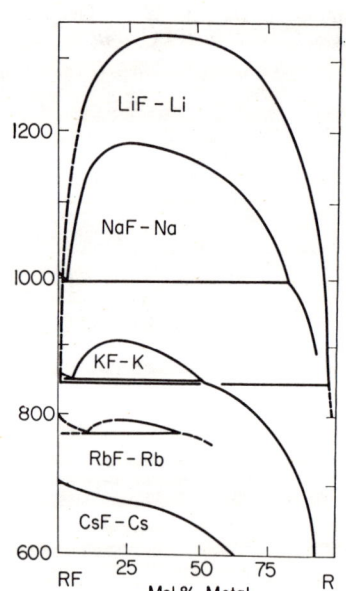

Fig. 1665.—Systems R–RF (R = Li, Na, K, Rb, Cs).

A. S. Dworkin, H. R. Bronstein, and M. A. Bredig, *J. Phys. Chem.*, **66**, 572–73 (1962).

(b) Three Substances

Al–Zn–AlCl$_3$

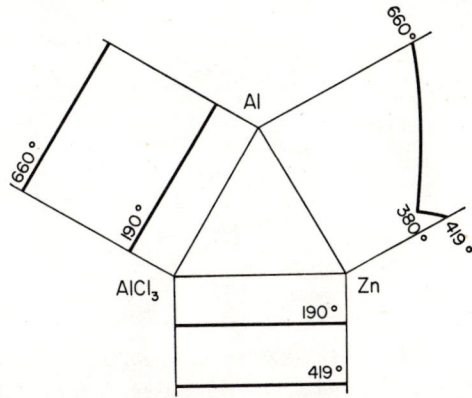

Fig. 1666.—System Zn–Al–AlCl$_3$.

A. P. Palkin and O. K. Belousov, *Zhur. Neorg. Khim.*, **2**, 1627 (1957).

Zn–ZnCl$_2$–AlCl$_3$

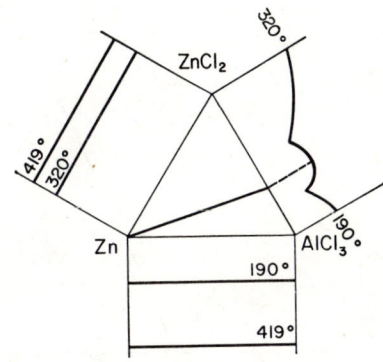

Fig. 1667.—System Zn–ZnCl$_2$–AlCl$_3$.

A. P. Palkin and O. K. Belousov, *Zhur. Neorg. Khim.*, **2**, 1627 (1957).

(c) Four Substances

Al–Zn–ZnCl$_2$–AlCl$_3$

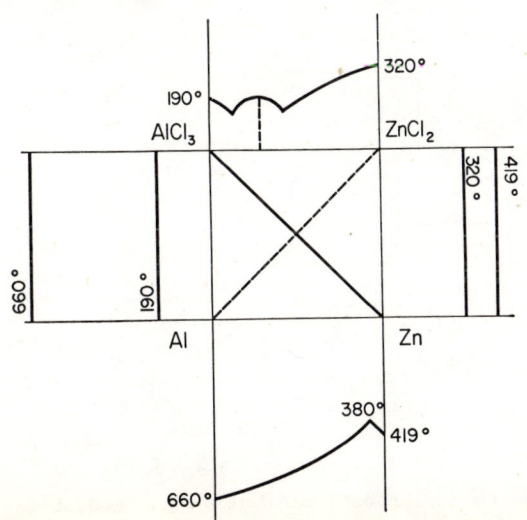

Cd–Zn–CdCl$_2$–ZnCl$_2$

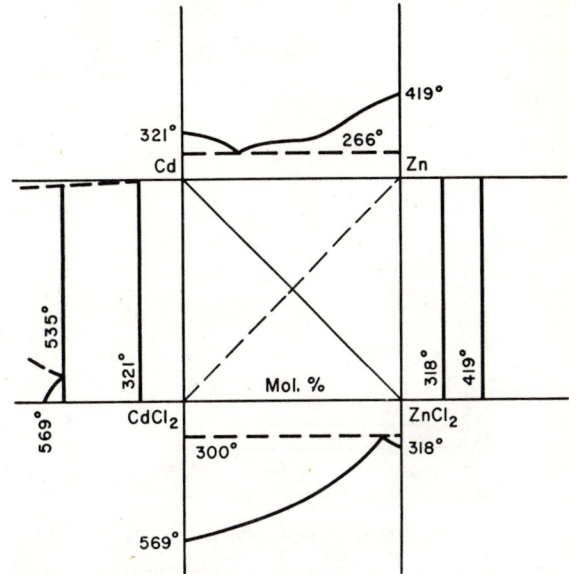

Fig. 1669.—System Cd–Zn–CdCl$_2$–ZnCl$_2$.

E. I. Elagina and A. P. Palkin, *Zhur. Neorg. Khim.*, **1** [5] 1044 (1956).

Mg–Pb–MgCl$_2$–PbCl$_2$

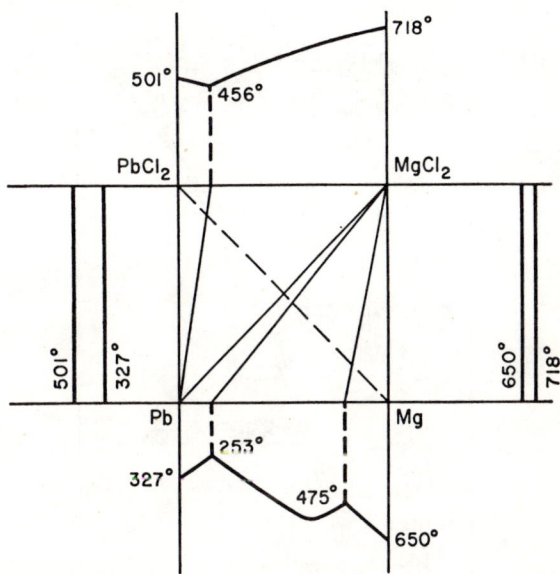

Fig. 1670.—System Mg–Pb–MgCl$_2$–PbCl$_2$.

A. P. Palkin and V. T. Redchenko, *Zhur. Neorg. Khim.*, **1** [1] 143 (1956).

Fig. 1668.—System Zn–ZnCl$_2$–Al–AlCl$_3$.

A. P. Palkin and O. K. Belousov, *Zhur. Neorg. Khim.*, **2**, 1627 (1959).

II. Fluorides with Metal Oxides

(a) Two Substances

CaF$_2$–CaO

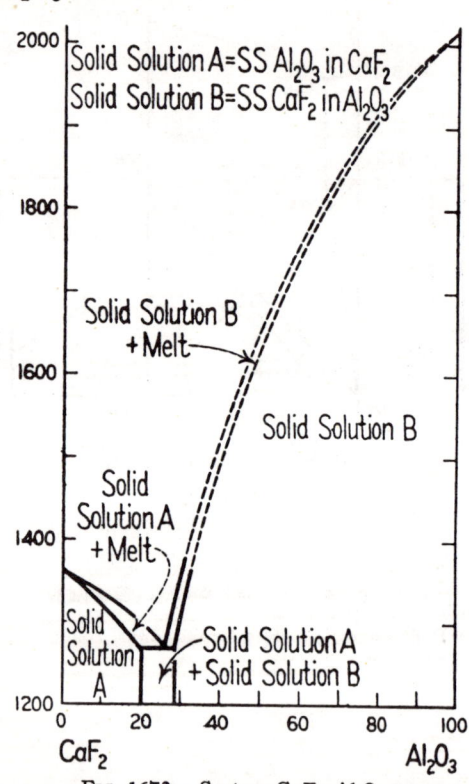

Fig. 1671.—System CaO–CaF$_2$.

Tryggve Bååk, *Acta Chem. Scand.*, 8 [9] 1727 (1954) (in English). See also P. P. Budnikov and S. G. Tresvyatskiĭ, *Doklady Akad. Nauk S.S.S.R.*, 89, 481 (1953).

CaF$_2$–MgO

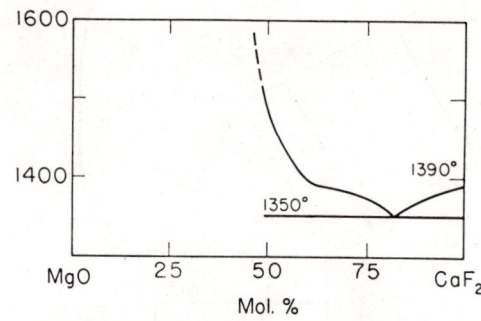

Fig. 1672.—System CaF$_2$–MgO.

P. P. Budnikov and S. G. Tresvyatskiĭ, *Ukrain. Khim. Zhur.*, 19, 555 (1953).

PbF$_2$–PbO

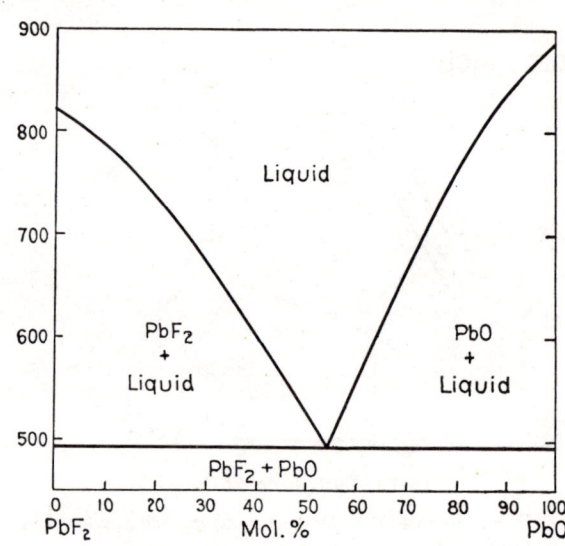

Fig. 1674.—System PbF$_2$–PbO.

C. Sandonnini, *Atti reale accad. sci.*, Torino, 22 [I] 959 (1914).

(b) Three Substances

CsF–Cs$_2$O–CrO$_3$

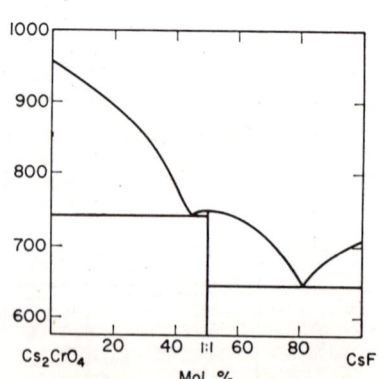

Fig. 1675.—System Cs$_2$CrO$_4$–CsF.

O. Schmitz-Dumont and Albert Weeg, *Z. anorg. Chem.*, 265, 147 (1951).

CaF$_2$–Al$_2$O$_3$

Fig. 1673.—System CaF$_2$–Al$_2$O$_3$.

Von P. Pascal, *Z. Elektrochem.*, 19, 611 (1913).

CsF–Cs₂O–MoO₃

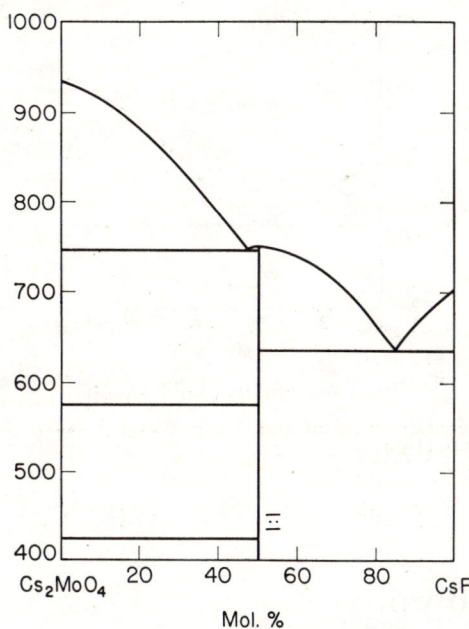

FIG. 1676.—System Cs₂MoO₄–CsF.

O. Schmitz-Dumont and Albert Weeg, *Z. anorg. Chem.*, **265**, 150 (1951).

CsF–Cs₂O–WO₃

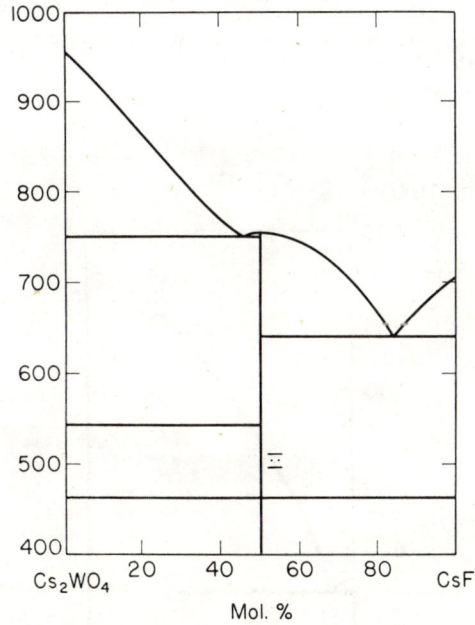

FIG. 1677.—System Cs₂WO₄–CsF.

O. Schmitz-Dumont and Albert Weeg, *Z. anorg. Chem.*, **265**, 152 (1951).

KF–K₂O–CrO₃

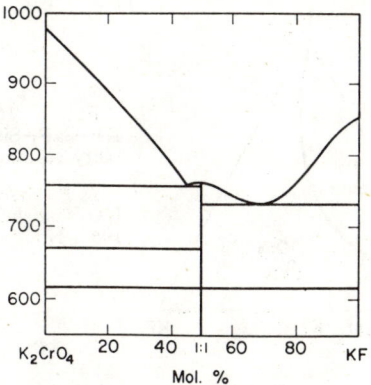

FIG. 1678.—System K₂CrO₄–KF.

O. Schmitz-Dumont and Albert Weeg, *Z. anorg. Chem.*, **265**, 147 (1951).

KF–K₂O–MoO₃

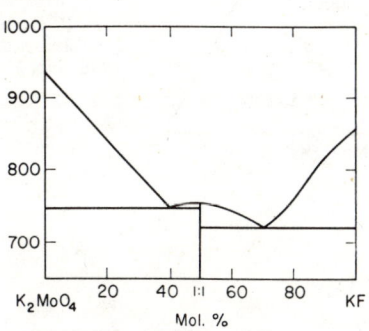

FIG. 1679.—System K₂MoO₄–KF. Inversion temperatures of K₂MoO₄ (475, 439, and 321°C.) as well as the inversion temperature of the compound (350°C.) are not shown.

O. Schmitz-Dumont and Albert Weeg, *Z. anorg. Chem.*, **265**, 149 (1951).

KF–K₂O–WO₃

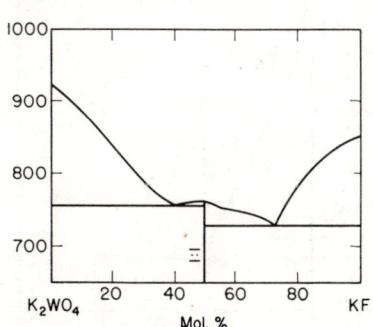

FIG. 1680.—System K₂WO₄–KF. Inversion temperature of K₂WO₄ (370°C.) not shown.

O. Schmitz-Dumont and Albert Weeg, *Z. anorg. Chem.*, **265**, 151 (1951).

KF–CaF$_2$–SiO$_2$

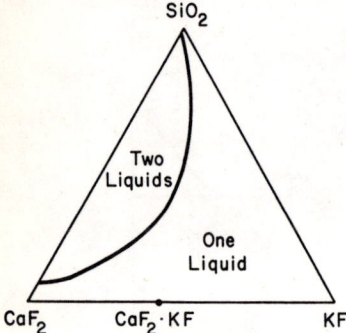

Fig. 1681.—System KF–CaF$_2$–SiO$_2$; immiscibility region.

Z. P. Ershova and Ya. I. Ol'shanskiĭ, *Geokhimiya*, 1958, No. 2, p. 148.

KF–MgF$_2$–SiO$_2$

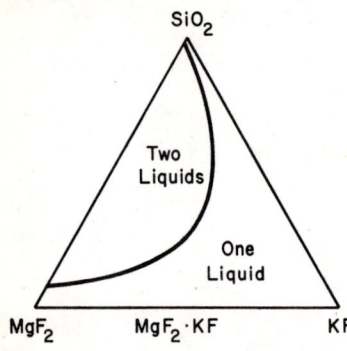

Fig. 1682.—System KF–MgF$_2$–SiO$_2$; immiscibility region.

Z. P. Ershova and Ya. I. Ol'shanskiĭ, *Geokhimiya*, 1958, No. 2, p. 148.

LiF–K$_2$O–TiO$_2$

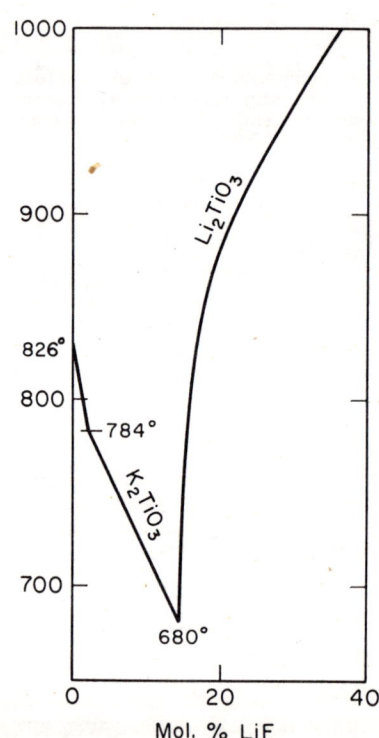

Fig. 1683.—System K$_2$TiO$_3$–2LiF.

I. N. Belyaev and N. P. Sigida, *Zhur. Obshcheĭ Khim.*, 26, 1557 (1956).

LiF–Li$_2$O–MoO$_3$

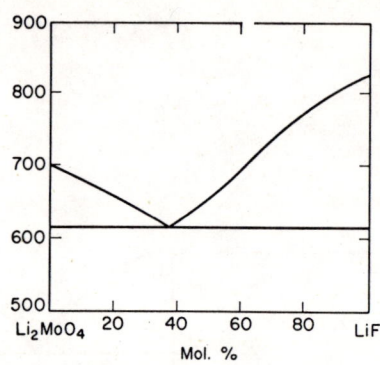

Fig. 1684.—System Li$_2$MoO$_4$–LiF.

O. Schmitz-Dumont and Albert Weeg, *Z. anorg. Chem.*, 265, 149 (1951).

LiF–Li$_2$O–WO$_3$

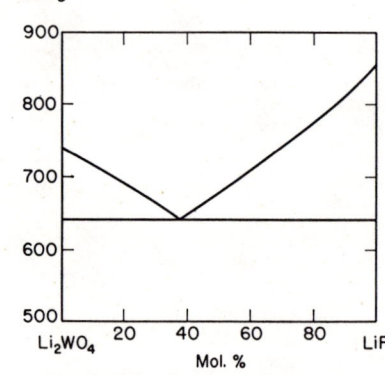

Fig. 1685.—System Li$_2$WO$_4$–LiF.

O. Schmitz-Dumont and Albert Weeg, *Z. anorg. Chem.*, 265, 151 (1951).

LiF–AlF$_3$–Al$_2$O$_3$

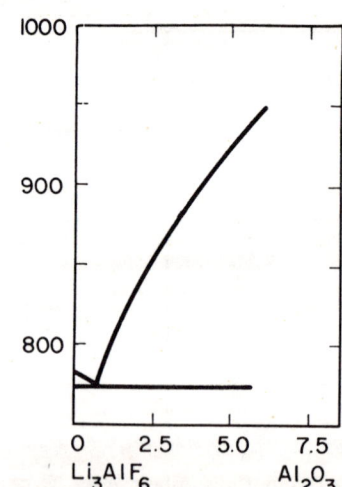

Fig. 1686.—System Li$_3$AlF$_6$–Al$_2$O$_3$.

V. P. Mashovets and V. I. Petrov, *Zhur. Prikl. Khim.*, 30, 1696 (1957). See also P. Drossbach, *Z. Elektrochem.*, 42 [2] 65 (1936).

NaF–Li₂O–TiO₂

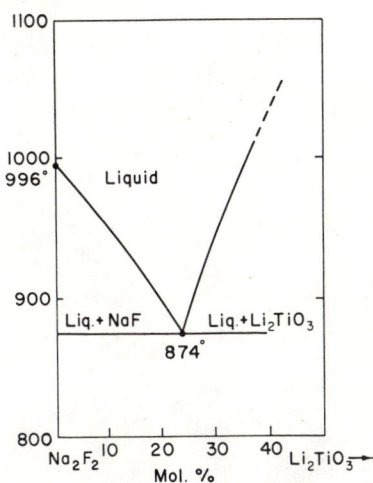

FIG. 1687.—System NaF–Li₂TiO₃.

I. N. Belyaev and N. P. Sigida, *Zhur. Neorg. Khim.*, **2** [5] 1121 (1957).

NaF–Na₂O–B₂O₃

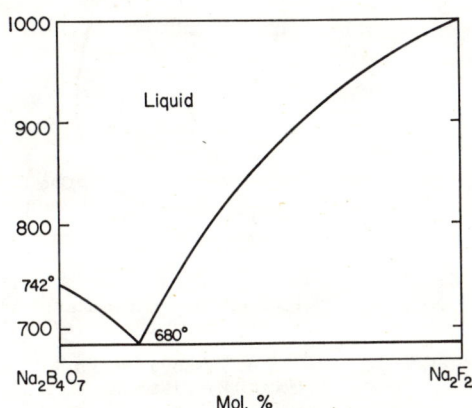

FIG. 1688.—System NaF–Na₂B₄O₇.

B. W. King, *J. Am. Ceram. Soc.*, **37** [5] 239 (1954).

NaF–Na₂O–SiO₂

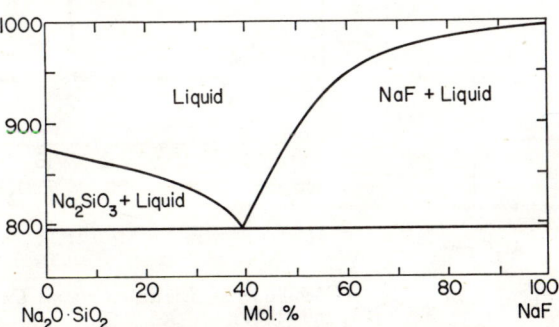

FIG. 1689.—System NaF–Na₂O·SiO₂.

H. S. Booth, B. A. Starrs, and M. J. Bahnsen, *J. Phys. Chem.*, **37**, 1106 (1933).

NaF–Na₂O–TiO₂

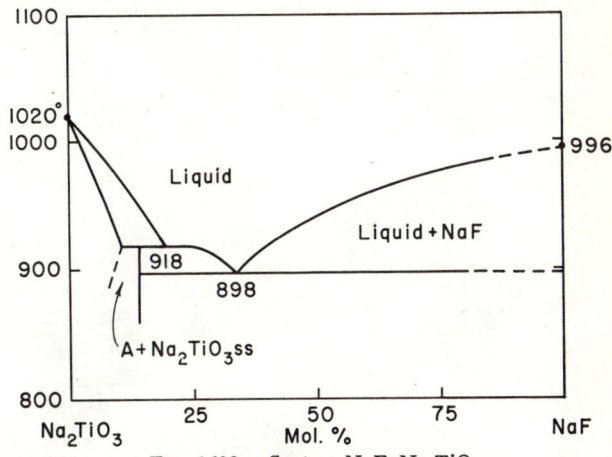

FIG. 1690.—System NaF–Na₂TiO₃.

I. N. Belyaev and N. P. Sigida, *Zhur. Neorg. Khim.*, **2** [5] 1121 (1957).

NaF–Na₂O–CrO₃

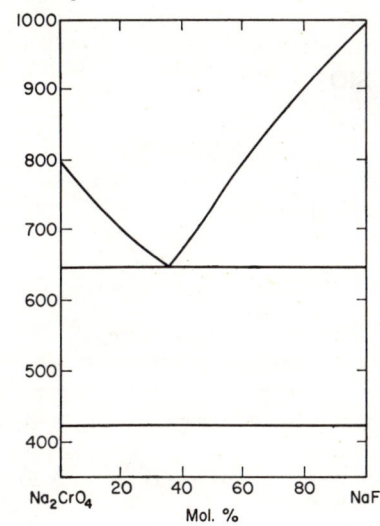

FIG. 1691.—System Na₂CrO₄–NaF.

O. Schmitz-Dumont and Albert Weeg, *Z. anorg. Chem.*, **265**, 147 (1951).

NaF–Na₂O–MoO₃

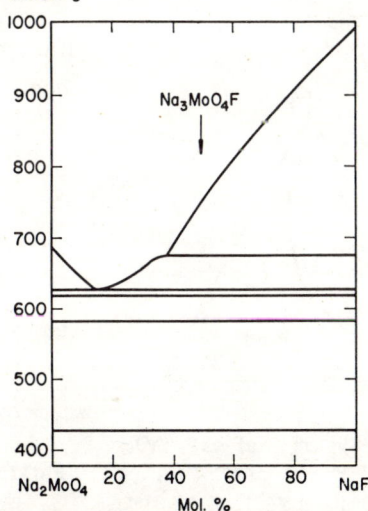

FIG. 1692.—System Na₂MoO₄–NaF.

O. Schmitz-Dumont and Albert Weeg, *Z. anorg. Chem.*, **265**, 149 (1951).

NaF–Na₂O–WO₃

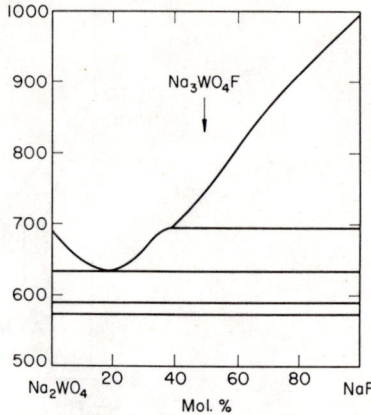

Fig. 1693.—System Na₂WO₄–NaF.

O. Schmitz-Dumont and Albert Weeg, *Z. anorg. Chem.*, **265**, 151 (1951).

NaF–BaF₂–SiO₂

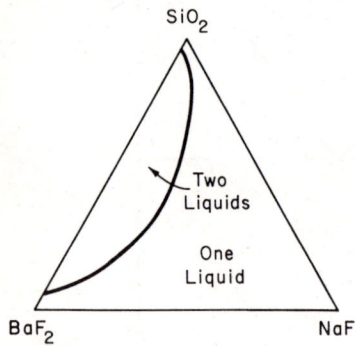

Fig. 1694.—System NaF–BaF₂–SiO₂; immiscibility region.

Z. P. Ershova and Ya. I. Ol'shanskiĭ, *Geokhimiya*, **1958**, No. 2, p. 150.

NaF–CaF₂–SiO₂

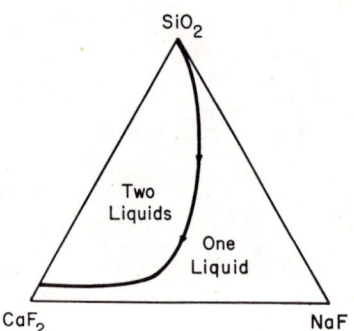

Fig. 1695.—System NaF–CaF₂–SiO₂; immiscibility region.

Z. P. Ershova and Ya. I. Ol'shanskiĭ, *Geokhimiya*, **1958**, No. 2, p. 146.

NaF–MgF₂–SiO₂

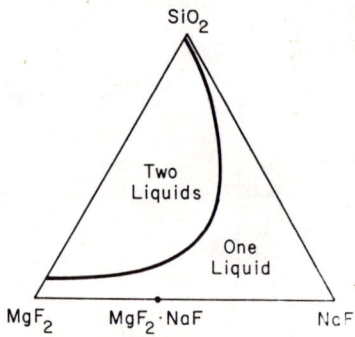

Fig. 1696.—System NaF–MgF₂–SiO₂; immiscibility region.

Z. P. Ershova and Ya. I. Ol'shanskiĭ, *Geokhimiya*, **1958**, No. 2, p. 146.

NaF–SrF₂–SiO₂

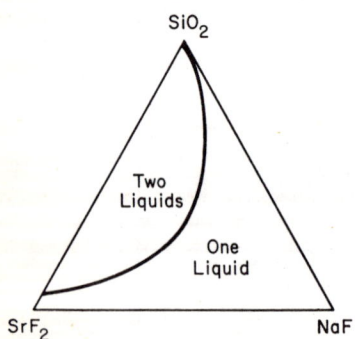

Fig. 1697.—System NaF–SrF₂–SiO₂; immiscibility region.

Z. P. Ershova and Ya. I. Ol'shanskiĭ, *Geokhimiya*, **1958**, No. 2, p. 150.

NaF–AlF₃–Al₂O₃

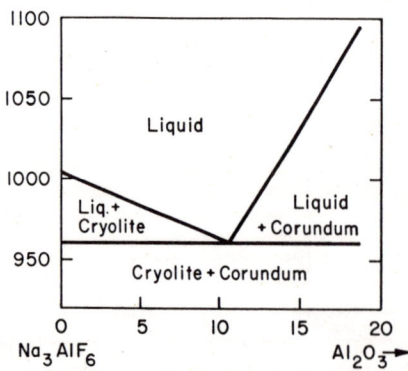

Fig. 1698.—System Na₃AlF₆–Al₂O₃; 0 to 18.5% Al₂O₃.

P. A. Foster, Jr., *J. Am. Ceram. Soc.*, **43** [2] 67 (1960). For diagram of NaF-rich corner of ternary system see P. A. Foster, Jr., *J. Am. Ceram. Soc.*, **45** [4] 148 (1962).

NaF–AlF₃–ZrO₂

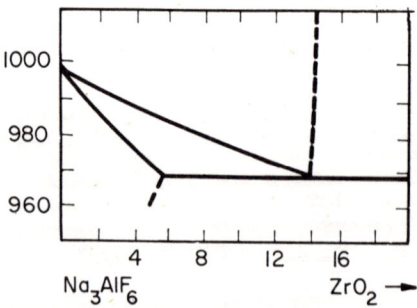

Fig. 1699.—System Na₃AlF₆–ZrO₂; cryolite region.

Hidehiko Kido, *Sci. Rept. Saitama Univ.*, **1A**, 166 (1954).

NaF–AlF$_3$–(RO, RO$_2$)

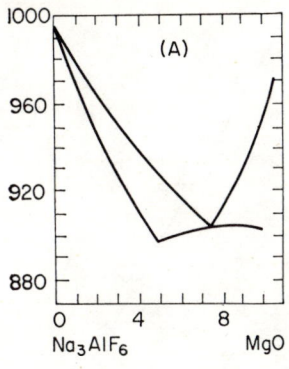
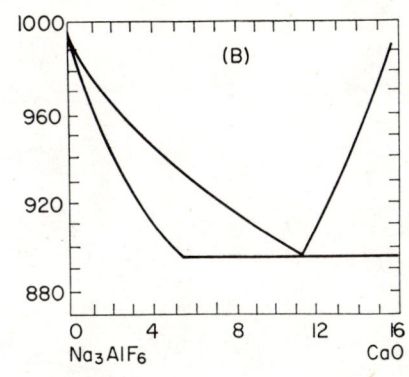
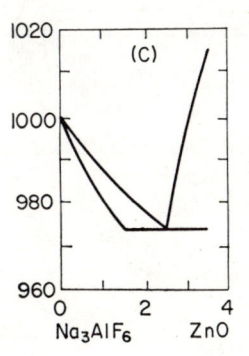
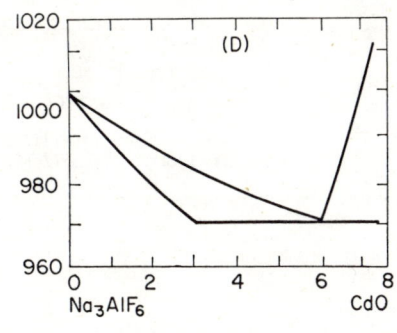
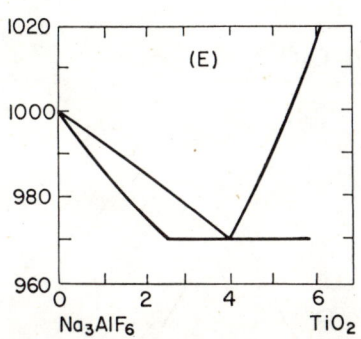

FIG. 1700.—NaF–AlF$_3$–(RO, RO$_2$). A = System Na$_3$AlF$_6$–MgO, B = System Na$_3$AlF$_6$–CaO, C = System Na$_3$AlF$_6$–ZnO, D = System Na$_3$AlF$_6$–CdO, E = System Na$_3$AlF$_6$–TiO$_2$.

Yasumasa Hayakawa and Hidehiko Kido, *Sci. Rept. Saitama Univ.*, **1**, 41 (1952).

RbF–Rb$_2$O–CrO$_3$

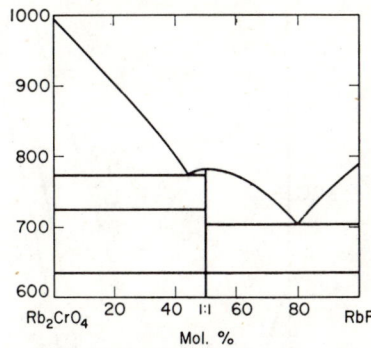

FIG. 1701.—System Rb$_2$CrO$_4$–RbF. Inversion temperature of Rb$_2$CrO$_4$ (552°C.) not shown.

O. Schmitz-Dumont and Albert Weeg, *Z. anorg. Chem.*, **265**, 147 (1951).

RbF–Rb$_2$O–MoO$_3$

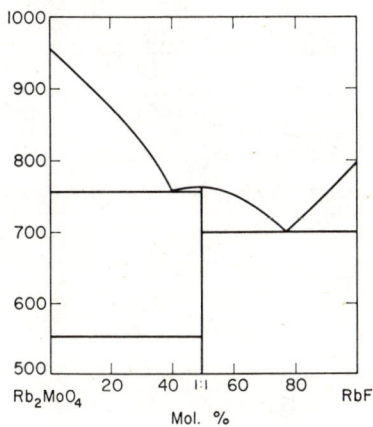

FIG. 1702.—System Rb$_2$MoO$_4$–RbF.

O. Schmitz-Dumont and Albert Weeg, *Z. anorg. Chem.*, **265**, 149 (1951).

RbF–Rb$_2$O–WO$_3$

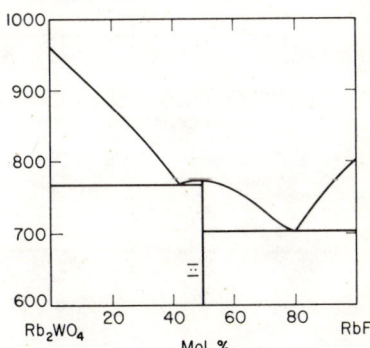

FIG. 1703.—System Rb$_2$WO$_4$–RbF.

O. Schmitz-Dumont and Albert Weeg, *Z. anorg. Chem.*, **265**, 151 (1951).

RF–Li$_2$O–TiO$_2$

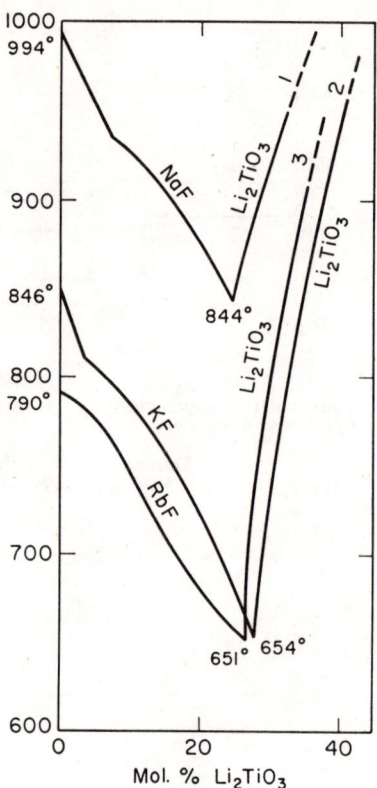

FIG. 1704.—Li$_2$O–TiO$_2$–(R)F. (1) 2NaF–Li$_2$TiO$_3$ (2); 2KF–Li$_2$TiO$_3$; (3) 2RbF–Li$_2$TiO$_3$.

I. N. Belyaev and N. P. Sigida, *Zhur. Obshcheĭ Khim.*, **26**, 1557 (1956).

CaF$_2$–Na$_2$O–SiO$_2$

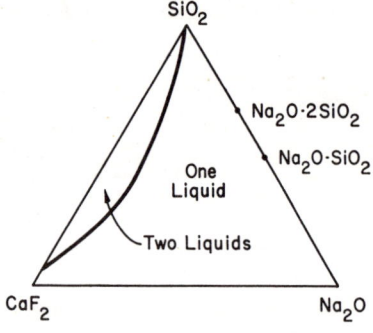

FIG. 1705.—System Na$_2$O–CaF$_2$–SiO$_2$; immiscibility region.

Z. P. Ershova and Ya. I. Ol'shanskiĭ *Geokhimiya*, 1958, No. 2, p. 145.

CaF$_2$–CaO–FeO

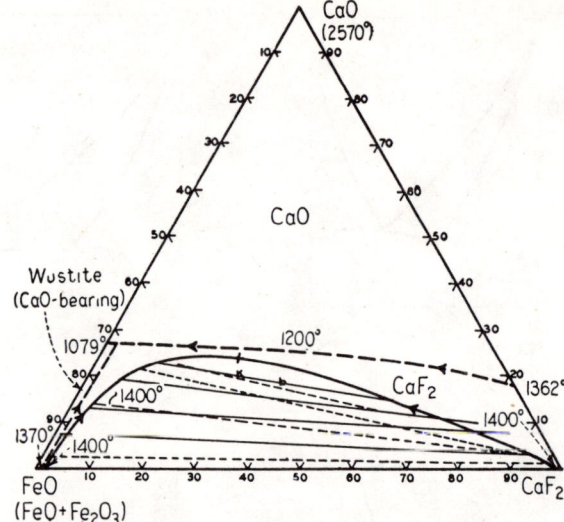

FIG. 1706.—System CaF$_2$–CaO–FeO.

W. Oelsen and H. Maetz, *Mitt. Kaiser-Wilhelm-Inst. Eisenforsch. Düsseldorf*, **23**, 195 (1941).

CaF$_2$–CaO–Al$_2$O$_3$

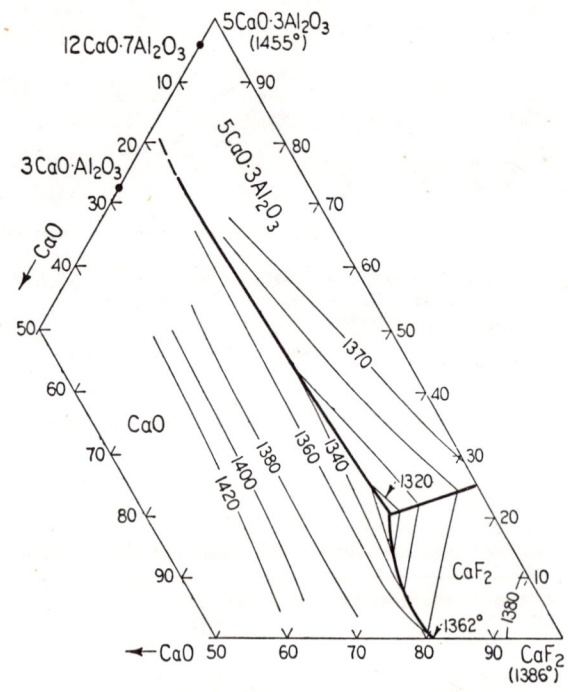

FIG. 1707.—System CaF$_2$–CaO–5CaO·3Al$_2$O$_3$.

W. Eitel, *Zement*, **27**, 30 (1938).

CaF₂–CaO–SiO₂

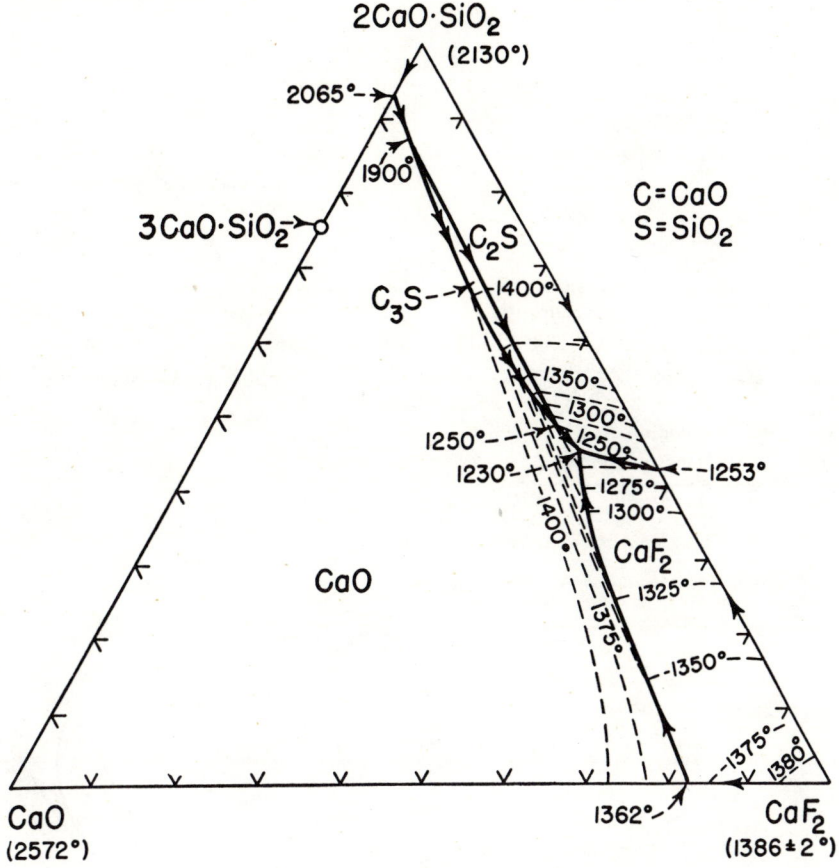

Fig. 1708.—System CaF_2–CaO–$2CaO \cdot SiO_2$.

W. Eitel, *Zement*, **27** [31] 469 (1938); *Z. angew. Mineral*, **1**, 272 (1938).

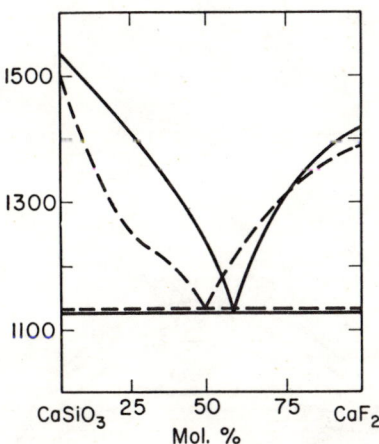

Fig. 1709.—System CaF_2–$CaSiO_3$. Dashed lines after B. Karandyéeff, *Z. anorg. Chem.*, **68**, 190 (1910).

Tryggve Bååk and Arne Ölander, *Acta Chem. Scand.*, **9** [8] 1351 (1955) (in English).

CaF₂–MgO–SiO₂

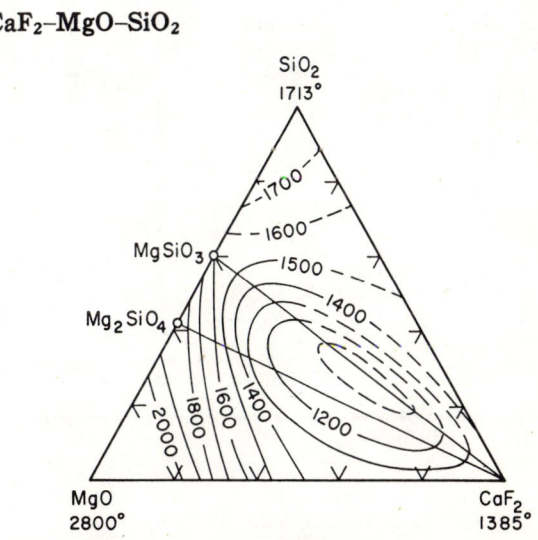

Fig. 1710.—System CaF_2–MgO–SiO_2; melting.

A. S. Berezhnoi, *Dopovidi Akad. Nauk Ukr. R.S.R.*, p. 250 (1951).

MgF$_2$–MgO–SiO$_2$

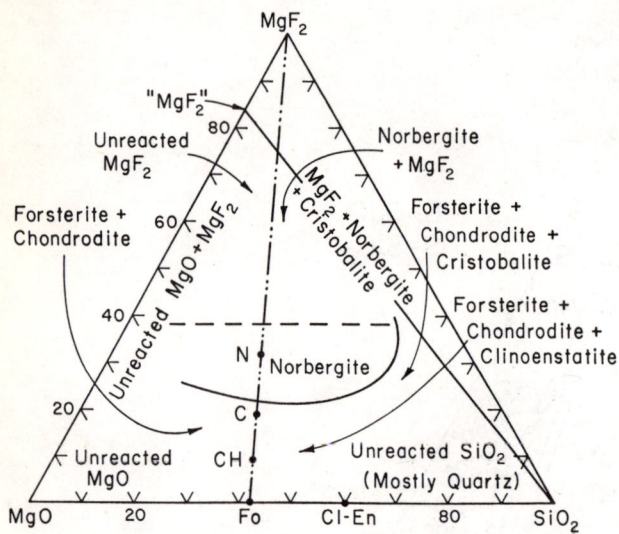

Fig. 1711.—System MgO–MgF$_2$–SiO$_2$; solid state reactions.

Takashi Fujii and Wilhelm Eitel, *Radex-Rundschau*, 1957, No. 1, p. 466.

PbF$_2$–PbO–As$_2$O$_5$

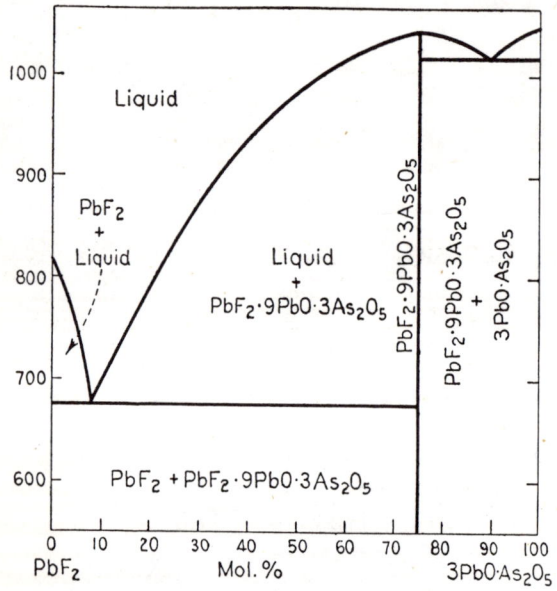

Fig. 1712.—System PbF$_2$–3PbO·As$_2$O$_5$.

M. Amadori, *Atti reale ist Veneto sci. (lettere ed arti)*, 22 [II] 200 (1915).

PbF$_2$–PbO–V$_2$O$_5$

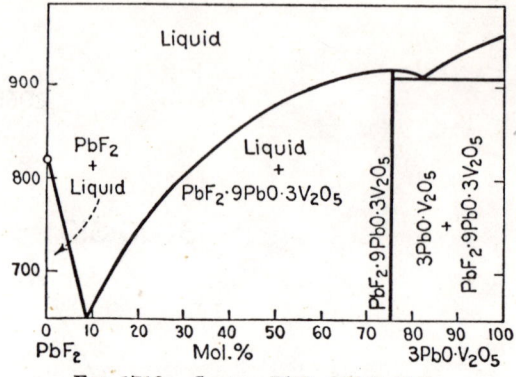

Fig. 1713.—System PbF$_2$–3PbO·V$_2$O$_5$.

M. Amadori, *Gazz. chim. ital.*, 49 [I] 44 (1919).

RF$_2$–RO–SiO$_2$

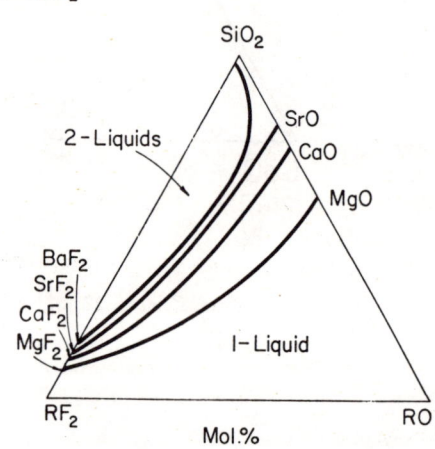

Fig. 1714.—System RF$_2$–RO–SiO$_2$.

Z. P. Ershova and Ya. I. Ol'shanskiĭ, *Geokhimiya*, 1957, No. 3, p. 220.

RF$_2$–Al$_2$O$_3$–SiO$_2$

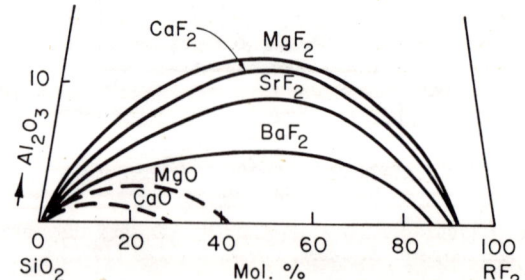

Fig. 1715.—System RF$_2$–Al$_2$O$_3$–SiO$_2$; immiscibility region.

Z. P. Ershova, *Geokhimiya*, 1957, No. 4, p. 303.

(c) Four Substances

KF–LiF–K₂SiO₃–Li₂SiO₃

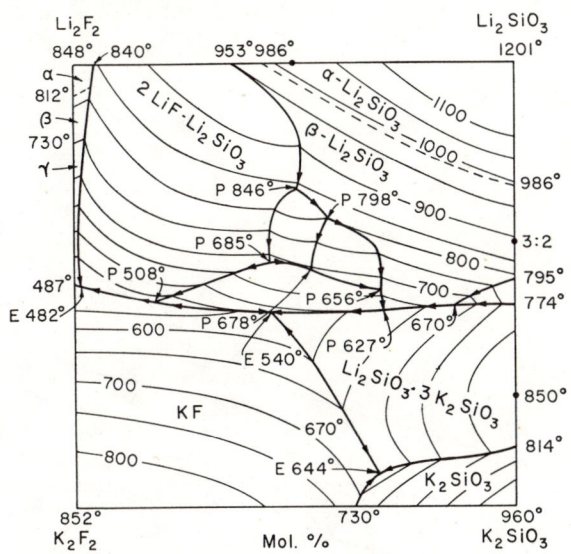

Fig. 1716.—System KF–K₂SiO₃–LiF–Li₂SiO₃.

A. G. Bergman and N. A. Bychkova-Shul'ga, *Zhur. Neorg. Khim.*, 2 [1] 188 (1957).

KF–NaF–KVO₃–NaVO₃

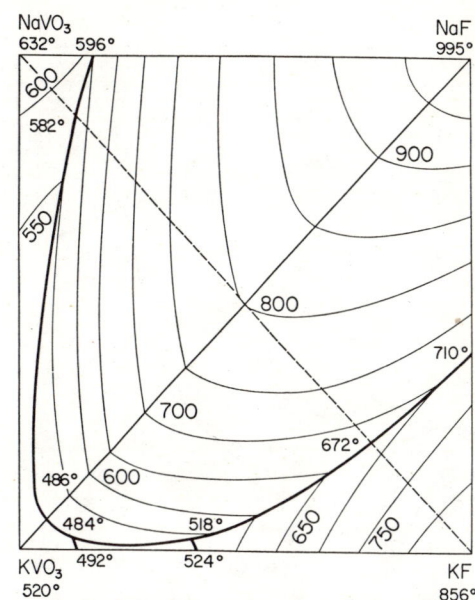

Fig. 1718.—System KF–KVO₃–NaF–NaVO₃.

M. A. Zakharchenko and A. G. Bergman, *Zhur. Neorg. Khim.*, 2, 881 (1957).

KF–NaF–K₂TiO₃–Na₂TiO₃

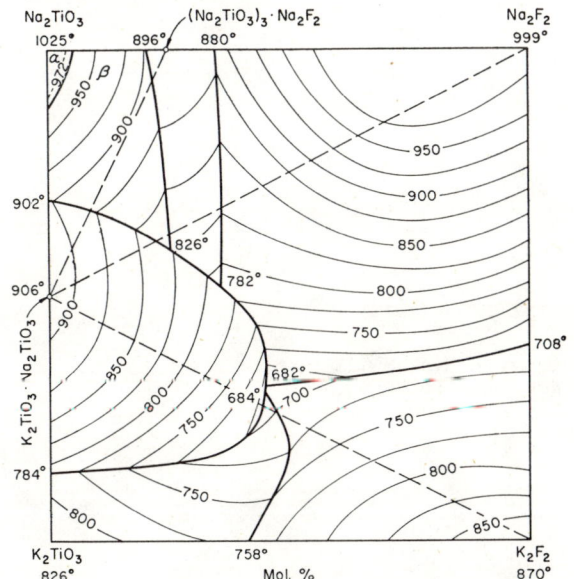

Fig. 1717.—System KF–K₂TiO₃–NaF–Na₂TiO₃; irreversible reciprocal.

M. L. Sholokhovich, *Zhur. Obshcheĭ Khim.*, 25, 1905 (1955).

KF–NaF–K₂MoO₄–Na₂MoO₄

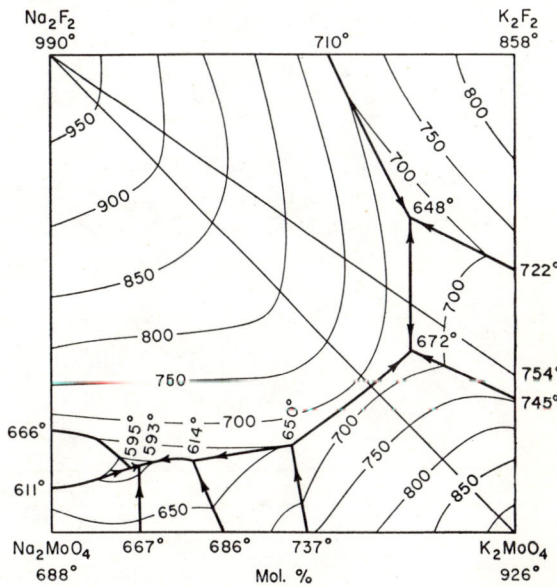

Fig. 1719.—System KF–K₂MoO₄–NaF–Na₂MoO₄; irreversible reciprocal.

Z. A. Mateiko and G. A. Bukhalova, *Zhur. Obshcheĭ Khim.*, 25, 1678 (1955).

KF–NaF–K$_2$WO$_4$–Na$_2$WO$_4$

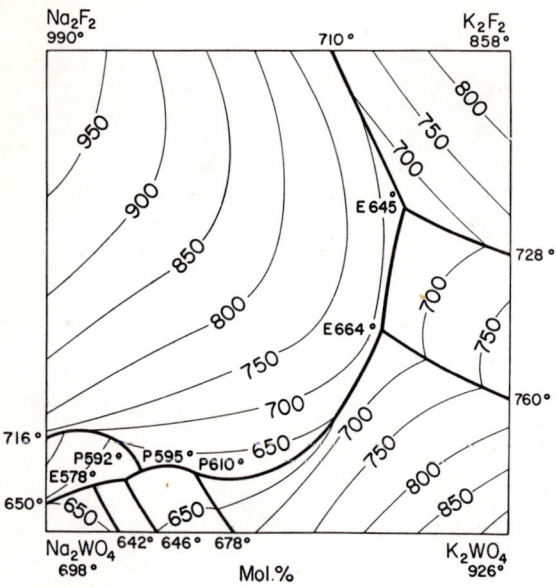

Fig. 1720.—System KF–K$_2$WO$_4$–NaF–Na$_2$WO$_4$.

Z. A. Mateĭko and G. A. Bukhalova, *Zhur. Neorg. Khim.*, **2**, 412 (1957).

LiF–NaF–Li$_2$TiO$_3$–Na$_2$TiO$_3$

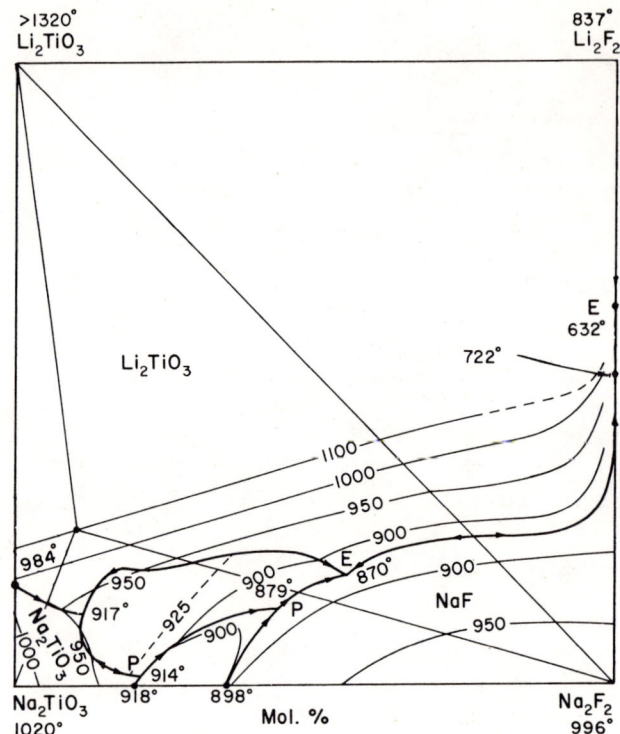

Fig. 1722.—System LiF–Li$_2$TiO$_3$–NaF–Na$_2$TiO$_3$.

I. N. Belyaev and N. P. Sigida, *Zhur. Neorg. Khim.*, **2** [5] 1125 (1957).

LiF–NaF–Li$_2$SiO$_3$–Na$_2$SiO$_3$

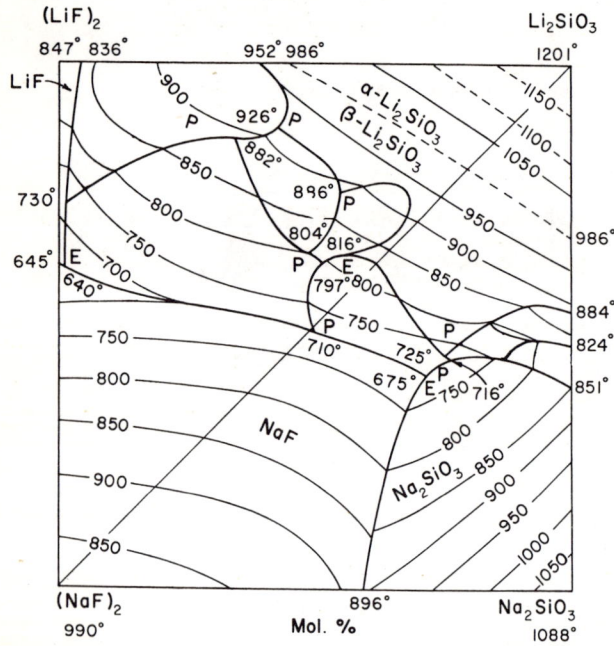

Fig. 1721.—System LiF–Li$_2$SiO$_3$–NaF–Na$_2$SiO$_3$.

I. N. Belyaev and N. P. Sigida, *Zhur. Neorg. Khim.*, **2** [5] 1126 (1957).

LiF–NaF–AlF$_3$–Al$_2$O$_3$

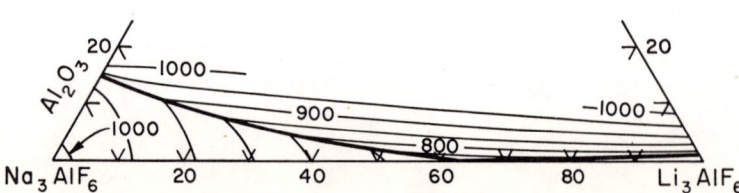

Fig. 1723.—System Li$_3$AlF$_6$–Na$_3$AlF$_6$–Al$_2$O$_3$.

V. P. Mashovets and V. I. Petrov, *Zhur. Prikl. Khim.*, **30**, 1697 (1957).

LiF–BaF$_2$–Li$_2$SiO$_3$–BaSiO$_3$

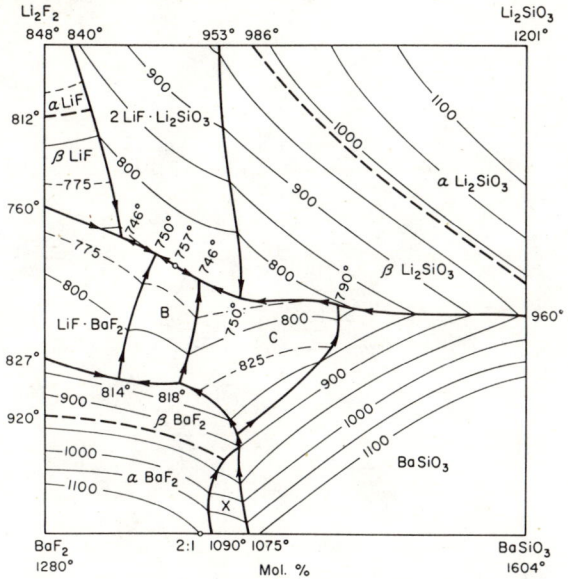

FIG. 1724.—System LiF–Li$_2$SiO$_3$–BaF$_2$–BaSiO$_3$; reciprocal.

N. A. Bychkova and A. G. Bergman, *Zhur. Obshcheĭ Khim.*, **26**, 647 (1956).

LiF–CaF$_2$–Li$_2$SiO$_3$–CaSiO$_3$

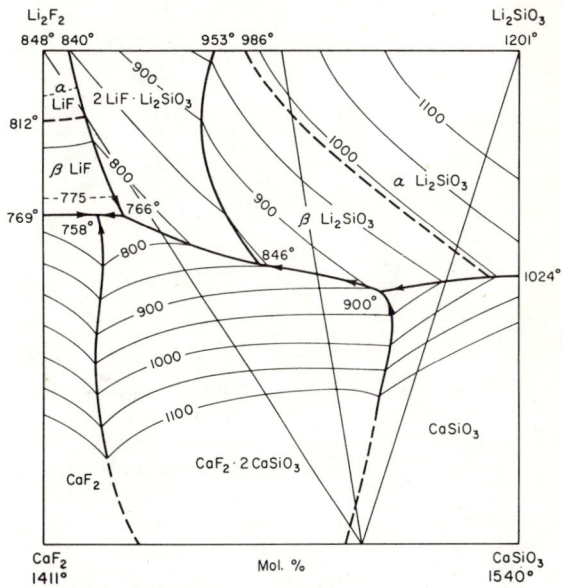

FIG. 1725.—System LiF–Li$_2$SiO$_3$–CaF$_2$–CaSiO$_3$; irreversible reciprocal.

A. G. Bergman and N. A. Bychkova, *Zhur. Obshcheĭ Khim.*, **25**, 1885 (1955).

NaF–Na$_2$O–MoO$_3$–WO$_3$

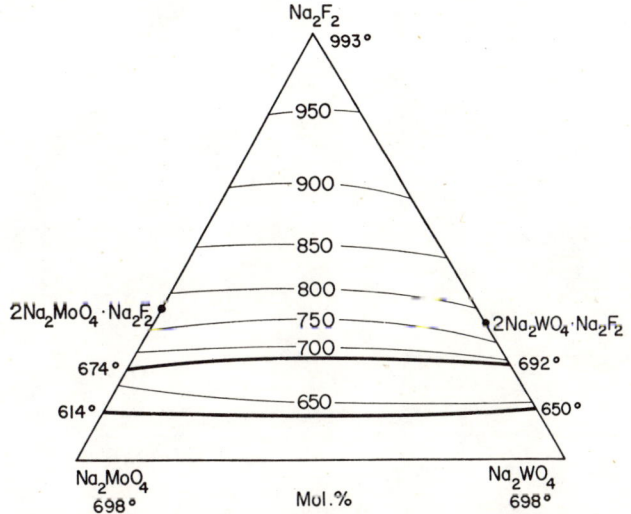

FIG. 1726.—System NaF–Na$_2$MoO$_4$–Na$_2$WO$_4$.

Z. A. Mateĭko and G. A. Bukhalova, *Zhur. Neorg. Khim.*, **4**, 1650 (1959).

NaF–CaF$_2$–AlF$_3$–Al$_2$O$_3$

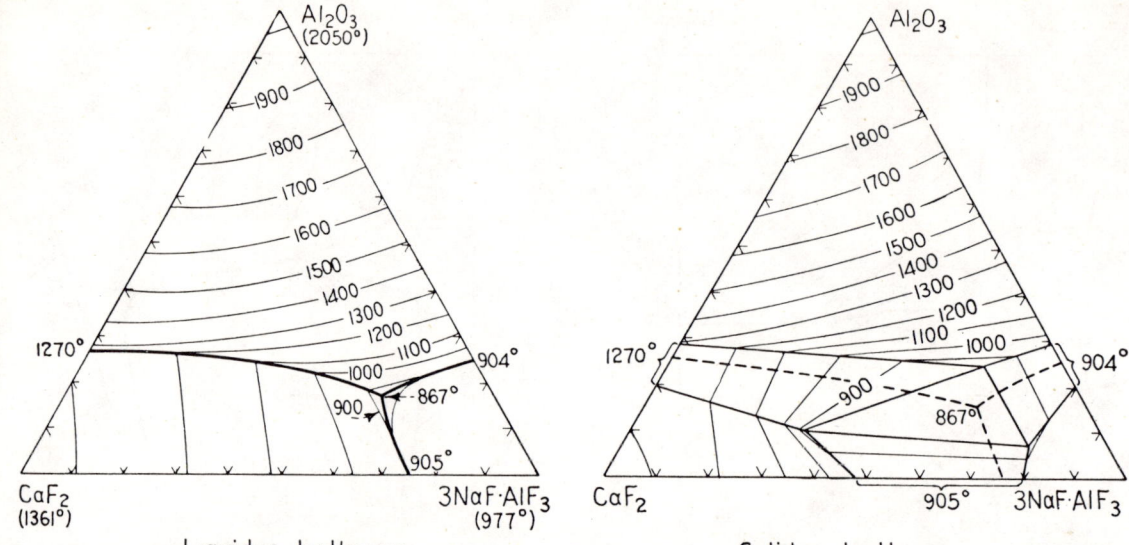

FIG. 1727.—System CaF$_2$–Al$_2$O$_3$–3NaF·AlF$_3$; isotherms of the liquidus and solidus.

Von P. Pascal, *Z. Electrochem.*, **19**, pp. 612, 613 (1913).

NaF–MgF$_2$–AlF$_3$–Al$_2$O$_3$

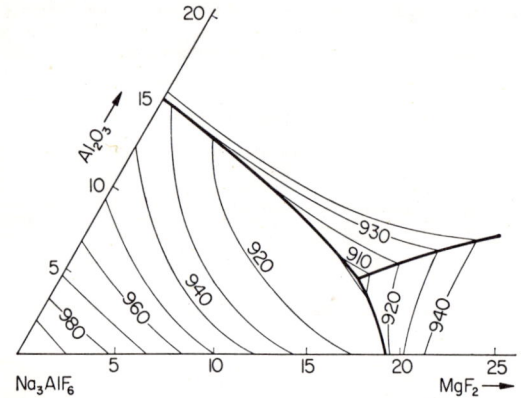

FIG. 1728.—System Na$_3$AlF$_6$–MgF$_2$–Al$_2$O$_3$.

E. Vatslavik and A. I. Belyaev, *Zhur. Neorg. Khim.*, **3**, 1044 (1958).

CaF$_2$–SiF$_4$–CaO–SiO$_2$

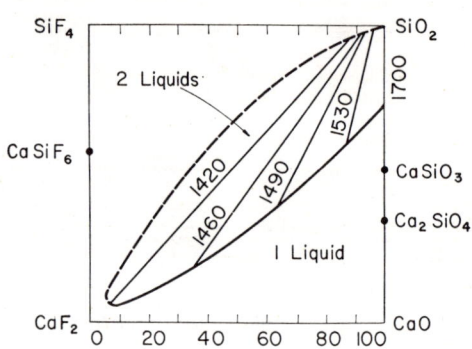

FIG. 1729.—System CaF$_2$–CaO–SiF$_4$–SiO$_2$; two-liquid region.

Ya. I. Ol'shanskiĭ, *Doklady Akad. Nauk S.S.S.R.* **114**, 1248 (1957).

III. Halides with Hydroxides

(a) Two Substances

KBr–KOH

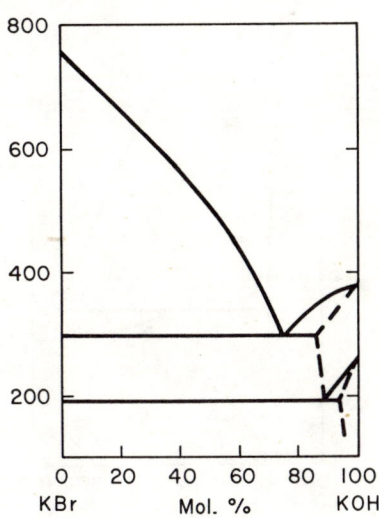

Fig. 1730.—System KBr–KOH.

Giuseppe Scarpa, *Atti reale accad. Lincei, Sez. I*, **24**, 746 (1915).

LiBr–LiOH

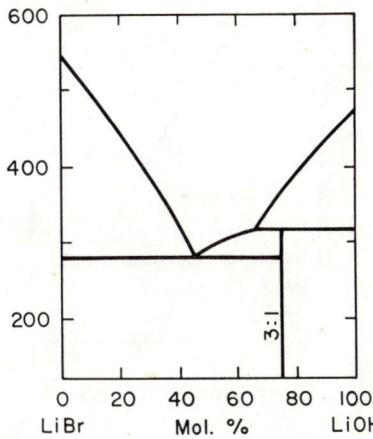

Fig. 1731.—System LiBr–LiOH.

Giuseppe Scarpa, *Atti reale accad. Lincei, Sez. II*, **24**, 480 (1915).

NaBr–NaOH

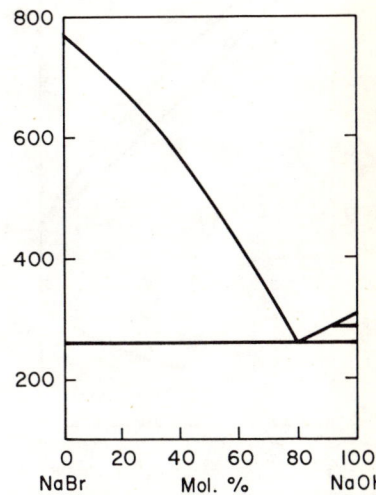

Fig. 1732.—System NaBr–NaOH.

Giuseppe Scarpa, *Atti reale accad. Lincei, Sez. I*, **24**, 961 (1915).

KCl–KOH

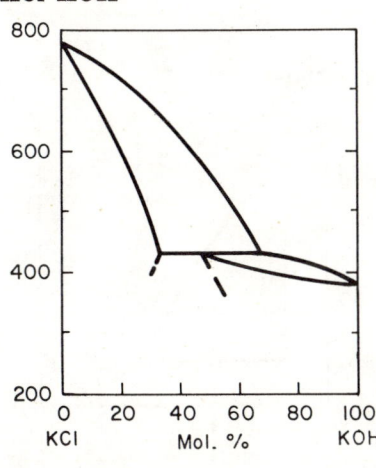

Fig. 1733.—System KCl–KOH.

Giuseppe Scarpa, *Atti reale accad. Lincei, Sez. I*, **24**, 744 (1915).

LiCl–LiOH

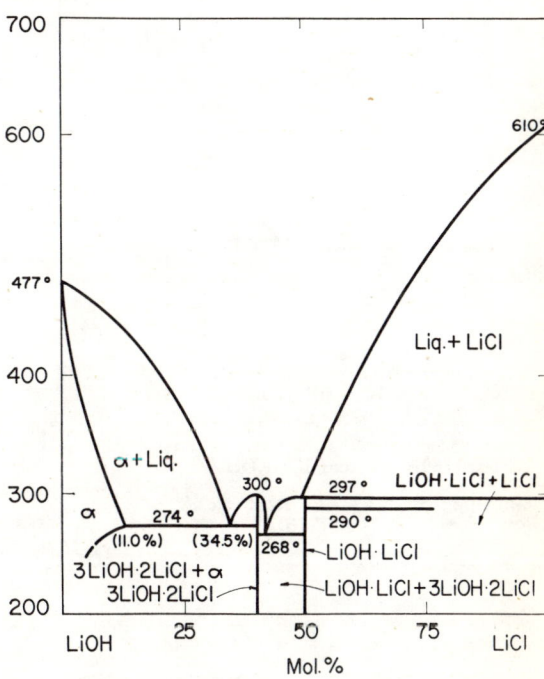

Fig. 1734.—System LiCl–LiOH.

N. A. Reshetnikov and G. M. Unzhakov, *Zhur. Neorg. Khim.*, **3**, 1435 (1958).

NaCl–NaOH

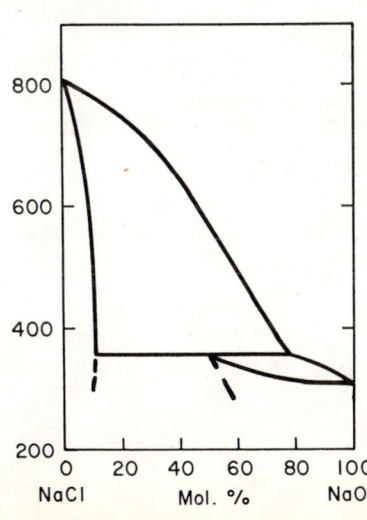

Fig. 1735.—System NaCl–NaOH.

Giuseppe Scarpa, *Atti reale accad. Lincei, Sez. I*, **24**, 957 (1915).

KF–KOH

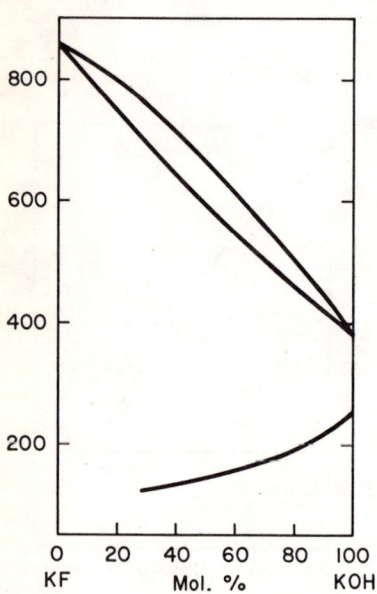

FIG. 1736.—System KF–KOH.

Giuseppe Scarpa, *Atti reale accad. Lincei, Sez. I*, **24**, 744 (1915).

LiF–LiOH

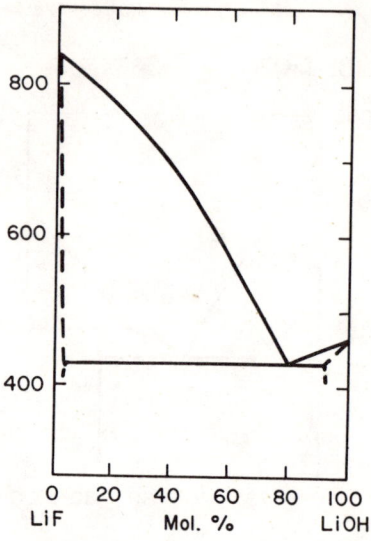

FIG. 1737.—System LiF–LiOH.

Giuseppe Scarpa, *Atti reale accad. Lincei, Sez. II*, **24**, 478 (1915).

NaF–NaOH

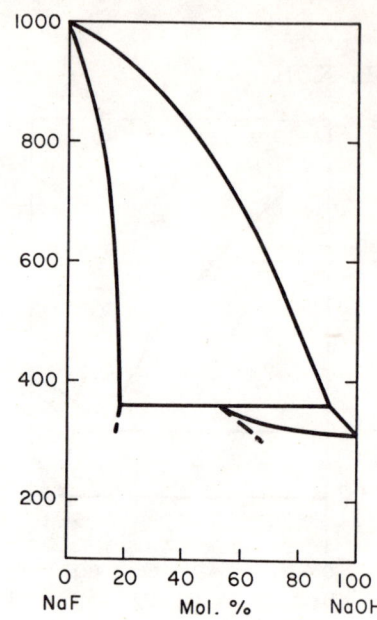

FIG. 1738.—System NaF–NaOH.

Giuseppe Scarpa, *Atti reale accad. Lincei, Sez. I*, **24**, 957 (1915).

KI–KOH

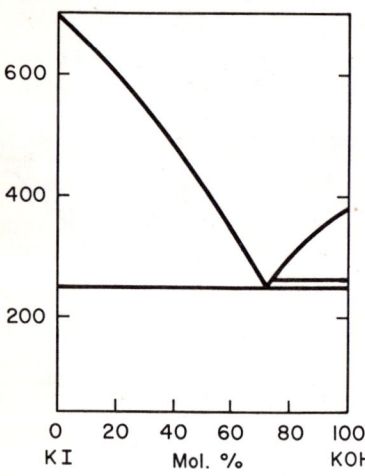

FIG. 1739.—System KI–KOH.

Giuseppe Scarpa, *Atti reale accad. Lincei, Sez. I*, **24**, 746 (1915).

LiI–LiOH

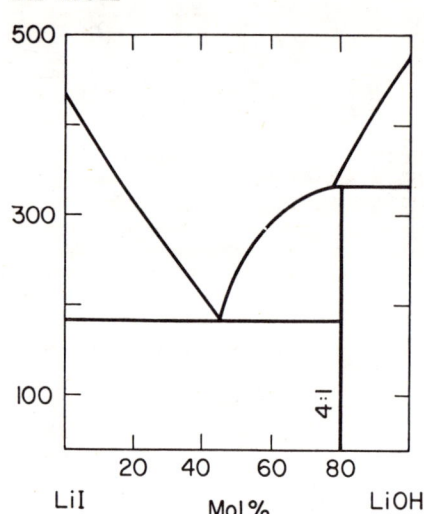

FIG. 1740.—System LiI–LiOH.

Giuseppe Scarpa, *Atti reale accad. Lincei, Sez. II*, **24**, 480 (1915).

NaI–NaOH

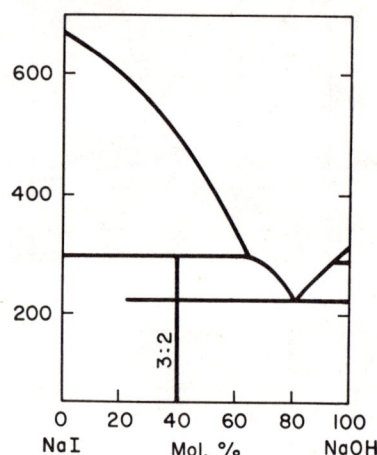

FIG. 1741.—System NaI–NaOH.

Giuseppe Scarpa, *Atti reale accad. Lincei, Sez. I*, **24**, 961 (1915).

IV. Halides with Nitrates

(a) Two Substances

AgBr–AgNO₃

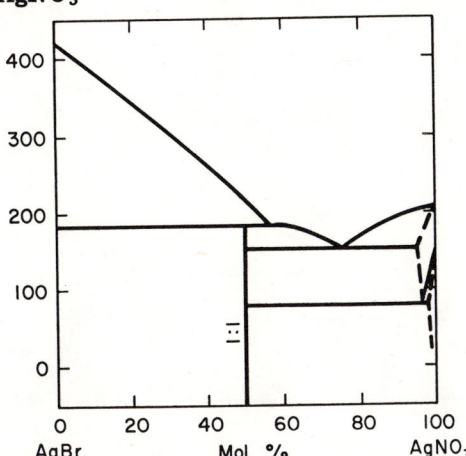

FIG. 1742.—System AgBr–AgNO₃.

Giuseppe Scarpa, *Atti reale accad. Lincei, Sez. II*, **22**, 458 (1913).

KBr–AgNO₃

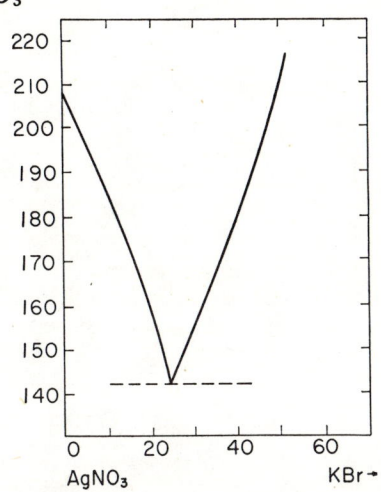

FIG. 1743.—System AgNO₃–KBr.

Iw. Kablukov, *Z. physik. Chem.*, **65**, 125 (1909).

AgCl–AgNO₃

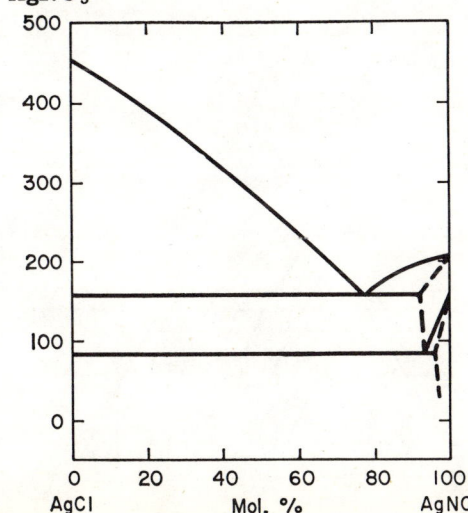

FIG. 1744.—System AgCl–AgNO₃.

Giuseppe Scarpa, *Atti reale accad. Lincei, Sez. II*, **22**, 457 (1913).

NH₄Cl–NH₄NO₃

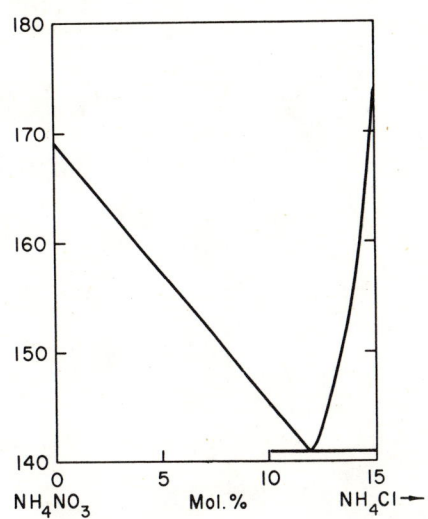

FIG. 1745.—System NH₄Cl–NH₄NO₃.

E. P. Perman, *J. Chem. Soc.*, **121**, 2475 (1922).

AgI–AgNO₃

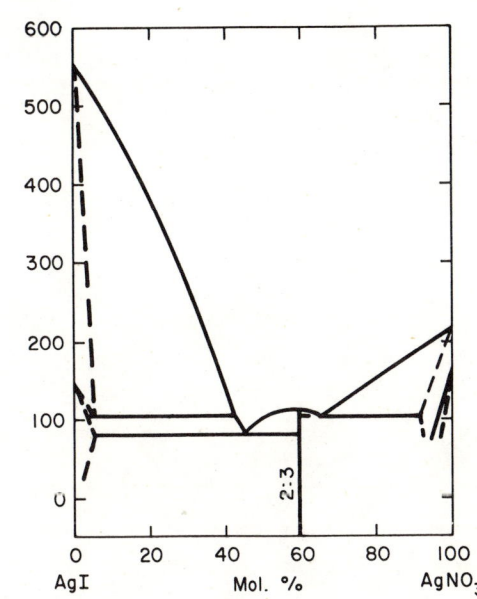

FIG. 1746.—System AgI–AgNO₃.

Giuseppe Scarpa, *Atti reale accad. Lincei, Sez. II*, **22**, 459 (1913).

TlI–TlNO₃

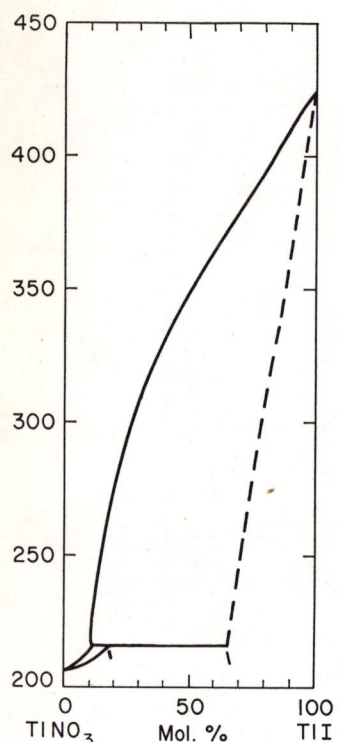

Fig. 1747.—System TlI–TlNO₃.

C. van Eyk, *Proc. Acad. Sci. Amsterdam*, **3**, 99 (1901).

(b) Three Substances

AgBr–AgCl–AgNO₃

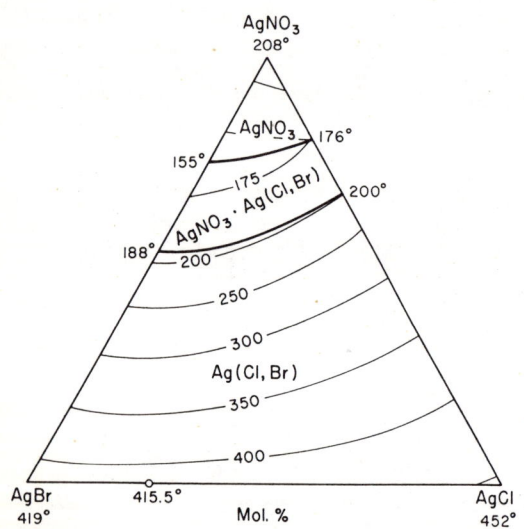

Fig. 1748.—System AgBr–AgCl–AgNO₃.

G. M. Lifshits, *Zhur. Obshcheĭ Khim.*, **25**, 2419 (1955).

AgBr–NaCl–KNO₃

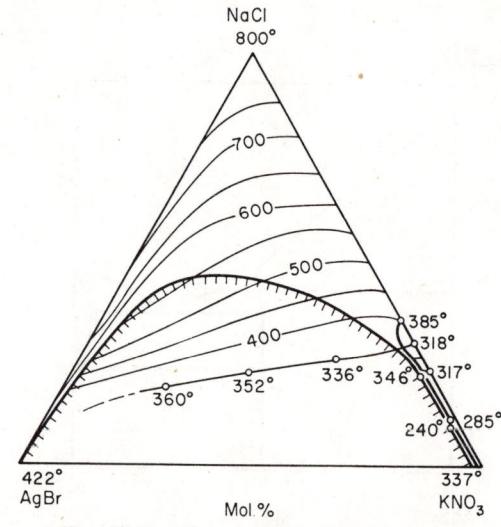

Fig. 1749.—System AgBr–KNO₃–NaCl.

N. S. Dombrovskaya and E. A. Alekseeva, *Zhur. Neorg. Khim.*, **1**, 2062 (1956).

KBr–KCl–KNO₃

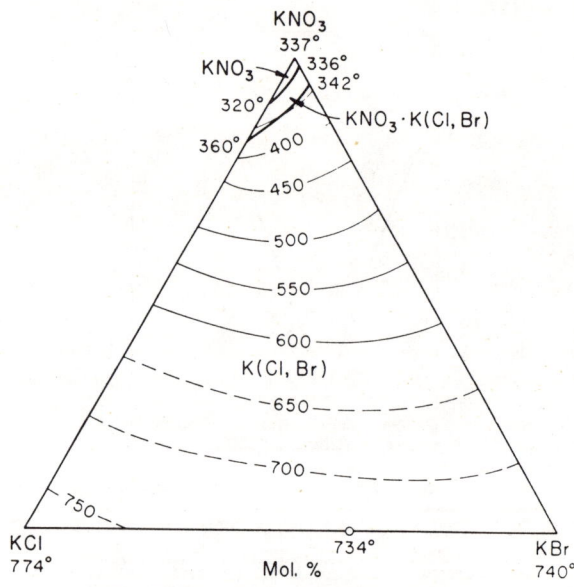

Fig. 1750.—System KBr–KCl–KNO₃.

G. M. Lifshits, *Zhur. Obshcheĭ Khim.*, **25**, 2417 (1955).

NaBr–KCl–AgNO₃

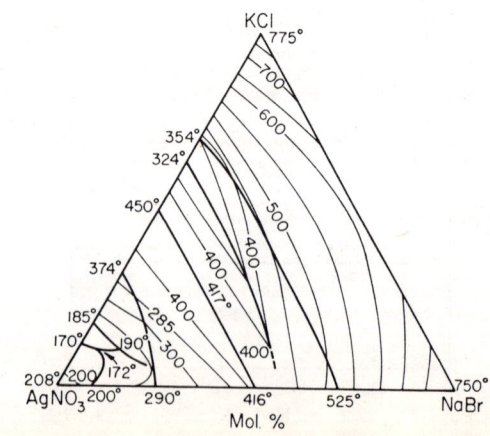

Fig. 1751.—System AgNO₃–KCl–NaBr.

N. S. Dombrovskaya and E. A. Alekseeva, *Zhur. Neorg. Khim.*, **1**, 2065 (1956).

(c) Four Substances

KBr–NaBr–KNO₃–NaNO₃

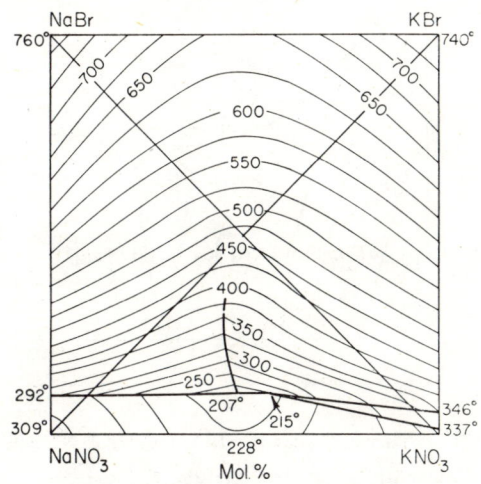

Fig. 1752.—System NaBr–NaNO₃–KBr–KNO₃.

R. N. Nyankovskaya, *Izvest. Sektora Fiz.-Khim. Anal., Inst. Obshcheĭ Neorg. Khim., Akad. Nauk S.S.S.R.*, **21**, 261 (1952).

CsCl–RbCl–CsNO₃–RbNO₃

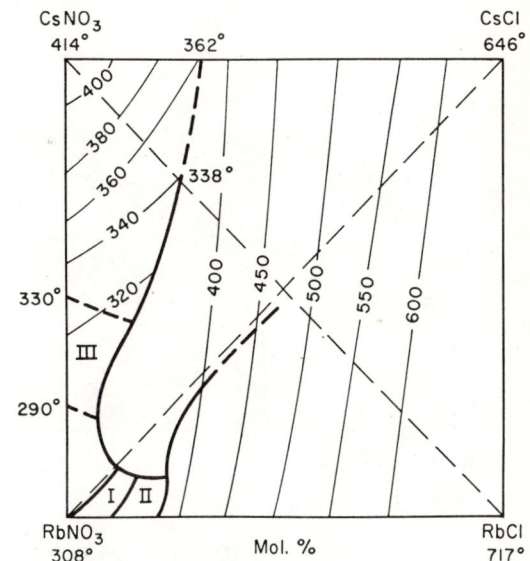

Fig. 1754.—System CsCl–CsNO₃–RbCl–RbNO₃. I = α-mRbNO₃·RbCl; II = β-mRbNO₃·nRbCl; III = homeomorphic transformation.

V. P. Blidin, *Izvest. Sektora Fiz.-Khim. Anal., Inst. Obshcheĭ Neorg. Inst., Akad. Nauk S.S.S.R.*, **23**, 237 (1953).

AgCl–LiCl–AgNO₃–LiNO₃

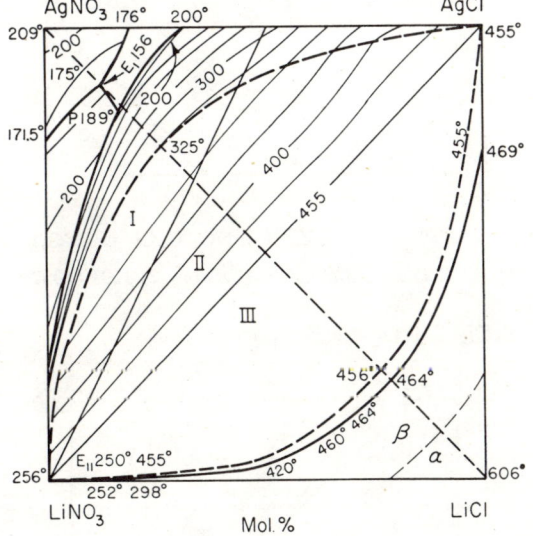

Fig. 1753.—System AgCl–AgNO₃–LiCl–LiNO₃; reciprocal.

M. A. Zakharchenko and A. G. Bergman, *Sbornik Stateĭ Obshcheĭ Khim., Akad. Nauk S.S.S.R.*, **1**, 131 (1953).

KCl–NaCl–KNO₃–NaNO₃

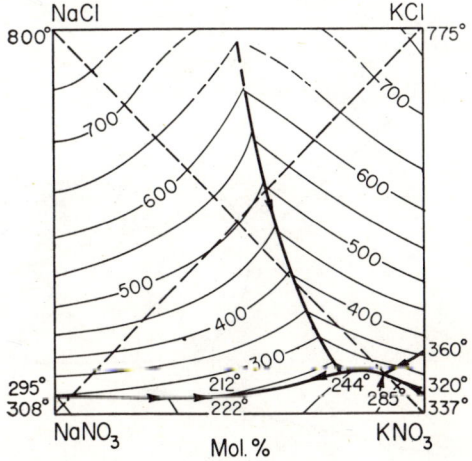

Fig. 1755.—System NaCl–NaNO₃–KCl–KNO₃.

R. N. Nyankovskaya, *Izvest. Sektora Fiz.-Khim. Anal., Inst. Obshcheĭ Neorg. Khim., Akad. Nauk S.S.S.R.*, **21**, 259 (1952).

KCl–NH₄Cl–KNO₃–NH₄NO₃

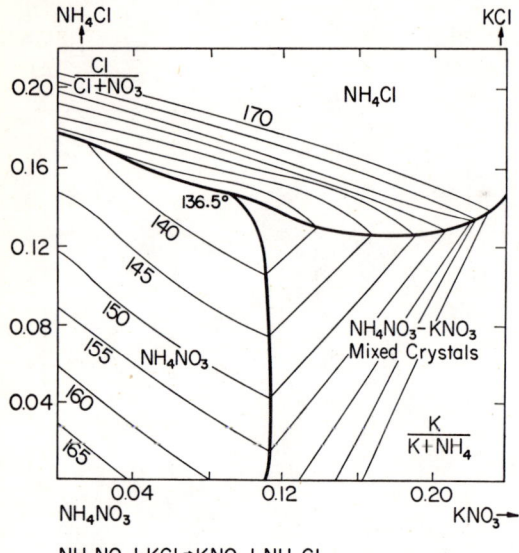

Fig. 1756.—System KCl–KNO₃–NH₄Cl–NH₄NO₃.

E. P. Perman and H. L. Saunders, *J. Chem. Soc.*, **123**, 846 (1923).

KCl–SrCl₂–KNO₃–Sr(NO₃)₂

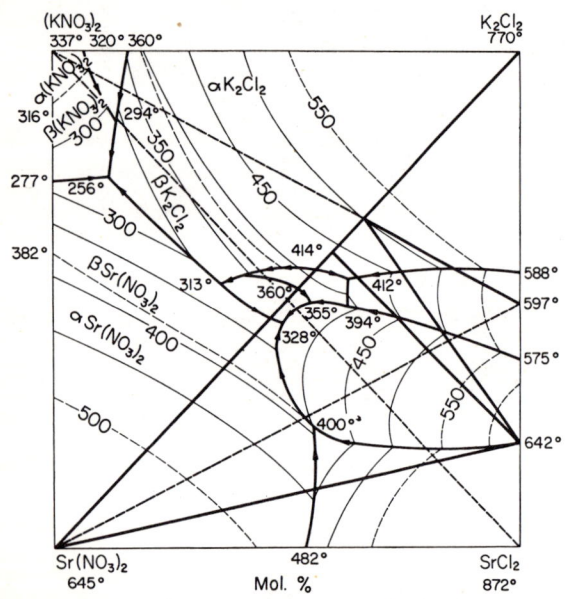

Fig. 1757.—System KCl–KNO₃–SrCl₂–Sr(NO₃)₂.

M. V. Tokareva and A. G. Bergman, *Zhur. Neorg. Khim.*, **2**, 1896 (1957).

LiCl–NH₄Cl–LiNO₃–NH₄NO₃

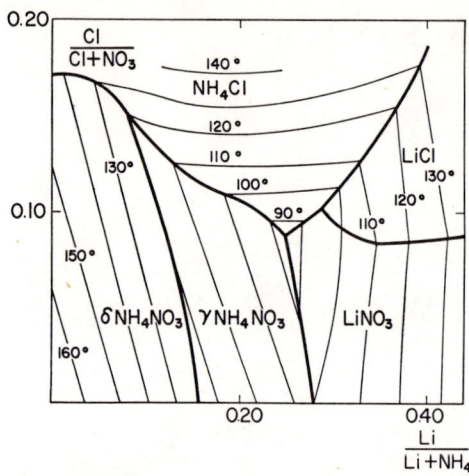

Fig. 1758.—System LiCl–LiNO₃–NH₄Cl–NH₄NO₃.

E. P. Perman and W. R. Harrison, *J. Chem. Soc.*, **125 II**, 1710 (1924).

LiCl–CaCl₂–LiNO₃–Ca(NO₃)₂

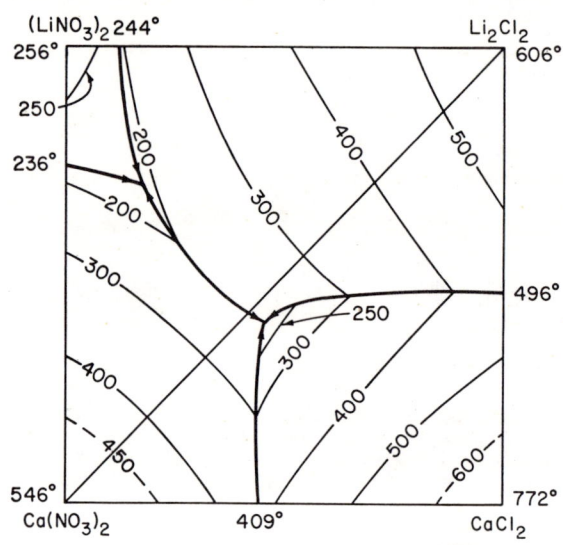

Fig. 1759.—System LiCl–LiNO₃–CaCl₂–Ca(NO₃)₂.

M. V. Tokareva, *Zhur. Neorg. Khim.*, **2**, 1591 (1957).

LiCl–SrCl₂–LiNO₃–Sr(NO₃)₂

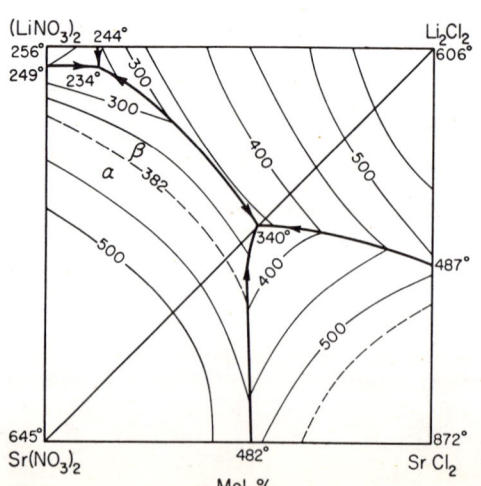

Fig. 1760.—System (LiCl)₂–(LiNO₃)₂–SrCl₂–Sr(NO₃)₂.

M. V. Tokareva and A. G. Bergman, *Zhur. Neorg. Khim.*, **1**, 2574 (1956).

NaCl–NH$_4$Cl–NaNO$_3$–NH$_4$NO$_3$

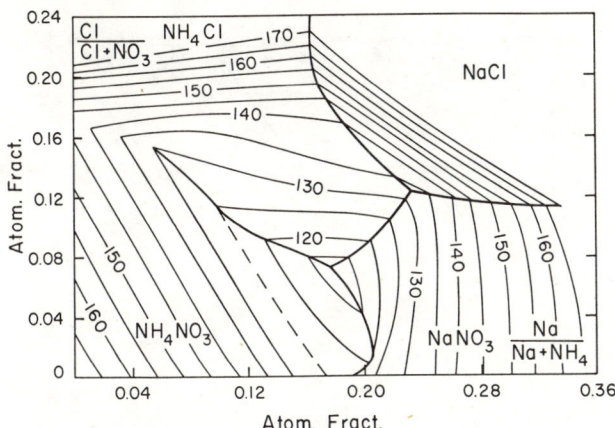

Fig. 1761.—System NaCl–NaNO$_3$–NH$_4$Cl–NH$_4$NO$_3$.

E. P. Perman, *J. Chem. Soc.*, **121**, 2478 (1922).

NaCl–BaCl$_2$–NaNO$_3$–Ba(NO$_3$)$_2$

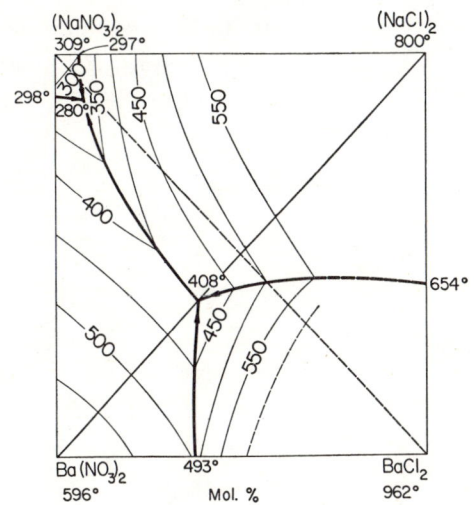

Fig. 1762.—System NaCl–NaNO$_3$–BaCl$_2$–Ba(NO$_3$)$_2$.

E. I. Speranskaya, *Izvest. Sektora Fiz.-Khim. Anal., Inst. Obshcheĭ Neorg. Khim., Akad. Nauk S.S.S.R.*, **24**, 218 (1954).

NaCl–CaCl$_2$–NaNO$_3$–Ca(NO$_3$)$_2$

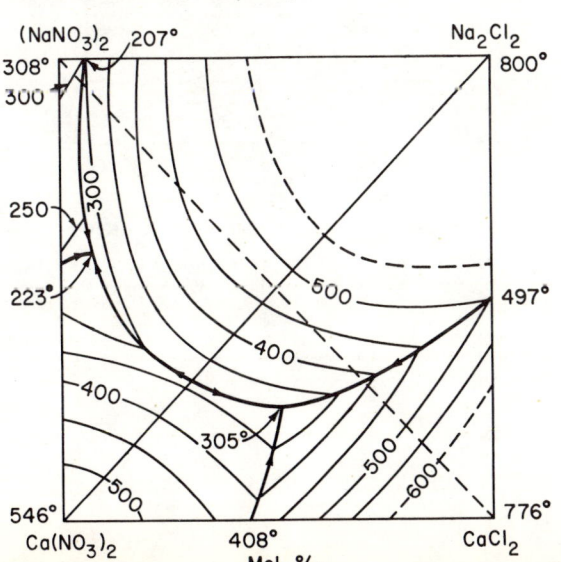

Fig. 1763.—System NaCl–NaNO$_3$–CaCl$_2$–Ca(NO$_3$)$_2$.

M. V. Tokareva, A. G. Bergman, and S. S. Kayalova, *Zhur. Neorg. Khim.*, **3**, 1911 (1958).

NaCl–SrCl$_2$–NaNO$_3$–Sr(NO$_3$)$_2$

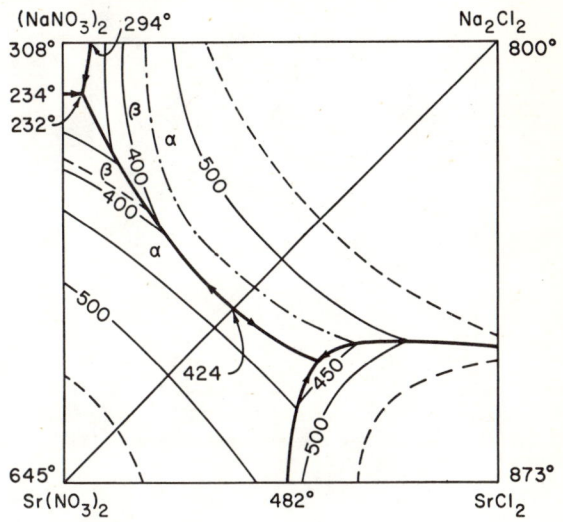

Fig. 1764.—System NaCl–NaNO$_3$–SrCl$_2$–Sr(NO$_3$)$_2$.

M. V. Tokareva, A. G. Bergman, and S. S. Kayalova, *Zhur. Neorg. Khim.*, **3**, 1913 (1958).

BaCl$_2$–CaCl$_2$–Ba(NO$_3$)$_2$–Ca(NO$_3$)$_2$

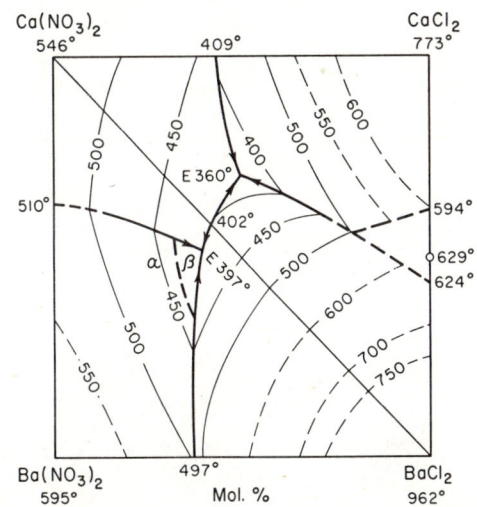

Fig. 1765.—System BaCl$_2$–Ba(NO$_3$)$_2$–CaCl$_2$–Ca(NO$_3$)$_2$.

A. G. Bergman and M. V. Tokareva, *Zhur. Neorg. Khim.*, **2** [8] 1890 (1957).

KF–NaF–KNO₃–NaNO₃

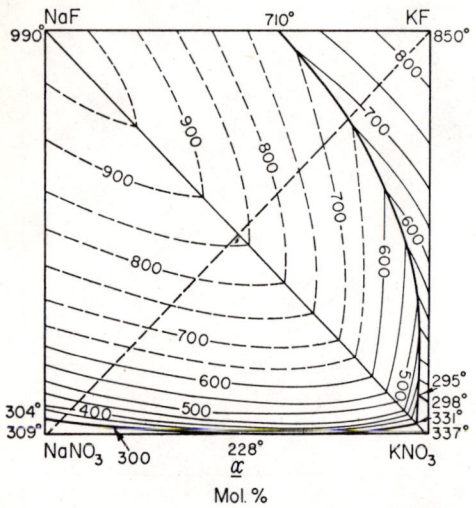

Fig. 1766.—System KF–KNO₃–NaF–NaNO₃.

R. N. Nyankovskaya and A. G. Bergman, *Izvest. Sektora Fiz.-Khim. Anal., Inst. Obshcheĭ Neorg. Khim., Akad. Nauk S.S.S.R.*, **21**, 250 (1952).

AgI–NaI–AgNO₃–NaNO₃

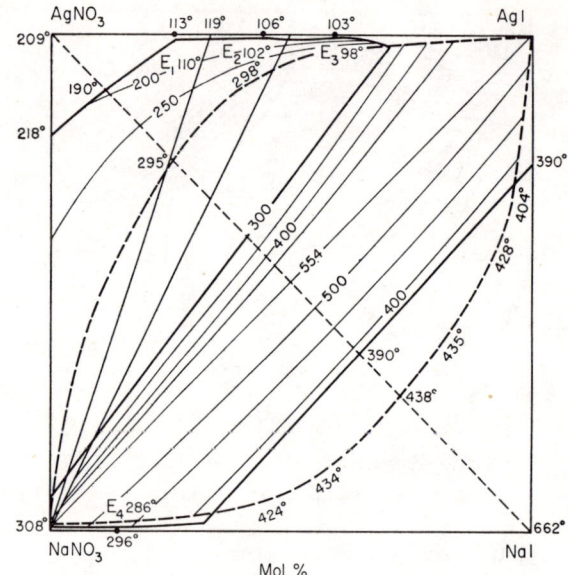

Fig. 1768.—System AgI–AgNO₃–NaI–NaNO₃,; reciprocal.

M. A. Zakharchenko and A. G. Bergman, *Zhur. Obshcheĭ Khim.*, **25**, 867 (1955).

AgI–KI–AgNO₃–KNO₃

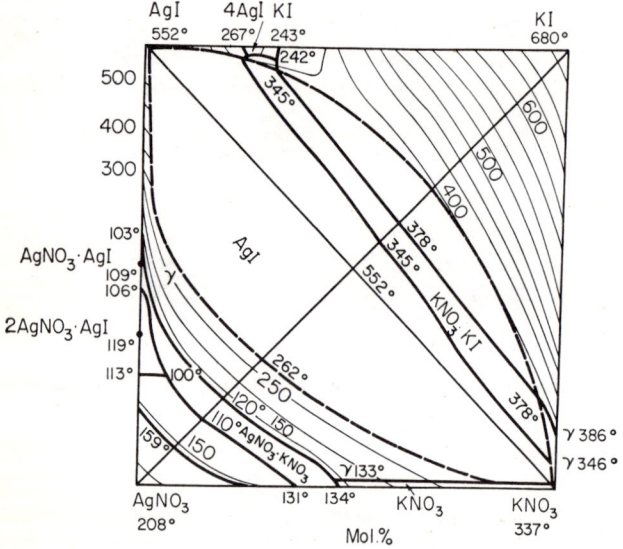

Fig. 1767.—System AgI–AgNO₃–KI–KNO₃.

N. S. Dombrovskaya and Z. A. Koloskova, *Izvest. Sektora Fiz.-Khim. Anal., Inst. Obshcheĭ Neorg. Khim., Akad. Nauk S.S.S.R.*, **22**, 183 (1953).

KI–NaI–KNO₃–NaNO₃

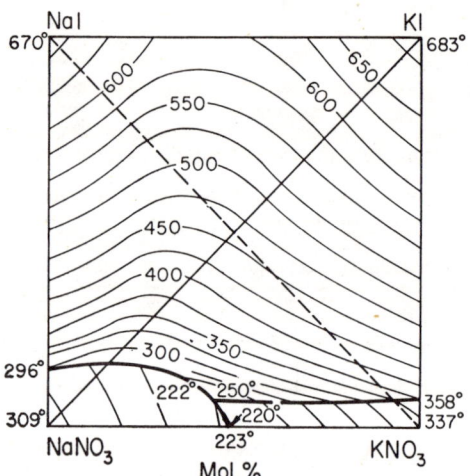

Fig. 1769.—System NaI–NaNO₃–KI–KNO₃.

R. N. Nyankovskaya, *Izvest. Sektora Fiz.-Khim. Anal., Inst. Obshcheĭ Neorg. Khim., Akad. Nauk S.S.S.R.*, **21**, 261 (1952).

V. Halides with Sulfates

(a) Two Substances

KBr–K₂SO₄

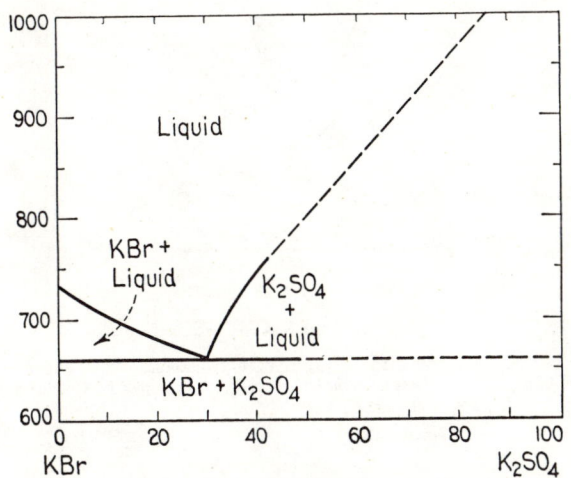

Fig. 1770.—System KBr–K₂SO₄.

R. F. Rea, *J. Am. Ceram. Soc.*, **21** [3] 100 (1938).

NaBr–Na₂SO₄

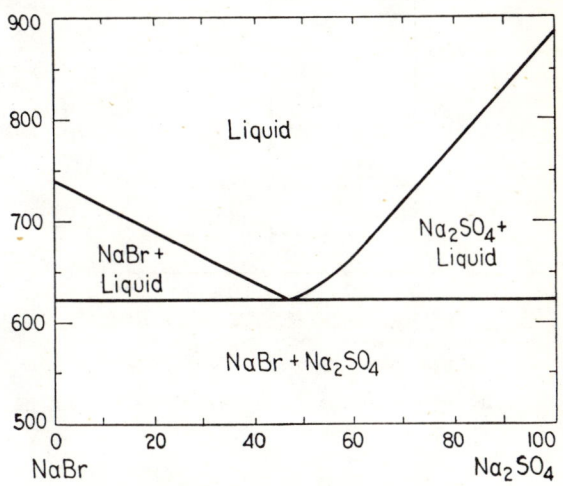

Fig. 1771.—System NaBr–Na₂SO₄.

R. F. Rea, *J. Am. Ceram. Soc.*, **21** [3] 100 (1938).

KCl–K₂SO₄

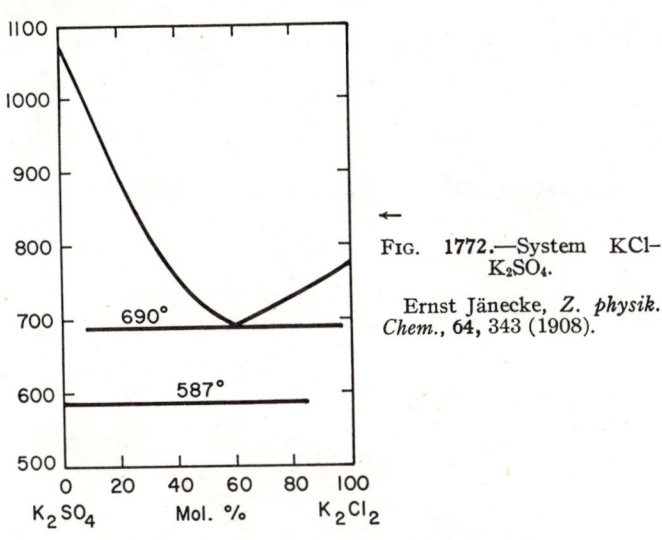

Fig. 1772.—System KCl–K₂SO₄.

Ernst Jänecke, *Z. physik. Chem.*, **64**, 343 (1908).

NaCl–Na₂SO₄

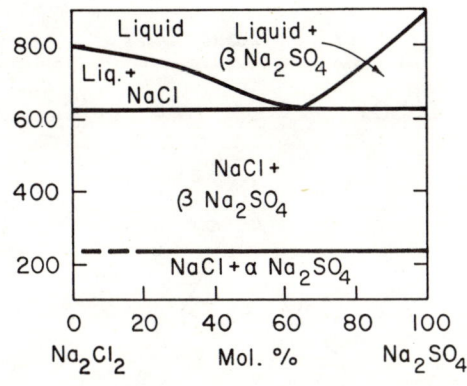

Fig. 1773.—System NaCl–Na₂SO₄.

P. I. Fedorov and K. A. Bol'shakov, *Zhur. Neorg. Khim.*, **4**, 893 (1959).

NaCl–CoSO₄

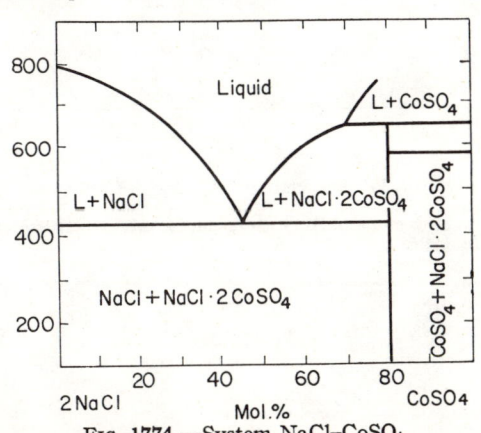

Fig. 1774.—System NaCl–CoSO₄.

P. I. Fedorov and K. A. Bol'shakov, *Zhur. Neorg. Khim.*, **4**, 893 (1959).

MgCl₂–MgSO₄

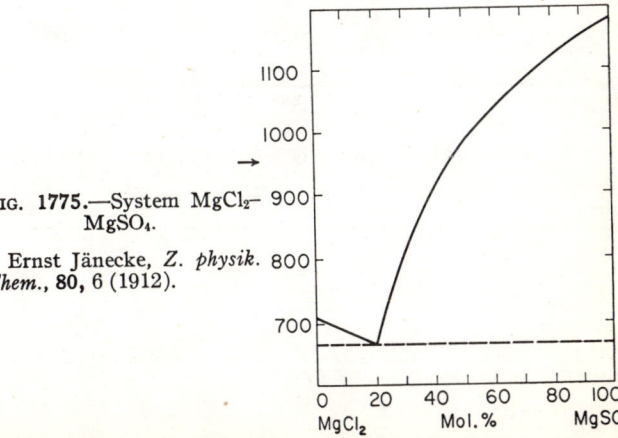

Fig. 1775.—System MgCl₂–MgSO₄.

Ernst Jänecke, *Z. physik. Chem.*, **80**, 6 (1912).

NaF–Na₂SO₄

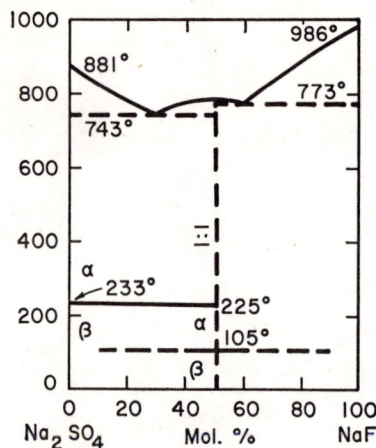

Fig. 1776.—System NaF–Na₂SO₄.

Adolph Wolters, *Neues Jahrb. Mineral., Geol.*, **30**, 64 (1910).

KCl–NaCl–Li₂SO₄

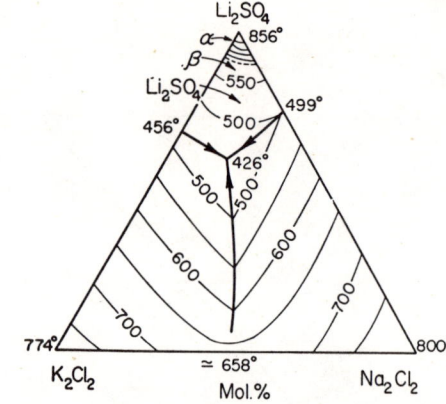

Fig. 1778.—System K_2Cl_2–Li_2SO_4–Na_2Cl_2.

E. K. Akopov and A. G. Bergman, *Izvest. Sektora Fiz.-Khim. Anal., Inst. Obshcheĭ Neorg. Khim., Akad. Nauk S.S.S.R.*, **25**, 266 (1954).

(b) Three Substances

KCl–LiCl–Li₂SO₄

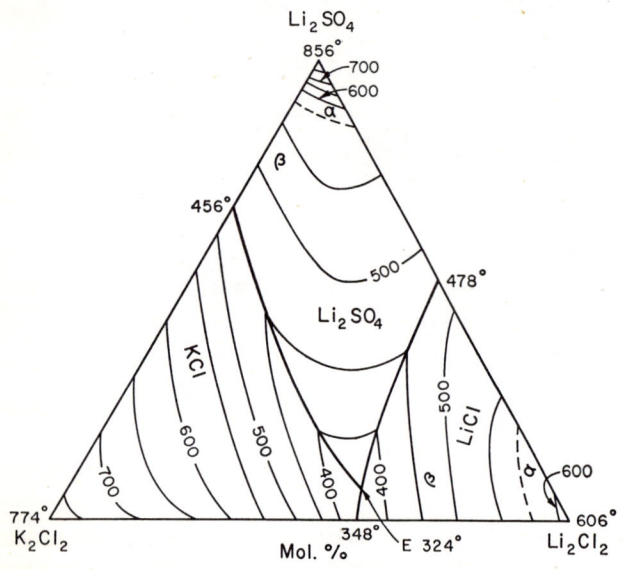

Fig. 1777.—System KCl–LiCl–Li₂SO₄.

E. K. Akopov and A. G. Bergman, *Zhur. Neorg. Khim.*, **2**, 386 (1957).

KCl–NaCl–CaSO₄

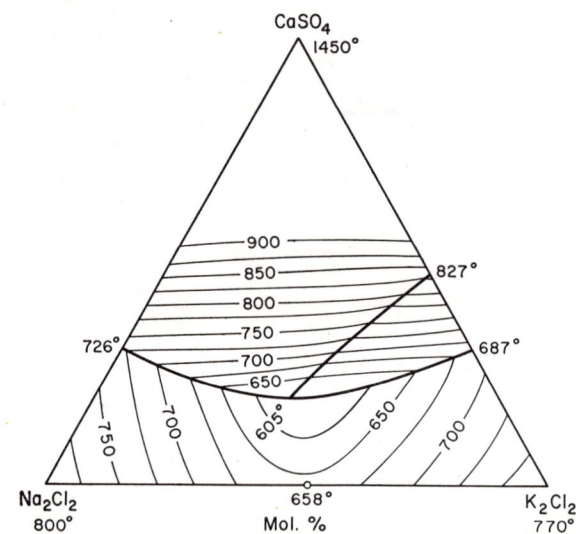

Fig. 1779.—System KCl–NaCl–CaSO₄.

V. V. Rubleva and A. G. Bergman, *Zhur. Obshcheĭ Khim.*, **26**, 654 (1956).

KCl–Li₂SO₄–Na₂SO₄

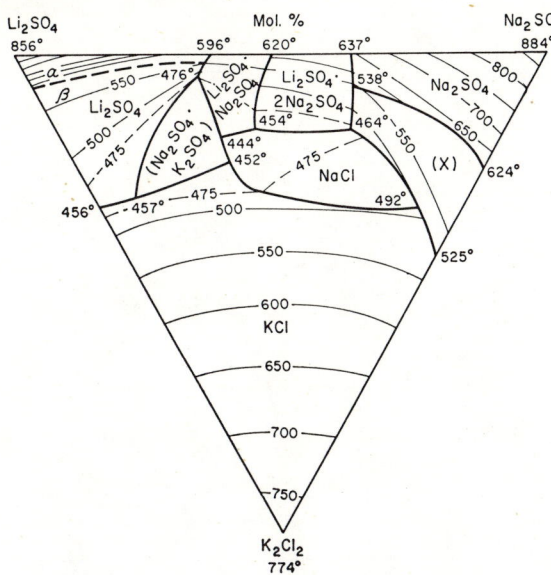

Fig. 1780.—System KCl–Li₂SO₄–Na₂SO₄.

E. K. Akopov and A. G. Bergman, *Zhur. Obshcheĭ Khim.*, **25**, 6 (1955).

NaCl–NaF–Na₂SO₄

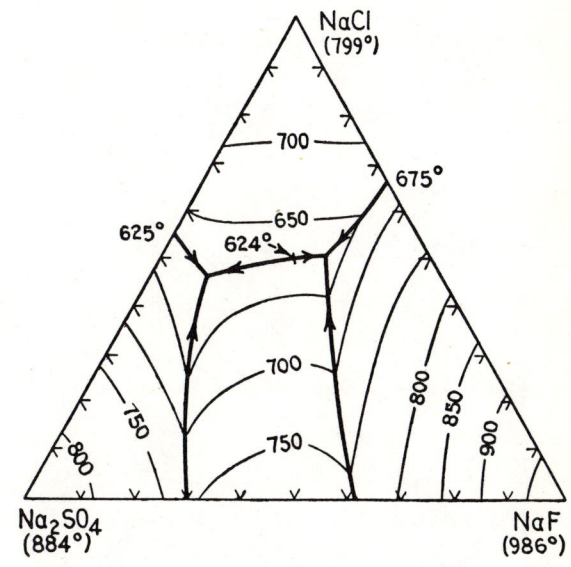

Fig. 1782.—System NaCl–NaF–Na₂SO₄.

L. A. H. Wolters, *Neues Jahrb. Mineral. Geol., Beiloge Bd.*, **30**, pp. 83, 87 (1910).

LiCl–NaCl–Li₂SO₄

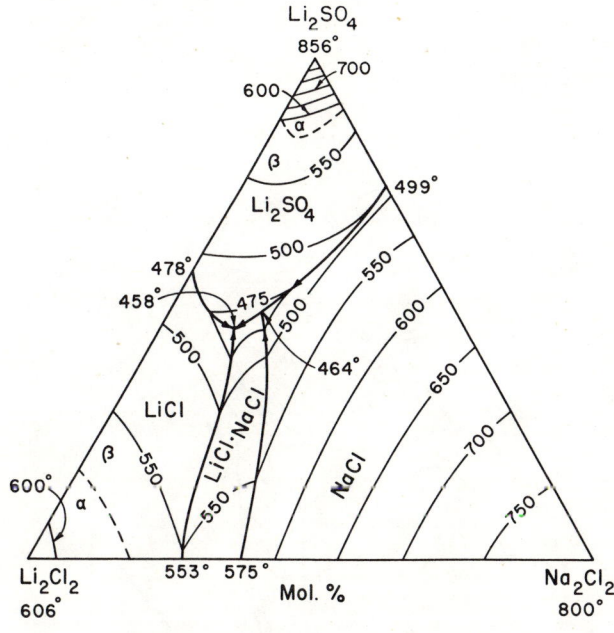

Fig. 1781.—System LiCl–NaCl–Li₂SO₄.

E. K. Akopov and A. G. Bergman, *Zhur. Neorg. Khim.*, **2**, 385 (1957).

LiF–BeF₂–Li₂SO₄

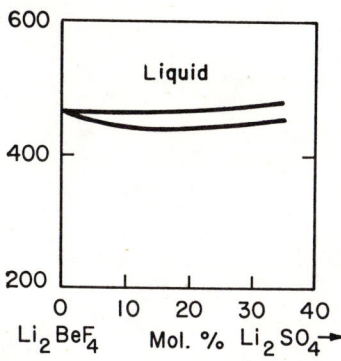

Fig. 1783.—System Li₂BeF₄–Li₂SO₄.

M. E. Levina, A. V. Novoselova, and Yu. P. Simanov, *Vestn. Mosk. Univ.*, **11**, Ser. Mat., Mekhan., Astron., Fiz. i Khim., **1**, 241 (1956).

NaF–AlF₃–Na₂SO₄

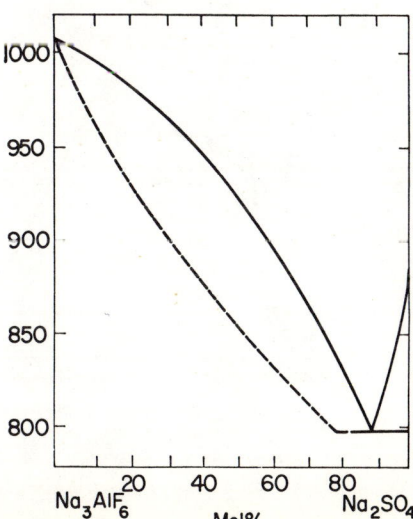

Fig. 1784.—System Na₃AlF₆–Na₂SO₄.

Kai Grjotheim, Tor Halvorsen, and Sigmund Urnes, *Can. J. Chem.*, **37**, 1171 (1959).

(c) *Four Substances*

AgCl–LiCl–Ag$_2$SO$_4$–Li$_2$SO$_4$

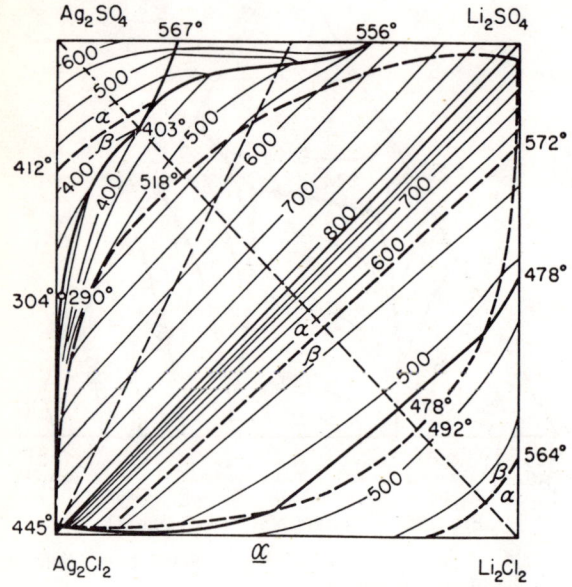

FIG. 1785.—System Ag$_2$Cl$_2$–Ag$_2$SO$_4$–Li$_2$Cl$_2$–Li$_2$SO$_4$.

D. S. Lesnykh and A. G. Bergman, *Zhur. Obshchei Khim.*, **23**, 378 (1953).

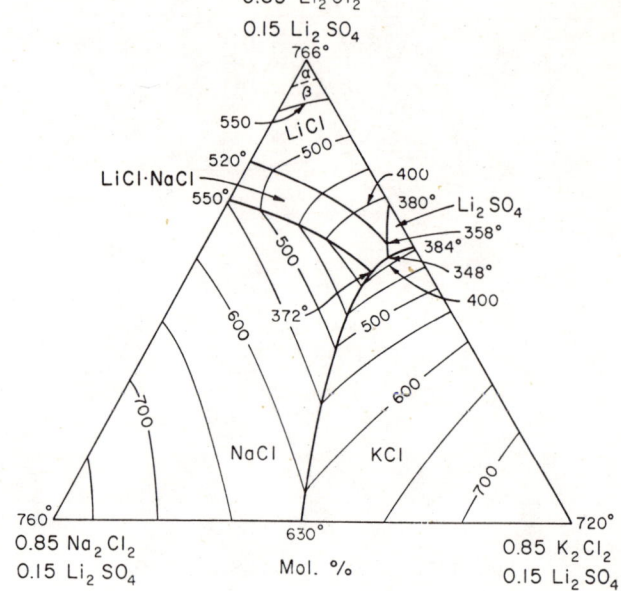

FIG. 1787.—System KCl–LiCl–NaCl–Li$_2$SO$_4$; plane with constant Li$_2$SO$_4$ at 15 mol. %.

E. K. Akopov and A. G. Bergman, *Zhur. Neorg. Khim.*, **2**, 391 (1957).

KCl–LiCl–NaCl–Li$_2$SO$_4$

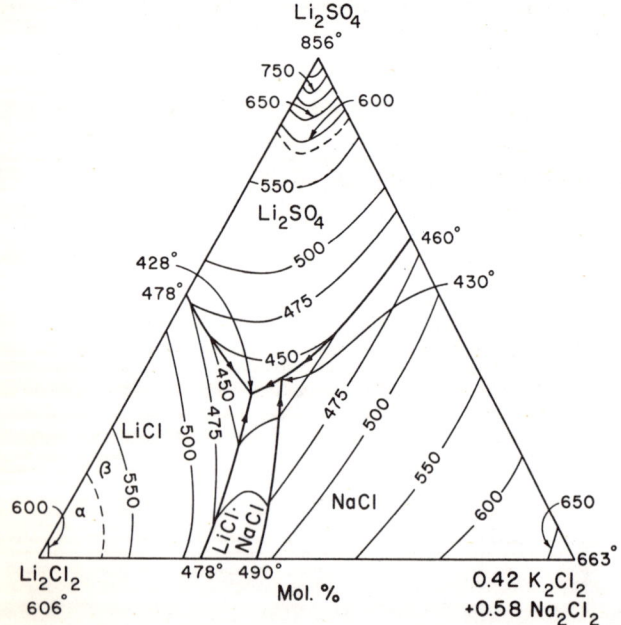

FIG. 1786.—System LiCl–Li$_2$SO$_4$–(0.42 KCl, 0.58 NaCl).

E. K. Akopov and A. G. Bergman, *Zhur. Neorg. Khim.*, **2**, 390 (1957).

KCl–LiCl–K$_2$SO$_4$–Li$_2$SO$_4$

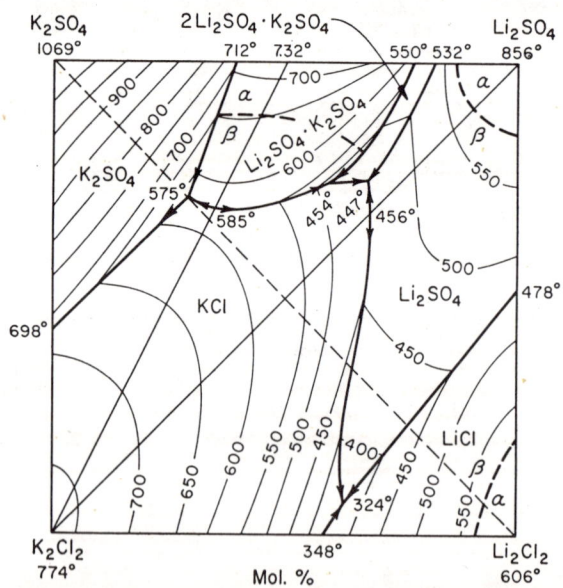

FIG. 1788.—System KCl–K$_2$SO$_4$–LiCl–Li$_2$SO$_4$; reciprocal.

E. K. Akopov and A. G. Bergman, *Doklady Akad. Nauk S.S.S.R.*, **102**, 82 (1955).

KCl–NaCl–K$_2$SO$_4$–Na$_2$SO$_4$

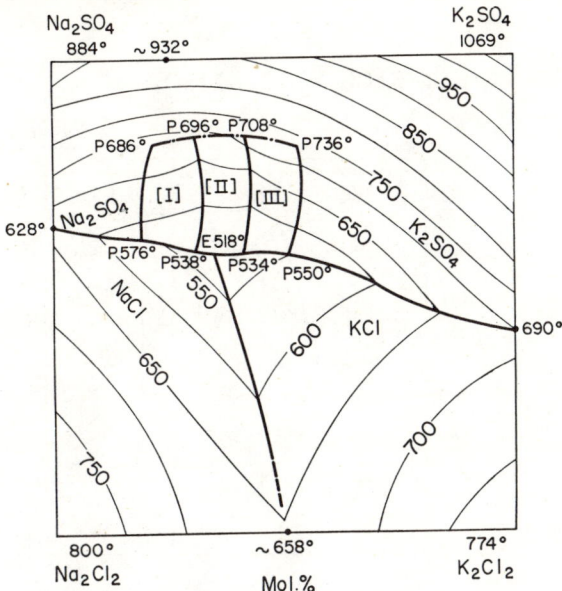

Fig. 1789.—System KCl–K$_2$SO$_4$–NaCl–Na$_2$SO$_4$. [I], [II], and [III] are unknown interior phases.

E. K. Akopov and A. G. Bergman, *Zhur. Neorg. Khim.*, **4** [7] 1655 (1959).

KCl–CaCl$_2$–K$_2$SO$_4$–CaSO$_4$

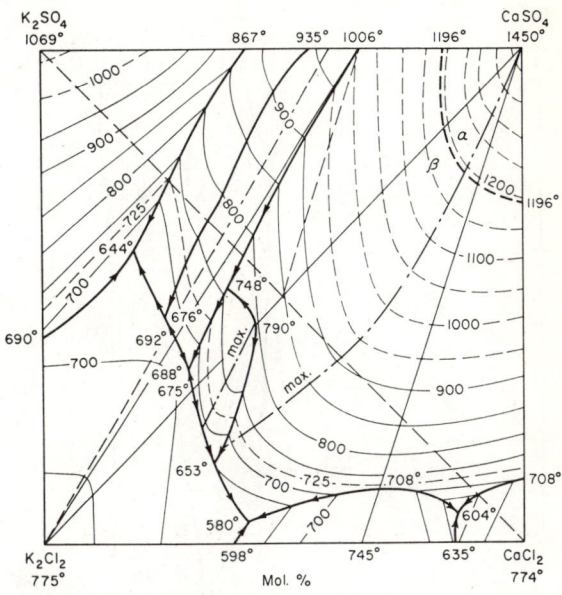

Fig. 1790.—System KCl–K$_2$SO$_4$–CaCl$_2$–CaSO$_4$; reciprocal.

M. S. Golubeva and A. G. Bergman, *Zhur. Obshcheĭ Khim.*, **26**, 330 (1956).

KCl–CdCl$_2$–K$_2$SO$_4$–CdSO$_4$

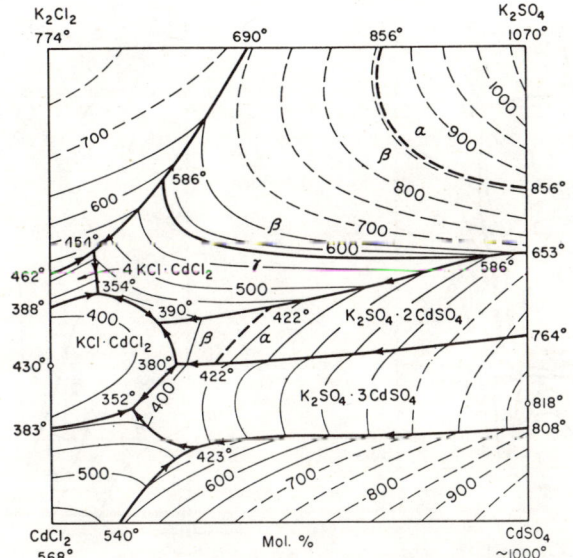

Fig. 1791.—System KCl–K$_2$SO$_4$–CdCl$_2$–CdSO$_4$; adiagonal reciprocal.

A. G. Bergman and E. L. Bakumskaya, *Zhur. Obshcheĭ Khim.*, **26**, 635 (1956).

KCl–CoCl₂–K₂SO₄–CoSO₄

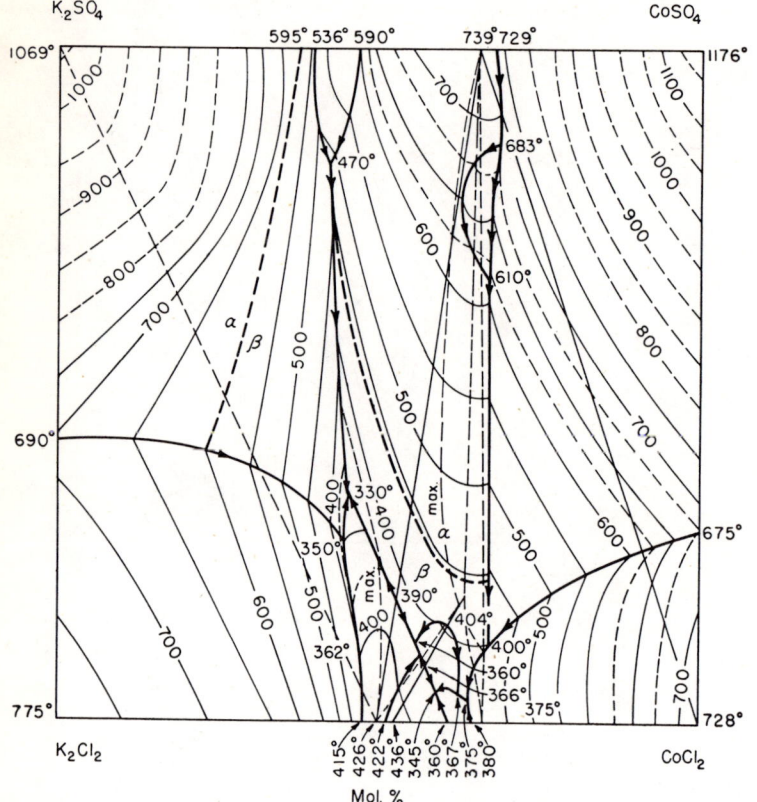

FIG. 1792.—System K₂Cl₂–K₂SO₄–CoCl₂–CoSO₄.

M. S. Golubeva and A. G. Bergman, *Doklady Akad. Nauk S.S.S.R.*, **89**, 690 (1953).

KCl–MgCl₂–K₂SO₄–MgSO₄

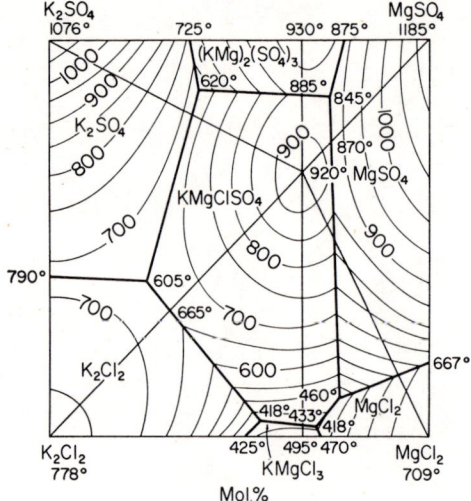

FIG. 1793.—System KCl–K₂SO₄–MgCl₂–MgSO₄.

Ernst Jänecke, *Z. physik. Chem.*, **80**, 11 (1912).

KCl–ZnCl₂–K₂SO₄–ZnSO₄

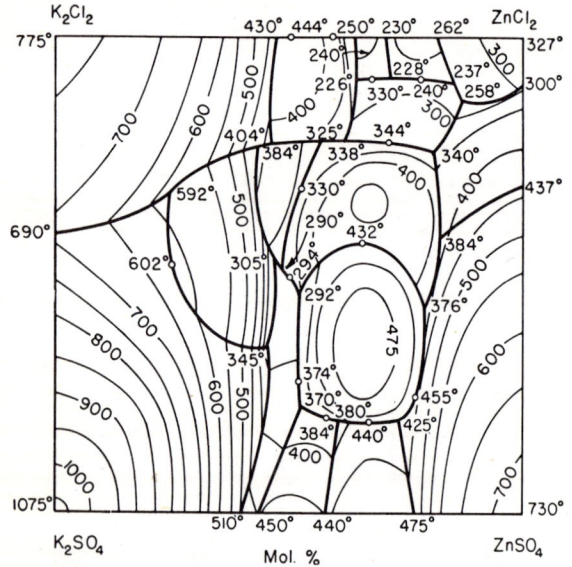

FIG. 1794.—System K₂Cl₂–K₂SO₄–ZnCl₂–ZnSO₄.

N. P. Luzhnaya, N. N. Evseeva, and I. P. Vereshchetina, *Zhur Neorg. Khim.*, **1** [7] 1499 (1956).

LiCl–NaCl–Li₂SO₄–Na₂SO₄

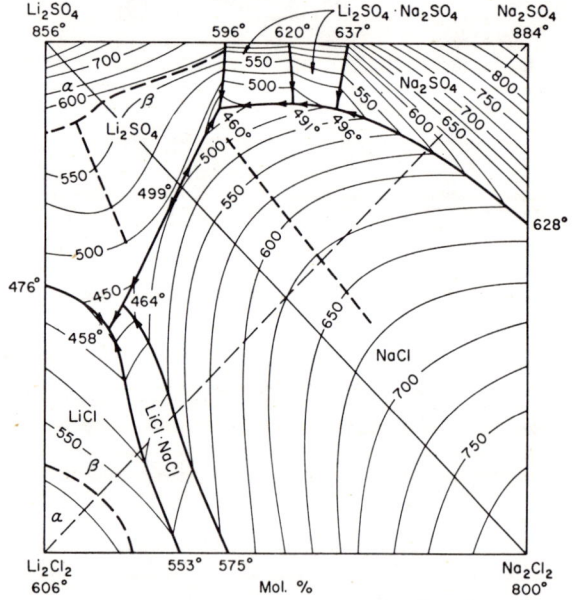

FIG. 1795.—System LiCl–Li₂SO₄–NaCl–Na₂SO₄; reciprocal.

E. K. Akopov and A. G. Bergman, *Doklady Akad. Nauk S.S.S.R.*, **102**, 82 (1955).

LiCl–TlCl–Li₂SO₄–Tl₂SO₄

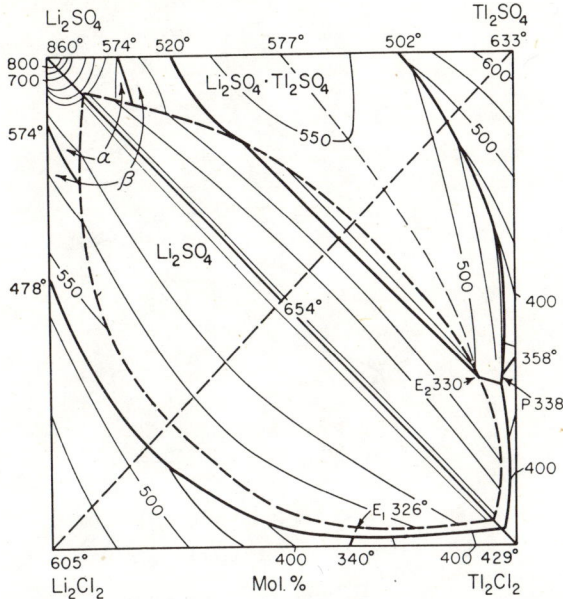

FIG. 1796.—System LiCl–Li₂SO₄–TlCl–Tl₂SO₄; reciprocal.

A. G. Bergman and M. L. Sholokhovich, *Zhur. Obshcheĭ Khim.*, **25**, 456 (1955).

LiCl–CaCl₂–Li₂SO₄–CaSO₄

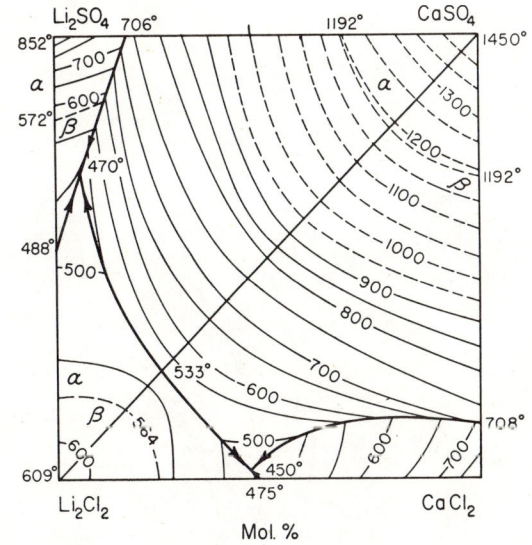

FIG. 1797.—System LiCl–Li₂SO₄–CaCl₂–CaSO₄; reciprocal.

M. S. Golubeva and A. G. Bergman, *Zhur. Obshcheĭ Khim.*, **25**, 462 (1955).

LiCl–CdCl₂–Li₂SO₄–CdSO₄

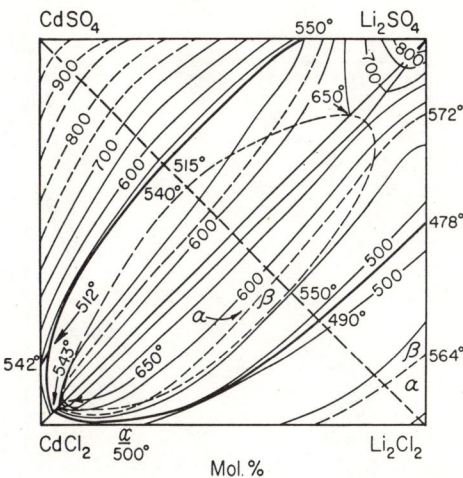

FIG. 1798.—System Li₂Cl₂–Li₂SO₄–CdCl₂–CdSO₄.

D. S. Lesnykh and A. G. Bergman, *Zhur. Obshcheĭ Khim.*, **23**, 542 (1953).

LiCl–CoCl₂–Li₂SO₄–CoSO₄

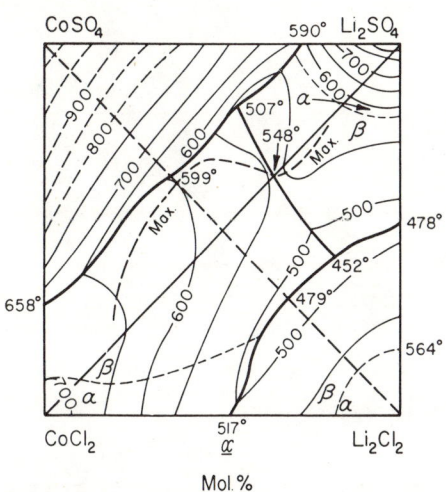

FIG. 1799.—System Li₂Cl₂–Li₂SO₄–CoCl₂–CoSO₄.

D. S. Lesnykh and A. G. Bergman, *Zhur. Obshcheĭ Khim.*, **23**, 900 (1953).

LiCl–SrCl₂–Li₂SO₄–SrSO₄

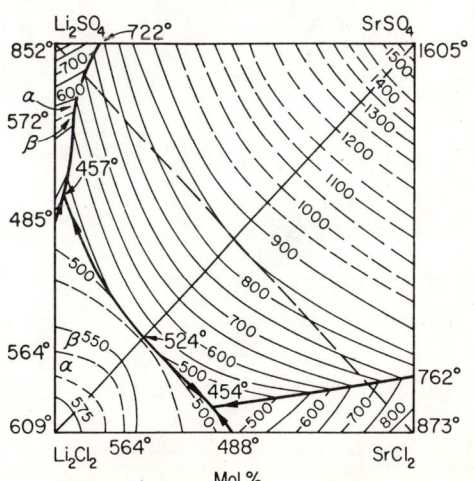

FIG. 1800.—System LiCl–Li₂SO₄–SrCl₂–SrSO₄; reciprocal.

M. S. Golubeva and A. G. Bergman, *Zhur. Obshcheĭ Khim.*, **25**, 461 (1955).

LiCl–ZnCl$_2$–Li$_2$SO$_4$–ZnSO$_4$

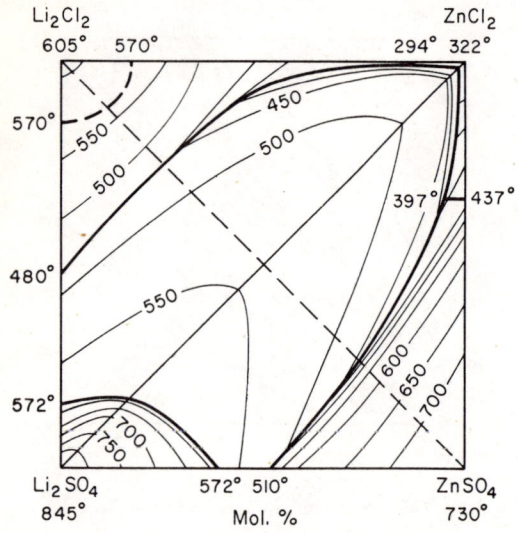

Fig. 1801.—System Li$_2$Cl$_2$–Li$_2$SO$_4$–ZnCl$_2$–ZnSO$_4$.

N. N. Evseeva and A. G. Bergman, *Zhur. Obshchei Khim.*, **21**, 1767 (1951).

NaCl–CoCl$_2$–Na$_2$SO$_4$–CoSO$_4$

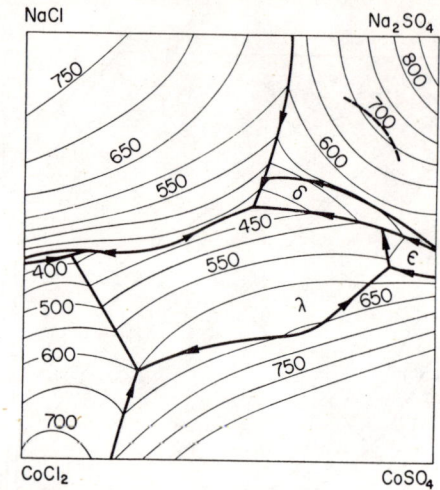

Fig. 1803.—System NaCl–Na$_2$SO$_4$–CoCl$_2$–CoSO$_4$.

P. I. Fedorov and K. A. Bol'shakov, *Zhur. Neorg. Khim.*, **4**, 895 (1959).

NaCl–CdCl$_2$–Na$_2$SO$_4$–CdSO$_4$

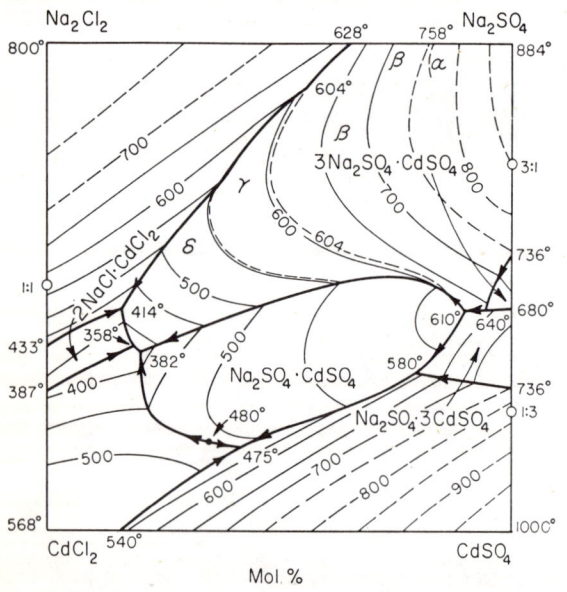

Fig. 1802.—System NaCl–Na$_2$SO$_4$–CdCl$_2$–CdSO$_4$; reciprocal.

A. G. Bergman and E. L. Bakumskaya, *Zhur. Obshchei Khim.*, **25**, 2412 (1955).

NaCl–ZnCl$_2$–Na$_2$SO$_4$–ZnSO$_4$

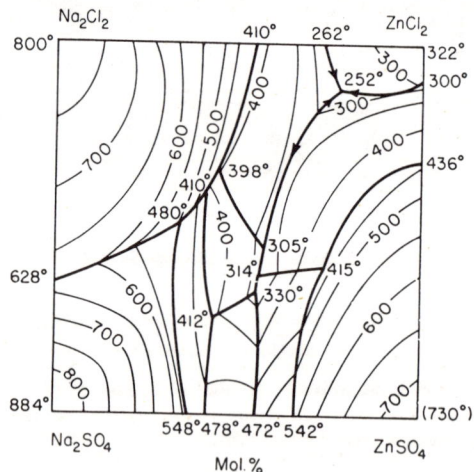

Fig. 1804.—System Na$_2$Cl$_2$–Na$_2$SO$_4$–ZnCl$_2$–ZnSO$_4$.

N. N. Evseeva and A. G. Bergman, *Izvest. Sektora Fiz.-Khim. Anal., Inst. Obshchei Neorg. Khim., Akad. Nauk S.S.S.R.*, **21**, 217 (1952).

TlCl–CdCl₂–Tl₂SO₄–CdSO₄

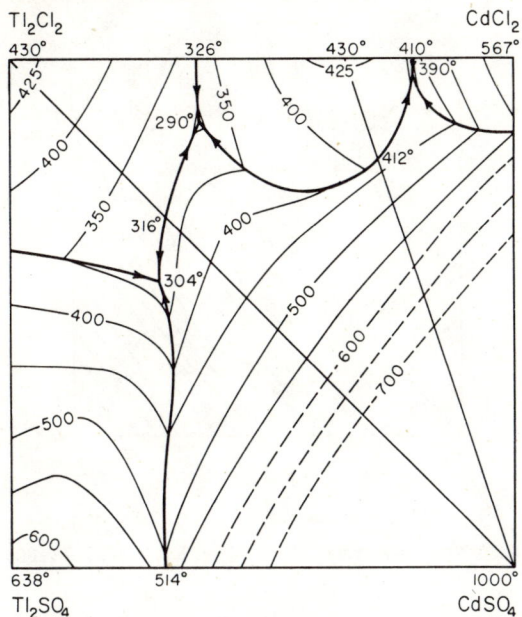

Fig. 1805.—System Tl₂Cl₂–Tl₂SO₄–CdCl₂–CdSO₄.

A. K. Sementsova, A. G. Bergman, and D. S. Lesnykh, *Zhur. Neorg. Khim.*, **1**, 168 (1956).

KF–PbF₂–K₂SO₄–PbSO₄

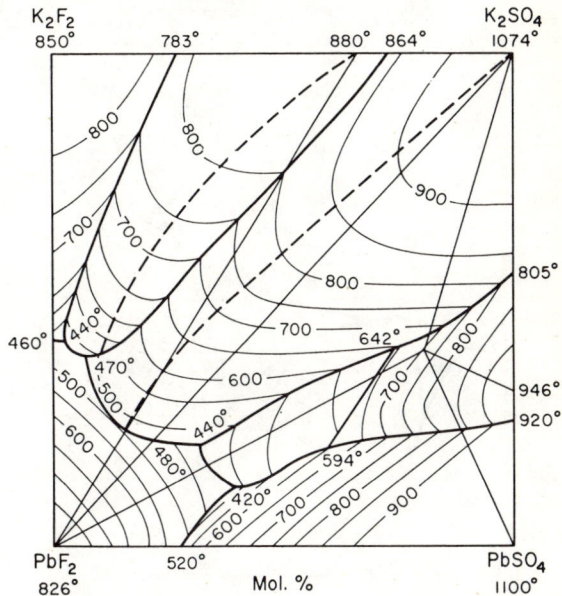

Fig. 1807.—System KF–K₂SO₄–PbF₂–PbSO₄; reciprocal.

V. A. Gladushchenko and A. G. Bergman, *Zhur. Obshchei Khim.*, **26**, 346 (1956).

KF–NaF–K₂SO₄–Na₂SO₄

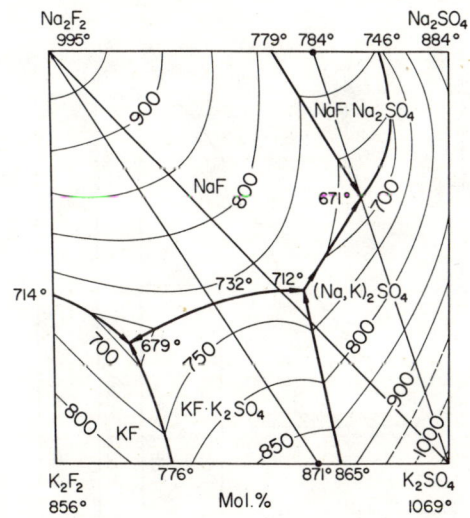

Fig. 1806.—System KF–K₂SO₄–NaF–Na₂SO₄.

A. G. Bergman and V. V. Rubleva, *Zhur. Neorg. Khim.*, **4** [1] 141 (1959).

LiF–NaF–Li₂SO₄–Na₂SO₄

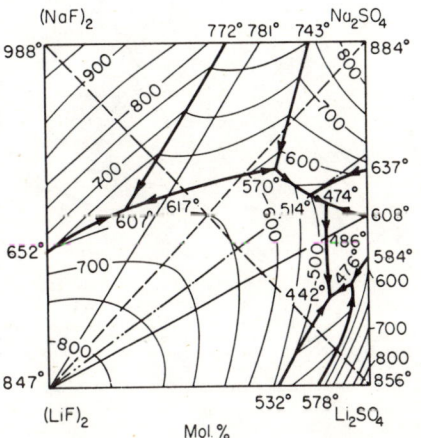

Fig. 1808.—System (LiF)₂–Li₂SO₄–(NaF)₂–Na₂SO₄.

E. I. Speranskaya and A. G. Bergman, *Izvest. Sektora Fiz.-Khim. Anal., Inst. Obshchei Neorg. Khim., Akad. Nauk S.S.S.R.*, **26**, 189 (1955).

NaF–RbF–Na$_2$SO$_4$–Rb$_2$SO$_4$

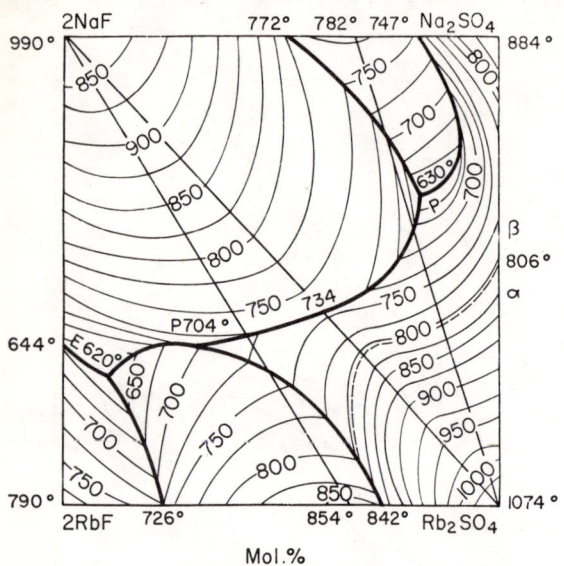

Fig. 1809.—System NaF–Na$_2$SO$_4$–RbF–Rb$_2$SO$_4$.

V. A. Gladushchenko and A. G. Bergman, *Zhur. Neorg. Khim.*, **3**, 1653 (1958).

KI–NaI–K$_2$SO$_4$–Na$_2$SO$_4$

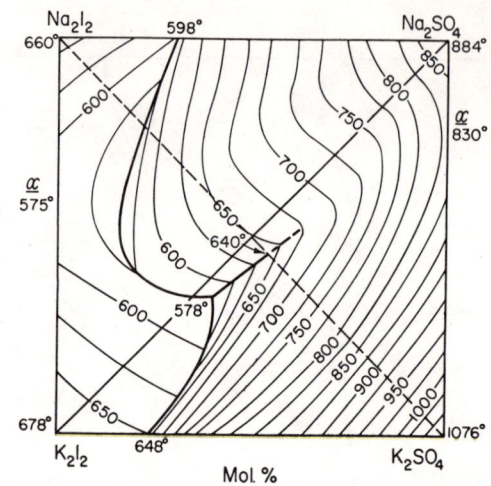

Fig. 1811.—System K$_2$I$_2$–K$_2$SO$_4$–Na$_2$I$_2$–Na$_2$SO$_4$.

R. N. Nyankovskaya, *Zhur. Neorg. Khim.*, **1**, 787 (1956).

NaF–PbF$_2$–Na$_2$SO$_4$–PbSO$_4$

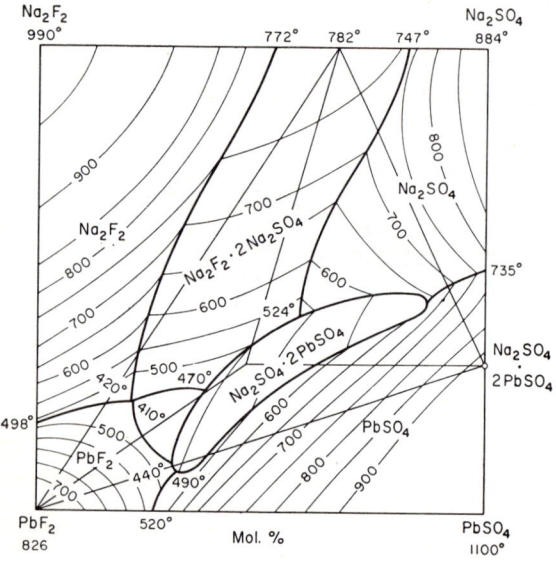

Fig. 1810.—System NaF–Na$_2$SO$_4$–PbF$_2$–PbSO$_4$; reciprocal

V. A. Gladushchenko and A. G. Bergman, *Zhur. Obshcheĭ Khim.*, **25**, 1656 (1955).

VI. Halides with Miscellaneous Oxides

(a) Two Substances

PbBr$_2$–PbO

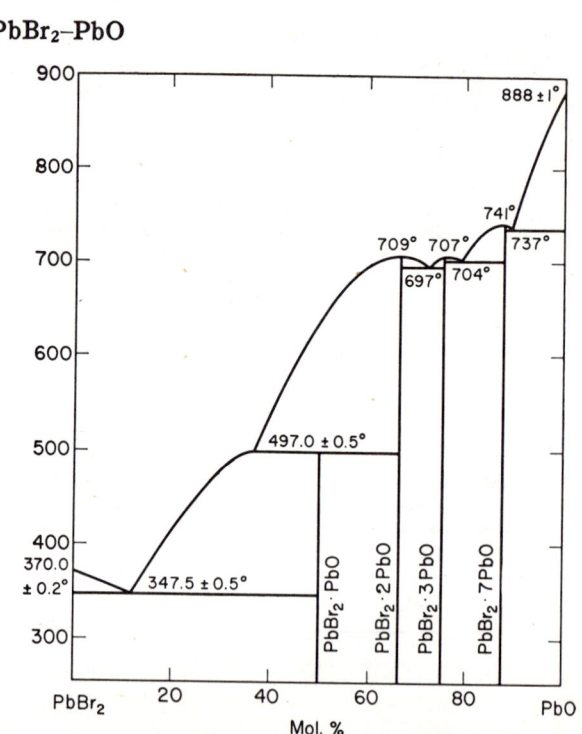

Fig. 1812.—System PbBr$_2$–PbO.

L. M. Knowles, *J. Chem. Phys.*, **19**, 1130 (1951).

KCl–K₂CO₃

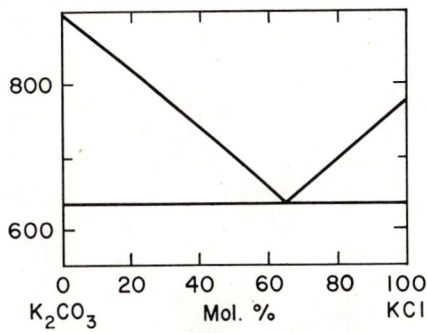

Fig. 1813.—System KCl–K$_2$CO$_3$.

M. Amadori, *Atti reale accad. Lincei, Sez. II*, **22**, 372 (1913).

KCl–KPO₃

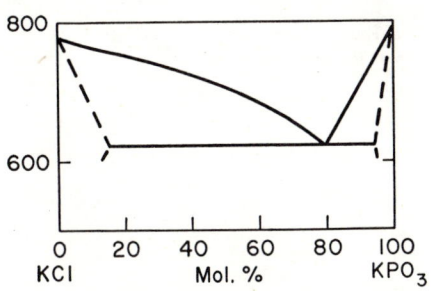

Fig. 1814.—System KCl–KPO$_3$.

M. Amadori, *Atti reale accad. Lincei, Sez. II*, **21**, 184 (1912).

KCl–K₃PO₄

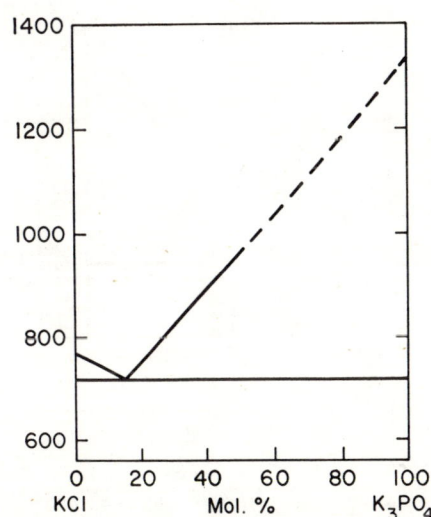

Fig. 1815.—System KCl–K$_3$PO$_4$.

M. Amadori, *Atti reale accad. Lincei, Sez. II*, **21**, 187 (1912).

KCl–K₄P₂O₇

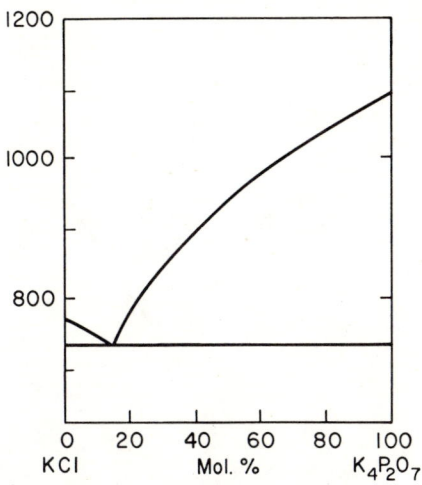

Fig. 1816.—System KCl–K$_4$P$_2$O$_7$.

M. Amadori, *Atti reale accad Lincei, Sez II*, **21**, 186 (1912)

KCl–K₂CrO₄

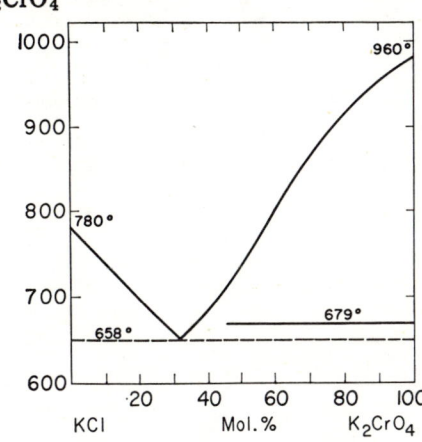

Fig 1817.—System KCl–K$_2$CrO$_4$.

S. Zhemchuzhnui, *Z. anorg. Chem.*, **57**, 270 (1908).

KCl–K₂Cr₂O₇

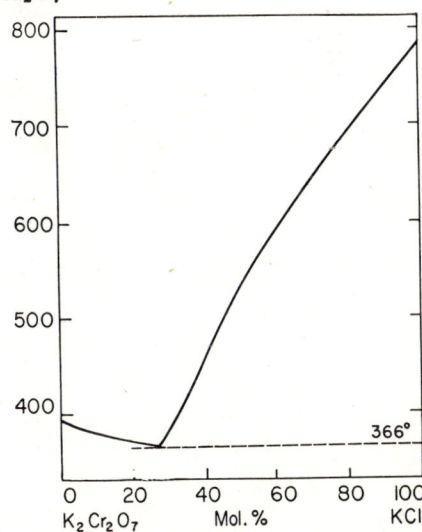

Fig 1818.—System KCl–K$_2$Cr$_2$O$_7$.

S. Zhemchuzhnui, *Z. anorg. Chem.*, **57**, 273 (1908).

LiCl–Li₂CO₃

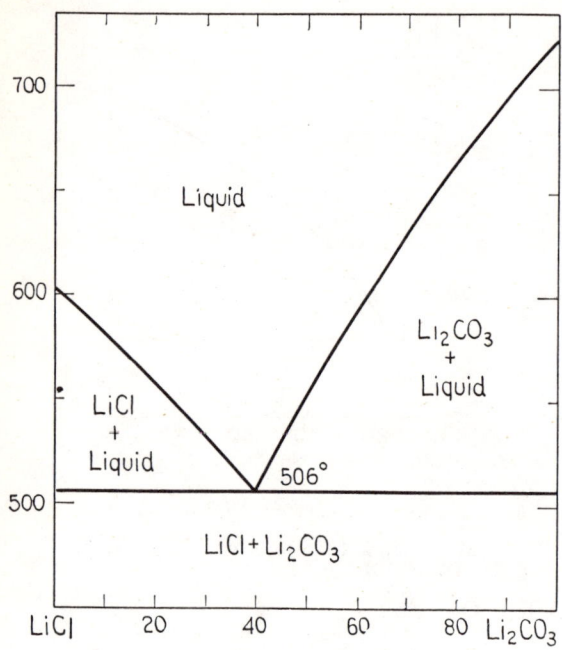

Fig. 1819.—System LiCl–Li₂CO₃.

Unpublished diagram, Metalloy Corporation, Minneapolis, Minn., 1944.

NaCl–Na₂CO₃

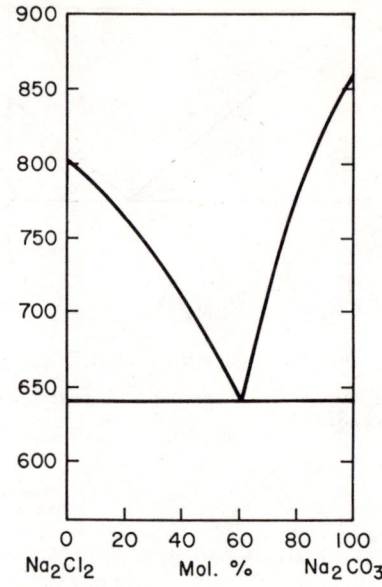

Fig. 1820.—System NaCl–Na₂CO₃.

Paul Niggli, *Z. anorg. u. allgem. Chem.*, **106**, 136 (1919).

PbCl₂–PbO

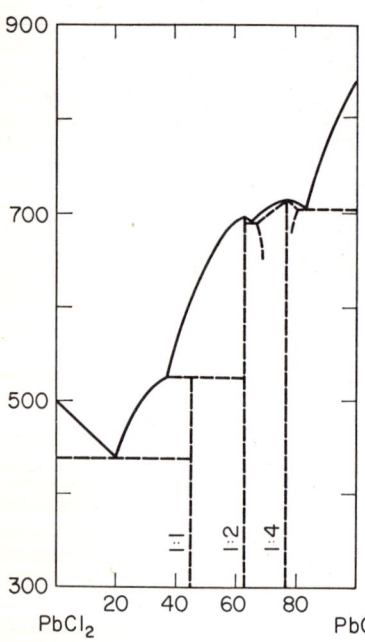

Fig. 1821.—System PbCl₂–PbO.

Rudolf Ruer, *Z. anorg. Chem.*, **49**, 372 (1906).

AlCl₃–POCl₃

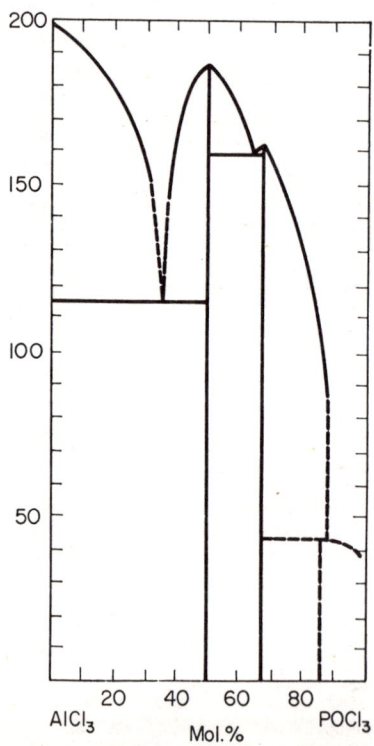

Fig. 1822.—System AlCl₃–POCl₃.

W. L. Groeneveld and A. P. Zuur, *Rec. Trav. Chim.*, **76**, 1006 (1957).

SbCl₅–POCl₃

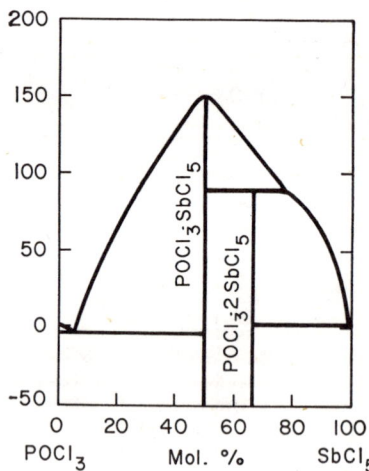

Fig. 1823.—System POCl₃–SbCl₅.

Geneviève Leman and Gabriel Tridot, *Compt. rend.*, **248**, 3440 (1959).

KF–K₂CO₃

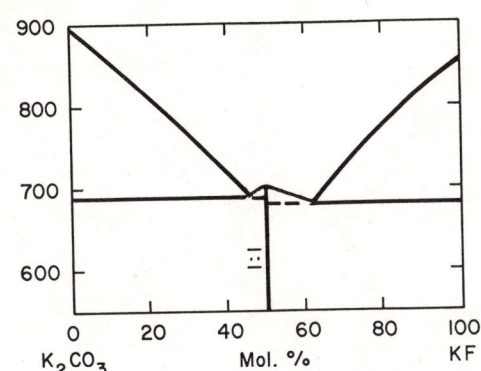

Fig. 1824.—System KF–K₂CO₃.

M. Amadori, *Atti reale accad. Lincei*, Sez. II, **22**, 369 (1913).

KF–K₄P₂O₇

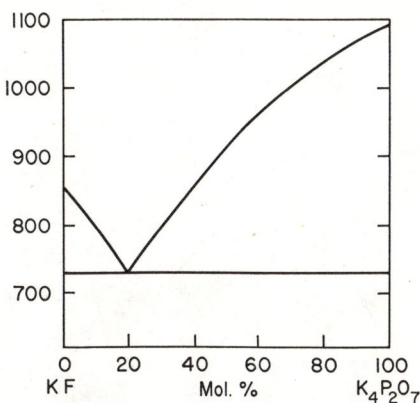

Fig. 1825.—System KF–K₄P₂O₇.

M. Amadori, *Atti reale accad. Lincei*, Sez. II, **21**, 692 (1912).

LiF–Li₂CO₃

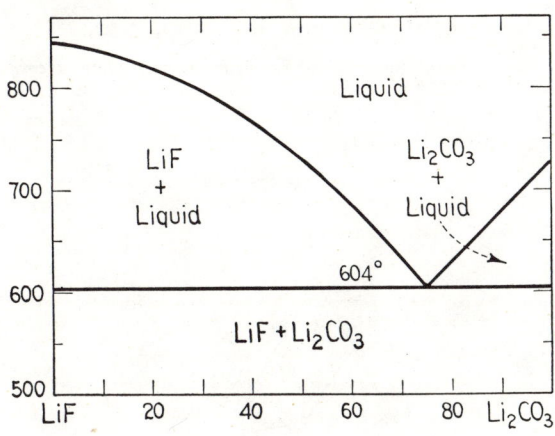

Fig. 1826.—System LiF–Li₂CO₃.

Unpublished diagram, Metalloy Corporation, Minneapolis, Minn., 1945.

NaF–Na₂CO₃

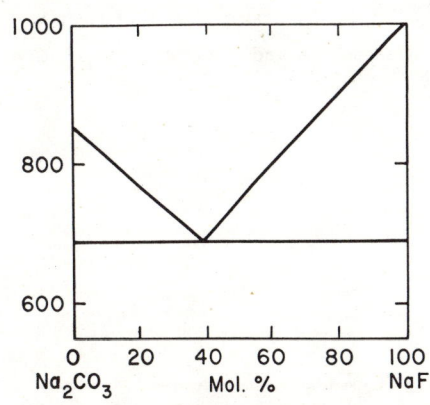

Fig. 1827.—System NaF–Na₂CO₃.

M. Amadori, *Atti reale accad. Lincei*, Sez. II, **22**, 368 (1913).

(b) Three Substances

KCl–KF–K₂CrO₄

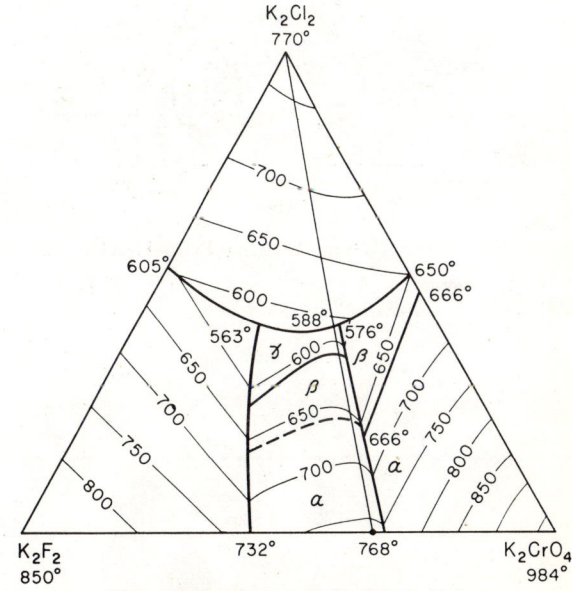

Fig. 1829.—System KCl–KF–K₂CrO₄.

I. S. Rassonskaya and A. G. Bergman, *Doklady Akad. Nauk S.S.S.R.*, **22**, 1092 (1952).

PbF₂–Pb₃(PO₄)₂

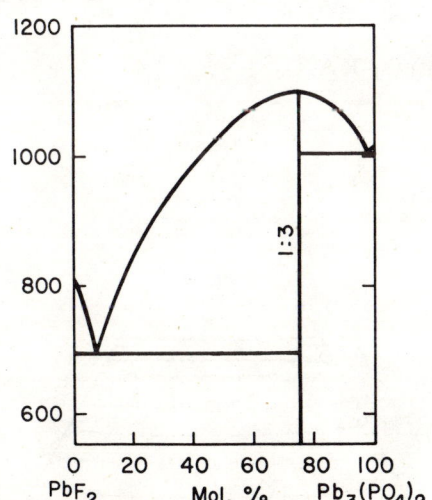

Fig. 1828.—System PbF₂–Pb₃(PO₄)₂.

M. Amadori, *Gazz. chim. ital.*, **49 I**, 44 (1919).

KCl–NaF–K₂CrO₄

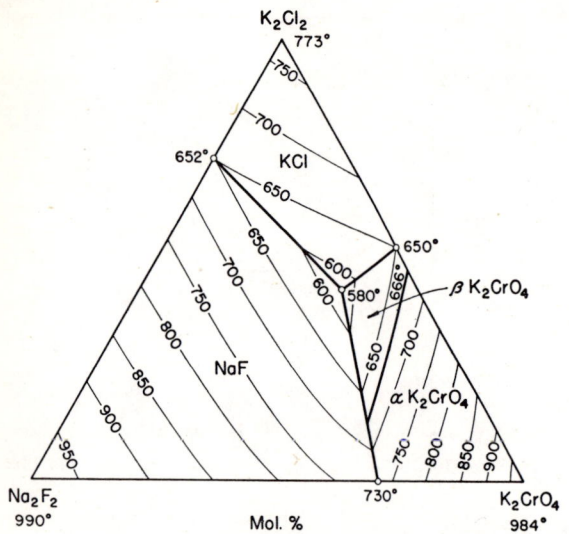

Fig. 1830.—System NaF–KCl–K₂CrO₄.

I. S. Rassonskaya and A. G. Bergman, *J. Gen. Chem. U.S.S.R.*, **23**, 18 (1953).

KCl–K₂CO₃–K₂SO₄

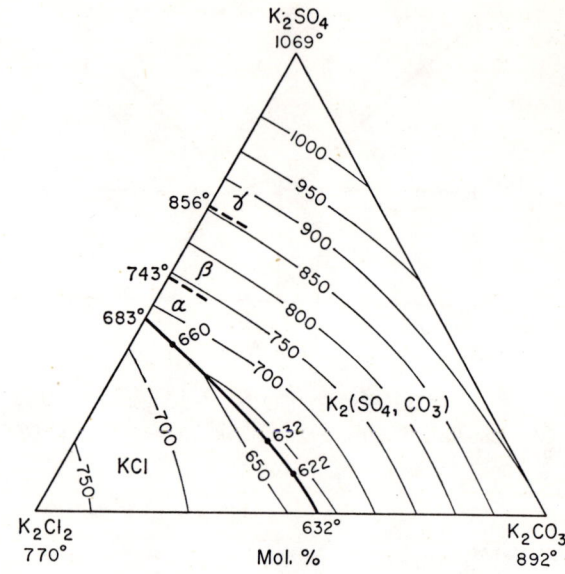

Fig. 1832.—System KCl–K₂CO₃–K₂SO₄.

A. G. Bergman and A. K. Sementsova, *Zhur. Neorg. Khim.*, **3** [2] 391 (1958).

KCl–Na₂CO₃–Na₂SO₄

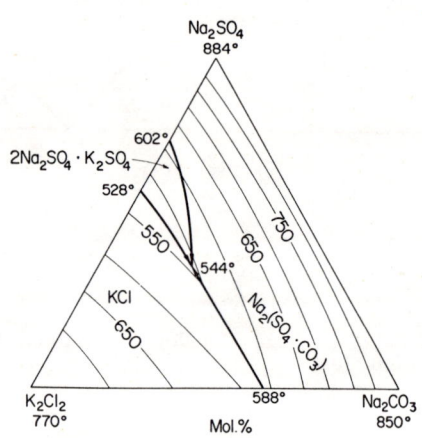

Fig. 1833.—System KCl–Na₂CO₃–Na₂SO₄.

A. G. Bergman and A. K. Sementsova, *Zhur. Neorg. Khim.*, **3** [2] 395 (1958).

KCl–KBO₂–K₂WO₄

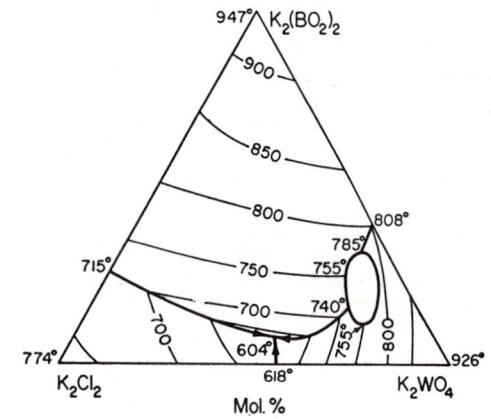

Fig. 1831.—System K₂Cl₂–K₂(BO₂)₂–K₂WO₄.

V. I. Posypaĭko, A. G. Bergman, and A. I. Kislova, *Zhur. Neorg. Khim.*, **1**, 2615 (1956).

KCl–K₂MoO₄–K₂WO₄

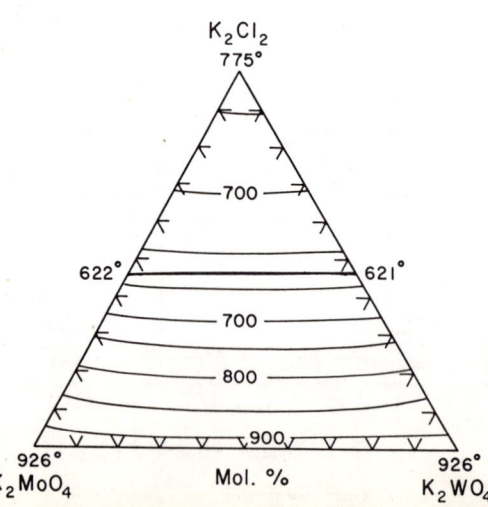

Fig. 1834.—System KCl–K₂MoO₄–K₂WO₄.

Z. A. Mateĭko and G. A. Bukhalova, *Zhur. Neorg. Khim.*, **3**, 1886 (1958).

KCl–K₂SO₄–Na₂CO₃

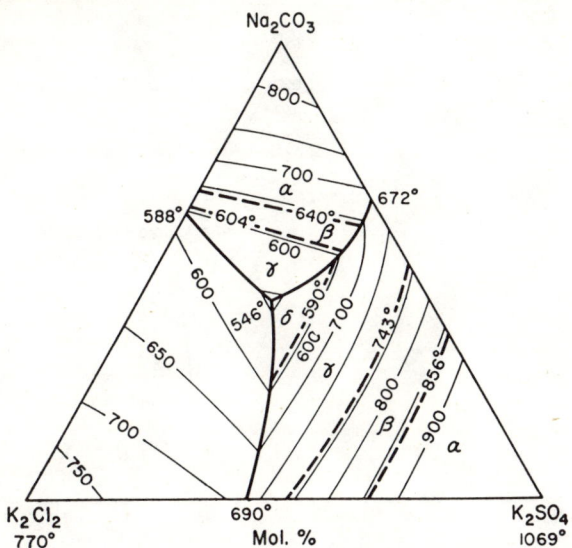

Fig. 1835.—System KCl–K₂SO₄–Na₂CO₃.

A. K. Sementsova and A. G. Bergman, *Zhur. Obshcheĭ Khim.*, **26**, 995 (1956).

KCl–K₂SO₄–K₂WO₄

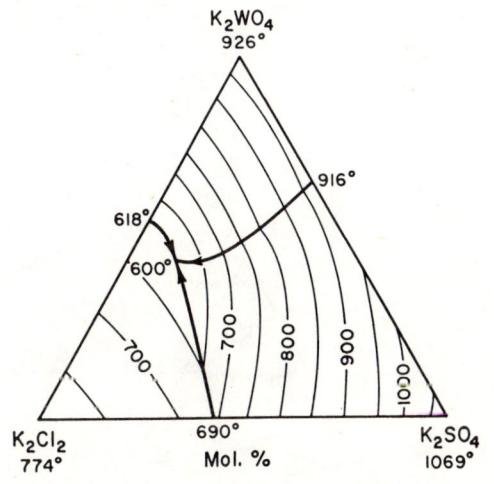

Fig. 1836.—System KCl–K₂SO₄–K₂WO₄.

A. G. Bergman, A. I. Kislova, and V. I. Posypaĭko, *Zhur. Obshcheĭ Khim.*, **25**, 14 (1955).

KCl–Li₂SO₄–Li₂WO₄

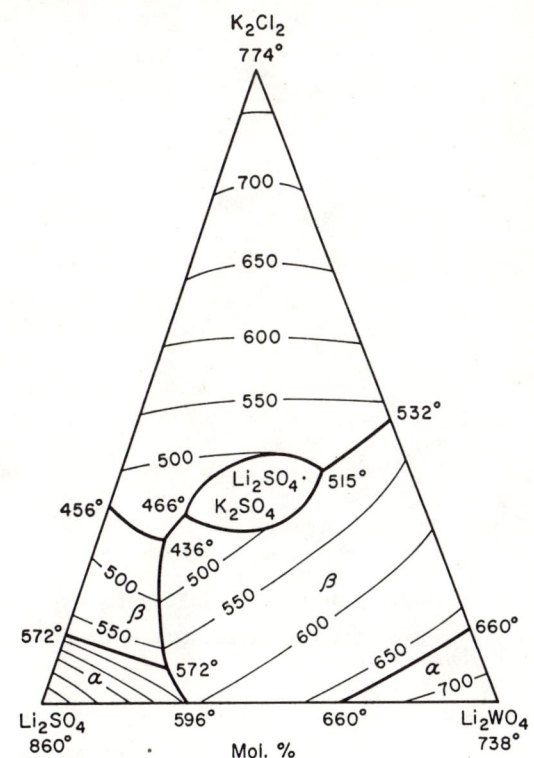

Fig. 1837.—System K₂Cl₂–Li₂SO₄–Li₂WO₄; reciprocal, quaternary.

A. G. Bergman, A. I. Kislova, and V. I. Posypaĭko, *Zhur. Obshcheĭ Khim.*, **24**, 1723 (1954).

LiCl–LiBO₂–Li₂SO₄

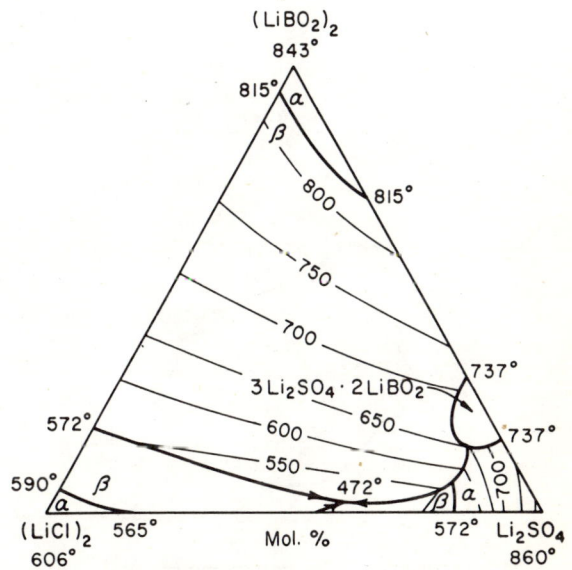

Fig. 1838.—System Li₂Cl₂–(LiBO₂)₂–Li₂SO₄.

V. I. Posypaĭko, A. I. Kislova, and A. G. Bergman, *Zhur. Neorg. Khim.*, **1**, 810 (1956).

LiCl–LiBO$_2$–Li$_2$WO$_4$

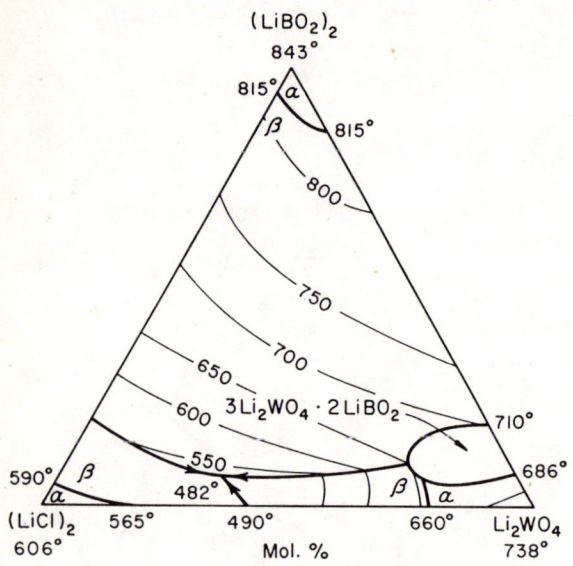

Fig. 1839.—System (LiCl)$_2$–(LiBO$_2$)$_2$–Li$_2$WO$_4$.

V. I. Posypaĭko, A. I. Kislova, and A. G. Bergman, *Zhur. Neorg. Khim.*, **1**, 818 (1956).

NaCl–NaF–K$_2$CrO$_4$

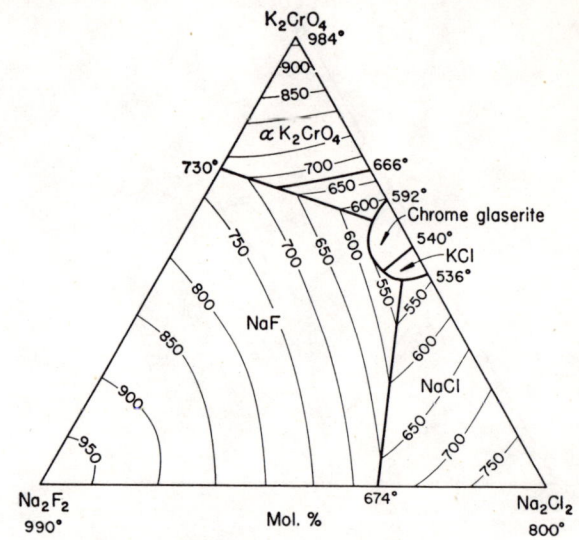

Fig. 1841.—System NaF–NaCl–K$_2$CrO$_4$.

I. S. Rassonskaya and A. G. Bergman, *J. Gen. Chem. U.S.S.R.*, **23**, 19 (1953).

LiCl–Li$_2$SO$_4$–Li$_2$WO$_4$

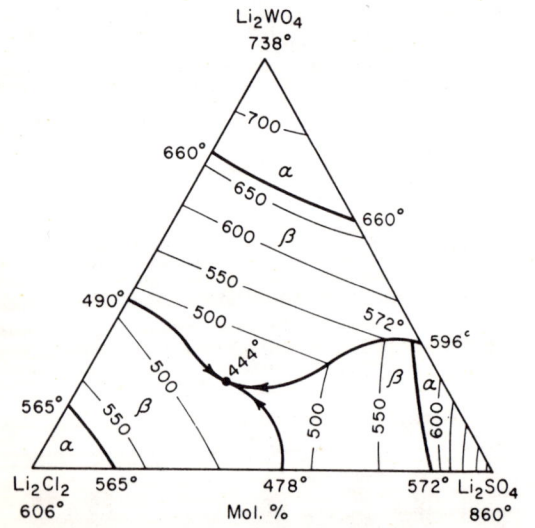

Fig. 1840.—System Li$_2$Cl$_2$–Li$_2$SO$_4$–Li$_2$WO$_4$.

A. G. Bergman, A. I. Kislova, and V. I. Posypaĭko, *Zhur. Obshchei Khim.*, **24**, 1939 (1954).

NaCl–NaF–Na$_2$CrO$_4$

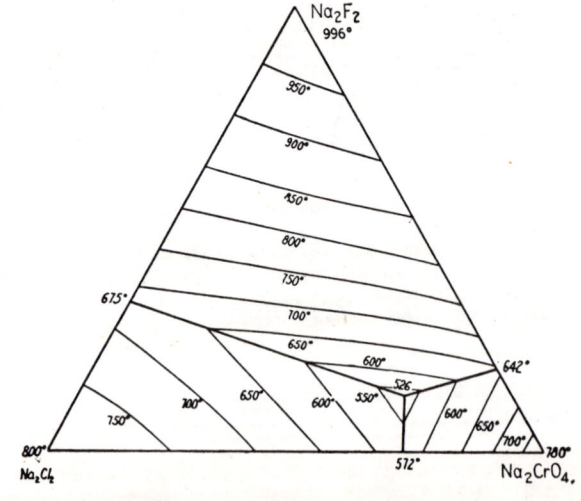

Fig. 1842.—System NaCl–NaF–Na$_2$O·CrO$_3$.

I. S. Rassonskaya and A. G. Bergman, *Compt. rend. acad. sci., U.R.S.S.*, **38**, 176 (1943).

NaCl–K₂CO₃–K₂SO₄

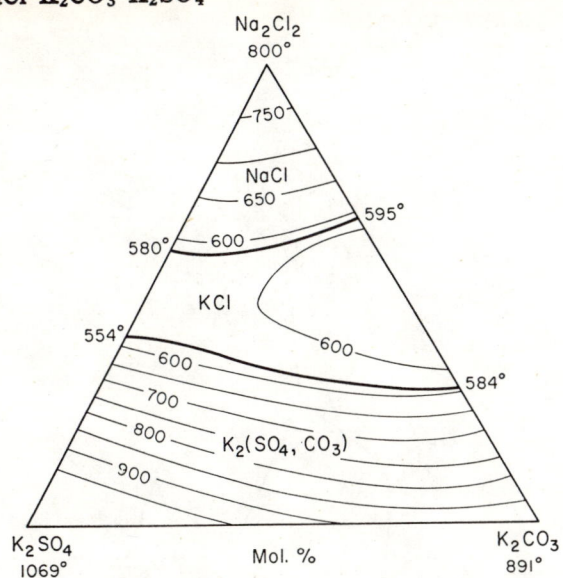

Fig. 1843.—System K₂CO₃–K₂SO₄–NaCl.

A. G. Bergman and A. K. Sementsova, *Zhur. Neorg. Khim.*, **3** [2] 401 (1958).

NaCl–Na₂CO₃–Na₂SO₄

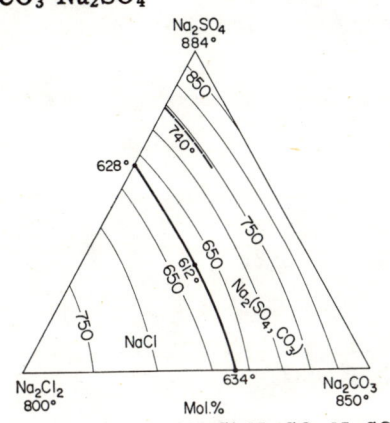

Fig. 1844.—System NaCl–Na₂CO₃–Na₂SO₄.

A. G. Bergman and A. K. Sementsova, *Zhur. Neorg. Khim.*, **3** [2] 388 (1958).

NaCl–Na₂MoO₄–Na₂WO₄

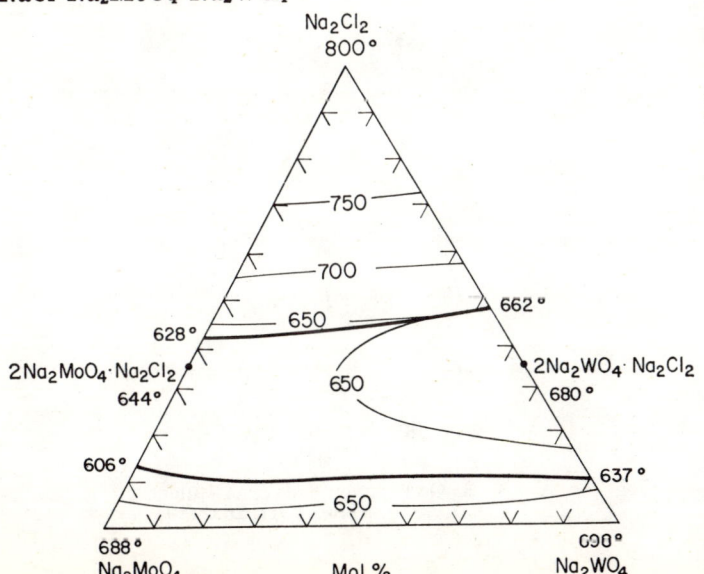

Fig. 1845.—System NaCl–Na₂MoO₄–Na₂WO₄.

Z. A. Mateĭko and G. A. Bukhalova, *Zhur. Neorg. Khim.*, **3**, 1884 (1958).

BaCl₂–BaO–TiO₂

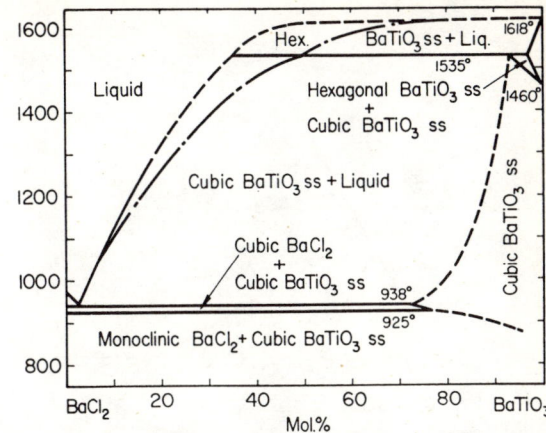

Fig. 1846.—System BaCl₂–BaTiO₃.

D. E. Rase and Rustum Roy, *J. Phys. Chem.*, **61**, 746 (1957).

BaCl₂–BaCO₃–BaTiO₃

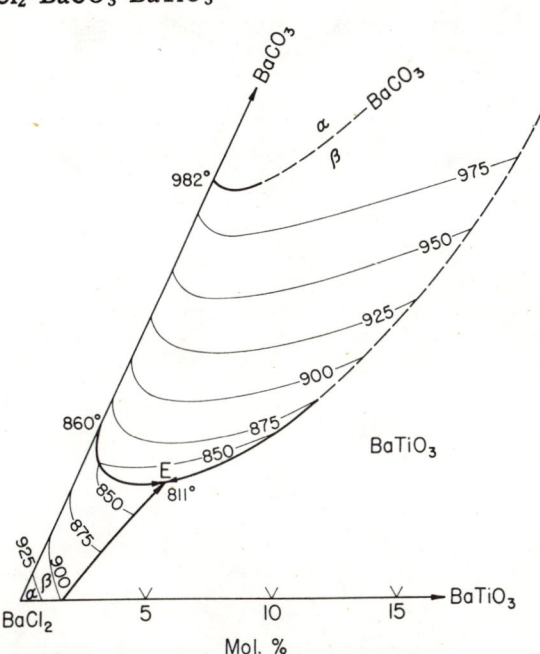

Fig. 1847.—System BaCl₂–BaCO₃–BaTiO₃. Eutectic, E at 11.25 mol. % BaCO₃, 9.25% BaTiO₃, 79.50% BaCl₂ and 811°C.

I. N. Belyaev and M. L. Sholokhovich, *Doklady Akad. Nauk S.S.S.R.*, **77** [1] 52 (1951).

KF–K$_2$CO$_3$–K$_2$SO$_4$

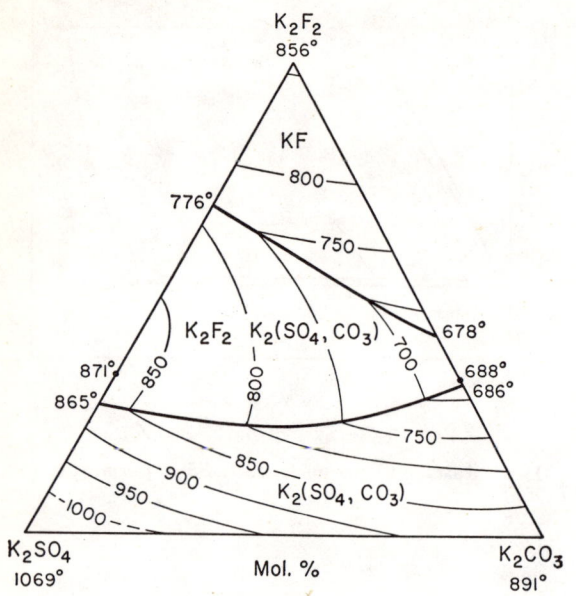

Fig. 1848.—System K$_2$CO$_3$–KF–K$_2$SO$_4$.

A. G. Bergman and V. V. Rubleva, *Zhur. Neorg. Khim.*, 2 [7] 1616 (1957).

KF–Na$_2$CO$_3$–K$_2$SO$_4$

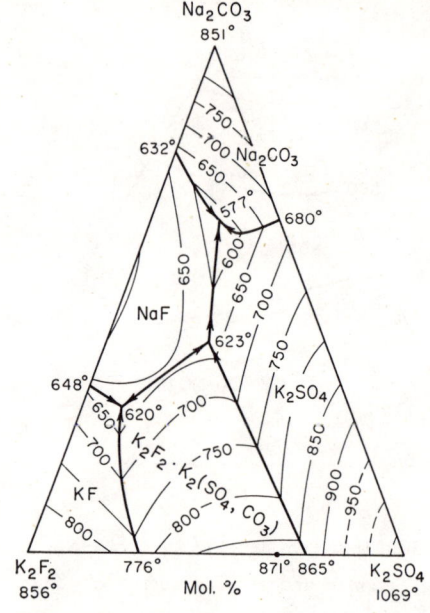

Fig. 1849.—System KF–K$_2$SO$_4$–Na$_2$CO$_3$.

A. G. Bergman and V. V. Rubleva, *Zhur. Neorg. Khim.*, 2 [11] 2636 (1957).

NaF–K$_2$CO$_3$–K$_2$SO$_4$

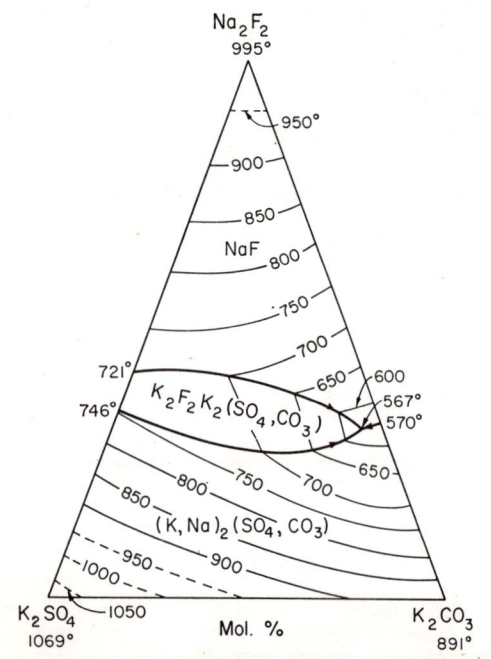

Fig. 1850.—System K$_2$CO$_3$–K$_2$SO$_4$–NaF.

A. G. Bergman and V. V. Rubleva, *Zhur. Neorg. Khim.*, 2 [11] 2626 (1957).

NaF–K$_2$CO$_3$–Na$_2$SO$_4$

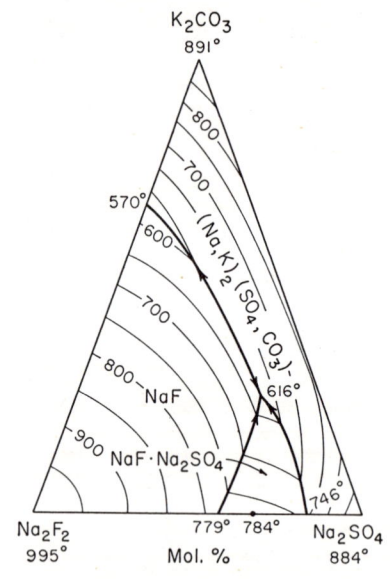

Fig. 1851.—System K$_2$CO$_3$–NaF–Na$_2$SO$_4$.

A. G. Bergman and V. V. Rubleva, *Zhur. Neorg. Khim.*, 2 [11] 2622 (1957).

NaF–Na$_2$CO$_3$–K$_2$SO$_4$

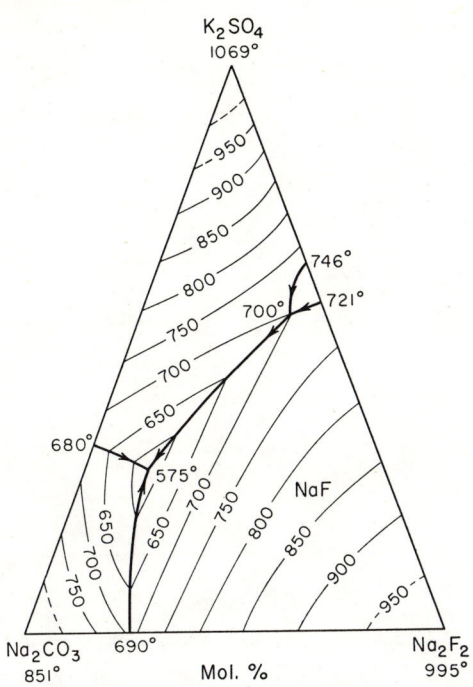

FIG. 1852.—System K$_2$SO$_4$–Na$_2$CO$_3$–NaF.

A. G. Bergman and V. V. Rubleva, *Zhur. Neorg. Khim.*, **2**, [11] 2631 (1957).

(c) *Four Substances*

KBr–NaBr–K$_2$CO$_3$–Na$_2$CO$_3$

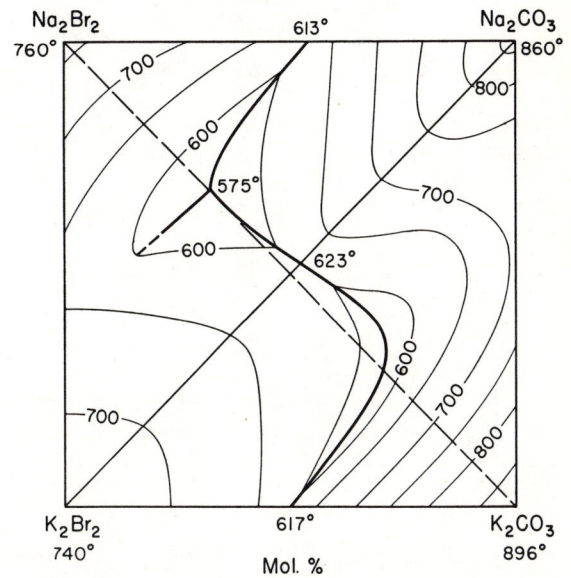

FIG. 1854.—K$_2$Br$_2$–Na$_2$CO$_3$–Na$_2$Br$_2$ and K$_2$Br$_2$–Na$_2$CO$_3$–K$_2$CO$_3$; reciprocal salt pairs, ternary systems.

R. N. Nyankovskaya, *Doklady Akad. Nauk S.S.S.R.*, **83**, 421 (1952).

NaF–Na$_2$CO$_3$–Na$_2$SO$_4$

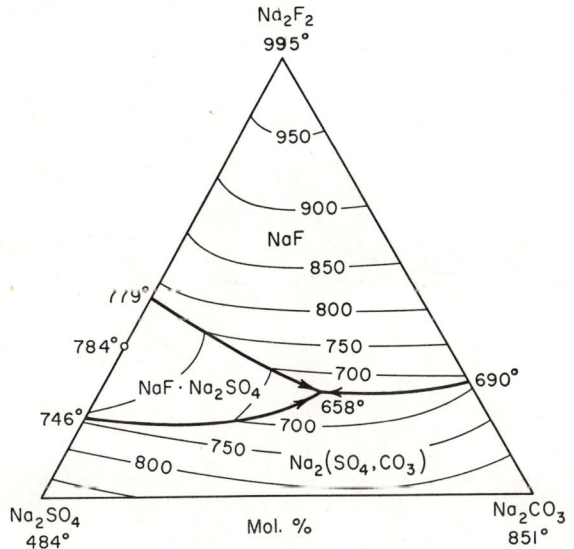

FIG. 1853.—System Na$_2$CO$_3$–NaF–Na$_2$SO$_4$.

A. G. Bergman and V. V. Rubleva, *Zhur. Neorg. Khim.*, **2** [7] 1612 (1957).

AgCl–LiCl–Ag$_2$CrO$_4$–Li$_2$CrO$_4$

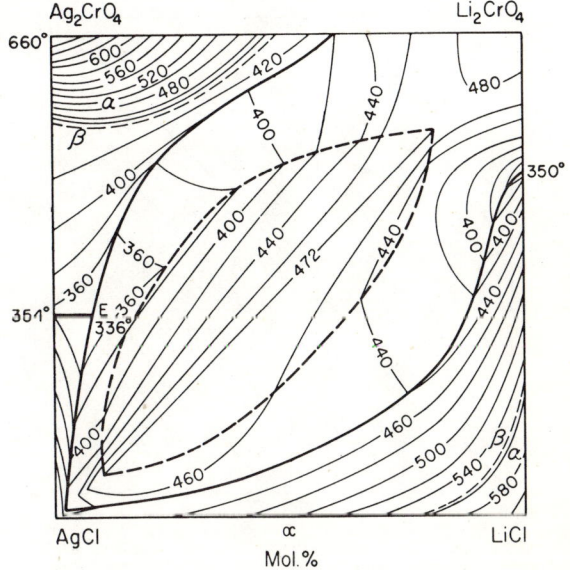

FIG. 1855.—System AgCl–Ag$_2$CrO$_4$–LiCl–Li$_2$CrO$_4$.

D. S. Lesnykh and A. G. Bergman, *Zhur. Fiz. Khim.*, **30**, 1961 (1956).

AgCl–LiCl–Ag₂WO₄–Li₂WO₄

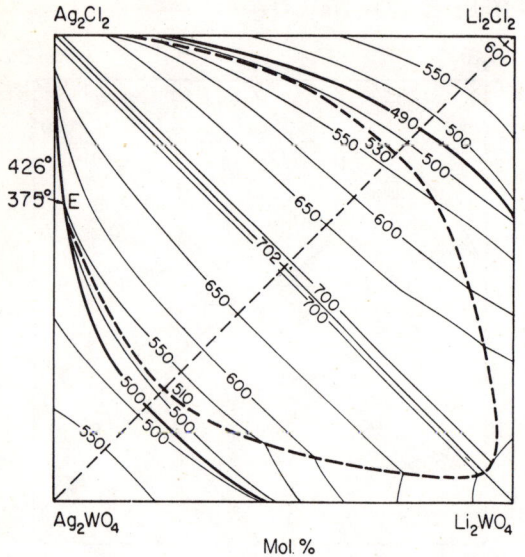

FIG. 1856.—System AgCl–Ag₂WO₄–LiCl–Li₂WO₄; reciprocal.

D. S. Lesnykh and A. G. Bergman, *Zhur. Obshcheĭ Khim.*, **26**, 1563 (1956).

KCl–NaCl–KVO₃–NaVO₃

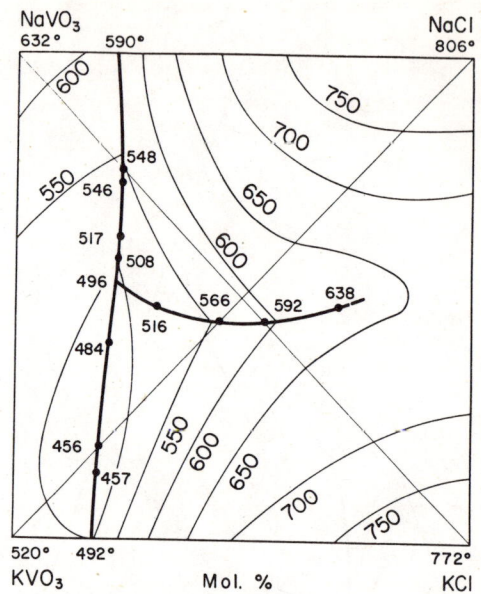

FIG. 1858.—System KCl–KVO₃–NaCl–NaVO₃.

M. A. Zakharchenko, *Zhur. Neorg. Khim.*, **2**, 2179 (1957).

KCl–NaCl–K₂CO₃–Na₂CO₃

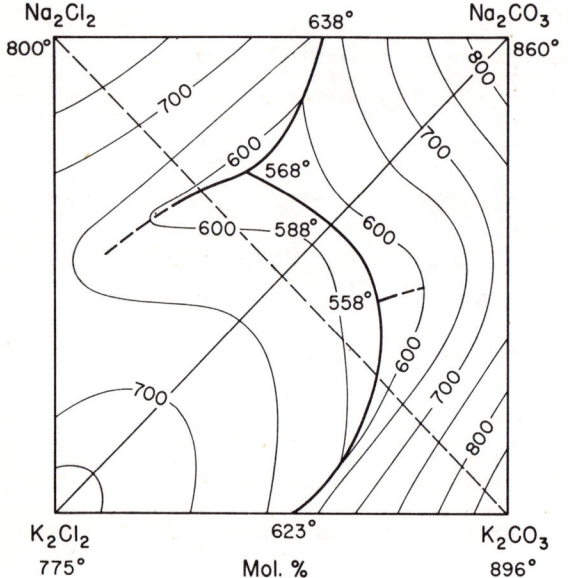

FIG. 1857.—K₂Cl₂–Na₂CO₃–Na₂Cl₂ and K₂Cl₂–Na₂CO₃–K₂CO₃; reciprocal salt pairs, ternary systems.

R. N. Nyankovskaya, *Doklady Akad. Nauk S.S.S.R.*, **83**, 420 (1952).

KCl–NaCl–K₂MoO₄–Na₂MoO₄

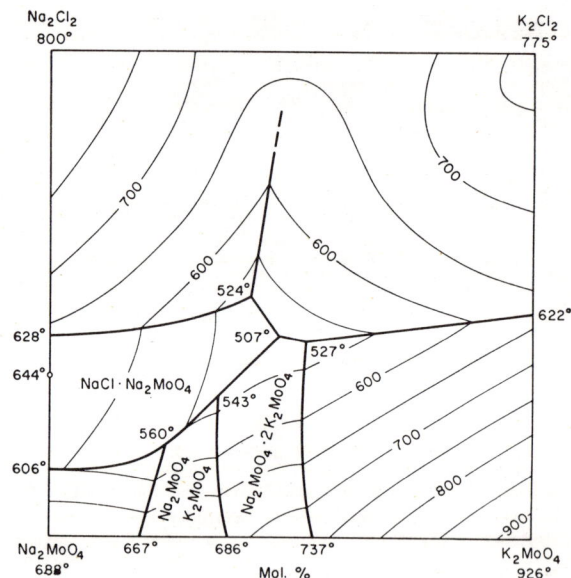

FIG. 1859.—System KCl–K₂MoO₄–NaCl–Na₂MoO₄; reciprocal.

G. A. Bukhalova and Z. A. Mateiko, *Zhur. Obshcheĭ Khim.*, **25**, 893 (1955).

KCl–NaCl–K₂WO₄–Na₂WO₄

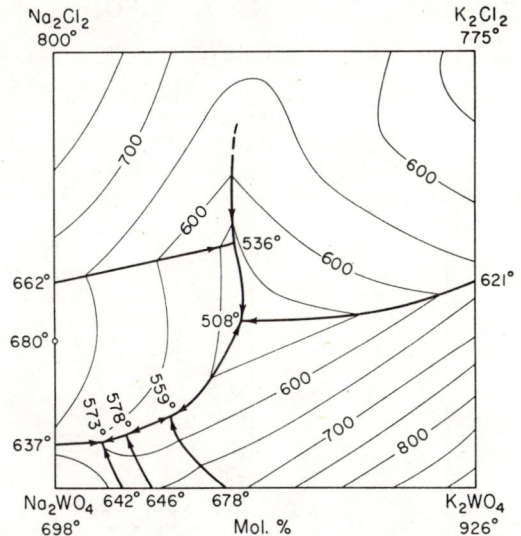

Fig. 1860.—System KCl–K₂WO₄–NaCl–Na₂WO₄; irreversible reciprocal.

G. A. Bukhalova and Z. A. Mateĭko, *Zhur. Obshcheĭ Khim.*, **26**, 2123 (1956).

NaCl–BaCl₂–Na₂CO₃–BaCO₃

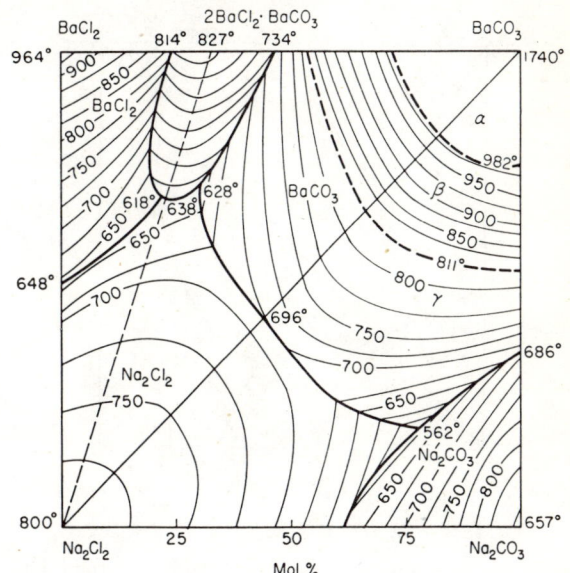

Fig. 1862.—System Na₂Cl₂–Na₂CO₃–BaCl₂–BaCO₃; reciprocal.

I. N. Belyaev and M. L. Sholokhovich, *Sbornik Stateĭ Obshcheĭ Khim., Akad. Nauk S.S.S.R.*, **1**, 141 (1953).

LiCl–CdCl₂–Li₂MoO₄–CdMoO₄

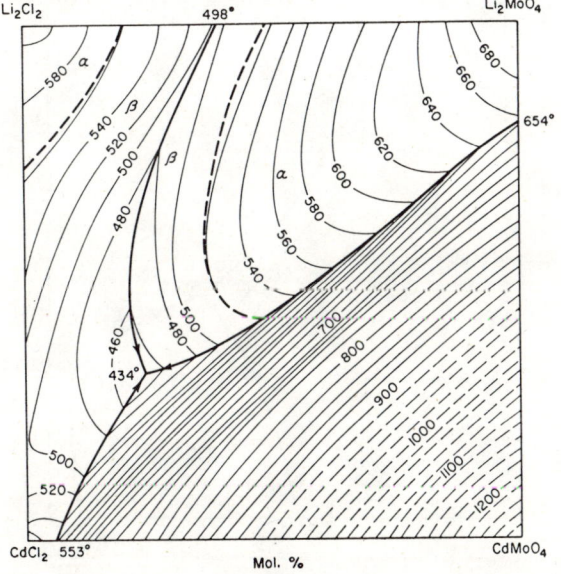

Fig. 1861.—System LiCl–Li₂MoO₄–CdCl₂–CdMoO₄; reciprocal.

D. S. Lesnykh, A. G. Bergman, and N. G. Bukun, *Zhur. Obshcheĭ Khim.*, **26**, 2676 (1956).

KF–NaF–K₂CO₃–Na₂CO₃

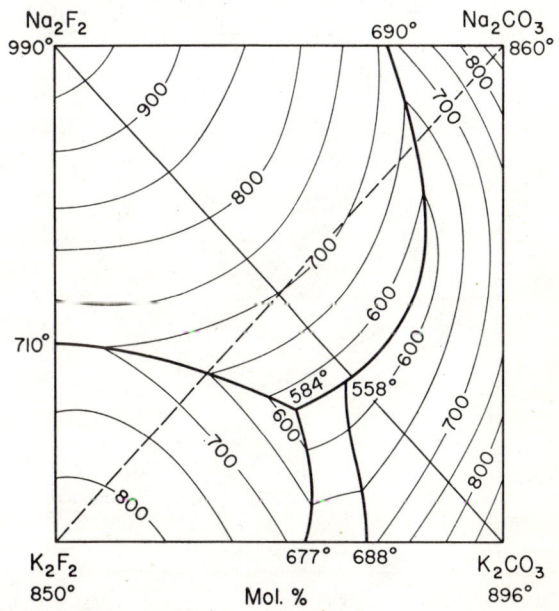

Fig. 1863.—K₂F₂–Na₂CO₃–Na₂F₂ and K₂F₂–Na₂CO₃–K₂CO₃; reciprocal salt pairs, ternary systems.

R. N. Nyankovskaya, *Doklady Akad. Nauk S.S.S.R.*, **83**, 420 (1952).

KF–NaF–K₂CrO₄–Na₂CrO₄

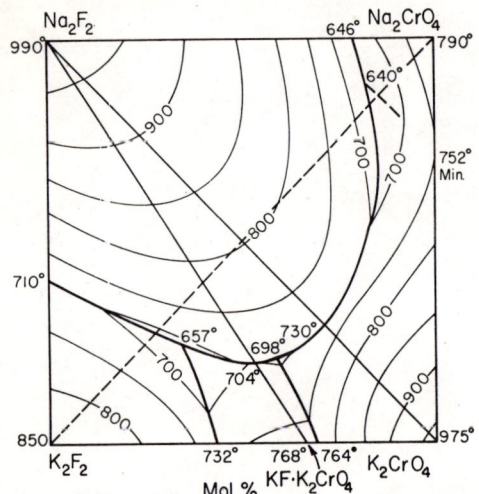

Fig. 1864.—System K₂F₂–K₂CrO₄–Na₂F₂–Na₂CrO₄.

E. P. Dergunov and A. G. Bergman, *Izvest. Sektora Fiz.-Khim. Anal., Inst. Obshcheĭ Neorg. Khim., Akad. Nauk S.S.S.R.*, **21**, 196 (1952).

(d) Six Substances

KF–NaF–K₂CO₃–Na₂CO₃–K₂SO₄–Na₂SO₄

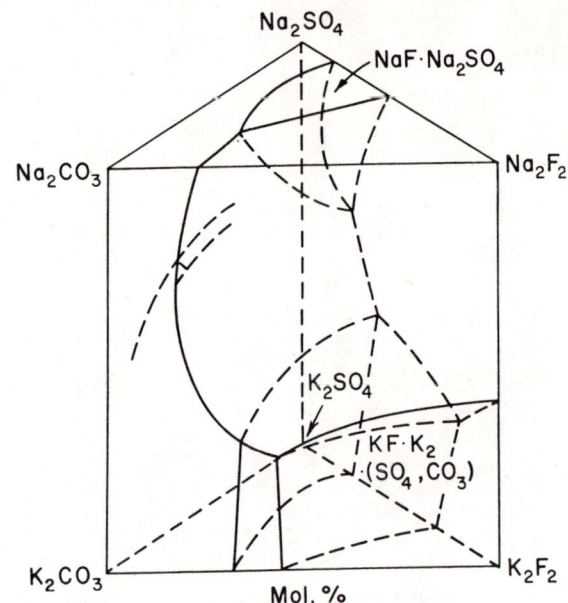

Fig. 1866.—System KF–K₂CO₃–K₂SO₄–NaF–Na₂CO₃–Na₂SO₄; schematic.

A. G. Bergman and V. V. Rubleva, *Zhur. Neorg. Khim.*, **3** [8] 1908 (1958).

KI–NaI–K₂CO₃–Na₂CO₃

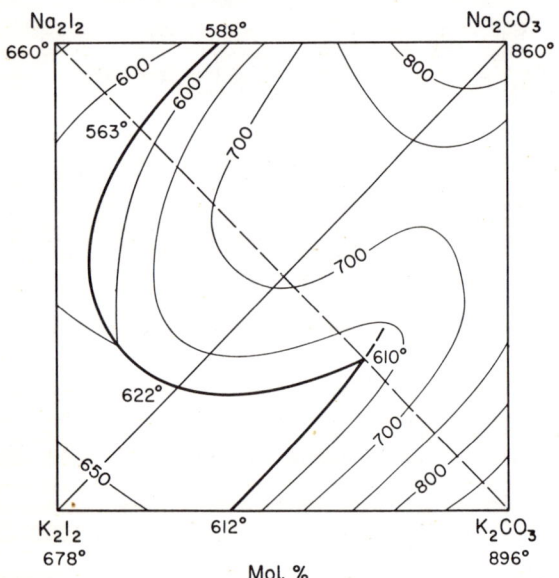

Fig. 1865.—K₂I₂–Na₂CO₃–Na₂I₂ and K₂I₂–Na₂CO₃–K₂CO₃; reciprocal salt pairs, ternary systems.

R. N. Nyankovskaya, *Doklady Akad. Nauk S.S.S.R.*, **83**, 421 (1952).

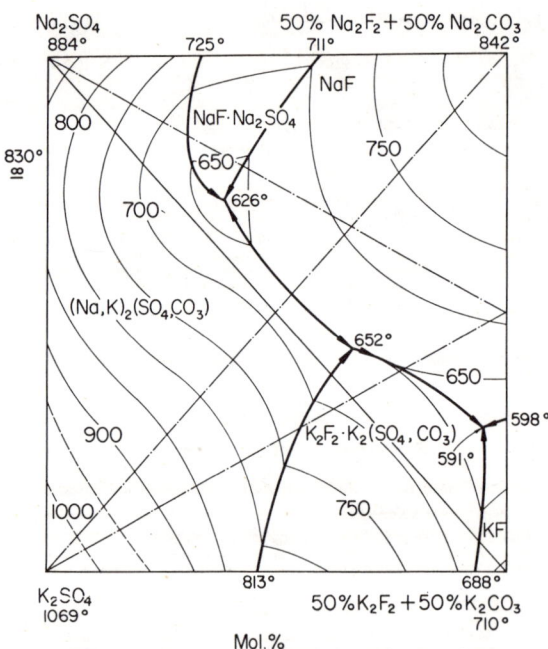

Fig. 1867.—System K₂SO₄–(50% KF, 50% K₂CO₃)–(50% NaF, 50% Na₂CO₃)–Na₂SO₄.

A. G. Bergman and V. V. Rubleva, *Zhur. Neorg. Khim.*, **3** [8] 1903 (1958).

VII. Halides with Miscellaneous Substances

(a) Two Substances

AgCl–Ag₂S

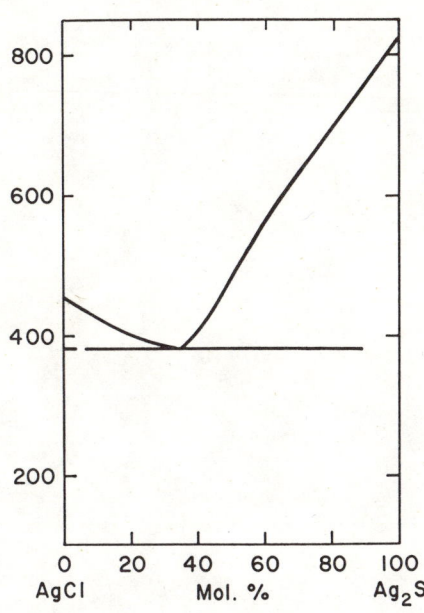

Fig. 1868.—System AgCl–Ag₂S.

Carlo Sandonnini, *Atti reale accad. Lincei, Sez. I*, **21**, 481 (1912).

NaCl–NaCN

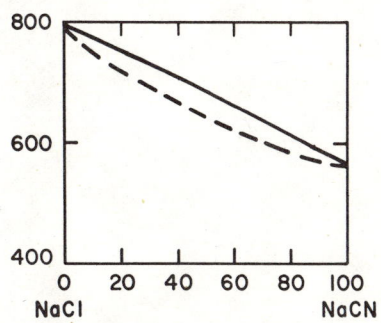

Fig. 1870.—System NaCl–NaCN.

Wilhelm Truthe, *Z. anorg. Chem.*, **76**, 138 (1912).

KCl–KCN

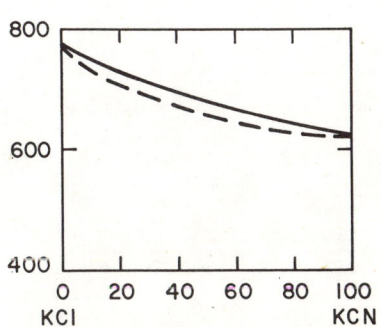

Fig. 1869.—System KCl–KCN.

Wilhelm Truthe, *Z. anorg. Chem.*, **76**, 138 (1912).

(b) Four Substances

KCl–NaCl–KCNS–NaCNS

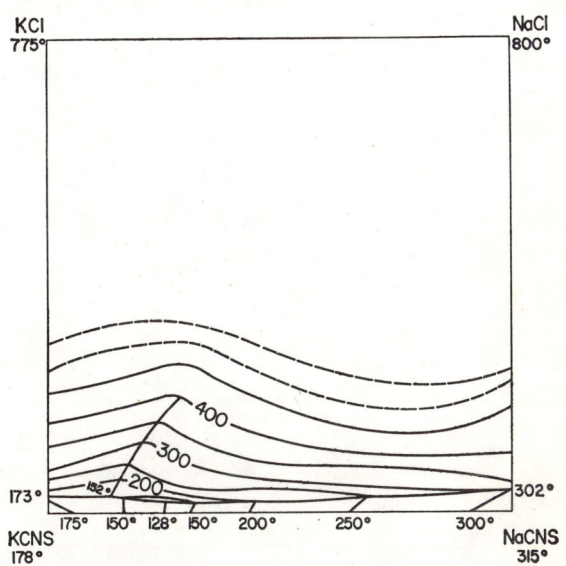

Fig. 1871.—System KCl–KCNS–NaCl–NaCNS.

A. F. Oparina and N. S. Dombrovskaya, *Zhur. Neorg. Khim.*, **3**, 420 (1958).

F. SYSTEMS CONTAINING CYANIDES, SULFIDES, ETC.

I. Cyanides Only

(a) Two Cyanides

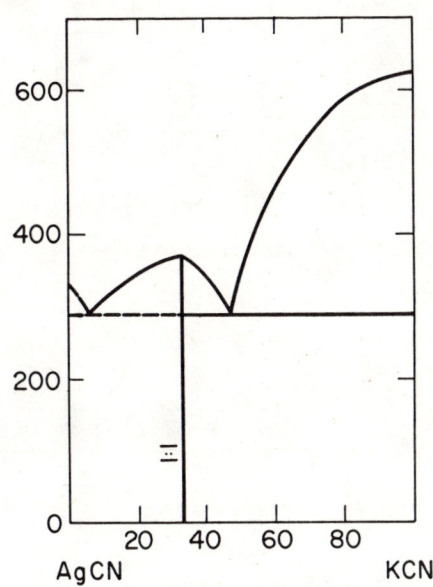

FIG. 1872.—System AgCN–KCN.
Wilhelm Truthe, Z. anorg. Chem., 76, 140 (1912).

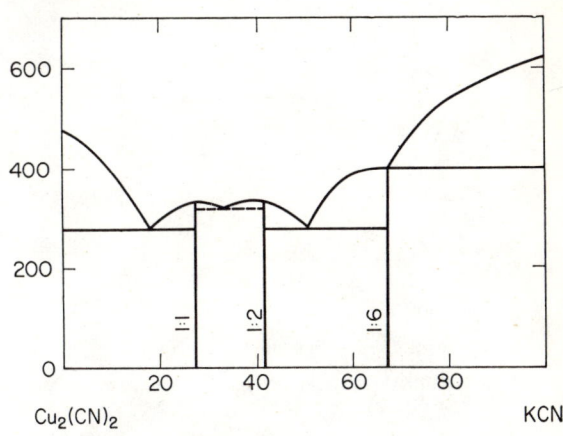

FIG. 1874.—System CuCN–KCN.
Wilhelm Truthe, Z. anorg. Chem., 76, 146 (1912).

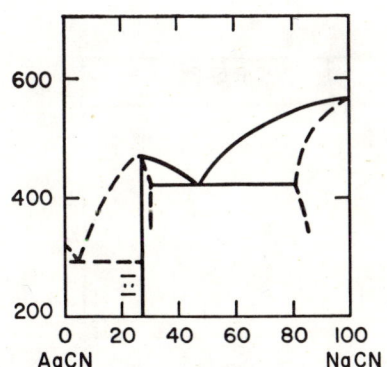

FIG. 1873.—System AgCN–NaCN.
Wilhelm Truthe, Z. anorg. Chem., 76, 143 (1912).

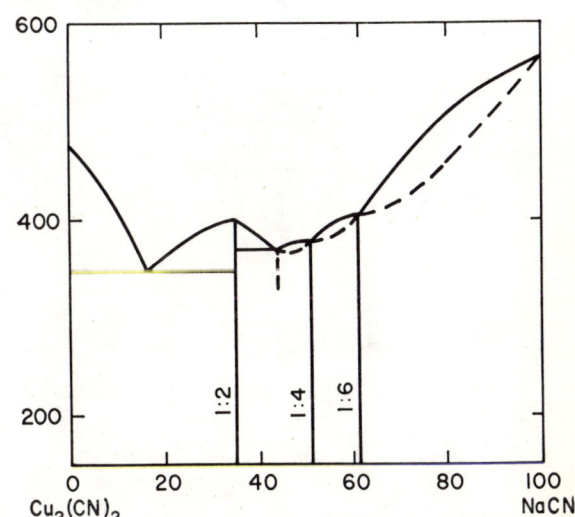

FIG. 1875.—System CuCN–NaCN.
Wilhelm Truthe, Z. anorg. Chem., 76, 152 (1912).

KCN–NaCN

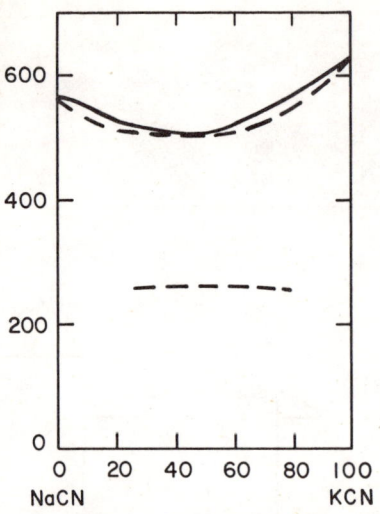

FIG. 1876.—System KCN–NaCN.

Wilhelm Truthe, *Z. anorg. Chem.*, **76**, 134 (1912).

KCN–Zn(CN)$_2$

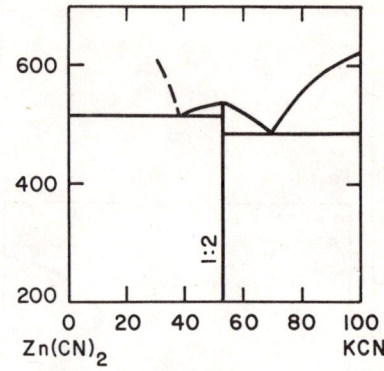

FIG. 1877.—System KCN–Zn(CN)$_2$.

Wilhelm Truthe, *Z. anorg. Chem.*, **76**, 156 (1912).

II. Metal with Sulfur

(a) Two Substances

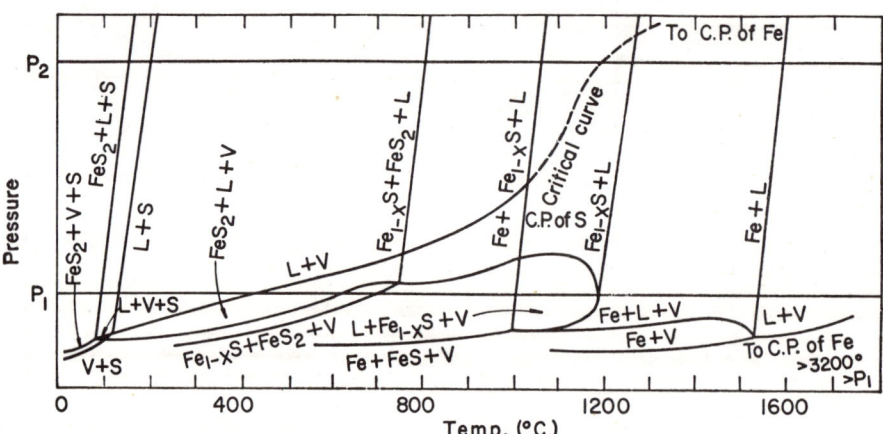

FIG. 1878.—System Fe–S; schematic p–T. Polymorphic changes have been neglected. P$_1$ is approximately 1 bar; P$_2$ is several thousand bars.

G. Kullerud and H. S. Yoder, *Econ. Geol.*, **54** [4] 559 (1959).

Fe–S (concl.)

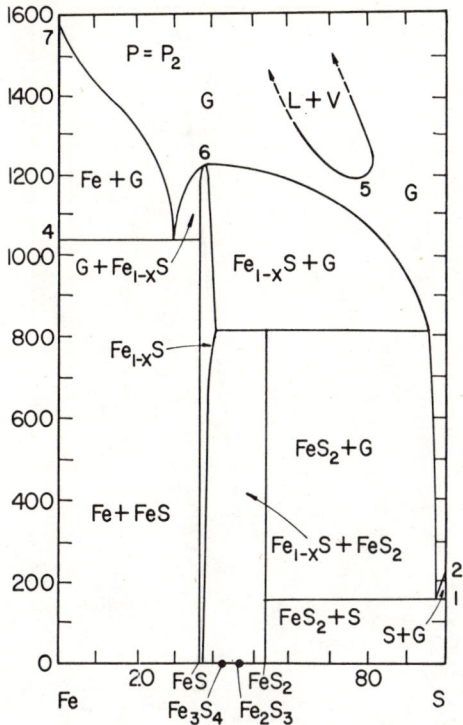

Fig. 1879.—System Fe–S; schematic temperature-composition diagram at the pressure P_2, several thousand bars. Polymorphic changes have been neglected, and the solubility of iron in sulfur has been exaggerated for clarity.

G. Kullerud and H. S. Yoder, *Econ. Geol.*, **54** [4] 562 (1959).

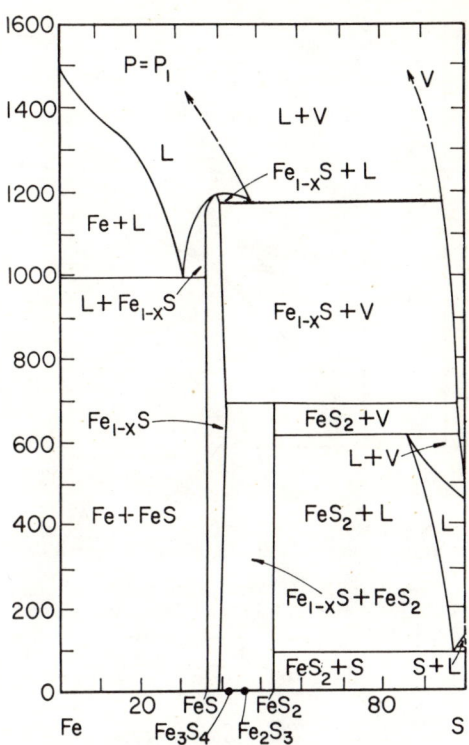

Fig. 1880.—System Fe–S; schematic temperature-composition diagram at the pressure P_1, approximately 1 atm. Polymorphic changes have been neglected, and the solubility of iron in sulfur has been exaggerated for clarity.

G. Kullerud and H. S. Yoder, *Econ. Geol.*, **54** [4] 561 (1959).

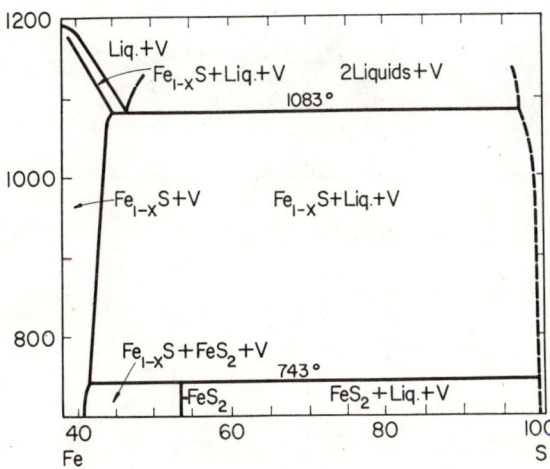

Fig. 1881.—System Fe–S above 700°C. Vapor, v, present throughout.

G. Kullerud, Ann. Rept. Director Geophys. Lab. 1960–61; in *Carnegie Inst. Washington Year Book*, **60**, 175 (1961).

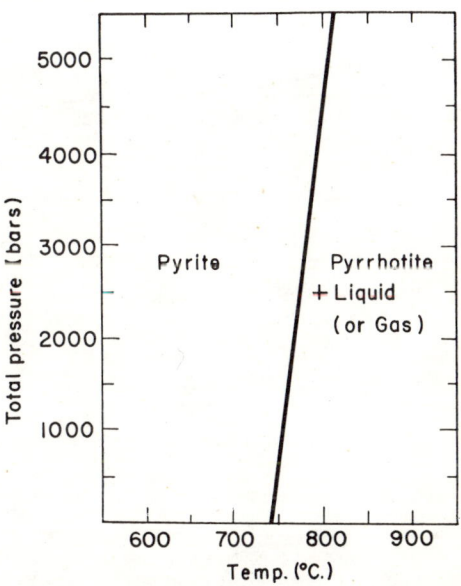

Fig. 1882.—System Fe–S. Univariant curve for the reaction $FeS_2 = Fe_{1-x}S + L$, which terminates at the point 743°C. and approximately 10 bars.

G. Kullerud and H. S. Yoder, *Econ. Geol.*, **54** [4] 554 (1959).

Ni–S

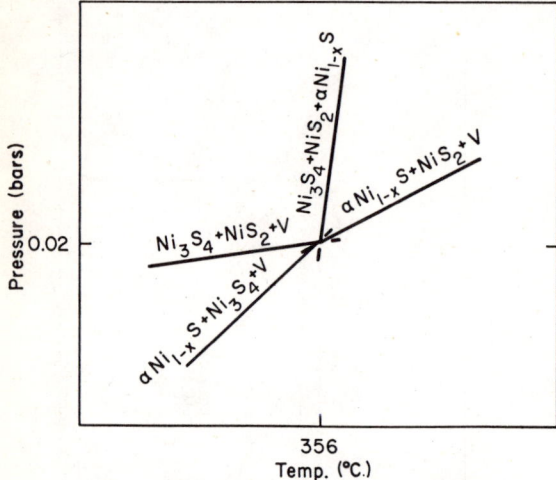

Fig. 1883.—System Ni–S; p–T, in the region of the invariant point: $\alpha Ni_{1-x}S$, Ni_3S_4, NiS_2, vapor.

G. Kullerud and R. A. Yund, Ann. Rept. Director Geophys. Lab. 1960–61; in *Carnegie Inst. Washington Year Book*, **60**, 176 (1961).

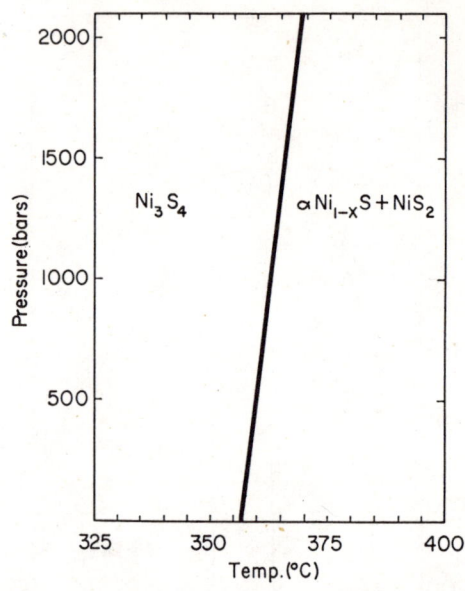

Fig. 1884.—System Ni–S; upper stability curve of polydymite. Univariant curve terminates at 356°C. and <0.002 bar.

G. Kullerud and R. A. Yund, Ann. Rept. Director Geophys. Lab. 1960–61; in *Carnegie Inst. Washington Year Book*, **60**, 177 (1961).

(b) Three Substances

As–Fe–S

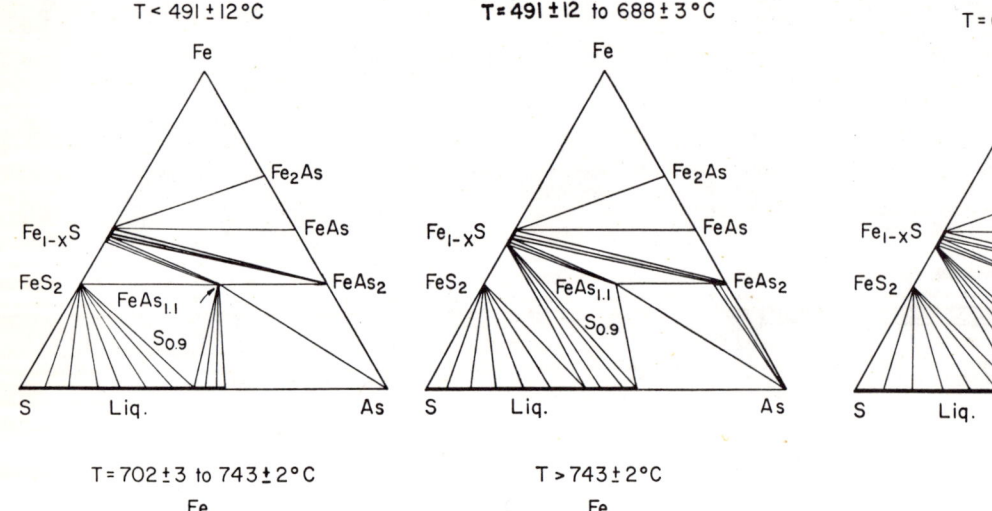

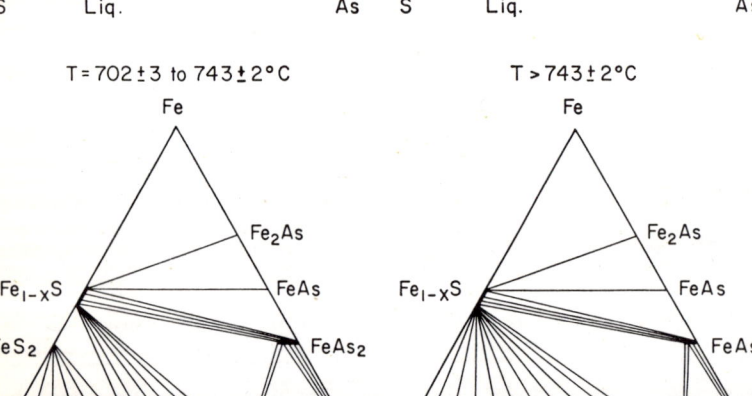

Fig. 1885.—System As–Fe–S; phase assemblages stable with vapor in specified temperature ranges.

L. A. Clark, *Econ. Geol.*, **55** [7] 1376 (1960).

(c) Four Substances

Co–Fe–CoS–FeS

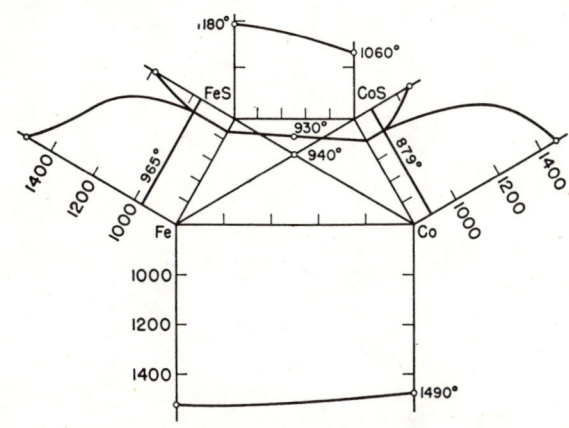

Fig. 1886.—System Fe–Co–CoS–FeS.

P. Asanti and E. J. Kohlmeyer, *Z. anorg. Chem.*, **265**, 93 (1951).

III. Sulfides Only

(a) Two Sulfides

Ag_2S–Tl_2S

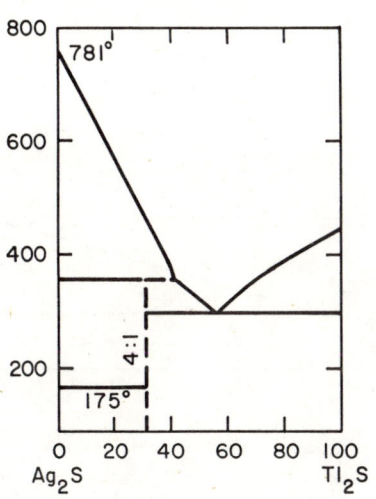

Fig. 1887.—System Ag_2S–Tl_2S.

Hans Huber, *Z. Anorg. u. allgem. Chem.*, **116**, 140 (1921).

Ag_2S–As_2S_3

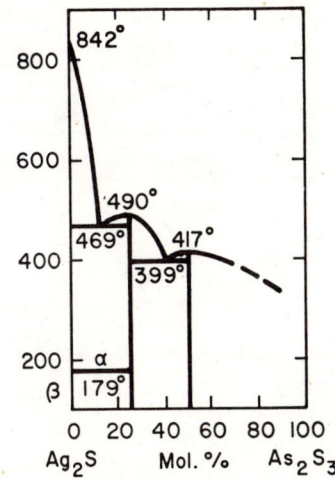

Fig. 1888.—System Ag_2S–As_2S_3.

F. M. Jaeger and H. S. van Klooster, *Z. anorg. Chem.*, **78**, 266 (1912).

Ag_2S–Sb_2S_3

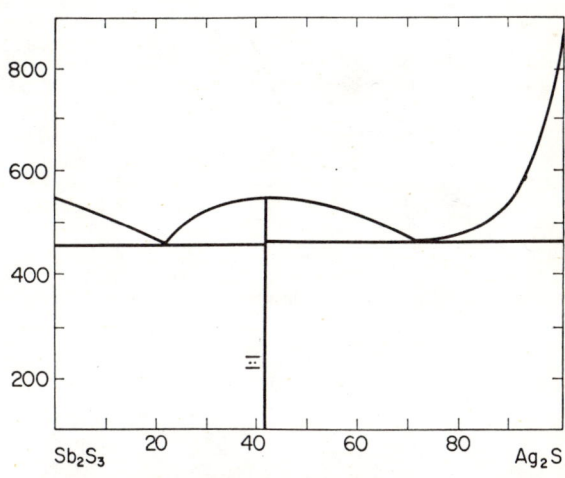

Fig. 1889.—System Ag_2S–Sb_2S_3.

Kosuke Konno, *Mem. Coll. Sci., Kyoto Imp. Univ.*, **4**, 53 (1919). See also F. M. Jaeger and H. S. van Klooster, *Z. anorg. Chem.*, **78**, 257 (1912).

Cu_2S–Ni_2S

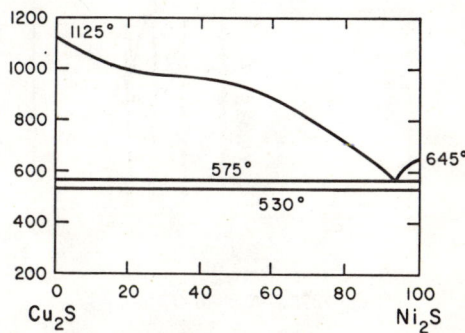

Fig. 1890.—System Cu_2S–Ni_2S.

K. Friedrich, *Z. Metall. u. Erzbergbau*, **1**, 164 (1914).

Cu_2S–Ni_3S_2

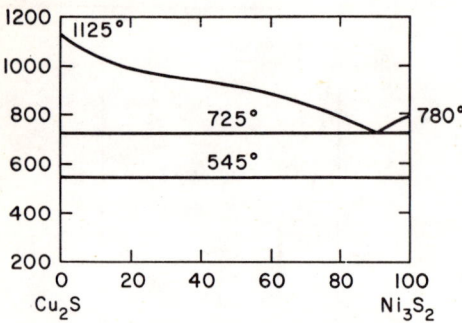

Fig. 1891.—System Cu_2S–Ni_3S_2.

K. Friedrich, *Z. Metall. u. Erzbergbau*, **1**, 162 (1914).

Tl_2S–PbS

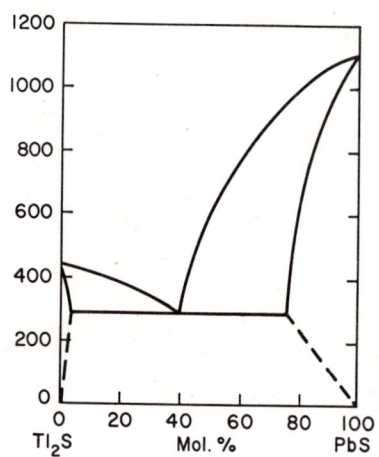

Fig. 1892.—System Tl_2S–PbS.

G. Canneri and L. Fernandes, *Atti reale accad. Lincei., Sez. I*, **1**, 674 (1925).

Tl_2S–As_2S_3

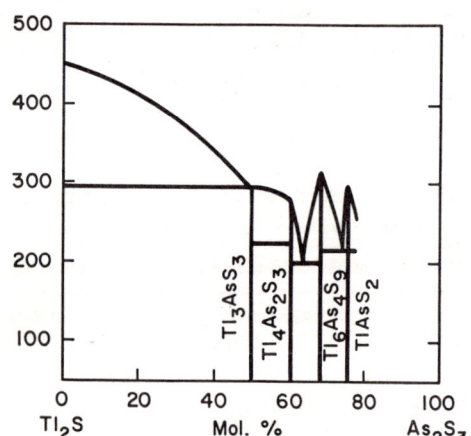

Fig. 1893.—System Tl_2S–As_2S_3.

G. Canneri and L. Fernandes, *Atti reale accad. Lincei, Sez. I*, **1**, 674 (1925).

BiS–Sb_2S_3

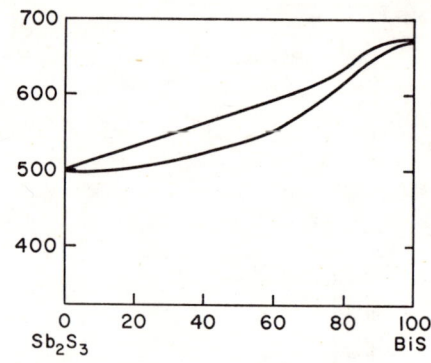

Fig. 1894.—System BiS–Sb_2S_3.

Yasuyo Takahashi, *Mem. Coll. Sci., Kyoto Imp. Univ.*, **4**, 49 (1919).

CoS–FeS

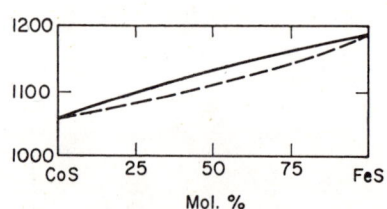

Fig. 1895.—System CoS–FeS.

P. Asanti and E. J. Kohlmeyer, *Z. anorg. Chem.*, **265**, 92 (1951).

FeS–SnS

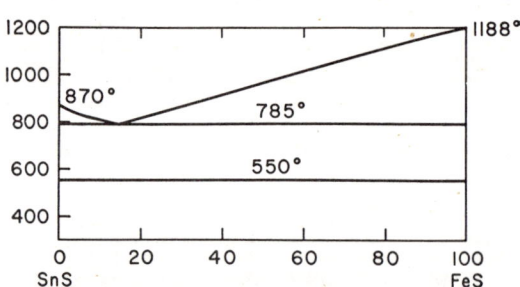

Fig. 1896.—System FeS–SnS.

Dipl.-Ing. Haan, *Z. Metall. u. Erzbergbau*, **10**, 832 (1913).

FeS–ZnS

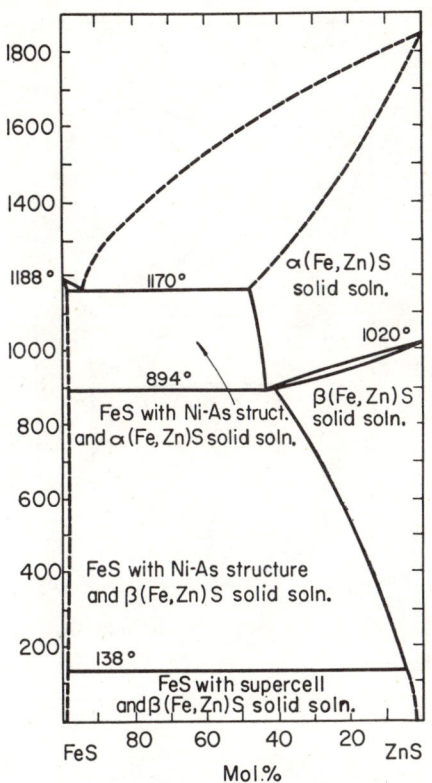

FIG. 1897.—System FeS–ZnS.

Gunnar Kullerud, *Norsk Geol. Tidsskr.*, **32**, 64 (1953).

PbS–Sb₂S₃

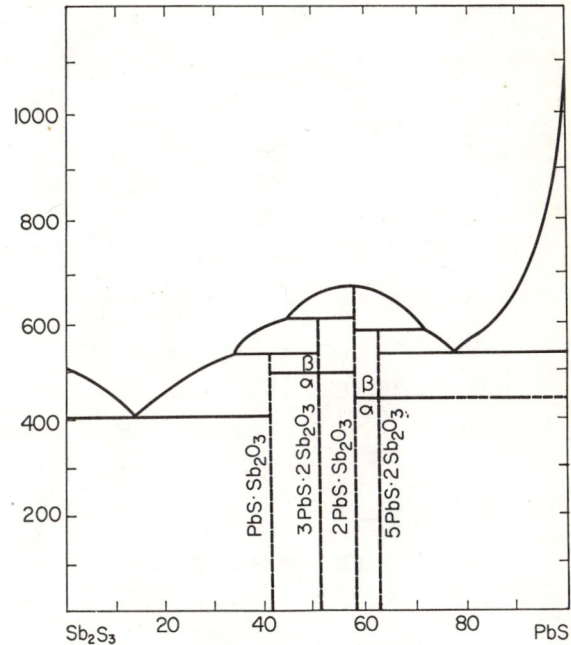

FIG. 1898.—System PbS–Sb₂S₃.

Daidzi Iitsuka, *Mem. Coll. Sci., Kyoto Imp. Univ.*, **4**, 61 (1920). See also F. M. Jaeger and H. S. van Klooster, *Z. anorg. Chem.*, **78**, 262 (1912).

(b) Three Sulfides

SnS–Sb₂S₃

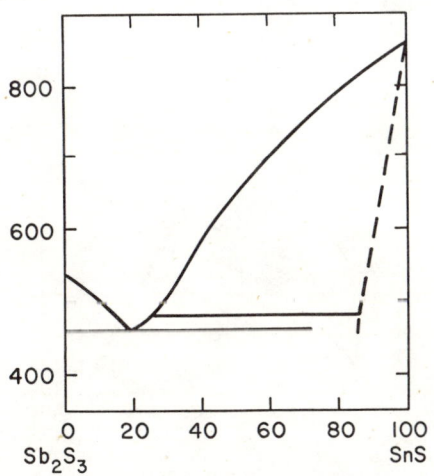

FIG. 1899.—System SnS–Sb₂S₃.

N. Parravano and P. DeCesaris, *Gazz. chim. ital.*, **42 II**, 6 (1912).

FeS–Ni₃S₂–Co₄S₃

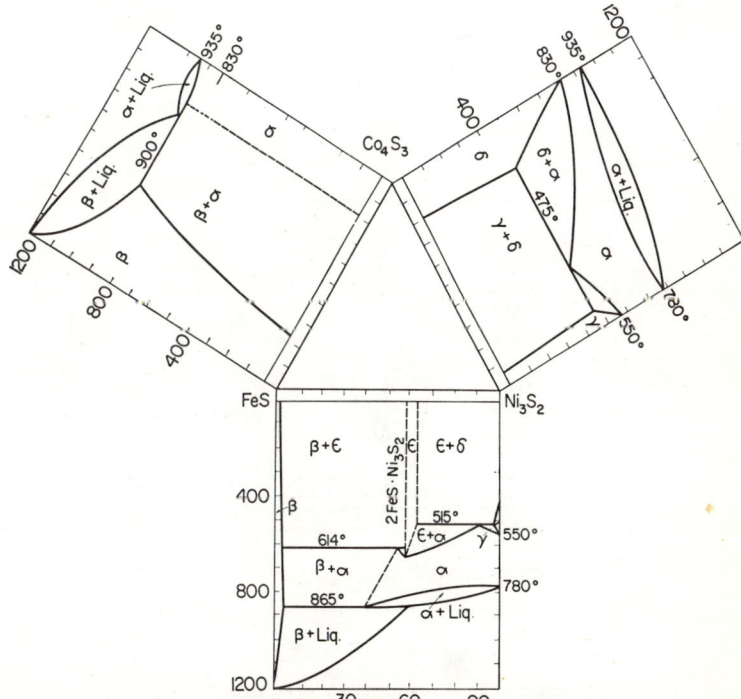

FIG. 1900.—System FeS–Ni₃S₂–Co₄S₃.

N. A. Anisheva and P. S. Kusakin, *Zhur. Neorg Khim.*, **3** [4] 916 (1958).

FeS–Ni$_3$S$_2$–Co$_4$S$_3$(concl.)

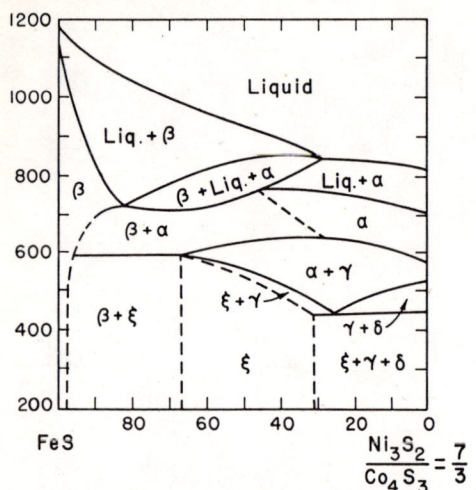

Fig. 1901.—System FeS–Ni$_3$S$_2$–Co$_4$S$_3$. Cut between FeS and Ni$_3$S$_2$/Co$_4$S$_3$ = 7/3 (See Fig. 1900).

N. A. Anisheva and P. S. Kusakin, *Zhur. Neorg. Khim.*, **3** [4] 917 (1958).

Sb$_2$S$_3$–Sb$_2$O$_3$

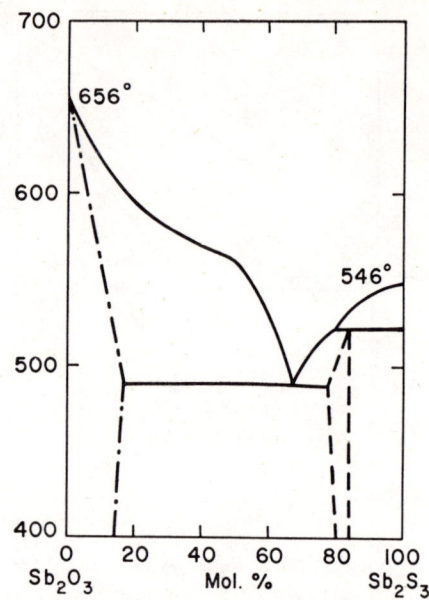

Fig. 1903.—System Sb$_2$O$_3$–Sb$_2$S$_3$.

E. Quercigh, *Atti reale accad. Lincei, Sez. I,* **21,** 418 (1912).

IV. Sulfides with Metal Oxides

(a) Two Substances

RS–RO

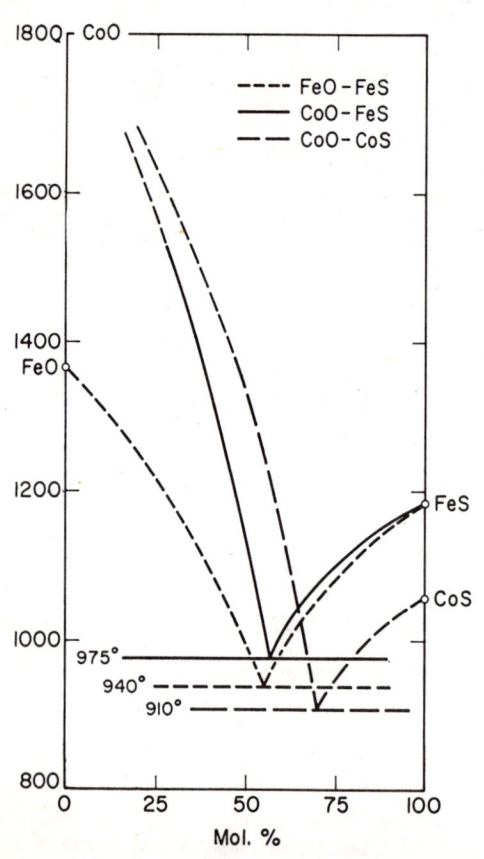

Fig. 1902.—Liquidus curves of systems FeO–FeS, CoO–FeS, CoO–CoS.

P. Asanti and E. J. Kohlmeyer, *Z. anorg. Chem.*, **265,** 94 (1951).

(b) Three Substances

Cu$_2$S–FeS–FeO

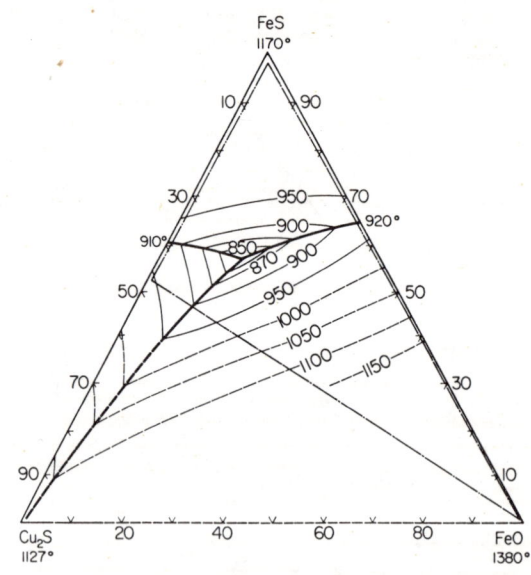

Fig. 1904.—System Cu$_2$S–FeO–FeS; partial liquidus.

Akira Yazawa and Mitsuo Kameda, *Technol. Rept. Tohoku Univ.*, **19,** 242 (1954).

FeS–FeO–SiO₂

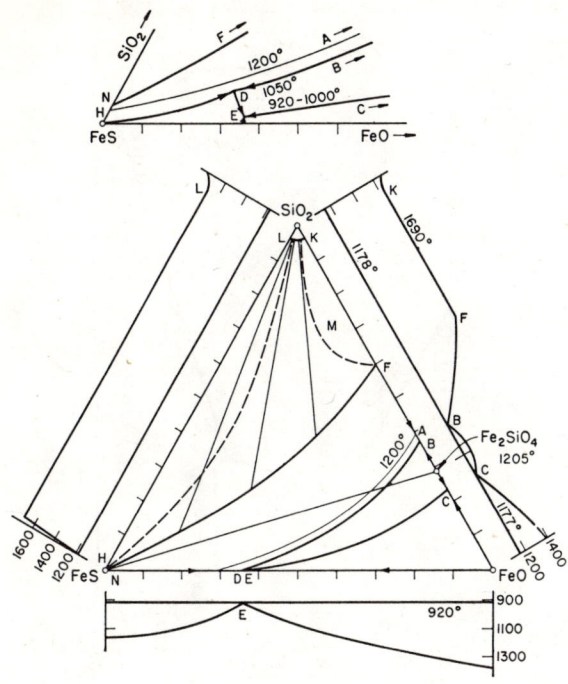

Fig. 1905.—System FeO–FeS–SiO₂.

Ya. I. Ol'shanskiĭ, *Doklady Akad. Nauk S.S.S.R.*, **70** [2] 246 (1950).

V. Miscellaneous

In–InP

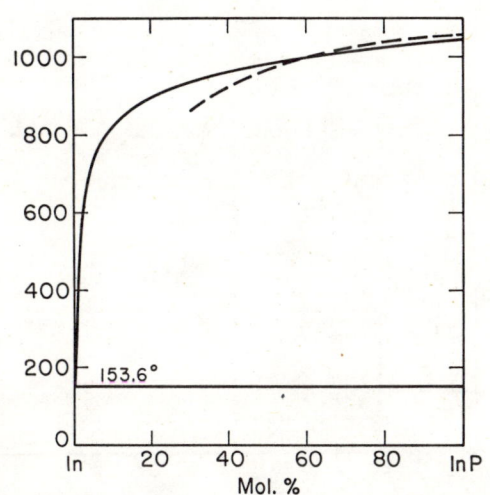

Fig. 1906.—System In–InP.

M. Shafer and K. Weiser, *J. Phys. Chem.*, **61**, 1425 (1957).

KCNS–NaCNS

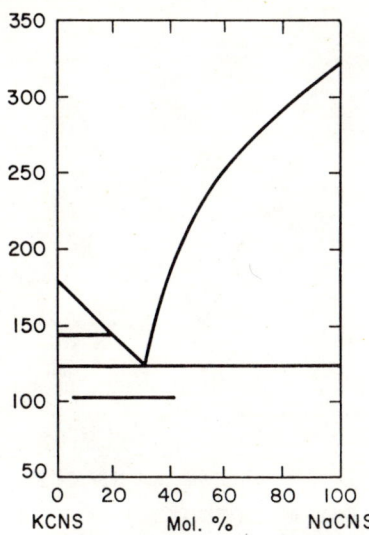

Fig. 1907.—System KCNS–NaCNS.

J. B. Wrzesnewsky, *Z. anorg. Chem.*, **74**, 99 (1912).

KCNS–NH₄CNS

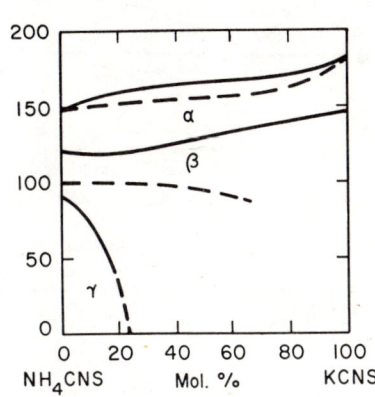

Fig. 1908.—System KCNS–NH₄CNS.

J. B. Wrzesnewsky, *Z. anorg. Chem.*, **74**, 102 (1912).

KCNS–RbCNS

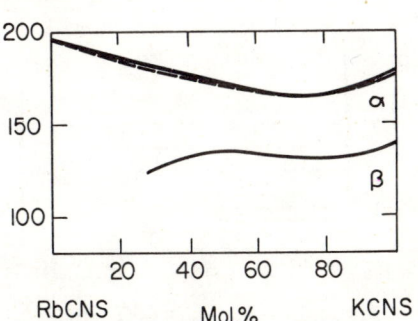

Fig. 1909.—System KCNS–RbCNS.

J. B. Wrzesnewsky, *Z. anorg. Chem.*, **74**, 105 (1912).

NaCNS–NaNO₃

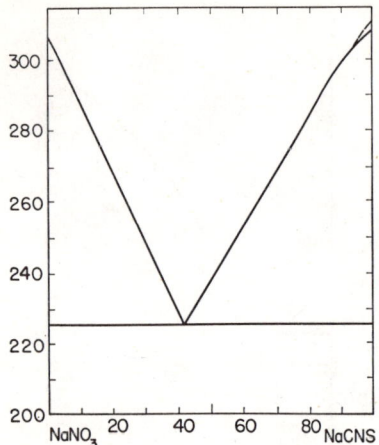

Fig. 1910.—System NaCNS–NaNO₃.

Blahoslav Stehlík, *Chem. Zvesti*, **10**, 534 (1956).

NH₄CNS–(NH₂)₂CS

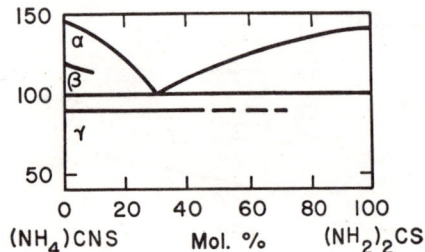

Fig. 1911.—System (NH₂)₂CS–(NH₄)CNS.

J. B. Wrzesnewsky, *Z. anorg. Chem.*, **74**, 104 (1912).

KNH₂–NaNH₂

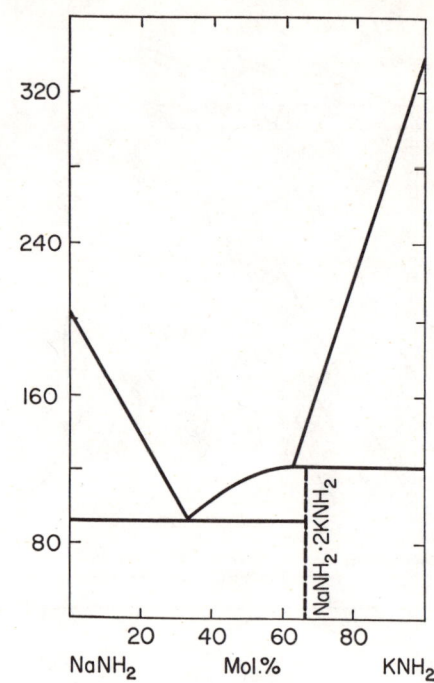

Fig. 1912.—System KNH₂–NaNH₂.

C. A. Kraus and E. J. Cuy, *J. Am. Chem. Soc.*, **45**, 713 (1923).

Na₂SO₄–Sb₂S₃

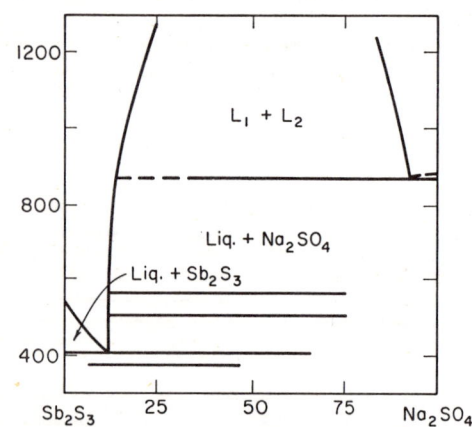

Fig. 1913.—System Na₂SO₄–Sb₂S₃.

K. A. Bol'shakov and P. I. Fedorov, *Izvest. Vysshikh Uchebn. Zavedeniĭ, Tsvetn. Met.*, **2** [2] 52 (1959).

Al₂O₃–Al₄C₃

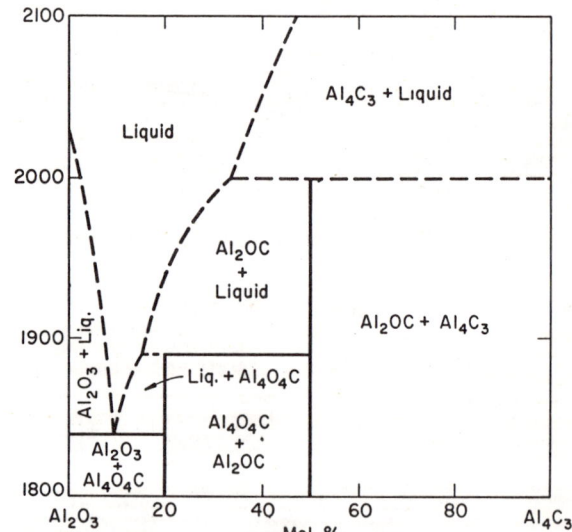

Fig. 1914.—System Al₂O₃–Al₄C₃.

L. M. Foster, G. Long, and M. S. Hunter, *J. Am. Ceram. Soc.*, **39** [1] 8 (1956).

G. SYSTEMS CONTAINING WATER

I. H_2O

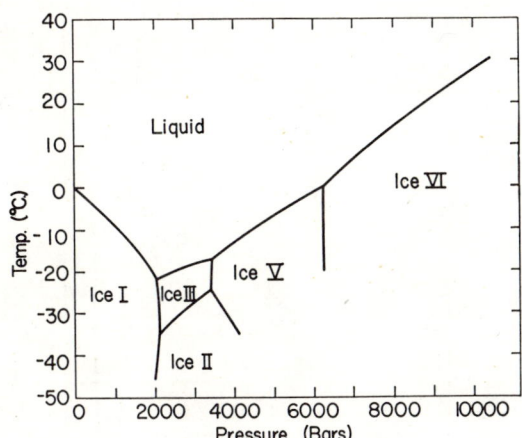

Fig. 1915.—System H_2O; p–T diagram between 0 and 10,000 bars.

The triple points are at:

°C	Bars	Coexisting Phases
−22.0	2072	III-L-I
−34.7	2128	II-III-I
−17.0	3465	V-III-L
−24.3	3445	V-II-III
0.16	6257	VI-V-L

Compiled by C. E. Weir of the National Bureau of Standards, after data by P. W. Bridgman. Triple point values obtained from I. C. T., **IV**, p. 11, 1928 (after Bridgman).

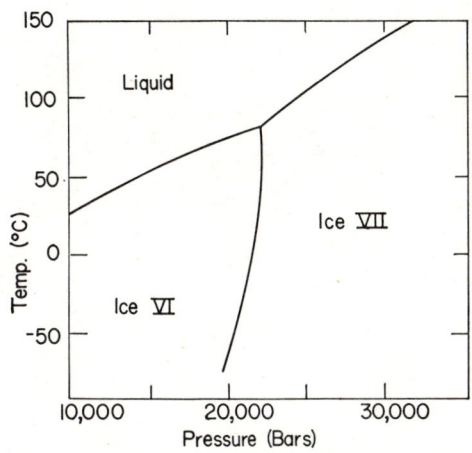

Fig. 1916.—System H_2O; p–T diagram between 10,000 and 30,000 bars. The triple point is 81.6°C and 21,970 bars.

P. W. Bridgman, *J. Chem. Phys.*, **5**, 965 (1937).

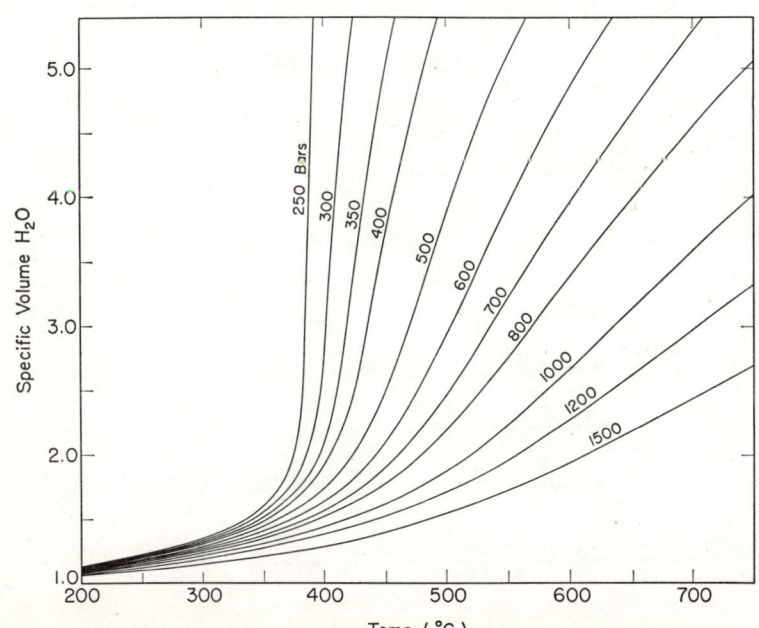

Fig. 1917.—System H_2O; isobaric curves for specific volume.

G. C. Kennedy, *Econ. Geol.*, **45** [7] 641 (1950).

II. One Metal Oxide with Water

H_2O_2–H_2O

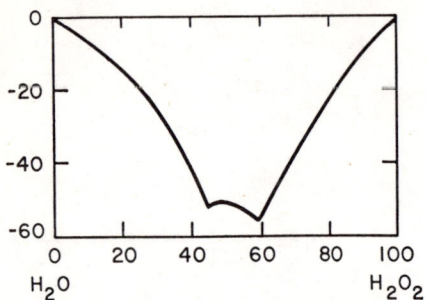

Fig. 1918.—System H_2O–H_2O_2.

O. Maass and O. W. Herzberg, *J. Am. Chem. Soc.*, **42**, 2570 (1920).

CaO–H_2O

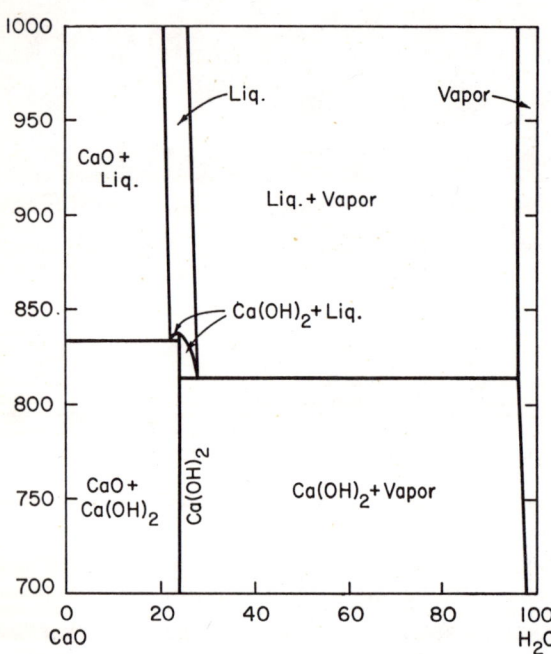

Fig. 1919.—System CaO–H_2O at 1000 bars pressure. Liq. = liquid.

P. J. Wyllie and O. F. Tuttle, *J. Am. Ceram. Soc.*, **42** [9] 449 (1959).

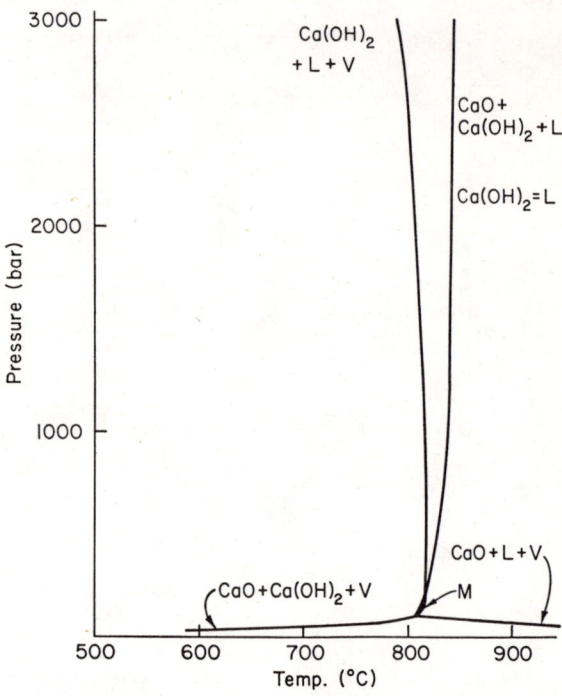

Fig. 1920.—System CaO–H_2O; p–T projection of univariant equilibria. L = liquid, V = vapor (almost pure H_2O). A curve $Ca(OH)_2$ = L, with restricted univariance, extends from the singular point, M, and is almost coincident with the curve CaO + $Ca(OH)_2$ + L.

P. J. Wyllie and O. F. Tuttle, *J. Am. Ceram. Soc.*, **42** [9] 449 (1959).

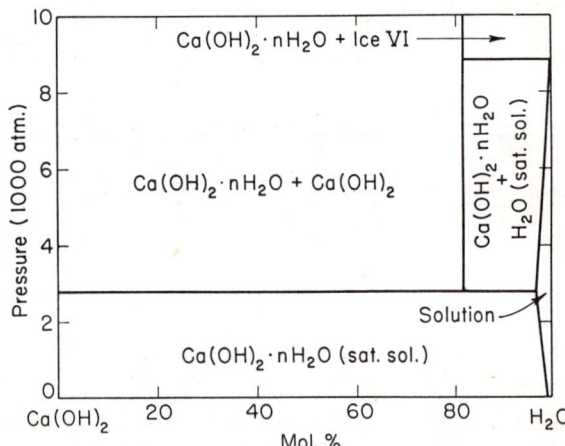

Fig. 1921.—System $Ca(OH)_2$–H_2O at 21°C.; proposed.

C. E. Weir, *J. Research Natl. Bur. Standards*, **54** [1] 40 (1955); RP2562.

Note: In the lower part of the diagram the field labeled $Ca(OH)_2 \cdot nH_2O$ (sat. sol.) should read $Ca(OH)_2$ + H_2O (sat. sol.).

CdO–H₂O

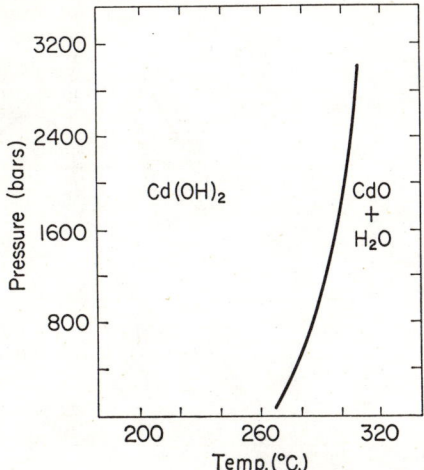

FIG. 1922.—System CdO–H₂O; p–T.

L. S. Dent Glasser and R. Roy, *J. Inorg. Nucl. Chem.*, **17**, 100 (1961).

MgO–H₂O

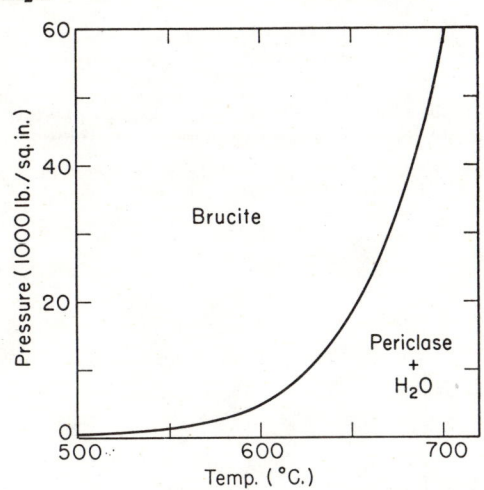

FIG. 1923.—System MgO–H₂O.

D. M. Roy and Rustum Roy, *Am. J. Sci.*, **255**, 579 (1957).

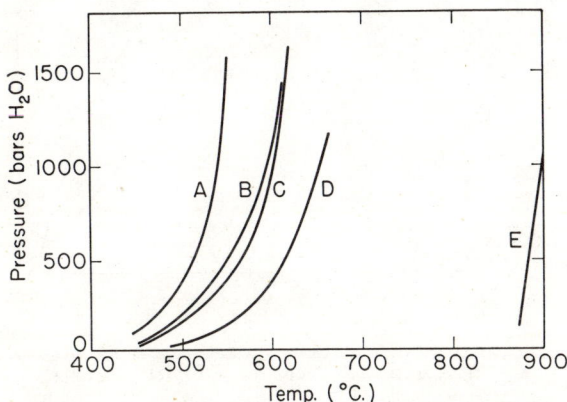

FIG. 1924.—System MgO–H₂O, by various workers. A = McDonald calculated; B = from vapor pressure measurement; C = McDonald calculated; D = Roy, Roy, and Osborn; E = Bowen and Tuttle. Field of brucite to the left of each curve; periclase plus water to the right.

G. C. Kennedy, *Am. J. Sci.*, **254**, 568 (1956).

MnO–H₂O

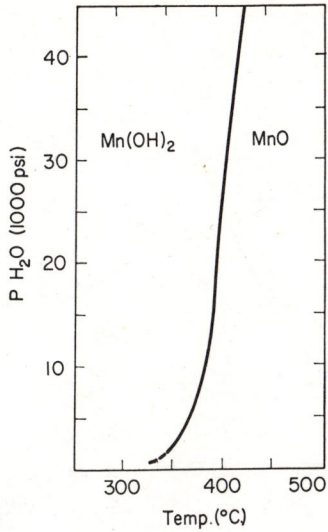

FIG. 1925.—System MnO–H₂O, pyrochroite-manganosite reaction.

Cyrus Klingsberg and Rustum Roy, *Am. Mineral.*, **44**, 829 (1959).

Al₂O₃–H₂O

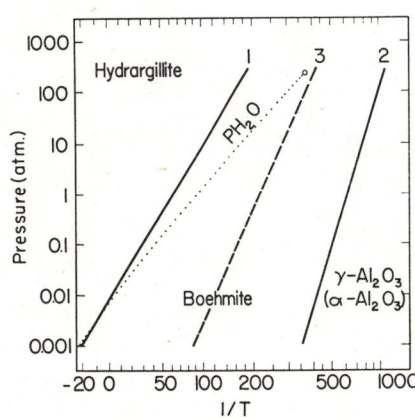

FIG. 1926.—System Al₂O₃–H₂O; p–T.

K. Torkar and H. Worel, *Monatsh. Chem.*, **88**, 743 (1957).

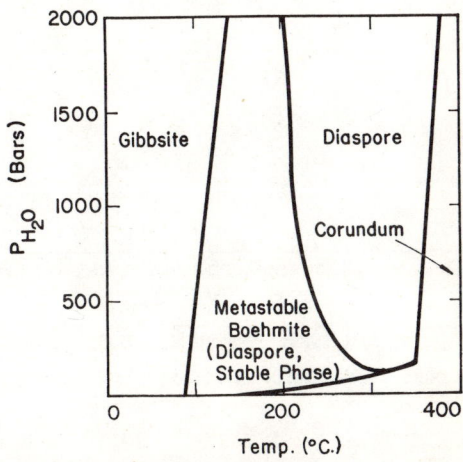

FIG. 1927.—System Al₂O₃–H₂O at low H₂O pressures.

G. C. Kennedy, *Am. J. Sci.*, **257**, 568 (1959).

Al_2O_3–H_2O (concl.)

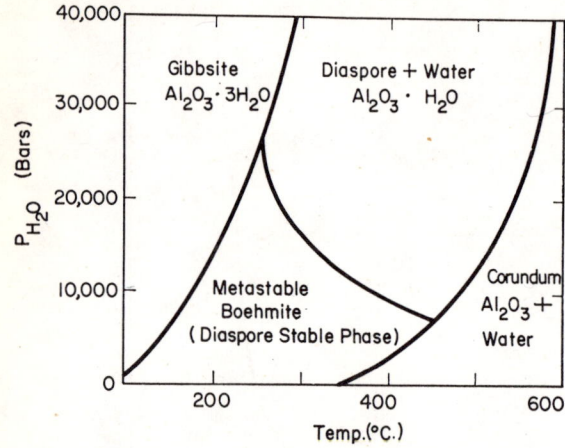

Fig. 1928.—System Al_2O_3–H_2O at high H_2O pressures.

G. C. Kennedy, *Am. J. Sci.*, **257**, 565 (1959).

B_2O_3–H_2O

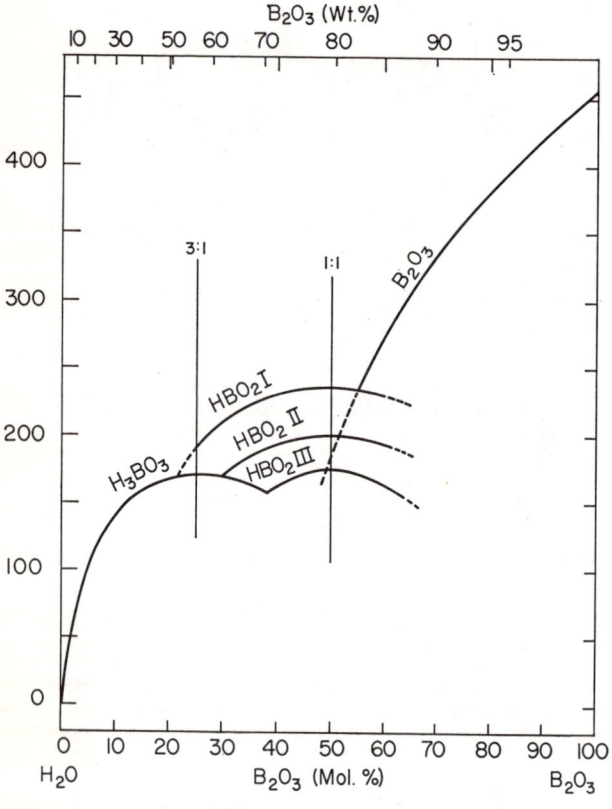

Fig. 1929.—System H_2O–B_2O_3.

F. C. Kracek, G. W. Morey, and H. E. Merwin, *Am. J. Sci.* 5th Ser., **35A**, 148 (1938).

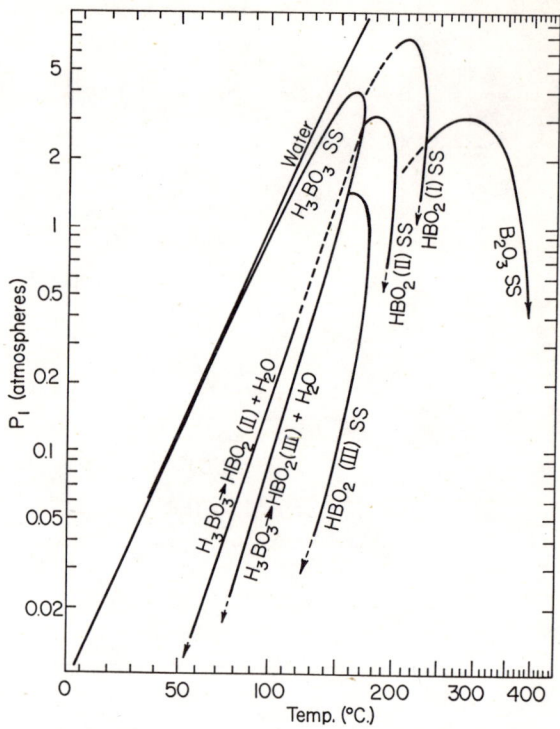

Fig. 1930.—System H_2O–B_2O_3; pressure-temperature diagram for vapor pressures of saturated solutions.

F. C. Kracek, G. W. Morey, and H. E. Merwin, *Am. J. Sci.* 5th Ser., **35A**, 162 (1938).

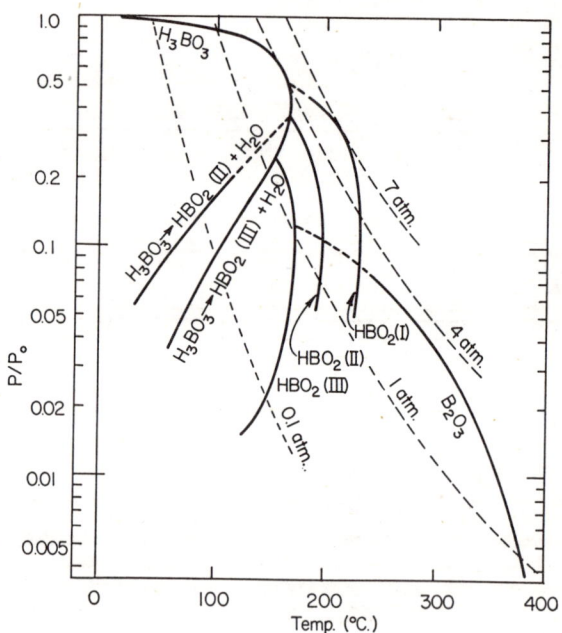

Fig. 1931.—System H_2O–B_2O_3; relative vapor pressure and temperature relations for saturated solutions.

F. C. Kracek, G. W. Morey, and H. E. Merwin, *Am. J. Sci.* 5th Ser., **35A**, 163 (1938).

Cr_2O_3–H_2O

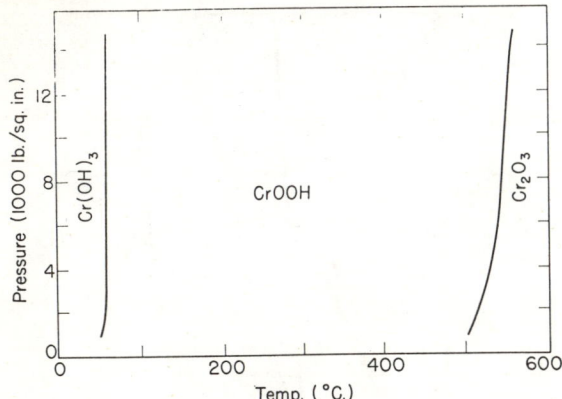

Fig. 1932.—System Cr_2O_3–H_2O.

M. W. Shafer and Rustum Roy, Z. anorg. u. allgem. Chem., 276, 275–288 (1954).

Ga_2O_3–H_2O

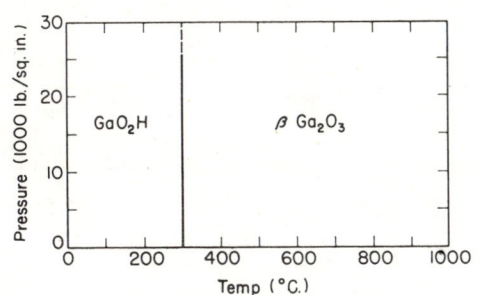

Fig. 1933.—System Ga_2O_3–H_2O.

V. G. Hill, Rustum Roy, and E. F. Osborn, J. Am Ceram. Soc., 35 [6] 136 (1952).

In_2O_3–H_2O

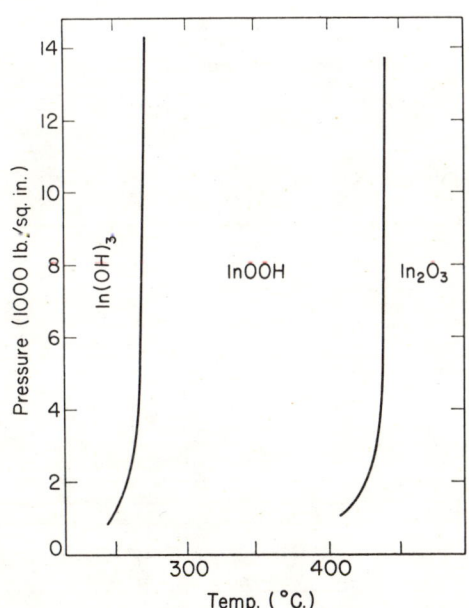

Fig. 1934.—System In_2O_3–H_2O.

Rustum Roy and M. W. Shafer, J. Phys. Chem., 58, 372–375 (1954).

La_2O_3–H_2O

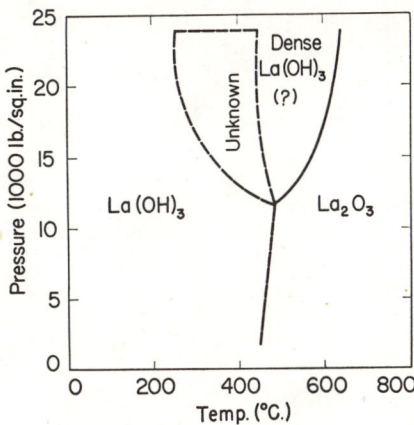

Fig. 1935.—System La_2O_3–H_2O; p–T.

M. W. Shafer and Rustum Roy, J. Am. Ceram. Soc., 42 [11] 567 (1959).

Mn_2O_3–H_2O

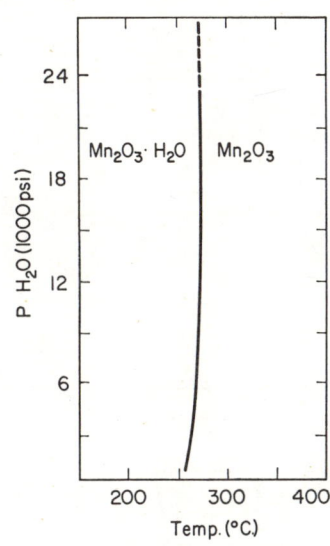

Fig. 1936.—System Mn_2O_3–H_2O; manganite-bixbyite reaction.

Cyrus Klingsberg and Rustum Roy, Am. Mineral., 44, 829 (1959).

Nd_2O_3–H_2O

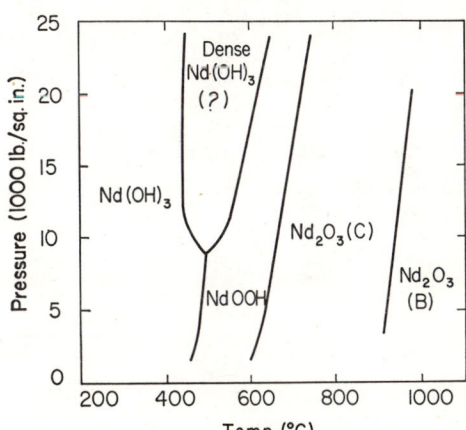

Fig. 1937.—System Nd_2O_3–H_2O; p–T.

M. W. Shafer and Rustum Roy, J. Am. Ceram. Soc., 42 [11] 567 (1959).

Sc_2O_3–H_2O

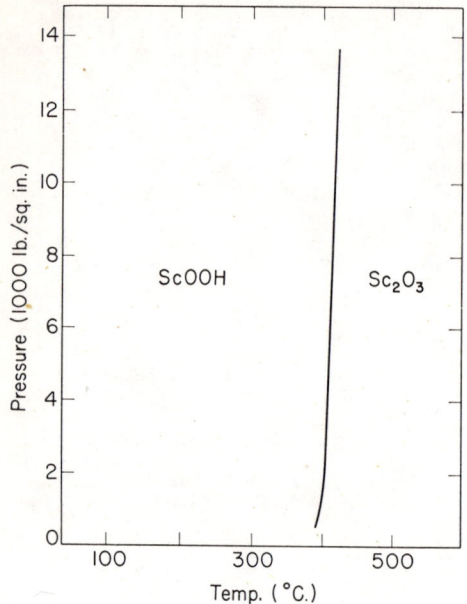

Fig. 1938.—System Sc_2O_3–H_2O.

M. W. Shafer and Rustum Roy, *Z. anorg. u. allgem. Chem.*, **276**, 275–288 (1954).

Sm_2O_3–H_2O

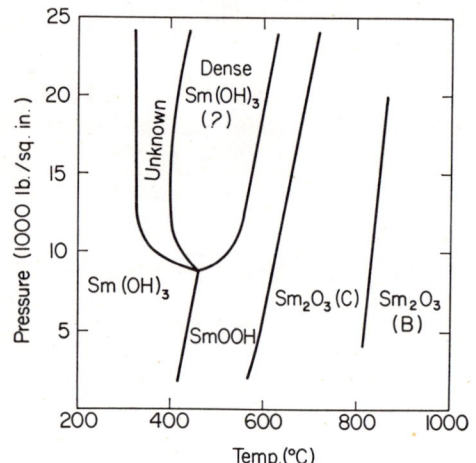

Fig. 1939.—System Sm_2O_3–H_2O; p–T.

M. W. Shafer and Rustum Roy, *J. Am. Ceram. Soc.*, **42** [11] 567 (1959).

Y_2O_3–H_2O

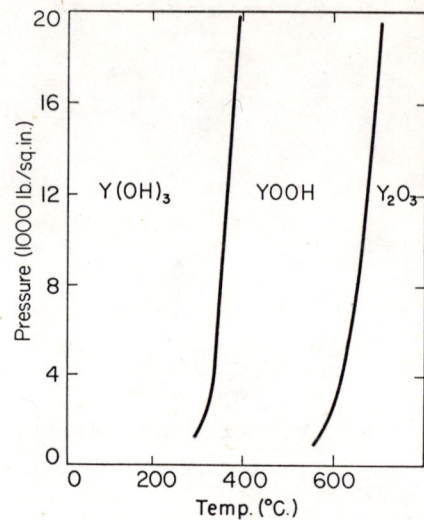

Fig. 1940.—System Y_2O_3–H_2O; p–T.

M. W. Shafer and Rustum Roy, *J. Am. Ceram. Soc.*, **42** [11] 567 (1959).

CO_2–H_2O

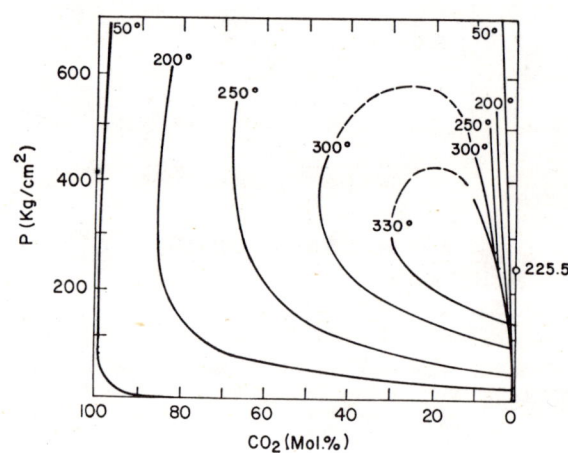

Fig. 1941.—System H_2O–CO_2.

N. I. Khitarov and S. D. Malinin, *Geokhimiya*, 1958, p. 679.

SiO_2–H_2O

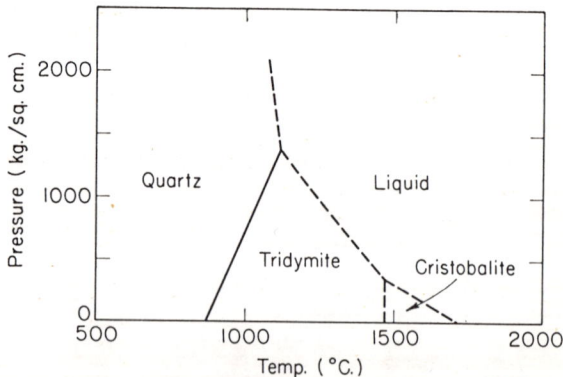

Fig. 1942.—System SiO_2–H_2O; tentative p-T diag. for silica-rich portion.

O. F. Tuttle and J. L. England, *Bull. Geol. Soc. Am.*, **66**, 150 (1955).

SiO₂–H₂O (cont.)

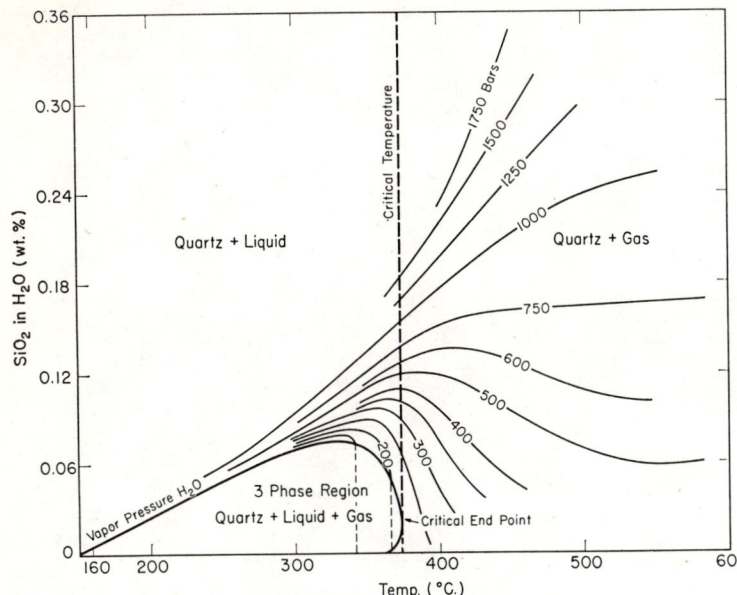

Fig. 1943.—System H_2O–SiO_2; isobaric solubility curves.

G. C. Kennedy, *Econ. Geol.*, **45** [7] 639 (1950).

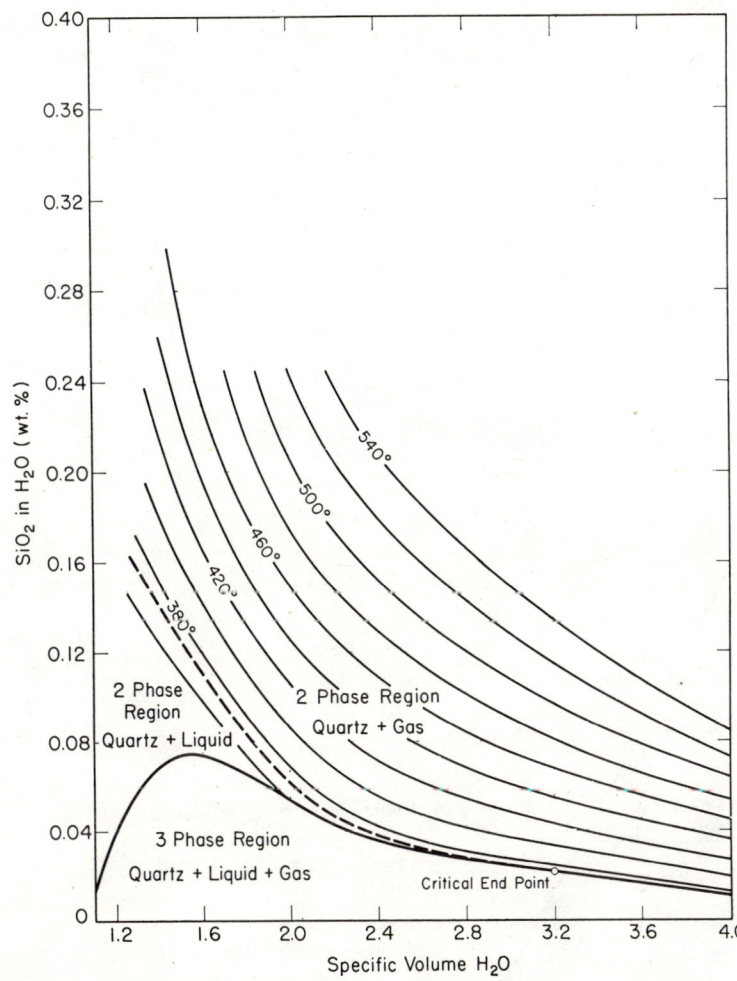

Fig. 1944.—System H_2O–SiO_2; isothermal solubility curves.

G. C. Kennedy, *Econ. Geol.*, **45** [7] 644 (1950).

SiO_2–H_2O (concl.)

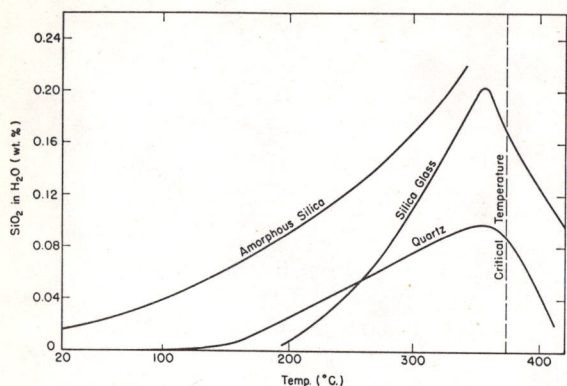

Fig. 1945.—System H_2O–SiO_2; relative solubility of various types of silica.

George C. Kennedy, *Econ. Geol.*, **45** [7] 652 (1950).

TiO_2–H_2O

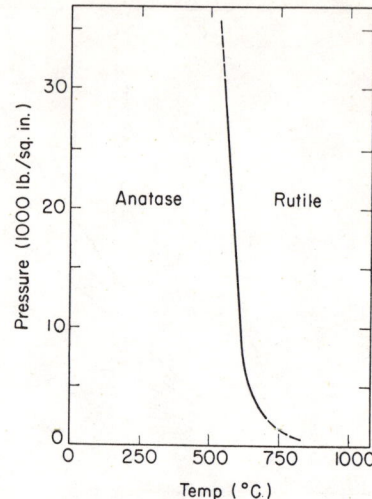

Fig. 1946.—System H_2O–TiO_2; conversion of anatase to rutile in the presence of water.

After K. C. Tu and E. F. Osborn, unpublished data.
E. F. Osborn, *J. Am. Ceram. Soc.*, **36** [5] 149 (1953).

III. Two Metal Oxides with Water

SeO_3–H_2O

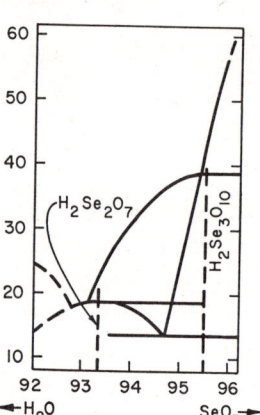

Fig. 1947.—System SeO_3–H_2O.

Karel Dostál and Martin Černohorský, *Chem. Listy*, **50**, 708 (1956).

K_2O–B_2O_3–H_2O

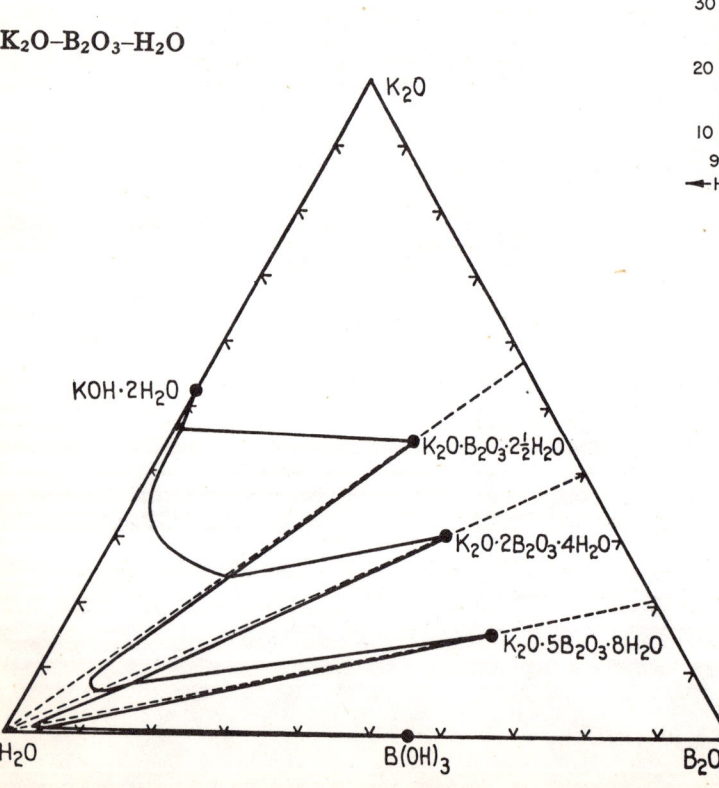

Fig. 1948.—System K_2O–B_2O_3–H_2O at 30°C.

M. Dukelski, *Z. anorg. Chem.*, **50**, 43 (1906).

K_2O–SiO_2–H_2O

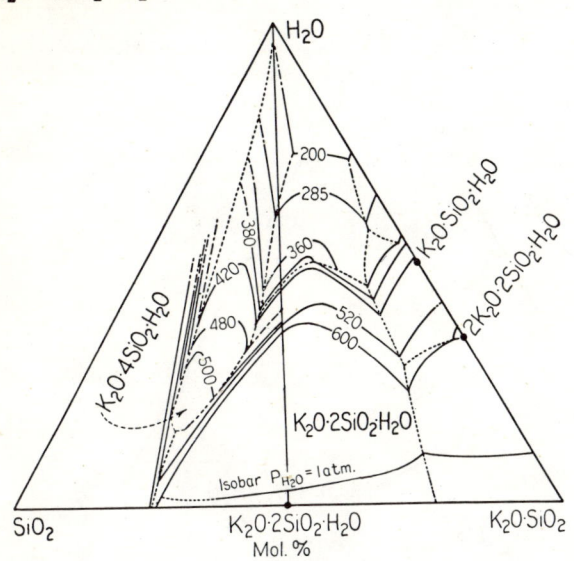

Fig. 1949.—Isothermal saturation curves for system SiO_2–K_2O·SiO_2–H_2O.

G. W. Morey, *J. Soc. Glass Technol.*, **6**, 25 (1922).

Li_2O–B_2O_3–H_2O

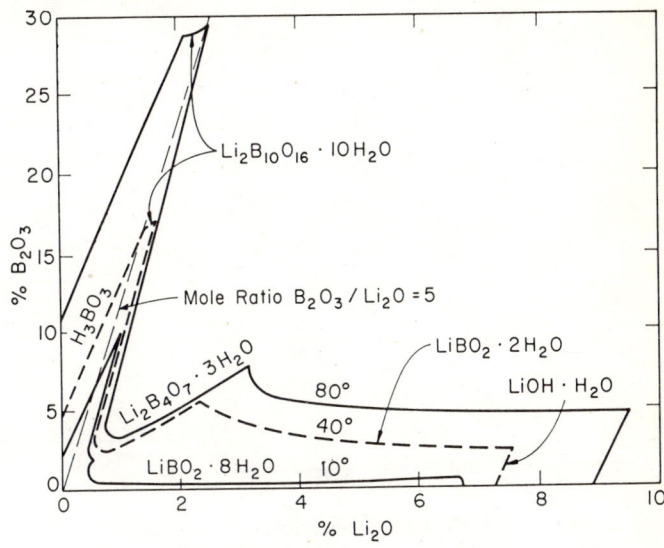

Fig. 1951.—System Li_2O–B_2O_3–H_2O.

W. T. Reburn and W. A. Gale, *J. Phys. Chem.*, **59**, 22 (1955).

K_2O–As_2O_5–H_2O

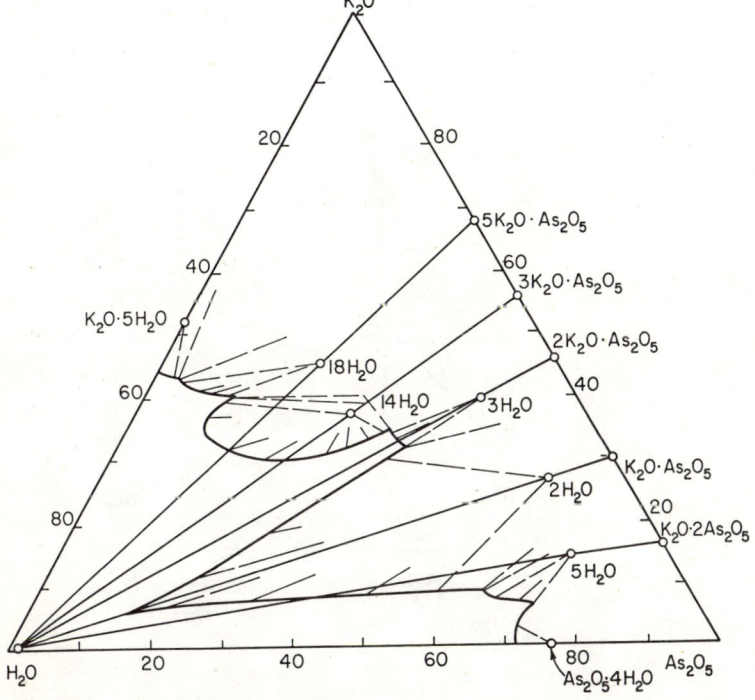

Fig. 1950.—System K_2O–As_2O_5–H_2O at 20°C.

Henri Guérin and Cyrille Duc-Maug, *Compt. rend.*, **240**, 2410 (1955).

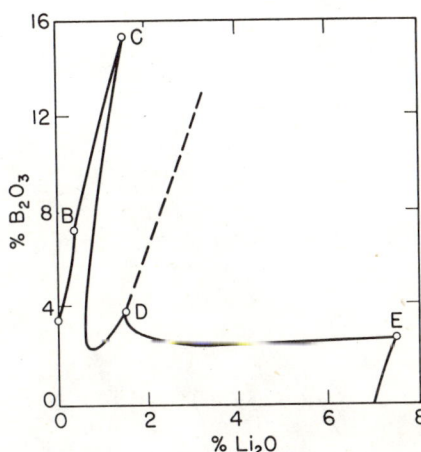

Fig. 1952.—System Li_2O–B_2O_3–H_2O, at 30°C. Point B = H_3BO_3, $5B_2O_3$·Li_2O·$10H_2O$, liquid; Point C = $5B_2O_3$·Li_2O·$10H_2O$, $2B_2O_3$·Li_2O·$4H_2O$, liquid; Point D = $2B_2O_3$·Li_2O·$4H_2O$, B_2O_3·Li_2O·$16H_2O$, liquid; Point E = B_2O_3·Li_2O·$16H_2O$, $LiOH$·H_2O, liquid.

A. P. Rollet and Roger Bouaziz, *Compt. rend.*, **240**, 1228 (1955).

Na₂O–PbO–H₂O

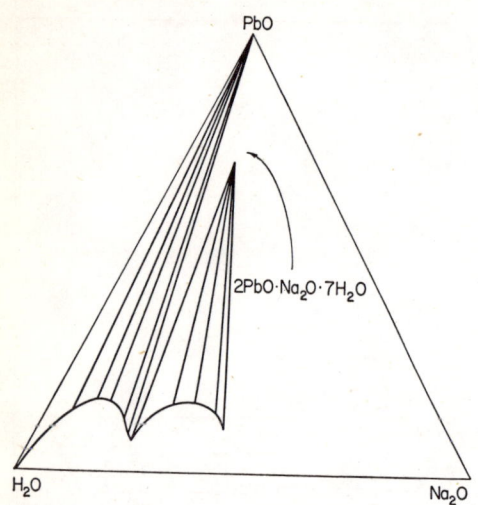

FIG. 1953.—System Na₂O–PbO–H₂O.

E. I. Sokolova and D. M. Chizhikov, *Zhur. Neorg. Khim.*, **2**, 1664 (1957).

Na₂O–B₂O₃–H₂O

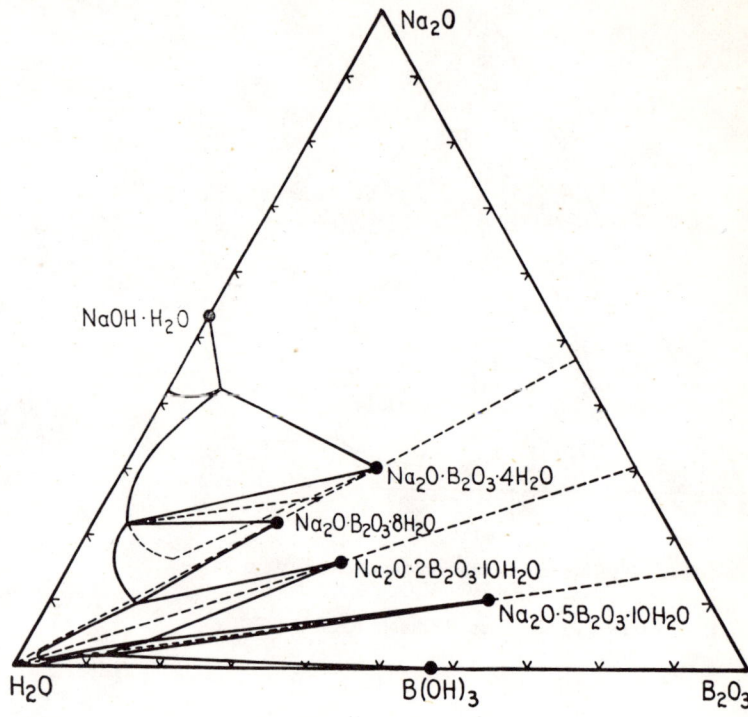

FIG. 1955.—System Na₂O–B₂O₃–H₂O at 30°C.

M. Dukelski, *Z. anorg. Chem.*, **50**, 47 (1906).

Na₂O–ZnO–H₂O

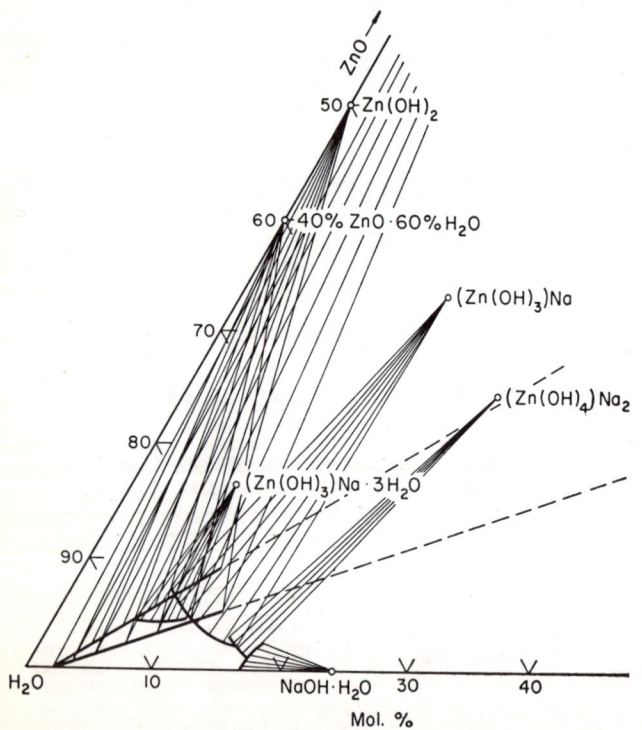

FIG. 1954.—System Na₂O–ZnO–H₂O.

V. G. Sochevanov, *Zhur. Obshcheĭ Khim.*, **22**, 1077 (1952).

Na₂O–Ga₂O₃–H₂O

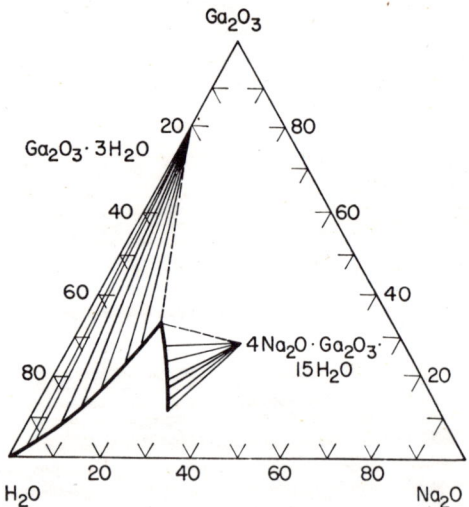

FIG. 1956.—System Na₂O–Ga₂O₃–H₂O at 18°C.

S. V. Gevorkyan and N. A. Gurovich, *Izvest. Akad. Nauk Arm. S.S.R., Ser. Khim. Nauk*, **10**, 388 (1957).

$Na_2O-Ga_2O_3-H_2O$ (concl.)

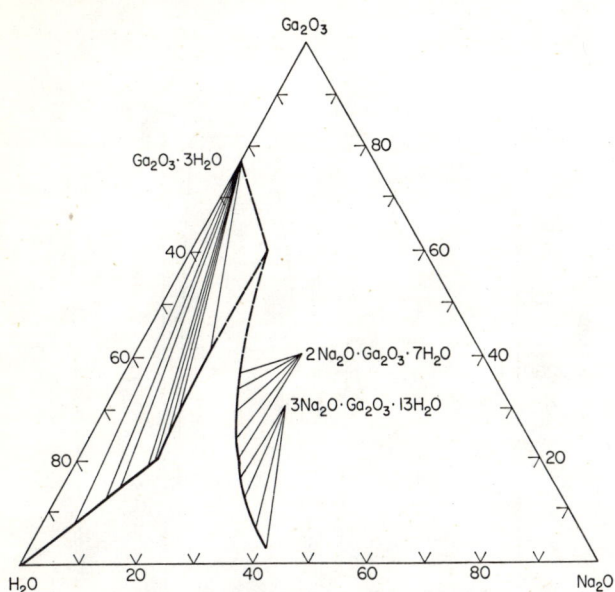

Fig. 1957.—System $Na_2O-Ga_2O_3-H_2O$ at 60°C.

S. V. Gevorkyan and N. A. Gurovich, *Izvest. Akad. Nauk Arm. S.S.R., Ser. Khim. Nauk*, **10**, 390 (1957).

$Na_2O-SiO_2-H_2O$

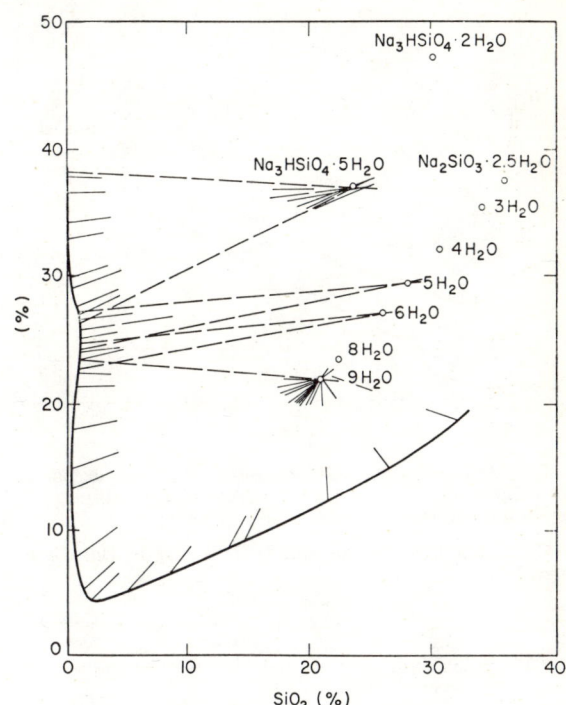

Fig. 1959.—System $H_2O-Na_2O-SiO_2$ at 10°C.

C. L. Baker and L. R. Jue, *J. Phys. Colloid Chem.*, **54**, 301 (1950).

$Na_2O-CO_2-H_2O$

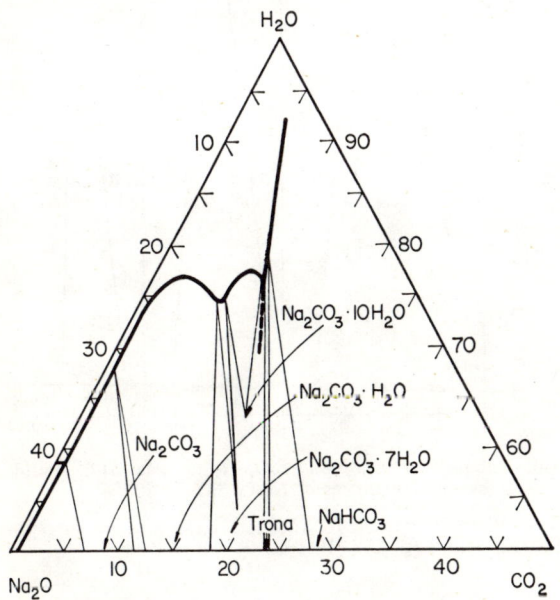

Fig. 1958.—System $Na_2O-CO_2-H_2O$ at 25°C.

F. A. Schimmel, Oak Ridge National Laboratory, Phase Diagrams for Nuclear Reactor Materials, R. E. Thoma, ed., ORNL-2548, p. 155 (1959).

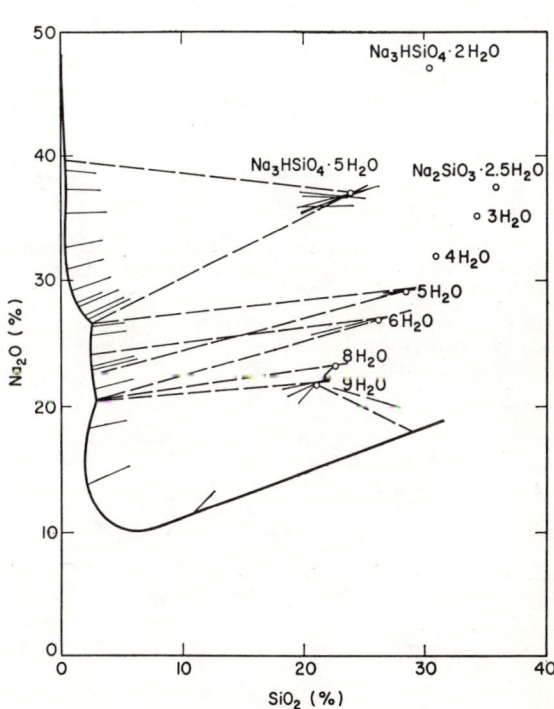

Fig. 1960.—System $H_2O-Na_2O-SiO_2$ at 31°C.

C. L. Baker and L. R. Jue, *J. Phys. Colloid Chem.*, **54**, 301 (1950).

Na_2O–SiO_2–H_2O (cont.)

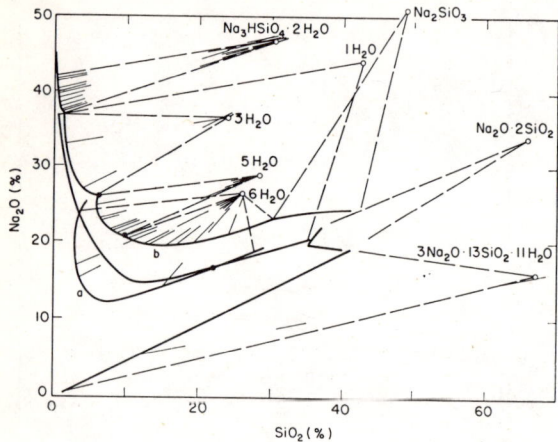

Fig. 1961.—System Na_2O–SiO_2–H_2O at 50°C. ●, points of isothermal invariance; ○, theoretical composition; —, field boundaries.

C. L. Baker, L. R. Jue, and J. H. Willis, *J. Am. Chem. Soc.*, **72**, 5378 (1950).

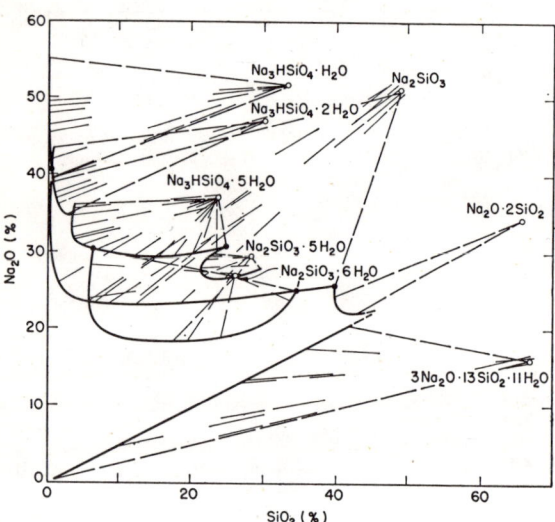

Fig. 1962.—System Na_2O–SiO_2–H_2O at 70°C. ●, points of isothermal invariance; ○, theoretical composition; — field boundaries.

C. L. Baker, L. R. Jue, and J. H. Willis, *J. Am. Chem. Soc.*, **72**, 5373 (1950).

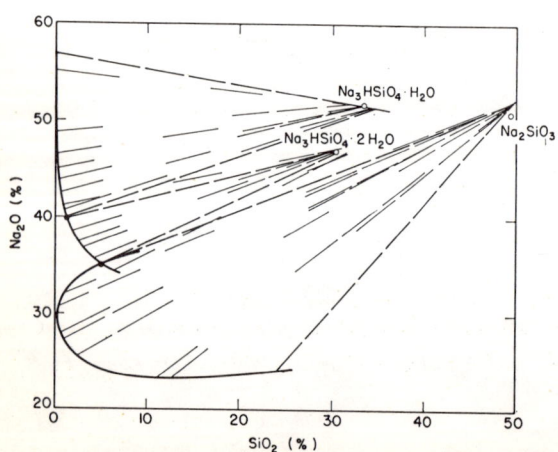

Fig. 1963.—System Na_2O–SiO_2–H_2O at 90°C.; alkaline salts only. ●, points of isothermal invariance; ○, theoretical composition; —, field boundaries.

C. L. Baker, L. R. Jue, and J. H. Willis, *J. Am. Chem. Soc.*, **72**, 5370 (1950).

Fig. 1964. See next page.

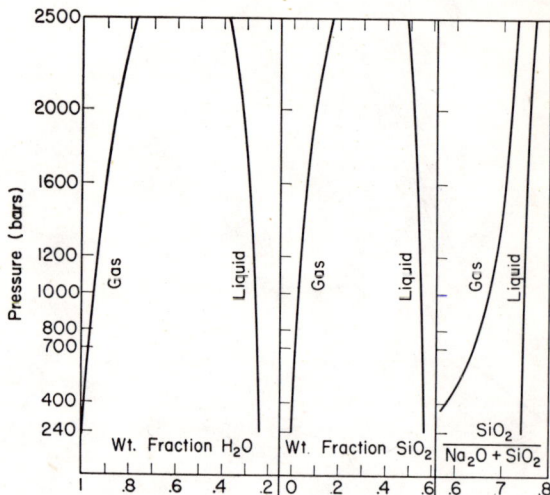

Fig. 1965.—System Na_2O–SiO_2–H_2O; gas and liquid saturation curves of quartz at 400°C.

G. W. Morey and J. M. Hesselgesser, *Am. J. Sci.*, Bowen volume, p. 361 (1952).

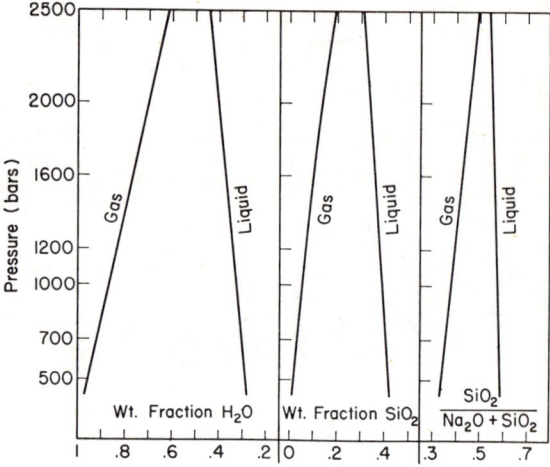

Fig. 1966.—System Na_2O–SiO_2–H_2O; gas and liquid saturation curves of $Na_2O \cdot SiO_2$ at 400°C.

G. W. Morey and J. M. Hesselgesser, *Am. J. Sci.*, Bowen volume, p. 367 (1952).

Na_2O–SiO_2–H_2O (cont.)

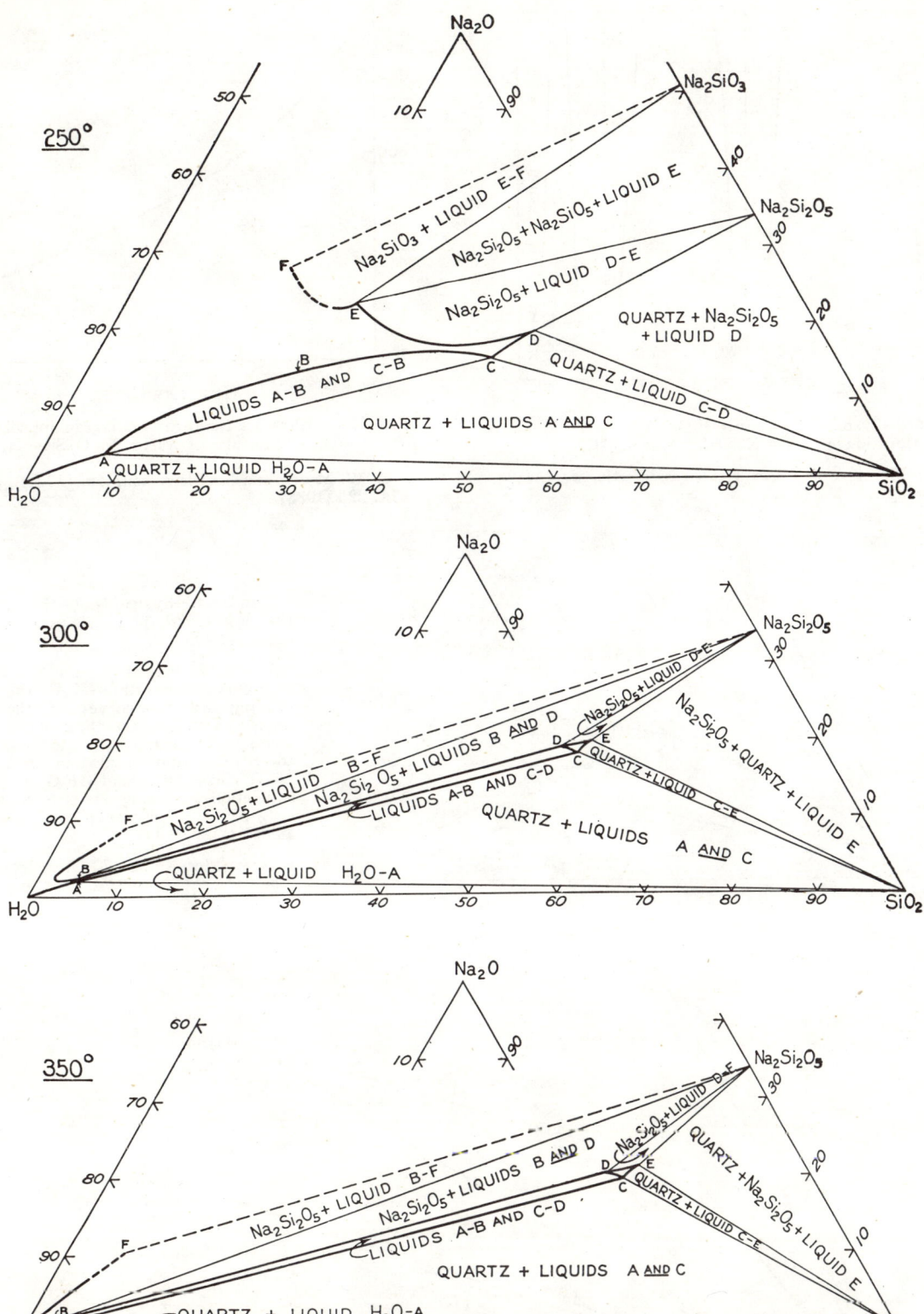

Fig. 1964.—Phase relations at 350°, 300°, and 250°C. in system Na_2O–SiO_2–H_2O.
O. F. Tuttle and I. Friedman, *J. Am. Chem. Soc.*, **70**, 923–25 (1948).

Na_2O–SiO_2–H_2O (cont.)

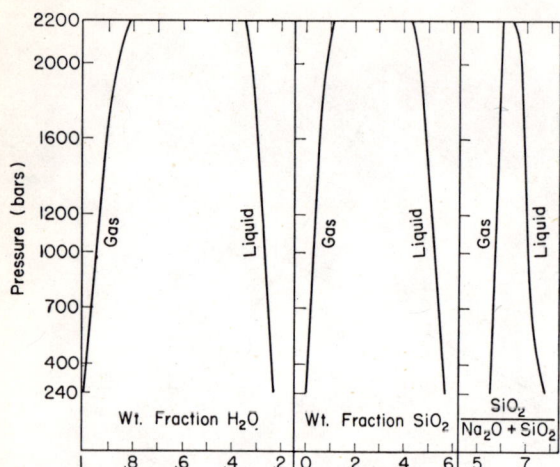

Fig. 1967.—System Na_2O–SiO_2–H_2O; gas and liquid saturation curves of $Na_2O \cdot 2SiO_2$ at 400°C.

G. W. Morey and J. M. Hesselgesser, *Am. J. Sci.*, Bowen volume, p. 363 (1952).

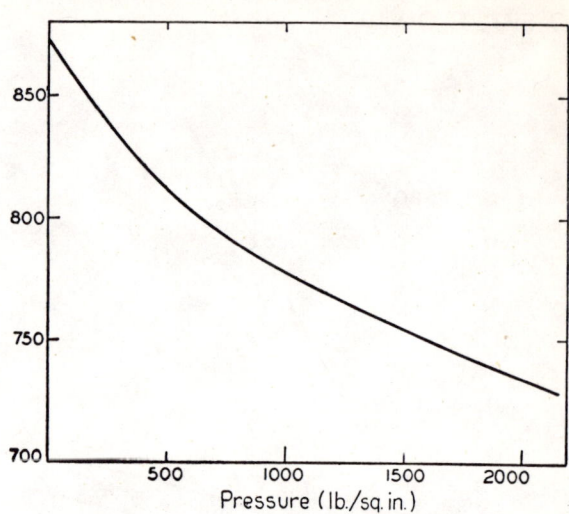

Fig. 1968.—Pressure-temperature diagram for solid-liquid-vapor equilibrium of system $Na_2O \cdot SiO_2$–H_2O.

G. W. Morey and Earl Ingerson, *Am. J. Sci.*, 5th Ser., **35A**, 225 (1938).

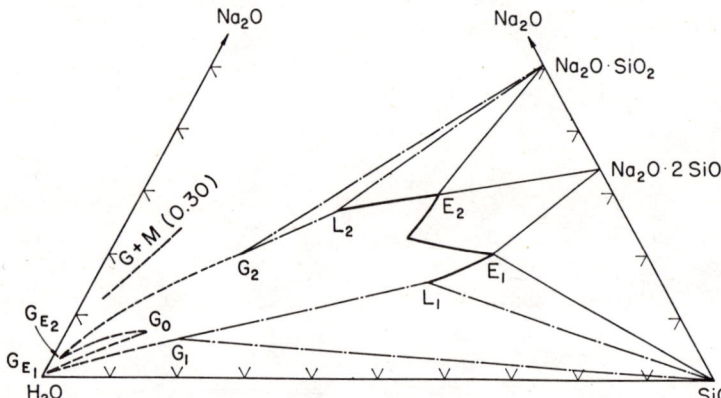

Fig. 1969.—System Na_2O–SiO_2–H_2O; the isothermal polybaric saturation curves at 400°C.

The curve E_1L_1 is the liquid saturation curve of quartz; GE_1G_1 the coexisting saturation curve; and the triangle $G_1L_1SiO_2$ is the three-phase triangle at 2500 bars. E_1E_2 and G_1G_2 are the gas and liquid saturation curves of $Na_2O \cdot 2SiO_2$. E_2L_2 and GE_2G_2 are the saturation curves of $Na_2O \cdot SiO_2$, and G_2L_2 $Na_2O \cdot SiO_2$ is the three-phase triangle at 2500 bars.

G. W. Morey and J. M. Hesselgesser, *Am. J. Sci.*, Bowen volume, p. 367 (1952).

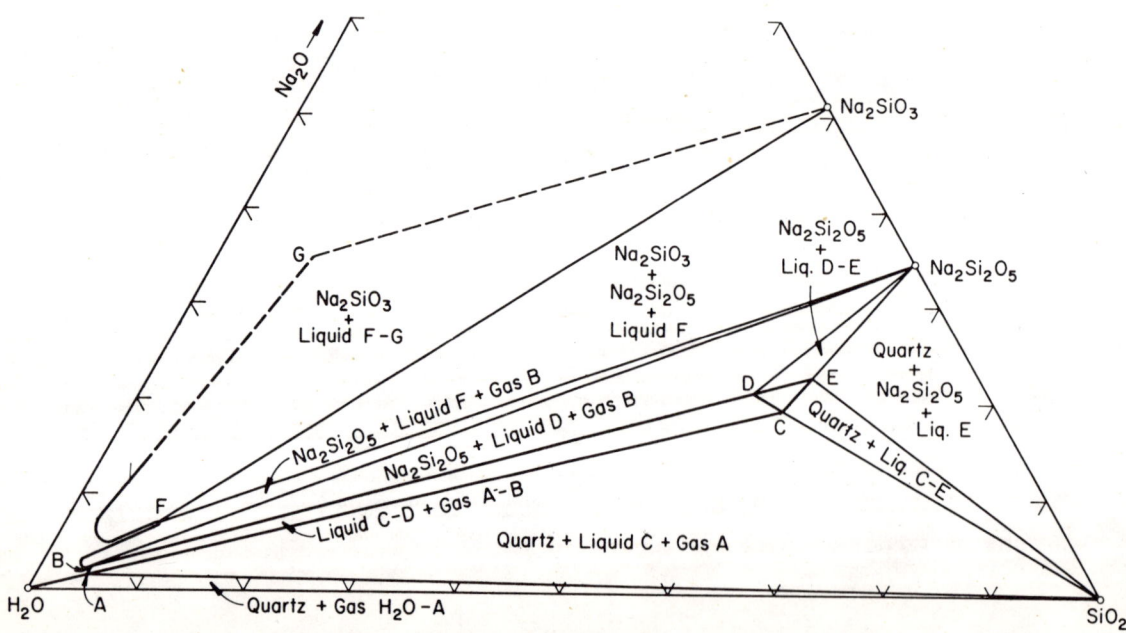

Fig. 1970.—System Na_2O–SiO_2–H_2O; polybaric saturation relations at 400°C.

Irving Friedman, *J. Am. Chem. Soc.*, **72**, 4572 (1950).

Na_2O–SiO_2–H_2O (concl.)

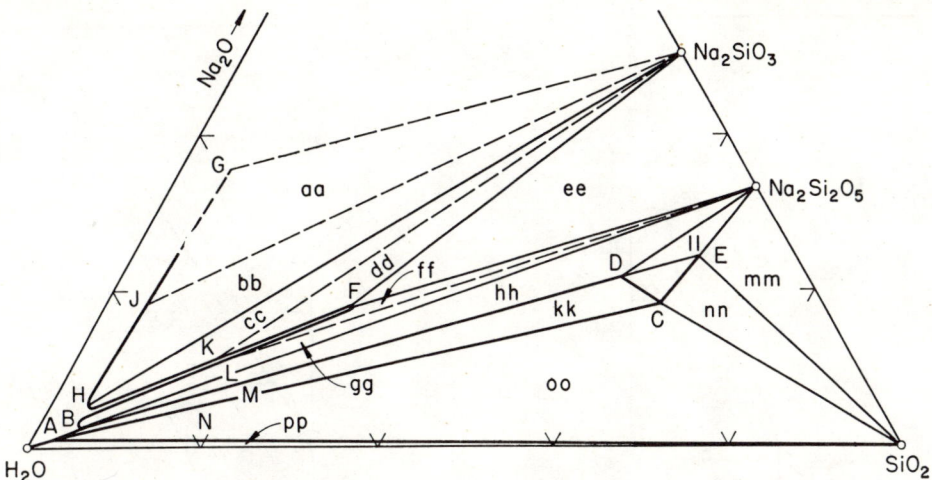

Fig. 1971.—System Na_2O–SiO_2–H_2O; polybaric saturation relations at 450°C.

aa, Na_2SiO_3 + liquid G-J
bb, Na_2SiO_3 + gas J-H
cc, Na_2SiO_3 + gas H-K
dd, Na_2SiO_3 + liquid K-F
ee, Na_2SiO_3 + $Na_2Si_2O_5$ + liquid F
ff, $Na_2Si_2O_5$ + liquid L-F
gg, $Na_2Si_2O_5$ + gas L-B
hh, $Na_2Si_2O_5$ + liquid D + gas B
kk, liquid C-D + gas A-B
ll, $Na_2Si_2O_5$ + liquid D-E
mm, quartz + $Na_2Si_2O_5$ + liquid E
nn, quartz + liquid C-E
oo, quartz + liquid C + gas A
pp, quartz + gas H_2O-A

Irving Friedman, *J. Am. Chem. Soc.*, **72**, 4573 (1950).

Na_2O–P_2O_5–H_2O

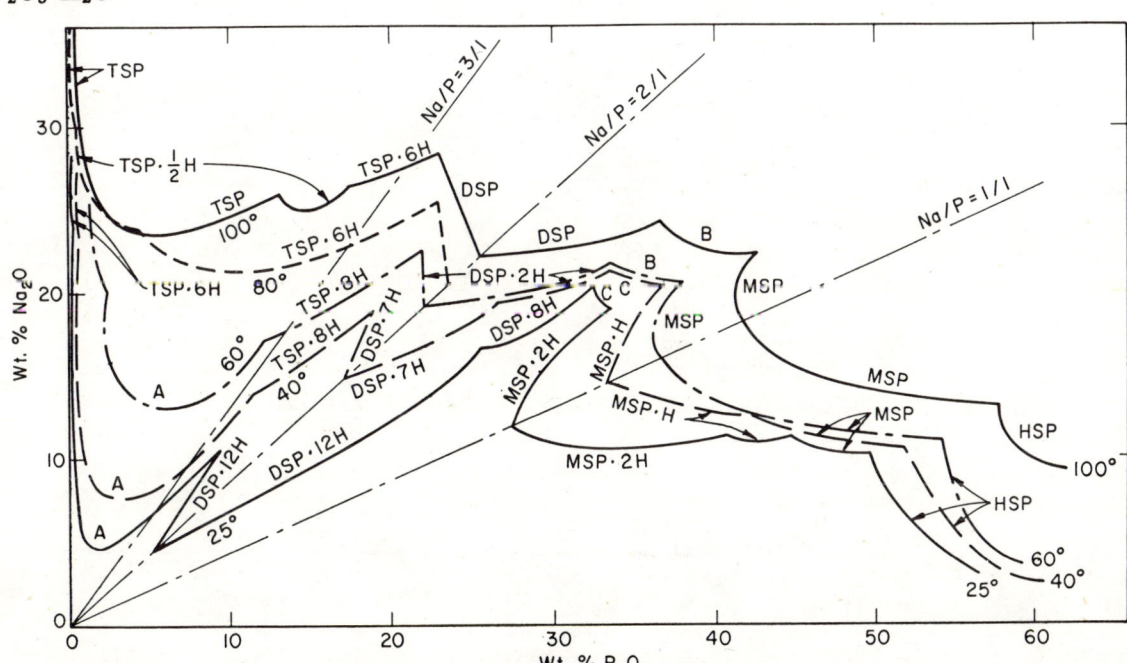

Fig. 1972.—System Na_2O–P_2O_5–H_2O. TSP = Na_3PO_4; DSP = Na_2HPO_4; MSP = NaH_2PO_4; HSP = $NaH_5(PO_4)_2$; A = $Na_3PO_4 \cdot 12H_2O \cdot 1/4 NaOH$; B = $Na_2HPO_4 \cdot NaH_2PO_4$; C = $Na_2HPO_4 \cdot 2NaH_2PO_4 \cdot 2H_2O$; H = H_2O.

Bernard Wendrow and K. A. Kobe, *Ind. Eng. Chem.*, **44** [6] 1444 (1952).

Na_2O–P_2O_5–H_2O (concl.)

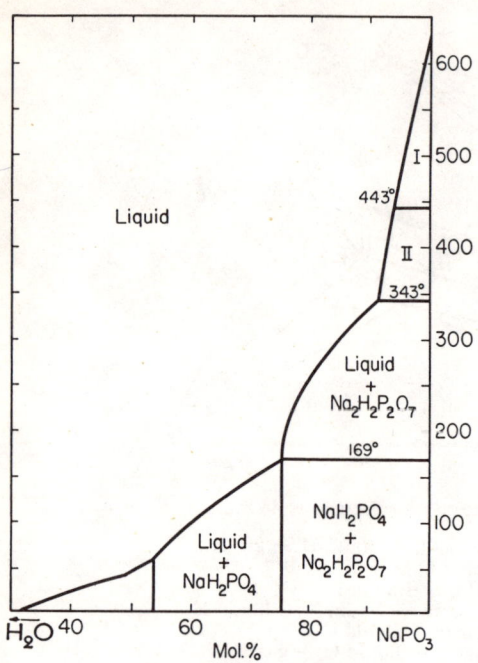

Fig. 1973.—System $NaPO_3$–H_2O.
G. W. Morey, *J. Am. Chem. Soc.*, **75**, 5795 (1953).

Na_2O–V_2O_5–H_2O

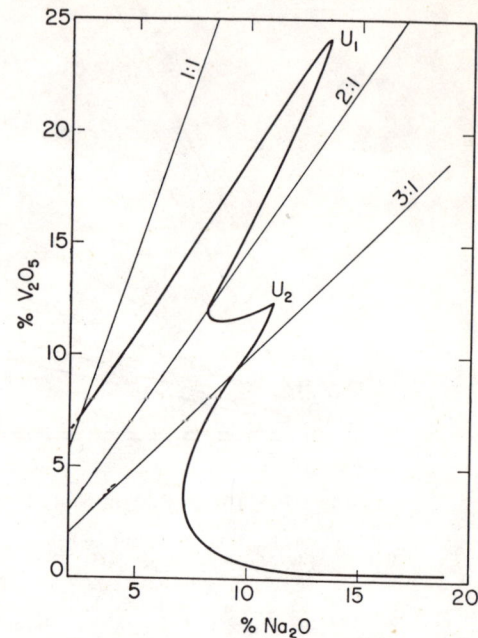

Fig. 1974.—Na_2O–V_2O_5–H_2O at 20°C.
Henrich Menzel and Gotthard Müller, *Z. anorg. u. allgem. Chem.*, **272**, 87 (1953).

Na_2O–UO_3–H_2O

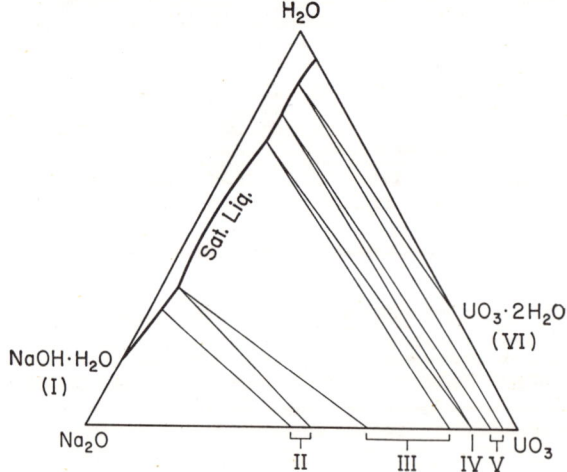

Fig. 1975.—System Na_2O–UO_3–H_2O at 50°C.; schematic.
J. E. Ricci and F. J. Loprest, *J. Am. Chem. Soc.*, **77**, 2126 (1955).

$BaO-Al_2O_3-H_2O$

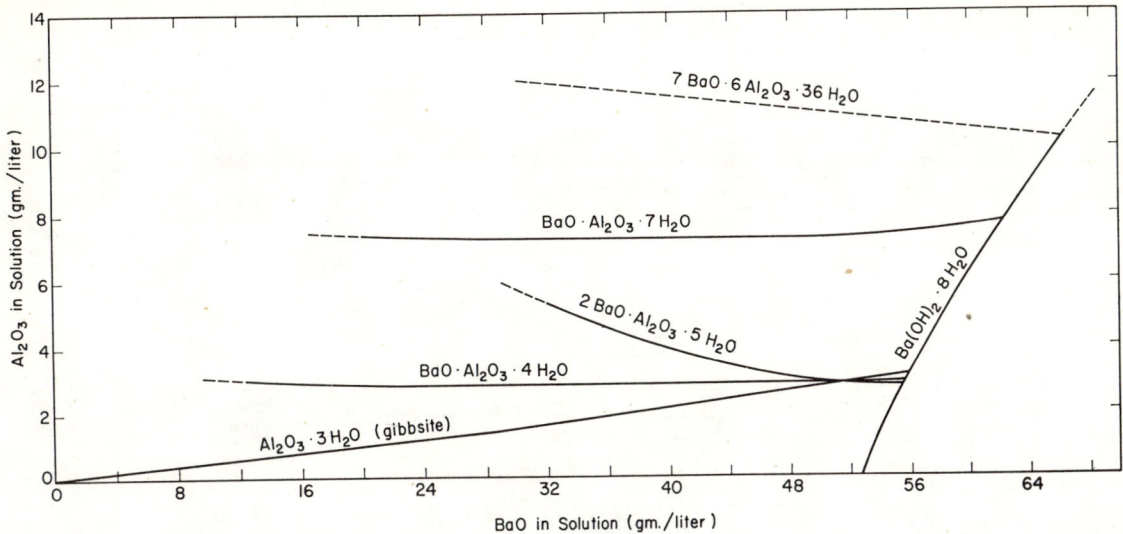

Fig. 1976.—System $BaO-Al_2O_3-H_2O$; concentration of barium aluminate solutions in stable or metastable equilibrium with solid phases at 30°C.

E. T. Carlson, T. J. Chaconas, and L. S. Wells, *J. Research Natl. Bur. Standards*, **45** [5] 396 (1950); RP 2149.

$BaO-P_2O_5-H_2O$

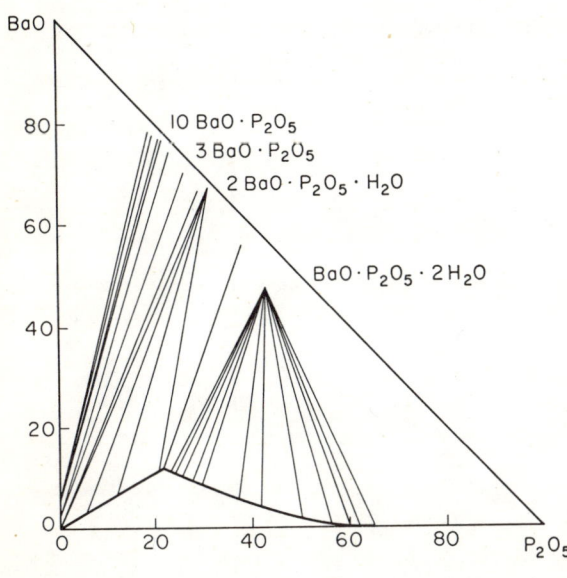

Fig. 1977.—System $BaO-P_2O_5-H_2O$ at 25°C.

André Artur, *Ann. Chim. (Paris)*, **10**, 974 (1955).

$CaO-Al_2O_3-H_2O$

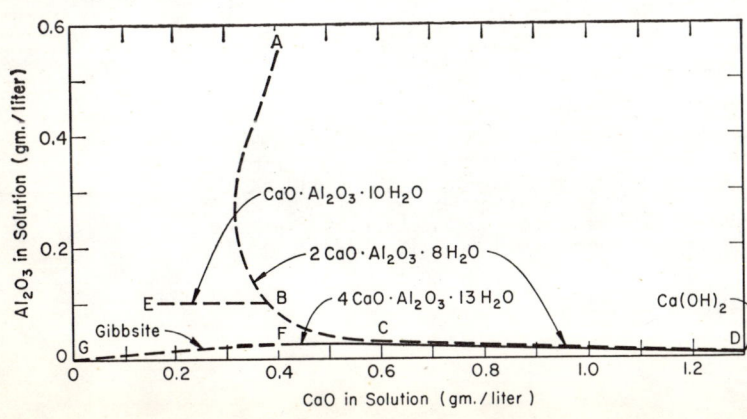

Fig. 1978.—System $CaO-Al_2O_3-H_2O$ at 1°C. ABCD = metastable equilibrium curve for $2CaO \cdot Al_2O_3 \cdot 8H_2O$, EB = metastable solubility curve for $CaO \cdot Al_2O_3 \cdot 10H_2O$, GF = (assumed) stable equilibrium curve for gibbsite, FCD = stable equilibrium curve for $4CaO \cdot Al_2O_3 \cdot 13H_2O$.

E. T. Carlson, *J. Research Natl. Bur. Standards*, **61** [1] 10 (1958); RP 2877.

CaO–Al$_2$O$_3$–H$_2$O (cont.)

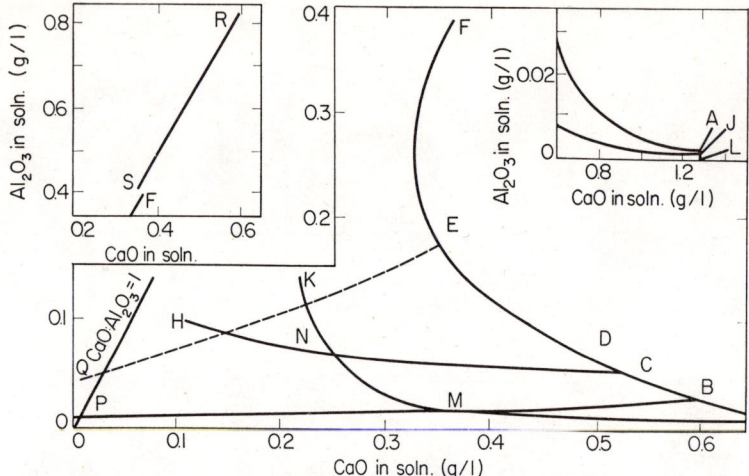

Fig. 1979.—System CaO–Al$_2$O$_3$–H$_2$O at 5°C. Insets are continuations on different scales.

Stable triple points: J = Ca(OH)$_2$, 3CaO·Al$_2$O$_3$·6H$_2$O, solution; M = 3CaO·Al$_2$O$_3$·6H$_2$O, gibbsite, solution.

Metastable triple points: A = Ca(OH)$_2$, 4CaO·Al$_2$O$_3$·19H$_2$O, soln.; B = 4CaO·Al$_2$O$_3$·19H$_2$O, gibbsite, soln.; C = 4CaO·Al$_2$O$_3$·19H$_2$O, CaO·Al$_2$O$_3$·10H$_2$O, soln.; D = 4CaO·Al$_2$O$_3$·19H$_2$O, 2CaO·Al$_2$O$_3$·8H$_2$O, soln.; E = 2CaO·Al$_2$O$_3$·8H$_2$O, alumina, soln.; H = CaO·Al$_2$O$_3$·10H$_2$O, alumina, soln.; K = 3CaO·Al$_2$O$_3$·6H$_2$O, alumina, soln.; N = 3CaO·Al$_2$O$_3$·6H$_2$O, CaO·Al$_2$O$_3$·10H$_2$O, soln.

F. G. Buttler and H. F. W. Taylor, *J. Chem. Soc.*, 2105 (1958).

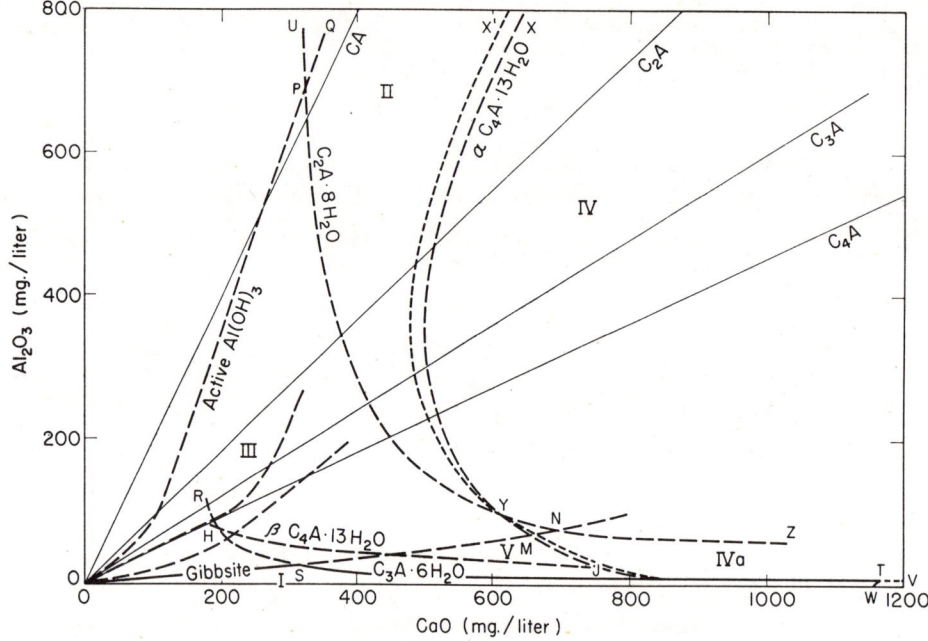

Fig. 1980.—System CaO–Al$_2$O$_3$–H$_2$O at 20°C.

Stable solubility curves: OS—Gibbsite; ST—C$_3$A·6H$_2$O; TW—Ca(OH)$_2$. Metastable solubility curves: OPQ—Boundary curve of active Al(OH)$_3$. The curve for the aged hydroxide approaches closer and closer to that of the gibbsite solubility; SMN—metastable extension of the gibbsite curve; SR—Metastable extension of the C$_3$A·6H$_2$O curve; UPYNZ—C$_2$A·8H$_2$O; XYMV—α C$_4$A·13H$_2$O; X'YV—precipitation curve of mixed hexagonal hydrates from supersaturated solutions; HJ—mixed crystals of hexagonal hydrates with β C$_4$A·13–14H$_2$O; A = Al$_2$O$_3$; C = CaO.

J. D'Ans and H. Eick, *Zement-Kalk-Gips*, 6 [6] 201 (1953).

CaO–Al$_2$O$_3$–H$_2$O (cont.)

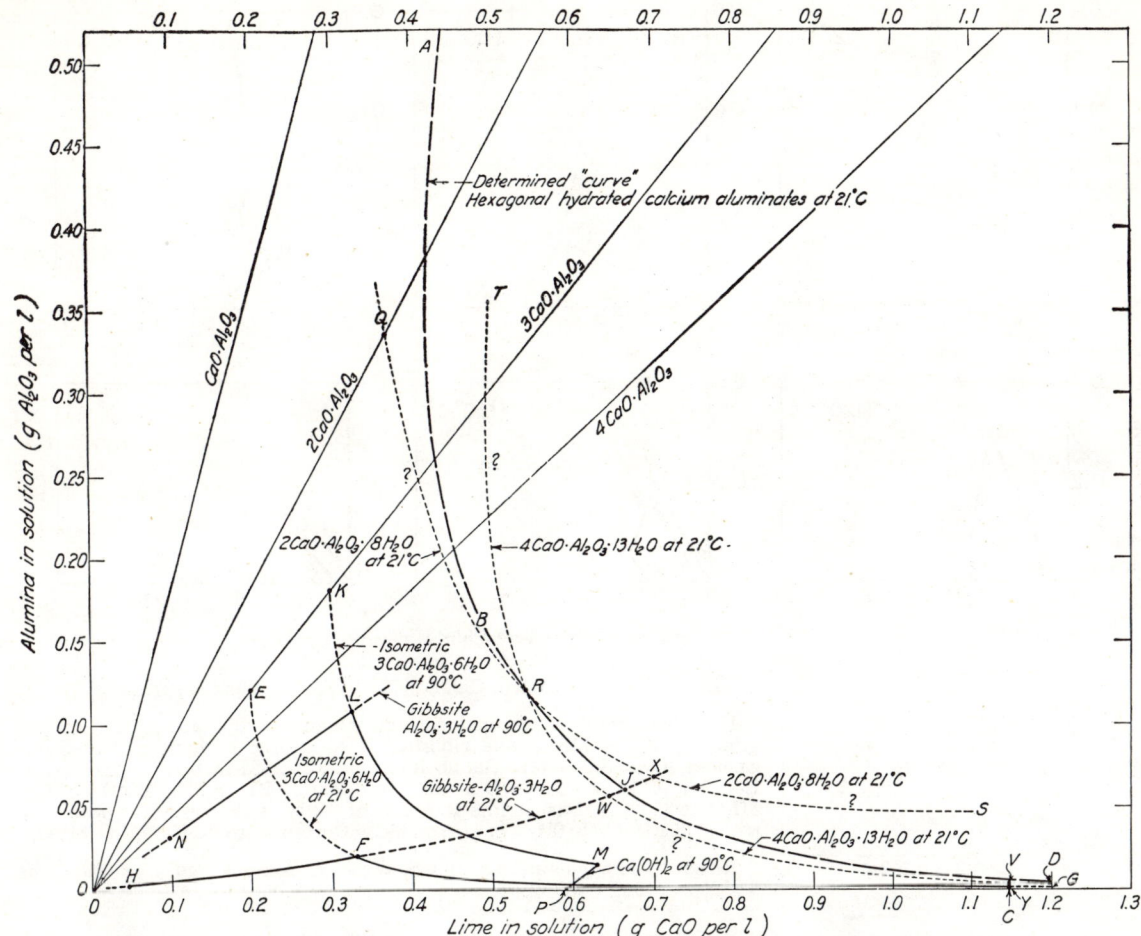

FIG. 1981.—System H$_2$O–CaO–Al$_2$O$_3$ at 21° and 90°C.; curves and points are as follows: *A-B-D*, metastable solubility curve for mixtures of hexagonal hydrates 2CaO·Al$_2$O$_3$·8H$_2$O and 4CaO·Al$_2$O$_3$·13H$_2$O at 21°C.; *Q-R-S*, suggested metastable solubility curve for hexagonal hydrate 2CaO·Al$_2$O$_3$·8H$_2$O at 21°C.; *Q*, approximate solubility of hexagonal hydrate 2CaO·Al$_2$O$_3$·8H$_2$O, apparently congruent in water at 21°C. (Bessey); *T-R-V*, suggested metastable solubility curve for hexagonal hydrate 4CaO·Al$_2$O$_3$·13H$_2$O at 21°C.; *R*, suggested metastable invariant point 2CaO·Al$_2$O$_3$·8H$_2$O–4CaO·Al$_2$O$_3$·13H$_2$O at 21°C.; *C*, solubility of coarsely crystalline Ca(OH)$_2$ in water at 21°C.; *V*, approximate metastable invariant point 4CaO·Al$_2$O$_3$·13H$_2$O–Ca(OH)$_2$ at 21°C.; *E-F-G*, solubility curve of isometric hydrate 3CaO·Al$_2$O$_3$·6H$_2$O at 21°C., along *E-F* the hydrate is metastable with respect to gibbsite and *F-G* is the stable section (see also Bessey); *Y*, stable invariant point 3CaO·Al$_2$O$_3$·6H$_2$O–Ca(OH)$_2$ at 21°C.; *Y-C*, solubility curve of Ca(OH)$_2$ at 21°C.; *H-F-J*, solubility curve of gibbsite Al$_2$O$_3$·3H$_2$O at 21°C., *H-F* is the stable section and *F-J* the metastable section with respect to 3CaO·Al$_2$O$_3$·6H$_2$O; *F*, stable invariant point 3CaO·Al$_2$O$_3$·6H$_2$O–Al$_2$O$_3$·3H$_2$O (gibbsite) at 21°C.; *W*, suggested metastable invariant point 4CaO·Al$_2$O$_3$·13H$_2$O–Al$_2$O$_3$·3H$_2$O (gibbsite) at 21°C.; *X*, possible metastable invariant point 2CaO·Al$_2$O$_3$·8H$_2$O–Al$_2$O$_3$·3H$_2$O (gibbsite) at 21°C.; *K-L-M*, solubility curve of isometric hydrate 3CaO·Al$_2$O$_3$·6H$_2$O at 90°C.; *M*, stable invariant point 3CaO·Al$_2$O$_3$·6H$_2$O–Ca(OH)$_2$ at 90°C.; *N-L*, solubility of Al$_2$O$_3$·3H$_2$O (gibbsite) at 90°C.; *L*, stable invariant point 3CaO·Al$_2$O$_3$·6H$_2$O–Al$_2$O$_3$·3H$_2$O at 90°C.; *P*, solubility of coarsely crystalline Ca(OH)$_2$ in water at 90°C. (Basset); *P-M*, solubility curve of Ca(OH)$_2$ at 90°C.

L. S. Wells, W. F. Clarke, and H. F. McMurdie, *J. Research Natl. Bur. Standards*, **30**, 103 (1943); R. P. 1539.

$CaO-Al_2O_3-H_2O$ (concl.)

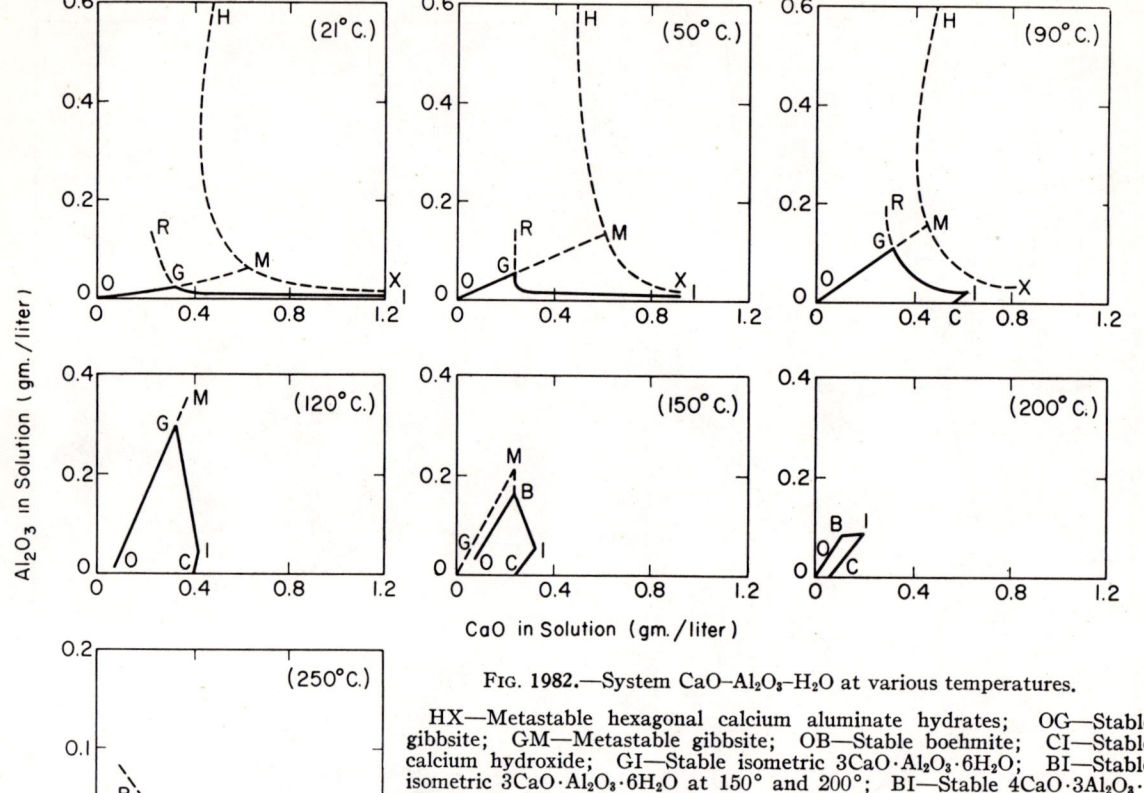

FIG. 1982.—System $CaO-Al_2O_3-H_2O$ at various temperatures.

HX—Metastable hexagonal calcium aluminate hydrates; OG—Stable gibbsite; GM—Metastable gibbsite; OB—Stable boehmite; CI—Stable calcium hydroxide; GI—Stable isometric $3CaO \cdot Al_2O_3 \cdot 6H_2O$; BI—Stable isometric $3CaO \cdot Al_2O_3 \cdot 6H_2O$ at 150° and 200°; BI—Stable $4CaO \cdot 3Al_2O_3 \cdot 3H_2O$ at 250°C.; Dashed line above BI at 250°C. indicates region where isometric $3CaO \cdot Al_2O_3 \cdot 6H_2O$ is metastable with respect to $4CaO \cdot 3Al_2O \cdot 3H_2O$.

Richard B. Peppler and Lansing S. Wells, *J. Research Natl. Bur. Standards*, **52** [2] 91 (1954); RP 2476.

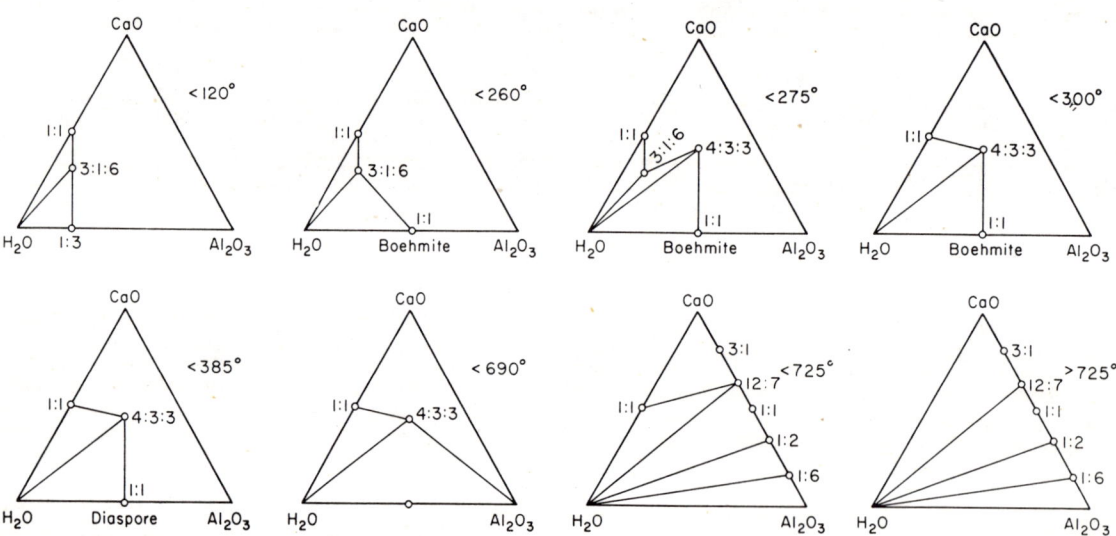

FIG. 1983.—System $CaO-Al_2O_3-H_2O$; temperature sequence of compatibility triangles drawn for a pressure of 10,000 p.s.i. of H_2O.

A. J. Majumdar and Rustum Roy, *J. Am. Ceram. Soc.*, **39** [12] 440 (1956).

$CaO-CO_2-H_2O$

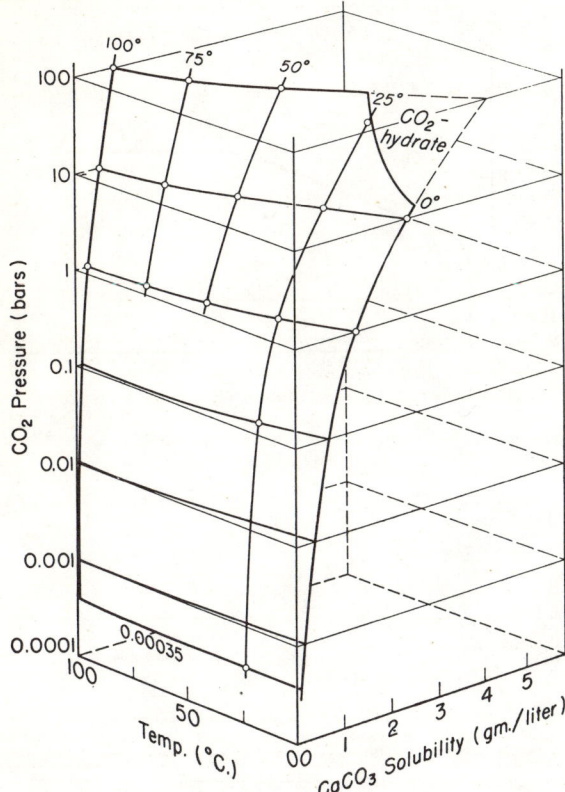

Fig. 1984.—System $CaCO_3-CO_2-H_2O$ for temperatures ranging from 0–100°C. and CO_2-pressures from 0.00035–100 bars.

John P. Miller, *Am. J. Sci.*, **250**, 192 (1952).

See also, G. T. Faust, *Am. Mineral.*, **34** [11/12] 810 (1949).

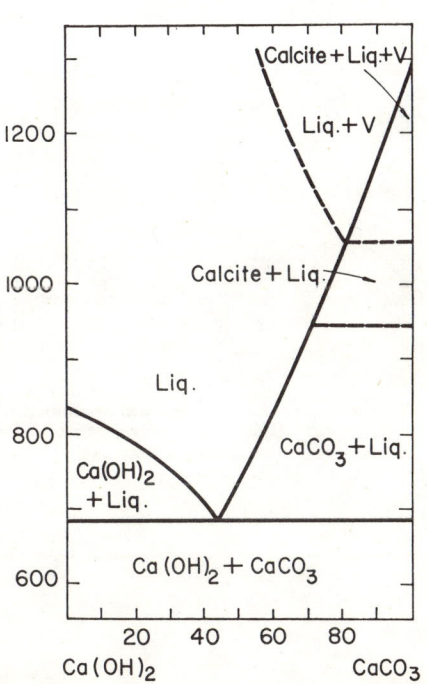

Fig. 1985.—System $Ca(OH)_2-CaCO_3$ at 1000 bars. Except where vapor is present, system is binary.

P. J. Wyllie and O. F. Tuttle, *J. Petrol.*, **1** [1] 19 (1960).

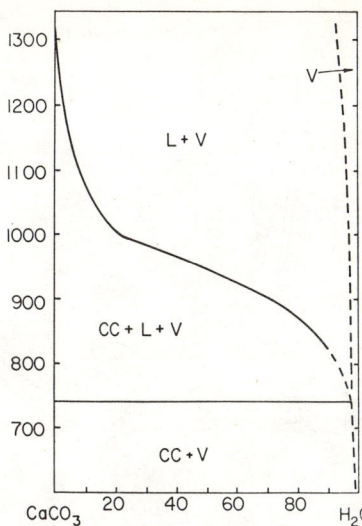

Fig. 1986.—System $CaCO_3-H_2O$ at 1000 bars pressure. CC = calcite, L = liquid, V = vapor. Where liquid and vapor coexist, system is not binary.

P. J. Wyllie and O. F. Tuttle, *J. Petrol.*, **1** [1] 20 (1960).

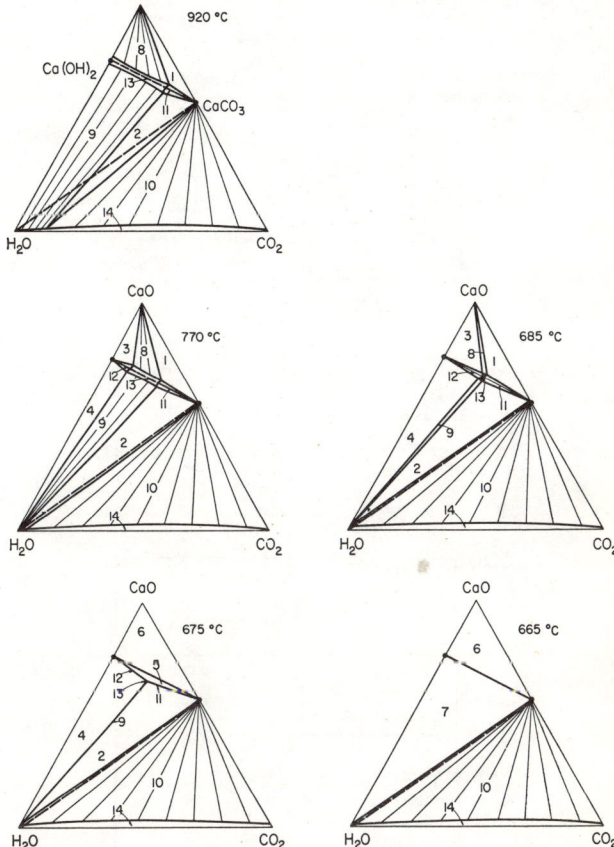

Fig. 1987.—System $CaO-CO_2-H_2O$; isothermal planes at 1000 bars. Vapor field enlarged to show phase relations.

Phase assemblages:

1 = $CaO + CaCO_3$ + liq.; 2 = $CaCO_3$ + liq. + vapor;
3 = $CaO + Ca(OH)_2$ + liq.; 4 = $Ca(OH)_2$ + liq. + vapor;
5 = $Ca(OH)_2 + CaCO_3$ + liq.; 6 = $CaO + Ca(OH)_2 + CaCO_3$;
7 = $Ca(OH)_2 + CaCO_3$ + vapor; 8 = CaO + liq.;
9 = liq. + vapor; 10 = $CaCO_3$ + vapor;
11 = $CaCO_3$ + liq.; 12 = $Ca(OH)_2$ + liq.; 13 = liq.;
14 = vapor.

P. J. Wyllie and O. F. Tuttle, *J. Petrol.*, **1** [1] 25 (1960).

CaO–CO$_2$–H$_2$O (concl.)

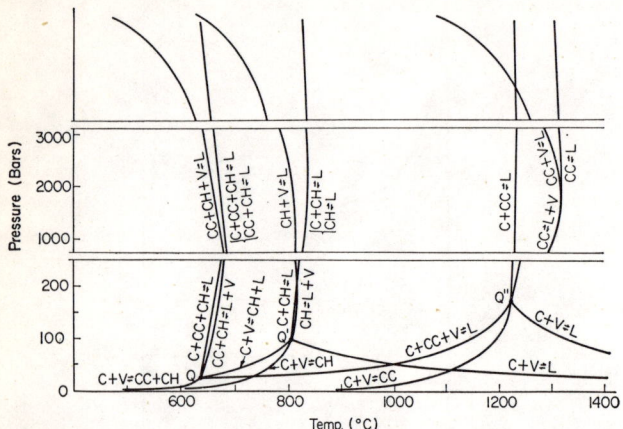

Fig. 1988.—System CaO–CO$_2$–H$_2$O; p–T projection of univariant curves. C = CaO; CC = CaCO$_3$; CH = Ca(OH)$_2$; L = liquid; V = vapor. Q, Q', Q'' are invariant points for the systems CaO–CO$_2$–H$_2$O, CaO–H$_2$O, and CaO–CO$_2$, respectively. High-pressure regions of diagram, especially, are schematic.

P. J. Wyllie and O. F. Tuttle, *J. Petrol.*, **1** [1] 39 (1960).

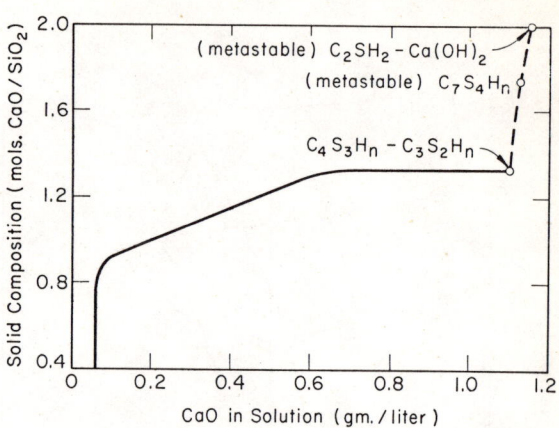

Fig. 1991.—System CaO–SiO$_2$–H$_2$O at 25°C. C = CaO, S = SiO$_2$, H = H$_2$O.

G. L. Kalousek and A. F. Prebus, *J. Am. Ceram. Soc.*, **41** [4] 130 (1958).

CaO–SiO$_2$–H$_2$O

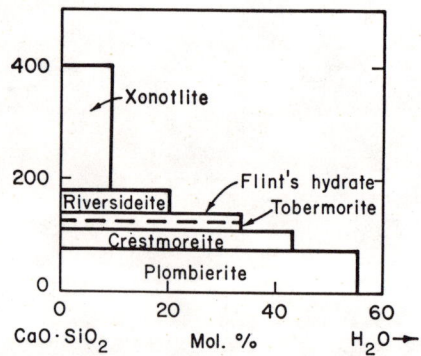

Fig. 1989.—System CaO·SiO$_2$–H$_2$O.

Gunnar Kullerud, *Norsk Geol. Tidsskr.*, **33**, 212 (1954).

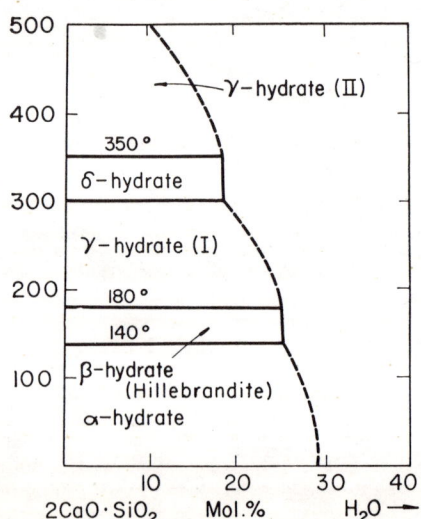

Fig. 1990.—System 2CaO·SiO$_2$–H$_2$O.

Gunnar Kullerud, *Norsk Geol. Tidsskr.*, **33**, 213 (1954).

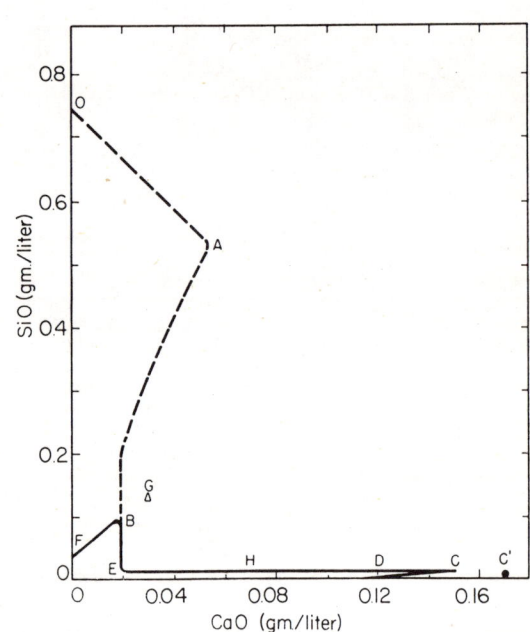

Fig. 1992.—System CaO–SiO$_2$–H$_2$O, at 180°C. OA = metastable silica gel; AB = metastable xonotlite; FB = stable alpha quartz; BEH = stable xonotlite; G = metastable gyrolite; HC = stable hillebrandite; CC' = metastable hillebrandite; CD = stable Ca(OH)$_2$.

R. B. Peppler, *J. Research Natl. Bur. Standards*, **54** [4] 208 (1955); RP2582.

CaO–SiO$_2$–H$_2$O (concl.)

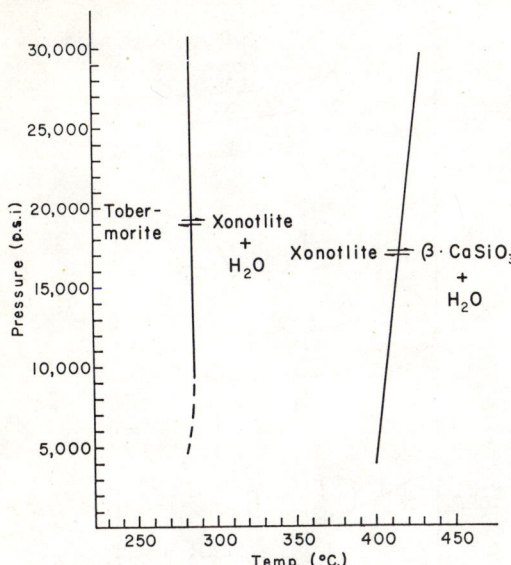

FIG. 1993.—System CaSiO$_3$–H$_2$O; univariant p–T curves for the reactions tobermorite = xonotlite + H$_2$O (or = xonotlite + truscottite + H$_2$O) and xonotlite = β-CaSiO$_3$ + H$_2$O.

D. A. Buckner, D. M. Roy, and Rustum Roy, *Am. J. Sci.*, **258**, 134 (1960).

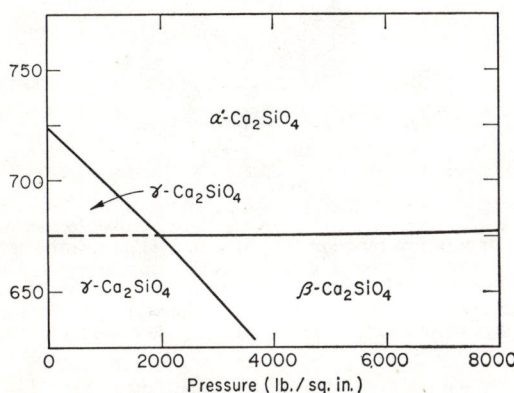

FIG. 1994.—Ca$_2$SiO$_4$; hypothetical p–T diagram.

D. M. Roy, *J. Am. Ceram. Soc.*, **41** [8] 293 (1958).

FeO–CO$_2$–H$_2$O

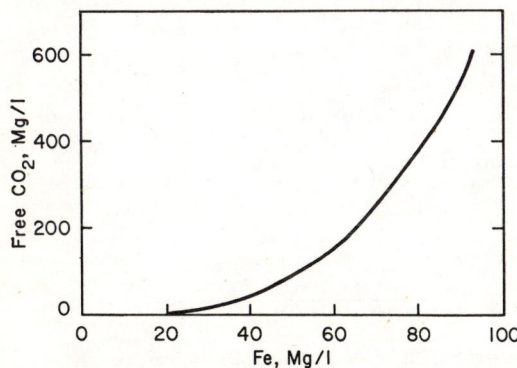

FIG. 1996.—System FeO–CO$_2$–H$_2$O at 19°C.

A. V. Kazakov, M. M. Tikhomirova, and V. I. Plotnikova, *Tr. Geol. Inst. Nauk Akad. S.S.S.R.*, No. 152, Geol. Ser., No. 64, 61 (1957).

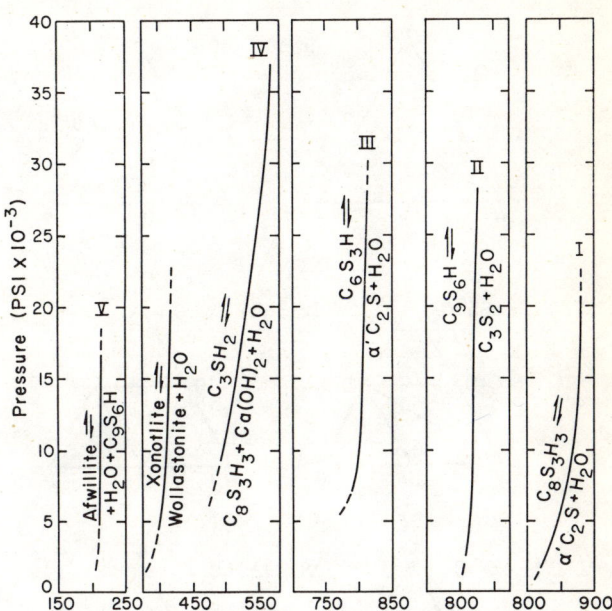

FIG. 1995.—System CaO–SiO$_2$–H$_2$O; p–T equilibrium curves. Xonotlite = wollastonite + H$_2$O, from Buckner and Roy.

D. M. Roy, *Am. Mineral.*, **43**, 1017 (1958).

MgO–Al$_2$O$_3$–H$_2$O

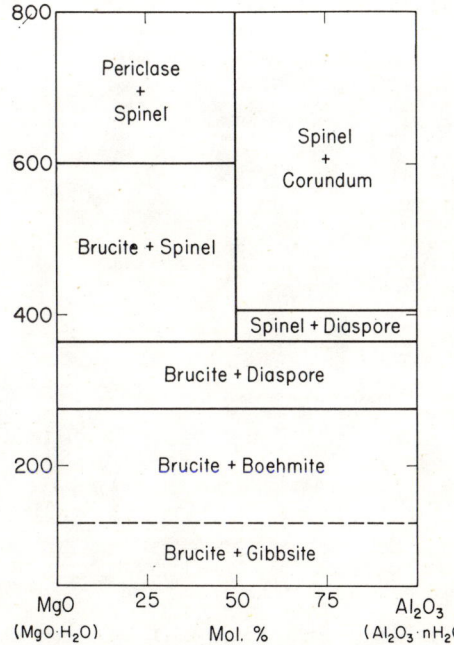

FIG. 1997.—System MgO–Al$_2$O$_3$–H$_2$O at 5000 psi water pressure.

Della M. Roy, Rustum Roy, and E. F. Osborn, *Am. J. Sci.*, **251** [5] 347 (1953).

MgO–Al₂O₂–H₂O (concl.)

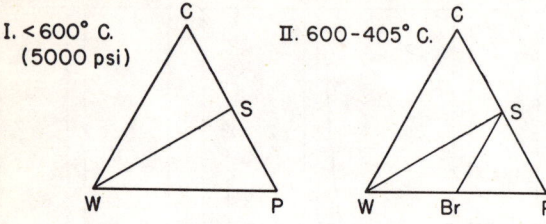

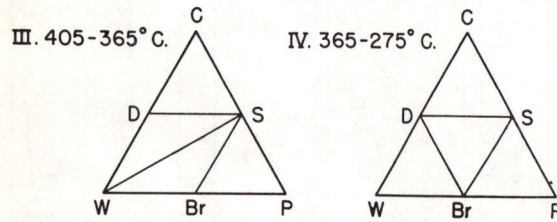

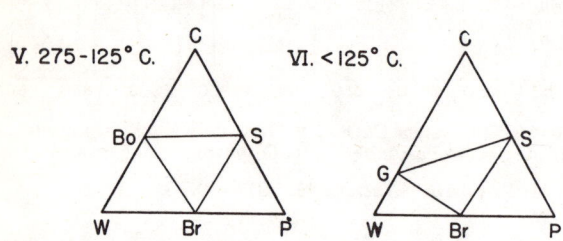

FIG. 1998.—System MgO–Al₂O₃–H₂O; composition triangles.

C—corundum, W—water, Bo—boehmite, S—spinel, Br—brucite, G—gibbsite, P—periclase, D—diaspore.
Della M. Roy, Rustum Roy, and E. F. Osborn, *Am. J. Sci.*, **251** [5] 351 (1953).

MgO–CO₂–H₂O

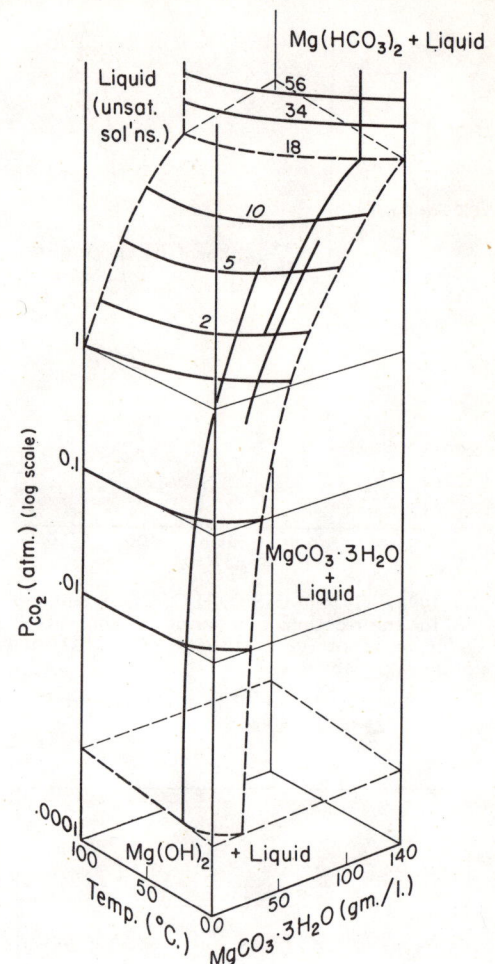

FIG. 1999.—System MgO–CO₂–H₂O; probable, within the temperature range of 0°C. and 100°C., and pressures of CO₂ ranging from 1×10^{-4} to 56 atmospheres. The stability field of magnesite is situated at higher temperatures than those shown here. The stability relations of lansfordite $MgCO_3 \cdot 5H_2O$, at very low temperatures have been ignored.

G. T. Faust, *Amer. Mineral.*, **34**, 806 (1949).

MgO–GeO₂–H₂O

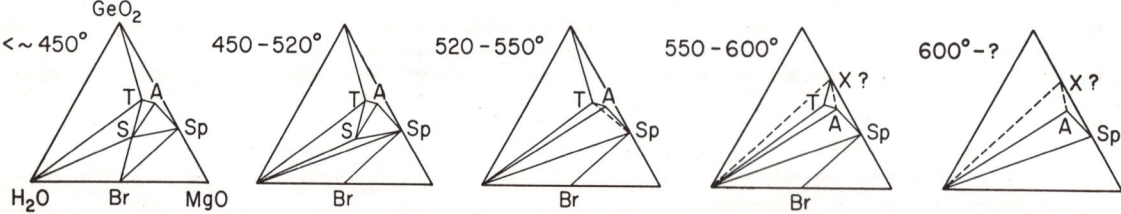

FIG. 2000.—System MgO–GeO₂–H₂O; compatibility triangles. T = talc; S = serpentine; Br = brucite; A = anthophyllite; Sp = spinel; X = unknown.

Della M. Roy and Rustum Roy, *Am. Mineral.*, **39**, 972 (1954).

MgO–SiO₂–H₂O

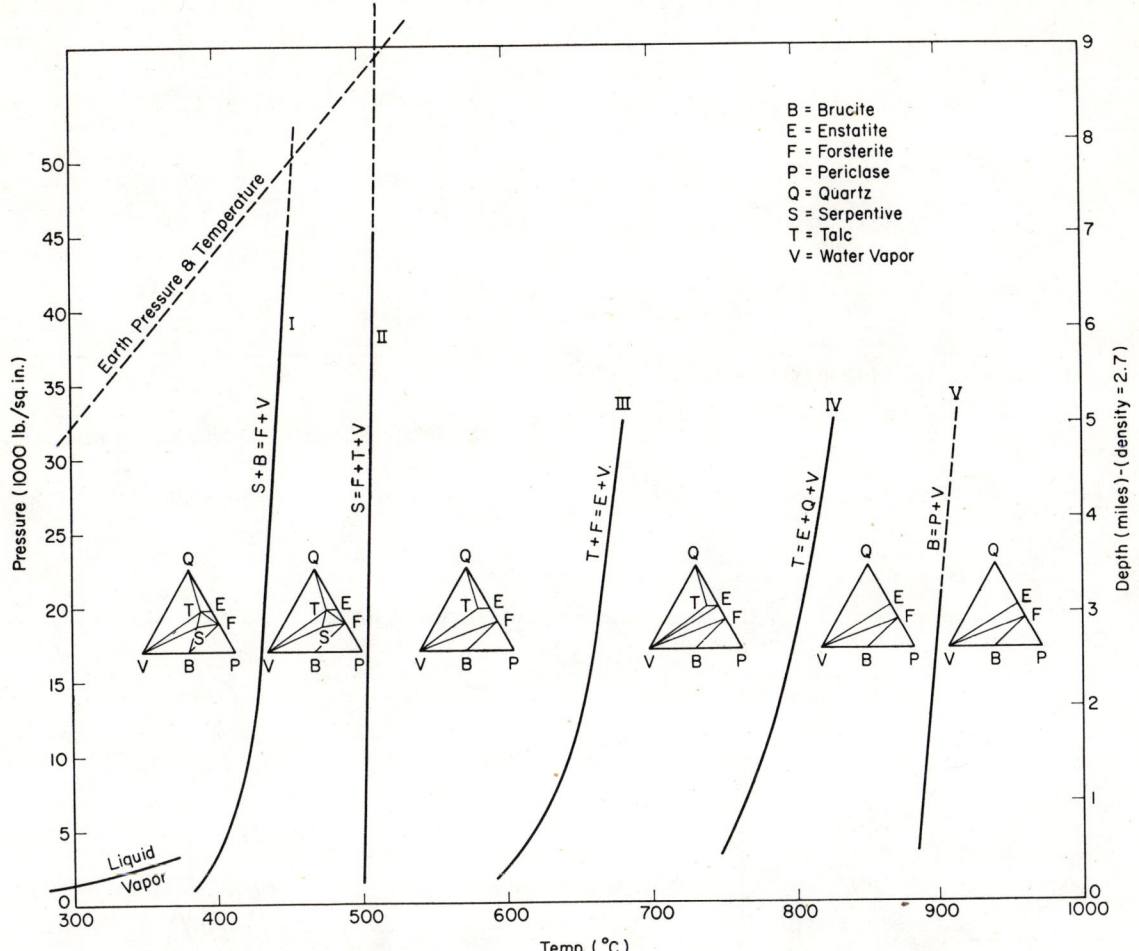

FIG. 2001.—System MgO–SiO₂–H₂O; pressure-temperature curves (I–V) of univariant equilibrium. Equation on each curve indicates the reaction to which curve refers. Triangular diagram on each divariant region between curves indicates, for all compositions, the stable assemblages under the range of p-t conditions represented by the region. Lower left—vapor-pressure curve of water ending at the critical temperature and pressure.

Upper left—curve of normal earth pressure and temperature assuming a temperature gradient of 55°C./mi. and a rock density of 2.7.

N. L. Bowen and O. F. Tuttle, *Bull. Geol. Soc. Am.*, **60**, 447 (1949).

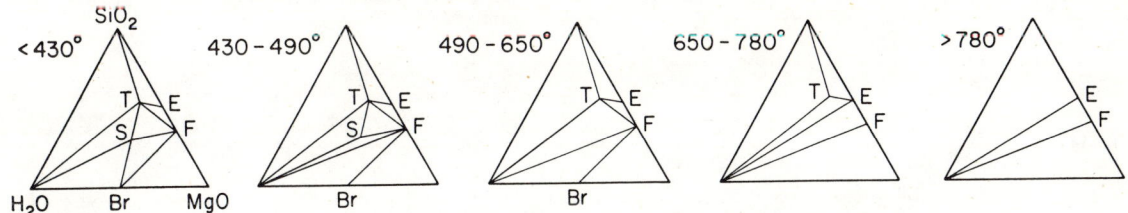

FIG. 2002.—System MgO–SiO₂–H₂O; compatibility triangles.

Mostly after N. L. Bowen and O. F. Tuttle, Fig. 2001, revised by Della M. Roy, Rustum Roy, and E. F. Osborn, *Am. J. Sci.*, **251**, 337–361 (1953).
T = talc, E = enstatite, F = forsterite, S = serpentine, Br = brucite.

Della M. Roy and Rustum Roy, *Am. Mineral.*, **39**, 972 (1954).

$MnO-P_2O_5-H_2O$

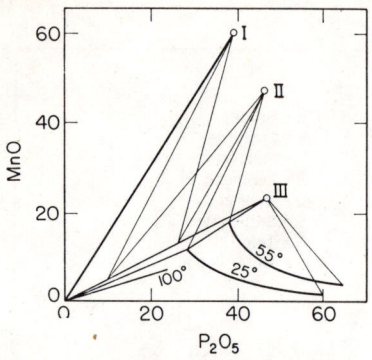

FIG. 2003.—System $MnO-P_2O_5-H_2O$.

I—$Mn_3(PO_4)_2$, II—$MnHPO_4$, III—$MnH_4(PO_4)_2 \cdot 3H_2O$. Data at 25° by Grube and Steche; at 55° by Zipfel; at 100° by Viardo.

A. A. Taperova and M. M. Isaeva, *Zhur. Prik. Khim.*, **22** [4] 343 (1949).

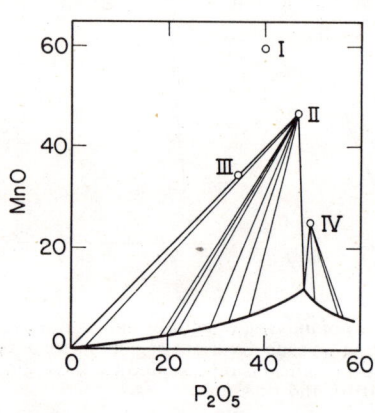

FIG. 2004.—System $MnO-P_2O_5-H_2O$, at 100°C.

I—$Mn_3(PO_4)_2$, II—$MnHPO_4$, III—$MnHPO_4 \cdot 3H_2O$, IV—$MnH_4(PO_4)_2 \cdot 2H_2O$.

A. A. Taperova and M. M. Isaeva, *Zhur. Prik. Khim.*, **22** [4] 347 (1949).

$ZnO-SiO_2-H_2O$

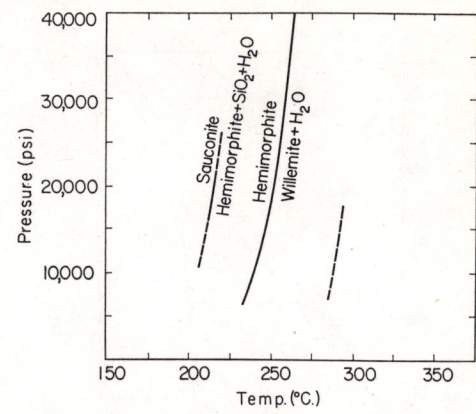

FIG. 2006.—System $ZnO-SiO_2-H_2O$; univariant p-T curves.

D. M. Roy and F. A. Mumpton, *Econ. Geol.*, **51**, 435 (1956).

$ZnO-CrO_3-H_2O$

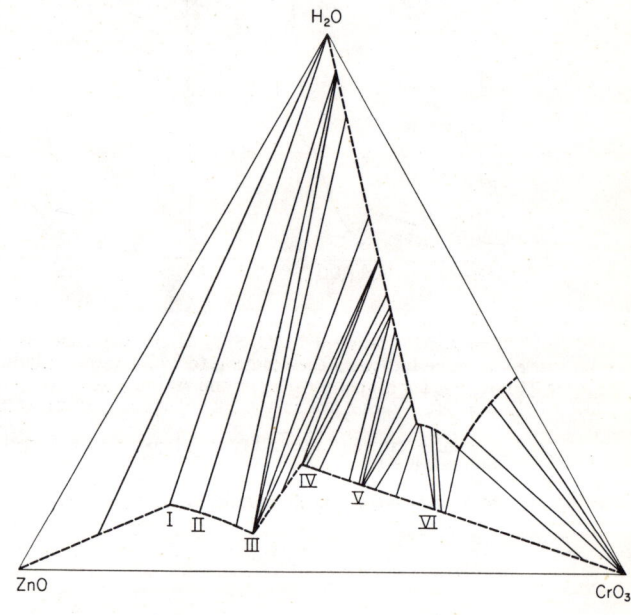

FIG. 2007.—System $ZnO-CrO_3-H_2O$ at 25°C.

A. E. Woodward, E. R. Allen, and R. H. Anderson, *J. Phys. Chem.*, **60**, 940 (1956).

$NiO-SiO_2-H_2O$

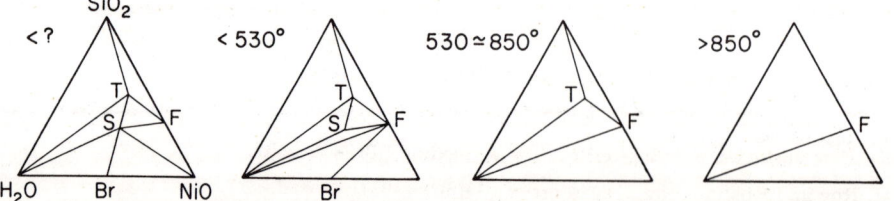

FIG. 2005.—System $NiO-SiO_2-H_2O$; compatibility triangles; T = talc; F = forsterite; S = serpentine; Br = brucite.

Della M. Roy and Rustum Roy, *Am. Mineral.*, **39**, 972 (1954).

Al_2O_3–Ga_2O_3–H_2O

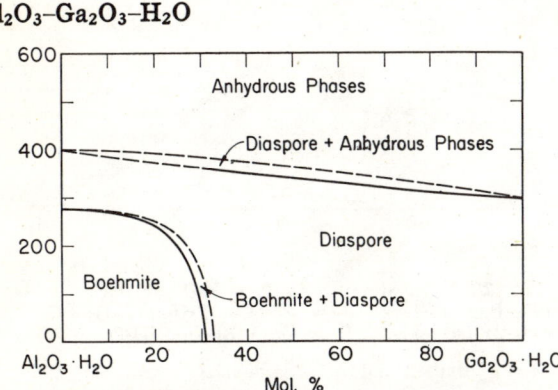

FIG. 2008.—System $Al_2O_3 \cdot H_2O$–$Ga_2O_3 \cdot H_2O$.

V. G. Hill, Rustum Roy, and E. F. Osborn, *J. Am. Ceram. Soc.*, **35** [6] 139 (1952).

Al_2O_3–SiO_2–H_2O

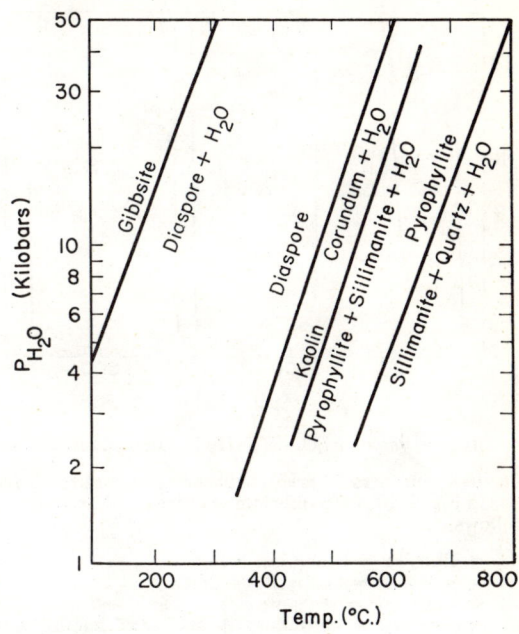

FIG. 2009.—System Al_2O_3–SiO_2–H_2O; v.p. of reactions.

G. C. Kennedy, *Am. J. Sci.*, **257**, 570 (1959).

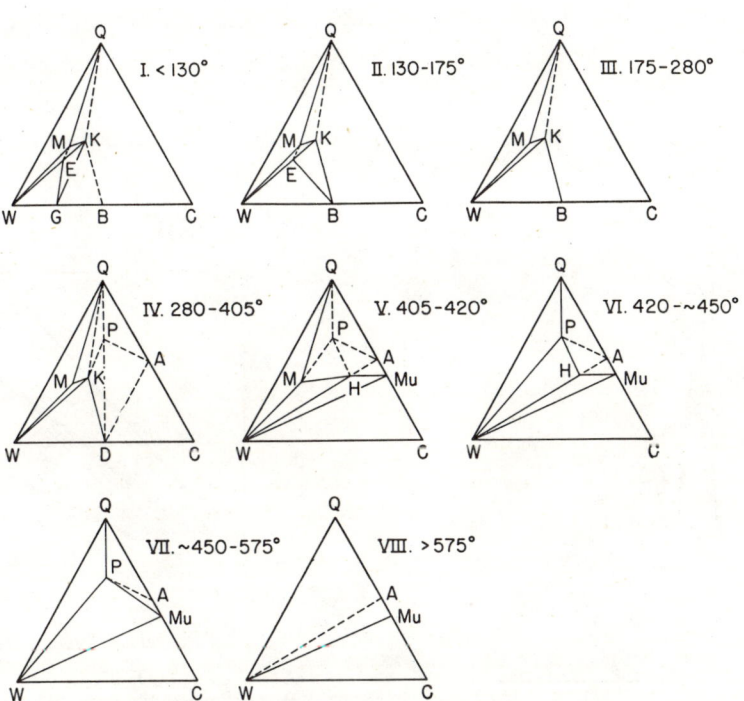

FIG. 2010.—System Al_2O_3–SiO_2–H_2O; compatibility triangles at about 10,000 psi water pressure.

Triangles bounded by solid lines have been experimentally determined. Dashed lines indicate probable relationships.

A = andalusite; B = boehmite; C = corundum; D = diaspore; E = endellite; G = gibbsite; H = hydralsite; K = kaolinite, nacrite, dickite, or halloysite; M = Al-montmorillonite; Mu = mullite; P = pyrophyllite; Q = quartz; W = water.

Rustum Roy and E. F. Osborn, *Am. Mineral.*, **39**, 875 (1954).

Al_2O_3–SiO_2–H_2O (concl.)

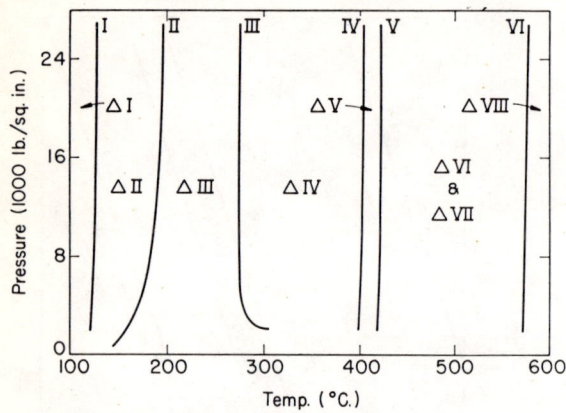

FIG. 2011.—System Al_2O_3–SiO_2–H_2O; univariant curves.

Triangle numbers in areas between curves refer to triangles in Fig. 2010. Equilibrium reactions along curves are as follows:

I = gibbsite = boehmite + H_2O
II = endellite = halloysite + H_2O
III = boehmite = diaspore
IV = diaspore = corundum + H_2O, and kaolinite = hydralsite + pyrophyllite + montmorillonite
V = montmorillonite = pyrophyllite + hydralsite + H_2O
VI = pyrophyllite = mullite (or andalusite) + quartz + H_2O

Rustum Roy and E. F. Osborn, *Am. Mineral.*, **39**, 876 (1954).

Cr_2O_3–P_2O_5–H_2O

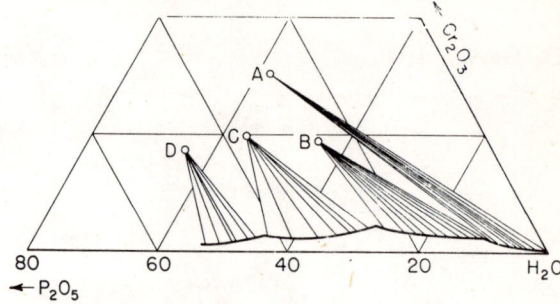

FIG. 2013.—System Cr_2O_3–P_2O_5–H_2O, at 0°C. A = $Cr_2O_3 \cdot P_2O_5 \cdot 12H_2O$; B = $2Cr_2O_3 \cdot 3P_2O_5 \cdot 51H_2O$; C = $Cr_2O_3 \cdot 2P_2O_5 \cdot 19H_2O$; D = $Cr_2O_3 \cdot 3P_2O_5 \cdot 18H_2O$.

R. F. Jameson and J. E. Salmon, *J. Chem. Soc.*, **1955**, p. 361.

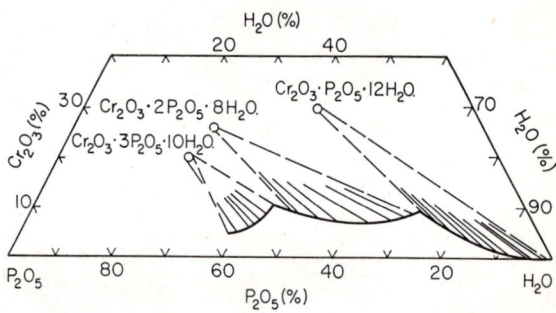

FIG. 2014.—System Cr_2O_3–P_2O_5–H_2O at 40°C.

R. F. Jameson and J. E. Salmon, *J. Chem. Soc.*, **1955**, p. 363.

Al_2O_3–SO_3–H_2O

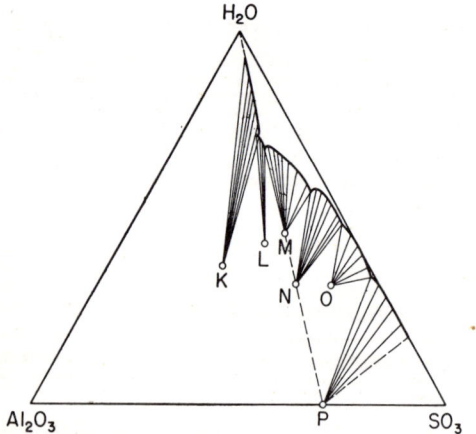

FIG. 2012.—System H_2O–Al_2O_3–SO_3 at 60°C.; K = $Al_2O_3 \cdot SO_3 \cdot 6H_2O$, L = $Al_2O_3 \cdot 2SO_3 \cdot 11H_2O$, M = $Al_2O_3 \cdot 3SO_3 \cdot 16H_2O$, N = $Al_2O_3 \cdot 3SO_3 \cdot 9H_2O$, O = $Al_2O_3 \cdot 6SO_3 \cdot 15H_2O$, P = $Al_2O_3 \cdot 3SO_3$.

Jack L. Henry and G. Brooks King, *J. Am. Chem. Soc.*, **72**, 1283 (1950).

Fe_2O_3–As_2O_5–H_2O

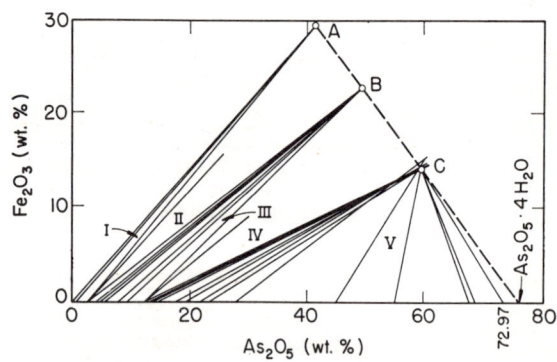

FIG. 2015.—System H_2O–Fe_2O_3–As_2O_5.

A = $Fe_2O_3 \cdot As_2O_5 \cdot 9H_2O$
B = $2Fe_2O_3 \cdot 3As_2O_5 \cdot 22H_2O$.
C = $Fe_2O_3 \cdot 3As_2O_5 \cdot 17H_2O$.
I = $Fe_2O_3 \cdot As_2O_5 \cdot 9H_2O$ + soln. (trace Fe_2O_3).
II = $Fe_2O_3 \cdot As_2O_5 \cdot 9H_2O$ + $2Fe_2O_3 \cdot 3As_2O_5 \cdot 22H_2O$ + soln. (As_2O_5 2.48%, Fe_2O_3 trace).
III = $2Fe_2O_3 \cdot 3As_2O_5 \cdot 22H_2O$ + soln.
IV = $2Fe_2O_3 \cdot 3As_2O_5 \cdot 22H_2O$ + $Fe_2O_3 \cdot 3As_2O_5 \cdot 17H_2O$ + soln. (As_2O_5 12.0%, Fe_2O_3 0.12%).
V = $Fe_2O_3 \cdot 3As_2O_5 \cdot 17H_2O$ + soln.

Takehiko Takahashi and Kumazo Sasaki, *J. Chem. Soc. Japan*, Ind. Chem. Sect., **53**, 384 (1950).

Fe_2O_3–P_2O_5–H_2O

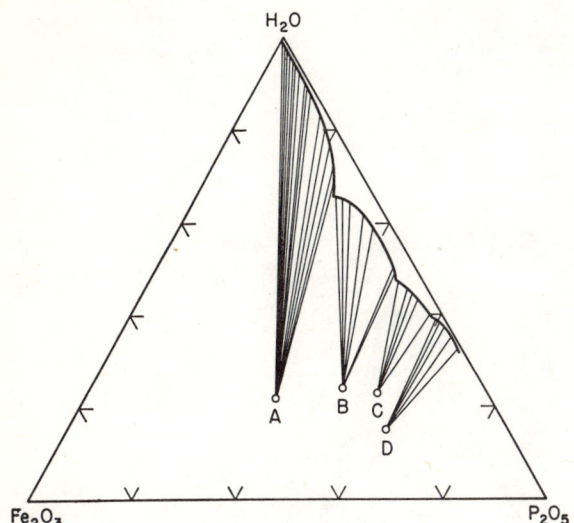

Fig. 2016.—System Fe_2O_3–P_2O_5–H_2O, at 25°C. A = $Fe_2O_3 \cdot P_2O_5 \cdot 5H_2O$; B = $Fe_2O_3 \cdot 2P_2O_5 \cdot 8H_2O$; C = $Fe_2O_3 \cdot 3P_2O_5 \cdot 10H_2O$; D = $Fe_2O_3 \cdot 3P_2O_5 \cdot 6H_2O$.

R. F. Jameson and J. E. Salmon, *J. Chem. Soc.*, **1954**, p. 29.

ThO_2–CrO_3–H_2O

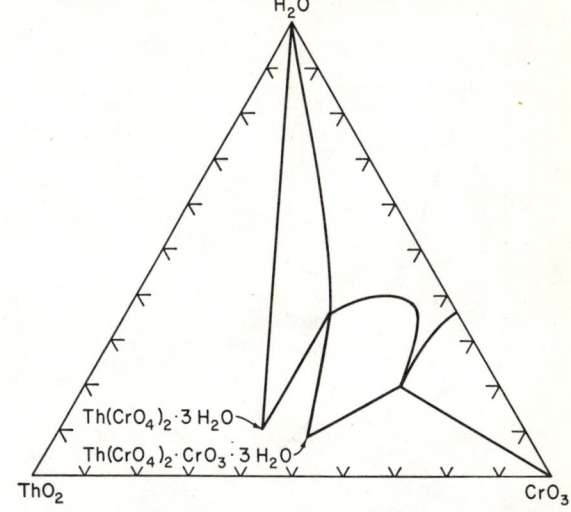

Fig. 2017.—System CrO_3–ThO_2–H_2O at 25°C.

H. T. S. Britton, *J. Chem. Soc.*, **123**, 1429 (1923). See also Oak Ridge National Laboratory, Phase Diagrams for Nuclear Reactor Materials, R. E. Thoma, ed., ORNL-2548, p. 199 (1959).

UO_2–P_2O_5–H_2O

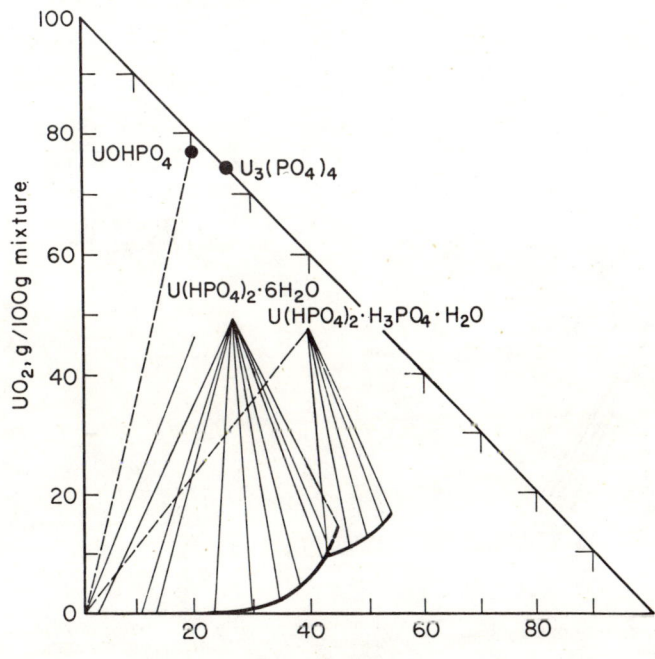

Fig. 2018.—System UO_2–P_2O_5–H_2O at 25°C.

J. M. Schreyer, *J. Am. Chem. Soc.*, **77**, 2972 (1955). See also Oak Ridge National Laboratory, Phase Diagrams of Nuclear Reactor Materials, R. E. Thoma, ed., ORNL-2548, p. 145 (1959).

IV. Three Metal Oxides with Water

$K_2O-Al_2O_3-SiO_2-H_2O$

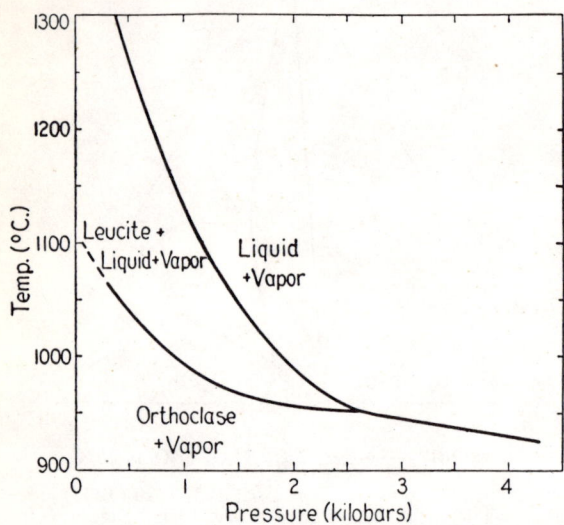

Fig. 2019.—Projection of system $K_2O \cdot Al_2O_3 \cdot 6SiO_2-H_2O$ on the temperature-pressure coordinate plane.

R. W. Goranson, *Am. J. Sci.*, 5th Ser., **35A**, 89 (1938).

$K_2O-CO_2-SiO_2-H_2O$

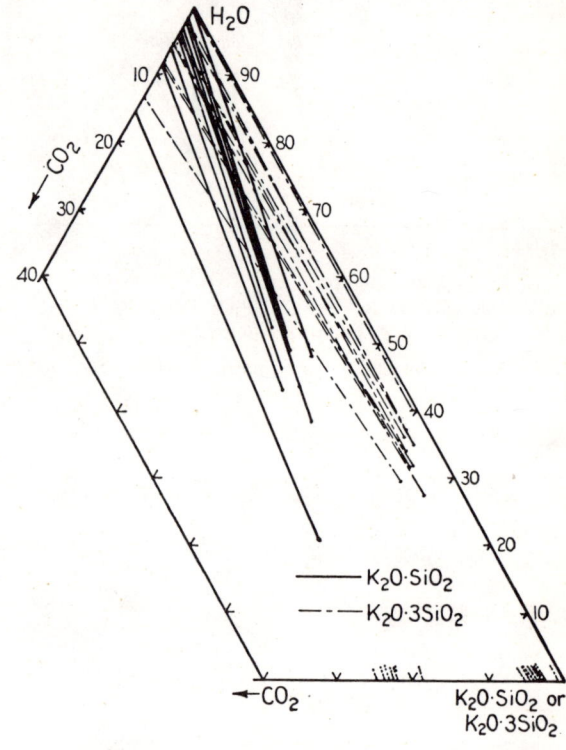

Fig. 2020.—System $H_2O-K_2O-SiO_2-CO_2$; compositions of some coexisting vapor and liquid phases; solid lines connect vapor and liquid in metasilicate melts; dot-dash lines, same in trisilicate melts.

G. W. Morey and Michael Fleischer, *Bull. Geol. Soc Am.*, **51**, Part II, 1052 (1940).

$Li_2O-CO_2-UO_3-H_2O$

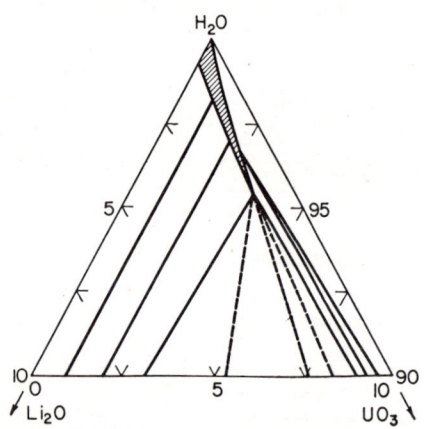

Fig. 2021.—System $Li_2O-UO_3-CO_2-H_2O$ at 250°C. and 1500 psi.

F. J. Loprest, W. L. Marshall, and C. H. Secoy, Oak Ridge National Laboratory, Phase Diagrams for Nuclear Reactor Materials, R. E. Thoma, ed., ORNL-2548, p. 196 (1959).

Na$_2$O–Al$_2$O$_3$–SiO$_2$–H$_2$O

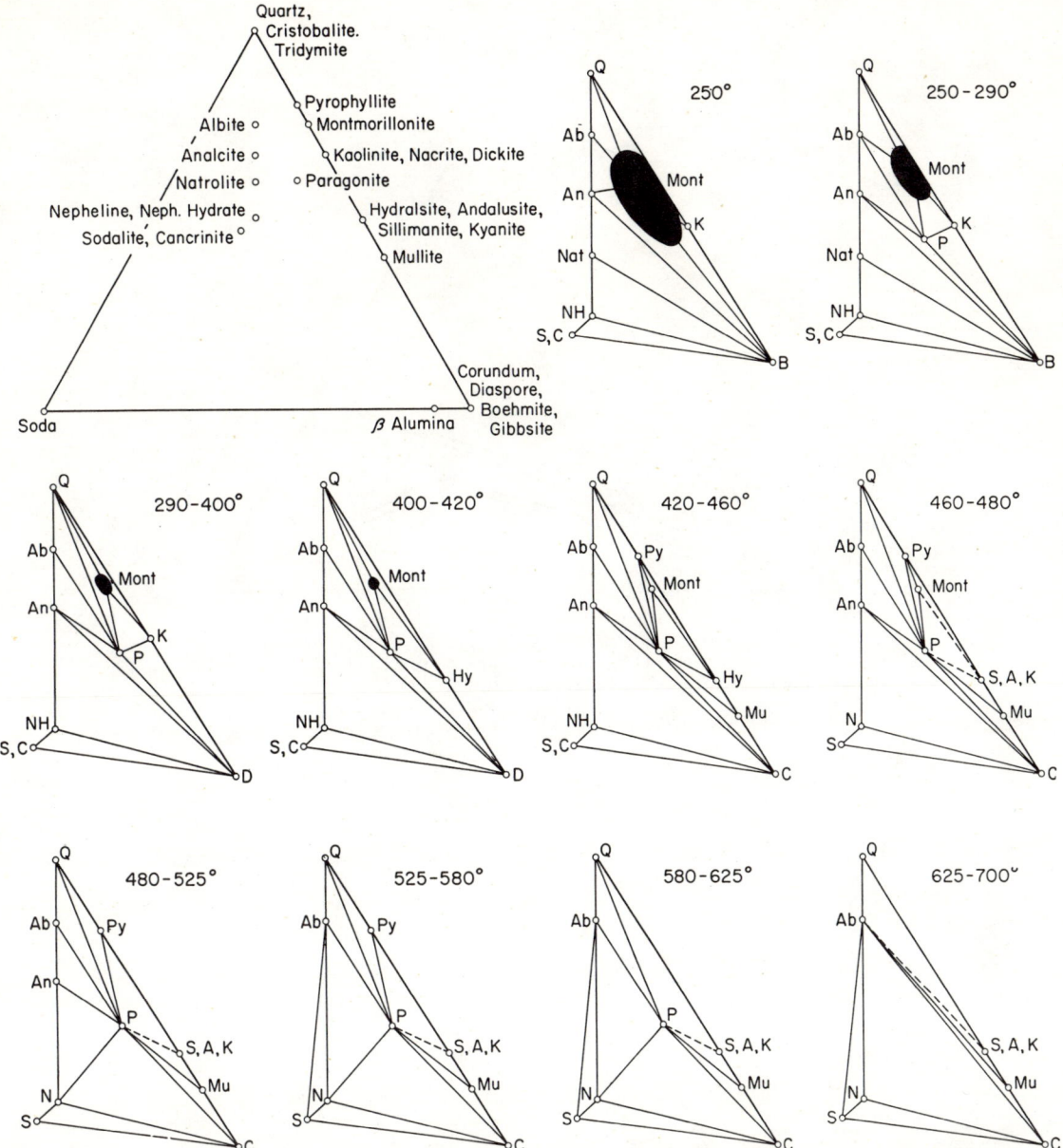

Fig. 2022.—System Na$_2$O–Al$_2$O$_3$–SiO$_2$–H$_2$O; compatibility triangles at 10,000 psi water pressure.

Abbreviations used in the smaller triangles represent the corresponding mineral phases indicated in the first composition triangle. Small triangles have been contorted to conserve space.

L. B. Sand, Rustum Roy, and E. F. Osborn, *Econ. Geol.*, **52** [2] 169–179 (1957).

Na₂O–Al₂O₃–SiO₂–H₂O (cont.)

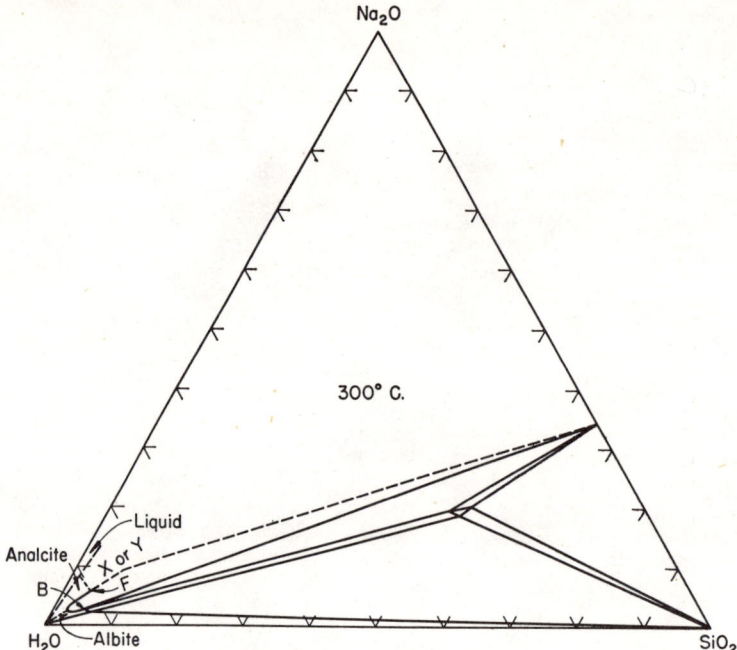

FIG. 2023.—System H_2O–Na_2O–Al_2O_3–SiO_2 at 300°C.; polybaric equilibrium relations. A projection on the ternary plane of the 1.2–1.7% Al_2O_3 section through the H_2O–Na_2O–Al_2O_3–SiO_2 tetrahedron at 300°C.

Irving Friedman, *J. Geol.*, **59**, 24 (1951).

FIG. 2024. See next page.

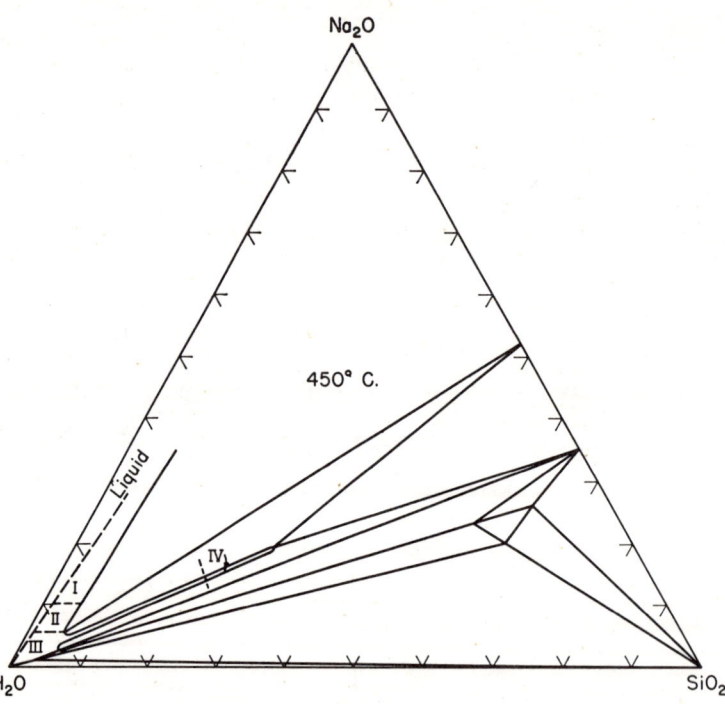

FIG. 2025.—System H_2O–Na_2O–Al_2O_3–SiO_2; polybaric equilibrium relations at 450°C. A section of the projection on the ternary plane of the 1.5–6% Al_2O_3 section through the H_2O–Na_2O–Al_2O_3–SiO_2 tetrahedron at 450°C.

Field I is occupied by phase X or Y, depending upon the SiO_2-Al_2O_3 ratio, the high-silica end favoring the stability of X. Field II is occupied at high H_2O pressure by X and at low pressure by a solution only. Field III is occupied at intermediate pressure by albite and at low pressure by a solution. Presumably, it will be occupied by analcite at very high pressures. Field IV is occupied by albite.

Irving Friedman, *J. Geol.*, **59**, 26 (1951).

Na_2O–Al_2O_3–SiO_2–H_2O (concl.)

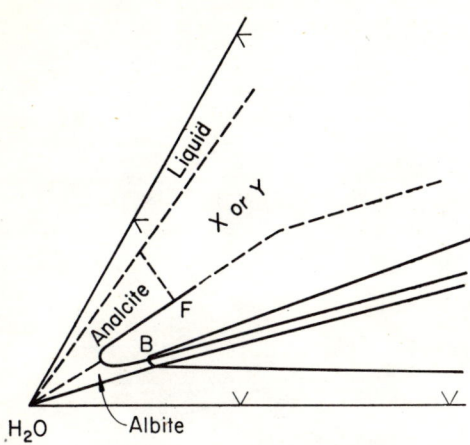

FIG. 2024.—H_2O–Na_2O–Al_2O_3–SiO_2; enlarged section of the H_2O corner of Fig. 2023.

Irving Friedman, *J. Geol.*, **59**, 24 (1951).

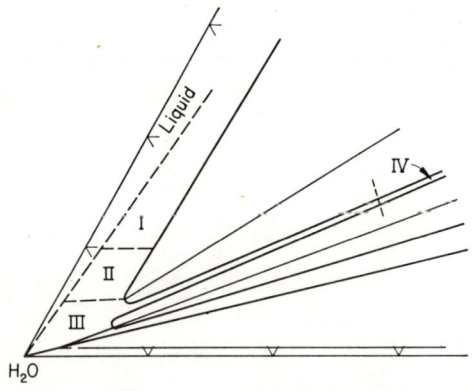

FIG. 2026.—H_2O–Na_2O–Al_2O_3–SiO_2; enlarged section of H_2O corner of Fig. 2025.

Irving Friedman, *J. Geol.*, **59**, 27 (1951).

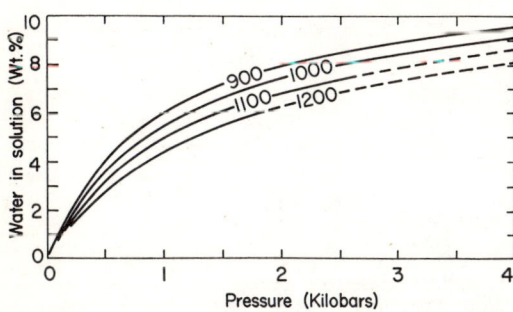

FIG. 2027.—System $Na_2O \cdot Al_2O_3 \cdot 6SiO_2$–$H_2O$.

R. W. Goranson, *Am. J. Sci.* 5th Ser., **35A**, 78 (1938).

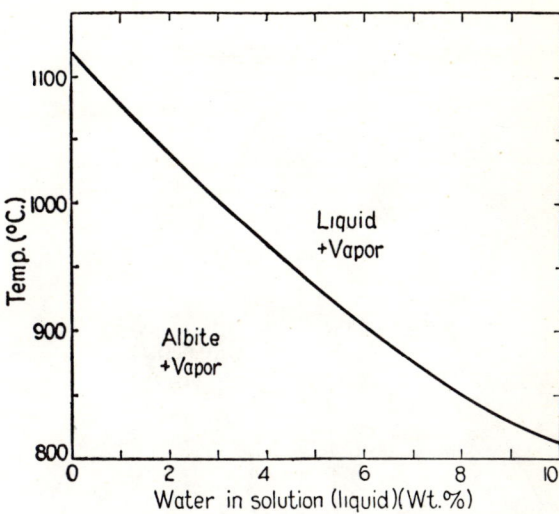

FIG. 2028.—Projection of 3-phase curve, albite + liquid + vapor, of system $Na_2O \cdot Al_2O_3 \cdot 6SiO_2$–$H_2O$ on temperature-concentration coordinate plane.

R. W. Goranson, *Am. J. Sci.*, 5th Ser., **35A**, 85 (1938).

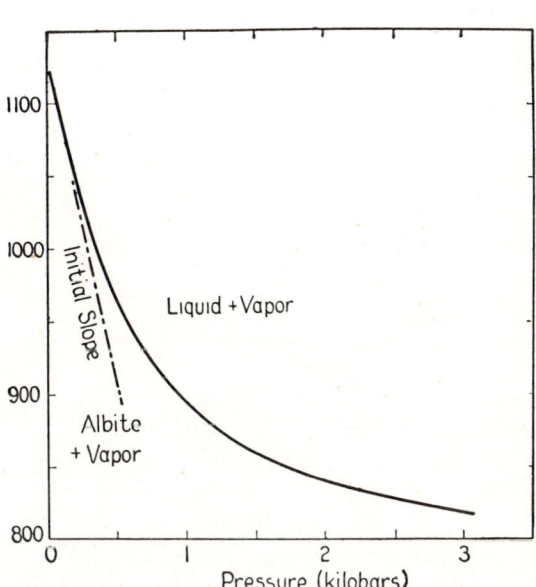

FIG. 2029.—Projection of 3-phase curve, albite + liquid + vapor, of system $Na_2O \cdot Al_2O_3 \cdot 6SiO_2$–$H_2O$ on temperature-pressure coordinate plane.

R. W. Goranson, *Am. J. Sci.*, 5th Ser., **35A**, 84 (1938).

Na_2O–CO_2–ThO_2–H_2O

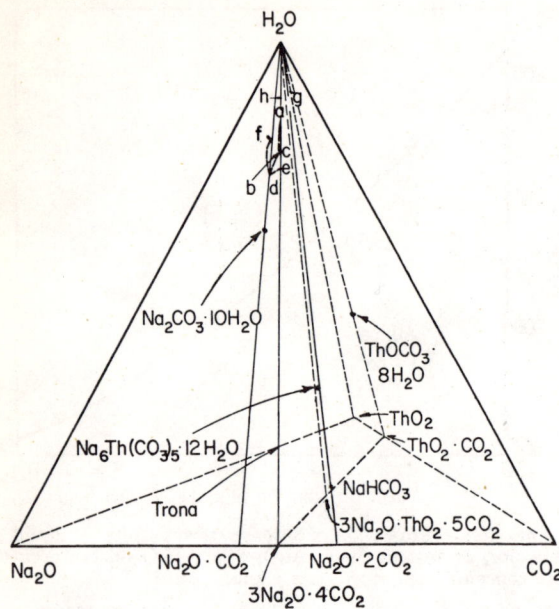

Fig. 2030.—System Na_2O–Th_2O–CO_2–H_2O at 25°C; partial. Equilibrium solid phases: a–b = z + $NaHCO_3$; b = z + $NaHCO_3$ + trona; b–c = $NaHCO_3$ + trona; b–d = z + trona; d = z + trona + $Na_2CO_3 \cdot 10H_2O$; d–e = trona + $Na_2CO_3 \cdot 10H_2O$; d–f = z + $Na_2CO_3 \cdot 10H_2O$; g–h = z + $ThOCO_3 \cdot H_2O$. z = $3Na_2O \cdot ThO_2 \cdot 5CO_2 \cdot 12H_2O$; trona = $Na_2CO_3 \cdot NaHCO_3 \cdot 2H_2O$.

F. A. Schimmel, Oak Ridge National Laboratory, Phase Diagrams of Nuclear Reactor Materials, R. E. Thoma, ed., ORNL-2548, p. 150 (1959).

Na_2O–CO_2–UO_3–H_2O

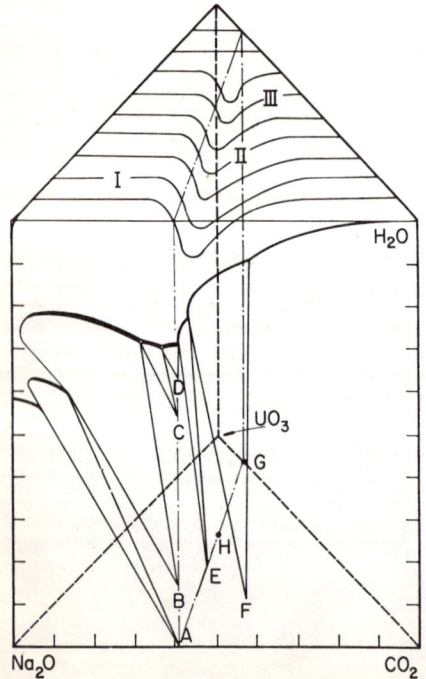

CaO–MgO–CO_2–H_2O

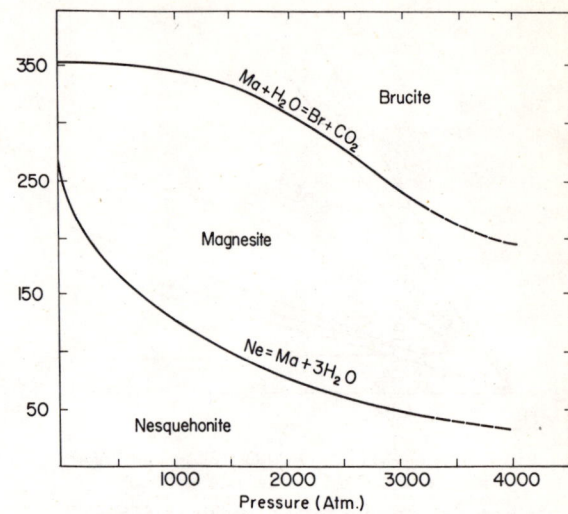

Fig. 2032.—System CaO–MgO–CO_2–H_2O.

H. Schloemer and R. Nacken, *Zement-Kalk-Gips*, 6 [10] 379 (1953).

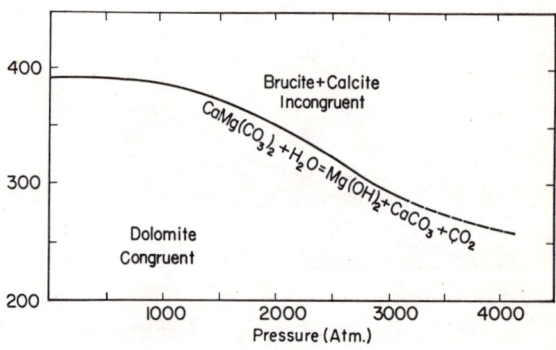

Fig. 2033.—System CaO–MgO–CO_2–H_2O.

H. Schloemer and R. Nacken, *Zement-Kalk-Gips*, 6 [10] 380 (1953).

Fig. 2031.—System Na_2O–UO_3–CO_2–H_2O at 26°C. Solutions in equilibrium with: I, sodium uranates; II, sodium uranyl tricarbonate; III, uranyl carbonate. A = Na_2CO_3, B = $Na_2CO_3 \cdot H_2O$, C = $Na_2CO_3 \cdot 7H_2O$, D = $Na_2CO_3 \cdot 10H_2O$, E = $Na_2CO_3 \cdot NaHCO_3 \cdot 2H_2O$, F = $NaHCO_3$, G = UO_2CO_3, H = $Na_4UO_2(CO_3)_3$. Entire upper face of the triangular prism represents pure water.

C. A. Blake, C. F. Coleman, K. B. Brown, D. G. Hill, R. S. Lowrie, and J. M. Schmitt, *J. Am. Chem. Soc.*, **78**, 5978 (1956). See also Oak Ridge National Laboratory, Phase Diagrams of Nuclear Reactor Materials, R. E. Thoma, ed., ORNL-2548, p. 148 (1959).

CaO–SrO–SiO$_2$–H$_2$O

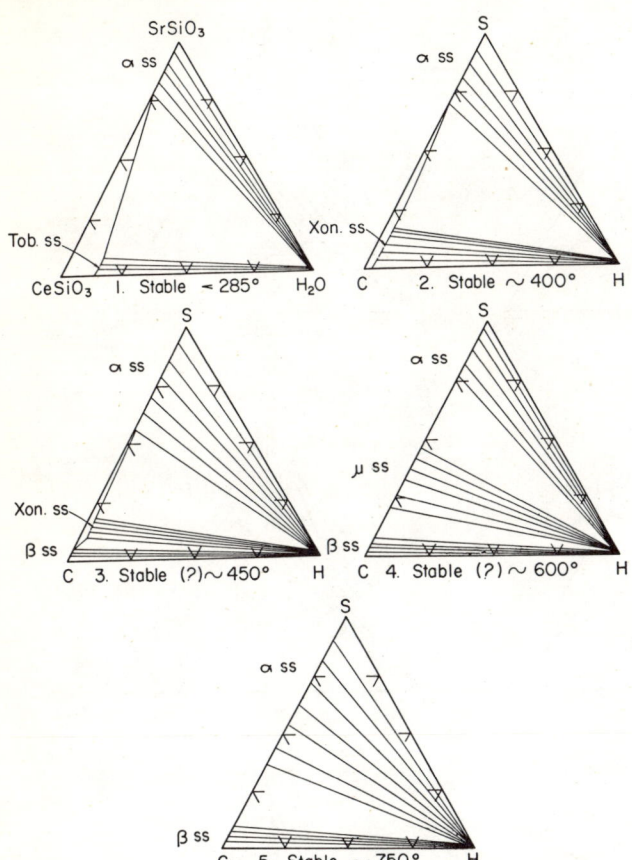

FIG. 2034.—System CaSiO$_3$–SrSiO$_3$–H$_2$O; probable compatibility triangles at a pressure of 10,000 psi. Tob ss = tobermorite solid solutions, Xon ss = xonotlite solid solutions, α ss = α(Sr, Ca)SiO$_3$ solid solutions, β ss = β(Ca,Sr)SiO$_3$ solid solutions.

D. A. Buckner, D. M. Roy, and Rustum Roy, *Am. J. Sci.*, **258**, 145 (1960).

NOTE: CeSiO$_3$ should read CaSiO$_3$.

FeO–Fe$_3$O$_4$–SiO$_2$–H$_2$O

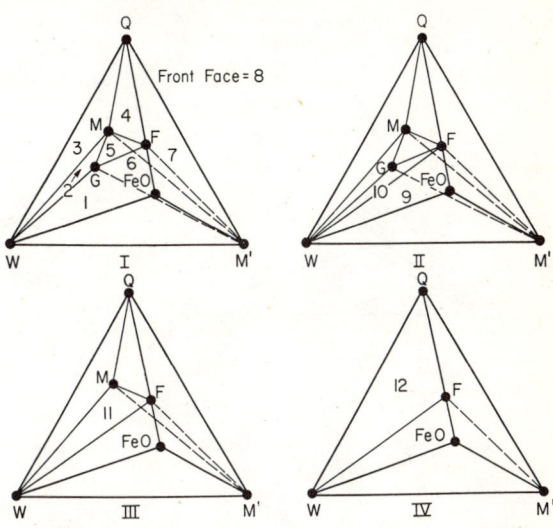

FIG. 2036.—System FeO–Fe$_3$O$_4$–SiO$_2$–H$_2$O; stable phase assemblages. I = temp. less than 250°C., II = 250°–470°C., III = 470°–480°C., IV = temp. greater than 480°C. Numbers refer to different three and four condensed phase assemblages. W = water, M' = magnetite Q = quartz, F = fayalite, M = minnesotaite, G = greenalite.

S. S. Flaschen and E. F. Osborn, *Econ. Geol.*, **52** [8] 934 (1957).

CaO–Al$_2$O$_3$–SiO$_2$–H$_2$O

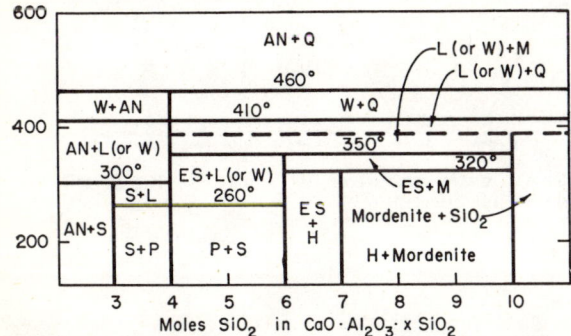

FIG. 2035.—CaO–Al$_2$O$_3$–SiO$_2$–H$_2$O. Phase relation among the calcium zeolites represented as a t–x isobaric (1000 atm.) projection on the anhydrous CaAl$_2$O$_4$–SiO$_2$ join. AN = anorthite, ES = epistilbite, H = heulandite, L = laumontite, M = mordenite, P = phillipsite, Q = quartz, S = scolecite, W = wairakite.

Mitsue Koizumi and Rustum Roy, *J. Geol.*, **68** [1] 50 (1960).

MgO–Al$_2$O$_3$–CO$_2$–H$_2$O

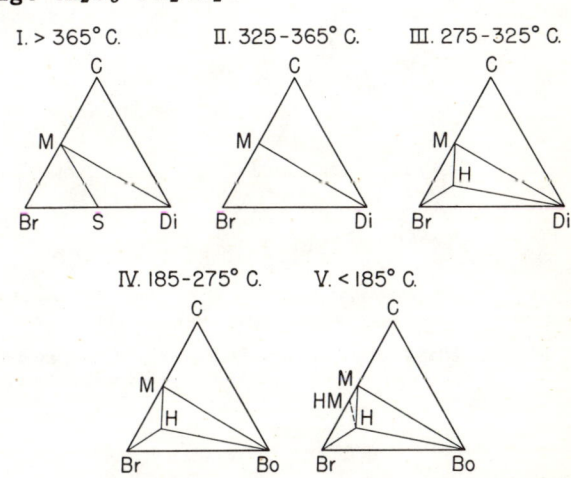

FIG. 2037.—System MgO–Al$_2$O$_3$–H$_2$O–CO$_2$; composition triangles are projections on the base of tetrahedra with H$_2$O at the top apex.

C—CO$_2$, Di—diaspore, S—spinel, Br—brucite, M—magnesite, H—hydrotalcite, HM—hydromagnesite, Bo—boehmite.

Della M. Roy, Rustum Roy, and E. F. Osborn, *Am. J. Sci.*, **251** [5] 352 (1953).

Fig. 2038A

MgO–Al$_2$O$_3$–SiO$_2$–H$_2$O

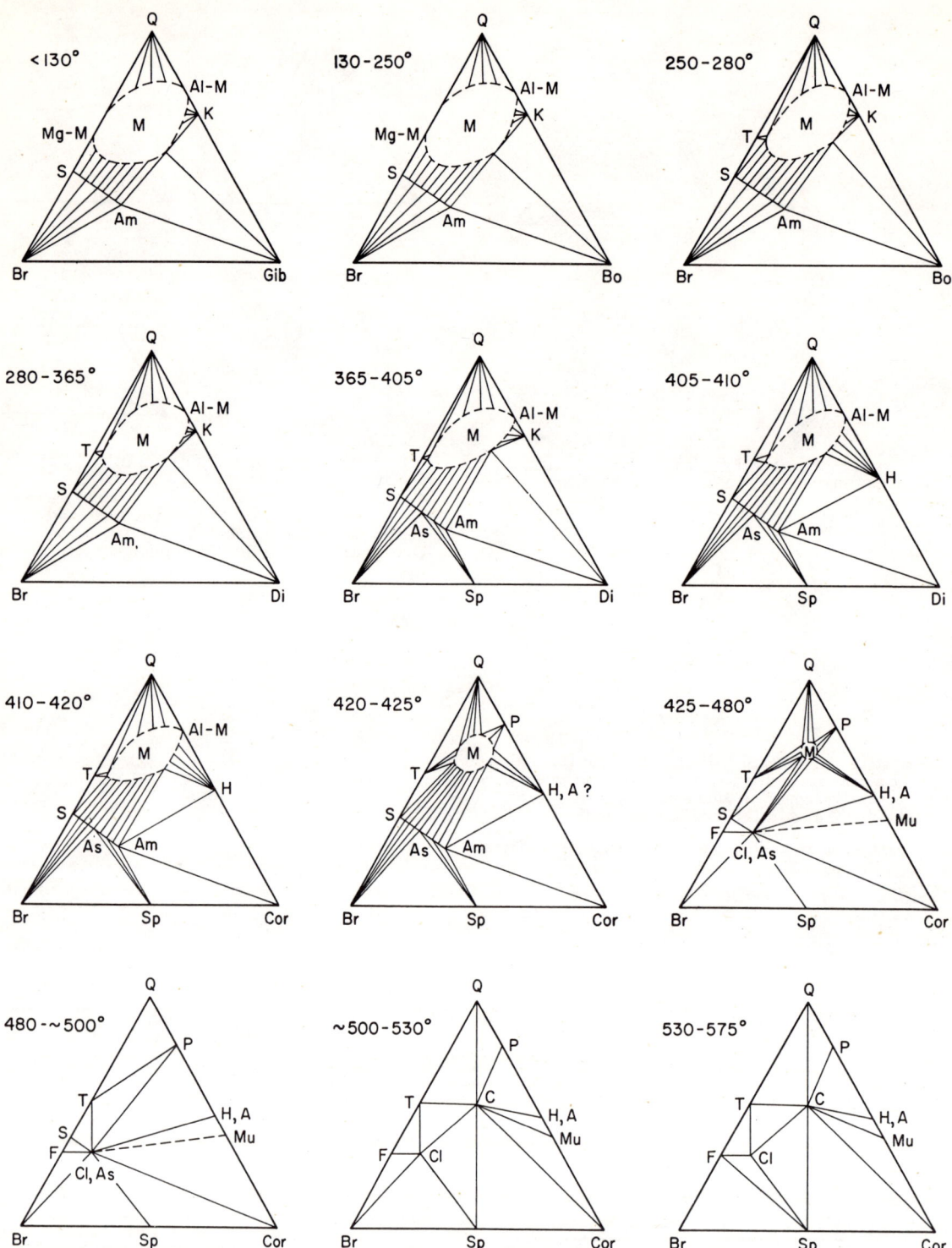

Fig. 2038A.—System MgO–Al$_2$O$_3$–SiO$_2$–H$_2$O; compatibility triangles at a water pressure of 10,000 psi. See Fig. 2038B on following page, for legend and reference.

MgO–Al₂O₃–SiO₂–H₂O

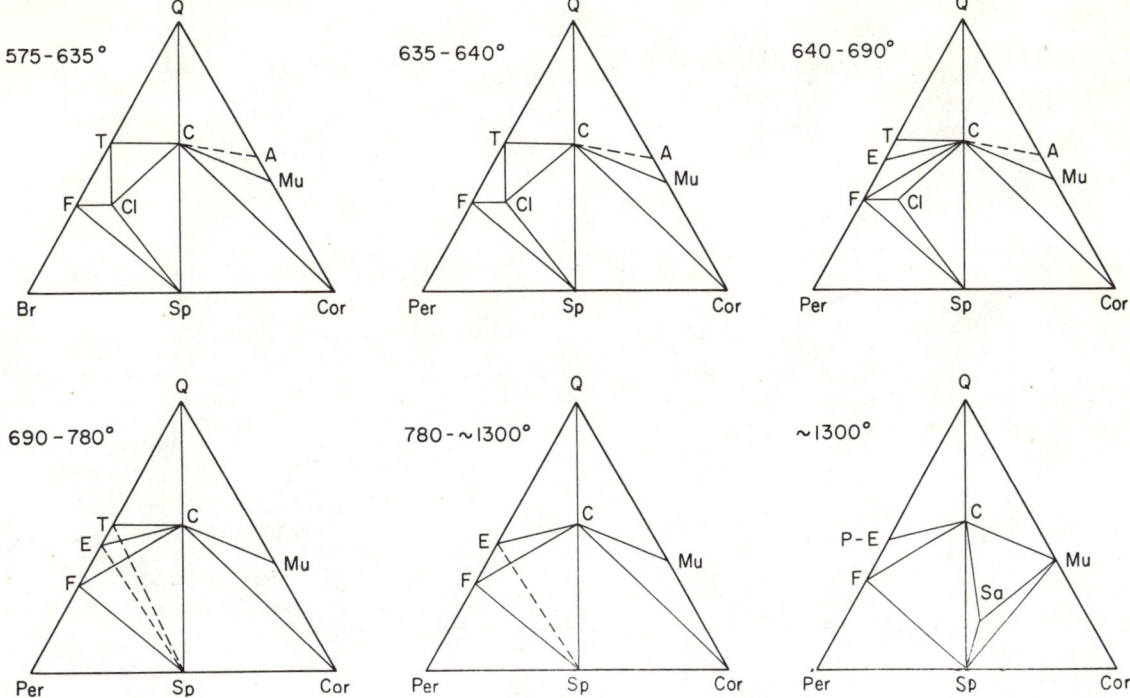

FIG. 2038B.—System MgO–Al₂O₃–SiO₂–H₂O; compatibility triangles at a water pressure of 10,000 psi.

A = andalusite; Al-M = aluminum montmorillonite; Am = amesite; As = aluminum serpentine; Bo = boehmite; Br = brucite; C = cordierite; Cl = clinochlore; Cor = corundum; Di = diaspore; E = enstatite; F = forsterite; Gib = gibbsite; H = hydralsite; K = kaolinite; M = montmorillonite; Mg-M = magnesian montmorillonite; Mu = mullite; P = pyrophyllite; P-E = proto-enstatite; Per = periclase; Q = quartz; S = serpentine; Sa = sapphirine; Sp = spinel; T = talc; Tr = tridymite.

Della M. Roy and Rustum Roy, *Am. Mineral.*, **40**, 162 (1955).

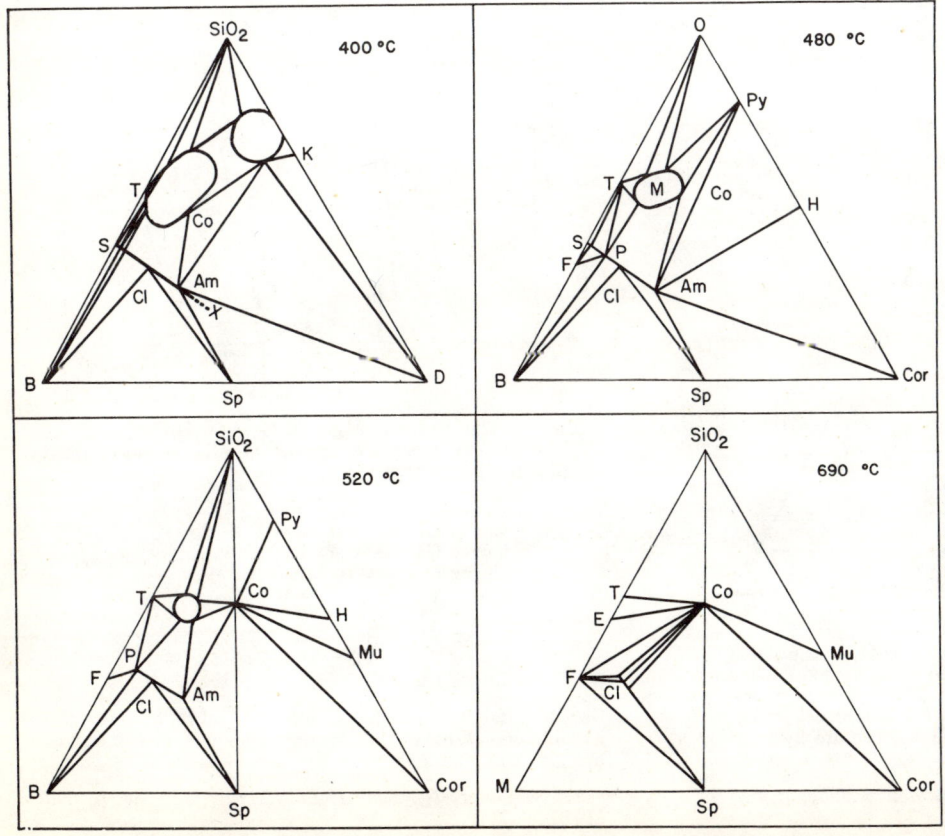

FIG. 2039.—System MgO–Al₂O₃–SiO₂–H₂O; isobaric (1000 atm.) compatibility triangles. Round quaternary solid solution areas are the various montmorillonoids. These revised data taken from Mumpton and Roy.

Am = amesite, B = brucite, Cl = clinochlore, Co = cordierite, Cor = corundum, D = diaspore, E = enstatite, F = forsterite, H = hydralsite, K = kaolinite, M = montmorillonoid, Mu = mullite, P = pyrope, Py = pyrophyllite, Q = quartz, S = chrysotile, Sp = spinel, T = talc.

B. W. Nelson and Rustum Roy, *Am. Mineral.*, **43**, 716 (1958).

MgO–Al$_2$O$_3$–SiO$_2$–H$_2$O (concl.)

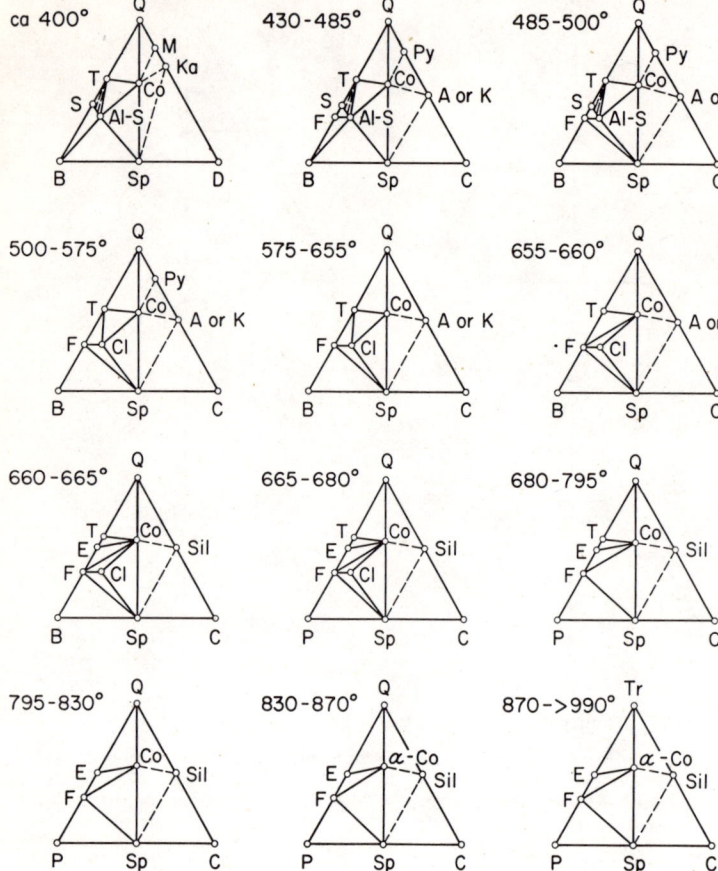

Fig. 2040.—System MgO–Al$_2$O$_3$–SiO$_2$–H$_2$O.

Assemblages stable in the presence of an excess of water vapor at significant temperature intervals and 15,000 psi. Abbreviations: A, andalusite; Al-S, aluminous serpentine; B, brucite; C, corundum; Cl, clinochlore; Co, cordierite; Cr, cristobalite; D, diaspore; E, enstatite; F, forsterite; K, kyanite; Ka, kaolinite; M, montmorillonite; Mu, mullite; P, periclase; Py, pyrophyllite; Q, quartz; S, serpentine; Sil, sillimanite; Sp, spinel; T, talc; Tr, tridymite; V, vapor.

H. S. Yoder, Jr., *Am. J. Sci.*, Bowen volume, p. 601 (1952).

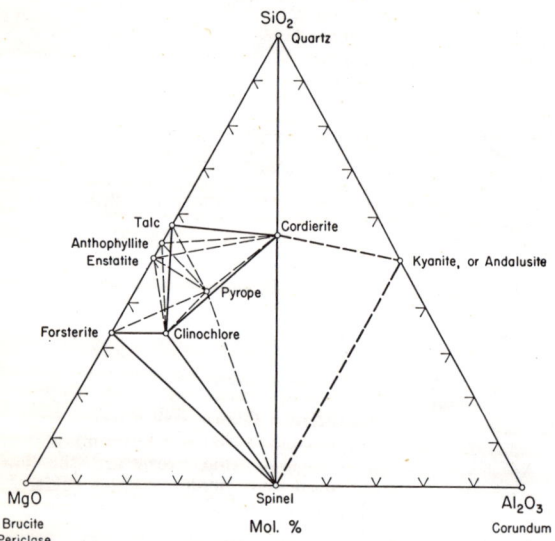

Fig. 2041.—System MgO–Al$_2$O$_3$–SiO$_2$–H$_2$O; assemblages in both excess-water regions (solid and heavy dashed lines) and water-deficient regions (light dashed lines) believed to be stable at ca. 600° and 15,000 psi in the system; Talc, anthophyllite, and clinochlore are hydrous minerals.

H. S. Yoder, Jr., *Am. J. Sci.*, Bowen volume, p. 613 (1952).

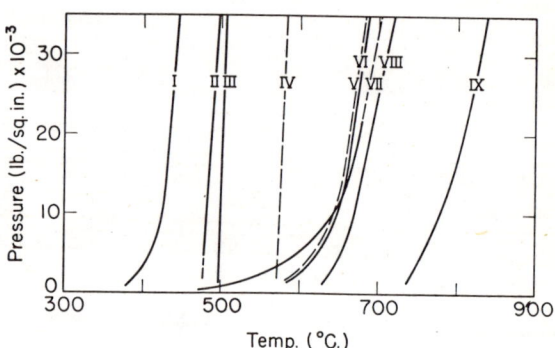

Fig. 2042.—System MgO–Al$_2$O$_3$–SiO$_2$–H$_2$O; some of the pressure-temperature curves (I–IX) of univariant equilibrium in the system:

Curve I.   S + B = F + V
Curve II.   B + Al-S = F + Sp + V
Curve III. S = F + T + V
Curve IV.   Py = Mu + Cr + V
Curve V.   Cl + T = F + Co + V
Curve VI.   F + T = E + V
Curve VII. B = P + V
Curve VIII. Cl = F + Co + Sp + V
Curve IX.   T = E + Q + V

See Fig. 2040 for code to mineral abbreviations; the assemblages stable in the divariant regions between curves are given in Fig. 2041.

H. S. Yoder, Jr., *Am. J. Sci.*, Bowen volume, p. 600 (1952).

$MgO-Al_2O_3-N_2O_5-H_2O$

V. Four Metal Oxides with Water

$K_2O-Na_2O-Al_2O_3-SiO_2-H_2O$

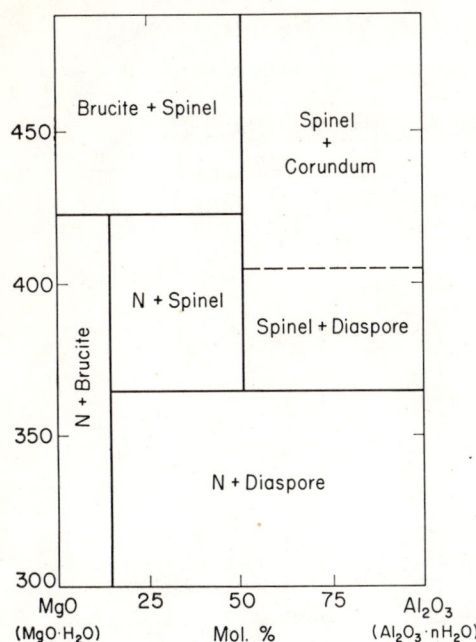

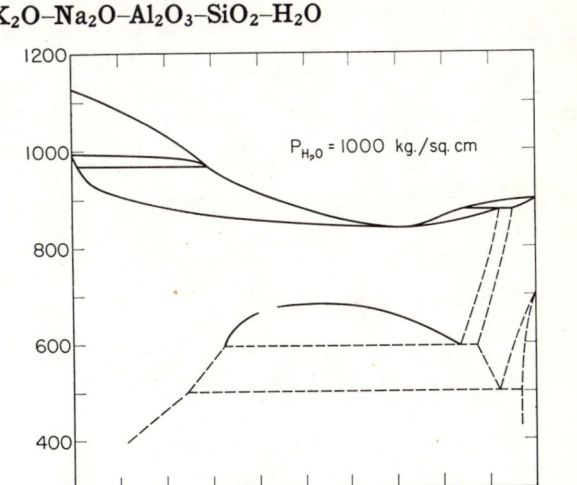

FIG. 2043.—$MgO-Al_2O_3-H_2O-N_2O_5$ at 10,000 psi. Small amount of nitrate ion in aqueous phase; N = nitrate-hydrotalcite phase.

Della M. Roy, Rustum Roy, and E. F. Osborn, *Am. J. Sci.*, **251** [5] 355 (1953).

FIG. 2044.—System $KAlSi_3O_8-NaAlSi_3O_8-H_2O$; isobaric equilibrium diagrams in dry melts and at 1,000 kg./cm.² (see modification by Wm. S. MacKenzie, Fig. 2045, and 2,000 kg./cm.² pressure of water).

N. L. Bowen and O. F. Tuttle, *J. Geol.*, **58** [5] 497 (1950).

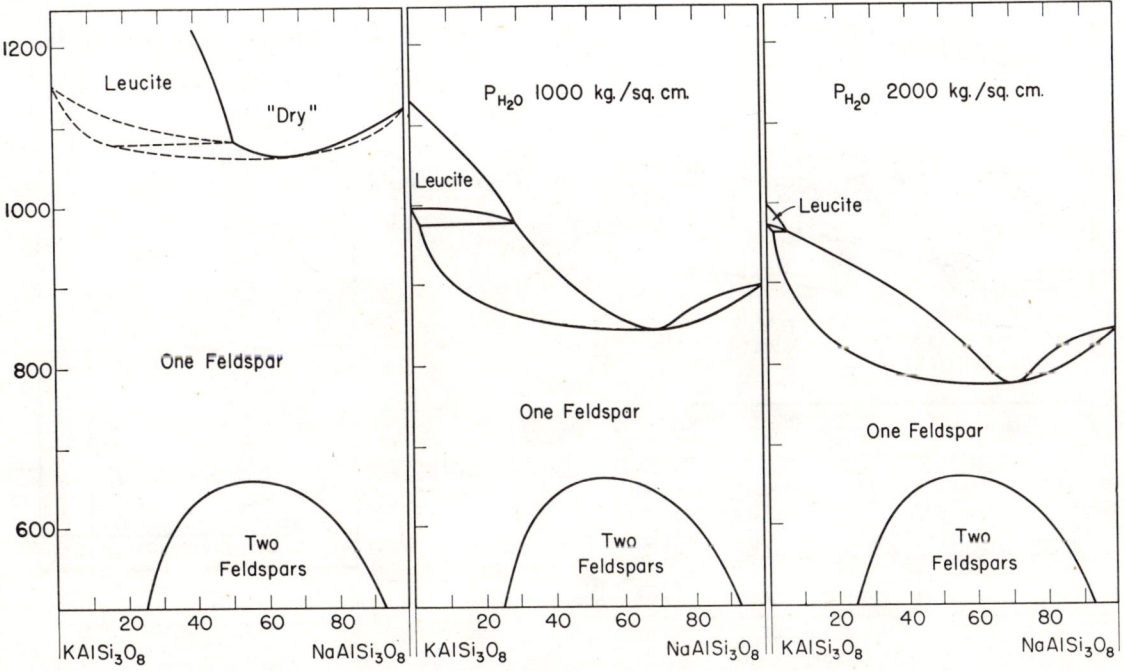

FIG. 2045.—System $KAlSi_3O_8-NaAlSi_3O_8$ at 1000 kg./cm.² pressure of water.

Modified from N. L. Bowen and O. F. Tuttle, *J. Geol.*, **58**, 497 (1950); (see Fig. 2044). Wm. S. MacKenzie, *Am. J. Sci.*, Bowen volume, p. 338 (1952).

Li_2O–Na_2O–Al_2O_3–SiO_2–H_2O

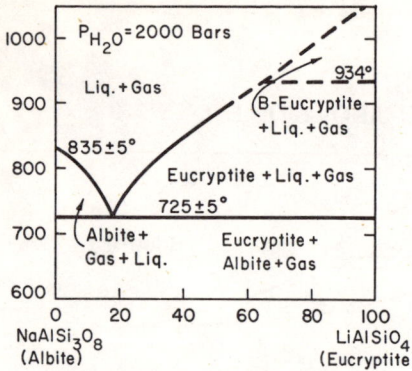

Fig. 2046.—System $LiAlSiO_4$–$NaAlSi_3O_8$ at 2000 bars H_2O pressure.

D. B. Stewart, *Intern. Geol. Congr.*, 21st., Copenhagen, 1960, p. 23 in *Rept. Session, Norden*.

Na_2O–MgO–Fe_2O_3–SiO_2–H_2O

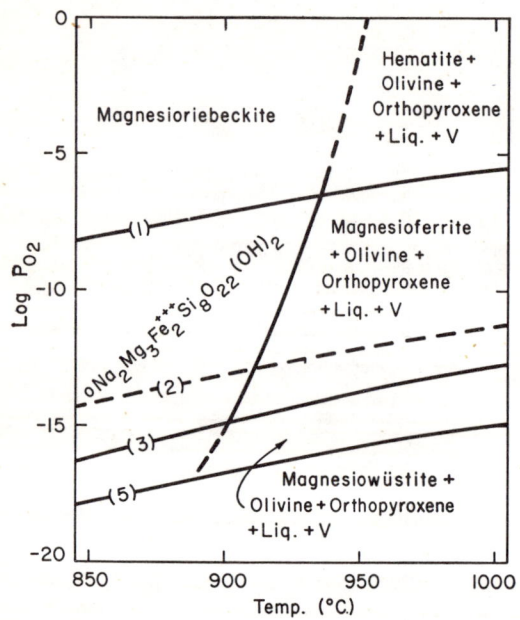

Fig. 2047.—System $Na_2O \cdot 3MgO \cdot Fe_2O_3 \cdot 8SiO_2$; P_{O_2}–T diagram with excess water at 2000 bars vapor pressure.

W. G. Ernst, *Geochim. et Cosmochim. Acta*, **19** [1] 25 (1960).

VI. Five Metal Oxides with Water

Na_2O–MgO–Al_2O_3–B_2O_3–SiO_2–H_2O

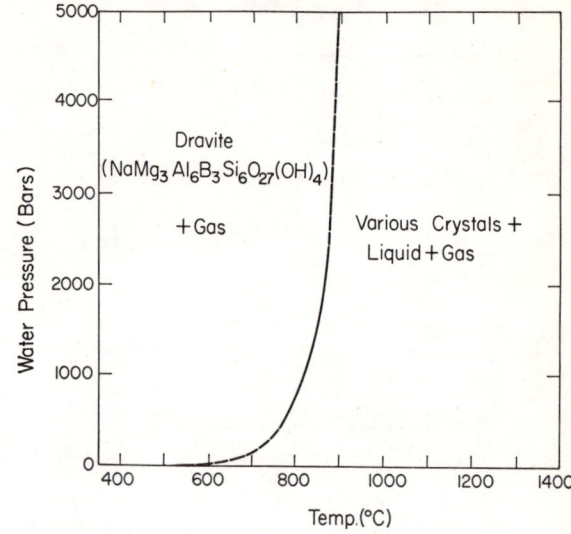

Fig. 2048.—System dravite–water; preliminary p–T diagram. At pressures up to 2000 bars the principal decomposition product is cordierite.

C. R. Robbins and H. S. Yoder, Jr., Ann. Rept. of the Director of the Geophysical Laboratory, 1961–1962, Carnegie Institution of Washington, Paper No. 1390, p. 108.

VII. Miscellaneous Substances with Water

(a) One Substance with Water

Na_2WO_4–H_2O

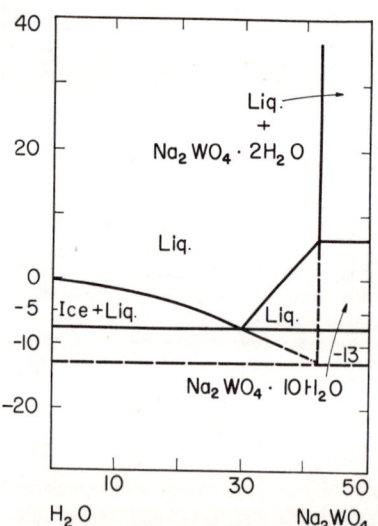

Fig. 2049.—System Na_2WO_4–H_2O.

R. L. Pilloton and G. E. Crawley, Jr., *J. Electrochem. Soc.*, **107** [2] 123 (1960).

RbOH–H₂O

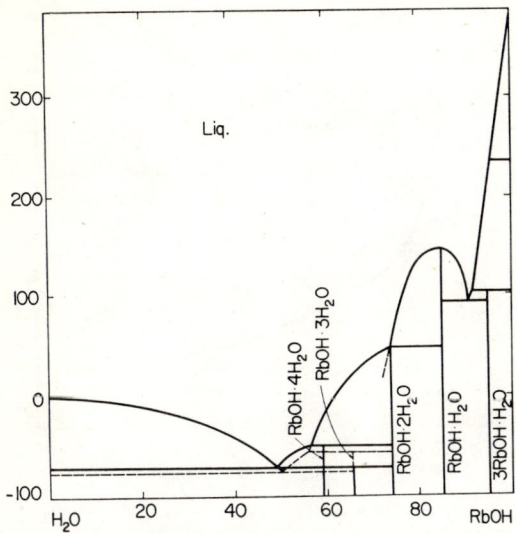

FIG. 2050.—System RbOH–H₂O.

A. P. Rollet, Roger Cohen-Adad, Maurice Michaud, and Aymond Tranquard, *Compt. rend.*, **246**, 3249 (1958).

(b) Two Substances with Water

K₂CO₃–Na₂CO₃–H₂O

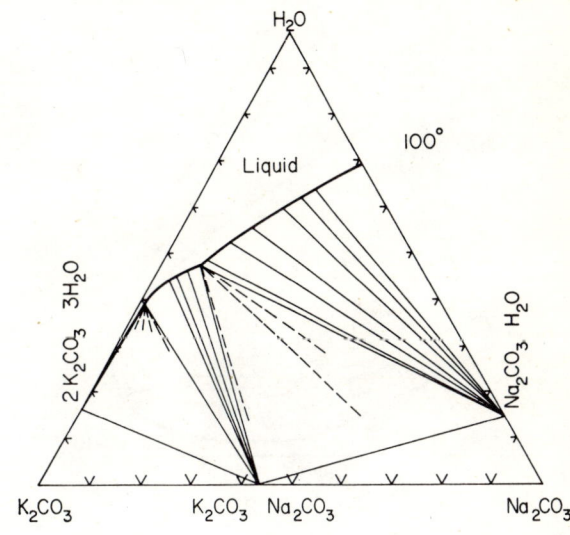

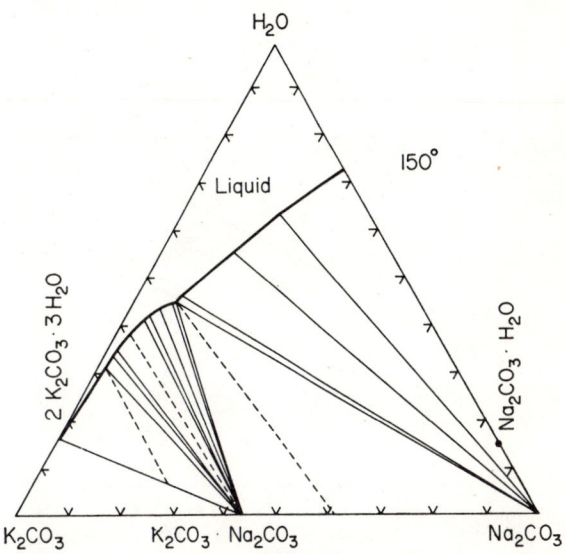

FIG. 2052.—Phase relations in system K₂CO₃–Na₂CO₃–H₂O at 100° and 150°C.

G. Ervin, Jr., A. L. Giorgi, and C. E. McCarthy, *J. Am. Chem. Soc.*, **66** [3] 384 (1944).

CaSO₄–H₂O

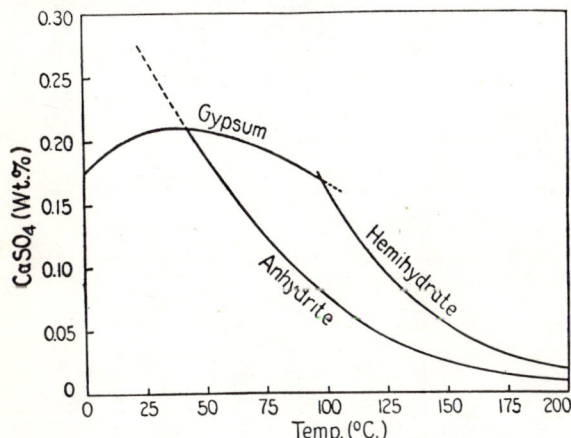

FIG. 2051.—System H₂O–CaSO₄; solubility of gypsum, hemihydrate, and anhydrite in water.

E. Posniak. *Am. J. Sci.*, 5th Ser., **35A**, 268 (1938).

$LiClO_4$–H_2O_2–H_2O

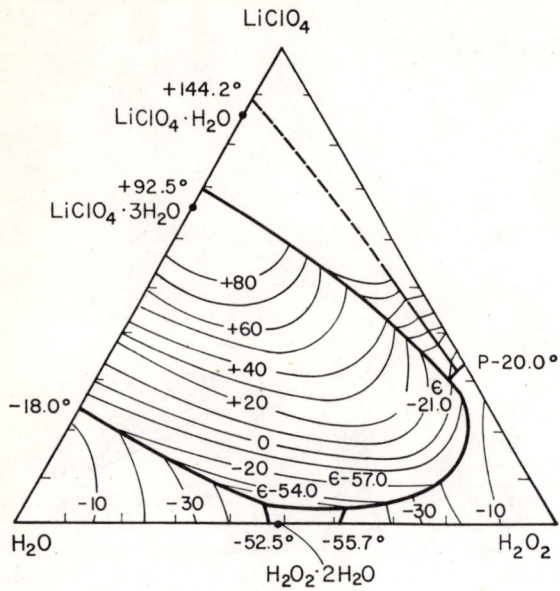

FIG. 2053.—System H_2O–H_2O_2–$LiClO_4$.

K. E. Mironov, M. Z. Pronina, and S. A. Tokareva, *Zhur. Neorg. Khim.*, **3**, 515 (1958).

$NaAlSi_3O_8$–NH_3–H_2O

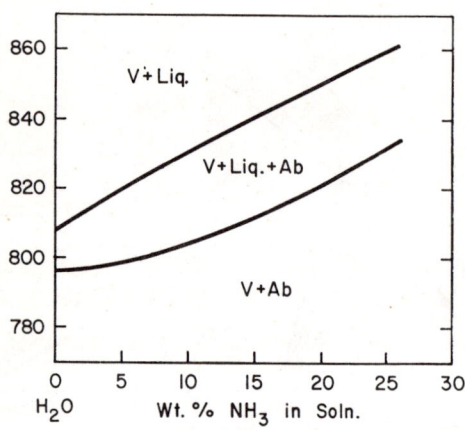

FIG. 2055.—System $NaAlSi_3O_8$–H_2O–NH_3 at 2750 bars pressure, determined for a 1:1 weight ratio of $NaAlSi_3O_8$ to total volatiles. Ab = albite; V = vapor.

P. J. Wyllie and O. F. Tuttle, *Am. J. Sci.*, **259**, 132 (1961).

Na_2CO_3–SiO_2–H_2O

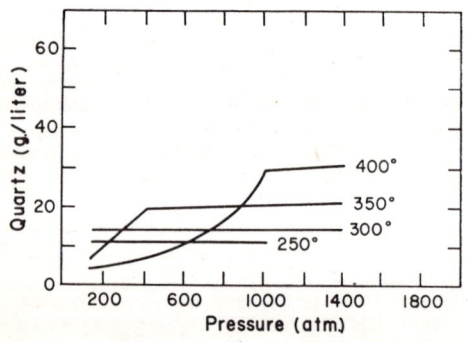

FIG. 2057.—H_2O–SiO_2–Na_2CO_3 (5% soln); solubility isotherms in relation to pressure.

V. P. Butuzov and L. V. Bryatov, *Kristallografiya*, **2**, 661 (1957).

$NaAlSi_3O_8$–HF–H_2O

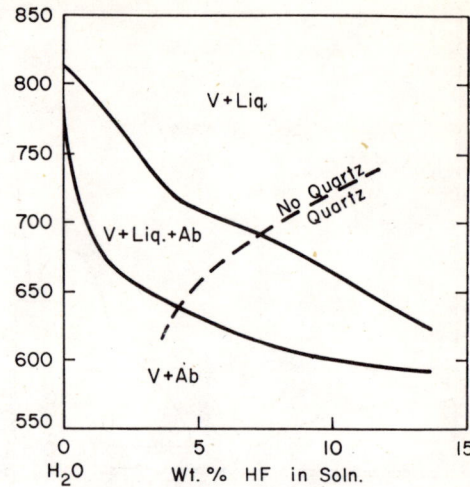

FIG. 2054.—System $NaAlSi_3O_8$–H_2O–HF at 2750 bars pressure determined for a 1:1 weight ratio of $NaAlSi_3O_8$ to total volatiles. Ab = albite; V = vapor.

P. J. Wyllie and O. F. Tuttle, *Am. J. Sci.*, **259**, 137 (1961).

$NaClO_4$–H_2O_2–H_2O

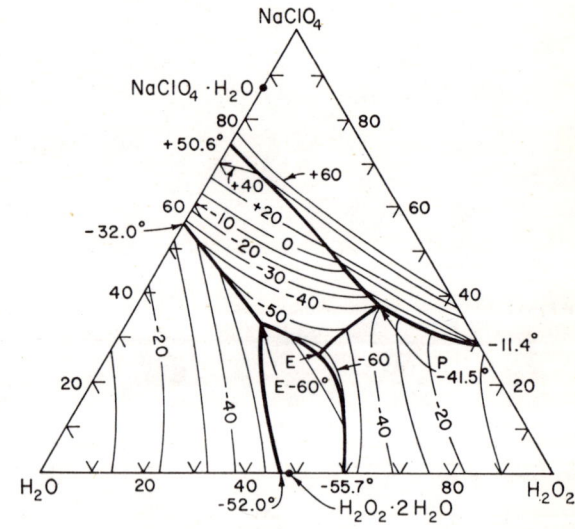

FIG. 2056.—System H_2O–H_2O_2–$NaClO_4$.

K. E. Mironov, M. Z. Pronina, and S. A. Tokareva, *Zhur. Neorg. Khim.*, **3**, 513 (1958).

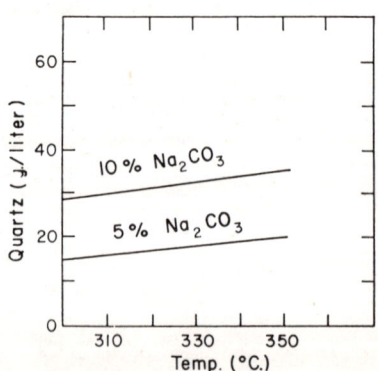

FIG. 2058.—H_2O–Na_2CO_3–SiO_2; relation between temperature and solubility.

V. P. Butuzov and L. V. Bryatov, *Kristallografiya*, **2**, 661 (1957).

Na_2CO_3–Na_2WO_4–H_2O

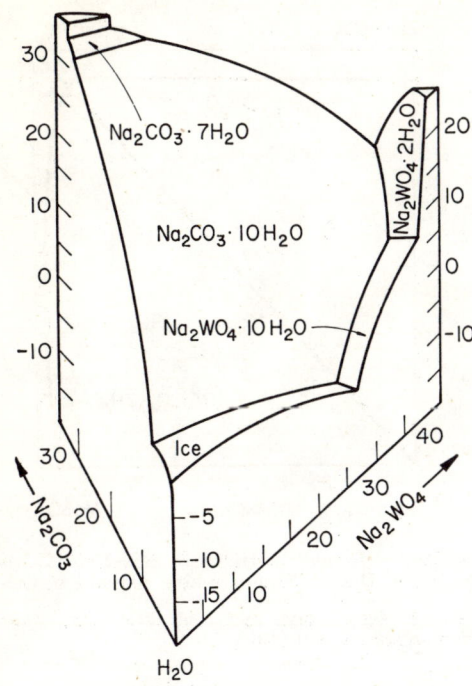

FIG. 2059.—System Na_2CO_3–Na_2WO_4–H_2O; crystallization surface.

R. L. Pilloton and G. E. Crawley, Jr., *J. Electrochem. Soc.*, **107** [2] 125 (1960).

CaO–$CaCl_2$–H_2O

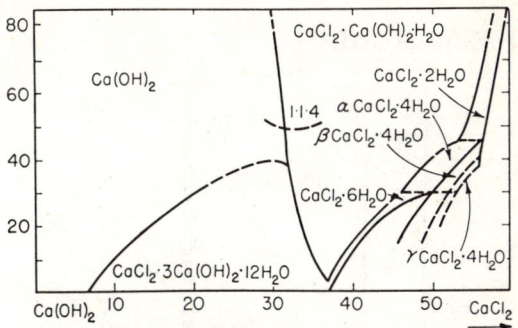

FIG. 2061.—System $Ca(OH)_2$–$CaCl_2$–H_2O.

S. Z. Makarov and I. I. Vol'nov, *Izvest. Sektora Fiz.-Khim. Anal., Inst. Obshcheĭ Neorg. Khim., Akad. Nauk S.S.S.R.*, **25**, 331 (1954).

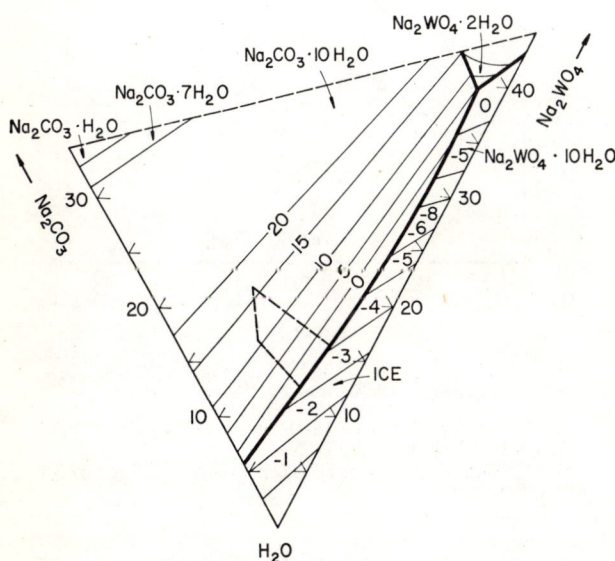

FIG. 2060.—System Na_2CO_3–Na_2WO_4–H_2O; liquidus projection.

R. L. Pilloton and G. E. Crawley, Jr., *J. Electrochem. Soc.*, **107** [2] 125 (1960).

$Pb(NO_3)_2$–HNO_3–H_2O

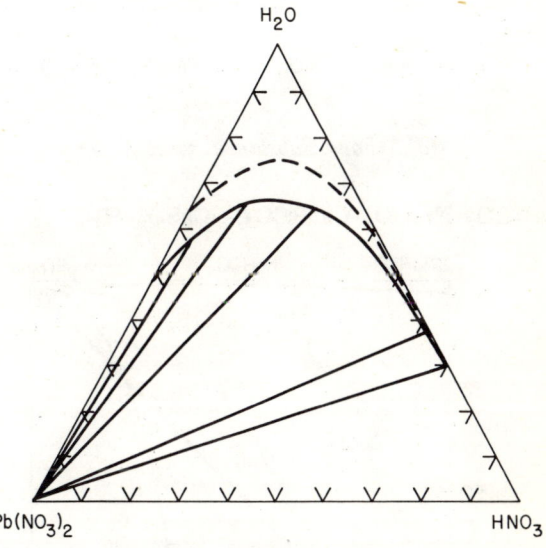

FIG. 2062.—System $Pb(NO_3)_2$–HNO_3–H_2O. Solid line at 80°C; dashed line at 26°C.

L. M. Ferris, *J. Chem. Eng. Data*, **5** [3] 242 (1960).

(c) Three Substances with Water

$CaO-Al_2O_3-SO_3-H_2O$

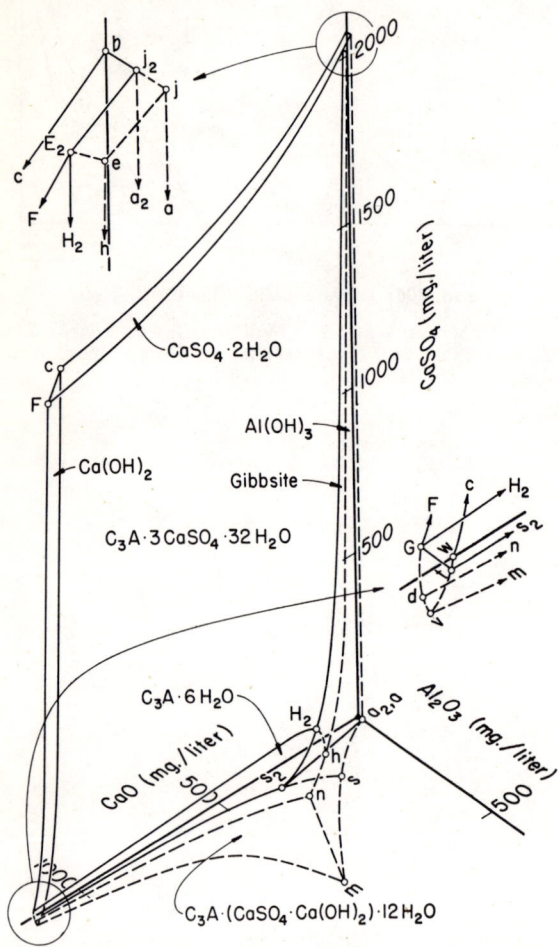

FIG. 2063.—System $CaO-Al_2O_3-CaSO_4-H_2O$ at 20°C. —— stable; - - - - metastable. A = Al_2O_3; C = CaO.

J. D'Ans and H. Eick, *Zement-Kalk-Gips*, **6** [6] 304 (1953).

(d) Four Substances with Water

$NaHCO_3-Na_2SO_4-Ca(HCO_3)_2-CaSO_4-H_2O$

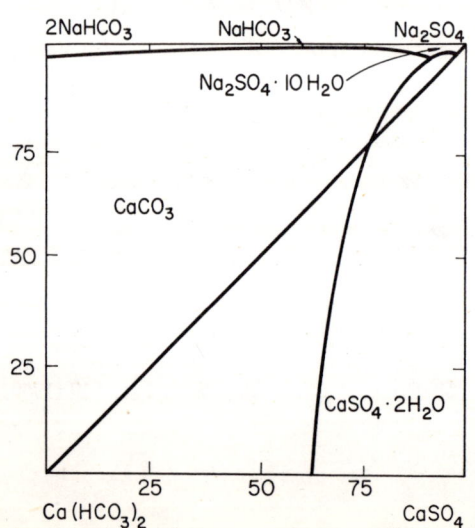

FIG. 2064.—System $NaHCO_3-Na_2SO_4-Ca(HCO_3)_2-CaSO_4-H_2O$ at 25°C. and 1 atm. pressure of CO_2.

Yu. P. Nikol'skaya and I. A. Moshkina, *Zhur. Neorg. Khim.*, **3**, 498 (1958).

$NaHCO_3-Na_2SO_4-Mg(HCO_3)_2-MgSO_4-H_2O$

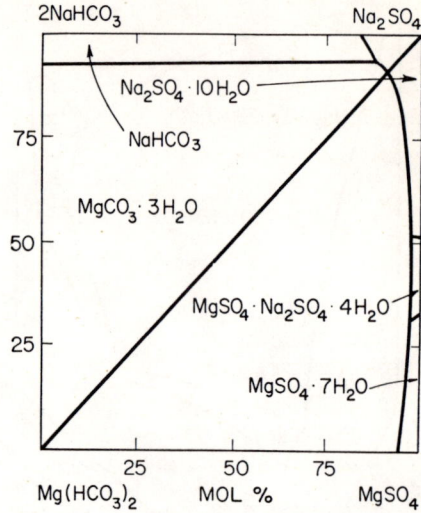

FIG. 2065.—System $NaHCO_3-Na_2SO_4-Mg(HCO_3)_2-MgSO_4-H_2O$ at 25°C. and 1 atm. pressure of CO_2.

Yu. P. Nikol'skaya and I. A. Moshkina, *Zhur. Neorg. Khim.*, **3**, 501 (1958).

$CaO-MgO-CO_2-SO_3-H_2O$

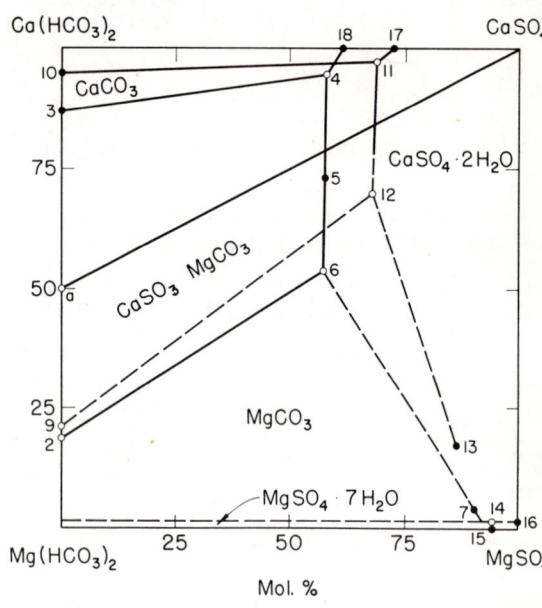

FIG. 2066.—System $Ca(HCO_3)_2-CaSO_4-Mg(HCO_3)_2-MgSO_4$ at 25°C. and $P(CO_2)$ = 1 atm.

Field 2-3-4-5-6-7-14-15-16-18 with 0% NaCl. Field 9-12-11-17-10-13 with 2.0% NaCl.

O. K. Yanat'eva, *Doklady Akad. Nauk S.S.S.R.*, **67** [3] 481 (1949).

Appendix I

Melting temperatures of the components are the basic reference points of phase diagrams. Nevertheless, in a recent comprehensive survey of previously published melting points of metal oxides, the author† observed wide variation in reported values for many of the oxides. An extracted portion of this survey is reproduced below.

MELTING POINTS OF THE METAL OXIDES†,‡

Oxide	Investigator	Date	Environment	Melting Point Original °C	Int. 1948†† °C
Ag_2O	F. C. Kracek	(1942)	......	191D	...
	Shun-ichiro Iijima	(1938)	......	230D	230D
Al_2O_3	E. Tiede & E. Birnbrauer	(1914)	Vacuum	1890	...
	R. F. Geller & P. J. Yavorsky	(1945)	(Air)	2000–2030	1994–2024
	O. Ruff & G. Lauschke	(1916)	Air at 7.5 mm Hg	2005	2005
	" " " "	"	Air at 7.7 mm Hg	2008	2008
	O. Weigel & F. Kaysser	(1931)	N_2 at 1 atm	2007	2009
	" " " "	"	Air	2010	2012
	" " " "	"	"	2005–2010	2007–2012
	" " " "	"	Reducing	2001	2003
	S. D. Mark, Jr.	(1959)	Neutral	2020	2020
	J. J. Diamond & S. J. Schneider	(1960)	Air	2025	2025
	R. F. Geller & E. N. Bunting	(1943)	Air	2035	2029
	W. A. Lambertson & F. H. Gunzel, Jr.	(1952)	He	2034	2034
	O. Ruff	(1913)	......	2010	2035
	E. N. Bunting	(1931)	(Air)	2045	2038
	R. H. McNally, R. I. Peters, & P. H. Ribbe	(1961)	Air, Ar, N_2	2043	2043
	O. Ruff & O. Goecke	(1911)	N_2	2020	2044
	S. M. Lang, F. P. Knudsen, C. L. Fillmore & R. S. Roth	(1956)	Ar	2049	2049
	H. v. Wartenberg, H. Lindy, & R. Jung	(1927)	Air	2055	2049
	C. W. Kanolt	(1914)	2 mm Hg; H_2	2050	2072*
As_2O_3	A. Smits and E. Beljaar	(1931)	66.1 mm Hg	312.3	312.3*
	H. V. Welsch & L. H. Duschak	(1915)	......	313	...
	E. R. Rushton & F. Daniels	(1926)	......	315	...
B_2O_3	F. C. Kracek, G. W. Morey, & H. E. Merwin	(1938)	(Air)	450	450*
	L. McCullock	(1937)	......	460–470	...
BaO	E. E. Schumacher	(1926)	H_2 at 0.2 atm	1923	1918
BeO	E. Tiede & E. Birnbrauer	(1914)	Vacuum	2400	...
	O. Ruff & G. Lauschke	(1916)	15 mm Hg	2410	2410
	S. M. Lang, F. P. Knudsen, C. L. Fillmore, & R. S. Roth	(1956)	Ar	2452	2452
	H. v. Wartenberg, H. J. Reusch & E. Saran	(1937)	Oxidizing	2520	2508
	H. v. Wartenberg & H. Werth	(1930)	Oxidizing	2570	2557
	Ya I. Ol'shanskiĭ	(1958)	N_2	2570	2570
	O. Ruff	(1913)	N_2 at 4–10 mm Hg	2525	2573
Bi_2O_3	L. Belladen	(1922)	......	817	...
	W. Guertler	(1903)	......	820	...
	G. Gattow & H. Schröder	(1962)	(Air)	824	824
	E. M. Levin & C. L. McDaniel	(1962)	Air	825	825*

† Samuel J. Schneider, "Compilation of the Melting Points of the Metal Oxides," NBS Monograph 68, Oct. 10, 1963, 31 pages, National Bureau of Standards, Washington, D. C.
‡ No data on the following oxides: Er_2O_3, Ho_2O_3, K_2O, Lu_2O_3, Na_2O, PdO_2, Pr_6O_{11}, Rb_2O, Rh_2O_3, RuO_4, Tb_4O_7, Tm_2O_3.
†† International Practical Temperature Scale of 1948 (text revision of 1960).
* Preferred value of those listed, based on adherence to criteria discussed in survey.

Oxide	Investigator	Date	Environment	Melting Point Original °C	Int. 1948†† °C
CaO	E. E. Schumacher	(1926)	H_2 at 0.2 atm	2576	2565
	C. W. Kanolt	(1914)	H_2	2572	2614*
	Ya I. Ol'shanskiĭ	(1958)	N_2	2620	2620
	R. C. Doman, J. B. Barr, N. R. McNally, & A. M. Alper	(1962)	......	2630	2630
CdO	R. S. Roth	(1961)	(Air)	>1500	>1500
CeO_2	O. Ruff	(1913)	......	1950	1973
	H. v. Wartenberg & W. Gurr	(1931)	Air	>2600	>2600
	F. Trombé	(1949)	......	2800	...
CoO	H. v. Wartenberg, H. J. Reusch, & E. Saran	(1937)	Oxidizing	1800	1795
	H. v. Wartenberg & E. Prophet	(1932)	Air	1810	1805
Cr_2O_3	O. Ruff	(1913)	N_2 at 30 mm Hg	1830–2080	1849–2107
			N_2 at 1 atm	1960	1983
	C. W. Kanolt	(1914)	Vacuum	1990	2011
	W. T. Wilde & W. J. Rees	(1943)	Air	2060	2053
	E. N. Bunting	(1931)	(Air)	2275	2266
	H. v. Wartenberg & H. J. Reusch	(1932)	Air	2275	2266
	R. H. McNally, R. I. Peters, & P. H. Ribbe	(1961)	N_2	2315	2315
			Air	2330	2330
	H. v. Wartenberg & K. Eckhardt	(1937)	Air	2435	2424
Cs_2O	M. E. Rengade	(1909)	N_2	490	...
Cu_2O	R. Ruer & M. Nakamoto	(1923)	N_2	1222	1222
	H. v. Wartenberg, H. J. Reusch, & E. Saran	(1937)	Oxidizing	1230	1229
	H. S. Roberts & F. H. Smyth	(1921)	0.6 mm Hg	1235	1236*
Dy_2O_3	L. G. Wisnyi & S. Pijanowski	(1956)	Either He, H_2 or vacuum	2340	2340
Eu_2O_3	L. G. Wisnyi & S. Pijanowski	(1956)	Either He, H_2 or vacuum	2050	2050
	S. J. Schneider	(1961)	Air	2240	2240*
FeO	J. Chipman & S. Marshall	(1940)	Slightly oxidizing	1369	1368
	R. Hay, D. D. Howat & J. White	(1932–33)	N_2 at 1 atm	1370	1368
	L. S. Darken & R. W. Gurry	(1946)	N_2 at 1 atm	1371	1369*
	N. L. Bowen & J. F. Schairer	(1932)	N_2-slightly oxidizing	1380	1382
Fe_3O_4	V. L. Moruzzi & M. W. Shafer	(1960)	Air	1591	1591
	L. S. Darken & R. W. Gurry	(1946)	O_2 at 1 atm	1583	1580*
			Air	1594	1591*
			O_2 at 0.0575 atm	1597	1594*
	J. W. Greig, E. Posnjak, H. E. Merwin, & R. B. Sosman	(1935)	Small O_2 pressure	1591	1594*
	H. v. Wartenberg & K. Eckhardt	(1937)	Air	1650	1647
Ga_2O_3	V. G. Hill, R. Roy, & E. F. Osborn	(1952)	......	1725	1725
	H. v. Wartenberg & H. J. Reusch	(1932)	Air	1740	1736
	S. J. Schneider & J. L. Waring	(1963)	Air	1795	1795*
Gd_2O_3	L. G. Wisnyi & S. Pijanowski	(1956)	Either He, H_2 or vacuum	2330	2330
	C. E. Curtis & J. R. Johnson	(1951)	Air	2350	2350
GeO_2	R. Schwarz, P. W. Schenk, & H. Giese	(1931)	Air	1115	1115*
	A. W. Laubengayer & B. S. Morton	(1932)	(Air)	1116	1116*
HfO_2	P. Clausing	(1932)	H_2	2774	2758
	S. D. Mark, Jr.	(1959)	Neutral	2770	2770
	F. Henning	(1925)	N_2 or H_2	2812	...
	C. E. Curtis, L. M. Doney, & J. R. Johnson	(1954)	......	2900	2900
In_2O_3	S. J. Schneider	(1961)	Air	1910	1910

Oxide	Investigator	Date	Environment	Melting Point Original °C	Int. 1948†† °C
IrO$_2$	E. H. P. Cordfunke & G. Meyer	(1962)	O$_2$ at 1 atm	1100D	1100D
La$_2$O$_3$	O. Ruff	(1913)	......	1840	1859
	W. A. Lambertson & F. H. Gunzel, Jr.	(1952)	H$_2$	2210	2210
	H. v. Wartenberg & H. J. Reusch	(1932)	Air	2315	2307
Li$_2$O	Handbook of Chemistry and Physics	(1956–7)	......	>1700	...
MgO	O. Ruff	(1913)	N$_2$ at 10–30 mm Hg	2120–2550	2150–2599
			N$_2$ at 1 atm	2250–2500	2285–2546
	K. K. Kelley	(1936)	......	2642	...
	R. H. McNally, R. I. Peters, & P. H. Ribbe	(1961)	N$_2$	2825	2825
	C. W. Kanolt	(1914)	CO and N$_2$ at 1 atm	2800	2852*
MnO	E. Tiede & E. Birnbrauer	(1914)	Vacuum	1650	...
	J. White, D. D. Howat, and R. Hay	(1933)	......	1785	1781
Mn$_3$O$_4$	H. v. Wartenberg & E. Prophet	(1932)	Air	1560	1557
	H. J. Van Hook & M. L. Keith	(1958)	Air	1562	1564*
	T. Ranganathan, B. E. MacKean, & A. Muan	(1962)	Air	1567	1567*
	H. v. Wartenberg, H. J. Reusch, & E. Saran	(1937)	Air	1590	1587
	H. v. Wartenberg & W. Gurr	(1931)	Air	1705	1701
MoO$_3$	T. Carnelley	(1878)	......	759	...
	E. Groschuff	(1908)	Air	791	...
	F. M. v. Jaeger & H. C. Germs	(1921)	Oxidizing	795	...
	F. Hoermann	(1929)	......	795	795
	G. D. Rieck	(1943)	Air	795	795
	L. A. Cosgrove & P. E. Snyder	(1953)	N$_2$ at 1 atm	795.4	795.4*
Nb$_2$O$_5$	G. Brauer	(1941)	O$_2$	1460	1458
	M. W. Shafer & R. Roy	(1958)	......	1465	1465
	M. Ibrahim, N. F. Bright & J. F. Rowland	(1962)	Air	1479	1479
	R. S. Roth & J. L. Waring	(1962)	Air	1485	1485*
	A. Reisman & F. Holtzberg	(1955)	O$_2$ + Air	1486	1486*
	R. S. Roth & J. L. Waring	(1961)	Air	1487	1487
	F. Holtzberg, A. Reisman, M. Berry, & M. Berkenblit	(1957)	(Air)	1491	1491*
	J. J. Diamond & S. J. Schneider	(1960)	Air	1496	1496
	R. S. Roth & L. W. Coughanour	(1955)	Air	1500	1500
	R. L. Orr	(1953)	......	1512	1512
	O. Ruff	(1913)	......	1520	1530
Nd$_2$O$_3$	W. A. Lambertson & F. H. Gunzel	(1952)	He	2272	2272
NiO	P. D. Merica & R. G. Waltenberg	(1925)	Vacuum	1552	...
			Air	1660	...
	H. v. Wartenberg & E. Prophet	(1932)	Air	1990	1984
OsO$_4$	H. v. Wartenberg	(1924)	≈11 mm Hg	40.1	...
	Handbook of Chemistry & Physics	(1956–7)	......	41	...
	K. K. Kelley	(1936)	......	56	...
P$_2$O$_5$	J. M. A. Hoeflake & M. F. C. Scheffer	(1926)	4600 mm Hg	569	569
PbO	L. Belladen	(1922)	......	870	...
	S. Hilpert & P. Weiller	(1909)	Air	876	...
	F. M. v. Jaeger & H. C. Germs	(1921)	Oxidizing	877	...
	R. Schenck & W. Rassbach	(1908)	Air	879	...
	R. F. Geller, A. S. Creamer, & E. N. Bunting	(1934)	Air	886	886*
	K. A. Krakau	(1936)	Air	886	886
	V. A. Kroll	(1912)	Air	888	...
	H. C. Cooper, L. I. Shaw, & N. E. Loomis	(1909)	(Air)	888	...
PtO$_2$	Handbook of Chemistry & Physics	(1956–7)	......	450	...
Re$_2$O$_7$	K. K. Kelley	(1936)	......	296	...
Sb$_2$O$_3$	W. B. Hincke	(1930)	8.5 mm Hg	655	655

Oxide	Investigator	Date	Environment	Melting Point Original °C	Int. 1948†† °C
Sc_2O_3	S. J. Schneider & J. L. Waring	(1963)	Air	>2405	>2405
SeO_2	Handbook of Chemistry & Physics	(1956–7)	......	340–350	...
SiO_2	R. Weitzel	(1921)	Air	1696	1691
	K. Endell & R. Rieke	(1913)	N_2	1685	1692
	J. White, D. D. Howat, & R. Hay	(1933)	......	1705	1701
	J. B. Ferguson & H. E. Merwin	(1918)	Air	1710	1720
	J. W. Greig	(1927)	Air	1713	1723*
	N. Zhirnova	(1934)	Air	1715	1728
	O. Ruff & G. Lauschke	(1916)	17 mm Hg	1850	1850
Sm_2O_3	L. G. Wisnyi & S. Pijanowski	(1956)	Either He, H_2 or vacuum	2300	2300
	C. E. Curtis & J. R. Johnson	(1951)	Air	2350	2350
SnO_2	O. Ruff	(1913)	......	1385	1391
			......	1625	1637
	V. J. Barczak & R. H. Insley	(1962)	......	1630	1630
SrO	E. E. Schumacher	(1926)	H_2 at 0.2 atm	2430	2420
Ta_2O_5	A. Reisman, F. Holtzberg, M. Berkenblit & M. Berry	(1956)	(Air)	1872	1872
	O. Ruff	(1913)	N_2 at reduced P	1875	1895
TeO_2	F. C. Kracek	(1942)	......	732.6	...
ThO_2	E. Tiede & E. Birnbrauer	(1914)	Vacuum	2000	...
	O. Ruff	(1913)	N_2 at reduced P	2425	2468
				2440	2483
				2470	2515
	F. Trombé	(1949)	......	3000	...
	O. Ruff, F. Ebert, & H. Woitinek	(1929)	......	3050	3030
	W. A. Lambertson & F. H. Gunzel, Jr.	(1952)	He	3220	3220*
TiO_2	W. O. Statton	(1951)	Vacuum	1720	1716
	H. v. Wartenberg & E. Prophet	(1932)	Air	1825	1820
	H. Sigurdson & S. S. Cole	(1949)	Oxidizing	1825	1825
	D. E. Rase & R. Roy	(1955)	(Air)	1830	...
	L. W. Coughanour & V. A. DeProsse	(1953)	Air	1839	1839
	P. D. S. St. Pierre	(1952)	Air	1840	1840
	J. J. Diamond & S. J. Schneider	(1960)	Air	1840	1840
	H. v. Wartenberg & W. Gurr	(1931)	Air	1850	1845
	S. M. Lang, C. L. Fillmore & L. H. Maxwell	(1952)	Air	1845	1845
	H. v. Wartenberg & K. Eckhardt	(1937)	Air	1855	1850
	G. Brauer & W. Littke	(1960)	O_2 at 300 torr + Ar at 460 torr	1840	1840
			O_2 at 500 torr + Ar at 260 torr	1860	1860
			O_2 at 600 torr + Ar at 160 torr	1870	1870
			O_2 at 760 torr	1870	1870
			O_2 at 1140 torr	1870	1870
Tl_2O_3	A. B. F. Duncan	(1929)	O_2 at 1 atm	717	717
UO_2	O. Ruff & O. Goecke	(1911)	N_2	2176	2208
	L. G. Wisnyi & S. Pijanowski	(1956)	Either He, H or vacuum	2760	2760
	T. C. Ehlert & J. L. Margrave	(1958)	Vacuum	2860	2860
	W. A. Lambertson & F. H. Gunzel, Jr.	(1952)	He	2878	2878*
V_2O_5	F. C. Kracek	(1942)	......	656	...
	T. Carnelley	(1878)	......	658	...
	O. A. Cook	(1947)	......	670	670
	V. V. Illarionov, R. P. Ozeron, & E. V. Kil'disheva	(1956)	......	672	672
	F. Holtzberg, A. Reisman, M. Berry, & M. Berkenblit	(1956)	Air + O_2	674	674*
	C. McDaniel	(1963)	Air	675	675*
	A. Burdese	(1957)	CO_2	685	685
	Handbook of Chemistry & Physics	(1956–7)	......	690	...

				Melting Point	
Oxide	Investigator	Date	Environment	Original °C	Int 1948†† °C
WO_3	F. M. v. Jaeger & H. C. Germs	(1921)	Oxidizing	1473	...
	F. Hoermann	(1929)	......	1473	1471
Y_2O_3	O. Ruff & G. Lauschke	(1916)	Air at 21.5 mm Hg	2410	2410
	O. Ruff	(1913)	N_2 at 15 mm Hg	2415	2458
ZnO	E. N. Bunting	(1930)	Air	1975	1969
ZrO_2	E. Tiede & E. Birnbrauer	(1914)	Vacuum	2430	...
	O. Ruff & G. Lauschke	(1916)	H_2 at 1 atm	2519	2519
			Air at 8.22 mm Hg	2563	2563
	O. Ruff	(1913)	N_2	2585	2636
	P. Clausing	(1932)	H_2	2677	2663
	E. Podszus	(1917)	Air	2677	...
			"	2727	...
	F. Henning	(1925)	H_2 and N_2	2687	...
	S. D. Mark, Jr.	(1959)	Neutral	2690	2690
	F. Trombé	(1949)	......	2700	...
	W. A. Lambertson & F. H. Gunzel, Jr.	(1952)	He	2710	2710
	N. Zhirnova	(1934)	Air	2715	2765
	C. E. Curtis, L. M. Doney, & J. R. Johnson	(1954)	......	2850	2850

Appendix 2

MOLECULAR WEIGHTS OF OXIDES

(Based on relative atomic weights 1961)

Oxide	Molecular Wt.	Oxide	Molecular Wt.	Oxide	Molecular Wt.
Ag_2O	231.739	MgO	40.311	Sb_2O_3	291.50
Al_2O_3	101.9612	MnO	70.9375	Sb_2O_5	323.50
Au_2O	409.933	MnO_2	86.9369	Sc_2O_3	137.910
Au_2O_3	441.932	Mn_2O_3	157.8744	SeO_2	110.96
As_2O_3	197.8414	Mn_3O_4	228.8119	SeO_3	126.96
As_2O_5	229.8402	MoO_2	127.94	SiO_2	60.085
B_2O_3	69.620	MoO_3	143.94	Sm_2O_3	348.70
BaO	153.34	Mo_2O_3	239.88	SnO	134.69
BeO	25.0116			SnO_2	150.69
Bi_2O_3	465.958	Na_2O	61.9790	SO_2	64.063
CaO	56.08	Nb_2O_5	265.809	SO_3	80.062
CdO	128.40	Nd_2O_3	336.48	SrO	103.62
CeO_2	172.12	NiO	74.71		
Ce_2O_3	328.24	Ni_2O_3	165.42	Ta_2O_5	441.893
CO	28.0106	Ni_3O_4	240.13	Tb_2O_3	365.846
CO_2	44.0100	NO	30.0061	Tb_4O_7	747.692
CoO	74.9326	NO_2	46.0055	TeO_2	159.60
Co_2O_3	165.8646	N_2O	44.0128	TeO_3	175.60
CrO_2	83.995			ThO_2	264.037
CrO_3	99.994	OsO	206.2	TiO	63.90
Cr_2O_3	151.990	OsO_2	222.2	TiO_2	79.90
Cs_2O	281.809	OsO_4	254.2	Ti_2O_3	143.80
CuO	79.54	Os_2O_3	428.4	Ti_3O_5	223.70
Cu_2O	143.08			Tl_2O	424.74
		PbO	223.19	Tl_2O_3	456.74
Dy_2O_3	373.00	PbO_2	239.19	Tm_2O_3	385.866
Er_2O_3	382.52	Pb_2O	430.38		
Eu_2O_3	351.92	Pb_2O_3	462.38		
		Pb_3O_4	685.57	UO	254.03
FeO	71.846	PdO	122.4	UO_2	270.03
Fe_2O_3	159.692	PdO_2	138.4	UO_3	286.03
Fe_3O_4	231.539	P_2O_5	141.9446	U_2O_5	556.06
Ga_2O_3	187.44	PrO_2	172.906	U_3O_7	826.09
Gd_2O_3	362.50	Pr_2O_3	329.812	U_3O_8	842.09
GeO_2	104.59	Pr_6O_{11}	1021.435		
		PtO	211.09		
H_2O	18.0153	PtO_2	227.09	VO	66.941
H_2O_2	34.0147			V_2O_3	149.882
HfO_2	210.49			V_2O_4	165.882
HgO	216.59	Rb_2O	186.94	V_2O_5	181.881
Hg_2O	417.18	ReO_2	218.2		
Ho_2O_3	377.858	ReO_3	234.2		
		Re_2O_3	420.4	WO_2	215.85
In_2O_3	277.64	Re_2O_7	484.4	WO_3	231.85
I_2O_5	333.8058	RhO	118.904	W_4O_{11}	911.39
IrO_2	224.2	RhO_2	134.904		
Ir_2O_3	432.4	Rh_2O_3	253.808		
		RuO	117.07	Yb_2O_3	394.08
K_2O	94.203	RuO_2	133.07	Y_2O_3	225.808
La_2O_3	325.82	RuO_3	149.07		
Li_2O	29.877	RuO_4	165.07	ZnO	81.37
Lu_2O_3	397.94	Ru_2O_3	250.14	ZrO_2	123.22

Author Index*

ADAMCZAK, R. L., 1451
ADAMS, L. H., 1003
ADAMSKY, R. F., 1203, 1205–06, 1608
AGAMAWI, Y. M., 768, 770–771, 954
AGASHKINA, G. D., 1305, 1309
AKOPOV, E. K., 1126–27, 1407, 1777–78, 1780–81, 1786–89, 1795
AKULKINA, L. M., 1404, 1416
ALBERMAN, K. B., 241
ALEKSEEVA, E. A., 1749, 1751
ALLEN, E. R., 2007
ALLEN, K. A., 1027, 1030
ALLEN, W. C., 42, 81, 587, 589–591, 987–990
ALPER, A. M., 229, 259, 262
AMADORI, M., 170, 195, 286, 290, 384, 387–389, 425–427, 1085–87, 1143, 1147–49, 1151–52, 1213, 1225, 1712–13, 1813–16, 1824–25, 1827–28
ANDERSEN, O., 266, 899
ANDERSON, J. A., 241
ANDERSON, R. H., 2007
ANDO, J., 613
ANDREWS, A. I., 783
ANISHEVA, N. A., 1900–01
ARAMAKI, S., 314
ARNOLD, H., 1400–01
ARTUR, A., 1977
ASANTI, P., 4, 255, 1886, 1895, 1902
ASKER, W. J., 1464
ATLAS, L. M., 28, 30–31, 33–35, 607
AURIOL, A., 204, 209, 215, 217, 219, 232, 548, 569–570, 575
AZZARIA, L. M., 1442

BÅÅK, T., 1671, 1709
BAHNSEN, M. J., 1689
BAĬKOVA, R. I., 181
BAKER, C. L., 1959–63
BAKUMSKAYA, E. L., 1128–29, 1135, 1791, 1802
BALZ, D., 1459
BANASHEK, E. I., 1645, 1647, 1649–50, 1655
BARBER, E. J., 1525
BARCZAK, V. J., 315
BARKOVA, G. V., 500, 545–547, 749–750, 812
BARR, J. B., 229
BARRETT, R. L., 665
BARTH, T. F. W., 507
BARTON, C. J., 1022, 1027, 1240, 1282–83, 1297–1298, 1321–22, 1337, 1445, 1449–50, 1454, 1456, 1460, 1468, 1474–76, 1479–81, 1489, 1492, 1494–95, 1498–99, 1501–03, 1507, 1522, 1527, 1530, 1532–34, 1536, 1538–42, 1544, 1550, 1558, 1565, 1569, 1573
BASKOW, A. V., 1448, 1455, 1463, 1484, 1488, 1493, 1504
BASMAJIAN, J. A., 555, 858
BAUER, A. A., 121-123
BECK, E., 1519
BELLANCA, A., 1099, 1116, 1130–33
BELOUSOV, O. K., 1666–68

BELOVA, Z. I., 1043, 1045, 1052, 1057, 1059–1062, 1067, 1079, 1082
BELYAEV, A. I., 1568, 1728
BELYAEV, I. N., 422–423, 438, 476, 500, 748–750, 792, 812, 1122, 1160–61, 1169, 1171, 1173–74, 1186, 1195, 1198, 1266, 1306, 1325, 1583–, 85, 1683, 1687, 1690, 1704, 1721–22, 1847, 1862
BENZ, R., 1278
BERAK, J., 272, 667, 669, 671, 725–728
BEREZHNAYA, V. T., 1551–52
BEREZHNOĬ, A. S., 164, 713–714, 773–774, 927–928, 930, 933–934, 1710
BERGMAN, A. G., 378, 418, 428, 437, 544, 785, 790–791, 1019–20, 1035–36, 1050, 1073–74, 1081, 1083, 1126–29, 1135, 1144–45, 1150, 1157–59, 1162–67, 1170, 1209, 1411, 1422, 1452–53, 1467, 1486, 1505, 1529, 1531, 1535, 1545, 1641–50, 1652, 1656–57, 1716, 1718, 1724–25, 1753, 1757, 1760, 1763–66, 1768, 1777–81, 1785–92, 1795–1802, 1804–10, 1829–33, 1835–44, 1848–53, 1855–56, 1861, 1864, 1866–67
BERKENBLIT, M., 173–174
BERNARDIS, G. B., 1609–10, 1630
BERRY, M., 173–174
BERRY, T. F., 987–990
BERUL, S. I., 1081, 1083
BEUSMAN, C., 1265, 1292
BINDER, H., 212
BIRCH, F., 514
BJORKLUND, C. W., 1296, 1316
BLAKE, C. A., 2031
BLAKELY, J. P., 1022, 1027, 1449, 1454, 1481, 1522, 1530, 1532–33, 1538, 1540–41, 1550, 1565, 1569
BLAKEY, R. C., 241
BLANKENSHIP, F. F., 1480, 1548
BLIDIN, V. P., 1754
BLOMEKE, J. O., 1521
BOEKE, H. E., 543, 1153
BOGATSKI, D. P., 16
BOGITCH, M. B., 1055
BOGUE, R. H., 480, 629, 827–829
BOL'SHAKOV, K. A., 1115, 1120, 1136, 1305, 1309, 1773–74, 1803, 1913
BONDAR, I. A., 549–550, 556–557
BOOTH, H. S., 1689
BORESKOV, G. K., 1137
BORNEFELD, H., 1447, 1472
BORZENKOVA, M. P., 1457
BOSTANDZHIYAN, A. K., 1411, 1422, 1590, 1653
BOTSCHWAR, A. A., 1599–1600
BOUAZIZ, R., 1952
BOWEN, N. L., 63, 80, 167, 266, 391–395, 408–415, 482, 484–485, 502–504, 508–510, 519–530, 588, 591–594, 599–601, 608, 634, 683–687, 795–796, 798–799, 833, 853–854, 973–974, 978–979, 2001, 2044
BOYD, F. R., 156, 267, 386, 513, 609
BRAND, H., 1185, 1194, 1207, 1264, 1412

BRATCHER, L. M., 1022, 1445, 1449–50, 1454, 1456, 1460, 1468, 1474, 1476, 1481, 1489, 1501–03, 1522, 1530, 1532–34, 1536, 1540–42, 1544, 1550, 1558, 1565, 1569
BRAUER, G., 3
BREDIG, M. A., 605, 1660–65
BREUSOV, O. N., 1446, 1516, 1622
BREWER, L., 1–2
BRIDGMAN, P. W., 1916
BRIGHT, N. F. H., 91–93, 244, 363
BRISI, C., 47–48, 54–55, 279
BRITTON, H. T. S., 2017
BRONSTEIN, H. R., 1661–63, 1665
BROWN, F. H., JR., 243, 271, 350, 369
BROWN, K. B., 2031
BROWNELL, W. E., 767
BROWNMILLER, L. T., 390, 480, 629
BRYATOV, L. V., 2057–58
BUCKNER, D. A., 621, 1993, 2034
BUDDINGTON, A. F., 923, 980
BUDNIKOV, P. P., 99–100, 181, 194, 225, 772, 1350, 1672
BUKHALOVA, G. A., 793, 1163, 1551–52, 1637, 1639, 1648, 1651, 1654, 1656, 1719–20, 1726, 1834, 1845, 1859–60
BUKUN, N. G., 1861
BUNK, A. J. H., 1259
BUNTING, E. N., 193, 280–281, 284, 299, 302, 309, 332, 404–405, 737–742, 758
BURDESE, A., 24, 47–48, 94–96, 320, 333
BURLEY, G., 1577, 1579
BUTTLER, F. G., 1979
BUTUZOV, V. P., 2057–58
BYCHKOVA, N. A., 437, 1724–25
BYCHKOVA-SHUL'GA, N. A., 1716

CALCAGNI, G., 1097–98, 1101, 1106–07, 1109–12, 1114, 1119, 1121
CALINGAERT, G., 1604
CANNERI, G., 1892–93
CARAPEZZA, M., 1099, 1116, 1130–33
CARLSON, E. T., 234, 1976, 1978
CARTER, P. T., 9, 487–493, 692–694
CARVETH, H. R., 1069
CASSEDANNE, J., 79
ČERNOHORSKÝ, M., 1947
CHACONAS, T. J., 1976
CHEN, W. T., 386
CHERNOV, R. V., 1238, 1279, 1317, 1336
CHERNYKH, V. I., 1457
CHIZHIKOV, D. M., 1953
CHUDINOVA, L. I., 1018
CLARK, G. L., 1051
CLARK, L. A., 1885
CLARK, S. P., JR., 1210–11, 1591
CLARKE, W. F., 1981
CLAUSSEN, W. F., 150
CLEEK, G. W., 207, 531–532, 559
COCCO, A., 282, 371, 584, 711
COHEN-ADAD, R., 2050
COLEMAN, C. F., 2031
COLEMAN, V. S., 1476, 1495

* All index listings are given by figure number.

Corbett, J. D., 1575
Coughanour, L. W., 270, 370, 372–373, 611–612, 673, 678, 729, 731, 939–940
Counts, W. E., 1471, 1513–14
Crawley, G. E., Jr., 2049, 2059–60
Creamer, A. S., 284, 572, 577, 579
Curtis, C. E., 236
Cuy, E. J., 1912

Dachille, F., 18, 148, 152, 157, 265. 719, 1442
Danilin, A. S., 120
D'Ans, J., 533–534, 1980, 2063
Darken, L. S., 8, 45, 70, 85–86
Darnell, A. J., 1658
Davis, H. M., 261
Davis, W. C., 1027, 1030
Deadmore, D. L., 1304, 1624–26
Dear, P. S., 755, 757
Dearborn, E. F., 97
de Carli, F., 276
DeCesaris, P., 1899
DeProsse, V. A., 270, 370
Dergunov, E. P., 1452–53, 1467, 1486, 1505, 1529, 1535, 1545, 1864
Desch, C. H., 231, 881
DeVries, R. C., 22, 113–115, 239, 553, 555, 660–663, 858, 895, 897–898, 903–904, 913–914, 1444, 1458
De Wys, E. C., 916–917, 921–922
D'Eye, R. W. M., 1512
Dietzel, A., 288, 441, 674–676, 746, 765
Diogenov, G. G., 1033, 1037–38
Doig, J. R., 121–123
Domagala, R. F., 25
Doman, R. C., 229, 259, 262
Dombrovskaya, N. S., 1134, 1634–35, 1640, 1749, 1751, 1767, 1871
Doney, L. M., 236
Doornbosch, H. J., 1587–89
Doroshenko, S. S., 1160–61
Dostál, K., 1947
Dryś, M., 297
Duckworth, W. H., 198, 273, 319, 374
Duc-Maug, C., 1950
Dudavskiĭ, I. F., 164
Dukelski, M., 1948, 1955
Dulin, F. H., 303
Durbin, E. A., 198, 273, 319, 374
Duwez, P., 243, 271, 350, 355, 368–369
Dworkin, A. S., 1665

Early, R. G., 1053
Eastman, E. D., 1607, 1611
Ebert, F., 242, 571, 582, 677, 730
Eckhardt, K., 356, 367
Egan, E. P., Jr., 318, 640, 642–644
Eggert, S., 1180, 1443
Ehrlich, P., 1232, 1271, 1272, 1313
Eichelberger, J. F., 1563
Eick, H., 1980, 2063
Eitel, W., 1007, 1010–11, 1707–08, 1711
Elagina, E. I., 1669
Elchardus, E., 1257, 1262
Elenevskaya, V. M., 1032
Elgersma, J., 1285

England, J. L., 156, 267, 386, 513, 609, 1942
Englert, W. J., 775
Epstein, L. F., 226
Ernst, W. G., 2047
Ershova, Z. P., 1681–82, 1694–97, 1705, 1714–15
Ervin, G., Jr., 2052
Eskola, P., 210, 296, 551, 620
Etter, D. E., 1469, 1518, 1549, 1554
Ettinger, V. J., 1077
Eubank, W. R., 827
Evans, P. E., 119
Evdokimova, K. A., 1167
Evseeva, N. N., 1123, 1794, 1801, 1804
Eyring, L., 19

Fackelmann, J. M., 121–123
Faust, G. T., 158–159, 246, 248–250, 419–421, 1999
Fedorov, P. I., 1115, 1120, 1136, 1305, 1309, 1773–74, 1803, 1913
Fedotieff, P. P., 1461, 1473, 1483, 1490, 1566
Fenn, E. M., 278
Ferguson, J. B., 603
Ferguson, R. E., 19
Fernandes, L., 1892–93
Ferris, L. M., 2062
Filbert, R. B., Jr., 1508–09
Fillmore, C. L., 23, 224, 226, 316–317, 349, 360, 578, 580–581, 583, 585
Fischer, J., 1524, 1528, 1592
Fischer, W. A., 26, 257
Flaschen, S. S., 424, 2036
Fleischer, M., 2020
Flint, E. P., 646, 649
Fonardzhyan, V. M., 1642
Ford, W. F., 38–39, 597, 626, 650, 655, 924–926
Forestier, H., 79
Foster, L. M., 1914
Foster, W. R., 631–632, 724, 763, 830–831, 841–842, 916–917, 921–922, 1140
Foster, P. A., Jr., 1698
Franco, R. R., 971–972
Friedman, H. A., 1465, 1477–80, 1485, 1487, 1496–98, 1506–07, 1517, 1526, 1538–39, 1547–48, 1553, 1556–57, 1560–62, 1564, 1571–72, 1574
Friedman, I., 1964, 1970–71, 2023–26
Friedrich, K., 1890–91
Fujii, T., 1711
Funk, R., 766

Galakhov, F. Ya., 218, 362, 459, 549–550, 556–557, 576, 734, 769
Gale, W. A., 1951
Gambino, R. J., 477
Garn, P. D., 424, 1182
Geach, G. A., 223
Gebhardt, E., 7, 17, 163
Gehman, W. G., 1658
Geller, R. F., 280–281, 284, 361, 404–405, 572–574, 577, 579, 737–742
Gentile, A. L., 631–632

Germs, H. C., 291–293, 751–753, 1139, 1154–56
Gevorkyan, S. V., 1956–57
Gielisse, P. J., 763
Gingerich, K., 3
Ginsberg, A. S., 1117–18
Giorgi, A. L., 2052
Gladushchenko, V. A., 1807, 1809–10
Glaeser, W., 1009
Glasser, F. P., 101, 235, 341, 614–619 651–654, 699–701
Glasser, L. S. D., 1922
Goldsmith, J. R., 843–844, 848–849, 1014
Golubeva, M. S., 1790, 1792, 1797, 1800
Goranson, R. W., 2019, 2027–29
Gorovits, N. N., 199
Goto Y., 61, 208
Graft, D. L., 1014
Grahmann, W., 1094–96, 1100, 1102–03
Grebenshchikov, I. V., 494, 497
Grebenshchikov, R. G., 1500, 1559
Greene, K. T., 478–479, 828–829
Greig, J. W., 136, 210, 266, 296, 507
Grena, C., 1280, 1319
Grieve, J., 88, 277, 695
Grimes, W. R., 1022, 1240, 1282–83, 1297–98, 1321–22, 1337, 1445, 1449–50, 1454, 1456, 1460, 1465, 1468, 1474, 1476, 1478–81, 1485, 1487, 1489, 1495–99, 1501–03, 1506–07, 1522, 1526–27, 1530, 1532–34, 1536–37, 1539–42, 1544, 1547–48, 1550, 1558, 1560–61, 1565, 1569–70 1572–73
Grjotheim, K., 1784
Groeneveld, W. L., 1822
Gromakov, S. D., 1072
Groschuff, E., 175, 1146
Grube, G., 1291
Grylicki, M., 857
Guérin, H., 1950
Gul'ko, N. V., 713–714, 773–774
Gulyanitskaya, Z. F., 285
Gummer, W. K., 832, 836–837
Gurovich, N. A., 1956–57
Gurry, R. W., 8, 45, 70
Guth, E. D., 19
Gutt, W., 245, 668

Haacke, A., 442
Haan, D.-I., 1896
Hachmeister, K., 1246, 1286, 1324, 1326–27
Haendler, H. M., 1615, 1617–1618
Hahn, W. C., Jr., 12–13
Halvorsen, T., 1784
Hamilton, E. H., 531–532
Hansen, W. C., 629
Hardenberg, H. V., 1285
Harker, R. I., 929, 1006, 1012–13
Harkins, W. D., 1051
Harkort, H.-J., 247, 670
Harris, L. A., 359, 1456, 1477, 1542
Harris, R. F., 1017
Harrison, D. E., 300, 568, 679, 942
Harrison, W. R., 1172, 1758
Hatch, R. A., 458
Hauser, G., 232, 548, 569–570, 575

HAY, R., 71, 275, 493, 688–689, 692–694
HAYAKAWA, Y., 1700
HEBERT, G. M., 1571, 1574
HEIDES, R. R., 178
HEINRICH, F., 166, 191
HENDRICKS, S. B., 158–159
HENNIG, H., 20, 107–108, 110–111, 125
HENRY, J. L., 2012
HERASYMENKO, P., 1312, 1375
HEROLD, P. G., 966–967
HERRMANN, G., 1181, 1202, 1250, 1252–56, 1371, 1385–87, 1389–90, 1394, 1582
HERZBERG, O. W., 1918
HESSELGESSER, J. M., 1965–67, 1969
HILL, D. G., 2031
HILL, V. G., 310, 1933, 2008
HILL, W. L., 158–159, 246, 248–250
HISSINK, D. H., 1040
HOERMANN, F., 177, 186–187, 200, 379, 439–440
HOFFMAN, M. V., 623, 703, 754
HOFFMANN, A., 26, 257
HOLMQUIST, S. B., 154, 168, 184
HOLTZBERG, F., 171, 173–174, 375
HORN, W. F., 776–777
HOTOP, W., 247, 670
HOUTGRAAF, H., 1391–92
HOWAT, D. D., 71
HOWELL, L. J., 1284, 1323, 1415
HOWELLS, W. J., 1168
HOWLAND, W. H., 226
HUBER, H., 1887
HUBER, K., 1328–31
HUDGENS, C. R., 1469, 1518, 1549, 1554, 1563
HUFFMAN, A. A., 1469, 1518, 1554
HUMMEL, F. A., 151, 180, 185, 222, 252, 268, 300, 305–308, 364, 443–445, 454–455, 460–474, 623, 679, 703–709, 717–718, 720, 732, 754, 756, 764, 775–777, 813–825, 942
HUNTER, M. S., 1914
HYTÖNEN, K., 889, 907–908, 911

IBRAHIM, M., 9, 244, 363, 487–492
IIDA, S., 49
IITSUKA, D., 1898
IKEDA, T., 861–862, 970
ILJINSKIĬ, W. P., 1483, 1490, 1566
ILLARIONOV, V. V., 105, 1137–38
ILLNER, K. W., 1009
IL'YASOV, I. I., 1208–09, 1411, 1422, 1590, 1641–44, 1646, 1652–53, 1657
INGERSON, E., 197, 406, 647–648, 759, 1968
INSLEY, H., 891–892, 1465, 1477–80, 1497–99, 1506–07, 1517, 1526–27, 1538–39, 1547, 1553, 1556–57, 1560–62, 1564, 1571–73
INSLEY, R. H., 315
IONOV, V. I., 1424
ISAEVA, M. M., 2003–04
IVANOV, A. I., 1267, 1360, 1419–20
IWASÉ, K., 932, 941, 955, 986

JAEGER, F. M., 291–293, 751–753, 1139, 1154–56, 1587–89, 1888

JÄGER, W., 41
JAMESON, R. F., 2013–14, 2016
JAMIESON, J. C., 1005
JÄNECKE, E., 1093, 1772, 1775, 1793
JEEVARATNAM, J., 235
JOHNSON, J. R., 236
JOHNSON, J. W., 1660, 1664
JOHNSTON, W. D., 178
JONES, L. V., 1469, 1518, 1554, 1563
JONKER, G. H., 565–566
JOST, E., 1328–31
JUE, L. R., 1959–63

KAHN, M., 1278
KABI, S. K., 493
KABLUKOV, Iw., 1743
KACHI, S., 61
KALOUSEK, G. L., 1991
KAMEDA, M., 1904
KAMENETSKIĬ, M. V., 1414
KAPYSHEV, A. G., 860
KARKHANAVALA, M. D., 89, 818–819
KARUTZ, I., 149
KARYAKIN, L. I., 164
KATNACK, F. L., 305–307, 707, 709
KAYALOVA, S. S., 1763–64
KAZAKOV, A. V., 1996
KEIHN, F. G., 262
KEITH, M. L., 72, 332, 715, 968–969
KELLNER, G., 1183–84, 1187–93, 1196–97
KELLOGG, H. H., 1284, 1323, 1415
KENNEDY, G. C., 1917, 1924, 1927–28, 1943–45, 2009
KHAKHLOVA, N. V., 1134
KHITAROV, N. I., 1941
KHITROV, V. A., 1034–1036
KIDO, H., 1699–1700
KIL'DISHEVA, E. V., 105
KIM, K. H., 308, 445, 460–464, 474, 764, 820–825
KING, B. W., 198, 273, 319, 374, 1688
KING, G. B., 2012
KIPARENKO, L. M., 1065
KIRKINA, D. F., 1510
KISLOVA, A. I., 785, 1158, 1162, 1165–66, 1831, 1836–40
KISSINGER, H. E., 1577, 1579
KITAMURA, T., 61
KLEMM, W., 1307
KLINGSBERG, C., 14–15, 1925, 1936
KLYUKINA, L. V., 1429–30
KNICK, R., 189
KNIGHT, M. A., 261
KNOWLES, L. M., 1812
KNUDSEN, F. P., 226, 317, 349, 360
KNUDSEN, R. P., 23
KOBE, K. A., 1972
KOFLER, A., 1047
KOHLMEYER, E. J., 4, 20, 107–108, 110–11, 125, 180, 255, 1880, 1895, 1902
KOIDE, S., 664
KOIZUMI, M., 2035
KOLOSKOVA, Z. A., 1767
KOMAREK, K., 1312, 1375
KOMISSAROVA, L. N., 1404, 1416, 1426
KONNO, K., 1889
KONOVALOV, P. F., 254
KÖPPEN, N., 441
KÖRBER, F., 1368, 1382

KORDES, E., 1141–42
KORENEV, Yu. M., 1466
KOROBKA, E. I., 1165, 1170
KOROTKOV, Yu. V., 1208
KORRENG, E., 1227, 1338, 1340, 1342–46
KORSHUNOV, B. G., 1424, 1431–32, 1434
KOSTENKO, T. G., 1233, 1335
KOSTROMIN, A. I., 1072
KOVALEV, F. V., 1408, 1427
KOWALCHIK, M., 11
KOZHEVNIKOV, G. N., 846
KRACEK, F. C., 155, 167, 182–183, 192, 365, 381–383, 391–395, 430–436 1929–31
KRAKAU, K. A., 495–496, 498–499
KRAUS, C. A., 1318, 1912
KROGER, C., 1009
KROGH-MOE, J., 160
KUBOTA, B., 5
KÜHNL, H., 1271
KULESHOV, I. M., 162, 176, 202–203
KULICHKINA, G. N., 1638
KULLERUD, G., 59, 1878–84, 1897, 1989–90
KURNAKOW, N. S., 1597
KUSAKIN, P. S., 846, 1900–01
KUSHAKOVSKIĬ, V. I., 99–100, 225
KUVYRKIN, O. N., 1622
KWESTROO, W., 552, 554, 565–566 622, 680–681, 778

LAFFITTE, P., 1257, 1262
LAMB, F. W., 1604
LAMBERTSON, W. A., 118, 349
LANDAU, B. S., 1465, 1478, 1485, 1487 1497, 1506–07, 1547, 1572
LANDER, J. J., 205
LANG, S. M., 23, 224, 226, 316–317, 349, 360–361, 573–574, 578, 580–581, 583, 585
LANTSBERRY, F. C. A. H., 1263, 1410
LEA, F. M., 231, 881, 943–947, 951–952
LEARY, J. A., 1278, 1296, 1316
LEHR, J. R., 318, 640, 642–644
LEHRMAN, A., 1070
LEIPOLD, M. H., 1140
LEMAN, G., 1823
LEONHARD, F., 477
LEONOV, Yu. S., 301
LESNYKH, D. S., 1785, 1798–99, 1805, 1855–56, 1861
LEVIN, E. M., 207, 211, 264, 321, 323, 325–330, 558–561
LEVINA, M. E., 1783
LIFSHITS, G. M., 1748, 1750
LINDSLEY, D. H., 32
LITVAKOVSKIĬ, A. A., 772
LOERPABEL, W., 571, 582, 677, 730
LÖFFLER, J., 533–534
LOH, E., 368
LONG, G., 1914
LOPREST, F. J., 1975, 2021
LOWRIE, R. S., 2031
LOWRY, T. M., 1053
LÖWY, S., 1204
LUZHNAYA, N. P., 1794

McCarthy, C. E., 2052
McCaughey, W. J., 665
McCreary, W. J., 1631
McDaniel, C., 323
McIntosh, A. B., 275, 688–689
McMullan, R. K., 1575
McMurdie, H. F., 207, 891–892, 1981
McNally, R. N., 229, 259, 262
McPherson, D. J., 25
McTaggart, G. D., 783
McVay, T. N., 1282, 1445, 1456, 1468, 1495–96, 1501, 1534, 1542, 1564
Maass, O., 1918
MacChesney, J. B., 90, 146–147
MacDonald, G. J. F., 514
Machin, J. S., 1304, 1624–26
Mackenzie, J. D., 150
MacKenzie, W. S., 2045
Maddocks, W. R., 197, 535–542, 690
Maetz, H., 595, 645, 1706
Majumdar, A. J., 238, 1983
Makarov, S. Z., 2061
Malakhova, A. Ya., 1084
Malinin, S. D., 1941
Margolin, H., 36, 104
Mark, S. D., Jr., 721–722
Markina, I. B., 1044, 1046, 1056, 1058
Markov, B. F., 1231, 1233, 1238, 1279, 1317, 1333, 1335–36
Markovskaya, N. F., 1091, 1125
Markowitz, M. M., 1017, 1049
Marotta, D., 1097–98, 1101, 1106–07 1109–11, 1114, 1119, 1121
Marshall, W. L., 2021
Martin, E., 40
Martin, F. S., 1512
Marzullo, S., 611–612, 673, 678, 729, 731, 939–940
Mashovets, V. P., 1546, 1686, 1723
Mason, B., 73
Massazza, F., 269, 294–295, 723
Mateĭko, Z. A., 793, 1719–20, 1726, 1834, 1845, 1859–60
Matignon, C., 1417
Matsumoto, K., 664
Mattern, J. A., 1451
Matthes, F., 1633, 1636
Maxwell, L. H., 224, 316, 573–574, 578
Mayer, S. W., 1658
Mazzetti, C., 276
Meadows, G. E., 1523
Meadows, R. E., 1456, 1501, 1542
Melent'ev, B. N., 985
Mellors, G. W., 1659
Meneghini, D., 1078
Menge, O., 1221–22, 1249, 1251, 1303, 1363, 1365–66, 1373, 1376
Menzel, H., 1974
Merwin, H. E., 188, 596, 603, 1929–31
Meshchaninova, L. V., 1404, 1416, 1426
Meyers, F., 1604
Michaud, M., 2050
Midgley, H. G., 800, 847, 953, 981, 999–1002
Miller, J. P., 1984
Mironov, K. E., 1186, 1195, 1198, 1266, 1306, 1325, 1583–85, 2053, 2056

Mirsoyapov, V. N., 1208
Miyashiro, A., 963
Modestova, T., 1606, 1628
Momin, A. C., 89
Mönkemeyer, K., 1593–94, 1601–02, 1612–13, 1620
Moore, R. E., 1456, 1479–80, 1498–99, 1501, 1507, 1527, 1534, 1537–38, 1542, 1558, 1573
Morey, G. W., 167, 172, 188, 197, 385–386, 391–395, 406, 482–486, 515–518, 647–648, 759, 852, 1929–31, 1949, 1965–69, 1973, 2020
Morgan, R. A., 222
Morgan, W. R., 478–479
Morozov, A. N., 251
Morozov, I. S., 1234–37, 1239, 1241–42, 1275–77, 1281, 1315, 1320, 1377, 1393, 1395, 1402–03, 1406, 1423–, 24, 1429–41
Moruzzi, V. L., 60
Moshkina, I. A., 2064–65
Muan, A., 10, 12–13, 27, 43–44, 46, 58, 64, 82–84, 87, 90, 102–103, 126–129, 131–135, 138–147, 237, 407, 501, 586, 598, 630, 656–658, 682, 696, 712, 760–762, 867, 871, 873–875
Mueller, M. H., 118, 349
Müller, G., 1974
Müller, H., 1105, 1113, 1124
Mumpton, F. A., 116–117, 124, 782, 784, 2006
Murad, A. B., 692–694
Murthy, M. K., 443–444, 454–455

Nacken, A. A., 1092, 1104
Nacken, R., 1088–90, 2032–33
Nadachowski, F., 857
Nelson, B. W., 2039
Nessle, G. J., 1449, 1522, 1538, 1540, 1569
Nesterova, A. K., 437, 748, 1169
Neuenschwander, E., 1328–31
Newkirk, T. F., 627–628
Newman, E. S., 672
Nielsen, J. W., 97
Niggli, P., 1016, 1820
Nikol'skaya, Yu. P., 2064–65
Nikonova, I. N., 1081
Nisioka, U., 931–932, 941, 955–956, 986, 1511
Novoselova, A. V., 1446, 1457, 1466, 1510, 1516, 1622, 1783
Nowotny, H., 766
Nurse, R. W., 668, 800, 847, 912, 953, 981, 999–1002
Nyankovskaya, R. N., 1752, 1755, 1766, 1769, 1811, 1854, 1857, 1863, 1865

Obolonchik, V. A., 1638
Obrowski, W., 7, 17, 163
Odell, F., 243, 271, 355
Oelsen, W., 595, 645, 1368, 1382, 1706
Okamoto, M., 1511
Ölander, A., 1709
Olds, L. E., 220, 311, 353

Ol'shanskiĭ, Ya. I., 6, 37, 985, 1681–82, 1694–97, 1705, 1714, 1729 1905
Oparina, A. F., 1871
Ordway, F., 238
Osborn, E. F., 82, 113–115, 138–145, 239, 260, 310, 407, 449–453, 456, 501, 586, 598, 602, 604, 606, 630, 635, 637, 651–654, 656, 660–663, 682, 696, 700–701, 712, 864–865, 867, 871, 873–875, 894–895, 987–898, 900–904, 906, 909–910, 913–915, 918, 1471, 1482, 1513–15, 1933, 1946, 1997–98, 2008, 2010–11, 2022, 2036–37, 2043
Otto, H. E., 216, 220–221, 227, 263, 311, 354
Ozerov, R. P., 105, 1137–38

Padurow, N. N., 366, 863
Paetsch, H. H., 288, 746
Page, R. A., 1263, 1410
Paladino, A. E., 65–68, 75–78
Palkin, A. P., 1224, 1341, 1666–70
Palkin, V. A., 1063, 1066
Palyura, I. P., 1341
Panchenko, I. D., 1231, 1233, 1333, 1335
Paping, H. A. M., 552, 554, 622
Parker, T. W., 943–947, 951–952
Parlenko, S. P., 1531
Parravano, N., 1899
Pascal, P. V., 1567, 1673, 1727
Pearson, J., 278
Peppler, R. B., 1982, 1992
Perman, E. P., 1168, 1172, 1745, 1756, 1758, 1761
Petrov, V. I., 1546, 1686, 1723
Phillips, B., 43–44, 46, 64, 87, 237, 657–658
Pilloton, R. L., 2049, 2059–60
Pish, G., 1563
Plato, W., 1614, 1619, 1621, 1623, 1629
Plotnikova, V. I., 1996
Plyushchev, V. E., 1044, 1046, 1056, 1058, 1091, 1125, 1404–05, 1408, 1416, 1426–27
Poch, W., 674–676
Pollard, A. J., 304
Popovskaya, N. P., 1075
Posnjak, E., 588, 592–594, 2051
Posypaĭko, V. I., 785, 1158, 1162 1831, 1836–40
Pozhitkova, S. A., 1405
Prebus, A. F., 1991
Prince, A. T., 882–887, 919–920, 957 982–984
Pröger, H., 1141–42
Prokopowicz, T. I., 813–817
Pronina, M. Z., 2053, 2056
Prophet, E., 52, 258, 274
Protsenko, P. I., 1043, 1045, 1050, 1052, 1057, 1059–62, 1064–65, 1067–68, 1071, 1073, 1075–76, 1079, 1082, 1084
Purt, G., 206
Puschin, N. A., 1204, 1448, 1455 1463, 1484, 1488, 1493, 1504

Quercigh, E., 1576, 1903

Rabenau, A., 659, 716
Rait, A. B., 688
Rait, J. R., 876–879, 880, 995–998
Rambach, F., 1226, 1228–29, 1260, 1287, 1299
Rang, H. J., 1391–92
Rankin, G. A., 596, 633, 636–637, 639
Rase, D. E., 115, 213, 303, 322, 564, 1846
Rassonskaya, I. S., 1074, 1157, 1159, 1829–30, 1841–42
Ravich, M. I., 1032
Rea, R. F., 1108, 1770–71
Reavis, J. G., 1296, 1316
Reburn, W. T., 1951
Reckhard, H., 161, 169, 201
Redchenko, V. T., 1670
Redman, J. D., 1492, 1494
Rees, W. J., 597, 626, 650, 655, 710, 924–926
Reinders, W., 1603
Reisman, A., 171, 173–174, 375, 1008
Reshetnikov, N. A., 1019–21, 1023–26, 1028–29, 1031, 1033, 1038, 1734
Reynolds, D. S., 246, 248–250
Rhinehammer, T. B., 1469, 1518, 1549, 1554, 1563
Ribbe, P. G., 259
Ricci, J. E., 1049, 1975
Richards, R. G., 29, 130
Ricker, R. W., 606
Rieck, G. D., 376
Ringwood, A. E., 702, 964–965
Roake, W. E., 1470, 1555
Robbins, C. R., 264, 321, 2048
Robertson, E. C., 514
Robin, J., 50, 51, 53, 56
Roedder, E., 397–403, 801
Rogers, N. E., 1469, 1518, 1549, 1554
Rollet, A. P., 165, 1952, 2050
Roth, B., 1328–31
Roth, R. S., 23, 211, 214, 226, 240, 253, 287, 312, 317, 324–331, 335, 337, 340, 342–352, 360, 370, 372–373, 562–563, 580–581, 583, 585, 611–612, 673, 678, 729, 731, 779, 781, 939–940
Rough, F. A., 121–123
Rowland, J. F., 244
Roy, D. M., 260, 456, 610, 1482, 1515, 1923, 1993–95, 1997–98, 2000, 2002, 2005–06, 2034, 2037–38, 2043
Roy, R., 14–15, 18, 22, 62, 113–115, 116–117, 124, 148, 152, 157, 179, 196, 213, 239, 260, 265, 298, 310, 314, 336, 357, 446–453, 456–457, 553, 564, 621, 660–63, 719, 780, 782, 784, 1444, 1458, 1471, 1482, 1513–15, 1846, 1922–23, 1925, 1932–40, 1983, 1993, 1997–98, 2000, 2002, 2005, 2008, 2010–11, 2022, 2034–35, 2037–39, 2043
Rozhkovskaya, L. V., 1641, 1644
Rubleva, V. V., 1068, 1779, 1806, 1848–53, 1866–67
Rüdel, W., 1291
Ruer, R., 1821
Ruff, O., 242, 571, 582, 677, 730

Sackmann, H., 1396–1401
St. Pierre, P. D. S., 641, 666, 958–962
Sakai, K., 1080
Saller, H. A., 121–123
Salmon, J. E., 2013–14, 2016
Samsonov, G. V., 1638
Sand, L. B., 2022
Sandonnini, C., 1176, 1178, 1201, 1214, 1216, 1218–19, 1230, 1243–45, 1247–48, 1261, 1268, 1288–89, 1293, 1295, 1300, 1308, 1332, 1351–54, 1356, 1359, 1361–62, 1364, 1367, 1369–70, 1372, 1374, 1378–81, 1383, 1388, 1578, 1580, 1586, 1605, 1627, 1632, 1674, 1868
Sarver, J. F., 151, 268, 358, 623, 703–709, 717–718, 720, 754, 756
Sasaki, K., 2015
Sastry, B. S. R., 180, 465–469
Sata, T., 935–938
Saunders, H. L., 1756
Sawaguchi, E., 747
Sawamoto, T., 664
Scarpa, G., 1176, 1268, 1293, 1308, 1347–48, 1578, 1730–33, 1735–42, 1744, 1746
Schaaf, H., 1400–01
Schairer, J. F., 63, 80, 408–415, 502–504, 508–510, 519–530, 588, 591, 594, 600–601, 634, 637, 683–687, 697–698, 786, 788, 795–796, 798–799, 802–811, 854–856, 864–872, 889, 907–908, 911, 915, 918, 971–977
Scheil, E., 1258–59
Schimmel, F. A., 1958, 2030
Schloemer, H., 2032–33
Schmidt, M. J. M., 1220, 1290, 1302, 1339, 1349, 1357–58
Schmitt, J. M., 2031
Schmitz-Dumont, O., 161, 169, 201, 1447, 1472, 1675–80, 1684–85, 1691–93, 1701–03
Schneider, S. J., 312, 331, 334–335, 337, 339, 340, 342, 344–345, 347–348, 351–352, 781
Scholze, H., 765
Schreyer, J. M., 2018
Schultz, J., 198, 273, 319, 374
Schürmann, E., 41
Schusterius, C., 863
Schwartz, R., 442, 475
Schwarz, R., 166, 191
Searcy, A. W., 1–2
Secoy, C. H., 2021
Segnit, E. R., 481, 624–625, 888, 890, 1464
Seifert, H.-J., 1232, 1272, 1313
Selivanov, V. G., 1462, 1491
Sementsova, A. K., 1167, 1805, 1832–33, 1835, 1843–44
Senderoff, S., 1659
Sennett, F. E., 611–612, 673, 678, 729, 731, 939–940
Sennett, P. S., 1615, 1617–18
Sense, K. A., 1508–09
Shafer, E. C., 357, 780
Shafer, M., 1906
Shafer, M. W., 60, 74, 115, 196, 1932, 1934–35, 1937–40
Shakhno, I. V., 1405, 1427

Shatillo, V. A., 1269, 1273, 1384, 1425
Shchemeleva, G. G., 1209, 1643
Shchetnikova, I. L., 1543
Shchirova, N. A., 1224
Sheil, R. J., 1282–83, 1298, 1321–22, 1337, 1456, 1474, 1501–02
Shevlin, T. S., 1140
Shevtsova, Z. N., 1429–30
Sheybany, H. A., 377
Shklover, L. P., 1044, 1046, 1056, 1058
Shlepov, V. K., 6, 37
Shmidt, N. E., 1074
Sholokhovich, M. L., 378, 428, 500, 544–547, 743, 749–750, 792, 812, 859, 1144–45, 1150, 1163–64, 1717, 1796, 1847, 1862
Shul'ga, N. A., 1637, 1639
Shumov, Yu. V., 860
Shvartsman, U. I., 1274, 1314
Sigida, N. P., 422–423, 438, 476, 1683, 1687, 1690, 1704, 1721–22
Simanov, Yu. P., 1446, 1457, 1466, 1510, 1516, 1783
Simonich, A. T., 1234, 1242, 1406, 1431–32
Sirchia, E., 269, 723
Skaliks, W., 1007, 1010–11
Smalley, R. G., 845
Smith, D. K., 238
Smith, J. V., 416, 512, 787, 789
Smits, A., 1285
Smolensky, S., 736
Smolina, L. P., 1018
Smothers, W. J., 966–967
Smyth, F. H., 1003
Snow, R. B., 42, 81, 587, 589–591, 733, 735, 987–990
Sochevanov, V. G., 1954
Sokolova, E. I., 1953
Sokolova, M. A., 1212
Solov'ev, S. P., 567
Sōmiya, S., 10, 58, 64, 126–129, 760–762
Sommer, R. C., 1284, 1323, 1415
Sosman, R. B., 21, 153
Speranskaya, E. I., 109, 112, 283, 285, 1762, 1808
Spitzyn, V., 162, 176, 202–203
Spivak, J., 834–835, 838–840
Stadelmaier, H., 1258–59
Starrs, B. A., 1689
Steger, A., 1581
Stehlik, B., 1910
Steierman, B. L., 572, 577, 579
Stender, V. V., 1462, 1491
Stephan, E., 242
Stewart, D. B., 2046
Stickland, A. E., 1607, 1611
Stone, L., 36, 104
Stone, P. E., 318, 640, 642–644
Stone, R. W., 1508–09
Stranski, I. N., 149
Strehlow, R. A., 1475, 1492, 1494, 1570
Strickler, D. W., 62, 179, 446–448
Strimple, J. H., 256
Studer, M., 1328–31
Stutterheim, N., 912
Subbarao, E. C., 252, 289

Sumarokova, T. N., 1606, 1628
Sumida, W. K., 28, 30–31, 33–35
Sun, I.-C, 1235–37, 1239, 1241, 1275–77, 1281, 1320, 1377, 1423
Surak, J. G., 1537
Swayze, M. A., 948–950, 991–994

Tait, D. B., 894, 906, 909
Takada, T., 61, 208
Takahashi, T., 2015
Takahashi, Y., 1894
Taperova, A. A., 2003–04
Taylor, H. F. W., 1979
Taylor, W. C., 396, 794, 797
Teltow, J., 1595
Thamer, B. J., 1523
Thoma, R. E., 1465, 1477–80, 1485, 1487, 1496–99, 1506–07, 1517, 1526–27, 1538–39, 1547–48, 1553, 1556–57, 1560–62, 1564, 1571–74
Thurmond, C. D., 11, 1607, 1611
Thwaite, R. D., 627–628
Tichelaar, G. W., 1259
Tieckelmann, H., 1451
Tien, T. Y., 185, 364, 470–473
Tikhomirova, M. M., 1996
Tikkanen, M. H., 228, 230
Tilley, C. E., 505–506, 511
Timofeeff, K., 1461, 1473
Tokareva, M. V., 1757, 1759–60, 1763–65
Tokareva, S. A., 2053, 2056
Toptygin, D. Ya., 1315, 1393, 1395, 1402–03, 1433, 1435, 1439, 1441
Torkar, K., 1926
Toropov, N. A., 362, 549–550, 556–557, 1543, 1559
Tranquard, A., 2050
Trapp, G., 1516
Treis, K., 1223, 1294, 1310, 1334, 1413
Tresvyatskiĭ, S. G., 99–100, 181, 190, 194, 225, 1350, 1672
Tridot, G., 1823
Trömel, G., 41, 246, 247, 670
Trumbore, F. A., 11
Truthe, W., 1869–70, 1872–77
Trzebiatowski, W., 297, 667, 669
Tsvetkov, A. I., 37
Tu, K. C., 1946
Tubandt, C., 1180, 1443
Tucker, P. A., 1469, 1518, 1549, 1554, 1563
Turkdogan, E. T., 197, 278, 535–542
Turnock, A. C., 32
Tuttle, O. F., 406, 416, 512, 759, 787, 833, 929, 1004, 1012–13, 1919–20, 1942, 1964, 1985–88, 2001, 2044, 2054–55

Ugai, Ya. A., 1269, 1273, 1384, 1421, 1425
Ugrinic, G., 558, 560–561
Unzhakov, G. M., 1026, 1734
Urazov, G. G., 109, 112, 285
Urnes, S., 1784
Ussow, A., 1039

Valentin, J., 1417
van Eyk, C., 1042, 1048, 1054, 1747
Van Hook, H. J., 72, 98
van Klooster, H. S., 380, 417, 429, 1888
Varicheva, V. I., 859
Vartbaronov, O. R., 418, 790–791
Vatslavik, E., 1568, 1728
Venevtsev, Yu. N., 567, 860
Verbitskaya, T. N., 567
Vereshchetina, I. P., 1794
Vilutis, N. I., 1021, 1023–25, 1028–29, 1031
Vogel, R., 40
Vogel, R. C., 1524, 1528, 1592
Vogt, T., 1520
Voitekhova, E. A., 120
Volkova, L. F., 1015
Vollbracht, L., 1391–92
Vol'nov, I. I., 2061
Volodin, P. L., 1350
Voronov, N. M., 120
Vortisch, E., 1270, 1301, 1311, 1355, 1409, 1418, 1428

Wagner, G., 212, 1459
Walsh, K. A., 1296, 1316
Ward, W. T., 1570
Waring, J. L., 214, 312, 321, 324, 331, 337, 339–340, 344, 351–352
Warshaw, I., 336
Wartenberg, H. v., 52, 258, 274, 356, 367
Washburn, E. W., 193
Watson, G. M., 1570
Weaver, B. S., 1027, 1030
Weaver, C. F., 1477, 1496, 1517, 1526, 1553, 1556–57, 1562, 1564, 1571, 1574
Webster, A. H., 91–93
Weeg, A., 1675–80, 1684–85, 1691–93, 1701–03
Weir, C. E., 1915, 1921
Weiser, K., 1906
Weiss, P., 1307
Welch, J. H., 245, 313, 668, 896, 905
Wells, L. S., 646, 649, 672, 1976, 1981–82
Wendolkowski, W. S., 1525
Wendrow, B., 1972
Wentrup, H., 338
Weymouth, J. H., 890
Wheeler, C. M., Jr., 1203, 1205–06, 1608, 1615, 1617–18

White, G. D., 1456, 1501, 1542
White, J., 29, 38–39, 57, 69, 71, 88, 130, 137, 275, 277, 688–689, 691, 695, 768, 770–771, 954
White, W. B., 18, 148, 152
Wickert, H., 441
Wilde, W. T., 710
Wilkerson, A. B., 1240, 1282, 1297, 1321, 1337
Willems, H. W. V., 519–530
Willis, J. H., 1961–63
Winternitz, P. F., 1049
Wisnyi, L. G., 233
Wittenberg, L. J., 1469, 1518, 1554, 1563
Wojciechowska, J., 667, 669, 671, 725–728
Wolters, A., 1776, 1782
Woodhouse, D., 57, 69, 137
Woodward, A. E., 2007
Worel, H., 1926
Wright, F. E., 633, 636–637, 639
Wrzesnewsky, J. B., 1596–98, 1616, 1907–09, 1911
Wurm, J. G., 204, 209, 215, 217, 219, 232, 548, 569–570, 575
Wylie, A. W., 1464
Wyllie, P. J., 1004, 1919–20, 1985–88, 2054–55

Yagi, K., 697–698
Yakel, H. L., Jr., 1477
Yanat'eva, O. K., 2066
Yarembash, E. I., 1457
Yavorsky, P. J., 572, 577, 579
Yazawa, A., 1904
Yoder, H. S., Jr., 638, 850–851, 855–856, 968–969, 975–977, 1878–80, 1882, 2040–42, 2048
Yosim, S. J., 1658
Yund, R. A., 59, 1883–84

Zakharchenko, G. H., 1179, 1199–1200
Zakharchenko, M. A., 1718, 1753, 1768, 1858
Zambonini, F., 744–745
Zawidzki, J., 1041
Zelikman, A. N., 199
Zhdanov, G. S., 567
Zhemchuzhnui, S. F., 1175, 1177, 1215, 1217, 1226, 1228–29, 1260, 1287, 1299, 1817–18
Zinov'ev, A. A., 1018
Zöfelt, B., 1141–42
Zuur A. P. 1822

System Index*

Ag–Cu–O–Pb 125	$AgNO_3$–KBr 1743	AlF_3–NaF–ZnO 1700
AgBr–AgCl 1593	$AgNO_3$–KCl–NaBr 1751	AlF_3–NaF–ZrO_2 1699
AgBr–AgCl–AgI 1633	$AgNO_3$–KNO_3 1039	AlF_3–RbF 1504
AgBr–AgCl–$AgNO_3$ 1748	$AgNO_3$–KNO_3–$TlNO_3$ 1063	Al_2O_3–Al 1–2
AgBr–AgI 1594	$AgNO_3$–$LiNO_3$–$RbNO_3$ 1065	Al_2O_3–Al_4C_3 1914
AgBr–$AgNO_3$ 1742	$AgNO_3$–NH_4NO_3 1041	Al_2O_3–AlF_3–CaF_2–NaF 1727
AgBr–$CdBr_2$ 1179	$AgNO_3$–$NaNO_3$ 1040	Al_2O_3–AlF_3–LiF 1686
AgBr–KBr 1175	$AgNO_3$–$NaNO_3$–$TlNO_3$ 1066	Al_2O_3–AlF_3–LiF–NaF 1723
AgBr–KNO_3–NaCl 1749	$AgNO_3$–$TlNO_3$ 1042	Al_2O_3–AlF_3–MgF_2–NaF 1728
AgBr–LiBr 1176	Ag_2S–AgCl 1868	Al_2O_3–AlF_3–NaF 1698
AgBr–NaBr 1177	Ag_2S–As_2S_3 1888	Al_2O_3–B_2O_3 308
AgBr–$PbBr_2$ 1180	Ag_2S–Sb_2S_3 1889	Al_2O_3–B_2O_3–H_2O–MgO–Na_2O–SiO_2 ....
AgBr–RbBr 1178	Ag_2S–Tl_2S 1887	 2048
AgCN–KCN 1872	Ag_2SO_4–AgCl–LiCl–Li_2SO_4 ... 1785	Al_2O_3–B_2O_3–Li_2O 445
AgCN–NaCN 1873	Ag_2SO_4–Ag_2MoO_4–Li_2MoO_4–Li_2SO_4	Al_2O_3–B_2O_3–Li_2O–SiO_2 820–825
AgCl–AgBr 1593	 1160	Al_2O_3–B_2O_3–SiO_2 763–765
AgCl–AgBr–AgI 1633	Ag_2SO_4–Ag_2MoO_4–Na_2MoO_4–Na_2SO_4 ..	Al_2O_3–BaO 206
AgCl–AgBr–$AgNO_3$ 1748	 1161	Al_2O_3–BaO–H_2O 1976
AgCl–Ag_2CrO_4–LiCl–Li_2CrO_4 .. 1855	Ag_2SO_4–K_2SO_4 1088	Al_2O_3–BaO–SiO_2 556–557
AgCl–AgI 1612	Ag_2SO_4–Li_2SO_4 1089	Al_2O_3–BeO 218
AgCl–$AgNO_3$ 1744	Ag_2SO_4–Na_2SO_4 1090	Al_2O_3–BeO–MgO 572
AgCl–$AgNO_3$–LiCl–$LiNO_3$ 1753	Ag_2WO_4–AgCl–LiCl–Li_2WO_4 .. 1856	Al_2O_3–BeO–SiO_2 576
AgCl–Ag_2S 1868	Al–$AlCl_3$–Zn 1666	Al_2O_3–BeO–ThO_2 577
AgCl–Ag_2SO_4–LiCl–Li_2SO_4 ... 1785	Al–$AlCl_3$–Zn–$ZnCl_2$ 1668	Al_2O_3–BeO–TiO_2 578
AgCl–Ag_2WO_4–LiCl–Li_2WO_4 .. 1856	Al–Al_2O_3 1–2	Al_2O_3–BeO–ZrO_2 579
AgCl–$BeCl_2$ 1220	Al–Cr–Fe–O 126–130	Al_2O_3–Bi_2O_3 327
AgCl–$BiCl_3$ 1212	Al–Fe–O 26–35	Al_2O_3–CO_2–H_2O–MgO 2037
AgCl–$CaCl_2$ 1221	Al–Fe–O–Si 131–136	Al_2O_3–CaF_2 1673
AgCl–CsCl 1213	Al–O 1–2	Al_2O_3–CaF_2–CaO 1707
AgCl–CuCl 1214	$AlBr_3$–BBr_3 1203	Al_2O_3–CaO 231–233
AgCl–KCl 1215	Al_4C_3–Al_2O_3 1914	Al_2O_3–CaO–Cr_2O_3 626
AgCl–LiCl 1216	$AlCl_3$–Al–Zn 1666	Al_2O_3–CaO–Cr_2O_3–MgO–SiO_2 .. 987–990
AgCl–$MgCl_2$ 1222	$AlCl_3$–Al–Zn–$ZnCl_2$ 1668	Al_2O_3–CaO–FeO–SiO_2 866–875
AgCl–NaCl 1217	$AlCl_3$–CsCl 1234	Al_2O_3–CaO–Fe_2O_3 627–629
AgCl–$PbCl_2$ 1223	$AlCl_3$–CsCl–$TaCl_5$ 1406	Al_2O_3–CaO–Fe_2O_3–K_2O 794
AgCl–RbCl 1218	$AlCl_3$–$FeCl_3$–$NbCl_5$ 1436	Al_2O_3–CaO–Fe_2O_3–MgO 876–879
AgCl–TlCl 1219	$AlCl_3$–$FeCl_3$–$TaCl_5$ 1437	Al_2O_3–CaO–Fe_2O_3–MgO–SiO_2 .991–1002
AgCl–$ZnCl_2$ 1224	$AlCl_3$–$FeCl_3$–$TiCl_4$ 1435	Al_2O_3–CaO–Fe_2O_3–Na_2O 827
Ag_2CrO_4–AgCl–LiCl–Li_2CrO_4 .. 1855	$AlCl_3$–KCl 1274	Al_2O_3–CaO–Fe_2O_3–SiO_2 .. 943–953
AgF–PbF_2 1443	$AlCl_3$–NaCl 1314	Al_2O_3–CaO–H_2O 1978–83
AgF–ZnF_2 1444	$AlCl_3$–NaCl–$NbCl_5$ 1431	Al_2O_3–CaO–H_2O–SO_3 2063
AgI–AgBr 1594	$AlCl_3$–NaCl–$TaCl_5$ 1432	Al_2O_3–CaO–H_2O–SiO_2 2035
AgI–AgBr–AgCl 1633	$AlCl_3$–$NbCl_5$–$TaCl_5$ 1438	Al_2O_3–CaO–K_2O 390
AgI–AgCl 1612	$AlCl_3$–$POCl_3$ 1822	Al_2O_3–CaO–K_2O–MgO–SiO_2 .. 973–974
AgI–$AgNO_3$ 1746	$AlCl_3$–$SeCl_4$ 1391	Al_2O_3–CaO–K_2O–Na_2O–SiO_2 .. 971–972
AgI–$AgNO_3$–KI–KNO_3 1767	$AlCl_3$–$TeCl_4$ 1392	Al_2O_3–CaO–K_2O–SiO_2 795–800
AgI–$AgNO_3$–NaI–$NaNO_3$ 1768	$AlCl_3$–$TiCl_4$ 1393	Al_2O_3–CaO–MgO 596
AgI–CuI 1576	$AlCl_3$–Zn–$ZnCl_2$ 1667	Al_2O_3–CaO–MgO–Na_2O–SiO_2 .. 975–981
AgI–HgI_2 1581	AlF_3–Al_2O_3–CaF_2–NaF 1727	Al_2O_3–CaO–MgO–SiO_2 880–923
AgI–KI 1577	AlF_3–Al_2O_3–LiF 1686	Al_2O_3–CaO–Na_2O 480
AgI–LiI 1578	AlF_3–Al_2O_3–LiF–NaF 1723	Al_2O_3–CaO–Na_2O–P_2O_5–SiO_2 985
AgI–NaI 1579	AlF_3–Al_2O_3–MgF_2–NaF 1728	Al_2O_3–CaO–Na_2O–SiO_2 .. 828–851
AgI–RbI 1580	AlF_3–Al_2O_3–NaF 1698	Al_2O_3–CaO–Na_2O–SiO_2–TiO_2 .. 982–984
Ag_2MoO_4–Ag_2SO_4–Li_2MoO_4–Li_2SO_4	AlF_3–CaF_2–NaF 1566–67	Al_2O_3–CaO–P_2O_5 640–644
.... 1160	AlF_3–CaO–NaF 1700	Al_2O_3–CaO–P_2O_5–SiO_2 958–962
Ag_2MoO_4–Ag_2SO_4–Na_2MoO_4–Na_2SO_4 ...	AlF_3–CdO–NaF 1700	Al_2O_3–CaO–SiO_2 630–639
.... 1161	AlF_3–CsF 1448	Al_2O_3–CaO–SiO_2–TiO_2 954–957
$AgNO_3$–AgBr 1742	AlF_3–KF 1461	Al_2O_3–CeO_2 356
$AgNO_3$–AgBr–AgCl 1748	AlF_3–KF–NaF 1536	Al_2O_3–Cr_2O_3 309
$AgNO_3$–AgCl 1744	AlF_3–LiF 1473	Al_2O_3–Cr_2O_3–FeO–Fe_2O_3 130
$AgNO_3$–AgCl–LiCl–$LiNO_3$ 1753	AlF_3–LiF–NaF 1546	Al_2O_3–Cr_2O_3–Fe_2O_3 126–129
$AgNO_3$–AgI 1746	AlF_3–MgF_2–NaF 1568	Al_2O_3–Cr_2O_3–MgO 710
$AgNO_3$–AgI–KI–KNO_3 1767	AlF_3–MgO–NaF 1700	Al_2O_3–Dy_2O_3 312
$AgNO_3$–AgI–NaI–$NaNO_3$ 1768	AlF_3–NaF 1490	Al_2O_3–Er_2O_3 312
$AgNO_3$–Ca(NO_3)$_2$ 1043	AlF_3–NaF–Na_2SO_4 1784	Al_2O_3–Eu_2O_3 312
$AgNO_3$–Cd(NO_3)$_2$–KNO_3 1064	AlF_3–NaF–TiO_2 1700	Al_2O_3–FeO 26

* All index listings are given by figure number.

$Al_2O_3-FeO-Fe_2O_3$ 29–30, 32	$AsBr_3-SbBr_3$ 1204	$BaBr_2-NaBr$ 1192
$Al_2O_3-FeO-Fe_2O_3-SiO_2$ 131–136	$AsBr_5-BBr_3$ 1206	$BaCO_3-BaCl_2-BaTiO_3$ 1847
$Al_2O_3-FeO-Fe_3O_4$ 31	AsI_3-PI_3 1587	$BaCO_3-BaCl_2-Na_2CO_3-NaCl$ ... 1862
$Al_2O_3-FeO-K_2O-SiO_2$ 801	AsI_3-SbI_3 1588	$BaCl_2-BaCO_3-BaTiO_3$ 1847
$Al_2O_3-FeO-MgO$ 680–681	As_2O_3 149	$BaCl_2-BaCO_3-Na_2CO_3-NaCl$ ... 1862
$Al_2O_3-FeO-MgO-SiO_2$ 963	$As_2O_5-Fe_2O_3-H_2O$ 2015	$BaCl_2-BaF_2$ 1621
$Al_2O_3-FeO-MnO$ 688	$As_2O_5-H_2O-K_2O$ 1950	$BaCl_2-BaF_2-CaCl_2-CaF_2$ 1656
$Al_2O_3-FeO-Na_2O-SiO_2$ 853–854	$As_2O_5-K_2O$ 170	$BaCl_2-BaF_2-KCl-KF$ 1645
$Al_2O_3-FeO-SiO_2$ 696–698	$As_2O_5-K_2O-Na_2O$ 384	$BaCl_2-BaF_2-LiCl-LiF$ 1647
$Al_2O_3-Fe_2O_3$ 27	$As_2O_5-Na_2O$ 195	$BaCl_2-BaF_2-NaCl-NaF$ 1650
$Al_2O_3-Fe_2O_3-Li_2O$ 446–448	$As_2O_5-PbF_2-PbO$ 1712	$BaCl_2-BaF_2-RbCl-RbF$ 1655
$Al_2O_3-Fe_2O_3-MgO$ 711	As_2O_5-PbO 286	$BaCl_2-Ba(NO_3)_2-CaCl_2-Ca(NO_3)_2$
$Al_2O_3-Fe_2O_3-SiO_2$ 766–767	$As_2S_3-Ag_2S$ 1888	 1765
$Al_2O_3-Ga_2O_3$ 310	$As_2S_3-Tl_2S$ 1893	$BaCl_2-Ba(NO_3)_2-NaCl-NaNO_3$ ... 1762
$Al_2O_3-Ga_2O_3-H_2O$ 2008		$BaCl_2-BaO-TiO_2$ 1846
$Al_2O_3-Gd_2O_3$ 312	BBr_3-AlBr_3 1203	$BaCl_2-BeCl_2$ 1349
$Al_2O_3-H_2O$ 1926–28	BBr_3-AsBr_5 1206	$BaCl_2-CaCl_2$ 1350
$Al_2O_3-H_2O-K_2O-Na_2O-SiO_2$	BBr_3-SnBr_4 1205	$BaCl_2-CaF_2-KCl$ 1637
.... 2044–45	BBr_3-SnI_4 1608	$BaCl_2-CaF_2-NaF$ 1639
$Al_2O_3-H_2O-K_2O-SiO_2$ 2019	$BF_3-KCl-KF$ 1638	$BaCl_2-CdCl_2$ 1351
$Al_2O_3-H_2O-Li_2O-Na_2O-SiO_2$ 2046	BF_3-KF 1462	$BaCl_2-KCl$ 1262
$Al_2O_3-H_2O-MgO$ 1997–98	$BF_3-KF-NaF$ 1537	$BaCl_2-KCl-MgCl_2$ 1417
$Al_2O_3-H_2O-MgO-N_2O_5$ 2043	BF_3-NaF 1491	$BaCl_2-KCl-NaCl$ 1409
$Al_2O_3-H_2O-MgO-SiO_2$ 2038–42	BF_3-UF_6 1524	$BaCl_2-KCl-SrCl_2$ 1418
$Al_2O_3-H_2O-Na_2O-SiO_2$ 2022–29	BF_5-UF_6 1528	$BaCl_2-LiCl$ 1289
$Al_2O_3-H_2O-SO_3$ 2012	B_2O_3 150	$BaCl_2-MgCl_2$ 1352
$Al_2O_3-H_2O-SiO_2$ 2009–11	$B_2O_3-Al_2O_3$ 308	$BaCl_2-MnCl_2$ 1353
$Al_2O_3-Ho_2O_3$ 312	$B_2O_3-Al_2O_3-H_2O-MgO-Na_2O-SiO_2$	$BaCl_2-NaCl$ 1301
$Al_2O_3-In_2O_3$ 312	 2048	$BaCl_2-NaCl-SrCl_2$ 1428
$Al_2O_3-K_2O-MgO-SiO_2$ 802–811	$B_2O_3-Al_2O_3-Li_2O$ 445	$BaCl_2-PbCl_2$ 1354
$Al_2O_3-K_2O-Na_2O-SiO_2$ 786–789	$B_2O_3-Al_2O_3-Li_2O-SiO_2$ 820–825	$BaCl_2-SrCl_2$ 1355
$Al_2O_3-K_2O-SiO_2$ 407–416	$B_2O_3-Al_2O_3-SiO_2$ 763–765	$BaCl_2-TlCl$ 1338
$Al_2O_3-La_2O_3$ 312	B_2O_3-BaO 207	$BaCl_2-ZnCl_2$ 1356
$Al_2O_3-Li_2O$ 179	$B_2O_3-BaO-PbO-TiO_2$ 859	$Ba(ClO_4)_2-NaClO_4$ 1018
$Al_2O_3-Li_2O-MgO-SiO_2$ 813–819	$B_2O_3-BaO-SiO_2$ 558–561	BaF_2-BaCl_2 1621
$Al_2O_3-Li_2O-SiO_2$ 449–459	$B_2O_3-Bi_2O_3$ 323, 327	$BaF_2-BaCl_2-CaCl_2-CaF_2$ 1656
$Al_2O_3-Li_2O-TiO_2$ 460–464	B_2O_3-CaO 234	$BaF_2-BaCl_2-KCl-KF$ 1645
$Al_2O_3-Lu_2O_3$ 312	$B_2O_3-CaO-Na_2O-SiO_2$ 852	$BaF_2-BaCl_2-LiCl-LiF$ 1647
Al_2O_3-MgO 259–260	$B_2O_3-CaO-SiO_2$ 646–649	$BaF_2-BaCl_2-NaCl-NaF$ 1650
$Al_2O_3-MgO-Na_2O-SiO_2$ 855–856	B_2O_3-CdO 252	$BaF_2-BaCl_2-RbCl-RbF$ 1655
$Al_2O_3-MgO-SiO_2$ 712	$B_2O_3-CdO-ZnO$ 679	$BaF_2-BaO-SiO_2$ 1714
$Al_2O_3-MgO-SiO_2-ZrO_2$ 966–967	B_2O_3-CoO 254	$BaF_2-BaSiO_3-LiF-Li_2SiO_3$ 1724
$Al_2O_3-MgO-TiO_2$ 713–714	$B_2O_3-CrO_3-K_2O$ 418	BaF_2-BeF_2 1510
Al_2O_3-MnO 275	$B_2O_3-CrO_3-K_2O-Na_2O$ 790–791	BaF_2-CaF_2-LiF 1551
$Al_2O_3-MnO-SiO_2$ 733–735	$B_2O_3-Cs_2O$ 160	BaF_2-KF 1455
$Al_2O_3-MnO-SiO_2-Y_2O_3$ 968–969	$B_2O_3-H_2O$ 1929–31	$BaF_2-LiF-MgF_2$ 1552
$Al_2O_3-Na_2O-SiO_2$ 501–514	$B_2O_3-H_2O-K_2O$ 1948	BaF_2-MgF_2 1511
$Al_2O_3-Nd_2O_3$ 312	$B_2O_3-H_2O-Li_2O$ 1951–52	$BaF_2-NaF-SiO_2$ 1694
$Al_2O_3-P_2O_5$ 318	$B_2O_3-H_2O-Na_2O$ 1955	BaF_2-UF_3 1512
$Al_2O_3-P_2O_5-SrO$ 756	$B_2O_3-K_2O$ 165	$Ba(NO_2)_2-Ba(NO_3)_2-KNO_2-KNO_3$
Al_2O_3-PbO 280	$B_2O_3-K_2O-Li_2O-WO_3$ 785	 1084
$Al_2O_3-PbO-SiO_2$ 737–739	$B_2O_3-K_2O-Na_2O$ 380	$Ba(NO_3)_2-BaCl_2-CaCl_2-Ca(NO_3)_2$
$Al_2O_3-Sc_2O_3$ 312	$B_2O_3-K_2O-P_2O_5$ 417	 1765
$Al_2O_3-SiO_2$ 313–314	$B_2O_3-La_2O_3$ 321	$Ba(NO_3)_2-BaCl_2-NaCl-NaNO_3$ ... 1762
$Al_2O_3-SiO_2-RF_2$ 1715	$B_2O_3-Li_2O$ 180	$Ba(NO_3)_2-Ba(NO_2)_2-KNO_2-KNO_3$
$Al_2O_3-SiO_2-SrO$ 755	$B_2O_3-Li_2O-Na_2O$ 429	 1084
$Al_2O_3-SiO_2-TiO_2$ 768–771	$B_2O_3-Li_2O-P_2O_5$ 470–473	$Ba(NO_3)_2-Ca(NO_3)_2$ 1060
$Al_2O_3-SiO_2-ZnO$ 758	$B_2O_3-Li_2O-SiO_2$ 465–469	$Ba(NO_3)_2-CsNO_3$ 1044
$Al_2O_3-SiO_2-ZrO_2$ 772	B_2O_3-MgO 261	$Ba(NO_3)_2-KNO_3$ 1049
$Al_2O_3-Sm_2O_3$ 312	B_2O_3-MnO 276	$Ba(NO_3)_2-KNO_3-NaNO_3$ 1073
$Al_2O_3-SnO_2$ 315	$B_2O_3-NaF-Na_2O$ 1688	$Ba(NO_3)_2-LiNO_3$ 1051
Al_2O_3-SrO 294–295	$B_2O_3-Na_2O$ 188	$Ba(NO_3)_2-RbNO_3$ 1056
$Al_2O_3-Ta_2O_5$ 319	$B_2O_3-Na_2O-SiO_2$ 515–518	$Ba(NO_3)_2-Sr(NO_3)_2$ 1061
$Al_2O_3-TiO_2$ 316	$B_2O_3-P_2O_5-SiO_2$ 775–777	$BaO-Al_2O_3$ 206
$Al_2O_3-TiO_2-ZrO_2$ 773–774	B_2O_3-PbO 281	$BaO-Al_2O_3-H_2O$ 1976
$Al_2O_3-Tm_2O_3$ 312	$B_2O_3-PbO-SiO_2$ 740–742	$BaO-Al_2O_3-SiO_2$ 556–557
$Al_2O_3-UO_2$ 317	$B_2O_3-PbO-TiO_2$ 743	$BaO-B_2O_3$ 207
$Al_2O_3-V_2O_5$ 320	$B_2O_3-SiO_2-ZnO$ 759	$BaO-B_2O_3-PbO-TiO_2$ 859
$Al_2O_3-Y_2O_3$ 311–312	$B_2O_3-ThO_2$ 322	$BaO-B_2O_3-SiO_2$ 558–561
$Al_2O_3-Yb_2O_3$ 312	B_2O_3-ZnO 300–301	$BaO-BaCl_2-TiO_2$ 1846
Al_2O_3-ZnO 299	$BaBr_2-KBr$ 1183	$BaO-BaF_2-SiO_2$ 1714
$As-Fe-S$ 1885	$BaBr_2-LiBr$ 1190	$BaO-BeO$ 204

BaO–BeO–La$_2$O$_3$..................548
BaO–Bi$_2$O$_3$....................326
BaO–CaO–MgO–SiO$_2$...........857
BaO–CaO–SiO$_2$...........549–551
BaO–CaO–SrO–TiO$_2$............858
BaO–CaO–TiO$_2$..........552–553
BaO–Fe$_2$O$_3$....................208
BaO–Fe$_2$O$_3$–Gd$_2$O$_3$.............562
BaO–Fe$_2$O$_3$–Na$_2$O.............477
BaO–Gd$_2$O$_3$–Nb$_2$O$_5$.............563
BaO–H$_2$O–P$_2$O$_5$...............1977
BaO–In$_2$O$_3$–La$_2$O$_3$–SrO–TiO$_2$–Y$_2$O$_3$.863
BaO–La$_2$O$_3$....................209
BaO–Li$_2$O–SiO$_2$...........437,441
BaO–MgO......................274
BaO–Na$_2$O–Nb$_2$O$_5$–TiO$_2$.......826
BaO–Na$_2$O–SiO$_2$...........478–479
BaO–Nb$_2$O$_5$....................214
BaO–NiO......................205
BaO–P$_2$O$_5$–TiO$_2$...............568
BaO–PbO–SnO$_2$–TiO$_2$.......860–861
BaO–PbO–TiO$_2$–ZrO$_2$...........862
BaO–SiO$_2$................210–211
BaO–SiO$_2$–TiO$_2$...............564
BaO–SnO$_2$....................212
BaO–SnO$_2$–TiO$_2$...............565
BaO–SrO–TiO$_2$...........554–555
BaO–TiO$_2$....................213
BaO–TiO$_2$–ZrO$_2$...........566–567
Ba(OH)$_2$–Sr(OH)$_2$...............1030
BaSO$_4$–K$_2$SO$_4$...................1094
BaSO$_4$–Na$_2$SO$_4$...................1112
BaSiO$_3$–BaF$_2$–LiF–Li$_2$SiO$_3$.......1724
BaTiO$_3$–BaCO$_3$–BaCl$_2$...........1847
BeCl$_2$–AgCl....................1220
BeCl$_2$–BaCl$_2$....................1349
BeCl$_2$–BeF$_2$....................1622
BeCl$_2$–CdCl$_2$....................1357
BeCl$_2$–LiCl....................1290
BeCl$_2$–NaCl....................1302
BeCl$_2$–PbCl$_2$....................1358
BeCl$_2$–TlCl....................1339
BeF$_2$–BaF$_2$....................1510
BeF$_2$–BeCl$_2$....................1622
BeF$_2$–CaF$_2$....................1513
BeF$_2$–CsF....................1446
BeF$_2$–KF...............1456–57
BeF$_2$–KF–NaF...............1534
BeF$_2$–LiF....................1469
BeF$_2$–LiF–Li$_2$SO$_4$...............1783
BeF$_2$–LiF–NaF...........1542–43
BeF$_2$–LiF–RbF...............1549
BeF$_2$–LiF–ThF$_4$...............1553
BeF$_2$–LiF–UF$_4$...............1554
BeF$_2$–MgF$_2$....................1514
BeF$_2$–NaF....................1482
BeF$_2$–NaF–RbF...........1558–59
BeF$_2$–NaF–ThF$_4$...............1562
BeF$_2$–NaF–UF$_4$...............1563
BeF$_2$–NaF–ZrF$_4$...............1564
BeF$_2$–PbF$_2$....................1515
BeF$_2$–RbF...............1500–01
BeF$_2$–SrF$_2$....................1516
BeF$_2$–ThF$_4$....................1517
BeF$_2$–ThF$_4$–UF$_4$...............1574
BeF$_2$–UF$_4$....................1518
BeO–Al$_2$O$_3$....................218
BeO–Al$_2$O$_3$–SiO$_2$...............576
BeO–Al$_2$O$_3$–ThO$_2$...............577
BeO–Al$_2$O$_3$–TiO$_2$...............578
BeO–Al$_2$O$_3$–ZrO$_2$...............579
BeO–BaO....................204

BeO–BaO–La$_2$O$_3$..................548
BeO–CaO....................215
BeO–CaO–La$_2$O$_3$...............570
BeO–CaO–SrO...............569
BeO–CaO–ZrO$_2$...............571
BeO–CeO$_2$–ZrO$_2$...........581–582
BeO–Cr$_2$O$_3$–ZrO$_2$...............580
BeO–La$_2$O$_3$....................219
BeO–La$_2$O$_3$–SrO...............575
BeO–MgO................216, 274
BeO–MgO–Al$_2$O$_3$...............572
BeO–MgO–ThO$_2$...............573
BeO–MgO–ZrO$_2$...............574
BeO–SiO$_2$....................222
BeO–SrO....................217
BeO–ThO$_2$....................223
BeO–TiO$_2$....................224
BeO–TiO$_2$–ZrO$_2$...........583–585
BeO–UO$_2$................225–226
BeO–Y$_2$O$_3$....................220
BeO–Yb$_2$O$_3$....................221
BeO–ZrO$_2$....................227
BeSO$_4$–K$_2$SO$_4$...................1095
Bi–BiCl$_3$....................1658
BiBr$_3$–PbBr$_2$....................1202
BiCl–AgCl....................1212
BiCl$_3$–Bi....................1658
BiCl$_3$–CuCl....................1255
BiCl$_3$–FeCl$_3$....................1394
BiCl$_3$–PbCl$_2$....................1385
BiCl$_3$–TlCl....................1347
BiCl$_3$–ZnCl$_2$....................1389
Bi$_2$O$_3$–Al$_2$O$_3$....................327
Bi$_2$O$_3$–B$_2$O$_3$................323,327
Bi$_2$O$_3$–BaO....................326
Bi$_2$O$_3$–CaO....................326
Bi$_2$O$_3$–CdO....................326
Bi$_2$O$_3$–CeO$_2$....................328
Bi$_2$O$_3$–CrO$_3$....................330
Bi$_2$O$_3$–Fe$_2$O$_3$....................327
Bi$_2$O$_3$–Ga$_2$O$_3$....................327
Bi$_2$O$_3$–GeO$_2$....................328
Bi$_2$O$_3$–La$_2$O$_3$....................327
Bi$_2$O$_3$–Li$_2$O....................325
Bi$_2$O$_3$–Lu$_2$O$_3$....................327
Bi$_2$O$_3$–MgO....................326
Bi$_2$O$_3$–Mn$_2$O$_3$....................327
Bi$_2$O$_3$–MoO$_3$....................330
Bi$_2$O$_3$–MoO$_3$–PbO..............744
Bi$_2$O$_3$–Nb$_2$O$_5$................324, 329
Bi$_2$O$_3$–NiO....................326
Bi$_2$O$_3$–P$_2$O$_5$....................329
Bi$_2$O$_3$–PbO....................326
Bi$_2$O$_3$–PbO–WO$_3$..............745
Bi$_2$O$_3$–RO....................326
Bi$_2$O$_3$–RO$_2$....................328
Bi$_2$O$_3$–RO$_3$....................330
Bi$_2$O$_3$–R$_2$O....................325
Bi$_2$O$_3$–R$_2$O$_3$....................327
Bi$_2$O$_3$–R$_2$O$_5$....................329
Bi$_2$O$_3$–Rb$_2$O....................325
Bi$_2$O$_3$–Sb$_2$O$_3$....................327
Bi$_2$O$_3$–SiO$_2$....................328
Bi$_2$O$_3$–Sm$_2$O$_3$....................327
Bi$_2$O$_3$–SnO$_2$....................328
Bi$_2$O$_3$–SrO....................326
Bi$_2$O$_3$–Ta$_2$O$_5$....................329
Bi$_2$O$_3$–TeO$_2$....................328
Bi$_2$O$_3$–TiO$_2$....................328
Bi$_2$O$_3$–V$_2$O$_5$....................329
Bi$_2$O$_3$–WO$_3$....................330
Bi$_2$O$_3$–ZnO....................326

Bi$_2$O$_3$–ZrO$_2$....................328
BiS–Sb$_2$S$_3$....................1894
Br$_2$–UF$_6$....................1592

C–Ti–O......................36
CCl$_4$–PbCl$_4$....................1396
CO$_2$–Al$_2$O$_3$–H$_2$O–MgO..........2037
CO$_2$–CaO................1003–05
CO$_2$–CaO–H$_2$O...........1984–88
CO$_2$–CaO–H$_2$O–MgO.........2032–33
CO$_2$–CaO–H$_2$O–MgO–SO$_3$.......2066
CO$_2$–CaO–MgO–SiO$_2$...........929
CO$_2$–FeO–H$_2$O...............1996
CO$_2$–H$_2$O....................1941
CO$_2$–H$_2$O–K$_2$O–SiO$_2$..........2020
CO$_2$–H$_2$O–Li$_2$O–UO$_3$..........2021
CO$_2$–H$_2$O–MgO...............1999
CO$_2$–H$_2$O–Na$_2$O...............1958
CO$_2$–H$_2$O–Na$_2$O–ThO$_2$.........2030
CO$_2$–H$_2$O–Na$_2$O–UO$_3$.........2031
CO$_2$–MgO....................1006
Ca–Cr–O....................37–39
Ca–Fe–O....................40–44
CaBr$_2$–KBr....................1184
CaBr$_2$–NaBr....................1193
CaCO$_3$................1003–05
CaCO$_3$–K$_2$CO$_3$...............1009
CaCO$_3$–K$_2$CO$_3$–Na$_2$CO$_3$.......1016
CaCO$_3$–Li$_2$CO$_3$...............1011
CaCO$_3$–MgCO$_3$...........1012–13
CaCO$_3$–MnCO$_3$...............1014
CaCl$_2$–AgCl....................1221
CaCl$_2$–BaCl$_2$....................1350
CaCl$_2$–BaCl$_2$–BaF$_2$–CaF$_2$.......1656
CaCl$_2$–BaCl$_2$–Ba(NO$_3$)$_2$–Ca(NO$_3$)$_2$.....
.........................1765
CaCl$_2$–CaF$_2$....................1623
CaCl$_2$–CaF$_2$–LiCl–LiF..........1648
CaCl$_2$–CaF$_2$–NaCl–NaF.........1651
CaCl$_2$–Ca(NO$_3$)$_2$–LiCl–LiNO$_3$....1759
CaCl$_2$–Ca(NO$_3$)$_2$–NaCl–NaNO$_3$...1763
CaCl$_2$–CaO–H$_2$O...............2061
CaCl$_2$–CaSO$_4$–KCl–K$_2$SO$_4$.......1790
CaCl$_2$–CaSO$_4$–LiCl–Li$_2$SO$_4$.......1797
CaCl$_2$–CdCl$_2$....................1359
CaCl$_2$–CsCl–KCl...............1404
CaCl$_2$–CsCl–NaCl...............1405
CaCl$_2$–CuCl....................1249
CaCl$_2$–KCl....................1263
CaCl$_2$–KCl–LiCl...............1408
CaCl$_2$–KCl–MgCl$_2$...........1419–20
CaCl$_2$–KCl–NaCl...............1410
CaCl$_2$–KCl–PbCl$_2$...............1421
CaCl$_2$–KCl–RbCl...............1416
CaCl$_2$–LaCl$_3$–NaCl...............1429
CaCl$_2$–LiCl....................1291
CaCl$_2$–LiCl–NaCl...............1426
CaCl$_2$–MgCl$_2$....................1360
CaCl$_2$–MnCl$_2$....................1361
CaCl$_2$–NaCl....................1303
CaCl$_2$–NaCl–NdCl$_3$...............1430
CaCl$_2$–NaCl–RbCl...............1427
CaCl$_2$–PbCl$_2$....................1362
CaCl$_2$–SnCl$_2$....................1363
CaCl$_2$–SrCl$_2$....................1364
CaCl$_2$–TlCl....................1340
CaCl$_2$–ZnCl$_2$....................1365
CaF$_2$–AlF$_3$–Al$_2$O$_3$–NaF.........1727
CaF$_2$–AlF$_3$–NaF...........1566–67
CaF$_2$–Al$_2$O$_3$....................1673
CaF$_2$–Al$_2$O$_3$–CaO...............1707
CaF$_2$–BaCl$_2$–BaF$_2$–CaCl$_2$.......1656

CaF_2–$BaCl_2$–KCl..............1637
CaF_2–$BaCl_2$–NaF....1639
CaF_2–BaF_2–LiF................1551
CaF_2–BeF_2......................1513
CaF_2–$CaCl_2$....................1623
CaF_2–$CaCl_2$–LiCl–LiF........1648
CaF_2–$CaCl_2$–NaCl–NaF........1651
CaF_2–CaI_2.....................1631
CaF_2–CaO.......................1671
CaF_2–CaO–FeO..................1706
CaF_2–CaO–SiF_4–SiO_2........1729
CaF_2–CaO–SiO_2........1708–09,1714
CaF_2–$CaSiO_3$–LiF–Li_2SiO_3......1725
CaF_2–KF–SiO_2.................1681
CaF_2–LiF.......................1470
CaF_2–LiF–MgF_2................1555
CaF_2–LiF–NaF..................1544
CaF_2–MgF_2....................1519
CaF_2–MgF_2–NaF...............1565
CaF_2–MgO......................1672
CaF_2–MgO–SiO_2...............1710
CaF_2–NaF......................1483
CaF_2–NaF–SiO_2...............1695
CaF_2–Na_2O–SiO_2.............1705
CaF_2–RbF......................1502
CaF_2–YF_3.....................1520
$Ca(HCO_3)_2$–$CaSO_4$–H_2O–$NaHCO_3$–
  Na_2SO_4......................2064
CaI_2–CaF_2.....................1631
$Ca(NO_3)_2$–$AgNO_3$..............1043
$Ca(NO_3)_2$–$BaCl_2$–$Ba(NO_3)_2$–$CaCl_2$.....
 1765
$Ca(NO_3)_2$–$Ba(NO_3)_2$............1060
$Ca(NO_3)_2$–$CaCl_2$–LiCl–$LiNO_3$.....1759
$Ca(NO_3)_2$–$CaCl_2$–NaCl–$NaNO_3$....1763
$Ca(NO_3)_2$–$CsNO_3$...............1045
$Ca(NO_3)_2$–$CsNO_3$–$RbNO_3$........1067
$Ca(NO_3)_2$–KNO_2................1079
$Ca(NO_3)_2$–KNO_3................1050
$Ca(NO_3)_2$–KNO_3–$LiNO_3$.........1070
$Ca(NO_3)_2$–KNO_3–$NaNO_3$.......1074
$Ca(NO_3)_2$–$LiNO_3$...............1052
$Ca(NO_3)_2$–$NaNO_2$...............1082
$Ca(NO_3)_2$–$NaNO_3$...............1050
$Ca(NO_3)_2$–$RbNO_3$...............1057
$Ca(NO_3)_2$–$Sr(NO_3)_2$...........1062
$Ca(NO_3)_2$–$TlNO_3$...............1059
CaO–AlF_3–NaF....................1700
CaO–Al_2O_3.................231–233
CaO–Al_2O_3–CaF_2................1707
CaO–Al_2O_3–Cr_2O_3...............626
CaO–Al_2O_3–Cr_2O_3–MgO–SiO_2..987–990
CaO–Al_2O_3–FeO–SiO_2........866–875
CaO–Al_2O_3–Fe_2O_3..........627–629
CaO–Al_2O_3–Fe_2O_3–K_2O..........794
CaO–Al_2O_3–Fe_2O_3–MgO......876–879
CaO–Al_2O_3–Fe_2O_3–MgO–SiO_2........
 991–1002
CaO–Al_2O_3–Fe_2O_3–Na_2O,........827
CaO–Al_2O_3–Fe_2O_3–SiO_2..........
 943–953
CaO–Al_2O_3–H_2O.............1978–83
CaO–Al_2O_3–H_2O–SO_3...........2063
CaO–Al_2O_3–H_2O–SiO_2..........2035
CaO–Al_2O_3–K_2O.................390
CaO–Al_2O_3–K_2O–MgO–SiO_2...973–974
CaO–Al_2O_3–K_2O–Na_2O–SiO_2...971–972
CaO–Al_2O_3–K_2O–SiO_2......795–800
CaO–Al_2O_3–MgO..................596
CaO–Al_2O_3–MgO–Na_2O–SiO_2.........
 975–981
CaO–Al_2O_3–MgO–SiO_2.......880–923

CaO–Al_2O_3–Na_2O................480
CaO–Al_2O_3–Na_2O–P_2O_5–SiO_2.....985
CaO–Al_2O_3–Na_2O–SiO_2......828–851
CaO–Al_2O_3–Na_2O–SiO_2–TiO_2........
 982–984
CaO–Al_2O_3–P_2O_5...........640–644
CaO–Al_2O_3–P_2O_5–SiO_2......958–962
CaO–Al_2O_3–SiO_2...........630–639
CaO–Al_2O_3–SiO_2–TiO_2.......954–957
CaO–B_2O_3.......................234
CaO–B_2O_3–Na_2O–SiO_2...........852
CaO–B_2O_3–SiO_2.............646–649
CaO–BaO–MgO–SiO_2................857
CaO–BaO–SiO_2...............549–551
CaO–BaO–SrO–TiO_2................858
CaO–BaO–TiO_2...............552–553
CaO–BeO...........................215
CaO–BeO–La_2O_3...................570
CaO–BeO–SrO.......................569
CaO–BeO–ZrO_2....................571
CaO–Bi_2O_3......................326
CaO–CO_2....................1003–05
CaO–CO_2–H_2O..............1984–88
CaO–CO_2–H_2O–MgO..........2032–33
CaO–CO_2–H_2O–MgO–SO_3........2066
CaO–CO_2–MgO–SiO_2..............929
CaO–$CaCl_2$–H_2O................2061
CaO–CaF_2........................1671
CaO–CaF_2–FeO...................1706
CaO–CaF_2–SiF_4–SiO_2..........1729
CaO–CaF_2–SiO_2........1708–09,1714
CaO–CoO...........................228
CaO–Cr–Cr_2O_3.....................37
CaO–CrO_3–Cr_2O_3.................38
CaO–Cr_2O_3........................39
CaO–Cr_2O_3–Fe_2O_3...............650
CaO–Cr_2O_3–MgO...................597
CaO–Cr_2O_3–MgO–SiO_2........924–926
CaO–Cr_2O_3–SiO_2............651–655
CaO–Fe–Fe_2O_3..................40–41
CaO–FeO............................42
CaO–FeO–Fe_2O_3................45–46
CaO–FeO–MgO–SiO_2...........864–865
CaO–FeO–P_2O_5....................595
CaO–FeO–SiO_2...............586–594
CaO–Fe_2O_3...................43–44
CaO–Fe_2O_3–Fe_3O_4.............47–48
CaO–Fe_2O_3–MgO–SiO_2.......927–928
CaO–Fe_2O_3–SiO_2............656–658
CaO–Ga_2O_3......................235
CaO–H_2O....................1919–21
CaO–H_2O–SiO_2.............1989–95
CaO–H_2O–SiO_2–SrO.............2034
CaO–HfO_2........................236
CaO–K_2O–SiO_2.............391–396
CaO–La_2O_3–ZrO_2...............659
CaO–Li_2O–SiO_2............437, 442
CaO–MgO...........................229
CaO–MgO–MnO–SiO_2–TiO_2........986
CaO–MgO–P_2O_5....................613
CaO–MgO–P_2O_5–SiO_2........935–938
CaO–MgO–SiO_2...............598–610
CaO–MgO–SiO_2–TiO_2........930–934
CaO–MgO–SnO_2....................611
CaO–MgO–SnO_2–TiO_2.........939–940
CaO–MgO–TiO_2....................612
CaO–MnO–SiO_2...............614–619
CaO–MnO–SiO_2–TiO_2.............941
CaO–MnO–SiO_2–WO_3..............942
CaO–Na_2O–P_2O_5.................486
CaO–Na_2O–SO_3–SiO_2........1141–42
CaO–Na_2O–SiO_2............481–485

CaO–Nb_2O_5......................244
CaO–NiO...........................230
CaO–P_2O_5..................245–250
CaO–P_2O_5–RO....................645
CaO–P_2O_5–SiO_2............665–671
CaO–P_2O_5–SrO...................623
CaO–SiO_2..................237–238
CaO–SiO_2–SrO..............620–621
CaO–SiO_2–TiO_2............660–663
CaO–SiO_2–ZnO..............624–625
CaO–SiO_2–ZrO_2................664
CaO–SnO_2–TiO_2...........673–676
CaO–SrO–TiO_2....................622
CaO–ThO_2–ZrO_2................677
CaO–TiO_2..................239–240
CaO–TiO_2–ZrO_2................678
CaO–UO_2........................241
CaO–V_2O_5.......................251
CaO–ZrO_2..................242–243
$CaSO_4$–$CaCl_2$–KCl–K_2SO_4........1790
$CaSO_4$–$CaCl_2$–LiCl–Li_2SO_4......1797
$CaSO_4$–$Ca(HCO_3)_2$–H_2O–$NaHCO_3$–
  Na_2SO_4......................2064
$CaSO_4$–H_2O...................2051
$CaSO_4$–KCl–NaCl.................1779
$CaSO_4$–K_2SO_4..................1096
$CaSO_4$–Li_2SO_4.................1105
$CaSO_4$–Na_2SO_4.................1113
$CaSO_4$–Rb_2SO_4.................1124
$CaSiO_3$–CaF_2–LiF–Li_2SiO_3.....1725
Ca_2SiO_4........................672
Cd–$CdCl_2$–Zn–$ZnCl_2$............1669
$CdBr_2$–AgBr.....................1179
$CdBr_2$–$CdCl_2$–$PbBr_2$–$PbCl_2$....1644
$CdBr_2$–CuBr.....................1181
$CdBr_2$–KBr......................1185
$CdBr_2$–KBr–NaBr.............1207–08
$CdBr_2$–NaBr.....................1194
$CdBr_2$–NaBr–$PbBr_2$.............1209
$CdBr_2$–$PbBr_2$..................1199
$CdBr_2$–$ZnBr_2$..................1200
$CdCl_2$–$BaCl_2$..................1351
$CdCl_2$–$BeCl_2$..................1357
$CdCl_2$–$CaCl_2$..................1359
$CdCl_2$–Cd–Zn–$ZnCl_2$............1669
$CdCl_2$–$CdBr_2$–$PbBr_2$–$PbCl_2$....1644
$CdCl_2$–CdF_2...................1626
$CdCl_2$–CdI_2–NaCl–NaI..........1652
$CdCl_2$–CdI_2–$PbCl_2$–PbI_2......1657
$CdCl_2$–$CdMoO_4$–LiCl–Li_2MoO_4...1861
$CdCl_2$–$CdSO_4$–KCl–K_2SO_4.......1791
$CdCl_2$–$CdSO_4$–LiCl–Li_2SO_4.....1798
$CdCl_2$–$CdSO_4$–NaCl–Na_2SO_4.....1802
$CdCl_2$–$CdSO_4$–TlCl–Tl_2SO_4.....1805
$CdCl_2$–CuCl.....................1250
$CdCl_2$–KCl......................1264
$CdCl_2$–KCl–NaCl.............1411–12
$CdCl_2$–KCl–$PbCl_2$..............1422
$CdCl_2$–LiF......................1624
$CdCl_2$–$MgCl_2$..................1366
$CdCl_2$–$MnCl_2$..................1367
$CdCl_2$–NH_4Cl..................1324
$CdCl_2$–NaCl.....................1304
$CdCl_2$–NaF......................1625
$CdCl_2$–$PbCl_2$..................1368
$CdCl_2$–$SnCl_2$..................1369
$CdCl_2$–$SrCl_2$..................1370
$CdCl_2$–TlCl.....................1341
$CdCl_2$–$ZnCl_2$..................1371
CdF_2–$CdCl_2$...................1626
CdF_2–NaF.......................1484
CdI_2–$CdCl_2$–NaCl–NaI..........1652

CdI_2–$CdCl_2$–$PbCl_2$–PbI_2 1657	$CoSO_4$–$CoCl_2$–$LiCl$–Li_2SO_4 1799	Cr_2O_3–Sm_2O_3 331
CdI_2–CuI 1582	$CoSO_4$–$CoCl_2$–$NaCl$–Na_2SO_4 .. 1803	Cr_2O_3–Tm_2O_3 331
CdI_2–HgI_2 1586	$CoSO_4$–K_2SO_4 1098	Cr_2O_3–V_2O_5 333
CdI_2–NaI–PbI_2 1590	$CoSO_4$–Li_2SO_4 1107	Cr_2O_3–Y_2O_3 331
$CdMoO_4$–$CdCl_2$–$LiCl$–Li_2MoO_4 ... 1861	$CoSO_4$–$NaCl$ 1774	Cr_2O_3–Yb_2O_3 331
$Cd(NO_3)_2$–$AgNO_3$–KNO_3 1064	$CoSO_4$–Na_2SO_4 1115	Cs–CsF 1665
$Cd(NO_3)_2$–$CsNO_3$–$TlNO_3$ 1068	$CoSO_4$–Na_2SO_4–$NiSO_4$ 1136	$CsCl$ 1210–11
$Cd(NO_3)_2$–KNO_3–$LiNO_3$ 1071	Cr–Al–Fe–O 126–130	$CsCl$–$AgCl$ 1213
$Cd(NO_3)_2$–KNO_3–$RbNO_3$ 1075	Cr–Ca–O 37–39	$CsCl$–$AlCl_3$ 1234
$Cd(NO_3)_2$–$LiNO_3$–$NaNO_3$ 1076	Cr–CaO–Cr_2O_3 37	$CsCl$–$AlCl_3$–$TaCl_5$ 1406
CdO–AlF_3–NaF 1700	Cr–Cr_2O_3 6	$CsCl$–$CaCl_2$–KCl 1404
CdO–B_2O_3 252	Cr–Fe–Mg–O 137	$CsCl$–$CaCl_2$–$NaCl$ 1405
CdO–B_2O_3–ZnO 679	Cr–Fe–O 57–58	$CsCl$–$CeCl_3$ 1235
CdO–Bi_2O_3 326	Cr–O 5–6	$CsCl$–$CsNO_3$–$RbCl$–$RbNO_3$ 1754
CdO–H_2O 1922	CrO_3–B_2O_3–K_2O 418	$CsCl$–$CuCl$ 1225
CdO–Nb_2O_5 253	CrO_3–B_2O_3–K_2O–Na_2O 790–791	$CsCl$–$HfCl_4$ 1239
$CdSO_4$–$CdCl_2$–KCl–K_2SO_4 1791	CrO_3–Bi_2O_3 330	$CsCl$–KCl 1226
$CdSO_4$–$CdCl_2$–$LiCl$–Li_2SO_4 ... 1798	CrO_3–CaO–Cr_2O_3 38	$CsCl$–$LaCl_3$ 1236
$CdSO_4$–$CdCl_2$–$NaCl$–Na_2SO_4 .. 1802	CrO_3–CsF–Cs_2O 1675	$CsCl$–$LiCl$ 1227
$CdSO_4$–$CdCl_2$–$TlCl$–Tl_2SO_4 ... 1805	CrO_3–H_2O–ThO_2 2017	$CsCl$–$MgCl_2$ 1231
$CdSO_4$–K_2SO_4 1097	CrO_3–H_2O–ZnO 2007	$CsCl$–$NaCl$ 1228
$CdSO_4$–K_2SO_4–Li_2SO_4 1128	CrO_3–KF–K_2O 1678	$CsCl$–$NbCl_5$ 1328
$CdSO_4$–K_2SO_4–Na_2SO_4 1129	CrO_3–K_2O 175	$CsCl$–$NdCl_3$ 1237
$CdSO_4$–Li_2SO_4 1106	CrO_3–K_2O–MoO_3 425	$CsCl$–$RbCl$ 1229
$CdSO_4$–Li_2SO_4–Na_2SO_4 1135	CrO_3–K_2O–Na_2O 387	$CsCl$–$TaCl_5$ 1242, 1330
$CdSO_4$–Na_2SO_4 1114	CrO_3–K_2O–P_2O_5 428	$CsCl$–$TiCl_3$ 1238
Ce–$CeCl_3$ 1659	CrO_3–K_2O–TiO_2 546	$CsCl$–$TlCl$ 1230
Ce–O 3	CrO_3–K_2O–WO_3 426	$CsCl$–UCl_4 1240
$CeCl_3$–Ce 1659	CrO_3–MoO_3–PbO 751	$CsCl$–VCl_2 1232
$CeCl_3$–$CsCl$ 1235	CrO_3–NaF–Na_2O 1691	$CsCl$–$ZnCl_2$ 1233
$CeCl_3$–KCl 1275	CrO_3–Na_2O–P_2O_5 544	$CsCl$–$ZrCl_4$ 1241
$CeCl_3$–KCl–$MgCl_2$ 1423	CrO_3–Na_2O–TiO_2 546	CsF–AlF_3 1448
$CeCl_3$–$MgCl_2$ 1377	CrO_3–PbO 291	CsF–BeF_2 1446
CeF_3–KF 1463	CrO_3–PbO–WO_3 752	CsF–CrO_3–Cs_2O 1675
CeF_3–LiF 1474	CrO_3–RbF–Rb_2O 1701	CsF–Cs 1665
CeF_3–NaF 1492	Cr_2O_3–Al_2O_3 309	CsF–Cs_2O–MoO_3 1676
CeF_3–NaF–ZrF_4 1570	Cr_2O_3–Al_2O_3–CaO 626	CsF–Cs_2O–WO_3 1677
CeO_2–Al_2O_3 356	Cr_2O_3–Al_2O_3–CaO–MgO–SiO_2 .. 987–990	CsF–LiF 1445
CeO_2–BeO–ZrO_2 581–582	Cr_2O_3–Al_2O_3–FeO–Fe_2O_3 130	CsF–UF_4 1449
CeO_2–Bi_2O_3 328	Cr_2O_3–Al_2O_3–Fe_2O_3 126–129	CsF–ZnF_2 1447
CeO_2–Cr_2O_3 356	Cr_2O_3–Al_2O_3–MgO 710	CsF–ZrF_4 1450
CeO_2–Fe_3O_4 356	Cr_2O_3–BeO–ZrO_2 580	$CsNO_3$–$Ba(NO_3)_2$ 1044
CeO_2–Mn_3O_4 356	Cr_2O_3–CaO 39	$CsNO_3$–$Ca(NO_3)_2$ 1045
CeO_2–ZrO_2 355	Cr_2O_3–CaO–Cr 37	$CsNO_3$–$Ca(NO_3)_2$–$RbNO_3$ 1067
ClF_3–UF_6 1525	Cr_2O_3–CaO–CrO_3 38	$CsNO_3$–$Cd(NO_3)_2$–$TlNO_3$ 1068
Co–CoS–Fe–FeS 1886	Cr_2O_3–CaO–Fe_2O_3 650	$CsNO_3$–$CsCl$–$RbCl$–$RbNO_3$ 1754
Co–Fe–O 49–50	Cr_2O_3–CaO–MgO 597	$CsNO_3$–$Sr(NO_3)_2$ 1046
Co–Mg–O 51–52	Cr_2O_3–CaO–MgO–SiO_2 924–926	Cs_2O–B_2O_3 160
Co–Ni–O 53	Cr_2O_3–CaO–SiO_2 651–655	Cs_2O–CrO_3–CsF 1675
Co–O 4	Cr_2O_3–CeO_2 356	Cs_2O–CsF–MoO_3 1676
Co–O–Zn 56	Cr_2O_3–Cr 6	Cs_2O–CsF–WO_3 1677
Co–V–O 54–55	Cr_2O_3–Dy_2O_3 331	Cs_2O–MoO_3 162
$CoCl_2$–$CoSO_4$–KCl–K_2SO_4 1792	Cr_2O_3–Er_2O_3 331	Cs_2O–TiO_2 161
$CoCl_2$–$CoSO_4$–$LiCl$–Li_2SO_4 .. 1799	Cr_2O_3–Eu_2O_3 331	Cs_2SO_4–$MgSO_4$ 1091
$CoCl_2$–$CoSO_4$–$NaCl$–Na_2SO_4 .. 1803	Cr_2O_3–FeO–Fe_2O_3–MgO 137	Cu–Ag–O–Pb 125
$CoCl_2$–$NaCl$ 1305	Cr_2O_3–Fe_2O_3 58	Cu–Cu_2O 7
CoO–B_2O_3 254	Cr_2O_3–Fe_2O_3–TiO_2 778	Cu–Fe–O 59
CoO–CaO 228	Cr_2O_3–Fe_3O_4–SiO_2 760–762	Cu–O 7
CoO–CoS 1902	Cr_2O_3–Ga_2O_3 340	$CuBr$–$CdBr_2$ 1181
CoO–Fe_2O_3 50	Cr_2O_3–Gd_2O_3 331	$CuBr$–$CuCl$ 1595
CoO–FeS 1902	Cr_2O_3–H_2O 1932	$CuCN$–KCN 1874
CoO–MgO 51–52	Cr_2O_3–H_2O–P_2O_5 2013–14	$CuCN$–$NaCN$ 1875
CoO–NiO 53	Cr_2O_3–Ho_2O_3 331	$CuCl$–$AgCl$ 1214
CoO–SiO_2 255	Cr_2O_3–In_2O_3 331	$CuCl$–$BiCl_3$ 1255
CoO–TiO_2 256	Cr_2O_3–La_2O_3 331	$CuCl$–$CaCl_2$ 1249
CoO–V_2O_5 55	Cr_2O_3–Lu_2O_3 331	$CuCl$–$CdCl_2$ 1250
CoO–ZnO 56	Cr_2O_3–MgO 262	$CuCl$–$CsCl$ 1225
CoS–Co–Fe–FeS 1886	Cr_2O_3–MgO–SiO_2 715	$CuCl$–$CuBr$ 1595
CoS–CoO 1902	Cr_2O_3–Nd_2O_3 331	$CuCl$–CuI 1613
CoS–FeS 1895	Cr_2O_3–R_2O_3 331	$CuCl$–$FeCl_3$ 1256
Co_4S_3–FeS–Ni_3S_2 1900–01	Cr_2O_3–Sc_2O_3 331	$CuCl$–KCl 1243
$CoSO_4$–$CoCl_2$–KCl–K_2SO_4 1792	Cr_2O_3–SiO_2 332	$CuCl$–$LiCl$ 1244

CuCl–MgCl$_2$ 1251	Fe–Cr–O 57–58	FeO–TiO$_2$ 88
CuCl–NH$_4$Cl 1246	Fe–Cu–O 59	FeO–ZrO$_2$ 257
CuCl–NaCl 1245	Fe–Fe$_2$O$_3$ 8	Fe$_2$O$_3$–Al$_2$O$_3$ 27
CuCl–PbCl$_2$ 1252	Fe–La–O 60–61	Fe$_2$O$_3$–Al$_2$O$_3$–CaO 627–629
CuCl–RbCl 1247	Fe–Li–O 62	Fe$_2$O$_3$–Al$_2$O$_3$–CaO–K$_2$O 794
CuCl–SnCl$_2$ 1253	Fe–Mg–O 63–69	Fe$_2$O$_3$–Al$_2$O$_3$–CaO–MgO 876–879
CuCl–TlCl 1248	Fe–Mg–O–Si 138–145	Fe$_2$O$_3$–Al$_2$O$_3$–CaO–MgO–SiO$_2$ .991–1002
CuCl–ZnCl$_2$ 1254	Fe–Mn–O 70–74	Fe$_2$O$_3$–Al$_2$O$_3$–CaO–Na$_2$O 827
CuI–AgI 1576	Fe–Ni–O 75–78	Fe$_2$O$_3$–Al$_2$O$_3$–CaO–SiO$_2$ 943–953
CuI–CdI$_2$ 1582	Fe–O 8–10	Fe$_2$O$_3$–Al$_2$O$_3$–Cr$_2$O$_3$ 126–129
CuI–CuCl 1613	Fe–O–Sc 79	Fe$_2$O$_3$–Al$_2$O$_3$–Cr$_2$O$_3$–FeO 130
Cu$_2$O–Cu 7	Fe–O–Si 80–87	Fe$_2$O$_3$–Al$_2$O$_3$–FeO 29–30, 32
Cu$_2$O–MgO 274	Fe–O–Si–Ti 146–147	Fe$_2$O$_3$–Al$_2$O$_3$–FeO–SiO$_2$ 131–136
Cu$_2$O–PbO 163	Fe–O–Ti 88–93	Fe$_2$O$_3$–Al$_2$O$_3$–Li$_2$O 446–448
Cu$_2$O–SiO$_2$ 164	Fe–O–V 94–96	Fe$_2$O$_3$–Al$_2$O$_3$–MgO 711
Cu$_2$S–FeO–FeS 1904	Fe–O–Y 97–98	Fe$_2$O$_3$–Al$_2$O$_3$–SiO$_2$ 766–767
Cu$_2$S–Ni$_2$S 1890	Fe–S 1878–82	Fe$_2$O$_3$–As$_2$O$_5$–H$_2$O 2015
Cu$_2$S–Ni$_3$S$_2$ 1891	FeCl$_2$–KCl 1265	Fe$_2$O$_3$–BaO 208
CuSO$_4$–K$_2$SO$_4$ 1099	FeCl$_2$–LiCl 1292	Fe$_2$O$_3$–BaO–Gd$_2$O$_3$ 562
CuSO$_4$–K$_2$SO$_4$–Na$_2$SO$_4$ 1130–33	FeCl$_3$–AlCl$_3$–NbCl$_5$ 1436	Fe$_2$O$_3$–BaO–Na$_2$O 477
CuSO$_4$–Na$_2$SO$_4$ 1116	FeCl$_3$–AlCl$_3$–TaCl$_5$ 1437	Fe$_2$O$_3$–Bi$_2$O$_3$ 327
	FeCl$_3$–AlCl$_3$–TiCl$_4$ 1435	Fe$_2$O$_3$–CaO 43–44
Dy$_2$O$_3$–Al$_2$O$_3$ 312	FeCl$_3$–BiCl$_3$ 1394	Fe$_2$O$_3$–CaO–Cr$_2$O$_3$ 650
Dy$_2$O$_3$–Cr$_2$O$_3$ 331	FeCl$_3$–CuCl 1256	Fe$_2$O$_3$–CaO–Fe 40–41
Dy$_2$O$_3$–Eu$_2$O$_3$ 335	FeCl$_3$–NH$_4$Cl 1327	Fe$_2$O$_3$–CaO–FeO 45–46
Dy$_2$O$_3$–Fe$_2$O$_3$ 337	FeCl$_3$–NaCl 1315	Fe$_2$O$_3$–CaO–Fe$_3$O$_4$ 47–48
Dy$_2$O$_3$–Ga$_2$O$_3$ 340	FeCl$_3$–NaCl–TiCl$_4$ 1433	Fe$_2$O$_3$–CaO–MgO–SiO$_2$ 927–928
Dy$_2$O$_3$–Gd$_2$O$_3$ 342	FeCl$_3$–NbCl$_5$–TaCl$_5$ 1440	Fe$_2$O$_3$–CaO–SiO$_2$ 656–658
Dy$_2$O$_3$–In$_2$O$_3$ 344	FeCl$_3$–NbCl$_5$–TiCl$_4$ 1439	Fe$_2$O$_3$–CoO 50
Dy$_2$O$_3$–La$_2$O$_3$ 345	FeCl$_3$–PbCl$_2$ 1386	Fe$_2$O$_3$–Cr$_2$O$_3$ 58
Dy$_2$O$_3$–Nd$_2$O$_3$ 348	FeCl$_3$–TiCl$_4$ 1395	Fe$_2$O$_3$–Cr$_2$O$_3$–FeO–MgO 137
Dy$_2$O$_3$–Sc$_2$O$_3$ 351	FeCl$_3$–TlCl 1348	Fe$_2$O$_3$–Cr$_2$O$_3$–TiO$_2$ 778
Dy$_2$O$_3$–Sm$_2$O$_3$ 352	FeCl$_3$–ZnCl$_2$ 1390	Fe$_2$O$_3$–Dy$_2$O$_3$ 337
	FeF$_2$–NaF 1485	Fe$_2$O$_3$–Er$_2$O$_3$ 337
Er$_2$O$_3$–Al$_2$O$_3$ 312	FeF$_3$–NaF 1493	Fe$_2$O$_3$–Eu$_2$O$_3$ 337
Er$_2$O$_3$–Cr$_2$O$_3$ 331	FeO–Al$_2$O$_3$ 26	Fe$_2$O$_3$–Fe 8
Er$_2$O$_3$–Eu$_2$O$_3$ 335	FeO–Al$_2$O$_3$–CaO–SiO$_2$ 866–875	Fe$_2$O$_3$–FeO–MgO 65–69
Er$_2$O$_3$–Fe$_2$O$_3$ 337	FeO–Al$_2$O$_3$–Cr$_2$O$_3$–Fe$_2$O$_3$ 130	Fe$_2$O$_3$–FeO–MgO–SiO$_2$ 138–145
Er$_2$O$_3$–Ga$_2$O$_3$ 340	FeO–Al$_2$O$_3$–Fe$_2$O$_3$ 29–30, 32	Fe$_2$O$_3$–FeO–MnO 70
Er$_2$O$_3$–Gd$_2$O$_3$ 342	FeO–Al$_2$O$_3$–Fe$_2$O$_3$–SiO$_2$ 131–136	Fe$_2$O$_3$–FeO–SiO$_2$ 82–84, 87
Er$_2$O$_3$–In$_2$O$_3$ 344	FeO–Al$_2$O$_3$–Fe$_3$O$_4$ 31	Fe$_2$O$_3$–FeO–TiO$_2$ 90–91
Er$_2$O$_3$–La$_2$O$_3$ 345	FeO–Al$_2$O$_3$–K$_2$O–SiO$_2$ 801	Fe$_2$O$_3$–Fe$_3$O$_4$ 10
Er$_2$O$_3$–Nd$_2$O$_3$ 348	FeO–Al$_2$O$_3$–MgO 680–681	Fe$_2$O$_3$–Fe$_3$O$_4$–Y$_2$O$_3$ 98
Er$_2$O$_3$–Sc$_2$O$_3$ 351	FeO–Al$_2$O$_3$–MgO–SiO$_2$ 963	Fe$_2$O$_3$–Ga$_2$O$_3$ 340
Er$_2$O$_3$–Sm$_2$O$_3$ 352	FeO–Al$_2$O$_3$–MnO 688	Fe$_2$O$_3$–Gd$_2$O$_3$ 336
Eu$_2$O$_3$–Al$_2$O$_3$ 312	FeO–Al$_2$O$_3$–Na$_2$O–SiO$_2$ 853–854	Fe$_2$O$_3$–Gd$_2$O$_3$–Nb$_2$O$_5$ 779
Eu$_2$O$_3$–Cr$_2$O$_3$ 331	FeO–Al$_2$O$_3$–SiO$_2$ 696–698	Fe$_2$O$_3$–H$_2$O–MgO–Na$_2$O–SiO$_2$ 2047
Eu$_2$O$_3$–Dy$_2$O$_3$ 335	FeO–CO$_2$–H$_2$O 1996	Fe$_2$O$_3$–H$_2$O–P$_2$O$_5$ 2016
Eu$_2$O$_3$–Er$_2$O$_3$ 335	FeO–CaF$_2$–CaO 1706	Fe$_2$O$_3$–Ho$_2$O$_3$ 337
Eu$_2$O$_3$–Fe$_2$O$_3$ 337	FeO–CaO 42	Fe$_2$O$_3$–In$_2$O$_3$ 337
Eu$_2$O$_3$–Ga$_2$O$_3$ 340	FeO–CaO–Fe$_2$O$_3$ 45–46	Fe$_2$O$_3$–K$_2$O–SiO$_2$ 419–421
Eu$_2$O$_3$–Gd$_2$O$_3$ 335	FeO–CaO–MgO–SiO$_2$ 864–865	Fe$_2$O$_3$–La$_2$O$_3$ 60–61
Eu$_2$O$_3$–Ho$_2$O$_3$ 335	FeO–CaO–P$_2$O$_5$ 595	Fe$_2$O$_3$–Li$_2$O 62
Eu$_2$O$_3$–In$_2$O$_3$ 334	FeO–CaO–SiO$_2$ 586–594	Fe$_2$O$_3$–Lu$_2$O$_3$ 337
Eu$_2$O$_3$–La$_2$O$_3$ 345	FeO–Cr$_2$O$_3$–Fe$_2$O$_3$–MgO 137	Fe$_2$O$_3$–MgO 64
Eu$_2$O$_3$–Ln$_2$O$_3$ 335	FeO–Cu$_2$S–FeS 1904	Fe$_2$O$_3$–Mn$_2$O$_3$ 73
Eu$_2$O$_3$–Lu$_2$O$_3$ 335	FeO–Fe$_2$O$_3$–MgO 65–69	Fe$_2$O$_3$–Na$_2$O 189
Eu$_2$O$_3$–Nd$_2$O$_3$ 348	FeO–Fe$_2$O$_3$–MgO–SiO$_2$ 138–145	Fe$_2$O$_3$–Na$_2$O–SiO$_2$ 519–530
Eu$_2$O$_3$–Sc$_2$O$_3$ 351	FeO–Fe$_2$O$_3$–MnO 70	Fe$_2$O$_3$–Nd$_2$O$_3$ 337
Eu$_2$O$_3$–Sm$_2$O$_3$ 352	FeO–Fe$_2$O$_3$–SiO$_2$ 82–84, 87	Fe$_2$O$_3$–P$_2$O$_5$ 338
Eu$_2$O$_3$–Tm$_2$O$_3$ 335	FeO–Fe$_2$O$_3$–TiO$_2$ 90–91	Fe$_2$O$_3$–PbO 282
Eu$_2$O$_3$–Yb$_2$O$_3$ 335	FeO–Fe$_3$O$_4$–H$_2$O–SiO$_2$ 2036	Fe$_2$O$_3$–R$_2$O$_3$ 337
	FeO–FeS 1902	Fe$_2$O$_3$–Sc$_2$O$_3$ 79
Fe–Al–Cr–O 126–130	FeO–FeS–SiO$_2$ 1905	Fe$_2$O$_3$–Sm$_2$O$_3$ 337
Fe–Al–O 26–35	FeO–K$_2$O–SiO$_2$ 397–398	Fe$_2$O$_3$–TiO$_2$ 89
Fe–Al–O–Si 131–136	FeO–MgO 63	Fe$_2$O$_3$–Tm$_2$O$_3$ 337
Fe–As–S 1885	FeO–MgO–SiO$_2$ 682–687	Fe$_2$O$_3$–V$_2$O$_5$ 96
Fe–Ca–O 40–48	FeO–MnO 71	Fe$_2$O$_3$–Y$_2$O$_3$ 97
Fe–CaO–Fe$_2$O$_3$ 40–41	FeO–MnO–SiO$_2$ 689–694	Fe$_2$O$_3$–Yb$_2$O$_3$ 337
Fe–Co–CoS–FeS 1886	FeO–MnO–TiO$_2$ 695	Fe$_3$O$_4$–Al$_2$O$_3$–FeO 31
Fe–Co–O 49–50	FeO–Na$_2$O–SiO$_2$ 487–492	Fe$_3$O$_4$–CaO–Fe$_2$O$_3$ 47–48
Fe–Cr–Mg–O 137	FeO–SiO$_2$ 80–81	Fe$_3$O$_4$–CeO$_2$ 356

Fe_3O_4–Cr_2O_3–SiO_2 760–762	GeO_2–MgO–NiO–SiO_2 964–965	H_2O–In_2O_3 1934
Fe_3O_4–FeO–H_2O–SiO_2 2036	GeO_2–MgO–SiO_2 717–719	H_2O–K_2CO_3–Na_2CO_3 2052
Fe_3O_4–Fe_2O_3 10	GeO_2–MgO–TiO_2 720	H_2O–K_2O–SiO_2 1949
Fe_3O_4–Fe_2O_3–Y_2O_3 98	GeO_2–MgO–ZnO 704	H_2O–La_2O_3 1935
Fe_3O_4–Mn_3O_4 72	GeO_2–Na_2O 190–191	H_2O–Mg(HCO$_3$)$_2$–MgSO$_4$–NaHCO$_3$–
FeS–Co–CoS–Fe 1886	GeO_2–PbO 283	Na_2SO_4 2065
FeS–CoO 1902	GeO_2–SiO_2 357	H_2O–MgO 1923–24
FeS–CoS 1895	GeO_2–TiO_2 358	H_2O–MgO–SiO_2 2001–02
FeS–Co$_4$S$_3$–Ni$_3$S$_2$ 1900–01		H_2O–MnO 1925
FeS–Cu$_2$S–FeO 1904	HF–H_2O–NaAlSi$_3$O$_8$ 2054	H_2O–MnO–P_2O_5 2003–04
FeS–FeO 1902	HF–NaF 1451	H_2O–Mn_2O_3 1936
FeS–FeO–SiO_2 1905	HNO$_3$–H_2O–Pb(NO$_3$)$_2$ 2062	H_2O–NH$_3$–NaAlSi$_3$O$_8$ 2055
FeS–SnS 1896	H_2O 1915–17	H_2O–Na_2CO_3–Na$_2$WO$_4$ 2059–60
FeS–ZnS 1897	H_2O–Al_2O_3 1926–28	H_2O–Na_2CO_3–SiO_2 2057–58
	H_2O–Al_2O_3–B_2O_3–MgO–Na_2O–SiO_2	H_2O–Na_2O–P_2O_5 1972–73
GaI–GaI$_3$ 1575	2048	H_2O–Na_2O–PbO 1953
GaI$_3$–GaI 1575	H_2O–Al_2O_3–BaO 1976	H_2O–Na_2O–SiO_2 1959–71
Ga_2O_3–Al_2O_3 310	H_2O–Al_2O_3–CO_2–MgO 2037	H_2O–Na_2O–UO_3 1975
Ga_2O_3–Al_2O_3–H_2O 2008	H_2O–Al_2O_3–CaO 1978–83	H_2O–Na_2O–V_2O_5 1974
Ga_2O_3–Bi_2O_3 327	H_2O–Al_2O_3–CaO–SO$_3$ 2063	H_2O–Na_2O–ZnO 1954
Ga_2O_3–CaO 235	H_2O–Al_2O_3–CaO–SiO_2 2035	H_2O–Na$_2$WO$_4$ 2049
Ga_2O_3–Cr_2O_3 340	H_2O–Al_2O_3–Ga_2O_3 2008	H_2O–Nd_2O_3 1937
Ga_2O_3–Dy_2O_3 340	H_2O–Al_2O_3–K_2O–Na_2O–SiO_2 ..2044–45	H_2O–NiO–SiO_2 2005
Ga_2O_3–Er_2O_3 340	H_2O–Al_2O_3–K_2O–SiO_2 2019	H_2O–P_2O_5–UO_2 2018
Ga_2O_3–Eu_2O_3 340	H_2O–Al_2O_3–Li_2O–Na_2O–SiO_2 2046	H_2O–RbOH 2050
Ga_2O_3–Fe_2O_3 340	H_2O–Al_2O_3–MgO 1997–98	H_2O–Sc_2O_3 1938
Ga_2O_3–Gd_2O_3 340	H_2O–Al_2O_3–MgO–N_2O_5 2043	H_2O–SeO_3 1947
Ga_2O_3–H_2O 1933	H_2O–Al_2O_3–MgO–SiO_2 2038–42	H_2O–SiO_2 1942–45
Ga_2O_3–H_2O–Na_2O 1956–57	H_2O–Al_2O_3–Na_2O–SiO_2 2022–29	H_2O–SiO_2–ZnO 2006
Ga_2O_3–Ho_2O_3 340	H_2O–Al_2O_3–SO$_3$ 2012	H_2O–Sm_2O_3 1939
Ga_2O_3–In_2O_3 340	H_2O–Al_2O_3–SiO_2 2009–11	H_2O–TiO_2 1946
Ga_2O_3–La_2O_3 340	H_2O–As_2O_5–Fe_2O_3 2015	H_2O–Y_2O_3 1940
Ga_2O_3–Lu_2O_3 340	H_2O–As_2O_5–K_2O 1950	H_2O_2–H_2O 1918
Ga_2O_3–Nd_2O_3 340	H_2O–B_2O_3 1929–31	H_2O_2–H_2O–LiClO$_4$ 2053
Ga_2O_3–P_2O_5–SiO_2 780	H_2O–B_2O_3–K_2O 1948	H_2O_2–H_2O–NaClO$_4$ 2056
Ga_2O_3–R_2O_3 340	H_2O–B_2O_3–Li_2O 1951–52	HfCl$_4$–CsCl 1239
Ga_2O_3–Sc_2O_3 ....339, 340	H_2O–B_2O_3–Na_2O 1955	HfCl$_4$–KCl 1281
Ga_2O_3–SiO_2 341	H_2O–BaO–P_2O_5 1977	HfCl$_4$–NaCl 1320
Ga_2O_3–Sm_2O_3 340	H_2O–CO_2 1941	HfF$_4$–NaF 1496
Ga_2O_3–Tm_2O_3 340	H_2O–CO_2–CaO 1984–88	HfO$_2$–CaO 236
Ga_2O_3–Y_2O_3 340	H_2O–CO_2–CaO–MgO 2032–33	HfO$_2$–MgO–ThO$_2$ 721–722
Ga_2O_3–Yb_2O_3 340	H_2O–CO_2–CaO–MgO–SO$_3$ 2066	HgBr$_2$–HgI$_2$ 1603
Gd_2O_3–Al_2O_3 312	H_2O–CO_2–FeO 1996	HgBr$_2$–KBr 1186
Gd_2O_3–BaO–Fe_2O_3 562	H_2O–CO_2–K_2O–SiO_2 2020	HgBr$_2$–NH$_4$Br 1198
Gd_2O_3–BaO–Nb_2O_5 563	H_2O–CO_2–Li_2O–UO_3 2021	HgBr$_2$–NaBr 1195
Gd_2O_3–Cr_2O_3 331	H_2O–CO_2–MgO 1999	HgBr$_2$–PbBr$_2$ 1201
Gd_2O_3–Dy_2O_3 342	H_2O–CO_2–Na_2O 1958	HgCl$_2$–KCl 1266
Gd_2O_3–Er_2O_3 342	H_2O–CO_2–Na_2O–ThO$_2$ 2030	HgCl$_2$–NH$_4$Cl 1325
Gd_2O_3–Eu_2O_3 335	H_2O–CO_2–Na_2O–UO_3 2031	HgCl$_2$–NaCl 1306
Gd_2O_3–Fe_2O_3 336	H_2O–CaCl$_2$–CaO 2061	HgI$_2$–AgI 1581
Gd_2O_3–Fe_2O_3–Nb_2O_5 779	H_2O–Ca(HCO$_3$)$_2$–CaSO$_4$–NaHCO$_3$–	HgI$_2$–CdI$_2$ 1586
Gd_2O_3–Ga_2O_3 340	Na_2SO_4 2064	HgI$_2$–HgBr$_2$ 1603
Gd_2O_3–Ho_2O_3 342	H_2O–CaO 1919–21	HgI$_2$–KI 1583
Gd_2O_3–In_2O_3 344	H_2O–CaO–SiO_2 1989–95	HgI$_2$–NH$_4$I 1585
Gd_2O_3–La_2O_3 345	H_2O–CaO–SiO_2–SrO 2034	HgI$_2$–NaI 1584
Gd_2O_3–Ln_2O_3 342	H_2O–CaSO$_4$ 2051	Ho$_2$O$_3$–Al_2O_3 312
Gd_2O_3–Lu_2O_3 342	H_2O–CdO 1922	Ho$_2$O$_3$–Cr_2O_3 331
Gd_2O_3–Nb_2O_5 343	H_2O–CrO$_3$–ThO$_2$ 2017	Ho$_2$O$_3$–Eu_2O_3 335
Gd_2O_3–Nd_2O_3 348	H_2O–CrO$_3$–ZnO 2007	Ho$_2$O$_3$–Fe_2O_3 337
Gd_2O_3–Sc_2O_3 351	H_2O–Cr_2O_3 1932	Ho$_2$O$_3$–Ga_2O_3 340
Gd_2O_3–Sm_2O_3 352	H_2O–Cr_2O_3–P_2O_5 2013–14	Ho$_2$O$_3$–Gd_2O_3 342
Gd_2O_3–Tm_2O_3 342	H_2O–FeO–Fe$_3$O$_4$–SiO_2 2036	Ho$_2$O$_3$–In_2O_3 344
Gd_2O_3–Yb_2O_3 342	H_2O–Fe_2O_3–MgO–Na_2O–SiO_2 ..2047	Ho$_2$O$_3$–La_2O_3 345
Ge–GeO$_2$ 11	H_2O–Fe_2O_3–P_2O_5 2016	Ho$_2$O$_3$–Nd_2O_3 348
Ge–O 11	H_2O–Ga_2O_3 1933	Ho$_2$O$_3$–Sc_2O_3 351
GeO$_2$ 151	H_2O–Ga_2O_3–Na_2O 1956–57	Ho$_2$O$_3$–Sm_2O_3 352
GeO_2–Bi_2O_3 328	H_2O–GeO$_2$–MgO 2000	
GeO_2–Ge 11	H_2O–HF–NaAlSi$_3$O$_8$ 2054	In–InP 1906
GeO_2–H_2O–MgO 2000	H_2O–HNO$_3$–Pb(NO$_3$)$_2$ 2062	In_2O_3–Al_2O_3 312
GeO_2–K_2O 166	H_2O–H_2O_2 1918	In_2O_3–BaO–La_2O_3–SrO–TiO_2–Y_2O_3 .863
GeO_2–Li_2O 181	H_2O–H_2O_2–LiClO$_4$ 2053	In_2O_3–Cr_2O_3 331
GeO_2–MgO 264–265	H_2O–H_2O_2–NaClO$_4$ 2056	In_2O_3–Dy_2O_3 344

In_2O_3–Er_2O_3	344	
In_2O_3–Eu_2O_3	334	
In_2O_3–Fe_2O_3	337	
In_2O_3–Ga_2O_3	340	
In_2O_3–Gd_2O_3	344	
In_2O_3–H_2O	1934	
In_2O_3–Ho_2O_3	344	
In_2O_3–La_2O_3	344	
In_2O_3–Lu_2O_3	344	
In_2O_3–Nd_2O_3	344	
In_2O_3–R_2O_3	344	
In_2O_3–Sc_2O_3	351	
In_2O_3–Sm_2O_3	344	
In_2O_3–Tm_2O_3	344	
In_2O_3–Y_2O_3	344	
In_2O_3–Yb_2O_3	344	
InP–In	1906	
K–KBr	1660	
K–KCl	1660	
K–KF	1660, 1665	
K–KI	1660	
KBO_2–KCl–K_2WO_4	1831	
KBO_2–K_2SO_4–$LiBO_2$–Li_2SO_4	1162	
KBr–AgBr	1175	
KBr–$AgNO_3$	1743	
KBr–$BaBr_2$	1183	
KBr–$CaBr_2$	1184	
KBr–$CdBr_2$	1185	
KBr–$CdBr_2$–NaBr	1207–08	
KBr–$HgBr_2$	1186	
KBr–K	1660	
KBr–K_2CO_3–NaBr–Na_2CO_3	1854	
KBr–KCl	1596	
KBr–KCl–KNO_3	1750	
KBr–KCl–TlBr–TlCl	1641	
KBr–KF	1597	
KBr–KF–KI	1634	
KBr–KF–NaBr–NaF	1640	
KBr–KI	1598	
KBr–KNO_3–NaBr–$NaNO_3$	1752	
KBr–KOH	1730	
KBr–K_2SO_4	1770	
KBr–LiBr	1182	
KBr–$MgBr_2$	1187	
KBr–$SrBr_2$	1188	
KCN–AgCN	1872	
KCN–CuCN	1374	
KCN–KCl	1869	
KCN–NaCN	1876	
KCN–$Zn(CN)_2$	1877	
KCNS–KCl–NaCNS–NaCl	1871	
KCNS–NH_4CNS	1908	
KCNS–NaCNS	1907	
KCNS–RbCNS	1909	
K_2CO_3–$CaCO_3$	1009	
K_2CO_3–$CaCO_3$–Na_2CO_3	1016	
K_2CO_3–H_2O–Na_2CO_3	2052	
K_2CO_3–KBr–NaBr–Na_2CO_3	1854	
K_2CO_3–KCl	1813	
K_2CO_3–KCl–Na_2CO_3–NaCl	1857	
K_2CO_3–KCl–K_2SO_4	1832	
K_2CO_3–KF	1824	
K_2CO_3–KF–K_2SO_4	1848	
K_2CO_3–KF–K_2SO_4–Na_2CO_3–NaF–Na_2SO_4	1866–67	
K_2CO_3–KF–Na_2CO_3–NaF	1863	
K_2CO_3–KI–Na_2CO_3–NaI	1865	
K_2CO_3–KNO_3	1085	
K_2CO_3–KOH–Na_2CO_3–NaOH	1034	
K_2CO_3–K_2SO_4–Na_2CO_3–Na_2SO_4	1167	
K_2CO_3–K_2SO_4–NaCl	1843	
K_2CO_3–K_2SO_4–NaF	1850	
K_2CO_3–Li_2CO_3	1007	
K_2CO_3–Li_2CO_3–Na_2CO_3	1015	
K_2CO_3–Na_2CO_3	1008	
K_2CO_3–NaF–Na_2SO_4	1851	
K_2CO_3–Nb_2O_5	171	
K_2CO_3–PbO–TiO_2	749	
K_2CO_3–Ta_2O_5	173	
K_2CO_3–V_2O_5	174	
KCl	1210–11	
KCl–AgCl	1215	
KCl–$AgNO_3$–NaBr	1751	
KCl–$AlCl_3$	1274	
KCl–BF_3–KF	1638	
KCl–$BaCl_2$	1262	
KCl–$BaCl_2$–BaF_2–KF	1645	
KCl–$BaCl_2$–CaF_2	1637	
KCl–$BaCl_2$–$MgCl_2$	1417	
KCl–$BaCl_2$–NaCl	1409	
KCl–$BaCl_2$–$SrCl_2$	1418	
KCl–$CaCl_2$	1263	
KCl–$CaCl_2$–$CaSO_4$–K_2SO_4	1790	
KCl–$CaCl_2$–CsCl	1404	
KCl–$CaCl_2$–LiCl	1408	
KCl–$CaCl_2$–$MgCl_2$	1419–20	
KCl–$CaCl_2$–NaCl	1410	
KCl–$CaCl_2$–$PbCl_2$	1421	
KCl–$CaCl_2$–RbCl	1416	
KCl–$CaSO_4$–NaCl	1779	
KCl–$CdCl_2$	1264	
KCl–$CdCl_2$–$CdSO_4$–K_2SO_4	1791	
KCl–$CdCl_2$–NaCl	1411–12	
KCl–$CdCl_2$–$PbCl_2$	1422	
KCl–$CeCl_3$	1275	
KCl–$CeCl_3$–$MgCl_2$	1423	
KCl–$CoCl_2$–$CoSO_4$–K_2SO_4	1792	
KCl–CsCl	1226	
KCl–CuCl	1243	
KCl–$FeCl_2$	1265	
KCl–$HfCl_4$	1281	
KCl–$HgCl_2$	1266	
KCl–K	1660	
KCl–KBO_2–K_2WO_4	1831	
KCl–KBr	1596	
KCl–KBr–KNO_3	1750	
KCl–KBr–TlBr–TlCl	1641	
KCl–KCN	1869	
KCl–KCNS–NaCNS–NaCl	1871	
KCl–K_2CO_3	1813	
KCl–K_2CO_3–K_2SO_4	1832	
KCl–K_2CO_3–Na_2CO_3–NaCl	1857	
KCl–K_2CrO_4	1817	
KCl–K_2CrO_4–KF	1829	
KCl–K_2CrO_4–NaF	1830	
KCl–$K_2Cr_2O_7$	1818	
KCl–KF	1614	
KCl–KI	1616	
KCl–KI–$PbCl_2$–PbI_2	1646	
KCl–K_2MoO_4–K_2WO_4	1834	
KCl–K_2MoO_4–NaCl–Na_2MoO_4	1859	
KCl–KNO_3–NH_4Cl–NH_4NO_3	1756	
KCl–KNO_3–NaCl–$NaNO_3$	1755	
KCl–KNO_3–$SrCl_2$–$Sr(NO_3)_2$	1757	
KCl–KOH	1733	
KCl–KPO_3	1814	
KCl–K_3PO_4	1815	
KCl–$K_4P_2O_7$	1816	
KCl–K_2SO_4	1772	
KCl–K_2SO_4–K_2WO_4	1836	
KCl–K_2SO_4–LiCl–Li_2SO_4	1788	
KCl–K_2SO_4–$MgCl_2$–$MgSO_4$	1793	
KCl–K_2SO_4–Na_2CO_3	1835	
KCl–K_2SO_4–NaCl–Na_2SO_4	1789	
KCl–K_2SO_4–$ZnCl_2$–$ZnSO_4$	1794	
KCl–KVO_3–NaCl–$NaVO_3$	1858	
KCl–K_2WO_4–NaCl–Na_2WO_4	1860	
KCl–$LaCl_3$	1276	
KCl–LiCl	1257	
KCl–LiCl–Li_2SO_4	1777	
KCl–LiCl–Li_2SO_4–NaCl	1786–87	
KCl–LiCl–NaCl	1407	
KCl–LiF	1615	
KCl–Li_2SO_4–Li_2WO_4	1837	
KCl–Li_2SO_4–NaCl	1778	
KCl–Li_2SO_4–Na_2SO_4	1780	
KCl–$MgCl_2$	1267	
KCl–$MgCl_2$–$NdCl_3$	1424	
KCl–$MnCl_2$	1268	
KCl–Na_2CO_3–Na_2SO_4	1833	
KCl–NaCl	1258–59	
KCl–NaCl–$PbCl_2$	1413	
KCl–NaCl–$TiCl_3$	1414	
KCl–NaCl–$ZrCl_4$	1415	
KCl–$NbCl_5$	1328–29	
KCl–$NdCl_3$	1277	
KCl–$PbCl_2$	1269	
KCl–$PbCl_2$–$ZnCl_2$	1425	
KCl–$PuCl_3$	1278	
KCl–RbCl	1260	
KCl–$SrCl_2$	1270	
KCl–$TaCl_5$	1330–31	
KCl–$TiCl_2$	1271	
KCl–$TiCl_3$	1279	
KCl–TlCl	1261	
KCl–UCl_4	1282	
KCl–VCl_2	1272	
KCl–VCl_3	1280	
KCl–$ZnCl_2$	1273	
KCl–$ZrCl_4$	1283–84	
K_2CrO_4–KCl	1817	
K_2CrO_4–KCl–KF	1829	
K_2CrO_4–KCl–NaF	1830	
K_2CrO_4–KF–Na_2CrO_4–NaF	1864	
K_2CrO_4–KNO_3–K_2SO_4	1157	
K_2CrO_4–KOH	1025	
K_2CrO_4–KOH–Li_2CrO_4–LiOH	1033	
K_2CrO_4–KOH–Na_2CrO_4–NaOH	1035	
K_2CrO_4–K_2SO_4	1146	
K_2CrO_4–NaCl–NaF	1841	
$K_2Cr_2O_7$–KCl	1818	
KF–AlF_3	1461	
KF–AlF_3–NaF	1536	
KF–BF_3	1462	
KF–BF_3–KCl	1638	
KF–BF_3–NaF	1537	
KF–$BaCl_2$–BaF_2–KCl	1645	
KF–BaF_2	1455	
KF–BeF_2	1456–57	
KF–BeF_2–NaF	1534	
KF–CaF_2–SiO_2	1681	
KF–CeF_3	1463	
KF–CrO_3–K_2O	1678	
KF–K	1660, 1665	
KF–KBr	1597	
KF–KBr–KI	1634	
KF–KBr–NaBr–NaF	1640	
KF–K_2CO_3	1824	
KF–K_2CO_3–K_2SO_4	1848	
KF–K_2CO_3–K_2SO_4–Na_2CO_3–NaF–Na_2SO_4	1866–67	
KF–K_2CO_3–Na_2CO_3–NaF	1863	
KF–KCl	1614	
KF–KCl–K_2CrO_4	1829	
KF–K_2CrO_4–Na_2CrO_4–NaF	1864	

KF–K$_2$MoO$_4$–NaF–Na$_2$MoO$_4$......1719	KNO$_3$–Ca(NO$_3$)$_2$–NaNO$_3$.........1074	K$_2$O–MoO$_3$–P$_2$O$_5$..............428
KF–KNO$_3$–NaF–NaNO$_3$........1766	KNO$_3$–Cd(NO$_3$)$_2$–LiNO$_3$..........1071	K$_2$O–MoO$_3$–PbO–TiO$_2$......749, 812
KF–K$_2$O–MoO$_3$................1679	KNO$_3$–Cd(NO$_3$)$_2$–RbNO$_3$..........1075	K$_2$O–MoO$_3$–TiO$_2$.............547
KF–K$_2$O–WO$_3$.................1680	KNO$_3$–KBr–KCl.................1750	K$_2$O–MoO$_3$–WO$_3$..............427
KF–KOH........................1736	KNO$_3$–KBr–NaBr–NaNO$_3$........1752	K$_2$O–Na$_2$O–P$_2$O$_5$..........385–386
KF–K$_4$P$_2$O$_7$....................1825	KNO$_3$–K$_2$CO$_3$..................1085	K$_2$O–Na$_2$O–SiO$_2$.............381–383
KF–K$_2$SO$_4$–Na$_2$CO$_3$.............1849	KNO$_3$–KCl–NH$_4$Cl–NH$_4$NO$_3$.....1756	K$_2$O–Na$_2$O–WO$_3$...............389
KF–K$_2$SO$_4$–NaF–Na$_2$SO$_4$........1806	KNO$_3$–KCl–NaCl–NaNO$_3$........1755	K$_2$O–Nb$_2$O$_5$...................171
KF–K$_2$SO$_4$–PbF$_2$–PbSO$_4$........1807	KNO$_3$–KCl–SrCl$_2$–Sr(NO$_3$)$_2$.....1757	K$_2$O–Nb$_2$O$_5$–Ta$_2$O$_5$............424
KF–K$_2$SiO$_3$–LiF–Li$_2$SiO$_3$.......1716	KNO$_3$–K$_2$CrO$_4$–K$_2$SO$_4$..........1157	K$_2$O–P$_2$O$_5$....................172
KF–K$_2$TiO$_3$–NaF–Na$_2$TiO$_3$......1717	KNO$_3$–KF–NaF–NaNO$_3$.........1766	K$_2$O–P$_2$O$_5$–PbO–TiO$_2$..........750
KF–KVO$_3$–NaF–NaVO$_3$........1718	KNO$_3$–KI–NaI–NaNO$_3$.........1769	K$_2$O–P$_2$O$_5$–SO$_3$...............428
KF–K$_2$WO$_4$–NaF–Na$_2$WO$_4$......1720	KNO$_3$–KNO$_2$....................1078	K$_2$O–P$_2$O$_5$–TiO$_2$............422–423
KF–LiF........................1452	KNO$_3$–KNO$_2$–NaNO$_2$–NaNO$_3$....1083	K$_2$O–P$_2$O$_5$–WO$_3$...............428
KF–LiF–MgF$_2$..................1531	KNO$_3$–KOH................1020–21	K$_2$O–PbO–SiO$_2$..............404–405
KF–LiF–NaF....................1529	KNO$_3$–K$_2$SO$_4$...................1143	K$_2$O–PbO–SiO$_2$–TiO$_2$............749
KF–LiF–RbF....................1530	KNO$_3$–K$_2$SO$_4$–LiNO$_3$–Li$_2$SO$_4$.....1163	K$_2$O–PbO–TiO$_2$–V$_2$O$_5$..........749
KF–LiF–UF$_4$...................1532	KNO$_3$–K$_2$SO$_4$–NH$_4$NO$_3$–(NH$_4$)$_2$SO$_4$....	K$_2$O–SO$_3$–TiO$_2$................546
KF–Li$_2$O–TiO$_2$.................1704	1168	K$_2$O–SiO$_2$..................167–168
KF–MgF$_2$......................1458	KNO$_3$–LiNO$_3$–NaNO$_3$............1069	K$_2$O–SiO$_2$–ZnO................406
KF–MgF$_2$–NaF..................1535	KNO$_3$–NaNO$_2$....................1080	K$_2$O–Ta$_2$O$_5$....................173
KF–MgF$_2$–SiO$_2$.................1682	KNO$_3$–NaNO$_3$....................1047	K$_2$O–TiO$_2$......................169
KF–NaF........................1453	KNO$_3$–NaNO$_3$–RbNO$_3$..........1072	K$_2$O–TiO$_2$–V$_2$O$_5$..............545
KF–NaF–RbF....................1533	KNO$_3$–NaOH..............1020, 1029	K$_2$O–TiO$_2$–WO$_3$................547
KF–NaF–UF$_4$...................1538	KNO$_3$–TlNO$_3$....................1048	K$_2$O–V$_2$O$_5$.....................174
KF–NaF–ZrF$_4$ 1539	K$_2$O–Al$_2$O$_3$–CaO................390	K$_2$O–WO$_3$......................177
KF–NiF$_2$......................1459	K$_2$O–Al$_2$O$_3$–CaO–Fe$_2$O$_3$..........794	KOH–KBr......................1730
KF–PbF$_2$–UF$_4$..................1540	K$_2$O–Al$_2$O$_3$–CaO–MgO–SiO$_2$...973–974	KOH–K$_2$CO$_3$–Na$_2$CO$_3$–NaOH....1034
KF–PbO–TiO$_2$...................749	K$_2$O–Al$_2$O$_3$–CaO–Na$_2$O–SiO$_2$...971–972	KOH–KCl......................1733
KF–RbF........................1454	K$_2$O–Al$_2$O$_3$–CaO–SiO$_2$........795–800	KOH–K$_2$CrO$_4$...................1025
KF–ThF$_4$......................1464	K$_2$O–Al$_2$O$_3$–FeO–SiO$_2$...........801	KOH–K$_2$CrO$_4$–Li$_2$CrO$_4$–LiOH...1033
KF–UF$_4$.......................1465	K$_2$O–Al$_2$O$_3$–H$_2$O–Na$_2$O–SiO$_2$...2044–45	KOH–K$_2$CrO$_4$–Na$_2$CrO$_4$–NaOH...1035
KF–ZnF$_2$......................1460	K$_2$O–Al$_2$O$_3$–H$_2$O–SiO$_2$..........2019	KOH–KF.......................1736
KF–ZrF$_4$......................1466	K$_2$O–Al$_2$O$_3$–MgO–SiO$_2$........802–811	KOH–KI.......................1739
KI–AgI........................1577	K$_2$O–Al$_2$O$_3$–Na$_2$O–SiO$_2$......786–789	KOH–KNO$_2$...................1019
KI–AgI–AgNO$_3$–KNO$_3$.........1767	K$_2$O–Al$_2$O$_3$–SiO$_2$............407–416	KOH–KNO$_3$................1020–21
KI–HgI$_2$......................1583	K$_2$O–As$_2$O$_5$....................170	KOH–K$_2$SO$_4$–NaOH–Na$_2$SO$_4$....1036
KI–K..........................1660	K$_2$O–As$_2$O$_5$–H$_2$O..............1950	KOH–LiOH.................1022–23
KI–KBr........................1598	K$_2$O–As$_2$O$_5$–Na$_2$O..............384	KOH–LiOH–NaOH..............1031
KI–KBr–KF.....................1634	K$_2$O–B$_2$O$_3$.....................165	KOH–NaNO$_2$...................1019
KI–K$_2$CO$_3$–Na$_2$CO$_3$–NaI........1865	K$_2$O–B$_2$O$_3$–CrO$_3$...............418	KOH–NaNO$_3$...................1020
KI–KCl........................1616	K$_2$O–B$_2$O$_3$–CrO$_3$–Na$_2$O......790–791	KOH–NaOH....................1024
KI–KCl–PbCl$_2$–PbI$_2$............1646	K$_2$O–B$_2$O$_3$–H$_2$O................1948	KPO$_3$–KCl....................1814
KI–KNO$_3$–NaI–NaNO$_3$.........1769	K$_2$O–B$_2$O$_3$–Li$_2$O–WO$_3$..........785	KPO$_3$–K$_2$SO$_4$..................1144
KI–KOH.......................1739	K$_2$O–B$_2$O$_3$–Na$_2$O...............380	KPO$_3$–K$_2$SO$_4$–LiPO$_3$–Li$_2$SO$_4$....1164
KI–K$_2$SO$_4$–NaI–Na$_2$SO$_4$.........1811	K$_2$O–B$_2$O$_3$–P$_2$O$_5$...............417	KPO$_3$–Li$_2$SO$_4$..................1145
K$_2$MoO$_4$–KCl–K$_2$WO$_4$...........1834	K$_2$O–CO$_2$–H$_2$O–SiO$_2$............2020	K$_3$PO$_4$–KCl....................1815
K$_2$MoO$_4$–KCl–NaCl–Na$_2$MoO$_4$...1859	K$_2$O–CaO–SiO$_2$...............391–396	K$_4$P$_2$O$_7$–KCl...................1816
K$_2$MoO$_4$–KF–NaF–Na$_2$MoO$_4$.....1719	K$_2$O–CrO$_3$......................175	K$_4$P$_2$O$_7$–KF....................1825
K$_2$MoO$_4$–K$_2$SO$_4$.................1147	K$_2$O–CrO$_3$–KF..................1678	K$_4$P$_2$O$_7$–K$_2$SO$_4$................428
K$_2$MoO$_4$–K$_2$SO$_4$–Li$_2$MoO$_4$–Li$_2$SO$_4$.....	K$_2$O–CrO$_3$–MoO$_3$................425	K$_2$SO$_4$–Ag$_2$SO$_4$................1088
...........................1165	K$_2$O–CrO$_3$–Na$_2$O................387	K$_2$SO$_4$–BaSO$_4$..................1094
KNH$_2$–NaNH$_2$..................1912	K$_2$O–CrO$_3$–P$_2$O$_5$................428	K$_2$SO$_4$–BeSO$_4$..................1095
KNO$_2$–Ba(NO$_2$)$_2$–Ba(NO$_3$)$_2$–KNO$_3$....	K$_2$O–CrO$_3$–TiO$_2$................546	K$_2$SO$_4$–CaCl$_2$–CaSO$_4$–KCl.......1790
...........................1084	K$_2$O–CrO$_3$–WO$_3$................426	K$_2$SO$_4$–CaSO$_4$..................1096
KNO$_2$–Ca(NO$_3$)$_2$................1079	K$_2$O–FeO–SiO$_2$...............397–398	K$_2$SO$_4$–CdCl$_2$–CdSO$_4$–KCl.......1791
KNO$_2$–KNO$_3$....................1078	K$_2$O–Fe$_2$O$_3$–SiO$_2$.............419–421	K$_2$SO$_4$–CdSO$_4$..................1097
KNO$_2$–KNO$_3$–NaNO$_2$–NaNO$_3$....1083	K$_2$O–GeO$_2$.....................166	K$_2$SO$_4$–CdSO$_4$–Li$_2$SO$_4$..........1128
KNO$_2$–KOH....................1019	K$_2$O–H$_2$O–SiO$_2$.................1949	K$_2$SO$_4$–CdSO$_4$–Na$_2$SO$_4$.........1129
KNO$_2$–NaNO$_2$...................1077	K$_2$O–KF–MoO$_3$..................1679	K$_2$SO$_4$–CoCl$_2$–CoSO$_4$–KCl.......1792
KNO$_2$–NaOH...................1019	K$_2$O–KF–WO$_3$...................1680	K$_2$SO$_4$–CoSO$_4$..................1098
KNO$_3$–AgBr–NaCl................1749	K$_2$O–LiF–TiO$_2$..................1683	K$_2$SO$_4$–CuSO$_4$..................1099
KNO$_3$–AgI–AgNO$_3$–KI..........1767	K$_2$O–Li$_2$O–MoO$_3$................379	K$_2$SO$_4$–CuSO$_4$–Na$_2$SO$_4$.......1130–33
KNO$_3$–AgNO$_3$...................1039	K$_2$O–Li$_2$O–P$_2$O$_5$................378	K$_2$SO$_4$–KBO$_2$–LiBO$_2$–Li$_2$SO$_4$.....1162
KNO$_3$–AgNO$_3$–Cd(NO$_3$)$_2$........1064	K$_2$O–Li$_2$O–P$_2$O$_5$–TiO$_2$............476	K$_2$SO$_4$–KBr....................1770
KNO$_3$–AgNO$_3$–TlNO$_3$..........1063	K$_2$O–Li$_2$O–SiO$_2$.............377, 437	K$_2$SO$_4$–K$_2$CO$_3$–KCl..............1832
KNO$_3$–Ba(NO$_2$)$_2$–Ba(NO$_3$)$_2$–KNO$_2$....	K$_2$O–Li$_2$O–TiO$_2$.................476	K$_2$SO$_4$–K$_2$CO$_3$–KF...............1866
...........................1084	K$_2$O–MgO–SiO$_2$..............399–403	K$_2$SO$_4$–K$_2$CO$_3$–KF–Na$_2$CO$_3$–NaF–
KNO$_3$–Ba(NO$_3$)$_2$................1049	K$_2$O–MoO$_3$......................176	Na$_2$SO$_4$...................1866–67
KNO$_3$–Ba(NO$_3$)$_2$–NaNO$_3$.......1073	K$_2$O–MoO$_3$–Na$_2$O................388	K$_2$SO$_4$–K$_2$CO$_3$–Na$_2$CO$_3$–Na$_2$SO$_4$...1167
KNO$_3$–Ca(NO$_3$)$_2$................1050	K$_2$O–MoO$_3$–Na$_2$O–P$_2$O$_5$..........792	K$_2$SO$_4$–K$_2$CO$_3$–NaCl............1843
KNO$_3$–Ca(NO$_3$)$_2$–LiNO$_3$........1070	K$_2$O–MoO$_3$–Na$_2$O–WO$_3$..........793	K$_2$SO$_4$–K$_2$CO$_3$–NaF.............1850

K_2SO_4–KCl ... 1772
K_2SO_4–KCl–K_2WO_4 ... 1836
K_2SO_4–KCl–LiCl–Li_2SO_4 ... 1788
K_2SO_4–KCl–$MgCl_2$–$MgSO_4$ ... 1793
K_2SO_4–KCl–Na_2CO_3 ... 1835
K_2SO_4–KCl–NaCl–Na_2SO_4 ... 1789
K_2SO_4–KCl–$ZnCl_2$–$ZnSO_4$ ... 1794
K_2SO_4–K_2CrO_4 ... 1146
K_2SO_4–K_2CrO_4–KNO_3 ... 1157
K_2SO_4–KF–Na_2CO_3 ... 1849
K_2SO_4–KF–NaF–Na_2SO_4 ... 1806
K_2SO_4–KF–PbF_2–$PbSO_4$ ... 1807
K_2SO_4–KI–NaI–Na_2SO_4 ... 1811
K_2SO_4–K_2MoO_4 ... 1147
K_2SO_4–K_2MoO_4–Li_2MoO_4–Li_2SO_4 ... 1165
K_2SO_4–KNO_3 ... 1143
K_2SO_4–KNO_3–$LiNO_3$–Li_2SO_4 ... 1163
K_2SO_4–KNO_3–NH_4NO_3–$(NH_4)_2SO_4$ ... 1168
K_2SO_4–KOH–NaOH–Na_2SO_4 ... 1036
K_2SO_4–KPO_3 ... 1144
K_2SO_4–KPO_3–$LiPO_3$–Li_2SO_4 ... 1164
K_2SO_4–$K_4P_2O_7$ ... 428
K_2SO_4–K_2TiO_3 ... 546
K_2SO_4–K_2WO_4 ... 1148
K_2SO_4–K_2WO_4–Li_2SO_4–Li_2WO_4 ... 1166
K_2SO_4–K_2WO_4–$PbSO_4$–$PbWO_4$ ... 1169
K_2SO_4–$LiPO_3$ ... 1145
K_2SO_4–Li_2SO_4 ... 1092
K_2SO_4–Li_2SO_4–Na_2SO_4 ... 1126
K_2SO_4–Li_2SO_4–Tl_2SO_4 ... 1127
K_2SO_4–$MgSO_4$ ... 1100
K_2SO_4–$MnSO_4$ ... 1101
K_2SO_4–Na_2CO_3–NaF ... 1852
K_2SO_4–Na_2SO_4 ... 1093
K_2SO_4–Na_2SO_4–$ZnSO_4$ ... 1134
K_2SO_4–$PbSO_4$ ... 1102
K_2SO_4–$SrSO_4$ ... 1103
K_2SO_4–V_2O_5 ... 1137
K_2SiO_3–KF–LiF–Li_2SiO_3 ... 1716
K_2TiO_3–KF–NaF–Na_2TiO_3 ... 1717
K_2TiO_3–K_2SO_4 ... 546
KVO_3–KCl–NaCl–$NaVO_3$ ... 1858
KVO_3–KF–NaF–$NaVO_3$ ... 1718
K_2WO_4–KBO_2–KCl ... 1831
K_2WO_4–KCl–K_2MoO_4 ... 1834
K_2WO_4–KCl–K_2SO_4 ... 1836
K_2WO_4–KCl–NaCl–Na_2WO_4 ... 1860
K_2WO_4–KF–NaF–Na_2WO_4 ... 1720
K_2WO_4–K_2SO_4 ... 1148
K_2WO_4–K_2SO_4–Li_2SO_4–Li_2WO_4 ... 1166
K_2WO_4–K_2SO_4–$PbSO_4$–$PbWO_4$ ... 1169

La–Fe–O ... 60–61
$LaCl_3$–$CaCl_2$–NaCl ... 1429
$LaCl_3$–CsCl ... 1236
$LaCl_3$–KCl ... 1276
La_2O_3–Al_2O_3 ... 312
La_2O_3–B_2O_3 ... 321
La_2O_3–BaO ... 209
La_2O_3–BaO–BeO ... 548
La_2O_3–BaO–In_2O_3–SrO–TiO_2–Y_2O_3 ... 863
La_2O_3–BeO ... 219
La_2O_3–BeO–CaO ... 570
La_2O_3–BeO–SrO ... 575
La_2O_3–Bi_2O_3 ... 327
La_2O_3–CaO–ZrO_2 ... 659
La_2O_3–Cr_2O_3 ... 331
La_2O_3–Dy_2O_3 ... 345
La_2O_3–Er_2O_3 ... 345
La_2O_3–Eu_2O_3 ... 345
La_2O_3–Fe_2O_3 ... 60–61
La_2O_3–Ga_2O_3 ... 340
La_2O_3–Gd_2O_3 ... 345
La_2O_3–H_2O ... 1935
La_2O_3–Ho_2O_3 ... 345
La_2O_3–In_2O_3 ... 344
La_2O_3–Ln_2O_3 ... 345
La_2O_3–Lu_2O_3 ... 345
La_2O_3–Lu_2O_3–Sm_2O_3 ... 781
La_2O_3–MgO–ZrO_2 ... 716
La_2O_3–Nd_2O_3 ... 345
La_2O_3–Sc_2O_3 ... 351
La_2O_3–Sm_2O_3 ... 345
La_2O_3–Tm_2O_3 ... 345
La_2O_3–Yb_2O_3 ... 345
La_2O_3–ZrO_2 ... 346
Li–Fe–O ... 62
Li–LiF ... 1665
$LiBO_2$–KBO_2–K_2SO_4–Li_2SO_4 ... 1162
$LiBO_2$–LiCl–Li_2SO_4 ... 1838
$LiBO_2$–LiCl–Li_2WO_4 ... 1839
$LiBO_2$–Li_2SO_4–Li_2WO_4 ... 1158
LiBr–AgBr ... 1176
LiBr–$BaBr_2$ ... 1190
LiBr–KBr ... 1182
LiBr–LiCl ... 1599
LiBr–LiF ... 1600
LiBr–LiOH ... 1731
LiBr–NaBr ... 1189
LiBr–$SrBr_2$ ... 1191
Li_2CO_3–$CaCO_3$ ... 1011
Li_2CO_3–K_2CO_3 ... 1007
Li_2CO_3–K_2CO_3–Na_2CO_3 ... 1015
Li_2CO_3–LiCl ... 1819
Li_2CO_3–LiF ... 1826
Li_2CO_3–$LiNO_3$ ... 1086
Li_2CO_3–Li_2SO_4 ... 1151
Li_2CO_3–Na_2CO_3 ... 1010
LiCl ... 1210
LiCl–AgCl ... 1216
LiCl–AgCl–Ag_2CrO_4–Li_2CrO_4 ... 1855
LiCl–AgCl–$AgNO_3$–$LiNO_3$ ... 1753
LiCl–AgCl–Ag_2SO_4–Li_2SO_4 ... 1785
LiCl–AgCl–Ag_2WO_4–Li_2WO_4 ... 1856
LiCl–$BaCl_2$ ... 1289
LiCl–$BaCl_2$–BaF_2–LiF ... 1647
LiCl–$BeCl_2$ ... 1290
LiCl–$CaCl_2$ ... 1291
LiCl–$CaCl_2$–CaF_2–LiF ... 1648
LiCl–$CaCl_2$–$Ca(NO_3)_2$–$LiNO_3$ ... 1759
LiCl–$CaCl_2$–$CaSO_4$–Li_2SO_4 ... 1797
LiCl–$CaCl_2$–KCl ... 1408
LiCl–$CaCl_2$–NaCl ... 1426
LiCl–$CdCl_2$–$CdMoO_4$–Li_2MoO_4 ... 1861
LiCl–$CdCl_2$–$CdSO_4$–Li_2SO_4 ... 1798
LiCl–$CoCl_2$–$CoSO_4$–Li_2SO_4 ... 1799
LiCl–CsCl ... 1227
LiCl–CuCl ... 1244
LiCl–$FeCl_2$ ... 1292
LiCl–KCl ... 1257
LiCl–KCl–K_2SO_4–Li_2SO_4 ... 1788
LiCl–KCl–Li_2SO_4 ... 1777
LiCl–KCl–Li_2SO_4–NaCl ... 1786–87
LiCl–KCl–NaCl ... 1407
LiCl–$LiBO_2$–Li_2SO_4 ... 1838
LiCl–$LiBO_2$–Li_2WO_4 ... 1839
LiCl–LiBr ... 1599
LiCl–Li_2CO_3 ... 1819
LiCl–LiF ... 1617
LiCl–LiF–$SrCl_2$–SrF_2 ... 1649
LiCl–$LiNO_3$–NH_4Cl–NH_4NO_3 ... 1758
LiCl–$LiNO_3$–$SrCl_2$–$Sr(NO_3)_2$ ... 1760
LiCl–LiOH ... 1734
LiCl–Li_2SO_4–Li_2WO_4 ... 1840
LiCl–Li_2SO_4–NaCl ... 1781
LiCl–Li_2SO_4–NaCl–Na_2SO_4 ... 1795
LiCl–Li_2SO_4–$SrCl_2$–$SrSO_4$ ... 1800
LiCl–Li_2SO_4–TlCl–Tl_2SO_4 ... 1796
LiCl–Li_2SO_4–$ZnCl_2$–$ZnSO_4$ ... 1801
LiCl–$MnCl_2$ ... 1293
LiCl–NH_4Cl ... 1286
LiCl–NaCl ... 1285
LiCl–$NbCl_5$ ... 1328
LiCl–$PbCl_2$ ... 1294
LiCl–$PuCl_3$ ... 1296
LiCl–RbCl ... 1287
LiCl–$SrCl_2$ ... 1295
LiCl–$TaCl_5$ ... 1330–31
LiCl–TlCl ... 1288
LiCl–UCl_3 ... 1297
LiCl–UCl_4 ... 1298
$LiClO_4$–H_2O–H_2O_2 ... 2053
$LiClO_4$–NH_4ClO_4 ... 1017
Li_2CrO_4–AgCl–Ag_2CrO_4–LiCl ... 1855
Li_2CrO_4–K_2CrO_4–KOH–LiOH ... 1033
Li_2CrO_4–LiOH–Na_2CrO_4–NaOH ... 1038
LiF–AlF_3 ... 1473
LiF–AlF_3–Al_2O_3 ... 1686
LiF–AlF_3–Al_2O_3–NaF ... 1723
LiF–AlF_3–NaF ... 1546
LiF–$BaCl_2$–BaF_2–LiCl ... 1647
LiF–BaF_2–$BaSiO_3$–Li_2SiO_3 ... 1724
LiF–BaF_2–CaF_2 ... 1551
LiF–BaF_2–MgF_2 ... 1552
LiF–BeF_2 ... 1469
LiF–BeF_2–Li_2SO_4 ... 1783
LiF–BeF_2–NaF ... 1542–43
LiF–BeF_2–RbF ... 1549
LiF–BeF_2–ThF_4 ... 1553
LiF–BeF_2–UF_4 ... 1554
LiF–$CaCl_2$–CaF_2–LiCl ... 1648
LiF–CaF_2 ... 1470
LiF–CaF_2–$CaSiO_3$–Li_2SiO_3 ... 1725
LiF–CaF_2–MgF_2 ... 1555
LiF–CaF_2–NaF ... 1544
LiF–$CdCl_2$ ... 1624
LiF–CeF_3 ... 1474
LiF–CsF ... 1445
LiF–KCl ... 1615
LiF–KF ... 1452
LiF–KF–K_2SiO_3–Li_2SiO_3 ... 1716
LiF–KF–MgF_2 ... 1531
LiF–KF–NaF ... 1529
LiF–KF–RbF ... 1530
LiF–KF–UF_4 ... 1532
LiF–K_2O–TiO_2 ... 1683
LiF–Li ... 1665
LiF–LiBr ... 1600
LiF–Li_2CO_3 ... 1826
LiF–LiCl ... 1617
LiF–LiCl–$SrCl_2$–SrF_2 ... 1649
LiF–Li_2O–MoO_3 ... 1684
LiF–Li_2O–WO_3 ... 1685
LiF–LiOH ... 1737
LiF–Li_2SO_4–NaF–Na_2SO_4 ... 1808
LiF–Li_2SiO_3–NaF–Na_2SiO_3 ... 1721
LiF–Li_2TiO_3–NaF–Na_2TiO_3 ... 1722
LiF–MgF_2 ... 1471
LiF–MgF_2–NaF ... 1545
LiF–NaCl ... 1618
LiF–NaF ... 1467
LiF–NaF–RbF ... 1541
LiF–NaF–UF_4 ... 1547
LiF–NaF–ZrF_4 ... 1548

LiF–PuF$_3$ 1475
LiF–RbF 1468
LiF–RbF–UF$_4$ 1550
LiF–ThF$_4$ 1478
LiF–ThF$_4$–UF$_4$ 1556–57
LiF–UF$_3$ 1476
LiF–UF$_4$ 1479
LiF–YF$_3$ 1477
LiF–ZnF$_2$ 1472
LiF–ZrF$_4$ 1480, 1509
LiI–AgI 1578
LiI–LiOH 1740
Li$_2$MoO$_4$–Ag$_2$MoO$_4$–Ag$_2$SO$_4$–Li$_2$SO$_4$ . . . .
. . . . 1160
Li$_2$MoO$_4$–CdCl$_2$–CdMoO$_4$–LiCl . . . 1861
Li$_2$MoO$_4$–K$_2$MoO$_4$–K$_2$SO$_4$–Li$_2$SO$_4$ . . . . . .
. . . . 1165
Li$_2$MoO$_4$–Li$_2$SO$_4$–Na$_2$MoO$_4$–Na$_2$SO$_4$ . . .
. . . . 1170
LiNO$_3$–AgCl–AgNO$_3$–LiCl 1753
LiNO$_3$–AgNO$_3$–RbNO$_3$ 1065
LiNO$_3$–Ba(NO$_3$)$_2$ 1051
LiNO$_3$–CaCl$_2$–Ca(NO$_3$)$_2$–LiCl 1759
LiNO$_3$–Ca(NO$_3$)$_2$ 1052
LiNO$_3$–Ca(NO$_3$)$_2$–KNO$_3$ 1070
LiNO$_3$–Cd(NO$_3$)$_2$–KNO$_3$ 1071
LiNO$_3$–Cd(NO$_3$)$_2$–NaNO$_3$ 1076
LiNO$_3$–KNO$_3$–K$_2$SO$_4$–Li$_2$SO$_4$ 1163
LiNO$_3$–KNO$_3$–NaNO$_3$ 1069
LiNO$_3$–Li$_2$CO$_3$ 1086
LiNO$_3$–LiCl–NH$_4$Cl–NH$_4$NO$_3$ 1758
LiNO$_3$–LiCl–SrCl$_2$–Sr(NO$_3$)$_2$ 1760
LiNO$_3$–LiOH–NaNO$_3$–NaOH 1037
LiNO$_3$–Li$_2$SO$_4$ 1149
Li$_2$O–Al$_2$O$_3$ 179
Li$_2$O–Al$_2$O$_3$–B$_2$O$_3$ 445
Li$_2$O–Al$_2$O$_3$–B$_2$O$_3$–SiO$_2$ 820–825
Li$_2$O–Al$_2$O$_3$–Fe$_2$O$_3$ 446–448
Li$_2$O–Al$_2$O$_3$–H$_2$O–Na$_2$O–SiO$_2$ . . . .
. . . . 2046
Li$_2$O–Al$_2$O$_3$–MgO–SiO$_2$ 813–819
Li$_2$O–Al$_2$O$_3$–SiO$_2$ 449–459
Li$_2$O–Al$_2$O$_3$–TiO$_2$ 460–464
Li$_2$O–B$_2$O$_3$ 180
Li$_2$O–B$_2$O$_3$–H$_2$O 1951–52
Li$_2$O–B$_2$O$_3$–K$_2$O–WO$_3$ 785
Li$_2$O–B$_2$O$_3$–Na$_2$O 429
Li$_2$O–B$_2$O$_3$–P$_2$O$_5$ 470–473
Li$_2$O–B$_2$O$_3$–SiO$_2$ 465–469
Li$_2$O–BaO–SiO$_2$ 437, 441
Li$_2$O–Bi$_2$O$_3$ 325
Li$_2$O–CO$_2$–H$_2$O–UO$_3$ 2021
Li$_2$O–CaO–SiO$_2$ 437, 442
Li$_2$O–Fe$_2$O$_3$ 62
Li$_2$O–GeO$_2$ 181
Li$_2$O–KF–TiO$_2$ 1704
Li$_2$O–K$_2$O–MoO$_3$ 379
Li$_2$O–K$_2$O–P$_2$O$_5$ 378
Li$_2$O–K$_2$O–P$_2$O$_5$–TiO$_2$ 476
Li$_2$O–K$_2$O–SiO$_2$ 377, 437
Li$_2$O–K$_2$O–TiO$_2$ 476
Li$_2$O–LiF–MoO$_3$ 1684
Li$_2$O–LiF–WO$_3$ 1685
Li$_2$O–MgO–SiO$_2$ 443–444
Li$_2$O–MnO 178
Li$_2$O–MoO$_3$ 186
Li$_2$O–NaF–TiO$_2$ 1687, 1704
Li$_2$O–Na$_2$O–MoO$_3$ 439
Li$_2$O–Na$_2$O–SiO$_2$ 430–437
Li$_2$O–Na$_2$O–SiO$_2$–TiO$_2$ 476
Li$_2$O–Na$_2$O–TiO$_2$ 438, 476
Li$_2$O–Na$_2$O–WO$_3$ 440

Li$_2$O–P$_2$O$_5$–TiO$_2$ 476
Li$_2$O–RbF–TiO$_2$ 1704
Li$_2$O–SiO$_2$ 182–184
Li$_2$O–SiO$_2$–TiO$_2$ 474
Li$_2$O–SiO$_2$–ZrO$_2$ 475
Li$_2$O–TiO$_2$ 185
Li$_2$O–WO$_3$ 187
LiOH–K$_2$CrO$_4$–KOH–Li$_2$CrO$_4$ 1033
LiOH–KOH 1022–23
LiOH–KOH–NaOH 1031
LiOH–LiBr 1731
LiOH–LiCl 1734
LiOH–Li$_2$CrO$_4$–Na$_2$CrO$_4$–NaOH . . 1038
LiOH–LiF 1737
LiOH–LiI 1740
LiOH–LiNO$_3$–NaNO$_3$–NaOH 1037
LiOH–NaOH 1026–27
LiPO$_3$–KPO$_3$–K$_2$SO$_4$–Li$_2$SO$_4$ 1164
LiPO$_3$–K$_2$SO$_4$ 1145
LiPO$_3$–Li$_2$SO$_4$ 1150
Li$_2$SO$_4$–AgCl–Ag$_2$SO$_4$–LiCl 1785
Li$_2$SO$_4$–Ag$_2$MoO$_4$–Ag$_2$SO$_4$–Li$_2$MoO$_4$ . . . .
. . . . 1160
Li$_2$SO$_4$–Ag$_2$SO$_4$ 1089
Li$_2$SO$_4$–BeF$_2$–LiF 1783
Li$_2$SO$_4$–CaCl$_2$–CaSO$_4$–LiCl 1797
Li$_2$SO$_4$–CaSO$_4$ 1105
Li$_2$SO$_4$–CdCl$_2$–CdSO$_4$–LiCl 1798
Li$_2$SO$_4$–CdSO$_4$ 1106
Li$_2$SO$_4$–CdSO$_4$–K$_2$SO$_4$ 1128
Li$_2$SO$_4$–CdSO$_4$–Na$_2$SO$_4$ 1135
Li$_2$SO$_4$–CoCl$_2$–CoSO$_4$–LiCl 1799
Li$_2$SO$_4$–CoSO$_4$ 1107
Li$_2$SO$_4$–KBO$_2$–K$_2$SO$_4$–LiBO$_2$ 1162
Li$_2$SO$_4$–KCl–K$_2$SO$_4$–LiCl 1788
Li$_2$SO$_4$–KCl–LiCl 1777
Li$_2$SO$_4$–KCl–LiCl–NaCl 1786–87
Li$_2$SO$_4$–KCl–Li$_2$WO$_4$ 1837
Li$_2$SO$_4$–KCl–NaCl 1778
Li$_2$SO$_4$–KCl–Na$_2$SO$_4$ 1780
Li$_2$SO$_4$–K$_2$MoO$_4$–K$_2$SO$_4$–Li$_2$MoO$_4$ . . . . .
. . . . 1165
Li$_2$SO$_4$–KNO$_3$–K$_2$SO$_4$–LiNO$_3$ 1163
Li$_2$SO$_4$–KPO$_3$ 1145
Li$_2$SO$_4$–KPO$_3$–K$_2$SO$_4$–LiPO$_3$ 1164
Li$_2$SO$_4$–K$_2$SO$_4$ 1092
Li$_2$SO$_4$–K$_2$SO$_4$–K$_2$WO$_4$–Li$_2$WO$_4$ . . 1166
Li$_2$SO$_4$–K$_2$SO$_4$–Na$_2$SO$_4$ 1126
Li$_2$SO$_4$–K$_2$SO$_4$–Tl$_2$SO$_4$ 1127
Li$_2$SO$_4$–LiBO$_2$–LiCl 1838
Li$_2$SO$_4$–LiBO$_2$–Li$_2$WO$_4$ 1158
Li$_2$SO$_4$–Li$_2$CO$_3$ 1151
Li$_2$SO$_4$–LiCl–Li$_2$WO$_4$ 1840
Li$_2$SO$_4$–LiCl–NaCl 1781
Li$_2$SO$_4$–LiCl–NaCl–Na$_2$SO$_4$ 1795
Li$_2$SO$_4$–LiCl–SrCl$_2$–SrSO$_4$ 1800
Li$_2$SO$_4$–LiCl–TlCl–Tl$_2$SO$_4$ 1796
Li$_2$SO$_4$–LiCl–ZnCl$_2$–ZnSO$_4$ 1801
Li$_2$SO$_4$–LiF–NaF–Na$_2$SO$_4$ 1808
Li$_2$SO$_4$–Li$_2$MoO$_4$–Na$_2$MoO$_4$–Na$_2$SO$_4$ . . .
. . . . 1170
Li$_2$SO$_4$–LiNO$_3$ 1149
Li$_2$SO$_4$–LiPO$_3$ 1150
Li$_2$SO$_4$–Li$_2$WO$_4$–PbSO$_4$–PbWO$_4$ . . . 1171
Li$_2$SO$_4$–MgSO$_4$ 1108
Li$_2$SO$_4$–MnSO$_4$ 1109
Li$_2$SO$_4$–Na$_2$SO$_4$ 1104
Li$_2$SO$_4$–PbSO$_4$ 1110
Li$_2$SO$_4$–SrSO$_4$ 1111
Li$_2$SiO$_3$–BaF$_2$–BaSiO$_3$–LiF 1724
Li$_2$SiO$_3$–CaF$_2$–CaSiO$_3$–LiF 1725
Li$_2$SiO$_3$–KF–K$_2$SiO$_3$–LiF 1716

Li$_2$SiO$_3$–LiF–NaF–Na$_2$SiO$_3$ 1721
Li$_2$TiO$_3$–LiF–NaF–Na$_2$TiO$_3$ 1722
Li$_2$WO$_4$–AgCl–Ag$_2$WO$_4$–LiCl 1856
Li$_2$WO$_4$–KCl–Li$_2$SO$_4$ 1837
Li$_2$WO$_4$–K$_2$SO$_4$–K$_2$WO$_4$–Li$_2$SO$_4$ . . . 1166
Li$_2$WO$_4$–LiBO$_2$–LiCl 1839
Li$_2$WO$_4$–LiBO$_2$–Li$_2$SO$_4$ 1158
Li$_2$WO$_4$–LiCl–Li$_2$SO$_4$ 1840
Li$_2$WO$_4$–Li$_2$SO$_4$–PbSO$_4$–PbWO$_4$ . . . 1171
Ln$_2$O$_3$–Ln$_2$'O$_3$ 347
Lu$_2$O$_3$–Al$_2$O$_3$ 312
Lu$_2$O$_3$–Bi$_2$O$_3$ 327
Lu$_2$O$_3$–Cr$_2$O$_3$ 331
Lu$_2$O$_3$–Eu$_2$O$_3$ 335
Lu$_2$O$_3$–Fe$_2$O$_3$ 337
Lu$_2$O$_3$–Ga$_2$O$_3$ 340
Lu$_2$O$_3$–Gd$_2$O$_3$ 342
Lu$_2$O$_3$–In$_2$O$_3$ 344
Lu$_2$O$_3$–La$_2$O$_3$ 345
Lu$_2$O$_3$–La$_2$O$_3$–Sm$_2$O$_3$ 781
Lu$_2$O$_3$–Nd$_2$O$_3$ 348
Lu$_2$O$_3$–Sc$_2$O$_3$ 351
Lu$_2$O$_3$–Sm$_2$O$_3$ 352

Mg–Co–O 51–52
Mg–Cr–Fe–O 137
Mg–Fe–O 63–69
Mg–Fe–O–Si 138–145
Mg–MgCl$_2$–Pb–PbCl$_2$ 1670
Mg–O–U 99–100
MgBr$_2$–KBr 1187
MgBr$_2$–NaBr 1196
MgCO$_3$ 1006
MgCO$_3$–CaCO$_3$ 1012–13
MgCl$_2$–AgCl 1222
MgCl$_2$–BaCl$_2$ 1352
MgCl$_2$–BaCl$_2$–KCl 1417
MgCl$_2$–CaCl$_2$ 1360
MgCl$_2$–CaCl$_2$–KCl 1419–20
MgCl$_2$–CdCl$_2$ 1366
MgCl$_2$–CeCl$_3$ 1377
MgCl$_2$–CeCl$_3$–KCl 1423
MgCl$_2$–CsCl 1231
MgCl$_2$–CuCl 1251
MgCl$_2$–KCl 1267
MgCl$_2$–KCl–K$_2$SO$_4$–MgSO$_4$ 1793
MgCl$_2$–KCl–NdCl$_3$ 1424
MgCl$_2$–Mg–Pb–PbCl$_2$ 1670
MgCl$_2$–MgSO$_4$ 1775
MgCl$_2$–MnCl$_2$ 1372
MgCl$_2$–NaCl 1307
MgCl$_2$–PbCl$_2$ 1373
MgCl$_2$–RbCl 1333
MgCl$_2$–SrCl$_2$ 1374
MgCl$_2$–TiCl$_2$ 1375
MgCl$_2$–TlCl 1342
MgCl$_2$–ZnCl$_2$ 1376
MgF$_2$–AlF$_3$–Al$_2$O$_3$–NaF 1728
MgF$_2$–AlF$_3$–NaF 1568
MgF$_2$–BaF$_2$ 1511
MgF$_2$–BaF$_2$–LiF 1552
MgF$_2$–BeF$_2$ 1514
MgF$_2$–CaF$_2$ 1519
MgF$_2$–CaF$_2$–LiF 1555
MgF$_2$–CaF$_2$–NaF 1565
MgF$_2$–KF 1458
MgF$_2$–KF–LiF 1531
MgF$_2$–KF–NaF 1535
MgF$_2$–KF–SiO$_2$ 1682
MgF$_2$–LiF 1471
MgF$_2$–LiF–NaF 1545
MgF$_2$–MgO–SiO$_2$ 1711, 1714

MgF_2–NaF	1486	
MgF_2–NaF–SiO_2	1696	
MgF_2–ThF_4	1521	
$Mg(HCO_3)_2$–H_2O–$MgSO_4$–$NaHCO_3$–Na_2SO_4	2065	
MgO–AlF_3–NaF	1700	
MgO–Al_2O_3	259–260	
MgO–Al_2O_3–B_2O_3–H_2O–Na_2O–SiO_2	2048	
MgO–Al_2O_3–BeO	572	
MgO–Al_2O_3–CO_2–H_2O	2037	
MgO–Al_2O_3–CaO	596	
MgO–Al_2O_3–CaO–Cr_2O_3–SiO_2	987–990	
MgO–Al_2O_3–CaO–Fe_2O_3	876–879	
MgO–Al_2O_3–CaO–Fe_2O_3–SiO_2	991–1002	
MgO–Al_2O_3–CaO–K_2O–SiO_2	973–974	
MgO–Al_2O_3–CaO–Na_2O–SiO_2	975–981	
MgO–Al_2O_3–CaO–SiO_2	880–923	
MgO–Al_2O_3–Cr_2O_3	710	
MgO–Al_2O_3–FeO	680–681	
MgO–Al_2O_3–FeO–SiO_2	963	
MgO–Al_2O_3–Fe_2O_3	711	
MgO–Al_2O_3–H_2O	1997–98	
MgO–Al_2O_3–H_2O–N_2O_5	2043	
MgO–Al_2O_3–H_2O–SiO_2	2038–42	
MgO–Al_2O_3–K_2O–SiO_2	802–811	
MgO–Al_2O_3–Li_2O–SiO_2	813–819	
MgO–Al_2O_3–Na_2O–SiO_2	855–856	
MgO–Al_2O_3–SiO_2	712	
MgO–Al_2O_3–SiO_2–ZrO_2	966–967	
MgO–Al_2O_3–TiO_2	713–714	
MgO–B_2O_3	261	
MgO–BaO	274	
MgO–BaO–CaO–SiO_2	857	
MgO–BeO	216, 274	
MgO–BeO–ThO_2	573	
MgO–BeO–ZrO_2	574	
MgO–Bi_2O_3	326	
MgO–CO_2	1006	
MgO–CO_2–CaO–H_2O	2032–33	
MgO–CO_2–CaO–H_2O–SO_3	2066	
MgO–CO_2–CaO–SiO_2	929	
MgO–CO_2–H_2O	1999	
MgO–CaF_2	1672	
MgO–CaF_2–SiO_2	1710	
MgO–CaO	229	
MgO–CaO–Cr_2O_3	597	
MgO–CaO–Cr_2O_3–SiO_2	924–926	
MgO–CaO–FeO–SiO_2	864–865	
MgO–CaO–Fe_2O_3–SiO_2	927–928	
MgO–CaO–MnO–SiO_2–TiO_2	986	
MgO–CaO–P_2O_5	613	
MgO–CaO–P_2O_5–SiO_2	935–938	
MgO–CaO–SiO_2	598–610	
MgO–CaO–SiO_2–TiO_2	930–934	
MgO–CaO–SnO_2	611	
MgO–CaO–SnO_2–TiO_2	939–940	
MgO–CaO–TiO_2	612	
MgO–CoO	51–52	
MgO–Cr_2O_3	262	
MgO–Cr_2O_3–FeO–Fe_2O_3	137	
MgO–Cr_2O_3–SiO_2	715	
MgO–Cu_2O	274	
MgO–FeO	63	
MgO–FeO–Fe_2O_3	65–69	
MgO–FeO–Fe_2O_3–SiO_2	138–145	
MgO–FeO–SiO_2	682–687	
MgO–Fe_2O_3	64	
MgO–Fe_2O_3–H_2O–Na_2O–SiO_2	2047	
MgO–GeO_2	264–265	
MgO–GeO_2–H_2O	2000	
MgO–GeO_2–NiO–SiO_2	964–965	
MgO–GeO_2–SiO_2	717–719	
MgO–GeO_2–TiO_2	720	
MgO–GeO_2–ZnO	704	
MgO–H_2O	1923–24	
MgO–H_2O–SiO_2	2001–02	
MgO–HfO_2–ThO_2	721–722	
MgO–K_2O–SiO_2	399–403	
MgO–La_2O_3–ZrO_2	716	
MgO–Li_2O–SiO_2	443–444	
MgO–MgF_2–SiO_2	1711, 1714	
MgO–MnO–SiO_2	699–701	
MgO–NiO	258	
MgO–NiO–SiO_2	702	
MgO–P_2O_5	272	
MgO–P_2O_5–SiO_2	725–728	
MgO–P_2O_5–SrO	703	
MgO–P_2O_5–ZnO	707–709	
MgO–SiO_2	266–268	
MgO–SiO_2–TiO_2	723	
MgO–SiO_2–ZnO	705–706	
MgO–SiO_2–ZrO_2	724	
MgO–SnO_2–TiO_2	729	
MgO–SrO	274	
MgO–Ta_2O_5	273	
MgO–ThO_2–ZrO_2	730	
MgO–TiO_2	269–270	
MgO–TiO_2–ZrO_2	731	
MgO–UO_2	99–100	
MgO–Y_2O_3	263	
MgO–ZrO_2	271	
$MgSO_4$–Cs_2SO_4	1091	
$MgSO_4$–H_2O–$Mg(HCO_3)_2$–$NaHCO_3$–Na_2SO_4	2065	
$MgSO_4$–KCl–K_2SO_4–$MgCl_2$	1793	
$MgSO_4$–K_2SO_4	1100	
$MgSO_4$–Li_2SO_4	1108	
$MgSO_4$–$MgCl_2$	1775	
$MgSO_4$–Na_2SO_4	1117–18	
$MgSO_4$–Rb_2SO_4	1125	
Mn–Fe–O	70–74	
Mn–O	12–15	
Mn–O–Si	101–103	
$MnCO_3$–$CaCO_3$	1014	
$MnCl_2$–$BaCl_2$	1353	
$MnCl_2$–$CaCl_2$	1361	
$MnCl_2$–$CdCl_2$	1367	
$MnCl_2$–KCl	1268	
$MnCl_2$–$LiCl$	1293	
$MnCl_2$–$MgCl_2$	1372	
$MnCl_2$–$NaCl$	1308	
$MnCl_2$–$PbCl_2$	1378	
$MnCl_2$–$SnCl_2$	1379	
$MnCl_2$–$SrCl_2$	1380	
$MnCl_2$–$ZnCl_2$	1381	
MnF_2	1442	
MnO–Al_2O_3	275	
MnO–Al_2O_3–FeO	688	
MnO–Al_2O_3–SiO_2	733–735	
MnO–Al_2O_3–SiO_2–Y_2O_3	968–969	
MnO–B_2O_3	276	
MnO–CaO–MgO–SiO_2–TiO_2	986	
MnO–CaO–SiO_2	614–619	
MnO–CaO–SiO_2–TiO_2	941	
MnO–CaO–SiO_2–WO_3	942	
MnO–FeO	71	
MnO–FeO–Fe_2O_3	70	
MnO–FeO–SiO_2	689–694	
MnO–FeO–TiO_2	695	
MnO–H_2O	1925	
MnO–H_2O–P_2O_5	2003–04	
MnO–Li_2O	178	
MnO–MgO–SiO_2	699–701	
MnO–Mn_3O_4	12	
MnO–Na_2O–SiO_2	493	
MnO–P_2O_5	278	
MnO–P_2O_5–ZnO	732	
MnO–SiO_2	101	
MnO–SiO_2–TiO_2	736	
MnO–TiO_2	277	
MnO_2–Mn_2O_3	15	
Mn_2O_3–Bi_2O_3	327	
Mn_2O_3–Fe_2O_3	73	
Mn_2O_3–H_2O	1936	
Mn_2O_3–MnO_2	15	
Mn_2O_3–Mn_3O_4	13–14	
Mn_3O_4–CeO_2	356	
Mn_3O_4–Fe_3O_4	72	
Mn_3O_4–MnO	12	
Mn_3O_4–Mn_2O_3	13–14	
Mn_3O_4–SiO_2	102–103	
$MnSO_4$–K_2SO_4	1101	
$MnSO_4$–Li_2SO_4	1109	
$MnSO_4$–Na_2SO_4	1119	
MoO_3–Bi_2O_3	330	
MoO_3–Bi_2O_3–PbO	744	
MoO_3–CrO_3–K_2O	425	
MoO_3–CrO_3–PbO	751	
MoO_3–CsF–Cs_2O	1676	
MoO_3–Cs_2O	162	
MoO_3–KF–K_2O	1679	
MoO_3–K_2O	176	
MoO_3–K_2O–Li_2O	379	
MoO_3–K_2O–Na_2O	388	
MoO_3–K_2O–Na_2O–P_2O_5	792	
MoO_3–K_2O–Na_2O–WO_3	793	
MoO_3–K_2O–P_2O_5	428	
MoO_3–K_2O–PbO–TiO_2	749, 812	
MoO_3–K_2O–TiO_2	547	
MoO_3–K_2O–WO_4	427	
MoO_3–LiF–Li_2O	1684	
MoO_3–Li_2O	186	
MoO_3–Li_2O–Na_2O	439	
MoO_3–NaF–Na_2O	1692	
MoO_3–NaF–Na_2O–WO_3	1726	
MoO_3–Na_2O	199	
MoO_3–Na_2O–P_2O_5	544	
MoO_3–Na_2O–PbO–TiO_2	750	
MoO_3–Na_2O–TiO_2	547	
MoO_3–Na_2O–WO_3	543	
MoO_3–PbO	292	
MoO_3–PbO–WO_3	753	
MoO_3–RbF–Rb_2O	1702	
MoO_3–Rb_2O	202	
MoO_3–WO_3	376	
N–O–Ti	104	
$(NH_2)_2CS$–$(NH_4)CNS$	1911	
NH_3–H_2O–$NaAlSi_3O_8$	2055	
NH_4Br–$HgBr_2$	1198	
NH_4CNS–$KCNS$	1908	
NH_4CNS–$(NH_2)_2CS$	1911	
NH_4Cl–$CdCl_2$	1324	
NH_4Cl–$CuCl$	1246	
NH_4Cl–$FeCl_3$	1327	
NH_4Cl–$HgCl_2$	1325	
NH_4Cl–KCl–KNO_3–NH_4NO_3	1756	
NH_4Cl–$LiCl$	1286	
NH_4Cl–$LiCl$–$LiNO_3$–NH_4NO_3	1758	
NH_4Cl–NH_4NO_3	1745	
NH_4Cl–NH_4NO_3–$NaCl$–$NaNO_3$	1761	
NH_4Cl–$ZnCl_2$	1326	
NH_4ClO_4–$LiClO_4$	1017	

NH_4I-HgI_2 1585
$NH_4NO_3-AgNO_3$ 1041
$NH_4NO_3-KCl-KNO_3-NH_4Cl$ 1756
$NH_4NO_3-KNO_3-K_2SO_4-(NH_4)_2SO_4$ ....
.... 1168
$NH_4NO_3-LiCl-LiNO_3-NH_4Cl$ 1758
$NH_4NO_3-NH_4Cl$ 1745
$NH_4NO_3-NH_4Cl-NaCl-NaNO_3$ ... 1761
$NH_4NO_3-(NH_4)_2SO_4-NaNO_3-Na_2SO_4$ ..
.... 1172
$NH_4NO_3-NaNO_3$ 1053
$NH_4NO_3-Pb(NO_3)_2$ 1055
$(NH_4)_2SO_4-KNO_3-K_2SO_4-NH_4NO_3$ ....
.... 1168
$(NH_4)_2SO_4-NH_4NO_3-NaNO_3-$
  Na_2SO_4 1172
$N_2O_5-Al_2O_3-H_2O-MgO$ 2043
$Na-NaBr$ 1661
$Na-NaCl$ 1662
$Na-NaF$ 1665
$Na-NaI$ 1663
$Na-O-V$ 105–106
$NaAlSi_3O_8-HF-H_2O$ 2054
$NaAlSi_3O_8-H_2O-NH_3$ 2055
$NaBr$ 1591
$NaBr-AgCl$ 1177
$NaBr-AgNO_3-KCl$ 1751
$NaBr-BaBr_2$ 1192
$NaBr-CaBr_2$ 1193
$NaBr-CdBr_2$ 1194
$NaBr-CdBr_2-KBr$ 1207–08
$NaBr-CdBr_2-PbBr_2$ 1209
$NaBr-HgBr_2$ 1195
$NaBr-KBr-KF-NaF$ 1640
$NaBr-KBr-KNO_3-NaNO_3$ 1752
$NaBr-K_2CO_3-KBr-Na_2CO_3$ 1854
$NaBr-LiBr$ 1189
$NaBr-MgBr_2$ 1196
$NaBr-Na$ 1661
$NaBr-NaCl-PbBr_2-PbCl_2$ 1643
$NaBr-NaCl-TlBr-TlCl$ 1642
$NaBr-NaF-NaI$ 1635
$NaBr-NaOH$ 1732
$NaBr-Na_2SO_4$ 1771
$NaBr-SrBr_2$ 1197
$NaCN-AgCN$ 1873
$NaCN-CuCN$ 1875
$NaCN-KCN$ 1876
$NaCN-NaCl$ 1870
$NaCNS-KCNS$ 1907
$NaCNS-KCNS-KCl-NaCl$ 1871
$NaCNS-NaNO_3$ 1910
$Na_2CO_3-BaCO_3-BaCl_2-NaCl$ ... 1862
$Na_2CO_3-CaCO_3-K_2CO_3$ 1016
$Na_2CO_3-H_2O-K_2CO_3$ 2052
$Na_2CO_3-H_2O-Na_2WO_4$ 2059–60
$Na_2CO_3-H_2O-SiO_2$ 2057–58
$Na_2CO_3-K_2CO_3$ 1008
$Na_2CO_3-K_2CO_3-KBr-NaBr$ 1854
$Na_2CO_3-K_2CO_3-KCl-NaCl$ 1857
$Na_2CO_3-K_2CO_3-KF-K_2SO_4-NaF-$
  Na_2SO_4 1866–67
$Na_2CO_3-K_2CO_3-KF-NaF$ 1863
$Na_2CO_3-K_2CO_3-KI-NaI$ 1865
$Na_2CO_3-K_2CO_3-KOH-NaOH$ 1034
$Na_2CO_3-K_2CO_3-K_2SO_4-Na_2SO_4$. 1167
$Na_2CO_3-K_2CO_3-Li_2CO_3$ 1015
$Na_2CO_3-KCl-K_2SO_4$ 1835
$Na_2CO_3-KCl-Na_2SO_4$ 1833
$Na_2CO_3-KF-K_2SO_4$ 1849
$Na_2CO_3-K_2SO_4-NaF$ 1852
$Na_2CO_3-Li_2CO_3$ 1010

Na_2CO_3-NaCl 1820
$Na_2CO_3-NaCl-Na_2SO_4$ 1844
Na_2CO_3-NaF 1827
$Na_2CO_3-NaF-Na_2SO_4$ 1853
$Na_2CO_3-NaNO_3$ 1087
$Na_2CO_3-PbO-TiO_2$ 749
$NaCl$ 1210, 1591
$NaCl-AgBr-KNO_3$ 1749
$NaCl-AgCl$ 1217
$NaCl-AlCl_3$ 1314
$NaCl-AlCl_3-NbCl_5$ 1431
$NaCl-AlCl_3-TaCl_5$ 1432
$NaCl-BaCO_3-BaCl_2-Na_2CO_3$ ... 1862
$NaCl-BaCl_2$ 1301
$NaCl-BaCl_2-BaF_2-NaF$ 1650
$NaCl-BaCl_2-Ba(NO_3)_2-NaNO_3$. 1762
$NaCl-BaCl_2-KCl$ 1409
$NaCl-BaCl_2-SrCl_2$ 1428
$NaCl-BeCl_2$ 1302
$NaCl-CaCl_2$ 1303
$NaCl-CaCl_2-CaF_2-NaF$ 1651
$NaCl-CaCl_2-Ca(NO_3)_2-NaNO_3$. 1763
$NaCl-CaCl_2-CsCl$ 1405
$NaCl-CaCl_2-KCl$ 1410
$NaCl-CaCl_2-LaCl_3$ 1429
$NaCl-CaCl_2-LiCl$ 1426
$NaCl-CaCl_2-NdCl_3$ 1430
$NaCl-CaCl_2-RbCl$ 1427
$NaCl-CaSO_4-KCl$ 1779
$NaCl-CdCl_2$ 1304
$NaCl-CdCl_2-CdI_2-NaI$ 1652
$NaCl-CdCl_2-CdSO_4-Na_2SO_4$ ... 1802
$NaCl-CdCl_2-KCl$ 1411–12
$NaCl-CoCl_2$ 1305
$NaCl-CoCl_2-CoSO_4-Na_2SO_4$ ... 1803
$NaCl-CoSO_4$ 1774
$NaCl-CsCl$ 1228
$NaCl-CuCl$ 1245
$NaCl-FeCl_3$ 1315
$NaCl-FeCl_3-TiCl_4$ 1433
$NaCl-HfCl_4$ 1320
$NaCl-HgCl_2$ 1306
$NaCl-KCNS-KCl-NaCNS$ 1871
$NaCl-K_2CO_3-KCl-Na_2CO_3$ 1857
$NaCl-K_2CO_3-K_2SO_4$ 1843
$NaCl-KCl$ 1258–59
$NaCl-KCl-K_2MoO_4-Na_2MoO_4$ ... 1859
$NaCl-KCl-KNO_3-NaNO_3$ 1755
$NaCl-KCl-K_2SO_4-Na_2SO_4$ 1789
$NaCl-KCl-KVO_3-NaVO_3$ 1858
$NaCl-KCl-K_2WO_4-Na_2WO_4$ 1860
$NaCl-KCl-LiCl$ 1407
$NaCl-KCl-LiCl-Li_2SO_4$ 1786–87
$NaCl-KCl-Li_2SO_4$ 1778
$NaCl-KCl-PbCl_2$ 1413
$NaCl-KCl-TiCl_3$ 1414
$NaCl-KCl-ZrCl_4$ 1415
$NaCl-K_2CrO_4-NaF$ 1841
$NaCl-LiCl$ 1285
$NaCl-LiCl-Li_2SO_4$ 1781
$NaCl-LiCl-Li_2SO_4-Na_2SO_4$ ... 1795
$NaCl-LiF$ 1618
$NaCl-MgCl_2$ 1307
$NaCl-MnCl_2$ 1308
$NaCl-NH_4Cl-NH_4NO_3-NaNO_3$ ... 1761
$NaCl-Na$ 1662
$NaCl-NaBr-PbBr_2-PbCl_2$ 1643
$NaCl-NaBr-TlBr-TlCl$ 1642
$NaCl-NaCN$ 1870
$NaCl-Na_2CO_3$ 1820
$NaCl-Na_2CO_3-Na_2SO_4$ 1844
$NaCl-Na_2CrO_4-NaF$ 1842

$NaCl-NaF$ 1619
$NaCl-NaF-Na_2SO_4$ 1782
$NaCl-NaF-SrCl_2-SrF_2$ 1654
$NaCl-NaI-PbCl_2-PbI_2$ 1653
$NaCl-Na_2MoO_4-Na_2WO_4$ 1845
$NaCl-NaNO_3-SrCl_2-Sr(NO_3)_2$. 1764
$NaCl-NaOH$ 1735
$NaCl-NaOH-Na_2SO_4$ 1032
$NaCl-Na_2SO_4$ 1773
$NaCl-Na_2SO_4-ZnCl_2-ZnSO_4$ ... 1804
$NaCl-NbCl_5$ 1328–29
$NaCl-NbCl_5-ZrCl_4$ 1434
$NaCl-NiCl_2$ 1309
$NaCl-PbCl_2$ 1310
$NaCl-PuCl_3$ 1316
$NaCl-RbCl$ 1299
$NaCl-SrCl_2$ 1311
$NaCl-TaCl_5$ 1330–31
$NaCl-TiCl_2$ 1312
$NaCl-TiCl_3$ 1317
$NaCl-TlCl$ 1300
$NaCl-UCl_3$ 1318
$NaCl-UCl_4$ 1321
$NaCl-VCl_2$ 1313
$NaCl-VCl_3$ 1319
$NaCl-ZrCl_4$ 1322–23
$NaClO_4-Ba(ClO_4)_2$ 1018
$NaClO_4-H_2O-H_2O_2$ 2056
$Na_2CrO_4-K_2CrO_4-KF-NaF$ 1864
$Na_2CrO_4-K_2CrO_4-KOH-NaOH$ ... 1035
$Na_2CrO_4-Li_2CrO_4-LiOH-NaOH$. 1038
$Na_2CrO_4-NaCl-NaF$ 1842
$Na_2CrO_4-NaNO_3-Na_2SO_4$ 1159
NaF 1591
$NaF-AlF_3$ 1490
$NaF-AlF_3-Al_2O_3$ 1698
$NaF-AlF_3-Al_2O_3-CaF_2$ 1727
$NaF-AlF_3-Al_2O_3-LiF$ 1723
$NaF-AlF_3-Al_2O_3-MgF_2$ 1728
$NaF-AlF_3-CaF_2$ 1566–67
$NaF-AlF_3-CaO$ 1700
$NaF-AlF_3-CdO$ 1700
$NaF-AlF_3-KF$ 1536
$NaF-AlF_3-LiF$ 1546
$NaF-AlF_3-MgF_2$ 1568
$NaF-AlF_3-MgO$ 1700
$NaF-AlF_3-Na_2SO_4$ 1784
$NaF-AlF_3-TiO_2$ 1700
$NaF-AlF_3-ZnO$ 1700
$NaF-AlF_3-ZrO_2$ 1699
$NaF-BF_3$ 1491
$NaF-BF_3-KF$ 1537
$NaF-B_2O_3-Na_2O$ 1688
$NaF-BaCl_2-BaF_2-NaCl$ 1650
$NaF-BaCl_2-CaF_2$ 1639
$NaF-BaF_2-SiO_2$ 1694
$NaF-BeF_2$ 1482
$NaF-BeF_2-KF$ 1534
$NaF-BeF_2-LiF$ 1542–43
$NaF-BeF_2-RbF$ 1558–59
$NaF-BeF_2-ThF_4$ 1562
$NaF-BeF_2-UF_4$ 1563
$NaF-BeF_2-ZrF_4$ 1564
$NaF-CaCl_2-CaF_2-NaCl$ 1651
$NaF-CaF_2$ 1483
$NaF-CaF_2-LiF$ 1544
$NaF-CaF_2-MgF_2$ 1565
$NaF-CaF_2-SiO_2$ 1695
$NaF-CdCl_2$ 1625
$NaF-CdF_2$ 1484
$NaF-CeF_3$ 1492
$NaF-CeF_3-ZrF_4$ 1570

NaF–CrO$_3$–Na$_2$O..............1691	NaF–ZrF$_4$..............1499, 1509	Na$_2$O–Al$_2$O$_3$–CaO–P$_2$O$_5$–SiO$_2$......985
NaF–FeF$_2$..............1485	NaHCO$_3$–Ca(HCO$_3$)$_2$–CaSO$_4$–H$_2$O–	Na$_2$O–Al$_2$O$_3$–CaO–SiO$_2$.......828–851
NaF–FeF$_3$..............1493	Na$_2$SO$_4$..............2064	Na$_2$O–Al$_2$O$_3$–CaO–SiO$_2$–TiO$_2$..982–984
NaF–HF..............1451	NaHCO$_3$–H$_2$O–Mg(HCO$_3$)$_2$–MgSO$_4$–	Na$_2$O–Al$_2$O$_3$–FeO–SiO$_2$......853–854
NaF–HfF$_4$..............1496	Na$_2$SO$_4$..............2065	Na$_2$O–Al$_2$O$_3$–H$_2$O–K$_2$O–SiO$_2$...2044–45
NaF–KBr–KF–NaBr..............1640	NaI..............1591	Na$_2$O–Al$_2$O$_3$–H$_2$O–Li$_2$O–SiO$_2$......2046
NaF–K$_2$CO$_3$–KF–K$_2$SO$_4$–Na$_2$CO$_3$–	NaI–AgI..............1579	Na$_2$O–Al$_2$O$_3$–H$_2$O–SiO$_2$......2022–29
Na$_2$SO$_4$..............1866–67	NaI–AgI–AgNO$_3$–NaNO$_3$.......1768	Na$_2$O–Al$_2$O$_3$–K$_2$O–SiO$_2$......786–789
NaF–K$_2$CO$_3$–KF–Na$_2$CO$_3$........1863	NaI–CdCl$_2$–CdI$_2$–NaCl..........1652	Na$_2$O–Al$_2$O$_3$–MgO–SiO$_2$......855–856
NaF–K$_2$CO$_3$–K$_2$SO$_4$..............1850	NaI–CdI$_2$–PbI$_2$..............1590	Na$_2$O–Al$_2$O$_3$–SiO$_2$..........501–514
NaF–K$_2$CO$_3$–Na$_2$SO$_4$..............1851	NaI–HgI$_2$..............1584	Na$_2$O–As$_2$O$_5$..............195
NaF–KCl–K$_2$CrO$_4$..............1830	NaI–K$_2$CO$_3$–KI–Na$_2$CO$_3$..........1865	Na$_2$O–As$_2$O$_5$–K$_2$O..............384
NaF–K$_2$CrO$_4$–KF–Na$_2$CrO$_4$......1864	NaI–KI–KNO$_3$–NaNO$_3$..........1769	Na$_2$O–B$_2$O$_3$..............188
NaF–K$_2$CrO$_4$–NaCl..............1841	NaI–KI–K$_2$SO$_4$–Na$_2$SO$_4$..........1811	Na$_2$O–B$_2$O$_3$–CaO–SiO$_2$..............852
NaF–KF..............1453	NaI–Na..............1663	Na$_2$O–B$_2$O$_3$–CrO$_3$–K$_2$O.......790–791
NaF–KF–K$_2$MoO$_4$–Na$_2$MoO$_4$..1719	NaI–NaBr–NaF..............1635	Na$_2$O–B$_2$O$_3$–H$_2$O..............1955
NaF–KF–KNO$_3$–NaNO$_3$........1766	NaI–NaCl–PbCl$_2$–PbI$_2$..........1653	Na$_2$O–B$_2$O$_3$–K$_2$O..............380
NaF–KF–K$_2$SO$_4$–Na$_2$SO$_4$........1806	NaI–NaOH..............1741	Na$_2$O–B$_2$O$_3$–Li$_2$O..............429
NaF–KF–K$_2$TiO$_3$–Na$_2$TiO$_3$......1717	Na$_2$MoO$_4$–Ag$_2$MoO$_4$–Ag$_2$SO$_4$–Na$_2$SO$_4$...	Na$_2$O–B$_2$O$_3$–NaF..............1688
NaF–KF–KVO$_3$–NaVO$_3$.........1718	1161	Na$_2$O–B$_2$O$_3$–SiO$_2$..........515–518
NaF–KF–K$_2$WO$_4$–Na$_2$WO$_4$.....1720	Na$_2$MoO$_4$–KCl–K$_2$MoO$_4$–NaCl...1859	Na$_2$O–BaO–Fe$_2$O$_3$..............477
NaF–KF–LiF..............1529	Na$_2$MoO$_4$–KF–K$_2$MoO$_4$–NaF.....1719	Na$_2$O–BaO–Nb$_2$O$_5$–TiO$_2$..........826
NaF–KF–MgF$_2$..............1535	Na$_2$MoO$_4$–Li$_2$MoO$_4$–Li$_2$SO$_4$–Na$_2$SO$_4$....	Na$_2$O–BaO–SiO$_2$......437, 478–479
NaF–KF–RbF..............1533	1170	Na$_2$O–CO$_2$–H$_2$O..............1958
NaF–KF–UF$_4$..............1538	Na$_2$MoO$_4$–NaCl–Na$_2$WO$_4$..........1845	Na$_2$O–CO$_2$–H$_2$O–ThO$_2$..........2030
NaF–KF–ZrF$_4$..............1539	Na$_2$MoO$_4$–Na$_2$SO$_4$–PbMoO$_4$–PbSO$_4$...	Na$_2$O–CO$_2$–H$_2$O–UO$_3$..........2031
NaF–K$_2$SO$_4$–Na$_2$CO$_3$..............1852	1173	Na$_2$O–CaF$_2$–SiO$_2$..............1705
NaF–LiF..............1467	NaNH$_2$–KNH$_2$..............1912	Na$_2$O–CaO–P$_2$O$_5$..............486
NaF–LiF–Li$_2$SO$_4$–Na$_2$SO$_4$........1808	NaNO$_2$–Ca(NO$_3$)$_2$..............1082	Na$_2$O–CaO–SO$_3$–SiO$_2$..........1141–42
NaF–LiF–Li$_2$SiO$_3$–Na$_2$SiO$_3$......1721	NaNO$_2$–KNO$_2$..............1077	Na$_2$O–CaO–SiO$_2$..........481–485
NaF–LiF–Li$_2$TiO$_3$–Na$_2$TiO$_3$.....1722	NaNO$_2$–KNO$_2$–KNO$_3$–NaNO$_3$....1083	Na$_2$O–CrO$_3$–K$_2$O..............387
NaF–LiF–MgF$_2$..............1545	NaNO$_2$–KNO$_3$..............1080	Na$_2$O–CrO$_3$–NaF..............1691
NaF–LiF–RbF..............1541	NaNO$_2$–KOH..............1019	Na$_2$O–CrO$_3$–P$_2$O$_5$..............544
NaF–LiF–UF$_4$..............1547	NaNO$_2$–NaNO$_3$..............1081	Na$_2$O–CrO$_3$–TiO$_2$..............546
NaF–LiF–ZrF$_4$..............1548	NaNO$_2$–NaOH..............1019, 1028	Na$_2$O–FeO–SiO$_2$..........487–492
NaF–Li$_2$O–TiO$_2$..........1687, 1704	NaNO$_3$–AgI–AgNO$_3$–NaI.......1768	Na$_2$O–Fe$_2$O$_3$..............189
NaF–MgF$_2$..............1486	NaNO$_3$–AgNO$_3$..............1040	Na$_2$O–Fe$_2$O$_3$–H$_2$O–MgO–SiO$_2$.....2047
NaF–MgF$_2$–SiO$_2$..............1696	NaNO$_3$–AgNO$_3$–TlNO$_3$..........1066	Na$_2$O–Fe$_2$O$_3$–SiO$_2$..........519–530
NaF–MoO$_3$–Na$_2$O–WO$_3$..........1726	NaNO$_3$–BaCl$_2$–Ba(NO$_3$)$_2$–NaCl...1762	Na$_2$O–Ga$_2$O$_3$–H$_2$O..........1956–57
NaF–Na..............1665	NaNO$_3$–Ba(NO$_3$)$_2$–KNO$_3$..........1073	Na$_2$O–GeO$_2$..............190–191
NaF–NaBr–NaI..............1635	NaNO$_3$–CaCl$_2$–Ca(NO$_3$)$_2$–NaCl...1763	Na$_2$O–H$_2$O–P$_2$O$_5$..............1972–73
NaF–Na$_2$CO$_3$..............1827	NaNO$_3$–Ca(NO$_3$)$_2$..............1050	Na$_2$O–H$_2$O–PbO..............1953
NaF–Na$_2$CO$_3$–Na$_2$SO$_4$..............1853	NaNO$_3$–Ca(NO$_3$)$_2$–KNO$_3$..........1074	Na$_2$O–H$_2$O–SiO$_2$..........1959–71
NaF–NaCl..............1619	NaNO$_3$–Cd(NO$_3$)$_2$–LiNO$_3$..........1076	Na$_2$O–H$_2$O–UO$_3$..............1975
NaF–NaCl–Na$_2$CrO$_4$..............1842	NaNO$_3$–KBr–KNO$_3$–NaBr.......1752	Na$_2$O–H$_2$O–V$_2$O$_5$..............1974
NaF–NaCl–Na$_2$SO$_4$..............1782	NaNO$_3$–KCl–KNO$_3$–NaCl.......1755	Na$_2$O–H$_2$O–ZnO..............1954
NaF–NaCl–SrCl$_2$–SrF$_2$..............1654	NaNO$_3$–KF–KNO$_3$–NaF........1766	Na$_2$O–K$_2$O–MoO$_3$..............388
NaF–Na$_2$O–MoO$_3$..............1692	NaNO$_3$–KI–KNO$_3$–NaI..........1769	Na$_2$O–K$_2$O–MoO$_3$–P$_2$O$_5$..........792
NaF–Na$_2$O–SiO$_2$..............1689	NaNO$_3$–KNO$_2$–KNO$_3$–NaNO$_2$...1083	Na$_2$O–K$_2$O–MoO$_3$–WO$_3$..........793
NaF–Na$_2$O–TiO$_2$..............1690	NaNO$_3$–KNO$_3$..............1047	Na$_2$O–K$_2$O–P$_2$O$_5$..........385–386
NaF–Na$_2$O–WO$_3$..............1693	NaNO$_3$–KNO$_3$–LiNO$_3$..............1069	Na$_2$O–K$_2$O–SiO$_2$..........381–383
NaF–NaOH..............1738	NaNO$_3$–KNO$_3$–RbNO$_3$..............1072	Na$_2$O–K$_2$O–WO$_3$..............389
NaF–Na$_2$SO$_4$..............1776	NaNO$_3$–KOH..............1020	Na$_2$O–Li$_2$O–MoO$_3$..............439
NaF–Na$_2$SO$_4$–PbF$_2$–PbSO$_4$........1810	NaNO$_3$–LiNO$_3$–LiOH–NaOH.....1037	Na$_2$O–Li$_2$O–SiO$_2$..........430–437
NaF–Na$_2$SO$_4$–RbF–Rb$_2$SO$_4$........1809	NaNO$_3$–NH$_4$Cl–NH$_4$NO$_3$–NaCl.....1761	Na$_2$O–Li$_2$O–SiO$_2$–TiO$_2$..............476
NaF–NiF$_2$..............1487	NaNO$_3$–NH$_4$NO$_3$..............1053	Na$_2$O–Li$_2$O–TiO$_2$..........438, 476
NaF–PbF$_2$..............1488	NaNO$_3$–NH$_4$NO$_3$–(NH$_4$)$_2$SO$_4$–Na$_2$SO$_4$..	Na$_2$O–Li$_2$O–WO$_3$..............440
NaF–PbF$_2$–UF$_4$..............1569	1172	Na$_2$O–MnO–SiO$_2$..............493
NaF–PbO–TiO$_2$..............750	NaNO$_3$–NaCNS..............1910	Na$_2$O–MoO$_3$..............199
NaF–PuF$_3$..............1494	NaNO$_3$–Na$_2$CO$_3$..............1087	Na$_2$O–MoO$_3$–NaF..............1692
NaF–RbF..............1481	NaNO$_3$–NaCl–SrCl$_2$–Sr(NO$_3$)$_2$....1764	Na$_2$O–MoO$_3$–NaF–WO$_3$..........1726
NaF–RbF–UF$_4$..............1560	NaNO$_3$–Na$_2$CrO$_4$–Na$_2$SO$_4$..........1159	Na$_2$O–MoO$_3$–P$_2$O$_5$..............544
NaF–RbF–ZrF$_4$..............1561	NaNO$_3$–NaNO$_2$..............1081	Na$_2$O–MoO$_3$–PbO–TiO$_2$..........750
NaF–SiO$_2$–SrF$_2$..............1697	NaNO$_3$–NaOH..............1020	Na$_2$O–MoO$_3$–TiO$_2$..............547
NaF–ThF$_4$..............1497	NaNO$_3$–Na$_2$SO$_4$..............1152	Na$_2$O–MoO$_3$–WO$_3$..............543
NaF–ThF$_4$–UF$_4$..............1571	NaNO$_3$–TlNO$_3$..............1054	Na$_2$O–NaF–SiO$_2$..............1689
NaF–ThF$_4$–ZrF$_4$..............1572	Na$_2$O–Al$_2$O$_3$–B$_2$O$_3$–H$_2$O–MgO–SiO$_2$...	Na$_2$O–NaF–TiO$_2$..............1690
NaF–UF$_3$..............1495	2048	Na$_2$O–NaF–WO$_3$..............1693
NaF–UF$_4$..............1498	Na$_2$O–Al$_2$O$_3$–CaO..............480	Na$_2$O–Nb$_2$O$_5$..............196
NaF–UF$_4$–ZrF$_4$..............1573	Na$_2$O–Al$_2$O$_3$–CaO–Fe$_2$O$_3$..........827	Na$_2$O–P$_2$O$_5$..............197
NaF–ZnF$_2$..............1489	Na$_2$O–Al$_2$O$_3$–CaO–K$_2$O–SiO$_2$...971–972	Na$_2$O–P$_2$O$_5$–PbO–TiO$_2$..........750
	Na$_2$O–Al$_2$O$_3$–CaO–MgO–SiO$_2$..975–981	Na$_2$O–P$_2$O$_5$–SO$_3$..............544

Na_2O–P_2O_5–SiO_2 535–542	Na_2SO_4–Li_2MoO_4–Li_2SO_4–Na_2MoO_4 1170	Nb_2O_5–K_2O–Ta_2O_5 424
Na_2O–P_2O_5–TiO_2 545	Na_2SO_4–Li_2SO_4 1104	Nb_2O_5–Na_2O 196
Na_2O–P_2O_5–WO_3 544	Na_2SO_4–$MgSO_4$ 1117–18	Nb_2O_5–PbO 287
Na_2O–PbO–SiO_2 494–499	Na_2SO_4–$MnSO_4$ 1119	Nb_2O_5–SiO_2 363
Na_2O–PbO–SiO_2–TiO_2 749	Na_2SO_4–NH_4NO_3–$(NH_4)_2SO_4$–$NaNO_3$ 1172	Nb_2O_5–Ta_2O_5 375
Na_2O–PbO–TiO_2–V_2O_5 749	Na_2SO_4–$NaBr$ 1771	Nb_2O_5–TiO_2 372
Na_2O–PbO–TiO_2–WO_3 750	Na_2SO_4–Na_2CO_3–$NaCl$ 1844	Nb_2O_5–ZnO 304
Na_2O–PbO–WO_3 500	Na_2SO_4–Na_2CO_3–NaF 1853	Nb_2O_5–ZrO_2 373
Na_2O–SO_3–TiO_2 546	Na_2SO_4–$NaCl$ 1773	$NdCl_3$–$CaCl_2$–$NaCl$ 1430
Na_2O–SO_3–V_2O_5 1140	Na_2SO_4–$NaCl$–NaF 1782	$NdCl_3$–$CsCl$ 1237
Na_2O–SiO_2 . . . . 168, 192	Na_2SO_4–$NaCl$–$NaOH$ 1032	$NdCl_3$–KCl 1277
Na_2O–SiO_2–TiO_2 531–532	Na_2SO_4–$NaCl$–$ZnCl_2$–$ZnSO_4$ 1804	$NdCl_3$–KCl–$MgCl_2$ 1424
Na_2O–SiO_2–ZrO_2 533–534	Na_2SO_4–Na_2CrO_4–$NaNO_3$ 1159	Nd_2O_3–Al_2O_3 312
Na_2O–Ta_2O_5 198	Na_2SO_4–NaF 1776	Nd_2O_3–Cr_2O_3 331
Na_2O–TiO_2 193–194	Na_2SO_4–NaF–PbF_2–$PbSO_4$ 1810	Nd_2O_3–Dy_2O_3 348
Na_2O–TiO_2–V_2O_5 545	Na_2SO_4–NaF–RbF–Rb_2SO_4 1809	Nd_2O_3–Er_2O_3 348
Na_2O–TiO_2–WO_3 547	Na_2SO_4–Na_2MoO_4–$PbMoO_4$–$PbSO_4$ 1173	Nd_2O_3–Eu_2O_3 348
Na_2O–V_2O_5 106	Na_2SO_4–$NaNO_3$ 1152	Nd_2O_3–Fe_2O_3 337
Na_2O–WO_3 200	Na_2SO_4–$Na_4P_2O_7$ 544	Nd_2O_3–Ga_2O_3 340
$NaOH$–K_2CO_3–KOH–Na_2CO_3 . . . 1034	Na_2SO_4–Na_2TiO_3 546	Nd_2O_3–Gd_2O_3 348
$NaOH$–K_2CrO_4–KOH–Na_2CrO_4 . . 1035	Na_2SO_4–Na_2WO_4 . . . . 1122, 1153	Nd_2O_3–H_2O 1937
$NaOH$–KNO_2 1019	Na_2SO_4–Na_2WO_4–$PbSO_4$–$PbWO_4$ 1174	Nd_2O_3–Ho_2O_3 348
$NaOH$–KNO_3 . . . . 1020, 1029	Na_2SO_4–$NiSO_4$ 1120	Nd_2O_3–In_2O_3 344
$NaOH$–KOH 1024	Na_2SO_4–$PbSO_4$ 1121–22	Nd_2O_3–La_2O_3 345
$NaOH$–KOH–K_2SO_4–Na_2SO_4 1036	Na_2SO_4–$PbWO_4$ 1122	Nd_2O_3–Ln_2O_3 348
$NaOH$–KOH–$LiOH$ 1031	Na_2SO_4–Sb_2S_3 1913	Nd_2O_3–Lu_2O_3 348
$NaOH$–Li_2CrO_4–$LiOH$–Na_2CrO_4 . . 1038	Na_2SO_4–V_2O_5 1138	Nd_2O_3–Sc_2O_3 351
$NaOH$–$LiNO_3$–$LiOH$–$NaNO_3$ 1037	Na_2SO_4–$ZnSO_4$ 1123	Nd_2O_3–Sm_2O_3 348
$NaOH$–$LiOH$ 1026–27	Na_2SiO_3–LiF–Li_2SiO_3–NaF 1721	Nd_2O_3–Tm_2O_3 348
$NaOH$–$NaBr$ 1732	Na_2TiO_3–KF–K_2TiO_3–NaF 1717	Nd_2O_3–UO_2 349
$NaOH$–$NaCl$ 1735	Na_2TiO_3–LiF–Li_2TiO_3–NaF 1722	Nd_2O_3–Yb_2O_3 348
$NaOH$–$NaCl$–Na_2SO_4 1032	Na_2TiO_3–Na_2SO_4 546	Nd_2O_3–ZrO_2 350
$NaOH$–NaF 1738	$NaVO_3$–KCl–KVO_3–$NaCl$ 1858	Ni–Co–O 53
$NaOH$–NaI 1741	$NaVO_3$–KF–KVO_3–NaF 1718	Ni–Fe–O 75–78
$NaOH$–$NaNO_2$ 1019, 1028	$NaVO_3$–V_2O_5 105	Ni–O 16
$NaOH$–$NaNO_3$ 1020	Na_2WO_4–H_2O 2049	Ni–S 1883–84
$Na_4P_2O_7$–Na_2SO_4 544	Na_2WO_4–H_2O–Na_2CO_3 2059–60	$NiCl_2$–$NaCl$ 1309
Na_2SO_4–Ag_2MoO_4–Ag_2SO_4–Na_2MoO_4 1161	Na_2WO_4–KCl–K_2WO_4–$NaCl$ 1860	NiF_2–KF 1459
Na_2SO_4–Ag_2SO_4 1090	Na_2WO_4–KF–K_2WO_4–NaF 1720	NiF_2–NaF 1487
Na_2SO_4–AlF_3–NaF 1784	Na_2WO_4–$NaCl$–Na_2MoO_4 1845	NiO–BaO 205
Na_2SO_4–$BaSO_4$ 1112	Na_2WO_4–Na_2SO_4 . . . . 1122, 1153	NiO–Bi_2O_3 326
Na_2SO_4–$Ca(HCO_3)_2$–$CaSO_4$–H_2O–$NaHCO_3$ 2064	Na_2WO_4–Na_2SO_4–$PbSO_4$–$PbWO_4$ 1174	NiO–CaO 230
Na_2SO_4–$CaSO_4$ 1113	Na_2WO_4–$PbSO_4$ 1122	NiO–CoO 53
Na_2SO_4–$CdCl_2$–$CdSO_4$–$NaCl$ 1802	Na_2WO_4–$PbWO_4$ 1122	NiO–GeO_2–MgO–SiO_2 964–965
Na_2SO_4–$CdSO_4$ 1114	$NbCl_5$–$AlCl_3$–$FeCl_3$ 1436	NiO–H_2O–SiO_2 2005
Na_2SO_4–$CdSO_4$–K_2SO_4 1129	$NbCl_5$–$AlCl_3$–$NaCl$ 1431	NiO–MgO 258
Na_2SO_4–$CdSO_4$–Li_2SO_4 1135	$NbCl_5$–$AlCl_3$–$TaCl_5$ 1438	NiO–MgO–SiO_2 702
Na_2SO_4–$CoCl_2$–$CoSO_4$–$NaCl$ 1803	$NbCl_5$–$CsCl$ 1328	NiO–V_2O_5 279
Na_2SO_4–$CoSO_4$ 1115	$NbCl_5$–$FeCl_3$–$TaCl_5$ 1440	Ni_2S–Cu_2S 1890
Na_2SO_4–$CoSO_4$–$NiSO_4$ 1136	$NbCl_5$–$FeCl_3$–$TiCl_4$ 1439	Ni_3S_2–Co_4S_3–FeS 1900–01
Na_2SO_4–$CuSO_4$ 1116	$NbCl_5$–KCl 1328–29	Ni_3S_2–Cu_2S 1891
Na_2SO_4–$CuSO_4$–K_2SO_4 1130–33	$NbCl_5$–$LiCl$ 1328	$NiSO_4$–$CoSO_4$–Na_2SO_4 1136
Na_2SO_4–H_2O–$Mg(HCO_3)_2$–$MgSO_4$–$NaHCO_3$ 2065	$NbCl_5$–$NaCl$ 1328–29	$NiSO_4$–Na_2SO_4 1120
Na_2SO_4–K_2CO_3–KF–K_2SO_4–Na_2CO_3–NaF 1866–67	$NbCl_5$–$NaCl$–$ZrCl_4$ 1434	
Na_2SO_4–K_2CO_3–K_2SO_4–Na_2CO_3 . . 1167	$NbCl_5$–$RbCl$ 1328–29	O–Ag–Cu–Pb 125
Na_2SO_4–K_2CO_3–NaF 1851	$NbCl_5$–$TaCl_5$–$TiCl_4$ 1441	O–Al 1–2
Na_2SO_4–KCl–K_2SO_4–$NaCl$ 1789	$NbCl_5$–$TiCl_4$ 1402	O–Al–Cr–Fe 126–130
Na_2SO_4–KCl–Li_2SO_4 1780	Nb_2O_5–BaO 214	O–Al–Fe 26–35
Na_2SO_4–KCl–Na_2CO_3 1833	Nb_2O_5–BaO–Gd_2O_3 563	O–Al–Fe–Si 131–136
Na_2SO_4–KF–K_2SO_4–NaF 1806	Nb_2O_5–BaO–Na_2O–TiO_2 826	O–C–Ti 36
Na_2SO_4–KI–K_2SO_4–NaI 1811	Nb_2O_5–Bi_2O_3 . . . . 324, 329	O–Ca–Cr 37–39
Na_2SO_4–KOH–K_2SO_4–$NaOH$ 1036	Nb_2O_5–CaO 244	O–Ca–Fe 40–48
Na_2SO_4–K_2SO_4 1093	Nb_2O_5–CdO 253	O–Ce 3
Na_2SO_4–K_2SO_4–Li_2SO_4 1126	Nb_2O_5–Fe_2O_3–Gd_2O_3 779	O–Co 4
Na_2SO_4–K_2SO_4–$ZnSO_4$ 1134	Nb_2O_5–Gd_2O_3 343	O–Co–Fe 49–50
Na_2SO_4–$LiCl$–Li_2SO_4–$NaCl$ 1795	Nb_2O_5–K_2CO_3 171	O–Co–Mg 51–52
Na_2SO_4–LiF–Li_2SO_4–NaF 1808	Nb_2O_5–K_2O 171	O–Co–Ni 53
		O–Co–V 54–55
		O–Co–Zn 56
		O–Cr 5–6
		O–Cr–Fe 57–58
		O–Cr–Fe–Mg 137

O–Cu............................7	P_2O_5–K_2O–PbO–TiO_2............750	$PbCl_2$–$SrCl_2$....................1383
O–Cu–Fe........................59	P_2O_5–K_2O–SO_3...................428	$PbCl_2$–TlCl....................1343
O–Fe.........................8–10	P_2O_5–K_2O–TiO_2..............422–423	$PbCl_2$–$ZnCl_2$....................1384
O–Fe–La......................60–61	P_2O_5–K_2O–WO_3...................428	$PbCl_4$–CCl_4....................1396
O–Fe–Li.........................62	P_2O_5–Li_2O–TiO_2..................476	$PbCl_4$–$SiCl_4$....................1397
O–Fe–Mg......................63–69	P_2O_5–MgO.......................272	$PbCl_4$–$SnCl_4$....................1398
O–Fe–Mg–Si..................138–145	P_2O_5–MgO–SiO_2............725–728	$PbCl_4$–$TiCl_4$....................1399
O–Fe–Mn......................70–74	P_2O_5–MgO–SrO...................703	$PbCrO_4$–$PbSO_4$..................1154
O–Fe–Ni......................75–78	P_2O_5–MgO–ZnO...............707–709	PbF_2–AgF.......................1443
O–Fe–Sc.........................79	P_2O_5–MnO.......................278	PbF_2–As_2O_5–PbO...............1712
O–Fe–Si......................80–87	P_2O_5–MnO–ZnO...................732	PbF_2–BeF_2....................1515
O–Fe–Si–Ti..................146–147	P_2O_5–MoO_3–Na_2O................544	PbF_2–KF–K_2SO_4–$PbSO_4$........1807
O–Fe–Ti......................88–93	P_2O_5–Na_2O....................197	PbF_2–KF–UF_4..................1540
O–Fe–V.......................94–96	P_2O_5–Na_2O–PbO–TiO_2...........750	PbF_2–NaF......................1488
O–Fe–Y.......................97–98	P_2O_5–Na_2O–SO_3.................544	PbF_2–NaF–Na_2SO_4–$PbSO_4$......1810
O–Ge............................11	P_2O_5–Na_2O–SiO_2............535–542	PbF_2–NaF–UF_4.................1569
O–Mg–U......................99–100	P_2O_5–Na_2O–TiO_2................545	PbF_2–$PbBr_2$....................1605
O–Mn.........................12–15	P_2O_5–Na_2O–WO_3.................544	PbF_2–$PbCl_2$....................1627
O–Mn–Si....................101–103	P_2O_5–PbO.......................288	PbF_2–PbI_2....................1632
O–N–Ti.........................104	P_2O_5–PbO–SiO_2..................746	PbF_2–PbO......................1674
O–Na–V.....................105–106	P_2O_5–SiO_2....................364	PbF_2–PbO–V_2O_5................1713
O–Ni............................16	P_2O_5–SrO–ZnO...................754	PbF_2–$Pb_3(PO_4)_2$...............1828
O–Pb.........................17–18	P_2O_5–ZnO...................305–307	PbF_2–UF_4....................1522
O–Pb–Sb....................107–112	$POCl_3$–$AlCl_3$...................1822	PbI_2–$CdCl_2$–CdI_2–$PbCl_2$......1657
O–Pr............................19	$POCl_3$–$SbCl_5$...................1823	PbI_2–CdI_2–NaI................1590
O–Sb............................20	Pb–Ag–Cu–O.....................125	PbI_2–KCl–KI–$PbCl_2$............1646
O–Si............................21	Pb–Mg–$MgCl_2$–$PbCl_2$...........1670	PbI_2–NaCl–NaI–$PbCl_2$..........1653
O–Si–Ti....................113–115	Pb–O..........................17–18	PbI_2–$PbBr_2$....................1606
O–Th–U.....................116–118	Pb–O–Sb....................107–112	PbI_2–$PbBr_2$–$PbCl_2$............1636
O–Ti............................22	Pb–PbO..........................17	PbI_2–$PbCl_2$....................1628
O–U.............................23	$PbBr_2$–AgBr.....................1180	PbI_2–PbF_2....................1632
O–U–Zr.....................119–124	$PbBr_2$–$BiBr_3$...................1202	$PbMoO_4$–Na_2MoO_4–Na_2SO_4–Pb-
O–V.............................24	$PbBr_2$–$CdBr_2$...................1199	SO_4.........................1173
O–Zr............................25	$PbBr_2$–$CdBr_2$–$CdCl_2$–$PbCl_2$......1644	$PbMoO_4$–$PbSO_4$..................1155
	$PbBr_2$–$CdBr_2$–NaBr..............1209	$Pb(NO_3)_2$–HNO_3–H_2O..........2062
PI_3–AsI_3.....................1587	$PbBr_2$–$HgBr_2$...................1201	$Pb(NO_3)_2$–NH_4NO_3..............1055
PI_3–SbI_3.....................1589	$PbBr_2$–NaBr–NaCl–$PbCl_2$.......1643	PbO............................148
P_2O_5.......................158–159	$PbBr_2$–$PbCl_2$...................1604	PbO–Al_2O_3......................280
P_2O_5–Al_2O_3...................318	$PbBr_2$–$PbCl_2$–PbI_2.............1636	PbO–Al_2O_3–SiO_2............737–739
P_2O_5–Al_2O_3–CaO.............640–644	$PbBr_2$–PbF_2....................1605	PbO–As_2O_5......................286
P_2O_5–Al_2O_3–CaO–Na_2O–SiO_2.....985	$PbBr_2$–PbI_2....................1606	PbO–As_2O_5–PbF_2...............1712
P_2O_5–Al_2O_3–CaO–SiO_2.......958–962	$PbBr_2$–PbO.....................1812	PbO–B_2O_3.......................281
P_2O_5–Al_2O_3–SrO...............756	$PbCl_2$–AgCl.....................1223	PbO–B_2O_3–BaO–TiO_2............859
P_2O_5–B_2O_3–K_2O...............417	$PbCl_2$–$BaCl_2$...................1354	PbO–B_2O_3–SiO_2............740–742
P_2O_5–B_2O_3–Li_2O...........470–473	$PbCl_2$–$BeCl_2$...................1358	PbO–B_2O_3–TiO_2................743
P_2O_5–B_2O_3–SiO_2...........775–777	$PbCl_2$–$BiCl_3$...................1385	PbO–BaO–SnO_2–TiO_2........860–861
P_2O_5–BaO–H_2O.................1977	$PbCl_2$–$CaCl_2$...................1362	PbO–BaO–TiO_2–ZrO_2............862
P_2O_5–BaO–TiO_2.................568	$PbCl_2$–$CaCl_2$–KCl...............1421	PbO–Bi_2O_3......................326
P_2O_5–Bi_2O_3...................329	$PbCl_2$–$CdBr_2$–$CdCl_2$–$PbBr_2$......1644	PbO–Bi_2O_3–MoO_3................744
P_2O_5–CaO...................245–250	$PbCl_2$–$CdCl_2$...................1368	PbO–Bi_2O_3–WO_3................745
P_2O_5–CaO–FeO...................595	$PbCl_2$–$CdCl_2$–CdI_2–PbI_2........1657	PbO–CrO_3.......................291
P_2O_5–CaO–MgO...................613	$PbCl_2$–$CdCl_2$–KCl...............1422	PbO–CrO_3–MoO_3.................751
P_2O_5–CaO–MgO–SiO_2.......935–938	$PbCl_2$–CuCl.....................1252	PbO–CrO_3–WO_3.................752
P_2O_5–CaO–Na_2O.................486	$PbCl_2$–$FeCl_3$...................1386	PbO–Cu_2O.......................163
P_2O_5–CaO–RO....................645	$PbCl_2$–KCl.....................1269	PbO–Fe_2O_3......................282
P_2O_5–CaO–SiO_2..............665–671	$PbCl_2$–KCl–KI–PbI_2............1646	PbO–GeO_2.......................283
P_2O_5–CaO–SrO...................623	$PbCl_2$–KCl–NaCl................1413	PbO–H_2O–Na_2O.................1953
P_2O_5–CrO_3–K_2O................428	$PbCl_2$–KCl–$ZnCl_2$..............1425	PbO–K_2CO_3–TiO_2...............749
P_2O_5–CrO_3–Na_2O...............544	$PbCl_2$–LiCl.....................1294	PbO–KF–TiO_2....................749
P_2O_5–Cr_2O_3–H_2O.............2013–14	$PbCl_2$–Mg–$MgCl_2$–Pb..........1670	PbO–K_2O–MoO_3–TiO_2.......749, 812
P_2O_5–Fe_2O_3...................338	$PbCl_2$–$MgCl_2$...................1373	PbO–K_2O–P_2O_5–TiO_2...........750
P_2O_5–Fe_2O_3–H_2O..............2016	$PbCl_2$–$MnCl_2$...................1378	PbO–K_2O–SiO_2.............404–405
P_2O_5–Ga_2O_3–SiO_2...............780	$PbCl_2$–NaBr–NaCl–$PbBr_2$.......1643	PbO–K_2O–SiO_2–TiO_2............749
P_2O_5–H_2O–MnO................2003–04	$PbCl_2$–NaCl.....................1310	PbO–K_2O–TiO_2–V_2O_5...........749
P_2O_5–H_2O–Na_2O.............1972–73	$PbCl_2$–NaCl–NaI–PbI_2.........1653	PbO–MoO_3.......................292
P_2O_5–H_2O–UO_2................2018	$PbCl_2$–$PbBr_2$...................1604	PbO–MoO_3–Na_2O–TiO_2...........750
P_2O_5–K_2O....................172	$PbCl_2$–$PbBr_2$–PbI_2.............1636	PbO–MoO_3–WO_3.................753
P_2O_5–K_2O–Li_2O................378	$PbCl_2$–PbF_2....................1627	PbO–Na_2CO_3–TiO_2..............749
P_2O_5–K_2O–Li_2O–TiO_2..........476	$PbCl_2$–PbI_2....................1628	PbO–NaF–TiO_2...................750
P_2O_5–K_2O–MoO_3................428	$PbCl_2$–PbO.....................1821	PbO–Na_2O–P_2O_5–TiO_2..........750
P_2O_5–K_2O–MoO_3–Na_2O..........792	$PbCl_2$–RbCl.....................1334	PbO–Na_2O–SiO_2............494–499
P_2O_5–K_2O–Na_2O............385–386	$PbCl_2$–$SnCl_2$...................1382	PbO–Na_2O–SiO_2–TiO_2..........749

PbO–Na$_2$O–TiO$_2$–V$_2$O$_5$.............749	RbCl–KCl....................1260	Sb–O–Pb.................107–112
PbO–Na$_2$O–TiO$_2$–WO$_3$..........750	RbCl–LiCl....................1287	SbBr$_3$–AsBr$_3$............1204
PbO–Na$_2$O–WO$_3$..................500	RbCl–MgCl$_2$...............1333	SbBr$_3$–SbCl$_3$...........1609
PbO–Nb$_2$O$_5$............................287	RbCl–NaCl...................1299	SbBr$_3$–SbI$_3$.............1610
PbO–P$_2$O$_5$..............................288	RbCl–NbCl$_5$........1328–29	SbCl$_3$–SbBr$_3$...........1609
PbO–P$_2$O$_5$–SiO$_2$................746	RbCl–PbCl$_2$.................1334	SbCl$_3$–SbI$_3$.............1630
PbO–Pb.....................................17	RbCl–Rb.....................1664	SbCl$_5$–POCl$_3$...........1823
PbO–PbBr$_2$..........................1812	RbCl–TaCl$_5$.........1330–31	SbI$_3$–AsI$_3$...............1588
PbO–PbCl$_2$..........................1821	RbCl–TiCl$_3$.................1336	SbI$_3$–PI$_3$................1589
PbO–PbF$_2$...........................1674	RbCl–TlCl....................1332	SbI$_3$–SbBr$_3$.............1610
PbO–PbF$_2$–V$_2$O$_5$.................1713	RbCl–UCl$_3$.................1337	SbI$_3$–SbCl$_3$.............1630
PbO–PbSO$_4$.........................1139	RbCl–ZnCl$_2$................1335	Sb$_2$O$_3$–Bi$_2$O$_3$................327
PbO–Sb$_2$O$_3$.................108–109	RbF–AlF$_3$...................1504	Sb$_2$O$_3$–PbO...........108–109
PbO–Sb$_2$O$_4$............................110	RbF–BaCl$_2$–BaF$_2$–RbCl...1655	Sb$_2$O$_3$–Sb$_2$S$_3$.............1903
PbO–Sb$_2$O$_5$....................111–112	RbF–BeF$_2$..............1500–01	Sb$_2$O$_4$–PbO....................110
PbO–SiO$_2$..............................284	RbF–BeF$_2$–LiF.............1549	Sb$_2$O$_5$–PbO...............111–112
PbO–SnO$_2$.............................285	RbF–BeF$_2$–NaF......1558–59	Sb$_2$S$_3$–Ag$_2$S...............1889
PbO–SrO–TiO$_2$–ZrO$_2$..........970	RbF–CaF$_2$..................1502	Sb$_2$S$_3$–BiS.................1894
PbO–Ta$_2$O$_5$...........................289	RbF–CrO$_3$–Rb$_2$O.........1701	Sb$_2$S$_3$–Na$_2$SO$_4$...........1913
PbO–TiO$_2$–V$_2$O$_5$...................748	RbF–KF......................1454	Sb$_2$S$_3$–PbS................1898
PbO–TiO$_2$–ZrO$_2$....................747	RbF–KF–LiF................1530	Sb$_2$S$_3$–Sb$_2$O$_3$..............1903
PbO–V$_2$O$_5$............................290	RbF–KF–NaF..............1533	Sb$_2$S$_3$–SnS................1899
PbO–WO$_3$.............................293	RbF–LiF.....................1468	Sc–Fe–O........................79
PbO$_2$.....................................152	RbF–LiF–NaF.............1541	Sc$_2$O$_3$–Al$_2$O$_3$...............312
Pb$_3$(PO$_4$)$_2$–PbF$_2$..................1828	RbF–LiF–UF$_4$.............1550	Sc$_2$O$_3$–Cr$_2$O$_3$..............331
PbS–Sb$_2$S$_3$..........................1898	RbF–Li$_2$O–TiO$_2$..........1704	Sc$_2$O$_3$–Dy$_2$O$_3$..............351
PbS–Tl$_2$S.............................1892	RbF–MoO$_3$–Rb$_2$O.......1702	Sc$_2$O$_3$–Er$_2$O$_3$...............351
PbSO$_4$–KF–K$_2$SO$_4$–PbF$_2$......1807	RbF–NaF....................1481	Sc$_2$O$_3$–Eu$_2$O$_3$..............351
PbSO$_4$–K$_2$SO$_4$..................1102	RbF–NaF–Na$_2$SO$_4$–Rb$_2$SO$_4$...1809	Sc$_2$O$_3$–Fe$_2$O$_3$................79
PbSO$_4$–K$_2$SO$_4$–K$_2$WO$_4$–PbWO$_4$....1169	RbF–NaF–UF$_4$............1560	Sc$_2$O$_3$–Ga$_2$O$_3$.........339, 340
PbSO$_4$–Li$_2$SO$_4$...................1110	RbF–NaF–ZrF$_4$...........1561	Sc$_2$O$_3$–Gd$_2$O$_3$..............351
PbSO$_4$–Li$_2$SO$_4$–Li$_2$WO$_4$–PbWO$_4$...1171	RbF–Rb.................1664–65	Sc$_2$O$_3$–H$_2$O................1938
PbSO$_4$–NaF–Na$_2$SO$_4$–PbF$_2$......1810	RbF–Rb$_2$O–WO$_3$.........1703	Sc$_2$O$_3$–Ho$_2$O$_3$..............351
PbSO$_4$–Na$_2$MoO$_4$–Na$_2$SO$_4$–Pb-MoO$_4$....1173	RbF–ThF$_4$..................1505	Sc$_2$O$_3$–In$_2$O$_3$................351
PbSO$_4$–Na$_2$SO$_4$............1121–22	RbF–UF$_4$....................1506	Sc$_2$O$_3$–La$_2$O$_3$..............351
PbSO$_4$–Na$_2$SO$_4$–Na$_2$WO$_4$–Pb-WO$_4$....1174	RbF–ZnF$_2$..................1503	Sc$_2$O$_3$–Ln$_2$O$_3$..............351
PbSO$_4$–Na$_2$WO$_4$.................1122	RbF–ZrF$_4$...............1507–09	Sc$_2$O$_3$–Lu$_2$O$_3$..............351
PbSO$_4$–PbCrO$_4$..................1154	RbI–AgI.....................1580	Sc$_2$O$_3$–Nd$_2$O$_3$..............351
PbSO$_4$–PbMoO$_4$..................1155	RbI–Rb......................1664	Sc$_2$O$_3$–Sm$_2$O$_3$.............351
PbSO$_4$–PbO..........................1139	RbNO$_3$–AgNO$_3$–LiNO$_3$...1065	Sc$_2$O$_3$–Tm$_2$O$_3$..............351
PbSO$_4$–PbWO$_4$...................1156	RbNO$_3$–Ba(NO$_3$)$_2$........1056	Sc$_2$O$_3$–Y$_2$O$_3$................351
PbWO$_4$–K$_2$SO$_4$–K$_2$WO$_4$–PbSO$_4$....1169	RbNO$_3$–Ca(NO$_3$)$_2$........1057	Sc$_2$O$_3$–Yb$_2$O$_3$..............351
PbWO$_4$–Li$_2$SO$_4$–Li$_2$WO$_4$–PbSO$_4$...1171	RbNO$_3$–Ca(NO$_3$)$_2$–CsNO$_3$...1067	SeCl$_4$–AlCl$_3$...............1391
PbWO$_4$–Na$_2$SO$_4$.................1122	RbNO$_3$–Cd(NO$_3$)$_2$–KNO$_3$....1075	SeO$_3$–H$_2$O.................1947
PbWO$_4$–Na$_2$SO$_4$–Na$_2$WO$_4$–Pb-SO$_4$....1174	RbNO$_3$–CsCl–CsNO$_3$–RbCl....1754	Si–Al–Fe–O............131–136
PbWO$_4$–Na$_2$WO$_4$...............1122	RbNO$_3$–KNO$_3$–NaNO$_3$....1072	Si–Fe–Mg–O...........138–145
PbWO$_4$–PbSO$_4$....................1156	RbNO$_3$–Sr(NO$_3$)$_2$..........1058	Si–Fe–O.....................80–87
Pr–O..19	Rb$_2$O–Bi$_2$O$_3$................325	Si–Fe–O–Ti............146–147
PuCl$_3$–KCl...........................1278	Rb$_2$O–CrO$_3$–RbF............1701	Si–Mn–O...............101–103
PuCl$_3$–LiCl...........................1296	Rb$_2$O–MoO$_3$.................202	Si–O...........................21
PuCl$_3$–NaCl.........................1316	Rb$_2$O–MoO$_3$–RbF..........1702	Si–O–Ti.................113–115
PuF$_3$–LiF.............................1475	Rb$_2$O–RbF–WO$_3$............1703	Si–SiO$_2$........................21
PuF$_3$–NaF............................1494	Rb$_2$O–TiO$_2$...................201	SiCl$_4$–PbCl$_4$................1397
	Rb$_2$O–WO$_3$...................203	SiCl$_4$–SnCl$_4$................1400
Rb–RbBr..............................1664	RbOH–H$_2$O................2050	SiCl$_4$–TiCl$_4$................1401
Rb–RbCl..............................1664	Rb$_2$SO$_4$–CaSO$_4$.............1124	SiF$_4$–CaF$_2$–CaO–SiO$_2$........1729
Rb–RbF..........................1664–65	Rb$_2$SO$_4$–MgSO$_4$............1125	SiO$_2$....................153–157
Rb–RbI................................1664	Rb$_2$SO$_4$–NaF–Na$_2$SO$_4$–RbF....1809	SiO$_2$–Al$_2$O$_3$............313–314
RbBr–AgBr..........................1178		SiO$_2$–Al$_2$O$_3$–B$_2$O$_3$........763–765
RbBr–Rb..............................1664	S–As–Fe....................1885	SiO$_2$–Al$_2$O$_3$–B$_2$O$_3$–H$_2$O–MgO–Na$_2$O....2048
RbCNS–KCNS......................1909	S–Fe.....................1878–82	
RbCl.............................1210–11	S–Ni....................1883–84	SiO$_2$–Al$_2$O$_3$–B$_2$O$_3$–Li$_2$O.....820–825
RbCl–AgCl...........................1218	SO$_3$–Al$_2$O$_3$–CaO–H$_2$O....2063	SiO$_2$–Al$_2$O$_3$–BaO........556–557
RbCl–BaCl$_2$–BaF$_2$–RbF.......1655	SO$_3$–Al$_2$O$_3$–H$_2$O..........2012	SiO$_2$–Al$_2$O$_3$–BeO............576
RbCl–CaCl$_2$–KCl.................1416	SO$_3$–CO$_2$–CaO–H$_2$O–MgO....2066	SiO$_2$–Al$_2$O$_3$–CaO........630–639
RbCl–CaCl$_2$–NaCl................1427	SO$_3$–CaO–Na$_2$O–SiO$_2$....1141–42	SiO$_2$–Al$_2$O$_3$–CaO–Cr$_2$O$_3$–MgO....987–990
RbCl–CsCl............................1229	SO$_3$–K$_2$O–P$_2$O$_5$...............428	SiO$_2$–Al$_2$O$_3$–CaO–FeO....866–875
RbCl–CsCl–CsNO$_3$–RbNO$_3$.....1754	SO$_3$–K$_2$O–TiO$_2$...............546	SiO$_2$–Al$_2$O$_3$–CaO–Fe$_2$O$_3$....943–953
RbCl–CuCl...........................1247	SO$_3$–Na$_2$O–P$_2$O$_5$.............544	SiO$_2$–Al$_2$O$_3$–CaO–Fe$_2$O$_3$–MgO....991–1002
	SO$_3$–Na$_2$O–TiO$_2$..............546	SiO$_2$–Al$_2$O$_3$–CaO–H$_2$O....2035
	SO$_3$–Na$_2$O–V$_2$O$_5$............1140	SiO$_2$–Al$_2$O$_3$–CaO–K$_2$O....795–800
	Sb–O...........................20	

SiO_2–Al_2O_3–CaO–K_2O–MgO...973–974
SiO_2–Al_2O_3–CaO–K_2O–Na_2O.........971–972
SiO_2–Al_2O_3–CaO–MgO......880–923
SiO_2–Al_2O_3–CaO–MgO–Na_2O.........975–981
SiO_2–Al_2O_3–CaO–Na_2O......828–851
SiO_2–Al_2O_3–CaO–Na_2O–P_2O_5.....985
SiO_2–Al_2O_3–CaO–Na_2O–TiO_2.........982–984
SiO_2–Al_2O_3–CaO–P_2O_5......958–962
SiO_2–Al_2O_3–CaO–TiO_2......954–957
SiO_2–Al_2O_3–FeO............696–698
SiO_2–Al_2O_3–FeO–Fe_2O_3......131–136
SiO_2–Al_2O_3–FeO–K_2O.........801
SiO_2–Al_2O_3–FeO–MgO.........963
SiO_2–Al_2O_3–FeO–Na_2O......853–854
SiO_2–Al_2O_3–Fe_2O_3..........766–767
SiO_2–Al_2O_3–H_2O............2009–11
SiO_2–Al_2O_3–H_2O–K_2O..........2019
SiO_2–Al_2O_3–H_2O–K_2O–Na_2O...2044–45
SiO_2–Al_2O_3–H_2O–Li_2O–Na_2O...2046
SiO_2–Al_2O_3–H_2O–MgO......2038–42
SiO_2–Al_2O_3–H_2O–Na_2O......2022–29
SiO_2–Al_2O_3–K_2O...........407–416
SiO_2–Al_2O_3–K_2O–MgO......802–811
SiO_2–Al_2O_3–K_2O–Na_2O......786–789
SiO_2–Al_2O_3–Li_2O.........449–459
SiO_2–Al_2O_3–Li_2O–MgO.....813–819
SiO_2–Al_2O_3–MgO.............712
SiO_2–Al_2O_3–MgO–Na_2O......855–856
SiO_2–Al_2O_3–MgO–ZrO_2......966–967
SiO_2–Al_2O_3–MnO..........733–735
SiO_2–Al_2O_3–MnO–Y_2O_3......968–969
SiO_2–Al_2O_3–Na_2O........501–514
SiO_2–Al_2O_3–PbO............737–739
SiO_2–Al_2O_3–RF_2..............1715
SiO_2–Al_2O_3–SrO.............755
SiO_2–Al_2O_3–TiO_2..........768–771
SiO_2–Al_2O_3–ZnO.............758
SiO_2–Al_2O_3–ZrO_2............772
SiO_2–B_2O_3–BaO...........558–561
SiO_2–B_2O_3–CaO...........646–649
SiO_2–B_2O_3–CaO–Na_2O..........852
SiO_2–B_2O_3–Li_2O..........465–469
SiO_2–B_2O_3–Na_2O.........515–518
SiO_2–B_2O_3–P_2O_5..........775–777
SiO_2–B_2O_3–PbO...........740–742
SiO_2–B_2O_3–ZnO..............759
SiO_2–BaF_2–BaO.............1714
SiO_2–BaF_2–NaF.............1694
SiO_2–BaO.............210–211
SiO_2–BaO–CaO.............549–551
SiO_2–BaO–CaO–MgO............857
SiO_2–BaO–Li_2O..........437, 441
SiO_2–BaO–Na_2O...........478–479
SiO_2–BaO–TiO_2..............564
SiO_2–BeO.....................222
SiO_2–Bi_2O_3....................328
SiO_2–CO_2–CaO–MgO............929
SiO_2–CO_2–H_2O–K_2O...........2020
SiO_2–CaF_2–CaO.........1708–09, 1714
SiO_2–CaF_2–CaO–SiF_4..........1729
SiO_2–CaF_2–KF................1681
SiO_2–CaF_2–MgF..............1710
SiO_2–CaF_2–NaF..............1695
SiO_2–CaF_2–Na_2O............1705
SiO_2–CaO..................237–238
SiO_2–CaO–Cr_2O_3.........651–655
SiO_2–CaO–Cr_2O_3–MgO......924–926
SiO_2–CaO–FeO.............586–594

SiO_2–CaO–FeO–MgO........864–865
SiO_2–CaO–Fe_2O_3............656–658
SiO_2–CaO–Fe_2O_3–MgO......927–928
SiO_2–CaO–H_2O............1989–95
SiO_2–CaO–H_2O–SrO............2034
SiO_2–CaO–K_2O...........391–396
SiO_2–CaO–Li_2O...........437, 442
SiO_2–CaO–MgO............598–610
SiO_2–CaO–MgO–MnO–TiO_2......986
SiO_2–CaO–MgO–P_2O_5......935–938
SiO_2–CaO–MgO–TiO_2......930–934
SiO_2–CaO–MnO...........614–619
SiO_2–CaO–MnO–TiO_2...........941
SiO_2–CaO–MnO–WO_3............942
SiO_2–CaO–Na_2O...........481–485
SiO_2–CaO–Na_2O–SO_3......1141–42
SiO_2–CaO–P_2O_5.........665–671
SiO_2–CaO–SrO.............620–621
SiO_2–CaO–TiO_2...........660–663
SiO_2–CaO–ZnO............624–625
SiO_2–CaO–ZrO_2.................664
SiO_2–CoO......................255
SiO_2–Cr_2O_3...................332
SiO_2–Cr_2O_3–Fe_3O_4.........760–762
SiO_2–Cr_2O_3–MgO..............715
SiO_2–Cu_2O....................164
SiO_2–FeO...................80–81
SiO_2–FeO–Fe_2O_3.........82–84, 87
SiO_2–FeO–Fe_2O_3–MgO......138–145
SiO_2–FeO–Fe_3O_4–H_2O.........2036
SiO_2–FeO–FeS................1905
SiO_2–FeO–K_2O............397–398
SiO_2–FeO–MgO............682–687
SiO_2–FeO–MnO............689–694
SiO_2–FeO–Na_2O...........487–492
SiO_2–Fe_2O_3–H_2O–MgO–Na_2O....2047
SiO_2–Fe_2O_3–K_2O..........419–421
SiO_2–Fe_2O_3–Na_2O..........519–530
SiO_2–Ga_2O_3...................341
SiO_2–Ga_2O_3–P_2O_5.............780
SiO_2–GeO_2....................357
SiO_2–GeO_2–MgO...........717–719
SiO_2–GeO_2–MgO–NiO.....964–965
SiO_2–H_2O.................1942–45
SiO_2–H_2O–K_2O...............1949
SiO_2–H_2O–MgO............2001–02
SiO_2–H_2O–Na_2CO_3..........2057–58
SiO_2–H_2O–Na_2O...........1959–71
SiO_2–H_2O–NiO...............2005
SiO_2–H_2O–ZnO..............2006
SiO_2–KF–MgF_2..............1682
SiO_2–K_2O................167–168
SiO_2–K_2O–Li_2O..........377, 437
SiO_2–K_2O–MgO............399–403
SiO_2–K_2O–Na_2O...........381–383
SiO_2–K_2O–PbO............404–405
SiO_2–K_2O–PbO–TiO_2..........749
SiO_2–K_2O–ZnO................406
SiO_2–Li_2O................182–184
SiO_2–Li_2O–MgO...........443–444
SiO_2–Li_2O–Na_2O..........430–437
SiO_2–Li_2O–Na_2O–TiO_2........476
SiO_2–Li_2O–TiO_2...............474
SiO_2–Li_2O–ZrO_2..............475
SiO_2–MgF_2–MgO..........1711, 1714
SiO_2–MgF_2–NaF.............1696
SiO_2–MgO..................266–268
SiO_2–MgO–MnO...........699–701
SiO_2–MgO–NiO................702
SiO_2–MgO–P_2O_5..........725–728
SiO_2–MgO–TiO_2...............723
SiO_2–MgO–ZnO............705–706

SiO_2–MgO–ZrO_2................724
SiO_2–MnO....................101
SiO_2–MnO–Na_2O..............493
SiO_2–MnO–TiO_2...............736
SiO_2–Mn_3O_4................102–103
SiO_2–NaF–Na_2O.............1689
SiO_2–NaF–SrF_2..............1697
SiO_2–Na_2O................168, 192
SiO_2–Na_2O–P_2O_5..........535–542
SiO_2–Na_2O–PbO...........494–499
SiO_2–Na_2O–PbO–TiO_2..........749
SiO_2–Na_2O–TiO_2.........531–532
SiO_2–Na_2O–ZrO_2..........533–534
SiO_2–Nb_2O_5....................363
SiO_2–P_2O_5....................364
SiO_2–P_2O_5–PbO...............746
SiO_2–PbO.....................284
SiO_2–(R_2O,RO)................365
SiO_2–Si........................21
SiO_2–SrF_2–SrO..............1714
SiO_2–SrO.....................296
SiO_2–SrO–ZrO_2................757
SiO_2–ThO_2....................359
SiO_2–ThO_2–ZrO_2..............782
SiO_2–TiO_2................113–114
SiO_2–TiO_2–Ti_2O_3..............115
SiO_2–TiO_2–ZrO_2...............783
SiO_2–UO_2....................360
SiO_2–UO_2–ZrO_2...............784
SiO_2–ZnO.....................302
SiO_2–ZrO_2................361–362
Sm_2O_3–Al_2O_3.................312
Sm_2O_3–Bi_2O_3.................327
Sm_2O_3–Cr_2O_3.................331
Sm_2O_3–Dy_2O_3.................352
Sm_2O_3–Er_2O_3.................352
Sm_2O_3–Eu_2O_3.................352
Sm_2O_3–Fe_2O_3.................337
Sm_2O_3–Ga_2O_3.................340
Sm_2O_3–Gd_2O_3.................352
Sm_2O_3–H_2O..................1939
Sm_2O_3–Ho_2O_3.................352
Sm_2O_3–In_2O_3.................344
Sm_2O_3–La_2O_3.................345
Sm_2O_3–La_2O_3–Lu_2O_3..........781
Sm_2O_3–Ln_2O_3.................352
Sm_2O_3–Lu_2O_3.................352
Sm_2O_3–Nd_2O_3.................348
Sm_2O_3–Sc_2O_3.................351
Sm_2O_3–Tm_2O_3.................352
Sm_2O_3–Yb_2O_3.................352
$SnBr_4$–BBr_3..................1205
$SnCl_2$–$CaCl_2$.................1363
$SnCl_2$–$CdCl_2$.................1369
$SnCl_2$–CuCl...................1253
$SnCl_2$–$MnCl_2$.................1379
$SnCl_2$–$PbCl_2$.................1382
$SnCl_2$–TlCl...................1344
$SnCl_2$–$ZnCl_2$.................1387
$SnCl_4$–$PbCl_4$.................1398
$SnCl_4$–$SiCl_4$.................1400
SnF_2–UF_4....................1523
SnI_4–BBr_3...................1608
SnO_2–Al_2O_3..................315
SnO_2–BaO.....................212
SnO_2–BaO–PbO–TiO_2......860–861
SnO_2–BaO–TiO_2...............565
SnO_2–Bi_2O_3..................328
SnO_2–CaO–MgO.................611
SnO_2–CaO–MgO–TiO_2......939–940
SnO_2–CaO–TiO_2...........673–676
SnO_2–MgO–TiO_2...............729

SnO_2–PbO 285	$TaCl_5$–$AlCl_3$–$NbCl_5$ 1438	$TiCl_4$–$SiCl_4$ 1401
SnO_2–TiO_2 366	$TaCl_5$–$CsCl$ 1242, 1330	$TiCl_4$–$TaCl_5$ 1403
SnS–FeS 1896	$TaCl_5$–$FeCl_3$–$NbCl_5$ 1440	TiO_2–AlF_3–NaF 1700
SnS–Sb_2S_3 1899	$TaCl_5$–KCl 1330–31	TiO_2–Al_2O_3 316
$SrBr_2$–KBr 1188	$TaCl_5$–$LiCl$ 1330–31	TiO_2–Al_2O_3–BeO 578
$SrBr_2$–$LiBr$ 1191	$TaCl_5$–$NaCl$ 1330–31	TiO_2–Al_2O_3–CaO–Na_2O–SiO_2 ...
$SrBr_2$–$NaBr$ 1197	$TaCl_5$–$NbCl_5$–$TiCl_4$ 1441	 982–984
$SrBr_2$–UBr_3 1607	$TaCl_5$–$RbCl$ 1330–31	TiO_2–Al_2O_3–CaO–SiO_2 ... 954–957
$SrCl_2$–$BaCl_2$ 1355	$TaCl_5$–$TiCl_4$ 1403	TiO_2–Al_2O_3–Li_2O 460–464
$SrCl_2$–$BaCl_2$–KCl 1418	Ta_2O_5–Al_2O_3 319	TiO_2–Al_2O_3–MgO 713–714
$SrCl_2$–$BaCl_2$–$NaCl$ 1428	Ta_2O_5–Bi_2O_3 329	TiO_2–Al_2O_3–SiO_2 768–771
$SrCl_2$–$CaCl_2$ 1364	Ta_2O_5–K_2CO_3 173	TiO_2–Al_2O_3–ZrO_2 773–774
$SrCl_2$–$CdCl_2$ 1370	Ta_2O_5–K_2O 173	TiO_2–B_2O_3–BaO–PbO 859
$SrCl_2$–KCl 1270	Ta_2O_5–K_2O–Nb_2O_5 424	TiO_2–B_2O_3–PbO 743
$SrCl_2$–KCl–KNO_3–$Sr(NO_3)_2$ ... 1757	Ta_2O_5–MgO 273	TiO_2–$BaCl_2$–BaO 1846
$SrCl_2$–$LiCl$ 1295	Ta_2O_5–Na_2O 198	TiO_2–BaO 213
$SrCl_2$–$LiCl$–LiF–SrF_2 1649	Ta_2O_5–Nb_2O_5 375	TiO_2–BaO–CaO 552–553
$SrCl_2$–$LiCl$–$LiNO_3$–$Sr(NO_3)_2$. 1760	Ta_2O_5–PbO 289	TiO_2–BaO–CaO–SrO 858
$SrCl_2$–$LiCl$–Li_2SO_4–$SrSO_4$ ... 1800	Ta_2O_5–ZrO_2 374	TiO_2–BaO–In_2O_3–La_2O_3–SrO–Y_2O_3 . 863
$SrCl_2$–$MgCl_2$ 1374	$TeCl_4$–$AlCl_3$ 1392	TiO_2–BaO–Na_2O–Nb_2O_5 826
$SrCl_2$–$MnCl_2$ 1380	TeO_2–Bi_2O_3 328	TiO_2–BaO–P_2O_5 568
$SrCl_2$–$NaCl$ 1311	Th–O–U 116–118	TiO_2–BaO–PbO–SnO_2 .. 860–861
$SrCl_2$–$NaCl$–NaF–SrF_2 1654	ThF_4–BeF_2 1517	TiO_2–BaO–PbO–ZrO_2 862
$SrCl_2$–$NaCl$–$NaNO_3$–$Sr(NO_3)_2$. 1764	ThF_4–BeF_2–LiF 1553	TiO_2–BaO–SiO_2 564
$SrCl_2$–$PbCl_2$ 1383	ThF_4–BeF_2–NaF 1562	TiO_2–BaO–SnO_2 565
$SrCl_2$–SrF_2 1629	ThF_4–BeF_2–UF_4 1574	TiO_2–BaO–SrO 554–555
$SrCl_2$–$TlCl$ 1345	ThF_4–KF 1464	TiO_2–BaO–ZrO_2 566–567
$SrCl_2$–$ZnCl_2$ 1388	ThF_4–LiF 1478	TiO_2–BeO 224
SrF_2–BeF_2 1516	ThF_4–LiF–UF_4 1556–57	TiO_2–BeO–ZrO_2 583–585
SrF_2–$LiCl$–LiF–$SrCl_2$ 1649	ThF_4–MgF_2 1521	TiO_2–Bi_2O_3 328
SrF_2–$NaCl$–NaF–$SrCl_2$ 1654	ThF_4–NaF 1497	TiO_2–CaO 239–240
SrF_2–NaF–SiO_2 1697	ThF_4–NaF–UF_4 1571	TiO_2–CaO–MgO 612
SrF_2–SiO_2–SrO 1714	ThF_4–NaF–ZrF_4 1572	TiO_2–CaO–MgO–MnO–SiO_2 ... 986
SrF_2–$SrCl_2$ 1629	ThF_4–RbF 1505	TiO_2–CaO–MgO–SiO_2 ... 930–934
SrI_2–UBr_3 1611	ThF_4–UF_4 1526	TiO_2–CaO–MgO–SnO_2 ... 939–940
$Sr(NO_3)_2$–$Ba(NO_3)_2$ 1061	ThO_2–Al_2O_3–BeO 577	TiO_2–CaO–MnO–SiO_2 941
$Sr(NO_3)_2$–$Ca(NO_3)_2$ 1062	ThO_2–B_2O_3 322	TiO_2–CaO–SiO_2 660–663
$Sr(NO_3)_2$–$CsNO_3$ 1046	ThO_2–BeO 223	TiO_2–CaO–SnO_2 673–676
$Sr(NO_3)_2$–KCl–KNO_3–$SrCl_2$ .. 1757	ThO_2–BeO–MgO 573	TiO_2–CaO–SrO 622
$Sr(NO_3)_2$–$LiCl$–$LiNO_3$–$SrCl_2$. 1760	ThO_2–CO_2–H_2O–Na_2O 2030	TiO_2–CaO–ZrO_2 678
$Sr(NO_3)_2$–$NaCl$–$NaNO_3$–$SrCl_2$. 1764	ThO_2–CaO–ZrO_2 677	TiO_2–CoO 256
$Sr(NO_3)_2$–$RbNO_3$ 1058	ThO_2–CrO_3–H_2O 2017	TiO_2–CrO_3–K_2O 546
SrO–Al_2O_3 294–295	ThO_2–HfO_2–MgO 721–722	TiO_2–CrO_3–Na_2O 546
SrO–Al_2O_3–P_2O_5 756	ThO_2–MgO–ZrO_2 730	TiO_2–Cr_2O_3–Fe_2O_3 778
SrO–Al_2O_3–SiO_2 755	ThO_2–SiO_2 359	TiO_2–Cs_2O 161
SrO–BaO–CaO–TiO_2 858	ThO_2–SiO_2–ZrO_2 782	TiO_2–FeO 88
SrO–BaO–In_2O_3–La_2O_3–TiO_2–Y_2O_3 . 863	ThO_2–TiO_2 367	TiO_2–FeO–Fe_2O_3 90–91
SrO–BaO–TiO_2 554–555	ThO_2–UO_2 118	TiO_2–FeO–MnO 695
SrO–BeO 217	ThO_2–ZrO_2 368	TiO_2–Fe_2O_3 89
SrO–BeO–CaO 569	Ti–C–O 36	TiO_2–GeO_2 358
SrO–BeO–La_2O_3 575	Ti–Fe–O 88–93	TiO_2–GeO_2–MgO 720
SrO–Bi_2O_3 326	Ti–Fe–O–Si 146–147	TiO_2–H_2O 1946
SrO–CaO–H_2O–SiO_2 2034	Ti–N–O 104	TiO_2–K_2CO_3–PbO 710
SrO–CaO–P_2O_5 623	Ti–O 22	TiO_2–KF–Li_2O 1704
SrO–CaO–SiO_2 620–621	Ti–O–Si 113–115	TiO_2–KF–PbO 749
SrO–CaO–TiO_2 622	Ti–TiO_2 22	TiO_2–K_2O 169
SrO–MgO 274	$TiCl_2$–KCl 1271	TiO_2–K_2O–LiF 1683
SrO–MgO–P_2O_5 703	$TiCl_2$–$MgCl_2$ 1375	TiO_2–K_2O–Li_2O 476
SrO–P_2O_5–ZnO 754	$TiCl_2$–$NaCl$ 1312	TiO_2–K_2O–Li_2O–P_2O_5 476
SrO–PbO–TiO_2–ZrO_2 970	$TiCl_3$–$CsCl$ 1238	TiO_2–K_2O–MoO_3 547
SrO–SiO_2 296	$TiCl_3$–KCl 1279	TiO_2–K_2O–MoO_3–PbO . 749, 812
SrO–SiO_2–SrF_2 1714	$TiCl_3$–KCl–$NaCl$ 1414	TiO_2–K_2O–P_2O_5 422–423
SrO–SiO_2–ZrO_2 757	$TiCl_3$–$NaCl$ 1317	TiO_2–K_2O–P_2O_5–PbO 750
SrO–TiO_2 297–298	$TiCl_3$–$RbCl$ 1336	TiO_2–K_2O–PbO–SiO_2 749
$Sr(OH)_2$–$Ba(OH)_2$ 1030	$TiCl_4$–$AlCl_3$ 1393	TiO_2–K_2O–PbO–V_2O_5 749
$SrSO_4$–K_2SO_4 1103	$TiCl_4$–$AlCl_3$–$FeCl_3$ 1435	TiO_2–K_2O–SO_3 546
$SrSO_4$–$LiCl$–Li_2SO_4–$SrCl_2$. 1800	$TiCl_4$–$FeCl_3$ 1395	TiO_2–K_2O–V_2O_5 545
$SrSO_4$–Li_2SO_4 1111	$TiCl_4$–$FeCl_3$–$NaCl$ 1433	TiO_2–K_2O–WO_3 547
	$TiCl_4$–$FeCl_3$–$NbCl_5$ 1439	TiO_2–Li_2O 185
$TaCl_5$–$AlCl_3$–$CsCl$ 1406	$TiCl_4$–$NbCl_5$ 1402	TiO_2–Li_2O–NaF 1687, 1704
$TaCl_5$–$AlCl_3$–$FeCl_3$ 1437	$TiCl_4$–$NbCl_5$–$TaCl_5$ 1441	TiO_2–Li_2O–Na_2O 438, 476
$TaCl_5$–$AlCl_3$–$NaCl$ 1432	$TiCl_4$–$PbCl_4$ 1399	TiO_2–Li_2O–Na_2O–SiO_2 476

TiO_2–Li_2O–P_2O_5 476	$TlNO_3$–$AgNO_3$–$NaNO_3$ 1066	UO_2–Nd_2O_3 349
TiO_2–Li_2O–RbF 1704	$TlNO_3$–$Ca(NO_3)_2$ 1059	UO_2–SiO_2 360
TiO_2–Li_2O–SiO_2 474	$TlNO_3$–$Cd(NO_3)_2$–$CsNO_3$ 1068	UO_2–SiO_2–ZrO_2 784
TiO_2–MgO 269–270	$TlNO_3$–KNO_3 1048	UO_2–ThO_2 118
TiO_2–MgO–SiO_2 723	$TlNO_3$–$NaNO_3$ 1054	UO_2–Y_2O_3 353
TiO_2–MgO–SnO_2 729	$TlNO_3$–TlI 1747	UO_2–ZrO_2 119–120
TiO_2–MgO–ZrO_2 731	Tl_2S–Ag_2S 1887	UO_3–CO_2–H_2O–Li_2O 2021
TiO_2–MnO 277	Tl_2S–As_2S_3 1893	UO_3–CO_3–H_2O–Na_2O 2031
TiO_2–MnO–SiO_2 736	Tl_2S–PbS 1892	UO_3–H_2O–Na_2O 1975
TiO_2–MoO_3–Na_2O 547	Tl_2SO_4–$CdCl_2$–$CdSO_4$–$TlCl$ 1805	
TiO_2–MoO_3–Na_2O–PbO 750	Tl_2SO_4–K_2SO_4–Li_2SO_4 1127	V–Co–O 54–55
TiO_2–Na_2CO_3–PbO 749	Tl_2SO_4–$LiCl$–Li_2SO_4–$TlCl$ 1796	V–Fe–O 94–96
TiO_2–NaF–Na_2O 1690	Tm_2O_3–Al_2O_3 312	V–Na–O 105–106
TiO_2–NaF–PbO 750	Tm_2O_3–Cr_2O_3 331	V–O 24
TiO_2–Na_2O 193–194	Tm_2O_3–Eu_2O_3 335	VCl_2–$CsCl$ 1232
TiO_2–Na_2O–P_2O_5 545	Tm_2O_3–Fe_2O_3 337	VCl_2–KCl 1272
TiO_2–Na_2O–P_2O_5–PbO 750	Tm_2O_3–Ga_2O_3 340	VCl_2–$NaCl$ 1313
TiO_2–Na_2O–PbO–SiO_2 749	Tm_2O_3–Gd_2O_3 342	VCl_3–KCl 1280
TiO_2–Na_2O–PbO–V_2O_5 749	Tm_2O_3–In_2O_3 344	VCl_3–$NaCl$ 1319
TiO_2–Na_2O–PbO–WO_3 750	Tm_2O_3–La_2O_3 345	V_2O_5–Al_2O_3 320
TiO_2–Na_2O–SO_3 546	Tm_2O_3–Nd_2O_3 348	V_2O_5–Bi_2O_3 329
TiO_2–Na_2O–SiO_2 531–532	Tm_2O_3–Sc_2O_3 351	V_2O_5–CaO 251
TiO_2–Na_2O–V_2O_5 545	Tm_2O_3–Sm_2O_3 352	V_2O_5–CoO 55
TiO_2–Na_2O–WO_3 547		V_2O_5–Cr_2O_3 333
TiO_2–Nb_2O_5 372	U–Mg–O 99–100	V_2O_5–Fe_2O_3 96
TiO_2–PbO–SrO–ZrO_2 970	U–O 23	V_2O_5–H_2O–Na_2O 1974
TiO_2–PbO–V_2O_5 748	U–O–Th 116–118	V_2O_5–K_2CO_3 174
TiO_2–PbO–ZrO_2 747	U–O–Zr 119–124	V_2O_5–K_2O 174
TiO_2–Rb_2O 201	UBr_3–$SrBr_2$ 1607	V_2O_5–K_2O–PbO–TiO_2 749
TiO_2–SiO_2 113–114	UBr_3–SrI_2 1611	V_2O_5–K_2O–TiO_2 545
TiO_2–SiO_2–Ti_2O_3 115	UCl_3–$LiCl$ 1297	V_2O_5–K_2SO_4 1137
TiO_2–SiO_2–ZrO_2 783	UCl_3–$NaCl$ 1318	V_2O_5–Na_2O 106
TiO_2–SnO_2 366	UCl_3–$RbCl$ 1337	V_2O_5–Na_2O–PbO–TiO_2 749
TiO_2–SrO 297–298	UCl_4–$CsCl$ 1240	V_2O_5–Na_2O–SO_3 1140
TiO_2–ThO_2 367	UCl_4–KCl 1282	V_2O_5–Na_2O–TiO_2 545
TiO_2–Ti 22	UCl_4–$LiCl$ 1298	V_2O_5–Na_2SO_4 1138
TiO_2–ZnO 303	UCl_4–$NaCl$ 1321	V_2O_5–$NaVO_3$ 105
TiO_2–ZrO_2 369–371	UF_3–BaF_2 1512	V_2O_5–NiO 279
Ti_2O_3–SiO_2–TiO_2 115	UF_3–LiF 1476	V_2O_5–PbF_2–PbO 1713
$TlBr$–KBr–KCl–$TlCl$ 1641	UF_3–NaF 1495	V_2O_5–PbO 290
$TlBr$–$NaBr$–$NaCl$–$TlCl$ 1642	UF_4–BeF_2 1518	V_2O_5–PbO–TiO_2 748
$TlBr$–$TlCl$ 1601	UF_4–BeF_2–LiF 1554	
$TlBr$–TlI 1602	UF_4–BeF_2–NaF 1563	WO_3–B_2O_3–K_2O–Li_2O 785
$TlCl$–$AgCl$ 1219	UF_4–BeF_2–ThF_4 1574	WO_3–Bi_2O_3 330
$TlCl$–$BaCl_2$ 1338	UF_4–CsF 1449	WO_3–Bi_2O_3–PbO 745
$TlCl$–$BeCl_2$ 1339	UF_4–KF 1465	WO_3–CaO–MnO–SiO_2 942
$TlCl$–$BiCl_3$ 1347	UF_4–KF–LiF 1532	WO_3–CrO_3–K_2O 426
$TlCl$–$CaCl_2$ 1340	UF_4–KF–NaF 1538	WO_3–CrO_3–PbO 752
$TlCl$–$CdCl_2$ 1341	UF_4–KF–PbF_2 1540	WO_3–CsF–Cs_2O 1677
$TlCl$–$CdCl_2$–$CdSO_4$–Tl_2SO_4 1805	UF_4–LiF 1479	WO_3–KF–K_2O 1680
$TlCl$–$CsCl$ 1230	UF_4–LiF–NaF 1547	WO_3–K_2O 177
$TlCl$–$CuCl$ 1248	UF_4–LiF–RbF 1550	WO_3–K_2O–MoO_3 427
$TlCl$–$FeCl_3$ 1348	UF_4–LiF–ThF_4 1556–57	WO_3–K_2O–MoO_3–Na_2O 793
$TlCl$–KBr–KCl–$TlBr$ 1641	UF_4–NaF 1498	WO_3–K_2O–Na_2O 389
$TlCl$–KCl 1261	UF_4–NaF–PbF_2 1569	WO_3–K_2O–P_2O_5 428
$TlCl$–$LiCl$ 1288	UF_4–NaF–RbF 1560	WO_3–K_2O–TiO_2 547
$TlCl$–$LiCl$–Li_2SO_4–Tl_2SO_4 1796	UF_4–NaF–ThF_4 1571	WO_3–LiF–Li_2O 1685
$TlCl$–$MgCl_2$ 1342	UF_4–NaF–ZrF_4 1573	WO_3–Li_2O 187
$TlCl$–$NaBr$–$NaCl$–$TlBr$ 1642	UF_4–PbF_2 1522	WO_3–Li_2O–Na_2O 440
$TlCl$–$NaCl$ 1300	UF_4–RbF 1506	WO_3–MoO_3 376
$TlCl$–$PbCl_2$ 1343	UF_4–SnF_2 1523	WO_3–MoO_3–NaF–Na_2O 1726
$TlCl$–$RbCl$ 1332	UF_4–ThF_4 1526	WO_3–MoO_3–Na_2O 543
$TlCl$–$SnCl_2$ 1344	UF_4–ZrF_4 1527	WO_3–MoO_3–PbO 753
$TlCl$–$SrCl_2$ 1345	UF_6–BF_3 1524	WO_3–NaF–Na_2O 1693
$TlCl$–$TlBr$ 1601	UF_6–BF_5 1592	WO_3–Na_2O 200
$TlCl$–TlI 1620	UF_6–Br_2 1592	WO_3–Na_2O–P_2O_5 544
$TlCl$–$ZnCl_2$ 1346	UF_6–ClF_3 1525	WO_3–Na_2O–PbO 500
TlI–$TlBr$ 1602	UO_2–Al_2O_3 317	WO_3–Na_2O–PbO–TiO_2 750
TlI–$TlCl$ 1620	UO_2–BeO 225–226	WO_3–Na_2O–TiO_2 547
TlI–$TlNO_3$ 1747	UO_2–CaO 241	WO_3–PbO 293
$TlNO_3$–$AgNO_3$ 1042	UO_2–H_2O–P_2O_5 2018	WO_3–RbF–Rb_2O 1703
$TlNO_3$–$AgNO_3$–KNO_3 1063	UO_2–MgO 99–100	WO_3–Rb_2O 203

Y–Fe–O 97–98	$ZnCl_2$–$MnCl_2$ 1381	ZrF_4–KF–NaF 1539
YF_3–CaF_2 1520	$ZnCl_2$–NH_4Cl 1326	ZrF_4–LiF 1480, 1509
YF_3–LiF 1477	$ZnCl_2$–NaCl–Na_2SO_4–$ZnSO_4$ 1804	ZrF_4–LiF–NaF 1548
Y_2O_3–Al_2O_3 311–312	$ZnCl_2$–$PbCl_2$ 1384	ZrF_4–NaF 1499, 1509
Y_2O_3–Al_2O_3–MnO–SiO_2 968–969	$ZnCl_2$–RbCl 1335	ZrF_4–NaF–RbF 1561
Y_2O_3–BaO–In_2O_3–La_2O_3–SrO–TiO_2 .863	$ZnCl_2$–$SnCl_2$ 1387	ZrF_4–NaF–ThF_4 1572
Y_2O_3–BeO 220	$ZnCl_2$–$SrCl_2$ 1388	ZrF_4–NaF–UF_4 1573
Y_2O_3–Cr_2O_3 331	$ZnCl_2$–TlCl 1346	ZrF_4–RbF 1507–09
Y_2O_3–Fe_2O_3 97	ZnF_2–AgF 1444	ZrF_4–UF_4 1527
Y_2O_3–Fe_2O_3–Fe_3O_4 98	ZnF_2–CsF 1447	ZrO_2–AlF_3–NaF 1699
Y_2O_3–Ga_2O_3 340	ZnF_2–KF 1460	ZrO_2–Al_2O_3–BeO 579
Y_2O_3–H_2O 1940	ZnF_2–LiF 1472	ZrO_2–Al_2O_3–MgO–SiO_2 966–967
Y_2O_3–In_2O_3 344	ZnF_2–NaF 1489	ZrO_2–Al_2O_3–SiO_2 772
Y_2O_3–MgO 263	ZnF_2–RbF 1503	ZrO_2–Al_2O_3–TiO_2 773–774
Y_2O_3–Sc_2O_3 351	ZnO–AlF_3–NaF 1700	ZrO_2–BaO–PbO–TiO_2 862
Y_2O_3–UO_2 353	ZnO–Al_2O_3 299	ZrO_2–BaO–TiO_2 566–567
Y_2O_3–ZrO_2 354	ZnO–Al_2O_3–SiO_2 758	ZrO_2–BeO 227
Yb_2O_3–Al_2O_3 312	ZnO–B_2O_3 300–301	ZrO_2–BeO–CaO 571
Yb_2O_3–BeO 221	ZnO–B_2O_3–CdO 679	ZrO_2–BeO–CeO_2 581–582
Yb_2O_3–Cr_2O_3 331	ZnO–B_2O_3–SiO_2 759	ZrO_2–BeO–Cr_2O_3 580
Yb_2O_3–Eu_2O_3 335	ZnO–Bi_2O_3 326	ZrO_2–BeO–MgO 574
Yb_2O_3–Fe_2O_3 337	ZnO–CaO–SiO_2 624–625	ZrO_2–BeO–TiO_2 583–585
Yb_2O_3–Ga_2O_3 340	ZnO–CoO 56	ZrO_2–Bi_2O_3 328
Yb_2O_3–Gd_2O_3 342	ZnO–CrO_3–H_2O 2007	ZrO_2–CaO 242–243
Yb_2O_3–In_2O_3 344	ZnO–GeO_2–MgO 704	ZrO_2–CaO–La_2O_3 659
Yb_2O_3–La_2O_3 345	ZnO–H_2O–Na_2O 1954	ZrO_2–CaO–SiO_2 664
Yb_2O_3–Nd_2O_3 348	ZnO–H_2O–SiO_2 2006	ZrO_2–CaO–ThO_2 677
Yb_2O_3–Sc_2O_3 351	ZnO–K_2O–SiO_2 406	ZrO_2–CaO–TiO_2 678
Yb_2O_3–Sm_2O_3 352	ZnO–MgO–P_2O_5 707–709	ZrO_2–CeO_2 355
	ZnO–MgO–SiO_2 705–706	ZrO_2–FeO 257
Zn–Al–$AlCl_3$ 1666	ZnO–MnO–P_2O_5 732	ZrO_2–La_2O_3 346
Zn–Al–$AlCl_3$–$ZnCl_2$ 1668	ZnO–Nb_2O_5 304	ZrO_2–La_2O_3–MgO 716
Zn–$AlCl_3$–$ZnCl_2$ 1667	ZnO–P_2O_5 305–307	ZrO_2–Li_2O–SiO_2 475
Zn–Cd–$CdCl_2$–$ZnCl_2$ 1669	ZnO–P_2O_5–SrO 754	ZrO_2–MgO 271
Zn–Co–O 56	ZnO–SiO_2 302	ZrO_2–MgO–SiO_2 724
$ZnBr_2$–$CdBr_2$ 1200	ZnO–TiO_2 303	ZrO_2–MgO–ThO_2 730
$Zn(CN)_2$–KCN 1877	ZnS–FeS 1897	ZrO_2–MgO–TiO_2 731
$ZnCl_2$–AgCl 1224	$ZnSO_4$–KCl–K_2SO_4–$ZnCl_2$ 1794	ZrO_2–Na_2O–SiO_2 533–534
$ZnCl_2$–Al–$AlCl_3$–Zn 1668	$ZnSO_4$–K_2SO_4–Na_2SO_4 1134	ZrO_2–Nb_2O_5 373
$ZnCl_2$–$AlCl_3$–Zn 1667	$ZnSO_4$–LiCl–Li_2SO_4–$ZnCl_2$ 1801	ZrO_2–Nd_2O_3 350
$ZnCl_2$–$BaCl_2$ 1356	$ZnSO_4$–NaCl–Na_2SO_4–$ZnCl_2$ 1804	ZrO_2–PbO–SrO–TiO_2 970
$ZnCl_2$–$BiCl_3$ 1389	$ZnSO_4$–Na_2SO_4 1123	ZrO_2–PbO–TiO_2 747
$ZnCl_2$–$CaCl_2$ 1365	Zr–O 25	ZrO_2–SiO_2 361–362
$ZnCl_2$–Cd–$CdCl_2$–Zn 1669	Zr–O–U 119–124	ZrO_2–SiO_2–SrO 757
$ZnCl_2$–$CdCl_2$ 1371	$ZrCl_4$–CsCl 1241	ZrO_2–SiO_2–ThO_2 782
$ZnCl_2$–CsCl 1233	$ZrCl_4$–KCl 1283–84	ZrO_2–SiO_2–TiO_2 783
$ZnCl_2$–CuCl 1254	$ZrCl_4$–KCl–NaCl 1415	ZrO_2–SiO_2–UO_2 784
$ZnCl_2$–$FeCl_3$ 1390	$ZrCl_4$–NaCl 1322–23	ZrO_2–Ta_2O_5 374
$ZnCl_2$–KCl 1273	$ZrCl_4$–NaCl–$NbCl_5$ 1434	ZrO_2–ThO_2 368
$ZnCl_2$–KCl–K_2SO_4–$ZnSO_4$ 1794	ZrF_4–BeF_2–NaF 1564	ZrO_2–TiO_2 369–371
$ZnCl_2$–KCl–$PbCl_2$ 1425	ZrF_4–CeF_3–NaF 1570	ZrO_2–UO_2 119–120
$ZnCl_2$–LiCl–Li_2SO_4–$ZnSO_4$ 1801	ZrF_4–CsF 1450	ZrO_2–Y_2O_3 354
$ZnCl_2$–$MgCl_2$ 1376	ZrF_4–KF 1466	

DATE DUE			
e8/15			
GAYLORD			PRINTED IN U.S.A.

570128 (5-9-80)

```
                                    QD
                                    501
BE                                  .L59

504 57012 8

PHASE DIAGRAMS FOR CERAMIST
LEVIN, ERNEST
```

Property of
University Libraries
University of South Carolina